J. M. Sørensen
W. Arlt

LIQUID-LIQUID EQUILIBRIUM DATA COLLECTION

Ternary Systems

DECHEMA

Chemistry Data Series
Vol. V, Part 2

Published by DECHEMA
Deutsche Gesellschaft für Chemisches Apparatewesen

Editors: Dieter Behrens, Reiner Eckermann

Liquid-Liquid Equilibrium Data Collection

2

Ternary Systems

Tables, Diagrams and Model Parameters

J. M. Sørensen*, W. Arlt**

* Instituttet for Kemiteknik, Danmarks Tekniske Højskole, Lyngby, Denmark

** Lehrstuhl Technische Chemie B, Universität Dortmund, Federal Republic of Germany

© 1980 DECHEMA Deutsche Gesellschaft für Chemisches Apparatewesen
6000 Frankfurt/Main, Federal Republic of Germany

ISBN 3-921 567-18-1

Printed by Schön & Wetzel GmbH, Frankfurt/Main, F. R. Germany

PREFACE OF AUTHORS

This work consists of three Parts. Part 1 deals with binary liquid-liquid equilibria and Parts 2 and 3 with ternary liquid-liquid equilibria.

The correlation and prediction of liquid-liquid equilibria (LLE) using models for liquid phase non-idealities is a subject within chemical engineering which is not yet mastered quantitatively. Model inadequacies may be one explanation for this. However, this work will show that the results may be quite good with existing models if they are used properly, i. e. if the parameter estimation procedure as well as the data base are chosen carefully.

In addition to an extensive collection of experimental binary, ternary, and quaternary data, this work contains NRTL and UNIQUAC parameters reduced from these data. For each ternary system (Parts 2 and 3) we give a specific set of NRTL and UNIQUAC parameters which emphasizes the distribution ratio of the solute at small concentrations. This makes semi-quantitative extraction calculations possible for many systems. We also include a table of UNIQUAC parameters for the A-B interaction reduced from a large number of systems with components A, B, and any third component. These parameters will usually be better for predicting multicomponent LLE than parameters based on VLE-data.

The work leading to these three books began in 1977 at Instituttet for Kemiteknik, Lyngby, Denmark, where Aa. Fredenslund started a project called "A Group Contribution Method for Predicting Liquid-Liquid Equilibria". An initial step in this project was to investigate how existing molecular models (e.g. NRTL and UNIQUAC) behave in LLE calculations. For this purpose a parameter estimation procedure and a large data base were established.

The above project was close to research plans at Dortmund University, and a collaboration was initiated in 1978 with the purpose of jointly preparing this publication. The data collection was extended, and the authors have profited very much by the experience of J. Gmehling and U. Onken who started Volume I of this series ("Vapor-Liquid Equilibrium Data Collection").

Simultaneously the parameter estimation procedure was further refined under daily guidance of Aa. Fredenslund and P. Rasmussen to whom the authors are very grateful.

The parameter estimation from ternary LLE-data involves many numerical problems. The authors wish to thank M. L. Michelsen (Instituttet for Kemiteknik, Lyngby) for providing the final version of the estimation procedure. Finally, we dedicate special thanks to T. Magnussen (DECHEMA, Frankfurt/Main) for many fruitful discussions and help and to colleagues and students in Lyngby and Dortmund for various kinds of assistance.

The authors hope that the data collection and parameters of this work will be of use to the industry. We also hope that the availability of the large amount of data will facilitate the testing of new models at universities.

The authors

PREFACE OF EDITORS

Subjects of the DECHEMA Chemistry Data Series are the physical and thermo-dynamic property data of chemical compounds and mixtures essentially for the fluid state covering PVT data, heat capacity, and entropy data, phase equili-brium data, transport and interfacial tension data.

The main purpose is to provide chemists and engineers with data for process design and development. For computer based calculations in process design appropriate correlation methods and accurate data must be used. These are only in some cases available in open literature. For that reason the most urgent requirement regarding the publication of data is to offer critically evaluated and reliable data. This will be the goal of the series.

DECHEMA gives an opportunity to authors especially from universities to publish not only their theoretical results, but also their measured or compiled data, most often a large amount.

After that a successful group contribution method for the prediction of vapor-liquid equilibria (UNIFAC) had been presented to the scientific community several years ago, the needs for a similar treatment of liquid-liquid equilibria led to a cooperation between the Dechema Data Compiler Development Group and Professor Fredenslund at the Instituttet for Kemiteknik in Lyngby, who has much experience in this field.

During this cooperation, J. M. Sørensen and W. Arlt, who is already co-author of Volume I of the Series, have collected the mutual solubility data of more than 2000 binary, ternary and quaternary mixtures of organic liquids.

This compilation is now being published as Volume V of this series, in 3 parts giving not only measured data but also evaluated correlation constants and recommended values. We hope that this gives an instrument that will allow the users to solve their problems considerably more easily aid more quickly than before.

Frankfurt/Main, October 1980

Dieter Behrens
Reiner Eckermann

CONTENTS
Vol. V, Part 2

LIST OF SYMBOLS

Symbols only used in the model expressions are not explained here.

a	activity or model parameter
$\triangle G$	molar Gibbs energy of mixing
n	number of moles
q	constant in the UNIQUAC model
r	constant in the UNIQUAC model
R	the gas constant
T	temperature
x	liquid mole fraction

Greek symbols

α	NRTL non-randomness parameter
γ	activity coefficient

Superscripts

I, II	phases

Subscripts

1,2,3,4	components
i	component
j	component or phase
k	tie line
ℓ	data set

INTRODUCTION

Parts 2 and 3 of the liquid-liquid equilibrium (LLE) data collection deal with multicomponent systems. The purposes are:

a. to report experimental ternary and quaternary LLE data from the literature within the range of components and conditions given in Section 1. We do not report how the experimental data were measured.

b. to provide "specific" UNIQUAC and NRTL parameters based on data for each individual ternary system. If several references report data for the same ternary system at temperatures between 20 °C and 30 °C, the data are combined to give only one set of parameters. Specific parameters are presented for all ternary systems of type 1 or 2 (see Section 1) with five experimental tie lines or more if these cover the immiscibility region reasonably well.

c. to provide "common" UNIQUAC parameters valid at temperatures between 20 °C and 30 °C. The common parameters for the AB interaction are the same in the systems ABC, ABD and any other system which contains components A and B. Data for nearly all ternary systems of type 1 or 2 between 20 °C and 30 °C were used in the estimation of common parameters.

The specific or common parameters may be used to represent smoothed values of the experimental ternary data from which they are reduced. Computed binodal curves are shown on figures. Binodal curves in tabular form may be generated with a computer program listed in this work.

The common UNIQUAC parameters have been used to predict the reported quaternary LLE data. The results are discussed in Section 13.

1. Ternary and Quaternary LLE Data Collection

All the listed data are based on information obtained from original papers. Papers in which the data are not presented numerically (i. e. only graphically) have not been included.

Types of Data

The ternary and quaternary data collection contains experimental tie line data measured at constant temperature. Only tie line data are included, i. e. all concentrations must be known in the two (or three) coexisting liquid phases. Data for binodal curves without knowledge of distribution ratios are not included. Only data sets with at least three tie lines are included. Each ternary system belongs to one of the six types shown in **Figure 1** (Sørensen et al., 1979a).

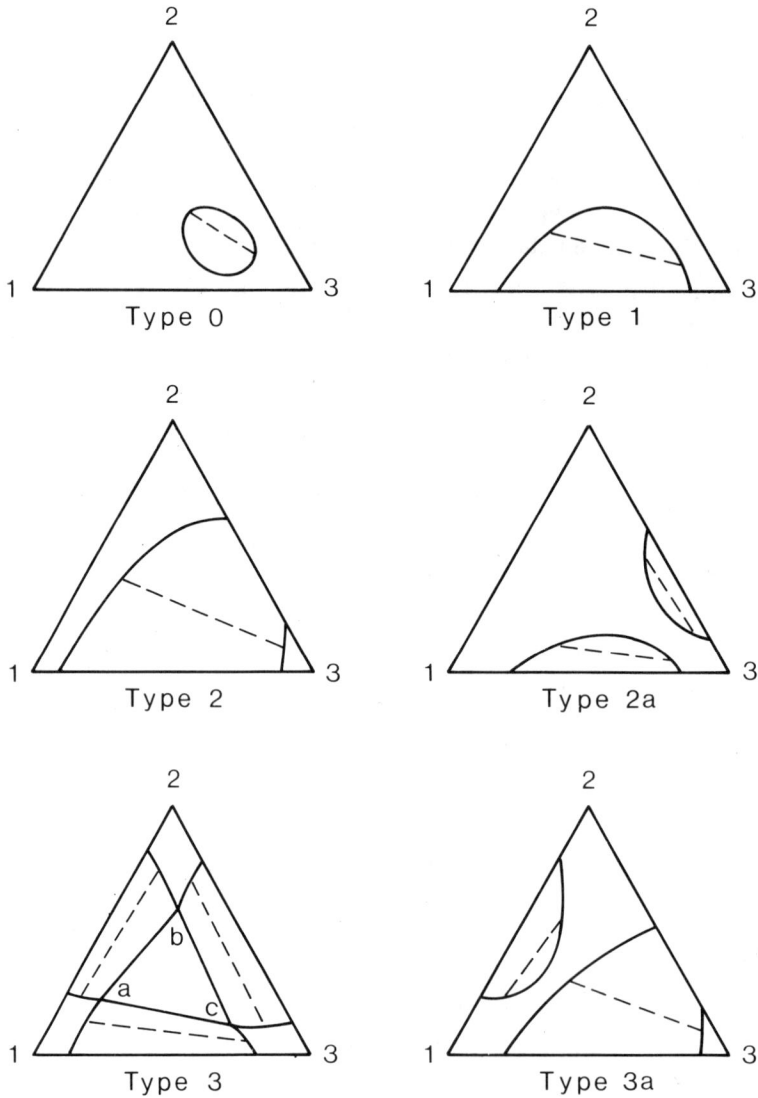

Figure 1 Types of ternary liquid-liquid equilibria

The dashed lines connecting two points on a binodal curve are tie lines giving the concentrations of two liquid phases in equilibrium. Type 3 (but not type 3a) may have three coexisting liquid phases (point a, b, and c). Each quaternary system belongs to one of the four types shown in **Figure 2.**

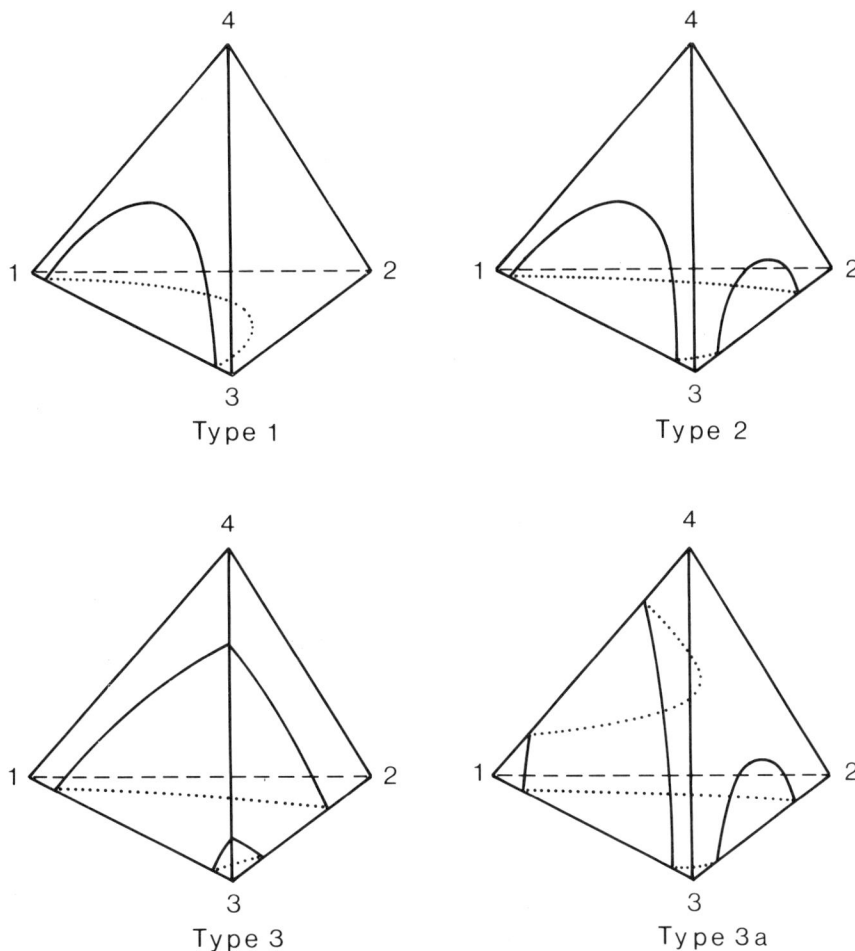

Figure 2 Types of quaternary liquid-liquid equilibria

Ternary Data Obtained by Graphical Interpolation

We present tie line data, i. e. the equilibrium concentrations in both phases. Many investigators report a binodal curve in tabular form together with co-existing concentrations of only one component. In those cases we have determined the remaining concentrations by graphical interpolation. Such references are in the data collection denoted by "GRAPH. INTERPOL.".

Types of Components

The components are non-electrolytes except for organic acids and bases. The components consist of two or more of the following atoms: C, H, Br, Cl, F, N, O, and S. Polymers and components with normal boiling points below 0 °C are not included.

Purity of Components

Liquid-liquid equilibria are often very sensitive to even small impurities in the components. A systematic attempt to specify the component purity has not been made in this work. However, some references which were obviously based on impure components have been excluded.

Temperature and Pressure

The temperature range is roughly 0–150 °C. In certain cases data beyond this range are included.

The data collection only contains low pressure data, i. e. below approximately 20 bar. The pressure is not registered in the data collection since it has very little influence on liquid-liquid equilibria under moderate conditions. In fact, the pressure is most often not even reported in the original publications.

Exclusion of Erroneous Data

Data which are obviously erroneous — most often when the sum of the concentrations in a given phase differs significantly from 100 mole per cent — are excluded from the data collection.

2. Thermodynamic Conditions of Multicomponent Liquid-Liquid Equilibria

The thermodynamic conditions of multicomponent liquid-liquid equilibria cannot be illustrated graphically as simply as for binary systems, but the criteria of equilibrium are the same.

The **necessary and sufficient** criterion of equilibrium is that the molar Gibbs energy of mixing $\triangle G$ for the system is minimum.

A **necessary but not sufficient** criterion is that the activity a_i for each component must be the same in the two phases:

$$a_i^I = a_i^{II} \qquad i = 1,2,...,N \text{ (components)}$$

As explained in Part I, p. XII, this "isoactivity" criterion does not help us in distinguishing between a maximum, an inflection point, a local minimum, and a global minimum of $\triangle G$ for the system.

Binary liquid-liquid equilibrium compositions calculated using the parameters of Part 1 correspond to one and only one solution of the isoactivity criterion, see Part 1, p. XII.

If the parameters of Parts 1, 2, and 3 are used to predict multicomponent LLE, there may be multiple solutions to the isoactivity criterion. The only correct solution is that corresponding to the global minimum of $\triangle G$ for the system. It is easy to discard the solutions corresponding to maxima and inflection points by a check for convexity of the predicted $\triangle G$-surface at the points found by the isoactivity criterion, see (Sørensen et al., 1979b). The problem is thus the distinction between global and local minima. It is very time-consuming to check whether a minimum is global or local. This requires the knowledge of the predicted $\triangle G$ at all points of the concentration space. We have therefore tried to reduce the risk of multiple solutions to the isoactivity criterion. This risk is influenced by the numerical size of the model parameters. Parameters with relatively large numerical values make the isoactivity equations more unlinear thus increasing the risk of multiple solutions. We have therefore introduced a constraint into the objective function used in the reduction of the experimental ternary data of Parts 2 and 3, see Section 7. This constraint tends to lower the numerical size of the estimated parameters. This can be done without seriously increasing the residual of the objective function because the model parameters are mutually very correlated. According to the experience of the authors this almost eliminates the risk of multiple solutions to the isoactivity criterion.

3. Comparison between Different References

Classical thermodynamics offers no method to evaluate the quality of a given set of tie line data as no true consistency test exists for LLE. We have instead, where possible, compared data from different references. Where data for the same system have been reported at temperatures between 20 °C and 30 °C in more than one reference, we have plotted the experimental distribution ratios in one diagram.

4. Models for Liquid-Liquid Equilibria

The two models used in this work are NRTL (Renon and Prausnitz, 1968) and UNIQUAC (Abrams and Prausnitz, 1975).

The NRTL Equation

$$\frac{G^E}{RT} = \sum_i x_i \frac{\sum_j \tau_{ji} G_{ji} x_j}{\sum_k G_{ki} x_k} \quad i,j,k, = 1,2,...,N \text{ (components)}$$

$$\tau_{ji} = \frac{g_{ji} - g_{ii}}{RT} \qquad G_{ji} = \exp(-\alpha_{ji} \tau_{ji})$$

$$g_{ji} = g_{ij} \qquad \alpha_{ji} = \alpha_{ij}$$

The UNIQUAC Equation

$$\frac{G^E}{RT} = \frac{G^E(\text{combinatorial})}{RT} + \frac{G^E(\text{residual})}{RT}$$

$$\frac{G^E(\text{combinatorial})}{RT} = \sum_i x_i ln \frac{\Phi_i}{x_i} + \frac{z}{2} \sum_i q_i x_i ln \frac{\Theta_i}{\Phi_i}$$

$$\frac{G^E(\text{residual})}{RT} = - \sum_i q_i x_i ln \sum_j \Theta_j \tau_{ji} \quad i,j = 1,2,...,N \text{ (components)}$$

$$\Theta_i = \frac{x_i q_i}{\sum_j x_j q_j} \qquad \Phi_i = \frac{x_i r_i}{\sum_j x_j r_j}$$

$$\tau_{ji} = \exp(- \frac{u_{ji} - u_{ii}}{RT}) \qquad u_{ji} = u_{ij} \qquad z = 10$$

Throughout this work we define the parameters as follows:

For NRTL: $a_{ij} = (g_{ij} - g_{jj})/R$ Kelvin

For UNIQUAC: $a_{ij} = (u_{ij} - u_{jj})/R$ Kelvin

Note: In the Vapor-Liquid Equilibrium Data Collection of this series (Gmehling, Onken and Arlt, ab 1977) the parameter unit is cal/mole.

5. Computation of Multicomponent Liquid-Lquid Equilibria

Multicomponent LLE compositions may be calculated using the above mentioned models. At a given temperature and a given total composition, we find the two sets of compositions $(n_1^I, n_2^I,...,n_N^I)$ and $(n_1^{II}, n_2^{II},...,n_N^{II})$, for which ΔG for the system is minimum. The function

$$(n^I + n^{II}) \Delta G = n^I \Delta G^I + n^{II} \Delta G^{II}$$

may be minimized under the constraints.

$$n_i^I + n_i^{II} = n_i \quad i = 1,2,...,N \text{ (components)}$$

where n_i is the total number of moles of component i. This procedure corresponds to the **necessary and sufficient** condition of equilibrium mentioned in Section 2.

It is computationally more adequate to use the **necessary but not sufficient** criterion

$$a_i^I = a_i^{II} \quad i = 1,2,...,N$$

still under the constraints

$$n_i^I + n_i^{II} = n_i \quad i = 1,2,...,N$$

We thus have 2N equations with 2N unknowns. Instead of specifying the total composition, we may specify the concentration of N-2 components in one of the phases and then solve

$$a_i^I = a_i^{II} \quad i = 1,2,...,N$$

under the constraints

$$\Sigma_i x_i^I = 1 \text{ and } \Sigma_i x_i^{II} = 1 \quad i = 1,2,...,N$$

When LLE compositions are computed using the isoactivity criterion, it is essential to perform a check for convexity of the $\triangle G$-surface at the predicted phases in order to avoid maxima and inflection points. It may thus be ensured that the solution corresponds to a minimum, either global or local.

6. Computation of Binodal Curves for Ternary Systems

Using a given set of UNIQUAC or NRTL parameters a binodal curve for a ternary system is constructed by establishing a series of tie lines by the method of the previous section. If these tie lines are spaced throughout the two-phase region, the entire binodal curve is readily drawn.

In practice, we first establish the tie line with no solute, i.e. on the base line of a triangular diagram. We next predict a tie line a little above the base line. This is done by specifying the concentration of one of the components in one of the phases. For example, the concentration of the solute, i.e. component (2), may be fixed at 2 mole per cent in phase I. The five unknowns x_1^I, x_3^I, x_1^{II}, x_2^{II}, and x_3^{II} are then found by solving the five equations

$$a_i^I = a_i^{II} \qquad\qquad i = 1,2,3$$
$$x_1^I + 0.02 + x_3^I = 1$$
$$x_1^{II} + x_2^{II} + x_3^{II} = 1$$

using Newton-Raphson iteration. We continue in this way until the plait point (type 1) or a side line (type 2) is reached. For each calculated tie line we check for convexity as explained in the previous section. A computer program for the binodal curve construction is listed in this work.

Further description of the binodal curve construction is given in (Fredenslund et al., 1980).

7. Parameter Estimation from Ternary Liquid-Liquid Equilibrium Data

The purpose of a parameter estimation is to find model parameters from experimental equilibrium data.

Since a ternary system is described by six UNIQUAC or NRTL parameters, we need at least six equations to estimate these. The minimum experimental information is two tie lines, each giving rise to three isoactivity equations. Most often a ternary data set contains more than two experimental tie lines at a given temperature. Therefore the parameter estimation involves the minimization of an objective function. The experimental tie lines are not reproduced exactly by the estimated parameters. Instead a smoothed fit is obtained. The result of the fit depends on the choice of objective function. Let us consider the activity objective function

$$F_a = \sum_l \sum_k \sum_i (a_{ikl}^I - a_{ikl}^{II})^2 \qquad\qquad (1)$$

and the concentration objective function

$$F_x = \sum_l \sum_k \min \sum_i \sum_j (x_{ijkl} - \hat{x}_{ijkl})^2 \qquad\qquad (2)$$

a = activity obtained directly from the model by insertion of the experimental concentrations

x = experimental concentration

$\hat{x}$ = concentration of the predicted tie line lying closest to the considered experimental tie line

i = 1,2,3 (components)

j = I, II (phases)

k = 1,2,...,M (tie lines)

l = 1,2,...,ND (data sets)

We start the parameter estimation using F_a since this requires no qualified guess at the parameters. After convergence we shift to F_x which is in agreement with our goal, i.e. a fit to the experimental concentrations.

The computation of F_x is rather time consuming since it requires a number of liquid-liquid equilibrium computations for each current parameter estimate. In practice, we construct a series of tie lines such as explained in the previous section and then interpolate in these in order to find the closest predicted tie line for each experimental one. The estimation procedure is described in (Sørensen, 1980 and Magnussen, 1980).

In Section 2 we mentioned that the risk of multiple solutions to the isoactivity criterion is dependent upon the numerical size of the model parameters. Therefore, the following "penalty" term is added to F_a and F_x:

$$Q \sum_n p_n^2$$

$Q =$ constant

$p_n =$ parameter value

$n = 1,2,...,NP$ (parameters)

Q is chosen empirically such that we get relatively small parameter values without increasing the minimum of (1) and (2) considerably. The penalty term also has the advantage that the minimum of the objective function becomes sharper, thus promoting the convergence.

One more term concerning distribution ratios is added to F_x, see Section 10.

8. Specific Parameters

"Specific" parameters are UNIQUAC or NRTL parameters fitted individually to each ternary system. If several references report data fo a system ABC at temperatures between 20 °C and 30 °C, the data are combined to give only one set of parameters. For data outside this temperature range a specific set of parameters is given for each data set. Specific parameters have been estimated for all ternary systems of type 1 or 2 with five experimental tie lines or more if these cover the immiscibility region reasonably well.

9. Common Parameters

„Common" parameters for the AB interaction are UNIQUAC parameters which may be used in LLE calculations for any mixture containing components A and B. The common parameters for the AB interaction are the same in the systems ABC, ABD, and any other system which contains components A and B, while the specific parameters for the AB interaction are different in the systems ABC and ABD. Data for nearly all ternary systems of type 1 and 2 between 20 °C and 30 °C were used in the estimation of common parameters. Theoretically the common parameters should be estimated in one parameter estimation; but since we have about 600 ternary data sets between 20 °C and 30 °C, the estimation has to take place in steps following a certain hierarchy. The parameters were estimated for one pair of components at time. The component pairs were given priorities according to the number of systems in which they appear. One exception to this rule was made: We did not use ternary data to establish common parameters for those binaries which are reported in Part 1 between 20 ° C and 30 °C. The parameters of Part 1 (closest to 25 °C) were used as common parameters.

Example: Estimation of common parameters for water-N,N-dimethylaniline. This binary is involved in two ternary data sets together with acetic acid and furfural respectively.

(1) Water

(2) N,N-dimethylaniline

(3a) Acetic acid

(3b) Furfural

The 1-3a parameters were estimated at an earlier stage since water and acetic acid are present in many ternary systems. The 1-3b parameters are known from Part 1. The 1-2, 2-3a, and 2-3b parameters are estimated at this stage. The 2-3a and 2-3b binaries are only involved in the two data sets used here. The parameters for these binaries are therefore automatically "common".

10. Distribution Ratios

The objective function F_x of Section 7 minimizes the absolute deviations between experimental and calculated mole fractions. Errors on small concentrations give only a small contribution to F_x even if these errors are relatively large. Since we are interested in a correct reproduction of the distribution ratio of the solute at small concentrations, we add the following term to F_x:

$$\sum_l \left[ln\left(\frac{\hat{\gamma}_{Sl\infty}^{I}}{\hat{\gamma}_{Sl\infty}^{II}} \cdot D_{Sl\infty} \right) \right]^2$$

$l = 1,2,...,ND$ (data sets)

$\hat{\gamma}_{Sl\infty}^{I}$ is the predicted activity coefficient of the solute S at infinite dilution in phase I of data set number l. $D_{Sl\infty}$ is the wanted ratio between the solute concentrations in phase I and phase II at infinite dilution and must be given in the input to the estimation program. This contribution to the objective function will vanish if the predicted distribution ratio is equal to $D_{Sl\infty}$ since $\hat{\gamma}_{Sl\infty}^{I}/\hat{\gamma}_{Sl\infty}^{II}$ is the reciprocal of the predicted distribution ratio.

By fixing the distribution ratio at infinite dilution to a wanted value, we have been able to improve the predicted distribution ratios at small solute concentrations considerably. This has been done for the specific as well as for the common parameters.

11. Prediction of Multicomponent Liquid-Liquid Equilibria

(Continued from Part 1, p. XV)

In Part 1, p. XV we described how parameters for miscible binaries may be obtained from different sources:

a. From ternary LLE.

b. From VLE.

c. From a UNIFAC table based on VLE.

d. From ternary LLE at another temperature (i.e. temperature extrapolation).

In Part 1, p. XVI we showed predicted results according to points b and c. We shall now show predictions following points a and d, and one more parameter source will be demonstrated:

e. A UNIFAC table based on LLE (Magnussen et al., 1980).

12. Examples of Predictions

Predictions following Sections 11e and 11a will now be performed for the system water(1)-ethanol(2)-benzene(3) at 45 °C.

Prediction Using UNIFAC Based on LLE

From Part 1 of this work we can get UNIQUAC parameters for water-benzene at 45 °C. Using a UNIFAC table based on LLE (Magnussen et al., 1980), we can calculate UNIQUAC parameters for the two miscible component pairs. **Figure 3** shows the predicted results compared with experimental data at 45 °C.

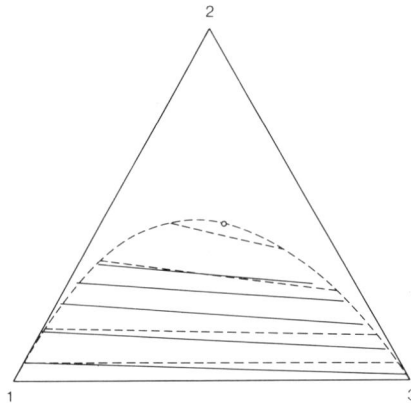

Figure 3. The system water(1)-ethanol(2)-benzene(3) at 45 °C

   - - - - - -   Predicted binodal curve and tie lines using UNIFAC (LLE)
      O   Predicted plait point using UNIFAC (LLE)
   — — —   Experimental tie lines (Morachevskii et al., 1958)

The quality of the prediction is considerably improved compared to UNIFAC based on VLE, see Part 1, p. XVII.

Prediction Using Ternary LLE Data

The parameters for water-benzene at 45 °C are taken from Part 1. The parameters for the miscible component pairs are taken from the list of common UNIQUAC parameters presented in Part 3. The predicted results are shown on **Figure 4.**

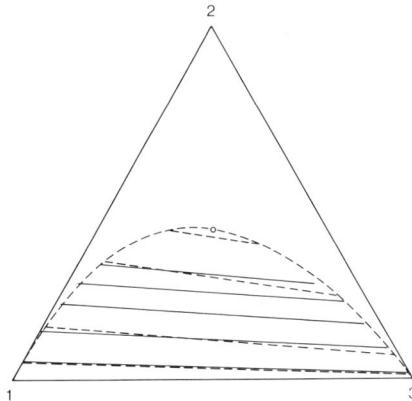

Figure 4. The system water(1)-ethanol(2)-benzene(3) at 45 °C

  - - - - - - Predicted binodal curve and tie lines using UNIQUAC with common parameters

  O Predicted plait point using UNIQUAC with common parameters

  — — — — Experimental tie lines (Morachevskii et al., 1958)

The slope of the tie lines near the base line is improved on Figure 4 compared to Figure 3.

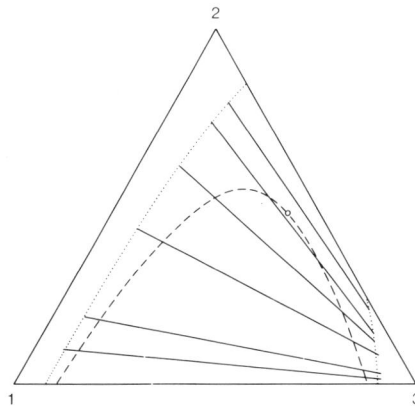

Figure 5. The system hexane(1)-methylcyclopentane(2)-aniline(3) at 25 °C

  - - o - - Predicted binodal curve and plait point using UNIQUAC with parameters valid at 45 °C

  . . . . . . Predicted binodal curve using UNIQUAC with (1)-(2) parameters valid at 45 °C and the remaining parameters valid at 25 °C

  — — — — Experimental tie lines (Darwent and Winkler, 1943)

Temperature Extrapolation

A prediction following Section 11d will be shown for the system hexane(1)-methylcyclopentane(2)-aniline(3). Specific parameters valid at 45 °C are available. These parameters may be used to predict equilibrium data for the system at 25 °C but the result is poor. An improved prediction is obtained if the (1)-(3) and (2)-(3) parameters are replaced by parameters valid at 25 °C. Such parameters may be found in Part 1 of this work.

Figure 5 shows predicted binodals compared with experimental tie lines at 25 °C.

13. Prediction of Quaternary Systems

In order to test the common UNIQUAC parameters we have used them to predict the reported quaternary systems. These were not used in the estimation of the common parameters. In those cases where common parameters are not available for a certain component pair, we have constructed the missing parameters from a table of UNIFAC parameters based on LLE data (Magnussen et al., 1980). The results is:

$$100 \sum_{l} \sqrt{\sum_{k} \min \sum_{i} \sum_{j} (x_{ijkl} - \hat{x}_{ijkl})^2 / 8M} / ND = 2.83 \text{ mole pct.}$$

$i = 1,2,3,4$ (components)
$j = I, II$ (phases)
$k = 1,2,...,M$ (tie lines)
$l = 1,2,...,ND$ (data sets)

The results are presented in detail in (Sørensen, 1980).

14. Conclusion

The NRTL and UNIQUAC models may often be used to smoothen experimental ternary LLE data within the accuracy of the measurements. However, for systems with a complicated distribution ratio for the solute as a function of concentration, e.g. many systems with 1- or 2-propanol, the reproduction of the distribution ratios is less quantitative.

The models may also be used to predict multicomponent LLE data from binary and ternary information. The parameters should always be based on data resembling the data to be predicted as much as possible, i.e. parameters based on LLE data are preferable to parameters based on VLE data. Normally the common UNIQUAC parameters are to be preferred for the prediction of multicomponent systems near room temperature.

REFERENCES TO LITERATURE

Abrams, D. S. and J. M. Prausnitz, AIChE J. 21 (1975) 116

Darwent, B. B. and C. A. Winkler, J. Phys. Chem. 47 (1943) 442

Fredenslund, Aa., M. L. Michelsen, and J. M. Sørensen, 2nd International Conference on Phase Equilibria and Fluid Properties in the Chemical Industry, Berlin (West), 1980

Gmehling, J., U. Onken, and W. Arlt, "Vapor-Liquid Equilibrium Data Collection", DECHEMA Chemistry Data Series, Frankfurt, ab 1977

Magnussen, T., P. Rasmussen, and Aa. Fredenslund, to be published 1980

Magnussen, T., Ph. D. thesis, Instituttet for Kemiteknik, Lyngby, 1980

Morachevskii, A. G. and V. P. Belousov, Vestn. Leningr. Univ. Ser. Mat. Fiz. Khim. 13,4 (1958) 117

Renon, H., and J. M. Prausnitz, AIChE J. 14 (1968) 135

Sørensen, J. M., T. Magnussen, P. Rasmussen, and Aa. Fredenslund, Fluid Phase Equilibria 2 (1979) 297

Sørensen, J. M., T. Magnussen, P. Rasmussen, and Aa. Fredenslund, **ibid,** 3 (1979) 47

Sørensen, J. M., Ph. D. thesis, Instituttet for Kemiteknik, Lyngby, 1980

GUIDE TO TABLES, FIGURES, LIST OF COMMON UNIQUAC PARAMETERS, AND INDEX

1. Guide to Ternary Tables

Order of Succession of Systems

The systems are ordered according to the chemical formulae of the components. That is, the order follows increasing numbers of C- and H-atoms in the molecules and similarly for the remaining elements in alphabetical order. Components with identical formulae are ordered alphabetically according to their names. If several references exist for a given system, the references are ordered chronologically. Often the same data are published in several references. We only report one of these, usually the original.

Order of Succession of Components

Within each system the three components are positioned as on Figure 1 of the introduction.

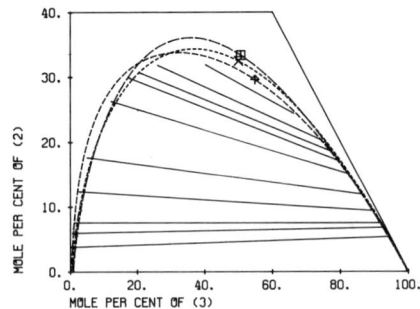

```
a---  (1) C10H22     DECANE
      ----------------------------------------
      (2) C8H18O     1-OCTANOL
      ----------------------------------------
      (3) C4H10O3    DIETHYLENE GLYCOL

b---  ITSIKSON L.B., BORUKHOVA M.S.
      ZH.PRIKL.KHIM.(LENINGRAD) 42(1969)1363          ----c

      TEMPERATURE = 30.0 DEG C      TYPE OF SYSTEM = 1

      EXPERIMENTAL TIE LINES IN MOLE PCT
      ----------------------------------------
            LEFT PHASE              RIGHT PHASE
        (1)     (2)     (3)      (1)     (2)     (3)

d---  95.792   3.808   0.401    0.151   5.364  94.485
      93.365   5.969   0.666    0.379   6.794  92.827
      91.099   7.573   1.328    0.456   7.470  92.075
      85.015  12.349   2.636    0.687   9.342  89.971
      77.285  17.632   5.084    1.230  11.847  86.923
      61.539  26.274  12.187    2.332  15.030  82.638
      52.654  29.883  17.463    3.527  17.127  79.345
      48.871  30.728  20.400    3.936  18.491  77.574
      42.171  31.807  26.022    5.237  20.287  74.476
      27.860  31.867  40.274    9.324  24.361  66.315

      SPECIFIC MODEL PARAMETERS IN KELVIN
      ----------------------------------------
                 UNIQUAC              NRTL(ALPHA=.2)      ----f
e---  I J    AIJ        AJI        AIJ        AJI

      1 2   482.83    -254.49     1291.7    -642.82
      1 3   333.80      98.934    1170.4    1440.7
      2 3   142.93     -86.902     439.71   -185.06

g---  R1 = 7.1974   R2 = 6.1519   R3 = 4.0013
      Q1 = 6.016    Q2 = 5.212    Q3 = 3.568

      MEAN DEV. BETWEEN CALC. AND EXP. CONC. IN MOLE PCT
      ----------------------------------------
h---  UNIQUAC (SPECIFIC PARAMETERS)      1.12
      NRTL    (SPECIFIC PARAMETERS)      0.56
      UNIQUAC (COMMON PARAMETERS)        1.19
```

a. Component Names

The nomenclature used in this work follows "Handbook of Chemistry and Physics", 58th ed., The Chemical Rubber Co., Ohio 1977-78. With some exceptions this is the same as the recommended IUPAC nomenclature. We do, however, make one small exception to the Chemical Rubber nomenclature: We use the expression "perfluoro" when all the hydrogen atoms in a molecule or atom group are substituted by fluorine atoms.

b. Literature Citation

The journal names follow Chemical Abstracts Service Source Index, 1907-1974 Cumulative, or later editions.

c. Type of System

See Figure 1 of the introduction

d. Experimental Data

If the statement "GRAPH. INTERPOL." appears after the text "EXPERIMEN-TAL TIE LINES IN MOLE PCT" the original publication contained a binodal curve in tabular form together with coexisting concentrations of only one component. In this case we have determined the remaining concentrations by graphical interpolation.

e. Specific Model Parameters

The specific parameters are fitted individually to each data set. Yet if several references report data for the same system at temperatures between 20 °C and 30 °C, the data are combined to give only one set of parameters. Specific parameters are presented only for systems of type 1 or 2 with at least five experimental tie lines and only if these cover the immiscibility region reasonably well.

f. α of the NRTL Model

$\alpha = 0.2$ is used throughout this work

g. r and q of the UNIQUAC Model

r and q are pure-component parameters. If parameters of this work are combined with parameters from other sources (e.g. for prediction purposes), the r- and q-values must be the same. This work and Volume I of the same series (Gmehling, Onken, and Arlt, ab 1977) use the same r- and q-values. A computer program for converting UNIQUAC parameters based on one set of r- and q-values to UNIQUAC parameters valid for a different set of r- and q-values is found on the pages following this guide.

h. Mean deviation between calculated and experimental concentrations.

The mean deviation between calculated and experimental concentrations is expressed in mole pct. as

$$100 \sqrt{\sum_k \min \sum_i \sum_j (x_{ijk} - \hat{x}_{ijk})^2 / 6M}$$

x = experimental concentration

$\hat{x}$ = concentration of the calculated tie line lying closest to the considered experimental tie line

i = 1,2,3 (components)

j = I,II (phases)

k = 1,2,...,M (tie lines)

The deviation is given for specific (see point **e**) as well as for common parameters. The common parameters for the AB interaction are the same in all systems with components A and B. Common parameters have been established for nearly all the reported ternary systems between 20 °C and 30 °C. The common parameters are listed in Part 3 before the index to the ternary systems.

2. Guide to Ternary Figures

Calculated data are shown for systems of type 1 or 2 if specific or common parameters have been established. Where experimental data for the same system have been reported at temperatures between 20 °C and 30 °C in more than one reference, the lower figure contains the experimental distribution ratios from all the references.

3. Guide to Quaternary Tables

The layout is equivalent to that for ternary systems with the exception that the systems are ordered chronologically. Within each system the components are positioned as on Figure 2 of the introduction.

4. Guide to List of Common UNIQUAC Parameters

For each pair of parameters the data base is indicated. Parameters based on binary data have been taken directly from Part 1. An equilibrium calculation will reproduce the binary immiscibility gap quantitatively. This is often not the case if the parameters for an immiscible binary have been reduced from ternary data.

5. Guide to Index

The alphabetical component index for ternary systems includes in addition to the nomenclature used in this work — the nomenclature used in Volume I of this series (Gmehling, Onken, and Arlt, ab 1977) and other trivial names.

COMPUTER PROGRAMS

Two computer programs are listed in the following:

a) UNIQUAC r- and q-conversion and b) Binodal Curve Construction.
The programs are written in FORTRAN IV G. Sample inputs and outputs are shown.

a. UNIQUAC r- and q-conversion

UNIQUAC interaction parameters taken from other sources, e.g. constructed from a UNIFAC table, should not be used directly together with parameters from this work unless they are based on the same r- and q-values.

If interaction parameters corresponding to a given set of r- and q-values are available, a new set of parameters corresponding to our set of r- and q-values may be found by a computer program listed in the following. This work and Volume I of the same series (Gmehling, Onken, and Arlt, ab 1977) use the same r- and q-values.

b. Binodal Curve Construction

All the computed ternary LLE data in this work shown on figures. Binodal curves in tabular form may be generated by a computer program listed in the following. With a given set of UNIQUAC or NRTL parameters the program establishes a grid of tie lines covering the whole immiscibility region. The distance between the computed tie lines is specified by the user. The program is listed with permission by M. L. Michelsen (Instituttet for Kemiteknik, Lyngby).

```
C
C ***********************************************************************
C *
C * UNIQUAC R- AND Q-CONVERSION BASED ON ACTIVITY COEFFICIENT
C * AT INFINITE DILUTION
C *
C ***********************************************************************
C
C    INPUT
C
C    A. 2 CARDS: ROLD,QOLD,RNEW,QNEW,TEXT   (4F10.,40A1)
C
C        ROLD = OLD R-VALUE
C        QOLD = OLD Q-VALUE
C        RNEW = NEW R-VALUE
C        QNEW = NEW Q-VALUE
C        TEXT = COMPONENT NAME
C               THE TWO CARDS ARE ONE FOR EACH COMPONENT
C
C    B. 1 CARD: TC,A12OLD,A21OLD   (3F10.)
C
C        TC     = TEMPERATURE, DEG C
C        A12OLD = OLD UNIQUAC INTERACTION PARAMETER 1-2 IN KELVIN
C        A21OLD = OLD UNIQUAC INTERACTION PARAMETER 2-1 IN KELVIN
C                 A12OLD AND A21OLD CORRESPOND TO ROLD AND QOLD
C
      LOGICAL*1 TEXT
      DIMENSION ROLD(2),QOLD(2),RNEW(2),QNEW(2),XLOLD(2),XLNEW(2),
     *TEXT(2,40)
      READ(5,100)(ROLD(I),QOLD(I),RNEW(I),QNEW(I),(TEXT(I,J),J=1,40),
     *I=1,2)
100   FORMAT(4F10.1,40A1)
      READ(5,100)TC,A12,A21
      WRITE(6,909)
909   FORMAT(1H1)
      WRITE(6,910)
910   FORMAT(' UNIQUAC R- AND Q-CONVERSION BASED ON ACTIVITY COEFFICIENT
     *',/,' AT INFINITE DILUTION',/)
      WRITE(6,912)(I,(TEXT(I,J),J=1,40),ROLD(I),QOLD(I),RNEW(I),QNEW(I),
     *I=1,2)
912   FORMAT(/,48X,'ROLD',3X,'QOLD',4X,'RNEW',3X,'QNEW',/,
     *2(/,' (',I1,') ',40A1,2(F8.4,F7.3)))
      WRITE(6,914)TC
914   FORMAT(/,' TEMPERATURE = ',F4.1,' DEG C')
      WRITE(6,915)
915   FORMAT(/,' OLD INTERACTION PARAMETERS IN KELVIN:')
      WRITE(6,916)A12,A21
916   FORMAT(/,4X,'A(1,2)OLD',6X,'A(2,1)OLD',//,3X,G12.5,3X,G12.5)
      TK=TC+273.15
      DO 1 I=1,2
      XLOLD(I)=5.*(ROLD(I)-QOLD(I))-(ROLD(I)-1.)
1     XLNEW(I)=5.*(RNEW(I)-QNEW(I))-(RNEW(I)-1.)
      TAU12=EXP(-A12/TK)
      TAU21=EXP(-A21/TK)
      R=ROLD(1)/ROLD(2)
      GAM1=ALOG(R)+5.*QOLD(1)*ALOG(QOLD(1)/QOLD(2)/R)+XLOLD(1)-
     *R*XLOLD(2)+QOLD(1)*(1.-ALOG(TAU21)-TAU12)
      R=1./R
      GAM2=ALOG(R)+5.*QOLD(2)*ALOG(QOLD(2)/QOLD(1)/R)+XLOLD(2)-
```

```
      *R*XLOLD(1)+QOLD(2)*(1.-ALOG(TAU12)-TAU21)
       R=RNEW(1)/RNEW(2)
       F2=GAM1-ALOG(R)-5.*QNEW(1)*ALOG(QNEW(1)/QNEW(2)/R)-XLNEW(1)+
      *R*XLNEW(2)
       R=1./R
       F1=GAM2-ALOG(R)-5.*QNEW(2)*ALOG(QNEW(2)/QNEW(1)/R)-XLNEW(2)+
      *R*XLNEW(1)
       F1=1.-F1/QNEW(2)
       F2=1.-F2/QNEW(1)
10     TAU21=EXP(F2-TAU12)
       F=ALOG(TAU12)+TAU21-F1
       DF=TAU21-1./TAU12
       DTAU12=F/DF
       TAU12=TAU12+DTAU12
       IF(ABS(DTAU12).GT.1.E-4)GOTO 10
       A12NEW=ALOG(TAU12)*(-TK)
       A21NEW=ALOG(TAU21)*(-TK)
       WRITE(6,815)
815    FORMAT(/,' NEW INTERACTION PARAMETERS IN KELVIN:')
       WRITE(6,816)A12NEW,A21NEW
816    FORMAT(/,4X,'A(1,2)NEW',6X,'A(2,1)NEW',//,3X,G12.5,3X,G12.5)
       WRITE(6,909)
       STOP
       END
```

```
*************************************************************************
*
* UNIQUAC R- AND Q-CONVERSION BASED ON ACTIVITY COEFFICIENT
* AT INFINITE DILUTION
*
* SAMPLE INPUT
*
*************************************************************************


4.7817    3.6      3.1878    2.4      BENZENE
.92       1.4      .92       1.4      WATER
25.       1056.1   96.272


*************************************************************************
*
* UNIQUAC R- AND Q-CONVERSION BASED ON ACTIVITY COEFFICIENT
* AT INFINITE DILUTION
*
* SAMPLE OUTPUT
*
*************************************************************************
```

UNIQUAC R- AND Q-CONVERSION BASED ON ACTIVITY COEFFICIENT
AT INFINITE DILUTION

	ROLD	QOLD	RNEW	QNEW
(1) BENZENE	4.7817	3.600	3.1878	2.400
(2) WATER	0.9200	1.400	0.9200	1.400

TEMPERATURE = 25.0 DEG C

OLD INTERACTION PARAMETERS IN KELVIN:

A(1,2)OLD	A(2,1)OLD
1056.1	96.272

NEW INTERACTION PARAMETERS IN KELVIN:

A(1,2)NEW	A(2,1)NEW
860.85	368.97

```
C
C  ***********************************************************************
C  *
C  * BINODAL CURVE CONSTRUCTION
C  *
C  ***********************************************************************
C
C  INPUT
C
C  A. 1 CARD: MODEL,TC,STEP,Y1,Y3  (I1,9X,4F10.)
C
C     MODEL = 1, UNIQUAC
C           = 2, NRTL
C     TC    = TEMPERATURE, DEG C
C     STEP  = STEP LENGTH USED IN THE BINODAL CURVE CONSTRUCTION,
C             MOLE FRACTION.  0.05>STEP>0.01.  IF STEP IS EQUALED
C             TO ZERO, IT IS AUTOMATICALLY SET TO 0.02
C     Y1    = INITIAL GUESS AT THE SOLUBILITY OF COMPONENT (1)
C             IN COMPONENT (3), MOLE PCT. MAY BE EQUALED TO ZERO
C             EXCEPT FOR HIGHLY UNSYMMETRIC SYSTEMS
C     Y3    = THE SAME AS Y1, BUT (3) IN (1)
C
C  B. 3 CARDS: R,Q,TEXT  (2F10.,40A1)
C
C     R    = UNIQUAC R-VALUE, NOT USED FOR NRTL
C     Q    = UNIQUAC Q-VALUE, NOT USED FOR NRTL
C     TEXT = COMPONENT NAME
C            THE FIRST CARD IS FOR COMPONENT (1) ETC. THE COMPONENTS
C            MUST BE POSITIONED AS ON FIGURE 1 OF THE INTRODUCTION
C
C  C. 1 CARD: A(1,2),A(2,1),ALPHA(1,2)  (3F10.)
C
C     A(1,2)     = UNIQUAC OR NRTL INTERACTION PARAMETER 1-2 IN KELVIN
C     A(2,1)     = UNIQUAC OR NRTL INTERACTION PARAMETER 2-1 IN KELVIN
C     ALPHA(1,2) = NON-RANDOMNESS PARAMETER 1-2 FOR NRTL,
C                  NOT USED FOR UNIQUAC. IF ALPHA(1,2) IS EQUALED
C                  TO ZERO, IT IS AUTOMATICALLY SET TO 0.2
C
C  D. 1 CARD: A(1,2),A(3,1),ALPHA(1,3)  (3F10.)
C
C  E. 1 CARD: A(2,3),A(3,2),ALPHA(2,3)  (3F10.)
C
      IMPLICIT REAL*8 (A-H,O-Z)
      LOGICAL*1 TEXT
      COMMON Q(3),R(3),PAR(3,3),TAU(3,3),G(3,3),ALP(3,3),TCK
      DIMENSION Y(6),YOLD(4),DY(4),DYOLD(4),DMAT(4,4),TEXT(3,40),P(3,3)
      READ(5,799)MODEL,TC,STEP,Y1,Y3
799   FORMAT(I1,9X,4F10.1)
      READ(5,800)(R(I),Q(I),(TEXT(I,J),J=1,40),I=1,3)
800   FORMAT(2F10.1,40A1)
      READ(5,801)P(1,2),P(2,1),ALP(1,2),P(1,3),P(3,1),ALP(1,3),P(2,3),
     *P(3,2),ALP(2,3)
801   FORMAT(3F10.1)
      IF(DABS(ALP(1,2)).LT.1.D-10)ALP(1,2)=.2
      IF(DABS(ALP(1,3)).LT.1.D-10)ALP(1,3)=.2
      IF(DABS(ALP(2,3)).LT.1.D-10)ALP(2,3)=.2
```

```
      WRITE(6,909)
909   FORMAT(1H1)
      IF(MODEL.EQ.1)WRITE(6,910)
910   FORMAT(' BINODAL CURVE CONSTRUCTION USING UNIQUAC',/)
      IF(MODEL.EQ.2)WRITE(6,911)
911   FORMAT(' BINODAL CURVE CONSTRUCTION USING NRTL',/)
      IF(MODEL.EQ.1)WRITE(6,912)(I,(TEXT(I,J),J=1,40),R(I),Q(I),I=1,3)
912   FORMAT(/,49X,'R',7X,'Q',/,3(/,' (',I1,') ',40A1,F8.4,F7.3))
      IF(MODEL.EQ.2)WRITE(6,913)(I,(TEXT(I,J),J=1,40),I=1,3)
913   FORMAT(/,3(/,' (',I1,') ',40A1))
      WRITE(6,914)TC
914   FORMAT(/,' TEMPERATURE = ',F4.1,' DEG C')
      TCK=TC+273.15D0
      WRITE(6,915)
915   FORMAT(/,' INTERACTION PARAMETERS:')
      I1=1
      I2=2
      I3=3
      IF(MODEL.EQ.1)WRITE(6,916)I1,I2,P(1,2),P(2,1),I1,I3,P(1,3),P(3,1),
     *I2,I3,P(2,3),P(3,2)
916   FORMAT(/,'  I J',5X,'A(I,J)',6X,'A(J,I)',/,10X,'KELVIN',6X,
     *'KELVIN',/,3(/,1X,2I2,3X,2G12.5))
      IF(MODEL.EQ.2)WRITE(6,917)I1,I2,P(1,2),P(2,1),ALP(1,2),I1,I3,
     *P(1,3),P(3,1),ALP(1,3),I2,I3,P(2,3),P(3,2),ALP(2,3)
917   FORMAT(/,'  I J',5X,'A(I,J)',6X,'A(J,I)',7X,'ALPHA',/,10X,
     *'KELVIN',6X,'KELVIN',/,3(/,1X,2I2,3X,2G12.5,F9.4))
      ALP(2,1)=ALP(1,2)
      ALP(3,1)=ALP(1,3)
      ALP(3,2)=ALP(2,3)
      DO 120 I=1,3
      P(I,I)=0.D0
      ALP(I,I)=0.D0
      DO 120 J=1,3
      TAU(I,J)=P(I,J)/TCK
      G(I,J)=DEXP(-ALP(I,J)*TAU(I,J))
120   PAR(I,J)=DEXP(-TAU(I,J))
      IF(STEP.LT.1.D-10)STEP=.02D0
      WRITE(6,918)STEP
918   FORMAT(/,' STEP LENGTH USED IN THE BINODAL CURVE CONSTRUCTION = ',
     *2PF4.1,' MOLE PCT')
      WRITE(6,919)Y1,Y3
919   FORMAT(/,' INITIAL GUESS AT MUTUAL SOLUBILITIES OF COMPONENTS (1)
     *AND (3):',//,' (1) IN (3):',F5.1,' MOLE PCT',/,' (3) IN (1):',
     *F5.1,' MOLE PCT')
      WRITE(6,909)
      WRITE(6,920)
920   FORMAT(' COMPUTED BINODAL CURVE:',//,9X,'LEFT PHASE',17X,
     *'RIGHT PHASE',/,2(5X,'(1)      (2)      (3)   '),/)
      DO 1 I=1,4
      DO 1 J=1,4
1     DMAT(I,J)=0.D0
      IRUND=200
      IC=0
      NOLD=2
      N=0
      Y(1)=1.D0-Y3/100.D0
      Y(2)=0.D0
      Y(3)=Y1/100.D0
      Y(4)=0.D0
```

```
C
C CALCULATE TIE LINE N+1
C
12      CALL SOLVE(MODEL,Y,DY,NOLD,NEW,NITER,N)
        IF(NITER.LE.10)GOTO 16
C
C CONVERGENCE FAILURE
C
        IF(N.GT.0)GOTO 19
        WRITE(6,902)
902     FORMAT(' THE BASE LINE CALCULATION DID NOT CONVERGE IN 10 ITERATIO
       *NS')
        WRITE(6,901)
901     FORMAT(' TRY AN INITIAL GUESS AT BASE LINE CONCENTRATIONS Y1 AND Y
       *3',/,' IF THIS DOES NOT HELP, PERHAPS THE BASE LINE PARAMETERS COR
       *RESPOND',/,' TO ONLY ONE LIQUID PHASE')
        WRITE(6,925)
925     FORMAT(' ARE THE COMPONENTS POSITIONED AS ON FIGURE 1 OF THE INTRO
       *DUCTION?')
        GOTO 3
19      IF(IHALF.LT.5)GOTO 20
        WRITE(6,900)
900     FORMAT(' THE NEXT TIE LINE CALCULATION DID NOT CONVERGE IN 10 ITER
       *ATIONS')
        GOTO 3
20      IHALF=IHALF+1
        ST=ST/2.D0
        GOTO 17
C
C CONVERGENCE OBTAINED
C
16      IF(DABS(Y(1)-Y(3))+DABS(Y(2)-Y(4)).GT.1.D-5)GOTO 21
        IF(N.GT.0)GOTO 19
        WRITE(6,903)
903     FORMAT(' THE CALCULATED CONCENTRATIONS ON THE BASE LINE ARE IDENTI
       *CAL IN THE TWO PHASES')
        WRITE(6,901)
        WRITE(6,925)
        GOTO 3
21      N=N+1
        IHALF=0
        Y(5)=1.D0-Y(1)-Y(2)
        Y(6)=1.D0-Y(3)-Y(4)
        WRITE(6,899)Y(1),Y(2),Y(5),Y(3),Y(4),Y(6)
899     FORMAT(1X,2(2P3F8.3,3X))
        IF(IC.EQ.1)GOTO 3
        IF(IC.EQ.2.AND.Y(1).LT.1.D-10)GOTO 3
C
C FIND COEFFICIENTS OF THIRD DEGREE POLYNOMIUM TO EXTRAPOLATE
C BINODAL CURVE
C
        DYMAX=DABS(DY(NEW))
        DO 4 I=1,4
4       DY(I)=DY(I)/DYMAX
        IF(N.EQ.1)GOTO 5
        STAP=DABS(Y(NEW)-YOLD(NEW))
        IF(DY(NEW)*DYOLD(NEW).GT.0.D0)GOTO 6
        DO 7 I=1,4
7       DY(I)=-DY(I)
6       IF(NEW.EQ.NOLD)GOTO 8
```

```
            RR=DY(NEW)/DYOLD(NEW)
            DO 9 I=1,4
9           DYOLD(I)=DYOLD(I)*RR
8           DO 10 I=1,4
            Z=(YOLD(I)-Y(I))/STAP
            DMAT(I,3)=(3.D0*Z+2.D0*DY(I)+DYOLD(I))/STAP
10          DMAT(I,4)=(2.D0*Z+DY(I)+DYOLD(I))/STAP**2
5           ST=RUND(Y(NEW),DY(NEW),STEP,IRUND)
            DO 18 I=1,4
            DMAT(I,1)=Y(I)
            DMAT(I,2)=DY(I)
            YOLD(I)=Y(I)
18          DYOLD(I)=DY(I)
C
C INITIAL GUESS AT NEXT TIE LINE
C
17          DO 11 I=1,4
            Y(I)=DMAT(I,4)
            DO 11 J=1,3
11          Y(I)=ST*Y(I)+DMAT(I,4-J)
            IF(IHALF.GT.0)GOTO 12
C
C CHECK FOR END OF BINODAL CURVE
C
            CALL TERM(Y,DMAT,IC,NEW)
            NOLD=NEW
            IF(IC.EQ.0.OR.IC.EQ.2)GOTO 12
            IF(IC.EQ.1)GOTO 21
            IF(IC.EQ.-2)WRITE(6,898)
898         FORMAT(' COMPONENTS (1) AND (2) ARE PREDICTED TO BE IMMISCIBLE')
            WRITE(6,925)
            IF(IC.EQ.-1)WRITE(6,897)
897         FORMAT(' PLAIT POINT CALCULATION DID NOT CONVERGE')
3           WRITE(6,909)
            STOP
            END
C
C
            SUBROUTINE SOLVE(MODEL,Y,DY,NOLD,NEW,NITER,N)
            IMPLICIT REAL*8(A-H,O-Z)
            DIMENSION Y1(3),Y2(3),ACT1(3),ACT2(3),DACT1(3,3),DACT2(3,3),
           *Y(6),DY(4),AMAT(3,5),INO(3)
            NITER=0
C
C CONVERGE THE TIE LINE CORRESPONDING TO Y(NOLD) AND
C FIND THE DERIVATIVES OF THE CONCENTRATIONS WITH RESPECT TO Y(NOLD)
C
11          NITER=NITER+1
            IF(NITER.GT.10)RETURN
            DO 2 I=1,4
2           IF(Y(I).LT.0.D0)Y(I)=0.D0
            DO 3 I=1,2
            Y1(I)=Y(I)
3           Y2(I)=Y(I+2)
            IF(MODEL.NE.1)GOTO 4
            CALL UNIQ(Y1,ACT1,DACT1)
            CALL UNIQ(Y2,ACT2,DACT2)
            GOTO 5
4           CALL NRTL(Y1,ACT1,DACT1)
            CALL NRTL(Y2,ACT2,DACT2)
```

```
5       J=0
        DO 6 I=1,4
        IF(I.EQ.NOLD)GOTO 6
        J=J+1
        INO(J)=I
6       CONTINUE
        DO 7 I=1,3
        DO 7 J=1,2
        AMAT(I,J)=DACT1(I,J)-DACT1(I,3)
7       AMAT(I,J+2)=DACT2(I,3)-DACT2(I,J)
        DO 8 I=1,3
        AMAT(I,5)=AMAT(I,NOLD)
        DO 9 J=1,3
9       AMAT(I,J)=AMAT(I,INO(J))
8       AMAT(I,4)=ACT1(I)-ACT2(I)
        CALL GAUSL(3,5,3,2,AMAT)
        RES=0.D0
        DO 10 I=1,3
        Y(INO(I))=Y(INO(I))-AMAT(I,4)
        DY(INO(I))=-AMAT(I,5)
10      RES=RES+AMAT(I,4)**2
        IF(RES.GT.1.D-10)GOTO 11
C
C CHECK FOR UNSTABLE PHASES
C
        CALL GMIX(Y1,ACT1,DACT1,ICVEX)
        IF(ICVEX.EQ.-1)WRITE(6,900)
900     FORMAT(' LEFT PHASE UNSTABLE IN THE FOLLOWING TIE LINE')
        CALL GMIX(Y2,ACT2,DACT2,ICVEX)
        IF(ICVEX.EQ.-1)WRITE(6,901)
901     FORMAT(' RIGHT PHASE UNSTABLE IN THE FOLLOWING TIE LINE')
C
C FIND NEW, THE NUMBER OF THE CONCENTRATION WITH GREATEST DERIVATIVE
C
        DY(NOLD)=1.D0
        NEW=NOLD
        DYMAX=1.D0
        DO 12 I=1,4
        IF(DABS(DY(I)).LE.DYMAX)GOTO 12
        NEW=I
        DYMAX=DABS(DY(I))
12      CONTINUE
        RETURN
        END
C
C
        SUBROUTINE GMIX(X,ACT,DACT,ICVEX)
        IMPLICIT REAL*8(A-H,O-Z)
        DIMENSION X(3),DG(2),DDG(2,2),ACT(3),DACT(3,3)
C
C CHECK FOR STABILITY OF EACH PHASE
C
        NK=3
        ICVEX=1
        X(3)=1.D0-X(1)-X(2)
        IRETUR=0
        DO 2 I=1,NK
2       IF(X(I).LT.1.D-10)IRETUR=1
        IF(IRETUR.EQ.1)RETURN
        DO 5 I=1,NK
```

```
        DO 5 J=1,NK
5       DACT(I,J)=DACT(I,J)/ACT(I)
        DO 20 I=2,NK
        II=I-1
        DO 20 J=2,NK
        JJ=J-1
20      DDG(II,JJ)=DACT(I,J)-DACT(1,J)-DACT(I,1)+DACT(1,1)
        DET=DDG(1,1)*DDG(2,2)-DDG(2,1)*DDG(2,1)
        IF(DET.LE.0.D0.OR.DDG(1,1).LE.0.D0.OR.DDG(2,2).LE.0.D0)ICVEX=-1
        RETURN
        END
C
C
        FUNCTION RUND(Y,DY,S,IRUND)
        IMPLICIT REAL*8(A-H,O-Z)
C
C FIND ROUND VALUE FOR CONCENTRATION STEP
C
        X=Y+S*DY+1.D-8*DY**2
        IX=IRUND*X
        Z=DFLOAT(IX)/IRUND-Y
        RUND=DABS(Z)
        RETURN
        END
C
C
        SUBROUTINE TERM(Y,DMAT,ICOND,NEW)
        IMPLICIT REAL*8(A-H,O-Z)
        DIMENSION Y(6),DMAT(4,4),A(4)
C
C CHECK FOR END OF BINODAL CURVE
C
        IF(Y(1).LT.1.D-10.OR.Y(3).LT.1.D-10)GOTO 1
        IF(Y(1)+Y(2).GT.1.D0.OR.Y(3)+Y(4).GT.1.D0)GOTO 2
        IF(Y(1)+Y(2)-.01D0.LT.Y(3)+Y(4).AND.Y(1)-.01D0.LT.Y(3))GOTO 3
        RETURN
1       ICOND=2
        DS=DMAT(1,1)/(DMAT(1,1)-Y(1))
        DO 5 I=1,4
5       Y(I)=DMAT(I,1)+DS*(Y(I)-DMAT(I,1))
        Y(1)=0.D0
        NEW=1
        RETURN
2       ICOND=-2
        RETURN
3       ICOND=1
        ND=2+NEW
        IF(ND.GT.4)ND=ND-4
        DO 6 I=1,4
6       A(I)=DMAT(NEW,I)-DMAT(ND,I)
        DS=0.D0
        NITER=0
7       NITER=NITER+1
        IF(NITER.LE.10)GOTO 8
        ICOND=-1
        RETURN
8       F=((A(4)*DS+A(3))*DS+A(2))*DS+A(1)
        DF=(3.D0*A(4)*DS+2.D0*A(3))*DS+A(2)
        DF=-F/DF
        DS=DS+DF
```

```
      IF(DABS(DF).GT.1.D-6)GOTO 7
      DO 9 I=1,4
      Y(I)=DMAT(I,4)
      DO 9 J=1,3
9     Y(I)=Y(I)*DS+DMAT(I,4-J)
      RETURN
      END
C
C
      SUBROUTINE UNIQ(X,ACT,DACT)
      IMPLICIT REAL*8(A-H,O-Z)
      COMMON Q(3),R(3),PAR(3,3),TAU(3,3),G(3,3),ALP(3,3),TCK
      DIMENSION X(3),ACT(3),DACT(3,3),THETA(3),PHI(3),THS(3),
     *QI(3),QIX(3),RI(3),PARA(3,3),PARB(3,3),GAM(3),QID(3)
C
C CALCULATION OF ACTIVITIES AND DERIVATIVES OF ACTIVITIES WITH RESPECT
C TO CONCENTRATIONS USING THE UNIQUAC EQUATION
C
      NK=3
      X(3)=1.D0-X(1)-X(2)
      IF(X(3).LT.0.D0)X(3)=0.D0
      NCOR=5
      THETS=0.D0
      PHS=0.D0
      DO 10 I=1,NK
      THETA(I)=Q(I)*X(I)
      THETS=THETS+THETA(I)
      PHI(I)=R(I)*X(I)
10    PHS=PHS+PHI(I)
      DO 20 I=1,NK
      THETA(I)=THETA(I)/THETS
      PHI(I)=PHI(I)/PHS
      RI(I)=R(I)/PHS
      QIX(I)=Q(I)/THETS
      QI(I)=RI(I)/QIX(I)
20    QID(I)=1.D0-QI(I)
      DO 30 I=1,NK
      THS(I)=0.D0
      DO 30 J=1,NK
30    THS(I)=THS(I)+PAR(J,I)*THETA(J)
      DO 40 I=1,NK
      GA=1.D0-RI(I)
      VAL=DLOG(QI(I)**NCOR*THS(I))
      GB=NCOR*QID(I)+VAL-1.D0
      DO 45 J=1,NK
      PARA(I,J)=PAR(I,J)/THS(J)
      PARB(I,J)=PARA(I,J)*THETA(J)
45    GB=GB+PARB(I,J)
      GAM(I)=DEXP(GA-Q(I)*GB)*RI(I)
40    ACT(I)=X(I)*GAM(I)
      DO 50 I=1,NK
      DO 50 J=I,NK
      PSUM=1.D0-PARA(I,J)-PARA(J,I)
      DO 55 K=1,NK
55    PSUM=PSUM+PARA(I,K)*PARB(J,K)
      PSUM=PSUM-NCOR*QID(I)*QID(J)
      DACT(I,J)=Q(I)*QIX(J)*PSUM+(1.D0-RI(I))*(1.D0-RI(J))
50    DACT(J,I)=DACT(I,J)
      DO 60 I=1,NK
      DO 60 J=1,NK
```

```
      DACT(I,J)=ACT(I)*DACT(I,J)
      IF (J.EQ.I) DACT(I,J)=DACT(I,J)+GAM(I)
60    CONTINUE
      RETURN
      END
C
C
      SUBROUTINE NRTL(X,ACT,DACT)
      IMPLICIT REAL*8 (A-H,O-Z)
      COMMON Q(3),R(3),PAR(3,3),TAU(3,3),G(3,3),ALP(3,3),TCK
      DIMENSION X(3),ACT(3),DACT(3,3),GAM(3),G1(3,3),TAU1(3,3),S(3,3),
     *G2(3,3),A1(3),B1(3)
C
C CALCULATION OF ACTIVITIES AND DERIVATIVES OF ACTIVITIES WITH RESPECT
C TO CONCENTRATIONS USING THE NRTL EQUATION
C
      NK=3
      X(3)=1.D0-X(1)-X(2)
      IF(X(3).LT.0.D0)X(3)=0.D0
      DO 20 I=1,NK
      AA=0.D0
      BB=0.D0
      DO 30 J=1,NK
      Z=G(J,I)*X(J)
      AA=AA+Z
30    BB=BB+Z*TAU(J,I)
      A1(I)=AA
      B1(I)=BB
      GAM(I)=BB/AA
      DO 20 J=1,NK
      G1(J,I)=G(J,I)/AA
      TAU1(J,I)=TAU(J,I)-GAM(I)
      G2(J,I)=G1(J,I)*TAU1(J,I)
20    S(J,I)=X(I)*G2(J,I)
      DO 40 I=1,NK
      DO 50 J=1,NK
50    GAM(I)=GAM(I)+S(I,J)
      GAM(I)=DEXP(GAM(I))
40    ACT(I)=X(I)*GAM(I)
      DO 60 I=1,NK
      DO 60 J=I,NK
      SUM=G2(J,I)+G2(I,J)
      DO 65 K=1,NK
65    SUM=SUM-G1(I,K)*S(J,K)-G1(J,K)*S(I,K)
      DACT(I,J)=SUM
60    DACT(J,I)=SUM
      DO 70 I=1,NK
      DO 70 J=1,NK
      DACT(I,J)=DACT(I,J)*ACT(I)
      IF(J.EQ.I)DACT(I,I)=DACT(I,I)+GAM(I)
70    CONTINUE
      RETURN
      END
C
C
      SUBROUTINE GAUSL(ND,NCOL,N,NS,A)
      IMPLICIT REAL*8(A-H,O-Z)
      DIMENSION A(ND,NCOL)
```

```
C
C GAUSL SOLVES A*X=B, WHERE A IS N*N AND B IS N*NS, BY GAUSSIAN
C ELIMINATION WITH PARTIAL PIVOTING. THE MATRIX (OR VECTOR) B
C IS PLACED ADJACENT TO A IN COLUMNS N+1 TO N+NS.
C A IS DESTROYED, AND THE RESULTING MATRIX X REPLACES B
C
      N1=N+1
      NT=N+NS
      IF(N.EQ.1)GOTO 50
C
C START ELIMINATION
C
      DO 10 I=2,N
      IP=I-1
      I1=IP
      X=DABS(A(I1,I1))
      DO 11 J=I,N
      IF(DABS(A(J,I1)).LT.X)GOTO 11
      X=DABS(A(J,I1))
      IP=J
11    CONTINUE
      IF(IP.EQ.I1)GOTO 13
C
C ROW INTERCHANGE
C
      DO 12 J=I1,NT
      X=A(I1,J)
      A(I1,J)=A(IP,J)
12    A(IP,J)=X
13    DO 10 J=I,N
      IF(DABS(A(I1,I1)).LT.1.D-10)A(I1,I1)=1.D0
      X=A(J,I1)/A(I1,I1)
      DO 10 K=I,NT
10    A(J,K)=A(J,K)-X*A(I1,K)
C
C ELIMINATION FINISHED, NOW BACKSUBSTITUTION
C
50    DO 20 IP=1,N
      I=N1-IP
      DO 20 K=N1,NT
      IF(DABS(A(I,I)).LT.1.D-10)A(I,I)=1.D0
      A(I,K)=A(I,K)/A(I,I)
      IF(I.EQ.1)GOTO 20
      I1=I-1
      DO 25 J=1,I1
25    A(J,K)=A(J,K)-A(I,K)*A(J,I)
20    CONTINUE
      RETURN
      END
```

```
****************************************************************************
*
* BINODAL CURVE CONSTRUCTION
*
* SAMPLE INPUT
*
****************************************************************************


1          30.
7.1974     6.016      DECANE
6.1519     5.212      1-OCTANOL
4.0013     3.568      DIETHYLENE GLYCOL
482.83     -254.49
333.80     98.934
142.93     -86.902


****************************************************************************
*
* BINODAL CURVE CONSTRUCTION
*
* SAMPLE OUTPUT
*
****************************************************************************


BINODAL CURVE CONSTRUCTION USING UNIQUAC


                                              R       Q

(1) DECANE                                 7.1974  6.016
(2) 1-OCTANOL                              6.1519  5.212
(3) DIETHYLENE GLYCOL                      4.0013  3.568

TEMPERATURE = 30.0 DEG C

INTERACTION PARAMETERS:

  I J     A(I,J)      A(J,I)
          KELVIN      KELVIN

   1 2    482.83     -254.49
   1 3    333.80      98.934
   2 3    142.93     -86.902

STEP LENGTH USED IN THE BINODAL CURVE CONSTRUCTION =  2.0 MOLE PCT

INITIAL GUESS AT MUTUAL SOLUBILITIES OF COMPONENTS (1) AND (3):

(1) IN (3):  0.0 MOLE PCT
(3) IN (1):  0.0 MOLE PCT
```

COMPUTED BINODAL CURVE:

LEFT PHASE			RIGHT PHASE		
(1)	(2)	(3)	(1)	(2)	(3)
99.124	0.0	0.876	0.306	0.0	99.694
98.129	0.867	1.004	0.441	2.000	97.559
96.824	1.986	1.189	0.619	4.000	95.381
94.958	3.552	1.490	0.849	6.000	93.151
92.500	5.551	1.949	1.089	7.664	91.247
90.500	7.129	2.371	1.233	8.538	90.228
88.500	8.664	2.836	1.346	9.165	89.489
86.500	10.159	3.341	1.437	9.646	88.917
84.500	11.615	3.885	1.515	10.043	88.442
82.500	13.034	4.466	1.586	10.396	88.019
80.500	14.417	5.083	1.655	10.728	87.617
78.500	15.766	5.734	1.725	11.056	87.220
76.500	17.080	6.420	1.797	11.389	86.813
74.500	18.360	7.140	1.875	11.735	86.390
72.500	19.607	7.893	1.958	12.097	85.945
70.500	20.820	8.680	2.048	12.479	85.473
68.500	21.999	9.501	2.145	12.882	84.972
66.500	23.143	10.357	2.252	13.309	84.439
64.500	24.253	11.247	2.368	13.760	83.872
62.500	25.326	12.174	2.495	14.236	83.269
60.500	26.362	13.138	2.634	14.737	82.629
58.500	27.360	14.140	2.786	15.265	81.949
56.500	28.318	15.182	2.952	15.820	81.228
54.500	29.235	16.265	3.134	16.402	80.464
52.500	30.108	17.392	3.334	17.012	79.654
50.500	30.936	18.564	3.553	17.651	78.796
48.500	31.716	19.784	3.793	18.319	77.887
46.500	32.445	21.055	4.058	19.017	76.925
44.500	33.120	22.380	4.350	19.745	75.905
42.500	33.738	23.762	4.673	20.504	74.823
40.500	34.295	25.205	5.030	21.294	73.675
38.500	34.786	26.714	5.427	22.117	72.456
36.500	35.206	28.294	5.869	22.971	71.160
34.500	35.551	29.949	6.362	23.859	69.779
32.500	35.811	31.689	6.915	24.779	68.306
30.500	35.982	33.518	7.538	25.731	66.731
28.500	36.052	35.448	8.243	26.715	65.042
26.500	36.012	37.488	9.045	27.729	63.226
24.500	35.849	39.651	9.964	28.770	61.267
22.500	35.548	41.952	11.024	29.833	59.143
20.500	35.090	44.410	12.258	30.911	56.831
18.500	34.452	47.048	13.711	31.992	54.297
16.500	33.604	49.896	15.443	33.057	51.499
15.965	33.336	50.699	15.965	33.336	50.699

TABLES AND FIGURES

$CCl_4\text{-}C_2H_4O_2$

EXP.TIE LINE —— CALC.BINODAL ---- CALC.PLAIT P.

UNIQ(SP) □ NRTL(SP) ---- ✦ UNIQ(CO) ---- ✗

MOLE PER CENT OF (3)

MOLE PER CENT OF (2)

EXP. DISTR. RATIO —— UNIQ(SP) ◇ NRTL(SP) ---- UNIQ(CO) ----
CALC.DISTR.RATIO

MOLE PER CENT OF (2) IN RIGHT PHASE

DISTRIBUTION RATIO FOR (2)

(1) C6H15N	AMINE, TRIETHYL
(2) CCL4	METHANE, TETRACHLORO
(3) C2H4O2	ACETIC ACID

PLEKHOTKIN V.F., MARKUZIN N.P.
FIZ.KHIM.SVOISTVA RASTVOROV,LENINGRAD,EDITOR:A.I.RUSANOV
(1964)12

TEMPERATURE = 20.0 DEG C   TYPE OF SYSTEM = 1

EXPERIMENTAL TIE LINES IN MOLE PCT

	LEFT PHASE			RIGHT PHASE	
(1)	(2)	(3)	(1)	(2)	(3)
93.400	0.0	6.600	41.000	0.0	59.000
91.400	2.000	6.600	41.200	0.900	57.900
88.700	4.600	6.700	41.300	2.000	56.700
86.100	7.100	6.800	41.400	3.400	55.200
82.000	11.100	6.900	41.500	5.300	53.200
77.000	14.400	8.600	41.800	7.100	51.100
72.700	17.200	10.100	42.300	8.900	48.800
69.800	19.200	11.000	42.300	10.400	47.300
65.800	20.900	13.300	42.500	11.700	45.800
64.200	21.600	14.200	42.500	13.500	44.000
62.000	22.200	15.800	42.500	14.700	42.800
58.000	22.900	19.100	43.400	16.700	39.900

SPECIFIC MODEL PARAMETERS IN KELVIN

I	J	UNIQUAC AIJ	AJI	NRTL(ALPHA=.2) AIJ	AJI
1	2	269.73	-226.60	-303.24	90.369
1	3	766.22	-159.96	1162.9	-143.69
2	3	251.77	-34.635	350.19	75.515

R1 = 5.0118   R2 = 3.3900   R3 = 2.2024
Q1 = 4.256   Q2 = 2.910   Q3 = 2.072

MEAN DEV. BETWEEN CALC. AND EXP. CONC. IN MOLE PCT

UNIQUAC (SPECIFIC PARAMETERS)	1.03
NRTL (SPECIFIC PARAMETERS)	0.69
UNIQUAC (COMMON PARAMETERS)	2.02

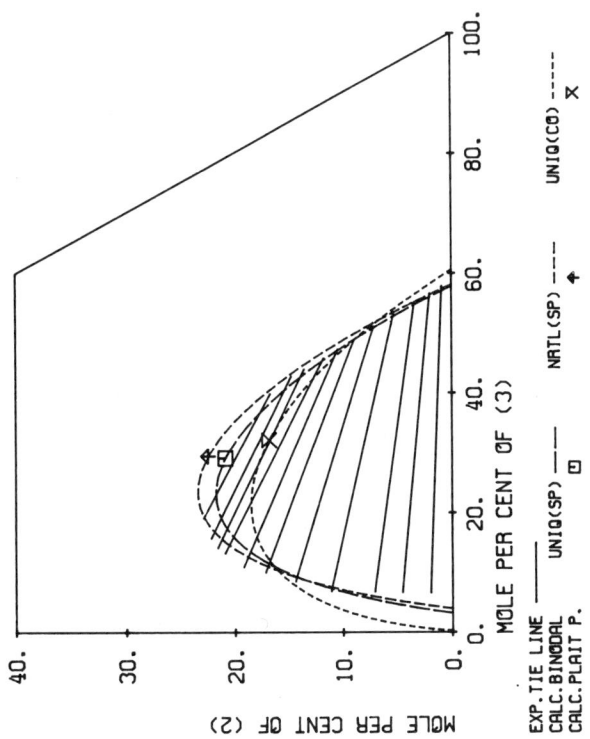

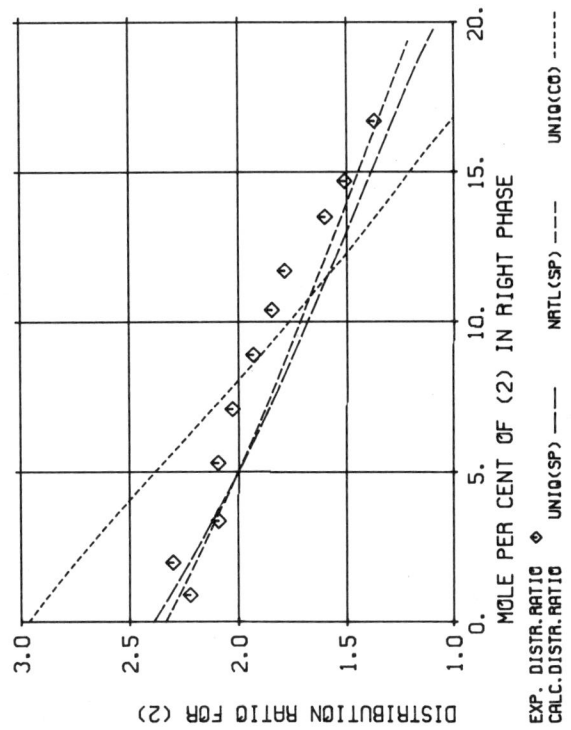

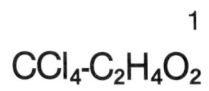

$CCl_4-C_2H_4O_2$

(1) H2O	WATER	
(2) C2H4O2	ACETIC ACID	
(3) CCL4	METHANE, TETRACHLORO	

KRISHNAMURTY V.V.G.; MURTI P.S., VENKATA RAO C.
J.SCI.IND.RES. 12B(1953)583

TEMPERATURE = 27.5 DEG C   TYPE OF SYSTEM = 1

EXPERIMENTAL TIE LINES IN MOLE PCT

	LEFT PHASE			RIGHT PHASE	
(1)	(2)	(3)	(1)	(2)	(3)
87.168	12.802	0.030	0.0	4.485	95.515
77.140	22.679	0.181	0.0	8.501	91.499
69.513	29.978	0.509	0.0	11.880	88.120
55.587	42.460	1.953	0.759	17.082	82.159
52.296	45.181	2.523	1.472	20.531	77.997

SPECIFIC MODEL PARAMETERS IN KELVIN

		UNIQUAC		NRTL(ALPHA=.2)	
I J	AIJ	AJI		AIJ	AJI
1 2	-238.97	-48.556		185.08	-374.61
1 3	525.57	361.16		2102.8	1267.2
2 3	29.822	60.383		473.17	-92.441

R1 = 0.9200   R2 = 2.2024   R3 = 3.3900
Q1 = 1.400   Q2 = 2.072   Q3 = 2.910

MEAN DEV. BETWEEN CALC. AND EXP. CONC. IN MOLE PCT

UNIQUAC (SPECIFIC PARAMETERS)   0.87
NRTL   (SPECIFIC PARAMETERS)   0.78
UNIQUAC (COMMON PARAMETERS)   1.12

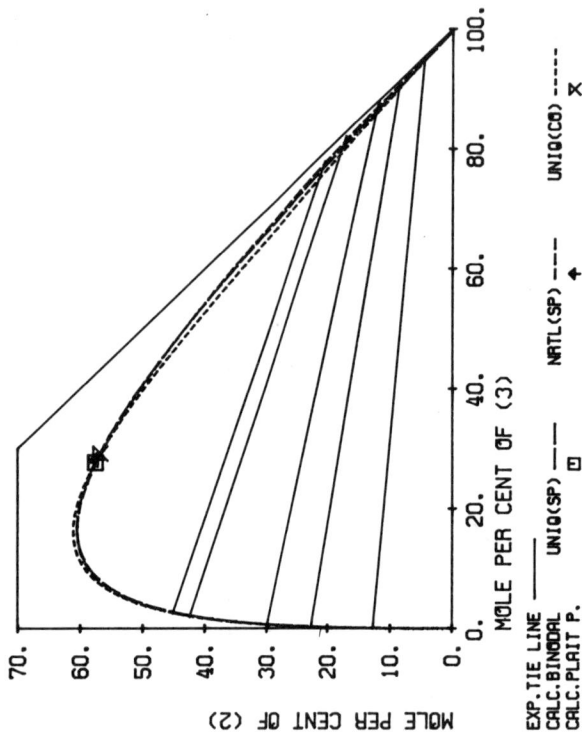

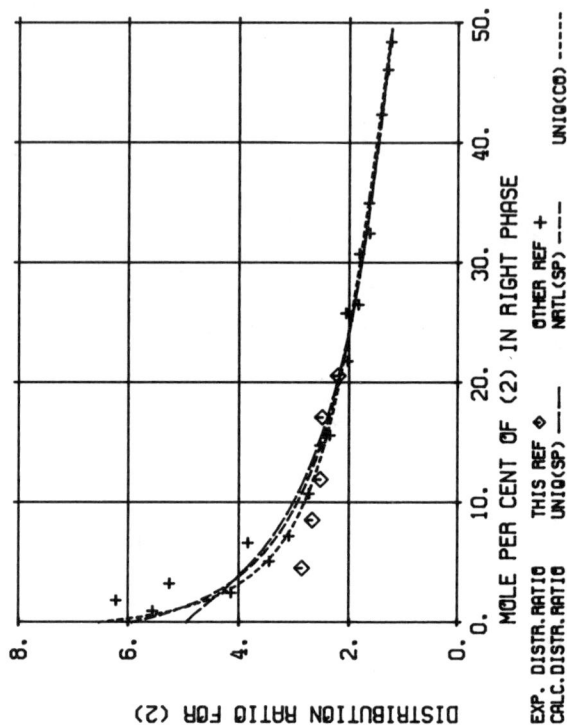

MOLE PER CENT OF (2)

MOLE PER CENT OF (3)

EXP.TIE LINE ——   UNIQ(SP) —— ▢   NRTL(SP) —— ✦   UNIQ(CO) ----- ✕
CALC.BINODAL
CALC.PLAIT P.

DISTRIBUTION RATIO FOR (2)

MOLE PER CENT OF (2) IN RIGHT PHASE

EXP. DISTR.RATIO ◇   THIS REF ◇   OTHER REF +   UNIQ(CO) -----
CALC.DISTR.RATIO   UNIQ(SP) ——   NRTL(SP) ——

$CCl_4\text{-}C_2H_4O_2$

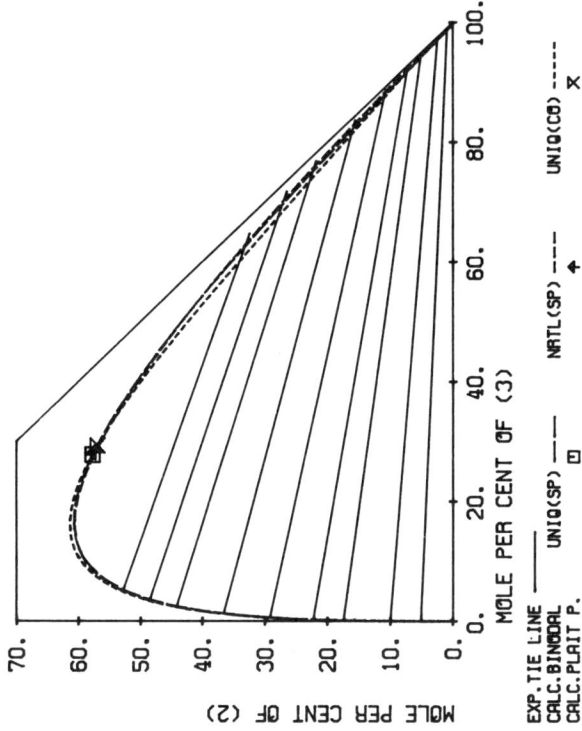

EXP.TIE LINE
CALC.BINODAL   UNIQ(SP) --□--   NRTL(SP) --♠--   UNIQ(C6) --✕--
CALC.PLAIT P.

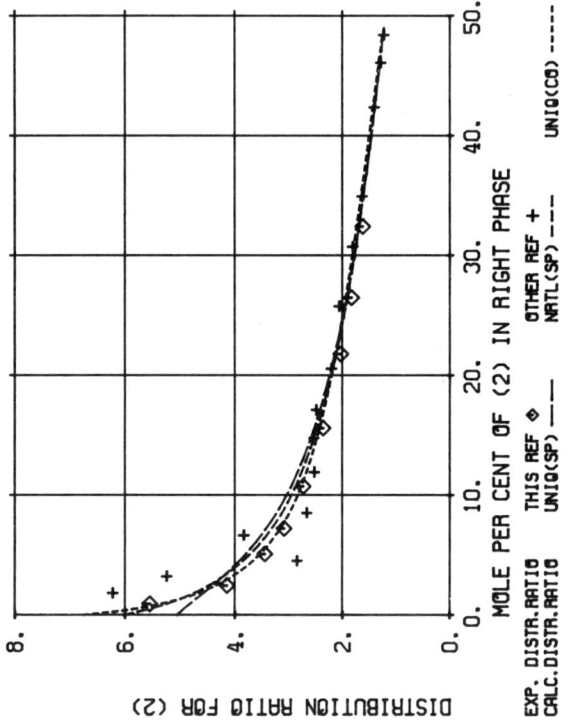

EXP. DISTR.RATIO   THIS REF ◇   OTHER REF +   UNIQ(C6) -----
CALC.DISTR.RATIO   UNIQ(SP) ----   NRTL(SP) ---

(1) H2O	WATER
(2) C2H4O2	ACETIC ACID
(3) CCL4	METHANE,TETRACHLORO

PRINCE R.G.H., HUNTER T.G.
CHEM.ENG.SCI. 6(1957)245

TEMPERATURE = 25.0 DEG C   TYPE OF SYSTEM = 1

EXPERIMENTAL TIE LINES IN MOLE PCT

	LEFT PHASE			RIGHT PHASE	
(1)	(2)	(3)	(1)	(2)	(3)
94.873	5.088	0.039	0.170	0.916	98.915
89.896	10.030	0.074	0.168	2.419	97.413
32.393	17.429	0.178	0.243	5.053	94.699
77.442	22.233	0.325	0.325	7.180	92.495
70.156	29.187	0.657	0.556	10.695	88.749
62.143	36.578	1.279	0.767	15.557	83.676
53.412	44.123	2.465	1.315	21.747	75.938
47.939	48.469	3.592	1.757	26.445	71.798
41.883	52.743	5.374	2.790	32.399	64.812

SPECIFIC MODEL PARAMETERS IN KELVIN

		UNIQUAC			NRTL(ALPHA=.2)	
I J		AIJ	AJI		AIJ	AJI
1 2		-238.97	-48.556		185.08	-374.61
1 3		525.57	851.16		2102.8	1267.2
2 3		29.822	60.388		473.17	-92.441

R1 = 0.9200   R2 = 2.2024   R3 = 3.3900
Q1 = 1.400   Q2 = 2.072   Q3 = 2.910

MEAN DEV. BETWEEN CALC. AND EXP. CONC. IN MOLE PCT

UNIQUAC (SPECIFIC PARAMETERS)	0.73
NRTL (SPECIFIC PARAMETERS)	0.59
UNIQUAC (COMMON PARAMETERS)	0.86

CCl$_4$-C$_2$H$_4$O$_2$

EXP.TIE LINE ———   UNIQ(SP) ——— ☐   NRTL(SP) ——— ✦   UNIQ(CO) ——— ✕
CALC.BINODAL
CALC.PLAIT P.

MOLE PER CENT OF (3)

MOLE PER CENT OF (2)

DISTRIBUTION RATIO FOR (2)

MOLE PER CENT OF (2) IN RIGHT PHASE

EXP. DISTR.RATIO ◇   THIS REF   OTHER REF +   UNIQ(CO) -----
CALC.DISTR.RATIO   UNIQ(SP) ———   NRTL(SP) ———

(1) H2O   WATER
(2) C2H4O2   ACETIC ACID
(3) CCL4   METHANE,TETRACHLORO

FUSE K., IGUCHI A.
KAGAKU KOGAKU 34(1970)1001

TEMPERATURE = 25.0 DEG C   TYPE OF SYSTEM = 1

EXPERIMENTAL TIE LINES IN MOLE PCT (GRAPH.INTERPOL.)

	LEFT PHASE			RIGHT PHASE	
(1)	(2)	(3)	(1)	(2)	(3)
88.724	11.157	0.119	0.421	1.792	97.787
83.004	16.798	0.198	0.666	3.196	96.139
74.203	25.378	0.418	1.135	5.617	92.243
61.713	37.261	1.026	2.280	14.729	82.991
42.377	53.023	4.600	3.735	25.733	70.532
38.137	55.713	6.150	4.489	30.700	64.811
34.651	57.439	7.910	5.190	34.935	59.875
27.621	60.101	12.277	6.331	42.326	51.292
26.395	59.955	13.650	7.372	46.073	46.555
23.852	60.064	16.084	8.307	48.383	43.310

SPECIFIC MODEL PARAMETERS IN KELVIN

I	J	UNIQUAC AIJ	AJI	NRTL(ALPHA=.2) AIJ	AJI
1	2	-238.97	-49.556	185.08	-374.61
1	3	525.57	861.16	2102.8	1267.2
2	3	29.822	60.388	473.17	-92.441

R1 = 0.9200   R2 = 2.2024   R3 = 3.3900
Q1 = 1.400   Q2 = 2.072   Q3 = 2.910

MEAN DEV. BETWEEN CALC. AND EXP. CONC. IN MOLE PCT

UNIQUAC (SPECIFIC PARAMETERS)   0.64
NRTL   (SPECIFIC PARAMETERS)   0.66
UNIQUAC (COMMON PARAMETERS)   0.90

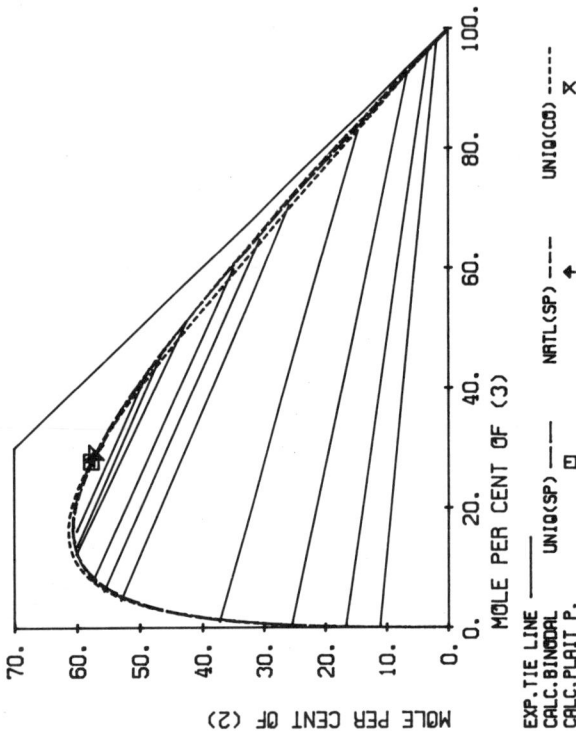

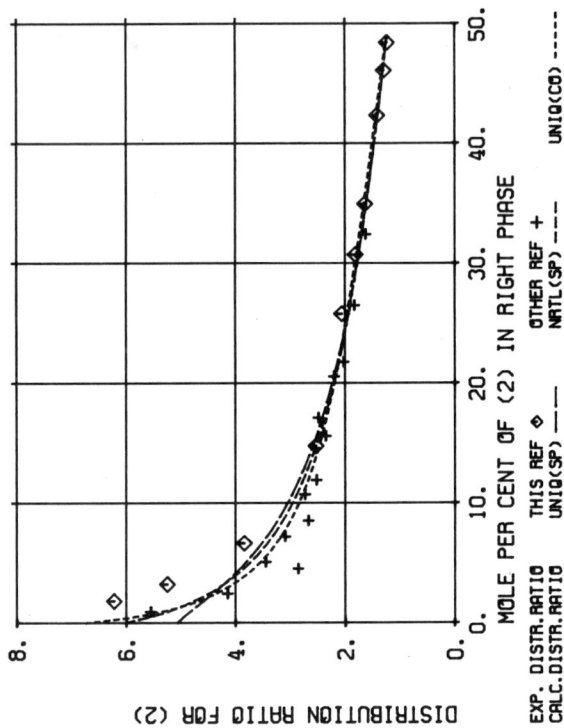

CCl$_4$-C$_2$H$_6$O

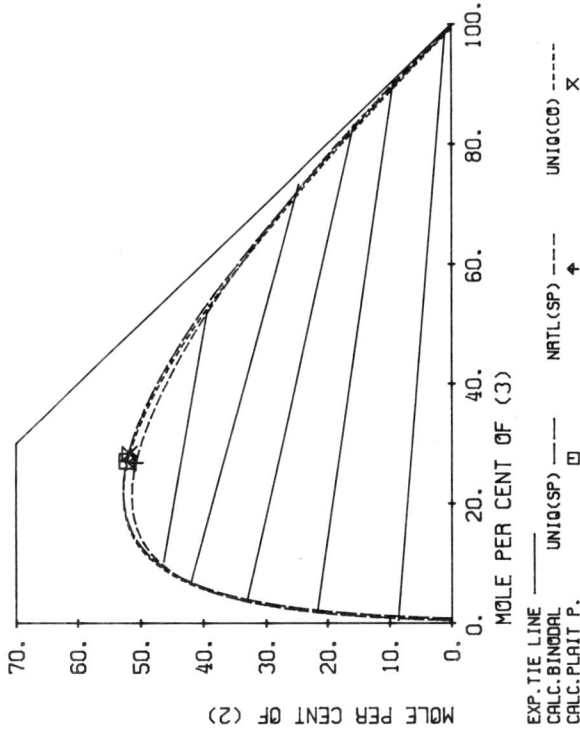

MOLE PER CENT OF (3)

MOLE PER CENT OF (2)

EXP.TIE LINE ——— UNIQ(SP) ⊡ NRTL(SP) ---- ▲ UNIQ(CO) ----- ✕
CALC.BINODAL
CALC.PLAIT P.

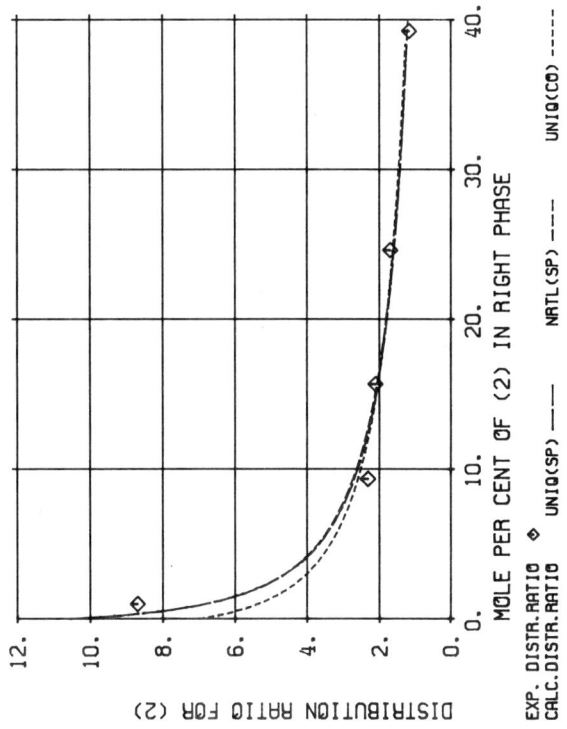

MOLE PER CENT OF (2) IN RIGHT PHASE

DISTRIBUTION RATIO FOR (2)

EXP. DISTR.RATIO ◇ UNIQ(SP) ——— NRTL(SP) ---- UNIQ(CO) -----
CALC.DISTR.RATIO

(1) C3H8O3   GLYCEROL
(2) C2H6O   ETHANOL
(3) CCL4   METHANE,TETRACHLORO
                                    .

MCDONALD H.J., KLUENDER A.F., LANE R.W.
J.PHYS.CHEM. 46(1942)946

TEMPERATURE =   25.0 DEG C   TYPE OF SYSTEM = 1

EXPERIMENTAL TIE LINES IN MOLE PCT

	LEFT PHASE			RIGHT PHASE	
(1)	(2)	(3)	(1)	(2)	(3)
90.782	8.643	0.575	0.994	98.840
76.472	21.687	1.840	0.312	9.348	90.340
53.028	33.053	3.919	1.036	15.679	83.285
51.176	41.964	6.860	2.052	24.620	73.328
43.293	46.377	10.330	8.428	39.239	52.333

SPECIFIC MODEL PARAMETERS IN KELVIN

		UNIQUAC		NRTL(ALPHA=.2)	
I J	AIJ	AJI	AIJ	AJI
1 2	-9.5920	-274.21	-181.37	-468.95
1 3	300.09	-329.12	1039.6	1426.6
2 3	74.791	-165.69	545.76	-497.24

R1 = 3.5857   R2 = 2.1055   R3 = 3.3900
Q1 = 3.060   Q2 = 1.972   Q3 = 2.910

MEAN DEV. BETWEEN CALC. AND EXP. CONC. IN MOLE PCT

UNIQUAC (SPECIFIC PARAMETERS)   1.22
NRTL   (SPECIFIC PARAMETERS)   1.09
UNIQUAC (COMMON PARAMETERS)   1.38

CCl$_4$-C$_2$H$_6$O

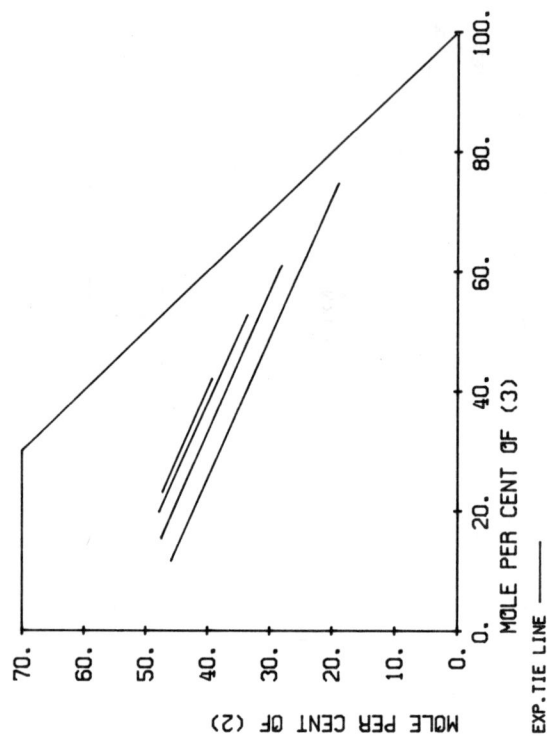

(1) H2O   WATER
----
(2) C2H6O   ETHANOL
----
(3) CCL4   METHANE,TETRACHLORO

BONNER W.D.
J.PHYS.CHEM. 14(1910)738

TEMPERATURE =   0.0 DEG C   TYPE OF SYSTEM = 1

EXPERIMENTAL TIE LINES IN MOLE PCT (GRAPH.INTERPOL.)

          LEFT PHASE   RIGHT PHASE
     (1)   (2)   (3)   (1)   (2)   (3)

   42.518   45.823   11.659   6.240   18.977   74.783
   37.142   47.429   15.429   10.869   28.099   61.032
   32.397   47.759   19.844   13.742   33.532   52.726
   29.595   47.244   23.162   18.654   39.277   42.069

MOLE PER CENT OF (2)

MOLE PER CENT OF (3)

EXP.TIE LINE

Chart 1 (ternary diagram):

Y axis: MOLE PER CENT OF (2) — 70., 60., 50., 40., 30., 20., 10., 0.

X axis: MOLE PER CENT OF (3) — 0., 20., 40., 60., 80., 100.

Legend:
EXP.TIE LINE	UNIQ(SP)	UNIQ(CO)
CALC.BINODAL	NRTL(SP)	
CALC.PLAIT P.		

Chart 2 (distribution diagram):

Y axis: DISTRIBUTION RATIO FOR (2) — 4.0, 3.5, 3.0, 2.5, 2.0, 1.5

X axis: MOLE PER CENT OF (2) IN RIGHT PHASE — 0., 10., 20., 30.

Legend:
EXP.DISTR.RATIO	UNIQ(SP)	UNIQ(CO)
CALC.DISTR.RATIO	NRTL(SP)	

(1) CCL4   METHANE, TETRACHLORO
(2) C3H6O   2-PROPANONE
(3) H2O   WATER

BUCHANAN R.H.
IND.ENG.CHEM. 44(1952)2449

TEMPERATURE = 30.0 DEG C   TYPE OF SYSTEM = 1

EXPERIMENTAL TIE LINES IN MOLE PCT (GRAPH.INTERPOL.)

LEFT PHASE			RIGHT PHASE		
(1)	(2)	(3)	(1)	(2)	(3)
97.826	2.090	0.084	0.028	1.066	98.906
89.959	9.801	0.240	0.032	3.447	96.521
88.391	11.292	0.317	0.034	3.884	96.081
67.812	29.146	3.042	0.056	8.209	91.736
62.979	33.194	3.827	0.059	9.141	90.800
50.966	43.629	5.405	0.090	12.012	87.899
39.686	53.525	6.789	0.189	14.956	84.854
32.153	57.861	9.987	0.286	17.626	82.037
23.479	60.132	16.388	0.468	21.776	77.756
18.418	59.892	21.690	0.844	25.028	74.128

SPECIFIC MODEL PARAMETERS IN KELVIN

	UNIQUAC		NRTL(ALPHA=.2)	
I J	AIJ	AJI	AIJ	AJI
1 2	368.93	-141.50	770.01	-359.03
1 3	1003.8	595.35	1516.6	2074.8
2 3	463.04	-105.87	382.16	227.04

R1 = 3.3900   R2 = 2.5735   R3 = 0.9200
Q1 = 2.910   Q2 = 2.336   Q3 = 1.400

MEAN DEV. BETWEEN CALC. AND EXP. CONC. IN MOLE PCT

UNIQUAC (SPECIFIC PARAMETERS)   0.46
NRTL   (SPECIFIC PARAMETERS)   0.55
UNIQUAC (COMMON PARAMETERS)   0.92

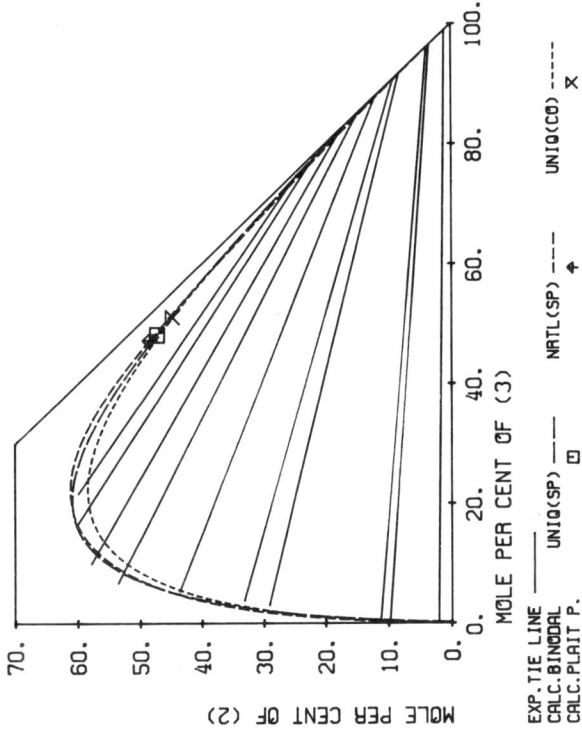

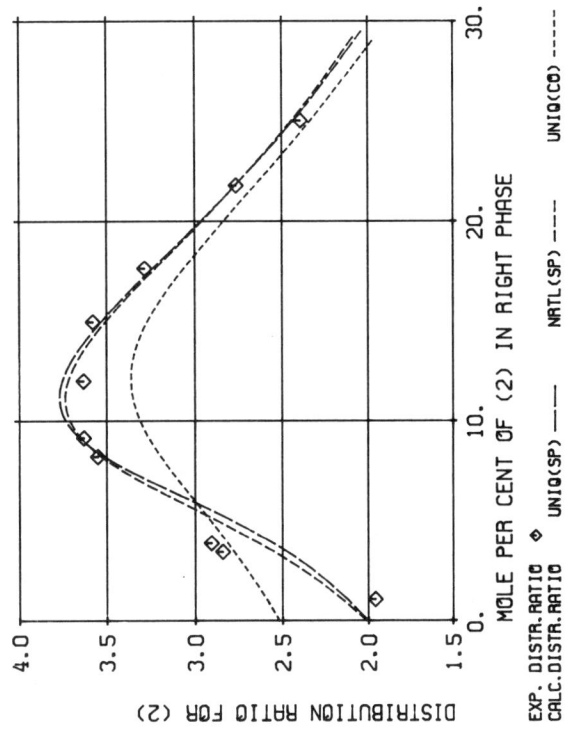

CCl$_4$-C$_3$H$_6$O$_2$

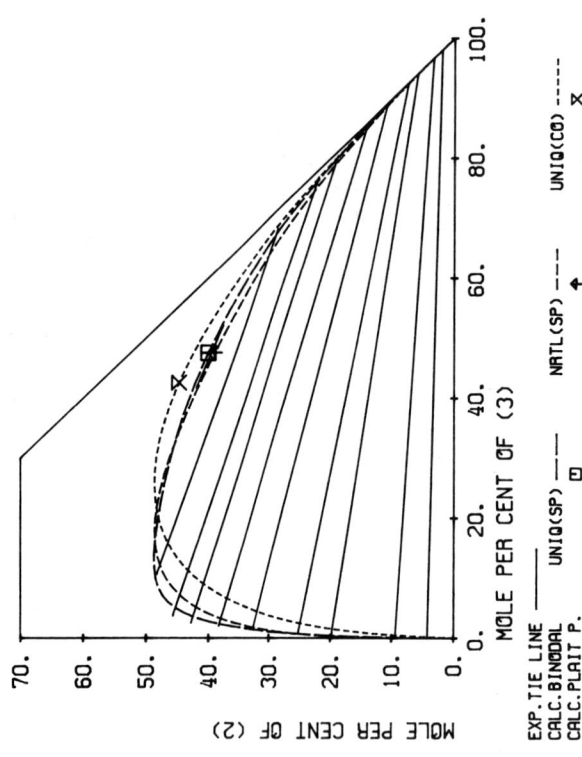

MOLE PER CENT OF (3)

MOLE PER CENT OF (2)

EXP.TIE LINE ———   UNIQ(SP) ☐   NRTL(SP) ✦   UNIQ(CO) -----
CALC.BINODAL   X
CALC.PLAIT P.   NRTL(SP) ✦

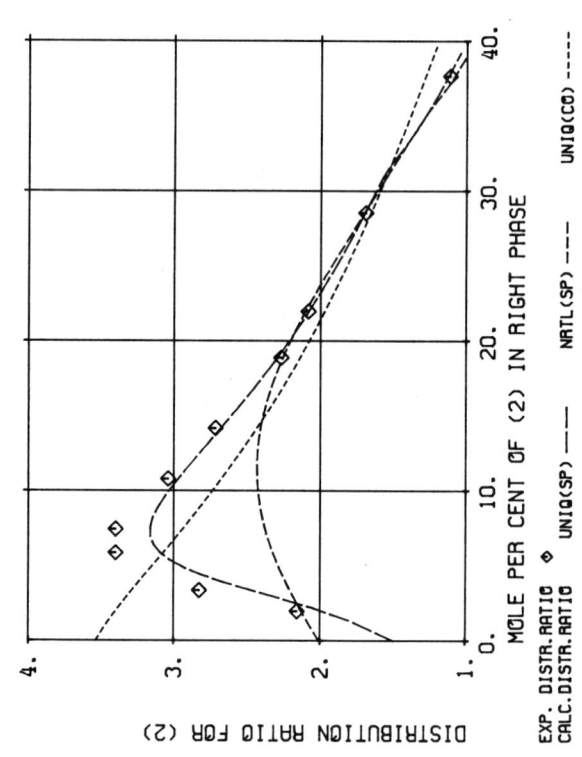

MOLE PER CENT OF (2) IN RIGHT PHASE

DISTRIBUTION RATIO FOR (2)

EXP. DISTR.RATIO ◇   UNIQ(SP) ◆   NRTL(SP) ———   UNIQ(CO) -----
CALC.DISTR.RATIO

---

(1) CCL4   METHANE,TETRACHLORO

(2) C3H6O2   PROPANOIC ACID

(3) H2O   WATER

IGUCHI A., FUSE K.
KAGAKU KOGAKU 36(1972)673

TEMPERATURE = 25.0 DEG C   TYPE OF SYSTEM = 1

EXPERIMENTAL TIE LINES IN MOLE PCT (GRAPH.INTERPOL.)

	LEFT PHASE			RIGHT PHASE	
(1)	(2)	(3)	(1)	(2)	(3)
95.776	4.224	0.0	0.025	1.956	98.019
90.560	9.440	0.0	0.052	3.344	96.604
79.417	19.822	0.760	0.083	5.835	94.081
74.040	25.223	0.737	0.102	7.426	92.473
65.831	32.772	1.397	0.126	10.781	89.093
59.635	38.358	2.007	0.291	14.144	85.564
54.619	42.801	2.580	0.731	18.866	80.403
50.533	45.727	3.740	1.205	21.934	76.861
41.068	48.294	10.638	3.246	28.520	68.234
16.331	41.941	41.729	0.192	37.651	52.157

SPECIFIC MODEL PARAMETERS IN KELVIN

		UNIQUAC		NRTL(ALPHA=.2)	
I	J	AIJ	AJI	AIJ	AJI
1	2	525.16	-250.92	693.53	-405.96
1	3	1470.0	1021.2	2588.4	2152.5
2	3	810.88	-225.50	361.86	86.533

R1 = 3.3900   R2 = 2.8768   R3 = 0.9200
Q1 = 2.910   Q2 = 2.612   Q3 = 1.400

MEAN DEV. BETWEEN CALC. AND EXP. CONC. IN MOLE PCT

UNIQUAC (SPECIFIC PARAMETERS)   0.41
NRTL (SPECIFIC PARAMETERS)   1.42
UNIQUAC (COMMON PARAMETERS)   2.20

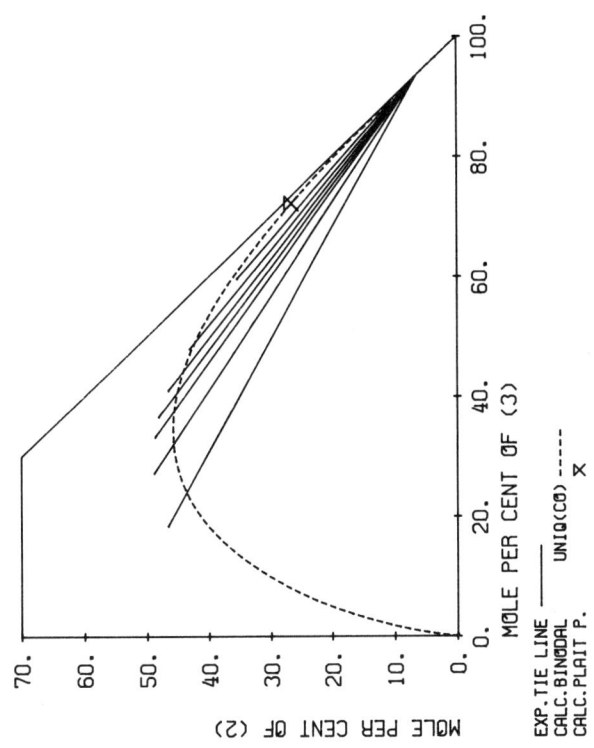

MOLE PER CENT OF (3)

MOLE PER CENT OF (2)

EXP.TIE LINE ———
CALC.BINODAL   - - - -
CALC.PLAIT P. ✕

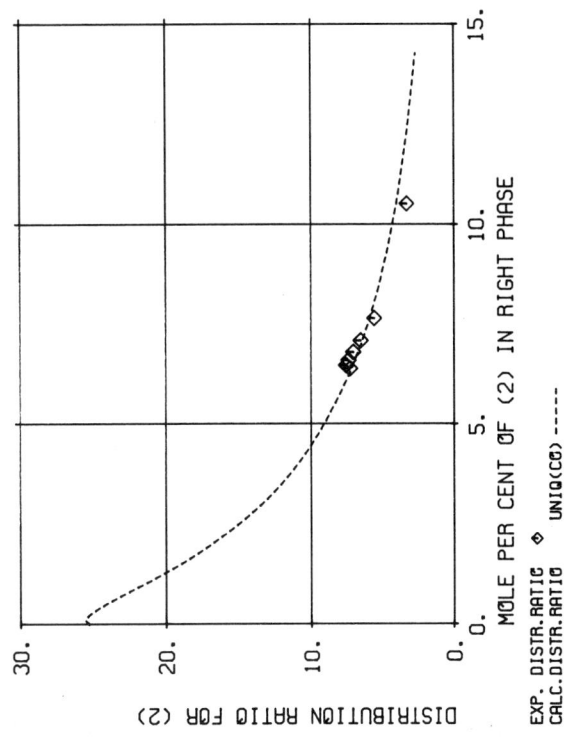

MOLE PER CENT OF (2) IN RIGHT PHASE

DISTRIBUTION RATIO FOR (2)

EXP. DISTR.RATIO ◇   UNIQ(C0) - - - -
CALC.DISTR.RATIO

(1) CCL4   METHANE,TETRACHLORO
----
(2) C3H8O   1-PROPANOL
----
(3) H2O   WATER

DENZLER C.G.
J.PHYS.CHEM. 49(1945)358

TEMPERATURE = 20.0 DEG C   TYPE OF SYSTEM = 1

EXPERIMENTAL TIE LINES IN MOLE PCT

      LEFT PHASE   RIGHT PHASE
   (1)   (2)   (3)   (1)   (2)   (3)

34.977   46.594   18.429   0.035   6.387   93.578
23.937   48.838   27.225   0.033   6.467   93.495
18.077   48.672   33.251   0.041   6.583   93.371
15.260   48.075   35.665   0.052   6.795   93.153
12.383   46.534   41.083   0.059   7.083   92.858
 8.948   43.065   47.987   0.100   7.643   92.257
 4.948   35.416   59.636   0.327 10.507   89.166

MEAN DEV. BETWEEN CALC. AND EXP. CONC. IN MOLE PCT

UNIQUAC (COMMON PARAMETERS)   2.01

CCl_4-C_3H_8O

(1) CCL4   METHANE,TETRACHLORO

(2) C3H8O   2-PROPANOL

(3) H2O   WATER

IZMAILOV N.A., FRANKE A.K.
ZH.FIZ.KHIM. 29(1955)263

TEMPERATURE = 25.0 DEG C   TYPE OF SYSTEM = 1

EXPERIMENTAL TIE LINES IN MOLE PCT

LEFT PHASE			RIGHT PHASE		
(1)	(2)	(3)	(1)	(2)	(3)
98.479	1.521	0.0	0.013	3.681	96.306
96.005	3.995	0.0	0.026	5.373	94.601
89.635	9.567	0.798	0.041	6.749	93.210
70.749	23.740	5.511	0.114	8.896	90.990
56.812	31.156	12.022	0.160	9.869	89.970
43.737	37.152	19.111	0.207	10.649	89.144
33.492	39.782	26.726	0.302	11.446	88.252
19.658	39.433	40.855	0.486	12.916	86.598
14.461	37.273	48.266	0.755	14.742	84.503

SPECIFIC MODEL PARAMETERS IN KELVIN

		UNIQUAC		NRTL(ALPHA=.2)	
I	J	AIJ	AJI	AIJ	AJI
1	2	336.80	-85.423	703.29	-183.39
1	3	656.78	248.47	385.90	983.89
2	3	75.985	68.463	-244.79	964.61

R1 = 3.3900   R2 = 2.7791   R3 = 0.9200
Q1 = 2.910   Q2 = 2.508   Q3 = 1.400

MEAN DEV. BETWEEN CALC. AND EXP. CONC. IN MOLE PCT

UNIQUAC (SPECIFIC PARAMETERS)   1:38
NRTL   (SPECIFIC PARAMETERS)   1:21
UNIQUAC (COMMON PARAMETERS)   3:38

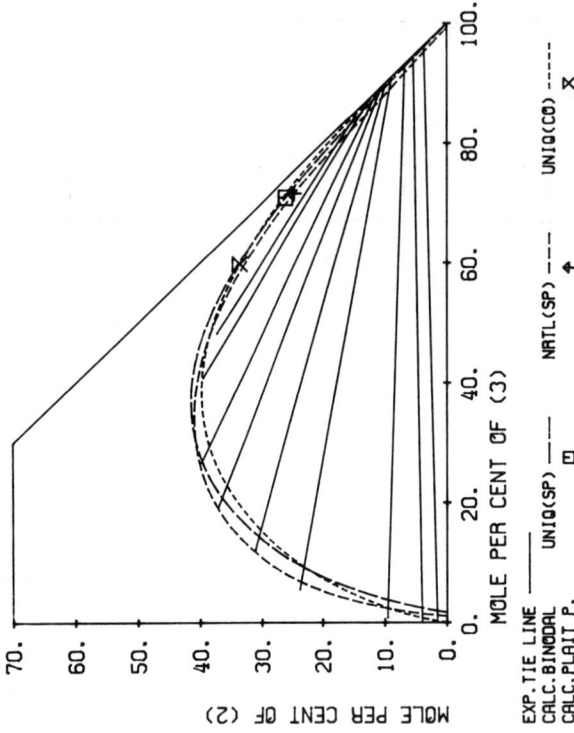

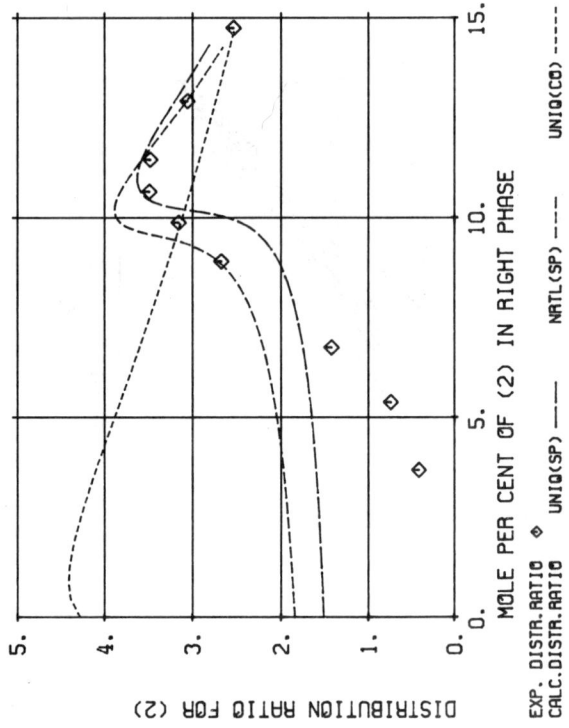

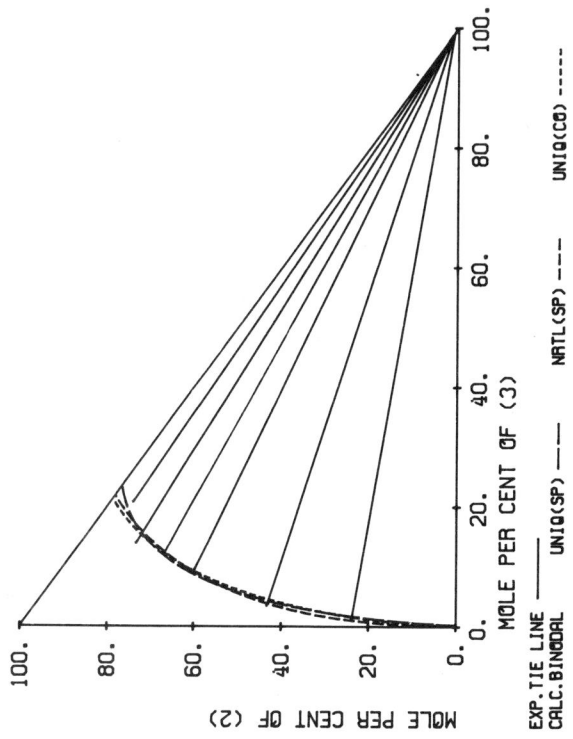

MØLE PER CENT ØF (3)

MØLE PER CENT ØF (2)

EXP.TIE LINE   UNIQ(SP) ————   NRTL(SP) ————   UNIQ(CO) -----
CALC.BINØDAL

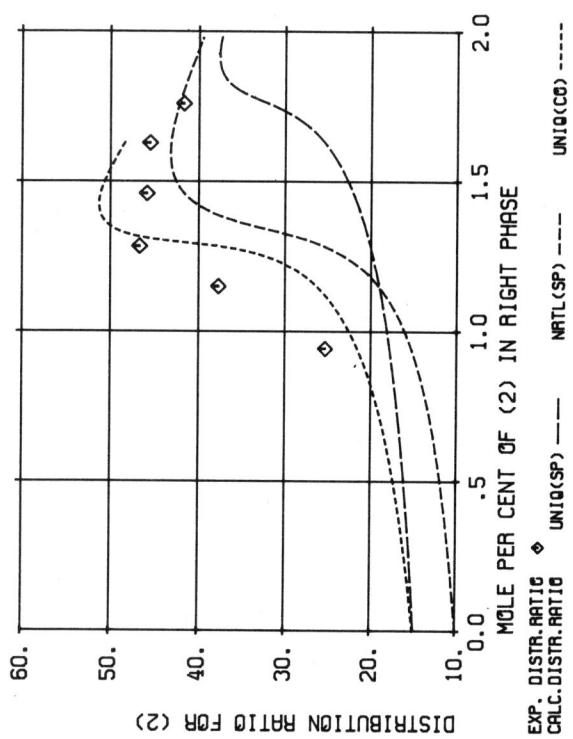

MØLE PER CENT ØF (2) IN RIGHT PHASE

DISTRIBUTIØN RATIØ FØR (2)

EXP. DISTR.RATIØ ◇ UNIQ(SP)   NRTL(SP) ————   UNIQ(CO) -----
CALC.DISTR.RATIØ

(1) CCL4   METHANE,TETRACHLORO

(2) C5H4O2   FURFURAL

(3) H2O   WATER

KRUPATKIN I.L., GLAGOLEVA M.F.
ZH.PRIKL.KHIM.(LENINGRAD) 42(1969)1526

TEMPERATURE = 25.0 DEG C   TYPE OF SYSTEM = 2

EXPERIMENTAL TIE LINES IN MOLE PCT

	LEFT PHASE		RIGHT PHASE	
(1)	(2)	(3)	(1)	(2)	(3)

(1)	(2)	(3)	(1)	(2)	(3)
75.060	23.786	1.153	0.074	0.942	98.985
53.271	43.284	3.446	0.087	1.149	98.765
30.726	59.820	9.454	0.075	1.283	98.642
21.075	66.856	12.070	0.037	1.459	98.504
5.286	73.991	20.723	0.013	1.627	98.360
12.830	73.283	13.887	0.025	1.758	98.216

SPECIFIC MODEL PARAMETERS IN KELVIN

		UNIQUAC		NRTL(ALPHA=.2)	
I J	AIJ	AJI		AIJ	AJI
1 2	39.512	144.99		427.61	103.56
1 3	918.49	633.38		1527.6	1577.5
2 3	210.12	50.877		48.559	1164.6

R1 = 3.3900   R2 = 3.1680   R3 = 0.9200
Q1 = 2.910   Q2 = 2.484   Q3 = 1.400

MEAN DEV. BETWEEN CALC. AND EXP. CONC. IN MOLE PCT

UNIQUAC (SPECIFIC PARAMETERS)   0.55
NRTL   (SPECIFIC PARAMETERS)   0.59
UNIQUAC (COMMON PARAMETERS)   0.65

CCl_4-$C_6H_{11}NO$

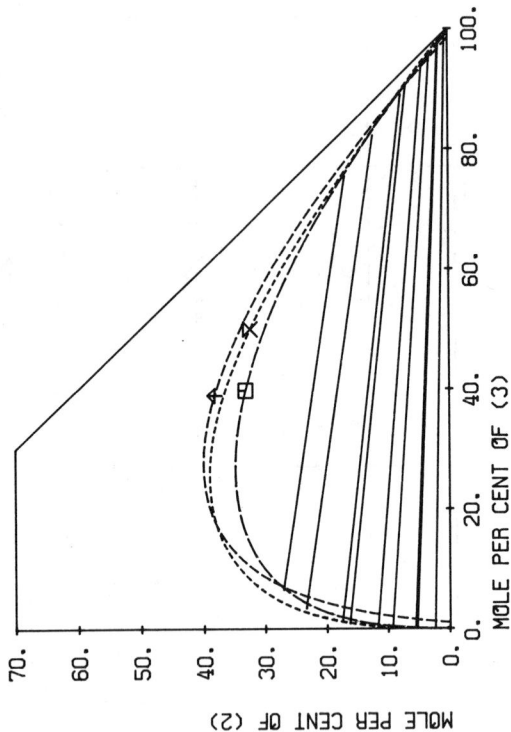

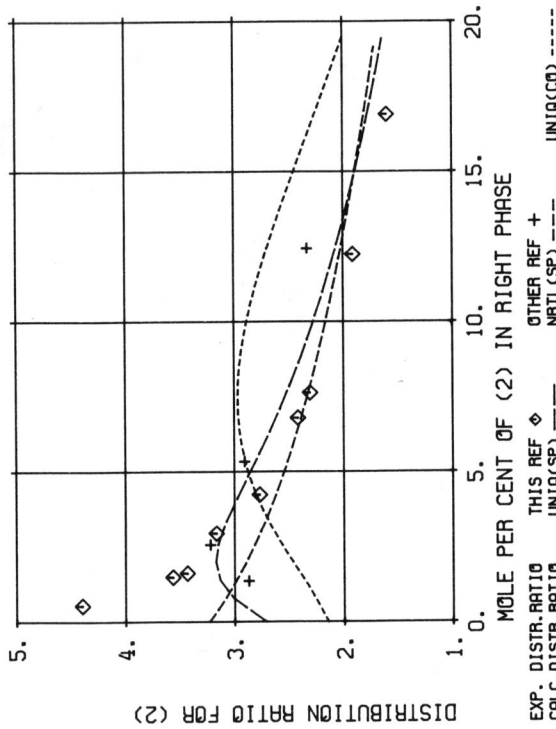

(1) H2O   WATER

(2) C6H11NO   HEXANOIC ACID,6-AMINO,LACTAM

(3) CCL4   METHANE,TETRACHLORO

MORACHEVSKII A.G., SABNIN V.E.
ZH.PRIKL.KHIM.(LENINGRAD) 33(1960)1775

TEMPERATURE = 20.0 DEG C   TYPE OF SYSTEM = 1

EXPERIMENTAL TIE LINES IN MOLE PCT

	LEFT PHASE			RIGHT PHASE	
(1)	(2)	(3)	(1)	(2)	(3)
97.585	2.389	0.026	0.0	0.543	99.457
94.640	5.315	0.045	0.0	1.489	98.511
94.374	5.580	0.046	0.0	1.624	98.376
90.553	9.324	0.123	0.841	2.945	96.214
88.076	11.732	0.191	1.663	4.237	94.099
82.718	16.399	0.883	2.460	6.788	90.752
31.253	17.590	1.157	3.249	7.628	89.123
73.142	23.408	3.450	5.494	12.244	82.262
66.378	27.047	6.575	7.585	16.906	75.509

SPECIFIC MODEL PARAMETERS IN KELVIN

		UNIQUAC		NRTL(ALPHA=.2)	
I J	AIJ	AJI		AIJ	AJI
1 2	-171.16	-79.783		449.77	-461.21
1 3	424.36	782.69		911.14	779.34
2 3	-46.065	51.435		144.92	54.695

R1 = 0.9200   R2 = 4.6106   R3 = 3.3900
Q1 = 1.400   Q2 = 3.724   Q3 = 2.910

MEAN DEV. BETWEEN CALC. AND EXP. CONC. IN MOLE PCT

UNIQUAC (SPECIFIC PARAMETERS)   1.05
NRTL   (SPECIFIC PARAMETERS)   1.51
UNIQUAC (COMMON PARAMETERS)   2.19

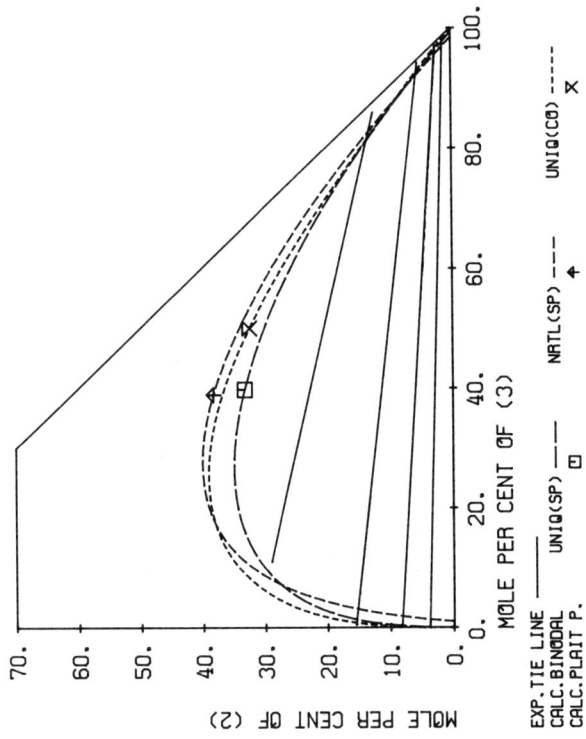

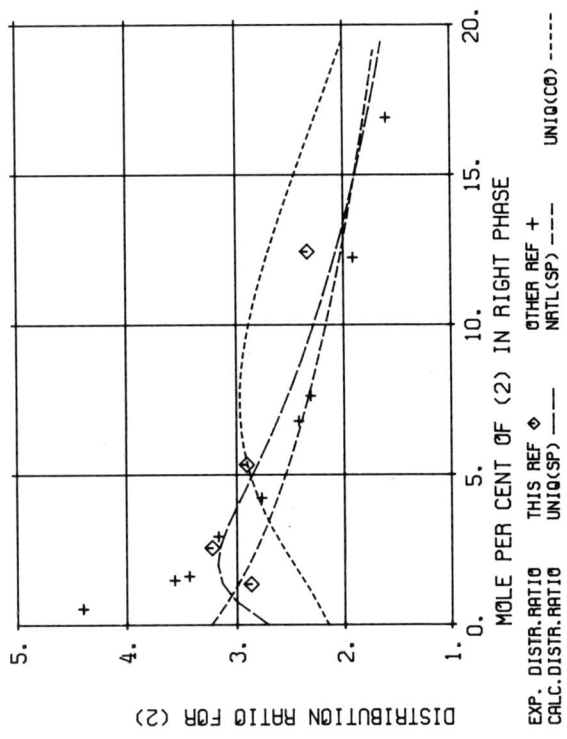

(1) H2O   WATER
----
(2) C6H11NO   HEXANOIC ACID,6-AMINO,LACTAM
----
(3) CCL4   METHANE,TETRACHLORO

TETTAMANTI K., NOGRADI M., SAWINSKY J.
PERIOD.POLYTECH.,CHEM.ENG. 4(1960)201

TEMPERATURE = 20.0 DEG C   TYPE OF SYSTEM = 1

EXPERIMENTAL TIE LINES IN MOLE PCT (GRAPH.INTERPOL.)

        LEFT PHASE   RIGHT PHASE
   (1)   (2)   (3)   (1)   (2)   (3)

 99.992   0.0   0.008   0.085   0.0   99.915
 96.090   3.882   0.028   0.170   1.352   98.478
 91.694   8.252   0.054   0.254   2.559   97.187
 83.940   15.519   0.540   0.419   5.340   94.240
 59.925   28.995   11.079   1.547   12.442   86.011

SPECIFIC MODEL PARAMETERS IN KELVIN

              UNIQUAC   NRTL(ALPHA=.2)
 I J   AIJ   AJI   AIJ   AJI
----
 1 2   -171.16   -79.783   448.77   -461.21
 1 3   424.36   732.69   911.14   779.34
 2 3   -46.065   61.435   144.92   54.695

 R1 = 0.9200   R2 = 4.6106   R3 = 3.3900
 Q1 = 1.400   Q2 = 3.724   Q3 = 2.910

MEAN DEV. BETWEEN CALC. AND EXP. CONC. IN MOLE PCT

UNIQUAC (SPECIFIC PARAMETERS)   1.89
NRTL   (SPECIFIC PARAMETERS)   1.96
UNIQUAC (COMMON PARAMETERS)   1.11

CCl₄-C₇F₁₆

(1) C7F16   HEPTANE, PERFLUORO
(2) C8F16O   OCTANE, 1,8-OXY, PERFLUORO
(3) CCL4   METHANE, TETRACHLORO

KYLE B.G., REED T.M.
J.CHEM.ENG.DATA 5(1960)266

TEMPERATURE = 30.0 DEG C   TYPE OF SYSTEM = 2

EXPERIMENTAL TIE LINES IN MOLE PCT

	LEFT PHASE			RIGHT PHASE	
(1)	(2)	(3)	(1)	(2)	(3)
64.049	0.0	35.951	3.508	0.0	96.492
60.829	5.293	33.878	3.840	0.354	95.806
56.740	9.758	33.501	3.731	0.751	95.518
51.574	13.234	35.192	3.250	0.906	95.844
44.983	16.928	38.090	2.206	0.971	96.823
37.170	23.939	38.891	2.455	1.776	95.769
30.579	30.353	39.068	1.998	2.458	95.544
21.476	37.302	41.222	1.886	3.078	95.036
14.357	45.247	40.397	1.075	4.010	94.915
9.351	47.492	42.558	0.774	4.249	94.977
4.354	52.498	43.149	0.344	4.738	94.918
0.0	56.297	43.703	0.0	5.143	94.852

SPECIFIC MODEL PARAMETERS IN KELVIN

		UNIQUAC		NRTL(ALPHA=.2)	
I J	AIJ	AJI		AIJ	AJI
1 2	-1.7634	-0.60995		-60.800	-45.923
1 3	148.54	43.149		-39.586	1049.0
2 3	133.66	43.892		-134.15	1110.7

R1 = 7.8645   R2 = 9.3279   R3 = 3.3900
Q1 = 7.360   Q2 = 7.600   Q3 = 2.910

MEAN DEV. BETWEEN CALC. AND EXP. CONC. IN MOLE PCT

UNIQUAC (SPECIFIC PARAMETERS)   0.58
NRTL   (SPECIFIC PARAMETERS)   0.72
UNIQUAC (COMMON PARAMETERS)   0.79

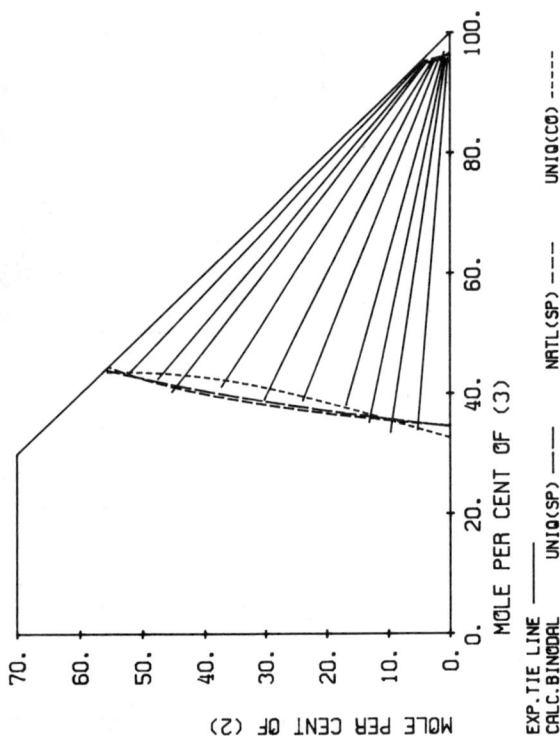

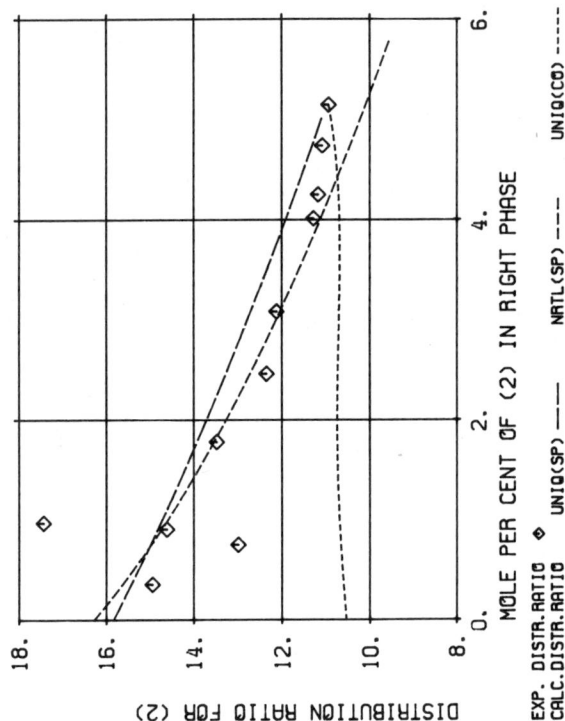

$$CCl_4\text{-}C_{10}H_{14}N_2$$

MOLE PER CENT OF (3)

MOLE PER CENT OF (2)

EXP.TIE LINE —— UNIQ(SP) —— ☐ NRTL(SP) ---- ✦ UNIQ(CO) ---- ✕
CALC.BINODAL
CALC.PLAIT P.

MOLE PER CENT OF (2) IN RIGHT PHASE

DISTRIBUTION RATIO FOR (2)

EXP. DISTR.RATIO ◇ UNIQ(SP) —— NRTL(SP) ---- UNIQ(CO) ----
CALC.DISTR.RATIO

(1) CCL4   METHANE,TETRACHLORO
----
(2) C10H14N2   NICOTINE
----
(3) H2O   WATER

FOWLER R.T., NOBLE R.A.S.
J.APPL.CHEM. 4(1954)546

TEMPERATURE = 25.0 DEG C   TYPE OF SYSTEM = 1

EXPERIMENTAL TIE LINES IN MOLE PCT

          LEFT PHASE   RIGHT PHASE
      (1)   (2)   (3)   (1)   (2)   (3)

    98.197   1.803   0.0   0.012   0.022   99.966
    70.808   23.442   5.750   0.066   1.457   98.477
    59.620   32.300   8.080   0.194   4.609   95.196
    53.059   37.352   9.588   0.479   9.254   90.268
    44.027   40.153   15.820   2.008   16.205   81.787

SPECIFIC MODEL PARAMETERS IN KELVIN

                UNIQUAC   NRTL(ALPHA=.2)
 I J   AIJ   AJI   AIJ   AJI
 1 2 -480.12   548.41   -337.83   -641.02
 1 3   1111.6   1042.0   893.50   1011.4
 2 3   5.1139   -244.34   -65.392   77.825

 R1 = 3.3900   R2 = 6.4898   R3 = 0.9200
 Q1 = 2.910   Q2 = 4.621   Q3 = 1.400

MEAN DEV. BETWEEN CALC. AND EXP. CONC. IN MOLE PCT

UNIQUAC (SPECIFIC PARAMETERS)   1.09
NRTL   (SPECIFIC PARAMETERS)   0.92
UNIQUAC (COMMON PARAMETERS)   6.61

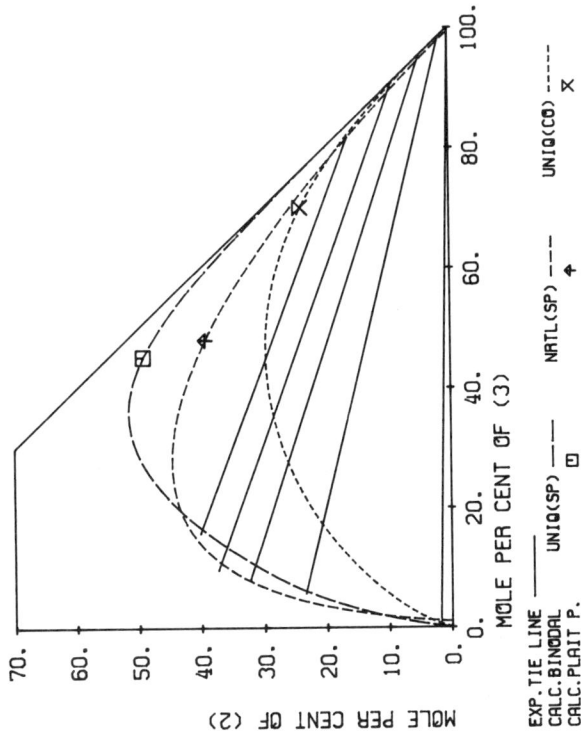

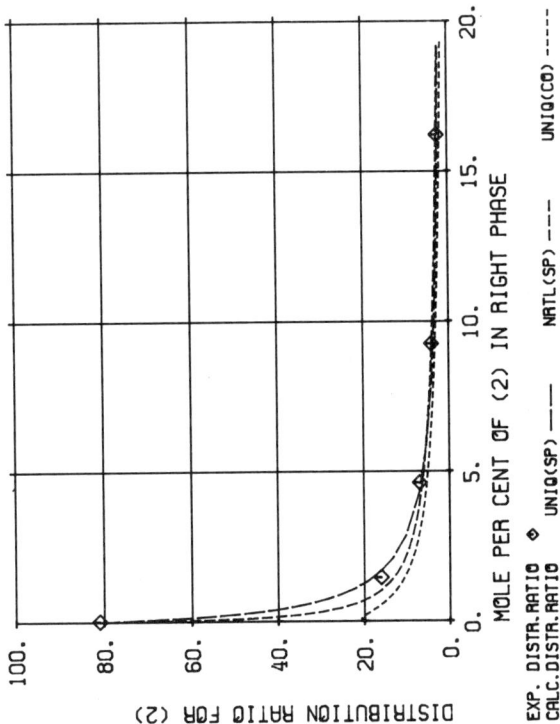

CHCl$_3$-CH$_2$O$_2$

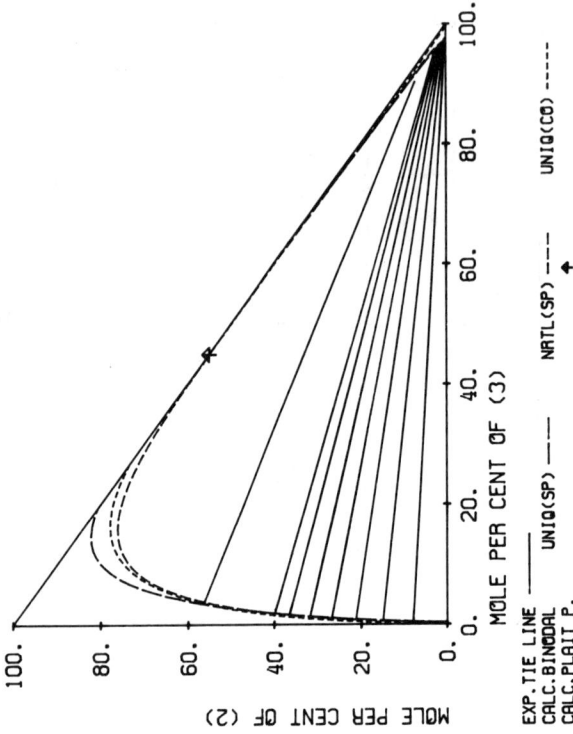

MOLE PER CENT OF (3)

MOLE PER CENT OF (2)

EXP.TIE LINE ——— UNIQ(SP) ——— NRTL(SP) ——— UNIQ(CO) -----
CALC.BINODAL
CALC.PLAIT P. ◆

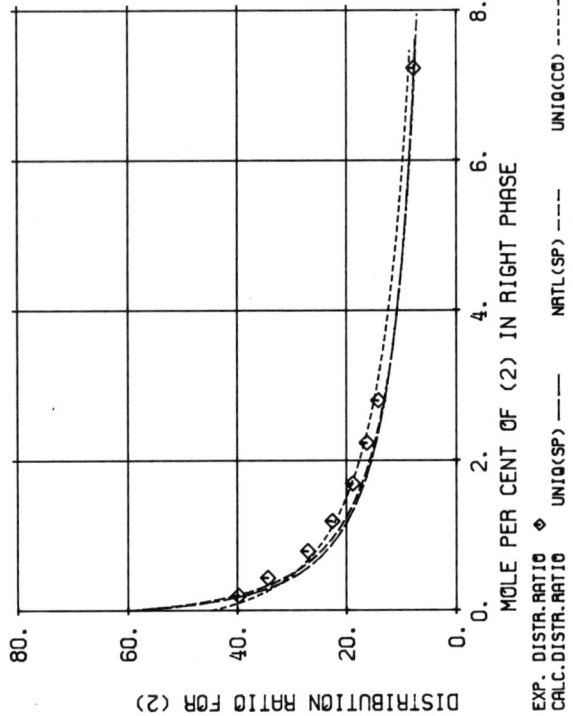

MOLE PER CENT OF (2) IN RIGHT PHASE

DISTRIBUTION RATIO FOR (2)

EXP. DISTR.RATIO ◇ UNIQ(SP) ——— NRTL(SP) ——— UNIQ(CO) -----
CALC.DISTR.RATIO

(1) H2O   WATER
(2) CH2O2   FORMIC ACID
(3) CHCL3   METHANE,TRICHLORO

OTHMER D.F., PING LIANG KU
J.CHEM.ENG.DATA 5(1960)42

TEMPERATURE = 25.0 DEG C   TYPE OF SYSTEM = 1

EXPERIMENTAL TIE LINES IN MOLE PCT (GRAPH.INTERPOL.)

	LEFT PHASE			RIGHT PHASE	
(1)	(2)	(3)	(1)	(2)	(3)
91.720	7.780	0.499	2.163	0.196	97.641
84.467	14.853	0.680	2.173	0.432	97.395
77.889	21.258	0.853	2.181	0.785	97.034
72.114	26.855	1.031	2.195	1.187	96.618
66.947	31.794	1.259	2.207	1.687	96.106
61.974	36.499	1.527	2.224	2.233	95.543
58.382	39.875	1.742	2.242	2.799	94.959
39.963	56.274	3.763	2.368	7.230	90.402

SPECIFIC MODEL PARAMETERS IN KELVIN

		UNIQUAC		NRTL(ALPHA=.2)	
I J		AIJ	AJI	AIJ	AJI
1 2		-227.67	-245.14	-240.56	-381.59
1 3		268.42	489.31	1391.1	555.11
2 3		141.78	285.60	434.73	245.64

R1 = 0.9200   R2 = 1.5280   R3 = 2.8700
Q1 = 1.400   Q2 = 1.532   Q3 = 2.410

MEAN DEV. BETWEEN CALC. AND EXP. CONC. IN MOLE PCT

UNIQUAC (SPECIFIC PARAMETERS)   0.32
NRTL   (SPECIFIC PARAMETERS)   0.34
UNIQUAC (COMMON PARAMETERS)   1.13

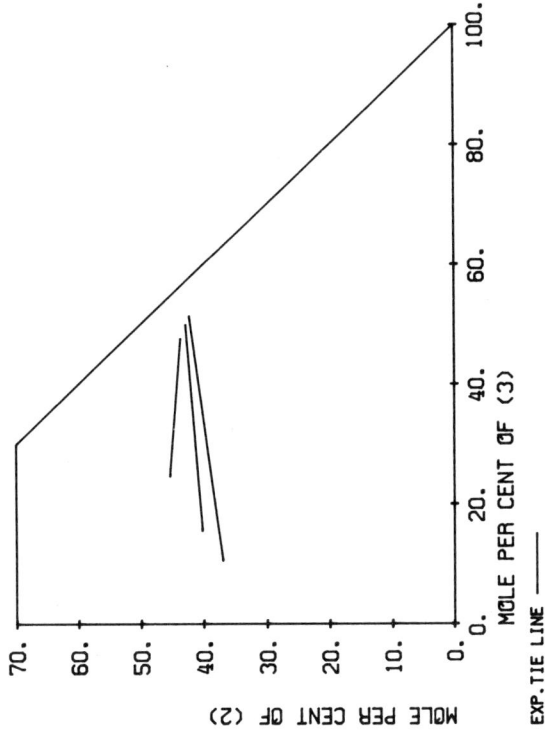

MOLE PER CENT OF (3)

EXP.TIE LINE ——

MOLE PER CENT OF (2)

(1) CHCL3   METHANE,TRICHLORO

(2) CH4O   METHANOL

(3) H2O   WATER

BONNER W.D.
J.PHYS.CHEM. 14(1910)738

TEMPERATURE =   0.0 DEG C   TYPE OF SYSTEM = 1

EXPERIMENTAL TIE LINES IN MOLE PCT (GRAPH.INTERPOL.)

LEFT PHASE			RIGHT PHASE		
(1)	(2)	(3)	(1)	(2)	(3)
52.457	36.945	10.598	6.363	42.198	51.439
44.230	40.184	15.586	7.252	42.780	49.969
30.026	45.331	24.643	8.767	43.586	47.647

CHCl$_3$-C$_2$H$_4$O$_2$

(1) CHCL3   METHANE, TRICHLORO

(2) C2H4O2   ACETIC ACID

(3) H2O   WATER

WRIGHT A.
PROC.ROY.SOC.LONDON 49(1891)174

TEMPERATURE = 18.0 DEG C   TYPE OF SYSTEM = 1

EXPERIMENTAL TIE LINES IN MOLE PCT

	LEFT PHASE			RIGHT PHASE	
(1)	(2)	(3)	(1)	(2)	(3)
93.786	0.0	6.214	0.128	0.0	99.872
93.492	1.968	4.541	0.147	2.047	97.807
85.970	6.892	7.138	0.137	6.104	93.759
80.251	11.759	7.990	0.224	9.250	90.525
81.116	14.053	4.831	0.353	10.528	89.119
74.883	18.732	6.385	0.607	13.688	85.705
61.380	27.028	11.592	1.751	21.037	77.212
50.736	32.560	16.704	3.546	27.139	69.315
47.186	34.444	18.370	4.381	28.925	66.693

SPECIFIC MODEL PARAMETERS IN KELVIN

		UNIQUAC		NRTL(ALPHA=.2)	
I	J	AIJ	AJI	AIJ	AJI
1	2	267.80	-110.57	295.10	-142.06
1	3	355.89	546.80	488.90	1779.9
2	3	100.20	-85.939	-55.868	176.02

R1 = 2.8700   R2 = 2.2024   R3 = 0.9200
Q1 = 2.410   Q2 = 2.072   Q3 = 1.400

MEAN DEV. BETWEEN CALC. AND EXP. CONC. IN MOLE PCT

UNIQUAC (SPECIFIC PARAMETERS)   1.01
NRTL   (SPECIFIC PARAMETERS)   0.95

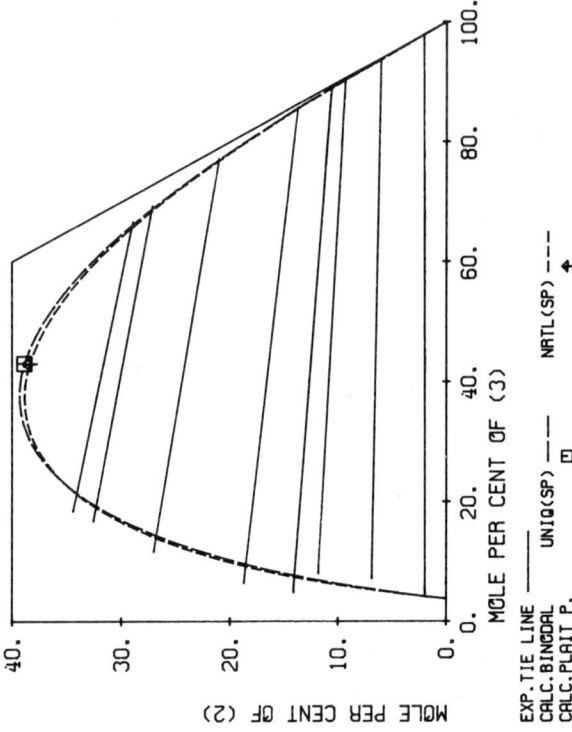

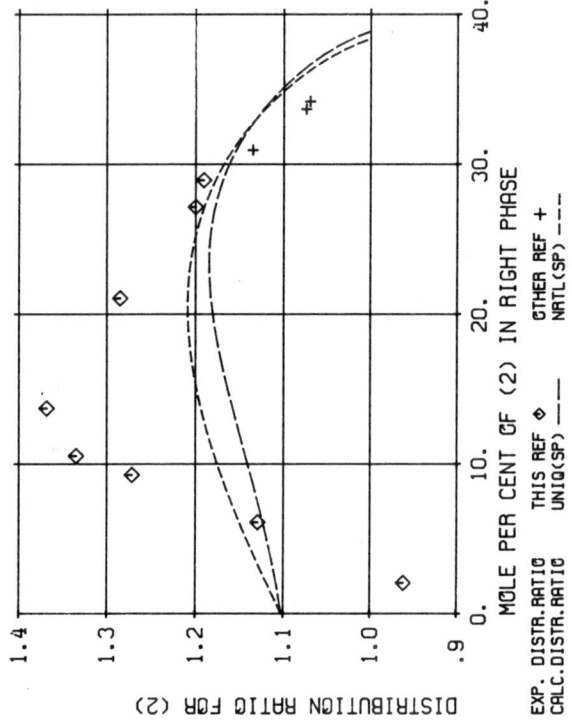

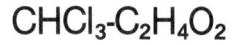

$CHCl_3$-$C_2H_4O_2$

(1) CHCL3   METHANE, TRICHLORO
(2) C2H4O2   ACETIC ACID
(3) H2O   WATER

WRIGHT A.
PROC.ROY.SOC.LONDON 50(1892)375

TEMPERATURE = 18.0 DEG C   TYPE OF SYSTEM = 1

EXPERIMENTAL TIE LINES IN MOLE PCT

	LEFT PHASE			RIGHT PHASE	
(1)	(2)	(3)	(1)	(2)	(3)
42.886	35.106	22.003	5.640	30.927	63.433
34.024	35.171	29.804	8.640	33.677	57.684
29.671	36.579	33.750	10.367	34.197	55.436

SPECIFIC MODEL PARAMETERS IN KELVIN

		UNIQUAC		NRTL(ALPHA=.2)	
I	J	AIJ	AJI	AIJ	AJI
1	2	267.80	-110.57	295.10	-142.06
1	3	355.89	546.80	488.90	1779.9
2	3	100.20	-35.939	-55.868	176.02

R1 = 2.8700 R2 = 2.2024 R3 = 0.9200
Q1 = 2.410   Q2 = 2.072   Q3 = 1.400

MEAN DEV. BETWEEN CALC. AND EXP. CONC. IN MOLE PCT

UNIQUAC (SPECIFIC PARAMETERS)   0.99
NRTL   (SPECIFIC PARAMETERS)   0.92

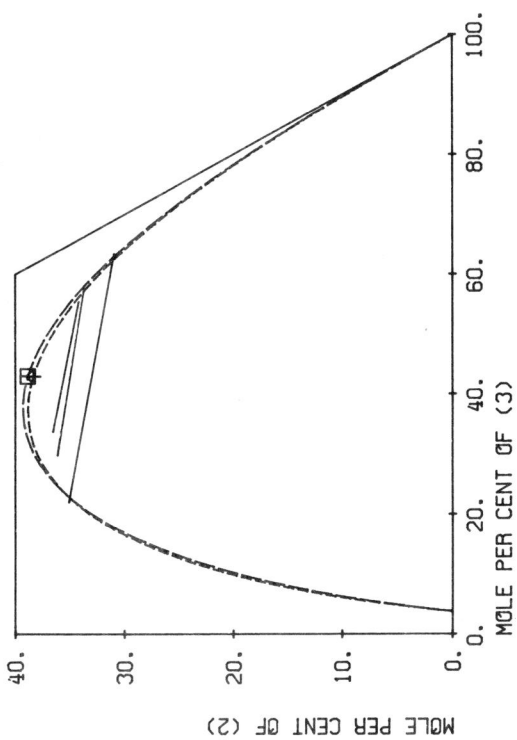

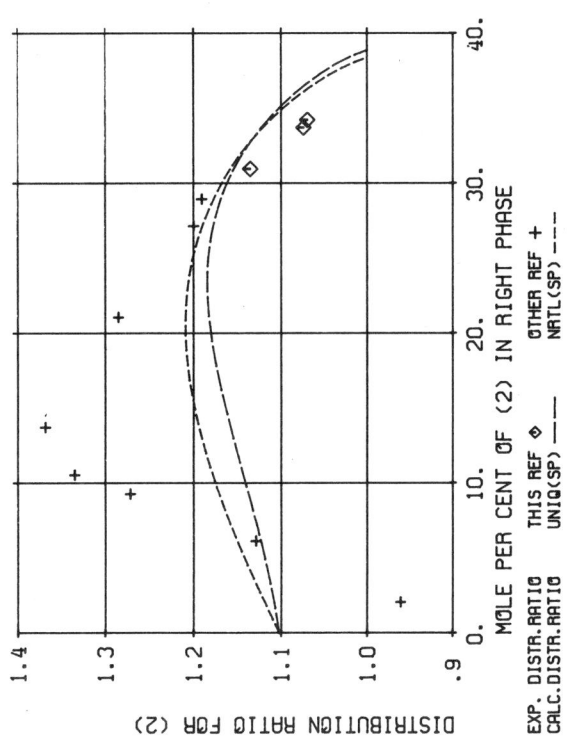

MOLE PER CENT OF (2)

MOLE PER CENT OF (3)

EXP.TIE LINE ——   UNIQ(SP) —— □   NRTL(SP) ---- ⧫
CALC.BINODAL
CALC.PLAIT P.

DISTRIBUTION RATIO FOR (2)

MOLE PER CENT OF (2) IN RIGHT PHASE

EXP.DISTR.RATIO   THIS REF ◇   OTHER REF +
CALC.DISTR.RATIO UNIQ(SP) —— NRTL(SP) ----

CHCl$_3$-C$_2$H$_4$O$_2$

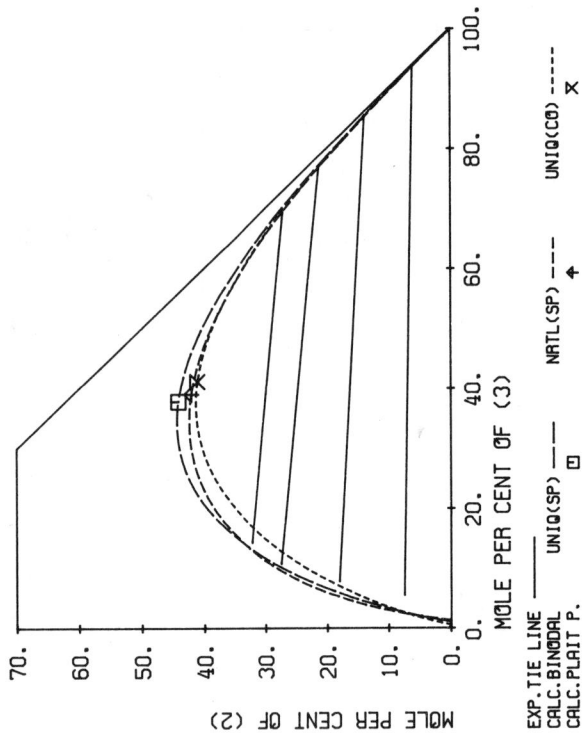

MOLE PER CENT OF (3)

MOLE PER CENT OF (2)

EXP.TIE LINE ——   UNIQ(SP) ⊟   NRTL(SP) ——   UNIQ(CO) ——
CALC.BINODAL ——
CALC.PLAIT P.   ⊀

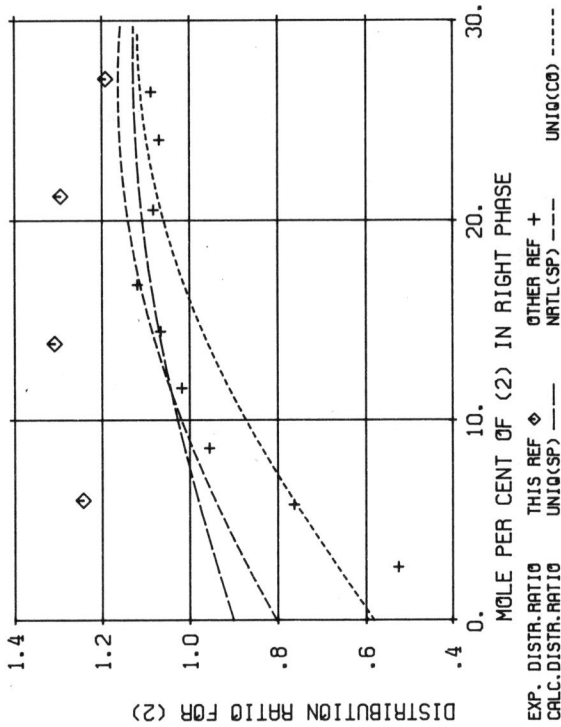

MOLE PER CENT OF (2) IN RIGHT PHASE

DISTRIBUTION RATIO FOR (2)

EXP. DISTR.RATIO ◇   THIS REF ◇   OTHER REF +   UNIQ(CO) ——
CALC.DISTR.RATIO   UNIQ(SP) ——   NRTL(SP) +

(1) CHCL3   METHANE,TRICHLORO
(2) C2H4O2   ACETIC ACID
(3) H2O   WATER

BRANCKER A.V., HUNTER T.G., NASH A.W.
J.PHYS.CHEM. 44(1940)683

TEMPERATURE = 25.0 DEG C   TYPE OF SYSTEM = 1

EXPERIMENTAL TIE LINES IN MOLE PCT

	LEFT PHASE			RIGHT PHASE	
	(1)	(2)	(3)	(1)	(2)	(3)
87.064	7.470	5.466	0.191	6.008	93.801
74.063	13.111	7.826	0.531	13.839	85.630
61.775	27.479	10.746	1.679	21.222	77.099
53.445	32.274	14.281	3.413	27.088	69.499

SPECIFIC MODEL PARAMETERS IN KELVIN

I J		UNIQUAC		NRTL(ALPHA=.2)	
		AIJ	AJI	AIJ	AJI
1 2	-146.59	54.545		-525.92	57.819
1 3	659.85	379.05		759.74	1518.5
2 3	-121.08	-108.23		-364.94	-98.812

R1 = 2.8700   R2 = 2.2024   R3 = 0.9200
Q1 = 2.410   Q2 = 2.072   Q3 = 1.400

MEAN DEV. BETWEEN CALC. AND EXP. CONC. IN MOLE PCT

UNIQUAC (SPECIFIC PARAMETERS)   1.72
NRTL   (SPECIFIC PARAMETERS)   1.78
UNIQUAC (COMMON PARAMETERS)   1.89

CHCl$_3$-C$_2$H$_4$O$_2$

(1) CHCL3   METHANE,TRICHLORO

(2) C2H4O2   ACETIC ACID

(3) H2O   WATER

OTHMER D.F., PING LIANG KU
J.CHEM.ENG.DATA 5(1960)42

TEMPERATURE = 25.0 DEG C   TYPE OF SYSTEM = 1

EXPERIMENTAL TIE LINES IN MOLE PCT (GRAPH.INTERPOL.)

| | LEFT PHASE | | | RIGHT PHASE | |
(1)	(2)	(3)	(1)	(2)	(3)
99.472	0.0	0.528	0.122	0.0	99.878
97.823	1.392	0.784	0.268	2.649	97.083
94.174	4.417	1.408	0.348	5.784	93.867
89.534	8.223	2.243	0.455	8.610	90.934
85.034	11.823	3.092	0.651	11.611	87.738
80.706	15.400	3.894	0.889	14.438	84.672
76.464	18.771	4.765	1.154	16.773	82.068
71.988	22.220	5.792	1.546	20.530	77.924
67.304	25.694	7.002	2.687	24.037	73.276
62.902	28.761	8.337	3.566	26.449	69.985

SPECIFIC MODEL PARAMETERS IN KELVIN

| | | UNIQUAC | | NRTL(ALPHA=.2) | |
I	J	AIJ	AJI	AIJ	AJI
1	2	-146.59	54.545	-525.92	57.819
1	3	659.85	379.05	759.74	1518.5
2	3	-121.08	-108.23	-364.94	-98.812

R1 = 2.8700   R2 = 2.2024   R3 = 0.9200
Q1 = 2.410   Q2 = 2.072   Q3 = 1.400

MEAN DEV. BETWEEN CALC. AND EXP. CONC. IN MOLE PCT

UNIQUAC (SPECIFIC PARAMETERS)   0.95
NRTL   (SPECIFIC PARAMETERS)   0.86
UNIQUAC (COMMON PARAMETERS)   1.24

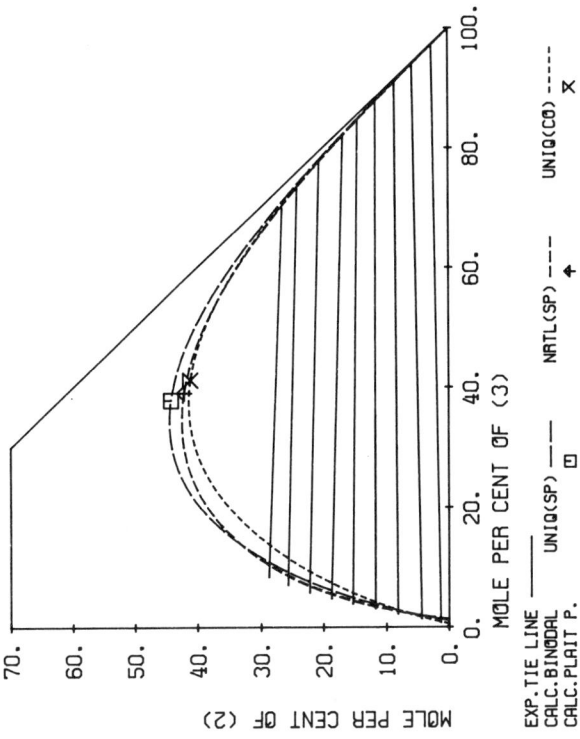

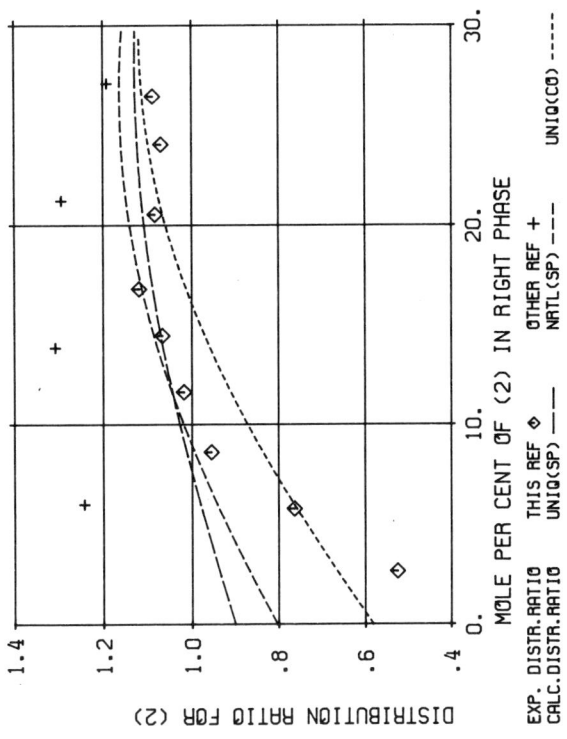

CHCl$_3$-C$_2$H$_5$ClO

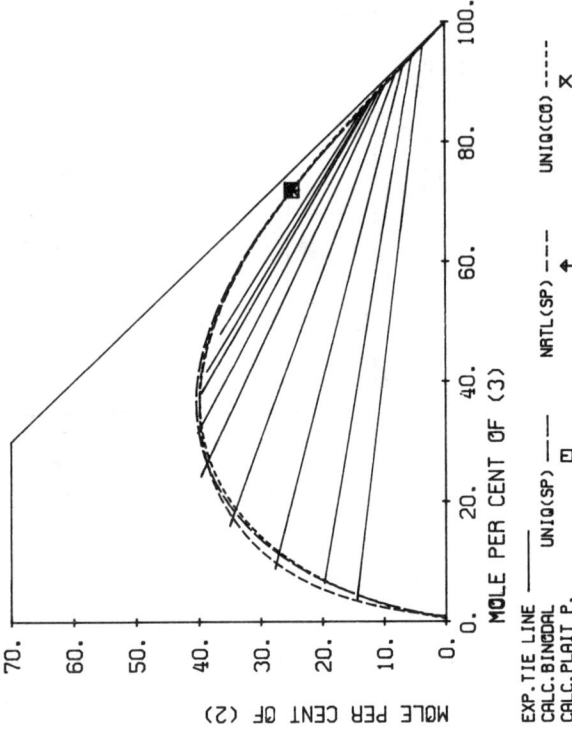

MOLE PER CENT OF (3)

MOLE PER CENT OF (2)

EXP.TIE LINE ——————   UNIQ(SP) —————— ▦   NRTL(SP) —————— ♦   UNIQ(CO) ------
CALC.BINODAL
CALC.PLAIT P.

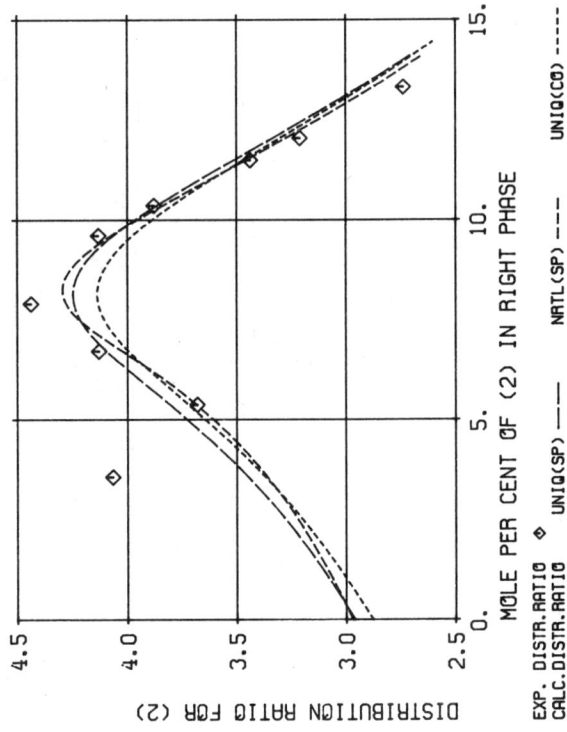

DISTRIBUTION RATIO FOR (2)

MOLE PER CENT OF (2) IN RIGHT PHASE

EXP. DISTR.RATIO ◇   UNIQ(SP) ——————   NRTL(SP) ——————   UNIQ(CO) ------
CALC.DISTR.RATIO

(1) CHCL3   METHANE,TRICHLORO
(2) C2H5CLO   ETHANOL,2-CHLORO
(3) H2O   WATER

ABABI V., POPA A., MIHAILA GH.
AN.STIINT.UNIV.AL.I.CUZA IASI. 9(1963)233

TEMPERATURE = 20.5 DEG C   TYPE OF SYSTEM = 1

EXPERIMENTAL TIE LINES IN MOLE PCT

	LEFT PHASE			RIGHT PHASE	
(1)	(2)	(3)	(1)	(2)	(3)
81.850	14.486	3.665	0.240	3.562	96.198
73.848	19.729	6.424	0.327	5.363	94.310
63.432	27.719	8.849	0.398	6.712	92.890
49.021	35.060	15.919	0.511	7.899	91.589
36.116	39.642	24.242	0.662	9.595	89.744
28.345	40.151	31.505	0.762	10.351	88.887
22.407	39.475	38.119	0.919	11.486	87.595
19.661	38.634	41.705	1.001	12.037	86.962
15.535	36.445	48.020	1.200	13.310	85.490

SPECIFIC MODEL PARAMETERS IN KELVIN

		UNIQUAC		NRTL(ALPHA=.2)	
I J	AIJ	AJI		AIJ	AJI
1 2	231.52	-70.165		655.36	-295.84
1 3	768.82	285.16		884.09	1313.0
2 3	-32.396	165.21		-333.52	1018.6

R1 = 2.8700   R2 = 2.6698   R3 = 0.9200
Q1 = 2.410   Q2 = 2.392   Q3 = 1.400

MEAN DEV. BETWEEN CALC. AND EXP. CONC. IN MOLE PCT

UNIQUAC (SPECIFIC PARAMETERS)   0.70
NRTL   (SPECIFIC PARAMETERS)   0.50
UNIQUAC (COMMON PARAMETERS)   0.73

CHCl$_3$-C$_2$H$_6$O

(1) CHCL3   METHANE,TRICHLORO
----
(2) C2H6O   ETHANOL
----
(3) H2O   WATER

KHANINA E.P.,BEREGOVYKH V.V., PAVLENKO T.G., TIMOFEEV V.S.
ZH.FIZ.KHIM. 52(1978)1558

TEMPERATURE = 20.0 DEG C   TYPE OF SYSTEM = 1

EXPERIMENTAL TIE LINES IN MOLE PCT

          LEFT PHASE   RIGHT PHASE
      (1)   (2)   (3)   (1)   (2)   (3)

   97.564   1.784   0.652   0.097   4.187   95.716
   92.926   5.181   1.893   0.102   7.507   92.392
   82.735   12.584   4.681   0.106   10.788   89.106
   74.395   17.953   7.652   0.092   13.728   86.180
   64.403   23.618   11.980   0.092   14.135   85.773
   54.236   28.671   17.093   0.206   14.817   84.977
   40.806   33.604   25.590   0.401   15.524   84.076
   34.972   34.222   30.806   0.505   16.608   82.887
   27.105   34.340   38.555   0.694   17.615   81.691
   22.637   34.383   42.980   1.194   19.105   79.701

SPECIFIC MODEL PARAMETERS IN KELVIN

            UNIQUAC   NRTL(ALPHA=.2)
 I J   AIJ   AJI   AIJ   AJI
----
 1 2   504.88   -126.98   1047.2   -385.18
 1 3   610.76   253.86   836.10   2424.7
 2 3 -129.94   331.69   90.875   214.44

R1 = 2.8700   R2 = 2.1055   R3 = 0.9200
Q1 = 2.410   Q2 = 1.972   Q3 = 1.400

MEAN DEV. BETWEEN CALC. AND EXP. CONC. IN MOLE PCT

UNIQUAC (SPECIFIC PARAMETERS)   1.12
NRTL   (SPECIFIC PARAMETERS)   0.59
UNIQUAC (COMMON PARAMETERS)   4.66

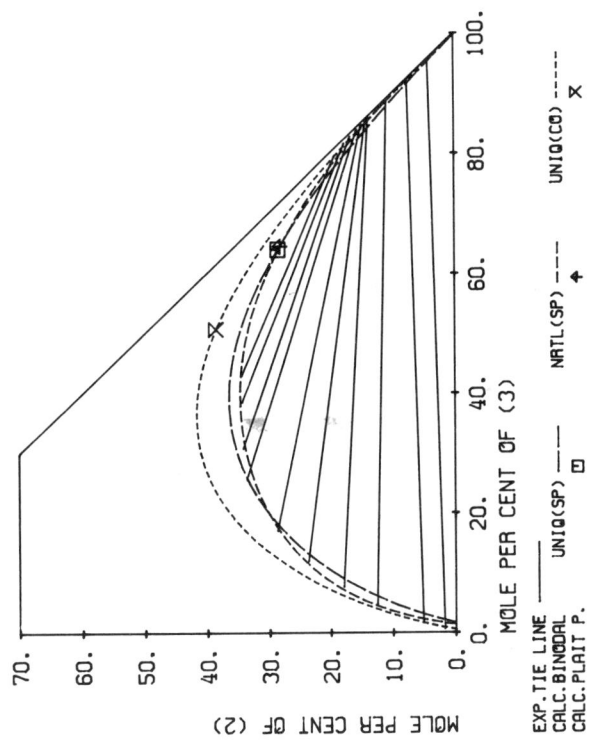

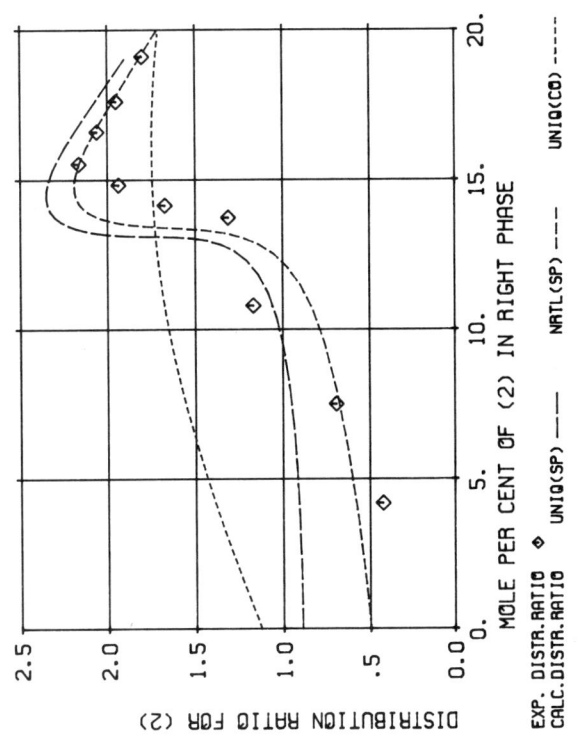

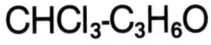

CHCl₃-C₃H₆O

(1) CHCL3   METHANE,TRICHLORO

(2) C3H6O   2-PROPANONE

(3) H2O   WATER

BONNER W.D.
J.PHYS.CHEM. 14(1910)738

TEMPERATURE =   0.0 DEG C   TYPE OF SYSTEM = 1

EXPERIMENTAL TIE LINES IN MOLE PCT (GRAPH.INTERPOL.)

LEFT PHASE			RIGHT PHASE		
(1)	(2)	(3)	(1)	(2)	(3)
16.073	62.219	21.703	0.257	13.286	86.457
15.533	61.738	22.729	0.400	16.803	82.796
14.657	61.141	24.202	0.508	19.504	79.989
15.090	61.437	23.473	0.508	19.504	79.989
14.255	60.554	25.190	0.618	21.696	77.686
13.892	60.206	25.901	0.845	24.799	74.356
12.814	58.981	28.205	1.079	27.080	71.841
11.237	56.890	31.873	1.784	31.054	67.162
6.606	47.674	45.721	4.292	40.326	55.382

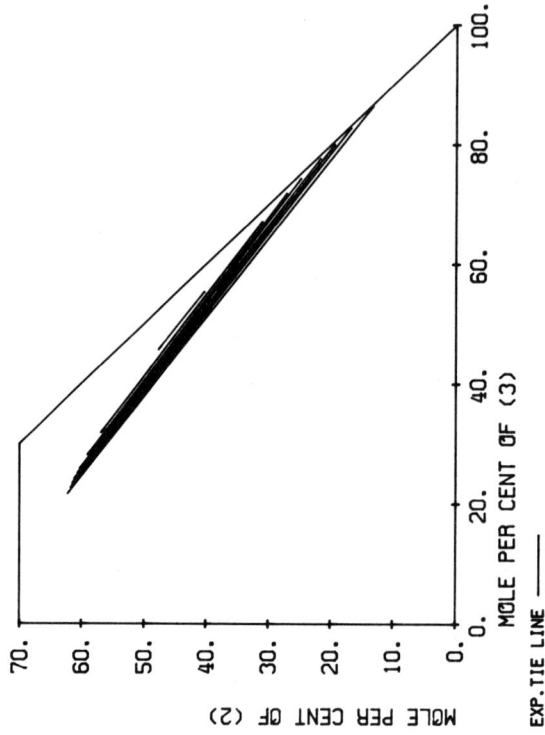

CHCl$_3$-C$_3$H$_6$O

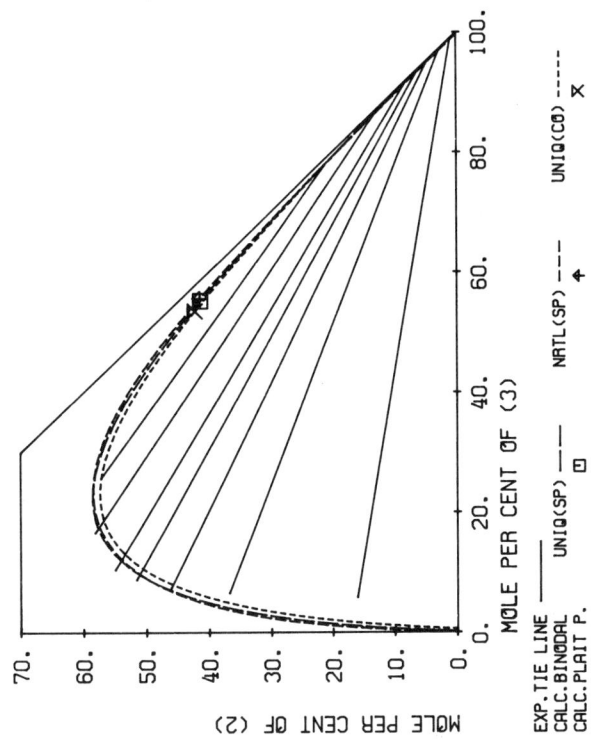

MOLE PER CENT OF (3)

MOLE PER CENT OF (2)

EXP.TIE LINE ——————   UNIQ(SP) ▣   NRTL(SP) ———— ✦   UNIQ(CO) ----- ✗
CALC.BINODAL ———————
CALC.PLAIT P. ————————

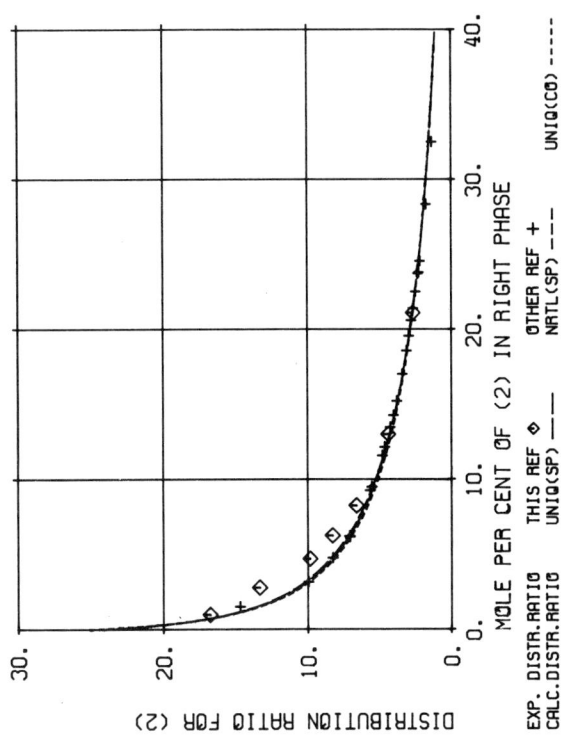

MOLE PER CENT OF (2) IN RIGHT PHASE

DISTRIBUTION RATIO FOR (2)

EXP. DISTR.RATIO ◇   THIS REF ◇   OTHER REF +   UNIQ(CO) -----
CALC.DISTR.RATIO ——————   UNIQ(SP) ———   NRTL(SP) ——

(1) CHCL3   METHANE,TRICHLORO
(2) C3H5O   2-PROPANONE
(3) H2O   WATER

BRANCKER A.V., HUNTER T.G., NASH A.W.
J.PHYS.CHEM. 44(1940)683

TEMPERATURE = 25.0 DEG C   TYPE OF SYSTEM = 1

EXPERIMENTAL TIE LINES IN MOLE PCT

	LEFT PHASE			RIGHT PHASE	
(1)	(2)	(3)	(1)	(2)	(3)
78.175	16.069	5.756	0.155	0.959	93.886
56.677	36.813	6.510	0.194	2.761	97.045
46.506	46.068	7.426	0.253	4.683	95.064
39.641	51.603	8.756	0.279	6.229	93.492
34.608	54.963	10.424	0.326	8.236	91.437
25.193	58.112	16.695	0.416	12.983	86.601
17.059	57.103	25.838	1.037	21.079	77.884

SPECIFIC MODEL PARAMETERS IN KELVIN

		UNIQUAC		NRTL(ALPHA=.2)	
I	J	AIJ	AJI	AIJ	AJI
1	2	112.04	-178.71	-5.0568	-315.17
1	3	1041.0	-311.36	1411.8	1116.7
2	3	312.58	-55.970	40.753	529.20

R1 = 2.8700   R2 = 2.5735   R3 = 0.9200
Q1 = 2.410   Q2 = 2.336   Q3 = 1.400

MEAN DEV. BETWEEN CALC. AND EXP. CONC. IN MOLE PCT

UNIQUAC (SPECIFIC PARAMETERS)   1.36
NRTL   (SPECIFIC PARAMETERS)   1.53
UNIQUAC (COMMON PARAMETERS)   1.49

CHCl₃-C₃H₆O

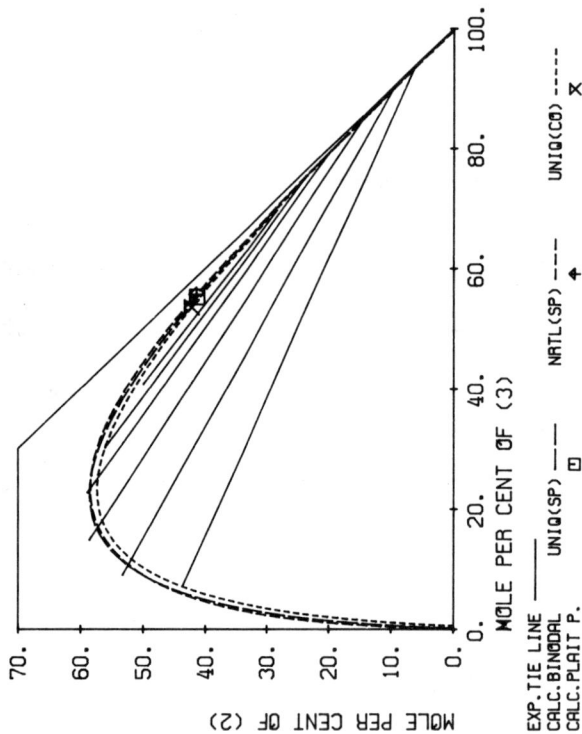

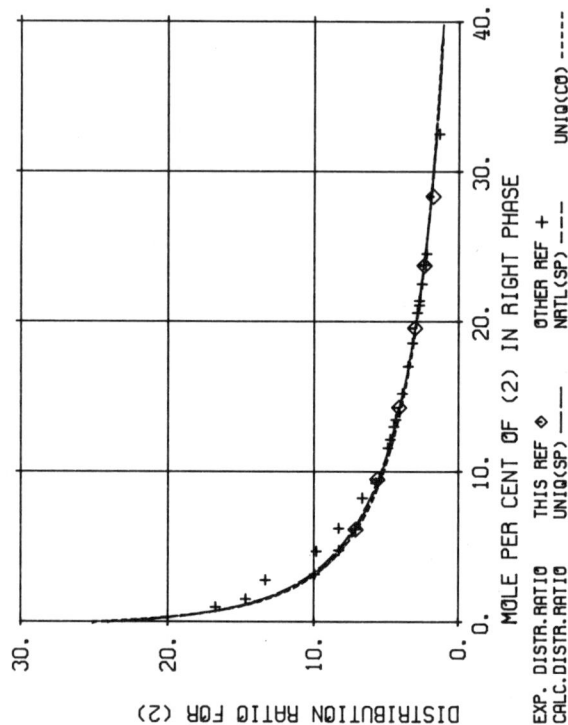

(1) CHCL3   METHANE,TRICHLORO

(2) C3H6O   2-PROPANONE

(3) H2O   WATER

BANCROFT W.D., HUBARD S.D.,
J.AM.CHEM.SOC. 64(1942)347

TEMPERATURE = 25.0 DEG C   TYPE OF SYSTEM = 1

EXPERIMENTAL TIE LINES IN MOLE PCT

	LEFT PHASE			RIGHT PHASE	
(1)	(2)	(3)	(1)	(2)	(3)
49.218	43.671	7.111	0.121	6.135	93.745
37.680	53.279	9.041	0.184	9.468	90.348
26.606	58.575	14.819	0.322	14.244	85.434
18.364	58.902	22.734	0.666	19.543	79.790
13.714	56.095	30.192	1.153	23.703	75.144
9.253	49.956	40.791	1.942	28.320	69.738

SPECIFIC MODEL PARAMETERS IN KELVIN

		UNIQUAC		NRTL(ALPHA=.2)	
I J		AIJ	AJI	AIJ	AJI
1 2		112.04	-178.71	-5.0563	-315.17
1 3		1041.0	311.36	1411.8	1116.7
2 3		312.58	-55.970	40.758	529.20

R1 = 2.8700   R2 = 2.5735   R3 = 0.9200
Q1 = 2.410   Q2 = 2.336   Q3 = 1.400

MEAN DEV. BETWEEN CALC. AND EXP. CONC. IN MOLE PCT

UNIQUAC (SPECIFIC PARAMETERS)   0.89
NRTL   (SPECIFIC PARAMETERS)   1.00
UNIQUAC (COMMON PARAMETERS)   1.11

CHCl$_3$-C$_3$H$_6$O

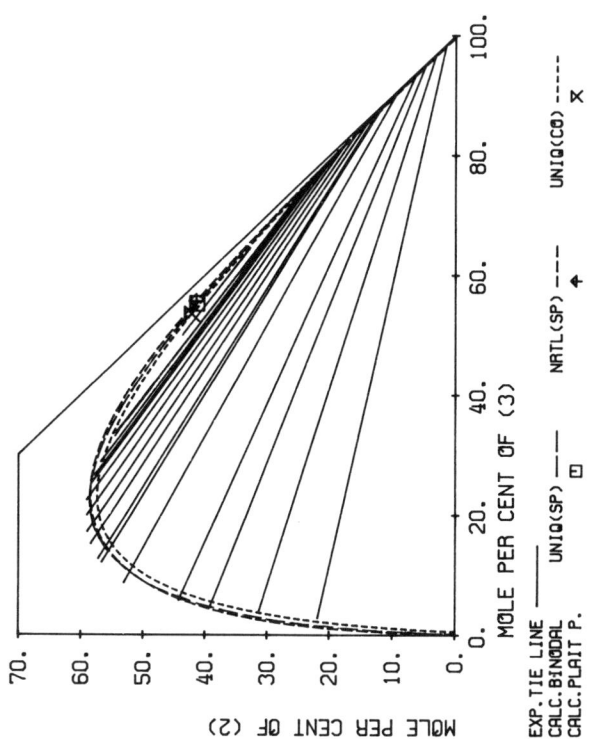

MOLE PER CENT OF (3)

MOLE PER CENT OF (2)

EXP.TIE LINE ——— CALC.BINODAL ——— UNIQ(SP) ——— ◻ NRTL(SP) ——— ✦ UNIQ(CO) ----- ✗
CALC.PLAIT P.

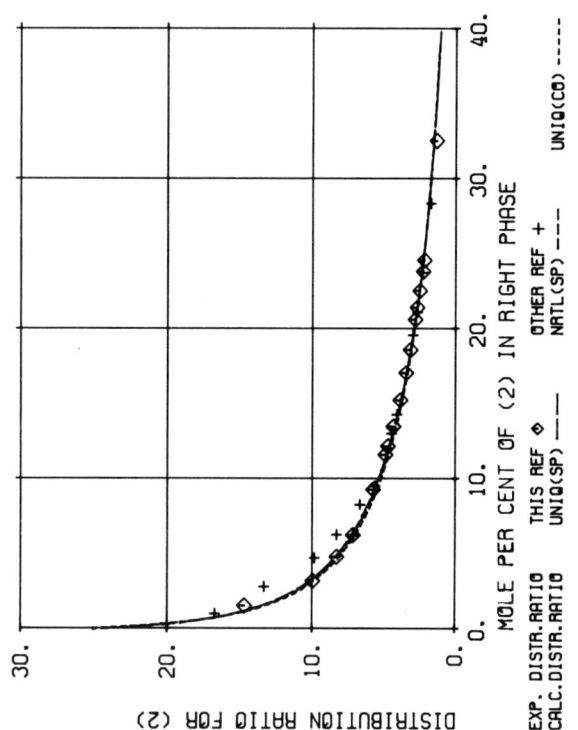

MOLE PER CENT OF (2) IN RIGHT PHASE

DISTRIBUTION RATIO FOR (2)

EXP. DISTR.RATIO ◇ THIS REF OTHER REF +
CALC.DISTR.RATIO UNIQ(SP) ——— NRTL(SP) ——— UNIQ(CO) -----

(1) CHCL3   METHANE,TRICHLORO
(2) C3H6O   2-PROPANONE
(3) H2O   WATER

REINDERS W., DE MINJER C.H.
RECL.TRAV.CHIM.PAYS-BAS. 66(1947)573

TEMPERATURE = 25.0 DEG C   TYPE OF SYSTEM = 1

EXPERIMENTAL TIE LINES IN MOLE PCT

	LEFT PHASE			RIGHT PHASE	
(1)	(2)	(3)	(1)	(2)	(3)
75.305	21.833	2.861	0.141	1.485	98.373
64.898	31.358	3.744	0.163	3.146	96.691
56.250	39.217	4.533	0.168	4.737	95.094
50.037	44.205	5.758	0.173	6.193	93.634
38.249	53.085	8.666	0.183	9.239	90.577
31.311	56.625	12.063	0.191	11.562	88.246
30.105	57.204	12.692	0.213	12.125	87.663
26.463	58.387	15.150	0.258	13.445	86.297
23.879	58.944	17.176	0.327	15.185	84.487
21.057	59.009	19.934	0.423	17.003	82.574
18.528	58.997	22.476	0.522	18.566	80.912
16.503	58.335	25.162	0.770	20.573	78.652
15.669	57.738	26.592	0.874	21.374	77.752
14.365	56.769	28.866	0.937	22.496	76.566
12.325	54.542	33.134	1.154	23.778	75.067
11.812	54.070	34.119	1.218	24.532	74.250
6.442	43.634	49.924	2.810	32.493	64.697

SPECIFIC MODEL PARAMETERS IN KELVIN

		UNIQUAC		NRTL(ALPHA=.2)	
I	J	AIJ	AJI	AIJ	AJI
1	2	112.04	-178.71	-5.0568	-315.17
1	3	1041.0	-311.36	1411.8	1116.7
2	3	312.58	-55.970	40.758	529.20

R1 = 2.8700   R2 = 2.5735   R3 = 0.9200
Q1 = 2.410   Q2 = 2.336   Q3 = 1.400

MEAN DEV. BETWEEN CALC. AND EXP. CONC. IN MOLE PCT

UNIQUAC (SPECIFIC PARAMETERS)   0.83
NRTL   (SPECIFIC PARAMETERS)   0.92
UNIQUAC (COMMON PARAMETERS)   1.20

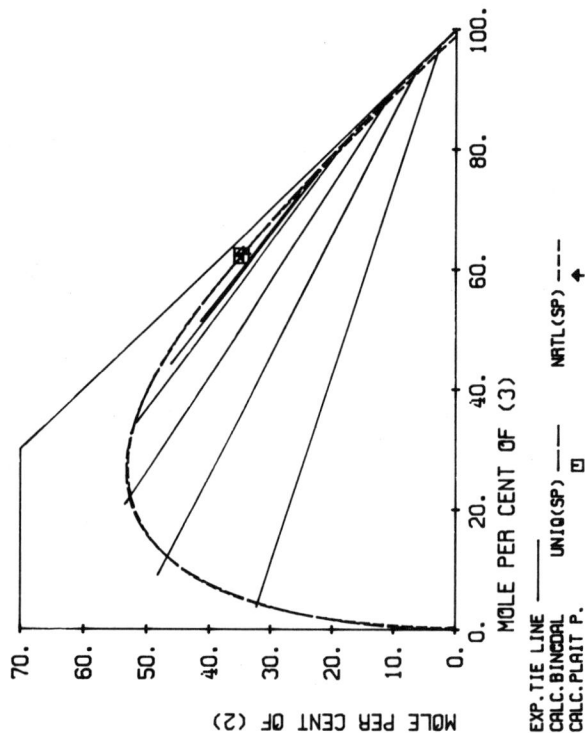

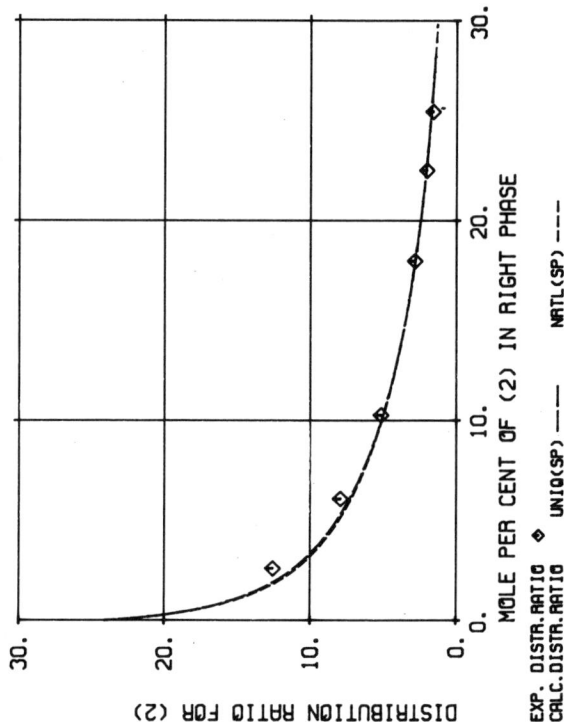

CHCl₃-C₃H₆O

(1) CHCL3   METHANE, TRICHLORO

(2) C3H6O   2-PROPANONE

(3) H2O   WATER

REINDERS W., DE MINJER C.H.
RECL.TRAV.CHIM.PAYS-BAS. 66(1947)573

TEMPERATURE = 60.0 DEG C   TYPE OF SYSTEM = 1

EXPERIMENTAL TIE LINES IN MOLE PCT

	LEFT PHASE			RIGHT PHASE	
(1)	(2)	(3)	(1)	(2)	(3)
63.952	32.326	3.722	0.145	2.578	97.278
42.806	48.232	8.962	0.155	6.066	93.779
25.825	53.420	20.755	0.263	10.254	89.483
14.002	51.642	34.356	0.852	17.956	81.192
9.655	46.148	44.197	1.659	22.489	75.851
7.320	41.375	51.305	2.274	25.421	72.305

SPECIFIC MODEL PARAMETERS IN KELVIN

		UNIQUAC		NRTL(ALPHA=.2)	
I	J	AIJ	AJI	AIJ	AJI
1	2	323.01	-292.79	229.76	-477.19
1	3	1529.7	386.73	1485.0	905.59
2	3	263.09	-25.362	-54.081	753.74

R1 = 2.8700   R2 = 2.5735   R3 = 0.9200
Q1 = 2.410   Q2 = 2.336   Q3 = 1.400

MEAN DEV. BETWEEN CALC. AND EXP. CONC. IN MOLE PCT

UNIQUAC (SPECIFIC PARAMETERS)   0.91
NRTL   (SPECIFIC PARAMETERS)   1.05

(1) CHCL3   METHANE,TRICHLORO
(2) C3H6O2   PROPANOIC ACID
(3) H2O   WATER

IGUCHI A.; FUSE K.
KAGAKU KOGAKU 36(1972)673

TEMPERATURE = 25.0 DEG C   TYPE OF SYSTEM = 1

EXPERIMENTAL TIE LINES IN MOLE PCT (GRAPH.INTERPOL.)

	LEFT PHASE			RIGHT PHASE	
(1)	(2)	(3)	(1)	(2)	(3)
91.083	7.644	1.273	0.127	1.411	98.462
82.844	15.315	1.841	0.130	2.293	97.577
74.197	22.860	2.943	0.150	3.189	96.660
63.729	31.268	5.003	0.155	4.368	95.477
55.603	36.500	7.897	0.160	5.475	94.365
44.097	41.142	14.761	0.281	7.311	92.407
39.507	41.529	18.965	0.387	8.393	91.220
33.627	40.259	26.114	0.494	9.042	90.464
22.889	36.155	40.957	0.999	11.334	87.666
16.157	31.714	52.129	2.930	15.447	81.623

SPECIFIC MODEL PARAMETERS IN KELVIN

		UNIQUAC		NRTL(ALPHA=.2)	
I	J	AIJ	AJI	AIJ	AJI
1	2	417.94	-228.61	1804.1	-880.74
1	3	1243.3	257.67	1943.3	2207.7
2	3	176.22	-20.420	476.94	73.994

R1 = 2.8700   R2 = 2.8768   R3 = 0.9200
Q1 = 2.410   Q2 = 2.612   Q3 = 1.400

MEAN DEV. BETWEEN CALC. AND EXP. CONC. IN MOLE PCT

UNIQUAC (SPECIFIC PARAMETERS)   1.83
NRTL   (SPECIFIC PARAMETERS)   0.52
UNIQUAC (COMMON PARAMETERS)   2.83

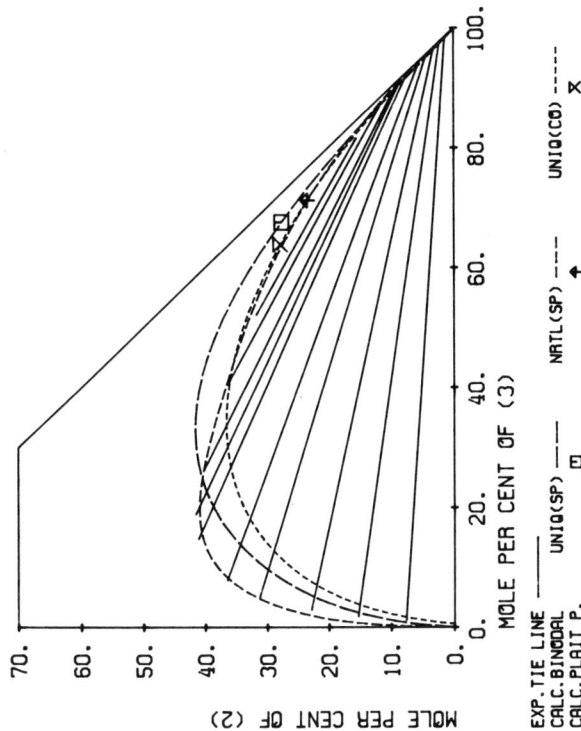

MOLE PER CENT OF (3)

MOLE PER CENT OF (2)

EXP.TIE LINE ———   UNIQ(SP) □   NRTL(SP) ◆   UNIQ(CO) ----- x
CALC.BINODAL
CALC.PLAIT P.

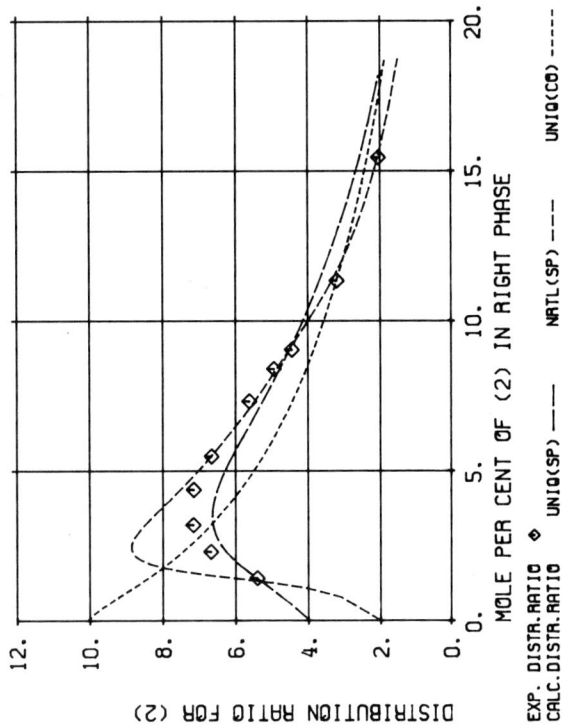

MOLE PER CENT OF (2) IN RIGHT PHASE

DISTRIBUTION RATIO FOR (2)

EXP. DISTR.RATIO ◆   UNIQ(SP) ◇   NRTL(SP) -----   UNIQ(CO) -----
CALC.DISTR.RATIO

CHCl$_3$-C$_3$H$_8$O

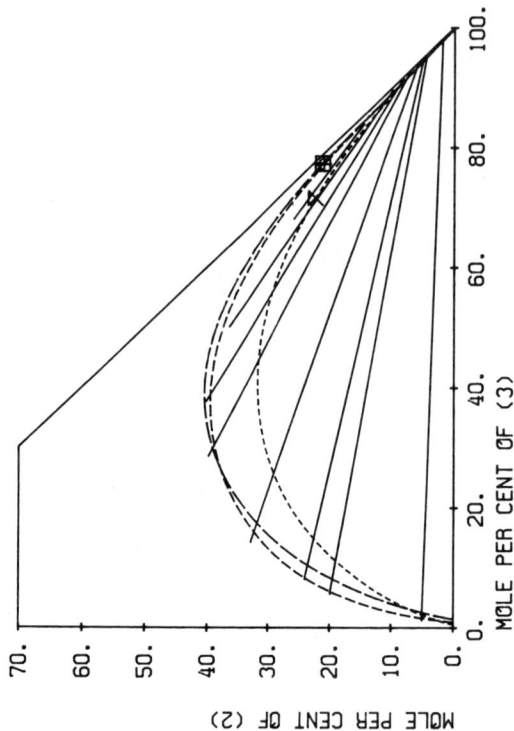

EXP.TIE LINE   UNIQ(SP) ☐   NRTL(SP) ----   UNIQ(CO) -----
CALC.BINODAL ----
CALC.PLAIT P.   ✕

MOLE PER CENT OF (3)

MOLE PER CENT OF (2)

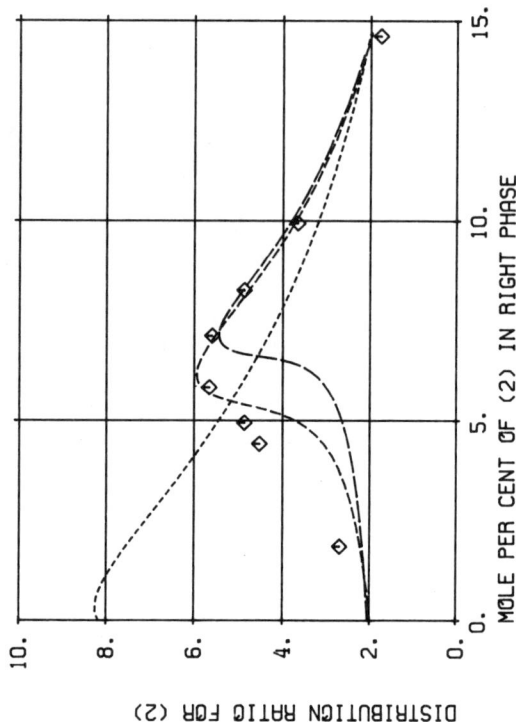

EXP. DISTR.RATIO ◇   UNIQ(SP) ◇   NRTL(SP) ----   UNIQ(CO) -----
CALC.DISTR.RATIO

MOLE PER CENT OF (2) IN RIGHT PHASE

DISTRIBUTION RATIO FOR (2)

(1) CHCL3   METHANE,TRICHLORO

(2) C3H8O   2-PROPANOL

(3) H2O   WATER

IZMAILOV N.A., FRANKE A.K.
ZH.FIZ.KHIM. 29(1955)263

TEMPERATURE = 25.0 DEG C   TYPE OF SYSTEM = 1

EXPERIMENTAL TIE LINES IN MOLE PCT

	LEFT PHASE			RIGHT PHASE	
(1)	(2)	(3)	(1)	(2)	(3)
93.742	4.980	1.278	0.063	1.851	98.086
74.364	19.985	5.651	0.117	4.424	95.458
67.895	24.033	8.073	0.136	4.949	94.915
52.806	32.969	14.225	0.156	5.834	94.010
31.630	39.804	28.566	0.178	7.122	92.701
22.232	40.139	37.628	0.274	8.257	91.470
13.465	36.439	50.095	0.514	9.941	89.545
6.046	25.881	68.073	1.438	14.625	83.937

SPECIFIC MODEL PARAMETERS IN KELVIN

		UNIQUAC		NRTL(ALPHA=.2)	
I J	AIJ	AJI		AIJ	AJI
1 2	276.43	-42.811		955.11	-362.40
1 3	629.70	284.71		868.79	1297.7
2 3	-0.14886	169.40		-354.02	1132.2

R1 = 2.8700   R2 = 2.7791   R3 = 0.9200
Q1 = 2.410   Q2 = 2.508   Q3 = 1.400

MEAN DEV. BETWEEN CALC. AND EXP. CONC. IN MOLE PCT

UNIQUAC (SPECIFIC PARAMETERS)   1.17
NRTL (SPECIFIC PARAMETERS)   0.69
UNIQUAC (COMMON PARAMETERS)   3.39

(1) CHCL3   METHANE, TRICHLORO
----
(2) C5H4O2   FURFURAL
----
(3) H2O   WATER

KRUPATKIN I.L., GLAGOLEVA M.F.:
ZH.PRIKL.KHIM.(LENINGRAD) 42(1969)1526

TEMPERATURE = 25.0 DEG C   TYPE OF SYSTEM = 2

EXPERIMENTAL TIE LINES IN MOLE PCT

          LEFT PHASE   RIGHT PHASE
     (1)   (2)   (3)   (1)   (2)   (3)

  72.718   20.126   7.157   0.108   0.279   99.614
  49.201   41.867   8.933   0.086   0.639   99.276
  32.491   57.409   10.101   0.063   0.940   98.997
  23.406   63.880   12.714   0.048   1.103   98.850
  14.755   70.578   14.667   0.032   1.352   98.616
   6.834   74.681   18.485   0.016   1.634   98.350

SPECIFIC MODEL PARAMETERS IN KELVIN

              UNIQUAC   NRTL(ALPHA=.2)
  I J   AIJ   AJI   AIJ   AJI

  1 2 -107.05   58.669   -193.99   2.0775
  1 3   382.44   300.38   525.79   1017.8
  2 3   260.51   35.961   107.92   1054.3

  R1 = 2.8700   R2 = 3.1680   R3 = 0.9200
  Q1 = 2.410   Q2 = 2.484   Q3 = 1.400

MEAN DEV. BETWEEN CALC. AND EXP. CONC. IN MOLE PCT

UNIQUAC (SPECIFIC PARAMETERS)   0.38
NRTL   (SPECIFIC PARAMETERS)   0.64
UNIQUAC (COMMON PARAMETERS)   1.11

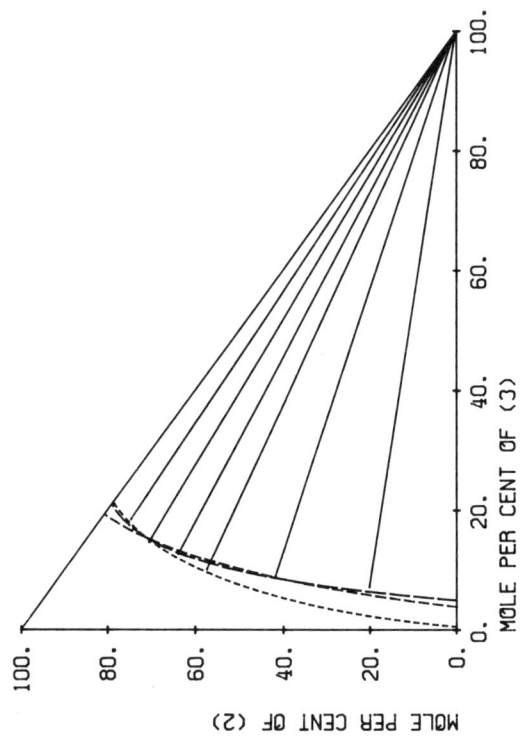

MOLE PER CENT OF (2)

MOLE PER CENT OF (3)

EXP.TIE LINE ———
CALC.BINODAL ———   UNIQ(SP) ———   NRTL(SP) ----   UNIQ(CO) -----

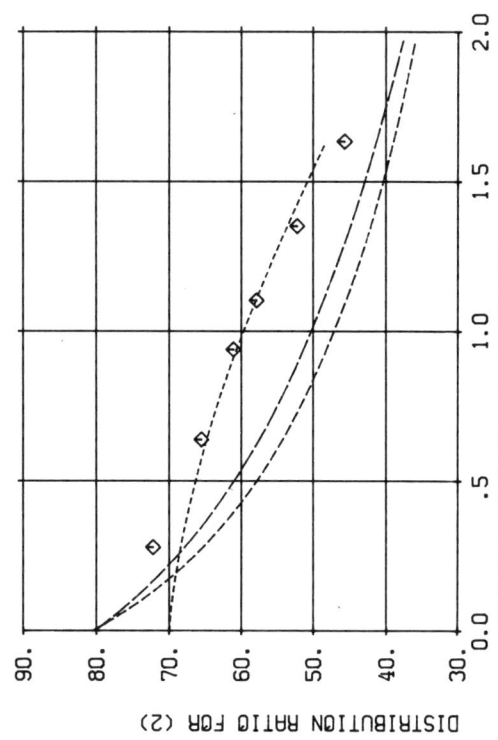

DISTRIBUTION RATIO FOR (2)

MOLE PER CENT OF (2) IN RIGHT PHASE

EXP. DISTR.RATIO ◇
CALC.DISTR.RATIO ———   UNIQ(SP) ◇   NRTL(SP) ———   UNIQ(CO) -----

CHCl$_3$-C$_6$H$_7$N

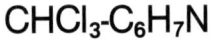

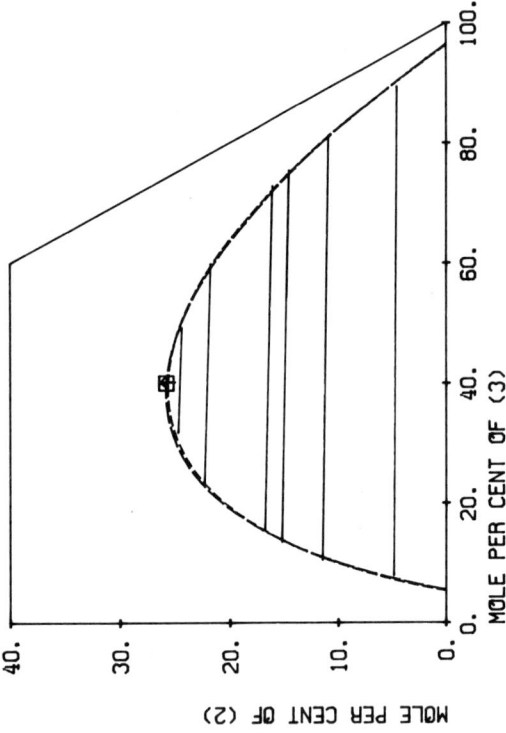

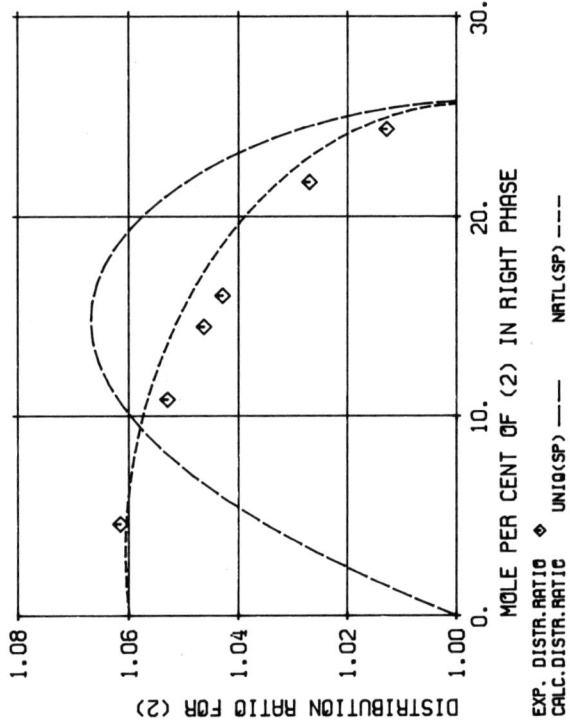

(1) C7H16   HEPTANE,
(2) CHCL3   METHANE,TRICHLORO
(3) C6H7N   ANILINE

PALATNIK L.S.,ET AL.
ZH.FIZ.KHIM. 33(1959)1939

TEMPERATURE = 18.0 DEG C   TYPE OF SYSTEM = 1

EXPERIMENTAL TIE LINES IN MOLE PCT

	LEFT PHASE			RIGHT PHASE	
(1)	(2)	(3)	(1)	(2)	(3)
87.228	4.887	7.885	5.957	4.604	89.439
78.111	11.410	10.479	7.996	10.836	81.168
71.361	15.147	13.492	10.036	14.477	75.487
67.904	16.714	15.382	11.064	16.027	72.908
54.390	22.311	23.299	18.386	21.725	59.889
43.554	24.689	31.758	26.306	24.375	49.318

SPECIFIC MODEL PARAMETERS IN KELVIN

		UNIQUAC		NRTL(ALPHA=.2)	
I J		AIJ	AJI	AIJ	AJI
1 2		-39.351	-21.790	-47.408	-239.22
1 3		277.36	54.002	520.61	676.90
2 3		-211.00	220.80	-293.74	32.530

R1 = 5.1742   R2 = 2.8700   R3 = 3.7165
Q1 = 4.396   Q2 = 2.410   Q3 = 2.816

MEAN DEV. BETWEEN CALC. AND EXP. CONC. IN MOLE PCT

UNIQUAC (SPECIFIC PARAMETERS)   0.29
NRTL   (SPECIFIC PARAMETERS)   0.25

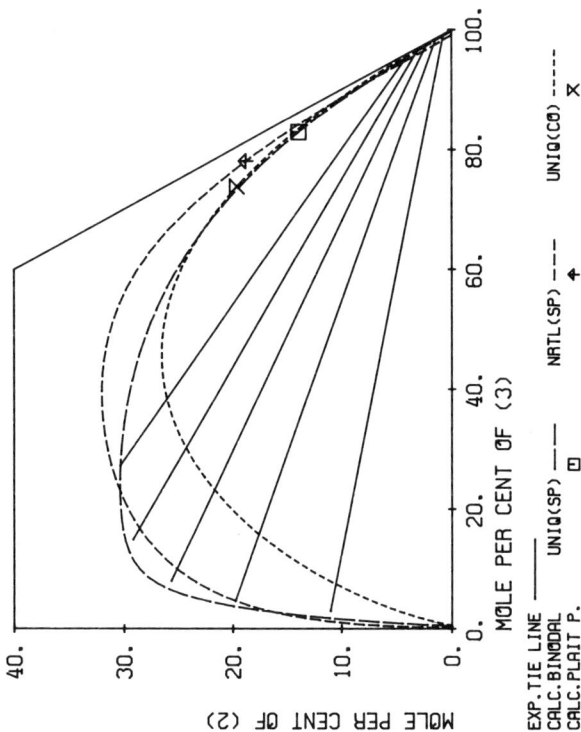

MOLE PER CENT OF (2)

MOLE PER CENT OF (3)

EXP.TIE LINE ———   UNIQ(SP) ——— ▣   NRTL(SP) ——— ✛   UNIQ(CO) - - - - ✕
CALC.BINODAL - - - - -
CALC.PLAIT P.

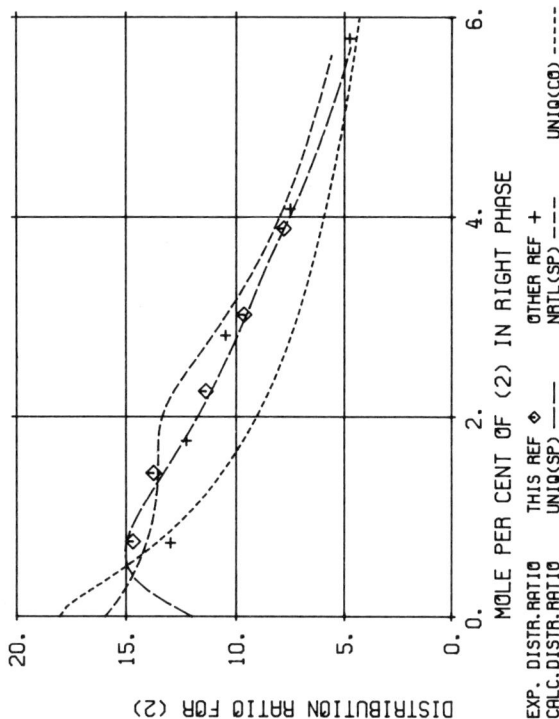

DISTRIBUTION RATIO FOR (2)

MOLE PER CENT OF (2) IN RIGHT PHASE

EXP. DISTR.RATIO   THIS REF ◇   OTHER REF +
CALC.DISTR.RATIO   UNIQ(SP) - - -   NRTL(SP) ———   UNIQ(CO) - - - - -

(1) CHCL3   METHANE, TRICHLORO
----
(2) C6H11NO   HEXANOIC ACID,6-AMINO,LACTAM
----
(3) H2O   WATER

KUDRYAVTSEVA G.I., KRUTIKOVA A.D.
ZH.PRIKL.KHIM.(LENINGRAD) 26(1953)1190

TEMPERATURE = 20.0 DEG C   TYPE OF SYSTEM = 1

EXPERIMENTAL TIE LINES IN MOLE PCT (GRAPH.INTERPOL.)

	LEFT PHASE			RIGHT PHASE	
(1)	(2)	(3)	(1)	(2)	(3)
86.101	11.008	2.892	0.163	0.750	99.087
75.594	19.747	4.659	0.189	1.436	98.376
66.416	25.664	7.920	0.215	2.253	97.532
55.943	29.156	14.901	0.230	3.022	96.748
42.443	30.354	27.203	0.250	3.882	95.868

SPECIFIC MODEL PARAMETERS IN KELVIN

		UNIQUAC		NRTL (ALPHA=.2)	
I J	AIJ	AJI	AIJ	AIJ	AJI
1 2	369.52	-227.56		-617.98	286.05
1 3	857.89	263.80		1504.9	869.92
2 3	1132.6	-250.83		-475.04	1141.7

R1 = 2.8700   R2 = 4.6106   R3 = 0.9200
Q1 = 2.410   Q2 = 3.724   Q3 = 1.400

MEAN DEV. BETWEEN CALC. AND EXP. CONC. IN MOLE PCT

UNIQUAC (SPECIFIC PARAMETERS)   0.45
NRTL (SPECIFIC PARAMETERS)   0.73
UNIQUAC (COMMON PARAMETERS)   3.57

CHCl$_3$-C$_6$H$_{11}$NO

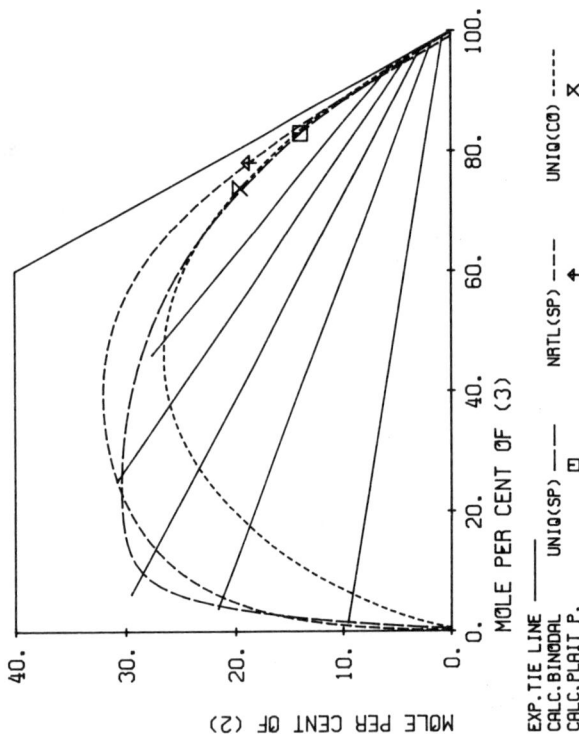

MOLE PER CENT OF (3) — MOLE PER CENT OF (2)

EXP.TIE LINE   UNIQ(SP) □   NRTL(SP) ✦   UNIQ(CO) ✕
CALC.BINODAL
CALC.PLAIT P.

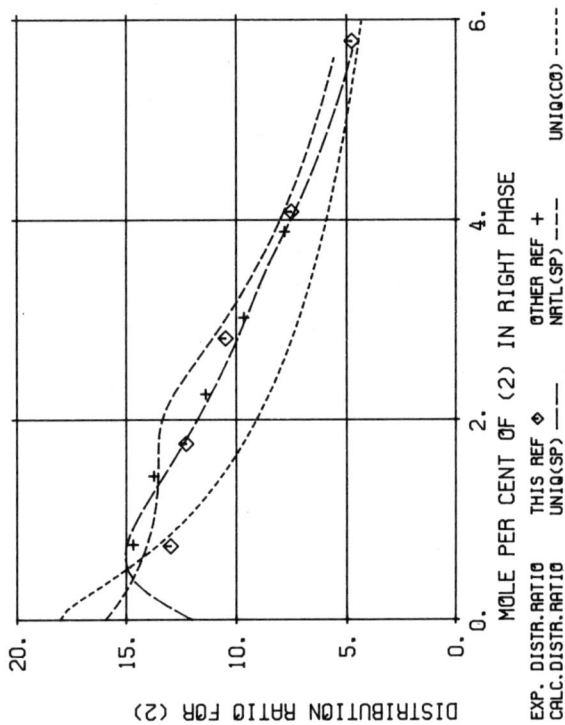

MOLE PER CENT OF (2) IN RIGHT PHASE — DISTRIBUTION RATIO FOR (2)

EXP. DISTR.RATIO   THIS REF ◇   OTHER REF +   UNIQ(CO) ----
CALC.DISTR.RATIO   UNIQ(SP) ---   NRTL(SP) ---

---

(1) CHCL3   METHANE,TRICHLORO

(2) C6H11NO   HEXANOIC ACID,6-AMINO,LACTAM

(3) H2O   WATER

TETTAMANTI K., NOGRADI M., SAWINSKY J.
PERIOD.POLYTECH.,CHEM.ENG. 4(1960)201

TEMPERATURE = 20.0 DEG C   TYPE OF SYSTEM = 1

EXPERIMENTAL TIE LINES IN MOLE PCT (GRAPH.INTERPOL.)

| | LEFT PHASE | | | RIGHT PHASE | |
(1)	(2)	(3)	(1)	(2)	(3)
99.538	0.0	0.462	0.125	0.0	99.875
88.969	9.534	1.497	0.160	0.734	99.106
74.604	21.593	3.803	0.208	1.759	98.033
64.320	29.502	6.177	0.254	2.814	96.931
44.062	30.692	25.246	0.305	4.082	95.613
26.330	27.589	46.081	0.370	5.786	93.844

SPECIFIC MODEL PARAMETERS IN KELVIN

I J	UNIQUAC AIJ	AJI	NRTL(ALPHA=.2) AIJ	AJI
1 2	369.52	-227.56	-617.98	286.05
1 3	857.89	-253.80	1504.9	869.92
2 3	1132.6	-250.83	-475.04	1141.7

R1 = 2.8700   R2 = 4.6106   R3 = 0.9200
Q1 = 2.410   Q2 = 3.724   Q3 = 1.400

MEAN DEV. BETWEEN CALC. AND EXP. CONC. IN MOLE PCT

UNIQUAC (SPECIFIC PARAMETERS)   0.57
NRTL (SPECIFIC PARAMETERS)   1.44
UNIQUAC (COMMON PARAMETERS)   3.55

CH_2Cl_2-CH_4O

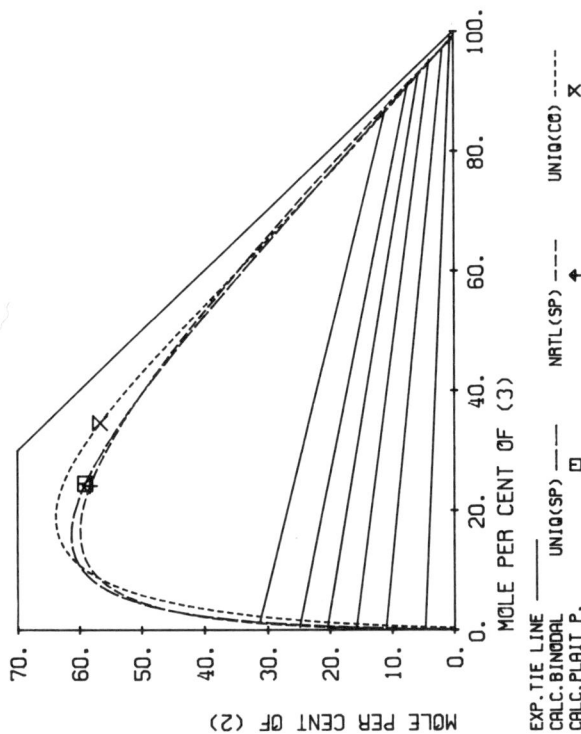

MOLE PER CENT OF (3)

MOLE PER CENT OF (2)

EXP.TIE LINE ———   UNIQ(SP) ——— ⊟   NRTL(SP) ——— ✦   UNIQ(CO) ----- ✕
CALC.BINODAL
CALC.PLAIT P.

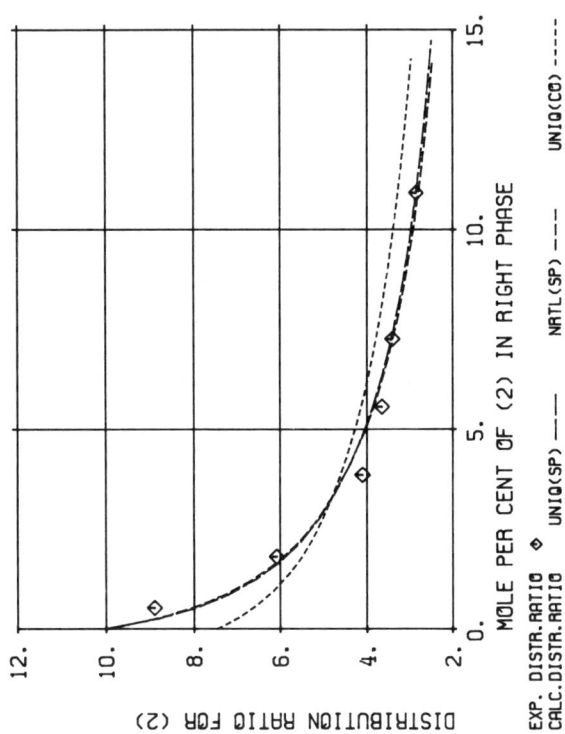

DISTRIBUTION RATIO FOR (2)

MOLE PER CENT OF (2) IN RIGHT PHASE

EXP. DISTR.RATIO ◇   UNIQ(SP) ———   NRTL(SP) ———   UNIQ(CO) -----
CALC.DISTR.RATIO

(1) H2O   - WATER
(2) CH4O   METHANOL
(3) CH2CL2   METHANE,DICHLORO

KHANINA E.P., PAVLENKO T.G., MALYSHEVA O.A., TIMOFEEV V.S.
ZH.FIZ.KHIM. 52(1978)1558

TEMPERATURE = 20.0 DEG C   TYPE OF SYSTEM = 1

EXPERIMENTAL TIE LINES IN MOLE PCT

	LEFT PHASE			RIGHT PHASE	
(1)	(2)	(3)	(1)	(2)	(3)
95.209	4.680	0.110	0.468	0.526	99.005
88.851	11.034	0.116	1.383	1.814	96.803
83.901	15.810	0.289	0.913	3.852	95.234
79.125	20.427	0.449	1.350	5.567	93.082
74.593	24.892	0.515	1.335	7.259	91.406
67.324	31.303	1.373	2.157	10.915	86.928

SPECIFIC MODEL PARAMETERS IN KELVIN

		UNIQUAC		NRTL(ALPHA=.2)	
I	J	AIJ	AJI	AIJ	AJI
1	2	-119.40	-250.23	-61.678	-419.22
1	3	528.46	-666.46	1561.7	833.45
2	3	190.90	-14.145	527.83	-282.00

R1 = 0.9200   R2 = 1.4311   R3 = 2.2564
Q1 = 1.400   Q2 = 1.432   Q3 = 1.988

MEAN DEV. BETWEEN CALC. AND EXP. CONC. IN MOLE PCT

UNIQUAC (SPECIFIC PARAMETERS)   0.25
NRTL   (SPECIFIC PARAMETERS)   0.25
UNIQUAC (COMMON PARAMETERS)   0.74

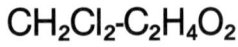

CH$_2$Cl$_2$-C$_2$H$_4$O$_2$

(1) H2O   WATER

(2) C2H4O2   ACETIC ACID

(3) CH2CL2   METHANE,DICHLORO

CASARICO A.
ANN.CHIM.(ROME) 41(1951)199

TEMPERATURE = 19.0 DEG C   TYPE OF SYSTEM = 1

EXPERIMENTAL TIE LINES IN MOLE PCT

LEFT PHASE			RIGHT PHASE		
(1)	(2)	(3)	(1)	(2)	(3)
96.460	3.008	0.532	2.308	0.928	96.764
94.848	4.505	0.647	2.534	0.940	96.526
94.475	4.849	0.676	2.759	0.966	96.275
59.438	29.425	11.138	11.814	24.454	63.732

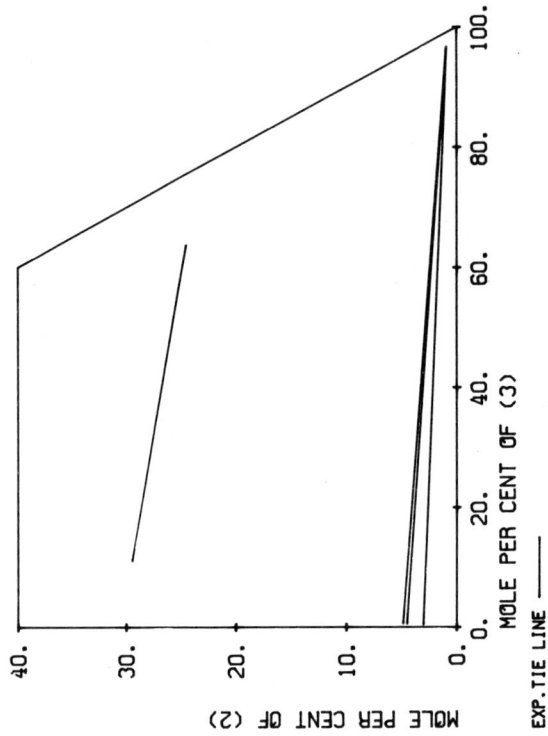

MOLE PER CENT OF (2)

MOLE PER CENT OF (3)

EXP.TIE LINE   UNIQ(SP) — □   NRTL(SP) — ⧫   UNIQ(C̄O) — ⊠
CALC.BINODAL
CALC.PLAIT P.

DISTRIBUTION RATIO FOR (2)

MOLE PER CENT OF (2) IN RIGHT PHASE

EXP. DISTR.RATIO ◇ THIS REF   + OTHER REF   UNIQ(C̄O) -----
CALC.DISTR.RATIO   UNIQ(SP)   --- NRTL(SP)

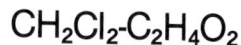

(1) H2O   WATER
(2) C2H4O2   ACETIC ACID
(3) CH2CL2   METHANE, DICHLORO

SABININ V.E., KIYA-OGLU N.V., GORICHNINA V.P.
ZH.PRIKL.KHIM.(LENINGRAD) 43(1970)1776

TEMPERATURE = 20.0 DEG C   TYPE OF SYSTEM = 1

EXPERIMENTAL TIE LINES IN MOLE PCT

	LEFT PHASE			RIGHT PHASE	
(1)	(2)	(3)	(1)	(2)	(3)
94.668	4.852	0.480	0.929	2.508	96.563
87.344	11.628	1.028	2.240	11.018	86.742
81.671	16.668	1.662	3.058	16.777	80.164
68.915	25.383	5.702	5.846	23.299	70.855

SPECIFIC MODEL PARAMETERS IN KELVIN

		UNIQUAC		NRTL(ALPHA=.2)	
I J	AIJ	AJI		AIJ	AJI
1 2	-89.211	-81.695		-77.014	-188.15
1 3	524.55	591.00		1232.9	751.37
2 3	-48.951	71.304		-232.69	102.66

R1 = 0.9200   R2 = 2.2024   R3 = 2.2564
Q1 = 1.400   Q2 = 2.072   Q3 = 1.988

MEAN DEV. BETWEEN CALC. AND EXP. CONC. IN MOLE PCT

UNIQUAC (SPECIFIC PARAMETERS)   1.43
NRTL   (SPECIFIC PARAMETERS)   1.03
UNIQUAC (COMMON PARAMETERS)   2.06

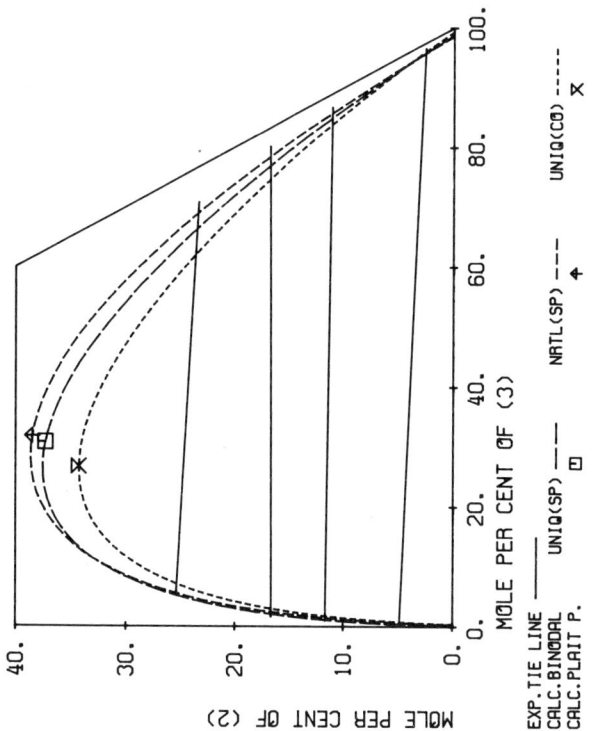

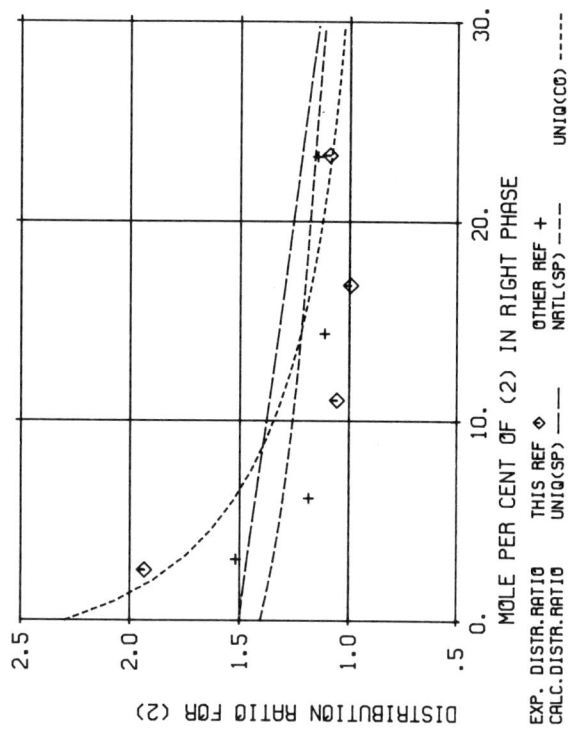

CH₂Cl₂-C₂H₄O₂

(1) H2O   WATER
(2) C2H4O2   ACETIC ACID
(3) CH2CL2   METHANE, DICHLORO

SABNIN V.E., KIYA-OGLU N.V., GORICHNINA V.P.
ZH.PRIKL.KHIM.(LENINGRAD) 43(1970)1776

TEMPERATURE = 30.0 DEG C   TYPE OF SYSTEM = 1

EXPERIMENTAL TIE LINES IN MOLE PCT

	LEFT PHASE			RIGHT PHASE	
(1)	(2)	(3)	(1)	(2)	(3)
94.648	4.630	0.722	1.386	3.050	95.564
91.724	7.278	0.998	2.273	6.136	91.591
81.703	15.976	2.321	4.354	14.369	81.277
64.884	25.612	8.504	8.559	23.231	68.210

SPECIFIC MODEL PARAMETERS IN KELVIN

		UNIQUAC		NRTL(ALPHA=.2)	
I J	AIJ	AJI	AIJ	AJI
1 2	-89.211	-81.695	-77.014	-188.15
1 3	524.55	591.00	1232.9	751.37
2 3	-48.951	71.304	-232.69	102.66

R1 = 0.9200   R2 = 2.2024   R3 = 2.2564
Q1 = 1.400   Q2 = 2.072   Q3 = 1.988

MEAN DEV. BETWEEN CALC. AND EXP. CONC. IN MOLE PCT

UNIQUAC (SPECIFIC PARAMETERS)   0.83
NRTL   (SPECIFIC PARAMETERS)   0.60
UNIQUAC (COMMON PARAMETERS)   1.86

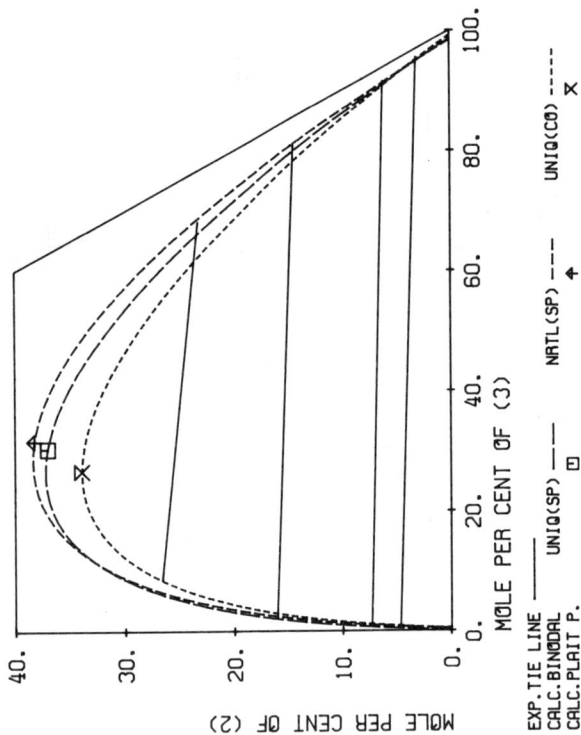

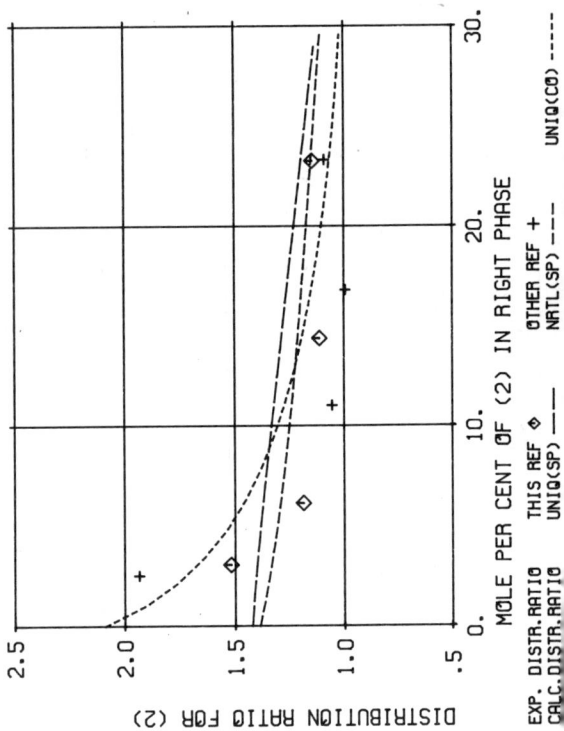

MOLE PER CENT OF (2)
MOLE PER CENT OF (3)

EXP.TIE LINE ——   UNIQ(SP) □   NRTL(SP) +   UNIQ(CO) X
CALC.BINODAL
CALC.PLAIT P.

DISTRIBUTION RATIO FOR (2)
MOLE PER CENT OF (2) IN RIGHT PHASE

EXP. DISTR.RATIO ◇   THIS REF ◇   OTHER REF +   UNIQ(CO)
CALC. DISTR.RATIO   UNIQ(SP)   NRTL(SP)

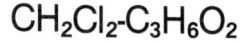

$CH_2Cl_2\text{-}C_3H_6O_2$

MOLE PER CENT OF (2)

MOLE PER CENT OF (3)

EXP.TIE LINE   UNIQ(SP) ———   NRTL(SP) ---△---   UNIQ(CO) ---✕---
CALC.BINODAL
CALC.PLAIT P.

DISTRIBUTION RATIO FOR (2)

MOLE PER CENT OF (2) IN RIGHT PHASE

EXP. DISTR.RATIO ◇   THIS REF ◇   OTHER REF +   UNIQ(CO) -----
CALC.DISTR.RATIO   UNIQ(SP) ——   NRTL(SP) ---

(1) CH2CL2   METHANE, DICHLORO

(2) C3H6O2   PROPANOIC ACID

(3) H2O   WATER

SABININ V.E., KIYA-OGLU N.V., GORICHNINA V.P.
ZH.PRIKL.KHIM.(LENINGRAD) 43(1970)1776

TEMPERATURE = 20.0 DEG C   TYPE OF SYSTEM = 1

EXPERIMENTAL TIE LINES IN MOLE PCT

	LEFT PHASE			RIGHT PHASE	
(1)	(2)	(3)	(1)	(2)	(3)
84.630	13.544	1.826	0.459	2.050	97.491
67.242	27.532	5.225	0.958	3.944	95.098
41.824	35.495	22.681	1.634	7.183	91.182

SPECIFIC MODEL PARAMETERS IN KELVIN

		UNIQUAC		NRTL(ALPHA=.2)	
I	J	AIJ	AJI	AIJ	AJI
1	2	187.15	-117.37	-195.27	196.66
1	3	739.38	281.20	-1010.7	921.86
2	3	72.639	54.624	-409.02	1121.2

R1 = 2.2564   R2 = 2.8758   R3 = 0.9200
Q1 = 1.983   Q2 = 2.612   Q3 = 1.400

MEAN DEV. BETWEEN CALC. AND EXP. CONC. IN MOLE PCT

UNIQUAC (SPECIFIC PARAMETERS)	1.72
NRTL (SPECIFIC PARAMETERS)	1.46
UNIQUAC (COMMON PARAMETERS)	2.26

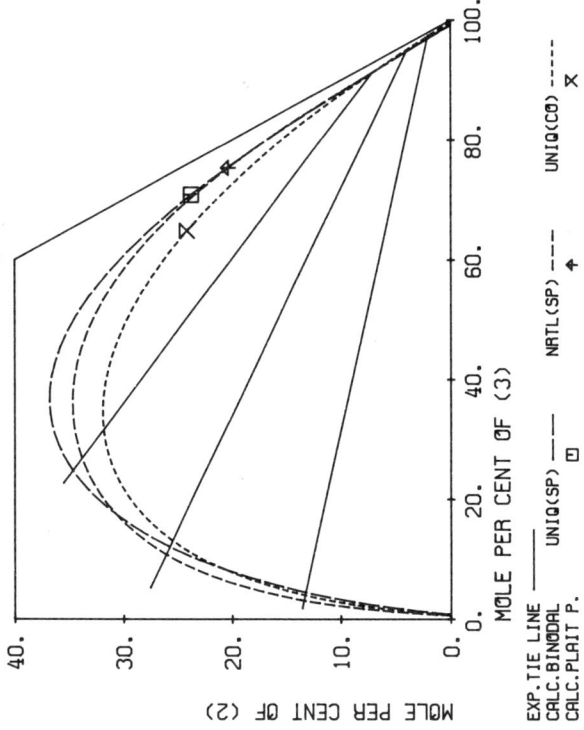

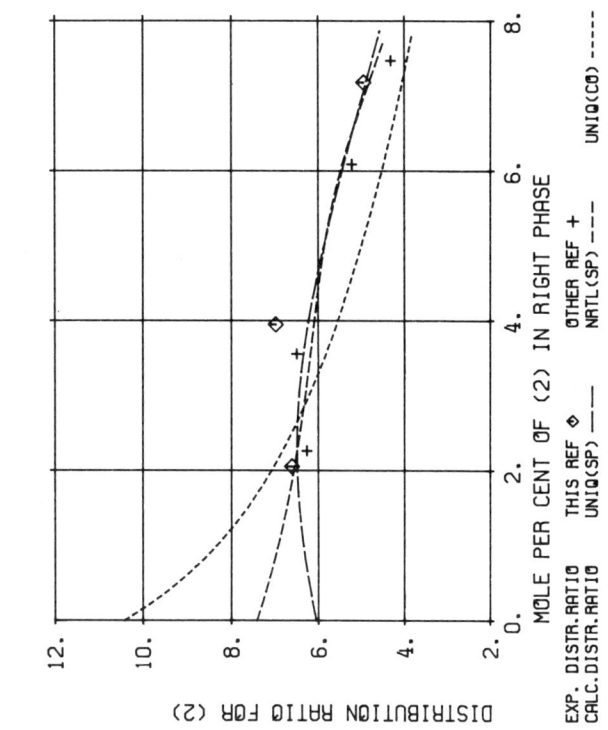

CH₂Cl₂-C₃H₆O₂

(1) CH2CL2   METHANE, DICHLORO
----
(2) C3H6O2   PROPANOIC ACID
----
(3) H2O   WATER

SABININ V.E., KIYA-OGLU N.V.; GORICHNINA V.P.
ZH.PRIKL.KHIM.(LENINGRAD) 43(1970)1776

TEMPERATURE = 30.0 DEG C   TYPE OF SYSTEM = 1

EXPERIMENTAL TIE LINES IN MOLE PCT

       LEFT PHASE   RIGHT PHASE
   (1)   (2)   (3)   (1)   (2)   (3)

 77.235 14.147   8.618   0.962   2.259 96.779
 66.830 23.088 10.082   1.329   3.554 95.117
 47.952 31.702 20.346   2.011   6.087 91.902
 30.877 32.258 36.865   2.472   7.472 90.056

SPECIFIC MODEL PARAMETERS IN KELVIN

          UNIQUAC   NRTL(ALPHA=.2)
 I J   AIJ   AJI   AIJ   AJI

 1 2 187.15   -117.37   -195.27   196.66
 1 3 739.38   281.20   1010.7   921.86
 2 3   72.639   54.624   -409.02   1121.2

 R1 = 2.2564   R2 = 2.8768   R3 = 0.9200
 Q1 = 1.988   Q2 = 2.612   Q3 = 1.400

MEAN DEV. BETWEEN CALC. AND EXP. CONC. IN MOLE PCT

UNIQUAC (SPECIFIC PARAMETERS)   1.26
NRTL   (SPECIFIC PARAMETERS)   1.25
UNIQUAC (COMMON PARAMETERS)   1.94

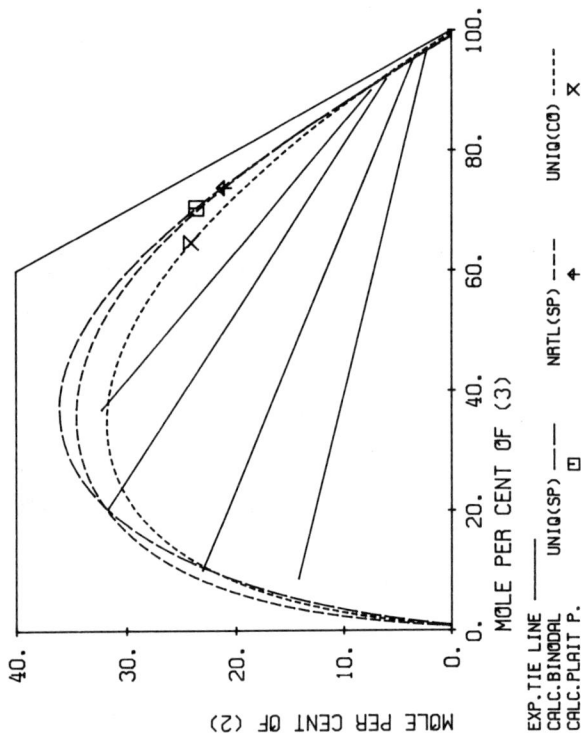

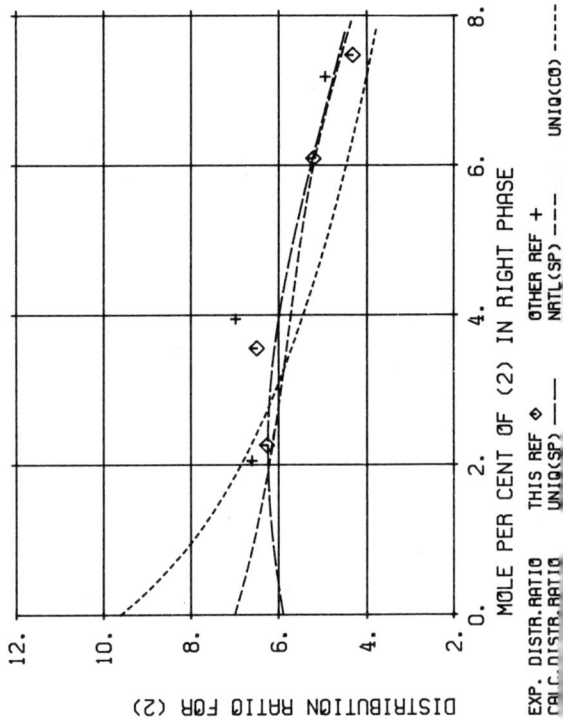

(1) CH2CL2   METHANE, DICHLORO

(2) C4H8O2   BUTANOIC ACID

(3) H2O   WATER

SABININ V.E., KIYA-OGLU N.V., GORICHNINA V.P.
ZH.PRIKL.KHIM.(LENINGRAD) 43(1970)1776

TEMPERATURE = 20.0 DEG C   TYPE OF SYSTEM = 1

EXPERIMENTAL TIE LINES IN MOLE PCT

	LEFT PHASE			RIGHT PHASE	
(1)	(2)	(3)	(1)	(2)	(3)
81.807	16.786	1.408	0.132	0.849	99.019
64.048	30.912	5.040	0.157	1.296	98.548
39.819	43.495	16.686	0.183	1.806	98.012
19.270	42.373	38.357	0.186	2.310	97.504

SPECIFIC MODEL PARAMETERS IN KELVIN

		UNIQUAC		NRTL(ALPHA=.2)	
I	J	AIJ	AJI	AIJ	AJI
1	2	59.826	-48.514	-15.680	-28.176
1	3	1028.6	587.50	1599.4	1127.2
2	3	80.222	81.344	-357.11	1320.6

R1 = 2.2564   R2 = 3.5512   R3 = 0.9200
Q1 = 1.988   Q2 = 3.152   Q3 = 1.400

MEAN DEV. BETWEEN CALC. AND EXP. CONC. IN MOLE PCT

UNIQUAC (SPECIFIC PARAMETERS)   1.61
NRTL   (SPECIFIC PARAMETERS)   1.25
UNIQUAC (COMMON PARAMETERS)   3.64

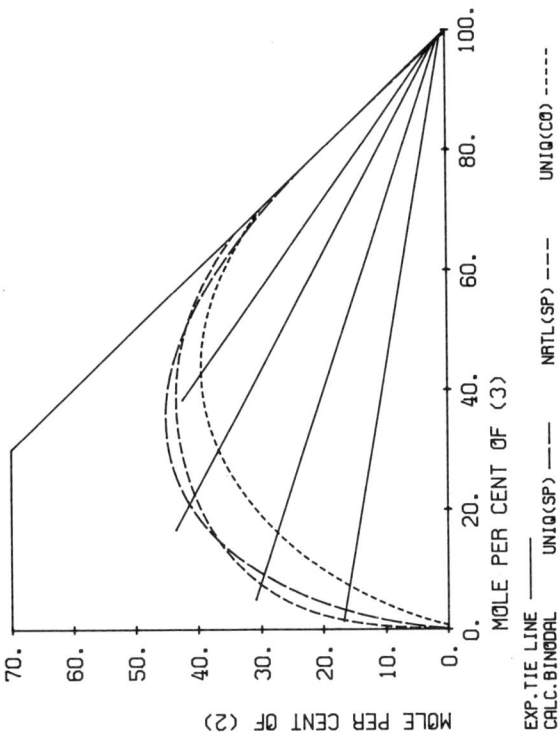

MOLE PER CENT OF (2)

MOLE PER CENT OF (3)

EXP.TIE LINE ——   UNIQ(SP) ——   NRTL(SP) ---   UNIQ(CO) ----
CALC.BINODAL

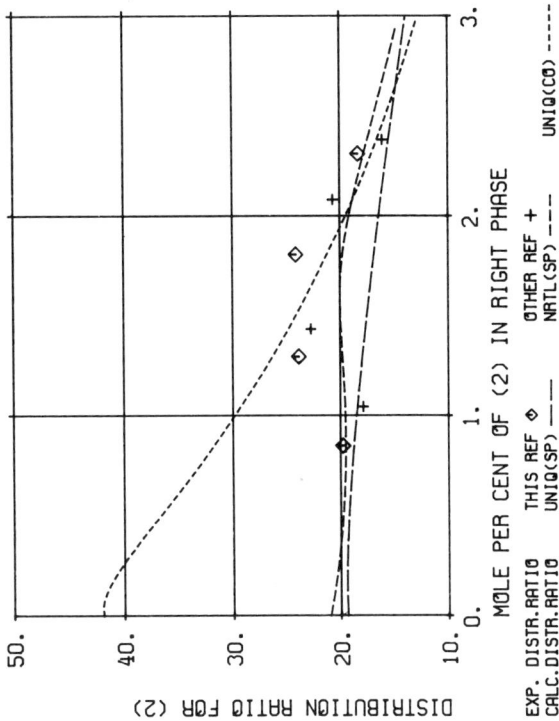

DISTRIBUTION RATIO FOR (2)

MOLE PER CENT OF (2) IN RIGHT PHASE

EXP. DISTR.RATIO ◇   THIS REF ◇   OTHER REF +   UNIQ(CO) ----
CALC.DISTR.RATIO   UNIQ(SP) ——   NRTL(SP) ---

CH₂Cl₂-C₄H₈O₂

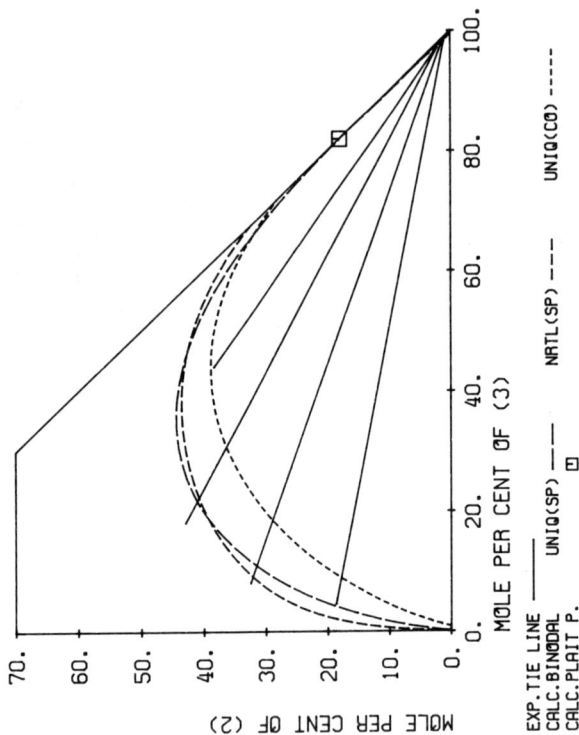

MOLE PER CENT OF (3)

MOLE PER CENT OF (2)

EXP.TIE LINE
CALC.BINODAL   UNIQ(SP) ----   NRTL(SP) ----   UNIQ(CO) -----
CALC.PLAIT P.   ⊡

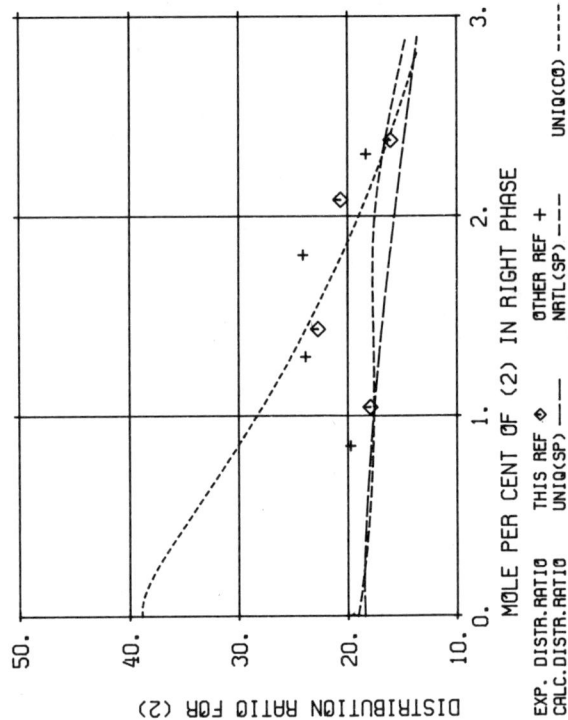

MOLE PER CENT OF (2) IN RIGHT PHASE

DISTRIBUTION RATIO FOR (2)

EXP. DISTR.RATIO ◇   THIS REF ◇   OTHER REF +
CALC.DISTR.RATIO   UNIQ(SP)   NRTL(SP) ----   UNIQ(CO) -----

(1) CH2CL2   METHANE,DICHLORO

(2) C4H8O2   BUTANOIC ACID

(3) H2O   WATER

SABININ V.E., KIYA-OGLU N.V., GORICHNINA V.P.
ZH.PRIKL.KHIM.(LENINGRAD) 43(1970)1775

TEMPERATURE = 30.0 DEG C   TYPE OF SYSTEM = 1

EXPERIMENTAL TIE LINES IN MOLE PCT

	LEFT PHASE			RIGHT PHASE	
(1)	(2)	(3)	(1)	(2)	(3)
76.706	18.716	4.578	0.609	1.043	98.348
59.394	32.554	8.051	0.618	1.433	97.949
38.953	42.930	18.117	0.680	2.081	97.239
17.685	38.270	44.046	0.712	2.379	95.909

SPECIFIC MODEL PARAMETERS IN KELVIN

		UNIQUAC		NRTL(ALPHA=.2)	
I J	AIJ	AJI		AIJ	AJI
1 2	59.826	-48.514		-15.680	-28.176
1 3	1028.6	587.50		1599.4	1127.2
2 3	80.222	81.344		-357.11	1320.6

R1 = 2.2564   R2 = 3.5512   R3 = 0.9200
Q1 = 1.988   Q2 = 3.152   Q3 = 1.400

MEAN DEV. BETWEEN CALC. AND EXP. CONC. IN MOLE PCT

UNIQUAC (SPECIFIC PARAMETERS)   1.85
NRTL (SPECIFIC PARAMETERS)   1.63
UNIQUAC (COMMON PARAMETERS)   3.27

CH_2Cl_2-$C_5H_4O_2$

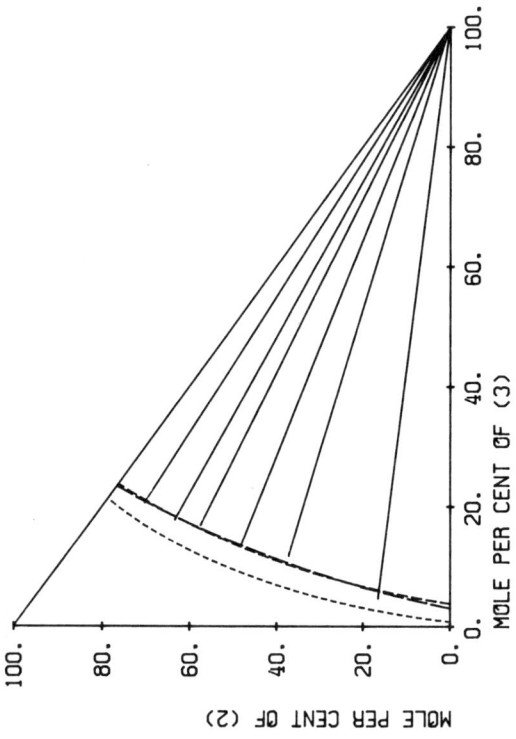

MOLE PER CENT OF (2)

MOLE PER CENT OF (3)

EXP.TIE LINE   UNIQ(SP) ———   NRTL(SP) —— ——   UNIQ(CO) - - - - -
CALC.BINODAL ———

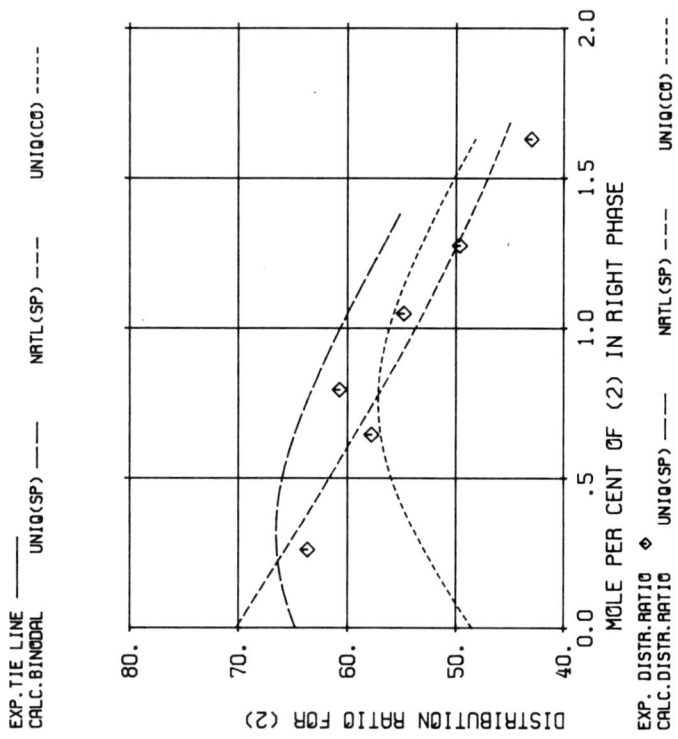

DISTRIBUTION RATIO FOR (2)

MOLE PER CENT OF (2) IN RIGHT PHASE

EXP. DISTR.RATIO   ◇   UNIQ(SP) ———   NRTL(SP) —— ——   UNIQ(CO) - - - - -
CALC.DISTR.RATIO

(1) CH2CL2   METHANE,DICHLORO

(2) C5H4O2   FURFURAL

(3) H2O   WATER

KRUPATKIN I.L., GLAGOLEVA M.F.
ZH.PRIKL.KHIM.(LENINGRAD) 42(1969)1525

TEMPERATURE = 25.0 DEG C   TYPE OF SYSTEM = 2

EXPERIMENTAL TIE LINES IN MOLE PCT

	LEFT PHASE			RIGHT PHASE	
(1)	(2)	(3)	(1)	(2)	(3)
78.812	16.544	4.644	0.392	0.260	99.349
51.077	37.198	11.724	0.331	0.644	99.025
38.269	48.200	13.531	0.288	0.794	98.918
25.636	57.479	16.885	0.246	1.048	98.706
18.862	63.353	17.785	0.203	1.275	98.522
9.114	70.280	20.607	0.068	1.630	98.302

SPECIFIC MODEL PARAMETERS IN KELVIN

		UNIQUAC		NRTL(ALPHA=.2)	
I J	AIJ	AJI		AIJ	AJI
1 2	32.277	-17.233		-204.28	150.91
1 3	484.44	313.16		546.27	953.55
2 3	127.56	150.71		11.332	1234.3

R1 = 2.2564   R2 = 3.1680   R3 = 0.9200
Q1 = 1.988   Q2 = 2.484   Q3 = 1.400

MEAN DEV. BETWEEN CALC. AND EXP. CONC. IN MOLE PCT

UNIQUAC (SPECIFIC PARAMETERS)   0.37
NRTL   (SPECIFIC PARAMETERS)   0.47
UNIQUAC (COMMON PARAMETERS)   1.79

CH$_2$Cl$_2$-C$_6$H$_{11}$NO

(1) CH2CL2   METHANE, DICHLORO

(2) C6H11NO   HEXANOIC ACID,6-AMINO,LACTAM

(3) H2O   WATER

KUDRYAVTSEVA G.I., KRUTIKOVA A.D.
ZH.PRIKL.KHIM.(LENINGRAD) 25(1953)1190

TEMPERATURE = 20.0 DEG C   TYPE OF SYSTEM = 1

EXPERIMENTAL TIE LINES IN MOLE PCT (GRAPH.INTERPOL.)

	LEFT PHASE			RIGHT PHASE	
(1)	(2)	(3)	(1)	(2)	(3)
87.864	7.169	4.967	0.441	1.136	98.423
75.990	13.977	10.034	0.470	1.922	97.608
62.201	18.435	19.364	0.520	2.686	96.794
50.159	21.319	28.522	0.597	3.413	95.990
39.154	22.329	38.517	0.691	4.021	95.289

SPECIFIC MODEL PARAMETERS IN KELVIN

		UNIQUAC		NRTL(ALPHA=.2)	
I J	AIJ	AJI		AIJ	AJI
1 2	239.27	-165.00		-784.54	34.081
1 3	610.69	474.19		762.69	1305.4
2 3	103.07	-136.30		-810.37	1147.5

R1 = 2.2564   R2 = 4.6106   R3 = 0.9200
Q1 = 1.988   Q2 = 3.724   Q3 = 1.400

MEAN DEV. BETWEEN CALC. AND EXP. CONC. IN MOLE PCT

UNIQUAC (SPECIFIC PARAMETERS)   0.18
NRTL   (SPECIFIC PARAMETERS)   0.23
UNIQUAC (COMMON PARAMETERS)   1.16

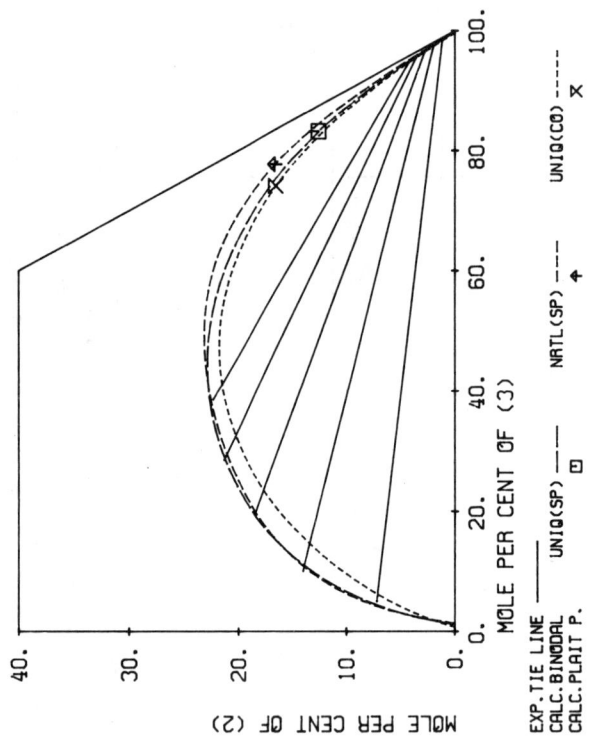

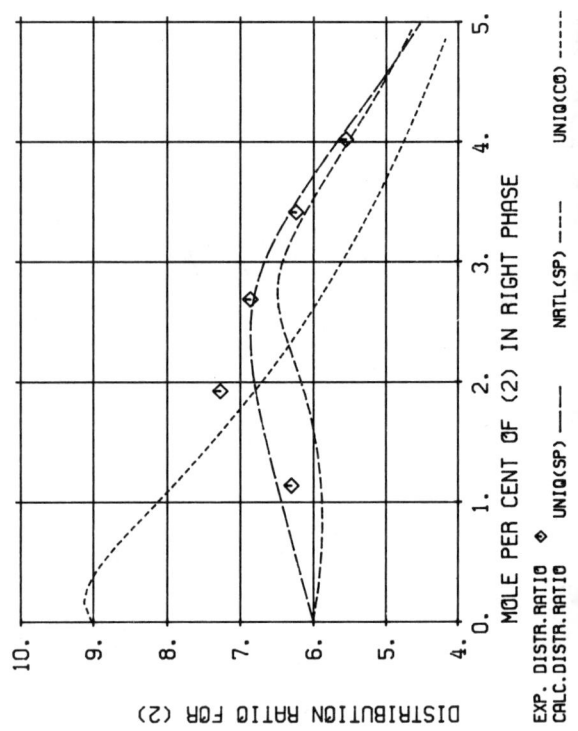

CH_2O_2-$C_2H_4Cl_2$

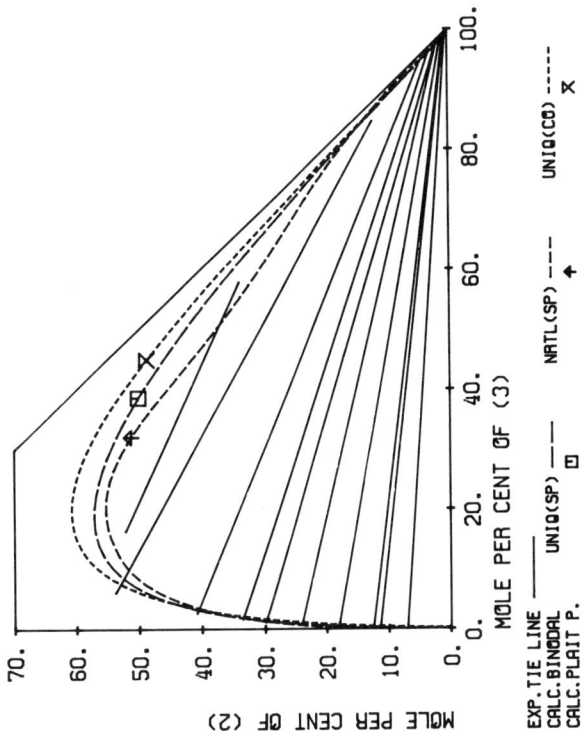

(1) H2O   WATER
(2) CH2O2   FORMIC ACID
(3) C2H4CL2   ETHANE,1,2-DICHLORO

UDOVENKO V.V., ALEXANDROVA L.P.
ZH.FIZ.KHIM. 32(1958)1889

TEMPERATURE = 30.0 DEG C   TYPE OF SYSTEM = 1

EXPERIMENTAL TIE LINES IN MOLE PCT

| | LEFT PHASE | | | RIGHT PHASE | |
(1)	(2)	(3)	(1)	(2)	(3)
92.613	7.018	0.369	0.055	0.380	99.566
88.168	11.396	0.436	0.273	0.590	99.137
87.037	12.521	0.442	0.164	0.697	99.139
81.251	18.032	0.717	0.272	1.102	98.625
75.005	24.093	0.902	0.488	1.784	97.729
68.956	29.941	1.103	0.541	1.900	97.558
64.824	33.580	1.596	0.805	3.106	96.089
56.519	40.864	2.617	1.167	4.630	94.203
40.185	53.751	6.064	3.000	12.183	84.817
31.519	52.180	16.301	8.258	33.806	57.936

SPECIFIC MODEL PARAMETERS IN KELVIN

| | | UNIQUAC | | NRTL(ALPHA=.2) | |
I	J	AIJ	AJI	AIJ	AJI
1	2	-408.32	-207.05	1479.2	-1024.5
1	3	416.13	791.02	2306.6	1817.3
2	3	-31.766	138.63	638.81	29.031

R1 = 0.9200   R2 = 1.5280   R3 = 2.9308
Q1 = 1.400   Q2 = 1.532   Q3 = 2.528

MEAN DEV. BETWEEN CALC. AND EXP. CONC. IN MOLE PCT

UNIQUAC (SPECIFIC PARAMETERS)   1.91
NRTL   (SPECIFIC PARAMETERS)   0.85
UNIQUAC (COMMON PARAMETERS)   2.94

MOLE PER CENT OF (3)

MOLE PER CENT OF (2)

EXP.TIE LINE   UNIQ(SP) ——□——   NRTL(SP) ——+——   UNIQ(CO) ——×——
CALC.BINODAL
CALC.PLAIT P.

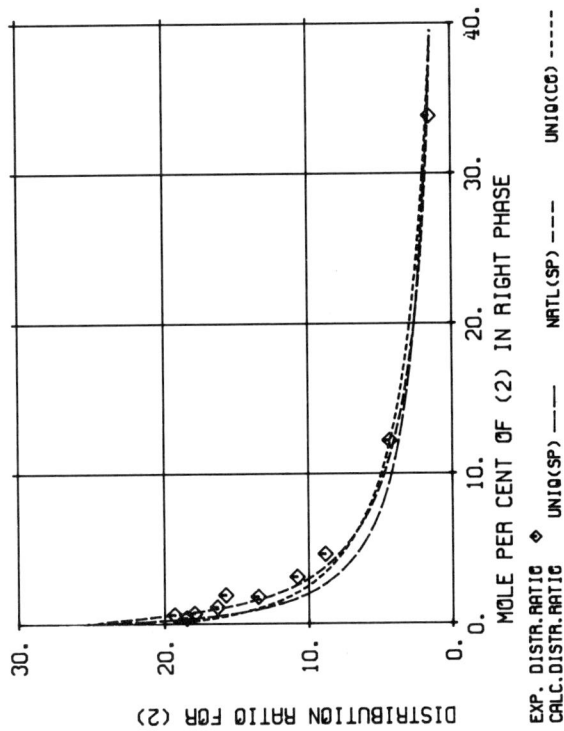

MOLE PER CENT OF (2) IN RIGHT PHASE

DISTRIBUTION RATIO FOR (2)

EXP. DISTR.RATIO ◇ UNIQ(SP)   NRTL(SP) ———   UNIQ(CO) -----
CALC.DISTR.RATIO

CH$_2$O$_2$-C$_2$H$_4$Cl$_2$

(1) H2O   WATER
(2) CH2O2   FORMIC ACID
(3) C2H4CL2 ETHANE,1,2-DICHLORO

UDOVENKO V.V., ALEXANDROVA L.P.
ZH.FIZ.KHIM. 32(1958)1889

TEMPERATURE = 45.0 DEG C   TYPE OF SYSTEM = 1

EXPERIMENTAL TIE LINES IN MOLE PCT

LEFT PHASE			RIGHT PHASE		
(1)	(2)	(3)	(1)	(2)	(3)
92.301	7.287	0.412	0.164	0.490	99.345
86.552	12.774	0.673	0.327	0.915	98.758
85.778	13.499	0.724	0.435	1.037	98.527
78.105	20.897	0.997	0.811	1.752	97.438
73.810	24.955	1.235	0.915	2.318	96.767
69.215	29.259	1.525	1.336	3.068	95.597
63.314	34.515	2.172	1.798	4.218	93.984
53.279	42.976	3.745	2.647	6.278	91.075
48.039	46.980	4.981	3.327	7.641	89.031
32.355	54.561	13.084	9.055	19.030	71.915
25.801	54.877	19.322	12.082	31.414	56.504

SPECIFIC MODEL PARAMETERS IN KELVIN

I J	UNIQUAC AIJ	AJI	NRTL(ALPHA=2) AIJ	AJI
1 2	-311.13	-207.68	1965.7	-1200.5
1 3	349.69	750.96	2522.9	1021.1
2 3	-25.561	233.75	650.36	9.5909

R1 = 0.9200   R2 = 1.5280   R3 = 2.9308
Q1 = 1.400   Q2 = 1.532   Q3 = 2.528

MEAN DEV. BETWEEN CALC. AND EXP. CONC. IN MOLE PCT
UNIQUAC (SPECIFIC PARAMETERS)   1.78
NRTL   (SPECIFIC PARAMETERS)   0.90

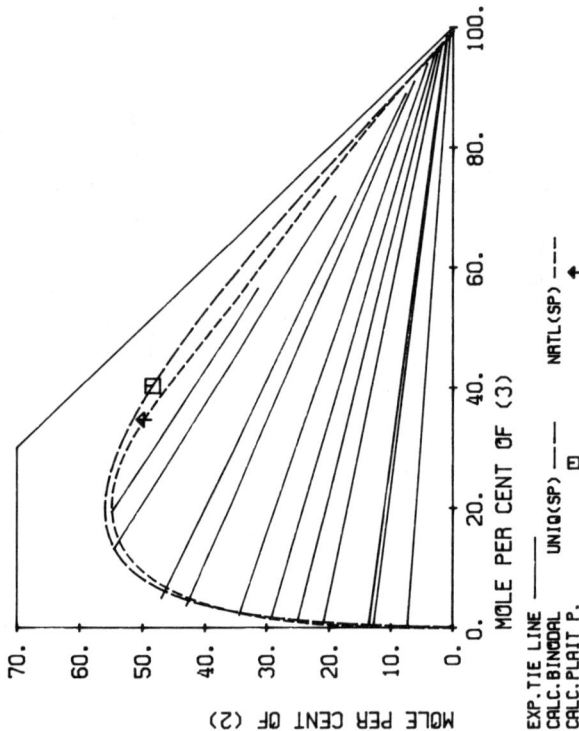

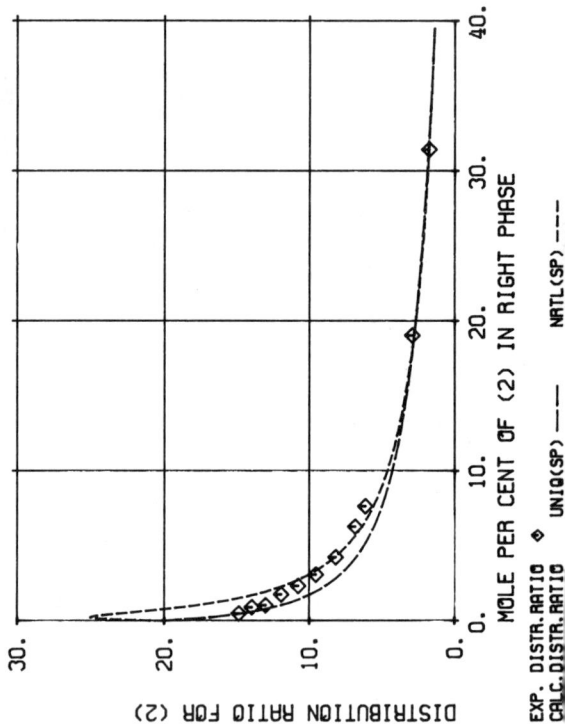

$CH_2O_2-C_2H_4Cl_2$

(1) H2O   WATER
(2) CH2O2   FORMIC ACID
(3) C2H4CL2   ETHANE,1,2-DICHLORO

UDOVENKO V.V., ALEXANDROVA L.P.
ZH.FIZ.KHIM. 32(1958)1889

TEMPERATURE = 60.0 DEG C   TYPE OF SYSTEM = 1

EXPERIMENTAL TIE LINES IN MOLE PCT

	LEFT PHASE			RIGHT PHASE	
(1)	(2)	(3)	(1)	(2)	(3)
93.134	6.499	0.366	0.328	0.577	99.095
89.914	9.639	0.447	0.436	0.885	98.679
85.929	13.395	0.676	0.435	1.435	98.131
78.131	20.744	1.125	0.915	2.430	96.655
73.185	25.216	1.599	1.335	3.121	95.543
66.558	31.189	2.252	1.797	4.315	93.888
63.742	33.597	2.660	2.202	5.150	92.648
52.045	43.146	4.809	3.472	7.877	88.651
48.726	45.549	5.725	4.534	10.445	85.021
44.676	48.246	7.079	10.044	16.223	73.733

SPECIFIC MODEL PARAMETERS IN KELVIN

		UNIQUAC		NRTL(ALPHA=.2)	
I	J	AIJ	AJI	AIJ	AJI
1	2	-212.29	-219.54	1159.7	-954.97
1	3	273.66	792.97	1463.7	933.16
2	3	12.474	254.33	784.28	-62.035

R1 = 0.9200   R2 = 1.5280   R3 = 2.9308
Q1 = 1.400   Q2 = 1.532   Q3 = 2.528

MEAN DEV. BETWEEN CALC. AND EXP. CONC. IN MOLE PCT

UNIQUAC (SPECIFIC PARAMETERS)   1.54
NRTL   (SPECIFIC PARAMETERS)   1.49

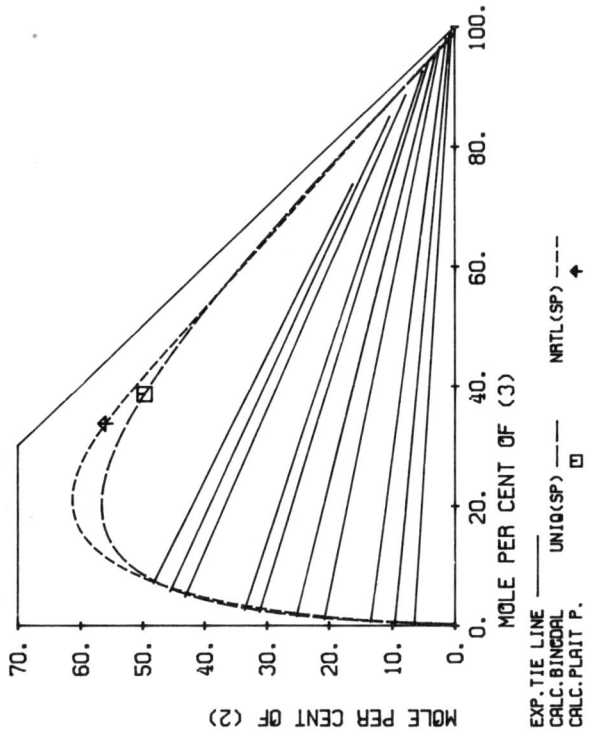

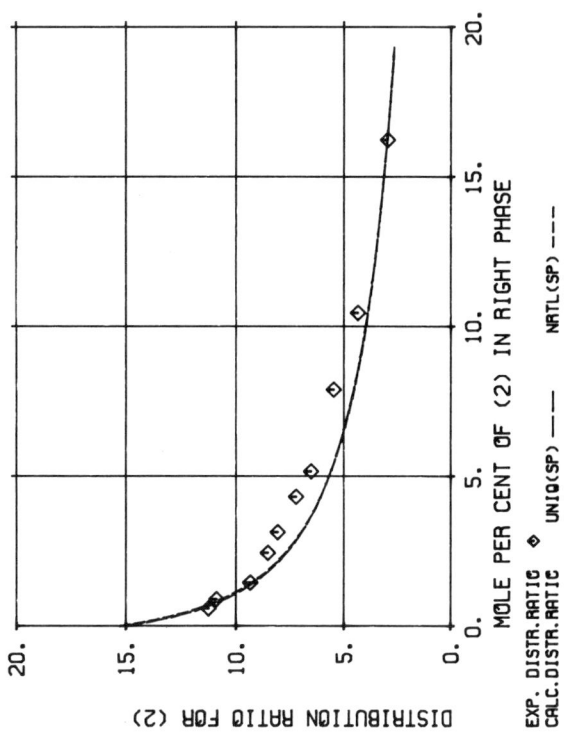

70.

60.

50.

40.

30.

20.

10.

0.

MOLE PER CENT OF (2)

0.   20.   40.   60.   80.   100.

MOLE PER CENT OF (3)

EXP.TIE LINE
CALC.BINODAL   UNIQ(SP) ---- ⊡
CALC.PLAIT P.   NRTL(SP) ---- ✦

20.

15.

10.

5.

0.

DISTRIBUTIION FOR RATIO (2)

0.   5.   10.   15.   20.

MOLE PER CENT OF (2) IN RIGHT PHASE

EXP. DISTR.RATIO   ◆ UNIQ(SP)   NRTL(SP) ---
CALC.DISTR.RATIO ----

CH_2O_2-$C_4H_8O_2$

(1) C4H8O2   ACETIC ACID,ETHYL ESTER
----
(2) CH2O2   FORMIC ACID
----
(3) H2O   WATER

RAMANA RAO M.V., SOMASUNDARA RAO K., VENKATA RAO C.
J.SCI.IND.RES. 20B(1961)379

TEMPERATURE = 35.0 DEG C   TYPE OF SYSTEM = 1

EXPERIMENTAL TIE LINES IN MOLE PCT

	LEFT PHASE			RIGHT PHASE	
(1)	(2)	(3)	(1)	(2)	(3)
76.630	2.836	20.533	2.063	1.073	96.863
68.803	5.599	25.599	2.219	2.102	95.679
63.528	7.733	28.739	2.438	3.320	94.242
57.470	9.677	32.853	2.631	4.350	93.018
50.708	11.471	37.820	3.001	5.557	91.442
36.949	14.147	43.904	4.076	8.205	87.719

SPECIFIC MODEL PARAMETERS IN KELVIN

		UNIQUAC		NRTL(ALPHA=.2)	
I	J	AIJ	AJI	AIJ	AJI
1	2	-250.46	-204.02	606.10	-613.60
1	3	-335.32	123.88	114.47	1280.2
2	3	-409.76	198.02	93.024	-358.31

R1 = 3.4786   R2 = 1.5280   R3 = 0.9200
Q1 = 3.116   Q2 = 1.532   Q3 = 1.400

MEAN DEV. BETWEEN CALC. AND EXP. CONC. IN MOLE PCT

UNIQUAC (SPECIFIC PARAMETERS)   0.35
NRTL   (SPECIFIC PARAMETERS)   0.29

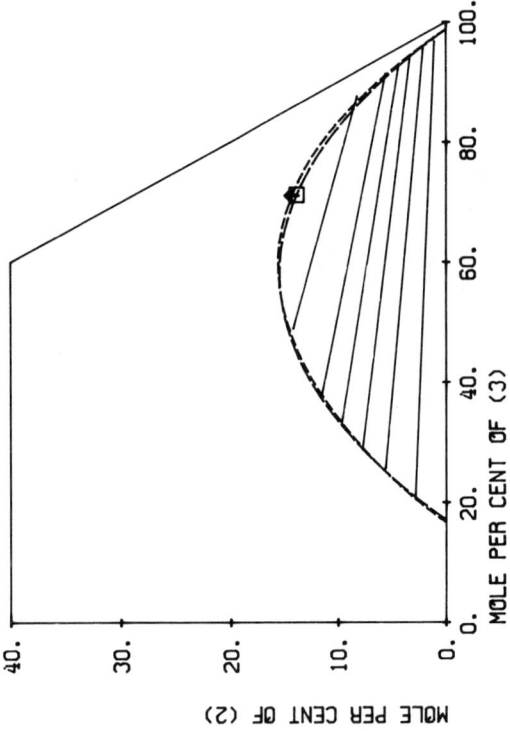

MOLE PER CENT OF (3)

MOLE PER CENT OF (2)

EXP.TIE LINE ———   UNIQ(SP) ——— ☐   NRTL(SP) ---- ♦
CALC.BINODAL
CALC.PLAIT P.

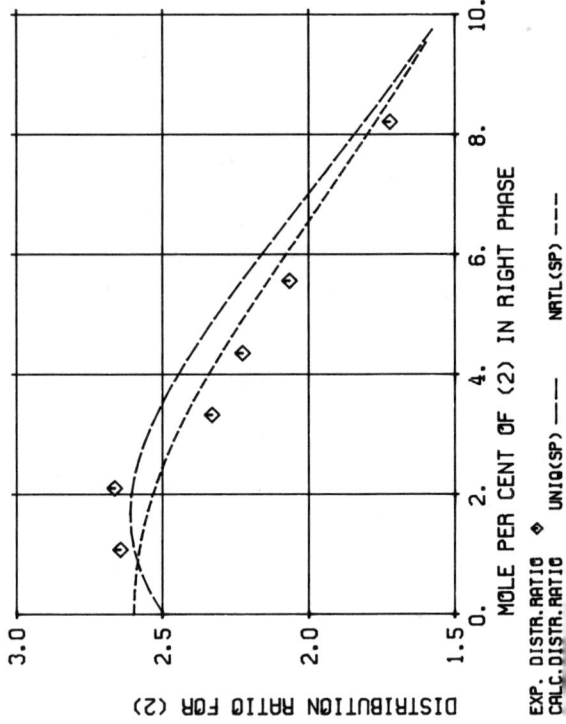

MOLE PER CENT OF (2) IN RIGHT PHASE

DISTRIBUTION RATIO FOR (2)

EXP. DISTR.RATIO ◇   UNIQ(SP) ☐   NRTL(SP) ----
CALC.DISTR.RATIO

CH₂O₂-C₄H₈O₂

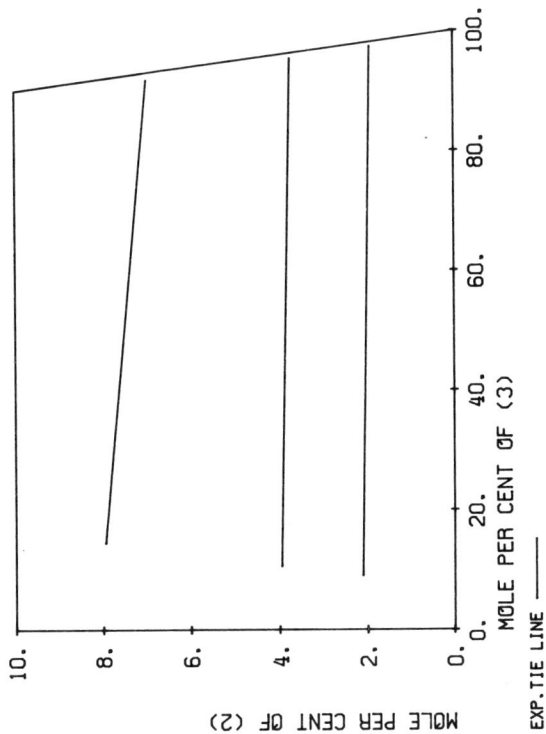

(1) C4H8O2   FORMIC ACID,PROPYL ESTER

(2) CH2O2   FORMIC ACID

(3) H2O   WATER

RIUS A., ALFONSO C.
AN.R.SOC.ESP.FIS.QUIM. 51B(1955)649

TEMPERATURE = 40.0 DEG C   TYPE OF SYSTEM = 1

EXPERIMENTAL TIE LINES IN MOLE PCT

	LEFT PHASE			RIGHT PHASE	
(1)	(2)	(3)	(1)	(2)	(3)
88.905	2.110	8.985	0.713	1.904	97.383
85.515	3.951	10.534	0.870	3.717	95.412
77.561	7.943	14.496	1.229	6.969	91.802

CH$_2$O$_2$-C$_5$H$_4$O$_2$

MOLE PER CENT OF (3)

MOLE PER CENT OF (2)

EXP.TIE LINE ——— UNIQ(SP) ——— ☐ NRTL(SP) ——— ✦ UNIQ(CO) ——— ✕
CALC.BINODAL
CALC.PLAIT P.

DISTRIBUTION RATIO FOR (2)

MOLE PER CENT OF (2) IN RIGHT PHASE

EXP. DISTR.RATIO ◇ UNIQ(SP) ——— NRTL(SP) ——— UNIQ(CO) -----
CALC.DISTR.RATIO

(1) C5H4O2   FURFURAL
(2) CH2O2   FORMIC ACID
(3) H2O   WATER

LANGFORD R.E., HERIC E.L.;
J.CHEM.ENG.DATA 17(1972)37

TEMPERATURE = 25.0 DEG C   TYPE OF SYSTEM = 1

EXPERIMENTAL TIE LINES IN MOLE PCT

	LEFT PHASE			RIGHT PHASE	
(1)	(2)	(3)	(1)	(2)	(3)
77.269	0.512	22.219	1.960	0.990	97.050
70.605	2.729	26.666	2.439	2.273	95.283
64.004	4.820	31.176	2.810	3.162	94.028
49.463	8.862	41.675	3.970	5.198	90.833
40.321	10.721	48.958	5.049	5.509	88.442
25.427	11.905	62.667	9.379	9.153	81.468

SPECIFIC MODEL PARAMETERS IN KELVIN

		UNIQUAC		NRTL(ALPHA=.2)	
I J		AIJ	AJI	AIJ	AJI
1 2		-333.21	-480.97	745.20	-517.04
1 3		110.78	190.59	22.522	1209.4
2 3		-627.56	136.29	-10.682	-492.26

R1 = 3.1680   R2 = 1.5280   R3 = 0.9200
Q1 = 2.484   Q2 = 1.532   Q3 = 1.400

MEAN DEV. BETWEEN CALC. AND EXP. CONC. IN MOLE PCT

UNIQUAC (SPECIFIC PARAMETERS)   0.82
NRTL   (SPECIFIC PARAMETERS)   0.98
UNIQUAC (COMMON PARAMETERS)   1.66

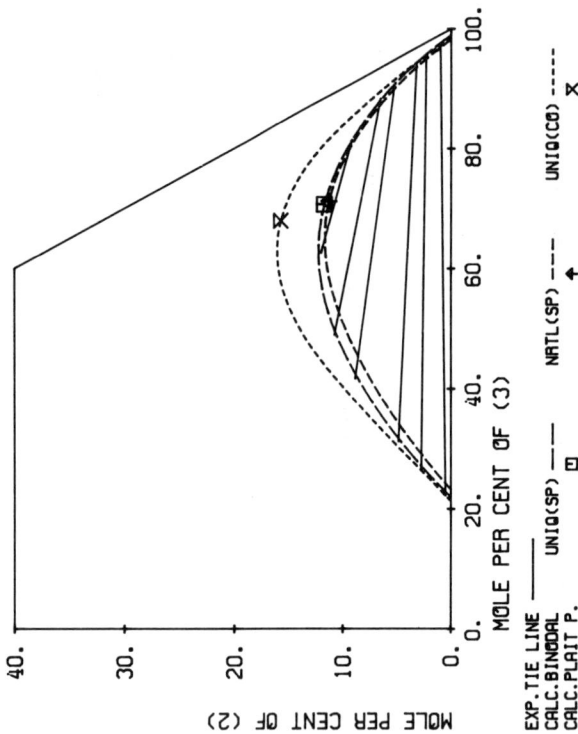

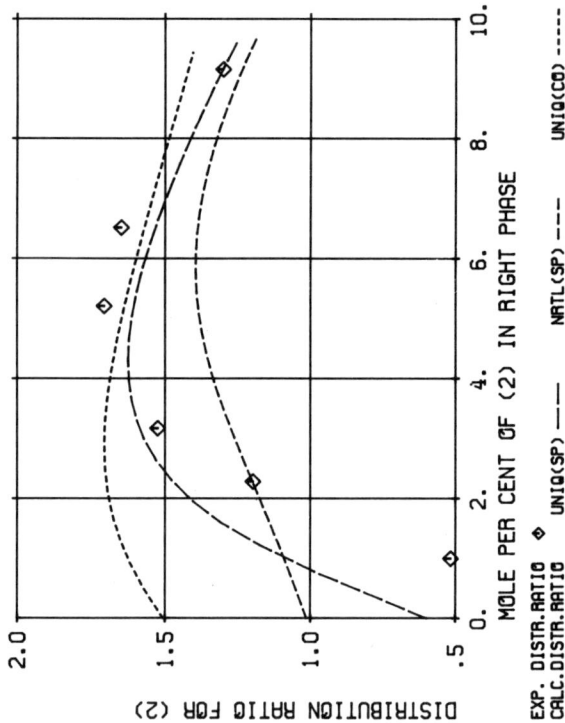

CH_2O_2-$C_5H_4O_2$

MOLE PER CENT OF (3)

MOLE PER CENT OF (2)

EXP.TIE LINE ———

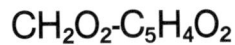

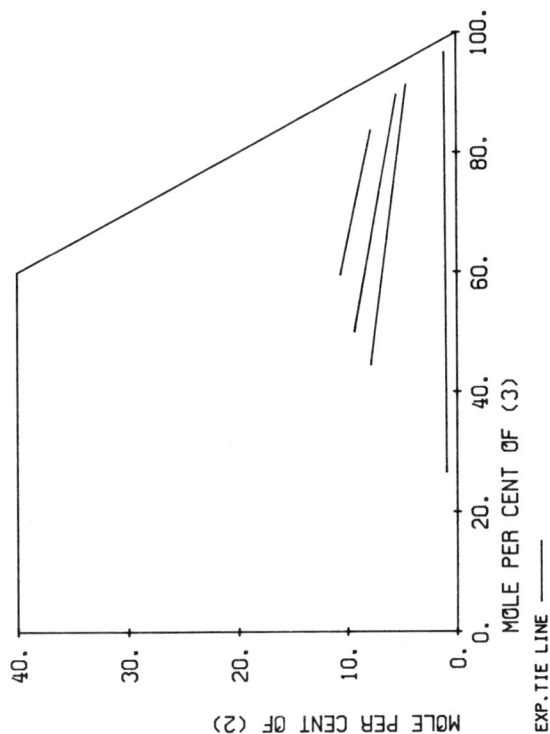

(1) C5H4O2   FURFURAL
(2) CH2O2   FORMIC ACID
(3) H2O   WATER

LANGFORD R.E., HERIC E.L.,
J.CHEM.ENG.DATA 17(1972)87

TEMPERATURE = 35.0 DEG C   TYPE OF SYSTEM = 1

EXPERIMENTAL TIE LINES IN MOLE PCT

LEFT PHASE			RIGHT PHASE		
(1)	(2)	(3)	(1)	(2)	(3)
72.436	0.976	26.589	2.234	1.090	96.676
47.575	7.850	44.575	4.098	4.595	91.307
40.663	9.331	50.007	4.829	5.471	89.700
29.883	10.590	59.527	8.488	7.811	83.701

CH_2O_2-$C_5H_{10}O_2$

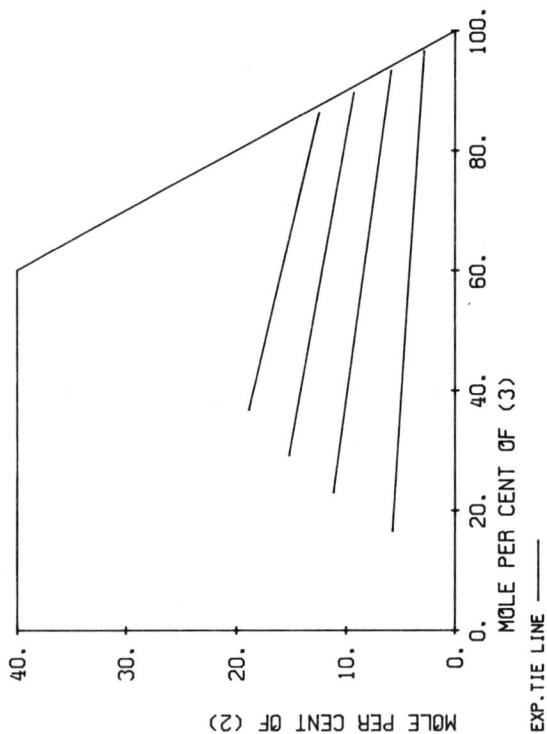

(1) C5H10O2   ACETIC ACID,PROPYL ESTER

(2) CH2O2   FORMIC ACID

(3) H2O   WATER

RAMANA RAO M.V., SOMASUNDARA RAO K., VENKATA RAO C.
J.SCI.IND.RES. 20B(1961)379

TEMPERATURE = 35.0 DEG C   TYPE OF SYSTEM = 1

EXPERIMENTAL TIE LINES IN MOLE PCT

LEFT PHASE			RIGHT PHASE		
(1)	(2)	(3)	(1)	(2)	(3)
77.755	5.726	16.518	0.528	2.845	96.627
65.902	11.145	22.952	0.734	5.858	93.408
55.685	15.163	29.152	0.987	9.270	89.743
44.439	18.850	36.712	1.213	12.424	86.363

MOLE PER CENT OF (3)

MOLE PER CENT OF (2)

EXP.TIE LINE ———

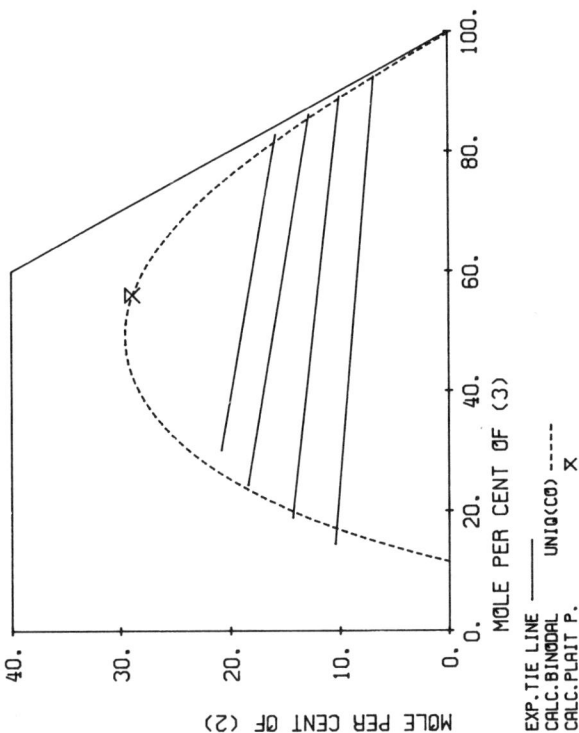

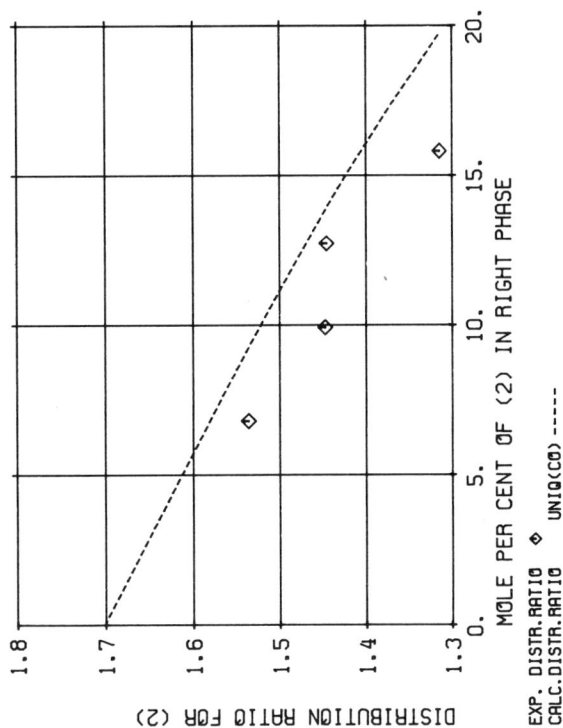

(1) C5H10O2   BUTANOIC ACID,METHYL ESTER
(2) CH2O2   FORMIC ACID
(3) H2O   WATER

RAMANA RAO M.V., DAKSHINAMURTY P., VENKATA RAO C.
INDIAN J.TECHNOL. 1(1963)220

TEMPERATURE = 30.0 DEG C   TYPE OF SYSTEM = 1

EXPERIMENTAL TIE LINES IN MOLE PCT

LEFT PHASE			RIGHT PHASE		
(1)	(2)	(3)	(1)	(2)	(3)
75.132	10.437	14.431	0.641	6.796	92.563
66.764	14.355	18.881	0.842	9.922	89.235
57.293	18.386	24.321	1.102	12.722	86.176
49.126	20.781	30.093	1.387	15.802	82.811

MEAN DEV. BETWEEN CALC. AND EXP. CONC. IN MOLE PCT
UNIQUAC (COMMON PARAMETERS)   0.93

CH$_2$O$_2$-C$_6$H$_6$

(1) H2O   WATER
(2) CH2O2   FORMIC ACID
(3) C6H6   BENZENE

UDOVENKO V.V., ALEKSANDROVA L.P.
ZH.FIZ.KHIM. 37(1963)52

TEMPERATURE = 30.0 DEG C   TYPE OF SYSTEM = 2

EXPERIMENTAL TIE LINES IN MOLE PCT

	LEFT PHASE			RIGHT PHASE	
(1)	(2)	(3)	(1)	(2)	(3)
90.689	9.232	0.079	0.043	0.203	99.753
82.268	17.614	0.118	0.052	0.423	99.525
73.926	25.912	0.162	0.065	0.795	99.141
67.808	31.879	0.313	0.086	1.198	98.715
41.159	57.409	1.432	0.149	4.737	95.114
19.927	76.342	3.732	0.819	12.035	87.146

SPECIFIC MODEL PARAMETERS IN KELVIN

		UNIQUAC		NRTL(ALPHA=.2)	
I	J	AIJ	AJI	AIJ	AJI
1	2	-276.64	-190.36	-429.88	-169.81
1	3	302.73	598.94	1187.0	1087.7
2	3	160.64	292.74	695.69	201.58

R1 = 0.9200   R2 = 1.5280   R3 = 3.1878
Q1 = 1.400   Q2 = 1.532   Q3 = 2.400

MEAN DEV. BETWEEN CALC. AND EXP. CONC. IN MOLE PCT

UNIQUAC (SPECIFIC PARAMETERS)   0.56
NRTL   (SPECIFIC PARAMETERS)   0.37
UNIQUAC (COMMON PARAMETERS)   0.74

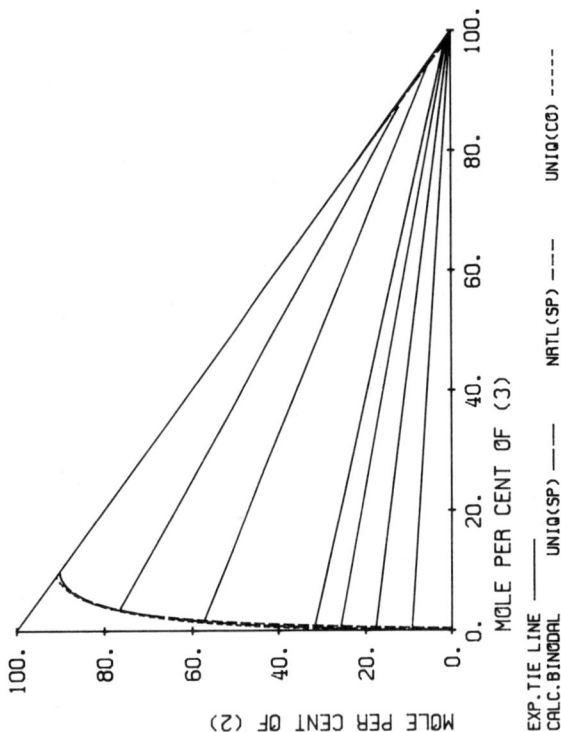

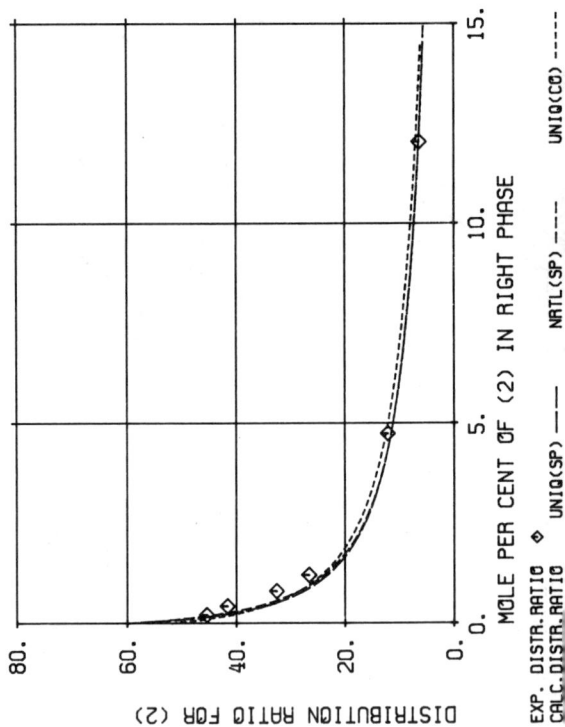

(1) H2O

(2) CH2O2   FORMIC ACID

(3) C6H6   BENZENE

UDOVENKO V.V., ALEKSANDROVA L.P.
ZH.FIZ.KHIM. 37(1963)52

TEMPERATURE = 45.0 DEG C   TYPE OF SYSTEM = 1

EXPERIMENTAL TIE LINES IN MOLE PCT

	LEFT PHASE			RIGHT PHASE	
(1)	(2)	(3)	(1)	(2)	(3)
90.067	9.853	0.080	0.043	0.271	99.686
83.436	16.419	0.145	0.087	0.491	99.423
72.424	27.278	0.298	0.129	1.047	98.824
62.119	37.328	0.553	0.172	1.984	97.844
40.942	57.304	1.753	0.212	5.369	94.420
28.022	68.817	3.161	0.417	8.725	90.858

SPECIFIC MODEL PARAMETERS IN KELVIN

		UNIQUAC		NRTL(ALPHA=.2)	
I	J	AIJ	AJI	AIJ	AJI
1	2	-240.29	-173.97	-347.62	-149.86
1	3	322.29	672.37	1314.6	1230.7
2	3	151.20	342.99	680.31	263.13

R1 = 0.9200   R2 = 1.5280   R3 = 3.1878
Q1 = 1.400   Q2 = 1.532   Q3 = 2.400

MEAN DEV. BETWEEN CALC. AND EXP. CONC. IN MOLE PCT

UNIQUAC (SPECIFIC PARAMETERS)   0.41
NRTL   (SPECIFIC PARAMETERS)   0.23

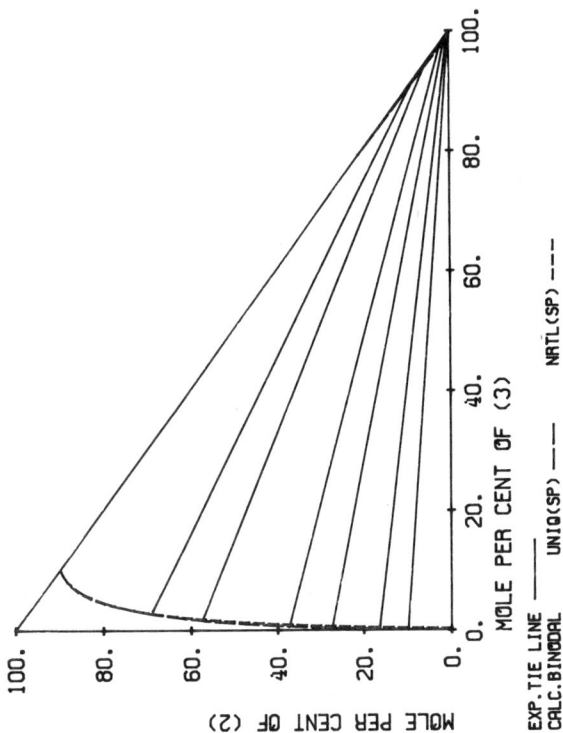

MOLE PER CENT OF (2)

MOLE PER CENT OF (3)

EXP.TIE LINE ——   UNIQ(SP) ——   NRTL(SP) ----
CALC.BINODAL ——

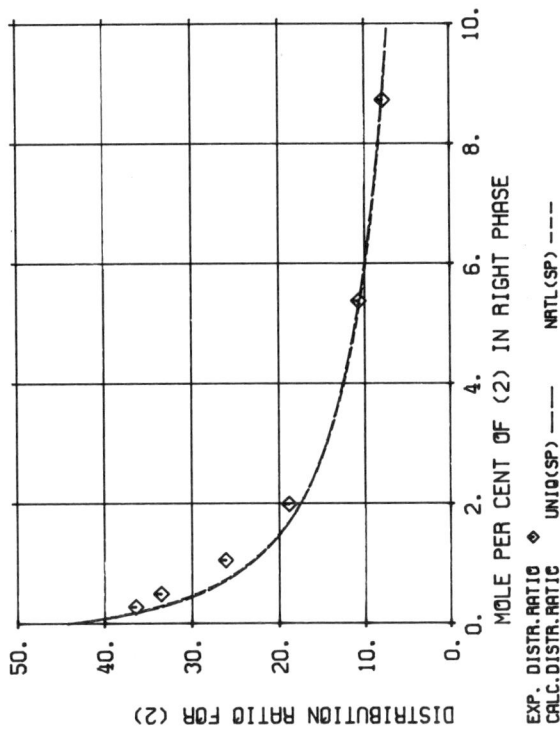

DISTRIBUTION RATIO FOR (2)

MOLE PER CENT OF (2) IN RIGHT PHASE

EXP. DISTR.RATIO ◆   UNIQ(SP) ——   NRTL(SP) ----
CALC.DISTR.RATIO

CH_2O_2-C_6H_6

(1) H2O   WATER
----
(2) CH2O2   FORMIC ACID
----
(3) C6H6   BENZENE

UDOVENKO V.V., ALEKSANDROVA L.P.
ZH.FIZ.KHIM. 37(1963)52

TEMPERATURE = 60.0 DEG C   TYPE OF SYSTEM = 1

EXPERIMENTAL TIE LINES IN MOLE PCT

	LEFT PHASE			RIGHT PHASE	
(1)	(2)	(3)	(1)	(2)	(3)
90.547	9.321	0.133	0.087	0.271	99.642
77.071	22.615	0.314	0.129	0.912	98.959
62.417	36.846	0.737	0.257	2.166	97.577
53.006	45.659	1.335	0.510	3.679	95.810
28.713	67.264	4.022	1.034	9.230	89.736

SPECIFIC MODEL PARAMETERS IN KELVIN

		UNIQUAC			NRTL(ALPHA=.2)	
I	J	AIJ	AJI		AIJ	AJI
1	2	-237.92	-200.69		-316.67	-214.51
1	3	299.94	621.54		1267.1	1132.7
2	3	135.81	355.14		663.75	262.67

R1 = 0.9200   R2 = 1.5280   R3 = 3.1878
Q1 = 1.400   Q2 = 1.532   Q3 = 2.400

MEAN DEV. BETWEEN CALC. AND EXP. CONC. IN MOLE PCT

UNIQUAC (SPECIFIC PARAMETERS)   0.58
NRTL   (SPECIFIC PARAMETERS)   0.36

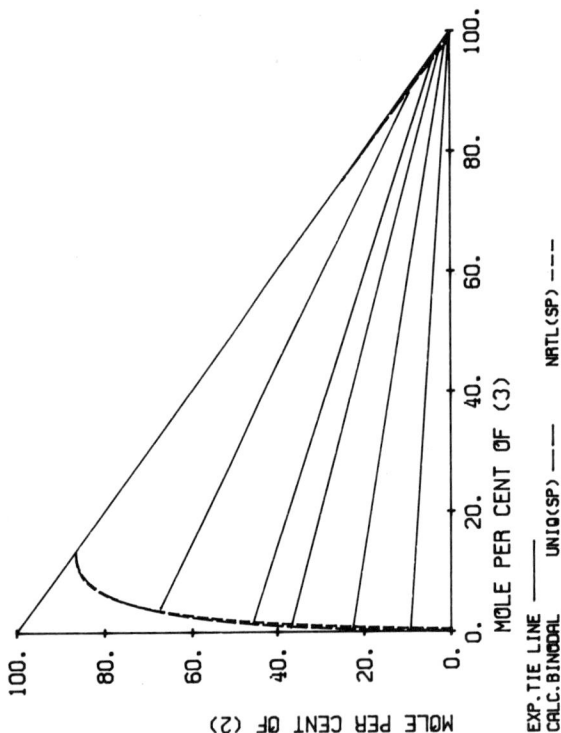

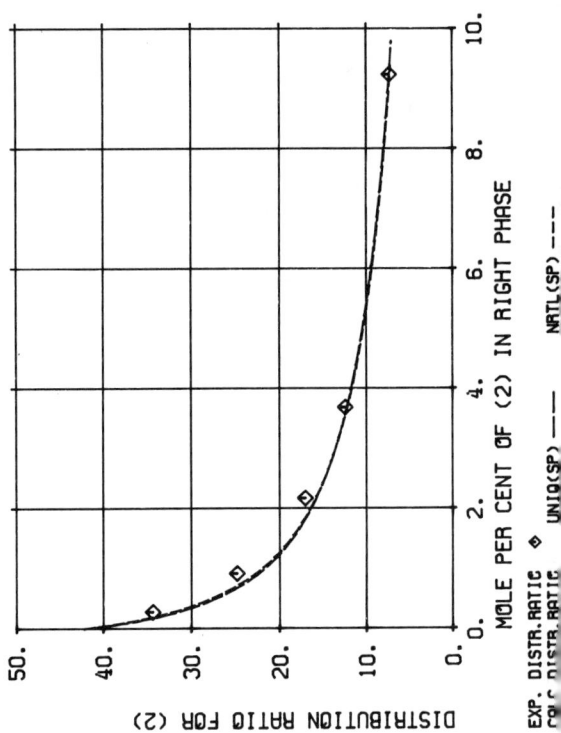

MOLE PER CENT OF (2)

MOLE PER CENT OF (3)

EXP.TIE LINE ——— UNIQ(SP) ———   NRTL(SP) ----
CALC.BINODAL

DISTRIBUTION RATIO FOR (2)

MOLE PER CENT OF (2) IN RIGHT PHASE

EXP. DISTR.RATIC ◊ UNIQ(SP)   NRTL(SP)
CALC. DISTR.RATIC

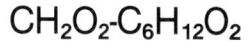

MOLE PER CENT OF (3)

MOLE PER CENT OF (2)

EXP.TIE LINE —— UNIQ(SP) ——□ NRTL(SP) ——+
CALC.BINODAL ——— CALC.PLAIT P. +

MOLE PER CENT OF (2) IN RIGHT PHASE

DISTRIBUTION RATIO FOR (2)

EXP. DISTR.RATIO ◇ UNIQ(SP) —— NRTL(SP) ———
CALC.DISTR.RATIO ◇

(1) C₆H₁₂O₂   ACETIC ACID, BUTYL ESTER
(2) CH₂O₂   FORMIC ACID
(3) H₂O   WATER

RAMANA RAO M.V., SOMASUNDARA RAO K., VENKATA RAO C.
J.SCI.IND.RES. 20B(1961)379

TEMPERATURE = 35.0 DEG C   TYPE OF SYSTEM = 1

EXPERIMENTAL TIE LINES IN MOLE PCT

	LEFT PHASE			RIGHT PHASE	
(1)	(2)	(3)	(1)	(2)	(3)
69.370	9.447	21.183	0.190	6.527	93.283
61.532	13.386	25.082	0.234	9.543	90.222
56.338	16.418	27.245	0.281	12.386	87.333
48.701	20.691	30.607	0.394	16.073	83.533
44.324	22.873	32.803	0.555	18.987	80.458
39.488	25.712	34.801	0.798	22.311	76.891

SPECIFIC MODEL PARAMETERS IN KELVIN

		UNIQUAC		NRTL(ALPHA=.2)	
I	J	AIJ	AJI	AIJ	AJI
1	2	40.307	-218.84	-192.37	287.47
1	3	174.70	404.39	103.89	2360.7
2	3	23.424	-542.18	-577.94	821.09

R1 = 4.8274   R2 = 1.5280   R3 = 0.9200
Q1 = 4.196   Q2 = 1.532   Q3 = 1.400

MEAN DEV. BETWEEN CALC. AND EXP. CONC. IN MOLE PCT

UNIQUAC (SPECIFIC PARAMETERS)   0.43
NRTL   (SPECIFIC PARAMETERS)   0.44

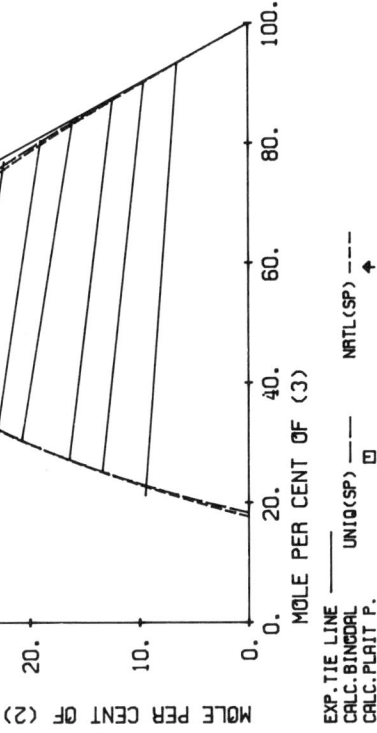

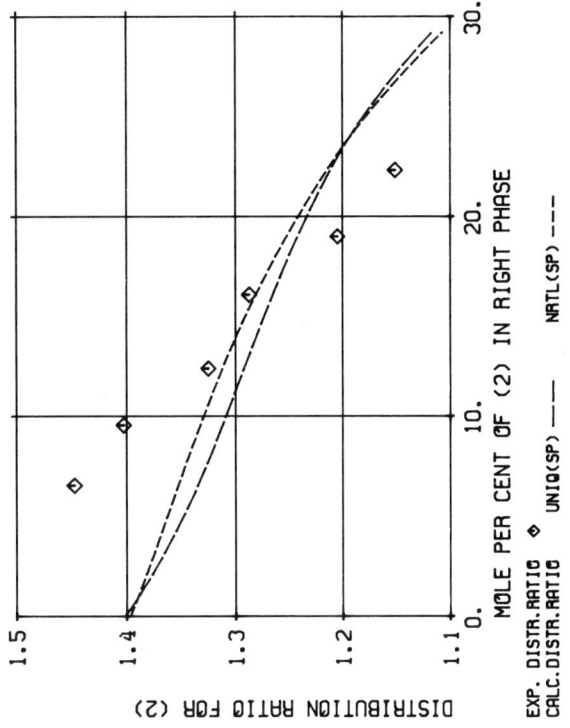

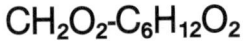

$CH_2O_2\text{-}C_6H_{12}O_2$

MOLE PER CENT OF (2)

MOLE PER CENT OF (3)

DISTRIBUTION RATIO FOR (2)

MOLE PER CENT OF (2) IN RIGHT PHASE

EXP.TIE LINE —— UNIQ(SP) ▣ NRTL(SP) ✦ UNIQ(CO) ✗
CALC.BINODAL
CALC.PLAIT P.

EXP. DISTR.RATIO ◇ UNIQ(SP) NRTL(SP) UNIQ(CO) -----
CALC.DISTR.RATIO

(1) C6H12O2   BUTANOIC ACID,ETHYL ESTER

(2) CH2O2   FORMIC ACID

(3) H2O   WATER

RAMANA RAO M.V., DAKSHINAMURTY P., VENKATA RAO C.
INDIAN J.TECHNOL. 1(1963)220

TEMPERATURE = 30.0 DEG C   TYPE OF SYSTEM = 1

EXPERIMENTAL TIE LINES IN MOLE PCT

	LEFT PHASE			RIGHT PHASE	
(1)	(2)	(3)	(1)	(2)	(3)
77.396	9.619	12.986	0.211	7.714	92.076
71.618	12.921	15.461	0.330	10.650	89.020
66.412	16.664	16.924	0.384	13.865	85.752
60.098	19.702	20.200	0.438	16.780	82.782
55.110	22.317	22.573	0.514	19.200	80.286
49.587	24.564	25.849	0.596	21.825	77.579

SPECIFIC MODEL PARAMETERS IN KELVIN

I J	UNIQUAC AIJ	AJI	NRTL(ALPHA=.2) AIJ	AJI
1 2	80.110	29.221	-355.95	188.77
1 3	512.09	89.214	254.62	1758.3
2 3	-252.11	259.66	-485.25	401.37

R1 = 4.8274   R2 = 1.5280   R3 = 0.9200
Q1 = 4.196   Q2 = 1.532   Q3 = 1.400

MEAN DEV. BETWEEN CALC. AND EXP. CONC. IN MOLE PCT

UNIQUAC (SPECIFIC PARAMETERS)   0.63
NRTL (SPECIFIC PARAMETERS)   0.67
UNIQUAC (COMMON PARAMETERS)   0.89

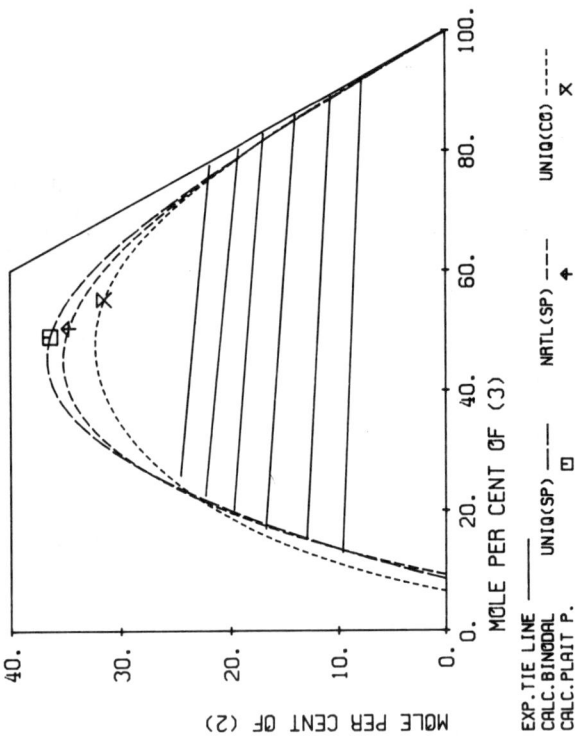

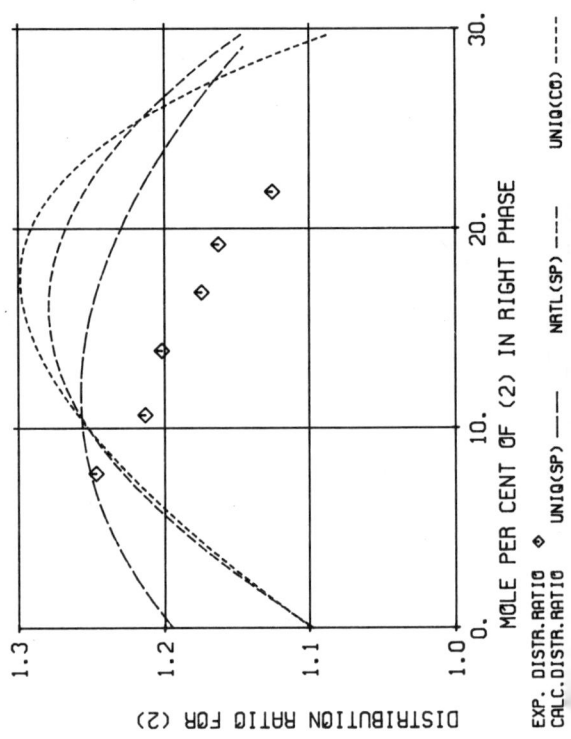

$CH_2O_2\text{-}C_6H_{12}O_2$

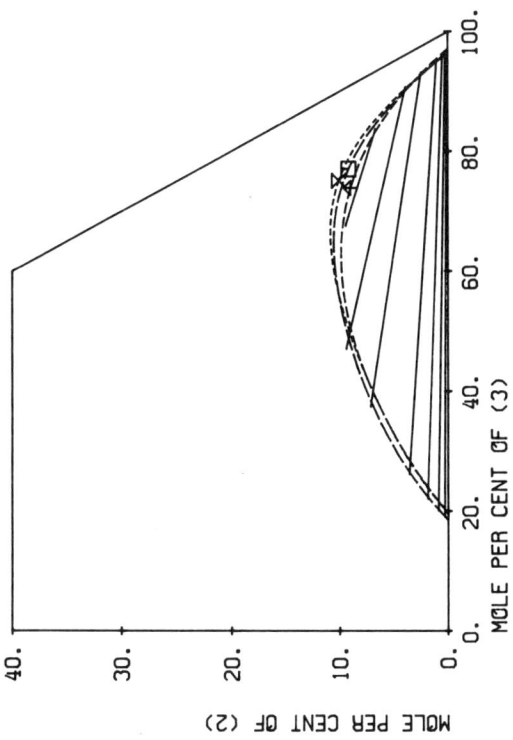

(1) C6H12O2   1,3-DIOXANE,4,4-DIMETHYL

(2) CH2O2   FORMIC ACID

(3) H2O   WATER

PUGACH L.M., OGORODNIKOV S.K., IDLIS G.S.
ZH.PRIKL.KHIM.(LENINGRAD) 47(1974)1856

TEMPERATURE = 20.0 DEG C   TYPE OF SYSTEM = 1

EXPERIMENTAL TIE LINES IN MOLE PCT

	LEFT PHASE			RIGHT PHASE	
(1)	(2)	(3)	(1)	(2)	(3)
80.369	0.0	19.631	3.590	0.0	96.410
80.630	0.317	19.054	3.504	0.093	96.402
79.047	0.895	20.058	3.553	0.253	96.194
76.035	1.882	22.083	3.517	0.531	95.952
70.306	3.556	26.138	3.717	1.064	95.219
55.518	7.138	37.344	4.347	2.471	93.182
43.594	9.399	47.007	5.349	3.894	90.757
23.015	9.433	67.552	9.631	6.659	83.709

SPECIFIC MODEL PARAMETERS IN KELVIN

		UNIQUAC			NRTL(ALPHA=.2)	
I J	AIJ	AJI		AIJ	AJI	
1 2	-134.72	68.382		564.57	-815.17	
1 3	604.70	-92.620		111.37	1003.1	
2 3	-165.07	133.22		-203.04	-586.07	

R1 = 4.5327   R2 = 1.5280   R3 = 0.9200
Q1 = 3.796   Q2 = 1.532   Q3 = 1.400

MEAN DEV. BETWEEN CALC. AND EXP. CONC. IN MOLE PCT

UNIQUAC (SPECIFIC PARAMETERS)   0.29
NRTL   (SPECIFIC PARAMETERS)   0.79
UNIQUAC (COMMON PARAMETERS)   0.50

EXP.TIE LINE ———   UNIQ(SP) ---- ⊡   NRTL(SP) ---- ✦   UNIQ(CO) ----- ⨯
CALC.BINODAL
CALC.PLAIT P.

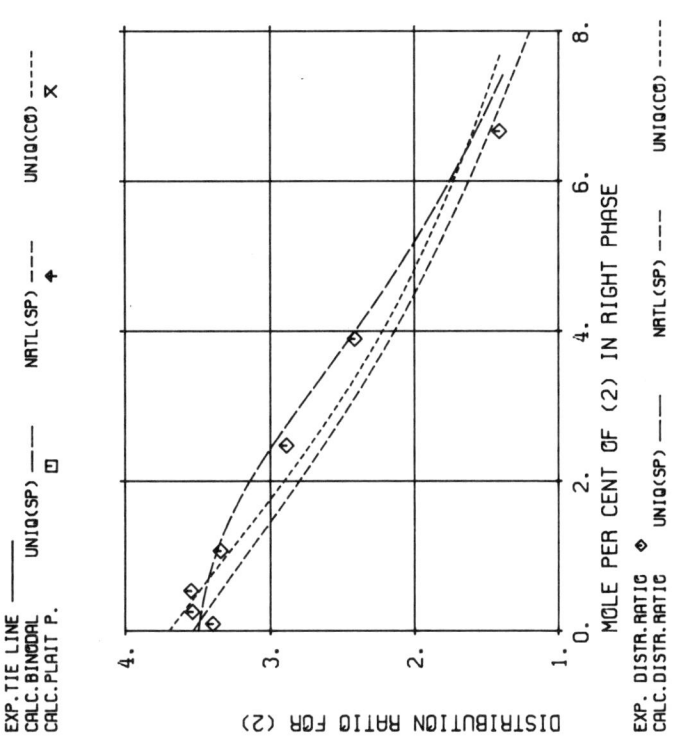

EXP. DISTR.RATIO ———   UNIQ(SP) ◇   NRTL(SP) ----   UNIQ(CO) -----
CALC.DISTR.RATIO

CH_2O_2-$C_6H_{14}O$

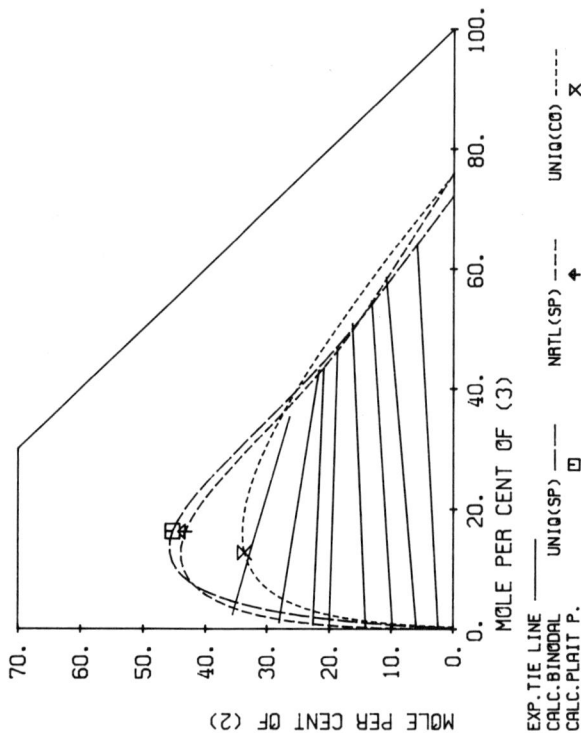

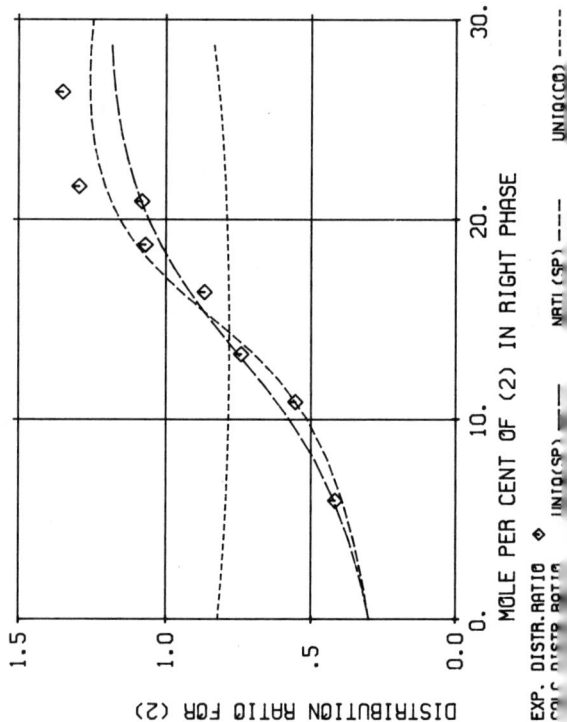

(1) H2O   WATER
(2) CH2O2   FORMIC ACID
(3) C6H14O   2-PENTANOL, 4-METHYL

RAJA RAO M., RAMAMURTY M., VENKATA RAO C.
CHEM.ENG.SCI. 8(1958)265

TEMPERATURE = 30.0 DEG C   TYPE OF SYSTEM = 1

EXPERIMENTAL TIE LINES IN MOLE PCT

| | LEFT PHASE | | | RIGHT PHASE | |
(1)	(2)	(3)	(1)	(2)	(3)
97.274	2.467	0.259	29.870	5.925	64.205
93.683	5.024	0.293	30.490	10.862	58.648
89.840	9.830	0.330	31.994	13.258	54.748
85.380	14.182	0.438	32.707	16.356	50.937
79.440	19.994	0.566	34.207	18.715	47.078
76.749	22.569	0.683	35.593	20.895	43.512
70.952	28.001	1.047	35.202	21.650	43.149
61.985	35.606	2.409	38.275	26.365	35.360

SPECIFIC MODEL PARAMETERS IN KELVIN

| | | UNIQUAC | | NRTL(ALPHA=.2) | |
I	J	AIJ	AJI	AIJ	AJI
1	2	648.37	-255.61	1471.2	-691.96
1	3	213.04	158.26	2695.3	38.258
2	3	78.063	189.59	862.87	-256.96

R1 = 0.9200   R2 = 1.5280   R3 = 4.8015
Q1 = 1.400   Q2 = 1.532   Q3 = 4.124

MEAN DEV. BETWEEN CALC. AND EXP. CONC. IN MOLE PCT

UNIQUAC (SPECIFIC PARAMETERS)   0.93
NRTL   (SPECIFIC PARAMETERS)   0.64
UNIQUAC (COMMON PARAMETERS)   3.16

$CH_2O_2\text{-}C_7H_{14}O_2$

(1) C7H14O2   ACETIC ACID,PENTYL ESTER

(2) CH2O2   FORMIC ACID

(3) H2O   WATER

KRISHNAMURTY R., JAYARAMA RAO G., VENKATA RAO C.
J.SCI.IND.RES. 21D(1962)282

TEMPERATURE = 28.0 DEG C   TYPE OF SYSTEM = 1

EXPERIMENTAL TIE LINES IN MOLE PCT

LEFT PHASE		RIGHT PHASE	
(1)	(2)	(3)	(1)	(2)	(3)

(1)	(2)	(3)	(1)	(2)	(3)
83.089	5.214	11.697	0.082	3.478	96.440
75.490	9.875	14.635	0.108	7.247	92.645
64.141	16.436	19.423	0.155	13.597	86.248
60.202	19.347	20.450	0.184	16.483	83.333
49.979	25.536	24.485	0.331	24.769	74.900

SPECIFIC MODEL PARAMETERS IN KELVIN

		UNIQUAC		NRTL(ALPHA=.2)	
I J	AIJ	AJI	AIJ	AJI
1 2	78.226	133.78	-344.28	703.87
1 3	484.91	81.892	287.41	2325.6
2 3	-237.12	351.14	-603.76	1005.8

R1 = 5.5018   R2 = 1.5280   R3 = 0.9200
Q1 = 4.736   Q2 = 1.532   Q3 = 1.400

MEAN DEV. BETWEEN CALC. AND EXP. CONC. IN MOLE PCT

UNIQUAC (SPECIFIC PARAMETERS)   0.54
NRTL   (SPECIFIC PARAMETERS)   0.31
UNIQUAC (COMMON PARAMETERS)   1.82

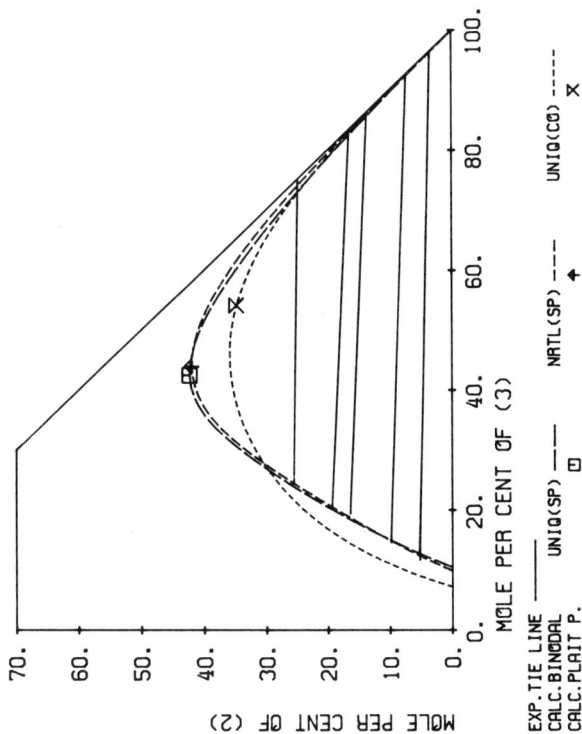

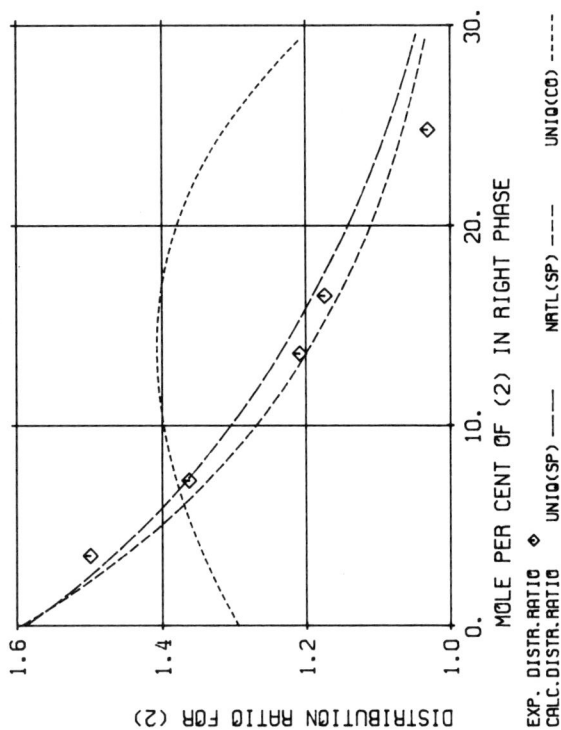

CH₂O₂-C₉H₁₀O₂
$CH_2O_2-C_9H_{10}O_2$

(1) H2O   WATER
(2) CH2O2   FORMIC ACID
(3) C9H1002   BENZOIC ACID,ETHYL ESTER

JAYA RAMA RAO G., VEKKATA RAO C.
J.SCI.IND.RES. 16B(1957)102

TEMPERATURE = 30.0 DEG C   TYPE OF SYSTEM = 1

EXPERIMENTAL TIE LINES IN MOLE PCT

	LEFT PHASE			RIGHT PHASE	
(1)	(2)	(3)	(1)	(2)	(3)
92.517	7.443	0.040	8.161	5.517	86.323
81.257	18.665	0.078	12.646	13.024	74.330
69.687	30.117	0.196	16.695	23.432	59.873
62.489	36.969	0.542	18.332	30.338	51.330

MEAN DEV. BETWEEN CALC. AND EXP. CONC. IN MOLE PCT
UNIQUAC (COMMON PARAMETERS)   1.25

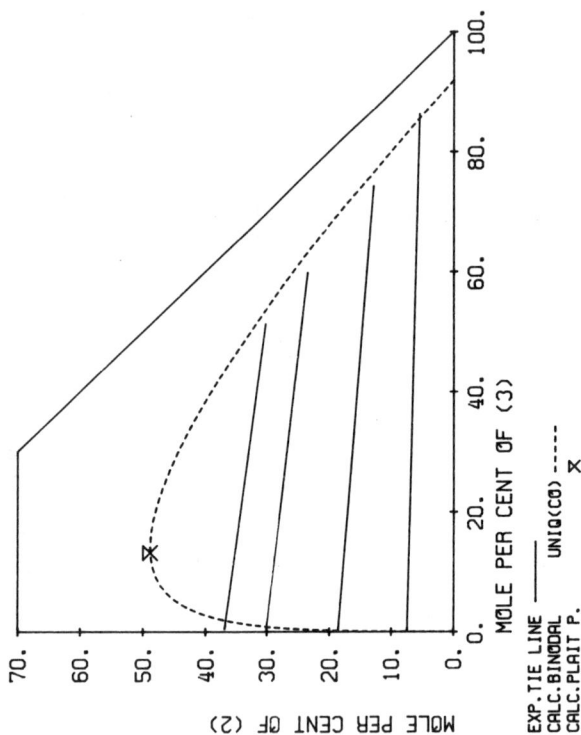

MOLE PER CENT OF (3)

MOLE PER CENT OF (2)

EXP.TIE LINE ———   UNIQ(CO) -----
CALC.BINODAL
CALC.PLAIT P.   X

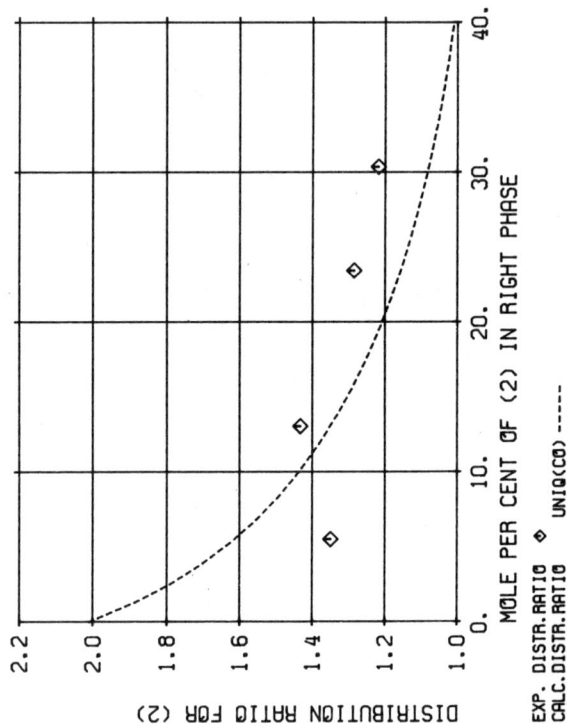

MOLE PER CENT OF (2) IN RIGHT PHASE

DISTRIBUTION RATIO FOR (2)

EXP. DISTR.RATIO ◇   UNIQ(CO) -----
CALC.DISTR.RATIO

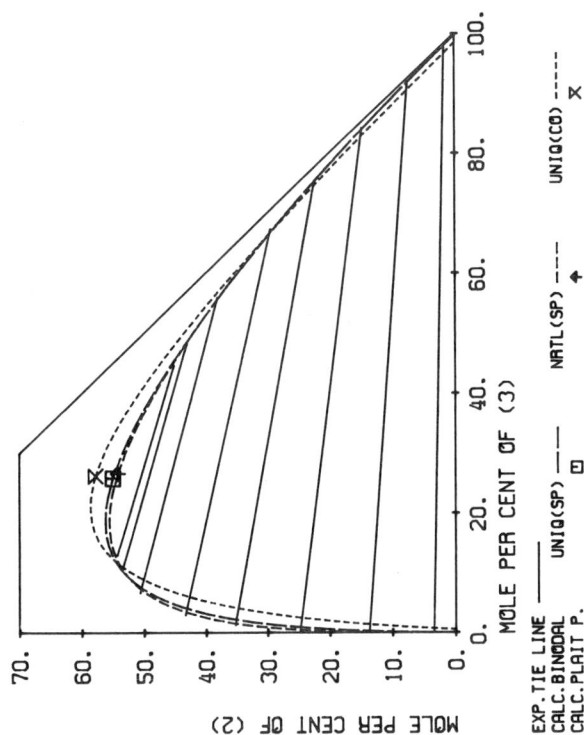

MOLE PER CENT OF (3)

MOLE PER CENT OF (2)

EXP.TIE LINE —— UNIQ(SP) ···· ⊡ NRTL(SP) ···· ✦ UNIQ(CO) ···· ☓
CALC.BINODAL
CALC.PLAIT P.

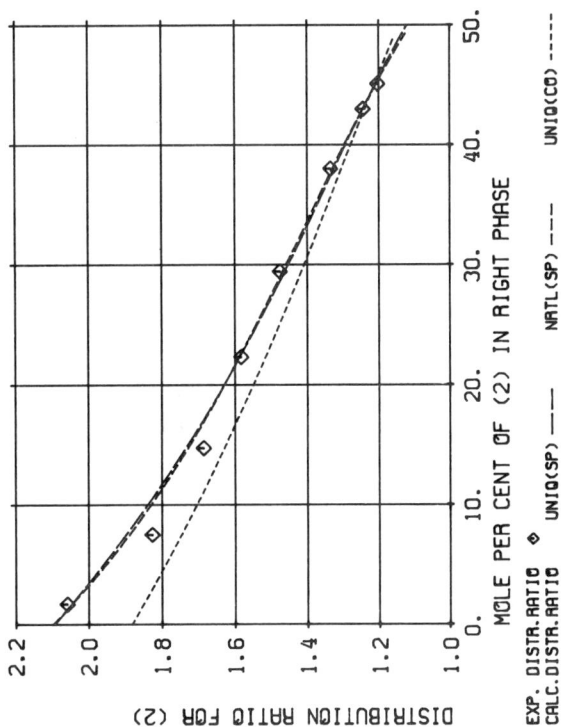

MOLE PER CENT OF (2) IN RIGHT PHASE

DISTRIBUTION RATIO FOR (2)

EXP. DISTR.RATIO ◇ UNIQ(SP) —— NRTL(SP) ---- UNIQ(CO) -----
CALC.DISTR.RATIO

(1) C8H10   BENZENE,ETHYL
(2) C3H6O   2-PROPANONE
(3) CH3NO   FORMIC ACID,AMIDE

BLANK M.G.
UKR.KHIM.ZH.(RUSS.ED.) 29(1963)1009

TEMPERATURE = 25.0 DEG C   TYPE OF SYSTEM = 1

EXPERIMENTAL TIE LINES IN MOLE PCT

	LEFT PHASE			RIGHT PHASE	
(1)	(2)	(3)	(1)	(2)	(3)
96.400	3.500	0.100	0.600	1.700	97.700
86.100	13.700	0.200	0.800	7.500	91.700
74.900	24.800	0.300	1.000	14.700	84.300
63.400	35.300	1.300	2.200	22.300	75.500
53.600	43.400	3.000	3.200	29.400	67.400
42.600	50.700	6.700	6.000	38.000	56.000
35.900	53.500	10.600	8.600	43.000	48.400
32.700	54.300	13.000	10.200	45.100	44.700

SPECIFIC MODEL PARAMETERS IN KELVIN

		UNIQUAC		NRTL(ALPHA=.2)	
I J		AIJ	AJI	AIJ	AJI
1 2		236.96	-194.19	151.11	-140.55
1 3		1398.0	-235.79	2615.9	1290.5
2 3		365.88	-136.19	307.14	-20.254

R1 = 4.5972   R2 = 2.5735   R3 = 1.6928
Q1 = 3.508   Q2 = 2.336   Q3 = 1.644

MEAN DEV. BETWEEN CALC. AND EXP. CONC. IN MOLE PCT

UNIQUAC (SPECIFIC PARAMETERS)   0.39
NRTL (SPECIFIC PARAMETERS)   0.21
UNIQUAC (COMMON PARAMETERS)   1.19

CH$_3$NO-C$_3$H$_6$O

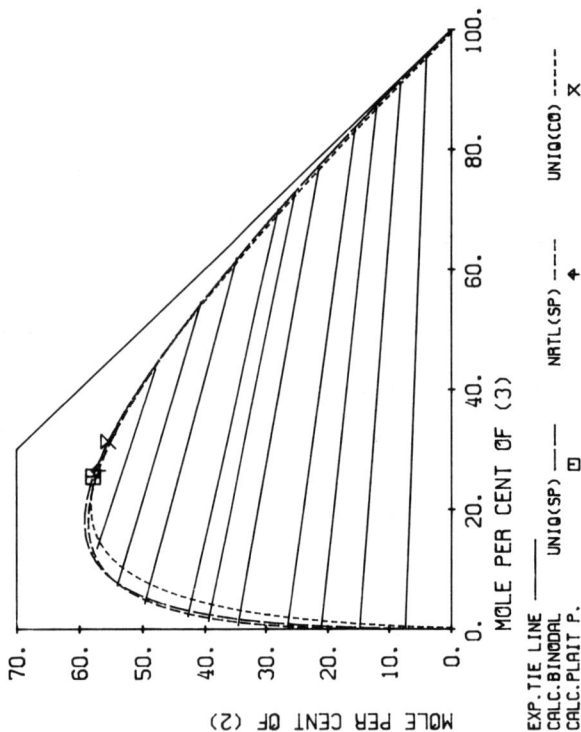

MOLE PER CENT OF (3)

MOLE PER CENT OF (2)

EXP.TIE LINE   UNIQ(SP) —— ⊟   NRTL(SP) ---- ✦   UNIQ(CO) -----
CALC.BINODAL
CALC.PLAIT P.   ✕

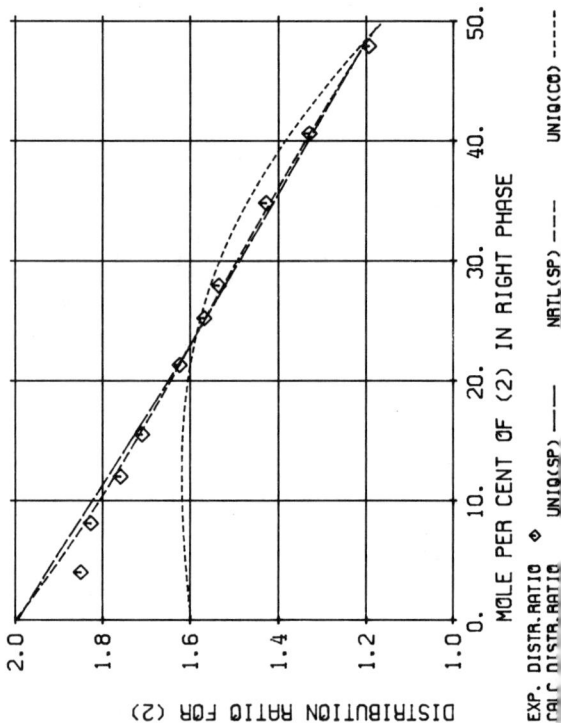

MOLE PER CENT OF (2) IN RIGHT PHASE

DISTRIBUTION RATIO FOR (2)

EXP. DISTR.RATIO ◇   UNIQ(SP) ——   NRTL(SP) ----   UNIQ(CO) -----
CALC. DISTR.RATIO

(1) C9H12   BENZENE,ISOPROPYL

(2) C3H6O   2-PROPANONE

(3) CH3NO   FORMIC ACID,AMIDE

BLANK M.G.
UKR.KHIM.ZH.(RUSS.ED.) 29(1963)1009

TEMPERATURE = 25.0 DEG C   TYPE OF SYSTEM = 1

EXPERIMENTAL TIE LINES IN MOLE PCT

	LEFT PHASE			RIGHT PHASE	
(1)	(2)	(3)	(1)	(2)	(3)
92.500	7.400	0.100	0.400	4.000	95.600
85.100	14.800	0.100	0.600	8.100	91.300
78.700	21.100	0.200	0.900	12.000	87.100
73.000	26.500	0.500	1.000	15.500	83.500
64.700	34.600	0.700	1.300	21.300	77.400
59.100	39.500	1.400	1.800	25.200	73.100
55.000	42.800	2.200	2.000	27.900	70.100
46.000	49.700	4.300	3.200	34.800	62.000
38.700	54.000	7.300	4.800	40.600	54.600
29.300	57.200	13.500	8.700	47.900	43.400

SPECIFIC MODEL PARAMETERS IN KELVIN

		UNIQUAC		NRTL(ALPHA=.2)	
I	J	AIJ	AJI	AIJ	AJI
1	2	242.03	-190.70	67.003	-41.274
1	3	1383.6	234.13	2718.3	1512.9
2	3	371.56	-144.38	346.14	-36.468

R1 = 5.2708   R2 = 2.5735   R3 = 1.6928
Q1 = 4.044   Q2 = 2.336   Q3 = 1.644

MEAN DEV. BETWEEN CALC. AND EXP. CONC. IN MOLE PCT

UNIQUAC (SPECIFIC PARAMETERS)   0.42
NRTL   (SPECIFIC PARAMETERS)   0.23
UNIQUAC (COMMON PARAMETERS)   1.68

$CH_3NO-C_5H_4O_2$

MASKHULIYA V.P., KRUPATKIN I.L.
FAZOVYE RAVNOVESIYA,NO.2,KALININ,EDITOR:I.KRUPATKIN
(1975)25

TEMPERATURE = 25.0 DEG C   TYPE OF SYSTEM = 1

EXPERIMENTAL TIE LINES IN MOLE PCT

	LEFT PHASE			RIGHT PHASE	
(1)	(2)	(3)	(1)	(2)	(3)
95.206	2.672	2.122	21.717	1.737	76.546
91.629	5.546	2.825	22.204	3.587	74.210
89.055	7.387	3.557	24.374	3.916	71.709
85.926	9.582	4.492	25.964	5.355	68.681
81.249	12.767	5.984	27.499	7.542	54.959
76.116	15.857	8.027	27.226	9.334	63.440
66.400	19.729	13.871	34.421	10.361	55.218

SPECIFIC MODEL PARAMETERS IN KELVIN

		UNIQUAC			NRTL(ALPHA=.2)	
I J	AIJ	AJI		AIJ	AJI	
1 2	-225.98	133.47		-111.99	86.578	
1 3	11.485	291.69		1056.0	93.190	
2 3	-27.702	236.73		-52.576	320.90	

R1 = 0.9200   R2 = 1.6928   R3 = 3.1680
Q1 = 1.400   Q2 = 1.644   Q3 = 2.484

MEAN DEV. BETWEEN CALC. AND EXP. CONC. IN MOLE PCT

UNIQUAC (SPECIFIC PARAMETERS)	0.96
NRTL (SPECIFIC PARAMETERS)	1.28
UNIQUAC (COMMON PARAMETERS)	1.12

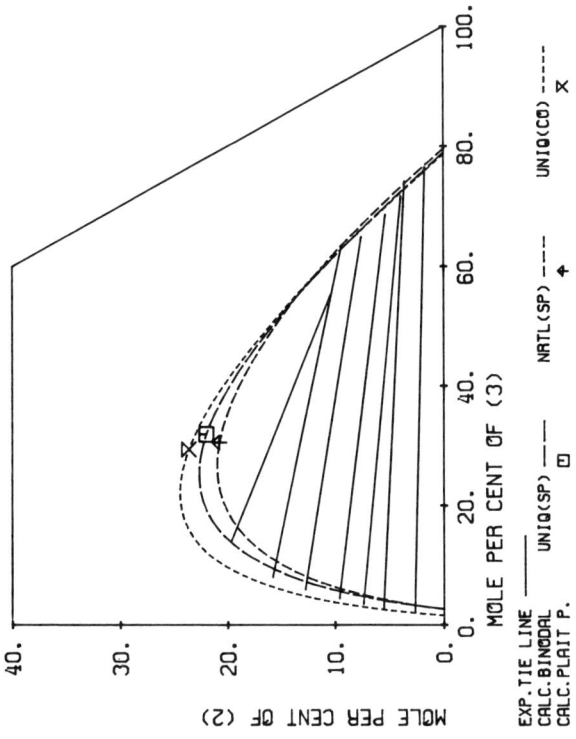

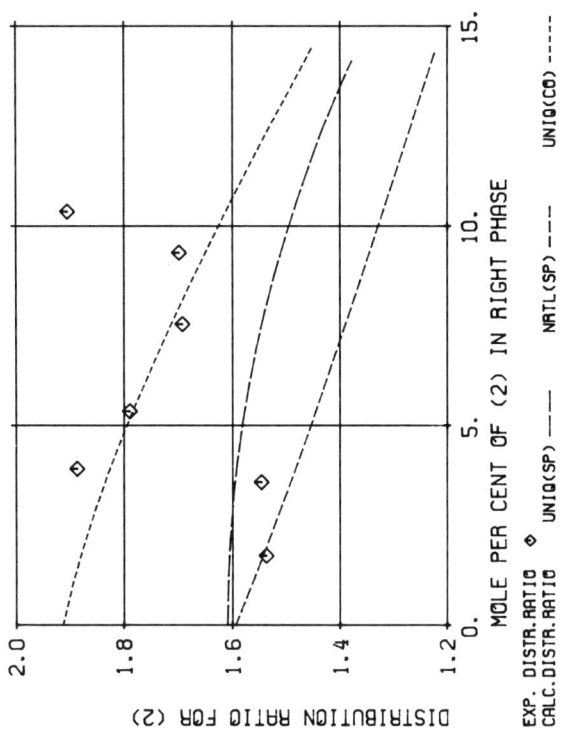

CH_3NO_2-$C_2H_4O_2$

(1) CH3NO2   METHANE,NITRO
----
(2) C2H4O2   ACETIC ACID
----
(3) H2O   WATER

SKRZEC A.E.; MURPHY N.F.
IND.ENG.CHEM. 46(1954)2245

TEMPERATURE = 26.7 DEG C   TYPE OF SYSTEM = 1

EXPERIMENTAL TIE LINES IN MOLE PCT (GRAPH.INTERPOL.)

            LEFT PHASE   RIGHT PHASE
        (1)   (2)   (3)   (1)   (2)   (3)

     93.873   0.0   6.127   3.039   0.0   96.961
     88.201   2.613   9.186   4.304   2.340   93.356
     81.648   4.595   13.757   5.130   4.172   90.698
     77.082   5.804   17.114   5.588   5.341   89.071
     74.787   6.551   18.662   6.064   6.319   87.617
     64.541   9.897   25.561   8.012   8.603   83.385

SPECIFIC MODEL PARAMETERS IN KELVIN

            UNIQUAC   NRTL(ALPHA=.2)
I J   AIJ   AJI   AIJ   AJI
----
1 2   -130.12   -110.53   706.31   -708.05
1 3   386.71   258.56   410.68   866.02
2 3   -333.09   102.04   67.480   -686.61

R1 = 2.0086   R2 = 2.2024   R3 = 0.9200
Q1 = 1.868   Q2 = 2.072   Q3 = 1.400

MEAN DEV. BETWEEN CALC. AND EXP. CONC. IN MOLE PCT

UNIQUAC (SPECIFIC PARAMETERS)   0.43
NRTL   (SPECIFIC PARAMETERS)   0.45
UNIQUAC (COMMON PARAMETERS)   0.72

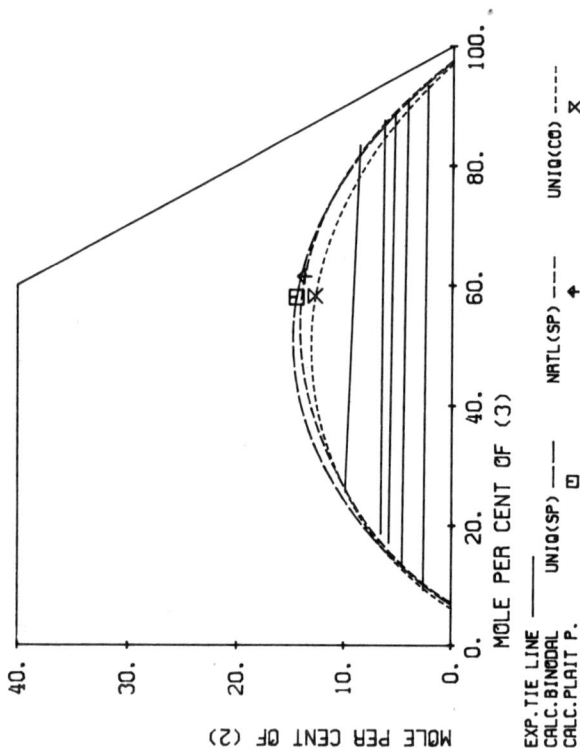

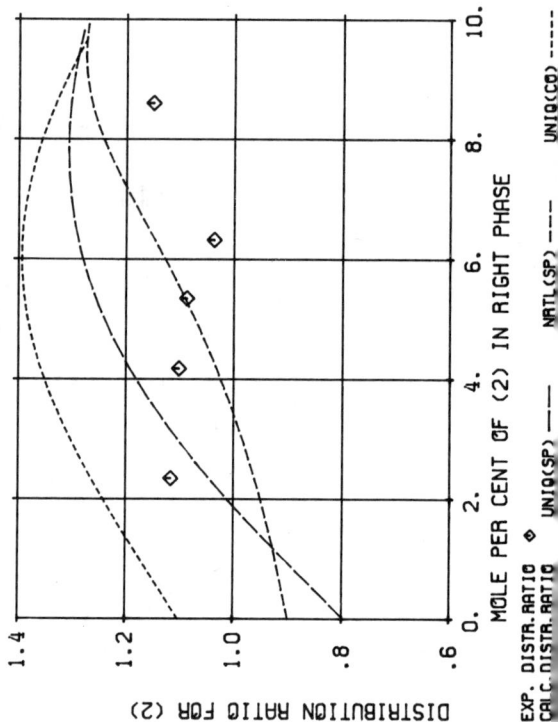

$CH_3NO_2-C_2H_6O_2$

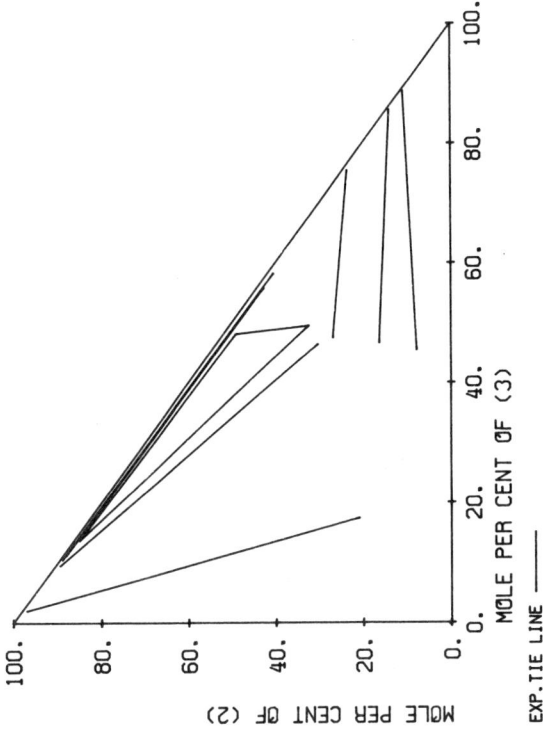

MOLE PER CENT OF (3)

MOLE PER CENT OF (2)

——— EXP.TIE LINE

(1) C12H26O   1-DODECANOL

(2) CH3NO2   METHANE,NITRO

(3) C2H6O2   1,2-ETHANEDIOL

MARKUZIN N.P., NIKANOROVA L.A.
ZH.OBSHCH.KHIM. 32(1962)3469

TEMPERATURE = 28.0 DEG C   TYPE OF SYSTEM = 3

EXPERIMENTAL TIE LINES IN MOLE PCT

	LEFT PHASE			UPPER PHASE			RIGHT PHASE	
(1)	(2)	(3)	(1)	(2)	(3)	(1)	(2)	(3)
46.698	7.864	45.437				0.368	10.839	89.793
37.095	16.277	46.628				0.504	13.835	85.662
25.632	26.847	47.522				0.981	23.447	75.572
18.075	32.414	49.510				2.829	49.045	48.126
			1.350	84.976	13.674	0.767	98.025	11.208
			1.640	42.558	55.802	0.632	88.891	10.473
			1.325	40.444	58.231			
23.257	30.298	46.445	1.039	89.505	9.456			
61.712	20.880	17.409	1.003	96.989	2.008			

CH$_3$NO$_2$-C$_6$H$_6$

(1) C6H12   CYCLOHEXANE

(2) C6H6   BENZENE

(3) CH3NO2   METHANE,NITRO

WECK H.I., HUNT H.
IND.ENG.CHEM. 46(1954)2521

TEMPERATURE = 25.0 DEG C   TYPE OF SYSTEM = 1

EXPERIMENTAL TIE LINES IN MOLE PCT

	LEFT PHASE			RIGHT PHASE	
(1)	(2)	(3)	(1)	(2)	(3)
82.036	11.651	6.314	4.696	3.533	91.771
72.127	18.856	9.017	6.957	9.883	83.161
56.282	25.345	17.373	9.371	14.722	75.907

MEAN DEV. BETWEEN CALC. AND EXP. CONC. IN MOLE PCT

UNIQUAC (COMMON PARAMETERS)   0.91

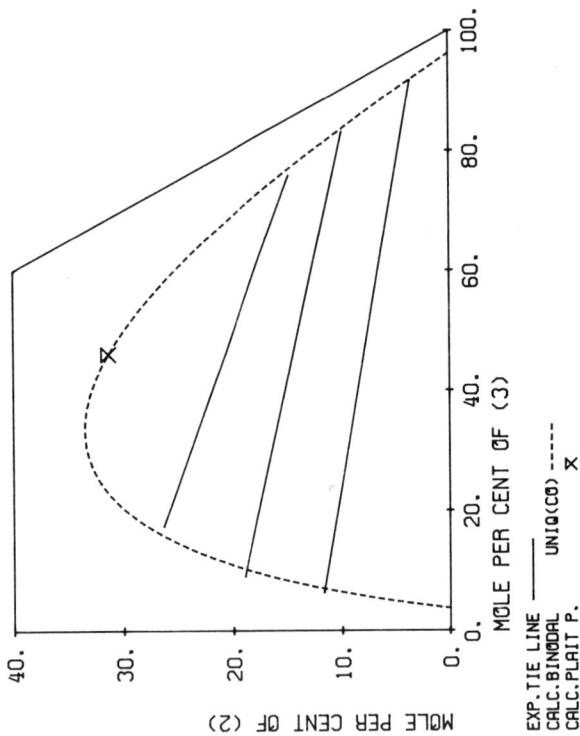

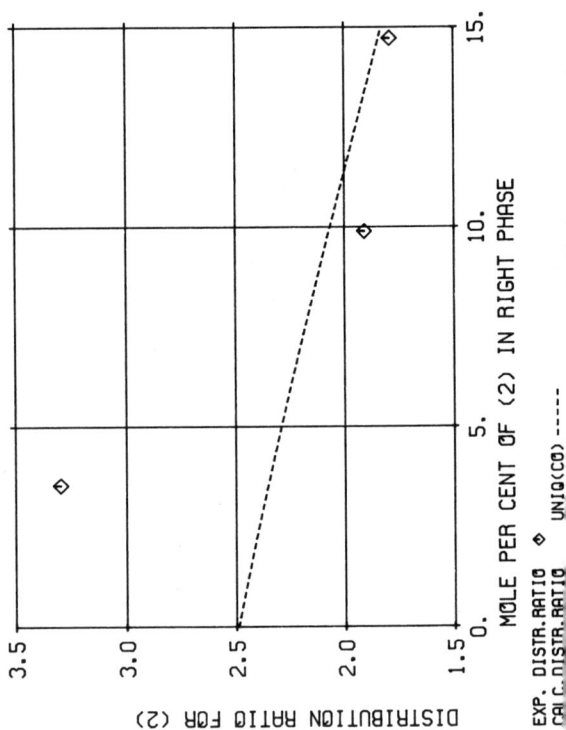

CH₃NO₂-C₆H₁₄O

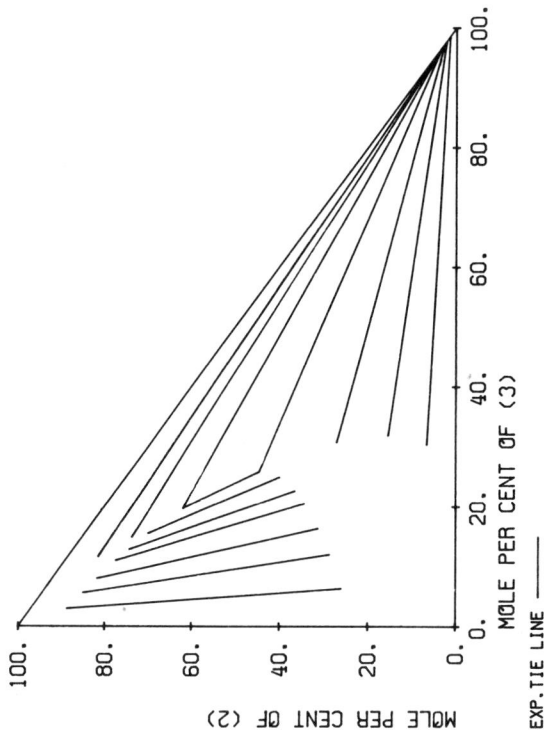

MOLE PER CENT OF (3)

MOLE PER CENT OF (2)

EXP.TIE LINE ————

(1) C6H14O   1-HEXANOL
----
(2) CH3NO2   METHANE,NITRO
----
(3) H2O   WATER

SAZONOV V.P., MARKUZIN N.P., FILIPPOV V.V.
ZH.PRIKL.KHIM.(LENINGRAD) 50(1977)1524

TEMPERATURE = 21.0 DEG C   TYPE OF SYSTEM = 3

EXPERIMENTAL TIE LINES IN MOLE PCT

LEFT PHASE			UPPER PHASE			RIGHT PHASE		
(1)	(2)	(3)	(1)	(2)	(3)	(1)	(2)	(3)
63.100	6.600	30.300	18.300	62.000	19.700	0.100	1.500	98.400
52.800	15.400	31.800	11.300	73.800	14.900	0.100	2.300	97.600
42.300	27.000	30.700	7.000	81.400	11.600	0.100	2.800	97.100
29.600	44.600	25.800	0.000	93.200	6.800	0.100	3.400	96.500
35.200	39.900	24.900	14.400	70.100	15.500	0.080	3.420	96.500
41.000	36.400	22.600	12.800	74.400	12.800	0.040	3.360	96.600
45.200	34.300	20.500	11.600	77.400	11.000	0.0	3.400	96.600
52.500	31.200	16.300	10.300	81.700	8.000			
59.400	28.600	12.000	9.500	84.900	5.600			
67.700	26.000	6.300	8.400	88.600	3.000			
76.800	23.200	0.0	7.500	92.500	0.0			

CH_3NO_2-$C_6H_{14}O$

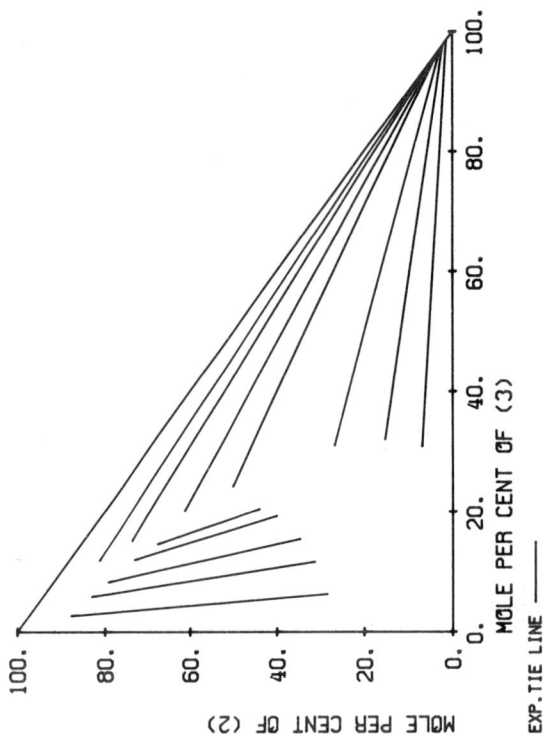

MOLE PER CENT OF (3) ————

MOLE PER CENT OF (2)

EXP.TIE LINE ————

(1) C6H14O	1-HEXANOL	
(2) CH3NO2	METHANE,NITRO	
(3) H2O	WATER	

SAZONOV V.P., MARKUZIN N.P., FILIPPOV V.V.
ZH.PRIKL.KHIM.(LENINGRAD) 50(1977)1524

TEMPERATURE = 23.0 DEG C   TYPE OF SYSTEM = 3A

EXPERIMENTAL TIE LINES IN MOLE PCT

LEFT PHASE			UPPER PHASE			RIGHT PHASE		
(1)	(2)	(3)	(1)	(2)	(3)	(1)	(2)	(3)
62.200	7.000	30.800	18.700	61.200	20.100	0.100	1.500	98.400
52.700	15.300	32.000	11.300	73.600	15.100	0.100	2.300	97.600
42.200	26.800	31.000	7.000	81.200	11.800	0.100	2.800	97.100
25.600	50.200	24.200	0.000	93.100	6.900	0.100	3.500	96.400
			35.500	44.000	20.500	0.100	3.440	96.500
17.900	67.500	14.600	40.700	40.000	19.300	0.060	3.470	96.500
15.000	73.000	12.000	49.900	34.600	15.500	0.030	3.470	96.500
12.500	79.200	8.300	57.000	31.300	11.700	0.0	3.500	96.500
11.100	83.000	5.900	65.200	28.500	6.300			
9.600	87.700	2.700	73.600	26.400	0.0			
8.200	91.800	0.0						

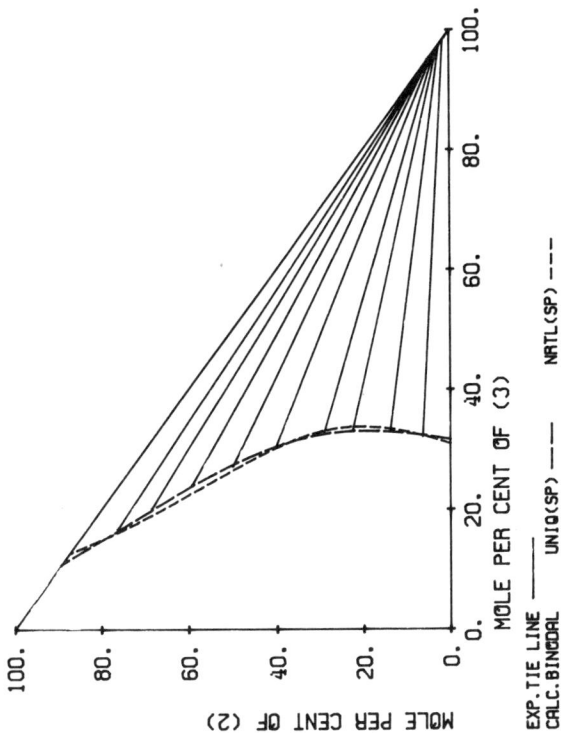

MOLE PER CENT OF (2)

MOLE PER CENT OF (3)

EXP.TIE LINE ———   UNIQ(SP) ———   NRTL(SP) ----
CALC.BINODAL

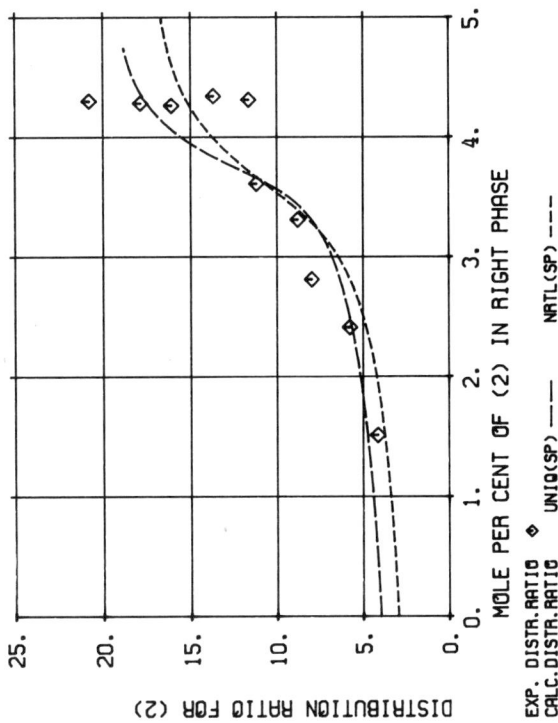

DISTRIBUTION RATIO FOR (2)

MOLE PER CENT OF (2) IN RIGHT PHASE

EXP. DISTR.RATIO ◇   UNIQ(SP) ———   NRTL(SP) ----
CALC.DISTR.RATIO

(2) CH3NO2   METHANE,NITRO

(3) H2O   WATER

SAZONOV V.P., MARKUZIN N.P., FILIPPOV V.V.
ZH.PRIKL.KHIM.(LENINGRAD) 50(1977)1524

TEMPERATURE = 40.0 DEG C   TYPE OF SYSTEM = 2

EXPERIMENTAL TIE LINES IN MOLE PCT

	LEFT PHASE			RIGHT PHASE	
(1)	(2)	(3)	(1)	(2)	(3)
61.600	6.300	32.100	0.090	1.510	98.400
53.000	13.800	33.200	0.090	2.410	97.500
44.500	22.400	33.100	0.090	2.810	97.100
38.400	29.100	32.500	0.090	3.310	96.600
29.500	40.300	30.200	0.090	3.610	96.300
11.900	68.500	20.100	0.040	4.260	95.700
7.000	76.300	16.700	0.020	4.280	95.700
0.000	89.300	10.700	0.0	4.300	95.700
22.800	50.000	27.200	0.090	4.310	95.600
16.900	59.300	23.800	0.060	4.340	95.600

SPECIFIC MODEL PARAMETERS IN KELVIN

		UNIQUAC		NRTL(ALPHA=.2)	
I	J	AIJ	AJI	AIJ	AJI
1	2	419.32	-18.011	669.26	82.182
1	3	74.291	333.35	-141.28	2192.2
2	3	353.78	204.68	327.23	774.33

R1 = 4.8031   R2 = 2.0086   R3 = 0.9200
Q1 = 4.132   Q2 = 1.868   Q3 = 1.400

MEAN DEV. BETWEEN CALC. AND EXP. CONC. IN MOLE PCT

UNIQUAC (SPECIFIC PARAMETERS)   0.19
NRTL   (SPECIFIC PARAMETERS)   0.56

CH₃NO₂-C₉H₂₀O

(1) C9H20O   1-NONANOL

(2) C10H8   NAPHTHALENE

(3) CH3NO2   METHANE, NITRO

SAZONOV V.P., CHERNYSHEVA M.F.
ZH.PRIKL.KHIM.(LENINGRAD) 51(1978)1019

TEMPERATURE = 20.0 DEG C   TYPE OF SYSTEM = 1

EXPERIMENTAL TIE LINES IN MOLE PCT

	LEFT PHASE			RIGHT PHASE	
(1)	(2)	(3)	(1)	(2)	(3)
82.952	0.0	17.048	1.821	0.0	98.179
75.898	5.077	19.025	1.912	2.252	95.836
67.232	10.243	22.525	2.677	4.934	92.389
60.109	14.206	25.685	3.643	7.551	88.806
55.510	17.239	27.250	4.836	10.211	84.953

SPECIFIC MODEL PARAMETERS IN KELVIN

		UNIQUAC		NRTL(ALPHA=.2)	
I J	AIJ	AJI	AIJ	AJI	
1 2	-52.301	94.409	-74.389	-265.26	
1 3	390.19	41.626	125.08	1069.0	
2 3	6.7120	86.049	-20.766	-147.08	

R1 = 6.8263   R2 = 4.9808   R3 = 2.0086
Q1 = 5.752   Q2 = 3.440   Q3 = 1.868

MEAN DEV. BETWEEN CALC. AND EXP. CONC. IN MOLE PCT

UNIQUAC (SPECIFIC PARAMETERS)   0.94
NRTL   (SPECIFIC PARAMETERS)   0.99
UNIQUAC (COMMON PARAMETERS)   1.05

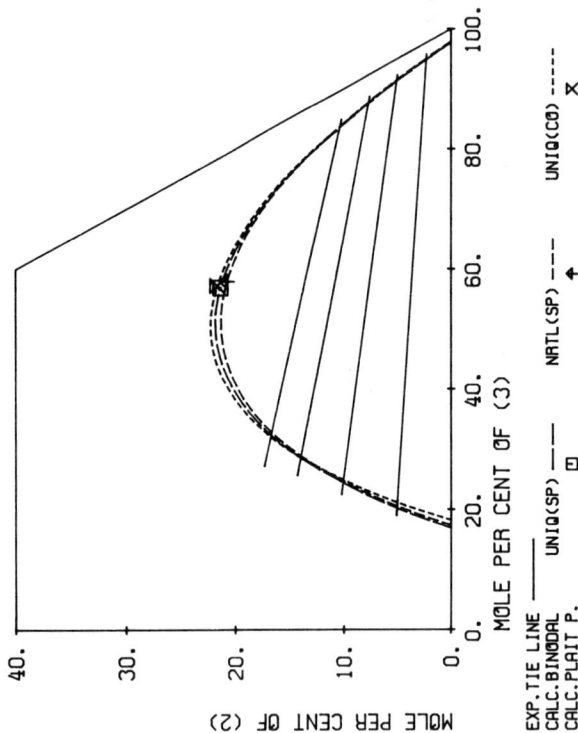

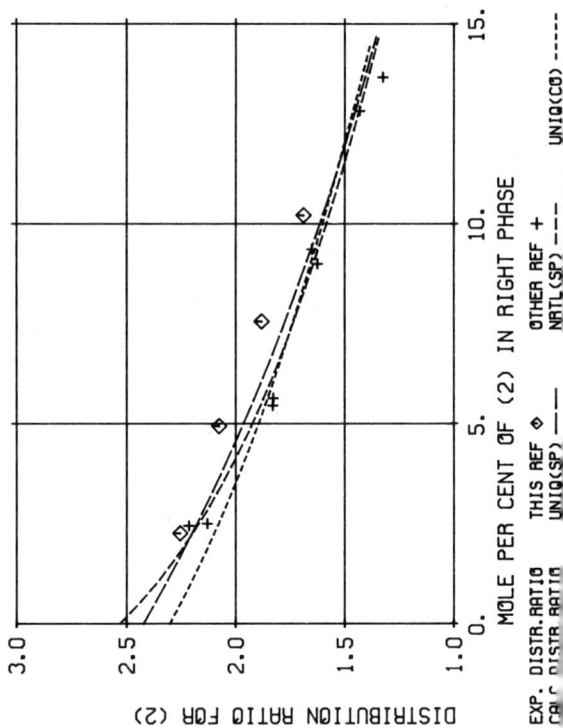

CH$_3$NO$_2$-C$_9$H$_{20}$O

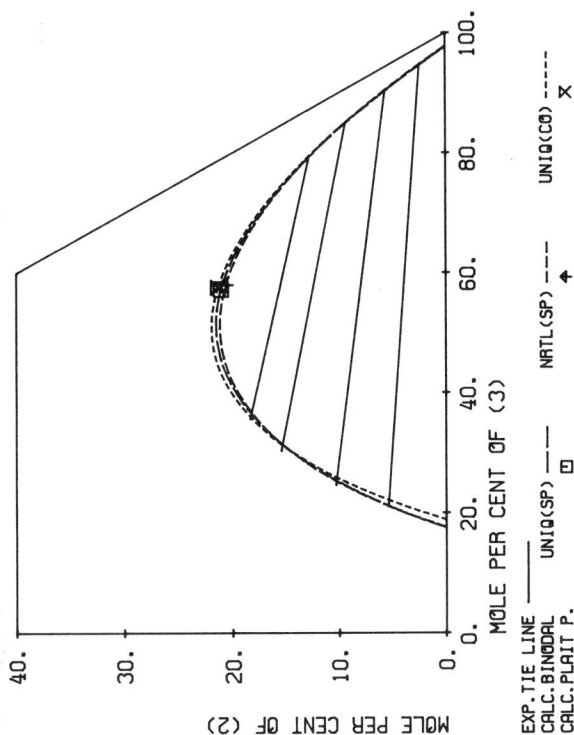

MOLE PER CENT OF (2)

40.

30.

20.

10.

0.

MOLE PER CENT OF (3)

0.   20.   40.   60.   80.   100.

EXP.TIE LINE ———   UNIQ(SP) ⊟   NRTL(SP) ——   UNIQ(CO) ——×
CALC.BINODAL
CALC.PLAIT P.

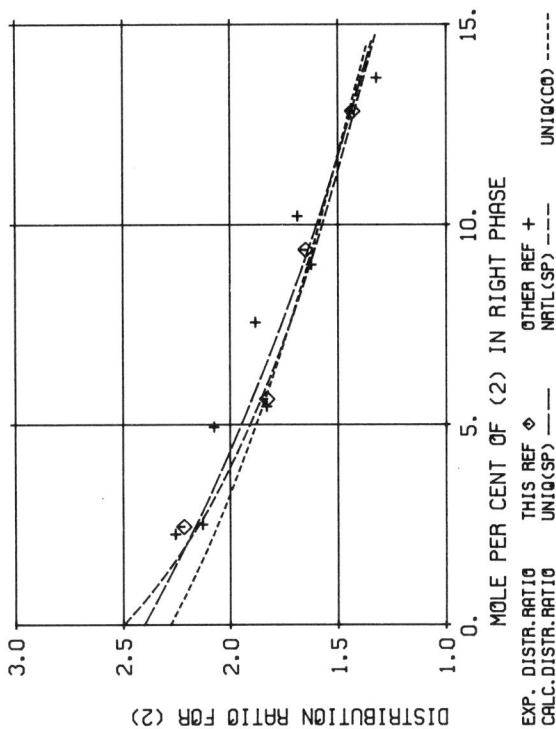

DISTRIBUTION RATIO FOR (2)

3.0

2.5

2.0

1.5

1.0

MOLE PER CENT OF (2) IN RIGHT PHASE

0.   5.   10.   15.

EXP. DISTR.RATIO   THIS REF ◇   OTHER REF +   UNIQ(CO) ——
CALC.DISTR.RATIO   UNIQ(SP) ——   NRTL(SP) ——

(2) C10H8   NAPHTHALENE

(3) CH3NO2   METHANE,NITRO

SAZONOV V.P., CHERNYSHEVA M.F.,
ZH.PRIKL.KHIM.(LENINGRAD) 51(1978)1019

TEMPERATURE = 23.0 DEG C   TYPE OF SYSTEM = 1

EXPERIMENTAL TIE LINES IN MOLE PCT

	LEFT PHASE			RIGHT PHASE	
(1)	(2)	(3)	(1)	(2)	(3)
81.430	0.0	18.570	2.268	0.0	97.732
73.563	5.401	21.036	2.934	2.438	94.628
65.101	10.289	24.609	3.867	5.626	90.507
54.197	15.452	30.351	5.963	9.361	84.676
45.964	18.374	35.662	3.160	12.820	79.019

SPECIFIC MODEL PARAMETERS IN KELVIN

		UNIQUAC		NRTL(ALPHA=.2)	
I	J	AIJ	AJI	AIJ	AJI
1	2	-52.301	94.409	-74.389	-265.26
1	3	390.19	41.626	125.08	1069.0
2	3	6.7120	86.049	-20.766	-147.08

R1 = 6.8263   R2 = 4.9808   R3 = 2.0085
Q1 = 5.752   Q2 = 3.440   Q3 = 1.868

MEAN DEV. BETWEEN CALC. AND EXP. CONC. IN MOLE PCT

UNIQUAC (SPECIFIC PARAMETERS)   0.39
NRTL   (SPECIFIC PARAMETERS)   0.33
UNIQUAC (COMMON PARAMETERS)   0.39

CH₃NO₂-C₉H₂₀O

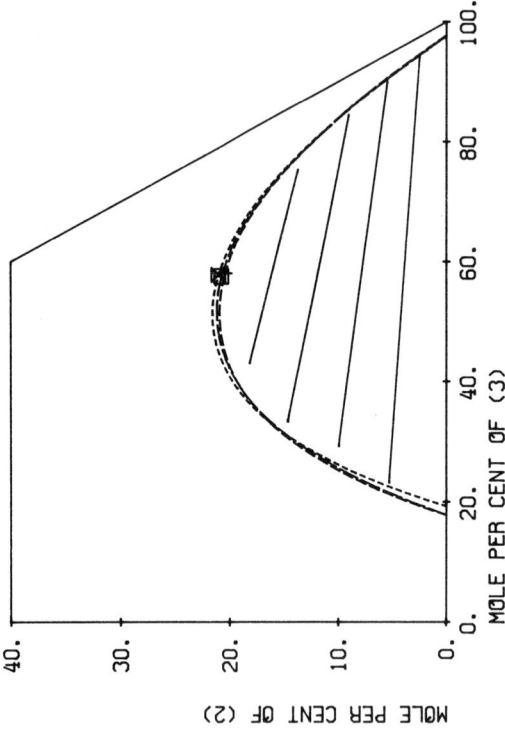

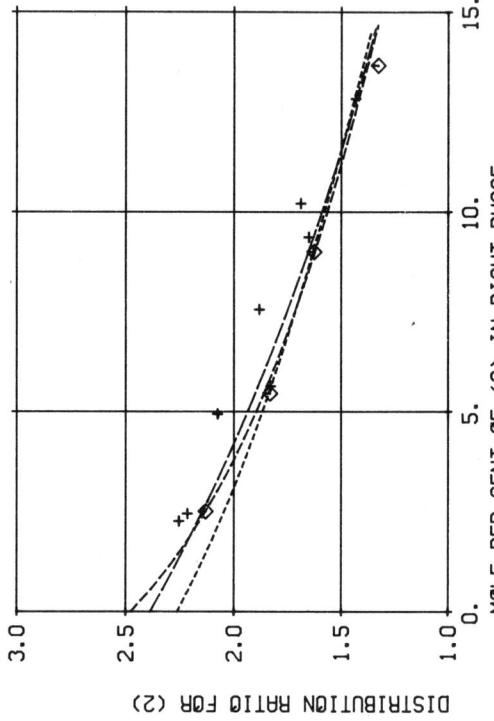

(1) C9H20O   1-NONANOL
(2) C10H8   NAPHTHALENE
(3) CH3NO2   METHANE,NITRO

SAZONOV V.P., CHERNYSHEVA M.F.
ZH.PRIKL.KHIM.(LENINGRAD) 51(1978)1019

TEMPERATURE = 24.9 DEG C   TYPE OF SYSTEM = 1

EXPERIMENTAL TIE LINES IN MOLE PCT

	LEFT PHASE			RIGHT PHASE	
(1)	(2)	(3)	(1)	(2)	(3)
80.680	0.0	19.320	2.766	0.0	97.234
71.495	5.325	23.180	3.219	2.500	94.282
60.767	9.967	29.266	4.508	5.446	90.046
52.088	14.610	33.302	6.480	8.989	84.531
38.843	18.110	43.047	10.880	13.666	75.453

SPECIFIC MODEL PARAMETERS IN KELVIN

		UNIQUAC		NRTL(ALPHA=.2)	
I	J	AIJ	AJI	AIJ	AJI
1	2	-52.301	94.409	-74.389	-265.26
1	3	390.19	41.626	125.08	1069.0
2	3	6.7120	86.049	-20.766	-147.08

R1 = 6.8263   R2 = 4.9808   R3 = 2.0086
Q1 = 5.752   Q2 = 3.440   Q3 = 1.868

MEAN DEV. BETWEEN CALC. AND EXP. CONC. IN MOLE PCT

UNIQUAC (SPECIFIC PARAMETERS)   0.87
NRTL   (SPECIFIC PARAMETERS)   0.96
UNIQUAC (COMMON PARAMETERS)   0.85

$CH_3NO_2-C_9H_{20}O$

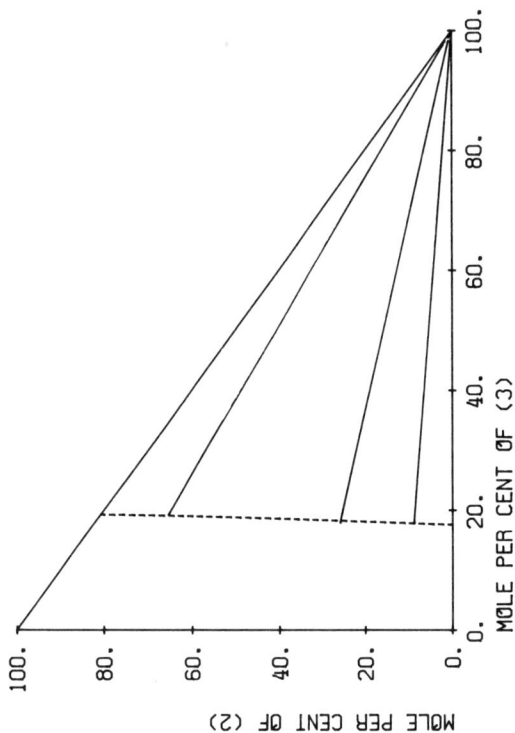

MOLE PER CENT OF (3)

MOLE PER CENT OF (2)

EXP.TIE LINE ——— UNIQ(CO) -----
CALC.BINODAL

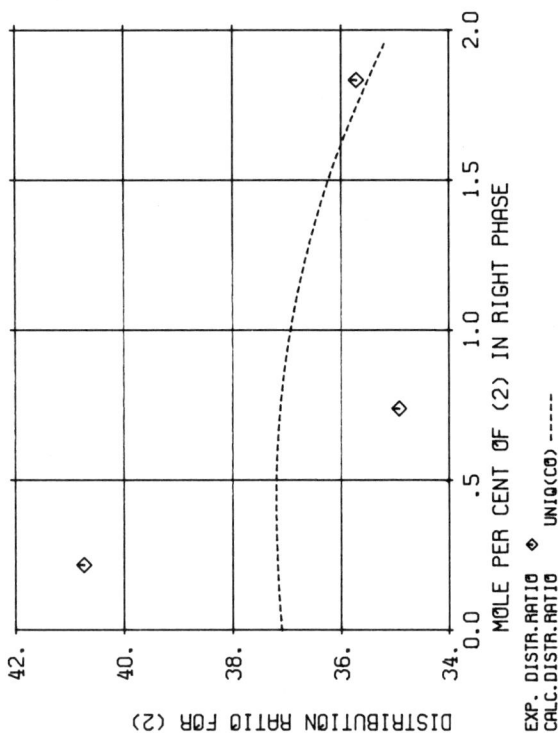

DISTRIBUTION RATIO FOR (2)

MOLE PER CENT OF (2) IN RIGHT PHASE

EXP. DISTR.RATIO ◇ UNIQ(CO) -----
CALC.DISTR.RATIO

(2) C9H20O   1-NONANOL
(3) CH3NO2   METHANE,NITRO

SAZONOV V.P., GROMAKOVSKAYA A.G., ZHURALEV E.F.
ZH.PRIKL.KHIM.(LENINGRAD) 50(1977)587

TEMPERATURE = 25.0 DEG C   TYPE OF SYSTEM = 2

EXPERIMENTAL TIE LINES IN MOLE PCT

	LEFT PHASE			RIGHT PHASE	
(1)	(2)	(3)	(1)	(2)	(3)
73.337	8.812	17.851	1.222	0.216	98.561
56.248	25.772	17.980	0.949	0.737	98.314
15.240	65.533	19.226	0.438	1.834	97.728

MEAN DEV. BETWEEN CALC. AND EXP. CONC. IN MOLE PCT

UNIQUAC (COMMON PARAMETERS)   0.29

CH₃NO₂-C₉H₂₀O

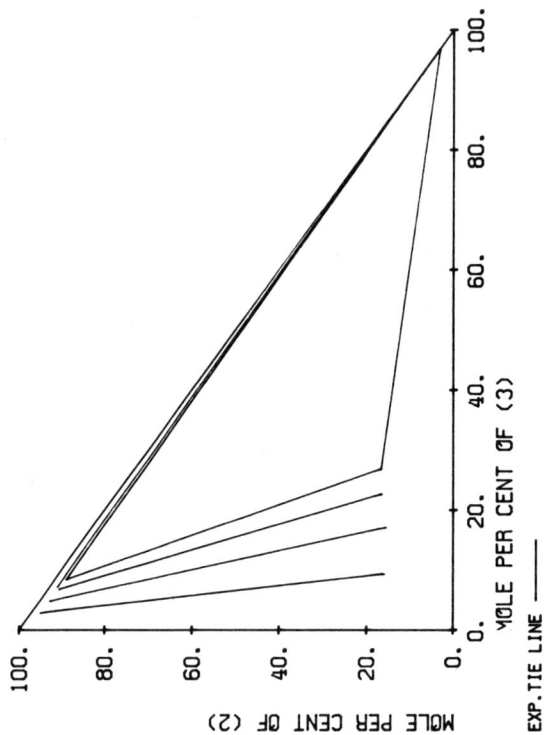

MOLE PER CENT OF (3)

EXP.TIE LINE ———

MOLE PER CENT OF (2)

(1) C9H20O   1-NONANOL
(2) CH3NO2   METHANE,NITRO
(3) H2O   WATER

SAZONOV V.P., CHERNYSHEVA M.F.,
ZH.PRIKL.KHIM.(LENINGRAD) 51(1978)1754

TEMPERATURE = 20.0 DEG C   TYPE OF SYSTEM = 3

EXPERIMENTAL TIE LINES IN MOLE PCT

	LEFT PHASE			UPPER PHASE			RIGHT PHASE	
(1)	(2)	(3)	(1)	(2)	(3)	(1)	(2)	(3)
74.982	0.0	25.018				0.001	0.0	99.999
56.577	16.615	26.808				0.0	3.209	96.791
						0.0	3.278	96.722
						0.000	3.381	96.619
60.836	16.532	22.632	2.819	88.880	8.301			
67.455	15.474	17.071	2.029	91.008	6.963			
74.884	15.840	9.276	0.0	93.532	6.468			
83.338	16.662	0.0	2.679	90.617	6.704			
			2.410	92.849	4.741			
			2.049	95.217	2.734			
			1.821	98.179	0.0			

$CH_3NO_2-C_9H_{20}O$

(1) C9H20O   1-NONANOL
----
(2) CH3NO2   METHANE,NITRO
----
(3) H2O   WATER

SAZONOV V.P., CHERNYSHEVA M.F.
ZH.PRIKL.KHIM.(LENINGRAD) 51(1978)1764

TEMPERATURE = 23.0 DEG C   TYPE OF SYSTEM = 3

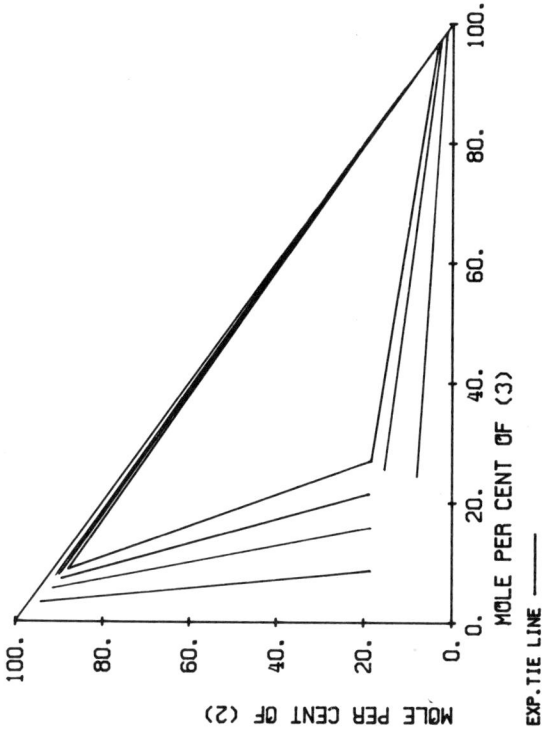

MOLE PER CENT OF (3)

MOLE PER CENT OF (2)

EXP.TIE LINE ———

EXPERIMENTAL TIE LINES IN MOLE PCT

	LEFT PHASE			UPPER PHASE			RIGHT PHASE	
(1)	(2)	(3)	(1)	(2)	(3)	(1)	(2)	(3)
74.982	0.0	25.018				0.001	0.0	99.999
67.628	8.053	24.319				0.0	1.309	98.691
59.289	15.316	25.395				0.0	2.635	97.365
54.848	18.229	26.923				0.0	3.347	96.653
						0.000	3.381	96.619
						0.0	3.485	96.515
						0.0	3.519	96.481
			3.463	87.851	8.686			
			2.403	89.968	7.629			
			1.770	90.977	7.253			
			0.0	93.224	6.776			
59.774	18.802	21.424	3.460	89.445	7.095			
65.854	18.405	15.742	3.057	91.504	5.439			
73.078	18.394	8.529	2.528	94.384	3.083			
81.619	18.381	0.0	2.268	97.732	0.0			

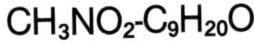

CH$_3$NO$_2$-C$_9$H$_{20}$O

MOLE PER CENT OF (3)

MOLE PER CENT OF (2)

EXP.TIE LINE ———

(1) C9H20O   1-NONANOL

(2) CH3NO2   METHANE,NITRO

(3) H2O   WATER

SAZONOV V.P., CHERNYSHEVA M.F.
ZH.PRIKL.KHIM.(LENINGRAD) 51(1978)1764

TEMPERATURE = 45.1 DEG C   TYPE OF SYSTEM = 3A

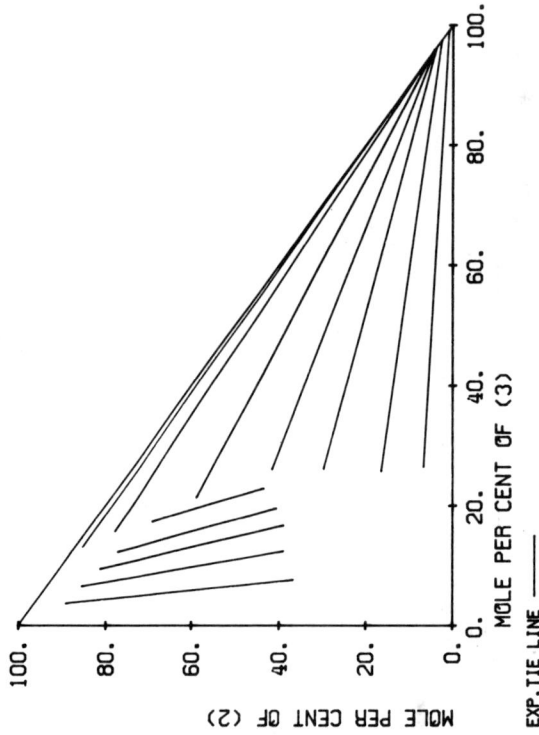

EXPERIMENTAL TIE LINES IN MOLE PCT

	LEFT PHASE			UPPER PHASE			RIGHT PHASE	
(1)	(2)	(3)	(1)	(2)	(3)	(1)	(2)	(3)
73.540	0.0	26.460				0.003	0.0	99.997
67.073	6.569	26.358				0.0	0.935	99.065
57.953	16.416	25.630				0.0	2.502	97.498
44.404	29.578	26.018				0.0	3.693	96.307
32.362	41.638	26.000	19.935	58.780	21.285	0.000	4.010	95.990
			6.547	77.857	15.596	0.0	4.117	95.883
			1.898	85.122	12.980	0.0	4.224	95.776
			0.0	88.196	11.804	0.0	4.367	95.633
33.580	43.572	22.848	13.481	69.246	17.273	0.0	4.548	95.452
39.849	40.623	19.527	10.566	77.244	12.190			
44.383	38.981	16.636	9.392	81.251	9.357			
48.734	38.971	12.296	8.060	85.450	6.490			
55.771	36.708	7.521	7.217	89.143	3.635			
63.006	36.994	0.0	6.645	93.355	0.0			

CH$_4$O-C$_2$HCl$_3$

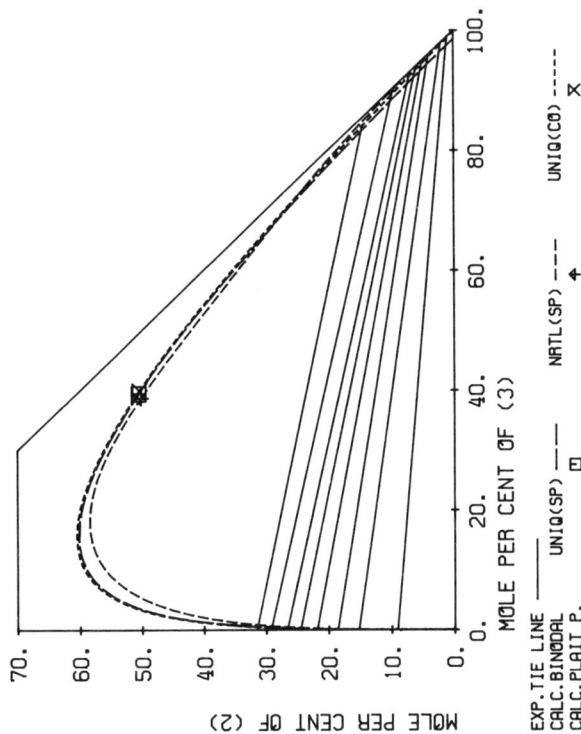

MOLE PER CENT OF (2)

MOLE PER CENT OF (3)

EXP.TIE LINE ——— UNIQ(SP) —□— NRTL(SP) —+— UNIQ(CO) ——×——
CALC.BINODAL
CALC.PLAIT P.

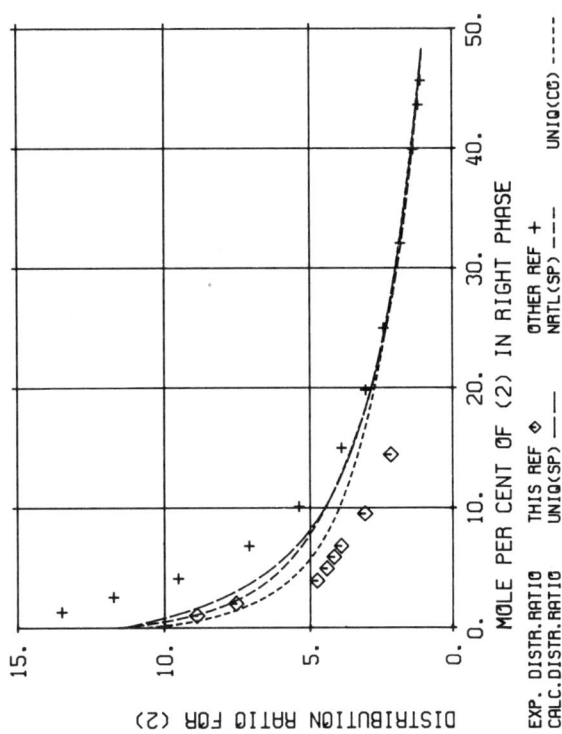

DISTRIBUTION RATIO FOR (2)

MOLE PER CENT OF (2) IN RIGHT PHASE

EXP. DISTR.RATIO ◇ THIS REF ◇ OTHER REF + UNIQ(CO) ——————
CALC.DISTR.RATIO UNIQ(SP) —— NRTL(SP) ———

(2) CH4O   METHANOL

(3) C2HCL3   ETHENE,TRICHLORO

KRISHNAMURTY V.V.G., MURTI P.S., VENKATA RAO C.
J.SCI.IND.RES. 12B(1953)583

TEMPERATURE = 27.5 DEG C   TYPE OF SYSTEM = 1

EXPERIMENTAL TIE LINES IN MOLE PCT

	LEFT PHASE			RIGHT PHASE	
(1)	(2)	(3)	(1)	(2)	(3)
	90.858	9.069	0.074	0.0	1.017	98.983
	84.559	15.255	0.186	0.0	2.019	97.981
	81.178	18.583	0.239	0.0	3.900	96.100
	77.892	21.831	0.277	0.0	4.935	95.065
	75.222	24.479	0.298	0.0	5.878	94.122
	73.077	26.587	0.337	0.0	6.807	93.193
	70.432	29.191	0.377	0.0	9.514	90.486
	68.081	31.482	0.435	1.283	14.432	84.285

SPECIFIC MODEL PARAMETERS IN KELVIN

		UNIQUAC		NRTL(ALPHA=.2)	
I J	AIJ	AJI	AIJ	AJI	
1 2	-610.17	-56.666	459.62	-517.73	
1 3	-868.91	570.62	2896.2	2509.0	
2 3	-64.077	117.49	501.29	106.50	

R1 = 0.9200   R2 = 1.4311   R3 = 3.3092
Q1 = 1.400   Q2 = 1.432   Q3 = 2.860

MEAN DEV. BETWEEN CALC. AND EXP. CONC. IN MOLE PCT

UNIQUAC (SPECIFIC PARAMETERS)   2.24
NRTL   (SPECIFIC PARAMETERS)   2.26
UNIQUAC (COMMON PARAMETERS)   1.74

CH_4O-C_2HCl_3

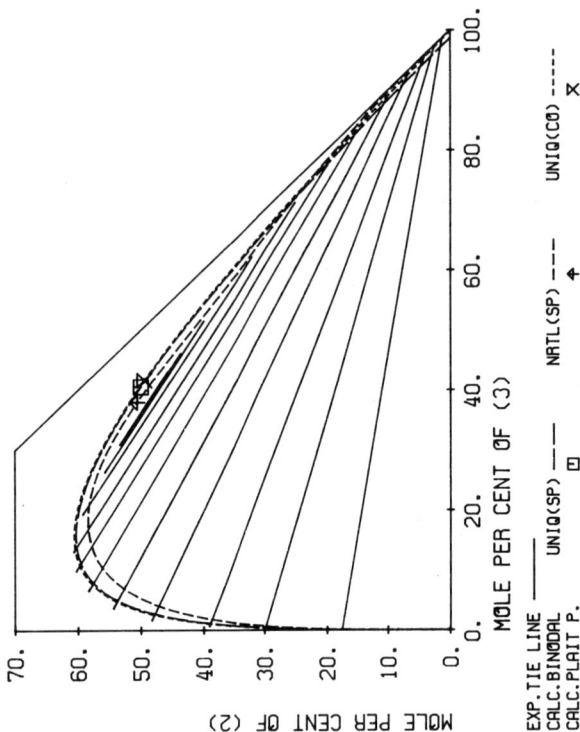

EXP.TIE LINE   UNIQ(SP) — □   NRTL(SP) — ✦   UNIQ(CO) — ✕
CALC.BINODAL
CALC.PLAIT P.

MOLE PER CENT OF (3)

MOLE PER CENT OF (2)

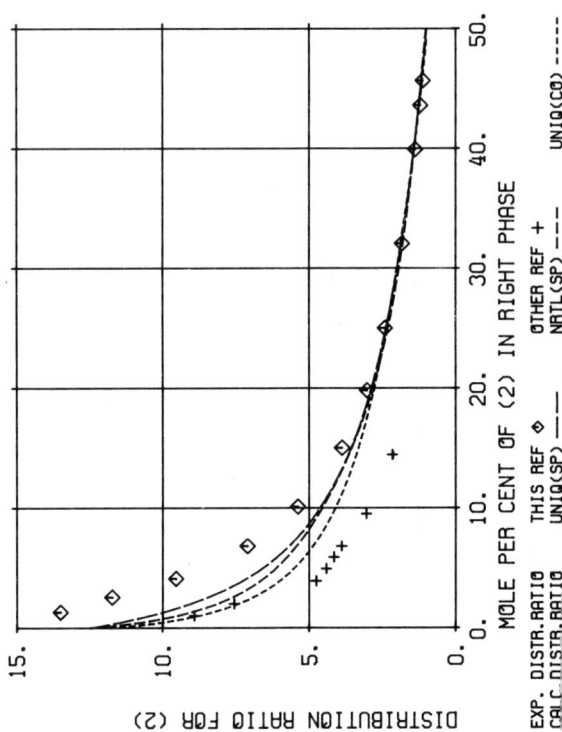

EXP. DISTR.RATIO ◇   THIS REF ◇   OTHER REF +   UNIQ(CO) -----
CALC.DISTR.RATIO   UNIQ(SP)   NRTL(SP) -----

MOLE PER CENT OF (2) IN RIGHT PHASE

DISTRIBUTION RATIO FOR (2)

(1) H2O   WATER
(2) CH4O   METHANOL
(3) C2HCL3   ETHENE,TRICHLORO

ROETHLIN S., CRUETZEN J.L., SCHULTZE G.R.
CHEM.ING.TECH. 29(1957)211

TEMPERATURE = 20.0 DEG C   TYPE OF SYSTEM = 1

EXPERIMENTAL TIE LINES IN MOLE PCT

	LEFT PHASE			RIGHT PHASE	
(1)	(2)	(3)	(1)	(2)	(3)
82.300	17.550	0.150	0.100	1.300	98.600
69.800	29.900	0.300	0.200	2.550	97.250
60.200	39.100	0.700	0.300	4.100	95.600
50.000	48.300	1.700	0.600	6.800	92.600
41.900	54.400	3.700	1.150	10.100	88.750
35.100	58.300	6.600	1.700	15.000	83.300
29.800	60.300	9.900	2.600	19.800	77.600
26.200	60.700	13.100	3.700	25.000	71.300
21.400	59.250	19.350	5.700	32.050	62.250
17.900	55.750	26.350	8.550	39.900	51.550
15.700	53.350	30.950	10.250	43.600	46.150
15.050	52.100	32.850	11.200	45.650	43.150

SPECIFIC MODEL PARAMETERS IN KELVIN

		UNIQUAC		NRTL(ALPHA=.2)	
I	J	AIJ	AJI	AIJ	AJI
1	2	-610.17	-56.666	459.62	-517.73
1	3	-868.91	570.62	2896.2	2509.0
2	3	-64.077	117.49	501.29	106.50

R1 = 0.9200   R2 = 1.4311   R3 = 3.3092
Q1 = 1.400   Q2 = 1.432   Q3 = 2.860

MEAN DEV. BETWEEN CALC. AND EXP. CONC. IN MOLE PCT

UNIQUAC (SPECIFIC PARAMETERS)   1.90
NRTL   (SPECIFIC PARAMETERS)   1.51
UNIQUAC (COMMON PARAMETERS)   2.64

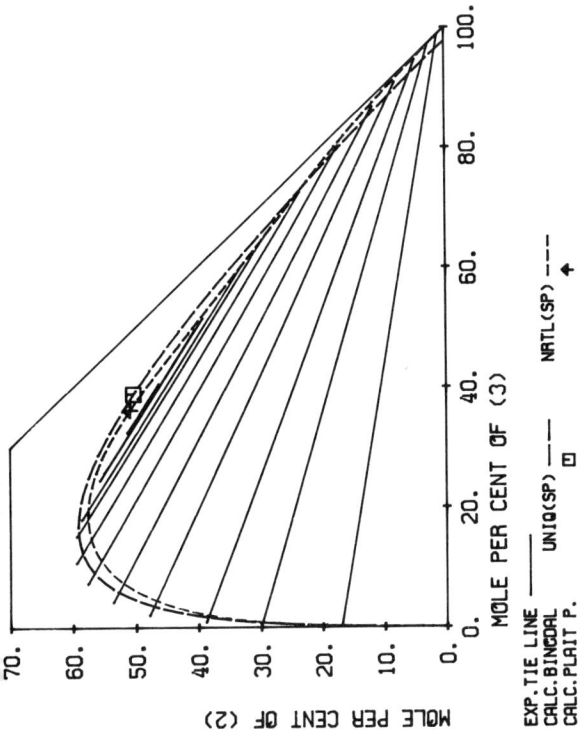

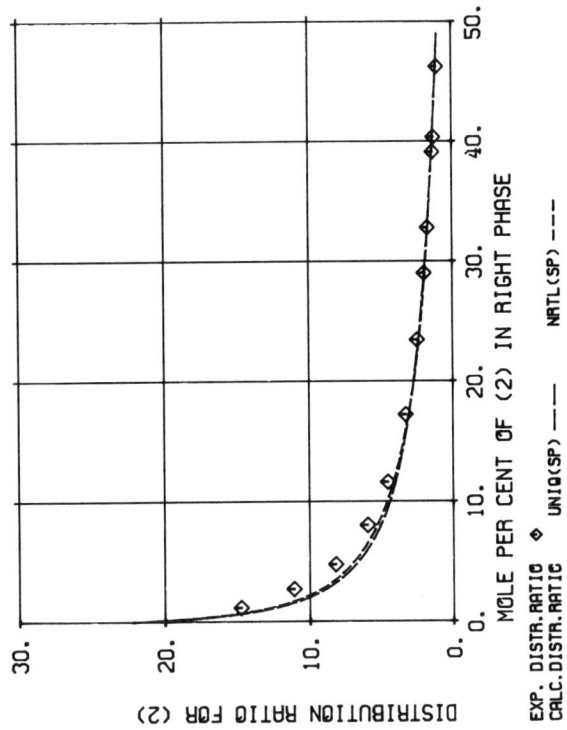

(2) CH4O   METHANOL
(3) C2HCL3   ETHENE,TRICHLORO

ROETHLIN S., CRUETZEN J.L., SCHULTZE G.R.
CHEM.ING.TECH. 29(1957)211

TEMPERATURE = 35.0 DEG C   TYPE OF SYSTEM = 1

EXPERIMENTAL TIE LINES IN MOLE PCT

	LEFT PHASE			RIGHT PHASE	
(1)	(2)	(3)	(1)	(2)	(3)
82.800	16.950	0.250	0.100	1.150	98.750
69.700	29.950	0.350	0.250	2.700	97.050
60.350	38.800	0.850	0.500	4.750	94.750
50.100	47.900	2.000	1.000	8.000	91.000
42.100	53.700	4.200	1.500	11.600	86.900
35.050	57.650	7.300	2.500	17.200	80.300
29.800	59.450	10.750	3.850	23.450	72.700
25.450	59.400	15.150	5.450	29.000	65.550
23.400	58.600	18.000	6.750	32.800	60.450
19.800	55.700	24.500	9.250	39.100	51.650
19.250	55.000	25.750	9.850	40.350	49.800
16.250	51.300	32.450	13.000	46.250	40.750

SPECIFIC MODEL PARAMETERS IN KELVIN

	UNIQUAC		NRTL(ALPHA=.2)	
I J	AIJ	AJI	AIJ	AJI
1 2	-329.16	-218.79	1106.9	-896.30
1 3	487.30	529.13	2567.1	1736.3
2 3	-48.748	307.55	580.45	68.663

R1 = 0.9200   R2 = 1.4311   R3 = 3.3092
Q1 = 1.400   Q2 = 1.432   Q3 = 2.860

MEAN DEV. BETWEEN CALC. AND EXP. CONC. IN MOLE PCT

UNIQUAC (SPECIFIC PARAMETERS)   1.81
NRTL   (SPECIFIC PARAMETERS)   1.15

CH$_4$O-C$_2$HCl$_3$

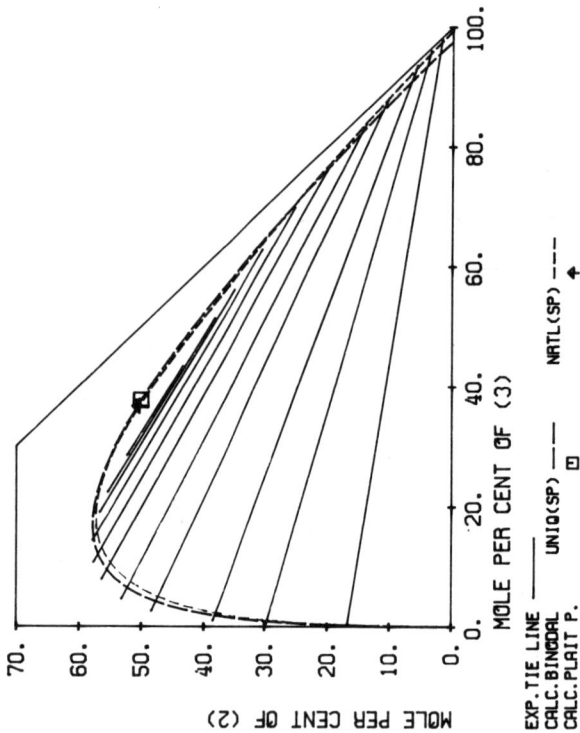

MOLE PER CENT OF (3)

MOLE PER CENT OF (2)

EXP.TIE LINE ———
CALC. BINODAL
CALC.PLAIT P.

UNIQ(SP) ——— ⊡
NRTL(SP) ----- ✦

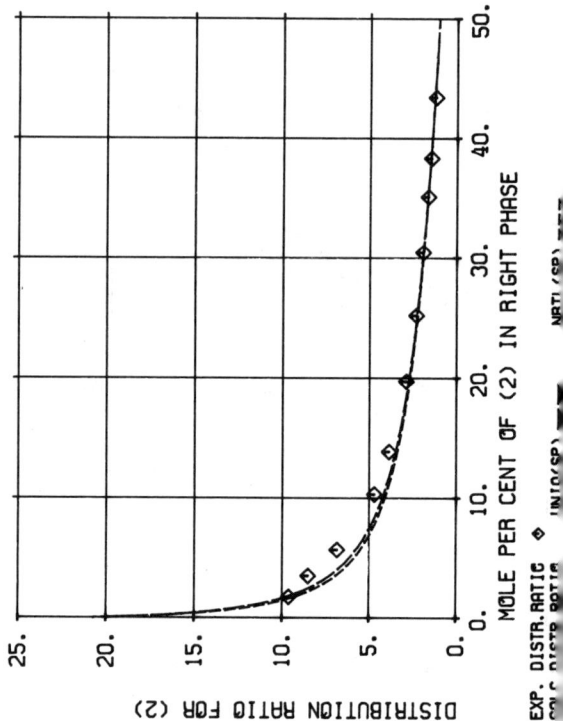

MOLE PER CENT OF (2) IN RIGHT PHASE

DISTRIBUTION RATIO FOR (2)

EXP. DISTR.RATIO ◇
CALC DISTR.RATIO

NRTL(SP)
UNIQ(SP)

(1) H2O   WATER
(2) CH4O   METHANOL
(3) C2HCL3   ETHENE,TRICHLORO

ROETHLIN S., CRUETZEN J.L., SCHULTZE G.R.
CHEM.ING.TECH. 29(1957)211

TEMPERATURE = 50.0 DEG C   TYPE OF SYSTEM = 1

EXPERIMENTAL TIE LINES IN MOLE PCT

LEFT PHASE			RIGHT PHASE		
(1)	(2)	(3)	(1)	(2)	(3)
82.900	16.800	0.300	0.200	1.750	98.050
69.750	29.750	0.500	0.450	3.500	96.050
60.450	38.550	1.000	0.700	5.650	93.650
49.100	48.350	2.550	1.550	10.300	88.150
42.300	53.150	4.550	2.300	13.850	83.850
35.800	56.350	7.850	3.500	19.700	76.800
31.700	57.650	10.650	4.900	25.200	69.900
27.900	57.800	14.300	6.700	30.450	62.850
24.350	56.650	19.000	8.700	35.100	56.200
22.300	55.350	22.350	10.250	38.300	51.450
19.200	52.300	28.500	13.150	43.400	43.450

SPECIFIC MODEL PARAMETERS IN KELVIN

		UNIQUAC		NRTL(ALPHA=.2)	
I J		AIJ	AJI	AIJ	AJI
1 2		-290.42	-244.04	967.60	-898.64
1 3		487.31	544.92	2192.2	1091.7
2 3		-54.939	330.71	552.43	29.269

R1 = 0.9200   R2 = 1.4311   R3 = 3.3092
Q1 = 1.400   Q2 = 1.432   Q3 = 2.860

MEAN DEV. BETWEEN CALC. AND EXP. CONC. IN MOLE PCT

UNIQUAC (SPECIFIC PARAMETERS)   1.70
NRTL   (SPECIFIC PARAMETERS)   1.57

$CH_4O\text{-}C_2H_4Cl_2$

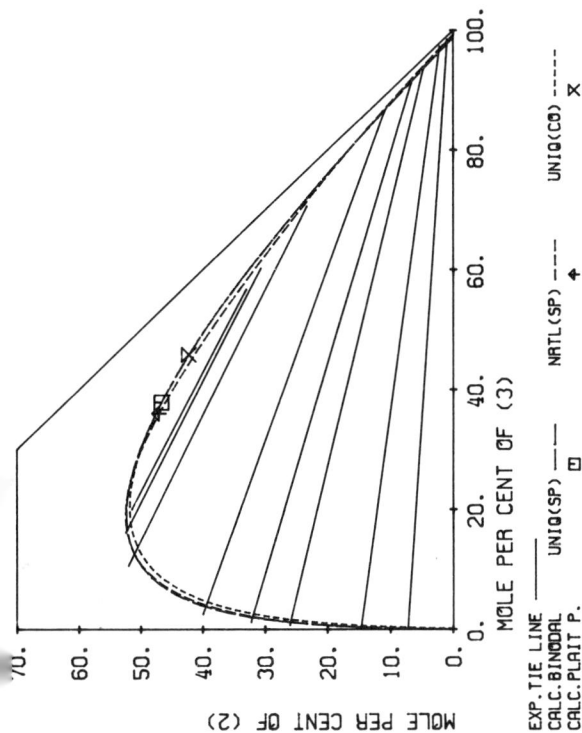

MOLE PER CENT OF (2)

MOLE PER CENT OF (3)

EXP.TIE LINE ——— UNIQ(SP) —— ▣ NRTL(SP) ———— ✦ UNIQ(CO) -----
CALC.BINODAL
CALC.PLAIT P.   ✕

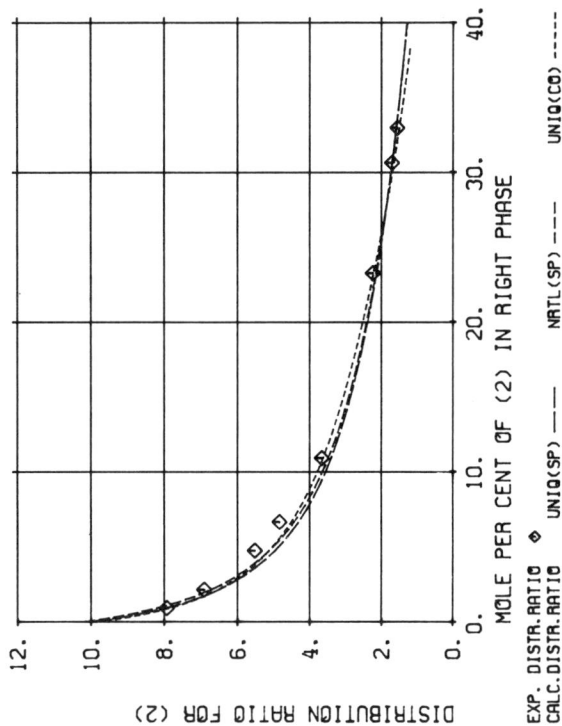

DISTRIBUTION RATIO FOR (2)

MOLE PER CENT OF (2) IN RIGHT PHASE

EXP. DISTR.RATIO ◇ UNIQ(SP) ——— NRTL(SP) ——— UNIQ(CO) -----
CALC.DISTR.RATIO

(2) CH4O   METHANOL

(3) C2H4CL2   ETHANE,1,2-DICHLORO

IZMAILOV N.A., FRANKE A.K.
ZH.FIZ.KHIM. 29(1955)620

TEMPERATURE = 25.0 DEG C   TYPE OF SYSTEM = 1

EXPERIMENTAL TIE LINES IN MOLE PCT

	LEFT PHASE			RIGHT PHASE	
(1)	(2)	(3)	(1)	(2)	(3)
92.456	7.272	0.272	0.544	0.917	98.540
84.939	14.690	0.371	0.539	2.121	97.340
73.190	26.068	0.743	1.574	4.720	93.705
66.536	32.276	1.187	2.061	6.664	91.275
57.531	39.954	2.515	2.488	10.912	86.600
37.414	52.010	10.576	6.097	23.263	70.640
31.381	52.453	16.166	9.079	30.629	60.291
28.519	51.468	20.013	10.279	32.963	56.758

SPECIFIC MODEL PARAMETERS IN KELVIN

		UNIQUAC		NRTL(ALPHA=.2)	
I	J	AIJ	AJI	AIJ	AJI
1	2	-177.22	-187.79	1020.1	-814.90
1	3	429.01	638.14	1790.4	944.01
2	3	-66.033	307.47	512.05	-18.927

R1 = 0.9200   R2 = 1.4311   R3 = 2.9308
Q1 = 1.400   Q2 = 1.432   Q3 = 2.528

MEAN DEV. BETWEEN CALC. AND EXP. CONC. IN MOLE PCT

UNIQUAC (SPECIFIC PARAMETERS)   0.90
NRTL   (SPECIFIC PARAMETERS)   0.61
UNIQUAC (COMMON PARAMETERS)   2.65

CH$_4$O-C$_3$H$_3$N

(1) C3H3N   PROPENOIC ACID,NITRILE

(2) CH4O   METHANOL

(3) H2O   WATER

NOVIKOVA K.E., KONDRATEVA N.M.
ZH.FIZ.KHIM. 39(1965)1432

TEMPERATURE = 20.0 DEG C   TYPE OF SYSTEM = 1

EXPERIMENTAL TIE LINES IN MOLE PCT

	LEFT PHASE			RIGHT PHASE	
(1)	(2)	(3)	(1)	(2)	(3)
85.166	3.014	11.820	2.842	3.027	94.131
81.925	5.169	12.906	3.194	5.395	91.411
76.601	7.788	15.611	3.648	8.933	87.420
67.689	11.768	20.544	4.347	11.436	84.217

MEAN DEV. BETWEEN CALC. AND EXP. CONC. IN MOLE PCT

UNIQUAC (COMMON PARAMETERS)   1.13

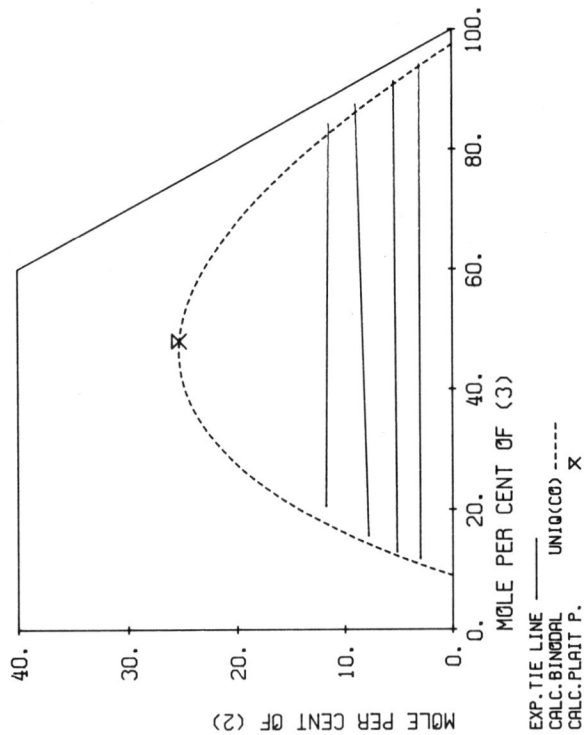

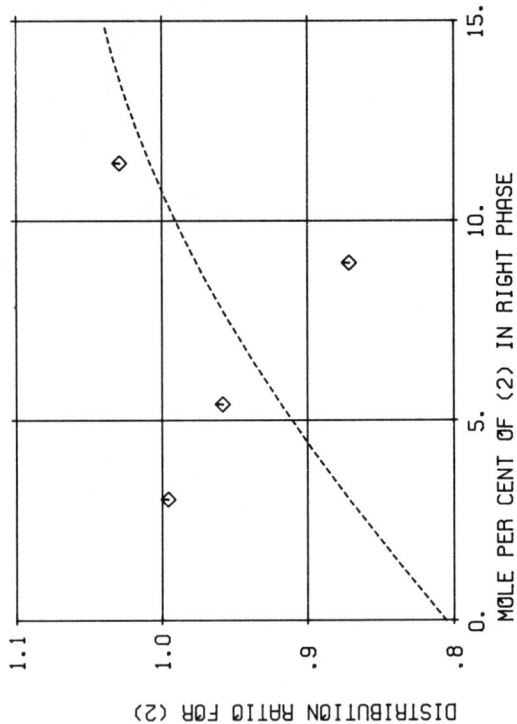

$$CH_4O\text{-}C_3H_6O_2$$

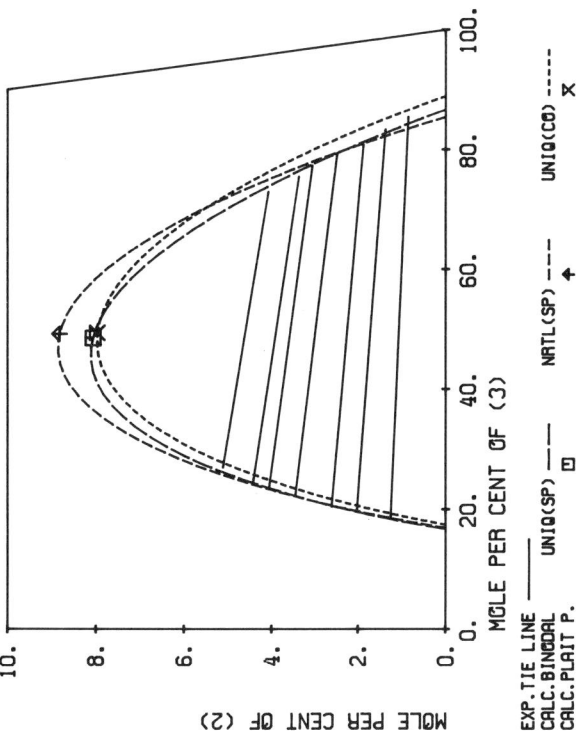

MOLE PER CENT OF (3)

MOLE PER CENT OF (2)

EXP.TIE LINE ——— UNIQ(SP) ⊟ NRTL(SP) ⊹ UNIQ(CG) ⊠
CALC.BINODAL ———
CALC.PLAIT P.

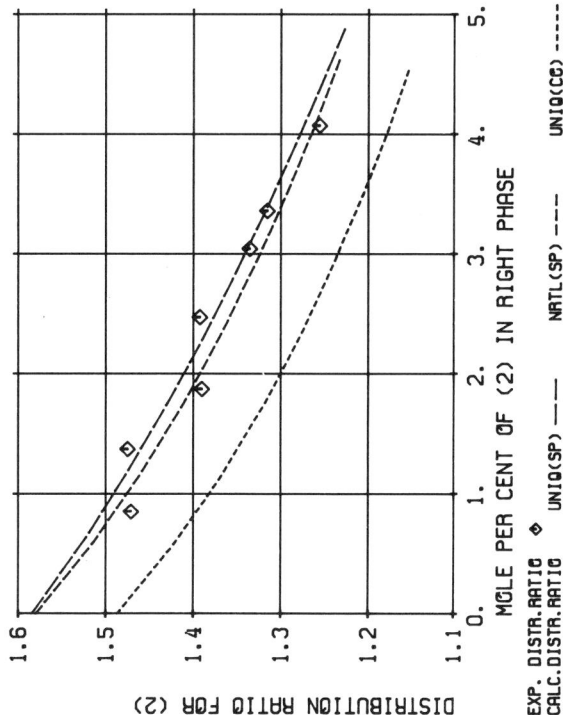

MOLE PER CENT OF (2) IN RIGHT PHASE

DISTRIBUTION RATIO FOR (2)

EXP. DISTR.RATIO ◇ UNIQ(SP) ——— NRTL(SP) ——— UNIQ(CG) -----
CALC.DISTR.RATIO

(2) C3H6O2   ACETIC ACID,METHYL ESTER
(3) C6H12   CYCLOHEXANE

SUGI H., NITTA T., KATAYAMA T.
J.CHEM.ENG.JPN. 9(1976)12

TEMPERATURE = 25.0 DEG C   TYPE OF SYSTEM = 1

EXPERIMENTAL TIE LINES IN MOLE PCT

| | LEFT PHASE | | | RIGHT PHASE | |
(1)	(2)	(3)	(1)	(2)	(3)
79.970	1.250	18.780	13.670	0.850	85.480
78.130	2.020	19.850	15.220	1.370	83.410
76.970	2.600	20.430	17.200	1.870	80.930
74.310	3.440	22.250	18.140	2.470	79.390
72.630	4.060	23.310	19.600	3.040	77.350
71.180	4.420	24.400	21.120	3.360	75.520
67.960	5.110	26.930	22.950	4.070	72.980

SPECIFIC MODEL PARAMETERS IN KELVIN

| | | UNIQUAC | | NRTL(ALPHA=.2) | |
I J	AIJ	AJI		AIJ	AJI
1 2	-16.965	-64.745		169.39	-580.60
1 3	-12.535	625.79		385.68	-467.08
2 3	-4.7990	-10.666		-234.39	-218.92

R1 = 1.4311   R2 = 2.8042   R3 = 4.0464
Q1 = 1.432   Q2 = 2.576   Q3 = 3.240

MEAN DEV. BETWEEN CALC. AND EXP. CONC. IN MOLE PCT

UNIQUAC (SPECIFIC PARAMETERS)   0.35
NRTL   (SPECIFIC PARAMETERS)   0.57
UNIQUAC (COMMON PARAMETERS)   1.22

CH$_4$O-C$_3$H$_7$NO$_2$

(1) C3H7NO2   PROPANE, 1-NITRO

(2) CH4O   METHANOL

(3) H2O   WATER

HANKINSON R.W., THOMPSON D.
J.CHEM.ENG.DATA 10(1965)18

TEMPERATURE = 25.0 DEG C   TYPE OF SYSTEM = 1

EXPERIMENTAL TIE LINES IN MOLE PCT

	LEFT PHASE			RIGHT PHASE	
(1)	(2)	(3)	(1)	(2)	(3)
82.446	11.475	6.079	0.682	12.587	86.731
67.051	21.884	11.066	0.740	14.728	84.532
48.319	34.840	16.841	0.859	18.854	80.287
36.062	40.141	23.797	1.106	22.075	75.819
25.874	43.671	30.454	1.570	27.131	71.298
21.104	43.889	35.006	2.271	31.804	65.925

SPECIFIC MODEL PARAMETERS IN KELVIN

		UNIQUAC		NRTL(ALPHA=.2)	
I	J	AIJ	AJI	AIJ	AJI
1	2	757.36	-183.64	836.34	-159.44
1	3	499.46	296.51	598.47	2472.6
2	3	-43.494	182.69	761.22	-69.325

R1 = 3.3573   R2 = 1.4311   R3 = 0.9200
Q1 = 2.943   Q2 = 1.432   Q3 = 1.400

MEAN DEV. BETWEEN CALC. AND EXP. CONC. IN MOLE PCT

UNIQUAC (SPECIFIC PARAMETERS)   0.62
NRTL   (SPECIFIC PARAMETERS)   0.65
UNIQUAC (COMMON PARAMETERS)   1.79

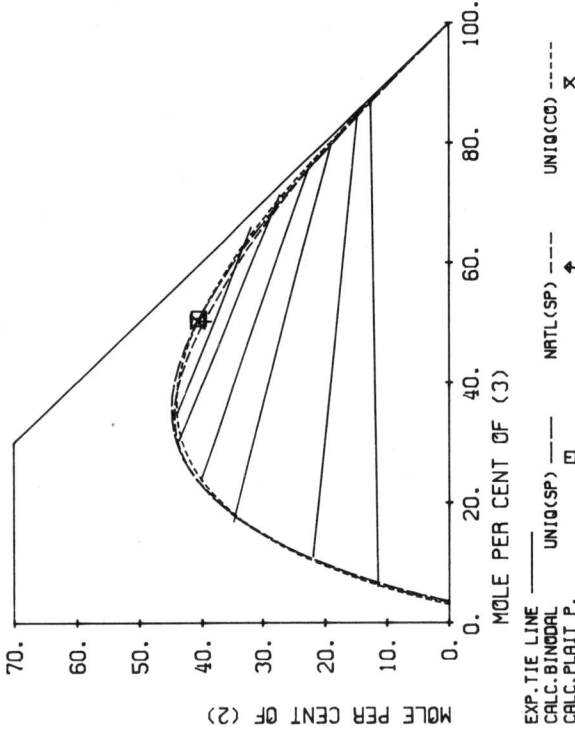

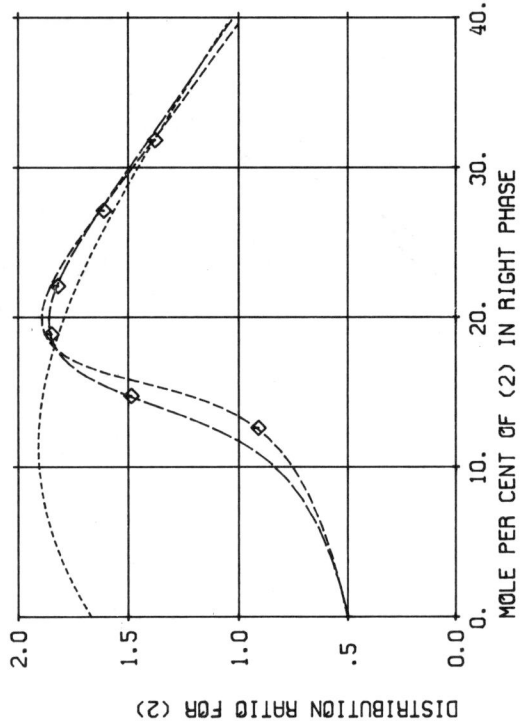

$$CH_4O\text{-}C_3H_8O$$

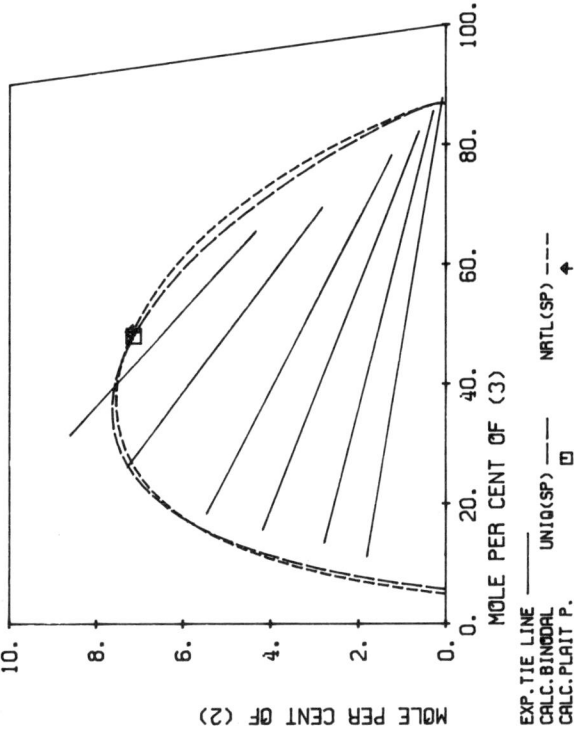

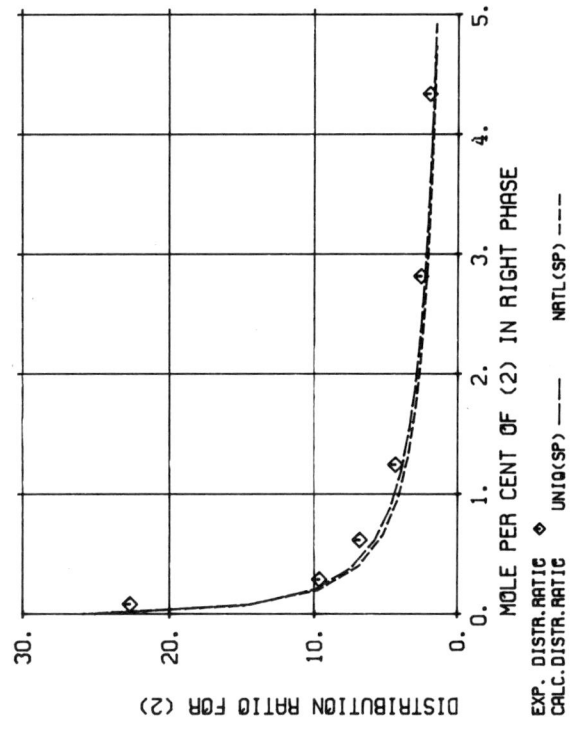

(2) C3H8O   2-PROPANOL
(3) C6H14   HEXANE

RADICE F.C., KNICKLE H.N.
J.CHEM.ENG.DATA 20(1975)371

TEMPERATURE = 5.0 DEG C   TYPE OF SYSTEM = 1

EXPERIMENTAL TIE LINES IN MOLE PCT

	LEFT PHASE			RIGHT PHASE	
(1)	(2)	(3)	(1)	(2)	(3)
88.637	0.0	11.363	10.361	0.0	89.639
86.909	1.801	11.291	12.192	0.079	87.729
83.674	2.766	13.560	14.151	0.287	85.562
80.096	4.172	15.732	17.191	0.613	82.196
76.138	5.447	18.415	20.508	1.244	73.248
66.472	7.279	26.249	27.668	2.814	69.519
59.931	8.592	31.477	30.140	4.334	65.526

SPECIFIC MODEL PARAMETERS IN KELVIN

		UNIQUAC		NRTL(ALPHA=.2)	
I J	AIJ	AJI		AIJ	AJI
1 2	29.648	-375.98		234.41	-1351.5
1 3	58.409	543.80		708.59	273.30
2 3	-108.29	-63.481		-889.36	-340.72

R1 = 1.4311   R2 = 2.7791   R3 = 4.4998
Q1 = 1.432   Q2 =. 2.508   Q3 = 3.856

MEAN DEV. BETWEEN CALC. AND EXP. CONC. IN MOLE PCT

UNIQUAC (SPECIFIC PARAMETERS)   1.27
NRTL   (SPECIFIC PARAMETERS)   1.40

CH₄O-C₃H₈O

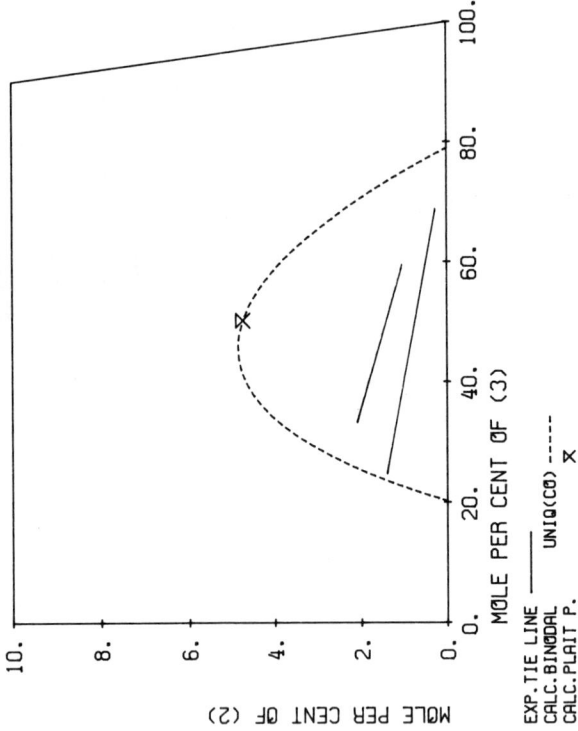

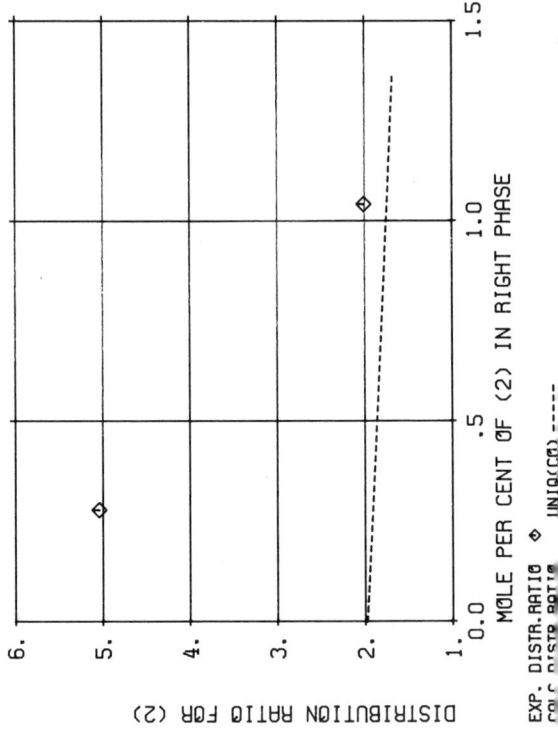

(1) CH4O   METHANOL
----
(2) C3H8O   2-PROPANOL
----
(3) C6H14   HEXANE

RADICE F.C., KNICKLE H.N.
J.CHEM.ENG.DATA 20(1975)371

TEMPERATURE = 25.0 DEG C   TYPE OF SYSTEM = 1

EXPERIMENTAL TIE LINES IN MOLE PCT

           LEFT PHASE   RIGHT PHASE
      (1)   (2)   (3)   (1)   (2)   (3)

   80.154   0.0   19.836   23.226   0.0   76.774
   73.831   1.396   24.773   30.917   0.277   68.806
   64.623   2.090   33.288   39.478   1.041   59.430

MEAN DEV. BETWEEN CALC. AND EXP. CONC. IN MOLE PCT

UNIQUAC (COMMON PARAMETERS)   3.16

CH$_4$O-C$_4$H$_4$Cl$_2$

(2) CH4O   METHANOL

(3) CH4CL2   1,3-BUTADIENE,2,3-DICHLORO

VOJTKO J., HRUSOVSKY M., KANALA A.
CHEM.ZVESTI 21(1967)443

TEMPERATURE = 20.0 DEG C   TYPE OF SYSTEM = 1

EXPERIMENTAL TIE LINES IN MOLE PCT

	LEFT PHASE			RIGHT PHASE	
(1)	(2)	(3)	(1)	(2)	(3)
91.527	8.457	0.016	0.467	2.624	96.909
81.886	18.030	0.084	0.783	5.134	94.084
67.258	32.557	0.185	1.082	7.875	91.043
55.222	44.213	0.565	1.433	10.159	88.408
45.818	52.783	1.399	1.910	14.769	83.321
38.024	59.427	2.549	2.269	17.341	80.390
24.135	68.185	7.680	3.306	25.784	70.910

SPECIFIC MODEL PARAMETERS IN KELVIN

I J	UNIQUAC		NRTL(ALPHA=.2)	
	AIJ	AJI	AIJ	AJI
1 2	-45.181	73.836	208.81	-47.899
1 3	350.15	805.25	1395.5	1029.7
2 3	16.272	434.77	582.35	102.99

R1 = 0.9200   R2 = 1.4311   R3 = 3.8166
Q1 = 1.400   Q2 = 1.432   Q3 = 3.424

MEAN DEV. BETWEEN CALC. AND EXP. CONC. IN MOLE PCT

UNIQUAC (SPECIFIC PARAMETERS)   0.97
NRTL   (SPECIFIC PARAMETERS)   1.12
UNIQUAC (COMMON PARAMETERS)   1.55

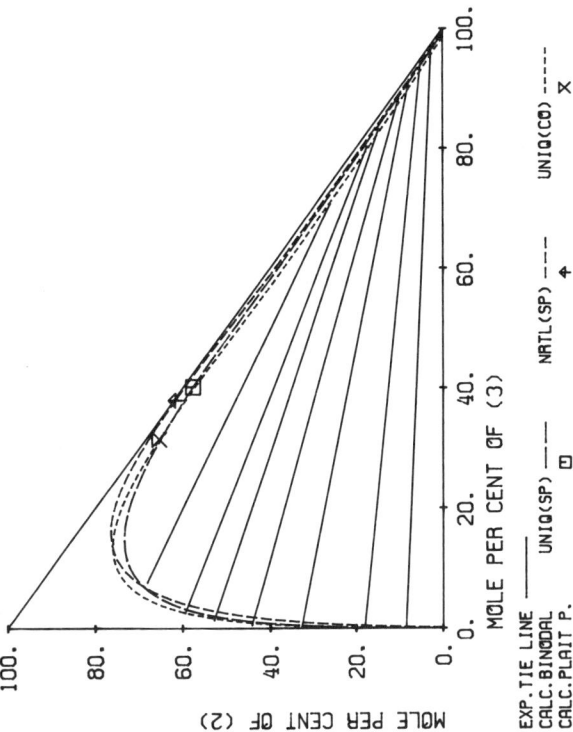

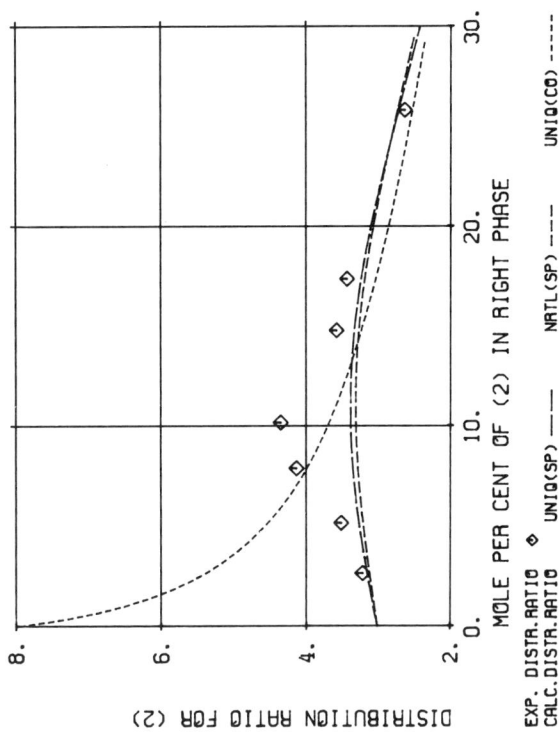

CH₄O-C₄H₆O₂

(1) H2O   WATER
(2) CH4O   METHANOL
(3) C4H6O2   ACETIC ACID,ETHENYL ESTER

NAKAMURA A.
KOGYO KAGAKU ZASSHI 71(1968)319

TEMPERATURE = 20.0 DEG C   TYPE OF SYSTEM = 1

EXPERIMENTAL TIE LINES IN MOLE PCT

	LEFT PHASE			RIGHT PHASE	
(1)	(2)	(3)	(1)	(2)	(3)
99.510	0.0	0.490	5.047	0.0	94.953
98.174	1.331	0.495	6.314	1.014	92.672
92.752	6.595	0.653	7.320	6.537	86.143
86.987	12.164	0.849	9.240	11.582	80.173
81.891	16.978	1.131	9.134	15.407	75.458
74.368	23.601	2.031	9.939	19.345	70.716
62.094	33.385	4.521	16.098	24.843	59.059

SPECIFIC MODEL PARAMETERS IN KELVIN

		UNIQUAC		NRTL(ALPHA=.2)	
I J	AIJ	AJI	AIJ	AJI	
1 2	-763.11	101.89	965.45	-635.70	
1 3	537.86	415.38	2285.6	552.64	
2 3	-189.69	-59.214	196.46	76.904	

R1 = 0.9200   R2 = 1.4311   R3 = 3.2485
Q1 = 1.400   Q2 = 1.432   Q3 = 2.904

MEAN DEV. BETWEEN CALC. AND EXP. CONC. IN MOLE PCT

UNIQUAC (SPECIFIC PARAMETERS)   1.94
NRTL   (SPECIFIC PARAMETERS)   1.98
UNIQUAC (COMMON PARAMETERS)   1.29

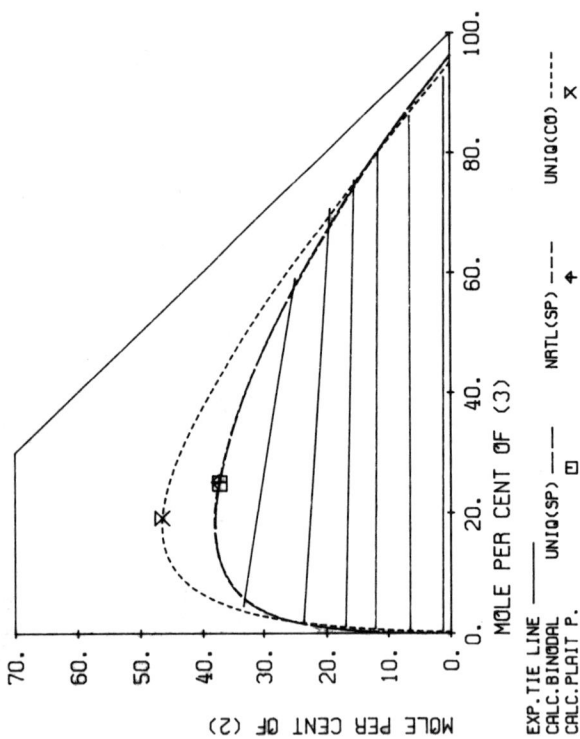

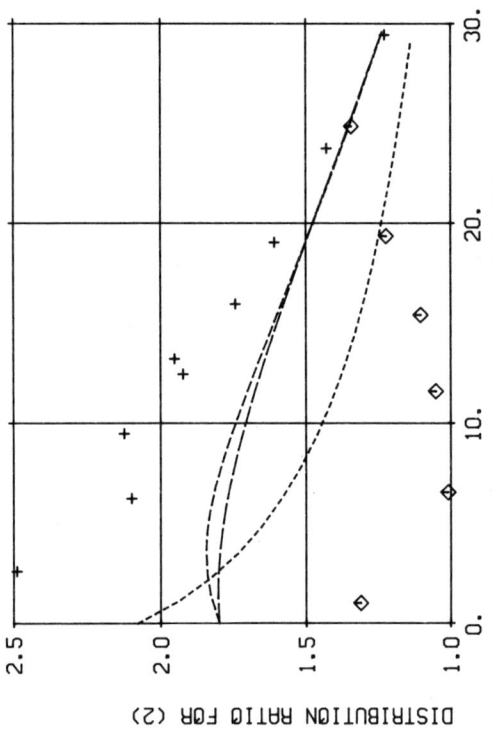

$CH_4O-C_4H_6O_2$

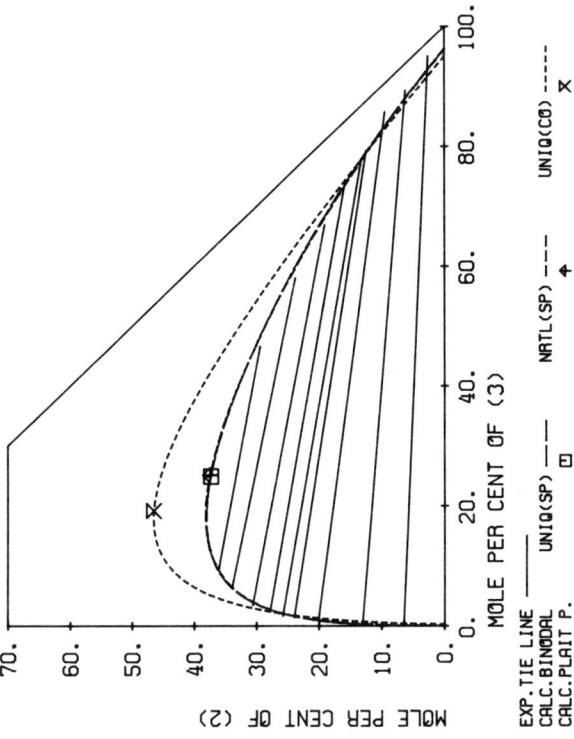

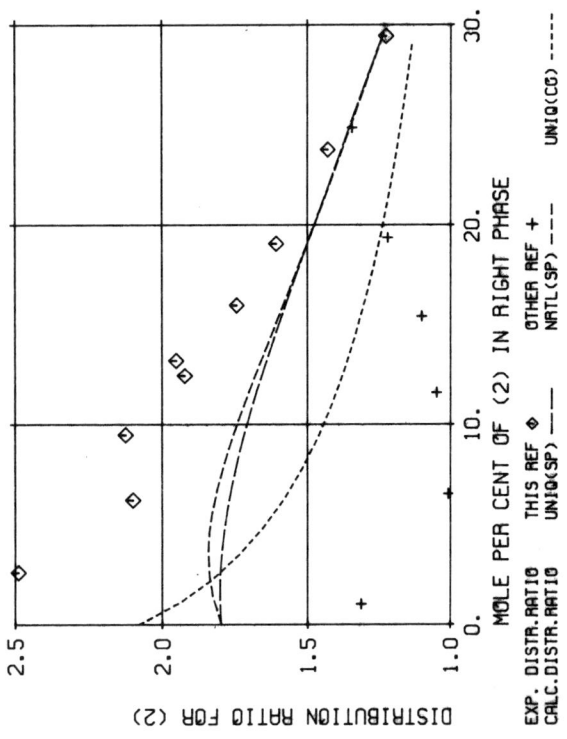

----

(2) CH4O   METHANOL

----

(3) C4H6O2   ACETIC ACID,ETHENYL ESTER

RUDAKOVSKAYA T.S., ET AL.
ZH.PRIKL.KHIM.(LENINGRAD) 41(1968)583

TEMPERATURE = 20.0 DEG C   TYPE OF SYSTEM = 1

EXPERIMENTAL TIE LINES IN MOLE PCT

	LEFT PHASE			RIGHT PHASE		
	(1)	(2)	(3)	(1)	(2)	(3)
	92.937	6.456	0.606	2.307	2.594	95.099
	86.292	13.048	0.660	4.425	6.220	89.355
	78.882	20.093	1.025	4.747	9.463	85.790
	74.656	23.917	1.427	8.192	12.437	79.371
	72.519	25.760	1.721	8.902	13.196	77.902
	69.691	27.842	2.467	10.888	15.961	73.152
	65.847	30.622	3.531	13.982	19.033	66.985
	59.595	33.933	6.472	18.238	23.738	58.024
	54.287	36.074	9.639	23.931	29.434	46.635

SPECIFIC MODEL PARAMETERS IN KELVIN

		UNIQUAC			NRTL(ALPHA=.2)	
I J		AIJ	AJI		AIJ	AJI
1 2		-763.11	101.89		965.45	-635.70
1 3		537.86	415.38		2285.6	552.64
2 3		-189.69	-59.214		196.46	75.904

R1 = 0.9200   R2 = 1.4311   R3 = 3.2485
Q1 = 1.400   Q2 = 1.432   Q3 = 2.904

MEAN DEV. BETWEEN CALC. AND EXP. CONC. IN MOLE PCT

UNIQUAC (SPECIFIC PARAMETERS)	1.45
NRTL (SPECIFIC PARAMETERS)	1.43
UNIQUAC (COMMON PARAMETERS)	3.14

CH₄O-C₄H₈O

(1) C6H12   CYCLOHEXANE

(2) C4H8O   FURAN, TETRAHYDRO

(3) CH4O   METHANOL

SUGI H., NITTA T., KATAYAMA T.
J.CHEM.ENG.JPN. 9(1976)12

TEMPERATURE = 25.0 DEG C   TYPE OF SYSTEM = 1

EXPERIMENTAL TIE LINES IN MOLE PCT

LEFT PHASE			RIGHT PHASE		
(1)	(2)	(3)	(1)	(2)	(3)
85.300	0.360	14.340	17.790	0.340	81.870
83.500	0.820	15.680	18.620	0.760	80.620
78.930	1.670	19.400	21.210	1.530	77.260
74.580	2.400	23.020	23.800	2.230	73.970
70.160	2.890	26.950	26.110	2.680	71.210

SPECIFIC MODEL PARAMETERS IN KELVIN

		UNIQUAC		NRTL(ALPHA=.2)	
I	J	AIJ	AJI	AIJ	AJI
1	2	-59.532	-130.17	-447.04	-336.27
1	3	618.50	13.935	445.78	396.55
2	3	-159.85	8.7418	-748.38	427.75

R1 = 4.0464   R2 = 2.9415   R3 = 1.4311
Q1 = 3.240   Q2 = 2.720   Q3 = 1.432

MEAN DEV. BETWEEN CALC. AND EXP. CONC. IN MOLE PCT

UNIQUAC (SPECIFIC PARAMETERS)   0.25
NRTL   (SPECIFIC PARAMETERS)   0.68
UNIQUAC (COMMON PARAMETERS)   1.43

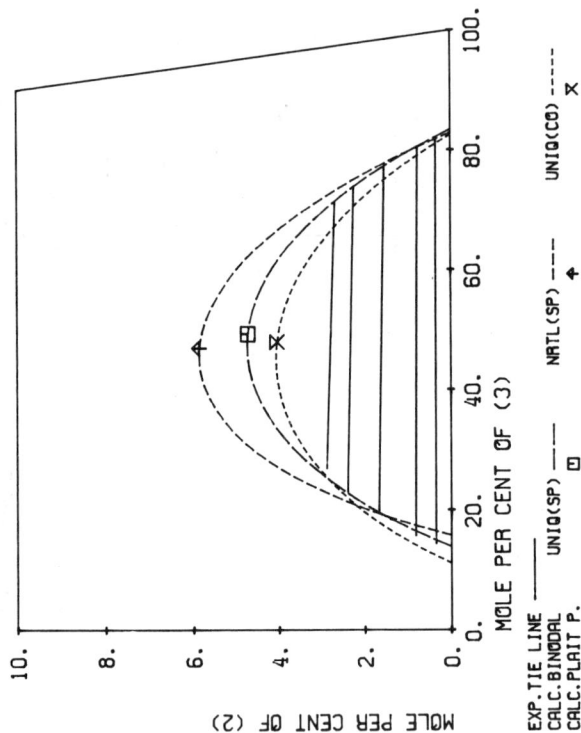

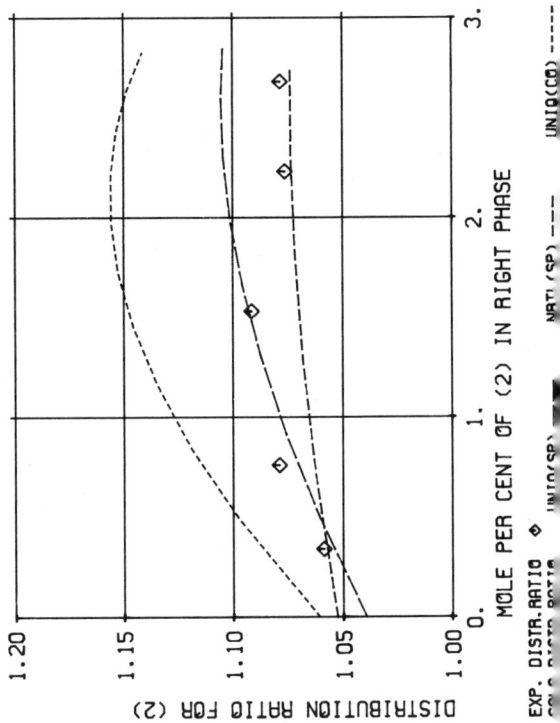

$CH_4O-C_4H_8O_2$

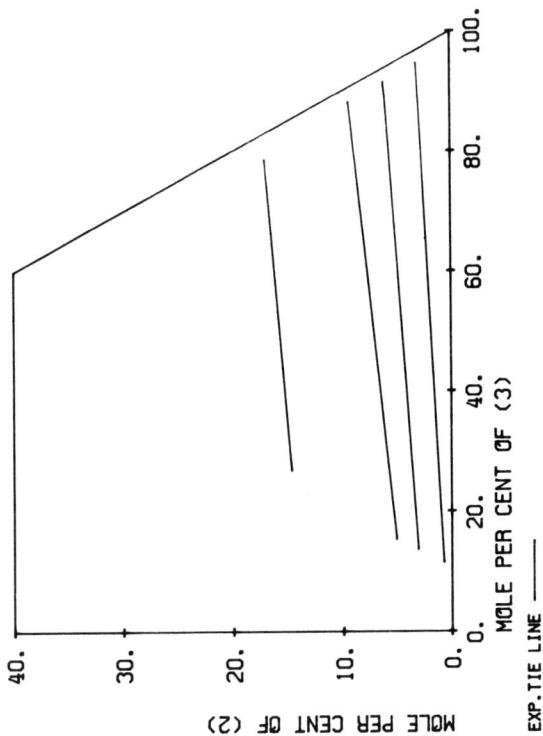

(1) C4H8O2   ACETIC ACID,ETHYL ESTER

(2) CH4O   METHANOL

(3) H2O   WATER

BEECH D.G., GLASSTONE S.
J.CHEM.SOC. (1938)67

TEMPERATURE =   0.0 DEG C   TYPE OF SYSTEM = 1

EXPERIMENTAL TIE LINES IN MOLE PCT

	LEFT PHASE			RIGHT PHASE	
(1)	(2)	(3)	(1)	(2)	(3)
87.760	0.746	11.495	2.153	3.179	94.668
83.240	3.116	13.643	2.202	6.184	91.614
79.578	5.117	15.305	2.435	9.390	88.174
58.522	14.684	26.794	4.337	17.051	78.612

MOLE PER CENT OF (2)

MOLE PER CENT OF (3)

EXP.TIE LINE

CH$_4$O-C$_4$H$_8$O$_2$

(1) C4H8O2   ACETIC ACID,ETHYL ESTER

(2) CH4O   METHANOL

(3) H2O   WATER

BEECH D.G., GLASSTONE S.
J.CHEM.SOC. (1938)67

TEMPERATURE = 20.0 DEG C   TYPE OF SYSTEM = 1

EXPERIMENTAL TIE LINES IN MOLE PCT

	LEFT PHASE			RIGHT PHASE	
(1)	(2)	(3)	(1)	(2)	(3)
83.144	2.392	14.464	1.782	2.573	95.646
75.803	6.979	17.218	1.918	5.400	92.683
67.427	11.056	21.518	2.173	7.520	90.307
64.481	12.783	22.735	2.523	9.913	87.564
50.765	17.496	31.739	3.725	14.343	81.932
39.100	20.663	40.237	5.019	18.178	75.803

SPECIFIC MODEL PARAMETERS IN KELVIN

		UNIQUAC		NRTL(ALPHA=.2)	
I J	AIJ	AJI		AIJ	AJI
1 2	-49.932	-118.89		561.00	-265.31
1 3	374.80	-126.62		149.88	1500.7
2 3	-340.40	103.49		230.22	-216.92

R1 = 3.4786   R2 = 1.4311   R3 = 0.9200
Q1 = 3.116   Q2 = 1.432   Q3 = 1.400

MEAN DEV. BETWEEN CALC. AND EXP. CONC. IN MOLE PCT

UNIQUAC (SPECIFIC PARAMETERS)   0.53
NRTL   (SPECIFIC PARAMETERS)   0.70
UNIQUAC (COMMON PARAMETERS)   1.17

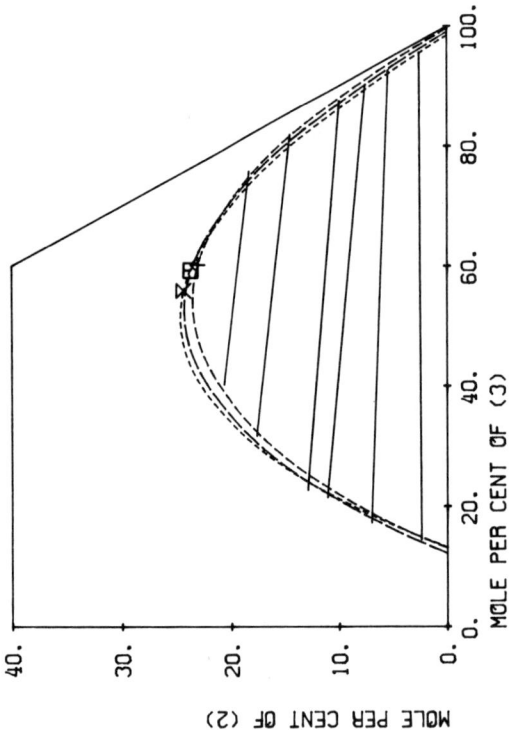

MOLE PER CENT OF (2)

MOLE PER CENT OF (3)

EXP.TIE LINE ——   UNIQ(SP) □   NRTL(SP) ——   UNIQ(CO) ——
CALC.BINODAL ——
CALC.PLAIT P.   ✕

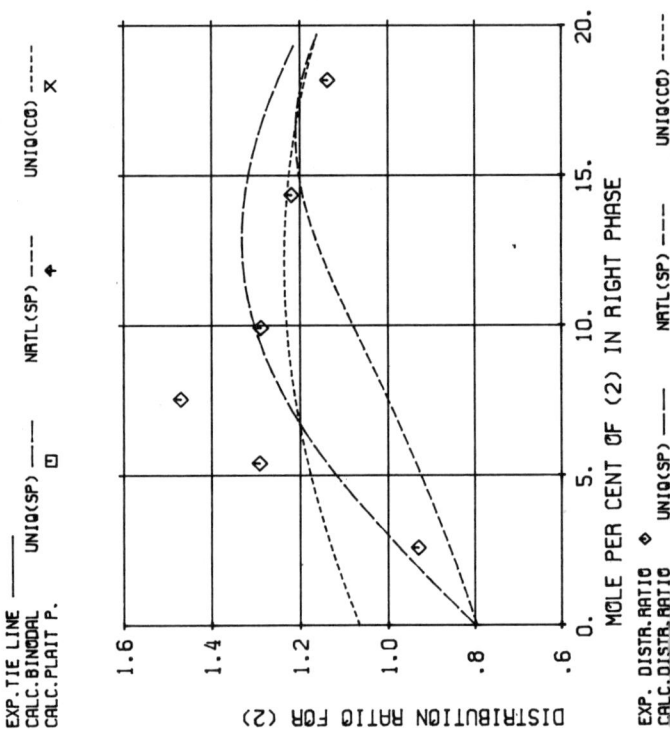

DISTRIBUTION RATIO FOR (2)

MOLE PER CENT OF (2) IN RIGHT PHASE

EXP. DISTR.RATIO ◇   UNIQ(SP) ◇   NRTL(SP) ——   UNIQ(CO) ——
CALC.DISTR.RATIO

CH$_4$O-C$_4$H$_8$O$_2$

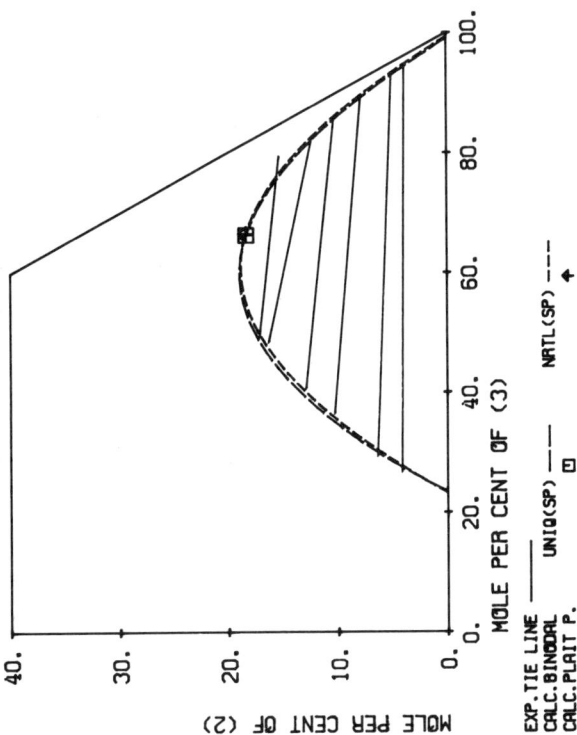

MOLE PER CENT OF (3)

MOLE PER CENT OF (2)

EXP.TIE LINE ——— UNIQ(SP) ▣ NRTL(SP) ‒‒‒
CALC.BINODAL
CALC.PLAIT P. ✦

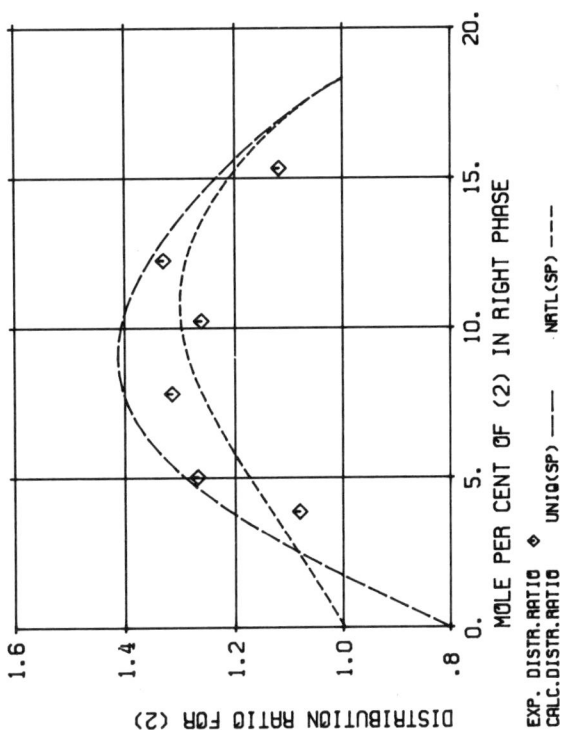

DISTRIBUTION RATIO FOR (2)

MOLE PER CENT OF (2) IN RIGHT PHASE

EXP. DISTR.RATIO ◇ UNIQ(SP) ——— NRTL(SP) ·····
CALC.DISTR.RATIO

(2) CH4O   METHANOL
(3) H2O   WATER

AKITA K., YOSHIDA F.
J.CHEM.ENG.DATA 8(1963)484

TEMPERATURE = 70.0 DEG C   TYPE OF SYSTEM = 1

EXPERIMENTAL TIE LINES IN MOLE PCT

	LEFT PHASE			RIGHT PHASE	
(1)	(2)	(3)	(1)	(2)	(3)
76.207	0.0	23.793	1.333	0.0	98.667
69.039	4.182	26.778	1.608	3.871	94.521
64.177	6.382	29.441	2.035	5.030	92.935
53.190	10.272	36.538	2.745	7.813	89.442
46.403	12.909	40.688	3.646	10.234	86.120
35.240	16.294	48.466	4.981	12.251	82.768
33.568	17.143	49.284	5.250	15.331	79.419

SPECIFIC MODEL PARAMETERS IN KELVIN

		UNIQUAC		NRTL(ALPHA=.2)	
I	J	AIJ	AJI	AIJ	AJI
1	2	-135.87	-234.67	638.54	-254.57
1	3	243.69	207.78	-20.116	1712.3
2	3	-477.25	116.23	271.48	-234.94

R1 = 3.4786   R2 = 1.4311   R3 = 0.9200
Q1 = 3.116   Q2 = 1.432   Q3 = 1.400

MEAN DEV. BETWEEN CALC. AND EXP. CONC. IN MOLE PCT

UNIQUAC (SPECIFIC PARAMETERS)   0.56
NRTL   (SPECIFIC PARAMETERS)   0.52

$CH_4O-C_4H_9NO$

(1) CH4O   METHANOL
----
(2) C4H9NO MORPHOLINE
----
(3) C7H16   HEPTANE

TAGLIAVINI G., ARICH G.
RIC.SCI. 28(1958)1902

TEMPERATURE = 18.0 DEG C   TYPE OF SYSTEM = 1

EXPERIMENTAL TIE LINES IN MOLE PCT

	LEFT PHASE			RIGHT PHASE	
(1)	(2)	(3)	(1)	(2)	(3)
91.100	0.0	8.900	10.450	0.0	89.550
87.720	3.610	8.670	10.350	0.740	88.910
74.280	16.580	9.140	9.920	3.820	85.260
72.120	18.530	9.350	9.340	4.230	86.430
58.040	31.190	10.770	8.580	8.950	82.470
44.860	41.610	13.530	8.230	15.670	76.100
37.110	47.440	15.450	7.770	20.460	71.770
28.450	51.930	19.620	7.330	26.830	65.840
21.120	53.060	25.820	6.440	34.610	58.950

SPECIFIC MODEL PARAMETERS IN KELVIN

		UNIQUAC		NRTL(ALPHA=.2)	
I	J	AIJ	AJI	AIJ	AJI
1	2	-239.49	325.72	-256.75	44.822
1	3	8.0479	680.19	553.56	422.41
2	3	-4.6706	165.21	456.90	58.160

R1 = 1.4311   R2 = 3.4740   R3 = 5.1742
Q1 = 1.432   Q2 = 2.796   Q3 = 4.396

MEAN DEV. BETWEEN CALC. AND EXP. CONC. IN MOLE PCT

UNIQUAC (SPECIFIC PARAMETERS)   0.40
NRTL   (SPECIFIC PARAMETERS)   0.50

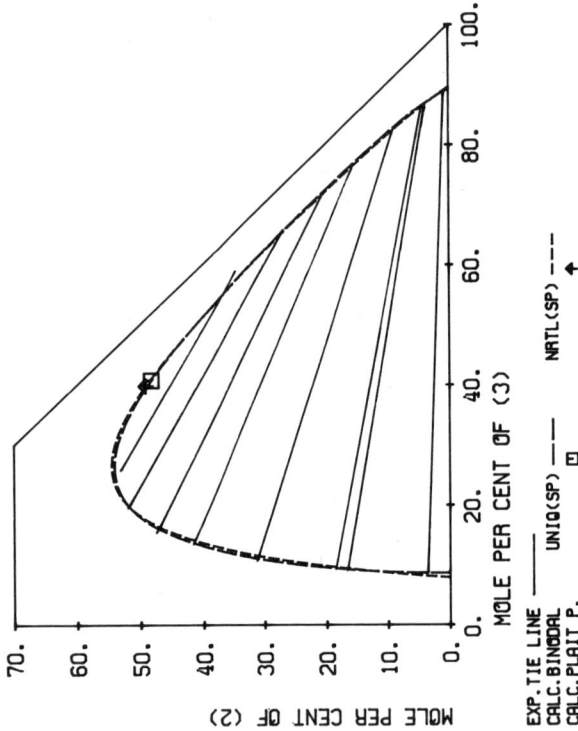

EXP.TIE LINE ———
CALC.BINODAL   UNIQ(SP) ☐   NRTL(SP) ----
CALC.PLAIT P. ◆

MOLE PER CENT OF (3)
MOLE PER CENT OF (2)

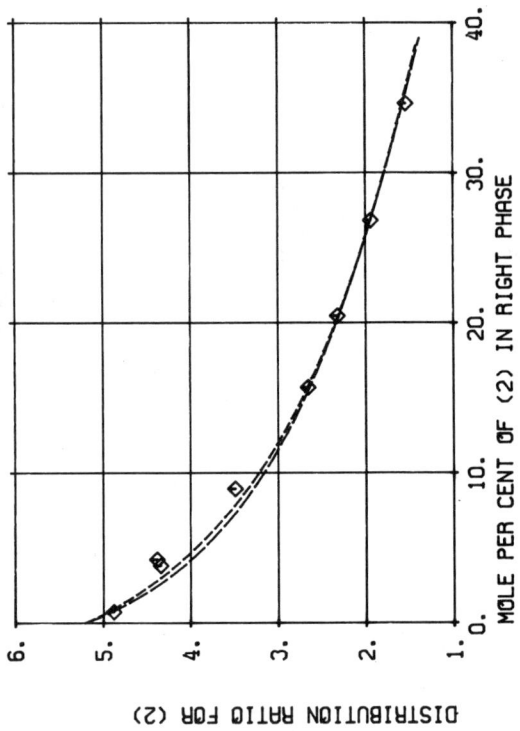

MOLE PER CENT OF (2) IN RIGHT PHASE
DISTRIBUTION RATIO FOR (2)

EXP. DISTR.RATIO ◇   UNIQ(SP)   NRTL(SP)

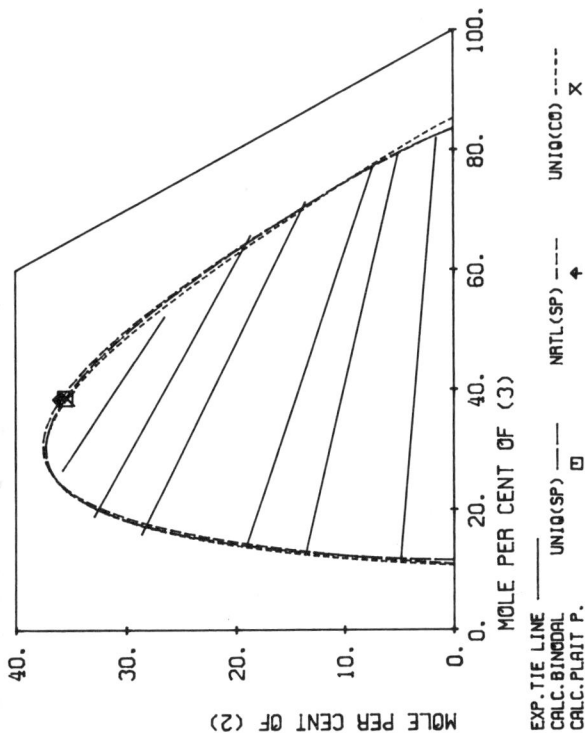

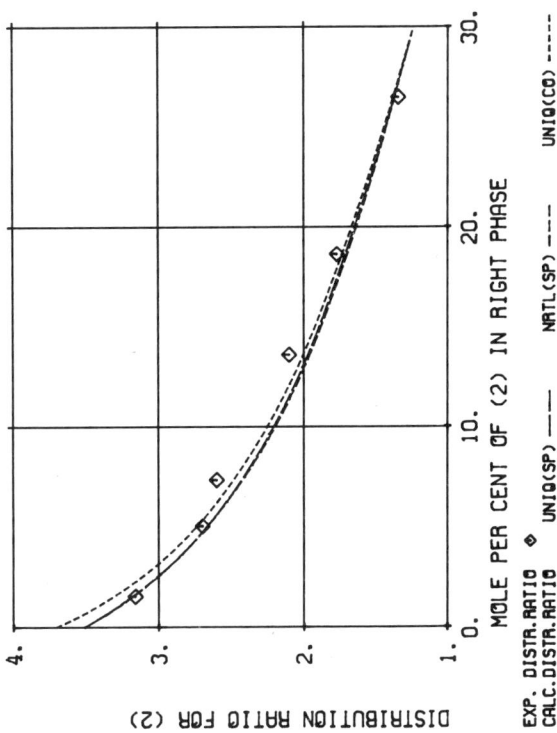

MOLE PER CENT OF (2)

MOLE PER CENT OF (3)

EXP.TIE LINE
CALC.BINODAL
CALC.PLAIT P.

UNIQ(SP) ——— □   NRTL(SP) ——— ✦   UNIQ(CO) ----- ✕

DISTRIBUTION RATIO FOR (2)

MOLE PER CENT OF (2) IN RIGHT PHASE

EXP. DISTR.RATIO ◇   UNIQ(SP) ———   NRTL(SP) ----   UNIQ(CO) -----
CALC.DISTR.RATIO

(2) C4H9NO   MORPHOLINE
(3) C7H16   HEPTANE

TAGLIAVINI G., ARICH G.
RIC.SCI. 28(1958)1902

TEMPERATURE = 30.0 DEG C   TYPE OF SYSTEM = 1

EXPERIMENTAL TIE LINES IN MOLE PCT

	LEFT PHASE			RIGHT PHASE	
(1)	(2)	(3)	(1)	(2)	(3)
88.350	0.0	11.650	16.810	0.0	83.190
83.070	4.830	12.100	16.510	1.530	81.960
73.620	13.570	12.810	15.790	5.030	79.180
67.030	19.060	13.910	15.500	7.340	77.160
55.440	28.590	15.970	15.050	13.600	71.350
48.190	32.880	18.930	15.630	18.610	65.760
37.570	35.750	26.680	21.300	26.480	52.220

SPECIFIC MODEL PARAMETERS IN KELVIN

		UNIQUAC		NRTL(ALPHA=.2)	
I	J	AIJ	AJI	AIJ	AJI
1	2	-241.57	207.67	-422.25	-2.2091
1	3	2.0824	649.07	565.80	345.19
2	3	2.7717	95.752	473.12	-223.06

R1 = 1.4311   R2 = 3.4740   R3 = 5.1742
Q1 = 1.432   Q2 = 2.796   Q3 = 4.396

MEAN DEV. BETWEEN CALC. AND EXP. CONC. IN MOLE PCT

UNIQUAC (SPECIFIC PARAMETERS)   0.64
NRTL (SPECIFIC PARAMETERS)   0.73
UNIQUAC (COMMON PARAMETERS)   0.90

CH₄O-C₄H₉NO

MOLE PER CENT OF (3)

MOLE PER CENT OF (2)

EXP.TIE LINE —— UNIQ(SP) —— ☐ NRTL(SP) —— ✦
CALC.BINODAL
CALC.PLAIT P.

DISTRIBUTION RATIO FOR (2)

MOLE PER CENT OF (2) IN RIGHT PHASE

EXP. DISTR.RATIO ◇ UNIQ(SP) —— NRTL(SP) ——
CALC. DISTR.RATIO

(1) CH4O   METHANOL
(2) C4H9NO   MORPHOLINE
(3) C7H16   HEPTANE

TAGLIAVINI G., ARICH G.
RIC.SCI. 28(1958)1902

TEMPERATURE = 40.0 DEG C   TYPE OF SYSTEM = 1

EXPERIMENTAL TIE LINES IN MOLE PCT

| | LEFT PHASE | | | RIGHT PHASE | |
	(1)	(2)	(3)	(1)	(2)	(3)
	84.860	0.0	15.140	25.360	0.0	74.640
	79.350	4.150	16.500	25.330	1.860	72.810
	73.150	8.760	18.090	26.580	4.540	63.880
	66.000	14.300	19.700	27.900	8.380	63.720
	59.680	18.360	21.960	31.200	12.240	56.560

SPECIFIC MODEL PARAMETERS IN KELVIN

| | | UNIQUAC | | NRTL(ALPHA=.2) | |
I	J	AIJ	AJI	AIJ	AJI
1	2	-89.653	-4.7683	-215.12	-360.63
1	3	0.35877	612.81	604.65	-250.35
2	3	132.88	8.1873	866.57	-610.47

R1 = 1.4311   R2 = 3.4740   R3 = 5.1742
Q1 = 1.432   Q2 = 2.796   Q3 = 4.396

MEAN DEV. BETWEEN CALC. AND EXP. CONC. IN MOLE PCT

UNIQUAC (SPECIFIC PARAMETERS)   0.38
NRTL (SPECIFIC PARAMETERS)   0.56

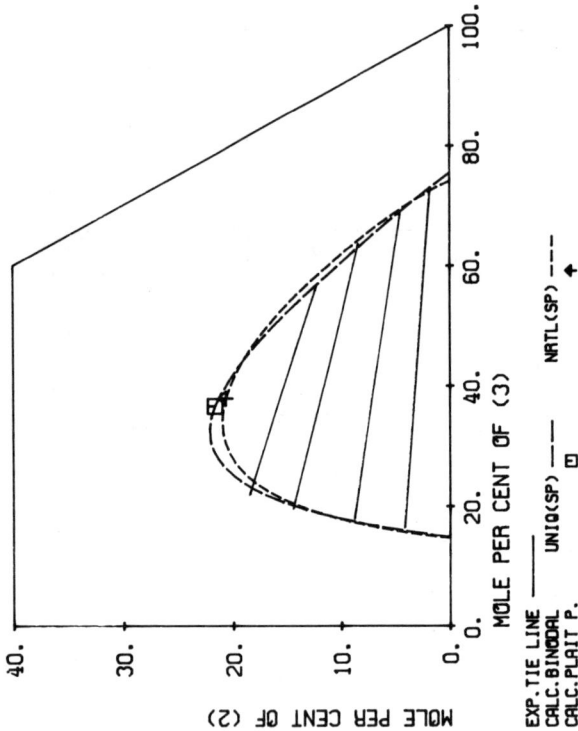

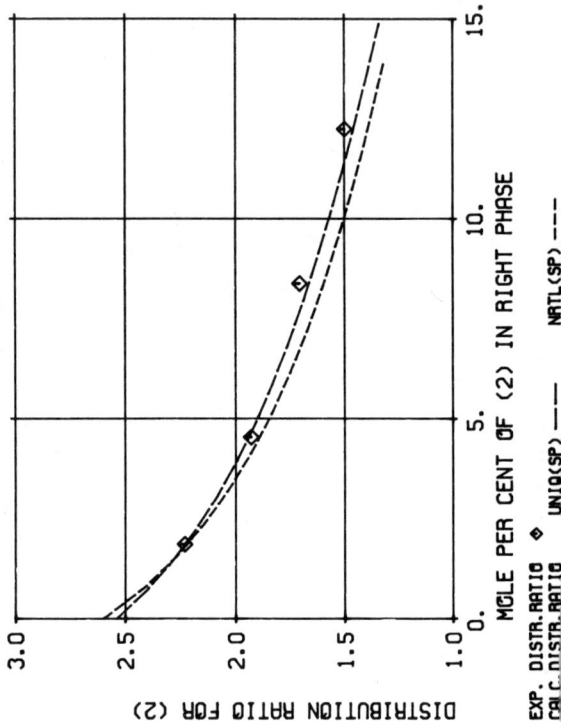

$CH_4O-C_4H_{10}O$

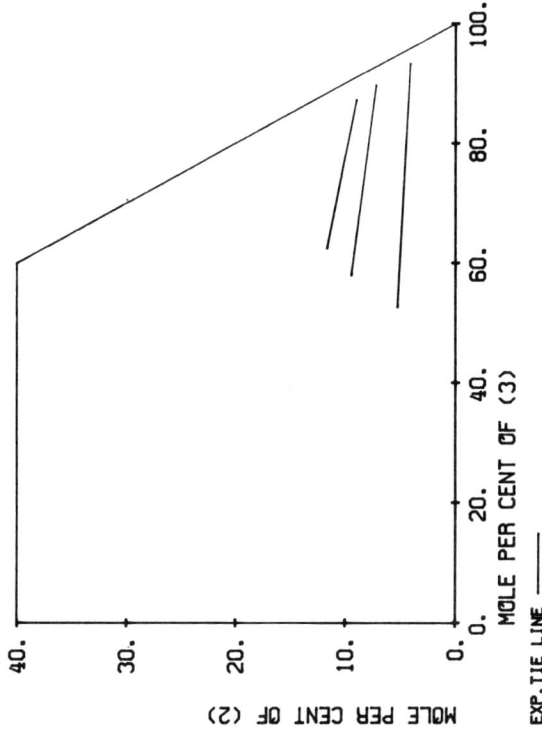

MOLE PER CENT OF (3)

MOLE PER CENT OF (2)

EXP.TIE LINE ⎯⎯⎯

(1) C4H10O   1-BUTANOL
----
(2) CH4O   METHANOL
----
(3) H2O   WATER

MUELLER A.J., PUGSLEY L.I., FERGUSON J.B.
J.PHYS.CHEM. 35(1931)1314

TEMPERATURE =   0.0 DEG C   TYPE OF SYSTEM = 1

EXPERIMENTAL TIE LINES IN MOLE PCT

      LEFT PHASE   RIGHT PHASE
  (1)   (2)   (3)   (1)   (2)   (3)

42.054   5.287   52.659   2.510 4.120 93.370
32.677   9.405   57.918   3.110 7.195 89.694
25.925 11.632   62.443   3.813 8.954 87.233

CH$_4$O-C$_4$H$_{10}$O

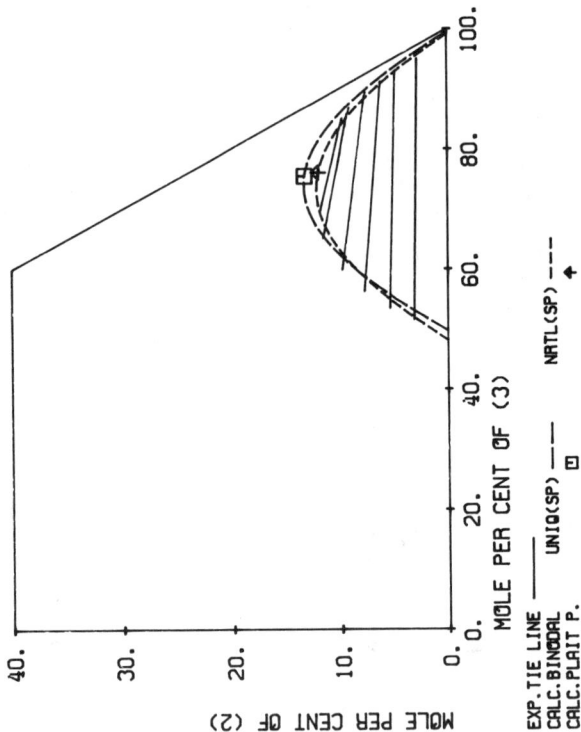

MOLE PER CENT OF (3)

MOLE PER CENT OF (2)

EXP.TIE LINE ———
CALC.BINODAL
CALC.PLAIT P.

UNIQ(SP) ——— ⊟

NRTL(SP) ——— ✦

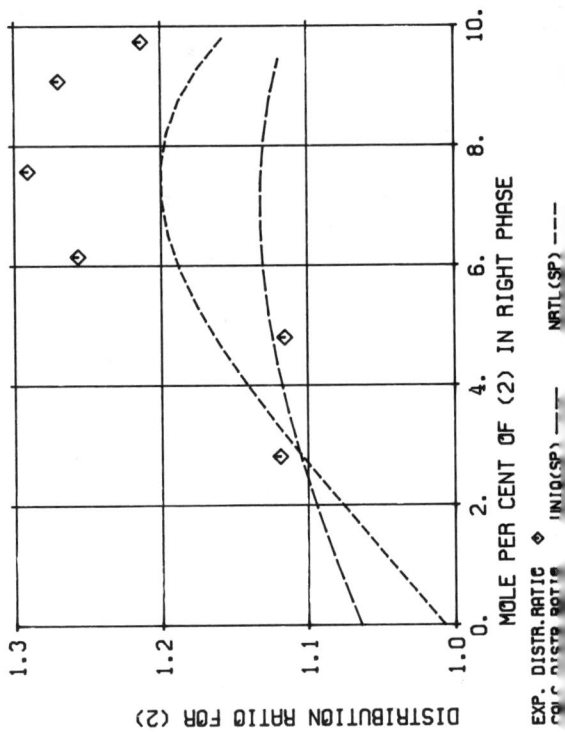

MOLE PER CENT OF (2) IN RIGHT PHASE

DISTRIBUTION RATIO FOR (2)

EXP. DISTR.RATIO ◆ UNIQ(SP) —— NRTL(SP) ——
CALC.DISTR.RATIO

(1) C4H10O   1-BUTANOL

(2) CH4O   METHANOL

(3) H2O   WATER

MUELLER A.J., PUGSLEY L.I., FERGUSON J.B.
J.PHYS.CHEM. 35(1931)1314

TEMPERATURE = 15.0 DEG C   TYPE OF SYSTEM = 1

EXPERIMENTAL TIE LINES IN MOLE PCT

	LEFT PHASE			RIGHT PHASE	
(1)	(2)	(3)	(1)	(2)	(3)
45.254	3.148	51.598	2.115	2.813	95.071
41.183	5.358	53.459	2.319	4.804	92.876
35.997	7.727	56.276	2.548	6.148	91.304
30.372	9.777	59.851	2.966	7.574	89.460
23.296	11.534	65.170	4.043	9.083	86.874
18.482	11.814	69.704	5.171	9.736	85.094

SPECIFIC MODEL PARAMETERS IN KELVIN

		UNIQUAC		NRTL(ALPHA=.2)	
I J	AIJ	AJI		AIJ	AJI
1 2	355.54	-164.09		706.80	-69.193
1 3	-82.688	443.56		-320.70	1664.7
2 3	-85.451	-321.92		100.23	-61.005

R1 = 3.4543   R2 = 1.4311   R3 = 0.9200
Q1 = 3.052   Q2 = 1.432   Q3 = 1.400

MEAN DEV. BETWEEN CALC. AND EXP. CONC. IN MOLE PCT

UNIQUAC (SPECIFIC PARAMETERS)   0.99
NRTL   (SPECIFIC PARAMETERS)   0.80

CH$_4$O-C$_4$H$_{10}$O

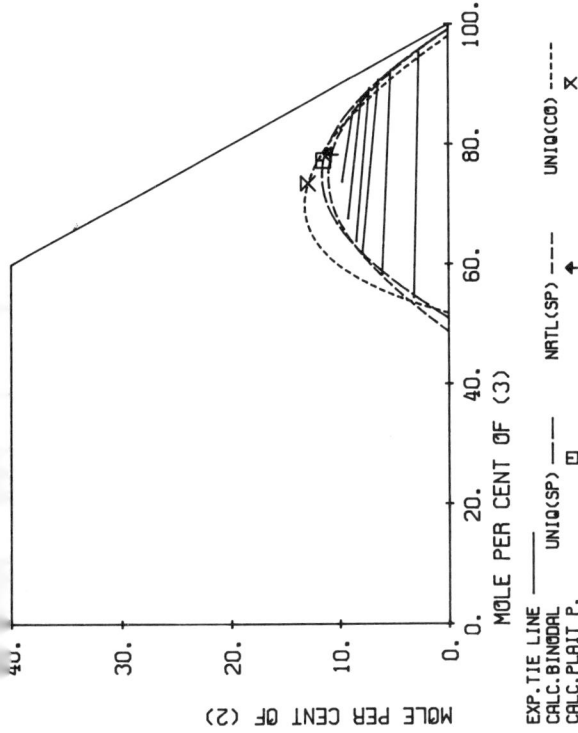

MOLE PER CENT OF (3)

MOLE PER CENT OF (2)

EXP.TIE LINE ——— UNIQ(SP) ☐ NRTL(SP) ✦ UNIQ(CC) -----⊠
CALC.BINODAL
CALC.PLAIT P.

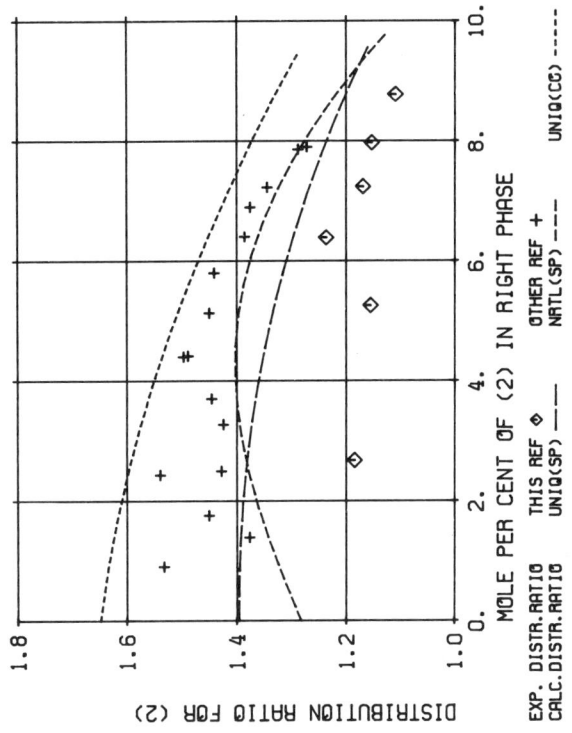

MOLE PER CENT OF (2) IN RIGHT PHASE

DISTRIBUTION RATIO FOR (2)

EXP. DISTR.RATIO ◇ THIS REF ◇ OTHER REF + UNIQ(CC) -----
CALC.DISTR.RATIO ——— UNIQ(SP) ——— NRTL(SP) ———

---

(2) CH4O   METHANOL

(3) H2O   WATER

MUELLER A.J., PUGSLEY L.I., FERGUSON J.B.
J.PHYS.CHEM. 35(1931)1314

TEMPERATURE = 30.0 DEG C   TYPE OF SYSTEM = 1

EXPERIMENTAL TIE LINES IN MOLE PCT

	LEFT PHASE			RIGHT PHASE	
(1)	(2)	(3)	(1)	(2)	(3)
42.416	3.165	54.419	1.917	2.673	95.410
35.667	6.067	58.266	2.298	5.254	92.447
30.419	7.907	61.674	2.790	6.391	90.819
27.704	8.458	63.838	3.327	7.240	89.434
23.195	9.178	67.627	4.139	7.969	87.892
16.439	9.731	73.830	6.120	8.777	85.103

SPECIFIC MODEL PARAMETERS IN KELVIN

		UNIQUAC		NRTL(ALPHA=.2)	
I J		AIJ	AJI	AIJ	AJI
1 2		320.76	-199.90	703.82	-178.09
1 3		-68.151	-396.99	-344.40	1756.8
2 3		-81.520	-319.75	45.740	-147.43

R1 = 3.4543   R2 = 1.4311   R3 = 0.9200
Q1 = 3.052   Q2 = 1.432   Q3 = 1.400

MEAN DEV. BETWEEN CALC. AND EXP. CONC. IN MOLE PCT

UNIQUAC (SPECIFIC PARAMETERS)   0.80
NRTL   (SPECIFIC PARAMETERS)   0.57
UNIQUAC (COMMON PARAMETERS)   2.41

CH$_4$O-C$_4$H$_{10}$O

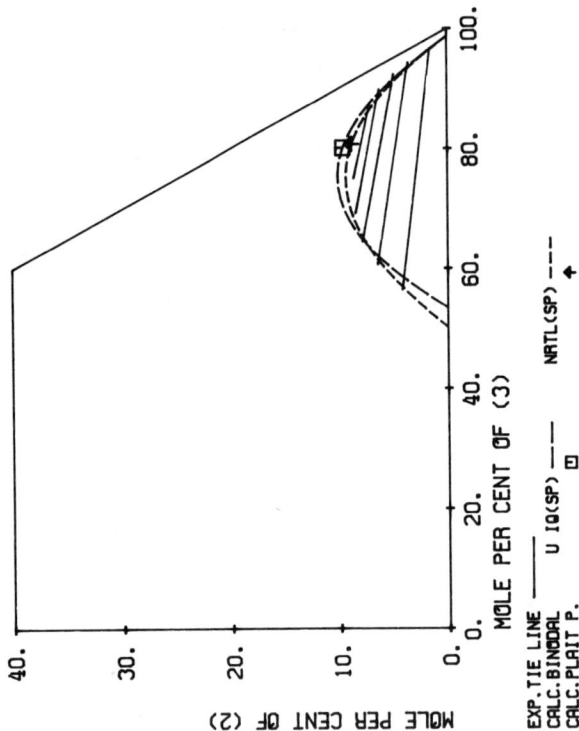

MOLE PER CENT OF (3)

MOLE PER CENT OF (2)

EXP.TIE LINE ———
CALC.BINODAL
CALC.PLAIT P.

U IQ(SP) ⊡ ———
NRTL(SP) - - - ↑

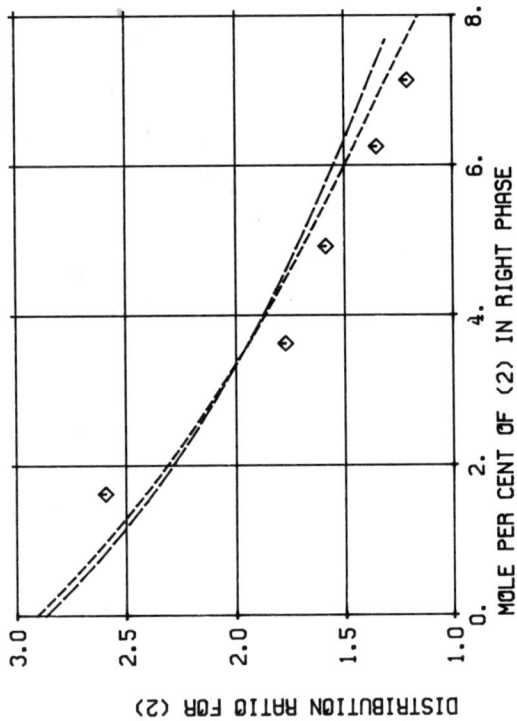

DISTRIBUTION RATIO FOR (2)

MOLE PER CENT OF (2) IN RIGHT PHASE

EXP. DISTR.RATIO ◇

(1) C4H10O   1-BUTANOL
(2) CH4O   METHANOL
(3) H2O   WATER

MUELLER A.J., PUGSLEY L.I., FERGUSON J.B.
J.PHYS.CHEM. 35(1931)1314   TYPE OF SYSTEM = 1

TEMPERATURE = 45.0 DEG C

EXPERIMENTAL TIE LINES IN MOLE PCT

	LEFT PHASE			RIGHT PHASE	
(1)	(2)	(3)	(1)	(2)	(3)
39.354	4.190	56.456	1.683	1.617	96.701
32.953	6.421	60.626	2.043	3.622	94.335
27.738	7.791	64.471	2.642	4.914	92.444
22.350	8.419	69.231	3.864	6.244	89.892
16.361	8.598	75.041	5.623	7.127	87.250

SPECIFIC MODEL PARAMETERS IN KELVIN

		UNIQUAC		NRTL(ALPHA=.2)	
I J	AIJ	AJI		AIJ	AJI
1 2	201.20	-285.19		225.09	-393.42
1 3	-63.331	372.64		-357.69	1740.5
2 3	-112.95	-233.81		-130.04	-229.65

R1 = 3.4543   R2 = 1.4311   R3 = 0.9200
Q1 = 3.052   Q2 = 1.432   Q3 = 1.400

MEAN DEV. BETWEEN CALC. AND EXP. CONC. IN MOLE PCT

UNIQUAC (SPECIFIC PARAMETERS)   0.62
NRTL   (SPECIFIC PARAMETERS)   0.44

$CH_4O-C_4H_{10}O$

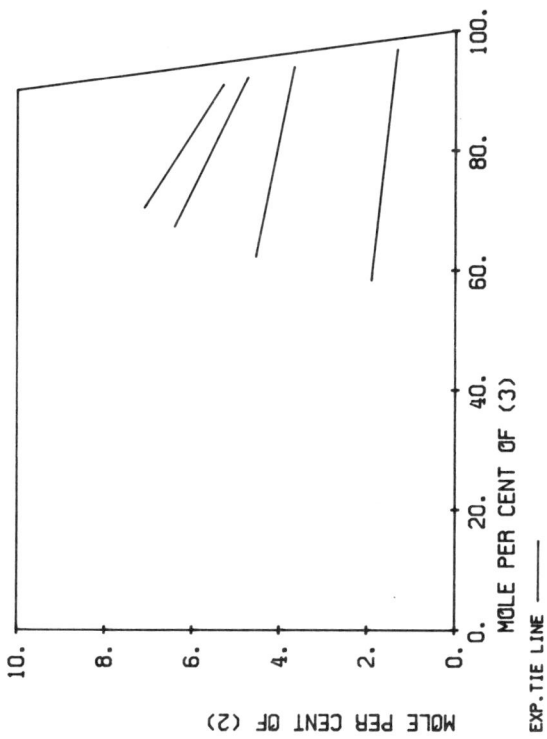

MOLE PER CENT OF (3)

EXP.TIE LINE ——

MOLE PER CENT OF (2)

(1) C4H10O   1-BUTANOL

(2) CH4O   METHANOL

(3) H2O   WATER

MUELLER A.J., PUGSLEY L.I., FERGUSON J.B.
J.PHYS.CHEM. 35(1931)1314

TEMPERATURE = 60.0 DEG C   TYPE OF SYSTEM = 1

EXPERIMENTAL TIE LINES IN MOLE PCT

	LEFT PHASE			RIGHT PHASE	
(1)	(2)	(3)	(1)	(2)	(3)
39.752	1.900	58.348	1.815	1.320	96.865
33.189	4.537	62.275	2.443	3.664	93.892
26.303	6.411	67.287	3.144	4.721	92.135
22.416	7.098	70.486	3.836	5.286	90.878

CH$_4$O-C$_4$H$_{10}$O

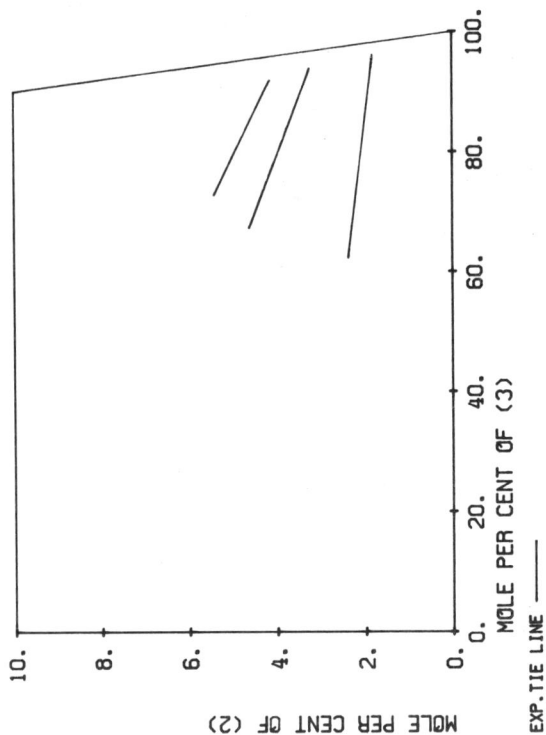

MOLE PER CENT OF (3)

MOLE PER CENT OF (2)

EXP.TIE LINE ——

(1) C4H10O   1-BUTANOL

(2) CH4O   METHANOL

(3) H2O   WATER

MUELLER A.J., PUGSLEY L.I., FERGUSON J.B.
J.PHYS.CHEM. 35(1931)1314

TEMPERATURE = 75.0 DEG C   TYPE OF SYSTEM = 1

EXPERIMENTAL TIE LINES IN MOLE PCT

	LEFT PHASE			RIGHT PHASE	
(1)	(2)	(3)	(1)	(2)	(3)
35.326	2.383	62.291	2.043	1.818	96.139
28.104	4.621	67.275	2.902	3.262	93.836
21.791	5.418	72.791	3.929	4.155	91.915

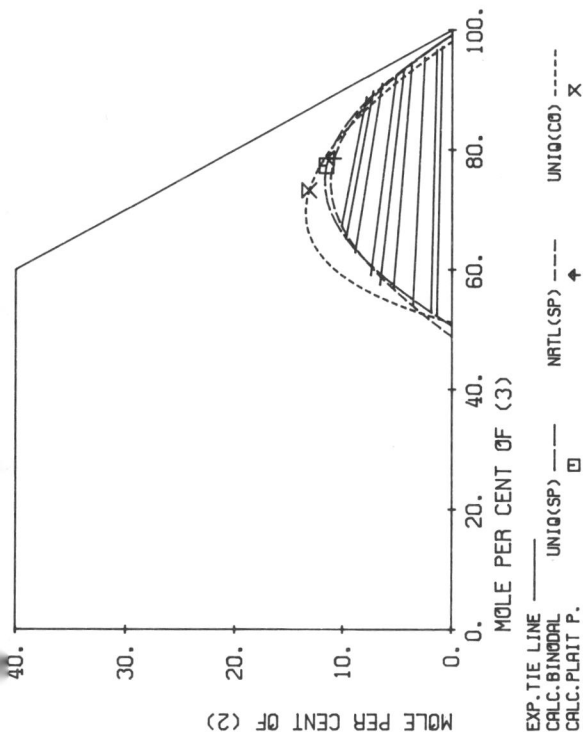

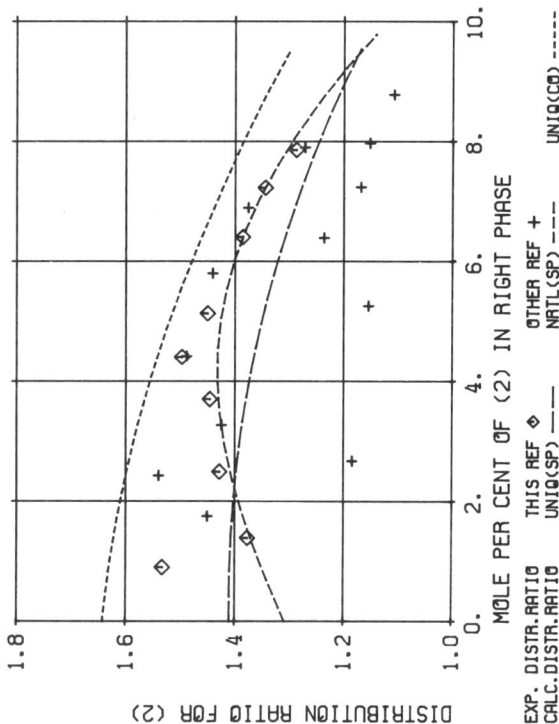

--- (2) CH4O   METHANOL ---

--- (3) H2O   WATER ---

PROCHAZKA O., SUSHKA J., PICK J.
COLLECT.CZECH.CHEM.COMMUN. 40(1975)781

TEMPERATURE = 25.0 DEG C   TYPE OF SYSTEM = 1

EXPERIMENTAL TIE LINES IN MOLE PCT

	LEFT PHASE		RIGHT PHASE	
(1)	(2)	(3)	(1)	(2)	(3)
46.201	1.377	52.421	1.891	0.899	97.210
45.106	1.905	52.989	1.898	1.384	96.718
42.414	3.566	54.020	2.026	2.496	95.479
37.581	5.348	57.071	2.158	3.697	94.145
35.913	6.588	57.499	2.198	4.402	93.400
33.506	7.444	59.050	2.354	5.133	92.513
28.215	8.871	62.914	2.494	6.403	91.103
25.098	9.712	65.190	2.928	7.226	89.846
22.555	10.115	67.329	3.311	7.856	88.834

SPECIFIC MODEL PARAMETERS IN KELVIN

		UNIQUAC		NRTL(ALPHA=.2)	
I J	AIJ	AJI	AIJ	AJI
1 2	320.76	-199.90	703.82	-178.09
1 3	-68.151	396.99	-344.40	1756.8
2 3	-81.520	-319.75	45.740	-147.43

R1 = 3.4543   R2 = 1.4311   R3 = 0.9200
Q1 = 3.052   Q2 = 1.432   Q3 = 1.400

MEAN DEV. BETWEEN CALC. AND EXP. CONC. IN MOLE PCT

UNIQUAC (SPECIFIC PARAMETERS)   0.54
NRTL   (SPECIFIC PARAMETERS)   0.42
UNIQUAC (COMMON PARAMETERS)   1.86

CH$_4$O-C$_4$H$_{10}$O

(1) C4H10O   1-BUTANOL
(2) CH4O   METHANOL
(3) H2O   WATER

PROCHAZKA O., SUSHKA J., PICK J.
COLLECT.CZECH.CHEM.COMMUN. 40(1975)781

TEMPERATURE = 45.0 DEG C   TYPE OF SYSTEM = 1

EXPERIMENTAL TIE LINES IN MOLE PCT

	LEFT PHASE			RIGHT PHASE	
(1)	(2)	(3)	(1)	(2)	(3)
43.177	1.457	55.366	1.538	0.890	97.572
40.954	2.191	56.854	1.489	1.247	97.264
37.844	3.850	58.306	1.719	2.289	95.992
33.473	5.272	61.255	1.927	3.359	94.714
27.668	7.448	64.884	2.234	4.857	92.909
24.326	8.405	67.269	2.659	5.770	91.571
22.329	8.725	68.946	2.935	6.150	90.915

SPECIFIC MODEL PARAMETERS IN KELVIN

		UNIQUAC		NRTL(ALPHA=.2)	
I	J	AIJ	AJI	AIJ	AJI
1	2	221.84	-184.91	259.86	-144.27
1	3	-100.56	441.67	-441.60	1912.0
2	3	-69.666	-231.74	-29.675	-174.04

R1 = 3.4543   R2 = 1.4311   R3 = 0.9200
Q1 = 3.052   Q2 = 1.432   Q3 = 1.400

MEAN DEV. BETWEEN CALC. AND EXP. CONC. IN MOLE PCT

UNIQUAC (SPECIFIC PARAMETERS)   0.58
NRTL   (SPECIFIC PARAMETERS)   0.42

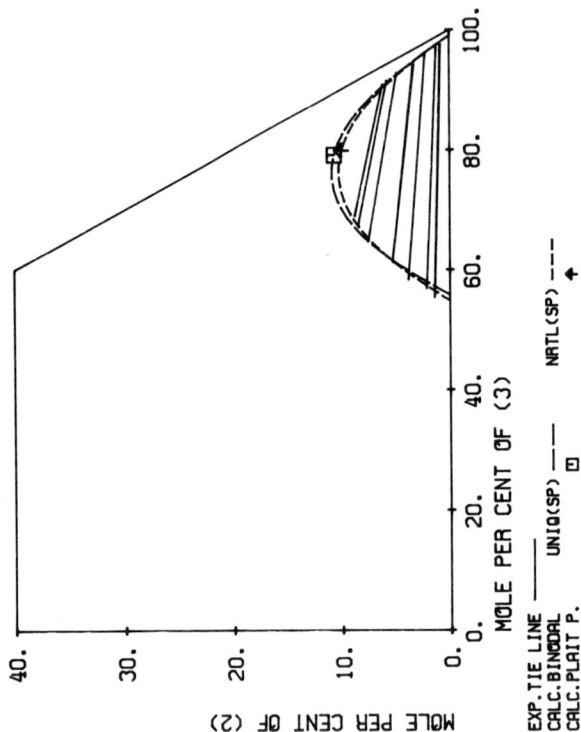

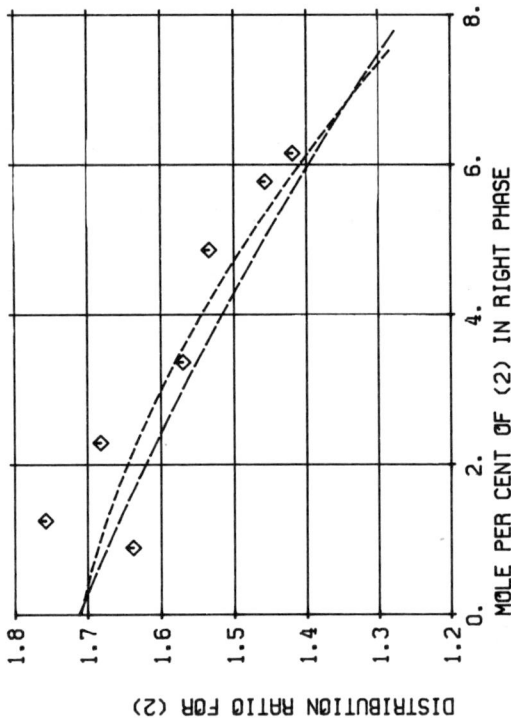

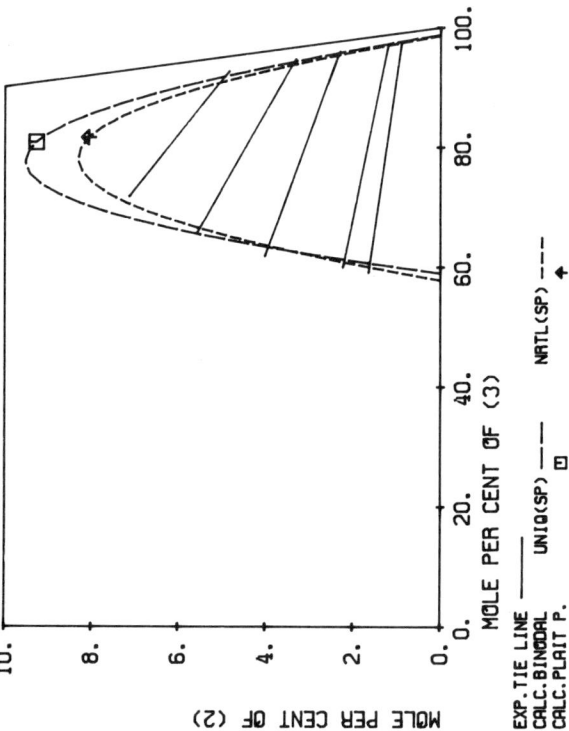

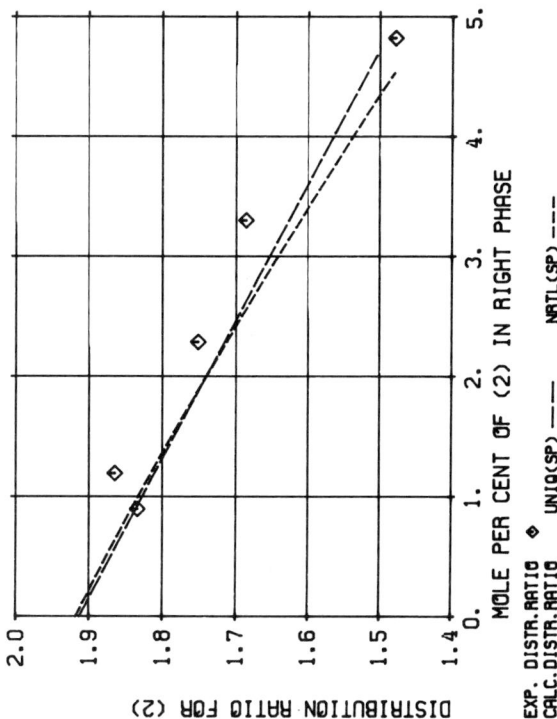

(2) CH4O   METHANOL

(3) H2O   WATER

PROCHAZKA O., SUSHKA J., PICK J.
COLLECT.CZECH.CHEM.COMMUN. 40(1975)781

TEMPERATURE = 65.0 DEG C   TYPE OF SYSTEM = 1

EXPERIMENTAL TIE LINES IN MOLE PCT

	LEFT PHASE			RIGHT PHASE	
(1)	(2)	(3)	(1)	(2)	(3)
39.309	1.635	59.056	1.619	0.892	97.489
37.836	2.222	59.941	1.623	1.192	97.185
34.176	4.006	61.818	1.691	2.287	96.022
28.826	5.566	65.608	1.982	3.302	94.716
21.107	7.127	71.767	2.435	4.820	92.744

SPECIFIC MODEL PARAMETERS IN KELVIN

		UNIQUAC		NRTL(ALPHA=.2)	
I J		AIJ	AJI	AIJ	AJI
1 2		204.47	-204.91	233.08	-227.88
1 3		-116.90	461.71	-488.99	1959.1
2 3		-79.411	-216.70	-116.03	-195.57

R1 = 3.4543   R2 = 1.4311   R3 = 0.9200
Q1 = 3.052   Q2 = 1.432   Q3 = 1.400

MEAN DEV. BETWEEN CALC. AND EXP. CONC. IN MOLE PCT

UNIQUAC (SPECIFIC PARAMETERS)   0.76
NRTL   (SPECIFIC PARAMETERS)   0.66

$CH_4O-C_4H_{10}O$

MOLE PER CENT OF (3)

MOLE PER CENT OF (2)

EXP.TIE LINE —— UNIQ(SP) —— NRTL(SP) ---- UNIQ(CO) ----
CALC.BINODAL   ⊡   +   ⨯
CALC.PLAIT P.

MOLE PER CENT OF (2) IN RIGHT PHASE

DISTRIBUTION RATIO FOR (2)

EXP. DISTR.RATIO   THIS REF ◇   OTHER REF +

(1) C4H10O   1-BUTANOL
(2) CH4O   METHANOL
(3) H2O   WATER

SUGI H., KATAYAMA T.
J.CHEM.ENG.JPN. 10(1977)400

TEMPERATURE = 25.0 DEG C   TYPE OF SYSTEM = 1

EXPERIMENTAL TIE LINES IN MOLE PCT

	LEFT PHASE			RIGHT PHASE	
(1)	(2)	(3)	(1)	(2)	(3)
45.310	2.540	52.150	1.740	1.750	96.510
42.690	3.740	53.570	1.770	2.430	95.800
40.940	4.660	54.400	2.070	3.270	94.660
35.560	6.580	57.860	2.420	4.420	93.160
31.300	8.360	60.340	2.820	5.800	91.380
26.430	9.490	64.080	3.320	6.900	89.780
21.890	10.050	68.060	4.300	7.900	87.800

SPECIFIC MODEL PARAMETERS IN KELVIN

I	J	UNIQUAC		NRTL(ALPHA=.2)	
		AIJ	AJI	AIJ	AJI
1	2	320.76	-199.90	703.82	-178.09
1	3	-68.151	396.99	-344.40	1756.8
2	3	-81.520	-319.75	45.740	-147.43

R1 = 3.4543   R2 = 1.4311   R3 = 0.9200
Q1 = 3.052   Q2 = 1.432   Q3 = 1.400

MEAN DEV. BETWEEN CALC. AND EXP. CONC. IN MOLE PCT

UNIQUAC (SPECIFIC PARAMETERS)   0.57
NRTL   (SPECIFIC PARAMETERS)   0.33
UNIQUAC (COMMON PARAMETERS)   1.72

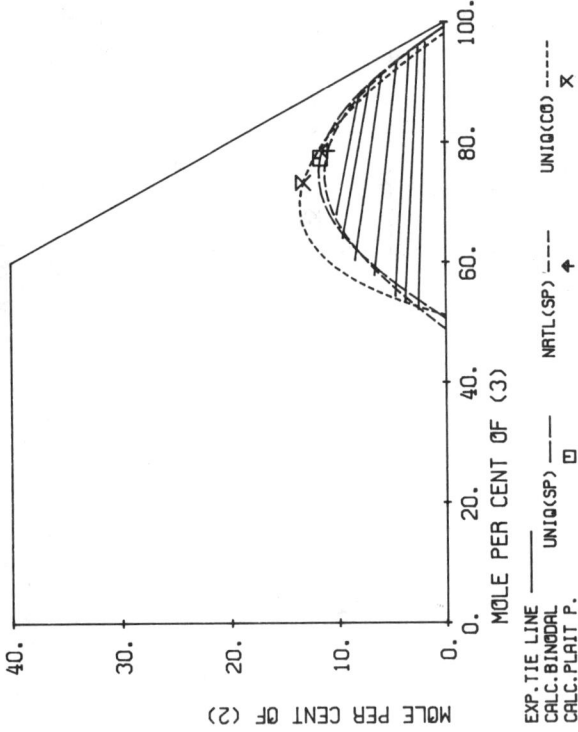

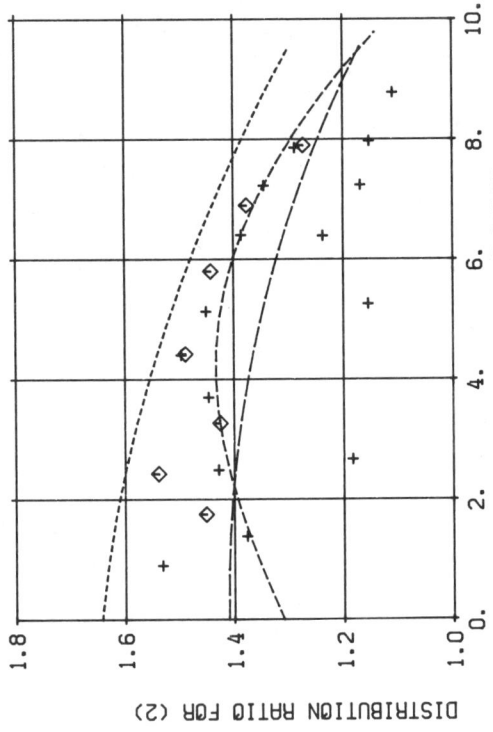

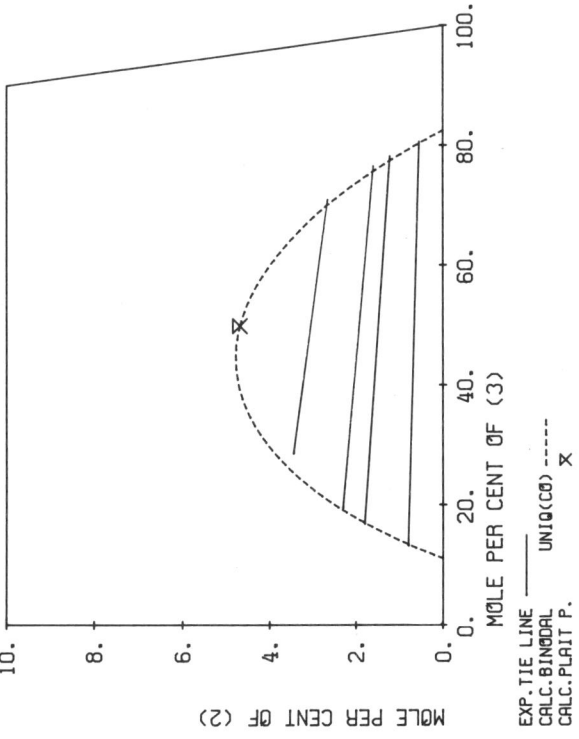

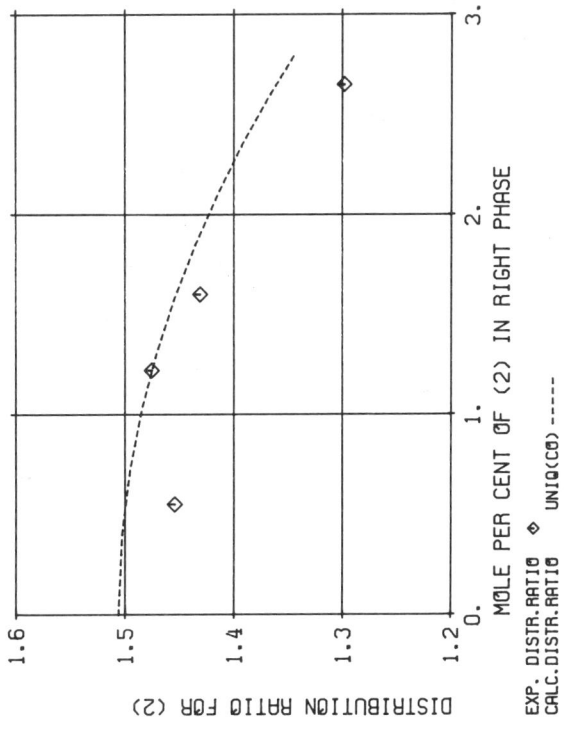

$CH_4O-C_4H_{10}O$

(2) C4H100   ETHER,DIETHYL

(3) CH40   METHANOL

SUGI H., NITTA T., KATAYAMA T.
J.CHEM.ENG.JPN. 9(1976)12

TEMPERATURE = 25.0 DEG C   TYPE OF SYSTEM = 1

EXPERIMENTAL TIE LINES IN MOLE PCT

	LEFT PHASE			RIGHT PHASE	
(1)	(2)	(3)	(1)	(2)	(3)
85.980	0.800	13.220	18.800	0.550	80.650
81.310	1.800	16.890	20.570	1.220	78.210
78.200	2.290	19.510	21.850	1.600	76.550
67.930	3.440	28.630	26.390	2.650	70.960

MEAN DEV. BETWEEN CALC. AND EXP. CONC. IN MOLE PCT

UNIQUAC (COMMON PARAMETERS)   0.63

MOLE PER CENT OF (2)

MOLE PER CENT OF (3)

EXP.TIE LINE ——— UNIQ(CO) -----
CALC.BINODAL
CALC.PLAIT P.   X

DISTRIBUTION RATIO FOR (2)

MOLE PER CENT OF (2) IN RIGHT PHASE

EXP. DISTR.RATIO ◇ UNIQ(CO) -----
CALC.DISTR.RATIO

CH$_4$O-C$_5$H$_8$

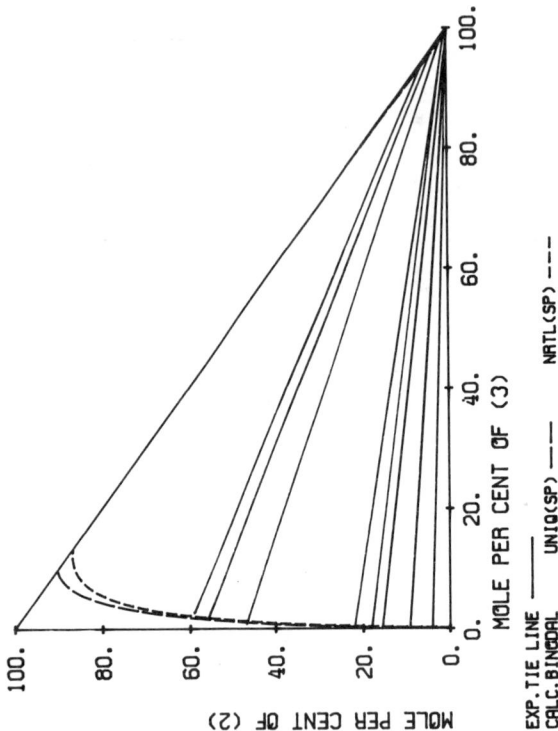

MOLE PER CENT OF (3)

MOLE PER CENT OF (2)

EXP.TIE LINE ——— UNIQ(SP) ——— NRTL(SP) ----

CALC.BINODAL

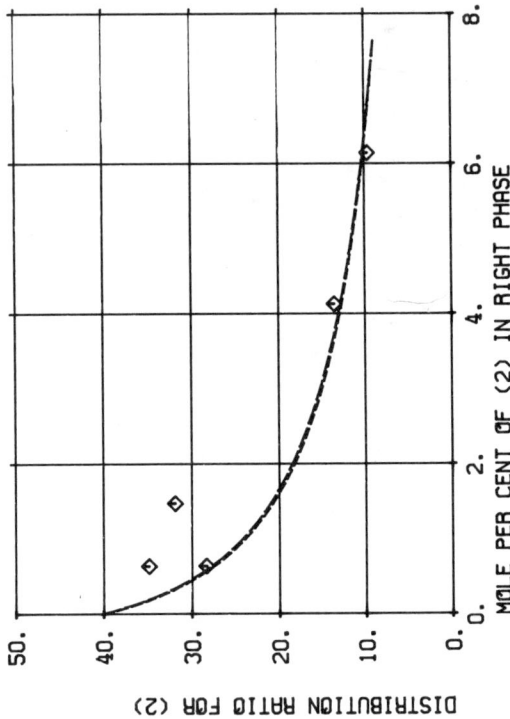

MOLE PER CENT OF (2) IN RIGHT PHASE

DISTRIBUTION RATIO FOR (2)

EXP. DISTR.RATIO ◇ UNIQ(SP) ——— NRTL (SP) ----

(1)	H2O	WATER
(2)	CH4O	METHANOL
(3)	C5H8	1,3-BUTADIENE,2-METHYL

LESTEVA T.M., OGORODNIKOV S.K., MOROZOVA V.I.
ZH.PRIKL.KHIM.(LENINGRAD) 39(1965)2134

TEMPERATURE = 15.0 DEG C   TYPE OF SYSTEM = 1

EXPERIMENTAL TIE LINES IN MOLE PCT

LEFT PHASE			RIGHT PHASE		
(1)	(2)	(3)	(1)	(2)	(3)
84.457	15.514	0.030	1.125	0.0	98.875
90.773	9.170	0.057	0.752	0.0	99.248
95.909	4.063	0.027	0.377	0.0	99.623
77.904	22.033	0.062	0.750	0.632	98.618
82.012	17.958	0.030	0.376	0.634	98.990
52.414	46.779	0.807	0.746	1.469	97.785
42.392	55.922	1.685	1.101	4.125	94.774
38.324	59.042	2.635	0.723	6.138	93.135

SPECIFIC MODEL PARAMETERS IN KELVIN

		UNIQUAC		NRTL(ALPHA=.2)	
I J		AIJ	AJI	AIJ	AJI
1 2		-292.22	-124.37	-433.50	11.073
1 3		301.02	733.18	1378.6	955.66
2 3		105.72	416.59	524.07	295.83

R1 = 0.9200   R2 = 1.4311   R3 = 3.3638
Q1 = 1.400   Q2 = 1.432   Q3 = 3.012

MEAN DEV. BETWEEN CALC. AND EXP. CONC. IN MOLE PCT

UNIQUAC (SPECIFIC PARAMETERS)	0.48
NRTL (SPECIFIC PARAMETERS)	0.47

$CH_4O-C_5H_8O_2$

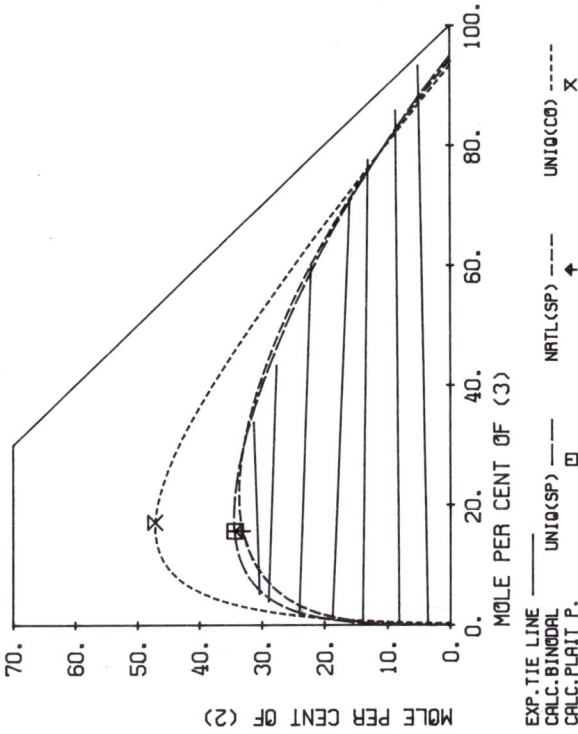

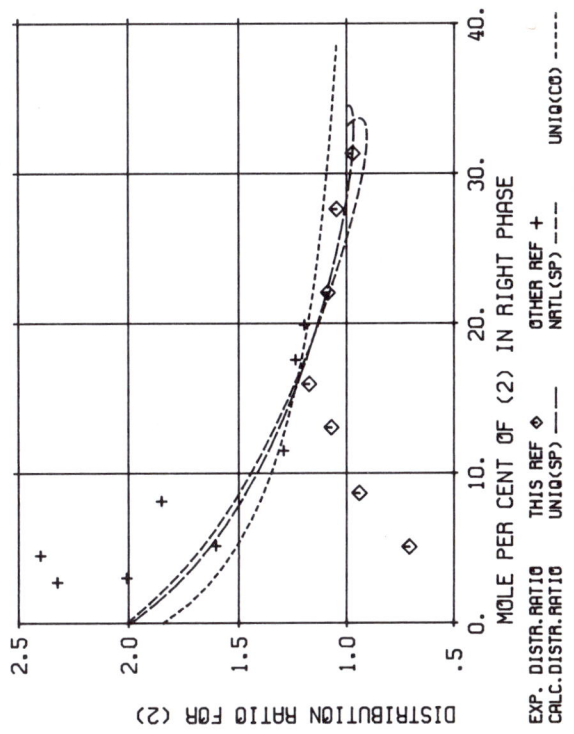

(2) CH4O   METHANOL

(3) C5H8O2   PROPENOIC ACID,2-METHYL,METHYL ESTER

KOOI J.
RECL.TRAV.CHIM.PAYS-BAS. 68(1949)34

TEMPERATURE = 25.0 DEG C   TYPE OF SYSTEM = 1

EXPERIMENTAL TIE LINES IN MOLE PCT

LEFT PHASE			RIGHT PHASE		
(1)	(2)	(3)	(1)	(2)	(3)
96.265	3.605	0.130	1.588	5.061	93.351
91.535	8.174	0.291	5.478	8.681	85.841
85.420	13.987	0.593	9.283	13.050	77.667
80.266	18.749	0.986	13.079	15.934	70.987
74.176	23.996	1.828	17.643	22.045	60.312
67.085	28.942	3.973	29.034	27.627	43.339
64.326	30.445	5.229	34.829	31.333	33.838

SPECIFIC MODEL PARAMETERS IN KELVIN

		UNIQUAC		NRTL(ALPHA=.2)	
I	J	AIJ	AJI	AIJ	AJI
1	2	-442.48	-122.69	-146.52	365.88
1	3	450.03	422.95	2585.4	476.27
2	3	-266.40	257.74	-139.42	600.55

R1 = 0.9200   R2 = 1.4311   R3 = 3.9215
Q1 = 1.400   Q2 = 1.432   Q3 = 3.564

MEAN DEV. BETWEEN CALC. AND EXP. CONC. IN MOLE PCT

UNIQUAC (SPECIFIC PARAMETERS)   1.19
NRTL (SPECIFIC PARAMETERS)   1.44
UNIQUAC (COMMON PARAMETERS)   4.19

CH$_4$O-C$_5$H$_8$O$_2$

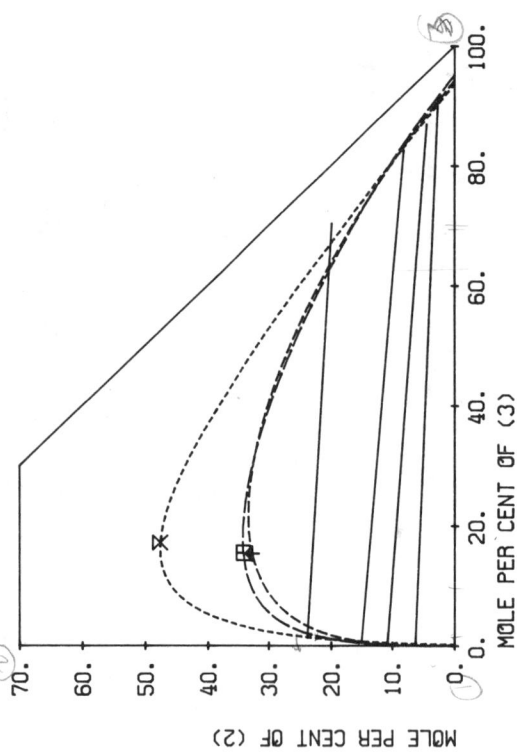

MOLE PER CENT OF (3)

MOLE PER CENT OF (2)

EXP.TIE LINE ——— CALC.BINODAL UNIQ(SP) ⊟ ——— NRTL(SP) —— ✦ UNIQ(CC) ✕ - - - - -
CALC.PLAIT P.

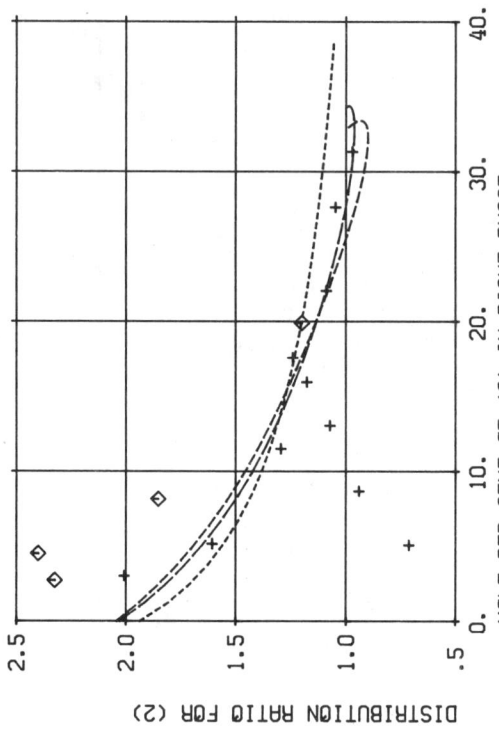

DISTRIBUTION RATIO FOR (2)

MOLE PER CENT OF (2) IN RIGHT PHASE

EXP. DISTR.RATIO ◇   THIS REF ◇   OTHER REF +
CALC. DISTR. RATIO   UNIQ(SP) ———   NRTL(SP) — — —   UNIQ(CC) - - - - -

(1) H2O   WATER
(2) CH4O   METHANOL
(3) C5H8O2   PROPENOIC ACID,2-METHYL,METHYL ESTER

FROLOV A.F., ET AL.
IZV.VYSSH.UCHEBN.ZAVED.KHIM.KHIM.TEKHNOL. 8(1965)570

TEMPERATURE = 20.0 DEG C   TYPE OF SYSTEM = 1

EXPERIMENTAL TIE LINES IN MOLE PCT

        LEFT PHASE   RIGHT PHASE
   (1)   (2)   (3)   (1)   (2)   (3)

99.718   0.0   0.282   6.324   0.0   93.676
93.320   6.296 0.384   6.060   2.708 91.231
88.762 10.801 0.437   8.501   4.499 87.000
84.379 15.087 0.534   9.180   8.150 82.670
74.816 23.838 1.346   9.610 19.894 70.496

SPECIFIC MODEL PARAMETERS IN KELVIN

            UNIQUAC   NRTL(ALPHA=.2)
I J   AIJ   AJI   AIJ   AJI
----
1 2 -442.48 -122.69   -146.52   365.88
1 3   450.03   422.95   2585.4   476.27
2 3 -266.40   257.74   -139.42   600.55

R1 = 0.9200   R2 = 1.4311   R3 = 3.9215
Q1 = 1.400   Q2 = 1.432   Q3 = 3.564

MEAN DEV. BETWEEN CALC. AND EXP. CONC. IN MOLE PCT

UNIQUAC (SPECIFIC PARAMETERS)   1.58
NRTL   (SPECIFIC PARAMETERS)   1.39
UNIQUAC (COMMON PARAMETERS)   1.15

$CH_4O-C_5H_8O_2$

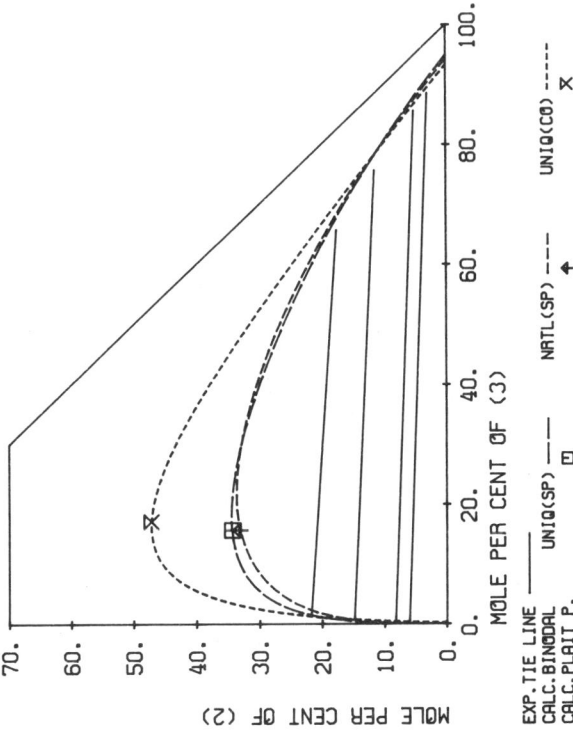

MOLE PER CENT OF (3)

MOLE PER CENT OF (2)

EXP.TIE LINE ———
CALC.BINODAL UNIQ(SP) ⊟ — — —   NRTL(SP) ◆ — — —   UNIQ(CO) ✕ -----
CALC.PLAIT P.

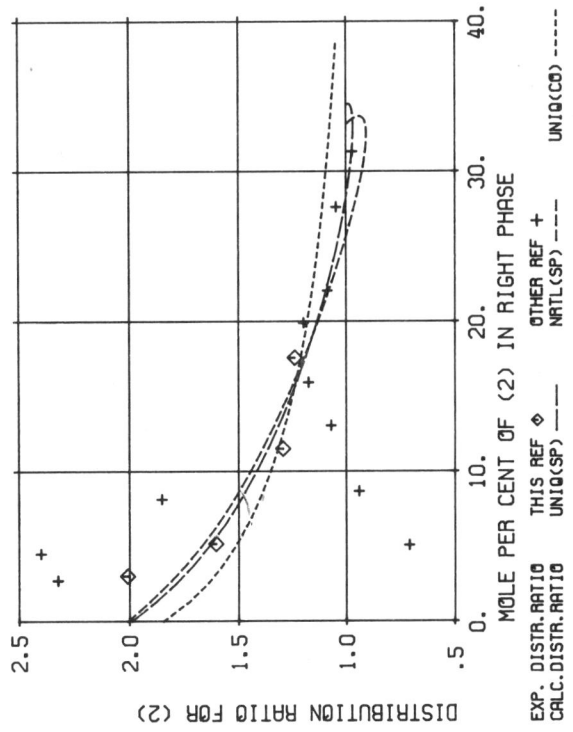

DISTRIBUTION RATIO FOR (2)

MOLE PER CENT OF (2) IN RIGHT PHASE

EXP. DISTR.RATIO   THIS REF ◇   OTHER REF +
CALC.DISTR.RATIO   UNIQ(SP) — — —   NRTL(SP) — — —   UNIQ(CO) -----

(2) CH4O   METHANOL
(3) C5H8O2   PROPENOIC ACID,2-METHYL,METHYL ESTER

CHUBAROV G.A., DANOV S.M., BROVKINA G.V., KUPRIYANOV T.V.
ZH.PRIKL.KHIM.(LENINGRAD) 51(1978)443

TEMPERATURE = 25.0 DEG C   TYPE OF SYSTEM = 1

EXPERIMENTAL TIE LINES IN MOLE PCT

	LEFT PHASE			RIGHT PHASE	
(1)	(2)	(3)	(1)	(2)	(3)
93.535	6.004	0.461	8.356	2.990	88.654
91.196	8.265	0.539	9.151	5.145	85.704
84.265	14.873	0.862	12.722	11.496	75.781
76.830	21.773	1.398	16.557	17.572	65.871

SPECIFIC MODEL PARAMETERS IN KELVIN

		UNIQUAC		NRTL(ALPHA=.2)	
I J	AIJ	AJI	AIJ	AJI	
1 2	-442.48	-122.69	-146.52	365.88	
1 3	450.03	422.95	2585.4	476.27	
2 3	-266.40	257.74	-139.42	600.55	

R1 = 0.9200   R2 = 1.4311   R3 = 3.9215
Q1 = 1.400   Q2 = 1.432   Q3 = 3.564

MEAN DEV. BETWEEN CALC. AND EXP. CONC. IN MOLE PCT

UNIQUAC (SPECIFIC PARAMETERS)   1.03
NRTL   (SPECIFIC PARAMETERS)   1.08
UNIQUAC (COMMON PARAMETERS)   1.19

CH_4O-$C_5H_8O_2$

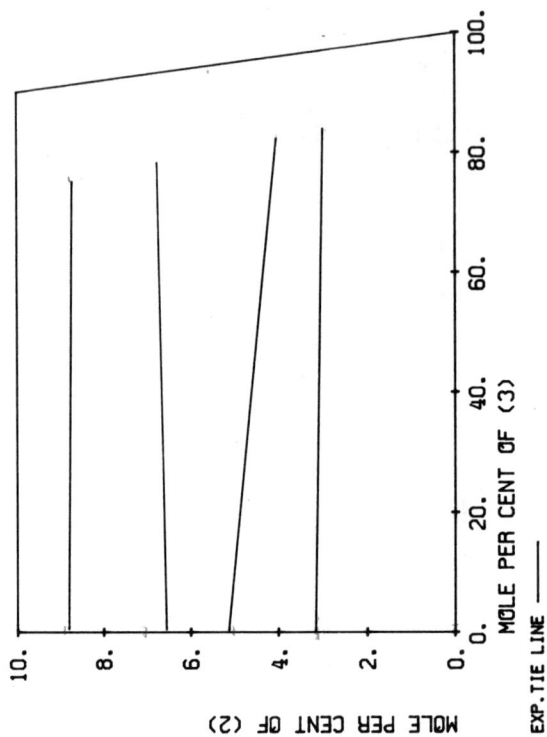

(1) H2O   WATER

(2) CH4O   METHANOL

(3) C5H8O2   PROPENOIC ACID,2-METHYL,METHYL ESTER

CHUBAROV G.A., DANOV S.M., BROVKINA G.V., KUPRIYANOV T.V.
ZH.PRIKL.KHIM.(LENINGRAD) 51(1978)443

TEMPERATURE = 70.0 DEG C   TYPE OF SYSTEM = 1

EXPERIMENTAL TIE LINES IN MOLE PCT

	LEFT PHASE			RIGHT PHASE	
(1)	(2)	(3)	(1)	(2)	(3)
96.451	3.164	0.384	13.090	2.999	83.911
94.412	5.130	0.458	13.635	4.035	82.330
92.933	6.554	0.512	15.011	6.752	78.237
90.600	8.808	0.591	16.170	8.713	75.117

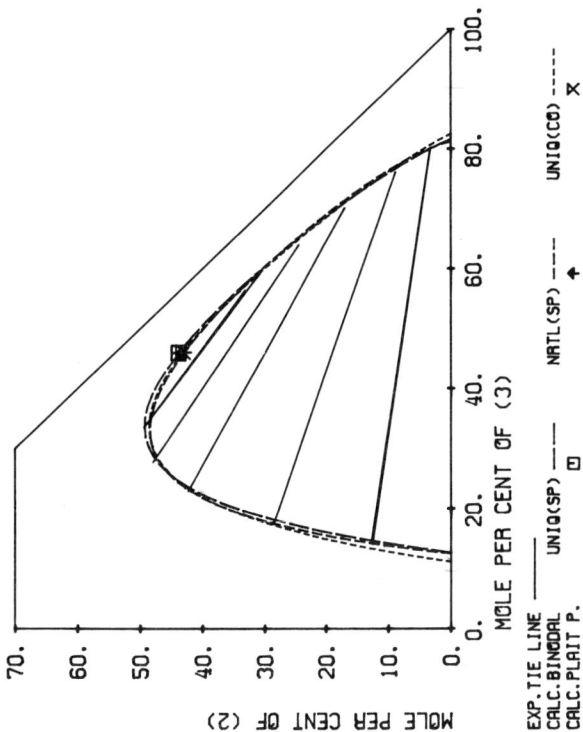

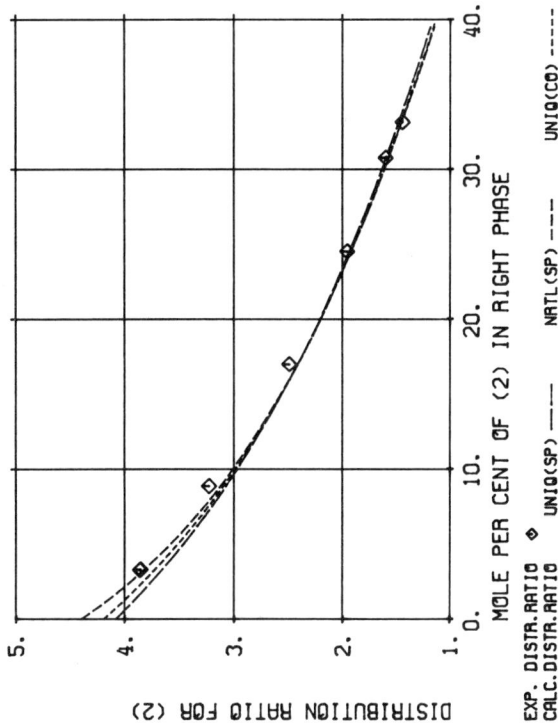

CH$_4$O-C$_5$H$_{10}$

(2) C5H10   CYCLOPENTANE

(3) CH4O   METHANOL

TAKEUCHI S., NITTA T., KATAYAMA T.
J.CHEM.ENG.JPN. 8(1975)248

TEMPERATURE = 25.0 DEG C   TYPE OF SYSTEM = 1

EXPERIMENTAL TIE LINES IN MOLE PCT

	LEFT PHASE			RIGHT PHASE	
(1)	(2)	(3)	(1)	(2)	(3)
87.090	0.0	12.910	17.520	0.0	82.480
73.340	12.580	14.080	16.620	3.260	80.120
72.590	12.710	14.700	16.590	3.300	80.110
54.020	28.610	17.370	14.990	8.860	76.150
34.760	42.300	22.940	12.850	16.970	70.180
24.280	47.930	27.790	11.540	24.500	63.960
17.080	49.360	33.560	10.010	30.750	59.240
16.500	48.050	35.450	10.470	33.130	56.400

SPECIFIC MODEL PARAMETERS IN KELVIN

		UNIQUAC		NRTL(ALPHA=.2)	
I	J	AIJ	AJI	AIJ	AJI
1	2	-15.774	-16.726	-97.419	-56.868
1	3	647.16	3.4857	544.82	318.02
2	3	485.41	-6.0894	361.38	234.33

R1 = 4.0464   R2 = 3.3720   R3 = 1.4311
Q1 = 3.240   Q2 = 2.700   Q3 = 1.432

MEAN DEV. BETWEEN CALC. AND EXP. CONC. IN MOLE PCT

UNIQUAC (SPECIFIC PARAMETERS)   0.51
NRTL   (SPECIFIC PARAMETERS)   0.58
UNIQUAC (COMMON PARAMETERS)   0.59

CH₄O-C₅H₁₀O₂

(1) H2O   WATER
(2) CH4O   METHANOL
(3) C5H10O2   PROPANOIC ACID,ETHYL ESTER

JAGANNADHA RAO R., VENKATA RAO C.
J.APPL.CHEM. 7(1957)435

TEMPERATURE = 30.0 DEG C   TYPE OF SYSTEM = 1

EXPERIMENTAL TIE LINES IN MOLE PCT

LEFT PHASE			RIGHT PHASE		
(1)	(2)	(3)	(1)	(2)	(3)
94.801	4.808	0.391	8.717	3.460	87.823
89.870	9.587	0.543	9.439	6.704	83.858
80.022	18.702	1.276	13.564	16.975	69.461
67.978	28.254	3.767	24.033	26.032	49.935

MEAN DEV. BETWEEN CALC. AND EXP. CONC. IN MOLE PCT

UNIQUAC (COMMON PARAMETERS)   0.95

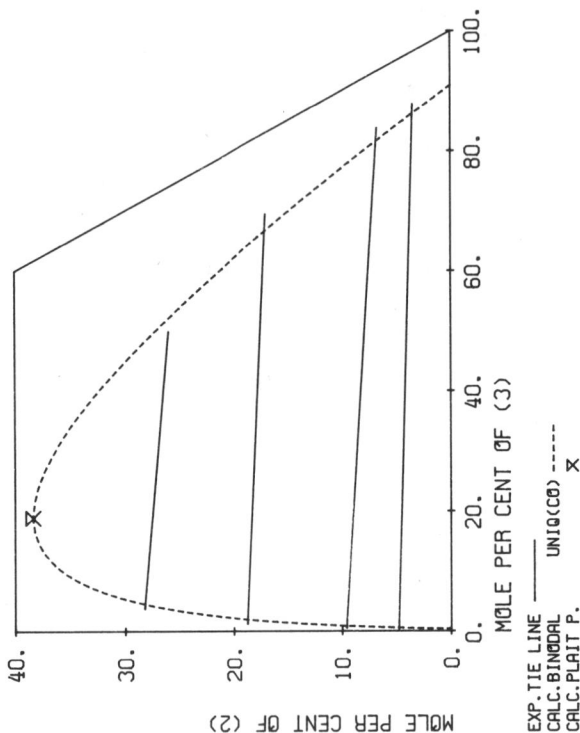

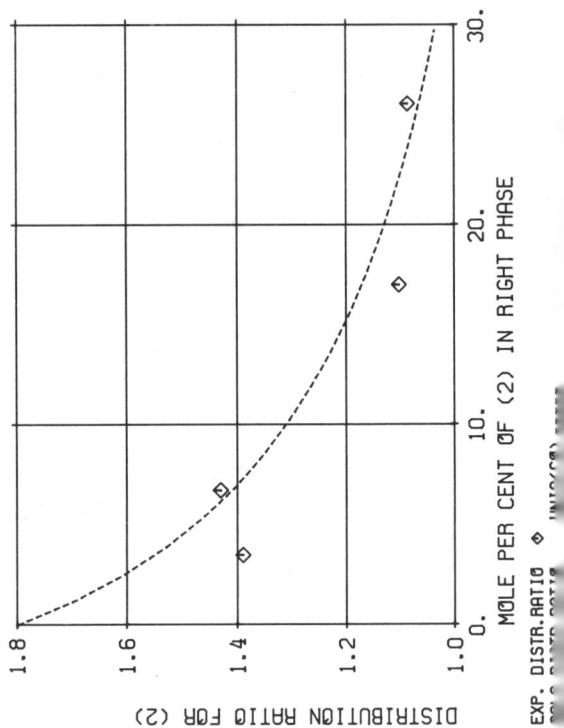

$CH_4O-C_5H_{12}O$

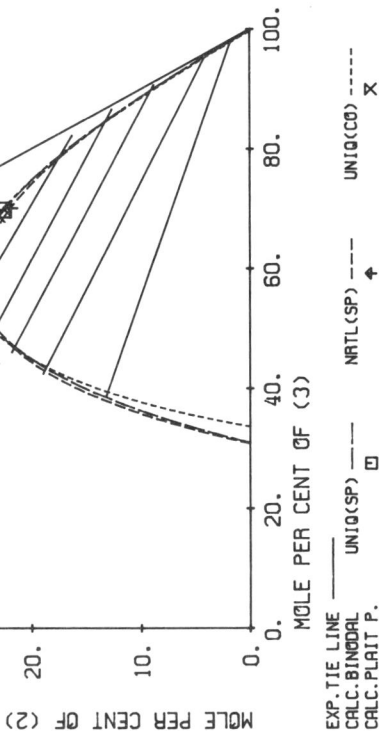

MOLE PER CENT OF (3)

MOLE PER CENT OF (2)

EXP.TIE LINE ——— UNIQ(SP) ——— NRTL(SP) ✦ UNIQ(CO) -----
CALC.BINODAL   ⊟   ✕
CALC.PLAIT P.

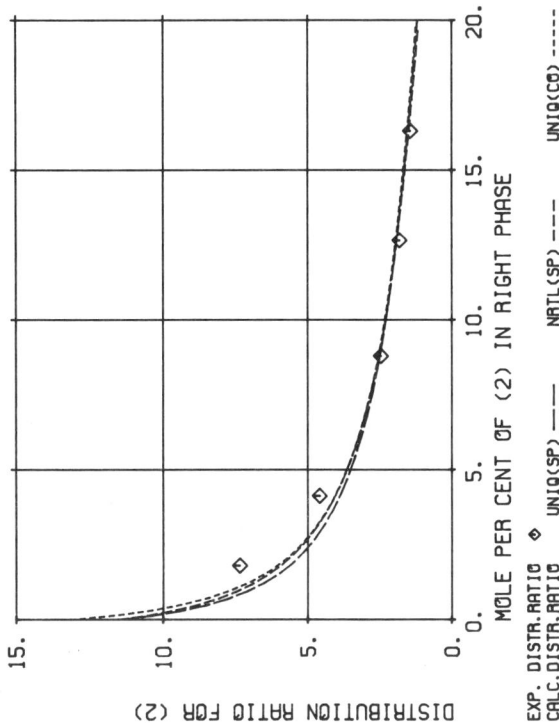

MOLE PER CENT OF (2) IN RIGHT PHASE

DISTRIBUTION FOR (2)

EXP. DISTR.RATIO ◇   UNIQ(SP) ———   NRTL(SP) ———   UNIQ(CO) -----
CALC.DISTR.RATIO

---

(2) CH4O   METHANOL
(3) H2O   WATER

LEBEDINSKAYA N.A., ET AL.
KHIM.PROM-ST.(MOSCOW)(1976)16

TEMPERATURE = 20.0 DEG C   TYPE OF SYSTEM = 1

EXPERIMENTAL TIE LINES IN MOLE PCT

	LEFT PHASE			RIGHT PHASE	
(1)	(2)	(3)	(1)	(2)	(3)
48.101	13.217	38.682	0.378	1.793	97.829
38.638	18.935	42.427	0.429	4.128	95.443
32.180	21.793	46.027	0.466	8.798	90.735
26.136	23.104	50.760	0.690	12.655	86.655
17.455	23.802	58.743	1.451	16.302	82.247

SPECIFIC MODEL PARAMETERS IN KELVIN

		UNIQUAC		NRTL(ALPHA=.2)	
I J	AIJ	AJI		AIJ	AJI
1 2	-109.19	-343.73		-508.67	-233.31
1 3	-69.052	310.11		-135.22	1874.2
2 3	-333.40	115.31		-245.85	155.92

R1 = 4.1287   R2 = 1.4311   R3 = 0.9200
Q1 = 3.592   Q2 = 1.432   Q3 = 1.400

MEAN DEV. BETWEEN CALC. AND EXP. CONC. IN MOLE PCT

UNIQUAC (SPECIFIC PARAMETERS)   0.81
NRTL   (SPECIFIC PARAMETERS)   0.78
UNIQUAC (COMMON PARAMETERS)   0.88

CH$_4$O-C$_6$H$_6$

MOLE PER CENT OF (3)

MOLE PER CENT OF (2)

EXP.TIE LINE	UNIQ(SP)	NRTL(SP)
CALC.BINODAL		
CALC.PLAIT P.		

MOLE PER CENT OF (2) IN RIGHT PHASE

DISTRIBUTION RATIO FOR (2)

EXP. DISTR.RATIO

---

(1) C7H16   HEPTANE
(2) C6H6   BENZENE
(3) CH4O   METHANOL

WITTRIG T.S.
THESIS ILLINOIS 1977

TEMPERATURE =   6.8 DEG C   TYPE OF SYSTEM = 1

EXPERIMENTAL TIE LINES IN MOLE PCT

	LEFT PHASE			RIGHT PHASE	
(1)	(2)	(3)	(1)	(2)	(3)
89.800	0.0	10.200	8.100	0.0	91.900
80.300	9.800	9.900	8.600	4.200	87.200
73.300	11.900	14.800	9.400	4.800	85.800
73.000	12.300	14.700	9.800	5.800	84.400
68.200	15.800	16.000	11.200	7.300	81.500
64.200	16.400	19.400	11.700	7.700	80.600
58.400	20.200	21.400	13.800	10.500	75.700

SPECIFIC MODEL PARAMETERS IN KELVIN

		UNIQUAC		NRTL(ALPHA=.2)	
I	J	AIJ	AJI	AIJ	AJI
1	2	55.852	-90.662	27.085	-266.12
1	3	689.14	8.8469	453.96	516.13
2	3	215.71	-71.223	-34.497	98.175

R1 = 5.1742   R2 = 3.1878   R3 = 1.4311
Q1 = 4.396   Q2 = 2.400   Q3 = 1.432

MEAN DEV. BETWEEN CALC. AND EXP. CONC. IN MOLE PCT

UNIQUAC (SPECIFIC PARAMETERS)   0.69
NRTL   (SPECIFIC PARAMETERS)   0.76

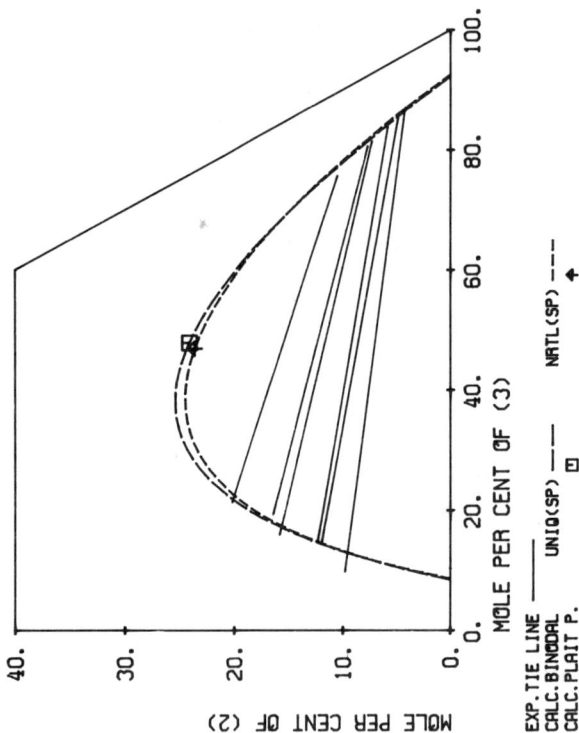

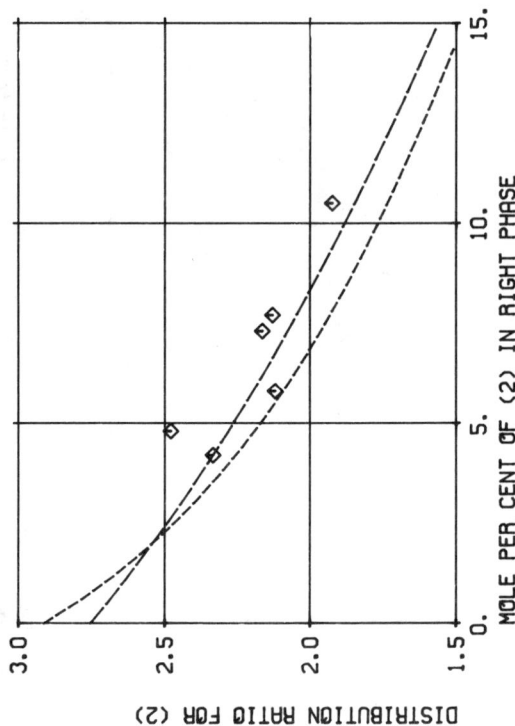

$CH_4O-C_6H_6$

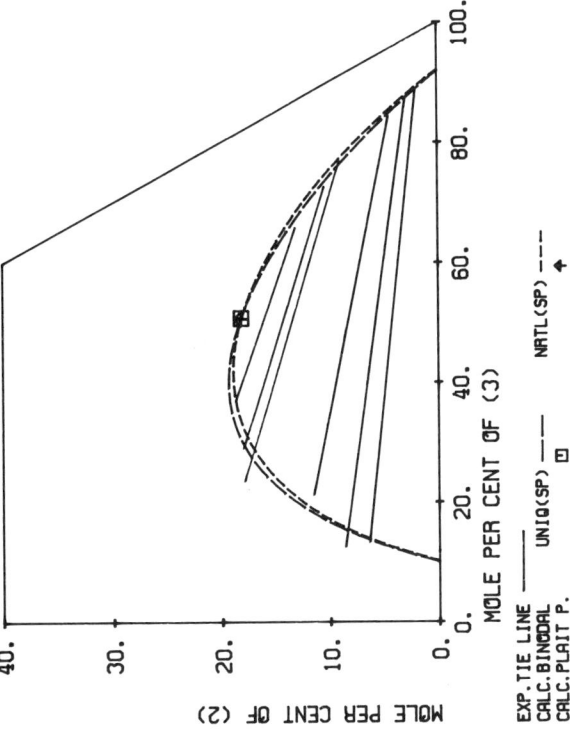

MOLE PER CENT OF (3)

MOLE PER CENT OF (2)

EXP.TIE LINE ——— UNIQ(SP) ⊞ NRTL(SP) —— ✦
CALC.BINODAL
CALC.PLAIT P.

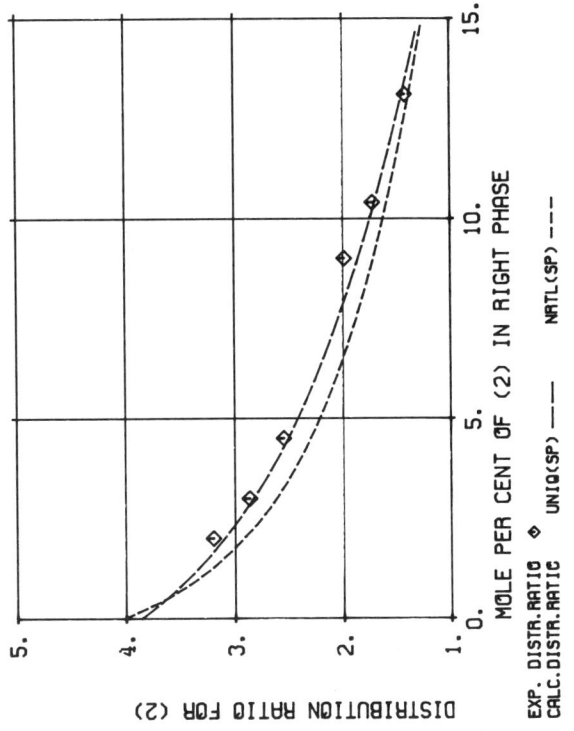

MOLE PER CENT OF (2) IN RIGHT PHASE

DISTRIBUTION RATIO FOR (2)

EXP.DISTR.RATIO ——— UNIQ(SP) ◈ NRTL(SP) ————
CALC.DISTR.RATIO

(2) C6H6   BENZENE

(3) CH4O   METHANOL

WITTRIG T.S.
THESIS ILLINOIS 1977

TEMPERATURE = 13.8 DEG C   TYPE OF SYSTEM = 1

EXPERIMENTAL TIE LINES IN MOLE PCT

	LEFT PHASE			RIGHT PHASE	
(1)	(2)	(3)	(1)	(2)	(3)
88.400	0.0	11.600	8.000	0.0	92.000
80.200	6.400	13.400	9.300	2.000	88.700
78.700	8.600	12.700	9.900	3.000	87.100
67.200	11.500	21.300	11.200	4.500	84.300
58.500	17.900	23.600	14.100	9.000	76.900
52.900	18.000	29.100	17.000	10.400	72.600
44.300	18.700	37.000	21.100	13.100	65.800

SPECIFIC MODEL PARAMETERS IN KELVIN

		UNIQUAC		NRTL(ALPHA=.2)	
I	J	AIJ	AJI	AIJ	AJI
1	2	92.194	-199.65	172.78	-521.56
1	3	678.79	9.9500	424.15	540.07
2	3	201.68	-143.65	-28.354	-61.370

R1 = 5.1742   R2 = 3.1878   R3 = 1.4311
Q1 = 4.396   Q2 = 2.400   Q3 = 1.432

MEAN DEV. BETWEEN CALC. AND EXP. CONC. IN MOLE PCT

UNIQUAC (SPECIFIC PARAMETERS)   0.74
NRTL   (SPECIFIC PARAMETERS)   0.87

CH₄O-C₆H₆

(1) C7H16   HEPTANE

(2) C6H6   BENZENE

(3) CH4O   METHANOL

WITTRIG T.S.
THESIS ILLINOIS 1977

TEMPERATURE = 32.8 DEG C   TYPE OF SYSTEM = 1

EXPERIMENTAL TIE LINES IN MOLE PCT

	LEFT PHASE			RIGHT PHASE	
(1)	(2)	(3)	(1)	(2)	(3)
78.800	0.0	21.200	11.700	0.0	88.300
70.300	4.800	24.900	14.800	2.100	83.100
66.800	7.200	26.000	14.900	3.300	81.800
63.700	7.900	28.400	16.800	4.000	79.200
60.900	8.800	30.300	18.800	4.600	76.600
49.200	8.800	42.000	23.700	7.000	69.300

SPECIFIC MODEL PARAMETERS IN KELVIN

		UNIQUAC		NRTL(ALPHA=.2)	
I J	AIJ	AJI		AIJ	AJI
1 2	-108.03	216.66		157.35	-590.50
1 3	639.76	3.4314		324.56	568.52
2 3	78.022	144.39		-125.53	-169.07

R1 = 5.1742   R2 = 3.1878   R3 = 1.4311
Q1 = 4.396   Q2 = 2.400   Q3 = 1.432

MEAN DEV. BETWEEN CALC. AND EXP. CONC. IN MOLE PCT

UNIQUAC (SPECIFIC PARAMETERS)   0.72
NRTL   (SPECIFIC PARAMETERS)   0.72

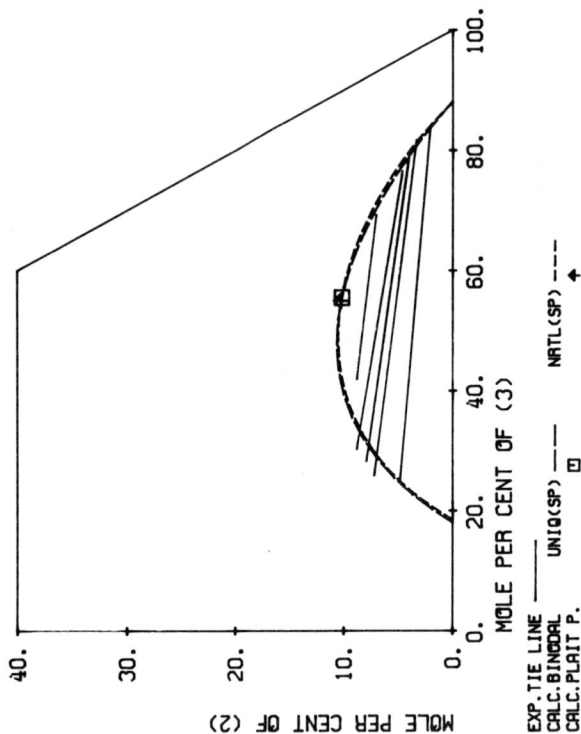

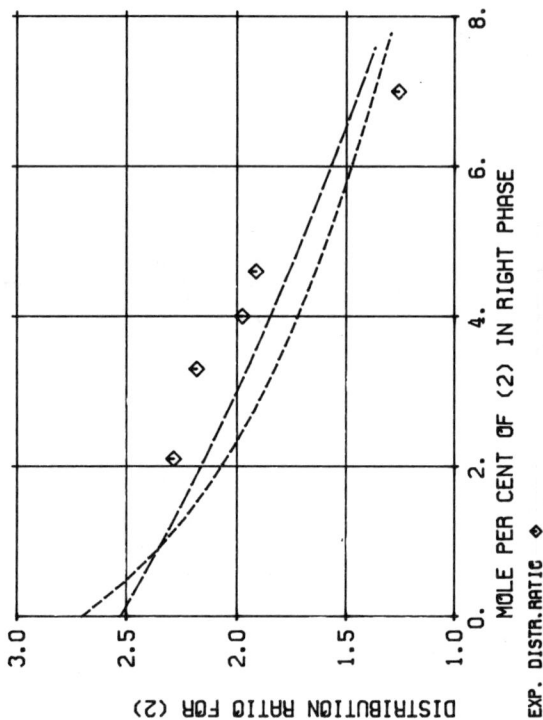

$CH_4O-C_6H_6$

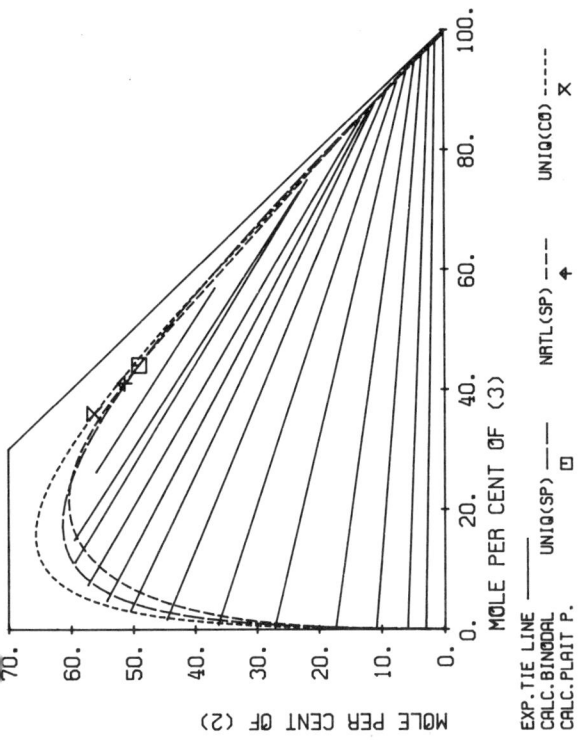

MOLE PER CENT OF (3)

MOLE PER CENT OF (2)

EXP.TIE LINE ——— UNIQ(SP) ——— NRTL(SP) ——— UNIQ(CO) ———
CALC.BINODAL □ ✦ ✕
CALC.PLAIT P.

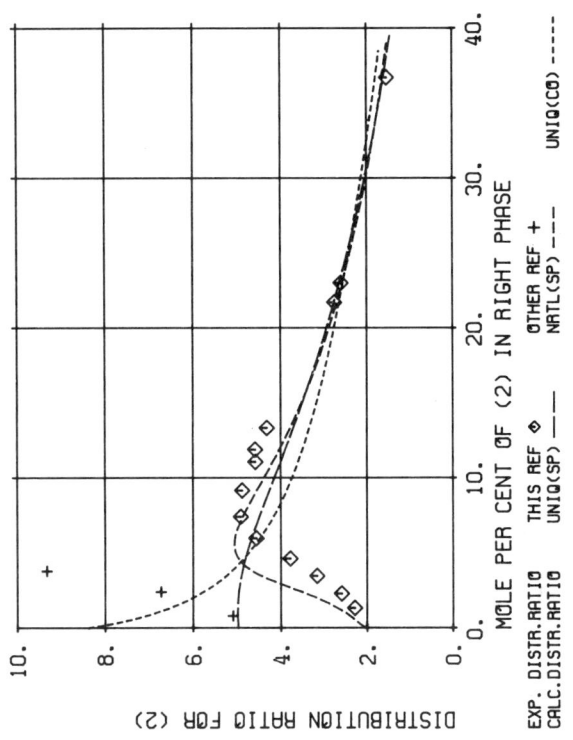

MOLE PER CENT OF (2) IN RIGHT PHASE

DISTRIBUTION RATIO FOR (2)

EXP. DISTR.RATIO THIS REF ◇ OTHER REF +
CALC.DISTR.RATIO UNIQ(SP) ——— NRTL(SP) ——— UNIQ(CO) ———

(2) CH4O   METHANOL
(3) C6H6   BENZENE

UDOVENKO V.V., MAZANKO T.F.
ZH.FIZ.KHIM. 37(1963)2324

TEMPERATURE = 30.0 DEG C   TYPE OF SYSTEM = 1

EXPERIMENTAL TIE LINES IN MOLE PCT

	LEFT PHASE			RIGHT PHASE	
(1)	(2)	(3)	(1)	(2)	(3)
96.914	3.027	0.059	0.215	1.328	98.457
94.033	5.894	0.073	0.426	2.278	97.296
89.043	10.881	0.075	0.634	3.446	95.920
82.463	17.379	0.158	0.838	4.595	94.567
72.481	27.123	0.395	1.037	5.950	93.013
62.875	36.310	0.816	1.232	7.388	91.380
53.700	44.600	1.700	1.219	9.136	89.646
46.399	50.494	3.106	1.402	11.040	87.557
40.860	54.319	4.821	1.592	11.860	86.548
35.236	57.341	7.423	1.772	13.284	84.945
29.389	59.386	11.225	3.305	21.677	75.019
25.521	59.495	14.985	3.978	22.975	73.048
17.791	55.993	26.215	6.370	36.710	56.920

SPECIFIC MODEL PARAMETERS IN KELVIN

		UNIQUAC		NRTL(ALPHA=.2)	
I	J	AIJ	AJI	AIJ	AJI
1	2	-284.24	266.87	1347.9	-693.61
1	3	315.18	753.90	2106.1	1391.5
2	3	-51.480	428.62	508.70	118.62

R1 = 0.9200   R2 = 1.4311   R3 = 3.1878
Q1 = 1.400   Q2 = 1.432   Q3 = 2.400

MEAN DEV. BETWEEN CALC. AND EXP. CONC. IN MOLE PCT

UNIQUAC (SPECIFIC PARAMETERS)   1.38
NRTL   (SPECIFIC PARAMETERS)   1.31
UNIQUAC (COMMON PARAMETERS)   2.96

CH$_4$O-C$_6$H$_6$

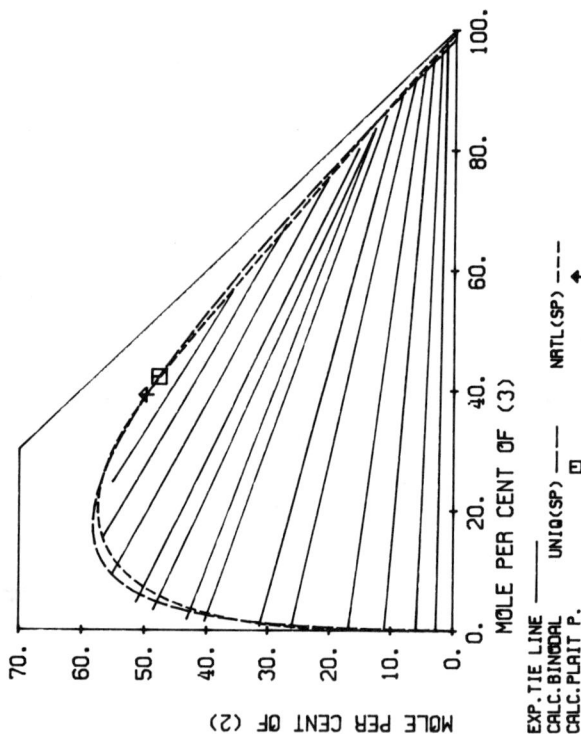

MOLE PER CENT OF (3)

MOLE PER CENT OF (2)

EXP.TIE LINE ———
CALC.BINODAL ——— □ UNIQ(SP) ——— □
CALC.PLAIT P. NRTL(SP) ——— ✦

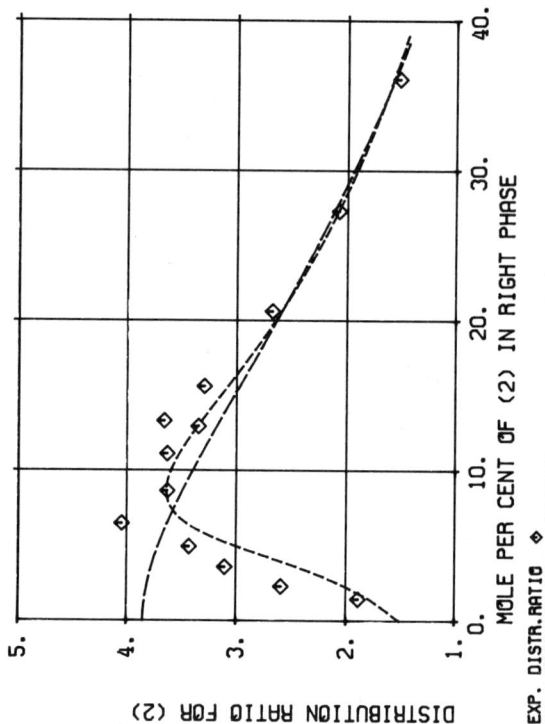

MOLE PER CENT OF (2) IN RIGHT PHASE

DISTRIBUTION RATIO FOR (2)

EXP. DISTR.RATIO ◇

(1) H2O   WATER
(2) CH4O   METHANOL
(3) C6H6   BENZENE

UDOVENKO V.V., MAZANKO T.F.
ZH.FIZ.KHIM. 37(1963)2324

TEMPERATURE = 45.0 DEG C   TYPE OF SYSTEM = 1

EXPERIMENTAL TIE LINES IN MOLE PCT

	LEFT PHASE			RIGHT PHASE	
(1)	(2)	(3)	(1)	(2)	(3)
97.208	2.733	0.059	0.429	1.446	98.126
94.033	5.894	0.073	0.639	2.274	97.088
88.842	11.033	0.126	0.843	3.557	95.600
82.920	16.843	0.236	1.457	4.914	93.629
73.325	26.111	0.564	2.051	6.459	91.490
67.876	31.303	0.821	2.421	8.621	88.958
58.158	40.300	1.543	3.158	11.097	85.745
54.960	43.169	1.871	3.501	12.903	83.596
48.032	48.562	3.406	3.868	13.267	82.866
44.010	51.327	4.662	4.178	15.590	80.232
35.608	55.180	9.213	4.396	20.595	75.009
27.961	56.619	15.421	5.198	27.277	67.525
20.265	55.071	24.664	7.585	36.075	56.340

SPECIFIC MODEL PARAMETERS IN KELVIN

		UNIQUAC		NRTL(ALPHA=.2)	
I J	AIJ	AJI		AIJ	AJI
1 2	-140.47	144.84		1303.5	-648.68
1 3	-331.19	583.44		2827.9	1190.2
2 3	-56.442	495.96		535.76	109.02

R1 = 0.9200   R2 = 1.4311   R3 = 3.1878
Q1 = 1.400   Q2 = 1.432   Q3 = 2.400

MEAN DEV. BETWEEN CALC. AND EXP. CONC. IN MOLE PCT

UNIQUAC (SPECIFIC PARAMETERS)   0.88
NRTL   (SPECIFIC PARAMETERS)   0.76

CH$_4$O-C$_6$H$_6$

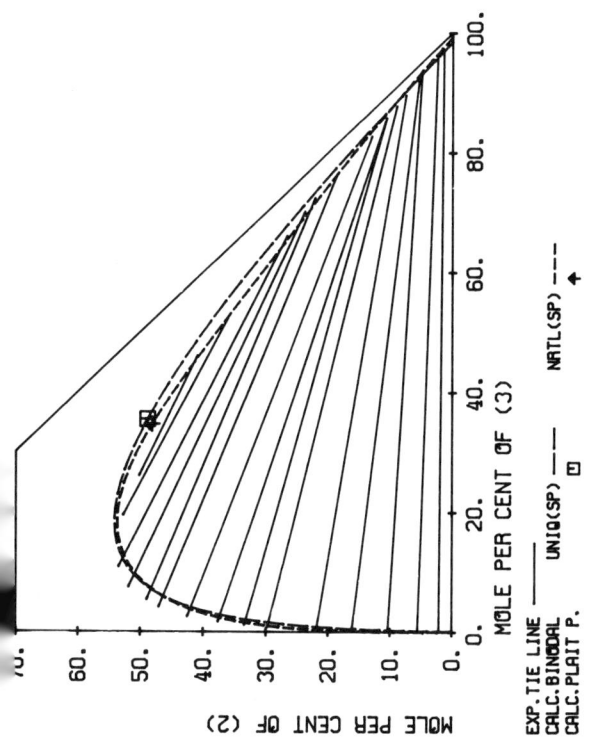

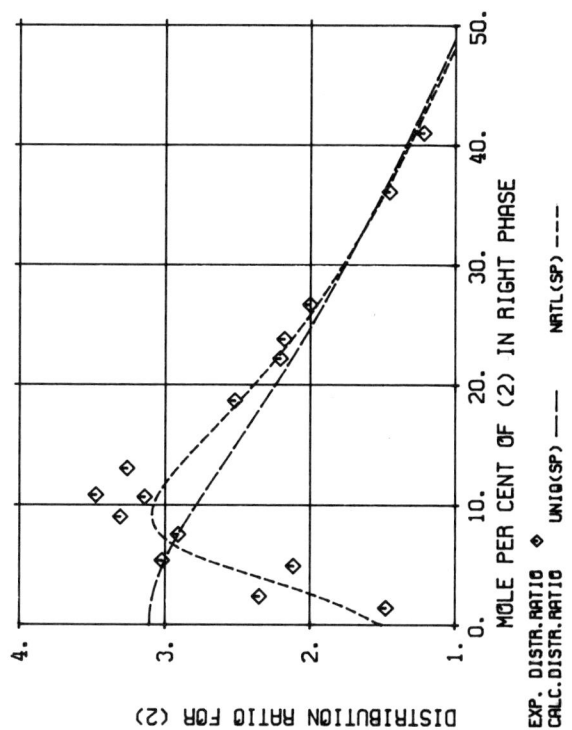

(2) CH4O	METHANOL	
(3) C6H6	BENZENE	

UDOVENKO V.V., MAZANKO T.F.
ZH.FIZ.KHIM. 37(1963)2324

TEMPERATURE = 60.0 DEG C   TYPE OF SYSTEM = 1

EXPERIMENTAL TIE LINES IN MOLE PCT

	LEFT PHASE			RIGHT PHASE	
(1)	(2)	(3)	(1)	(2)	(3)
97.810	2.120	0.070	1.277	1.436	97.287
94.283	5.596	0.121	1.270	2.380	96.350
89.471	10.379	0.150	1.662	4.907	93.431
83.602	16.137	0.262	2.065	5.342	92.593
77.667	21.868	0.465	2.835	7.514	89.651
69.351	29.777	0.872	3.200	8.995	87.805
65.330	33.440	1.230	3.551	10.649	85.800
60.584	37.638	1.778	3.928	10.823	85.248
54.883	42.501	2.616	4.248	13.027	82.725
48.774	47.061	4.166	5.162	18.657	76.181
45.601	48.997	5.402	5.383	22.198	72.419
40.665	51.863	7.472	5.997	23.801	70.202
35.721	53.459	10.819	7.171	26.686	66.143
27.813	52.707	19.479	11.510	36.104	52.386
23.734	50.202	26.063	12.864	40.988	46.147

SPECIFIC MODEL PARAMETERS IN KELVIN

	UNIQUAC			NRTL(ALPHA=.2)	
I J	AIJ	AJI		AIJ	AJI
1 2	-334.07	240.38		1356.6	-703.07
1 3	-283.21	633.85		3008.7	1161.3
2 3	-85.755	364.15		590.53	47.990

R1 = 0.9200   R2 = 1.4311   R3 = 3.1878
Q1 = 1.400   Q2 = 1.432   Q3 = 2.400

MEAN DEV. BETWEEN CALC. AND EXP. CONC. IN MOLE PCT

UNIQUAC	(SPECIFIC PARAMETERS)	1.21
NRTL	(SPECIFIC PARAMETERS)	0.76

CH$_4$O-C$_6$H$_6$

(1) H2O   WATER
(2) CH4O   METHANOL
(3) C6H6   BENZENE

MERTSLIN R.V., KAMAEVSKAYA L.A., NIKURASHINA N.I.
ZH.FIZ.KHIM. 40(1966)2539

TEMPERATURE = 26.0 DEG C   TYPE OF SYSTEM = 1

EXPERIMENTAL TIE LINES IN MOLE PCT (GRAPH.INTERPOL.)

	LEFT PHASE			RIGHT PHASE	
(1)	(2)	(3)	(1)	(2)	(3)
95.886	4.067	0.048	0.516	0.798	98.687
83.843	16.052	0.104	0.807	2.389	96.804
63.984	35.282	0.734	1.051	3.783	95.166

SPECIFIC MODEL PARAMETERS IN KELVIN

		UNIQUAC		NRTL(ALPHA=.2)	
I J	AIJ	AJI		AIJ	AJI
1 2	-284.24	266.87		1347.9	-693.61
1 3	315.18	753.90		2106.1	1391.5
2 3	-51.480	428.62		508.70	118.62

R1 = 0.9200   R2 = 1.4311   R3 = 3.1878
Q1 = 1.400   Q2 = 1.432   Q3 = 2.400

MEAN DEV. BETWEEN CALC. AND EXP. CONC. IN MOLE PCT

UNIQUAC (SPECIFIC PARAMETERS)   1.25
NRTL   (SPECIFIC PARAMETERS)   1.03
UNIQUAC (COMMON PARAMETERS)   1.58

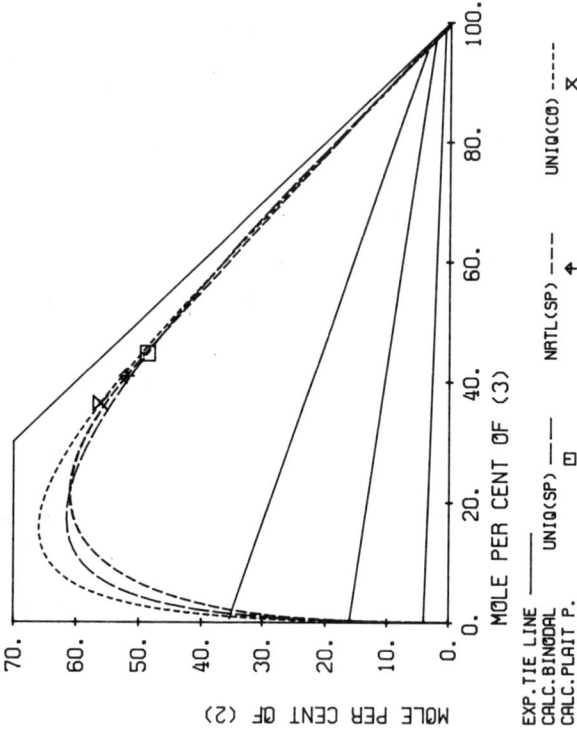

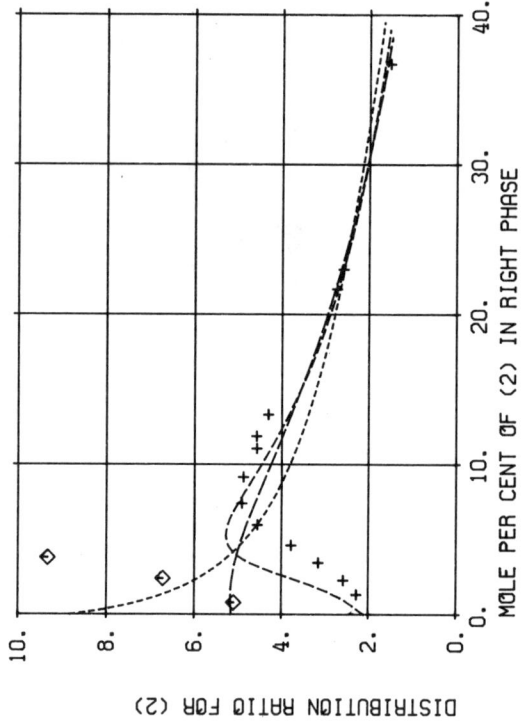

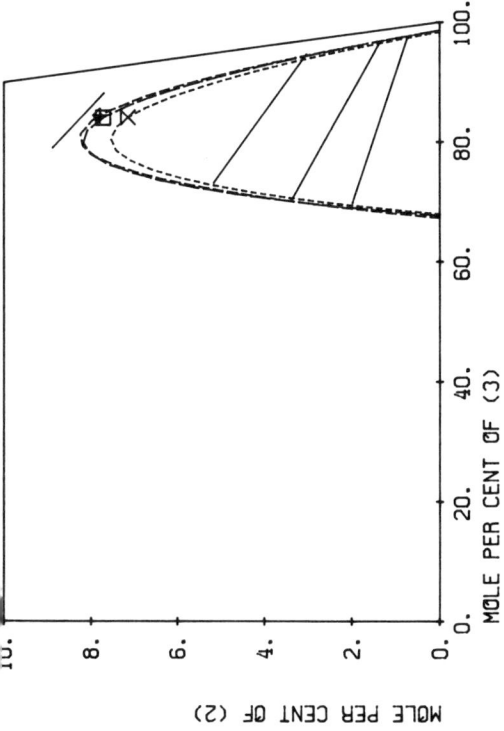

MOLE PER CENT OF (3)

MOLE PER CENT OF (2)

EXP.TIE LINE ——   UNIQ(SP) ☐   NRTL(SP) ----   UNIQ(CO) -----
CALC.BINODAL
CALC.PLAIT P.

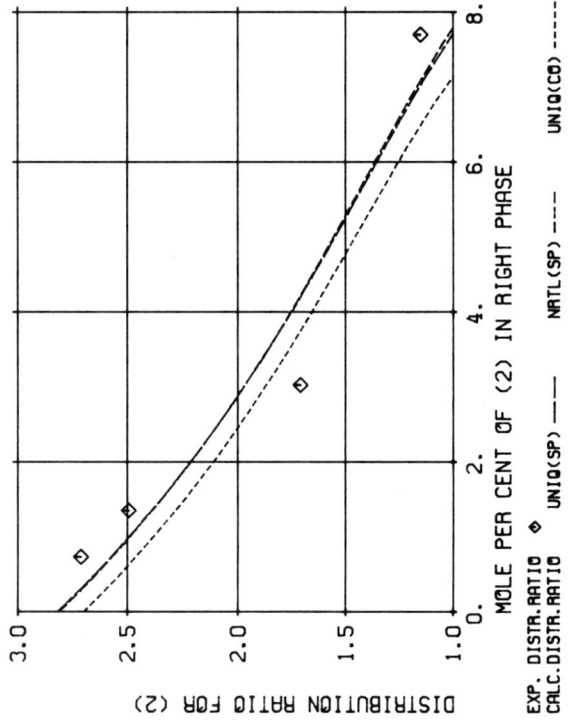

MOLE PER CENT OF (2) IN RIGHT PHASE

DISTRIBUTION RATIO FOR (2)

EXP. DISTR.RATIO ◇   UNIQ(SP) ◇   NRTL(SP) ----   UNIQ(CO) -----
CALC.DISTR.RATIO

(2) CH4O   METHANOL

(3) H2O   WATER

PRUTTON C.F., WALSH T.J., DESAI A.M.
IND.ENG.CHEM. 42(1950)1210

TEMPERATURE = 25.0 DEG C   TYPE OF SYSTEM = 1

EXPERIMENTAL TIE LINES IN MOLE PCT

	LEFT PHASE			RIGHT PHASE	
(1)	(2)	(3)	(1)	(2)	(3)
32.229	0.0	67.771	1.725	0.0	98.275
28.143	1.983	69.874	1.846	0.731	97.423
26.397	3.371	70.232	1.945	1.352	96.703
21.679	5.169	73.152	2.272	3.021	94.707
12.100	8.885	79.014	4.108	7.698	88.193

SPECIFIC MODEL PARAMETERS IN KELVIN

		UNIQUAC		NRTL(ALPHA=.2)	
I	J	AIJ	AJI	AIJ	AJI
1	2	-270.21	-80.219	-42.444	-33.676
1	3	-257.86	566.90	-567.84	1884.0
2	3	-235.52	129.16	55.212	44.631

R1 = 3.5517   R2 = 1.4311   R3 = 0.9200
Q1 = 2.680   Q2 = 1.432   Q3 = 1.400

MEAN DEV. BETWEEN CALC. AND EXP. CONC. IN MOLE PCT

UNIQUAC (SPECIFIC PARAMETERS)   0.45
NRTL   (SPECIFIC PARAMETERS)   0.44
UNIQUAC (COMMON PARAMETERS)   0.61

CH₄O-C₆H₁₀

(1) H2O   WATER
(2) CH4O   METHANOL
(3) C6H10   CYCLOHEXENE

WASHBURN E.R., GRAHAM C.L., ARNOLD G.B., TRANSUE L.F.
J.AM.CHEM.SOC. 62(1940)1454

TEMPERATURE = 25.0 DEG C   TYPE OF SYSTEM = 1

EXPERIMENTAL TIE LINES IN MOLE PCT (GRAPH.INTERPOL.)

	LEFT PHASE			RIGHT PHASE	
(1)	(2)	(3)	(1)	(2)	(3)
94.365	5.635	0.0	0.0	0.256	99.744
82.744	17.231	0.025	0.0	0.511	99.489
89.113	10.863	0.024	0.0	0.511	99.489
79.790	20.185	0.025	0.0	0.766	99.234
72.578	27.342	0.080	0.0	1.019	98.981
34.564	63.729	1.707	0.0	3.020	96.980
23.759	72.519	3.722	0.0	5.213	94.787

SPECIFIC MODEL PARAMETERS IN KELVIN

		UNIQUAC		NRTL(ALPHA=.2)	
I	J	AIJ	AJI	AIJ	AJI
1	2	-203.92	-59.166	-209.25	34.904
1	3	285.65	756.70	1559.9	1392.3
2	3	67.011	606.75	507.40	529.89

R1 = 0.9200   R2 = 1.4311   R3 = 3.8143
Q1 = 1.400   Q2 = 1.432   Q3 = 3.027

MEAN DEV. BETWEEN CALC. AND EXP. CONC. IN MOLE PCT

UNIQUAC (SPECIFIC PARAMETERS)   0.42
NRTL   (SPECIFIC PARAMETERS)   0.26
UNIQUAC (COMMON PARAMETERS)   0.39

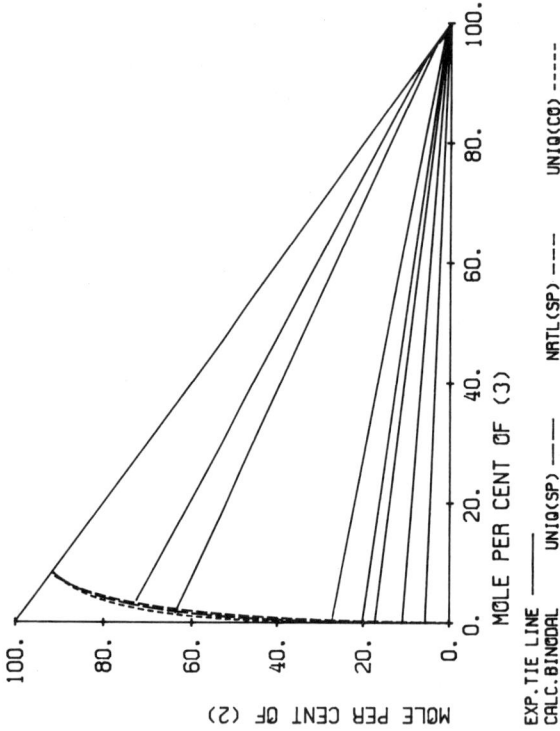

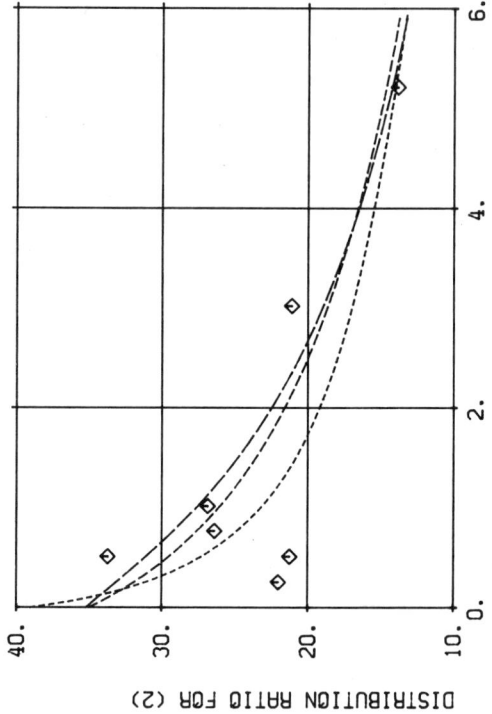

EXP.TIE LINE
CALC.BINODAL   UNIQ(SP) ----   NRTL(SP) ----   UNIQ(CO) ----

MOLE PER CENT OF (3)

MOLE PER CENT OF (2)

EXP. DISTR.RATIO ◇

DISTRIBUTION RATIO FOR (2)

MOLE PER CENT OF (2) IN RIGHT PHASE

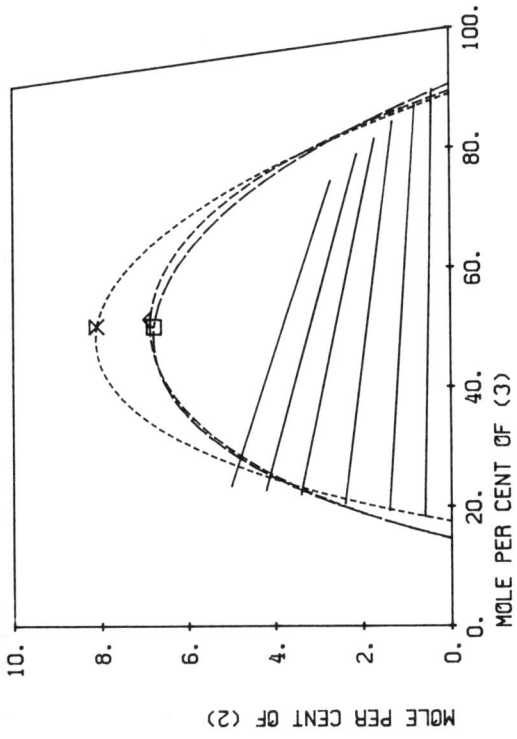

MOLE PER CENT OF (2)

MOLE PER CENT OF (3)

EXP.TIE LINE ———   UNIQ(SP) ——— ⊡   NRTL(SP) ——— ✦   UNIQ(CO) ----- ✕
CALC.BINODAL
CALC.PLAIT P.

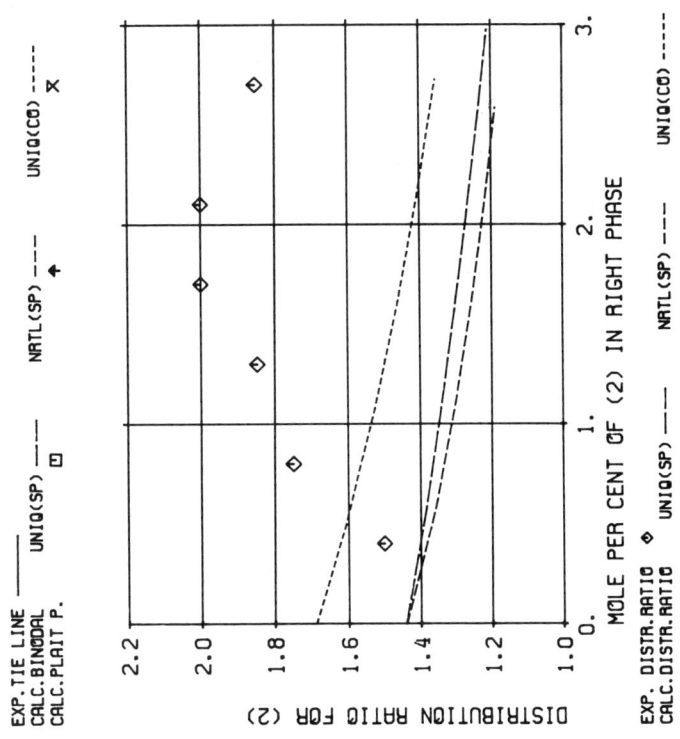

DISTRIBUTION RATIO FOR (2)

MOLE PER CENT OF (2) IN RIGHT PHASE

EXP. DISTR.RATIO ◇   UNIQ(SP) ———   NRTL(SP) ———   UNIQ(CO) -----
CALC.DISTR.RATIO

(2) C7H6O2   BENZOIC ACID
(3) C6H12   CYCLOHEXANE

SERGEEVA V.F., ET AL.
ZH.OBSHCH.KHIM. 41(1971)1895

TEMPERATURE = 25.0 DEG C   TYPE OF SYSTEM = 1

EXPERIMENTAL TIE LINES IN MOLE PCT

	LEFT PHASE			RIGHT PHASE	
(1)	(2)	(3)	(1)	(2)	(3)
82.100	0.0	17.900	7.500	0.0	92.500
81.100	0.600	18.300	9.900	0.400	89.700
79.300	1.400	19.300	12.200	0.800	87.000
77.100	2.400	20.500	14.400	1.300	84.300
74.600	3.400	22.700	16.800	1.700	81.500
73.100	4.200	22.700	19.000	2.100	78.900
71.600	5.000	23.400	22.900	2.700	74.400

SPECIFIC MODEL PARAMETERS IN KELVIN

		UNIQUAC		NRTL(ALPHA=.2)	
I	J	AIJ	AJI	AIJ	AJI
1	2	-8.3649	-72.360	322.69	-921.90
1	3	12.256	678.14	373.25	53.45
2	3	-46.742	-7.8853	-619.13	-362.75

R1 = 1.4311   R2 = 4.3230   R3 = 4.0464
Q1 = 1.432   Q2 = 3.344   Q3 = 3.240

MEAN DEV. BETWEEN CALC. AND EXP. CONC. IN MOLE PCT

UNIQUAC (SPECIFIC PARAMETERS)   1.31
NRTL   (SPECIFIC PARAMETERS)   1.57
UNIQUAC (COMMON PARAMETERS)   1.51

CH$_4$O-C$_6$H$_{12}$

MOLE PER CENT OF (2)

MOLE PER CENT OF (3)

EXP.TIE LINE ———   UNIQ(SP) ———   NRTL(SP) ———   UNIQ(CO) - - - - -
CALC.BINODAL
CALC.PLAIT P.

DISTRIBUTION RATIO FOR (2)

MOLE PER CENT OF (2) IN RIGHT PHASE

EXP. DISTR.RATIO ◇   UNIQ(SP) ———   NRTL(SP) ———   UNIQ(CO) - - - - -

(1) C6H12   CYCLOHEXANE
(2) C10H8   NAPHTHALENE
(3) CH4O   METHANOL

SERGEEVA V.F., ET AL.
ZH.OBSHCH.KHIM. 41(1971)1895

TEMPERATURE = 25.0 DEG C   TYPE OF SYSTEM = 1

EXPERIMENTAL TIE LINES IN MOLE PCT

LEFT PHASE			RIGHT PHASE		
(1)	(2)	(3)	(1)	(2)	(3)
92.500	0.0	7.500	17.900	0.0	82.100
87.800	2.200	10.000	18.600	0.700	80.700
84.100	3.500	12.400	19.300	1.400	79.300
79.800	4.600	15.600	20.100	2.000	77.900
74.700	5.700	19.600	21.400	2.600	76.000
68.400	6.700	24.900	21.900	3.200	74.900
61.200	7.300	31.500	24.400	4.000	71.600
54.100	7.800	38.100	26.300	4.600	69.100

SPECIFIC MODEL PARAMETERS IN KELVIN

		UNIQUAC		NRTL(ALPHA=.2)	
I	J	AIJ	AJI	AIJ	AJI
1	2	-11.799	-24.202	-647.20	-376.15
1	3	753.28	-8.1030	659.47	300.44
2	3	36.338	110.15	-681.65	725.01

R1 = 4.0464   R2 = 4.9808   R3 = 1.4311
Q1 = 3.240   Q2 = 3.440   Q3 = 1.432

MEAN DEV. BETWEEN CALC. AND EXP. CONC. IN MOLE PCT

UNIQUAC (SPECIFIC PARAMETERS)   0.29
NRTL   (SPECIFIC PARAMETERS)   0.99
UNIQUAC (COMMON PARAMETERS)   1.06

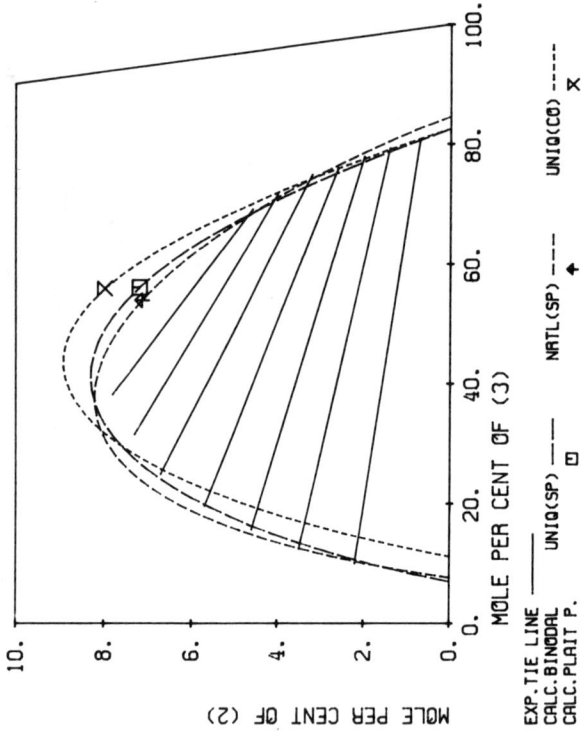

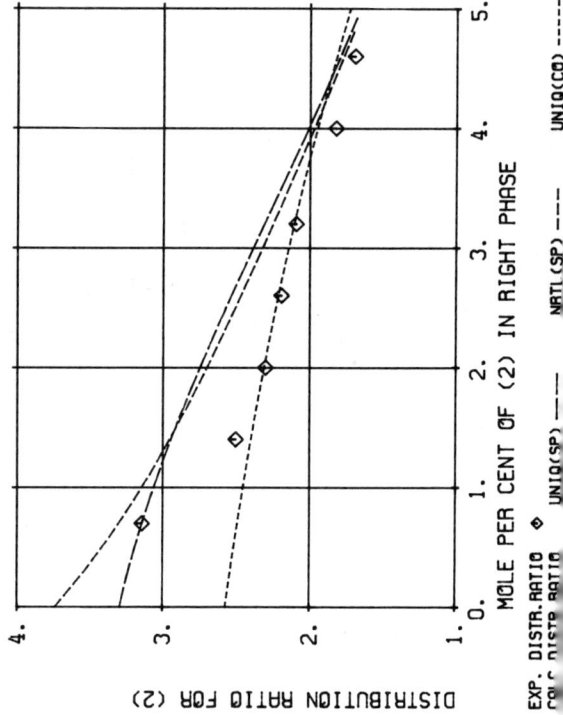

$$CH_4O\text{-}C_6H_{12}$$

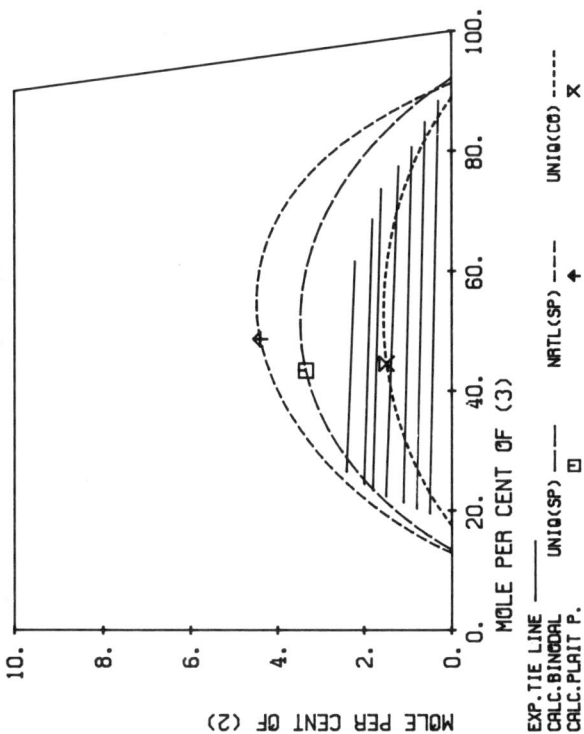

MOLE PER CENT OF (2)

MOLE PER CENT OF (3)

EXP.TIE LINE ———   UNIQ(SP) ——— ☐   NRTL(SP) ——— ✦   UNIQ(CB) ----- ✗
CALC.BINODAL ———
CALC.PLAIT P.

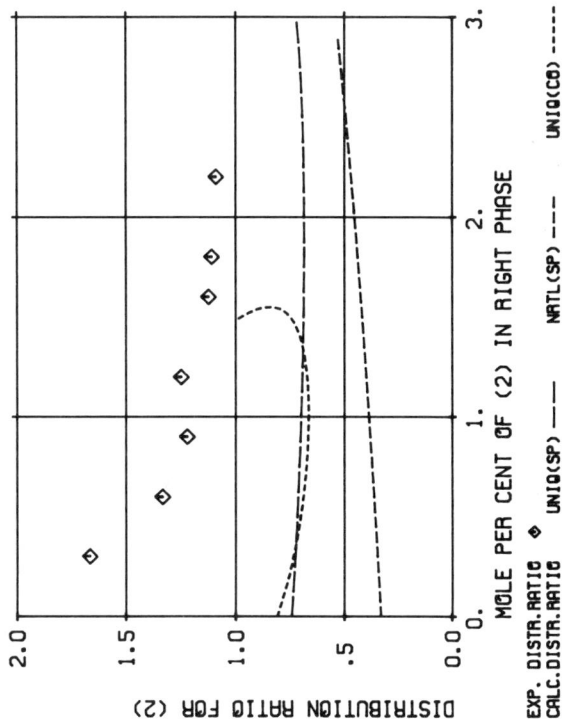

DISTRIBUTION RATIO FOR (2)

MOLE PER CENT OF (2) IN RIGHT PHASE

EXP. DISTR.RATIO ◆   UNIQ(SP) ———   NRTL(SP) -----   UNIQ(CB) -----
CALC.DISTR.RATIO

(2) C12H11N   AMINE,DIPHENYL
(3) C6H12   CYCLOHEXANE

SERGEEVA V.F., ET AL.
ZH.OBSHCH.KHIM. 41(1971)1895

TEMPERATURE = 25.0 DEG C   TYPE OF SYSTEM = 1

EXPERIMENTAL TIE LINES IN MOLE PCT

	LEFT PHASE			RIGHT PHASE	
(1)	(2)	(3)	(1)	(2)	(3)
82.100	0.0	17.900	7.500	0.0	92.500
80.000	0.500	19.500	11.300	0.300	88.400
78.900	0.800	20.300	14.500	0.600	84.900
77.600	1.100	21.300	18.400	0.900	80.700
76.200	1.500	22.300	21.300	1.200	77.500
74.900	1.800	23.300	24.700	1.600	73.700
73.600	2.000	24.400	29.500	1.800	68.700
71.200	2.400	26.400	36.100	2.200	61.700

SPECIFIC MODEL PARAMETERS IN KELVIN

		UNIQUAC		NRTL(ALPHA=.2)	
I	J	AIJ	AJI	AIJ	AJI
1	2	103.66	-179.61	873.57	-1245.0
1	3	14.019	704.77	379.39	578.07
2	3	-108.50	-75.541	-987.32	-856.11

R1 = 1.4311   R2 = 6.5760   R3 = 4.0464
Q1 = 1.432   Q2 = 4.636   Q3 = 3.240

MEAN DEV. BETWEEN CALC. AND EXP. CONC. IN MOLE PCT

UNIQUAC (SPECIFIC PARAMETERS)   1.39
NRTL   (SPECIFIC PARAMETERS)   2.32
UNIQUAC (COMMON PARAMETERS)   1.45

CH$_4$O-C$_6$H$_{12}$

(1) CH4O   METHANOL
----
(2) C18H36O2 OCTADECANOIC ACID
----
(3) C6H12   CYCLOHEXANE

SERGEEVA V.F., ESKARAEVA L.A.
ZH.OBSHCH.KHIM. 39(1969)731

TEMPERATURE = 40.0 DEG C   TYPE OF SYSTEM = 1

EXPERIMENTAL TIE LINES IN MOLE PCT

	LEFT PHASE			RIGHT PHASE	
(1)	(2)	(3)	(1)	(2)	(3)
72.620	0.0	27.380	24.510	0.0	75.490
71.340	0.060	28.600	32.790	0.010	67.200
66.100	2.850	31.050	32.810	0.530	66.660
65.310	3.130	31.560	33.000	0.710	66.290
57.280	3.340	39.380	36.340	0.880	62.280

SPECIFIC MODEL PARAMETERS IN KELVIN

I J	UNIQUAC		NRTL(ALPHA=.2)	
	AIJ	AJI	AIJ	AJI
1 2	-18.696	-65.942	-293.49	455.50
1 3	50.527	518.15	421.67	360.27
2 3	-32.129	-26.291	-350.76	1201.6

R1 = 1.4311   R2 =12.9928   R3 = 4.0464
Q1 = 1.432   Q2 =10.712   Q3 = 3.240

MEAN DEV. BETWEEN CALC. AND EXP. CONC. IN MOLE PCT

UNIQUAC (SPECIFIC PARAMETERS)   1.47
NRTL   (SPECIFIC PARAMETERS)   1.36

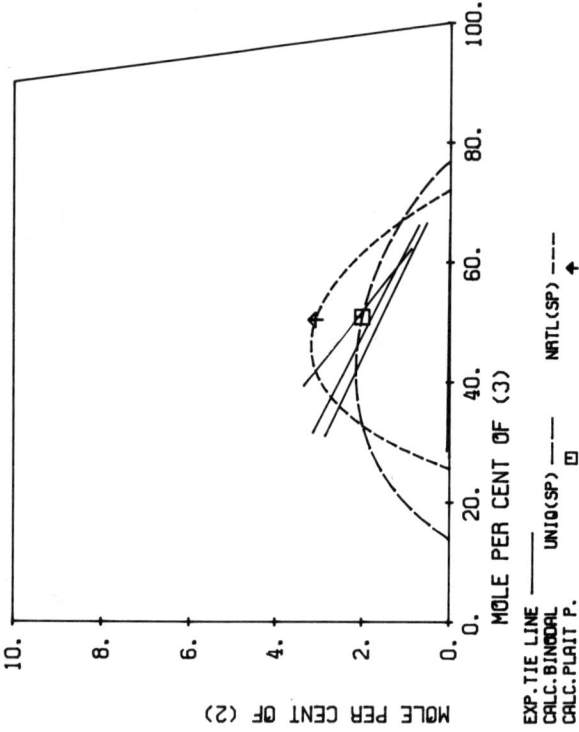

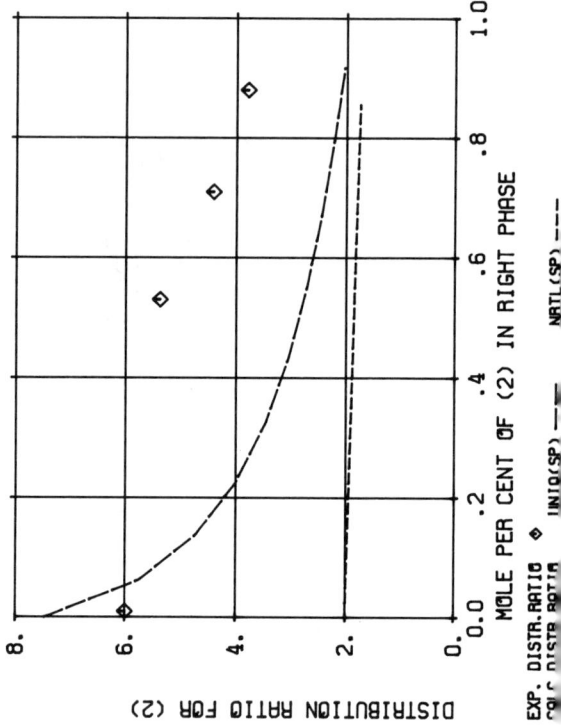

CH$_4$O-C$_6$H$_{12}$O$_2$

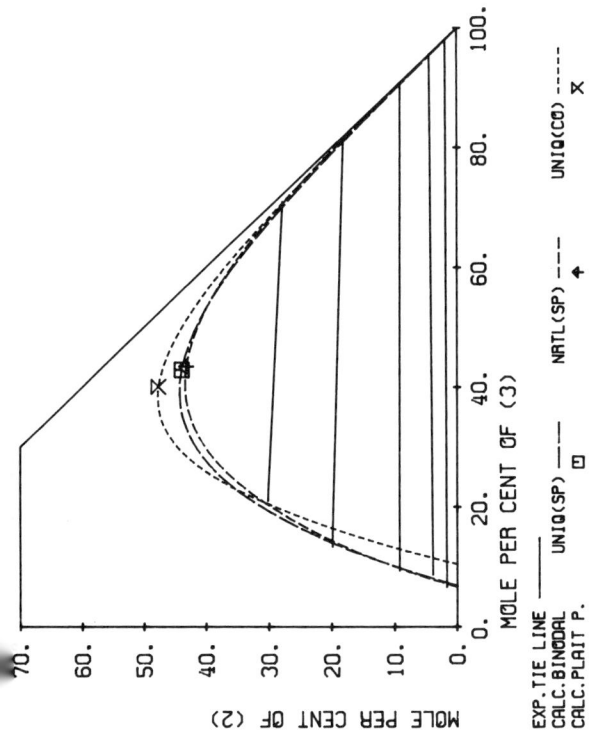

MOLE PER CENT OF (2)

MOLE PER CENT OF (3)

EXP.TIE LINE	UNIQ(SP) □	NRTL(SP) ✦	UNIQ(CO) ✕
CALC.BINODAL			
CALC.PLAIT P.			

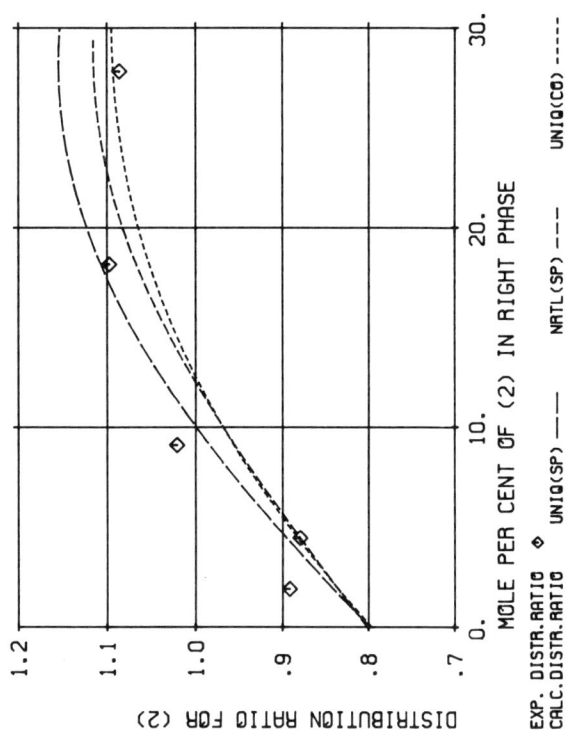

DISTRIBUTION RATIO FOR (2)

MOLE PER CENT OF (2) IN RIGHT PHASE

| EXP. DISTR.RATIO ◇ | UNIQ(SP) ◇ | NRTL(SP) | UNIQ(CO) |
| CALC.DISTR.RATIO | | | |

(1) CH1202   ACETIC ACID, BUTYL ESTER
(2) CH4O   METHANOL
(3) H2O   WATER

JAGANNADHA RAO R., VENKATA RAO C.
J.APPL.CHEM. 7(1957)435

TEMPERATURE = 30.0 DEG C   TYPE OF SYSTEM = 1

EXPERIMENTAL TIE LINES IN MOLE PCT

	LEFT PHASE			RIGHT PHASE		
	(1)	(2)	(3)	(1)	(2)	(3)
	91.701	1.689	6.610	0.111	1.894	97.995
	87.404	3.908	8.688	0.097	4.444	95.460
	81.357	9.285	9.358	0.134	9.092	90.774
	66.675	19.920	13.405	0.360	18.146	81.494
	48.777	30.203	21.020	1.192	27.813	70.995

SPECIFIC MODEL PARAMETERS IN KELVIN

		UNIQUAC		NRTL(ALPHA=.2)	
I J		AIJ	AJI	AIJ	AJI
1 2		48.236	-44.117	-386.02	418.40
1 3		531.76	116.68	404.00	2358.7
2 3		-221.05	-9.4739	-473.85	416.46

R1 = 4.8274   R2 = 1.4311   R3 = 0.9200
Q1 = 4.196   Q2 = 1.432   Q3 = 1.400

MEAN DEV. BETWEEN CALC. AND EXP. CONC. IN MOLE PCT

UNIQUAC (SPECIFIC PARAMETERS)   0.44
NRTL   (SPECIFIC PARAMETERS)   0.32
UNIQUAC (COMMON PARAMETERS)   1.50

CH$_4$O-C$_6$H$_{12}$O$_2$

(1) C6H1202   BUTANOIC ACID,ETHYL ESTER

(2) CH4O   METHANOL

(3) H2O   WATER

JAGANNADHA RAO R., VENKATA RAO C.
J.APPL.CHEM. 7(1957)435

TEMPERATURE = 30.0 DEG C   TYPE OF SYSTEM = 1

EXPERIMENTAL TIE LINES IN MOLE PCT

	LEFT PHASE			RIGHT PHASE	
(1)	(2)	(3)	(1)	(2)	(3)
92.019	1.355	6.626	0.095	1.951	97.954
88.449	3.947	7.604	0.113	4.447	95.440
77.612	12.273	10.115	0.135	10.668	89.197
66.140	21.457	12.403	0.359	17.700	81.941

MEAN DEV. BETWEEN CALC. AND EXP. CONC. IN MOLE PCT

UNIQUAC (COMMON PARAMETERS)   1.02

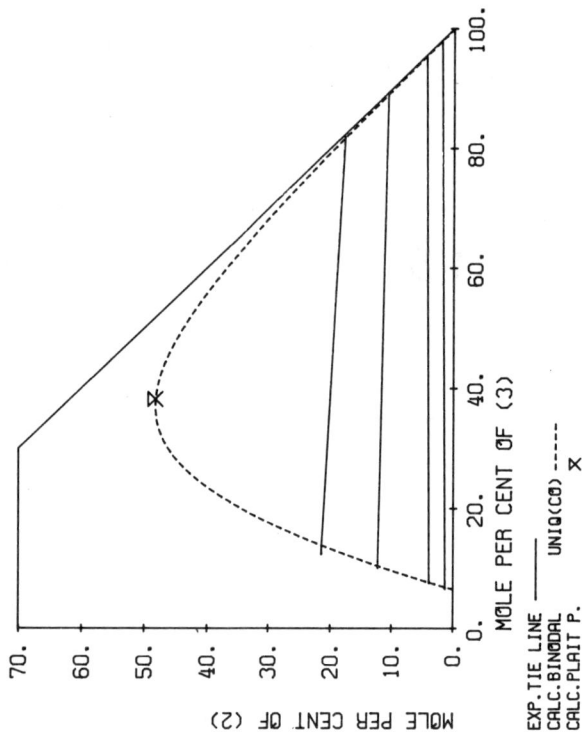

EXP.TIE LINE ——— UNIQ(CO) -----
CALC.BINODAL
CALC.PLAIT P.   ✕

MOLE PER CENT OF (2)
MOLE PER CENT OF (3)

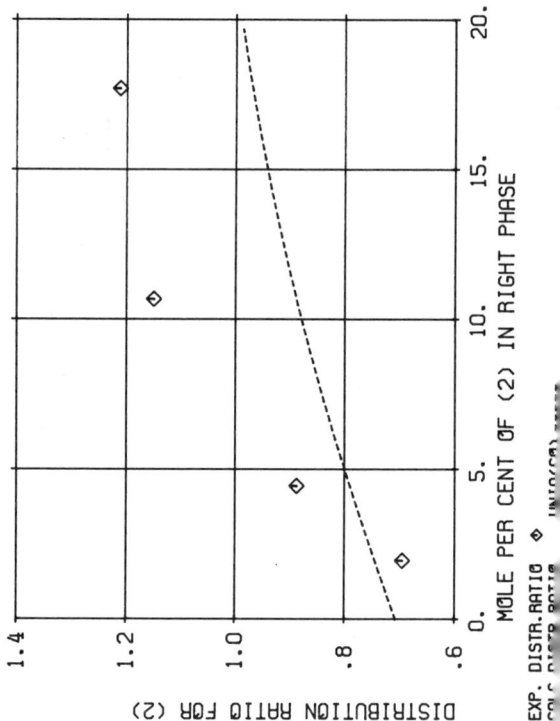

EXP. DISTR.RATIO   ◇
DISTRIBUTION RATIO FOR (2)
MOLE PER CENT OF (2) IN RIGHT PHASE

$CH_4O\text{-}C_6H_{14}$

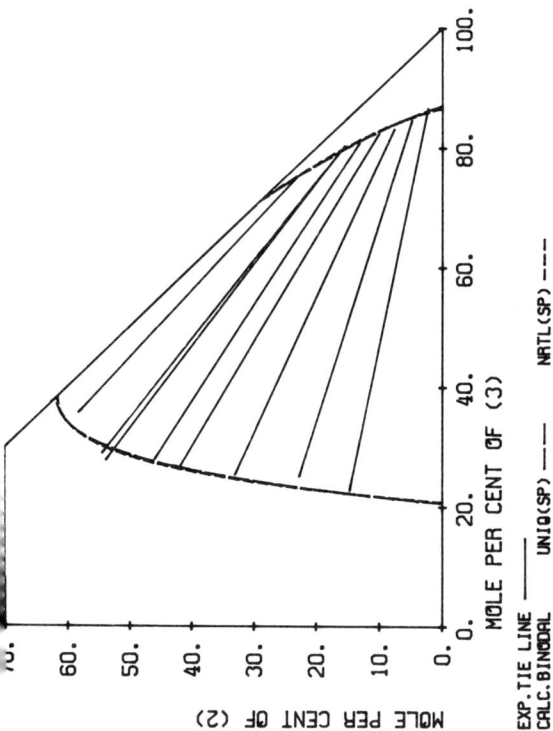

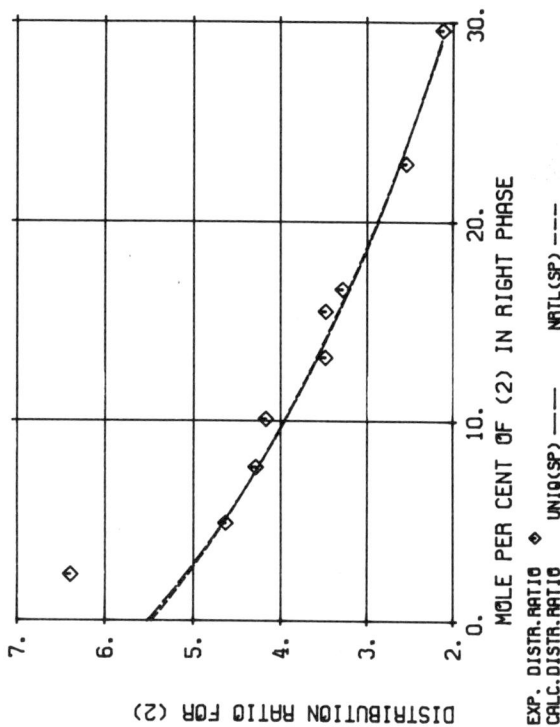

(2) C6H14   HEXANE

(3) CH4O   METHANOL

WITTRIG T.S.
THESIS ILLINOIS 1977

TEMPERATURE = 32.8 DEG C   TYPE OF SYSTEM = 2

EXPERIMENTAL TIE LINES IN MOLE PCT

	LEFT PHASE			RIGHT PHASE	
(1)	(2)	(3)	(1)	(2)	(3)
80.600	0.0	19.400	13.000	0.0	87.000
62.500	14.700	22.800	11.000	2.300	86.700
52.200	22.700	25.100	10.400	4.900	84.700
41.600	33.000	25.400	9.100	7.700	83.200
31.600	42.100	26.300	7.400	10.100	82.500
26.100	45.900	28.000	6.100	13.200	80.800
18.400	53.800	27.800	4.100	15.500	80.400
16.600	54.400	29.000	4.500	16.600	78.900
6.100	58.200	35.700	1.800	22.900	75.300
0.0	62.600	37.400	0.0	29.600	70.400

SPECIFIC MODEL PARAMETERS IN KELVIN

		UNIQUAC		NRTL(ALPHA=.2)	
I J		AIJ	AJI	AIJ	AJI
1 2		-71.593	57.909	-325.46	295.13
1 3		620.68	2.3547	305.35	560.90
2 3		547.69	-2.0154	258.44	475.42

R1 = 5.1742   R2 = 4.4998   R3 = 1.4311
Q1 = 4.396   Q2 = 3.856   Q3 = 1.432

MEAN DEV. BETWEEN CALC. AND EXP. CONC. IN MOLE PCT

UNIQUAC (SPECIFIC PARAMETERS)   0.70
NRTL   (SPECIFIC PARAMETERS)   0.72

CH₄O-C₆H₁₄O

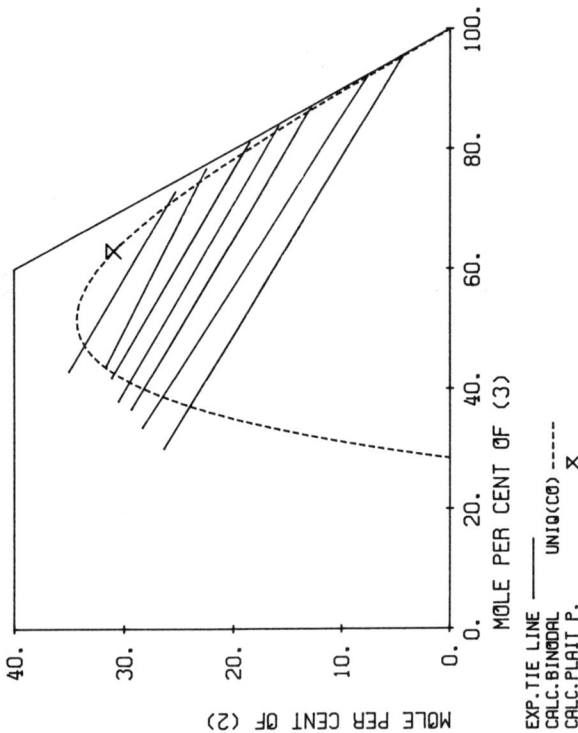

MOLE PER CENT OF (3)

MOLE PER CENT OF (2)

EXP.TIE LINE ——— UNIQ(C0) -----
CALC.BINODAL
CALC.PLAIT P.   x

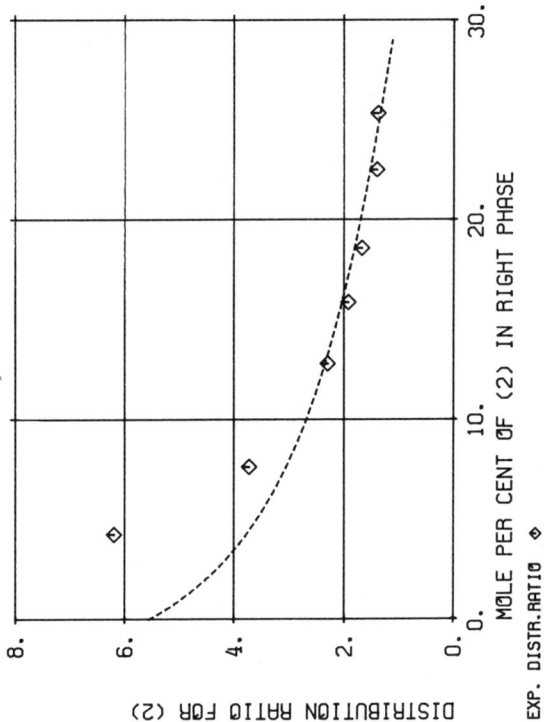

MOLE PER CENT OF (2) IN RIGHT PHASE

DISTRIBUTION RATIO FOR (2)

EXP. DISTR.RATIO ◇

(1) C6H14O   1-HEXANOL

(2) CH4O   METHANOL

(3) H2O   WATER

LEBEDINSKAYA N.A., ET AL.
KHIM.PROM-ST.(MOSCOW) (1976)16

TEMPERATURE = 20.0 DEG C   TYPE OF SYSTEM = 1

EXPERIMENTAL TIE LINES IN MOLE PCT

	LEFT PHASE			RIGHT PHASE	
(1)	(2)	(3)	(1)	(2)	(3)
43.495	26.395	30.109	0.110	4.262	95.628
38.074	28.332	33.593	0.150	7.614	92.235
34.040	29.319	36.641	0.195	12.778	87.026
31.550	30.490	37.959	0.220	15.875	83.906
27.032	31.131	41.838	0.306	18.571	81.123
24.563	31.630	43.807	0.725	22.501	76.774
22.113	35.002	42.885	1.759	25.320	72.921

MEAN DEV. BETWEEN CALC. AND EXP. CONC. IN MOLE PCT

UNIQUAC (COMMON PARAMETERS)   2.98

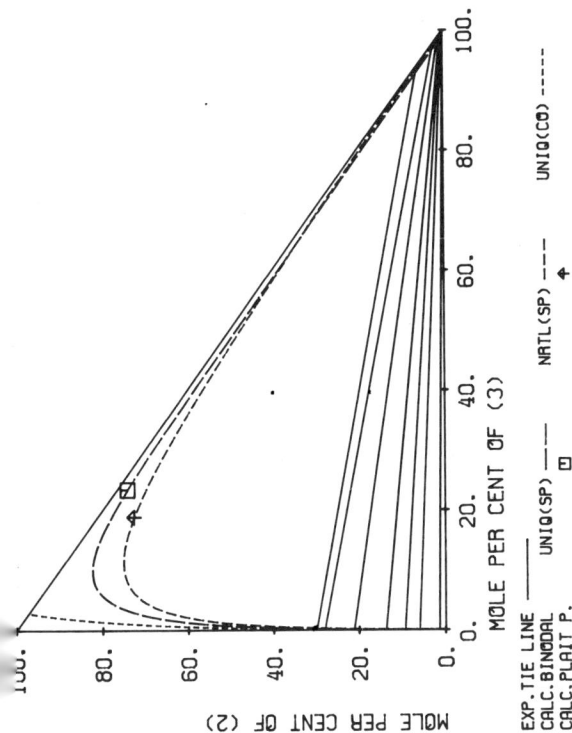

MOLE PER CENT OF (3)

MOLE PER CENT OF (2)

EXP.TIE LINE ——— UNIQ(SP) ☐ NRTL(SP) ——— ＋ UNIQ(CO) -----
CALC.BINODAL
CALC.PLAIT P.

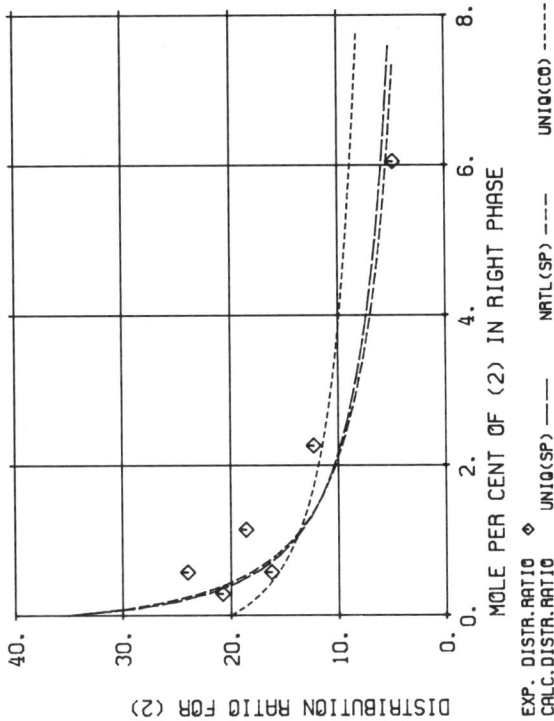

MOLE PER CENT OF (2) IN RIGHT PHASE

DISTRIBUTION RATIO FOR (2)

EXP. DISTR.RATIO ◇ UNIQ(SP) ——— NRTL(SP) ----- UNIQ(CO) -----
CALC.DISTR.RATIO

(2) CH4O   METHANOL

(3) C7H8   TOLUENE

MASON L.S., WASHBURN E.R.
J.AM.CHEM.SOC. 59(1937)2076

TEMPERATURE = 25.0 DEG C   TYPE OF SYSTEM = 1

EXPERIMENTAL TIE LINES IN MOLE PCT (GRAPH.INTERPOL.)

	LEFT PHASE			RIGHT PHASE	
(1)	(2)	(3)	(1)	(2)	(3)
98.568	1.422	0.010	0.255	0.0	99.745
94.043	5.945	0.012	0.305	0.286	99.408
86.272	13.703	0.026	0.330	0.572	99.098
90.693	9.291	0.016	0.330	0.572	99.093
78.812	21.128	0.060	0.369	1.138	98.492
71.992	27.893	0.115	0.447	2.259	97.294
70.081	29.781	0.138	0.660	6.044	93.296

SPECIFIC MODEL PARAMETERS IN KELVIN

		UNIQUAC			NRTL(ALPHA=.2)	
I J		AIJ	AJI		AIJ	AJI
1 2		-174.97	-288.54		-543.60	-385.81
1 3		311.77	666.30		1745.0	1241.7
2 3		121.24	167.48		818.50	-442.13

R1 = 0.9200   R2 = 1.4311   R3 = 3.9228
Q1 = 1.400   Q2 = 1.432   Q3 = 2.968

MEAN DEV. BETWEEN CALC. AND EXP. CONC. IN MOLE PCT

UNIQUAC (SPECIFIC PARAMETERS)   0.67
NRTL   (SPECIFIC PARAMETERS)   0.53
UNIQUAC (COMMON PARAMETERS)   0.77

CH₄O-C₇H₈O

(1) C7H8O   TOLUENE, 4-HYDROXY
----
(2) CH4O   METHANOL
----
(3) H2O   WATER

PRUTTON C.F., WALSH T.J., DESAI A.M.
IND.ENG.CHEM. 42(1950)1210

TEMPERATURE = 35.0 DEG C   TYPE OF SYSTEM = 1

EXPERIMENTAL TIE LINES IN MOLE PCT

	LEFT PHASE			RIGHT PHASE	
(1)	(2)	(3)	(1)	(2)	(3)
51.415	0.0	48.585	0.460	0.0	99.540
44.689	4.599	50.712	0.532	4.785	94.683
34.314	11.520	54.166	0.597	7.141	92.263
30.984	13.745	55.271	0.734	8.364	90.902
22.784	17.945	59.271	1.000	10.637	88.363
17.630	19.190	63.180	1.486	12.807	85.707
15.342	19.420	65.239	1.835	13.529	84.636

SPECIFIC MODEL PARAMETERS IN KELVIN

		UNIQUAC		NRTL(ALPHA=.2)	
I J		AIJ	AJI	AIJ	AJI
1 2		724.35	8.3651	840.48	102.41
1 3		-221.62	794.61	-371.32	2580.3
2 3		275.73	-85.345	199.99	293.15

R1 = 4.2867   R2 = 1.4311   R3 = 0.9200
Q1 = 3.248   Q2 = 1.432   Q3 = 1.400

MEAN DEV. BETWEEN CALC. AND EXP. CONC. IN MOLE PCT

UNIQUAC (SPECIFIC PARAMETERS)   0.59
NRTL   (SPECIFIC PARAMETERS)   0.40

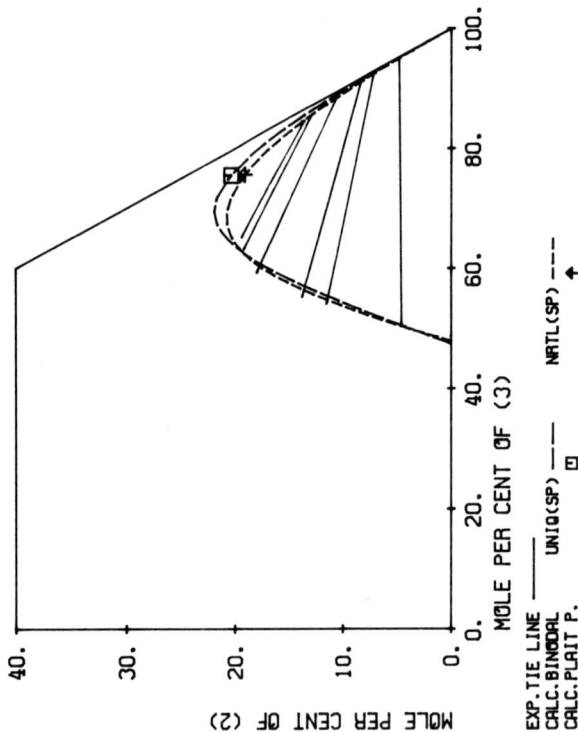

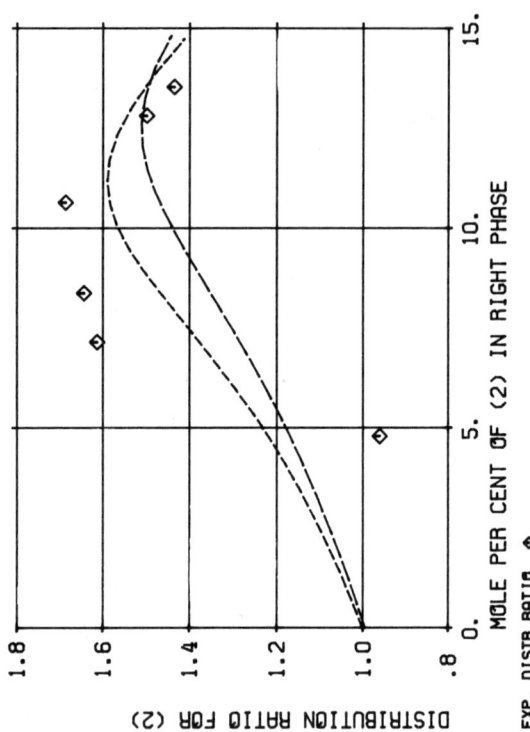

MOLE PER CENT OF (3)

MOLE PER CENT OF (2)

EXP.TIE LINE —— UNIQ(SP) ---- ⊟ NRTL(SP) ---- ♦
CALC.BINODAL
CALC.PLAIT P.

DISTRIBUTION RATIO FOR (2)

MOLE PER CENT OF (2) IN RIGHT PHASE

EXP. DISTR.RATIO ◇

$CH_4O-C_7H_{14}O_2$

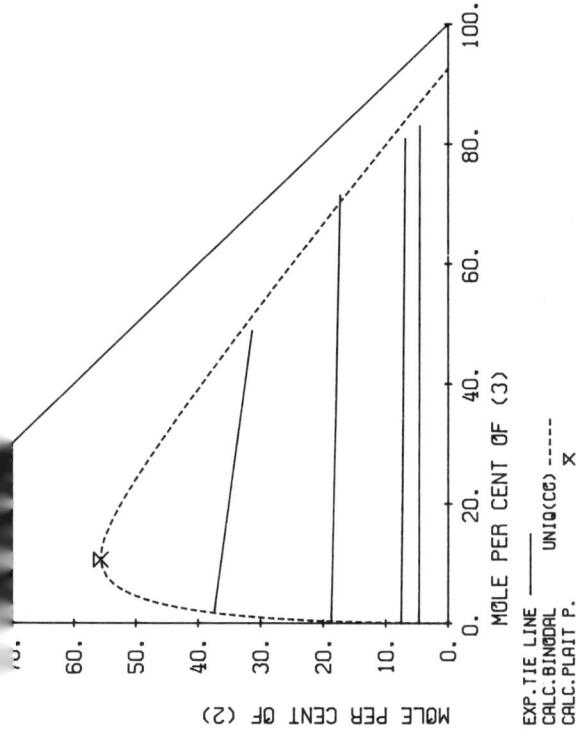

MOLE PER CENT OF (3)

EXP.TIE LINE ——— UNIQ(CC) -----
CALC.BINODAL
CALC.PLAIT P.   X

MOLE PER CENT OF (2)

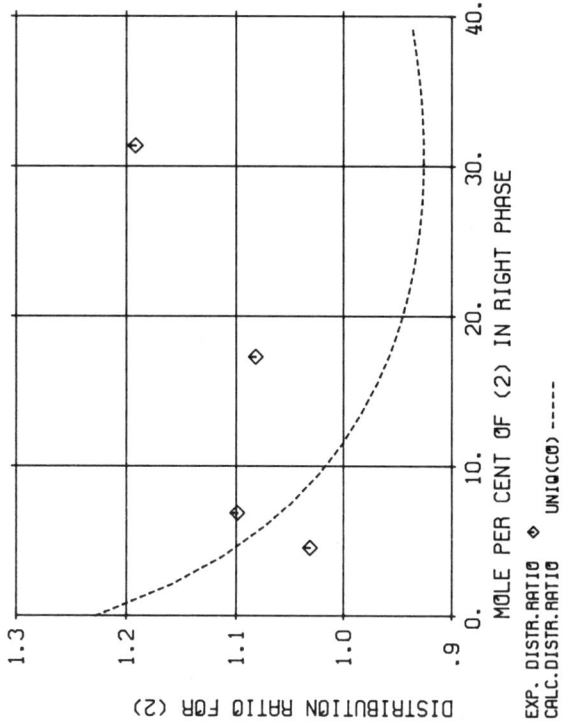

MOLE PER CENT OF (2) IN RIGHT PHASE

EXP. DISTR.RATIO ◇ UNIQ(CC) -----
CALC.DISTR.RATIO

DISTRIBUTION RATIO FOR (2)

(2) CH4O   METHANOL
(3) C7H14O2   ACETIC ACID,PENTYL ESTER

JAGANNADHA RAO R., VENKATA RAO C.
J.APPL.CHEM. 7(1957)435

TEMPERATURE = 30.0 DEG C   TYPE OF SYSTEM = 1

EXPERIMENTAL TIE LINES IN MOLE PCT

| | LEFT PHASE | | | RIGHT PHASE | |
	(2)	(3)	(1)	(2)	(3)
95.264	4.678	0.058	12.413	4.537	83.050
92.394	7.532	0.074	12.189	6.853	80.958
81.096	18.695	0.208	11.178	17.284	71.538
60.465	37.365	2.171	19.733	31.356	48.912

MEAN DEV. BETWEEN CALC. AND EXP. CONC. IN MOLE PCT

UNIQUAC (COMMON PARAMETERS)   2.26

138

CH₄O-C₇H₁₆O

(1) C7H16O   1-HEPTANOL

(2) CH4O   METHANOL

(3) H2O   WATER

LEBEDINSKAYA N.A., ET AL.
KHIM.PROM-ST.(MOSCOW) (1976)16

TEMPERATURE = 20.0 DEG C   TYPE OF SYSTEM = 1

EXPERIMENTAL TIE LINES IN MOLE PCT

	LEFT PHASE			RIGHT PHASE	
(1)	(2)	(3)	(1)	(2)	(3)
51.855	23.679	24.466	0.032	5.952	94.015
42.074	29.728	28.198	0.084	9.587	90.329
37.632	32.313	30.055	0.120	13.294	86.585
32.211	33.963	33.826	0.160	18.173	81.666
28.498	35.302	36.200	0.200	20.691	79.109
24.829	36.646	38.525	0.263	25.261	74.476
20.159	35.893	42.948	0.288	28.536	71.176

MEAN DEV. BETWEEN CALC. AND EXP. CONC. IN MOLE PCT

UNIQUAC (COMMON PARAMETERS)   2.51

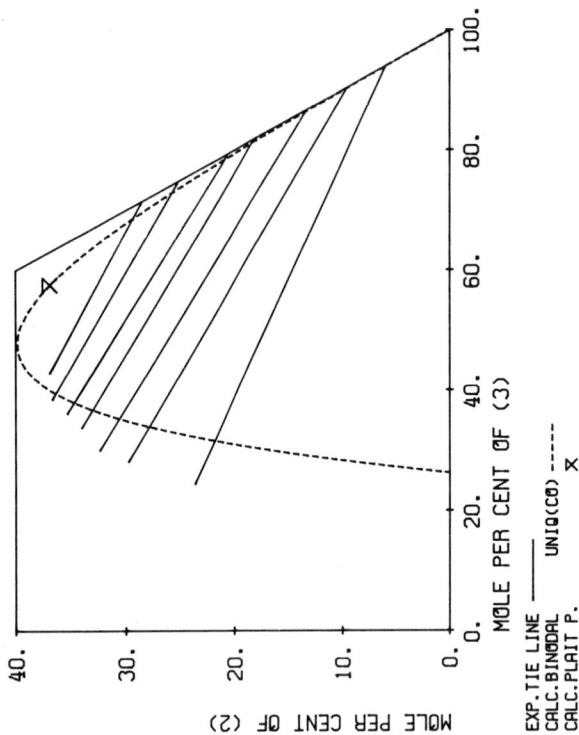

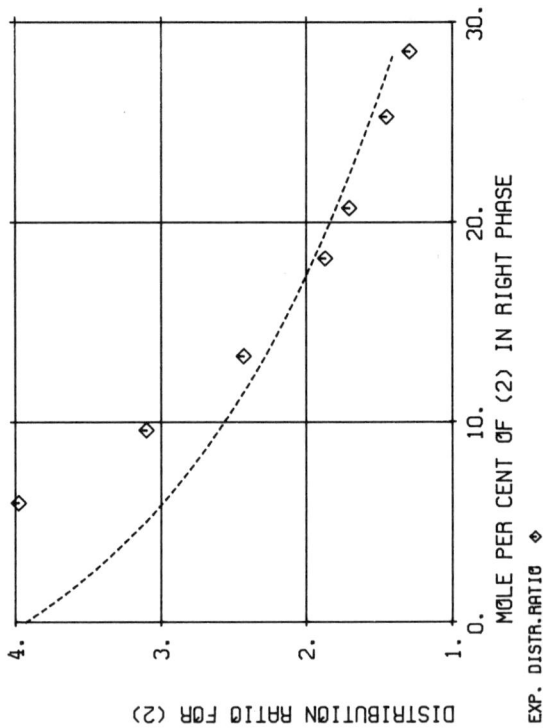

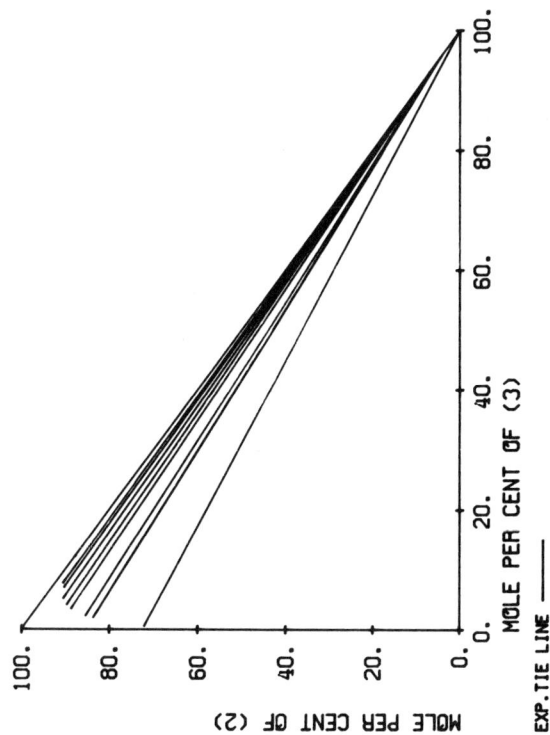

MOLE PER CENT OF (3) ⟶

EXP.TIE LINE

MOLE PER CENT OF (2)

(1) H2O   WATER
(2) CH4O   METHANOL
(3) C8H18   PENTANE,2,2,4-TRIMETHYL

BUCHOWSKI H., TEPEREK J.
ROCZ.CHEM. 33(1959)1093

TEMPERATURE = 18.0 DEG C   TYPE OF SYSTEM = 2

EXPERIMENTAL TIE LINES IN MOLE PCT

| | LEFT PHASE | | | RIGHT PHASE | |
(1)	(2)	(3)	(1)	(2)	(3)
27.395	72.002	0.603	0.0	0.0	100.000
14.400	83.633	1.966	0.000	4.151	95.849
12.337	85.351	2.312	0.0	4.485	95.515
7.875	88.665	3.460	0.000	6.782	93.217
6.144	89.708	4.148	0.000	8.060	91.940
4.359	90.455	5.187	0.000	8.690	91.310
2.703	90.248	7.050	0.000	10.543	89.457
1.695	90.552	7.753	0.0	15.237	84.763
0.000	87.862	12.138	0.0	17.729	82.271

CH₄O-C₈H₁₈

(1) H2O   WATER
----
(2) CH4O   METHANOL
----
(3) C8H18   PENTANE,2,2,4-TRIMETHYL

BUCHOWSKI H., TEPEREK J.
ROCZ.CHEM. 33(1959)1093

TEMPERATURE = 20.0 DEG C   TYPE OF SYSTEM = 2

EXPERIMENTAL TIE LINES IN MOLE PCT

        LEFT PHASE   RIGHT PHASE
----
  (1)   (2)   (3)   (1)   (2)   (3)

20.043   78.602   1.355   0.0   0.0   100.000
13.534   84.107   2.359   0.0   2.107   97.893
 7.634   88.212   4.154   0.0   5.808   94.192
 7.118   88.483   4.399   0.000   7.743   92.257
 5.884   89.105   5.011   0.0   8.376   91.624
 3.682   89.833   6.485   0.0   9.931   90.069
 2.540   89.615   7.845   0.000   11.750   88.250
 1.521   89.568   8.911   0.0   14.954   85.046
 1.099   89.504   9.397   0.0   15.800   84.200

MEAN DEV. BETWEEN CALC. AND EXP. CONC. IN MOLE PCT

UNIQUAC (COMMON PARAMETERS)   2.40

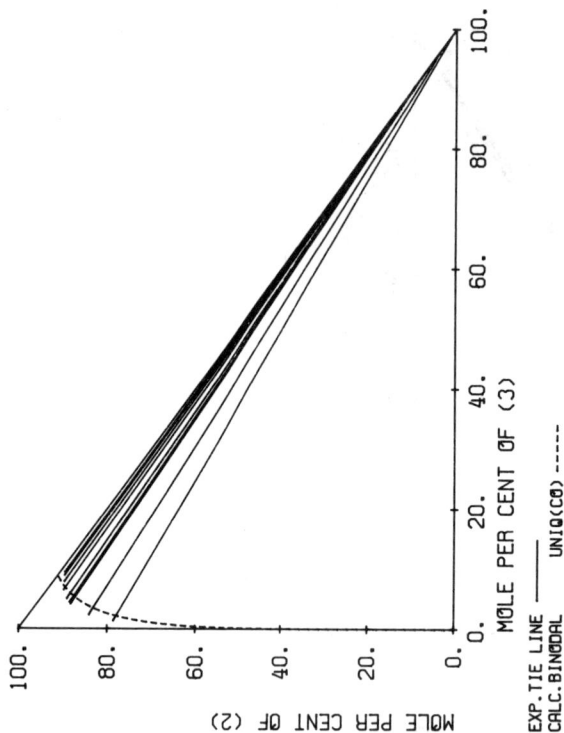

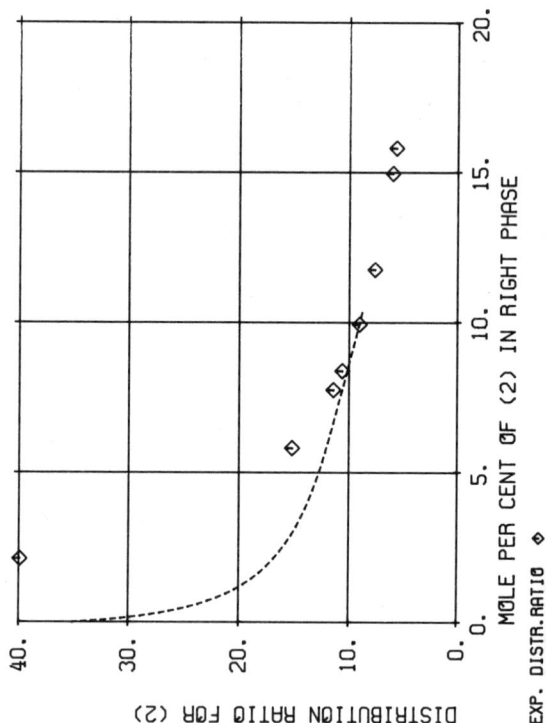

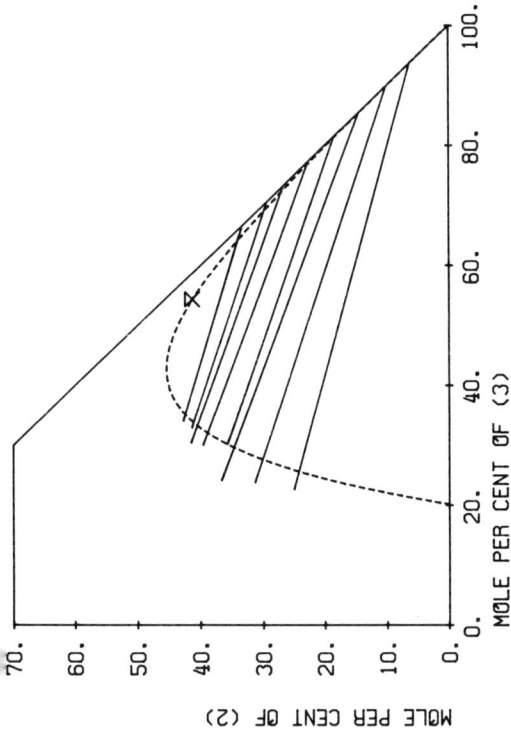

MOLE PER CENT OF (2)

MOLE PER CENT OF (3)

EXP.TIE LINE ———
CALC.BINODAL - - - - UNIQ(C0) - - - -
CALC.PLAIT P. X

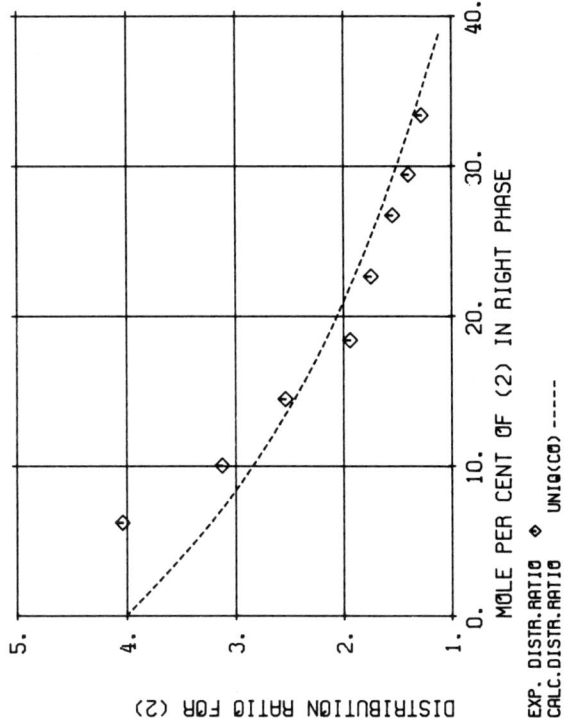

DISTRIBUTION RATIO FOR (2)

MOLE PER CENT OF (2) IN RIGHT PHASE

EXP. DISTR.RATIO ◇ UNIQ(C0) - - - -
CALC.DISTR.RATIO ◆

(2) CH4O   METHANOL

(3) H2O   WATER

LEBEDINSKAYA N.A., ET AL.
KHIM.PROM-ST.(MOSCOW) (1976)16

TEMPERATURE = 20.0 DEG C   TYPE OF SYSTEM = 1

EXPERIMENTAL TIE LINES IN MOLE PCT

LEFT PHASE			RIGHT PHASE		
(1)	(2)	(3)	(1)	(2)	(3)
52.274	25.026	22.700	0.015	6.194	93.792
44.850	31.329	23.822	0.030	10.018	89.952
39.151	36.643	24.206	0.062	14.435	85.503
34.031	35.727	30.242	0.079	18.399	81.521
30.357	39.572	30.071	0.115	22.620	77.265
28.102	41.473	30.426	0.169	26.720	73.111
25.645	41.366	32.990	0.224	29.423	70.353
23.140	42.738	34.122	0.265	33.369	66.366

MEAN DEV. BETWEEN CALC. AND EXP. CONC. IN MOLE PCT

UNIQUAC (COMMON PARAMETERS)   3.27

$CH_4O-C_{11}H_{10}$

(1) H2O	WATER
(2) CH4O	METHANOL
(3) C11H10	NAPHTHALENE, 1-METHYL

PRUTTON C.F., WALSH T.J., DESAI A.M.
IND.ENG.CHEM. 42(1950)1210

TEMPERATURE = 25.0 DEG C   TYPE OF SYSTEM = 1

EXPERIMENTAL TIE LINES IN MOLE PCT

	LEFT PHASE			RIGHT PHASE	
	(1)	(2)	(3)	(1)	(2)	(3)
100.000	0.0	0.0	0.0	0.0	100.000
72.736	27.264	0.0	0.0	4.291	95.709
60.432	39.568	0.0	0.0	5.115	94.885
36.683	62.736	0.581	0.0	7.523	92.477
29.376	69.280	1.344	0.0	10.217	89.783
24.010	73.565	2.425	0.0	12.070	87.930

SPECIFIC MODEL PARAMETERS IN KELVIN

		UNIQUAC		NRTL(ALPHA=.2)	
I J		AIJ	AJI	AIJ	AJI
1 2		-41.470	65.988	125.92	87.484
1 3		342.70	850.09	1294.7	1334.6
2 3		106.89	446.30	937.33	196.54

R1 = 0.9200   R2 = 1.4311   R3 = 5.7158
Q1 = 1.400   Q2 = 1.432   Q3 = 4.008

MEAN DEV. BETWEEN CALC. AND EXP. CONC. IN MOLE PCT

UNIQUAC (SPECIFIC PARAMETERS)	0.74
NRTL (SPECIFIC PARAMETERS)	0.61
UNIQUAC (COMMON PARAMETERS)	1.22

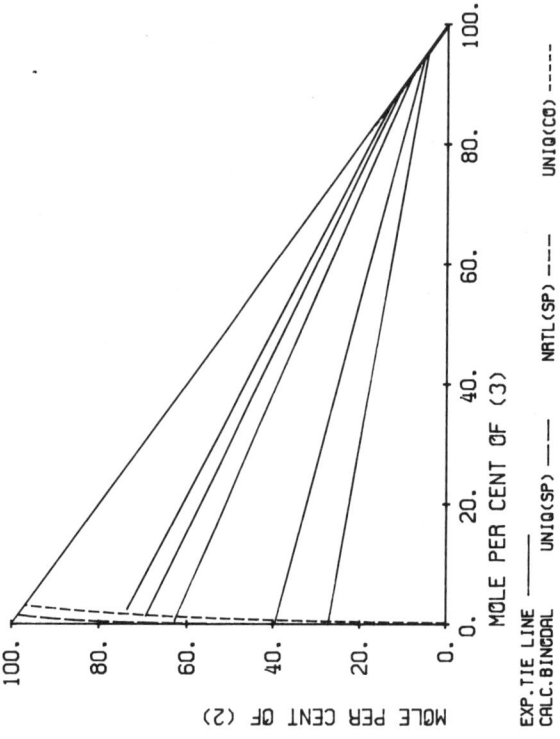

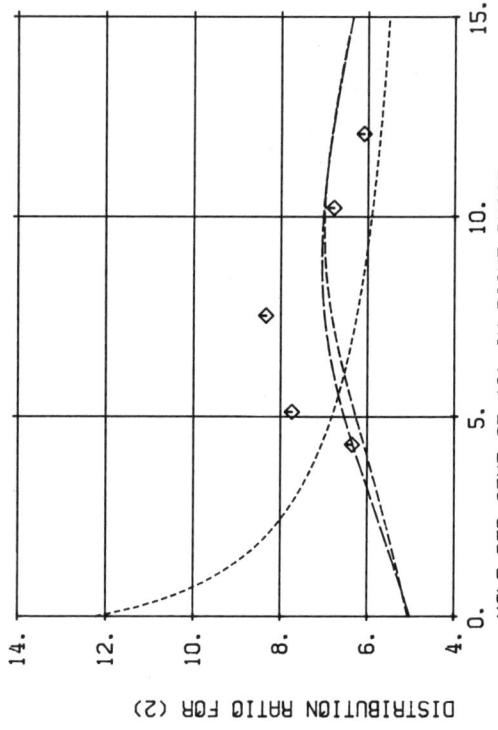

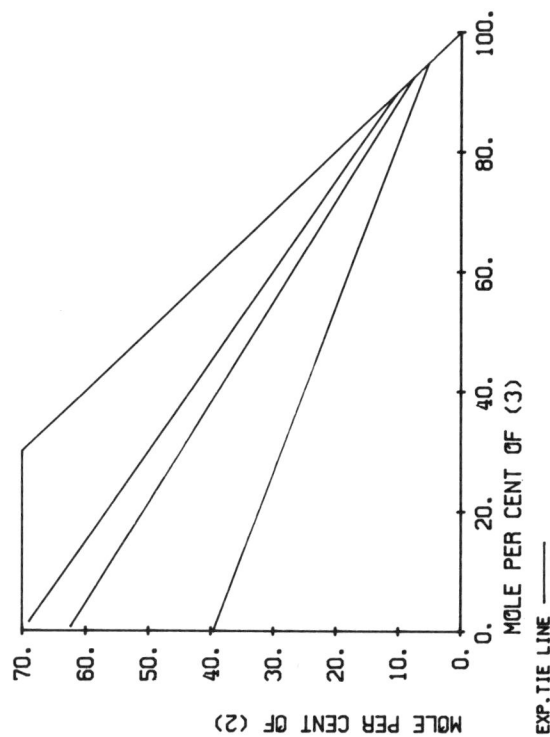

(1) H2O   WATER
----
(2) CH4O   METHANOL
----
(3) C11H10   NAPHTHALENE,1-METHYL

PRUTTON C.F., WALSH T.J., DESAI A.M.
IND.ENG.CHEM. 42(1950)1210

TEMPERATURE = 35.0 DEG C   TYPE OF SYSTEM = 1

EXPERIMENTAL TIE LINES IN MOLE PCT

        LEFT PHASE   RIGHT PHASE
    (1)   (2)   (3)   (1)   (2)   (3)

100.000   0.0   0.0   0.0   0.0   100.000
 60.432   39.568   0.0   0.0   5.115   94.885
 36.929   62.291   0.780   0.0   7.523   92.477
 29.586   68.852   1.562   0.0   10.217   89.783

CH$_4$O-C$_{12}$H$_{10}$O

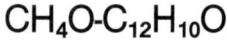

(1) H2O	WATER	
(2) CH4O	METHANOL	
(3) C12H10O	ETHER,DIPHENYL	

PURNELL J.H., BOWDEN S.T.
J.CHEM.SOC. (1954)539

TEMPERATURE = 25.0 DEG C   TYPE OF SYSTEM = 1

EXPERIMENTAL TIE LINES IN MOLE PCT

	LEFT PHASE			RIGHT PHASE	
(1)	(2)	(3)	(1)	(2)	(3)
81.691	18.309	0.0	0.0	1.053	98.947
70.302	29.698	0.0	0.0	3.610	96.390
53.202	46.755	0.043	0.0	5.579	94.421
42.556	57.180	0.264	0.0	7.484	92.516
31.624	67.531	0.845	0.0	11.554	88.446
22.626	75.747	1.627	0.0	20.021	79.979
13.461	82.926	3.613	0.0	30.106	69.894

SPECIFIC MODEL PARAMETERS IN KELVIN

		UNIQUAC		NRTL(ALPHA=.2)	
I J	AIJ	AJI		AIJ	AJI
1 2	-423.70	-74.877		-938.25	554.01
1 3	293.85	649.97		1381.1	1149.9
2 3	-25.771	408.82		807.56	-68.632

R1 = 0.9200   R2 = 1.4311   R3 = 6.2873
Q1 = 1.400   Q2 = 1.432   Q3 = 4.480

MEAN DEV. BETWEEN CALC. AND EXP. CONC. IN MOLE PCT

UNIQUAC (SPECIFIC PARAMETERS)	0.56
NRTL (SPECIFIC PARAMETERS)	0.54
UNIQUAC (COMMON PARAMETERS)	2.25

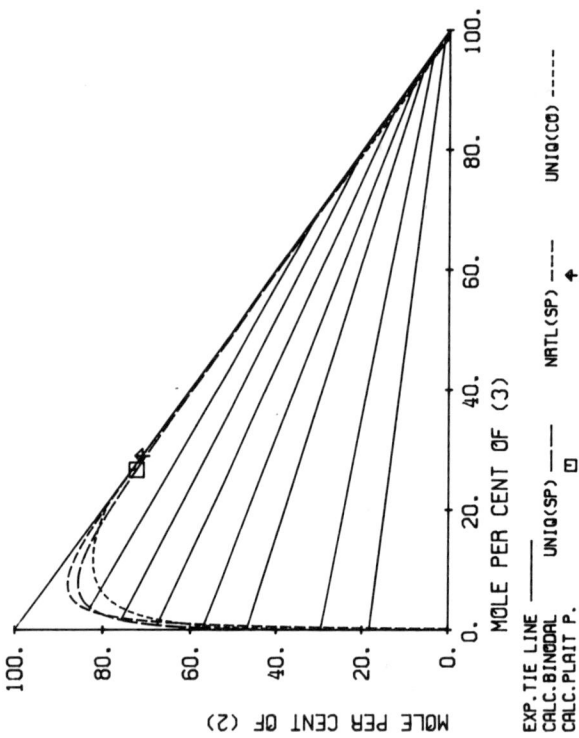

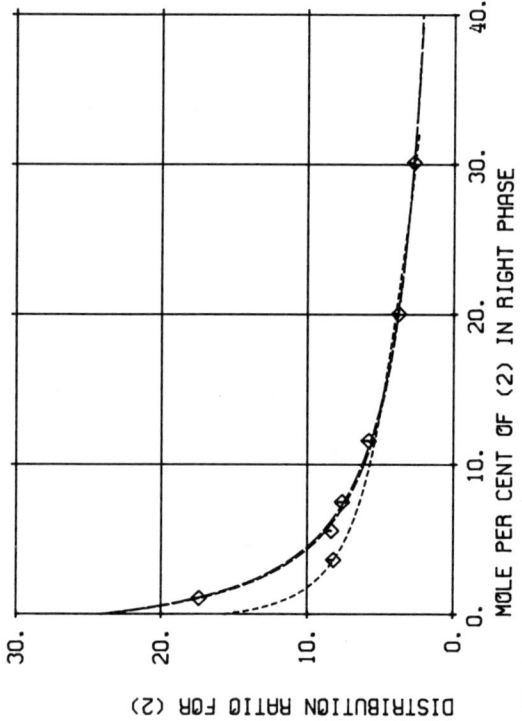

$CH_4O-C_{12}H_{24}$

MOLE PER CENT OF (3)

MOLE PER CENT OF (2)

EXP.TIE LINE ——— UNIQ(SP) —— □ NRTL(SP) ---- ↟
CALC.BINODAL
CALC.PLAIT P.

DISTRIBUTION RATIO FOR (2)

MOLE PER CENT OF (2) IN RIGHT PHASE

EXP. DISTR.RATIO —— ◇ UNIQ(SP) —— NRTL(SP) ----
CALC.DISTR.RATIO ◇

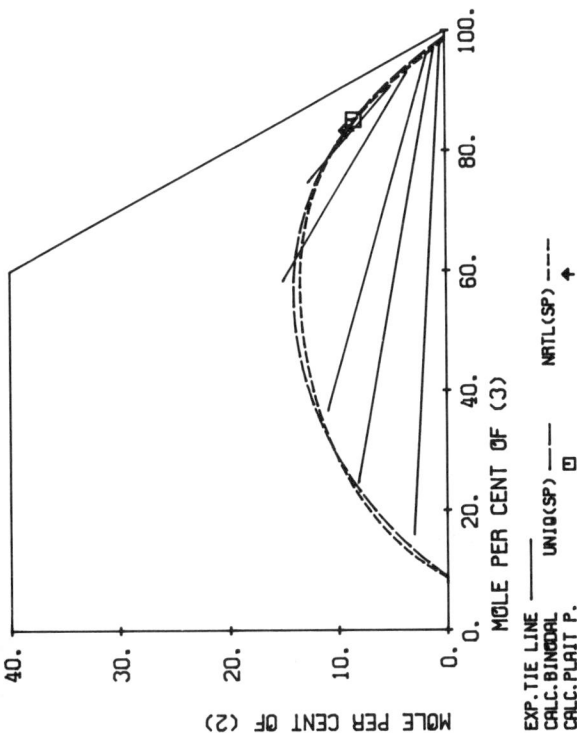

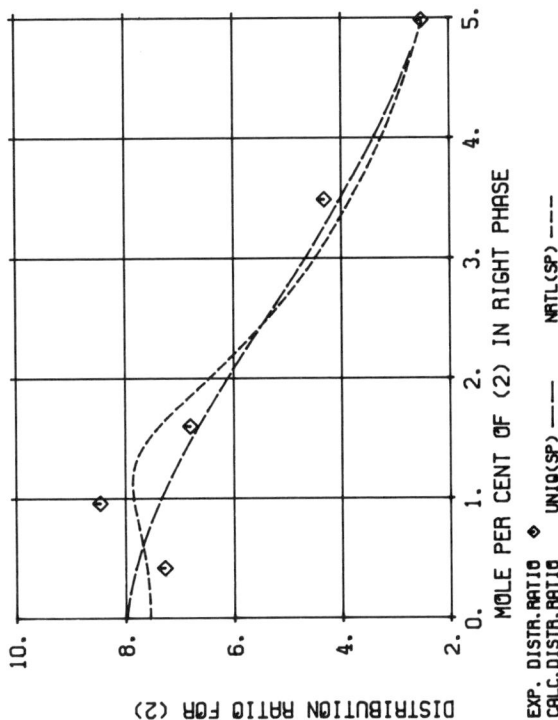

(2) C12H24O   DODECANE, 1,2-EPOXY
(3) CH4O   METHANOL

VOJTKO J., HRUSOVSKY M., FANCOVIC K., RATTAY V., HARGAS R.
CHEM.ZVESTI 27(1973)477

TEMPERATURE = 0.5 DEG C   TYPE OF SYSTEM = 1

EXPERIMENTAL TIE LINES IN MOLE PCT

	LEFT PHASE			RIGHT PHASE	
(1)	(2)	(3)	(1)	(2)	(3)
92.130	0.0	7.870	1.327	0.0	98.673
80.920	3.029	16.051	1.594	0.416	97.990
67.250	8.119	24.630	1.886	0.959	97.155
52.411	10.875	36.714	2.036	1.596	96.368
26.603	15.042	58.355	3.196	3.484	93.320
12.605	12.625	74.770	4.111	4.981	90.908

SPECIFIC MODEL PARAMETERS IN KELVIN

		UNIQUAC		NRTL(ALPHA=.2)	
I	J	AIJ	AJI	AIJ	AJI
1	2	20.602	-17.042	-764.86	-110.61
1	3	675.97	32.274	290.81	954.70
2	3	54.627	105.85	-833.77	1125.7

R1 = 8.3161   R2 = 8.3359   R3 = 1.4311
Q1 = 6.884   Q2 = 6.716   Q3 = 1.432

MEAN DEV. BETWEEN CALC. AND EXP. CONC. IN MOLE PCT

UNIQUAC (SPECIFIC PARAMETERS)   0.54
NRTL   (SPECIFIC PARAMETERS)   0.68

CS$_2$-C$_6$F$_{14}$

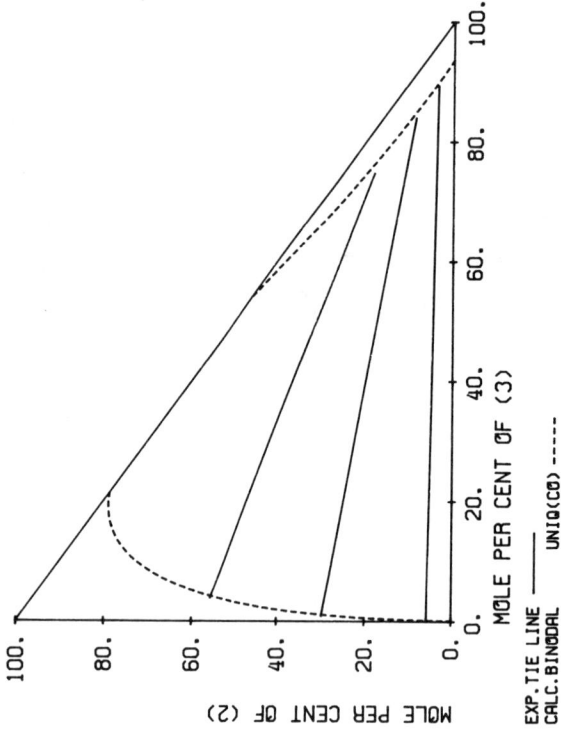

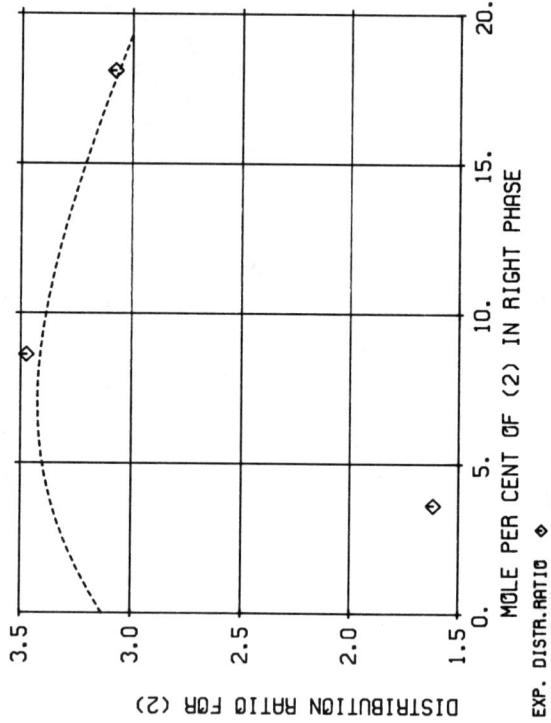

(1) CS2   CARBON DISULFIDE

(2) C6H14   HEXANE

(3) C6F14   HEXANE,PERFLUORO

PLISKIN I., TREYBAL R.E.
J.CHEM.ENG.DATA 11(1966)49

TEMPERATURE = 25.0 DEG C   TYPE OF SYSTEM = 1

EXPERIMENTAL TIE LINES IN MOLE PCT

LEFT PHASE			RIGHT PHASE		
(1)	(2)	(3)	(1)	(2)	(3)
99.887	0.0	0.113	6.373	0.0	93.627
93.994	5.823	0.183	6.939	3.606	89.455
69.134	29.924	0.942	7.410	8.614	83.976
40.440	55.672	3.888	7.188	18.098	74.714

MEAN DEV. BETWEEN CALC. AND EXP. CONC. IN MOLE PCT

UNIQUAC (COMMON PARAMETERS)   0.77

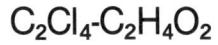

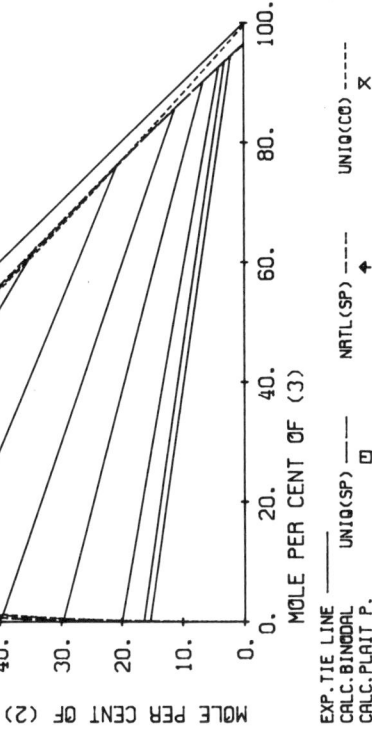

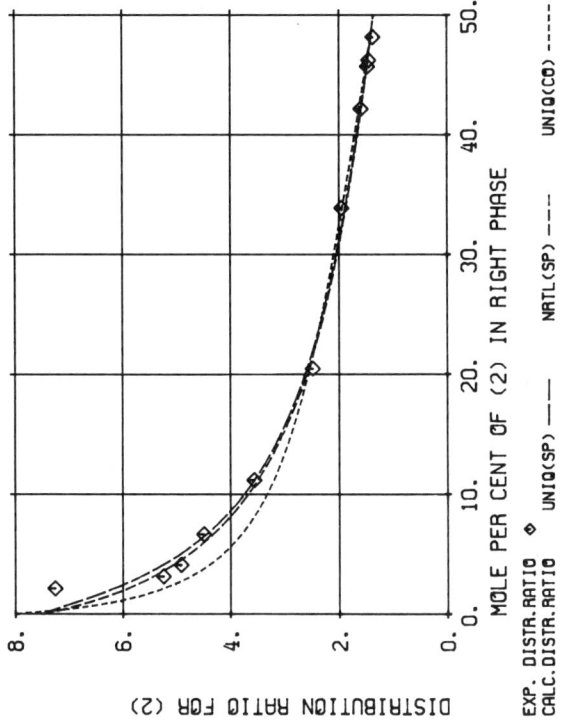

$C_2Cl_4\text{-}C_2H_4O_2$

EXP.TIE LINE   UNIQ(SP) ▭   CALC.BINODAL   NRTL(SP) ----- ◆   CALC.PLAIT P.   UNIQ(CO) ----- ✕

MOLE PER CENT OF (3)

MOLE PER CENT OF (2)

EXP. DISTR.RATIO ◇   UNIQ(SP) ◇   NRTL(SP) -----   UNIQ(CO) -----
CALC.DISTR.RATIO

MOLE PER CENT OF (2) IN RIGHT PHASE

DISTRIBUTION RATIO FOR (2)

---

(2) C2H4O2   ACETIC ACID

(3) C2CL4   ETHENE,TETRACHLORO

FUSE K., IGUCHI A.
KAGAKU KOGAKU 34(1970)1001

TEMPERATURE = 25.0 DEG C   TYPE OF SYSTEM = 1

EXPERIMENTAL TIE LINES IN MOLE PCT (GRAPH.INTERPOL.)

LEFT PHASE		RIGHT PHASE	
(1)	(2)	(3)	(1)	(2)	(3)

(1)	(2)	(3)	(1)	(2)	(3)
84.510	15.311	0.179	3.517	2.110	94.373
83.484	16.334	0.182	3.494	3.118	93.387
79.831	19.943	0.226	3.473	4.063	92.464
69.955	29.671	0.374	3.415	6.608	89.978
59.679	39.720	0.601	3.312	11.151	85.537
47.703	51.051	1.246	3.100	20.459	76.442
27.306	66.362	6.333	3.465	33.891	62.643
23.244	67.738	9.018	3.849	42.144	54.007
22.122	67.896	9.982	4.317	45.701	49.982
20.315	67.682	12.003	4.297	46.223	49.480
21.454	66.893	11.653	4.223	48.137	47.641

SPECIFIC MODEL PARAMETERS IN KELVIN

		UNIQUAC		NRTL(ALPHA=.2)	
I	J	AIJ	AJI	AIJ	AJI
1	2	-249.26	-57.329	-521.57	99.648
1	3	765.22	420.28	2278.1	555.92
2	3	46.855	108.20	511.81	-137.28

R1 = 0.9200   R2 = 2.2024   R3 = 3.8879
Q1 = 1.400   Q2 = 2.072   Q3 = 3.400

MEAN DEV. BETWEEN CALC. AND EXP. CONC. IN MOLE PCT

UNIQUAC (SPECIFIC PARAMETERS)   0.68
NRTL   (SPECIFIC PARAMETERS)   0.67
UNIQUAC (COMMON PARAMETERS)   2.03

C$_2$Cl$_4$-C$_3$H$_6$O

(1) C2CL4   ETHENE,TETRACHLORO

(2) C3H6O   2-PROPANONE

(3) H2O   WATER

RAJA RAO M., VENKATA RAO C.
TRANS.INDIAN INST.CHEM.ENGRS. 7(1954-55)78

TEMPERATURE = 30.0 DEG C   TYPE OF SYSTEM = 1

EXPERIMENTAL TIE LINES IN MOLE PCT

	LEFT PHASE			RIGHT PHASE	
(1)	(2)	(3)	(1)	(2)	(3)
93.135	5.988	0.877	0.012	3.085	96.903
85.961	13.204	0.835	0.025	5.906	94.069
78.708	20.106	1.186	0.065	8.790	91.144
69.836	29.057	1.106	0.096	11.727	88.176
62.172	36.446	1.382	0.115	14.014	85.871
55.013	41.766	3.221	0.150	16.402	83.449
47.530	47.701	4.769	0.154	18.325	81.520
39.534	53.406	7.061	0.177	21.002	78.821
27.754	58.510	13.736	0.345	25.172	74.483
14.714	57.710	27.577	0.754	30.229	69.018

SPECIFIC MODEL PARAMETERS IN KELVIN

		UNIQUAC		NRTL(ALPHA=.2)	
I	J	AIJ	AJI	AIJ	AJI
1	2	360.27	-119.46	483.32	-116.92
1	3	1504.4	605.60	2022.0	2382.4
2	3	343.18	-83.354	291.60	259.50

R1 = 3.8879   R2 = 2.5735   R3 = 0.9200
Q1 = 3.400   Q2 = 2.336   Q3 = 1.400

MEAN DEV. BETWEEN CALC. AND EXP. CONC. IN MOLE PCT

UNIQUAC (SPECIFIC PARAMETERS)   0.78
NRTL   (SPECIFIC PARAMETERS)   0.67
UNIQUAC (COMMON PARAMETERS)   1.17

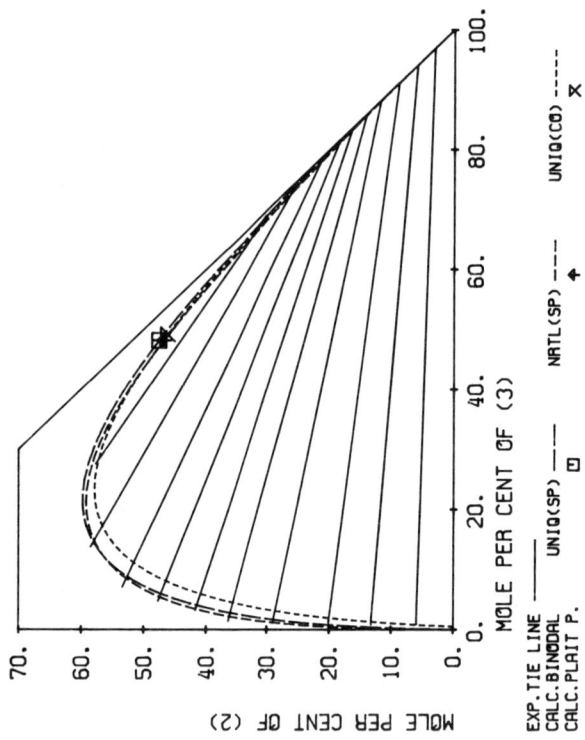

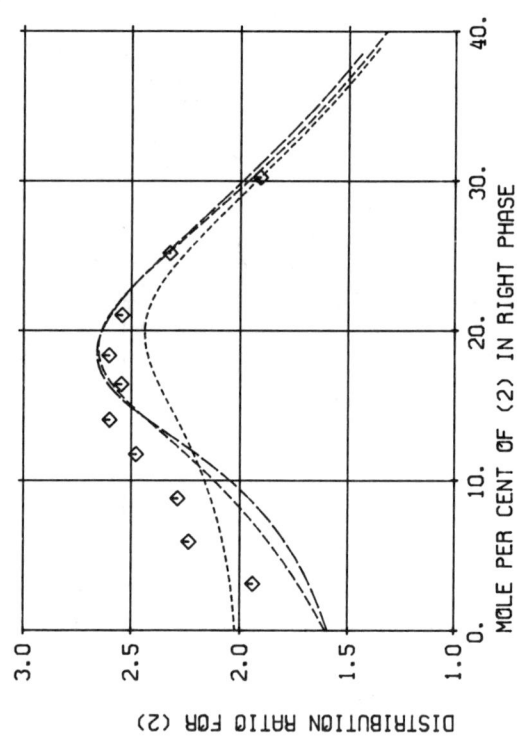

C_2Cl_4-$C_3H_6O_2$

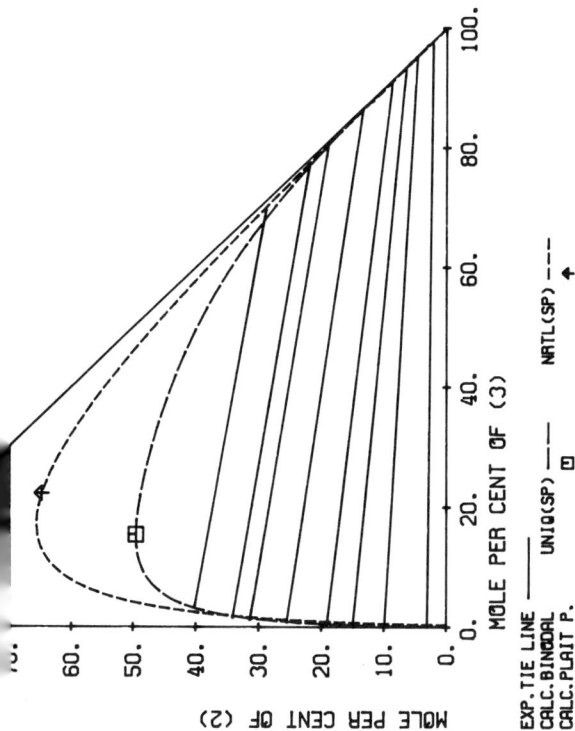

MOLE PER CENT OF (3)

MOLE PER CENT OF (2)

EXP.TIE LINE ——— UNIQ(SP) ——— ☐ NRTL(SP) ----
CALC.BINODAL
CALC.PLAIT P.

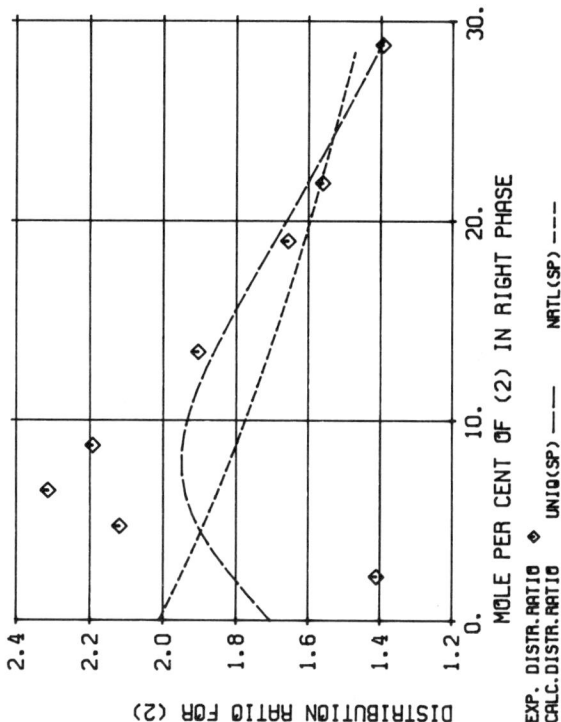

DISTRIBUTION RATIO FOR (2)

MOLE PER CENT OF (2) IN RIGHT PHASE

EXP. DISTR.RATIO ◇ UNIQ(SP) ——— NRTL(SP) ----
CALC. DISTR.RATIO

---

(2) C3H6O2 PROPANOIC ACID
---
(3) H2O WATER

RAJA RAO M., VENKATA RAO C.
J.APPL.CHEM. 6(1956)269

TEMPERATURE = 31.0 DEG C   TYPE OF SYSTEM = 1

EXPERIMENTAL TIE LINES IN MOLE PCT

	LEFT PHASE			RIGHT PHASE	
(1)	(2)	(3)	(1)	(2)	(3)
96.919	3.081	0.0	0.023	2.186	97.791
90.058	9.942	0.0	0.031	4.693	95.276
85.014	14.986	0.0	0.046	6.473	93.475
80.454	19.136	0.410	0.069	8.729	91.202
73.669	25.547	0.784	0.140	13.413	86.448
67.451	31.422	1.127	0.424	18.970	80.606
64.407	34.124	1.469	0.640	21.877	77.482
56.416	40.145	3.439	1.514	28.789	69.697

SPECIFIC MODEL PARAMETERS IN KELVIN

		UNIQUAC		NRTL(ALPHA=.2)	
I	J	AIJ	AJI	AIJ	AJI
1	2	427.55	-234.64	-285.27	193.24
1	3	1248.1	814.03	1173.1	1577.7
2	3	898.74	-285.01	483.68	-256.39

R1 = 3.8879   R2 = 2.8768   R3 = 0.9200
Q1 = 3.400   Q2 = 2.612   Q3 = 1.400

MEAN DEV. BETWEEN CALC. AND EXP. CONC. IN MOLE PCT

UNIQUAC (SPECIFIC PARAMETERS)   0.46
NRTL   (SPECIFIC PARAMETERS)   0.71

C₂Cl₄-C₃H₈O

(1) H2O   WATER
----
(2) C3H8O   2-PROPANOL
----
(3) C2CL4   ETHENE,TETRACHLORO

BERGELIN O., LOCKHART F.J., BROWN G.G.
TRANS.AM.INST.CHEM.ENG. 39(1943)173

TEMPERATURE = 25.0 DEG C   TYPE OF SYSTEM = 1

EXPERIMENTAL TIE LINES IN MOLE PCT

       LEFT PHASE   RIGHT PHASE
    (1)   (2)   (3)   (1)   (2)   (3)

  95.699   4.301   0.0   0.0   1.638   98.362
  95.223   4.777   0.0   0.0   1.908   98.092
  92.843   7.157   0.0   0.445   4.665   94.890
  92.847   7.153   0.013   0.445   4.665   94.890
  91.158   8.829   0.013   0.870   7.305   91.825
  90.762   9.225   0.013   0.868   7.801   91.332
  90.085   9.875   0.040   1.716   8.279   90.006
  88.527   11.376   0.097   1.685   10.980   87.335
  85.436   14.302   0.262   3.990   15.305   80.705
  84.032   15.620   0.348   3.205   15.848   80.947
  79.811   19.363   0.826   4.633   18.792   76.574
  74.906   23.506   1.587   5.339   19.429   75.232
  70.592   26.831   2.576   6.676   21.120   72.204
  61.338   33.089   5.573   8.492   24.391   67.117
  56.890   35.710   7.400   9.033   25.823   65.144
  54.560   36.925   8.515   10.136   27.542   62.322
  48.402   39.498   12.100   11.739   29.317   58.944
  45.174   40.240   14.586   14.665   32.231   53.104

SPECIFIC MODEL PARAMETERS IN KELVIN

              UNIQUAC   NRTL(ALPHA=.2)
  I J   AIJ   AJI   AIJ   AJI
----
  1 2 -1.4582   -112.24   319.11   -280.94
  1 3   335.30   702.78   1877.8   708.87
  2 3   -85.104   205.12   -100.12   328.53

  R1 = 0.9200   R2 = 2.7791   R3 = 3.8879
  Q1 = 1.400   Q2 = 2.508   Q3 = 3.400

MEAN DEV. BETWEEN CALC. AND EXP. CONC. IN MOLE PCT

  UNIQUAC (SPECIFIC PARAMETERS)   1.92
  NRTL   (SPECIFIC PARAMETERS)   1.72
  UNIQUAC (COMMON PARAMETERS)   1.57

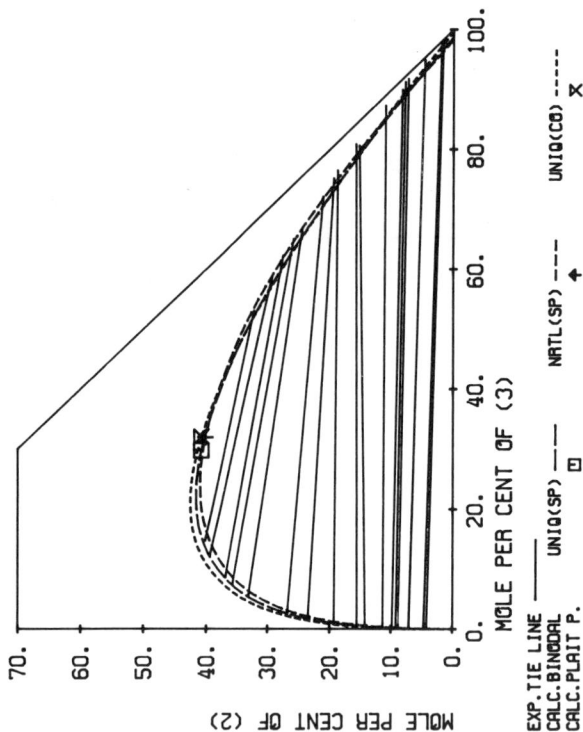

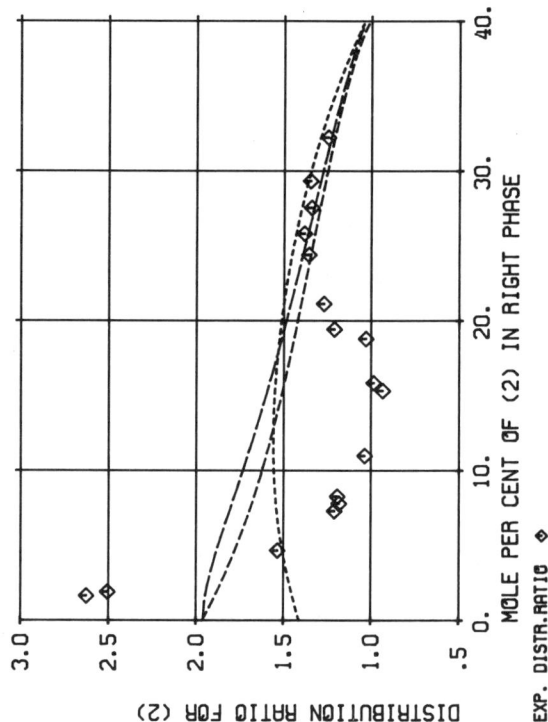

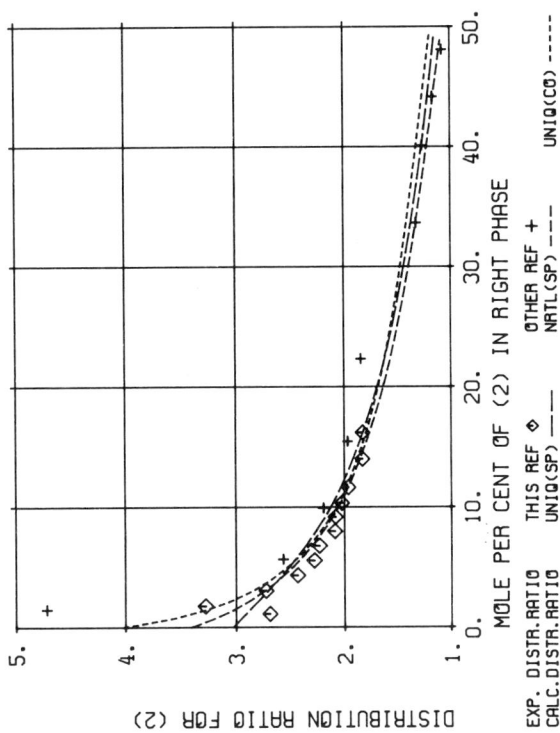

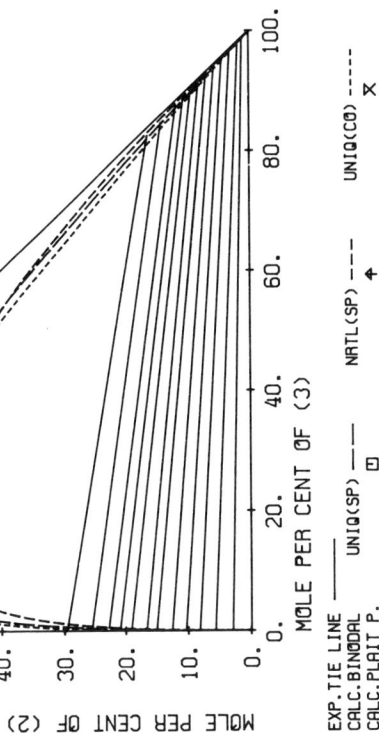

(2) C2H4O2   ACETIC ACID

(3) C2HCL3   ETHENE,TRICHLORO

KRISHNAMURTY V.V.G., MURTI P.S., VENKATA RAO C.
J.SCI.IND.RES. 12B(1953)583

TEMPERATURE = 30.0 DEG C   TYPE OF SYSTEM = 1

EXPERIMENTAL TIE LINES IN MOLE PCT

	LEFT PHASE			RIGHT PHASE	
(1)	(2)	(3)	(1)	(2)	(3)
97.067	2.918	0.015	0.0	1.088	98.912
94.306	5.679	0.016	0.0	1.734	98.266
91.794	8.190	0.016	0.0	3.013	96.987
89.589	10.377	0.034	0.0	4.274	95.726
87.384	12.563	0.053	0.0	5.518	94.482
84.822	15.048	0.130	0.0	6.745	93.255
83.173	16.635	0.192	0.0	7.955	92.045
80.643	19.056	0.301	0.0	9.149	90.851
78.699	20.886	0.415	0.0	10.327	89.673
76.691	22.773	0.536	0.679	11.612	87.709
73.816	25.533	0.651	1.331	13.978	84.691
69.570	29.471	0.959	1.314	16.164	82.522

SPECIFIC MODEL PARAMETERS IN KELVIN

		UNIQUAC		NRTL(ALPHA=.2)	
I	J	AIJ	AJI	AIJ	AJI
1	2	-290.66	-48.216	-198.29	-204.11
1	3	535.20	1034.4	2156.2	1566.1
2	3	79.159	-104.48	222.60	-247.48

R1 = 0.9200   R2 = 2.2024   R3 = 3.3092
Q1 = 1.400   Q2 = 2.072   Q3 = 2.860

MEAN DEV. BETWEEN CALC. AND EXP. CONC. IN MOLE PCT

UNIQUAC (SPECIFIC PARAMETERS)   0.36
NRTL   (SPECIFIC PARAMETERS)   0.22
UNIQUAC (COMMON PARAMETERS)   0.60

C$_2$HCl$_3$-C$_2$H$_4$O$_2$

(1) H2O   WATER
(2) C2H4O2   ACETIC ACID
(3) C2HCL3   ETHENE,TRICHLORO

FUSE K., IGUCHI A.
KAGAKU KOGAKU 34(1970)543

TEMPERATURE = 25.0 DEG C   TYPE OF SYSTEM = 1

EXPERIMENTAL TIE LINES IN MOLE PCT (GRAPH.INTERPOL.)

	LEFT PHASE			RIGHT PHASE	
(1)	(2)	(3)	(1)	(2)	(3)
92.964	6.827	0.209	9.918	1.448	88.634
85.323	14.398	0.278	9.691	5.621	84.688
77.793	21.743	0.463	9.458	9.912	80.629
68.559	30.345	1.096	9.159	15.424	75.417
56.050	41.013	2.937	8.785	22.314	68.900
51.646	44.441	3.913	8.664	33.660	57.675
41.310	50.740	7.950	8.763	40.082	51.155
37.695	51.661	10.644	9.850	44.160	45.990
33.021	51.902	15.077	13.981	48.071	37.948

SPECIFIC MODEL PARAMETERS IN KELVIN

		UNIQUAC		NRTL(ALPHA=.2)	
I	J	AIJ	AJI	AIJ	AJI
1	2	-290.66	-48.216	-198.29	-204.11
1	3	535.20	1034.4	2156.2	1565.1
2	3	79.159	-104.48	222.60	-247.48

R1 = 0.9200   R2 = 2.2024   R3 = 3.3092
Q1 = 1.400   Q2 = 2.072   Q3 = 2.860

MEAN DEV. BETWEEN CALC. AND EXP. CONC. IN MOLE PCT

UNIQUAC (SPECIFIC PARAMETERS)   3.75
NRTL   (SPECIFIC PARAMETERS)   3.54
UNIQUAC (COMMON PARAMETERS)   3.59

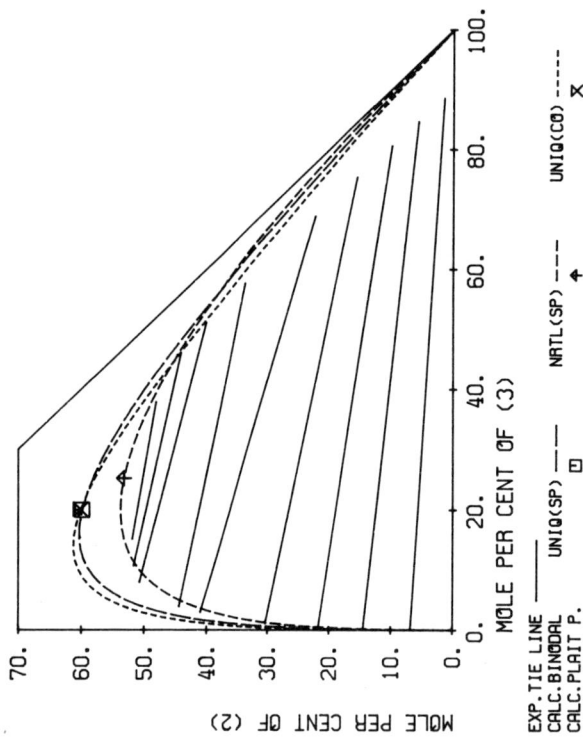

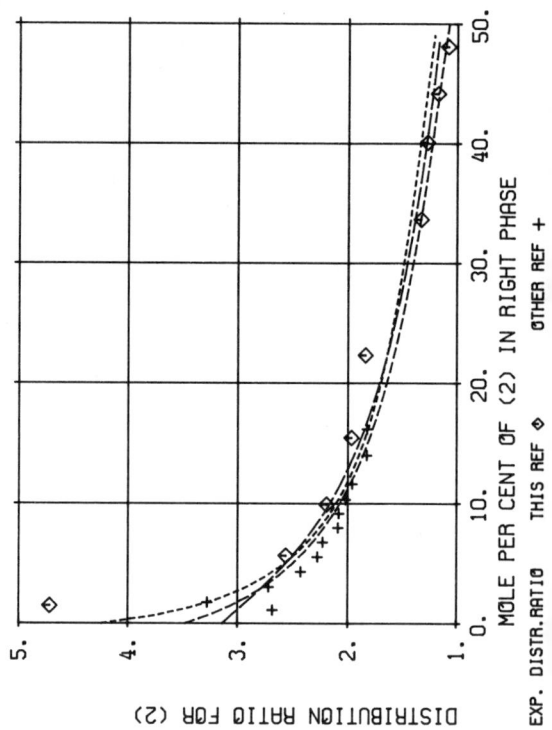

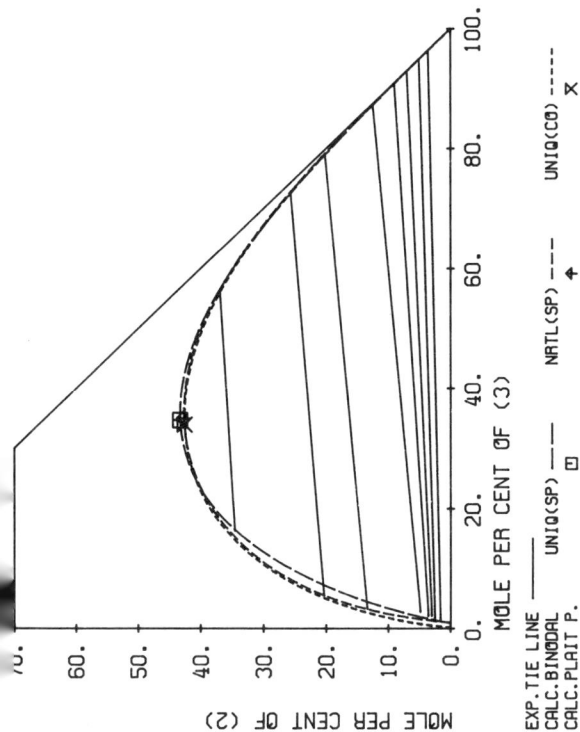

MOLE PER CENT OF (3)

MOLE PER CENT OF (2)

EXP.TIE LINE	——————	UNIQ(SP)	☐	NRTL(SP)	---	UNIQ(CO)	-----
CALC.BINODAL							✗
CALC.PLAIT P.					♦		

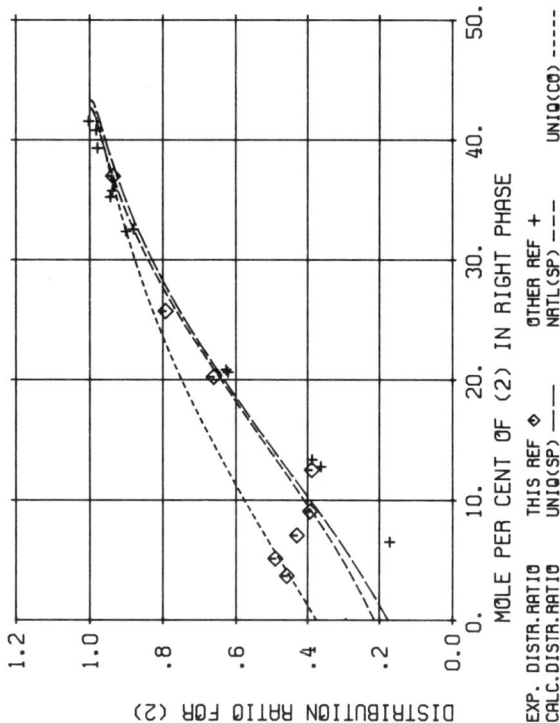

MOLE PER CENT OF (2) IN RIGHT PHASE

DISTRIBUTION RATIO FOR (2)

| EXP.DISTR.RATIO | ♦ | THIS REF + | | OTHER REF + | | | |
| CALC.DISTR.RATIO | | UNIQ(SP) ——— | | NRTL(SP) --- | | UNIQ(CO) ----- | |

(2) C2H6O   ETHANOL
(3) H2O   WATER

COLBURN A.P., PHILLIPS J.C.
TRANS.AM.INST.CHEM.ENG. 40(1944)333

TEMPERATURE =   25.0 DEG C   TYPE OF SYSTEM = 1

EXPERIMENTAL TIE LINES IN MOLE PCT

| | LEFT PHASE | | | RIGHT PHASE | |
(1)	(2)	(3)	(1)	(2)	(3)
96.904	1.672	1.425	0.073	3.653	96.275
96.089	2.494	1.417	0.089	5.091	94.820
94.876	3.019	2.105	0.107	7.027	92.866
94.347	3.555	2.098	0.126	9.027	90.847
92.394	4.850	2.756	0.165	12.485	87.350
83.391	13.381	3.228	0.555	20.215	79.230
74.235	20.375	5.390	1.387	25.714	72.899
49.174	34.608	16.218	6.922	36.964	56.114

SPECIFIC MODEL PARAMETERS IN KELVIN

| | | UNIQUAC | | NRTL(ALPHA=.2) | |
I	J	AIJ	AJI	AIJ	AJI
1	2	7.5913	-90.570	-1049.4	52.334
1	3	714.75	461.75	931.38	2434.4
2	3	-249.88	-148.10	-406.94	-959.95

R1 = 3.3092   R2 = 2.1055   R3 = 0.9200
Q1 = 2.860   Q2 = 1.972   Q3 = 1.400

MEAN DEV. BETWEEN CALC. AND EXP. CONC. IN MOLE PCT

UNIQUAC (SPECIFIC PARAMETERS)	0.53
NRTL (SPECIFIC PARAMETERS)	0.46
UNIQUAC (COMMON PARAMETERS)	0.73

C$_2$HCl$_3$-C$_2$H$_6$O

(1) C2HCL3   ETHENE,TRICHLORO
----
(2) C2H6O   ETHANOL
----
(3) H2O   WATER

REINDERS W., DE MINJER C.H.
RECL.TRAV.CHIM.PAYS-BAS. 66(1947)552

TEMPERATURE = 20.0 DEG C   TYPE OF SYSTEM = 1

EXPERIMENTAL TIE LINES IN MOLE PCT

      LEFT PHASE   RIGHT PHASE
    (1)   (2)   (3)   (1)   (2)   (3)

  97.452   1.118   1.430   0.051   6.480   93.460
  93.959   4.644   1.397   0.156   12.767   87.067
  93.436   5.172   1.392   0.157   13.338   86.495
  85.202   12.830   1.969   0.432   20.607   78.912
  84.970   13.064   1.965   0.483   20.817   78.701
  60.338   29.159   10.503   3.391   32.561   64.237
  60.797   28.658   10.545   3.560   32.561   63.879
  53.122   33.266   13.611   5.164   35.246   59.590
  52.442   33.571   13.987   5.539   35.768   58.693
  41.997   38.487   19.515   9.646   39.312   51.043
  36.131   40.090   23.779   13.029   40.808   46.163
  29.322   41.675   29.003   16.796   41.558   41.645

SPECIFIC MODEL PARAMETERS IN KELVIN

              UNIQUAC   NRTL(ALPHA=.2)
  I J   AIJ   AJI   AIJ   AJI

  1 2   7.5913   -90.570   -1049.4   52.334
  1 3 714.75   461.75   931.38   2434.4
  2 3 -249.88   -148.10   -406.94   -959.95

  R1 = 3.3092 R2 = 2.1055 R3 = 0.9200
  Q1 = 2.860   Q2 = 1.972   Q3 = 1.400

MEAN DEV. BETWEEN CALC. AND EXP. CONC. IN MOLE PCT

UNIQUAC (SPECIFIC PARAMETERS)   0.70
NRTL   (SPECIFIC PARAMETERS)   0.49
UNIQUAC (COMMON PARAMETERS)   0.95

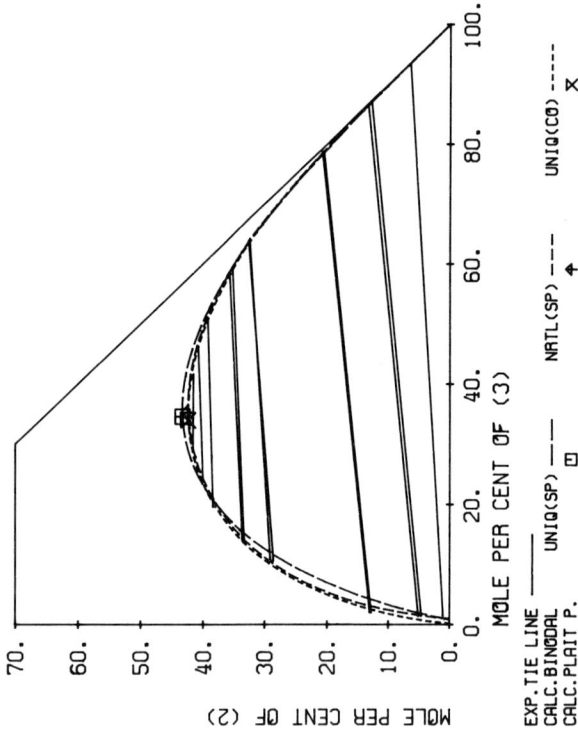

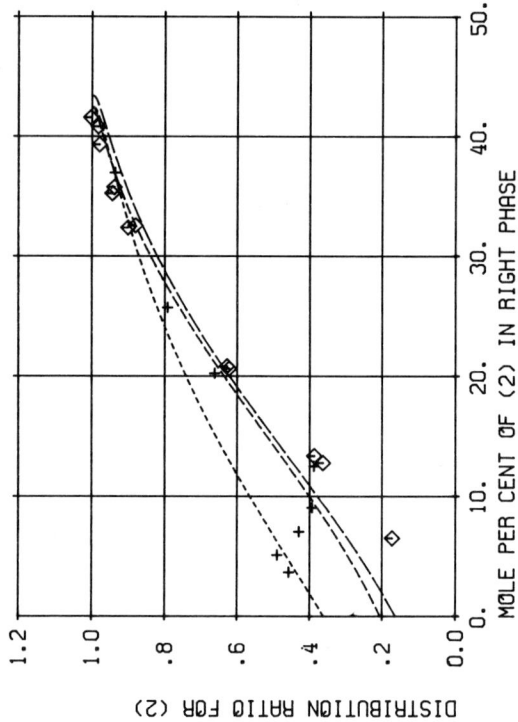

C$_2$HCl$_3$-C$_2$H$_6$O

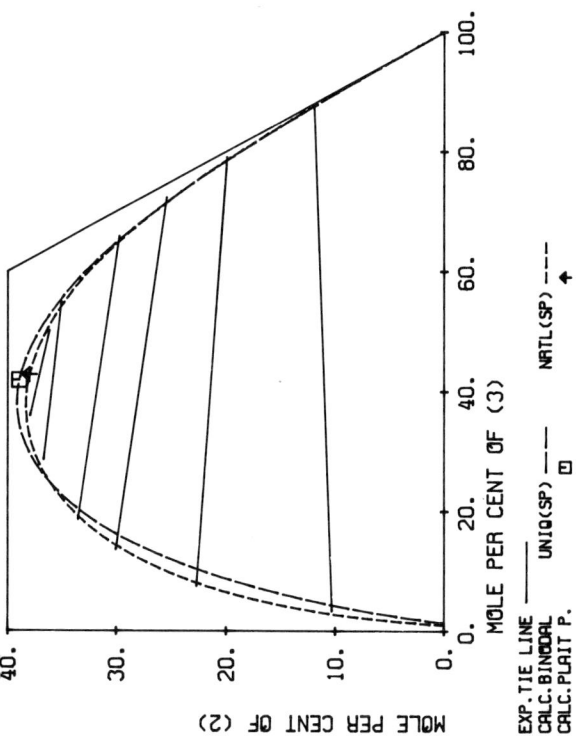

MOLE PER CENT OF (3)

MOLE PER CENT OF (2)

EXP.TIE LINE ——— UNIQ(SP) — ⊡ — NRTL(SP) – – –
CALC.BINODAL
CALC.PLAIT P. ✦

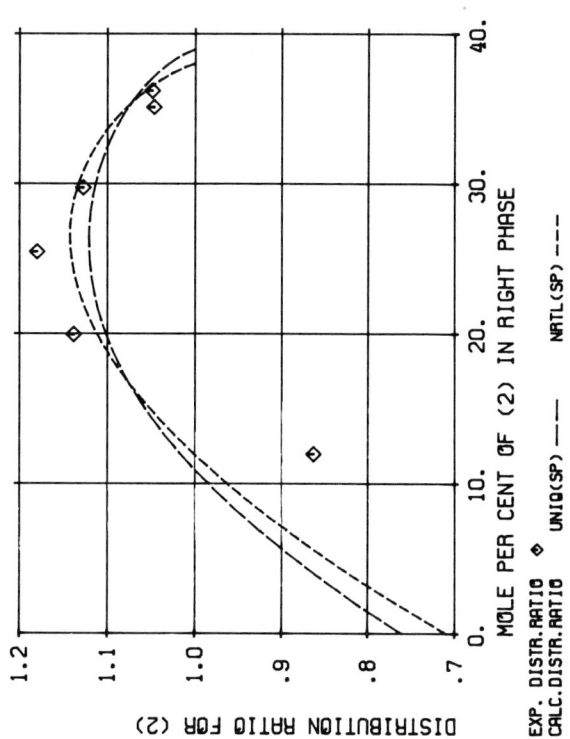

MOLE PER CENT OF (2) IN RIGHT PHASE

DISTRIBUTION RATIO FOR (2)

EXP. DISTR.RATIO ◊ UNIQ(SP) ——— NRTL(SP) – – –
CALC.DISTR.RATIO

(2) C2H6O   ETHANOL

(3) H2O   WATER

REINDERS W., DE MINJER C.H.
RECL.TRAV.CHIM.PAYS-BAS. 66(1947)552

TEMPERATURE = 67.0 DEG C   TYPE OF SYSTEM = 1

EXPERIMENTAL TIE LINES IN MOLE PCT

	LEFT PHASE		RIGHT PHASE	
(1)	(2)	(3)	(1) (2)	(3)
86.383	10.319	3.299	0.281 11.957	87.762
69.826	22.701	7.472	0.959 19.942	79.100
56.424	30.025	13.550	2.141 25.443	72.415
47.884	33.520	18.596	4.331 29.723	65.946
34.719	36.728	28.553	9.949 35.086	54.965
26.130	37.956	35.914	13.654 36.192	50.154

SPECIFIC MODEL PARAMETERS IN KELVIN

		UNIQUAC		NRTL(ALPHA=.2)	
I J		AIJ	AJI	AIJ	AJI
1 2		114.22	-165.92	-469.22	-137.36
1 3		742.70	432.30	1003.0	2212.9
2 3		-211.76	-54.337	-509.30	-83.304

R1 = 3.3092   R2 = 2.1055   R3 = 0.9200
Q1 = 2.860   Q2 = 1.972   Q3 = 1.400

MEAN DEV. BETWEEN CALC. AND EXP. CONC. IN MOLE PCT

UNIQUAC (SPECIFIC PARAMETERS)   0.88
NRTL   (SPECIFIC PARAMETERS)   0.56

C$_2$HCl$_3$-C$_4$H$_6$O$_2$

MOLE PER CENT OF (3)

MOLE PER CENT OF (2)

EXP.TIE LINE ——— UNIQ(SP) ----- NRTL(SP) ----- UNIQ(CO) -----
CALC.BINODAL

DISTRIBUTION RATIO FOR (2)

MOLE PER CENT OF (2) IN RIGHT PHASE

EXP. DISTR.RATIO ◇

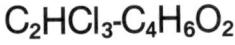

(1) C2HCL3   ETHENE,TRICHLORO
(2) C4H6O2   PROPENOIC ACID,2-METHYL
(3) H2O   WATER

FROLOV A.F., ET AL.
ZH.PRIKL.KHIM.(LENINGRAD) 39(1966)1805

TEMPERATURE = 20.0 DEG C   TYPE OF SYSTEM = 2

EXPERIMENTAL TIE LINES IN MOLE PCT

	LEFT PHASE			RIGHT PHASE	
(1)	(2)	(3)	(1)	(2)	(3)
79.539	20.055	0.406	0.017	0.666	99.317
67.641	31.585	0.774	0.017	0.932	99.050
45.241	52.084	2.675	0.019	1.522	98.459
30.988	63.656	5.356	0.021	1.885	98.094
11.388	70.252	18.360	0.023	2.636	97.342

SPECIFIC MODEL PARAMETERS IN KELVIN

		UNIQUAC		NRTL(ALPHA=.2)	
I J	AIJ	AJI		AIJ	AJI
1 2	-127.82	190.59		-133.74	170.70
1 3	1457.8	1042.1		1832.8	1473.2
2 3	284.04	65.418		79.415	946.49

R1 = 3.3092   R2 = 3.3197   R3 = 0.9200
Q1 = 2.860   Q2 = 3.060   Q3 = 1.400

MEAN DEV. BETWEEN CALC. AND EXP. CONC. IN MOLE PCT

UNIQUAC (SPECIFIC PARAMETERS)   0.89
NRTL   (SPECIFIC PARAMETERS)   1.09
UNIQUAC (COMMON PARAMETERS)   0.99

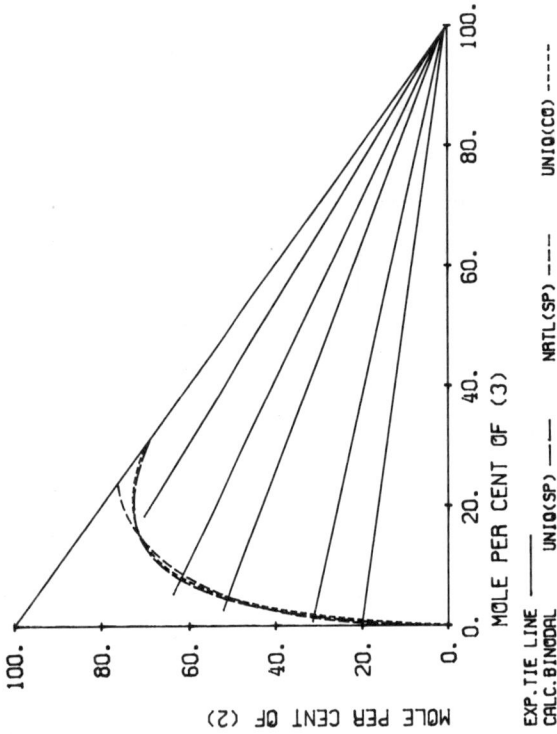

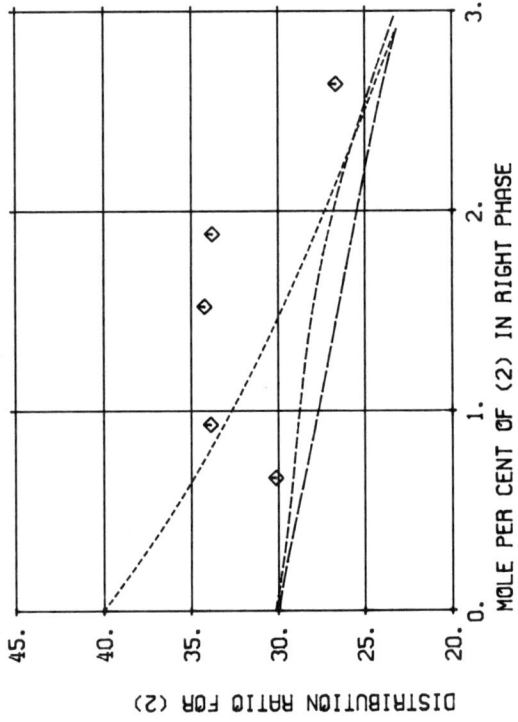

C_2HCl_3-C_4H_8O

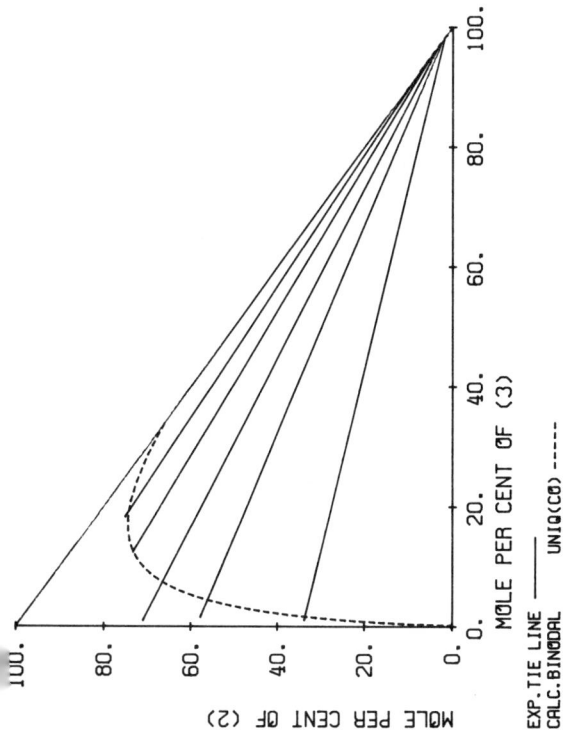

MOLE PER CENT OF (3)

MOLE PER CENT OF (2)

EXP.TIE LINE ———
CALC.BINODAL - - - - - UNIQ(CO) - - - - -

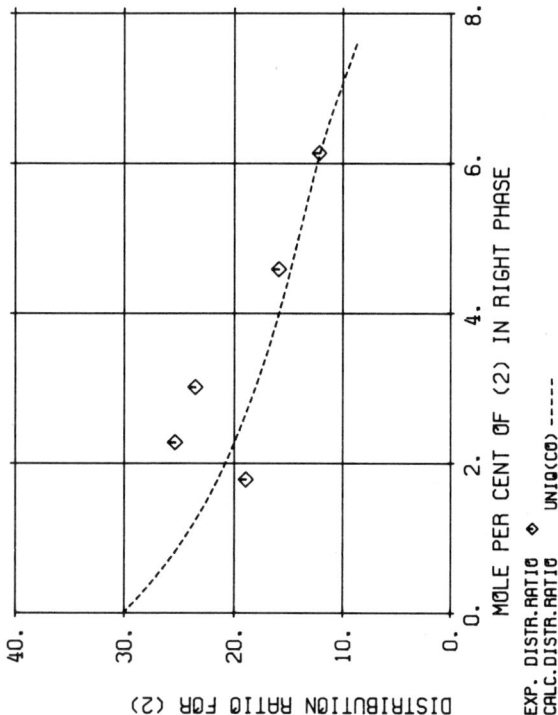

MOLE PER CENT OF (2) IN RIGHT PHASE

DISTRIBUTION RATIO FOR (2)

EXP. DISTR.RATIO ◇
CALC.DISTR.RATIO UNIQ(CO) - - - -

(2)	C4H8O	2-BUTANONE
(3)	H2O	WATER

NEWMAN M., HAYWORTH C.B., TREYBAL R.E.
IND.ENG.CHEM. 41(1949)2039

TEMPERATURE = 25.0 DEG C   TYPE OF SYSTEM = 2

EXPERIMENTAL TIE LINES IN MOLE PCT

LEFT PHASE			RIGHT PHASE		
(1)	(2)	(3)	(1)	(2)	(3)
65.149	33.750	1.101	0.055	1.782	98.163
40.638	57.774	1.588	0.053	2.275	97.672
27.948	70.976	1.076	0.048	3.013	96.939
14.218	73.327	12.455	0.034	4.584	95.382
6.590	75.064	18.346	0.029	6.136	93.835

MEAN DEV. BETWEEN CALC. AND EXP. CONC. IN MOLE PCT

UNIQUAC (COMMON PARAMETERS)   1.67

C$_2$HCl$_3$-C$_5$H$_4$O$_2$

(1) C2HCL3   ETHENE,TRICHLORO

(2) C5H4O2   FURFURAL

(3) H2O   WATER

RUCHAI N.S., RESHTO M.V., KHOLKIN YU.I.
ZH.PRIKL.KHIM.(LENINGRAD) 50(1977)2385

TEMPERATURE = 20.0 DEG C   TYPE OF SYSTEM = 2

EXPERIMENTAL TIE LINES IN MOLE PCT

	LEFT PHASE			RIGHT PHASE	
(1)	(2)	(3)	(1)	(2)	(3)
96.478	3.522	0.0	0.008	0.081	99.911
78.458	20.179	1.362	0.008	0.436	99.556
72.893	25.766	1.341	0.008	0.528	99.464
63.904	32.873	3.223	0.008	0.626	99.365
42.085	52.012	5.903	0.009	0.900	99.091
34.161	57.384	8.455	0.010	0.974	99.016

MEAN DEV. BETWEEN CALC. AND EXP. CONC. IN MOLE PCT

UNIQUAC (COMMON PARAMETERS)   0.25

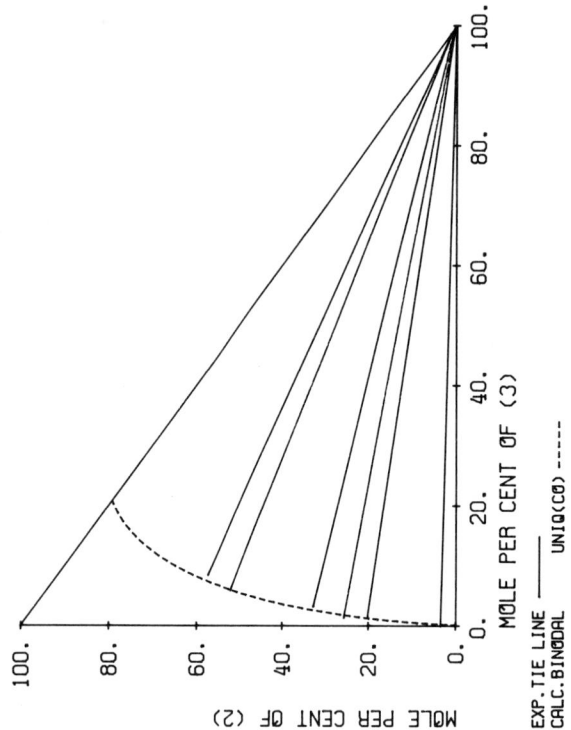

MOLE PER CENT OF (2)

MOLE PER CENT OF (3)

EXP.TIE LINE ———
CALC.BINODAL   UNIQ(C0) - - - - -

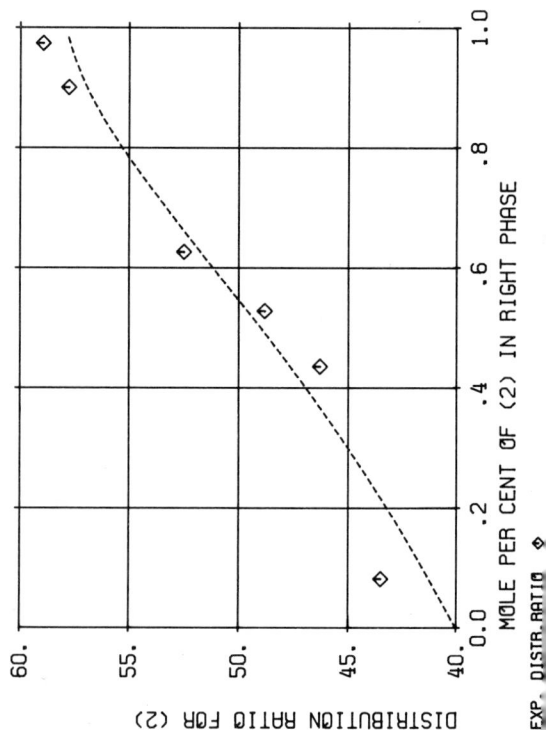

DISTRIBUTION RATIO FOR (2)

MOLE PER CENT OF (2) IN RIGHT PHASE

EXP. DISTR.RATIO ◇

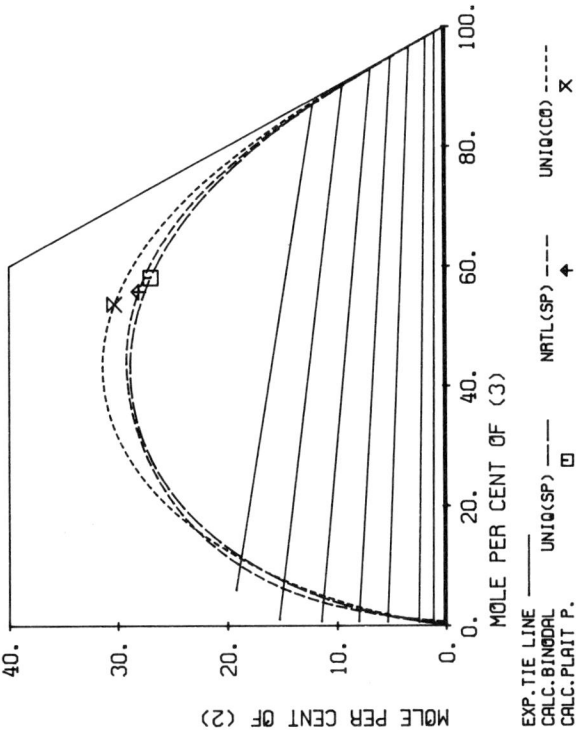

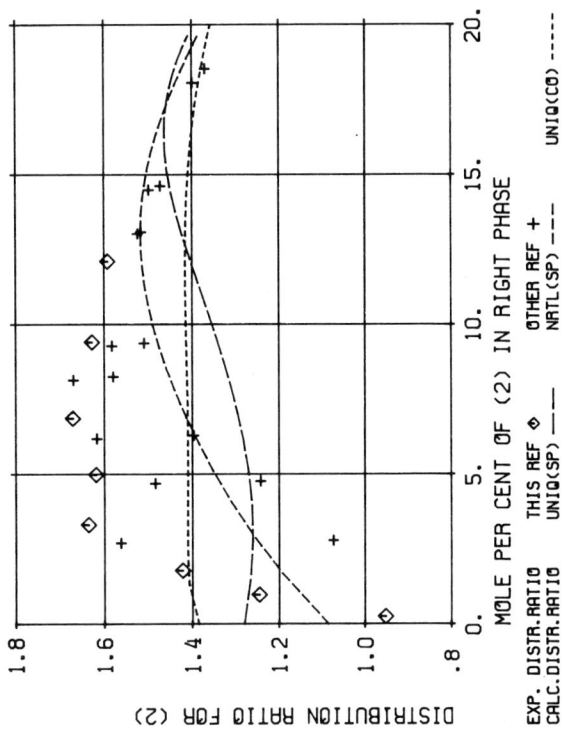

MOLE PER CENT OF (3)

MOLE PER CENT OF (2)

EXP.TIE LINE —— UNIQ(SP) ⊟ NRTL(SP) ▲ UNIQ(CO) ⊠
CALC.BINODAL
CALC.PLAIT P.

DISTRIBUTION RATIO FOR (2)

MOLE PER CENT OF (2) IN RIGHT PHASE

EXP. DISTR.RATIO ◇ THIS REF ◇ OTHER REF + UNIQ(CO) ----
CALC.DISTR.RATIO UNIQ(SP) —— NRTL(SP) +---

(1) C2HCL3   ETHENE,TRICHLORO
(2) C6H11NO   HEXANOIC ACID,6-AMINO,LACTAM
(3) H2O   WATER

TETTAMANTI K., NOGRADI M., SAWINSKY J.
PERIOD.POLYTECH.,CHEM.ENG. 4(1960)201

TEMPERATURE = 20.0 DEG C   TYPE OF SYSTEM = 1

EXPERIMENTAL TIE LINES IN MOLE PCT (GRAPH.INTERPOL.)

	LEFT PHASE			RIGHT PHASE	
(1)	(2)	(3)	(1)	(2)	(3)
99.419	0.0	0.581	0.014	0.0	99.986
99.189	0.231	0.580	0.017	0.242	99.741
98.209	1.211	0.580	0.026	0.972	99.002
96.854	2.532	0.614	0.041	1.781	98.179
93.944	5.409	0.648	0.082	3.306	96.612
91.217	8.066	0.717	0.145	4.983	94.873
87.787	11.465	0.749	0.223	6.857	92.920
83.628	15.311	1.061	0.346	9.403	90.251
74.755	19.262	5.983	0.735	12.100	87.165

SPECIFIC MODEL PARAMETERS IN KELVIN

I	J	UNIQUAC		NRTL(ALPHA=.2)	
		AIJ	AJI	AIJ	AJI
1	2	284.61	-173.08	-1743.8	-430.25
1	3	937.03	-808.07	978.70	2026.7
2	3	-121.04	26.069	-661.97	-1289.9

R1 = 3.3092   R2 = 4.6106   R3 = 0.9200
Q1 = 2.860   Q2 = 3.724   Q3 = 1.400

MEAN DEV. BETWEEN CALC. AND EXP. CONC. IN MOLE PCT

UNIQUAC (SPECIFIC PARAMETERS)   1.09
NRTL   (SPECIFIC PARAMETERS)   0.93
UNIQUAC (COMMON PARAMETERS)   1.16

C$_2$HCl$_3$-C$_6$H$_{11}$NO

(1) C2HCL3   ETHENE,TRICHLORO

(2) C6H11NO   HEXANOIC ACID,6-AMINO,LACTAM

(3) H2O   WATER

SEDMEROVA V.; NOVAK J.P.
CHEM.PRUM. 16/41(1966)270

TEMPERATURE = 20.0 DEG C   TYPE OF SYSTEM = 1

EXPERIMENTAL TIE LINES IN MOLE PCT

LEFT PHASE			RIGHT PHASE		
(1)	(2)	(3)	(1)	(2)	(3)
96.291	2.988	0.722	0.047	2.783	97.170
90.576	5.917	3.506	0.086	4.763	95.151
87.723	8.784	3.492	0.110	6.289	93.601
81.520	13.026	5.455	0.179	8.245	91.576
79.137	14.136	6.727	0.228	9.370	90.402
71.645	19.831	8.523	0.862	13.088	86.051
65.920	21.511	12.569	1.376	14.617	84.007
53.958	25.389	20.652	3.002	18.531	78.467

SPECIFIC MODEL PARAMETERS IN KELVIN

		UNIQUAC		NRTL(ALPHA=.2)	
I J		AIJ	AJI	AIJ	AJI
1 2		284.61	-173.08	-1743.8	-430.25
1 3		937.03	808.07	978.70	2025.7
2 3		-121.04	26.069	-661.97	-1289.9

R1 = 3.3092   R2 = 4.6106   R3 = 0.9200
Q1 = 2.860   Q2 = 3.724   Q3 = 1.400

MEAN DEV. BETWEEN CALC. AND EXP. CONC. IN MOLE PCT

UNIQUAC (SPECIFIC PARAMETERS)   0.77
NRTL   (SPECIFIC PARAMETERS)   0.64
UNIQUAC (COMMON PARAMETERS)   0.55

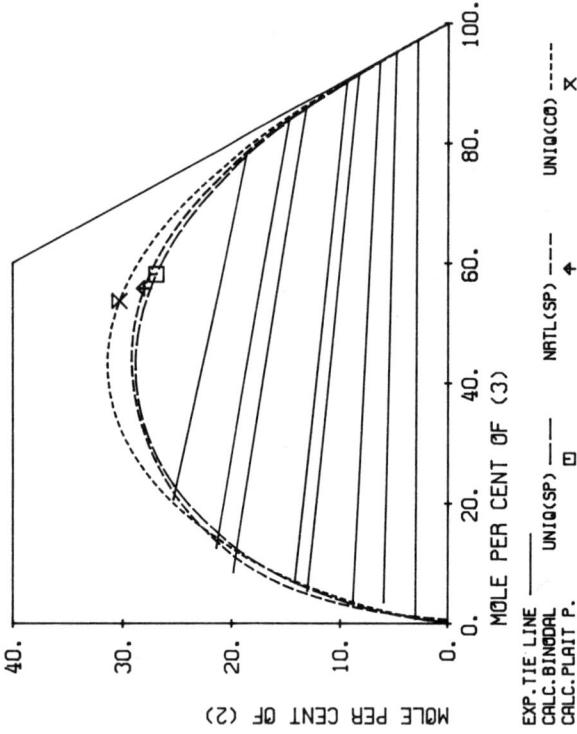

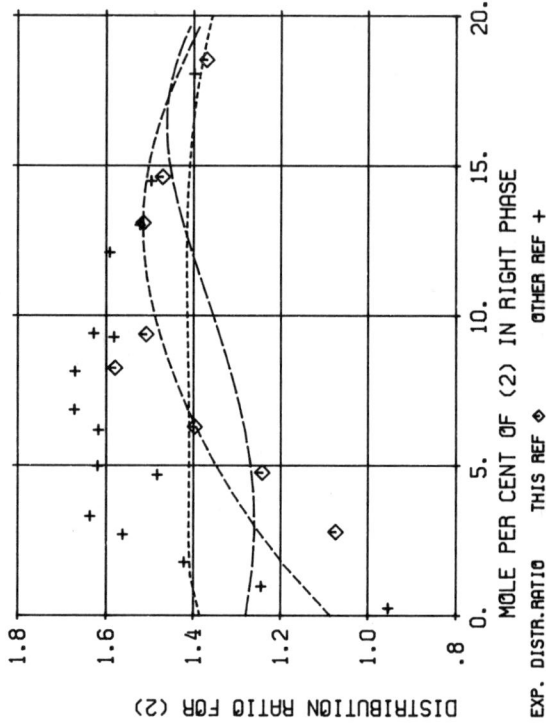

EXP.TIE LINE ———   UNIQ(SP) ——— ⊡   NRTL(SP) ——— ✦   UNIQ(C0) ----- ✕
CALC.BINODAL
CALC.PLAIT P.

MOLE PER CENT OF (3)

MOLE PER CENT OF (2)

MOLE PER CENT OF (2) IN RIGHT PHASE

DISTRIBUTION RATIO FOR (2)

EXP. DISTR.RATIO ◇   THIS REF ◇   OTHER REF +

C$_2$HCl$_3$-C$_6$H$_{11}$NO

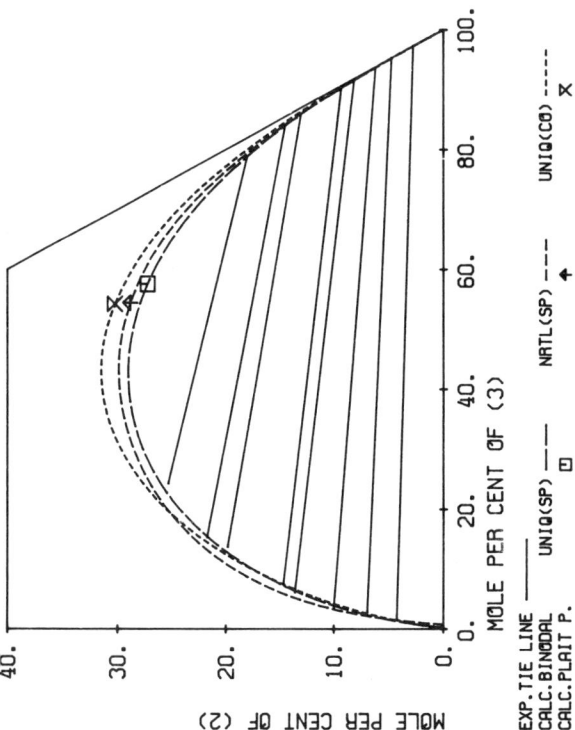

MOLE PER CENT OF (3)

MOLE PER CENT OF (2)

EXP.TIE LINE ———   UNIQ(SP) ⊟   NRTL(SP) ---- ♠   UNIQ(CO) ---- ✕
CALC.BINODAL
CALC.PLAIT P.

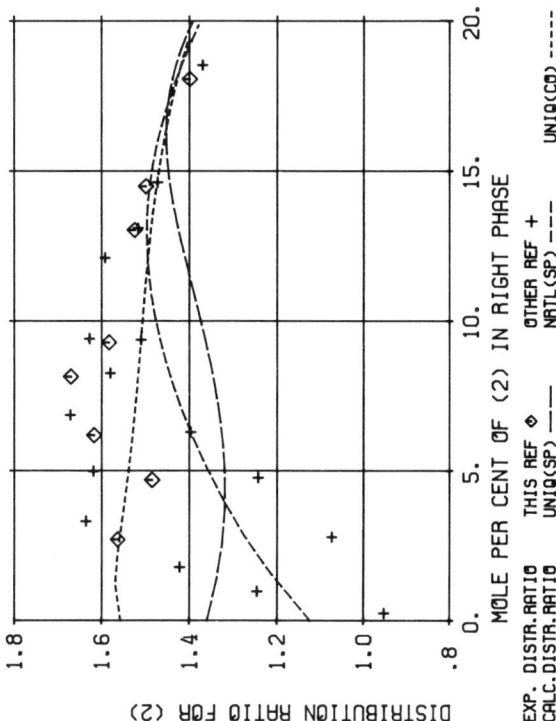

MOLE PER CENT OF (2) IN RIGHT PHASE

DISTRIBUTION RATIO FOR (2)

EXP. DISTR.RATIO   THIS REF ◇   OTHER REF +
CALC.DISTR.RATIO   UNIQ(SP) ———   NRTL(SP) ----   UNIQ(CO) ----

---

(2) C6H11NO   HEXANOIC ACID,6-AMINO,LACTAM

(3) H2O   WATER

SEDMEROVA V., NOVAK J.P.
CHEM.PRUM. 16/41(1966)270

TEMPERATURE = 30.0 DEG C   TYPE OF SYSTEM = 1

EXPERIMENTAL TIE LINES IN MOLE PCT

	LEFT PHASE			RIGHT PHASE	
(1)	(2)	(3)	(1)	(2)	(3)
94.350	4.218	1.432	0.047	2.699	97.254
90.227	6.954	2.818	0.086	4.688	95.226
86.523	9.991	3.486	0.110	6.176	93.715
80.320	13.585	6.095	0.278	8.131	91.591
77.969	14.680	7.350	0.374	9.276	90.351
66.457	19.836	13.706	1.134	13.022	85.844
63.021	21.699	15.280	1.585	14.489	83.925
50.480	25.270	24.251	3.223	18.059	78.717

SPECIFIC MODEL PARAMETERS IN KELVIN

I J	UNIQUAC		NRTL(ALPHA=.2)	
	AIJ	AJI	AIJ	AJI
1 2	284.61	-173.08	-1743.8	-430.25
1 3	937.03	808.07	978.70	2026.7
2 3	-121.04	26.069	-661.97	-1289.9

R1 = 3.3092   R2 = 4.6106   R3 = 0.9200
Q1 = 2.860   Q2 = 3.724   Q3 = 1.400

MEAN DEV. BETWEEN CALC. AND EXP. CONC. IN MOLE PCT

UNIQUAC (SPECIFIC PARAMETERS)   0.59
NRTL (SPECIFIC PARAMETERS)   0.73
UNIQUAC (COMMON PARAMETERS)   0.49

C$_2$HCl$_3$-C$_6$H$_{11}$NO

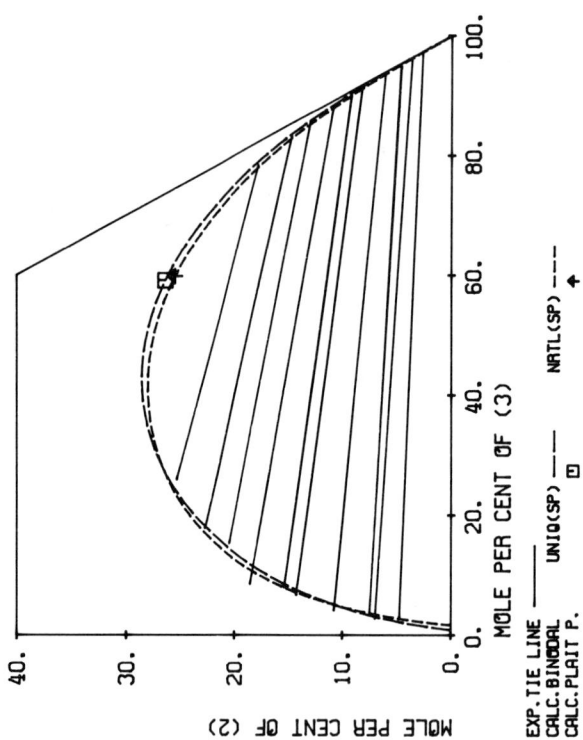

MOLE PER CENT OF (3)

MOLE PER CENT OF (2)

EXP.TIE LINE —— UNIQ(SP) ⊡   NRTL(SP) ---- ✦
CALC.BINODAL ----
CALC.PLAIT P.

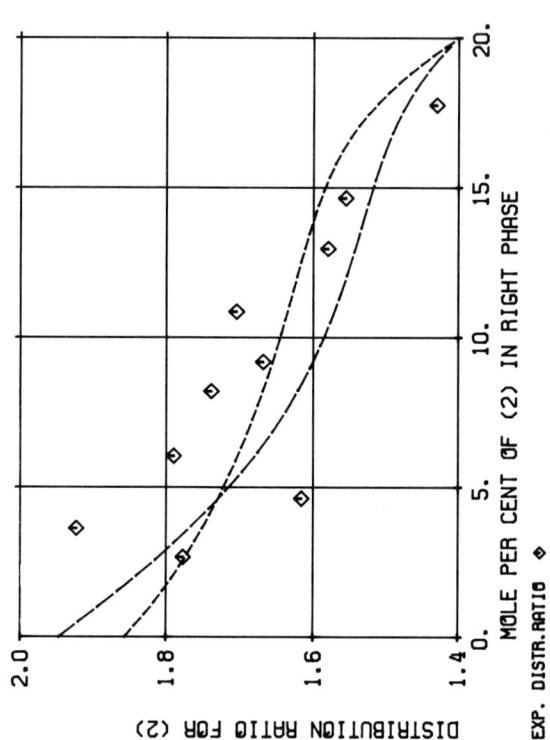

MOLE PER CENT OF (2) IN RIGHT PHASE

DISTRIBUTION RATIO FOR (2)

EXP. DISTR.RATIO ◇

(1) C2HCL3   ETHENE,TRICHLORO
----
(2) C6H11NO   HEXANOIC ACID,6-AMINO,LACTAM
----
(3) H2O   WATER

SEDMEROVA V.; NOVAK J.P.
CHEM.PRUM. 16/41(1966)270

TEMPERATURE = 40.0 DEG C   TYPE OF SYSTEM = 1

EXPERIMENTAL TIE LINES IN MOLE PCT

	LEFT PHASE			RIGHT PHASE	
(1)	(2)	(3)	(1)	(2)	(3)
92.447	4.726	2.827	0.063	2.660	97.277
90.227	6.954	2.818	0.082	3.618	96.300
89.037	7.464	3.499	0.103	4.618	95.279
85.046	10.800	4.154	0.182	6.036	93.782
79.033	14.241	6.726	0.400	8.193	91.407
76.714	15.321	7.965	0.520	9.181	90.299
72.951	18.509	8.539	0.802	10.858	88.340
64.223	20.470	15.307	1.410	12.956	85.635
59.383	22.794	17.823	2.090	14.650	83.260
48.681	25.380	25.940	3.823	17.756	78.421

SPECIFIC MODEL PARAMETERS IN KELVIN

I J	UNIQUAC		NRTL(ALPHA=.2)	
	AIJ	AJI	AIJ	AJI
1 2	292.37	-169.67	724.20	-656.16
1 3	724.60	-672.14	756.24	1799.6
2 3	-26.022	-58.173	42.957	-166.49

R1 = 3.3092   R2 = 4.6106   R3 = 0.9200
Q1 = 2.860   Q2 = 3.724   Q3 = 1.400

MEAN DEV. BETWEEN CALC. AND EXP. CONC. IN MOLE PCT

UNIQUAC (SPECIFIC PARAMETERS)   0.41
NRTL   (SPECIFIC PARAMETERS)   0.43

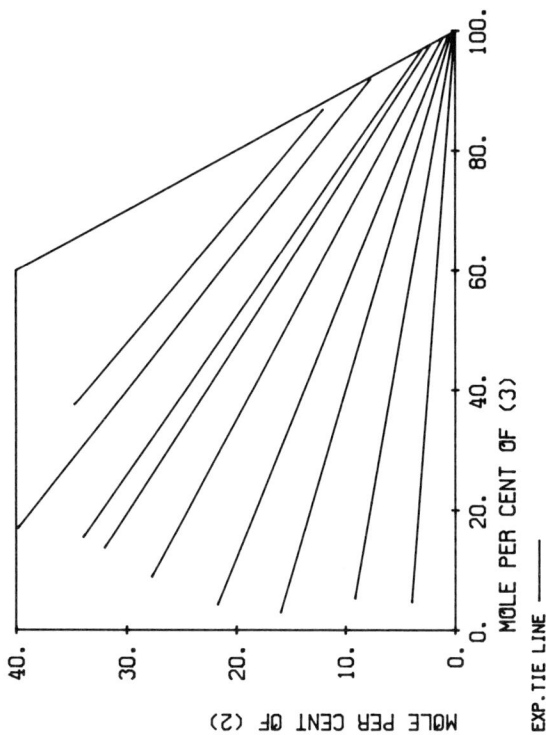

(1) C2HCL3   ETHENE,TRICHLORO

(2) C10H14N2   NICOTINE

(3) H2O   WATER

REILLY J., KELLEY D.F., OCONNOR M.
J.CHEM.SOC. (1941)275

TEMPERATURE = 17.0 DEG C   TYPE OF SYSTEM = 1

EXPERIMENTAL TIE LINES IN MOLE PCT

	LEFT PHASE			RIGHT PHASE	
(1)	(2)	(3)	(1)	(2)	(3)
91.396	3.939	4.664	0.014	0.055	99.932
85.612	9.121	5.267	0.011	0.151	99.838
81.099	15.948	2.952	0.017	0.353	99.630
74.074	21.708	4.218	0.014	0.557	99.429
63.407	27.725	8.868	0.024	1.125	98.851
54.306	32.012	13.682	0.041	2.239	97.721
50.559	33.922	15.520	0.069	3.064	96.868
43.171	39.839	16.990	0.357	7.706	91.937
27.671	34.727	37.602	1.069	12.021	86.910

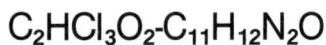

$C_2HCl_3O_2\text{-}C_{11}H_{12}N_2O$

MOLE PER CENT OF (3)

MOLE PER CENT OF (2)

EXP.TIE LINE ———

(1) C2HCL3O2   ACETIC ACID,TRICHLORO

(2) C11H12N2O   ANTIPYRINE

(3) H2O   WATER

KRUPATKIN I.L., ROZHENTSOVA E.P.
ZH.FIZ.KHIM. 45(1971)556

TEMPERATURE = 30.0 DEG C   TYPE OF SYSTEM = 0

EXPERIMENTAL TIE LINES IN MOLE PCT

LEFT PHASE			RIGHT PHASE		
(1)	(2)	(3)	(1)	(2)	(3)
32.194	20.338	47.468	0.582	0.101	99.316
34.435	14.283	51.283	0.644	0.112	99.244
26.399	7.093	66.508	1.038	0.115	98.847
22.463	23.341	54.196	0.299	0.189	99.512
4.964	9.612	85.424	0.299	0.210	99.491

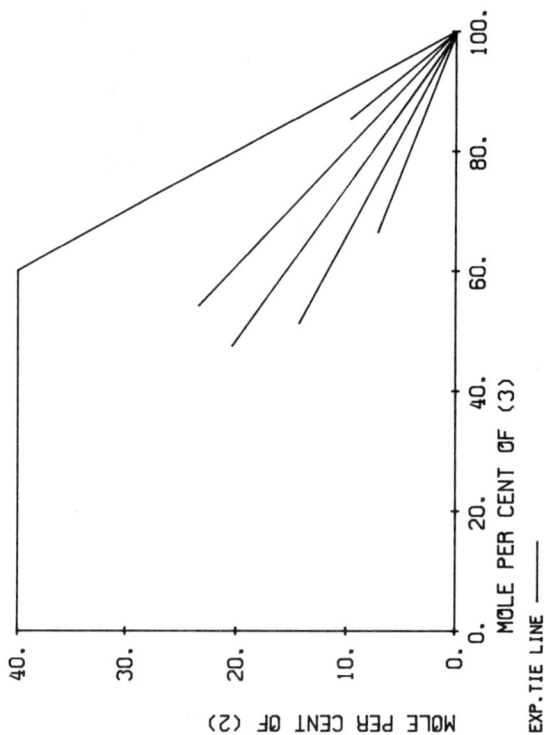

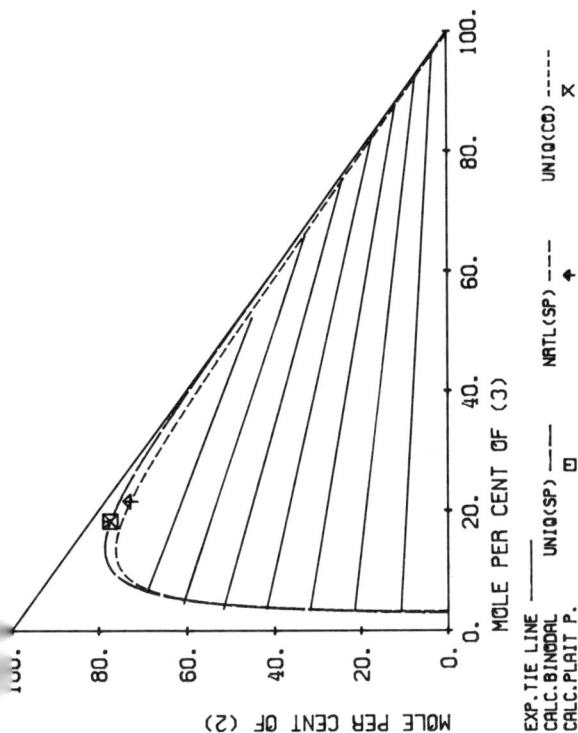

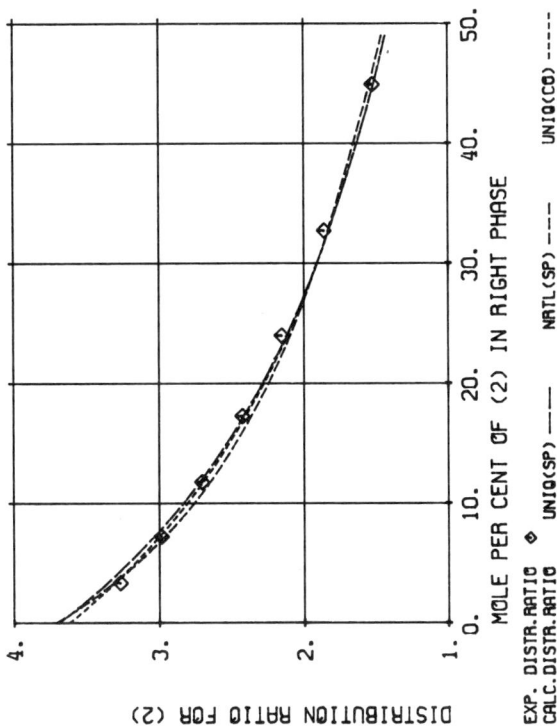

(2) C2H2BR4   ETHANE,1,1,2,2-TETRABROMO

(3) C4H10O3   DIETHYLENE GLYCOL

GARY L.H., CRICHTON J.S., FEILD R.
J.CHEM.ENG.DATA 3(1958)111

TEMPERATURE = 30.0 DEG C   TYPE OF SYSTEM = 1

EXPERIMENTAL TIE LINES IN MOLE PCT

	LEFT PHASE			RIGHT PHASE	
(1)	(2)	(3)	(1)	(2)	(3)
96.416	0.0	3.584	0.246	0.0	99.754
85.594	10.867	3.540	0.294	3.325	96.381
75.039	21.465	3.496	0.318	7.189	92.493
64.742	31.805	3.453	0.381	11.746	87.873
54.464	41.792	3.744	0.495	17.215	82.290
44.681	51.620	3.699	0.720	23.942	75.337
34.604	60.776	4.619	1.399	32.713	65.888
24.176	68.782	7.042	2.839	44.912	52.248

SPECIFIC MODEL PARAMETERS IN KELVIN

		UNIQUAC		NRTL(ALPHA=.2)	
I J	AIJ	AJI		AIJ	AJI
1 2	8.0875	27.927		-629.70	494.51
1 3	252.87	116.35		550.85	1305.1
2 3	127.01	-14.169		937.39	-371.82

R1 =16.1018   R2 = 4.6906   R3 = 4.0013
Q1 =12.848   Q2 = 3.784   Q3 = 3.568

MEAN DEV. BETWEEN CALC. AND EXP. CONC. IN MOLE PCT

UNIQUAC (SPECIFIC PARAMETERS)   0.61
NRTL   (SPECIFIC PARAMETERS)   0.42
UNIQUAC (COMMON PARAMETERS)   0.61

C$_2$H$_2$Cl$_2$O$_2$-C$_6$H$_5$Cl

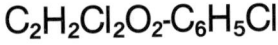

(1) H2O   WATER
(2) C2H2CL2O2 ACETIC ACID,DICHLORO
(3) C6H5CL   BENZENE,CHLORO

PEAKE J.S., THOMPSON K.E.
IND.ENG.CHEM. 44(1952)2439

TEMPERATURE = 25.0 DEG C   TYPE OF SYSTEM = 1

EXPERIMENTAL TIE LINES IN MOLE PCT (GRAPH.INTERPOL.)

	LEFT PHASE			RIGHT PHASE	
(1)	(2)	(3)	(1)	(2)	(3)
97.035	2.944	0.021	0.561	1.070	98.369
95.333	4.517	0.150	0.567	1.088	98.345
92.636	7.329	0.035	0.716	1.626	97.658
89.997	9.954	0.049	0.766	1.765	97.469
86.653	12.983	0.364	0.902	2.269	96.829
82.989	16.143	0.868	1.008	2.624	96.368
79.481	18.932	1.587	1.125	2.979	95.896
75.565	21.913	2.522	1.625	4.368	94.007

SPECIFIC MODEL PARAMETERS IN KELVIN

		UNIQUAC		NRTL(ALPHA=.2)	
I J	AIJ	AJI		AIJ	AJI
1 2	715.59	-343.46		592.30	-75.714
1 3	423.82	906.49		1145.7	999.01
2 3	-176.18	468.79		356.60	571.54

R1 = 0.9200   R2 = 3.3619   R3 = 3.8127
Q1 = 1.400   Q2 = 2.908   Q3 = 2.844

MEAN DEV. BETWEEN CALC. AND EXP. CONC. IN MOLE PCT

UNIQUAC (SPECIFIC PARAMETERS)   0.16
NRTL   (SPECIFIC PARAMETERS)   0.45
UNIQUAC (COMMON PARAMETERS)   0.16

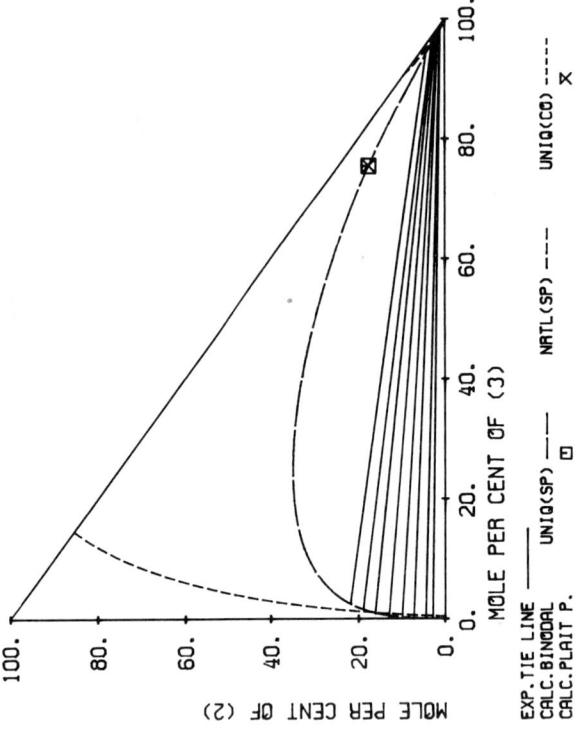

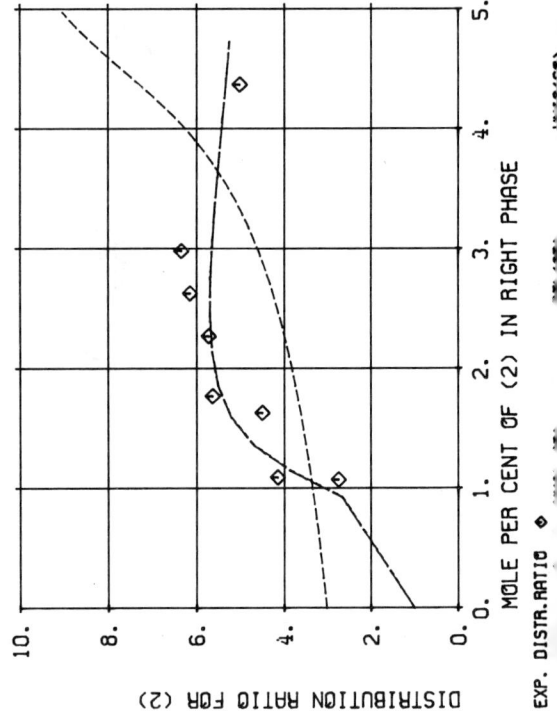

MOLE PER CENT OF (2)
MOLE PER CENT OF (3)

EXP.TIE LINE   UNIQ(SP) ▣   NRTL(SP) ----   UNIQ(CO) -----
CALC.BINODAL ----   X
CALC.PLAIT P.

DISTRIBUTION RATIO FOR (2)
MOLE PER CENT OF (2) IN RIGHT PHASE
EXP. DISTR.RATIO ◇

$C_2H_3Cl_3$-C_3H_6O

(2) C3H6O   2-PROPANONE
(3) H2O   WATER

TREYBAL R.E., WEBER L.D., DALEY J.F.
IND.ENG.CHEM. 38(1946)817

TEMPERATURE = 25.0 DEG C   TYPE OF SYSTEM = 1

EXPERIMENTAL TIE LINES IN MOLE PCT

	LEFT PHASE			RIGHT PHASE	
(1)	(2)	(3)	(1)	(2)	(3)
80.186	17.724	2.090	0.074	1.937	97.989
77.070	20.375	2.556	0.077	2.125	97.799
59.013	35.965	5.022	0.102	4.827	95.071
52.817	41.350	5.833	0.113	6.033	93.855
49.178	44.001	6.821	0.124	6.857	93.019
37.689	52.725	9.586	0.166	9.932	89.902
35.994	53.787	10.219	0.171	10.366	89.463
34.969	54.354	10.677	0.175	10.707	89.117
32.199	56.551	11.249	0.191	11.648	88.161
30.697	56.191	13.112	0.203	12.329	87.467
25.040	58.340	16.620	0.292	14.983	84.725
19.164	59.371	21.465	0.405	18.131	81.464
14.560	56.964	28.476	0.779	21.978	77.243
9.943	52.482	37.575	1.501	27.388	71.110

SPECIFIC MODEL PARAMETERS IN KELVIN

		UNIQUAC		NRTL(ALPHA=.2)	
I	J	AIJ	AJI	AIJ	AJI
1	2	168.08	-158.59	-198.14	21.091
1	3	1005.7	229.73	1300.9	1237.3
2	3	314.25	-55.813	9.0838	571.74

R1 = 3.5260   R2 = 2.5735   R3 = 0.9200
Q1 = 2.948   Q2 = 2.336   Q3 = 1.400

MEAN DEV. BETWEEN CALC. AND EXP. CONC. IN MOLE PCT

UNIQUAC (SPECIFIC PARAMETERS)   0.48
NRTL   (SPECIFIC PARAMETERS)   0.59
UNIQUAC (COMMON PARAMETERS)   0.75

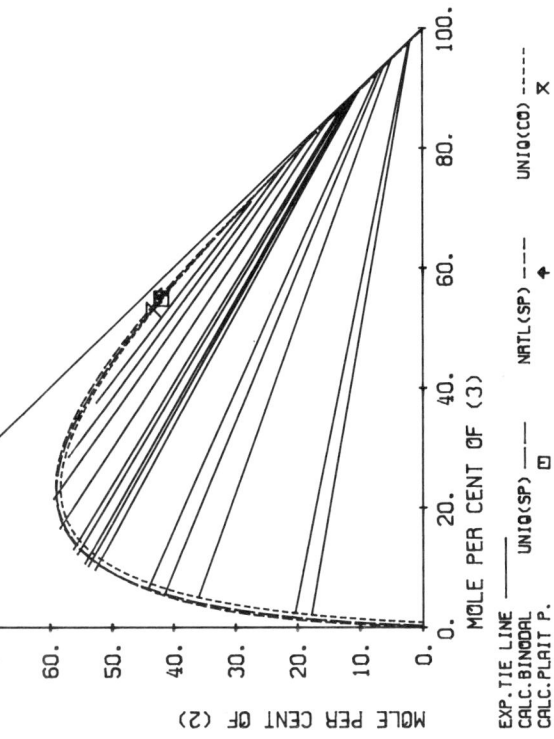

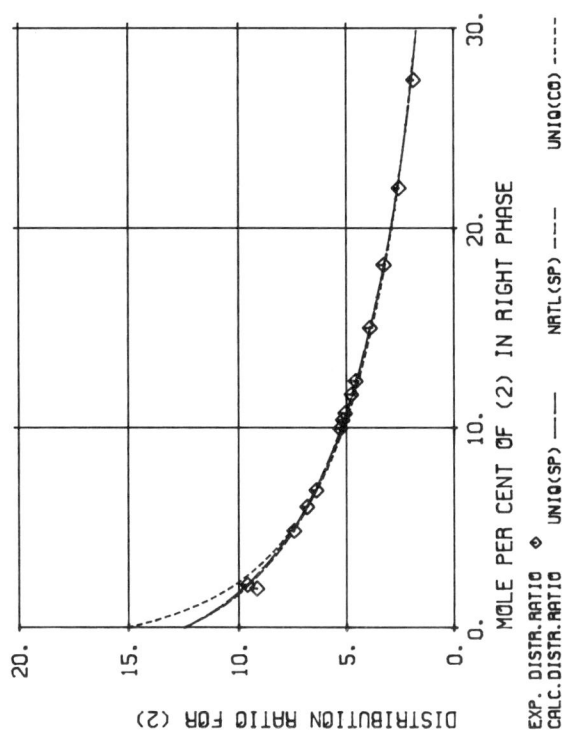

MOLE PER CENT OF (3)
MOLE PER CENT OF (2)

EXP.TIE LINE ———   UNIQ(SP) □   NRTL(SP) ———   UNIQ(CO) ✕
CALC.BINODAL
CALC.PLAIT P.

MOLE PER CENT OF (2) IN RIGHT PHASE
DISTRIBUTION RATIO FOR (2)

EXP. DISTR.RATIO ◇   UNIQ(SP) ———   NRTL(SP) ———   UNIQ(CO) ———
CALC.DISTR.RATIO

C₃H₃Cl₃-C₄H₈O

NEWMAN M., HAYWORTH C.B., TREYBAL R.E.
IND.ENG.CHEM. 41(1949)2039

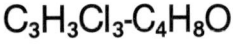

(1)	C2H3CL3	ETHANE,1,1,2-TRICHLORO
(2)	C4H8O	2-BUTANONE
(3)	H2O	WATER

TEMPERATURE = 25.0 DEG C   TYPE OF SYSTEM = 2

EXPERIMENTAL TIE LINES IN MOLE PCT

	LEFT PHASE			RIGHT PHASE	
(1)	(2)	(3)	(1)	(2)	(3)
90.145	9.855	0.0	0.056	0.258	99.686
70.165	26.562	3.272	0.051	0.725	99.224
51.008	43.421	5.571	0.043	1.574	98.384
36.781	55.773	7.446	0.033	2.482	97.484
23.102	64.814	12.084	0.024	3.536	96.440
10.149	70.684	19.167	0.017	5.254	94.729

SPECIFIC MODEL PARAMETERS IN KELVIN

	UNIQUAC		NRTL(ALPHA=.2)	
I J	AIJ	AJI	AIJ	AJI
1 2	-67.170	12.304	-145.28	-61.099
1 3	758.81	308.03	1062.7	1104.1
2 3	337.47	4.4505	88.755	856.59

R1 = 3.5260	R2 = 3.2479	R3 = 0.9200
Q1 = 2.943	Q2 = 2.876	Q3 = 1.400

MEAN DEV. BETWEEN CALC. AND EXP. CONC. IN MOLE PCT

UNIQUAC (SPECIFIC PARAMETERS)	0.34
NRTL (SPECIFIC PARAMETERS)	0.40
UNIQUAC (COMMON PARAMETERS)	0.35

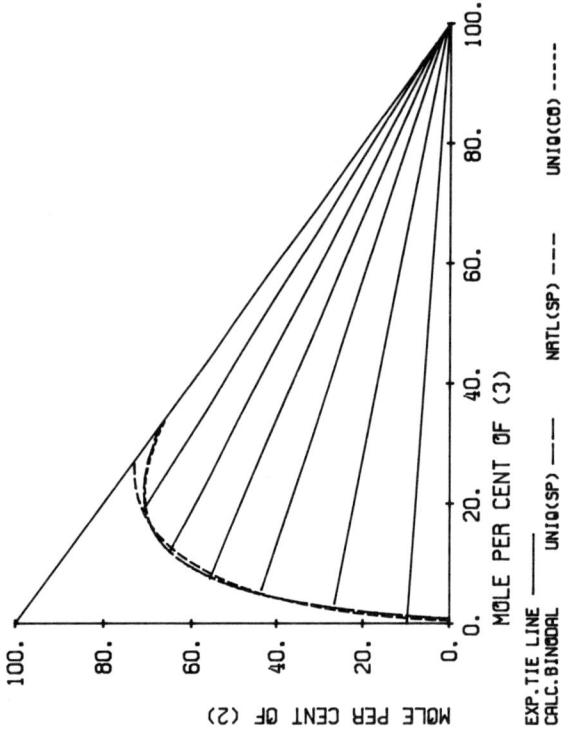

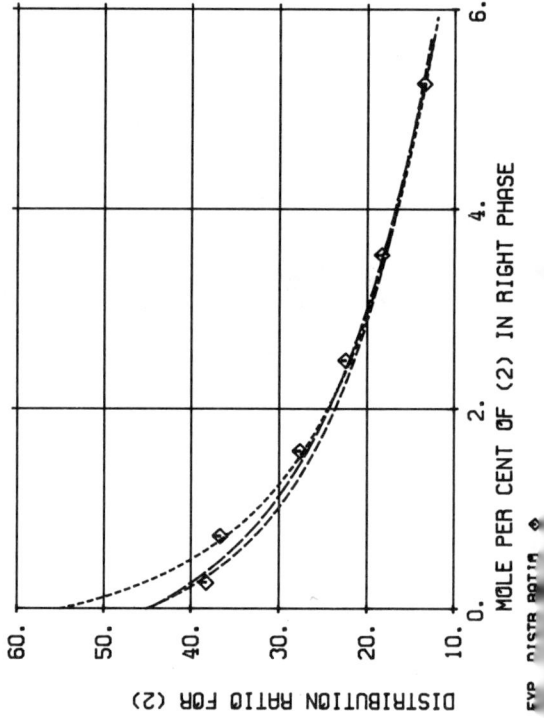

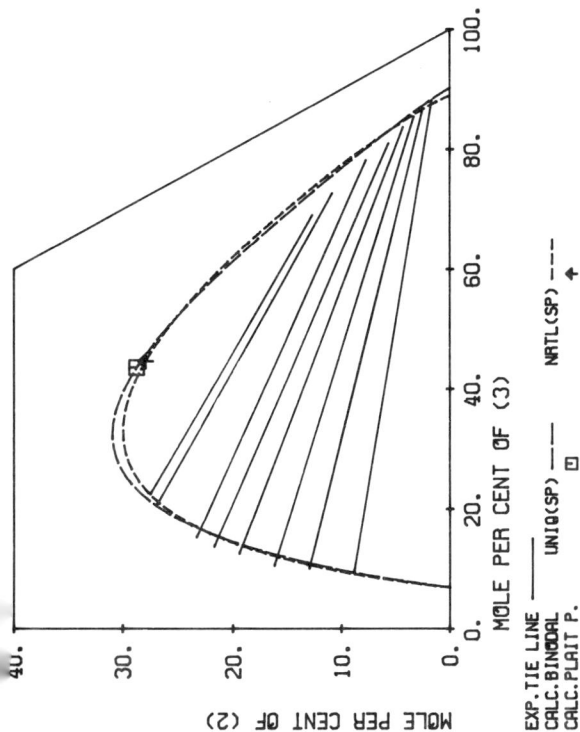

EXP.TIE LINE ———
CALC.BINODAL
CALC.PLAIT P.

UNIQ(SP) —— □
NRTL(SP) —— +

MOLE PER CENT OF (3)

MOLE PER CENT OF (2)

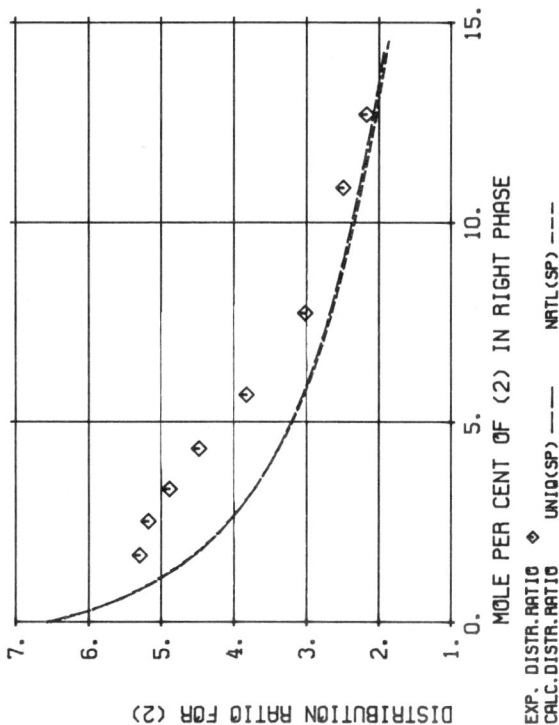

EXP. DISTR.RATIO ◇   UNIQ(SP) ——   NRTL(SP) ———
CALC.DISTR.RATIO

MOLE PER CENT OF (2) IN RIGHT PHASE

DISTRIBUTION RATIO FOR (2)

---

(2) C2H6O   ETHANOL
(3) C6H14   HEXANE

SUGI H., KATAYAMA T.
J.CHEM.ENG.JPN. 11(1978)167

TEMPERATURE = 40.0 DEG C   TYPE OF SYSTEM = 1

EXPERIMENTAL TIE LINES IN MOLE PCT

LEFT PHASE			RIGHT PHASE		
(1)	(2)	(3)	(1)	(2)	(3)
81.530	8.790	9.680	10.030	1.660	88.310
76.980	12.990	10.030	10.750	2.510	86.740
73.240	16.220	10.540	11.220	3.320	85.460
68.080	19.420	12.500	11.950	4.330	83.720
64.640	21.690	13.670	13.320	5.670	81.010
61.470	23.310	15.220	14.070	7.720	78.210
52.520	26.950	20.530	16.420	10.860	72.720
50.030	27.480	22.490	18.190	12.690	69.120

SPECIFIC MODEL PARAMETERS IN KELVIN

	UNIQUAC		NRTL(ALPHA=.2)	
I J	AIJ	AJI	AIJ	AJI
1 2	138.33	-238.27	225.23	-547.10
1 3	38.603	533.51	652.38	411.80
2 3	-12.562	231.51	248.66	0.82503

R1 = 1.8701   R2 = 2.1055   R3 = 4.4998
Q1 = 1.724   Q2 = 1.972   Q3 = 3.856

MEAN DEV. BETWEEN CALC. AND EXP. CONC. IN MOLE PCT

UNIQUAC (SPECIFIC PARAMETERS)   1.02
NRTL   (SPECIFIC PARAMETERS)   1.16

C$_2$H$_3$N-C$_3$H$_3$N

MOLE PER CENT OF (3)

MOLE PER CENT OF (2)

EXP.TIE LINE —————
CALC.BINODAL - - - -
CALC.PLAIT P.

UNIQ(SP) ⊡
NRTL(SP) ◆

UNIQ(CO) - - - -
⊠

DISTRIBUTION RATIO FOR (2)

MOLE PER CENT OF (2) IN RIGHT PHASE

EXP. DISTR.RATIO ◇ THIS REF ◇ OTHER REF +

(1) C3H3N   PROPENOIC ACID,NITRILE
(2) C2H3N   ACETIC ACID,NITRILE
(3) H2O   WATER

PROKHOROVA V.V., ET AL.
ZH.FIZ.KHIM. 38(1964)1488

TEMPERATURE = 25.0 DEG C   TYPE OF SYSTEM = 1

EXPERIMENTAL TIE LINES IN MOLE PCT

	LEFT PHASE		RIGHT PHASE	
	(2)	(3)	(2)	(3)
86.677	3.232	10.092	1.482	0.821	97.697
48.497	35.445	16.058	1.917	7.185	90.898
12.975	51.486	35.539	2.443	20.379	77.178

SPECIFIC MODEL PARAMETERS IN KELVIN

I J	UNIQUAC		NRTL(ALPHA=.2)	
	AIJ	AJI	AIJ	AJI
1 2	24.992	21.440	-27.096	95.447
1 3	326.12	228.14	316.42	993.78
2 3	140.19	115.83	76.406	590.18

R1 = 2.3144   R2 = 1.8701   R3 = 0.9200
Q1 = 2.052   Q2 = 1.724   Q3 = 1.400

MEAN DEV. BETWEEN CALC. AND EXP. CONC. IN MOLE PCT

UNIQUAC (SPECIFIC PARAMETERS)   1.17
NRTL   (SPECIFIC PARAMETERS)   1.19
UNIQUAC (COMMON PARAMETERS)   1.24

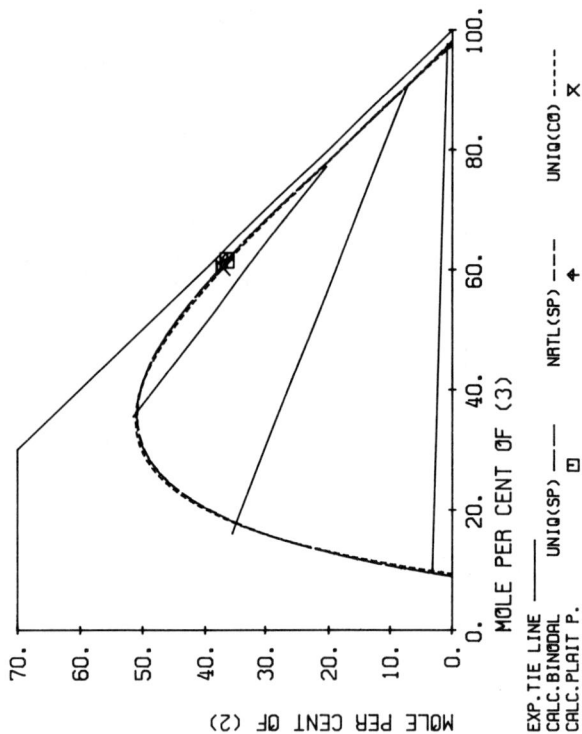

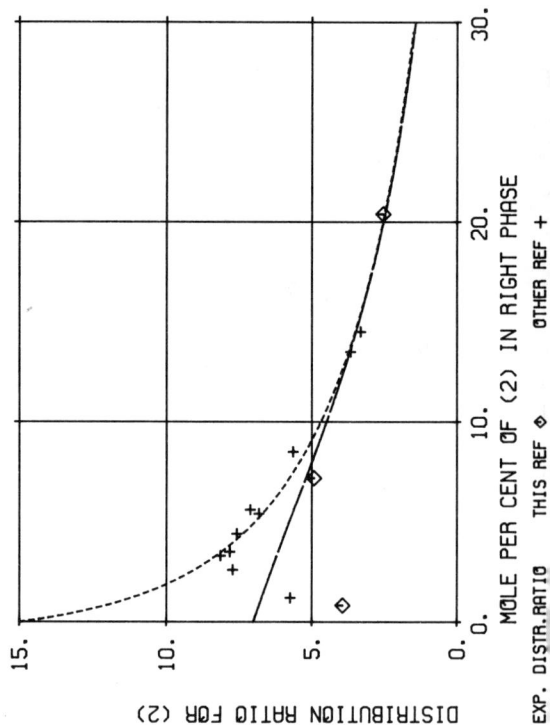

$C_2H_3N-C_3H_3N$

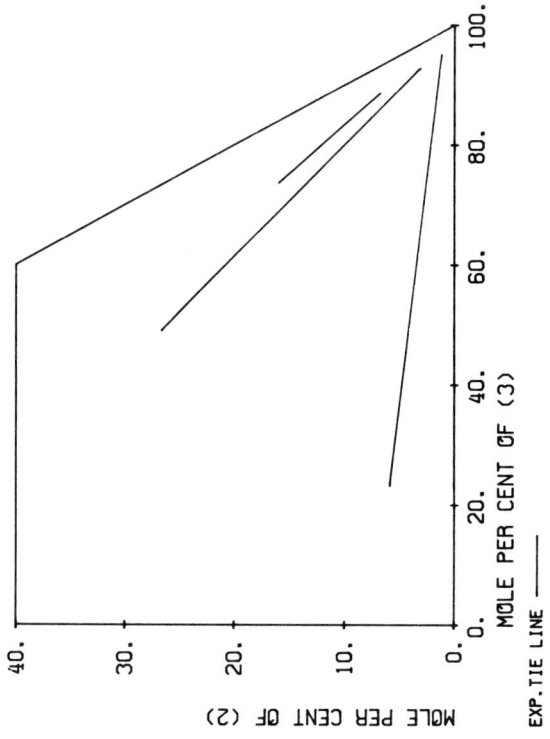

MOLE PER CENT OF (3)

EXP.TIE LINE

MOLE PER CENT OF (2)

(1) C3H3N   PROPENOIC ACID,NITRILE

(2) C2H3N   ACETIC ACID,NITRILE

(3) H2O   WATER

PROKHOROVA V.V., ET AL.
ZH.FIZ.KHIM. 38(1964)1488

TEMPERATURE = 70.0 DEG C   TYPE OF SYSTEM = 1

EXPERIMENTAL TIE LINES IN MOLE PCT

LEFT PHASE			RIGHT PHASE		
(1)	(2)	(3)	(1)	(2)	(3)
77.016	0.0	22.984	3.831	0.0	96.169
70.780	5.920	23.300	3.691	1.193	95.117
24.243	26.644	49.113	4.025	3.141	92.833
10.300	15.964	73.736	4.594	6.815	88.591

C_2H_3N-C_3H_3N

(1) C3H3N   PROPENOIC ACID,NITRILE

(2) C2H3N   ACETIC ACID,NITRILE

(3) H2O   WATER

VOLPICELLI G.
J.CHEM.ENG.DATA 13(1968)150

TEMPERATURE = 25.0 DEG C   TYPE OF SYSTEM = 1

EXPERIMENTAL TIE LINES IN MOLE PCT

	LEFT PHASE			RIGHT PHASE	
(1)	(2)	(3)	(1)	(2)	(3)
90.000	0.0	10.000	2.600	0.0	97.400
81.600	6.900	11.500	2.700	1.200	96.100
67.700	20.100	12.200	2.800	2.600	94.600
57.800	26.900	15.300	2.700	3.300	94.000
58.400	27.400	14.200	2.500	3.500	94.000
49.900	33.400	16.700	2.300	4.400	93.300
42.800	36.900	20.300	2.300	5.400	92.300
40.600	39.900	19.500	2.500	5.600	91.900
30.600	48.000	21.400	2.500	8.500	89.000
19.300	49.400	31.300	2.500	13.500	84.000
14.400	48.000	37.600	2.600	14.500	82.900

SPECIFIC MODEL PARAMETERS IN KELVIN

I	J	UNIQUAC		NRTL(ALPHA=.2)	
		AIJ	AJI	AIJ	AJI
1	2	24.992	21.440	-27.096	95.447
1	3	326.12	228.14	316.42	993.78
2	3	140.19	115.83	76.406	590.18

R1 = 2.3144   R2 = 1.8701   R3 = 0.9200
Q1 = 2.052   Q2 = 1.724   Q3 = 1.400

MEAN DEV. BETWEEN CALC. AND EXP. CONC. IN MOLE PCT

UNIQUAC (SPECIFIC PARAMETERS)   1.08
NRTL   (SPECIFIC PARAMETERS)   1.07
UNIQUAC (COMMON PARAMETERS)   0.90

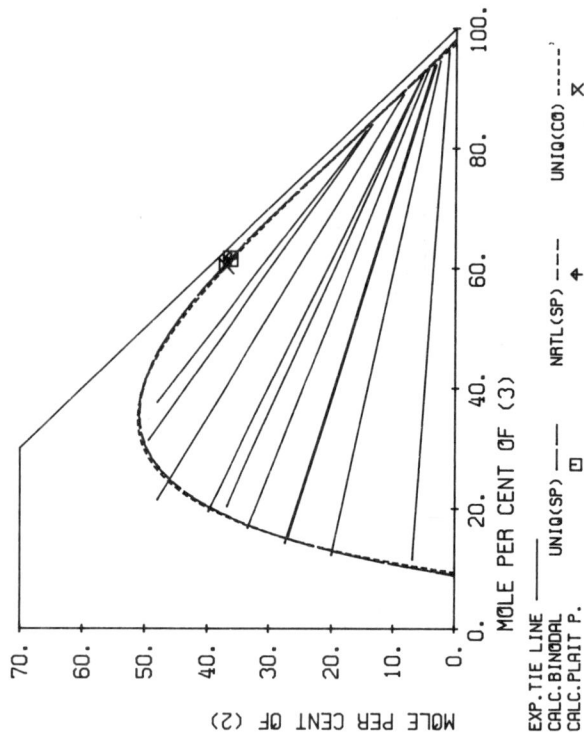

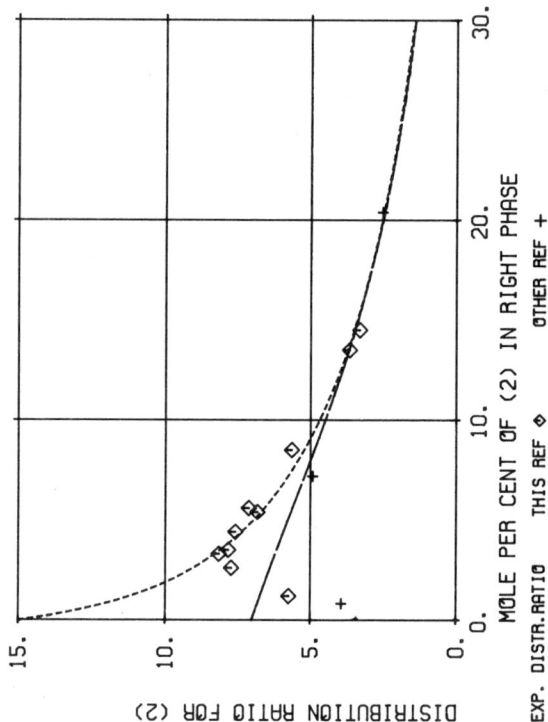

MOLE PER CENT OF (2)

MOLE PER CENT OF (3)

EXP.TIE LINE

CALC.BINODAL   UNIQ(SP) ——— ⊡   NRTL(SP) ——— ✦   UNIQ(CO) ----- ✕

CALC.PLAIT P.

DISTRIBUTION RATIO FOR (2)

MOLE PER CENT OF (2) IN RIGHT PHASE

EXP. DISTR.RATIO ◇   THIS REF ◇   OTHER REF +

MOLE PER CENT OF (2)

MOLE PER CENT OF (3)

EXP.TIE LINE ——— UNIQ(SP) ☐ NRTL(SP) ----
CALC.BINODAL
CALC.PLAIT P.

DISTRIBUTION RATIO FOR (2)

MOLE PER CENT OF (2) IN RIGHT PHASE

EXP. DISTR.RATIO ◇ UNIQ(SP) ——— NRTL(SP) ----
CALC.DISTR.RATIO

---

(2) C2H3N   ACETIC ACID,NITRILE
(3) H2O   WATER

VOLPICELLI G.
J.CHEM.ENG.DATA 13(1968)150

TEMPERATURE = 40.0 DEG C   TYPE OF SYSTEM = 1

EXPERIMENTAL TIE LINES IN MOLE PCT

LEFT PHASE			RIGHT PHASE		
(1)	(2)	(3)	(1)	(2)	(3)
88.500	0.0	11.500	2.800	0.0	97.200
75.000	10.000	15.000	3.000	1.000	96.000
76.300	9.800	13.900	2.500	1.100	96.400
78.000	10.300	11.700	2.800	1.700	95.600
70.500	14.200	15.300	2.700	1.700	95.600
62.300	21.000	16.700	3.300	2.400	94.300
60.000	21.900	18.100	3.200	2.900	94.300
49.700	29.600	20.700	2.900	4.100	93.000
47.500	31.200	21.300	2.600	4.600	92.800
47.200	30.300	22.500	3.000	5.100	91.900
42.700	34.500	22.800	2.800	5.600	91.600
28.000	40.400	31.600	3.000	8.700	88.300
27.300	41.400	31.300	3.200	9.200	87.600
24.600	42.700	32.700	3.400	9.600	87.000
22.200	43.200	34.600	2.600	9.600	87.800
21.800	41.500	36.700	3.400	11.700	84.900

SPECIFIC MODEL PARAMETERS IN KELVIN

		UNIQUAC		NRTL(ALPHA=.2)	
I	J	AIJ	AJI	AIJ	AJI
1	2	-1.5166	-39.879	-14.134	-95.855
1	3	343.80	183.69	316.65	909.03
2	3	44.540	190.77	-40.129	731.12

R1 = 2.3144   R2 = 1.8701   R3 = 0.9200
Q1 = 2.052   Q2 = 1.724   Q3 = 1.400

MEAN DEV. BETWEEN CALC. AND EXP. CONC. IN MOLE PCT

UNIQUAC (SPECIFIC PARAMETERS)   0.52
NRTL   (SPECIFIC PARAMETERS)   0.52

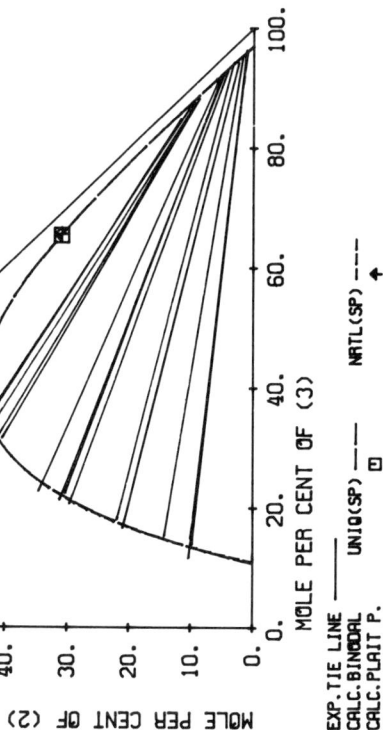

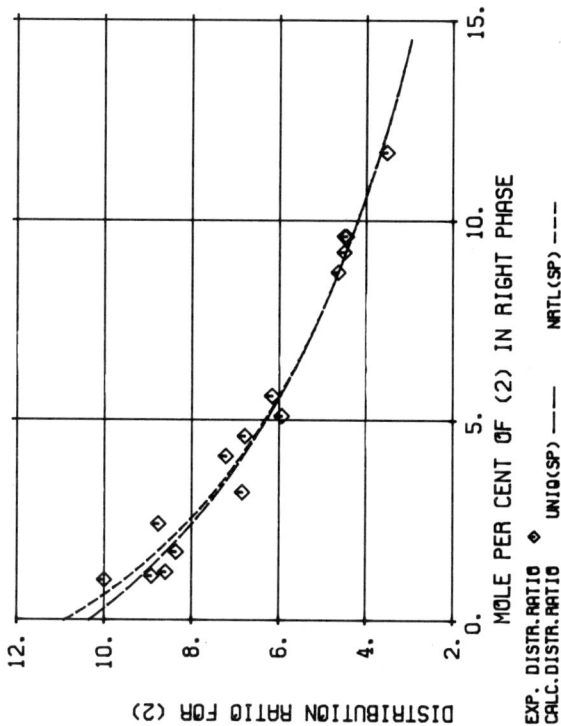

$C_2H_3N-C_3H_3N$

(1) C3H3N   PROPENOIC ACID,NITRILE
(2) C2H3N   ACETIC ACID,NITRILE
(3) H2O   WATER

VOLPICELLI G.
J.CHEM.ENG.DATA 13(1968)150

TEMPERATURE = 60.0 DEG C   TYPE OF SYSTEM = 1

EXPERIMENTAL TIE LINES IN MOLE PCT

	LEFT PHASE			RIGHT PHASE	
(1)	(2)	(3)	(1)	(2)	(3)
82.700	0.0	17.300	3.300	0.0	96.700
68.800	11.600	19.600	3.600	1.400	95.000
64.800	13.300	21.900	3.300	1.800	94.900
57.100	17.200	25.000	3.300	2.300	94.400
43.700	25.900	30.400	3.800	4.600	91.600
21.000	30.800	48.200	5.300	11.600	83.100

SPECIFIC MODEL PARAMETERS IN KELVIN

		UNIQUAC		NRTL(ALPHA=.2)	
I	J	AIJ	AJI	AIJ	AJI
1	2	-24.772	-76.230	70.273	-239.95
1	3	277.79	210.73	217.48	1023.5
2	3	-73.051	259.73	-199.99	845.19

R1 = 2.3144   R2 = 1.8701   R3 = 0.9200
Q1 = 2.052   Q2 = 1.724   Q3 = 1.400

MEAN DEV. BETWEEN CALC. AND EXP. CONC. IN MOLE PCT

UNIQUAC (SPECIFIC PARAMETERS)   0.50
NRTL   (SPECIFIC PARAMETERS)   0.48

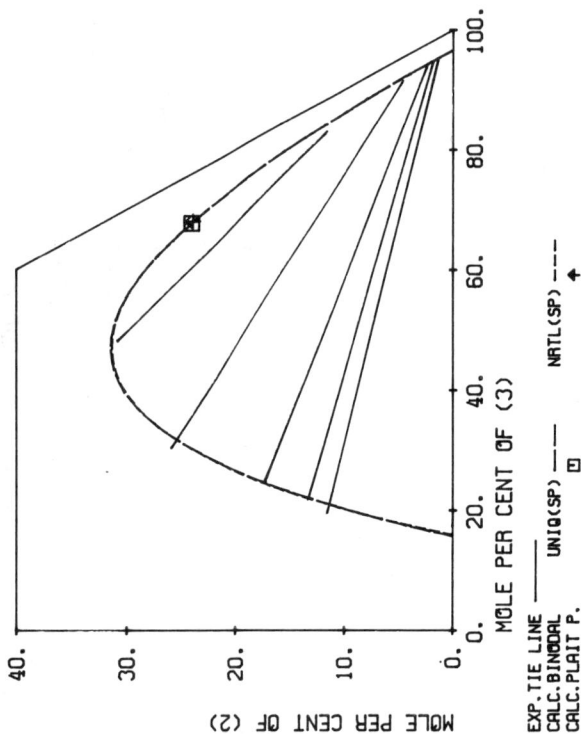

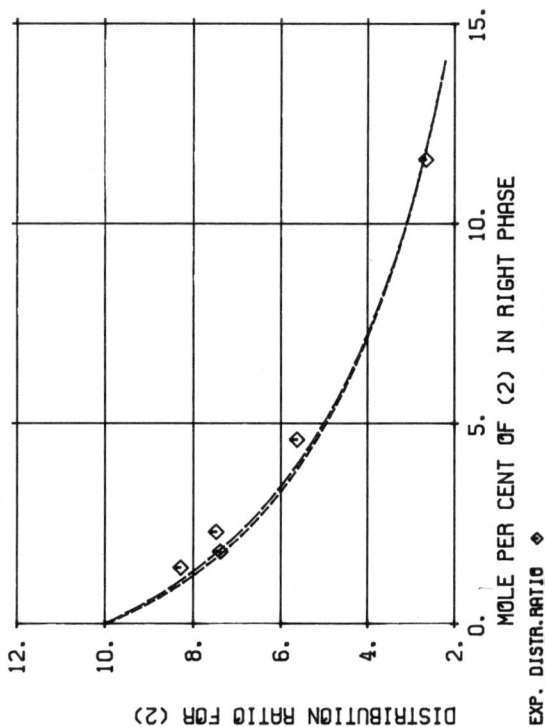

$C_2H_3N-C_4H_8O_2$

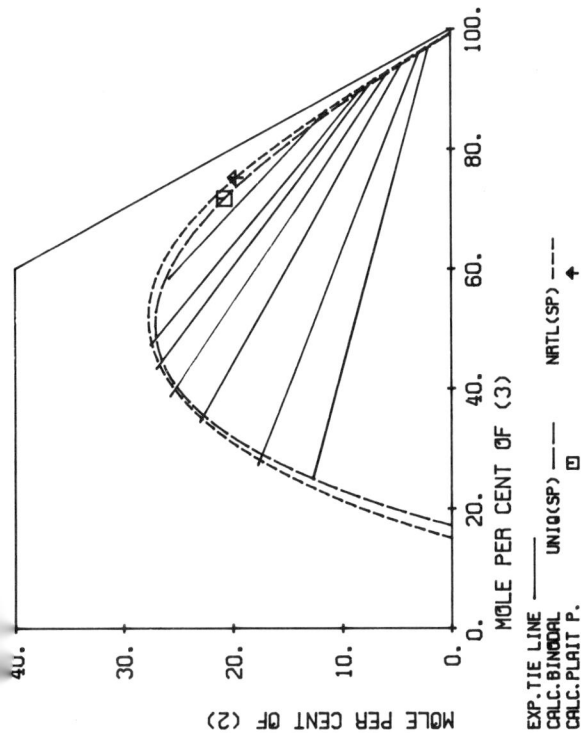

MOLE PER CENT OF (3)

MOLE PER CENT OF (2)

EXP.TIE LINE ——— UNIQ(SP) □ NRTL(SP) ——— ✦
CALC.BINODAL
CALC.PLAIT P.

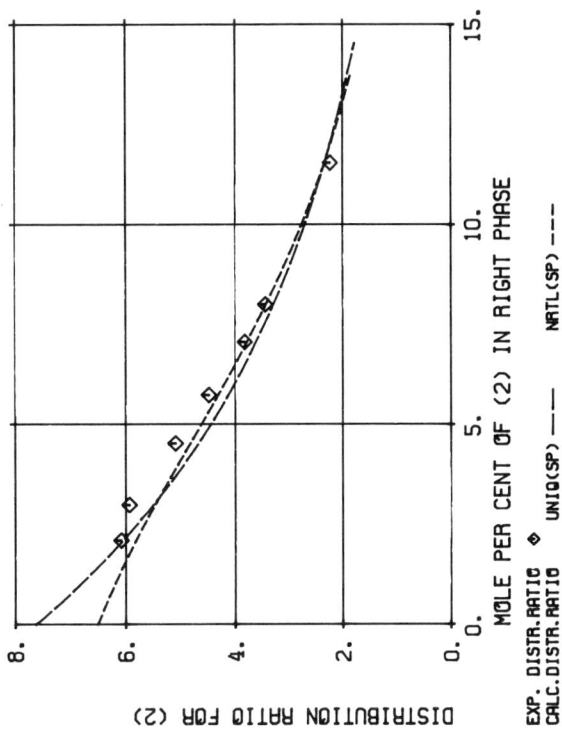

MOLE PER CENT OF (2) IN RIGHT PHASE

DISTRIBUTION RATIO FOR (2)

EXP. DISTR.RATIO ◇ UNIQ(SP) ◇
CALC.DISTR.RATIO ——— NRTL(SP) ———

(2) C2H3N   ACETIC ACID,NITRILE
(3) H2O   WATER

SUGI H., KATAYAMA T.
J.CHEM.ENG.JPN. 11(1978)167

TEMPERATURE = 60.0 DEG C   TYPE OF SYSTEM = 1

EXPERIMENTAL TIE LINES IN MOLE PCT

LEFT PHASE			RIGHT PHASE		
(1)	(2)	(3)	(1)	(2)	(3)
62.310	12.770	24.920	1.080	2.100	96.820
55.000	17.740	27.260	1.170	2.990	95.840
42.540	23.010	34.450	1.180	4.520	94.300
35.540	25.720	38.740	1.210	5.730	93.060
29.630	26.970	43.400	1.230	7.060	91.710
25.180	27.490	47.330	1.510	8.000	90.490
15.790	25.810	58.400	2.230	11.540	86.230

SPECIFIC MODEL PARAMETERS IN KELVIN

		UNIQUAC		NRTL(ALPHA=.2)	
I	J	AIJ	AJI	AIJ	AJI
1	2	-190.08	155.03	266.96	-178.44
1	3	-310.96	194.56	134.91	-1546.3
2	3	-182.01	317.08	-475.89	1235.4

R1 = 3.4786   R2 = 1.8701   R3 = 0.9200
Q1 = 3.116   Q2 = 1.724   Q3 = 1.400

MEAN DEV. BETWEEN CALC. AND EXP. CONC. IN MOLE PCT

UNIQUAC (SPECIFIC PARAMETERS)   0.94
NRTL   (SPECIFIC PARAMETERS)   0.38

C₂H₃N-C₆H₅Cl

(1) C6H5CL   BENZENE,CHLORO

(2) C2H3N   ACETIC ACID,NITRILE

(3) H2O   WATER

VENKATA SIVA RAMA RAO C. ET AL.
J.CHEM.ENG.DATA 23(1978)23

TEMPERATURE = 31.0 DEG C   TYPE OF SYSTEM = 1

EXPERIMENTAL TIE LINES IN MOLE PCT

	LEFT PHASE			RIGHT PHASE	
(1)	(2)	(3)	(1)	(2)	(3)
81.147	6.565	12.288	0.116	2.085	97.799
73.172	15.247	11.581	0.153	4.489	95.358
66.028	21.643	12.329	0.158	6.984	92.857
59.681	28.624	11.695	0.178	8.001	91.821
52.394	35.824	11.782	0.197	8.659	91.144
45.865	42.672	11.463	0.201	10.444	89.355
40.030	47.183	11.988	0.221	10.912	88.867

SPECIFIC MODEL PARAMETERS IN KELVIN

		UNIQUAC		NRTL(ALPHA=.2)	
I J	AIJ	AJI		AIJ	AJI
1 2	157.13	131.76		-12.646	548.56
1 3	226.88	246.74		172.47	1842.3
2 3	299.50	102.75		593.60	317.94

R1 = 3.8127   R2 = 1.8701   R3 = 0.9200
Q1 = 2.844   Q2 = 1.724   Q3 = 1.400

MEAN DEV. BETWEEN CALC. AND EXP. CONC. IN MOLE PCT

UNIQUAC (SPECIFIC PARAMETERS)   0.82
NRTL   (SPECIFIC PARAMETERS)   0.23

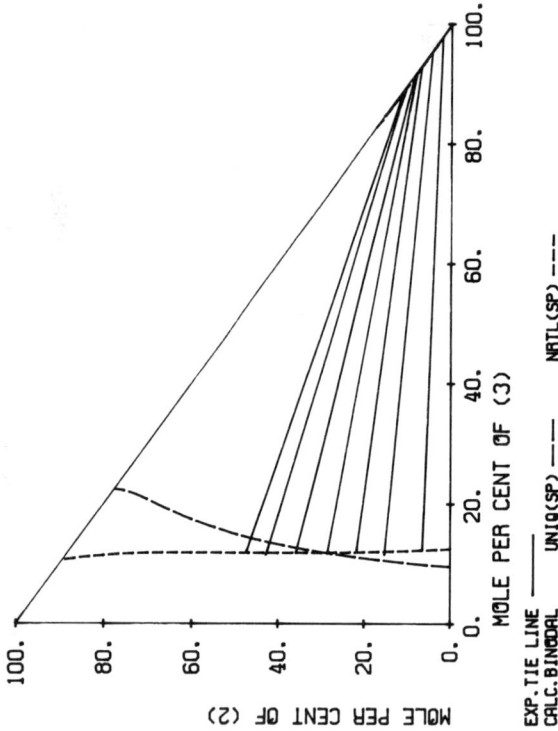

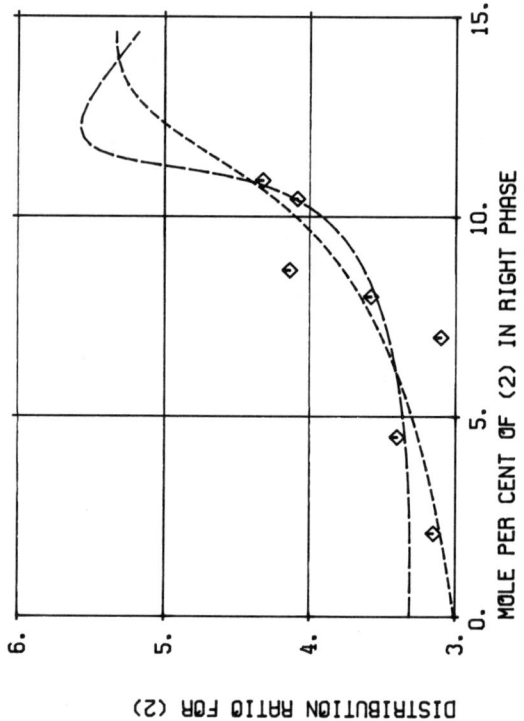

$C_2H_3N-C_6H_6$

```
------------------------------
(2) C6H6      BENZENE
------------------------------
(3) C2H3N     ACETIC ACID,NITRILE
------------------------------

HARTWIG G.M., HOOD G.C., MAYCOCK R.L.
J.PHYS.CHEM. 59(1955)52

TEMPERATURE =   25.0 DEG C      TYPE OF SYSTEM = 1

EXPERIMENTAL TIE LINES IN MOLE PCT
-----------------------------------------
        LEFT PHASE    (3)         RIGHT PHASE    (3)
    (1)     (2)     (3)         (1)     (2)     (3)
-----------------------------------------
  85.599   6.381   8.019       4.782   3.588  91.630
  76.395  12.322  11.283       6.229   7.869  85.902
  63.290  19.775  16.935       9.344  13.907  76.749

SPECIFIC MODEL PARAMETERS IN KELVIN
-----------------------------------------
            UNIQUAC              NRTL(ALPHA=.2)
I J     AIJ      AJI         AIJ      AJI
-----------------------------------------
1 2   -77.340   119.27      -82.539  -170.79
1 3   632.92     29.945     557.12    689.63
2 3    18.087    72.756    -239.57    215.35

R1 = 5.1742   R2 = 3.1878   R3 = 1.8701
Q1 = 4.396    Q2 = 2.400    Q3 = 1.724

MEAN DEV. BETWEEN CALC. AND EXP. CONC. IN MOLE PCT

UNIQUAC (SPECIFIC PARAMETERS)    0.45
NRTL   (SPECIFIC PARAMETERS)     0.44
UNIQUAC (COMMON PARAMETERS)      0.82
```

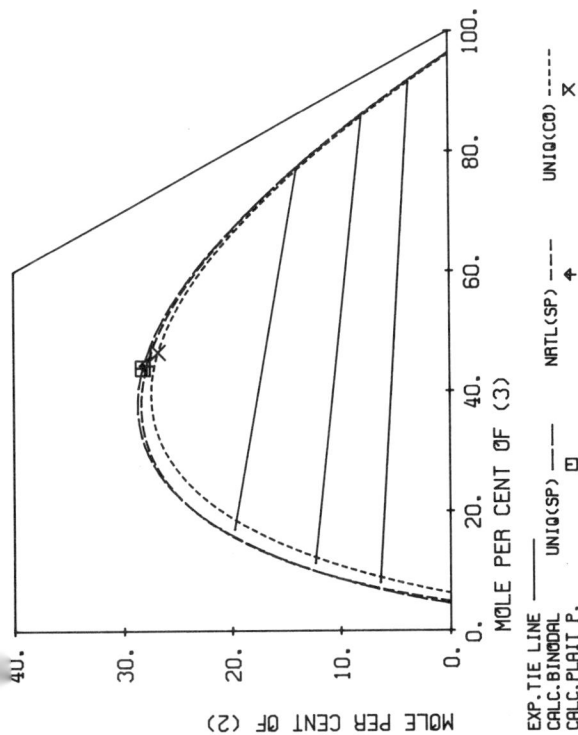

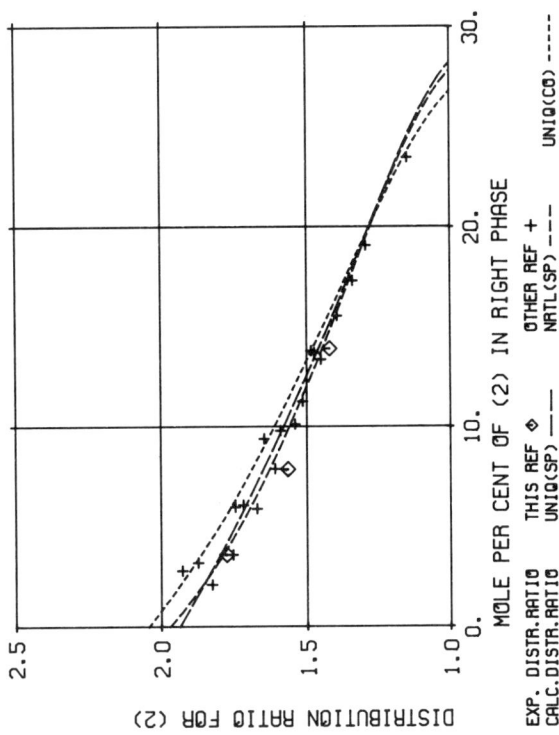

C_2H_3N-C_6H_6

(1) C7H16   HEPTANE

(2) C6H6   BENZENE

(3) C2H3N   ACETIC ACID,NITRILE

WERNER G., SCHUBERTH H.
J.PRAKT.CHEM. 4,31(1966)225

TEMPERATURE = 20.0 DEG C   TYPE OF SYSTEM = 1

EXPERIMENTAL TIE LINES IN MOLE PCT

	LEFT PHASE			RIGHT PHASE	
(1)	(2)	(3)	(1)	(2)	(3)
88.600	5.400	6.000	3.900	2.800	93.300
87.800	6.000	6.200	4.000	3.200	92.800
81.800	10.500	7.700	4.600	6.000	89.400
81.800	10.500	7.700	4.600	6.100	89.300
74.800	15.500	9.700	5.500	9.400	85.100
74.600	15.600	9.800	5.700	9.700	84.500
67.900	20.200	11.900	7.100	13.700	79.200
68.400	20.500	11.100	6.900	13.800	79.300

SPECIFIC MODEL PARAMETERS IN KELVIN

		UNIQUAC		NRTL(ALPHA=.2)	
I J	AIJ	AJI		AIJ	AJI
1 2	-77.340	119.27		-82.539	-170.79
1 3	632.92	29.945		557.12	689.63
2 3	18.087	72.756		-239.57	215.35

R1 = 5.1742   R2 = 3.1878   R3 = 1.8701
Q1 = 4.396   Q2 = 2.400   Q3 = 1.724

MEAN DEV. BETWEEN CALC. AND EXP. CONC. IN MOLE PCT

UNIQUAC (SPECIFIC PARAMETERS)   0.85
NRTL   (SPECIFIC PARAMETERS)   0.81
UNIQUAC (COMMON PARAMETERS)   1.66

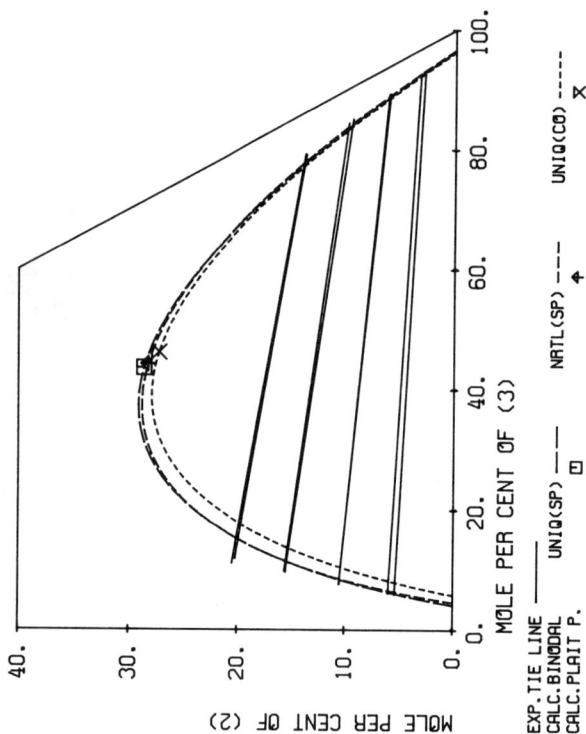

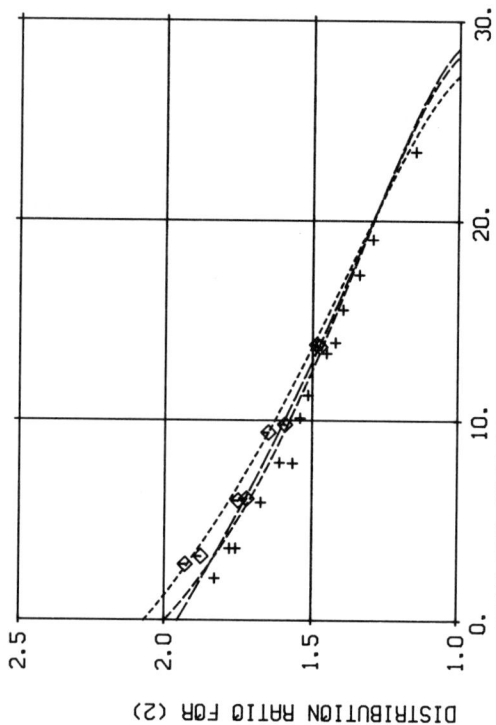

$C_2H_3N-C_6H_6$

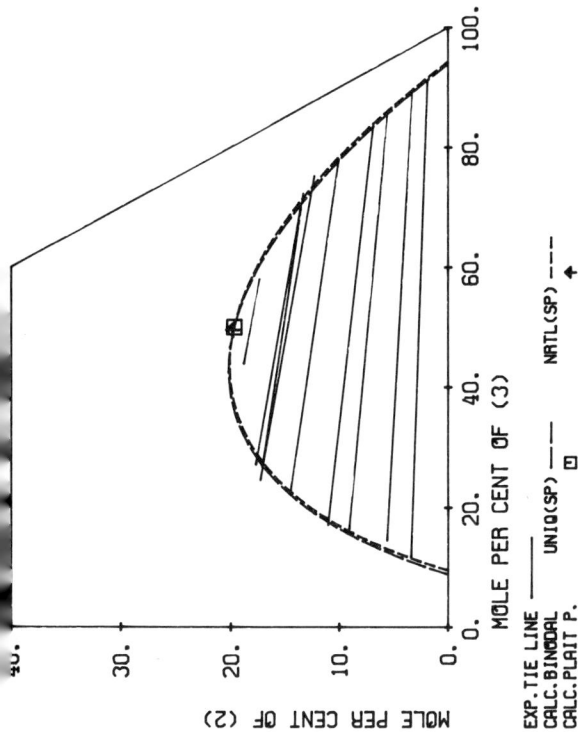

MOLE PER CENT OF (3)

MOLE PER CENT OF (2)

EXP.TIE LINE ——— UNIQ(SP) ☐ NRTL(SP) ——
CALC.BINODAL
CALC.PLAIT P. ▲

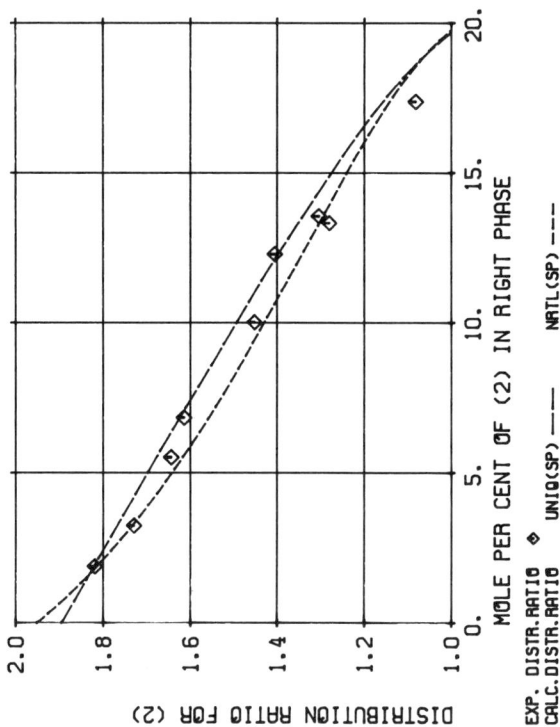

MOLE PER CENT OF (2) IN RIGHT PHASE

DISTRIBUTION RATIO FOR (2)

EXP. DISTR.RATIO ◇ UNIQ(SP) —— NRTL(SP) ——
CALC.DISTR.RATIO

---

(2) C6H6   BENZENE
---
(3) C2H3N   ACETIC ACID,NITRILE

PALMER D.A., SMITH B.D.
J.CHEM.ENG.DATA 17(1972)71

TEMPERATURE = 45.0 DEG C   TYPE OF SYSTEM = 1

EXPERIMENTAL TIE LINES IN MOLE PCT

	LEFT PHASE			RIGHT PHASE	
(1)	(2)	(3)	(1)	(2)	(3)
84.910	3.420	11.670	6.830	1.880	91.290
79.870	5.620	14.510	7.210	3.250	89.540
74.510	9.070	16.420	8.430	5.520	86.050
71.850	11.040	17.110	9.100	6.840	84.060
63.200	14.550	22.250	11.880	10.020	78.100
58.070	17.270	24.660	12.420	12.290	75.290
56.170	17.090	26.740	14.320	13.330	72.350
55.060	17.710	27.230	16.190	13.560	70.250
37.200	18.820	43.980	24.600	17.370	58.030

SPECIFIC MODEL PARAMETERS IN KELVIN

		UNIQUAC		NRTL(ALPHA=.2)	
I	J	AIJ	AJI	AIJ	AJI
1	2	-70.322	55.844	-126.34	-285.45
1	3	590.26	15.951	438.50	697.86
2	3	-17.008	71.450	-354.42	191.76

R1 = 5.1742   R2 = 3.1878   R3 = 1.8701
Q1 = 4.396   Q2 = 2.400   Q3 = 1.724

MEAN DEV. BETWEEN CALC. AND EXP. CONC. IN MOLE PCT

UNIQUAC (SPECIFIC PARAMETERS)   0.66
NRTL   (SPECIFIC PARAMETERS)   0.60

C₂H₃N-C₆H₆

(1) C7H16   HEPTANE
- - - - - - - - - - - - - - - - -
(2) C6H6   BENZENE
- - - - - - - - - - - - - - - - -
(3) C2H3N   ACETIC ACID,NITRILE
- - - - - - - - - - - - - - - - -

SETHY A., CULLINAN H.T.
AICHE J. 21(1975)571

TEMPERATURE = 25.0 DEG C   TYPE OF SYSTEM = 1

EXPERIMENTAL TIE LINES IN MOLE PCT

	LEFT PHASE			RIGHT PHASE	
(1)	(2)	(3)	(1)	(2)	(3)
93.710	0.0	6.290	3.810	0.0	96.190
88.960	3.840	7.200	4.420	2.100	93.480
86.070	6.310	7.620	5.140	3.590	91.270
80.100	9.890	10.010	6.000	5.910	88.090
76.690	12.730	10.580	6.080	7.900	86.020
71.280	15.600	13.120	7.230	10.120	82.650
68.970	17.050	13.980	7.240	11.260	81.500
64.650	19.410	15.940	8.490	13.370	78.140
60.600	21.700	17.700	9.840	15.540	74.620
56.190	23.190	20.620	10.800	17.290	71.910
52.340	24.670	22.990	12.380	19.060	68.560
40.650	26.950	32.400	18.550	23.460	57.990

SPECIFIC MODEL PARAMETERS IN KELVIN

		UNIQUAC		NRTL(ALPHA=.2)	
I J	AIJ	AJI	AIJ	AJI	
1 2	-77.340	119.27	-82.539	-170.79	
1 3	632.92	29.945	557.12	689.63	
2 3	18.087	72.756	-239.57	215.35	

R1 = 5.1742   R2 = 3.1878   R3 = 1.8701
Q1 = 4.396   Q2 = 2.400   Q3 = 1.724

MEAN DEV. BETWEEN CALC. AND EXP. CONC. IN MOLE PCT

UNIQUAC (SPECIFIC PARAMETERS)   0.45
NRTL   (SPECIFIC PARAMETERS)   0.40
UNIQUAC (COMMON PARAMETERS)   0.98

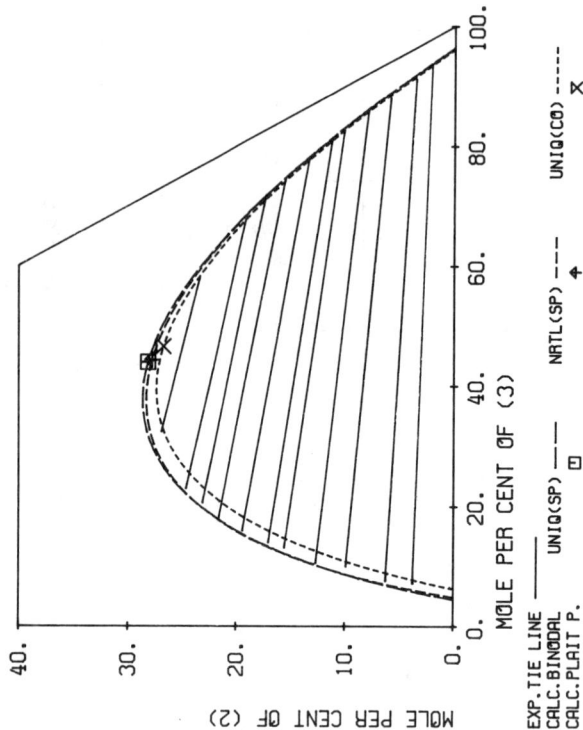

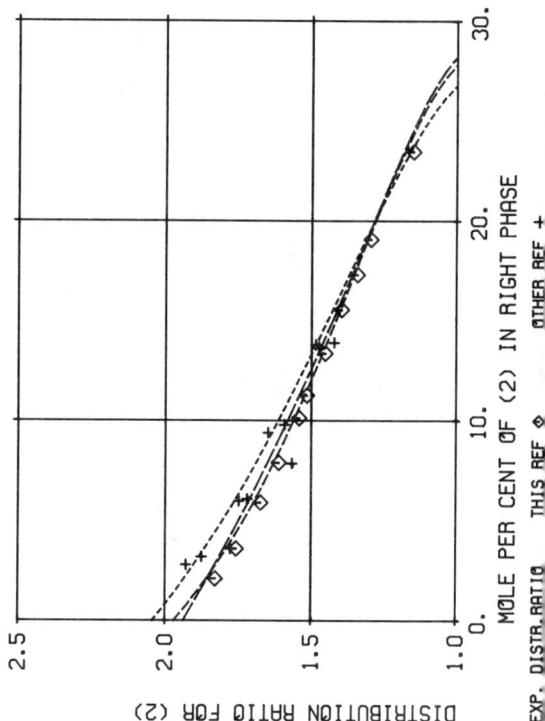

$C_2H_3N-C_6H_6$

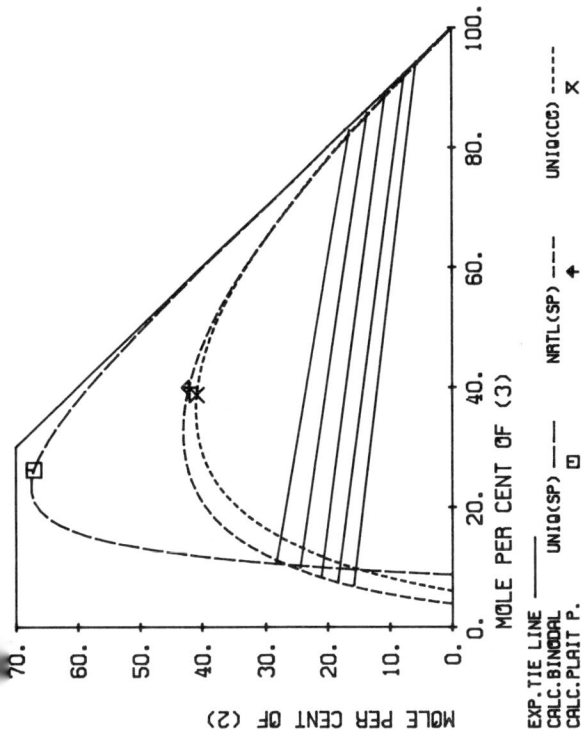

MOLE PER CENT OF (2)

MOLE PER CENT OF (3)

EXP.TIE LINE ———   UNIQ(SP) ☐   NRTL(SP) ♦   UNIQ(CC) ----
CALC.BINODAL
CALC.PLAIT P.

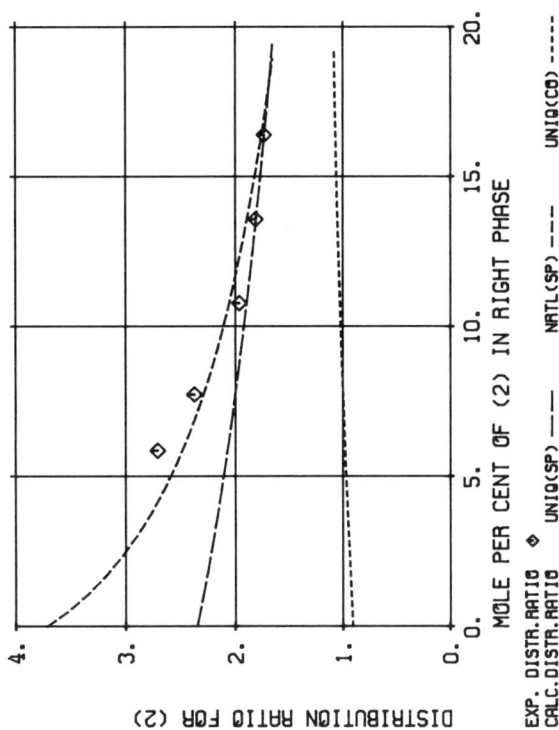

DISTRIBUTION RATIO FOR (2)

MOLE PER CENT OF (2) IN RIGHT PHASE

EXP. DISTR.RATIO ♦   UNIQ(SP) ———   NRTL(SP) ---   UNIQ(CC) ----
CALC.DISTR.RATIO

(2) C6H6   BENZENE

(3) C2H3N   ACETIC ACID,NITRILE

BONDARENKO M.F., ET AL.
KHIM.TEKHNOL.TOPL.MASEL (1972)4,8

TEMPERATURE = 30.0 DEG C   TYPE OF SYSTEM = 1

EXPERIMENTAL TIE LINES IN MOLE PCT

	LEFT PHASE			RIGHT PHASE	
(1)	(2)	(3)	(1)	(2)	(3)
77.188	15.827	6.984	0.393	5.843	93.765
74.006	18.335	7.659	0.442	7.719	91.839
70.626	21.096	8.278	0.519	10.759	88.722
65.559	24.545	9.896	0.599	13.567	85.834
60.984	28.370	10.646	0.682	16.381	82.937

SPECIFIC MODEL PARAMETERS IN KELVIN

		UNIQUAC			NRTL(ALPHA=.2)	
I J	AIJ	AJI		AIJ	AJI
1 2	7.5319	27.842		-412.18	-154.01
1 3	436.19	254.05		492.91	-1556.3
2 3	68.867	3.1813		167.74	-358.67

R1 =10.5694   R2 = 3.1878   R3 = 1.8701
Q1 = 8.716   Q2.= 2.400   Q3 = 1.724

MEAN DEV. BETWEEN CALC. AND EXP. CONC. IN MOLE PCT

UNIQUAC (SPECIFIC PARAMETERS)   0.93
NRTL   (SPECIFIC PARAMETERS)   0.38
UNIQUAC (COMMON PARAMETERS)   3.94

C₂H₃N-C₆H₆

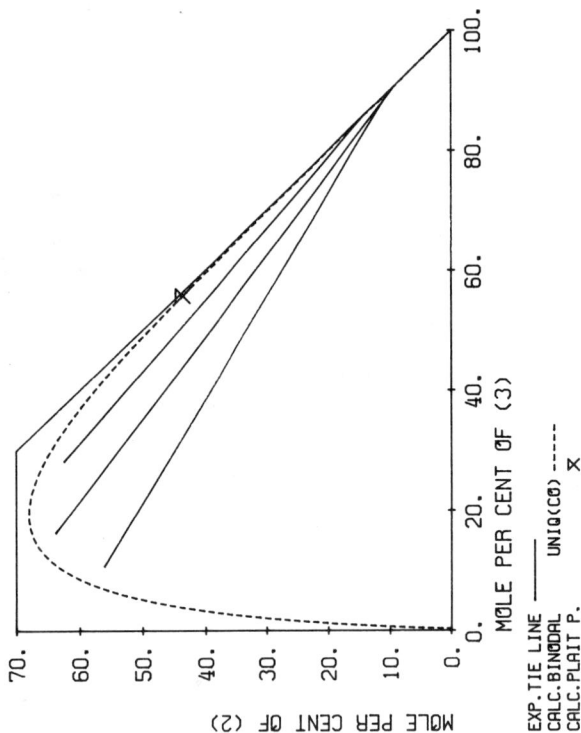

MOLE PER CENT OF (3)

MOLE PER CENT OF (2)

EXP.TIE LINE —————— UNIQ(C0) - - - - -
CALC.BINODAL
CALC.PLAIT P.   ✗

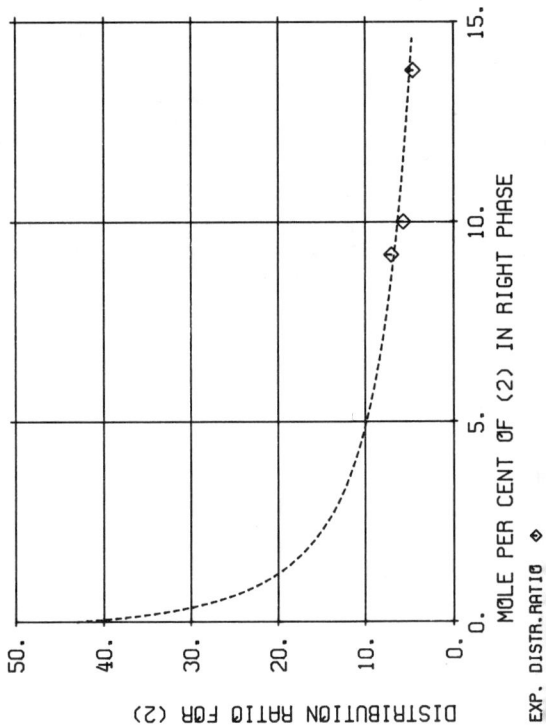

MOLE PER CENT OF (2) IN RIGHT PHASE

DISTRIBUTION RATIO FOR (2)

EXP. DISTR.RATIO ◇

(1) C6H6   BENZENE
(2) C2H3N   ACETIC ACID,NITRILE
(3) H2O   WATER

HARTWIG G.M., HOOD G.C., MAYCOCK R.L.
J.PHYS.CHEM. 59(1955)52

TEMPERATURE = 25.0 DEG C   TYPE OF SYSTEM = 1

EXPERIMENTAL TIE LINES IN MOLE PCT

	LEFT PHASE			RIGHT PHASE	
(1)	(2)	(3)	(1)	(2)	(3)
19.893	63.747	16.360	0.0	9.169	90.831
33.226	56.039	10.735	0.0	9.998	90.002
9.340	62.394	28.266	0.0	13.782	86.218

MEAN DEV. BETWEEN CALC. AND EXP. CONC. IN MOLE PCT

UNIQUAC (COMMON PARAMETERS)   3.34

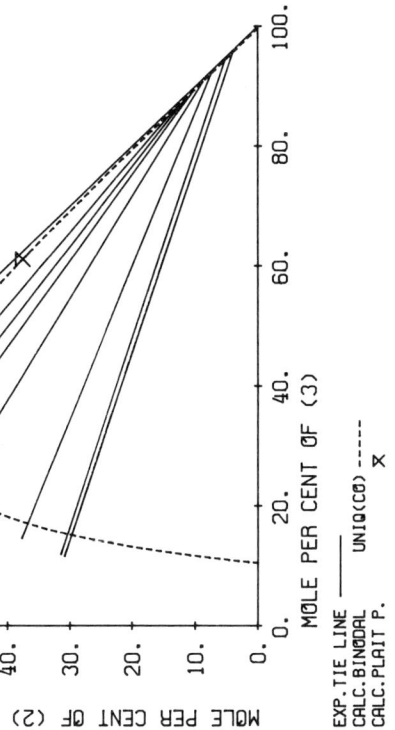

MOLE PER CENT OF (3)

MOLE PER CENT OF (2)

EXP.TIE LINE ——— UNIQ(CO) -----
CALC.BINODAL
CALC.PLAIT P.   x

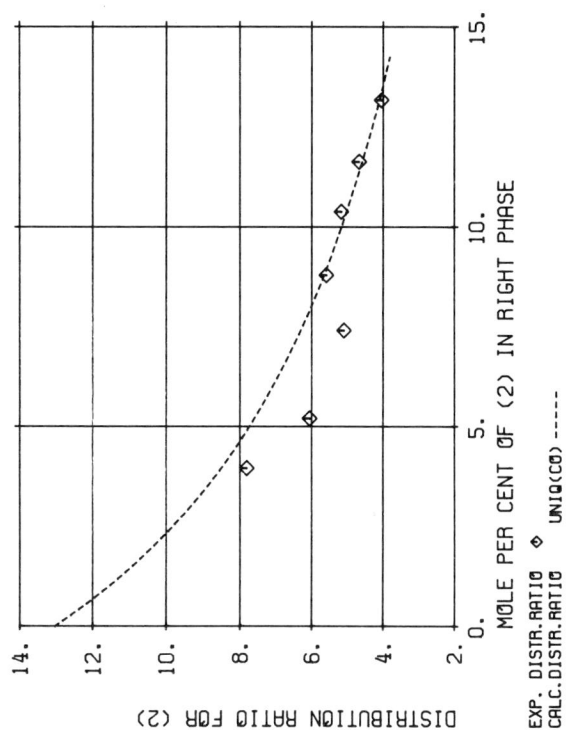

MOLE PER CENT OF (2) IN RIGHT PHASE

DISTRIBUTION RATIO FOR (2)

EXP. DISTR.RATIO ◇   UNIQ(CO) -----
CALC.DISTR.RATIO

(2) C2H3N   ACETIC ACID,NITRILE
(3) H2O   WATER

SUBBA RAO D. ET AL.
J.CHEM.ENG.DATA 24(1979)241

TEMPERATURE = 30.0 DEG C   TYPE OF SYSTEM = 1

EXPERIMENTAL TIE LINES IN MOLE PCT

	LEFT PHASE			RIGHT PHASE	
(1)	(2)	(3)	(1)	(2)	(3)
57.505	30.847	11.648	0.190	3.951	95.858
56.591	31.461	11.948	0.213	5.195	94.592
47.673	37.704	14.624	0.339	7.405	92.255
28.849	49.079	22.072	0.407	8.784	90.809
19.420	53.588	26.992	0.456	10.378	89.166
14.947	54.185	30.867	0.506	11.621	87.874
9.607	53.300	37.093	0.580	13.160	86.260

MEAN DEV. BETWEEN CALC. AND EXP. CONC. IN MOLE PCT

UNIQUAC (COMMON PARAMETERS)   1.51

C_2H_3N-$C_6H_{12}O_2$

(1) C6H12O2   ACETIC ACID,BUTYL ESTER

(2) C2H3N   ACETIC ACID,NITRILE

(3) H2O   WATER

VENKATA SIVA RAMA RAO C. ET AL.
J.CHEM.ENG.DATA 23(1978)23

TEMPERATURE = 31.0 DEG C   TYPE OF SYSTEM = 1

EXPERIMENTAL TIE LINES IN MOLE PCT

	LEFT PHASE			RIGHT PHASE	
(1)	(2)	(3)	(1)	(2)	(3)
83.007	6.894	10.099	0.157	0.445	99.398
79.516	9.575	10.909	0.158	0.805	99.037
71.994	16.208	11.798	0.160	1.811	98.029
58.899	26.547	14.554	0.178	2.799	97.023
51.900	33.217	14.883	0.182	4.646	95.172
47.111	38.675	14.214	0.202	5.709	94.090
38.094	44.076	17.831	0.205	7.057	92.738

SPECIFIC MODEL PARAMETERS IN KELVIN

		UNIQUAC		NRTL(ALPHA=.2)	
I J	AIJ	AJI		AIJ	AJI
1 2	-44.118	-74.912		-89.462	-106.81
1 3	416.61	168.23		240.52	1442.8
2 3	88.592	37.578		221.97	435.64

R1 = 4.8274   R2 = 1.8701   R3 = 0.9200
Q1 = 4.196   Q2 = 1.724   Q3 = 1.400

MEAN DEV. BETWEEN CALC. AND EXP. CONC. IN MOLE PCT

UNIQUAC (SPECIFIC PARAMETERS)   0.43
NRTL   (SPECIFIC PARAMETERS)   0.43

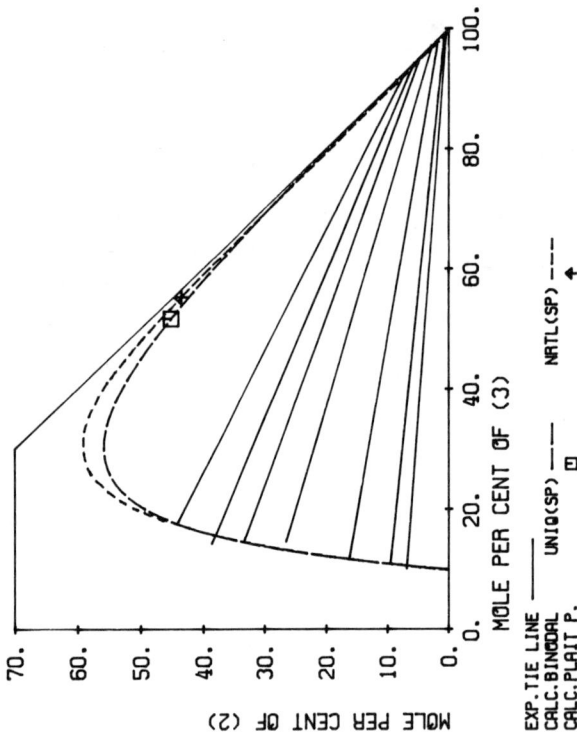

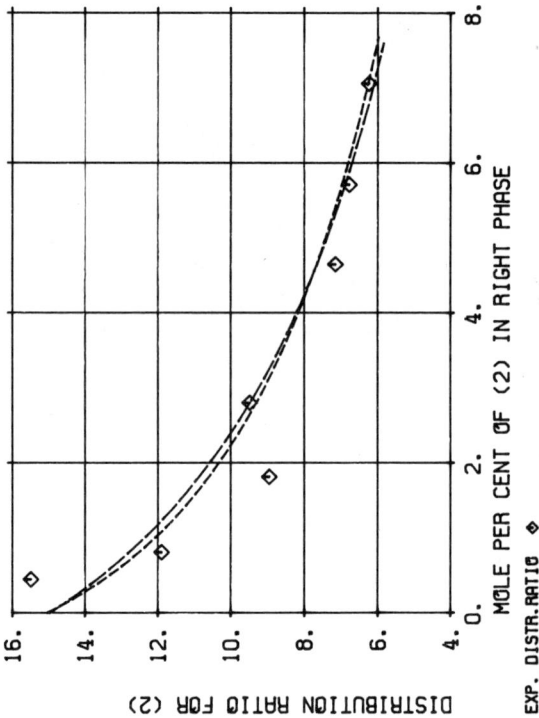

MOLE PER CENT OF (2)

MOLE PER CENT OF (3)

EXP.TIE LINE ———   UNIQ(SP) ——— ⊡   NRTL(SP) ––––– ♦
CALC.BINODAL
CALC.PLAIT P.

DISTRIBUTION RATIO FOR (2)

MOLE PER CENT OF (2) IN RIGHT PHASE

EXP. DISTR.RATIO ◇

$C_2H_3N-C_6H_{14}$

(2) C17H34O2   HEXADECANOIC ACID,METHYL ESTER

(3) C2H3N   ACETIC ACID,NITRILE

BARFORD R.A., BERTSCH R.J., ROTHBART H.L.
J.AM.OIL CHEM.SOC. 45(1968)141

TEMPERATURE = 20.0 DEG C   TYPE OF SYSTEM = 2

EXPERIMENTAL TIE LINES IN MOLE PCT

	LEFT PHASE			RIGHT PHASE	
(1)	(2)	(3)	(1)	(2)	(3)
79.636	8.222	12.142	5.015	0.342	94.643
66.803	13.698	19.499	4.644	0.575	94.781
37.317	22.447	40.237	4.514	0.711	94.775
11.188	26.091	62.721	1.825	1.916	96.259

SPECIFIC MODEL PARAMETERS IN KELVIN

		UNIQUAC		NRTL(ALPHA=.2)	
I	J	AIJ	AJI	AIJ	AJI
1	2	4.6241	11.727	-9.2861	-272.60
1	3	556.78	34.279	493.14	595.34
2	3	250.56	27.178	-527.33	1544.8

R1 = 4.4998   R2 =12.2458   R3 = 1.8701
Q1 = 3.856   Q2 =10.136   Q3 = 1.724

MEAN DEV. BETWEEN CALC. AND EXP. CONC. IN MOLE PCT

UNIQUAC (SPECIFIC PARAMETERS)   0.71
NRTL   (SPECIFIC PARAMETERS)   0.95
UNIQUAC (COMMON PARAMETERS)   1.04

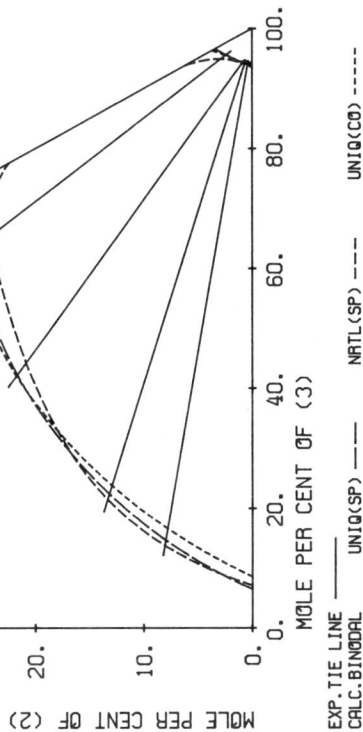

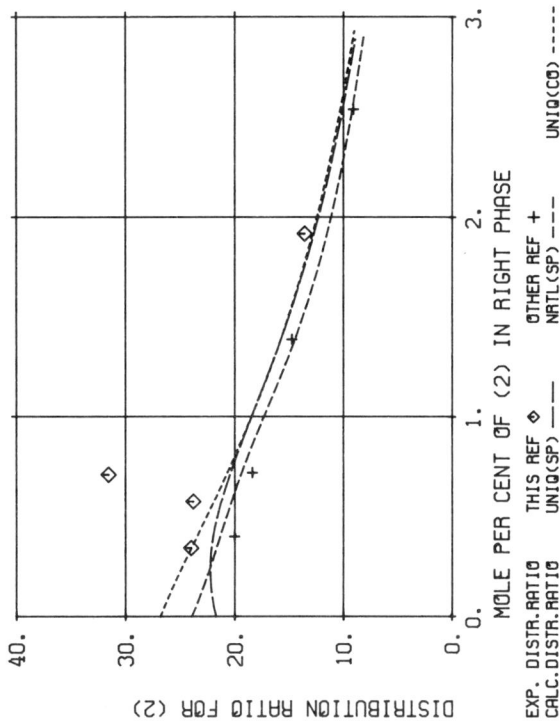

186

C₂H₃N-C₆H₁₄

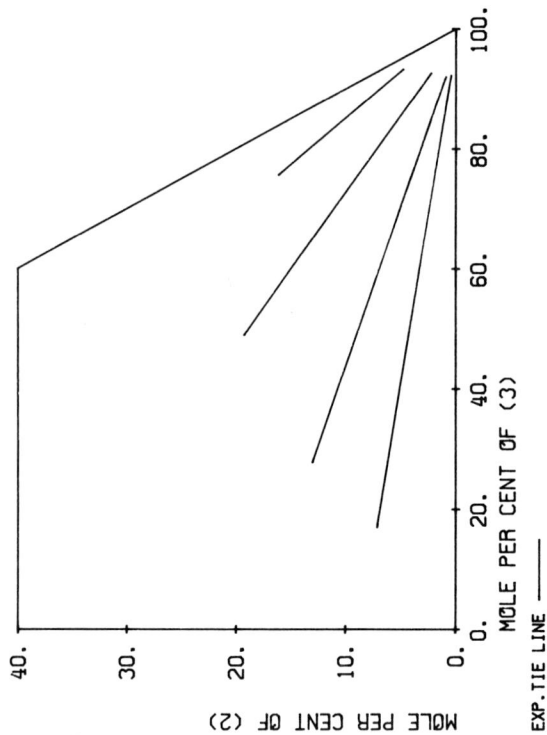

40.

30.

MOLE PER CENT OF (2)

20.

10.

0.

0.   20.   40.   60.   80.   100.

MOLE PER CENT OF (3)

EXP.TIE LINE ⎯⎯⎯⎯

(1) C6H14   HEXANE
(2) C17H34O2   HEXADECANOIC ACID,METHYL ESTER
(3) C2H3N   ACETIC ACID,NITRILE

BARFORD R.A., BERTSCH R.J., ROTHBART H.L.
J.AM.OIL CHEM.SOC. 45(1968)141

TEMPERATURE = 30.4 DEG C   TYPE OF SYSTEM = 2

EXPERIMENTAL TIE LINES IN MOLE PCT

	LEFT PHASE			RIGHT PHASE	
(1)	(2)	(3)	(1)	(2)	(3)
75.920	7.115	16.965	7.258	0.436	92.306
59.207	12.986	27.807	7.033	0.889	92.078
31.765	19.250	48.985	5.062	2.240	92.697
8.250	16.104	75.646	1.901	4.767	93.332

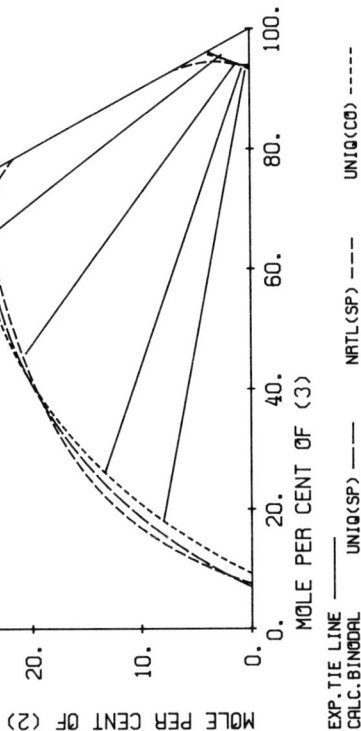

MOLE PER CENT OF (3)

MOLE PER CENT OF (2)

EXP.TIE LINE ———
CALC.BINODAL

UNIQ(SP) ———   NRTL(SP) ———   UNIQ(CO) -----

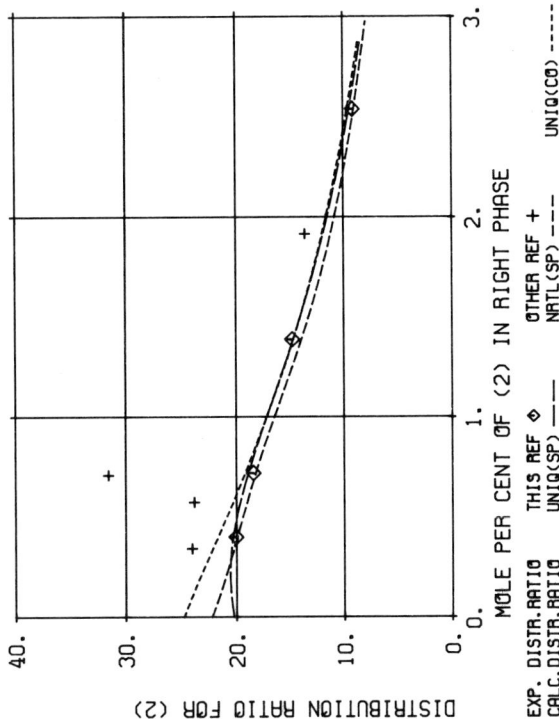

MOLE PER CENT OF (2) IN RIGHT PHASE

DISTRIBUTION RATIO FOR (2)

EXP. DISTR.RATIO   THIS REF ◇   OTHER REF +
CALC.DISTR.RATIO   UNIQ(SP) ———   NRTL(SP) ———   UNIQ(CO) -----

(2) C17H34O2	HEXADECANOIC ACID,METHYL ESTER
(3) C2H3N	ACETIC ACID,NITRILE

RUSLING J.F., BERTSCH R.J., BARFORD R.A., ROTHBART H.L.
J.CHEM.ENG.DATA 14(1969)169

TEMPERATURE = 25.0 DEG C   TYPE OF SYSTEM = 2

EXPERIMENTAL TIE LINES IN MOLE PCT

	LEFT PHASE			RIGHT PHASE	
(1)	(2)	(3)	(1)	(2)	(3)
73.935	7.990	18.075	6.625	0.399	92.976
60.516	13.257	26.227	5.896	0.721	93.383
33.428	20.540	46.032	4.679	1.388	93.933
10.307	23.098	66.595	1.770	2.538	95.692

SPECIFIC MODEL PARAMETERS IN KELVIN

		UNIQUAC		NRTL(ALPHA=.2)	
I J	AIJ	AJI		AIJ	AJI
1 2	4.6241	11.727		-9.2861	-272.60
1 3	556.78	34.279		493.14	595.34
2 3	250.56	27.178		-527.33	1544.8

R1 = 4.4998   R2 =12.2458   R3 = 1.8701
Q1 = 3.856   Q2 =10.136   Q3 = 1.724

MEAN DEV. BETWEEN CALC. AND EXP. CONC. IN MOLE PCT

UNIQUAC (SPECIFIC PARAMETERS)	0.60
NRTL (SPECIFIC PARAMETERS)	0.84
UNIQUAC (COMMON PARAMETERS)	0.55

C_2H_3N-C_6H_{14}

(1) C6H14   HEXANE

(2) C19H36O2   9-OCTADECENOIC ACID(CIS),METHYL ESTER

(3) C2H3N   ACETIC ACID,NITRILE

BARFORD R.A., BERTSCH R.J., ROTHBART H.L.
J.AM.OIL CHEM.SOC. 45(1963)141

TEMPERATURE = 20.0 DEG C   TYPE OF SYSTEM = 2

EXPERIMENTAL TIE LINES IN MOLE PCT

	LEFT PHASE			RIGHT PHASE	
	(1)	(3)	(1)	(2)	(3)
77.778	7.568	14.654	5.481	0.406	94.113
65.479	12.527	21.995	5.135	0.701	94.164
35.598	19.929	44.473	3.959	1.529	94.511
12.355	21.443	66.202	1.413	2.743	95.845

SPECIFIC MODEL PARAMETERS IN KELVIN

		UNIQUAC		NRTL(ALPHA=.2)	
I	J	AIJ	AJI	AIJ	AJI
1	2	10.242	9.1914	-77.716	-344.39
1	3	569.76	27.180	529.14	590.41
2	3	252.77	17.879	-628.17	1533.3

R1 = 4.4998   R2 =13.3625   R3 = 1.8701
Q1 = 3.856   Q2 =11.003   Q3 = 1.724

MEAN DEV. BETWEEN CALC. AND EXP. CONC. IN MOLE PCT

UNIQUAC (SPECIFIC PARAMETERS)   0.58
NRTL   (SPECIFIC PARAMETERS)   1.44
UNIQUAC (COMMON PARAMETERS)   0.84

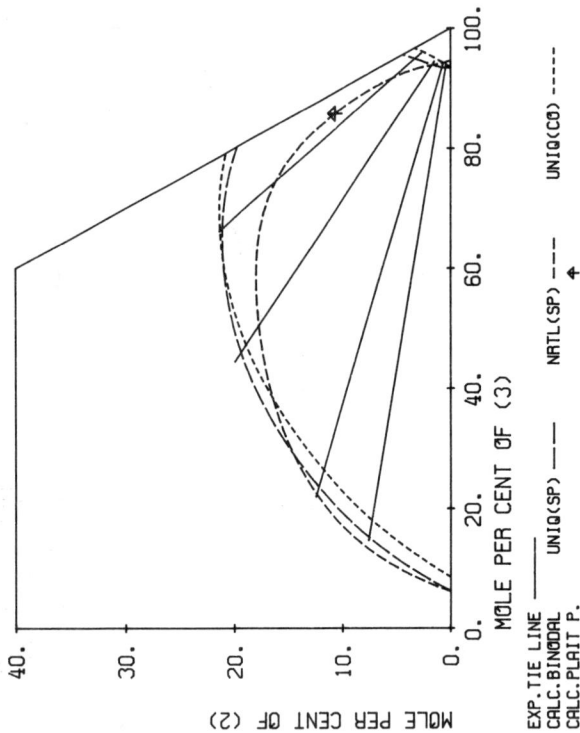

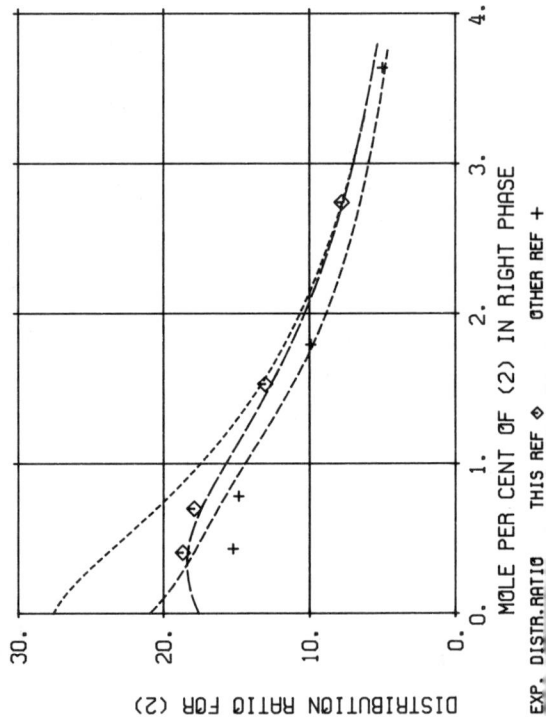

$C_2H_3N-C_6H_{14}$

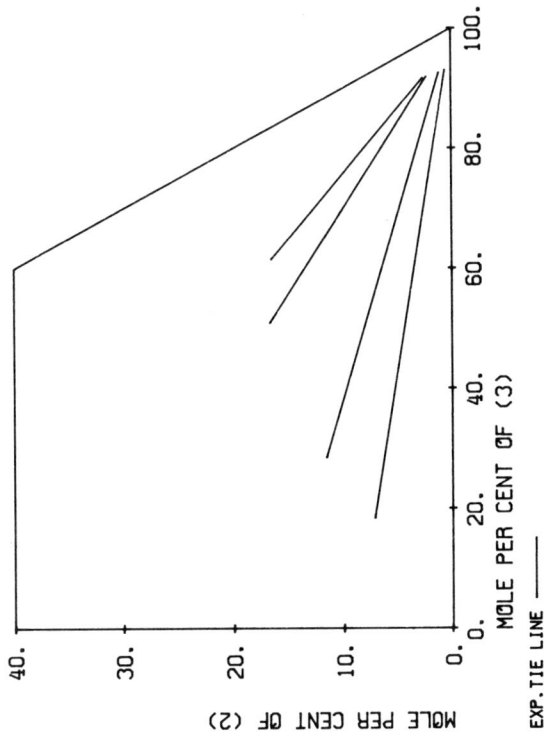

(1) C6H14   HEXANE
----
(2) C19H36O2   9-OCTADECENOIC ACID(CIS),METHYL ESTER
----
(3) C2H3N   ACETIC ACID,NITRILE

BARFORD R.A., BERTSCH R.J., ROTHBART H.L.
J.AM.OIL CHEM.SOC. 45(1968)141

TEMPERATURE = 30.4 DEG C   TYPE OF SYSTEM = 1

EXPERIMENTAL TIE LINES IN MOLE PCT

        LEFT PHASE   RIGHT PHASE
   (1)   (2)   (3)   (1)   (2)   (3)

 74.516   7.146 18.338   6.370   0.566 93.064
 59.913 11.555 28.532   6.235   1.119 92.647
 32.369 16.663 50.967   5.797   2.269 91.935
 21.950 16.531 61.518   5.659   2.611 91.730

C₂H₃N-C₆H₁₄

(1) C6H14   HEXANE
----
(2) C19H36O2 9-OCTADECENOIC ACID(CIS),METHYL ESTER
----
(3) C2H3N   ACETIC ACID,NITRILE

RUSLING J.F., BERTSCH R.J., BARFORD R.A., ROTHBART H.L.
J.CHEM.ENG.DATA 14(1969)169

TEMPERATURE = 25.0 DEG C   TYPE OF SYSTEM = 2

EXPERIMENTAL TIE LINES IN MOLE PCT

           LEFT PHASE   RIGHT PHASE
        (1)   (2)   (3)   (1)   (2)   (3)

      78.035   6.557   15.408   7.675   0.431   91.895
      66.064   11.637   22.299   7.605   0.784   91.611
      35.045   17.817   47.137   5.118   1.795   93.087
      11.099   18.322   70.580   1.781   3.640   94.579

SPECIFIC MODEL PARAMETERS IN KELVIN

               UNIQUAC   NRTL(ALPHA=.2)
  I J   AIJ   AJI   AIJ   AJI
----
  1 2   10.242   9.1914   -77.716   -344.39
  1 3 569.76   27.180   529.14   590.41
  2 3 252.77   17.879   -628.17   1533.3

  R1 = 4.4998   R2 =13.3625   R3 = 1.8701
  Q1 = 3.856   Q2 =11.003   Q3 = 1.724

MEAN DEV. BETWEEN CALC. AND EXP. CONC. IN MOLE PCT

UNIQUAC (SPECIFIC PARAMETERS)   0.72
NRTL   (SPECIFIC PARAMETERS)   1.28
UNIQUAC (COMMON PARAMETERS)   1.10

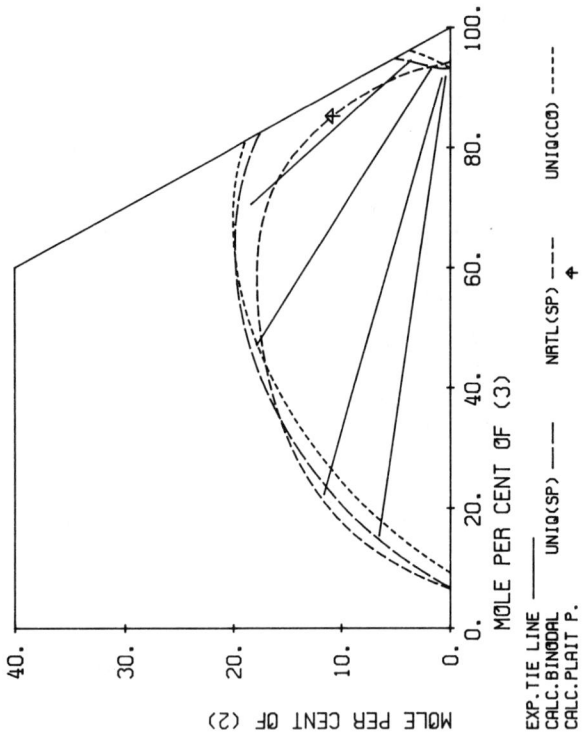

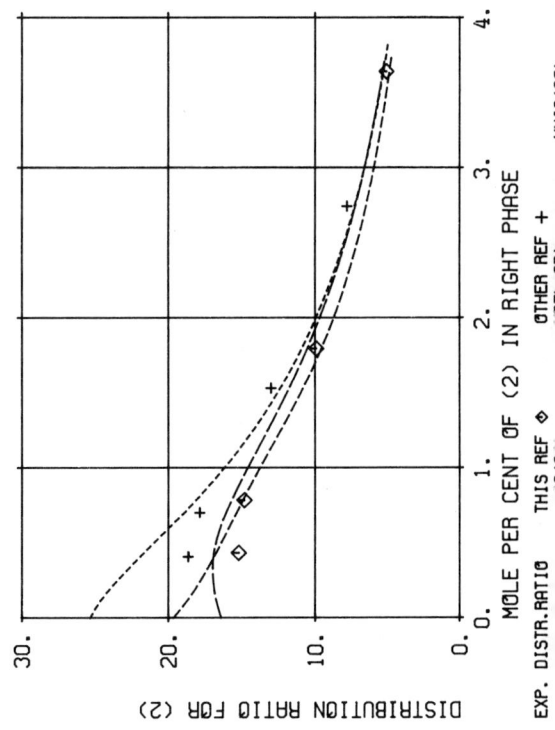

C$_2$H$_3$N-C$_7$H$_8$

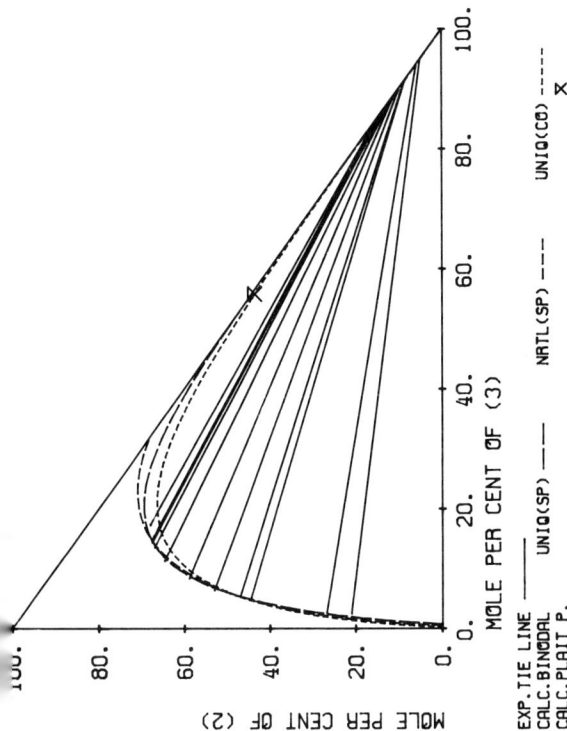

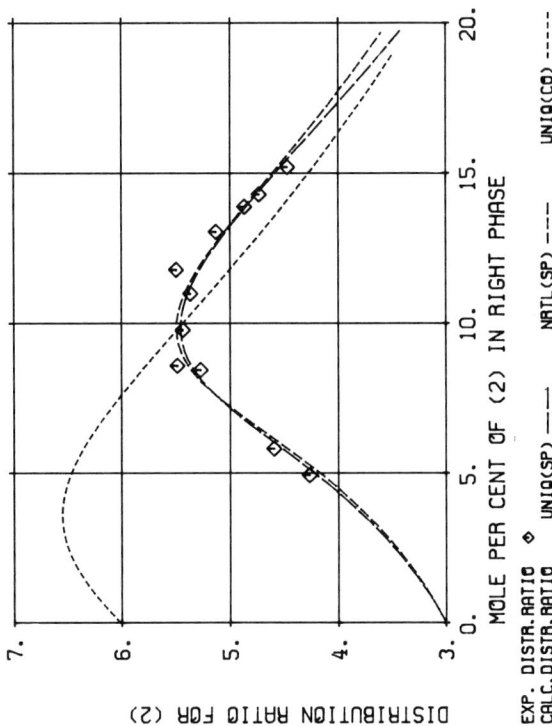

(2) C2H3N   ACETIC ACID,NITRILE

(3) H2O   WATER

SUBBA RAO D. ET AL.
J.CHEM.ENG.DATA 24(1979)241

TEMPERATURE = 30.0 DEG C   TYPE OF SYSTEM = 1

EXPERIMENTAL TIE LINES IN MOLE PCT

LEFT PHASE			RIGHT PHASE		
(1)	(2)	(3)	(1)	(2)	(3)
76.493	20.989	2.517	0.104	4.917	94.978
70.757	26.691	2.552	0.137	5.805	94.057
50.439	44.459	5.102	0.163	8.438	91.399
47.618	47.040	5.342	0.175	8.577	91.248
40.587	53.050	6.362	0.188	9.767	90.044
32.725	58.905	8.370	0.180	10.981	88.839
24.068	64.662	11.269	0.192	11.769	88.039
19.487	66.957	13.556	0.207	13.050	86.743
18.105	67.547	14.348	0.220	13.877	85.903
17.618	67.621	14.761	0.233	14.284	85.483
14.705	67.975	17.320	0.247	15.197	84.556

SPECIFIC MODEL PARAMETERS IN KELVIN

	UNIQUAC		NRTL(ALPHA=.2)	
I J	AIJ	AJI	AIJ	AJI
1 2	388.59	-121.25	440.17	-70.267
1 3	700.29	-438.76	1068.5	1703.7
2 3	197.12	101.10	295.31	451.76

R1 = 3.9228   R2 = 1.8701   R3 = 0.9200
Q1 = 2.968   Q2 = 1.724   Q3 = 1.400

MEAN DEV. BETWEEN CALC. AND EXP. CONC. IN MOLE PCT

UNIQUAC (SPECIFIC PARAMETERS)   0.25
NRTL   (SPECIFIC PARAMETERS)   0.32
UNIQUAC (COMMON PARAMETERS)   1.25

MOLE PER CENT OF (3)
MOLE PER CENT OF (2)

EXP.TIE LINE ——— UNIQ(SP) ——— NRTL(SP) ——— UNIQ(CO) -----
CALC.BINODAL
CALC.PLAIT P.   X

MOLE PER CENT OF (2) IN RIGHT PHASE
DISTRIBUTION RATIO FOR (2)

EXP. DISTR.RATIO ◇ UNIQ(SP) ——— NRTL(SP) ——— UNIQ(CO) -----
CALC.DISTR.RATIO

$C_2H_3N-C_7H_{14}O_2$

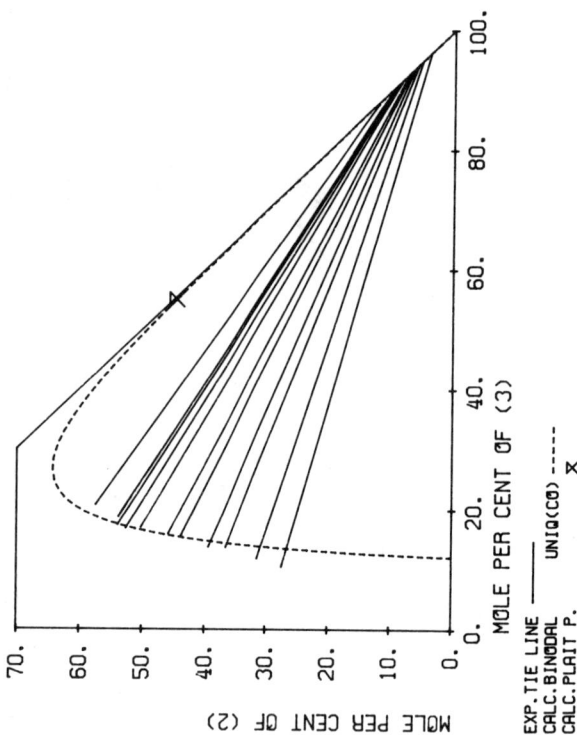

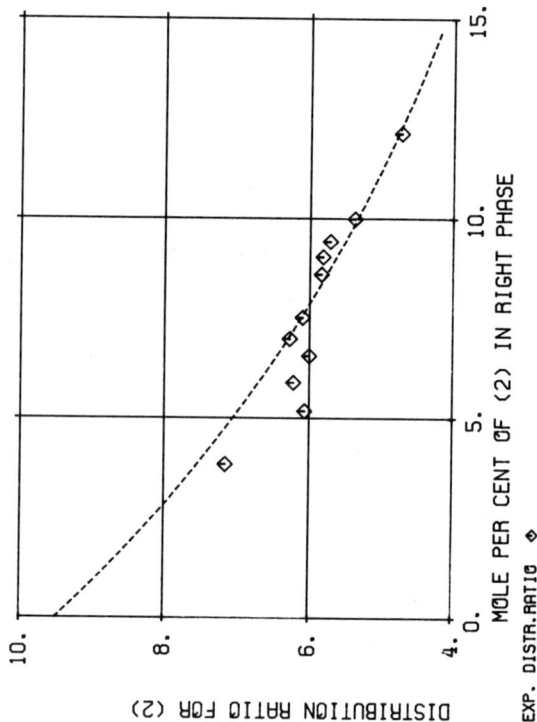

MOLE PER CENT OF (3)

MOLE PER CENT OF (2)

EXP.TIE LINE ———
CALC.BINODAL   UNIQ(CO) ------
CALC.PLAIT P.   X

DISTRIBUTION RATIO FOR (2)

MOLE PER CENT OF (2) IN RIGHT PHASE

EXP. DISTR. RATIO ◇

(1) C7H14O2   ACETIC ACID,3-METHYLBUTYL ESTER

(2) C2H3N   ACETIC ACID,NITRILE

(3) H2O   WATER

SUBBA RAO D. ET AL.
J.CHEM.ENG.DATA 24(1979)241

TEMPERATURE = 30.0 DEG C   TYPE OF SYSTEM = 1

EXPERIMENTAL TIE LINES IN MOLE PCT

	LEFT PHASE			RIGHT PHASE	
(1)	(2)	(3)	(1)	(2)	(3)
62.089	27.476	10.435	0.095	3.843	96.062
56.750	31.405	11.845	0.104	5.178	94.718
49.678	36.636	13.685	0.112	5.890	93.998
46.883	39.405	13.712	0.121	6.562	93.317
40.646	43.808	15.546	0.129	6.983	92.883
38.293	45.826	15.881	0.138	7.514	92.348
32.864	50.172	16.964	0.147	8.594	91.259
30.592	52.594	16.814	0.156	9.037	90.807
28.731	53.879	17.390	0.156	9.424	90.419
27.540	53.749	18.711	0.165	9.988	89.846
21.835	57.436	20.730	0.177	12.129	87.694

MEAN DEV. BETWEEN CALC. AND EXP. CONC. IN MOLE PCT
UNIQUAC (COMMON PARAMETERS)   0.72

$C_2H_3N-C_8H_{10}$

MOLE PER CENT OF (3)

MOLE PER CENT OF (2)

EXP.TIE LINE ——— UNIQ(SP) □ NRTL(SP) ----
CALC.BINODAL
CALC.PLAIT P.

MOLE PER CENT OF (2) IN RIGHT PHASE

DISTRIBUTION RATIO FOR (2)

EXP. DISTR.RATIO ◇ UNIQ(SP) ——— NRTL(SP) ----
CALC.DISTR.RATIO

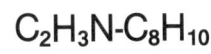

(1) C8H10   BENZENE,DIMETHYL(ISOMER NOT SPECIFIED)
(2) C2H3N   ACETIC ACID,NITRILE
(3) H2O   WATER

VENKATA SIVA RAMA RAO C. ET AL.
J.CHEM.ENG.DATA 23(1978)23

TEMPERATURE = 31.0 DEG C   TYPE OF SYSTEM = 1

EXPERIMENTAL TIE LINES IN MOLE PCT

| | LEFT PHASE | | | RIGHT PHASE | |
(1)	(2)	(3)	(1)	(2)	(3)
82.551	4.700	12.749	0.070	1.802	98.128
78.790	7.829	13.380	0.126	4.088	95.787
76.556	10.275	13.169	0.186	6.624	93.190
67.597	18.760	13.643	0.191	8.869	90.941
63.268	23.542	13.191	0.238	12.314	87.448
52.850	34.659	12.491	0.243	14.379	85.378
42.561	45.705	11.735	0.319	18.591	81.090

SPECIFIC MODEL PARAMETERS IN KELVIN

| | | UNIQUAC | | NRTL(ALPHA=.2) | |
I	J	AIJ	AJI	AIJ	AJI
1	2	215.30	62.307	-68.894	531.22
1	3	247.11	224.41	144.72	1768.4
2	3	207.13	72.080	1078.6	-21.915

R1 = 4.6578   R2 = 1.8701 * R3 = 0.9200
Q1 = 3.536   Q2 = 1.724   Q3 = 1.400

MEAN DEV. BETWEEN CALC. AND EXP. CONC. IN MOLE PCT

UNIQUAC (SPECIFIC PARAMETERS)   1.36
NRTL   (SPECIFIC PARAMETERS)   0.49

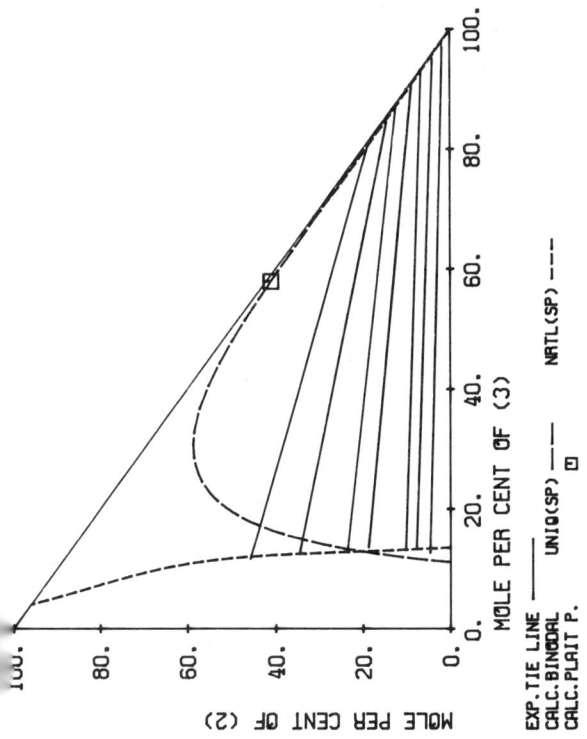

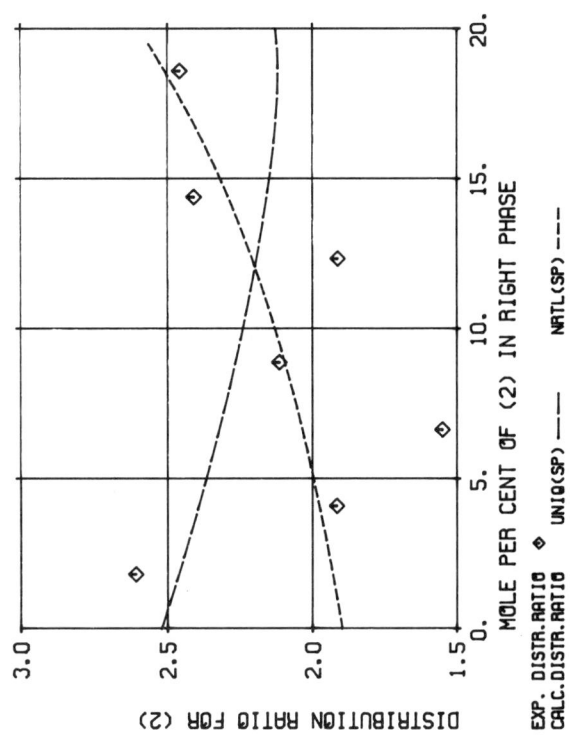

$C_2H_3N-C_8H_{10}$

(1) C15H32   PENTADECANE
(2) C8H10   BENZENE,ETHYL
(3) C2H3N   ACETIC ACID,NITRILE

BONDARENKO M.F., ET AL.
KHIM.TEKHNOL.TOPL.MASEL (1972)4,8

TEMPERATURE = 30.0 DEG C   TYPE OF SYSTEM = 1

EXPERIMENTAL TIE LINES IN MOLE PCT

	LEFT PHASE			RIGHT PHASE	
(1)	(2)	(3)	(1)	(2)	(3)
77.103	14.446	3.451	0.332	3.734	95.934
71.078	19.513	9.409	0.402	4.955	94.643
65.106	24.623	10.271	0.481	6.997	92.522
53.093	34.487	12.420	0.563	8.970	90.467
47.707	38.598	13.695	0.726	11.431	87.843

SPECIFIC MODEL PARAMETERS IN KELVIN

		UNIQUAC		NRTL(ALPHA=.2)	
I J	AIJ	AJI	AIJ	AJI	
1 2	231.53	-133.38	22.926	139.38	
1 3	529.84	163.00	344.56	1372.5	
2 3	261.63	1.8622	255.07	346.60	

R1 =10.5694   R2 = 4.5972   R3 = 1.8701
Q1 = 8.716   Q2 = 3.508   Q3 = 1.724

MEAN DEV. BETWEEN CALC. AND EXP. CONC. IN MOLE PCT

UNIQUAC (SPECIFIC PARAMETERS)   0.32
NRTL   (SPECIFIC PARAMETERS)   0.26
UNIQUAC (COMMON PARAMETERS)   0.43

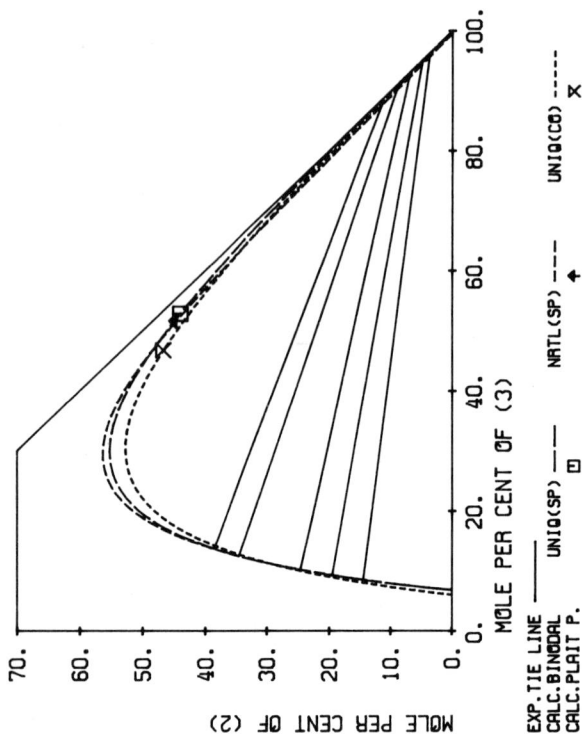

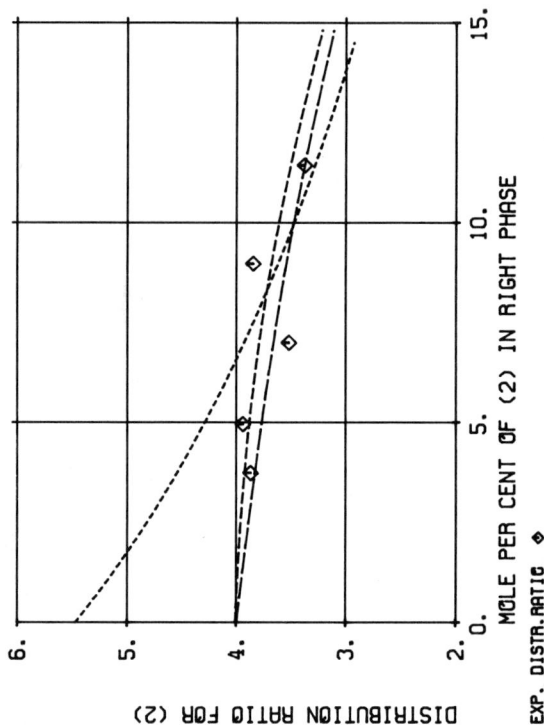

EXP.TIE LINE   UNIQ(SP) □   NRTL(SP) ---- ♦   UNIQ(CO) ----- ✕
CALC.BINODAL
CALC.PLAIT P.

MOLE PER CENT OF (3)
MOLE PER CENT OF (2)

MOLE PER CENT OF (2) IN RIGHT PHASE
DISTRIBUTION RATIO FOR (2)
EXP. DISTR.RATIO ♦

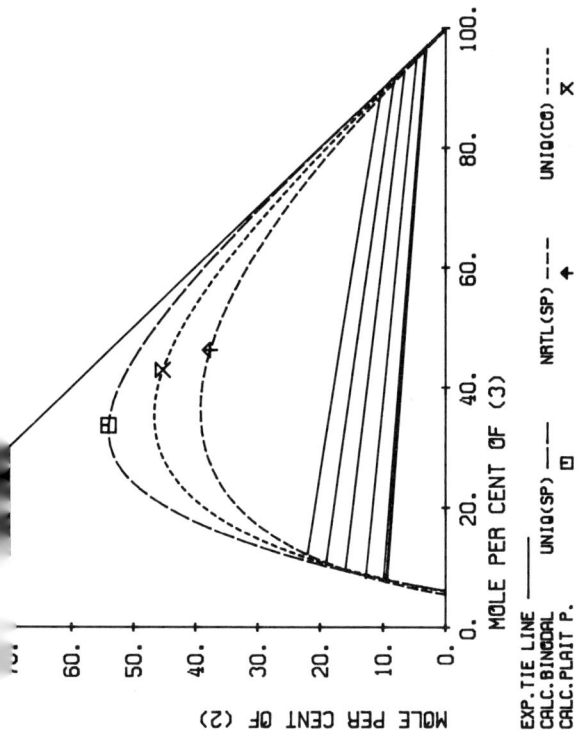

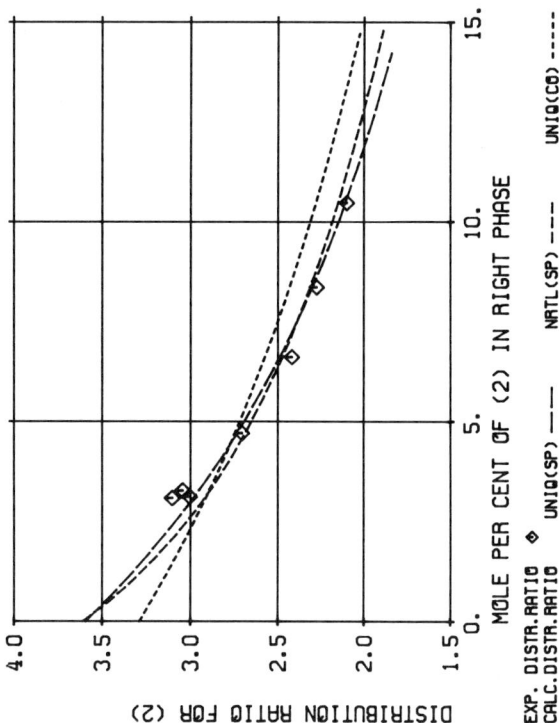

$C_2H_3N-C_{10}H_8$

(2) C10H8   NAPHTHALENE

(3) C2H3N   ACETIC ACID,NITRILE

BONDARENKO M.F., ET AL.
KHIM.TEKHNOL.TOPL.MASEL (1972)4,8

TEMPERATURE = 30.0 DEG C   TYPE OF SYSTEM = 1

EXPERIMENTAL TIE LINES IN MOLE PCT

	LEFT PHASE			RIGHT PHASE	
(1)	(2)	(3)	(1)	(2)	(3)
82.969	9.564	7.467	0.397	3.084	96.519
82.707	9.385	7.908	0.419	3.124	96.458
82.153	9.957	7.890	0.420	3.272	96.308
79.042	12.732	8.226	0.520	4.705	94.775
74.660	15.982	9.358	0.608	6.607	92.785
70.162	19.007	10.831	0.723	8.352	90.924
65.789	21.992	12.219	0.851	10.473	88.677

SPECIFIC MODEL PARAMETERS IN KELVIN

		UNIQUAC		NRTL(ALPHA=.2)	
I J	AIJ	AJI		AIJ	AJI
1 2	24.429	6.5148		-214.53	-164.37
1 3	516.21	215.43		401.55	1452.8
2 3	1.7316	123.80		15.992	-55.865

R1 =10.5694   R2 = 4.9808   R3 = 1.8701
Q1 = 8.716   Q2 = 3.440   Q3 = 1.724

MEAN DEV. BETWEEN CALC. AND EXP. CONC. IN MOLE PCT

UNIQUAC (SPECIFIC PARAMETERS)   0.41
NRTL   (SPECIFIC PARAMETERS)   0.16
UNIQUAC (COMMON PARAMETERS)   0.38

C₂H₃N-C₁₀H₁₄

(1) C15H32   PENTADECANE
----
(2) C10H14   BENZENE,BUTYL
----
(3) C2H3N   ACETIC ACID,NITRILE

BONDARENKO M.F., ET AL.
KHIM.TEKHNOL.TOPL.MASEL (1972)4,8

TEMPERATURE = 30.0 DEG C   TYPE OF SYSTEM = 1

EXPERIMENTAL TIE LINES IN MOLE PCT

         LEFT PHASE   RIGHT PHASE
     (1)   (2)   (3)   (1)   (2)   (3)

  65.924   22.562   11.513   0.249   2.826   96.925
  62.048   25.059   12.893   0.295   3.435   96.271
  51.318   33.106   15.575   0.326   4.845   94.829
  39.362   40.984   19.654   0.407   6.684   92.910

MEAN DEV. BETWEEN CALC. AND EXP. CONC. IN MOLE PCT

UNIQUAC (COMMON PARAMETERS)   0.24

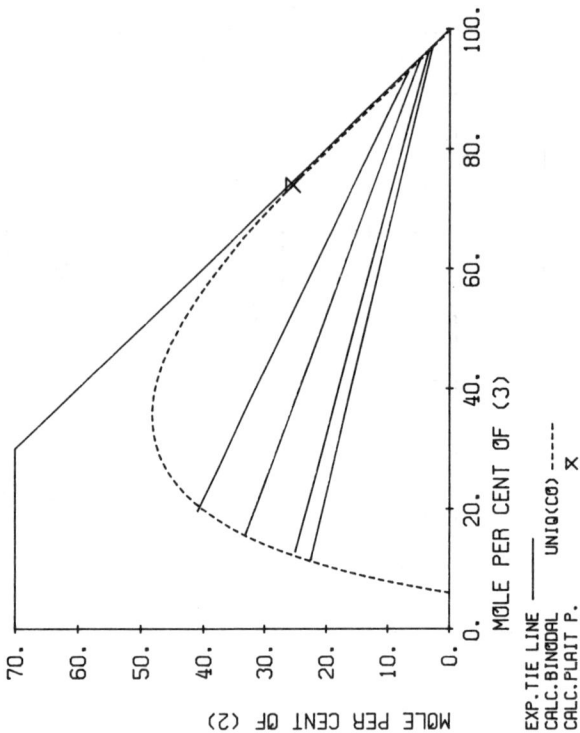

MOLE PER CENT OF (2)

MOLE PER CENT OF (3)

EXP.TIE LINE ──────   UNIQ(C0) -----
CALC.BINODAL
CALC.PLAIT P.   X

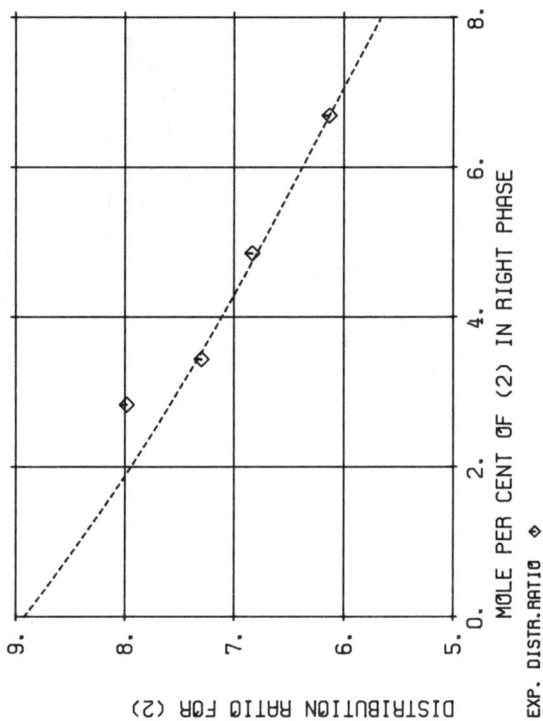

DISTRIBUTION RATIO FOR (2)

MOLE PER CENT OF (2) IN RIGHT PHASE

EXP. DISTR.RATIO ◇

$C_2H_3N-C_{14}H_{16}$

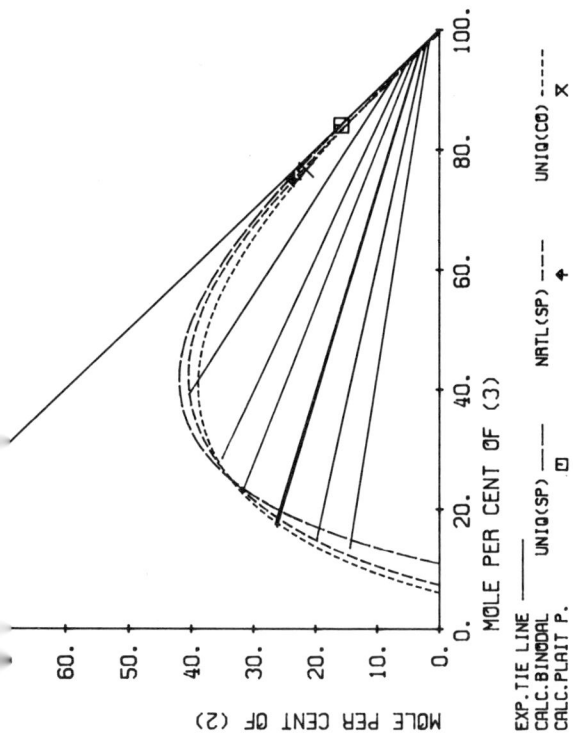

MOLE PER CENT OF (2)

MOLE PER CENT OF (3)

EXP.TIE LINE ——— UNIQ(SP) ——— 🔷 NRTL(SP) ——— ✦ UNIQ(CO) -----
CALC.BINODAL
CALC.PLAIT P.

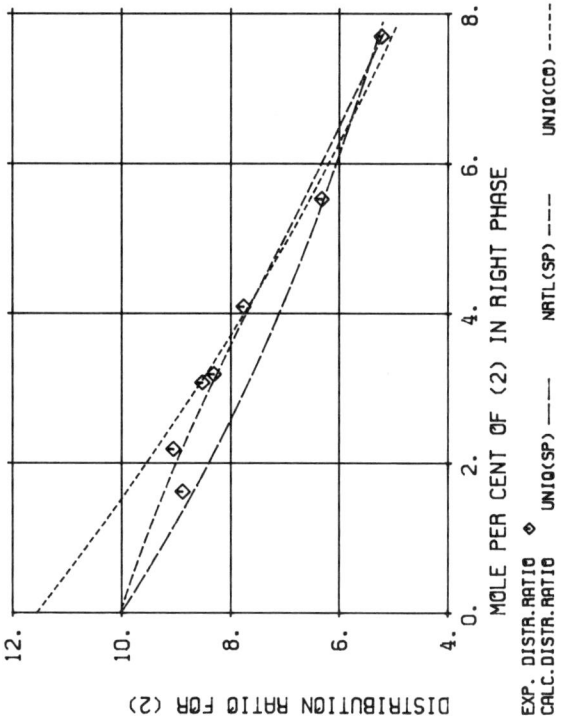

DISTRIBUTION RATIO FOR (2)

MOLE PER CENT OF (2) IN RIGHT PHASE

EXP. DISTR.RATIO ◆ UNIQ(SP) ——— NRTL(SP) ----- UNIQ(CO) -----
CALC.DISTR.RATIO

```
------------------------------
(2) C14H16     NAPHTHALENE,1-BUTYL
------------------------------
(3) C2H3N      ACETIC ACID,NITRILE
------------------------------

BONDARENKO M.F., ET AL.
KHIM.TEKHNOL.TOPL.MASEL (1972)4,8

TEMPERATURE = 30.0 DEG C    TYPE OF SYSTEM = 1

EXPERIMENTAL TIE LINES IN MOLE PCT
------------------------------
       LEFT PHASE              RIGHT PHASE
   (1)     (2)     (3)      (1)     (2)     (3)

 72.096  14.370  13.534    0.289   1.619  98.092
 65.238  19.758  15.004    0.316   2.184  97.500
 56.659  26.220  17.121    0.369   3.076  96.555
 56.024  26.496  17.480    0.370   3.187  96.442
 45.422  31.754  22.824    0.449   4.087  95.464
 36.561  34.967  28.472    0.564   5.528  93.908
 20.496  40.345  39.160    0.728   7.697  91.575

SPECIFIC MODEL PARAMETERS IN KELVIN
------------------------------
            UNIQUAC                 NRTL(ALPHA=.2)
  I J     AIJ       AJI           AIJ        AJI

  1 2   23.043     4.5991       75.687    -24.109
  1 3  410.25    230.83        326.67    1361.0
  2 3  119.01     98.880      -262.00    1025.4

R1 =10.5694   R2 = 7.7390   R3 = 1.8701
Q1 = 8.716    Q2 = 5.628    Q3 = 1.724

MEAN DEV. BETWEEN CALC. AND EXP. CONC. IN MOLE PCT
------------------------------
UNIQUAC (SPECIFIC PARAMETERS)    0.80
NRTL   (SPECIFIC PARAMETERS)    0.39
UNIQUAC (COMMON PARAMETERS)     0.54
```

$C_2H_3N-C_{14}H_{22}$

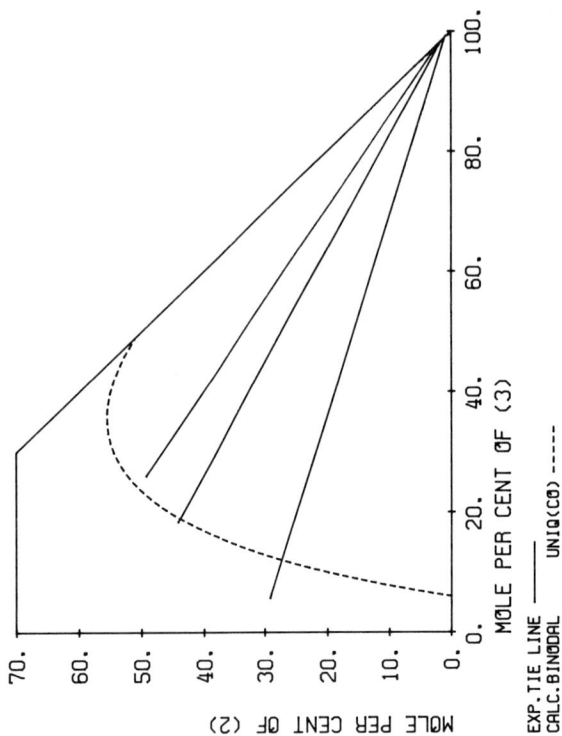

MOLE PER CENT OF (2)

MOLE PER CENT OF (3)

EXP.TIE LINE ——— UNIQ(CB) -----
CALC.BINODAL

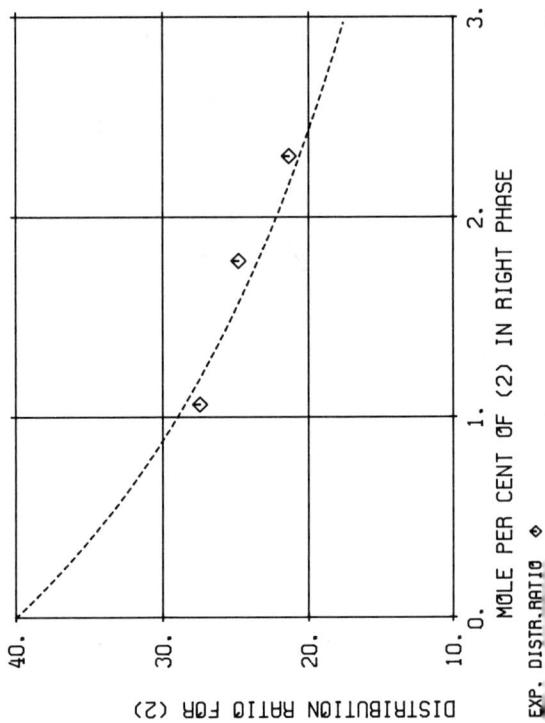

DISTRIBUTION RATIO FOR (2)

MOLE PER CENT OF (2) IN RIGHT PHASE

EXP. DISTR.RATIO ◇

(1) C 15H32   PENTADECANE
(2) C14H22   BENZENE,OCTYL
(3) C2H3N   ACETIC ACID,NITRILE

BONDARENKO M.F., ET AL.
KHIM.TEKHNOL.TOPL.MASEL (1972)4,8

TEMPERATURE = 30.0 DEG C   TYPE OF SYSTEM = 1

EXPERIMENTAL TIE LINES IN MOLE PCT

	LEFT PHASE			RIGHT PHASE	
(1)	(2)	(3)	(1)	(2)	(3)
65.100	29.165	5.734	0.223	1.062	98.715
37.480	44.169	18.352	0.166	1.780	98.055
24.757	49.239	26.004	0.147	2.304	97.549

MEAN DEV. BETWEEN CALC. AND EXP. CONC. IN MOLE PCT
UNIQUAC (COMMON PARAMETERS)   1.78

$C_2H_3N-C_{15}H_{32}$

MOLE PER CENT OF (3)

MOLE PER CENT OF (2)

EXP.TIE LINE ———   UNIQ(SP) ———   NRTL(SP) ———   UNIQ(CO) -----
CALC.BINODAL -----

DISTRIBUTION RATIO FOR (2)

MOLE PER CENT OF (2) IN RIGHT PHASE

EXP. DISTR.RATIO ◇   UNIQ(SP) ◇   NRTL(SP) ———   UNIQ(CO) -----
CALC. DISTR.RATIO

(2) C18H24   NAPHTHALENE,1-OCTYL

(3) C2H3N   ACETIC ACID,NITRILE

BONDARENKO M.F., ET AL.
KHIM.TEKHNOL.TOPL.MASEL (1972)4,8

TEMPERATURE = 30.0 DEG C   TYPE OF SYSTEM = 1

EXPERIMENTAL TIE LINES IN MOLE PCT

	LEFT PHASE			RIGHT PHASE	
(1)	(2)	(3)	(1)	(2)	(3)
74.222	13.850	11.928	0.219	0.386	99.395
63.307	23.322	13.371	0.201	0.639	99.160
52.342	30.410	17.249	0.183	0.879	98.938
43.215	36.622	20.163	0.164	1.086	98.751
30.654	43.791	25.556	0.145	1.393	98.461

SPECIFIC MODEL PARAMETERS IN KELVIN

		UNIQUAC		NRTL(ALPHA=.2)	
I	J	AIJ	AJI	AIJ	AJI
1	2	-32.568	32.073	-104.03	-4.2424
1	3	417.22	301.53	316.76	1287.2
2	3	242.06	49.645	-172.54	1183.0

R1 =10.5694   R2 =10.4366   R3 = 1.8701
Q1 = 8.716   Q2 = 7.788   Q3 = 1.724

MEAN DEV. BETWEEN CALC. AND EXP. CONC. IN MOLE PCT

UNIQUAC (SPECIFIC PARAMETERS)   0.21
NRTL (SPECIFIC PARAMETERS)   0.40
UNIQUAC (COMMON PARAMETERS)   0.43

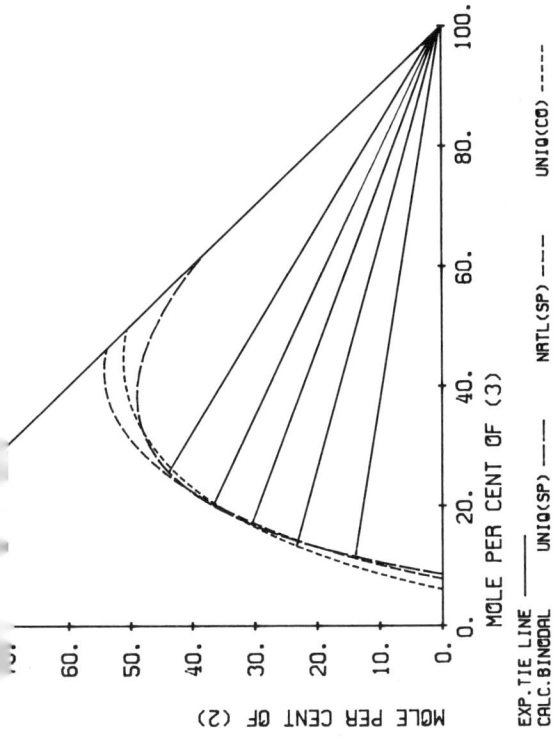

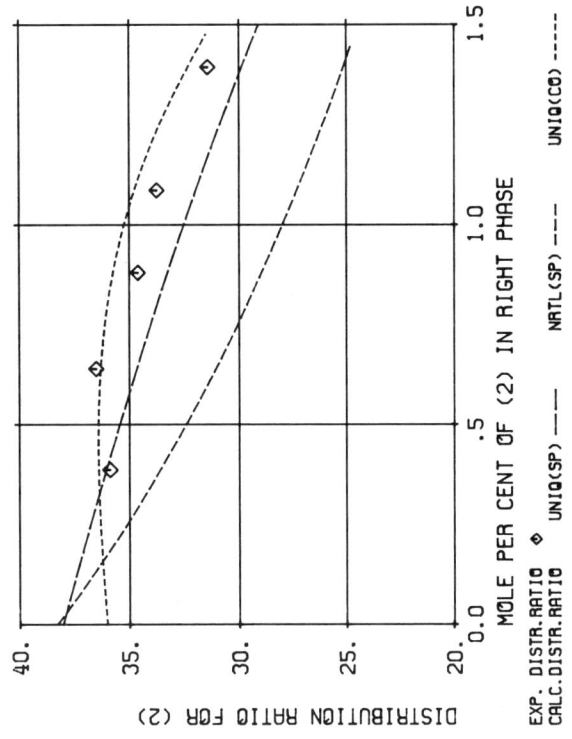

$C_2H_3N\text{-}C_{17}H_{34}O_2$

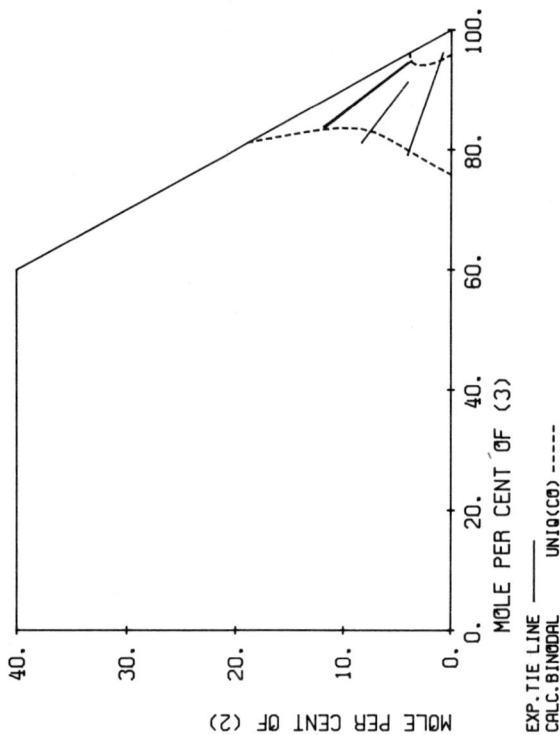

MOLE PER CENT OF (2)

MOLE PER CENT OF (3)

EXP.TIE LINE ——— UNIQ(CO) ------
CALC.BINODAL

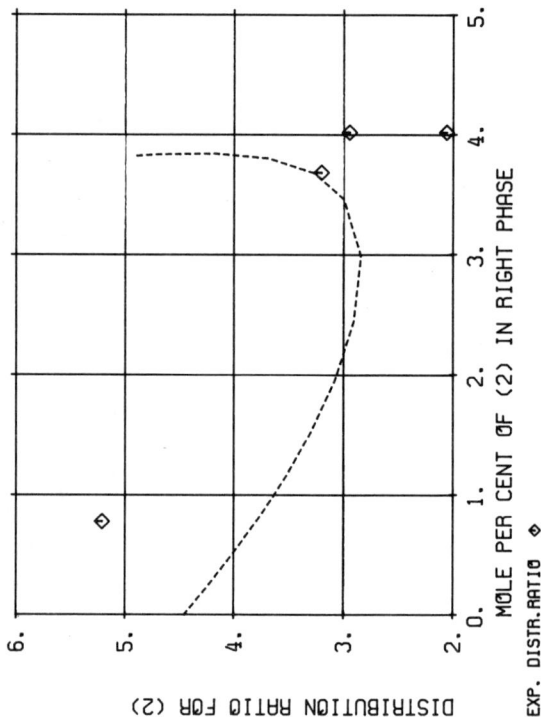

DISTRIBUTION RATIO FOR (2)

MOLE PER CENT OF (2) IN RIGHT PHASE

EXP. DISTR.RATIO ◇

(1) C17H34O2   HEXADECANOIC ACID,METHYL ESTER
(2) C19H36O2   9-OCTADECENOIC ACID(CIS),METHYL ESTER
(3) C2H3N   ACETIC ACID,NITRILE

RUSLING J.F., BERTSCH R.J., BARFORD R.A., ROTHBART H.L.
J.CHEM.ENG.DATA 14(1969)169

TEMPERATURE = 25.0 DEG C   TYPE OF SYSTEM = 2

EXPERIMENTAL TIE LINES IN MOLE PCT

	LEFT PHASE			RIGHT PHASE	
(1)	(2)	(3)	(1)	(2)	(3)
16.834	4.036	79.130	3.009	0.775	96.216
4.593	11.796	83.611	1.452	3.683	94.866
10.536	8.271	81.194	4.680	4.018	91.301
4.360	11.848	83.793	1.455	4.018	94.527

MEAN DEV. BETWEEN CALC. AND EXP. CONC. IN MOLE PCT
UNIQUAC (COMMON PARAMETERS)   0.92

C₂H₄Cl₂-C₂H₆O

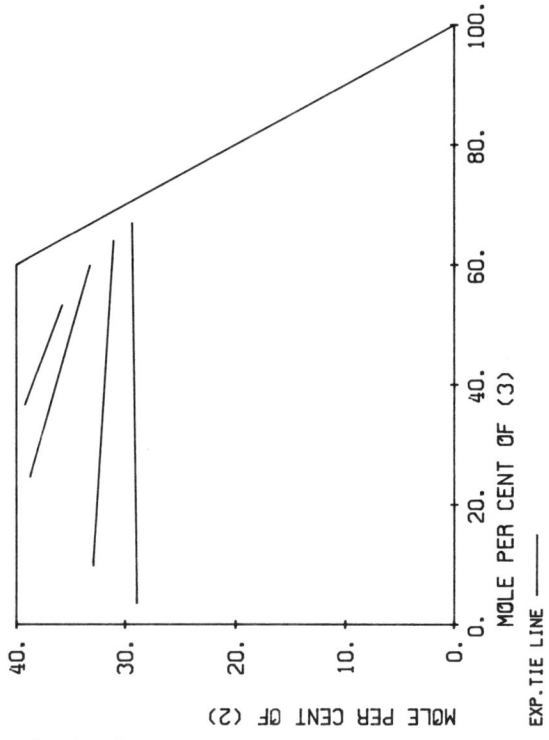

MOLE PER CENT OF (3) ──────

MOLE PER CENT OF (2)

EXP.TIE LINE ──────

(1) C2H4CL2   ETHANE,1,1-DICHLORO
----
(2) C2H6O   ETHANOL
----
(3) H2O   WATER

BONNER W.D.
J.PHYS.CHEM. 14(1910)738

TEMPERATURE =   0.0 DEG C   TYPE OF SYSTEM = 1

EXPERIMENTAL TIE LINES IN MOLE PCT (GRAPH.INTERPOL.)

LEFT PHASE			RIGHT PHASE		
(1)	(2)	(3)	(1)	(2)	(3)
67.490	28.924	3.586	3.661	29.361	66.978
57.278	32.916	9.807	4.971	31.031	63.998
36.592	38.729	24.679	6.962	33.201	59.837
24.140	39.206	36.655	10.969	35.744	53.287

C₂H₄Cl₂-C₂H₄O₂

DUBOVSKAYA A.S., KARAPETYANTS M.KH.
ZH.FIZ.KHIM. 44(1970)2299

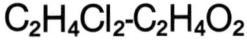

(1) H2O WATER
(2) C2H4O2 ACETIC ACID
(3) C2H4CL2 ETHANE, 1,2-DICHLORO

TEMPERATURE = 30.0 DEG C   TYPE OF SYSTEM = 1

EXPERIMENTAL TIE LINES IN MOLE PCT

	LEFT PHASE			RIGHT PHASE	
(1)	(2)	(3)	(1)	(2)	(3)
93.330	6.160	0.510	2.126	3.827	94.047
86.372	12.553	1.074	2.100	6.929	90.971
79.541	18.580	1.879	4.002	14.406	81.592
70.514	26.020	3.466	7.510	21.402	71.088
65.287	29.610	5.104	11.459	26.443	62.098
58.191	33.492	8.318	17.048	30.931	52.021

SPECIFIC MODEL PARAMETERS IN KELVIN

		UNIQUAC		NRTL(ALPHA=.2)	
I J	AIJ	AJI		AIJ	AJI
1 2	-16.853	-58.866		503.62	-531.48
1 3	642.80	702.87		1878.8	1023.2
2 3	-166.46	399.33		-62.672	80.132

R1 = 0.9200   R2 = 2.2024   R3 = 2.9308
Q1 = 1.400   Q2 = 2.072   Q3 = 2.528

MEAN DEV. BETWEEN CALC. AND EXP. CONC. IN MOLE PCT

UNIQUAC (SPECIFIC PARAMETERS)   0.94
NRTL (SPECIFIC PARAMETERS)   0.98
UNIQUAC (COMMON PARAMETERS)   0.78

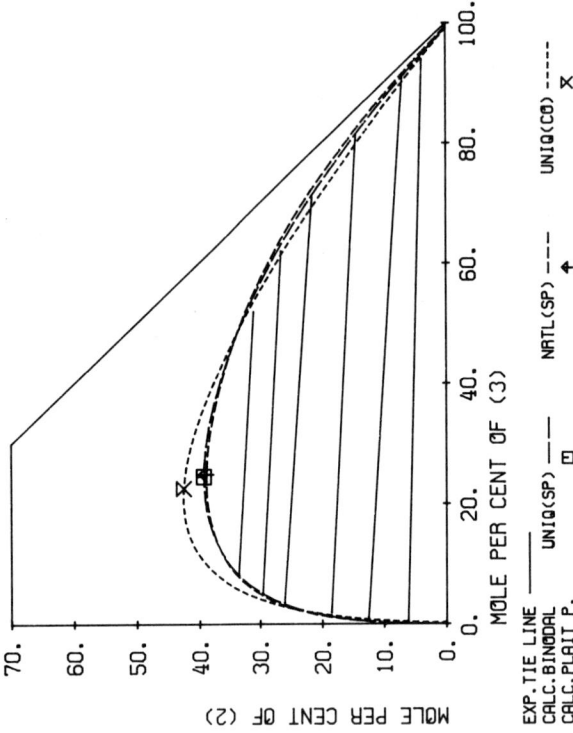

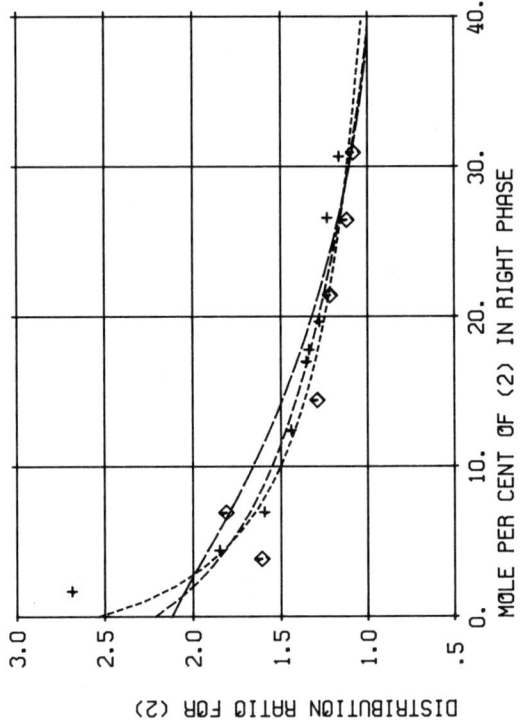

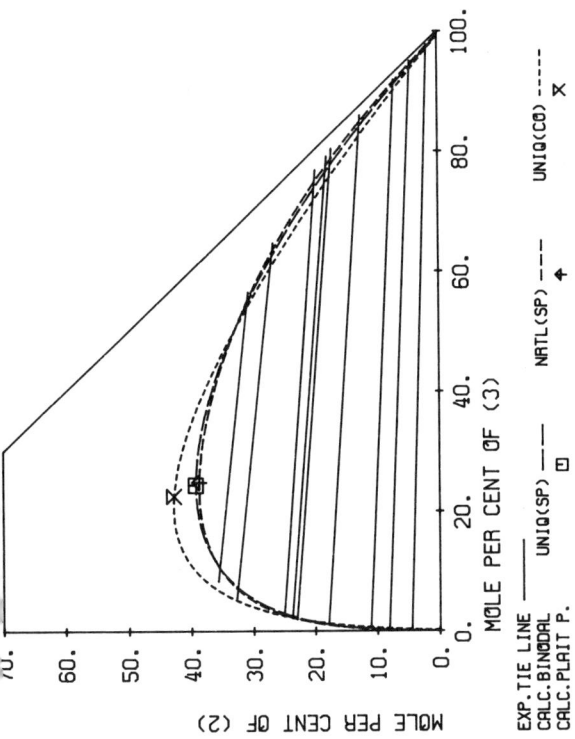

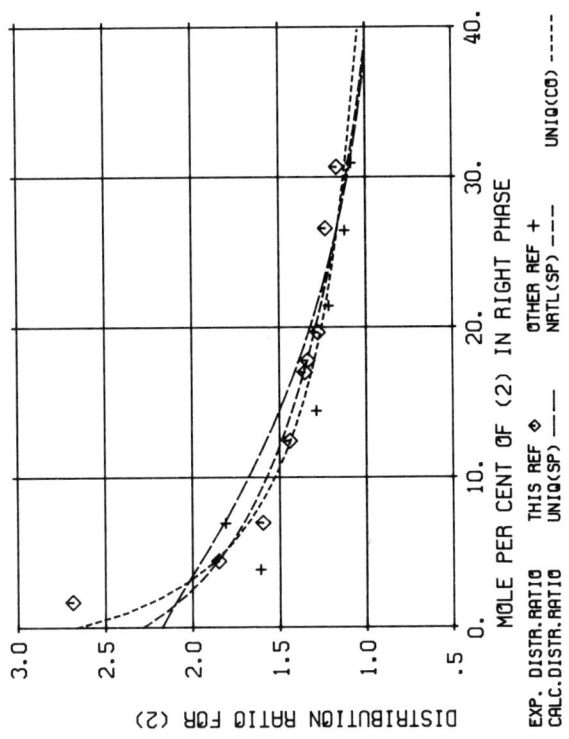

(2) C2H4O2   ACETIC ACID

(3) C2H4CL2   ETHANE,1,2-DICHLORO

FUSE K., IGUCHI A.
KAGAKU KOGAKU 34(1970)328

TEMPERATURE = 25.0 DEG C   TYPE OF SYSTEM = 1

EXPERIMENTAL TIE LINES IN MOLE PCT (GRAPH.INTERPOL.)

| | LEFT PHASE | | | RIGHT PHASE | |
(1)	(2)	(3)	(1)	(2)	(3)
95.207	4.508	0.285	0.543	1.679	97.778
91.511	8.116	0.373	0.537	4.385	95.077
88.386	11.100	0.513	1.059	6.958	91.983
80.998	17.880	1.122	1.547	12.405	86.048
75.039	23.009	1.952	2.507	16.996	80.498
74.119	23.779	2.102	2.985	17.776	79.239
72.458	25.104	2.438	3.440	19.638	76.922
61.745	32.763	5.492	8.612	26.569	64.819
56.009	35.722	8.269	12.772	30.652	56.576

SPECIFIC MODEL PARAMETERS IN KELVIN

| | | UNIQUAC | | NRTL(ALPHA=.2) | |
I J	AIJ	AJI		AIJ	AJI
1 2	-16.853	-58.866		503.62	-531.48
1 3	642.80	702.87		1878.8	1023.2
2 3	-166.46	399.33		-62.672	80.132

| R1 = 0.9200 | R2 = 2.2024 | R3 = 2.9308 |
| Q1 = 1.400 | Q2 = 2.072 | Q3 = 2.528 |

MEAN DEV. BETWEEN CALC. AND EXP. CONC. IN MOLE PCT

UNIQUAC (SPECIFIC PARAMETERS)	0.99
NRTL (SPECIFIC PARAMETERS)	0.77
UNIQUAC (COMMON PARAMETERS)	1.82

EXP.TIE LINE —— UNIQ(SP) —— ⊡ NRTL(SP) —— ✦ UNIQ(CO) —— ✕
CALC.BINODAL ——
CALC.PLAIT P.

EXP. DISTR.RATIO ◇ THIS REF ◇ OTHER REF + UNIQ(CO) ——
CALC.DISTR.RATIO   UNIQ(SP) —— NRTL(SP) ——

C₂H₄Cl₂-C₂H₅ClO

(1) C2H4CL2   ETHANE,1,2-DICHLORO
(2) C2H5CLO   ETHANOL,2-CHLORO
(3) H2O   WATER

ABABI V., POPA A., MIHAILA GH.
AN.STIINT.UNIV.AL.I.CUZA IASI. 9(1963)233

TEMPERATURE = 20.5 DEG C   TYPE OF SYSTEM = 1

EXPERIMENTAL TIE LINES IN MOLE PCT

	LEFT PHASE			RIGHT PHASE	
(1)	(2)	(3)	(1)	(2)	(3)
91.400	5.943	2.657	0.081	3.108	96.811
82.239	12.133	5.628	0.108	5.135	94.758
67.918	21.972	10.110	0.133	5.971	93.896
60.695	27.164	12.141	0.229	7.151	92.620
40.393	35.357	24.250	0.436	8.921	90.644
30.936	37.361	31.702	0.596	9.740	89.663
27.633	37.611	34.756	0.713	10.581	88.706
23.098	37.502	39.400	0.916	11.415	87.669
19.737	36.455	43.808	1.036	11.880	87.085
18.114	35.962	45.924	1.135	12.486	86.380

SPECIFIC MODEL PARAMETERS IN KELVIN

		UNIQUAC		NRTL(ALPHA=.2)	
I J		AIJ	AJI	AIJ	AJI
1 2		377.11	-130.18	933.32	-410.90
1 3		507.98	407.42	721.54	1872.5
2 3		-36.510	151.74	-308.15	892.70

R1 = 2.9308   R2 = 2.6698   R3 = 0.9200
Q1 = 2.528   Q2 = 2.392   Q3 = 1.400

MEAN DEV. BETWEEN CALC. AND EXP. CONC. IN MOLE PCT

UNIQUAC (SPECIFIC PARAMETERS)   0.63
NRTL   (SPECIFIC PARAMETERS)   0.47
UNIQUAC (COMMON PARAMETERS)   1.51

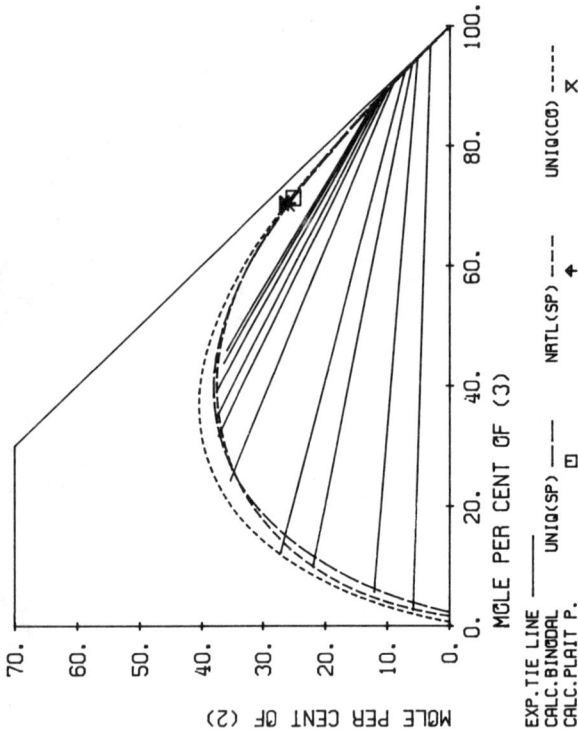

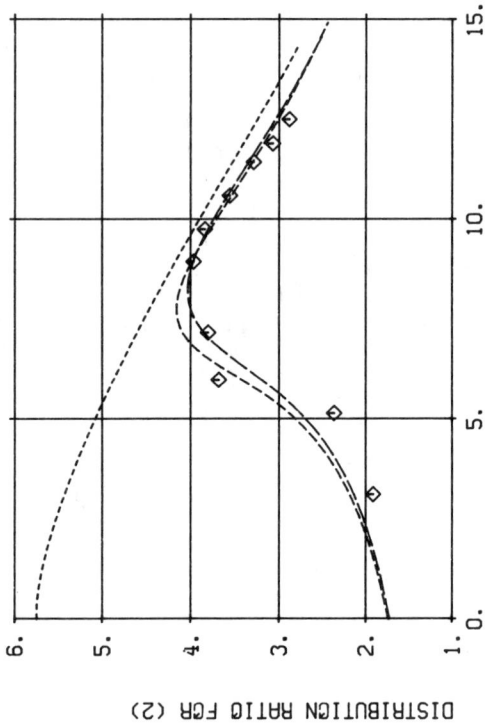

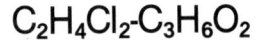

(2) C3H6O2   PROPANOIC ACID
(3) H2O   WATER

DUBOVSKAYA A.S., KARAPETYANTS M.KH.
ZH.FIZ.KHIM. 44(1970)2299

TEMPERATURE = 30.0 DEG C   TYPE OF SYSTEM = 1

EXPERIMENTAL TIE LINES IN MOLE PCT

	LEFT PHASE			RIGHT PHASE	
(1)	(2)	(3)	(1)	(2)	(3)
85.267	11.618	3.116	0.364	3.025	96.612
67.966	24.336	7.698	0.434	5.502	94.065
48.821	31.191	19.989	0.857	8.650	90.493
26.084	32.522	41.394	3.003	14.505	82.491
29.116	32.898	37.986	3.225	14.961	81.814
20.029	30.065	49.906	4.636	17.221	78.143

SPECIFIC MODEL PARAMETERS IN KELVIN

		UNIQUAC		NRTL(ALPHA=.2)	
I J	AIJ	AJI		AIJ	AJI
1 2	518.95	-256.17		1484.6	-833.88
1 3	920.01	237.44		1250.3	1719.6
2 3	374.14	-145.39		402.49	-79.164

R1 = 2.9308   R2 = 2.8768   R3 = 0.9200
Q1 = 2.528   Q2 = 2.612   Q3 = 1.400

MEAN DEV. BETWEEN CALC. AND EXP. CONC. IN MOLE PCT

UNIQUAC (SPECIFIC PARAMETERS)   0.86
NRTL   (SPECIFIC PARAMETERS)   0.73
UNIQUAC (COMMON PARAMETERS)   1.10

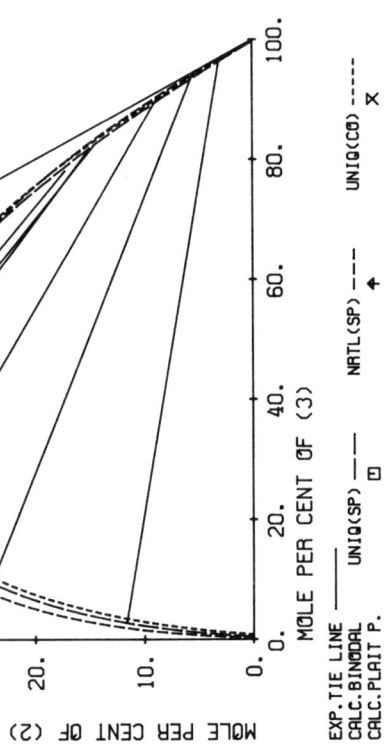

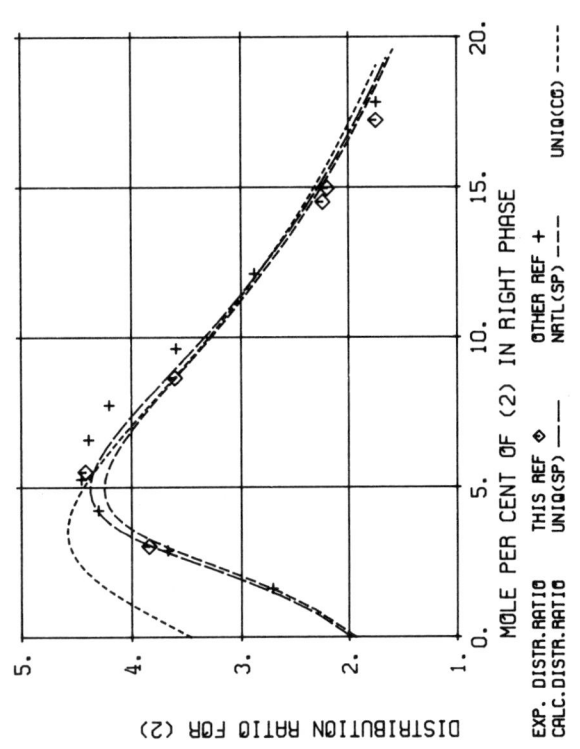

C$_2$H$_4$Cl$_2$-C$_3$H$_6$O$_2$

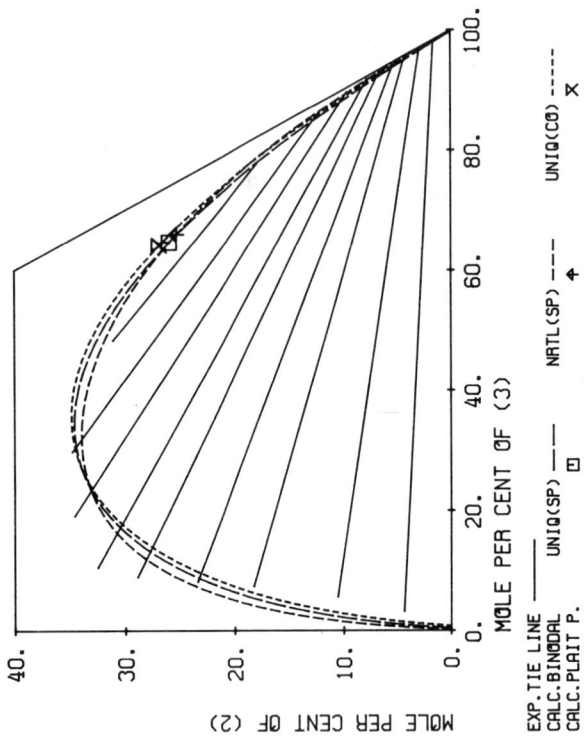

MOLE PER CENT OF (3)

MOLE PER CENT OF (2)

EXP.TIE LINE	UNIQ(SP)	NRTL(SP)	UNIQ(CB)
CALC.BINODAL			
CALC.PLAIT P.			

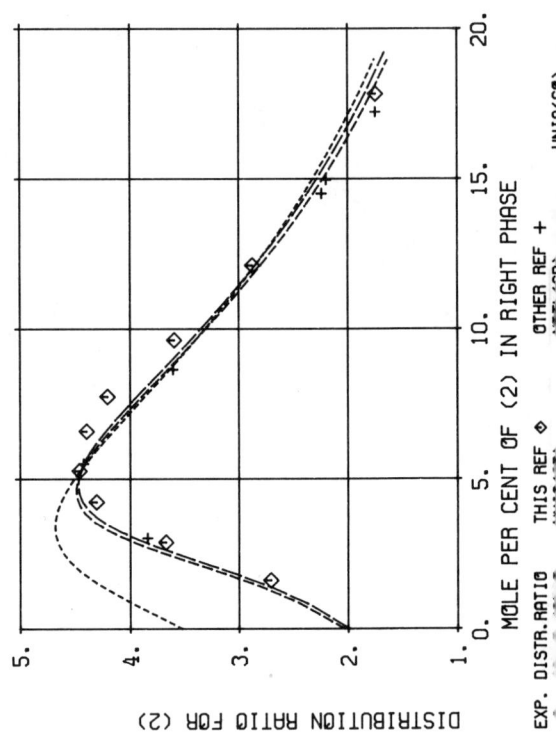

MOLE PER CENT OF (2) IN RIGHT PHASE

DISTRIBUTION RATIO FOR (2)

EXP. DISTR.RATIO ◇   THIS REF ◇   OTHER REF +

---

(1) C2H4CL2   ETHANE,1,2-DICHLORO

(2) C3H6O2   PROPANOIC ACID

(3) H2O   WATER

IGUCHI A., FUSE K.
KAGAKU KOGAKU 36(1972)673

TEMPERATURE = 25.0 DEG C   TYPE OF SYSTEM = 1

EXPERIMENTAL TIE LINES IN MOLE PCT (GRAPH.INTERPOL.)

	LEFT PHASE			RIGHT PHASE	
(1)	(2)	(3)	(1)	(2)	(3)
92.452	4.374	3.174	0.173	1.616	98.211
83.826	10.570	5.605	0.180	2.877	96.943
74.411	18.223	7.366	0.208	4.232	95.561
68.352	23.484	8.164	0.257	5.265	94.478
62.183	28.899	8.917	0.333	6.575	93.092
56.977	32.508	10.515	0.413	7.732	91.856
46.283	34.598	19.119	0.703	9.623	89.674
35.446	34.785	29.769	1.416	12.105	86.479
20.614	31.070	43.316	4.296	17.832	77.872

SPECIFIC MODEL PARAMETERS IN KELVIN

		UNIQUAC		NRTL(ALPHA=.2)	
I J	AIJ	AJI		AIJ	AJI
1 2	518.95	-256.17		1484.6	-833.88
1 3	920.01	-237.44		1250.3	1719.6
2 3	374.14	-145.39		402.49	-79.164

R1 = 2.9308	R2 = 2.8768	R3 = 0.9200
Q1 = 2.528	Q2 = 2.612	Q3 = 1.400

MEAN DEV. BETWEEN CALC. AND EXP. CONC. IN MOLE PCT

UNIQUAC (SPECIFIC PARAMETERS)	1.25
NRTL (SPECIFIC PARAMETERS)	1.17
UNIQUAC (COMMON PARAMETERS)	1.30

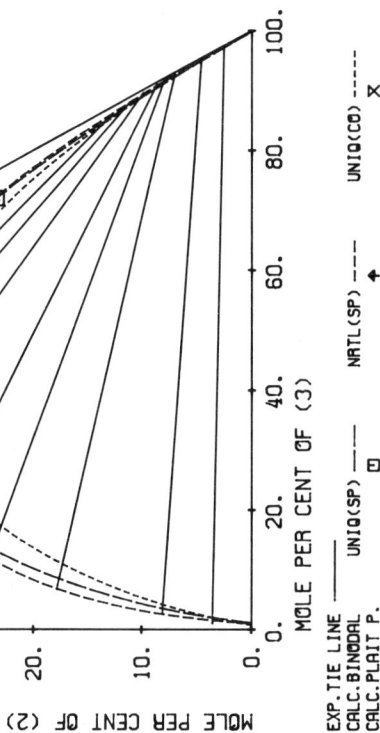

MOLE PER CENT OF (2)

MOLE PER CENT OF (3)

EXP.TIE LINE ———   UNIQ(SP)   ⊡   NRTL(SP) ———   UNIQ(CO) -----
CALC.BINODAL -----   ✕
CALC.PLAIT P. ⊡

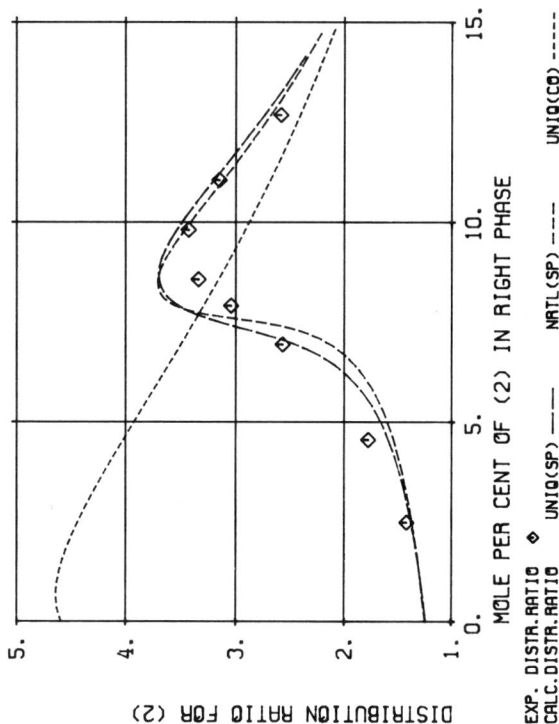

DISTRIBUTION RATIO FOR (2)

MOLE PER CENT OF (2) IN RIGHT PHASE

EXP. DISTR.RATIO ◇   UNIQ(SP) ◇   NRTL(SP) ———   UNIQ(CO) -----
CALC.DISTR.RATIO

(2) C3H8O   2-PROPANOL
(3) H2O   WATER

IZMAILOV N.A., FRANKE A.K.
ZH.FIZ.KHIM. 29(1955)120

TEMPERATURE = 25.0 DEG C   TYPE OF SYSTEM = 1

EXPERIMENTAL TIE LINES IN MOLE PCT

	LEFT PHASE			RIGHT PHASE	
(1)	(2)	(3)	(1)	(2)	(3)
95.387	3.540	1.074	0.175	2.492	97.333
89.293	8.106	2.601	0.244	4.554	95.202
75.412	17.860	6.728	0.300	6.941	92.759
63.959	24.064	11.977	0.328	7.909	91.763
53.173	28.572	18.256	0.354	8.565	91.081
36.816	33.604	29.579	0.548	9.807	89.645
26.636	34.828	38.536	0.801	11.055	88.143
18.141	32.806	49.052	1.230	12.678	86.092

SPECIFIC MODEL PARAMETERS IN KELVIN

		UNIQUAC		NRTL(ALPHA=.2)	
I J		AIJ	AJI	AIJ	AJI
1 2	408.35	-135.17	1108.0	-482.67
1 3	701.76	290.66	854.94	1640.1
2 3	-43.345	158.32	-452.09	1090.3

R1 = 2.9308   R2 = 2.7791   R3 = 0.9200
Q1 = 2.528   Q2 = 2.508   Q3 = 1.400

MEAN DEV. BETWEEN CALC. AND EXP. CONC. IN MOLE PCT

UNIQUAC (SPECIFIC PARAMETERS)   0.55
NRTL   (SPECIFIC PARAMETERS)   0.32
UNIQUAC (COMMON PARAMETERS)   2.40

C$_2$H$_4$Cl$_2$-C$_4$H$_8$O$_2$

(1) C2H4CL2   ETHANE,1,2-DICHLORO

(2) C4H8O2   BUTANOIC ACID

(3) H2O   WATER

DUBOVSKAYA A.S., KARAPETYANTS M.KH.
ZH.FIZ.KHIM. 44(1970)2299

TEMPERATURE = 30.0 DEG C   TYPE OF SYSTEM = 1

EXPERIMENTAL TIE LINES IN MOLE PCT

	LEFT PHASE			RIGHT PHASE	
(1)	(2)	(3)	(1)	(2)	(3)
16.638	38.970	44.392	0.179	2.239	97.582
48.676	35.756	15.567	0.179	2.239	97.582
76.716	17.147	6.137	0.179	2.239	97.582
85.018	9.782	5.201	0.179	2.239	97.582
5.446	29.026	65.528	0.160	2.286	97.555
42.902	37.752	19.346	0.160	2.286	97.555
45.127	37.390	17.482	0.160	2.286	97.555
58.375	31.001	10.624	0.160	2.286	97.555
60.540	29.707	9.753	0.160	2.286	97.555
30.453	41.006	28.541	0.181	2.485	97.334
48.323	36.116	15.561	0.181	2.485	97.334

SPECIFIC MODEL PARAMETERS IN KELVIN

		UNIQUAC		NRTL(ALPHA=.2)	
I	J	AIJ	AJI	AIJ	AJI
1	2	-6.2653	119.14	171.80	0.21285
1	3	793.19	601.44	1251.1	1511.5
2	3	-62.882	252.30	-445.69	1478.2

R1 = 2.9308   R2 = 3.5512   R3 = 0.9200
Q1 = 2.528   Q2 = 3.152   Q3 = 1.400

MEAN DEV. BETWEEN CALC. AND EXP. CONC. IN MOLE PCT

UNIQUAC (SPECIFIC PARAMETERS)   1.21
NRTL   (SPECIFIC PARAMETERS)   1.23
UNIQUAC (COMMON PARAMETERS)   1.91

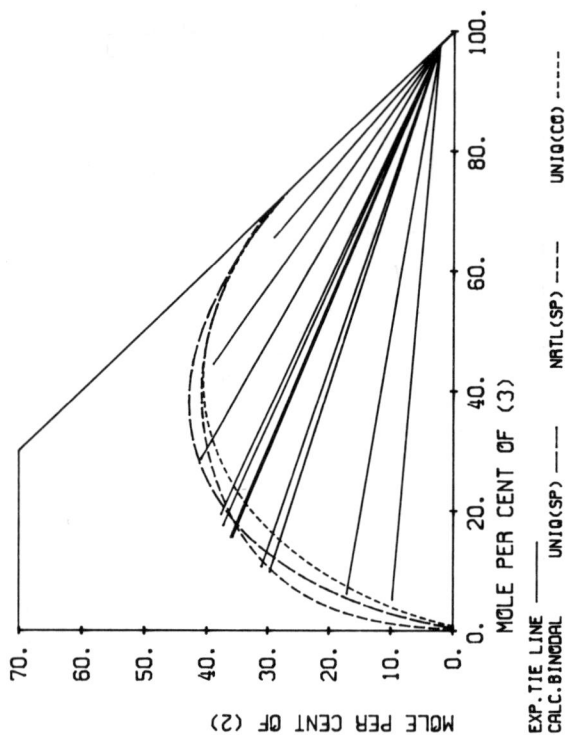

MOLE PER CENT OF (3)

MOLE PER CENT OF (2)

EXP.TIE LINE ——— UNIQ(SP) ———   NRTL(SP) -----
CALC.BINODAL   UNIQ(CO) -----

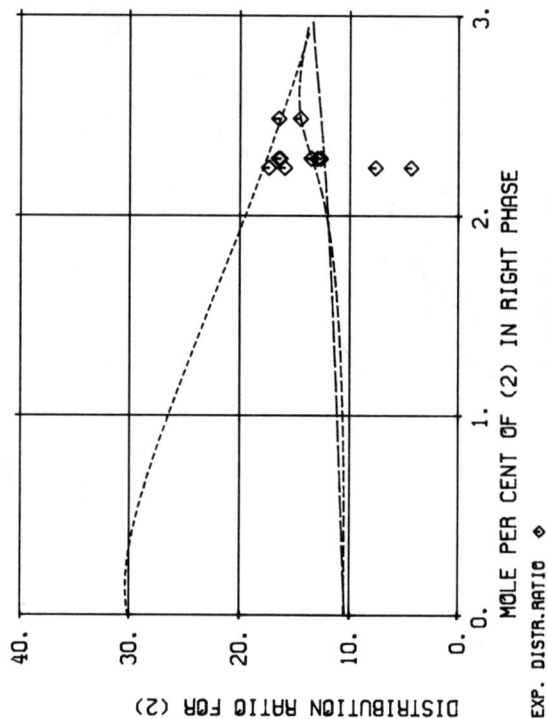

MOLE PER CENT OF (2) IN RIGHT PHASE

DISTRIBUTION RATIO FOR (2)

EXP. DISTR.RATIO ◇

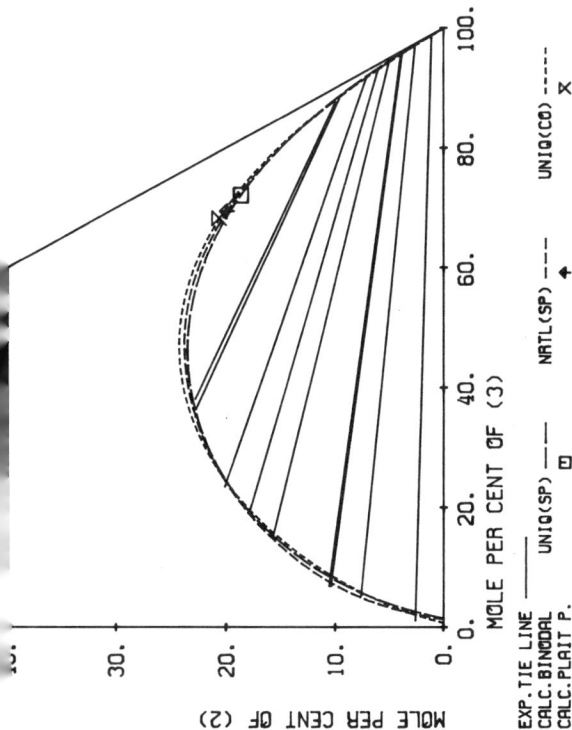

MOLE PER CENT OF (3)

MOLE PER CENT OF (2)

EXP.TIE LINE ——— UNIQ(SP) ▢ NRTL(SP) ——— + UNIQ(CO) -----
CALC.BINODAL
CALC.PLAIT P. ✕

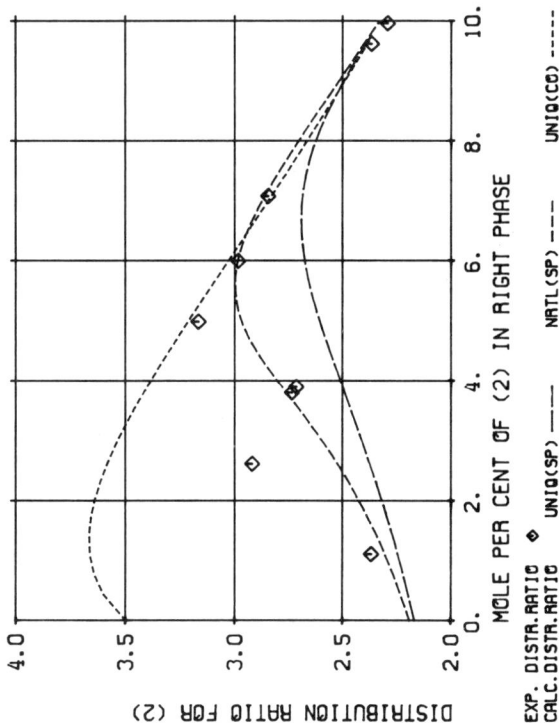

DISTRIBUTION RATIO FOR (2)

MOLE PER CENT OF (2) IN RIGHT PHASE

EXP. DISTR.RATIO ◇ UNIQ(SP) ◇ NRTL(SP) ——— UNIQ(CO) -----
CALC. DISTR.RATIO

(2) C6H11NO   HEXANOIC ACID,6-AMINO,LACTAM

(3) H2O   WATER

MORACHEVSKII A.G., SABININ V.E.
ZH.PRIKL.KHIM.(LENINGRAD) 33(1960)1775

TEMPERATURE = 20.0 DEG C   TYPE OF SYSTEM = 1

EXPERIMENTAL TIE LINES IN MOLE PCT

	LEFT PHASE			RIGHT PHASE	
(1)	(2)	(3)	(1)	(2)	(3)
96.297	2.610	1.093	0.155	1.102	98.743
87.071	7.614	5.314	0.188	2.610	97.203
82.754	10.398	6.848	0.243	3.800	95.958
82.582	10.569	6.849	0.244	3.893	95.864
69.388	15.761	14.851	0.420	4.979	94.601
63.030	17.867	19.103	0.712	5.989	93.300
56.418	20.131	23.451	1.034	7.077	91.889
40.883	22.764	36.353	2.193	9.616	88.191
39.108	22.842	38.050	2.378	9.954	87.668

SPECIFIC MODEL PARAMETERS IN KELVIN

		UNIQUAC		NRTL(ALPHA=.2)	
I J		AIJ	AJI	AIJ	AJI
1 2		286.60	-173.27	1157.7	-849.47
1 3		590.02	435.90	757.44	1429.4
2 3		-76.802	-8.0864	182.68	-342.09

R1 = 2.9308   R2 = 4.6106   R3 = 0.9200
Q1 = 2.528   Q2 = 3.724   Q3 = 1.400

MEAN DEV. BETWEEN CALC. AND EXP. CONC. IN MOLE PCT

UNIQUAC (SPECIFIC PARAMETERS)   0.44
NRTL   (SPECIFIC PARAMETERS)   0.29
UNIQUAC (COMMON PARAMETERS)   0.50

$C_2H_4O-C_4H_6O_2$

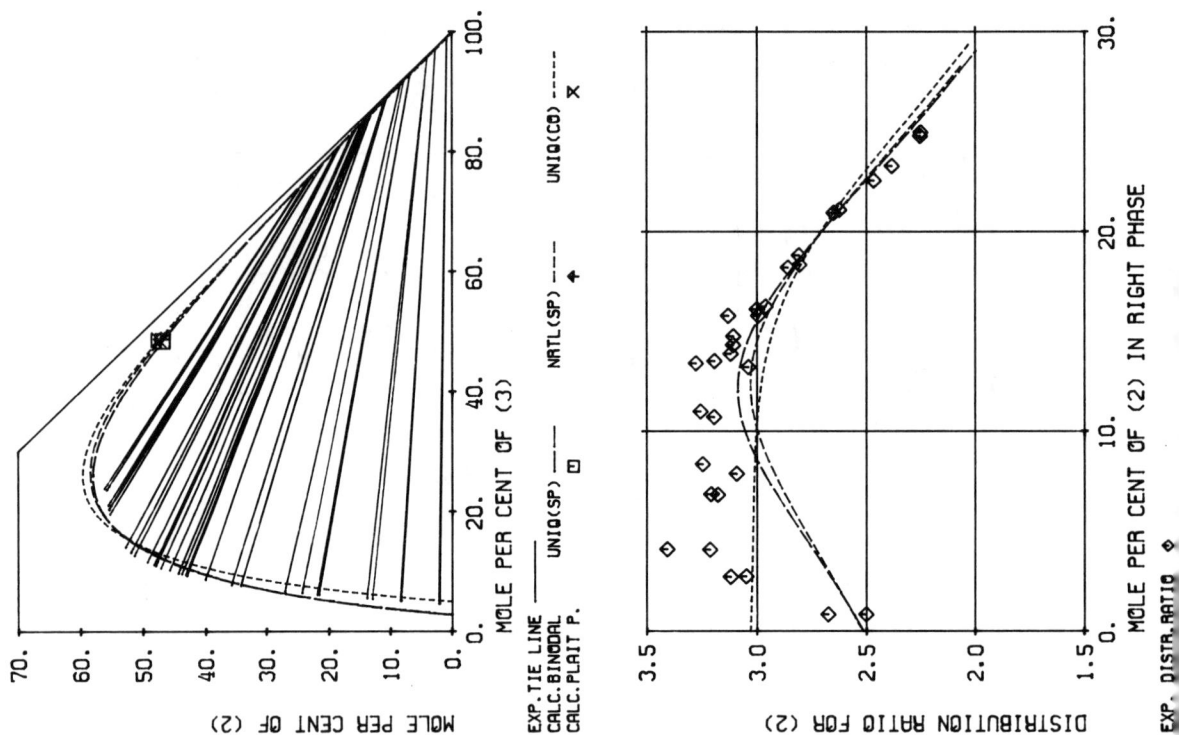

(1) C4H6O2   ACETIC ACID,ETHENYL ESTER
(2) C2H4O   ACETALDEHYDE
(3) H2O   WATER

PRATT H.R.C., GLOVER S.T.
TRANS.INST.CHEM.ENG. 24(1946)54

TEMPERATURE = 20.0 DEG C   TYPE OF SYSTEM = 1

EXPERIMENTAL TIE LINES IN MOLE PCT (GRAPH.INTERPOL.)

LEFT PHASE			RIGHT PHASE		
(1)	(2)	(3)	(1)	(2)	(3)
93.373	2.069	4.558	0.257	0.827	98.916
93.210	2.235	4.555	0.257	0.836	98.908
86.510	8.440	5.050	0.297	2.706	96.997
86.628	8.320	5.053	0.297	2.728	96.975
81.639	13.019	5.341	0.336	4.053	95.611
80.588	13.892	5.520	0.336	4.076	95.588
72.269	21.666	6.066	0.443	5.819	92.738
71.983	21.961	6.056	0.444	5.844	92.712
69.291	24.356	6.353	0.493	7.874	91.628
66.287	27.089	6.624	0.344	8.345	91.311
58.268	34.156	7.576	0.616	10.694	88.690
56.599	35.726	7.676	0.618	10.973	88.409
51.257	40.127	8.616	0.727	13.203	86.070
46.611	43.881	9.507	0.741	13.388	85.871
47.637	43.120	9.243	0.742	13.507	85.750
47.329	43.282	9.388	0.759	13.874	85.367
45.901	44.474	9.625	0.776	14.306	84.918
44.019	45.837	10.144	0.807	14.750	84.444
39.150	49.356	11.494	0.844	15.764	83.392
42.283	47.221	10.497	0.844	15.764	83.392
40.695	48.303	11.002	0.874	16.098	83.028
41.065	48.060	10.875	0.875	16.226	82.899
35.936	52.009	13.055	0.979	18.179	80.842
36.017	51.412	12.571	0.981	18.313	80.706
33.243	52.864	13.893	1.014	18.806	80.179
24.264	55.482	20.254	1.169	20.928	77.903
25.146	55.630	19.524	1.186	21.083	77.731
23.402	55.503	20.357	1.295	22.545	76.160
24.140	55.846	20.968	1.365	23.272	75.363
20.619	55.846	23.535	1.510	24.768	73.722
20.026	56.211	23.764	1.544	24.953	73.503

$C_2H_4O\text{-}C_4H_6O_2$

SPECIFIC MODEL PARAMETERS IN KELVIN

I J	UNIQUAC AIJ	AJI	NRTL(ALPHA=.2) AIJ	AJI
1 2	293.81	-94.756	276.67	-24.396
1 3	536.55	318.01	573.36	1661.2
2 3	329.64	-32.897	309.00	254.31

R1 = 3.2485   R2 = 1.8991   R3 = 0.9200
Q1 = 2.904   Q2 = 1.796   Q3 = 1.400

MEAN DEV. BETWEEN CALC. AND EXP. CONC. IN MOLE PCT

UNIQUAC (SPECIFIC PARAMETERS)   0.65
NRTL   (SPECIFIC PARAMETERS)   0.69
UNIQUAC (COMMON PARAMETERS)   0.86

C₂H₄O-C₄H₁₀O

(1) C4H100   ETHER,DIETHYL
----
(2) C2H4O   ACETALDEHYDE
----
(3) H2O   WATER

SUSKA J.
COLLECT.CZECH.CHEM.COMMUN. 44(1979)1999

TEMPERATURE = 15.1 DEG C   TYPE OF SYSTEM = 1

EXPERIMENTAL TIE LINES IN MOLE PCT

      LEFT PHASE   RIGHT PHASE
  (1)   (2)   (3)   (1)   (2)   (3)

76.700   15.500   7.800   2.500   8.300   89.200
63.700   27.400   8.900   3.300   17.500   79.200
56.600   32.700 10.700   4.500   23.800   71.700
46.100   40.600 13.300   5.700   29.500   64.800
37.300   46.100 16.600   7.000   34.600   58.400

SPECIFIC MODEL PARAMETERS IN KELVIN

              UNIQUAC   NRTL(ALPHA=.2)
 I J   AIJ   AJI   AIJ   AJI

 1 2   125.60   113.44   -496.49   557.56
 1 3   591.06   48.614   465.71   821.19
 2 3   384.47   -27.487   244.93   -108.02

R1 = 3.3949   R2 = 1.8991   R3 = 0.9200
Q1 = 3.016   Q2 = 1.796   Q3 = 1.400

MEAN DEV. BETWEEN CALC. AND EXP. CONC. IN MOLE PCT

UNIQUAC (SPECIFIC PARAMETERS)   0.68
NRTL   (SPECIFIC PARAMETERS)   0.62

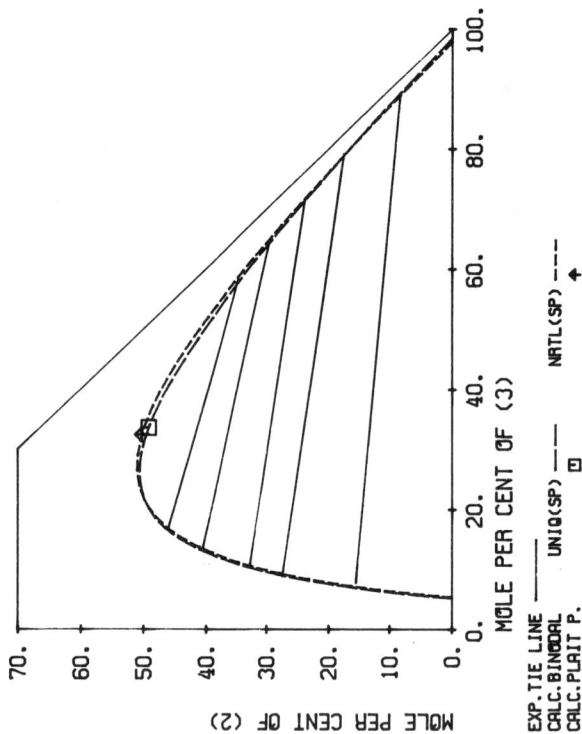

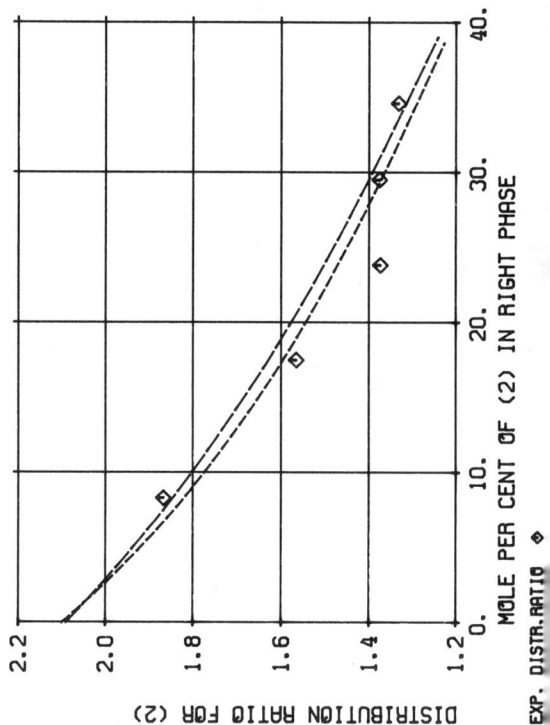

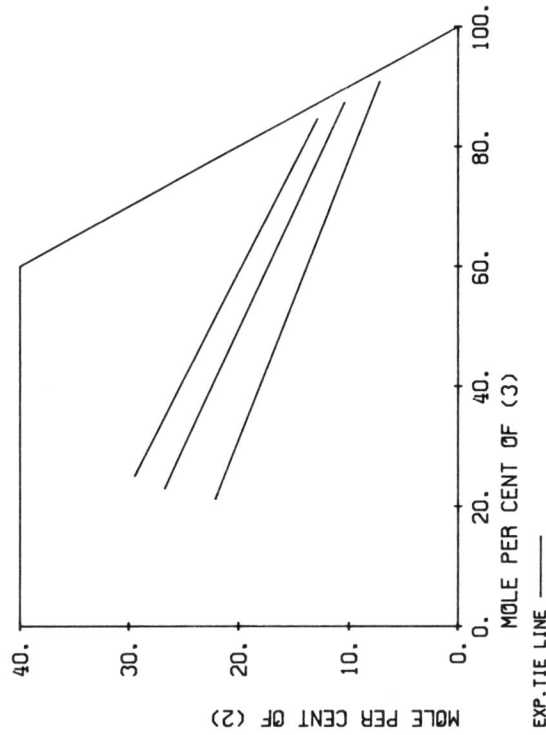

$C_2H_4O-C_5H_4O_2$

MOLE PER CENT OF (3)

MOLE PER CENT OF (2)

EXP.TIE LINE ————

(1) C5H4O2   FURFURAL

(2) C2H4O   ACETALDEHYDE

(3) H2O   WATER

OTHMER D.F., TOBIAS P.E.
IND.ENG.CHEM. 34(1942)690

TEMPERATURE = 16.0 DEG C   TYPE OF SYSTEM = 1

EXPERIMENTAL TIE LINES IN MOLE PCT (GRAPH.INTERPOL.)

LEFT PHASE			RIGHT PHASE		
(1)	(2)	(3)	(1)	(2)	(3)
56.620	22.064	21.316	1.959	7.196	90.845
50.310	26.689	23.001	2.296	10.372	37.332
45.444	29.434	25.122	2.556	12.828	84.615

C$_2$H$_4$O-C$_6$H$_6$

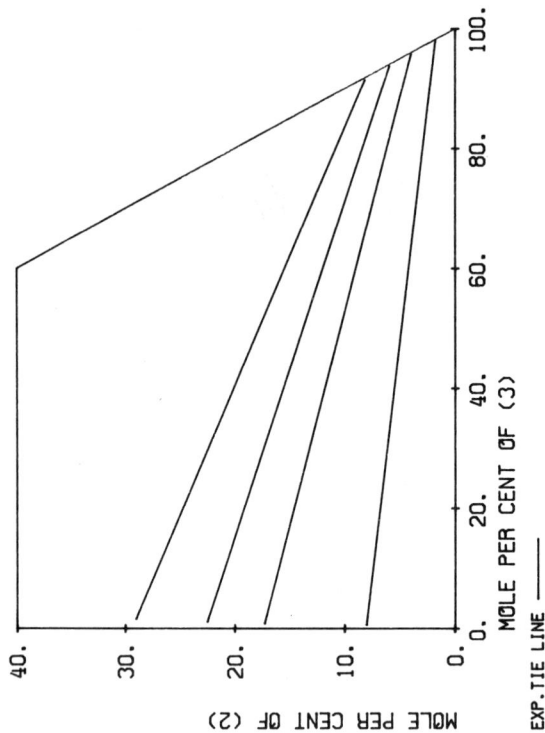

MOLE PER CENT OF (3)

MOLE PER CENT OF (2)

EXP.TIE LINE

(1) C6H6   BENZENE
(2) C2H4O   ACETALDEHYDE
(3) H2O   WATER

OTHMER D.F., TOBIAS P.E.
IND.ENG.CHEM. 34(1942)690

TEMPERATURE = 18.0 DEG C   TYPE OF SYSTEM = 1

EXPERIMENTAL TIE LINES IN MOLE PCT (GRAPH.INTERPOL.)

LEFT PHASE			RIGHT PHASE		
(1)	(2)	(3)	(1)	(2)	(3)
91.352	8.003	0.645	0.119	1.768	98.113
81.884	17.260	0.856	0.218	3.962	95.820
76.366	22.491	1.143	0.306	5.915	93.779
69.379	29.015	1.606	0.407	8.143	91.450

$C_2H_4O-C_7H_8$

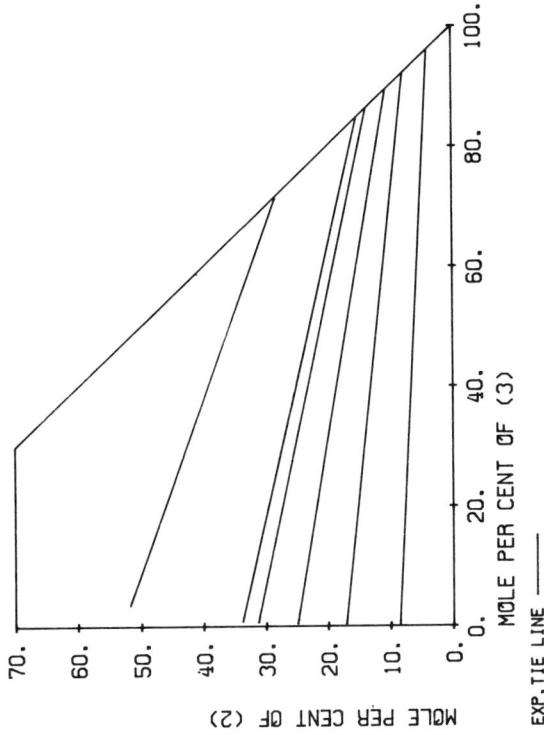

MOLE PER CENT OF (3)

MOLE PER CENT OF (2)

EXP.TIE LINE ———

(1) C7H8   TOLUENE
(2) C2H4O   ACETALDEHYDE
(3) H2O   WATER

OTHMER D.F.; TOBIAS P.E.
IND.ENG.CHEM. 34(1942)690

TEMPERATURE = 17.0 DEG C   TYPE OF SYSTEM = 1

EXPERIMENTAL TIE LINES IN MOLE PCT (GRAPH.INTERPOL.)

	LEFT PHASE			RIGHT PHASE	
(1)	(2)	(3)	(1)	(2)	(3)
91.177	8.575	0.249	0.030	3.892	96.078
82.505	17.086	0.409	0.045	7.743	92.212
74.519	24.989	0.492	0.060	10.476	89.464
67.997	31.289	0.714	0.081	13.650	86.269
65.336	33.828	0.836	0.092	15.152	84.756
44.661	51.770	3.568	0.275	28.283	71.443

$C_2H_4O_2$-$C_2H_6O_2$

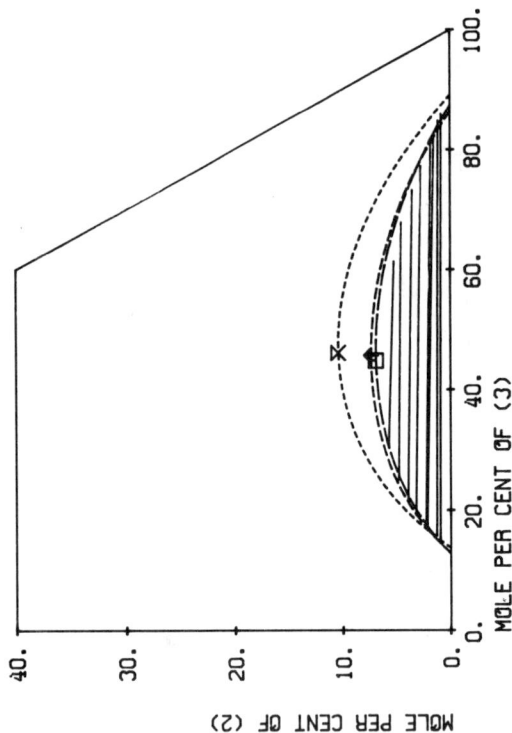

MOLE PER CENT OF (3)

MOLE PER CENT OF (2)

EXP.TIE LINE ——— UNIQ(SP) ☐ NRTL(SP) ✦ UNIQ(C6) ✕
CALC.BINODAL
CALC.PLAIT P.

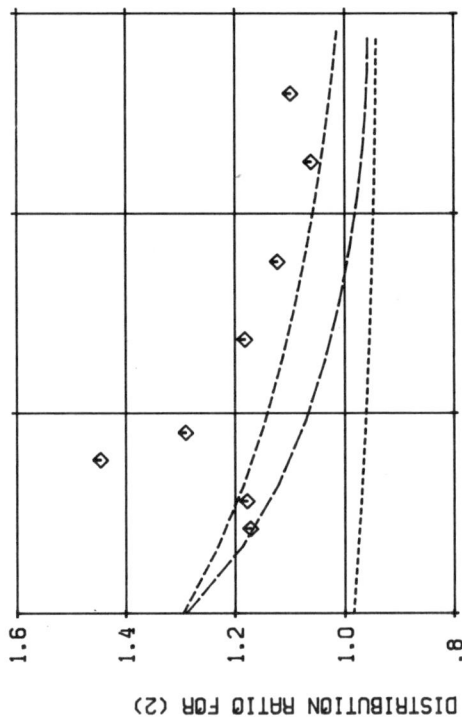

MOLE PER CENT OF (2) IN RIGHT PHASE

DISTRIBUTION RATIO FOR (2)

EXP. DISTR.RATIO ◇

(1) C2H6O2   1,2-ETHANEDIOL
(2) C2H4O2   ACETIC ACID
(3) C4H8O2   ACETIC ACID,ETHYL ESTER

ORELL A.
J.CHEM.ENG.DATA 12(1967)1

TEMPERATURE = 25.0 DEG C   TYPE OF SYSTEM = 1

EXPERIMENTAL TIE LINES IN MOLE PCT

| | LEFT PHASE | | | RIGHT PHASE | |
(1)	(2)	(3)	(1)	(2)	(3)
83.884	0.989	15.127	13.197	0.844	85.960
82.994	1.321	15.584	13.957	1.121	84.912
80.658	2.214	17.127	15.756	1.531	82.713
80.385	2.326	17.288	16.510	1.804	81.687
77.606	3.235	19.159	19.859	2.737	77.404
74.416	3.942	21.642	23.129	3.512	73.359
70.181	4.790	25.028	27.477	4.512	68.011
63.659	5.707	30.634	33.409	5.193	61.398

SPECIFIC MODEL PARAMETERS IN KELVIN

| | | UNIQUAC | | NRTL(ALPHA=.2) | |
I	J	AIJ	AJI	AIJ	AJI
1	2	87.971	-391.53	324.88	-838.73
1	3	72.350	295.47	462.97	431.49
2	3	-303.67	-123.43	-464.49	-396.64

R1 = 2.4088   R2 = 2.2024   R3 = 3.4786
Q1 = 2.248   Q2 = 2.072   Q3 = 3.116

MEAN DEV. BETWEEN CALC. AND EXP. CONC. IN MOLE PCT

UNIQUAC (SPECIFIC PARAMETERS)   0.49
NRTL   (SPECIFIC PARAMETERS)   0.77
UNIQUAC (COMMON PARAMETERS)   1.92

$C_2H_4O_2$-$C_4H_6O_2$

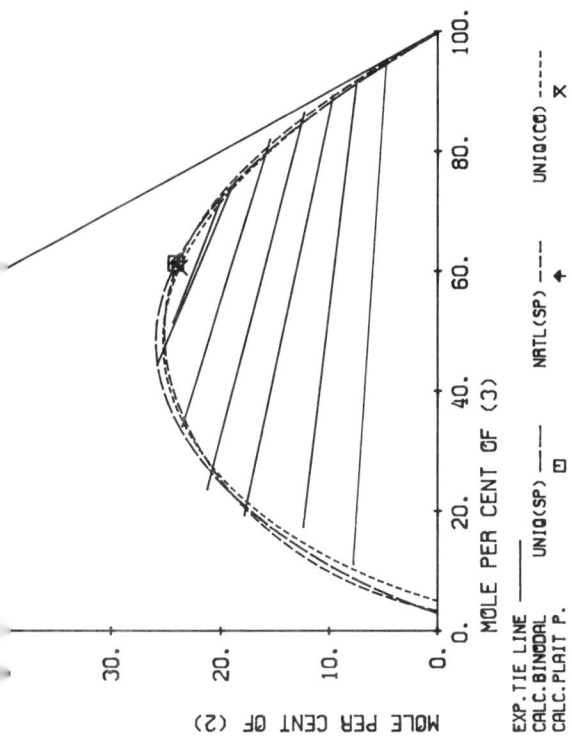

MOLE PER CENT OF (3)

MOLE PER CENT OF (2)

EXP.TIE LINE
CALC.BINODAL   UNIQ(SP) ----- ⊡   NRTL(SP) ----- ✦   UNIQ(CO) ----- ✗
CALC.PLAIT P.

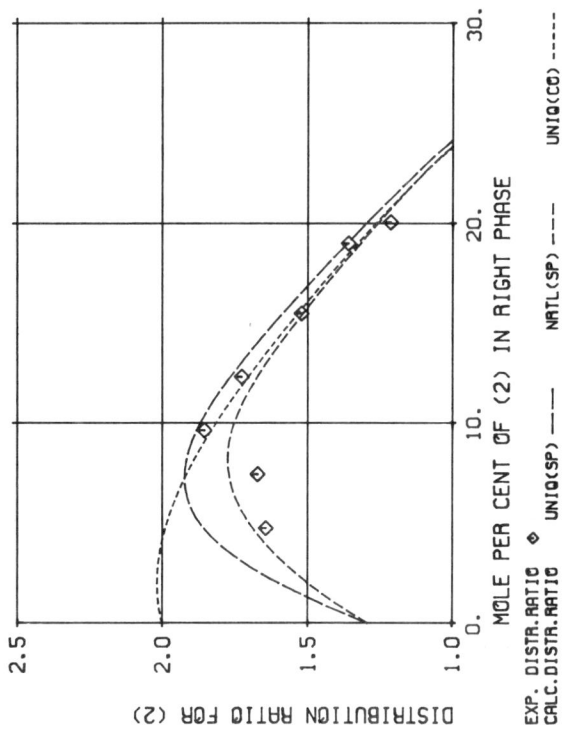

MOLE PER CENT OF (2) IN RIGHT PHASE

DISTRIBUTION RATIO FOR (2)

EXP. DISTR.RATIO   ◇ UNIQ(SP) -----   NRTL(SP) -----   UNIQ(CO) -----
CALC.DISTR.RATIO

(2) C2H4O2   ACETIC ACID

(3) H2O   WATER

TICHONOVA N.K., ET AL.
IZV.VYSSH.UCHEBN.ZAVED.KHIM.KHIM.TEKHNOL. 13(1970)175

TEMPERATURE = 20.0 DEG C   TYPE OF SYSTEM = 1

EXPERIMENTAL TIE LINES IN MOLE PCT

	LEFT PHASE			RIGHT PHASE	
(1)	(2)	(3)	(1)	(2)	(3)
99.524	0.0	0.476	0.425	0.0	99.575
81.177	7.776	11.047	0.617	4.727	94.656
70.226	12.422	17.352	0.754	7.427	91.818
62.817	17.852	19.330	1.004	9.619	89.377
55.120	21.259	23.621	1.170	12.301	86.530
42.412	23.546	34.042	2.530	15.468	82.002
29.907	25.824	44.269	7.089	18.993	73.918
24.220	24.366	51.414	8.316	20.048	71.637

SPECIFIC MODEL PARAMETERS IN KELVIN

		UNIQUAC		NRTL(ALPHA=.2)	
I J		AIJ	AJI	AIJ	AJI
1 2		-80.671	-166.88	-1479.4	-445.64
1 3		573.81	219.51	522.78	1639.1
2 3		-308.92	81.515	-787.64	-754.87

R1 = 3.2485 R2 = 2.2024 R3 = 0.9200
Q1 = 2.904   Q2 = 2.072   Q3 = 1.400

MEAN DEV. BETWEEN CALC. AND EXP. CONC. IN MOLE PCT

UNIQUAC (SPECIFIC PARAMETERS)   0.89
NRTL (SPECIFIC PARAMETERS)   0.98
UNIQUAC (COMMON PARAMETERS)   1.18

$C_2H_4O_2$-C_4H_8O

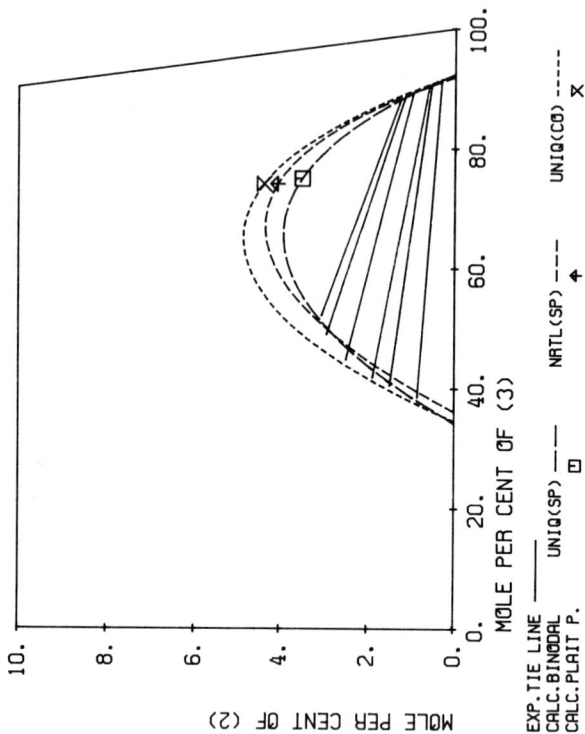

MOLE PER CENT OF (3)

MOLE PER CENT OF (2)

EXP.TIE LINE ——— UNIQ(SP) □ ——— NRTL(SP) ——+ UNIQ(CO) ----- ⋈
CALC.BINODAL -----
CALC.PLAIT P.

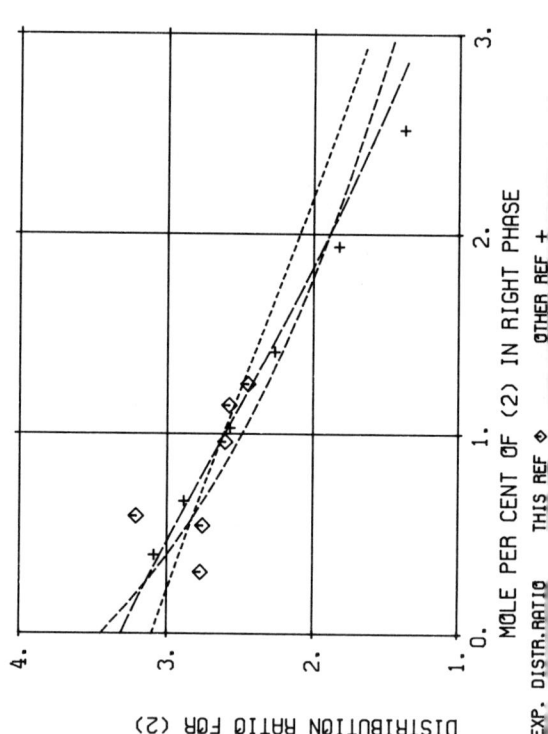

MOLE PER CENT OF (2) IN RIGHT PHASE

DISTRIBUTION RATIO FOR (2)

EXP. DISTR.RATIO ◇ THIS REF ◇   OTHER REF +

(1) C4H8O   2-BUTANONE
(2) C2H4O2   ACETIC ACID
(3) H2O   WATER

SKRZEC A.E., MURPHY N.F.
IND.ENG.CHEM. 46(1954)2245

TEMPERATURE = 26.7 DEG C   TYPE OF SYSTEM = 1

EXPERIMENTAL TIE LINES IN MOLE PCT (GRAPH.INTERPOL.)

	LEFT PHASE			RIGHT PHASE	
(1)	(2)	(3)	(1)	(2)	(3)
63.617	0.0	36.383	7.311	0.0	92.689
60.547	0.851	38.601	8.049	0.307	91.644
57.986	1.482	40.531	8.623	0.538	90.839
56.287	1.878	41.835	8.733	0.586	90.681
52.640	2.494	44.866	9.717	0.958	89.325
47.994	2.937	49.069	10.228	1.140	83.631
44.769	3.051	52.180	10.669	1.247	88.084

SPECIFIC MODEL PARAMETERS IN KELVIN

		UNIQUAC		NRTL(ALPHA=.2)	
I J		AIJ	AJI	AIJ	AJI
1 2		-254.13	-4.5537	518.52	-916.62
1 3		345.53	-2.0832	7.0045	895.00
2 3		-301.02	254.15	-453.71	-533.93

R1 = 3.2479   R2 = 2.2024   R3 = 0.9200
Q1 = 2.876   Q2 = 2.072   Q3 = 1.400

MEAN DEV. BETWEEN CALC. AND EXP. CONC. IN MOLE PCT

UNIQUAC (SPECIFIC PARAMETERS)   0.21
NRTL   (SPECIFIC PARAMETERS)   0.27
UNIQUAC (COMMON PARAMETERS)   0.36

C$_2$H$_4$O$_2$-C$_4$H$_8$O

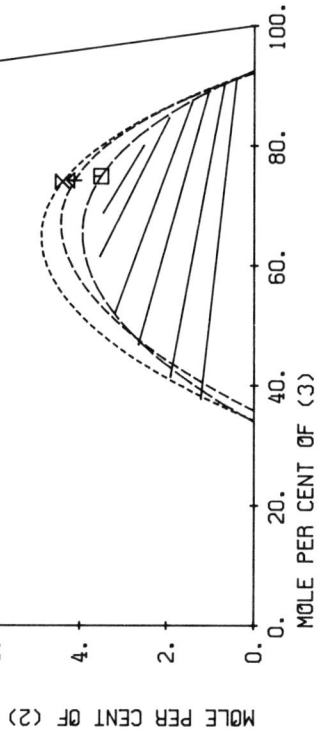

MOLE PER CENT OF (3)

MOLE PER CENT OF (2)

EXP.TIE LINE ———   UNIQ(SP) ——— □   NRTL(SP) — — +   UNIQ(CO) ----- ⚹
CALC.BINODAL
CALC.PLAIT P.

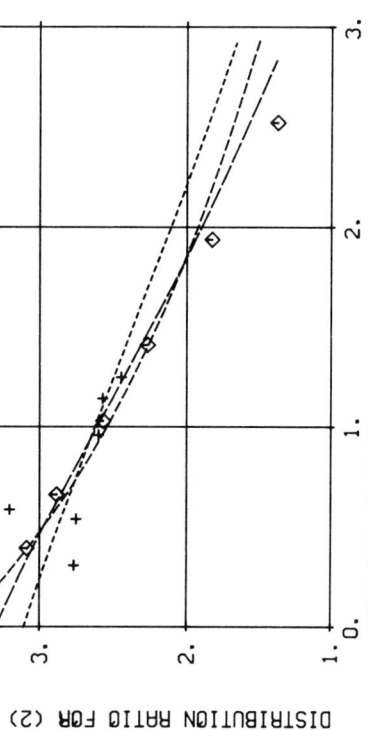

DISTRIBUTION RATIO FOR (2)

MOLE PER CENT OF (2) IN RIGHT PHASE

EXP. DISTR.RATIO ◇   THIS REF ◇   OTHER REF +   UNIQ(CO) -----
CALC.DISTR.RATIO   UNIQ(SP)   NRTL(SP)

(2) C2H4O2   ACETIC ACID

(3) H2O   WATER

IGUCHI A., FUSE K.
KAGAKU KOGAKU 35(1971)477

TEMPERATURE = 25.0 DEG C   TYPE OF SYSTEM = 1

EXPERIMENTAL TIE LINES IN MOLE PCT (GRAPH.INTERPOL.)

	LEFT PHASE			RIGHT PHASE	
(1)	(2)	(3)	(1)	(2)	(3)
61.025	1.210	37.765	8.220	0.392	91.388
56.633	1.911	41.456	8.910	0.662	90.428
50.454	2.644	46.902	9.988	1.029	88.984
44.785	3.194	52.021	11.225	1.410	87.366
34.895	3.534	61.571	13.371	1.936	84.693
27.693	3.460	68.848	17.430	2.521	80.049

SPECIFIC MODEL PARAMETERS IN KELVIN

	UNIQUAC			NRTL (ALPHA=.2)	
I J	AIJ	AJI		AIJ	AJI
1 2	-254.13	-4.5537		518.52	-916.62
1 3	345.53	-2.0882		7.0045	895.00
2 3	-301.02	254.15		-453.71	-533.93

R1 = 3.2479   R2 = 2.2024   R3 = 0.9200
Q1 = 2.876   Q2 = 2.072   Q3 = 1.400

MEAN DEV. BETWEEN CALC. AND EXP. CONC. IN MOLE PCT

UNIQUAC (SPECIFIC PARAMETERS)   0.23
NRTL   (SPECIFIC PARAMETERS)   0.45
UNIQUAC (COMMON PARAMETERS)   0.59

$C_2H_4O_2$-$C_4H_8O_2$

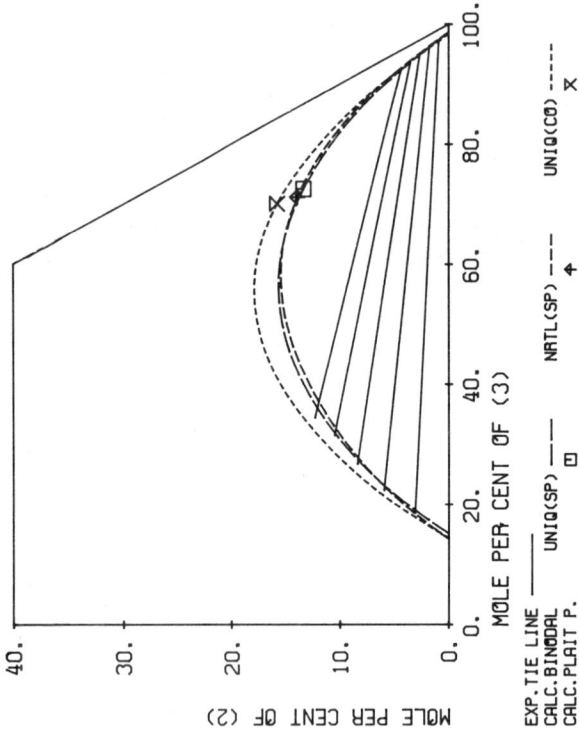

MOLE PER CENT OF (3)

MOLE PER CENT OF (2)

EXP.TIE LINE ——— UNIQ(SP) ☐ NRTL(SP) — ✦ UNIQ(CB) ----- ✗
CALC.BINODAL
CALC.PLAIT P.

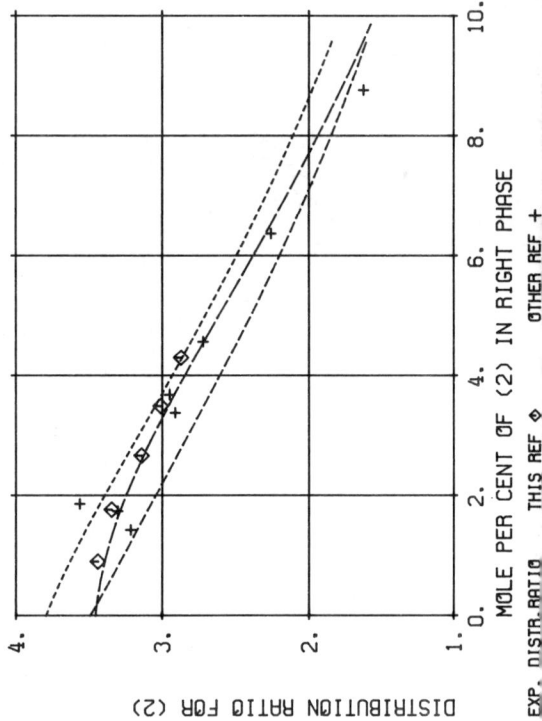

MOLE PER CENT OF (2) IN RIGHT PHASE

DISTRIBUTION RATIO FOR (2)

EXP. DISTR.RATIO ◇ THIS REF ◇ OTHER REF +

(1) C4H8O2   ACETIC ACID,ETHYL ESTER
(2) C2H4O2   ACETIC ACID
(3) H2O   WATER

SOHONI V.R., WARHADPANDE U.R.
IND.ENG.CHEM. 44(1952)1428

TEMPERATURE = 30.0 DEG C   TYPE OF SYSTEM = 1

EXPERIMENTAL TIE LINES IN MOLE PCT

	LEFT PHASE			RIGHT PHASE	
(1)	(2)	(3)	(1)	(2)	(3)
78.362	3.091	18.547	1.771	0.899	97.329
71.716	5.892	22.391	2.002	1.762	96.236
64.855	8.368	26.777	2.359	2.666	94.975
57.982	10.504	31.513	2.718	3.490	93.791
53.210	12.332	34.457	2.902	4.294	92.804

SPECIFIC MODEL PARAMETERS IN KELVIN

		UNIQUAC		NRTL(ALPHA=.2)	
I J	AIJ	AJI	AIJ	AIJ	AJI
1 2	-212.18	29.450	643.30	643.30	-702.57
1 3	376.94	97.519	166.36	166.36	1190.1
2 3	-259.46	188.77	-1.6825	-1.6825	-302.63

R1 = 3.4786   R2 = 2.2024   R3 = 0.9200
Q1 = 3.116   Q2 = 2.072   Q3 = 1.400

MEAN DEV. BETWEEN CALC. AND EXP. CONC. IN MOLE PCT

UNIQUAC (SPECIFIC PARAMETERS)   0.20
NRTL (SPECIFIC PARAMETERS)   0.30
UNIQUAC (COMMON PARAMETERS)   0.47

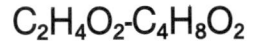

```
--------
(2) C2H4O2    ACETIC ACID
--------
(3) H2O       WATER
--------

GARNER F.H., ELLIS S.R.M.
CHEM.ENG.SCI. 2(1953)282

TEMPERATURE =   30.0 DEG C    TYPE OF SYSTEM =  1

EXPERIMENTAL TIE LINES IN MOLE PCT
--------
        LEFT PHASE                  RIGHT PHASE
        (1)      (2)      (3)       (1)      (2)      (3)
--------
71.981   4.572   23.447            1.805    1.423   96.772
70.917   5.723   23.360            1.817    1.733   96.449
56.066   9.824   34.110            2.396    3.375   94.229
26.617  14.203   59.180            6.931    8.756   84.313

SPECIFIC MODEL PARAMETERS IN KELVIN
--------
              UNIQUAC              NRTL(ALPHA=.2)
I  J     AIJ      AJI           AIJ       AJI
--------
1  2  -212.18    29.450        643.30    -702.57
1  3   376.94    97.519        166.36    1190.1
2  3  -259.46   188.77        -1.6825    -302.63

R1 =  3.4786    R2 =  2.2024    R3 =  0.9200
Q1 =  3.116     Q2 =  2.072     Q3 =  1.400

MEAN DEV. BETWEEN CALC. AND EXP. CONC. IN MOLE PCT
--------
UNIQUAC (SPECIFIC PARAMETERS)      0.68
NRTL   (SPECIFIC PARAMETERS)      0.89
UNIQUAC (COMMON PARAMETERS)       1.60
```

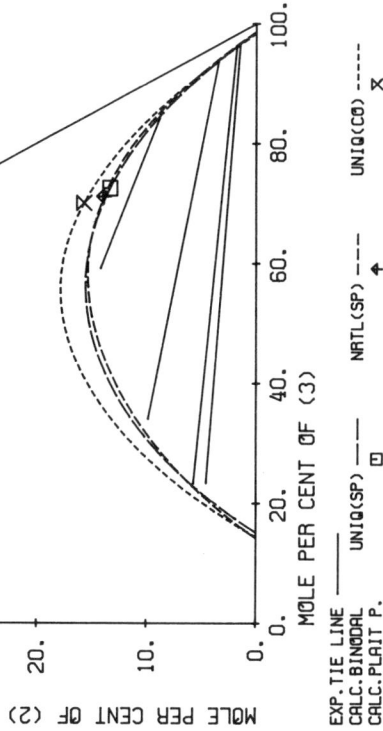

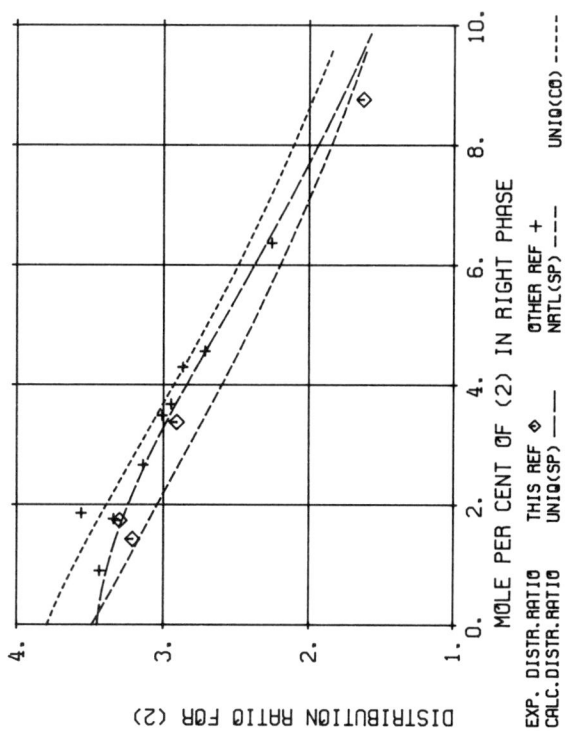

$C_2H_4O_2$-$C_4H_8O_2$

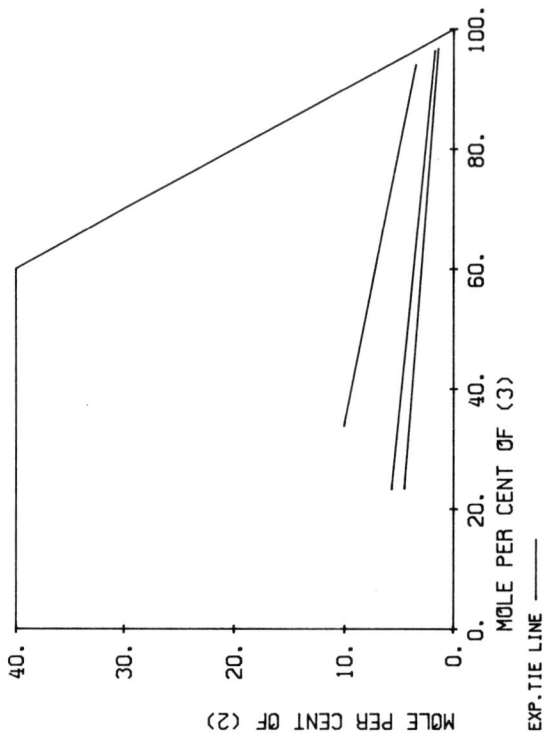

(1) C4H8O2   ACETIC ACID,ETHYL ESTER ---------

(2) C2H4O2   ACETIC ACID ---------

(3) H2O   WATER ---------

GARNER F.H., ELLIS S.R.M.
CHEM.ENG.SCI. 2(1953)282

TEMPERATURE = 40.0 DEG C   TYPE OF SYSTEM = 1

EXPERIMENTAL TIE LINES IN MOLE PCT

LEFT PHASE			RIGHT PHASE		
(1)	(2)	(3)	(1)	(2)	(3)
71.981	4.572	23.447	1.805	1.423	96.772
70.917	5.723	23.360	1.817	1.733	96.449
56.112	10.049	33.839	2.404	3.528	94.069

$C_2H_4O_2\text{-}C_4H_8O_2$

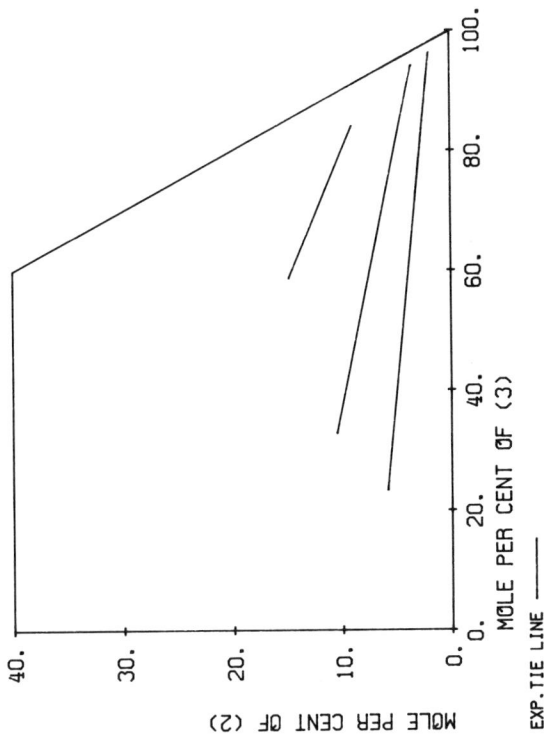

MOLE PER CENT OF (3)

MOLE PER CENT OF (2)

EXP.TIE LINE ⎯⎯⎯

(1) C4H8O2   ACETIC ACID,ETHYL ESTER
(2) C2H4O2   ACETIC ACID
(3) H2O   WATER

GARNER F.H., ELLIS S.R.M.
CHEM.ENG.SCI. 2(1953)282

TEMPERATURE = 50.0 DEG C   TYPE OF SYSTEM = 1

EXPERIMENTAL TIE LINES IN MOLE PCT

| | LEFT PHASE | | | RIGHT PHASE | |
(1)	(2)	(3)	(1)	(2)	(3)
70.811	5.838	23.351	1.823	1.873	96.304
2.404	3.528	94.069	2.400	3.451	94.149
56.744	10.460	32.796	2.404	3.528	94.069
26.477	14.823	58.695	6.960	8.970	84.071

$C_2H_4O_2$-$C_4H_8O_2$

```
(1) C4H8O2     ACETIC ACID, ETHYL ESTER
(2) C2H4O2     ACETIC ACID
(3) H2O        WATER

MIRADA LILLO R.; GONZALES TRIGO G.
AN.R.SOC.ESP.FIS.QUIM. 56B(1960)217

TEMPERATURE = 20.0 DEG C     TYPE OF SYSTEM = 1

EXPERIMENTAL TIE LINES IN MOLE PCT

        LEFT PHASE                    RIGHT PHASE
   (1)        (2)        (3)       (1)        (2)        (3)

 73.344      6.614     20.042     1.810      1.858     96.332
 60.426     10.846     28.728     2.412      3.677     93.911
 54.253     12.417     33.330     2.809      4.563     92.629
 42.727     14.393     42.879     3.848      6.364     89.787

SPECIFIC MODEL PARAMETERS IN KELVIN

              UNIQUAC                    NRTL(ALPHA=.2)
 I  J    AIJ        AJI              AIJ        AJI

 1  2  -212.18      29.450         643.30     -702.57
 1  3   376.94      97.519         166.36     1190.1
 2  3  -259.46     188.77          -1.6825    -302.63

 R1 = 3.4786    R2 = 2.2024    R3 = 0.9200
 Q1 = 3.116     Q2 = 2.072     Q3 = 1.400

MEAN DEV. BETWEEN CALC. AND EXP. CONC. IN MOLE PCT

 UNIQUAC (SPECIFIC PARAMETERS)      0.46
 NRTL   (SPECIFIC PARAMETERS)      0.66
 UNIQUAC (COMMON PARAMETERS)       0.66
```

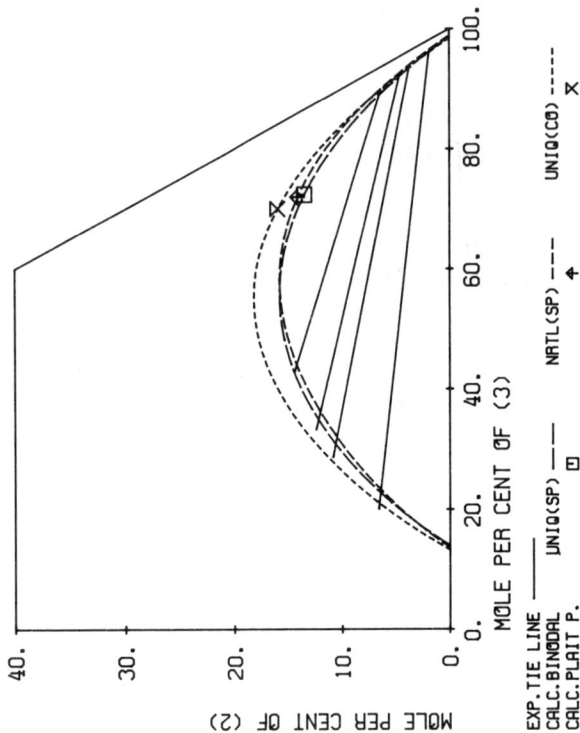

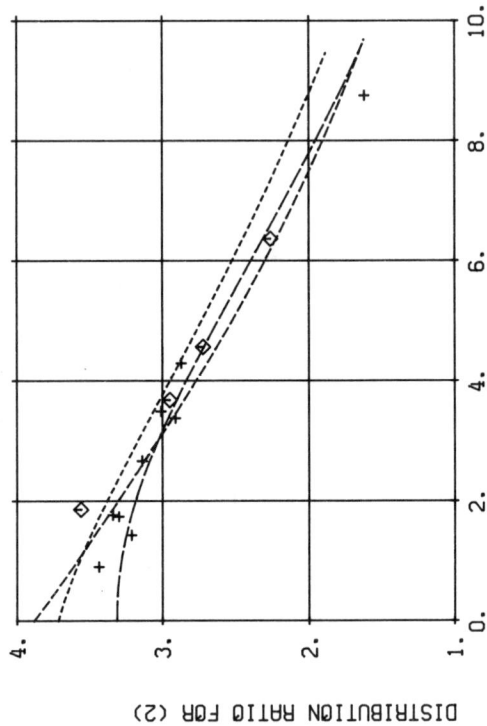

$C_2H_4O_2-C_4H_8O_2$

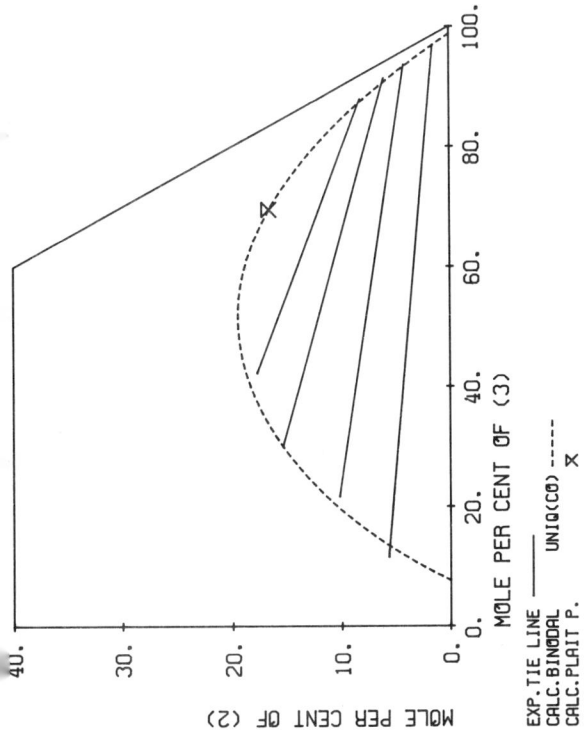

MOLE PER CENT OF (3)

MOLE PER CENT OF (2)

EXP.TIE LINE ———
CALC.BINODAL UNIQ(CO) -----
CALC.PLAIT P. ✕

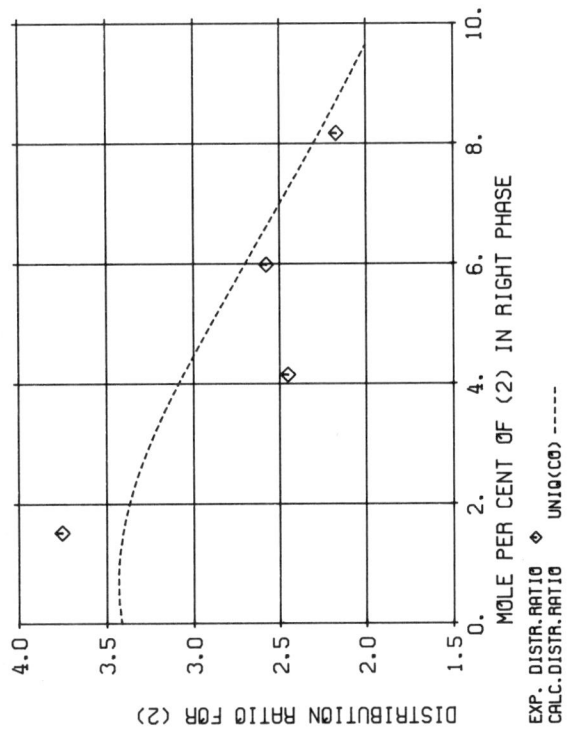

MOLE PER CENT OF (2) IN RIGHT PHASE

DISTRIBUTION RATIO FOR (2)

EXP. DISTR.RATIO ◇
CALC.DISTR.RATIO UNIQ(CO) -----

(2) C2H4O2   ACETIC ACID
(3) H2O   WATER

BOMSHTEIN A.L., TROFIMOV A.N., SERAFIMOV L.A.
ZH.PRIKL.KHIM.(LENINGRAD) 51(1978)1280

TEMPERATURE = 20.0 DEG C   TYPE OF SYSTEM = 1

EXPERIMENTAL TIE LINES IN MOLE PCT

	LEFT PHASE			RIGHT PHASE	
(1)	(2)	(3)	(1)	(2)	(3)
82.810	5.670	11.520	1.520	1.510	96.970
68.130	10.180	21.690	2.180	4.150	93.670
54.590	15.410	30.000	2.530	5.980	91.490
39.950	17.740	42.310	3.870	8.170	87.960

MEAN DEV. BETWEEN CALC. AND EXP. CONC. IN MOLE PCT

UNIQUAC (COMMON PARAMETERS)   0.51

$C_2H_4O_2$-$C_4H_{10}O$

(1) C4H10O   1-BUTANOL

(2) C2H4O2   ACETIC ACID

(3) H2O   WATER

SKRZEC A.E., MURPHY N.F.
IND.ENG.CHEM. 46(1954)2245

TEMPERATURE = 26.7 DEG C   TYPE OF SYSTEM = 1

EXPERIMENTAL TIE LINES IN MOLE PCT (GRAPH.INTERPOL.)

LEFT PHASE			RIGHT PHASE		
(1)	(2)	(3)	(1)	(2)	(3)
48.522	0.0	51.478	1.878	0.0	98.122
46.514	1.054	52.432	1.863	0.281	97.856
44.932	1.832	53.237	1.862	0.585	97.553
43.186	2.657	54.157	1.868	0.851	97.281
41.333	3.482	55.186	1.902	1.187	96.911
39.524	4.221	56.255	1.983	1.436	96.581
35.232	5.679	59.089	2.333	2.138	95.529
31.017	6.822	62.161	2.712	2.830	94.457

SPECIFIC MODEL PARAMETERS IN KELVIN

		UNIQUAC		NRTL(ALPHA=.2)	
I J	AIJ	AJI		AIJ	AJI
1 2	155.34	-211.40		192.64	-412.70
1 3	-30.037	311.03		-330.50	1601.7
2 3	-30.340	-170.32		-114.89	-182.85

R1 = 3.4543   R2 = 2.2024   R3 = 0.9200
Q1 = 3.052   Q2 = 2.072   Q3 = 1.400

MEAN DEV. BETWEEN CALC. AND EXP. CONC. IN MOLE PCT

UNIQUAC (SPECIFIC PARAMETERS)   0.16
NRTL   (SPECIFIC PARAMETERS)   0.17
UNIQUAC (COMMON PARAMETERS)   0.29

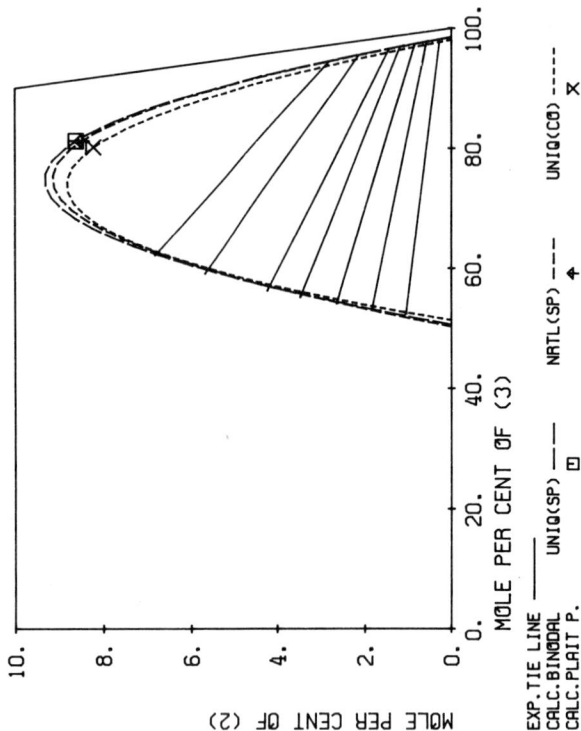

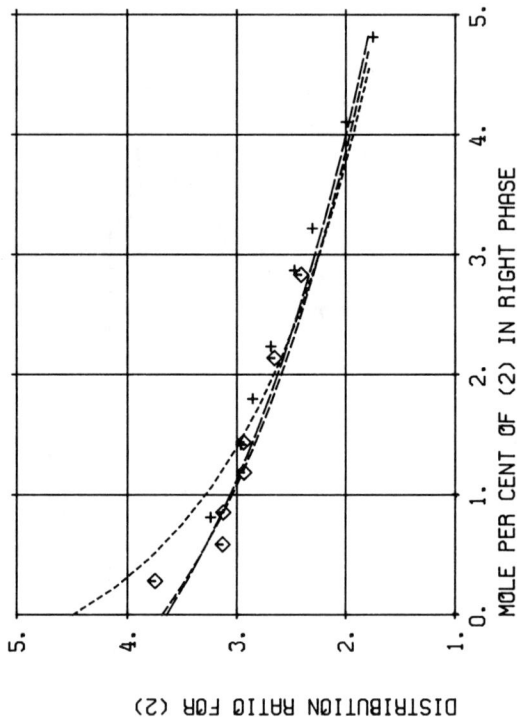

MOLE PER CENT OF (3)

MOLE PER CENT OF (2)

EXP.TIE LINE ———   UNIQ(SP) —— □   NRTL(SP) —— ✦   UNIQ(CO) ---- ✗
CALC.BINODAL
CALC.PLAIT P.

DISTRIBUTION RATIO FOR (2)

MOLE PER CENT OF (2) IN RIGHT PHASE

EXP. DISTR.RATIO ◇   THIS REF ◇   OTHER REF +

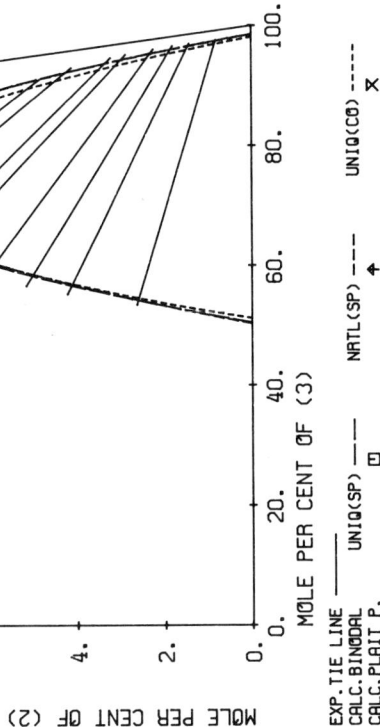

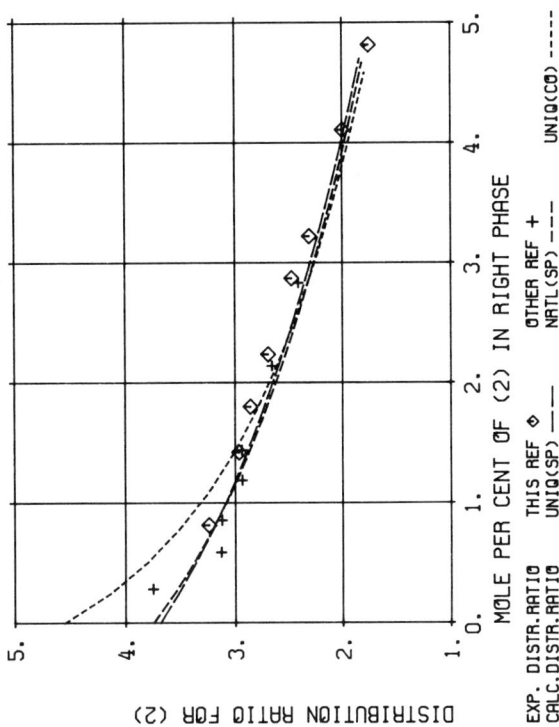

(2) C2H4O2	ACETIC ACID
(3) H2O	WATER

FUSE K., IGUCHI A.
KAGAKU KOGAKU 35(1971)801

TEMPERATURE = 25.0 DEG C   TYPE OF SYSTEM = 1

EXPERIMENTAL TIE LINES IN MOLE PCT (GRAPH.INTERPOL.)

LEFT PHASE			RIGHT PHASE		
(1)	(2)	(3)	(1)	(2)	(3)
43.930	2.624	53.446	1.529	0.809	97.661
40.553	4.200	55.247	1.551	1.418	97.032
38.251	5.130	56.619	1.564	1.796	96.640
35.303	6.014	58.683	1.664	2.236	96.101
30.617	7.088	62.295	1.888	2.867	95.245
28.597	7.424	63.979	2.078	3.218	94.705
23.598	8.207	68.195	2.819	4.105	93.076
20.784	8.418	70.798	4.032	4.813	91.156

SPECIFIC MODEL PARAMETERS IN KELVIN

		UNIQUAC		NRTL(ALPHA=.2)	
I J		AIJ	AJI	AIJ	AJI
1 2		155.34	-211.40	192.64	-412.70
1 3		-30.037	311.03	-330.50	-1601.7
2 3		-30.340	-170.32	-114.89	-182.85

F1 = 3.4543   R2 = 2.2024   R3 = 0.9200
Q1 = 3.052   Q2 = 2.072   Q3 = 1.400

MEAN DEV. BETWEEN CALC. AND EXP. CONC. IN MOLE PCT

UNIQUAC (SPECIFIC PARAMETERS)	0.54
NRTL (SPECIFIC PARAMETERS)	0.54
UNIQUAC (COMMON PARAMETERS)	0.95

C$_2$H$_4$O$_2$-C$_4$H$_{10}$O

(1) C4H10O   ETHER,DIETHYL
----
(2) C2H4O2   ACETIC ACID
----
(3) H2O   WATER

MAJOR C.J., SWENSON O.J.
IND.ENG.CHEM. 38(1946)834

TEMPERATURE = 25.0 DEG C   TYPE OF SYSTEM = 1

EXPERIMENTAL TIE LINES IN MOLE PCT

           LEFT PHASE   RIGHT PHASE
       (1)   (2)   (3)   (1)   (2)   (3)

    84.773   4.725   10.501   1.822   1.951   96.227
    74.591   9.462   15.947   2.076   3.844   94.080
    63.359   14.174   22.467   2.413   5.812   91.775
    52.281   18.123   29.596   3.047   7.907   89.046
    39.625   20.705   39.671   4.247   10.443   85.309
    28.229   21.048   50.723   6.553   13.315   80.132

SPECIFIC MODEL PARAMETERS IN KELVIN

              UNIQUAC   NRTL(ALPHA=.2)
 I J   AIJ   AJI   AIJ   AJI

 1 2 -212.01   26.114   955.63   -679.82
 1 3   493.70   127.55   374.70   1168.1
 2 3 -254.66   129.86   65.933   -58.806

R1 = 3.3949   R2 = 2.2024   R3 = 0.9200
Q1 = 3.016   Q2 = 2.072   Q3 = 1.400

MEAN DEV. BETWEEN CALC. AND EXP. CONC. IN MOLE PCT

UNIQUAC (SPECIFIC PARAMETERS)   0.31
NRTL   (SPECIFIC PARAMETERS)   0.30
UNIQUAC (COMMON PARAMETERS)   0.83

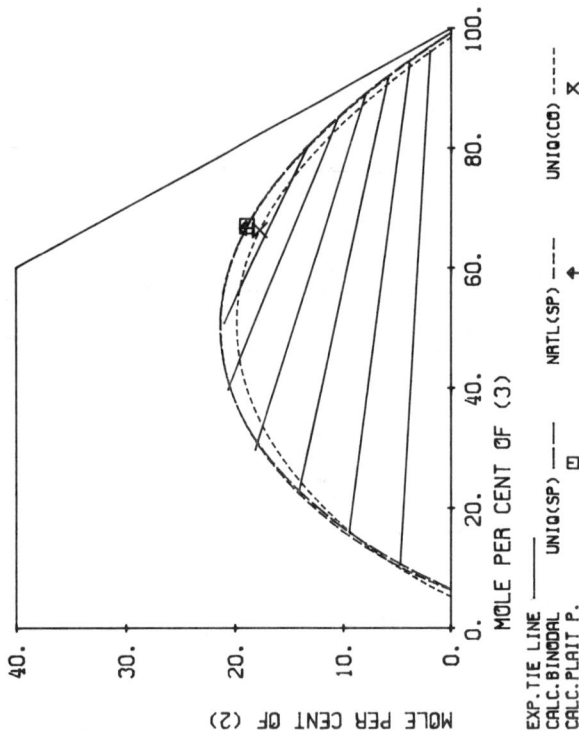

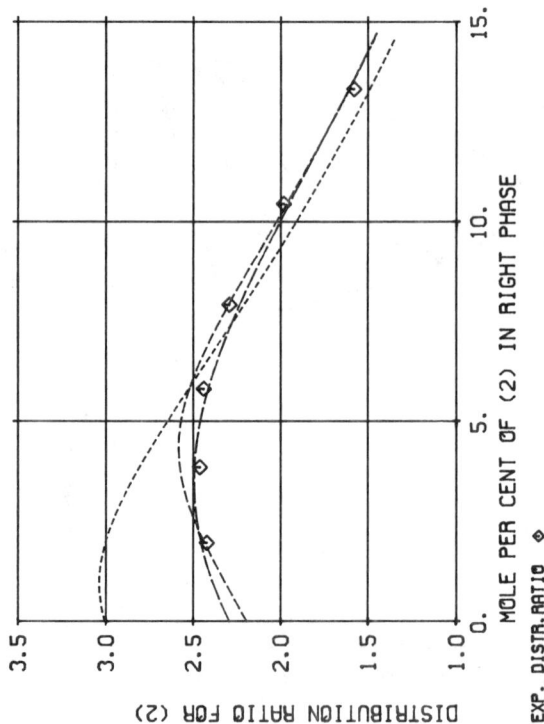

$C_2H_4O_2-C_4H_{10}O$

(2) C2H4O2   ACETIC ACID

(3) H2O   WATER

CASARICO A.
ANN.CHIM.(ROME) 41(1951)199

TEMPERATURE = 19.0 DEG C   TYPE OF SYSTEM = 1

EXPERIMENTAL TIE LINES IN MOLE PCT

	LEFT PHASE		RIGHT PHASE	
(1)	(2)	(3)	(1)	(2)	(3)
92.327	0.886	6.787	0.794	0.385	98.822
85.266	3.277	11.457	0.521	1.339	98.140
73.990	6.212	19.799	0.454	2.536	97.010
53.733	11.083	35.184	0.165	4.736	95.099
25.198	15.121	59.681	0.818	8.499	90.683

SPECIFIC MODEL PARAMETERS IN KELVIN

		UNIQUAC		NRTL(ALPHA=.2)	
I J	AIJ	AJI		AIJ	AJI
1 2	-339.02	-229.82		-860.01	-499.13
1 3	356.63	234.53		320.56	1548.0
2 3	-445.96	217.78		-1129.9	854.25

R1 = 3.3949   R2 = 2.2024   R3 = 0.9200
Q1 = 3.016   Q2 = 2.072   Q3 = 1.400

MEAN DEV. BETWEEN CALC. AND EXP. CONC. IN MOLE PCT

UNIQUAC (SPECIFIC PARAMETERS)   1.40
NRTL   (SPECIFIC PARAMETERS)   0.55

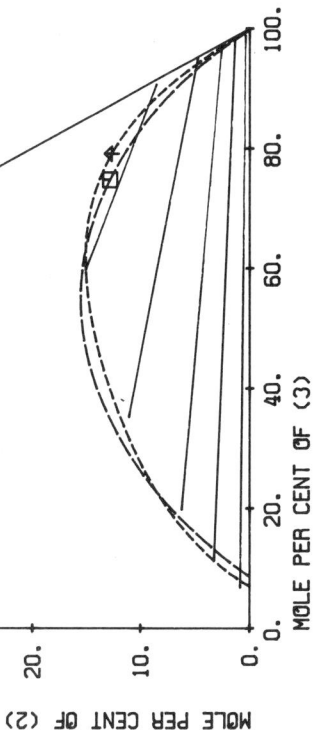

MOLE PER CENT OF (2)

MOLE PER CENT OF (3)

EXP.TIE LINE   UNIQ(SP) ▫   NRTL(SP) ---
CALC.BINODAL
CALC.PLAIT P.   ＋

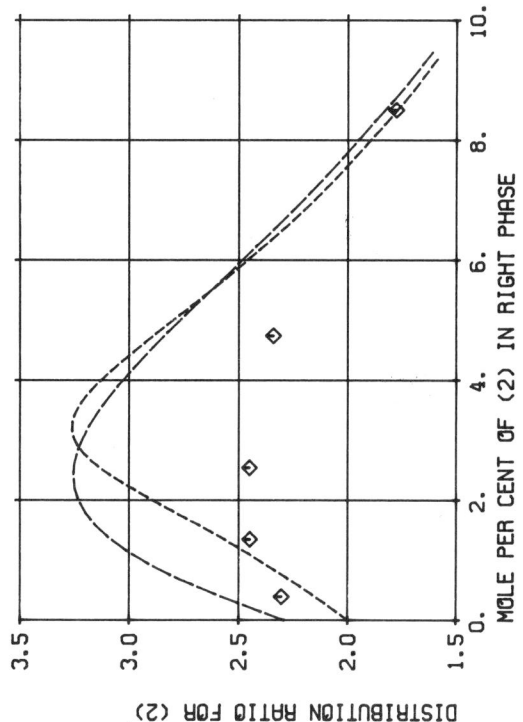

DISTRIBUTION RATIO FOR (2)

MOLE PER CENT OF (2) IN RIGHT PHASE

EXP. DISTR.RATIO ◇ UNIQ(SP) ---   NRTL(SP) ---
CALC.DISTR.RATIO

$C_2H_4O_2$-$C_5H_4O_2$

MOLE PER CENT OF (3)

MOLE PER CENT OF (2)

DISTRIBUTION RATIO FOR (2)

MOLE PER CENT OF (2) IN RIGHT PHASE

EXP.TIE LINE —— UNIQ(SP) ☐ NRTL(SP) ✦ UNIQ(CO) - - - ✕
CALC.BINODAL - - - -
CALC.PLAIT P.

(1) C5H4O2   FURFURAL
(2) C2H4O2   ACETIC ACID
(3) H2O   WATER

SKRZEC A.E., MURPHY N.F.
IND.ENG.CHEM. 46(1954)2245

TEMPERATURE = 26.7 DEG C   TYPE OF SYSTEM = 1

EXPERIMENTAL TIE LINES IN MOLE PCT (GRAPH.INTERPOL.)

	LEFT PHASE			RIGHT PHASE	
(1)	(2)	(3)	(1)	(2)	(3)
76.310	0.0	23.690	1.712	0.0	93.288
71.039	2.306	26.655	1.871	0.778	97.350
65.484	4.107	30.409	2.027	1.405	96.567
56.483	6.163	37.354	2.312	2.248	95.440
47.354	7.473	45.173	2.681	3.026	94.292
44.846	8.381	46.773	3.208	3.753	93.039
40.550	9.859	49.591	4.227	4.756	91.017
38.504	10.298	51.198	4.593	5.057	90.350

SPECIFIC MODEL PARAMETERS IN KELVIN

		UNIQUAC			NRTL (ALPHA=.2)	
I	J	AIJ	AJI		AIJ	AJI
1	2	342.99	-361.92		514.63	-757.65
1	3	149.81	110.29		28.810	1248.6
2	3	-127.49	-419.83		-250.83	-502.50

R1 = 3.1680   R2 = 2.2024   R3 = 0.9200
Q1 = 2.484   Q2 = 2.072   Q3 = 1.400

MEAN DEV. BETWEEN CALC. AND EXP. CONC. IN MOLE PCT

UNIQUAC (SPECIFIC PARAMETERS)   0.79
NRTL   (SPECIFIC PARAMETERS)   0.39
UNIQUAC (COMMON PARAMETERS)   1.89

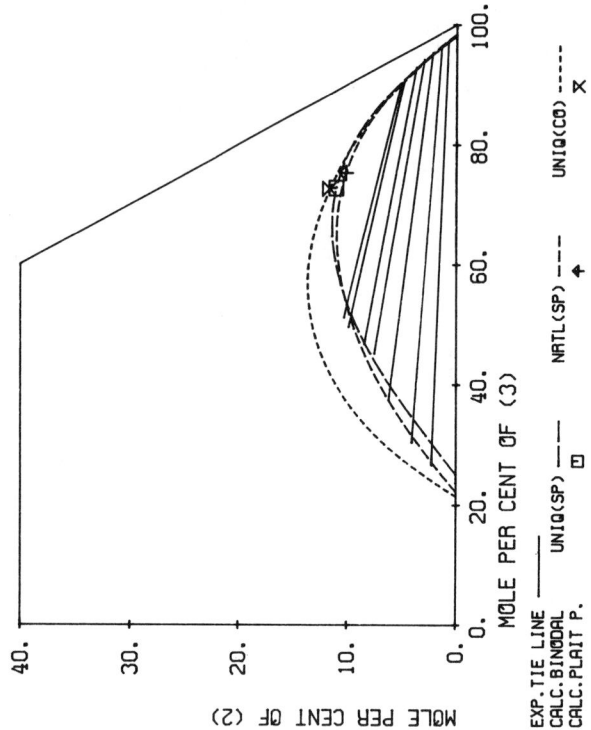

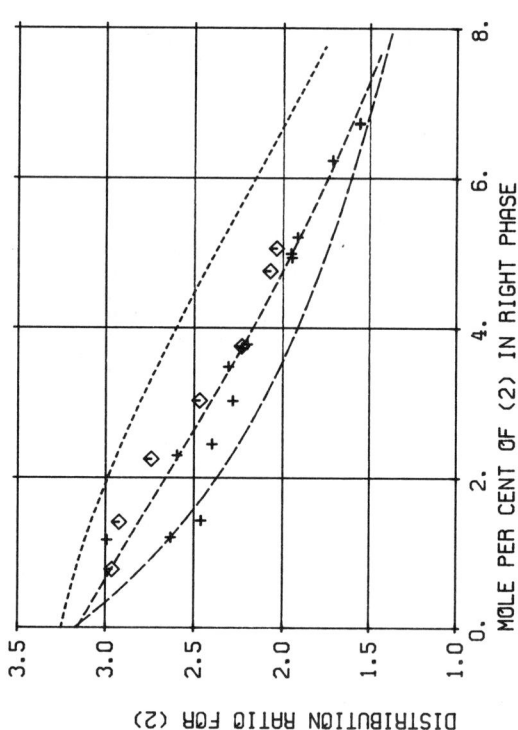

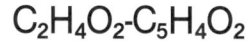

(2) C2H4O2   ACETIC ACID

(3) H2O   WATER

PEGORARO M.; GUGLIELMI G.
CHIM.IND.(MILAN) 37(1955)1035

TEMPERATURE = 25.0 DEG C   TYPE OF SYSTEM = 1

EXPERIMENTAL TIE LINES IN MOLE PCT

	LEFT PHASE			RIGHT PHASE	
(1)	(2)	(3)	(1)	(2)	(3)
67.091	3.507	29.402	2.131	1.425	96.444
53.088	6.896	40.017	2.767	3.021	94.211
49.391	8.324	42.285	2.785	3.732	93.483
40.228	9.616	50.156	3.928	4.936	91.136
30.840	10.708	58.453	4.916	6.236	88.849

SPECIFIC MODEL PARAMETERS IN KELVIN

		UNIQUAC		NRTL(ALPHA=.2)	
I J		AIJ	AJI	AIJ	AJI
1 2		342.99	-361.92	514.63	-757.65
1 3		149.81	110.29	28.810	1248.6
2 3		-127.49	-419.83	-250.83	-502.50

R1 = 3.1680   R2 = 2.2024   R3 = 0.9200
Q1 = 2.484   Q2 = 2.072   Q3 = 1.400

MEAN DEV. BETWEEN CALC. AND EXP. CONC. IN MOLE PCT

UNIQUAC (SPECIFIC PARAMETERS)   1.01
NRTL   (SPECIFIC PARAMETERS)   0.37
UNIQUAC (COMMON PARAMETERS)   2.07

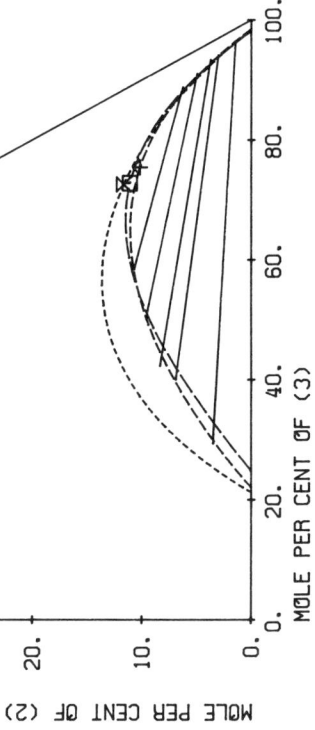

MOLE PER CENT OF (3)

MOLE PER CENT OF (2)

EXP.TIE LINE   UNIQ(SP) ——   NRTL(SP) ——   UNIQ(CO) ------
CALC.BINODAL
CALC.PLAIT P.

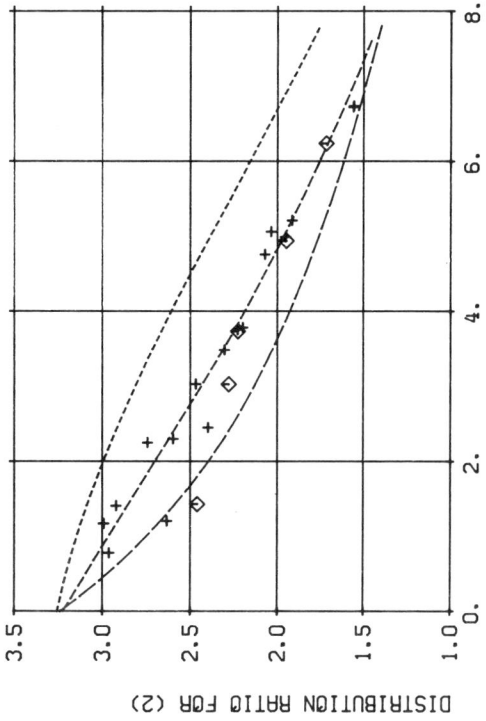

MOLE PER CENT OF (2) IN RIGHT PHASE

DISTRIBUTION RATIO FOR (2)

EXP. DISTR.RATIO   THIS REF ◊   OTHER REF +   UNIQ(CO) ------
CALC.DISTR.RATIO   UNIQ(SP) ——   NRTL(SP) ——

$C_2H_4O_2$-$C_5H_4O_2$

(1) C5H4O2   FURFURAL

(2) C2H4O2   ACETIC ACID

(3) H2O   WATER

HERIC E.L., RUTLEDGE R.M.
CAN.J.CHEM.ENG. 38(1960)46

TEMPERATURE = 25.0 DEG C   TYPE OF SYSTEM = 1

EXPERIMENTAL TIE LINES IN MOLE PCT

	LEFT PHASE			RIGHT PHASE	
(1)	(2)	(3)	(1)	(2)	(3)
67.622	3.491	28.887	1.937	1.167	96.896
58.683	5.966	35.351	2.458	2.297	95.245
49.809	8.031	42.160	3.228	3.480	93.292
39.787	9.740	50.472	4.441	4.987	90.572
28.138	10.498	61.364	6.873	6.721	86.407

SPECIFIC MODEL PARAMETERS IN KELVIN

		UNIQUAC		NRTL(ALPHA=.2)	
I J	AIJ	AJI		AIJ	AJI
1 2	342.99	-361.92		514.63	-757.65
1 3	149.81	110.29		28.810	1248.6
2 3	-127.49	-419.83		-250.83	-502.50

R1 = 3.1680   R2 = 2.2024   R3 = 0.9200
Q1 = 2.484   Q2 = 2.072   Q3 = 1.400

MEAN DEV. BETWEEN CALC. AND EXP. CONC. IN MOLE PCT

UNIQUAC (SPECIFIC PARAMETERS)   0.89
NRTL   (SPECIFIC PARAMETERS)   0.22
UNIQUAC (COMMON PARAMETERS)   1.86

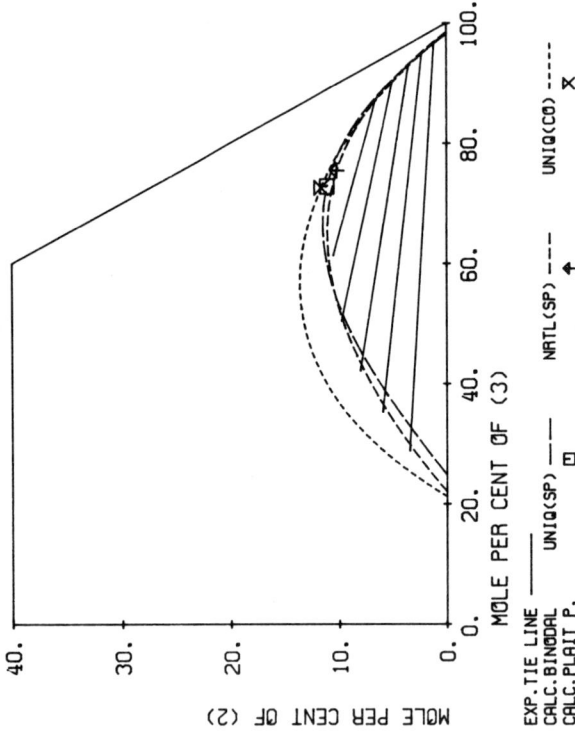

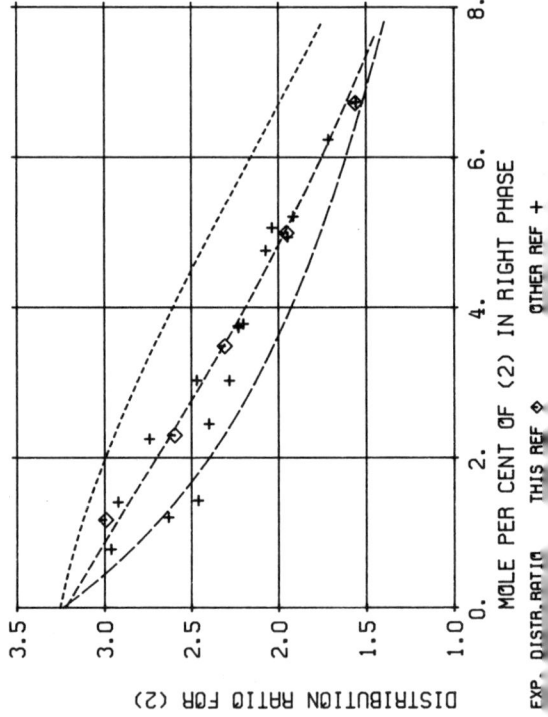

$C_2H_4O_2$-$C_5H_4O_2$

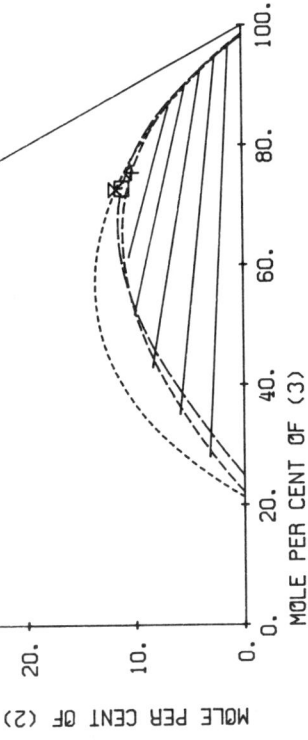

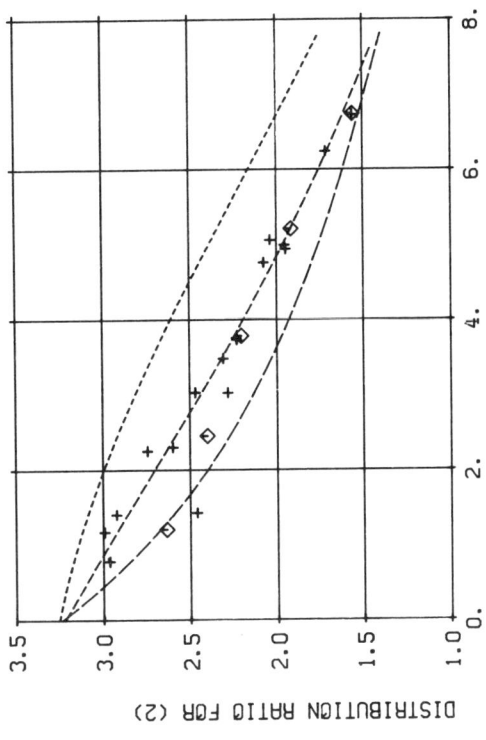

```
----|----------------------
(2) C2H4O2    ACETIC ACID
----|----------------------
(3) H2O       WATER
----|----------------------

HERIC E.L., RUTLEDGE R.M.
J.CHEM.ENG.DATA 5(1960)272

TEMPERATURE = 25.0 DEG C    TYPE OF SYSTEM = 1

EXPERIMENTAL TIE LINES IN MOLE PCT
----|----------------------------------
     LEFT PHASE              RIGHT PHASE
   (1)     (2)      (3)      (1)     (2)      (3)

 68.840   3.165   27.995   1.962   1.202   96.836
 59.022   5.875   35.103   2.514   2.449   95.037
 48.704   8.329   42.967   3.406   3.782   92.812
 38.270   9.975   51.755   4.695   5.207   90.098
 23.138  10.498   61.364   6.944   6.736   86.320

SPECIFIC MODEL PARAMETERS IN KELVIN
-----------------------------------------------
                 UNIQUAC              NRTL(ALPHA=.2)
 I J     AIJ       AJI          AIJ        AJI

 1 2   342.99    -361.92      514.63    -757.65
 1 3   149.81     110.29       28.810   1248.6
 2 3  -127.49    -419.83     -250.83    -502.50

R1 = 3.1680   R2 = 2.2024   R3 = 0.9200
Q1 = 2.484    Q2 = 2.072    Q3 = 1.400

MEAN DEV. BETWEEN CALC. AND EXP. CONC. IN MOLE PCT
--------------------------------------------------
UNIQUAC (SPECIFIC PARAMETERS)     0.91
NRTL    (SPECIFIC PARAMETERS)     0.27
UNIQUAC (COMMON PARAMETERS)       1.81
```

C$_2$H$_4$O$_2$-C$_5$H$_4$O$_2$

(1) C5H4O2   FURFURAL

(2) C2H4O2   ACETIC ACID

(3) H2O   WATER

HERIC E.L., RUTLEDGE R.M.
J.CHEM.ENG.DATA 5(1960)272

TEMPERATURE = 35.0 DEG C   TYPE OF SYSTEM = 1

EXPERIMENTAL TIE LINES IN MOLE PCT

	LEFT PHASE			RIGHT PHASE	
(1)	(2)	(3)	(1)	(2)	(3)
69.308	1.944	28.748	2.077	0.764	97.159
60.215	4.801	34.984	2.628	1.912	95.460
53.594	6.461	39.945	3.153	2.775	94.071
45.244	8.128	46.627	4.060	3.875	92.065
35.838	9.308	54.853	5.495	5.088	89.417

SPECIFIC MODEL PARAMETERS IN KELVIN

		UNIQUAC		NRTL(ALPHA=.2)	
I J	AIJ	AJI	AIJ	AJI	
1 2	-258.56	-1.6822	416.14	-665.47	
1 3	135.40	102.85	-22.801	1270.0	
2 3	-313.68	229.03	-205.88	-431.99	

R1 = 3.1680	R2 = 2.2024	R3 = 0.9200
Q1 = 2.484	Q2 = 2.072	Q3 = 1.400

MEAN DEV. BETWEEN CALC. AND EXP. CONC. IN MOLE PCT

UNIQUAC (SPECIFIC PARAMETERS)   0.78
NRTL   (SPECIFIC PARAMETERS)   0.43

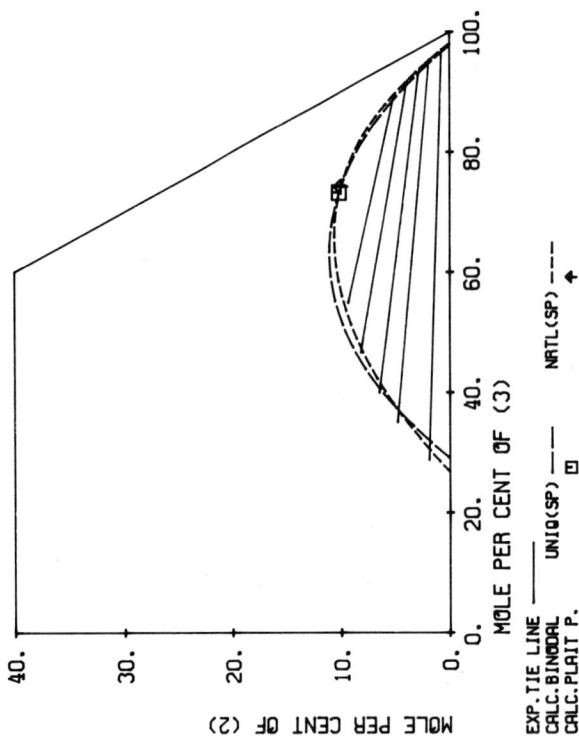

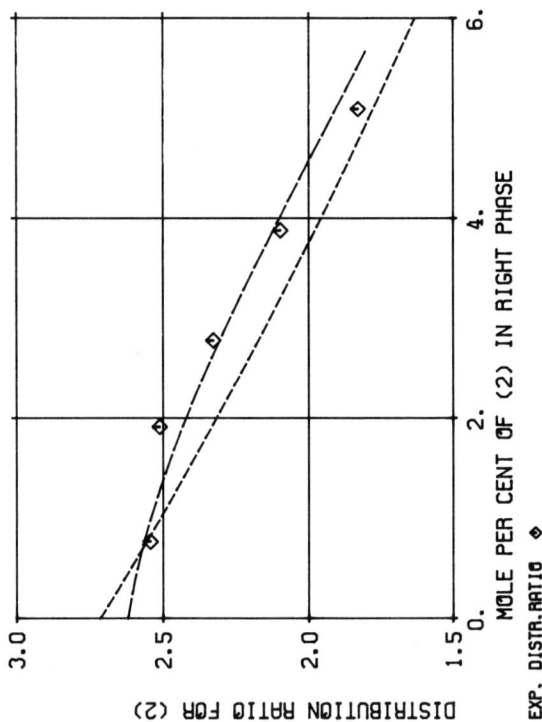

$C_2H_4O_2\text{-}C_5H_{10}O$

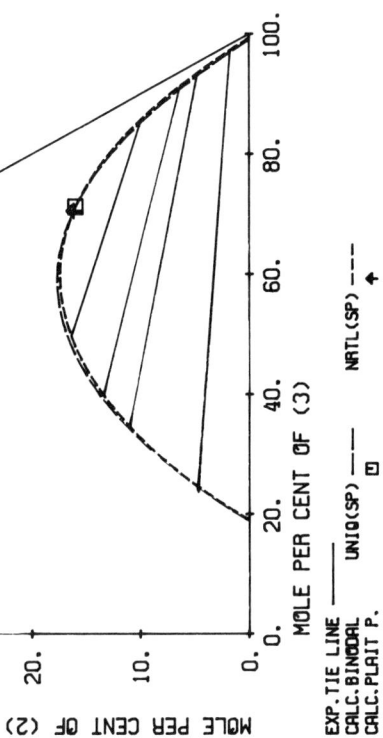

MOLE PER CENT OF (3)

MOLE PER CENT OF (2)

EXP.TIE LINE ——— UNIQ(SP) ⊡ NRTL(SP) ----
CALC.BINODAL
CALC.PLAIT P.   ↟

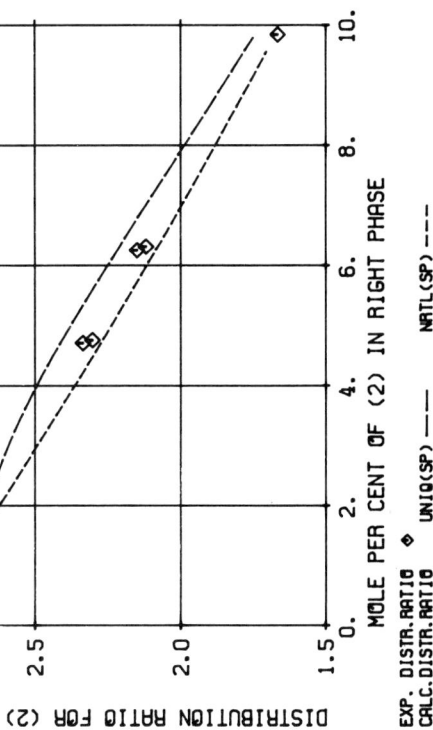

MOLE PER CENT OF (2) IN RIGHT PHASE

DISTRIBUTION RATIO FOR (2)

EXP. DISTR.RATIO   ◇   UNIQ(SP) ——— NRTL(SP) ----
CALC.DISTR.RATIO

---

(2) C2H4O2   ACETIC ACID
---
(3) H2O   WATER

HADDAD P.O., EDMISTER W.C.
J.CHEM.ENG.DATA 17(1972)275

TEMPERATURE = 70.0 DEG C   TYPE OF SYSTEM = 1

EXPERIMENTAL TIE LINES IN MOLE PCT

	LEFT PHASE			RIGHT PHASE	
(1)	(2)	(3)	(1)	(2)	(3)
71.192	4.770	24.038	1.180	1.735	97.085
71.611	4.684	23.705	1.180	1.745	97.074
54.406	10.996	34.598	1.798	4.707	93.495
54.555	10.962	34.483	1.826	4.758	93.416
46.686	13.440	39.874	2.267	6.252	91.482
46.784	13.376	39.841	2.297	6.313	91.390
33.868	16.405	49.728	4.146	9.841	86.013

SPECIFIC MODEL PARAMETERS IN KELVIN

		UNIQUAC		NRTL(ALPHA=.2)	
I J	AIJ	AJI		AIJ	AJI
1 2	-153.19	-85.952		444.69	-578.48
1 3	374.63	102.60		55.599	1648.4
2 3	-310.35	147.39		50.055	-300.88

R1 = 3.9223   R2 = 2.2024   R3 = 0.9200
Q1 = 3.416   Q2 = 2.072   Q3 = 1.400

MEAN DEV. BETWEEN CALC. AND EXP. CONC. IN MOLE PCT

UNIQUAC (SPECIFIC PARAMETERS)   0.26
NRTL   (SPECIFIC PARAMETERS)   0.18

C$_2$H$_4$O$_2$-C$_5$H$_{10}$O$_2$

(1) C5H10O2   ACETIC ACID,ISOPROPYL ESTER
(2) C2H4O2   ACETIC ACID
(3) H2O   WATER

HLAVATY K., LINEK J.
COLLECT.CZECH.CHEM.COMMUN. 38(1973)374

TEMPERATURE = 24.6 DEG C   TYPE OF SYSTEM = 1

EXPERIMENTAL TIE LINES IN MOLE PCT

	LEFT PHASE			RIGHT PHASE	
(1)	(2)	(3)	(1)	(2)	(3)
88.690	0.0	11.310	0.395	0.0	99.605
66.806	10.890	22.304	0.734	3.881	95.385
47.325	18.456	34.219	1.266	7.648	91.086
33.804	21.757	44.439	2.072	10.854	87.074
22.825	22.161	55.014	3.784	13.861	82.355
17.964	21.303	60.732	5.851	15.815	78.334

SPECIFIC MODEL PARAMETERS IN KELVIN

		UNIQUAC		NRTL(ALPHA=.2)	
I J		AIJ	AJI	AIJ	AJI
1 2		-193.47	-14.584	713.58	-561.27
1 3		-403.99	126.88	195.15	1601.4
2 3		-270.16	150.36	38.397	-11.587

R1 = 4.1522   R2 = 2.2024   R3 = 0.9200
Q1 = 3.652   Q2 = 2.072   Q3 = 1.400

MEAN DEV. BETWEEN CALC. AND EXP. CONC. IN MOLE PCT

UNIQUAC (SPECIFIC PARAMETERS)   0.26
NRTL   (SPECIFIC PARAMETERS)   0.28
UNIQUAC (COMMON PARAMETERS)   0.52

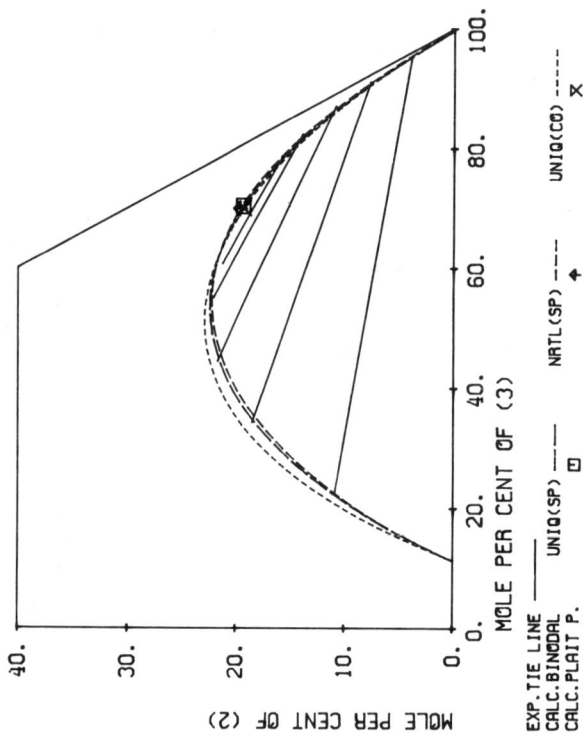

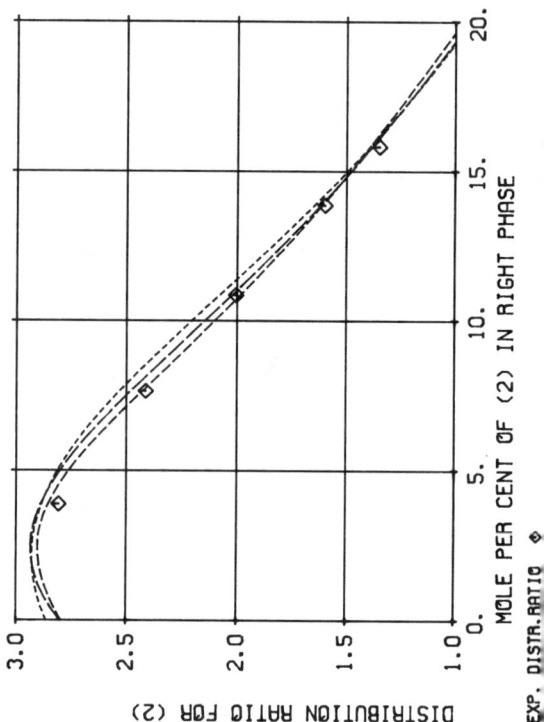

MOLE PER CENT OF (2)
MOLE PER CENT OF (3)

EXP.TIE LINE
CALC.BINODAL
CALC.PLAIT P.

UNIQ(SP)   NRTL(SP)   UNIQ(CO)

DISTRIBUTION RATIO FOR (2)
MOLE PER CENT OF (2) IN RIGHT PHASE
EXP. DISTR.RATIO

$C_2H_4O_2-C_5H_{10}O_2$

(2) C2H4O2   ACETIC ACID

(3) H2O   WATER

SITARAMA MURTY N.; SUBRAHMANYAM V., DAKSHINA MURTY P.
J.CHEM.ENG.DATA 11(1966)335

TEMPERATURE = 30.0 DEG C   TYPE OF SYSTEM = 1

EXPERIMENTAL TIE LINES IN MOLE PCT

	LEFT PHASE			RIGHT PHASE	
(1)	(2)	(3)	(1)	(2)	(3)
63.737	13.518	22.745	0.588	5.243	94.169
58.212	17.449	24.339	0.839	7.761	91.399
47.872	20.243	31.885	1.130	9.804	89.066
40.733	22.033	37.234	1.667	11.664	86.669
36.918	22.983	40.099	2.034	13.328	84.638
30.760	24.890	44.350	4.199	16.760	79.040

SPECIFIC MODEL PARAMETERS IN KELVIN

		UNIQUAC		NRTL(ALPHA=.2)	
I J	AIJ	AJI		AIJ	AJI
1 2	-0.23557	-194.93		260.18	-438.72
1 3	395.00	133.43		185.64	1757.6
2 3	-213.25	0.85389		29.117	-99.443

R1 = 4.1530	R2 = 2.2024	R3 = 0.9200
Q1 = 3.656	Q2 = 2.072	Q3 = 1.400

MEAN DEV. BETWEEN CALC. AND EXP. CONC. IN MOLE PCT

UNIQUAC (SPECIFIC PARAMETERS)	0.65
NRTL (SPECIFIC PARAMETERS)	0.57
UNIQUAC (COMMON PARAMETERS)	0.86

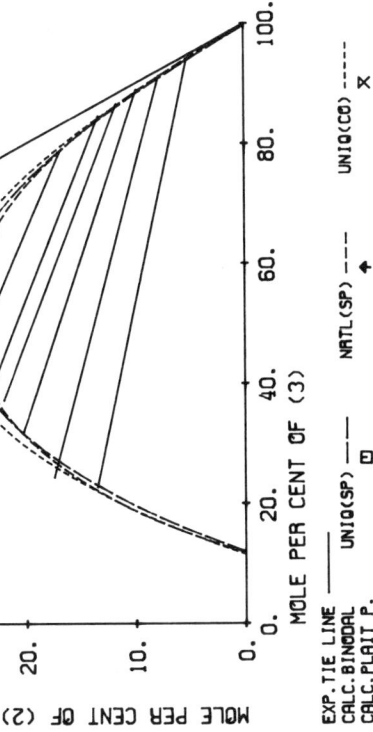

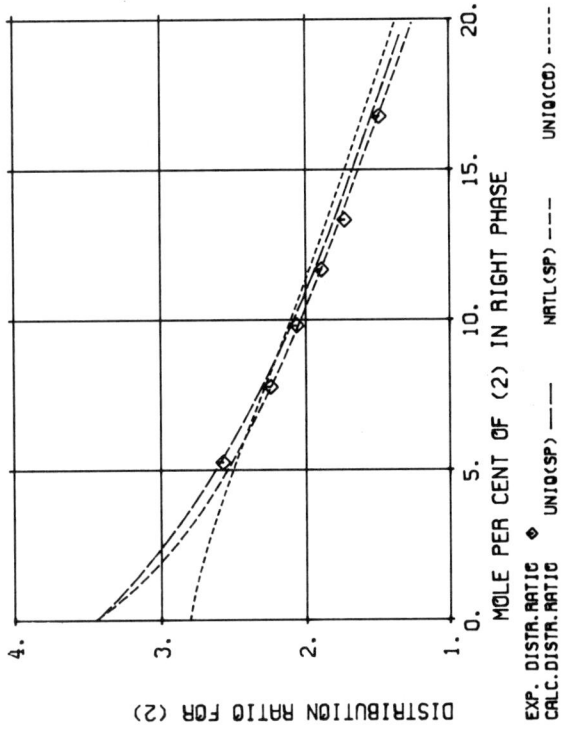

$C_2H_4O_2$-$C_5H_{10}O_2$

(1) C5H1002	PROPANOIC ACID, ETHYL ESTER
(2) C2H402	ACETIC ACID
(3) H20	WATER

JAYA RAMA RAO G., VENTAKA RAO C.
J.SCI.IND.RES. 14B(1955)444

TEMPERATURE = 28.0 DEG C   TYPE OF SYSTEM = 1

EXPERIMENTAL TIE LINES IN MOLE PCT

	LEFT PHASE			RIGHT PHASE	
(1)	(2)	(3)	(1)	(2)	(3)
83.799	4.224	11.977	0.476	1.758	97.766
72.496	10.384	17.121	0.615	4.150	95.235
62.882	15.385	21.733	0.887	6.447	92.666
49.444	20.235	30.322	1.302	9.281	89.416
38.234	23.550	38.216	2.167	12.166	85.666
27.822	24.373	47.805	3.762	15.368	80.870

SPECIFIC MODEL PARAMETERS IN KELVIN

		UNIQUAC		NRTL(ALPHA=.2)	
I	J	AIJ	AJI	AIJ	AJI
1	2	-163.95	3.7835	679.69	-511.28
1	3	-465.57	126.76	265.37	1605.3
2	3	-247.27	122.02	73.128	11.787

R1 = 4.1530   R2 = 2.2024   R3 = 0.9200
Q1 = 3.656   Q2 = 2.072   Q3 = 1.400

MEAN DEV. BETWEEN CALC. AND EXP. CONC. IN MOLE PCT

UNIQUAC (SPECIFIC PARAMETERS)	0.26
NRTL (SPECIFIC PARAMETERS)	0.21
UNIQUAC (COMMON PARAMETERS)	0.51

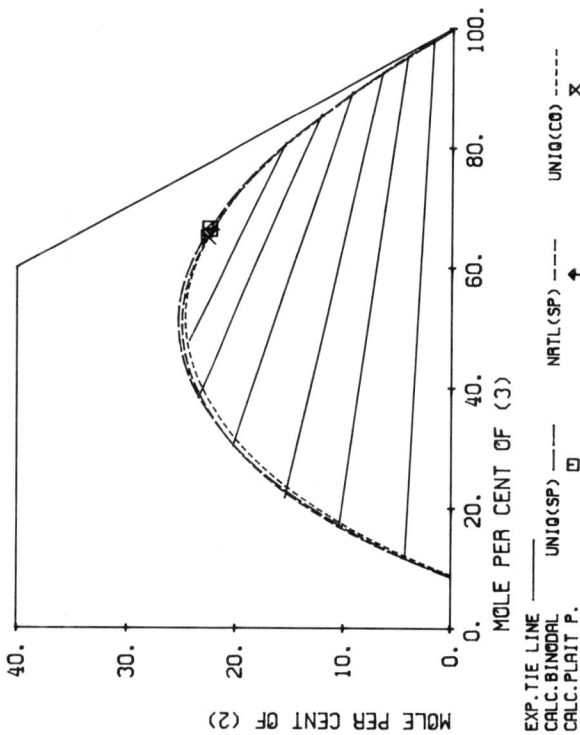

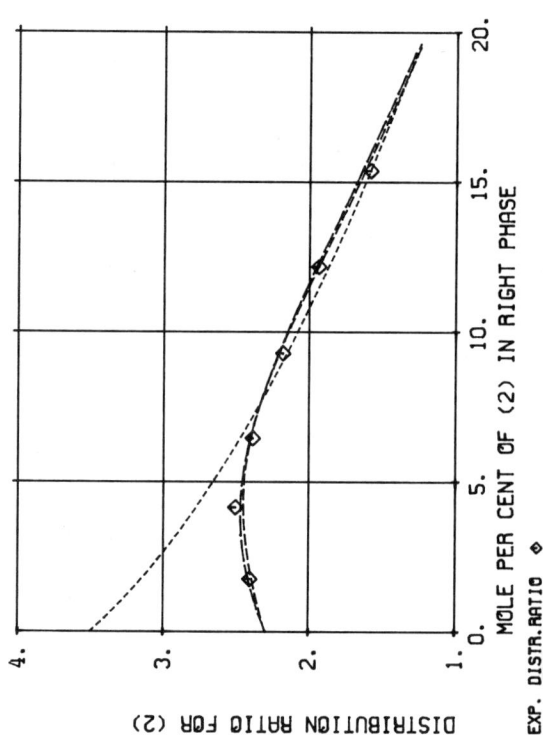

EXP.TIE LINE ———
CALC.BINODAL - - -
CALC.PLAIT P.

UNIQ(SP) ⊡   NRTL(SP) ✦   UNIQ(CO) - - - ✗

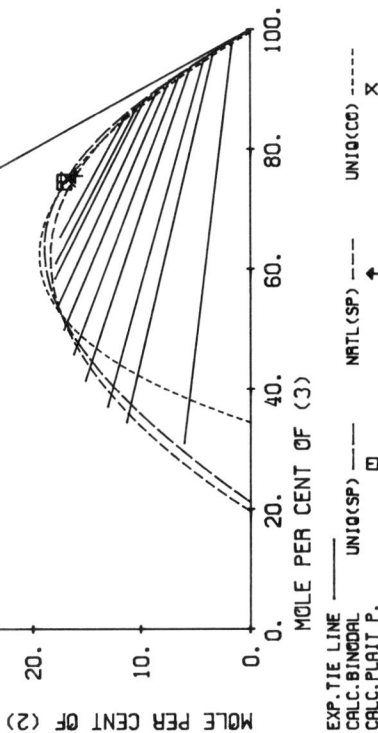

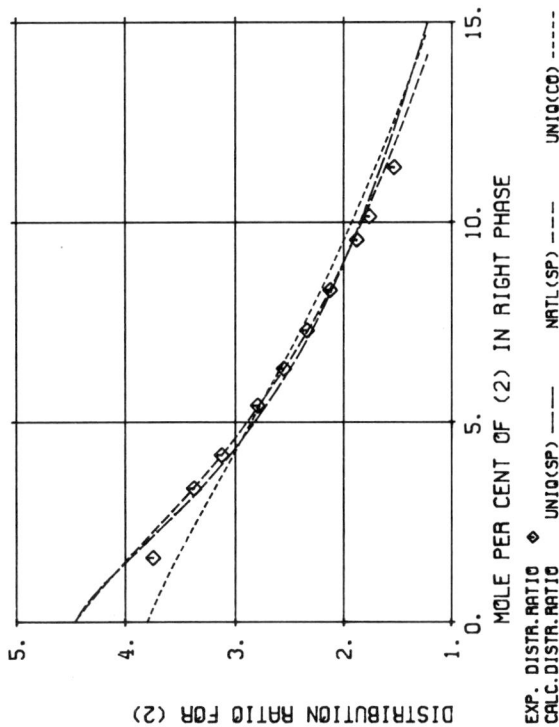

(2) C2H4O2   ACETIC ACID

(3) H2O   WATER

FUSE K., IGUCHI A.
KAGAKU KOGAKU 35(1971)801

TEMPERATURE = 25.0 DEG C   TYPE OF SYSTEM = 1

EXPERIMENTAL TIE LINES IN MOLE PCT (GRAPH.INTERPOL.)

	LEFT PHASE			RIGHT PHASE	
(1)	(2)	(3)	(1)	(2)	(3)
63.070	6.008	30.922	0.519	1.606	97.876
54.286	11.274	34.440	0.631	3.341	96.029
49.933	13.038	37.029	0.689	4.172	95.139
43.491	15.093	41.416	0.804	5.407	93.789
38.061	16.187	45.752	0.919	6.335	92.746
32.991	17.098	49.910	1.090	7.290	91.620
28.565	17.668	53.768	1.403	8.294	90.303
23.633	17.914	58.452	1.993	9.550	88.456
21.185	17.852	60.963	2.333	10.139	87.528
17.180	17.436	65.384	3.179	11.363	85.458

SPECIFIC MODEL PARAMETERS IN KELVIN

		UNIQUAC		NRTL(ALPHA=.2)	
I J	AIJ	AJI		AIJ	AJI
1 2	384.62	-318.92		610.35	-501.10
1 3	182.33	249.91		27.313	1807.9
2 3	17.453	-317.31		-66.424	-69.290

R1 = 4.1279   R2 = 2.2024   R3 = 0.9200
Q1 = 3.588   Q2 = 2.072   Q3 = 1.400

MEAN DEV. BETWEEN CALC. AND EXP. CONC. IN MOLE PCT

UNIQUAC (SPECIFIC PARAMETERS)   0.75
NRTL   (SPECIFIC PARAMETERS)   0.51
UNIQUAC (COMMON PARAMETERS)   1.69

C$_2$H$_4$O$_2$-C$_6$H$_5$Cl

(1)	H2O	WATER
(2)	C2H4O2	ACETIC ACID
(3)	C6H5CL	BENZENE,CHLORO

PEAKE J.S., THOMPSON K.E.
IND.ENG.CHEM. 44(1952)2439

TEMPERATURE = 25.0 DEG C   TYPE OF SYSTEM = 1

EXPERIMENTAL TIE LINES IN MOLE PCT (GRAPH.INTERPOL.)

| LEFT PHASE | | | RIGHT PHASE | | |
(1)	(2)	(3)	(1)	(2)	(3)
94.376	5.606	0.018	0.359	1.337	98.304
39.314	10.643	0.042	0.435	3.165	96.400
85.062	14.829	0.108	0.539	5.685	93.775
80.978	18.822	0.200	0.635	7.761	91.604
73.853	25.701	0.447	0.895	13.258	85.847
70.521	28.845	0.634	0.964	14.581	84.455
65.040	33.863	1.097	1.319	20.177	78.504
56.102	41.581	2.317	2.093	27.407	70.500

SPECIFIC MODEL PARAMETERS IN KELVIN

| | UNIQUAC | | NRTL(ALPHA=.2) | |
I J	AIJ	AJI	AIJ	AJI
1 2	-258.85	-117.14	-391.95	-216.63
1 3	488.62	813.20	1553.0	1145.1
2 3	44.178	-95.440	632.92	-577.80

R1 = 0.9200   R2 = 2.2024   R3 = 3.8127
Q1 = 1.400   Q2 = 2.072   Q3 = 2.844

MEAN DEV. BETWEEN CALC. AND EXP. CONC. IN MOLE PCT

UNIQUAC (SPECIFIC PARAMETERS)	0.48
NRTL (SPECIFIC PARAMETERS)	0.31
UNIQUAC (COMMON PARAMETERS)	0.64

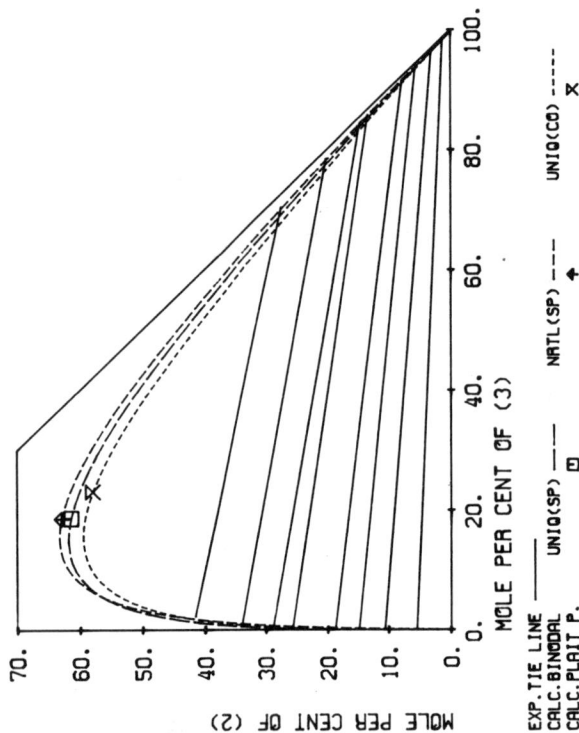

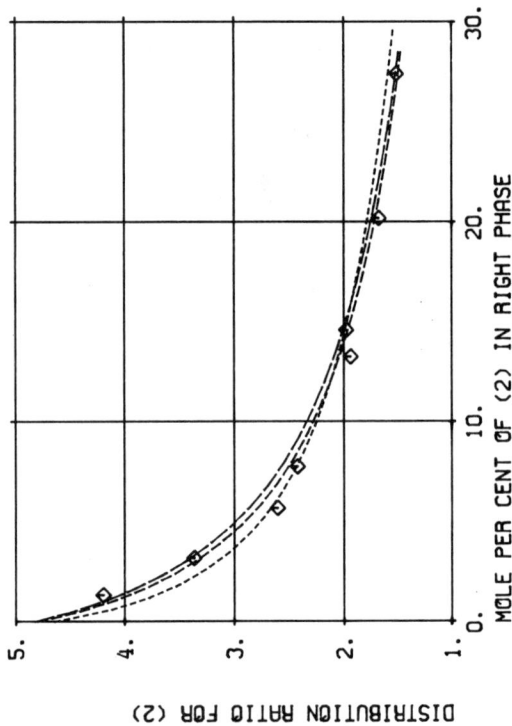

$C_2H_4O_2-C_6H_6$

MOLE PER CENT OF (3)

MOLE PER CENT OF (2)

EXP.TIE LINE ——— UNIQ(SP) ☐ NRTL(SP) ♦ UNIQ(CO) -----
CALC.BINODAL
CALC.PLAIT P. ⊠

MOLE PER CENT OF (2) IN RIGHT PHASE

DISTRIBUTION RATIO FOR (2)

EXP. DISTR.RATIO THIS REF ◇ OTHER REF + UNIQ(CO) -----
CALC.DISTR.RATIO UNIQ(SP) ——— NRTL(SP) ---

---

(2) C2H4O2   ACETIC ACID
(3) C6H6   BENZENE

WADDELL J.
J.PHYS.CHEM. 2(1898)233

TEMPERATURE = 25.0 DEG C   TYPE OF SYSTEM = 1

EXPERIMENTAL TIE LINES IN MOLE PCT

	LEFT PHASE			RIGHT PHASE	
(1)	(2)	(3)	(1)	(2)	(3)
96.932	3.023	0.045	0.087	0.597	99.316
94.307	5.612	0.081	0.259	1.681	98.060
89.252	10.595	0.153	0.641	3.976	95.383
84.234	15.501	0.265	1.059	6.606	92.335
80.157	19.335	0.508	1.428	8.821	89.751
76.988	22.365	0.648	1.751	10.881	87.368
72.659	26.307	1.034	2.109	13.026	84.865
62.109	35.202	2.688	3.186	19.718	77.096
55.054	40.505	4.441	4.102	25.441	70.457
53.412	41.661	4.927	4.316	26.835	68.850
46.778	45.852	7.370	7.943	34.121	57.936
42.152	47.788	10.060	8.966	36.158	54.876
30.823	50.438	18.739	18.303	45.594	36.103

SPECIFIC MODEL PARAMETERS IN KELVIN

		UNIQUAC		NRTL(ALPHA=.2)	
I J	AIJ	AJI		AIJ	AJI
1 2	-256.09	-129.95		-88.738	-325.86
1 3	-596.58	703.31		1804.8	924.96
2 3	-69.726	0.82389E-01		188.14	-154.94

R1 = 0.9200   R2 = 2.2024   R3 = 3.1878
Q1 = 1.400   Q2 = 2.072   Q3 = 2.400

MEAN DEV. BETWEEN CALC. AND EXP. CONC. IN MOLE PCT

UNIQUAC (SPECIFIC PARAMETERS)   0.61
NRTL   (SPECIFIC PARAMETERS)   0.62
UNIQUAC (COMMON PARAMETERS)   0.66

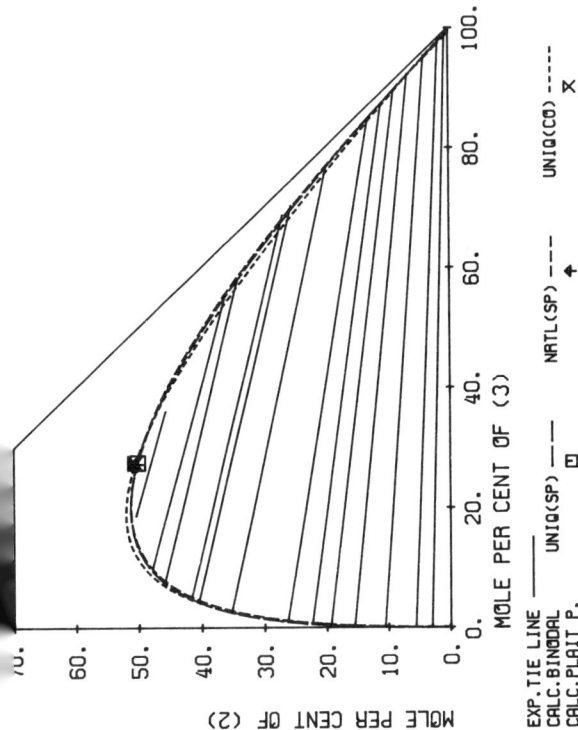

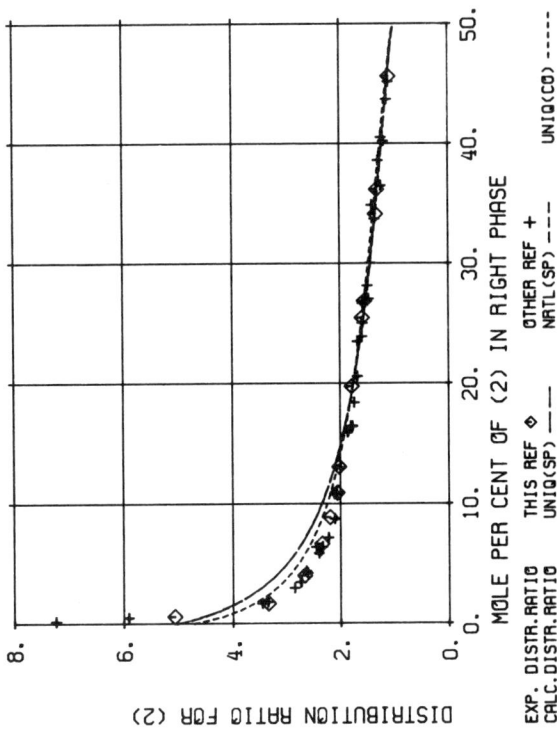

C$_2$H$_4$O$_2$-C$_6$H$_6$

MOLE PER CENT OF (3)

MOLE PER CENT OF (2)

EXP.TIE LINE ———

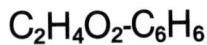

(1) H2O   WATER
(2) C2H4O2   ACETIC ACID
(3) C6H6   BENZENE

WADDELL J.
J.PHYS.CHEM. 2(1898)233

TEMPERATURE = 35.0 DEG C   TYPE OF SYSTEM = 1

EXPERIMENTAL TIE LINES IN MOLE PCT

LEFT PHASE			RIGHT PHASE		
(1)	(2)	(3)	(1)	(2)	(3)
94.249	5.588	0.162	0.345	1.603	98.052
76.306	22.939	0.756	2.404	11.189	86.408
35.228	49.079	15.694	19.481	43.832	36.688

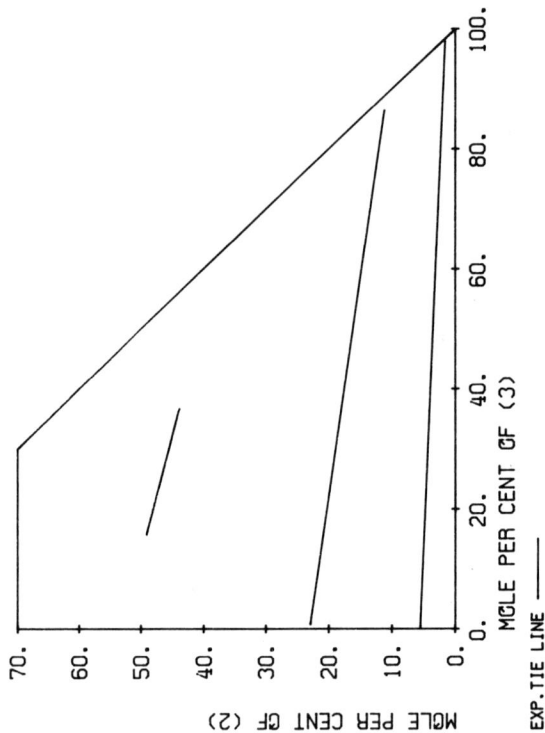

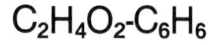

C$_2$H$_4$O$_2$-C$_6$H$_6$

MOLE PER CENT OF (3)

MOLE PER CENT OF (2)

EXP.TIE LINE ———	UNIQ(SP) ———	NRTL(SP) ———	UNIQ(CC) -----
CALC.BINODAL	□	✦	✕
CALC.PLAIT P.			

MOLE PER CENT OF (2) IN RIGHT PHASE

DISTRIBUTION RATIO FOR (2)

	THIS REF ◇	OTHER REF +	
EXP. DISTR.RATIO	UNIQ(SP) ———	NRTL(SP) ———	UNIQ(CC) -----
CALC.DISTR.RATIO			

---

(2) C2H4O2   ACETIC ACID

(3) C6H6   BENZENE

HAND D.B.
J.PHYS.CHEM. 34(1930)1961

TEMPERATURE = 25.0 DEG C   TYPE OF SYSTEM = 1

EXPERIMENTAL TIE LINES IN MOLE PCT

| | LEFT PHASE | | | RIGHT PHASE | |
(1)	(2)	(3)	(1)	(2)	(3)
98.577	1.414	0.010	0.004	0.195	99.801
93.876	6.072	0.053	0.173	1.811	98.016
88.926	10.958	0.116	0.471	4.197	95.332
69.060	29.619	1.320	1.647	16.426	81.928
66.327	32.010	1.663	2.042	18.378	79.580
58.883	38.135	2.982	2.802	23.899	73.299
56.450	39.905	3.645	3.360	27.039	69.601
40.236	49.333	10.431	7.123	34.866	58.011
37.290	50.109	12.601	9.113	38.602	52.285
35.091	50.563	14.345	10.717	40.512	48.771
30.219	50.243	19.538	15.487	45.149	39.364

SPECIFIC MODEL PARAMETERS IN KELVIN

| | UNIQUAC | | | NRTL(ALPHA=.2) | |
I J	AIJ	AJI		AIJ	AJI
1 2	-266.09	-129.95		-88.738	-325.86
1 3	596.58	703.31		1804.8	924.96
2 3	-69.726	0.82339E-01		188.14	-154.94

R1 = 0.9200   R2 = 2.2024   R3 = 3.1878
Q1 = 1.400   Q2 = 2.072   Q3 = 2.400

MEAN DEV. BETWEEN CALC. AND EXP. CONC. IN MOLE PCT

UNIQUAC (SPECIFIC PARAMETERS)   1.04
NRTL   (SPECIFIC PARAMETERS)   1.07
UNIQUAC (COMMON PARAMETERS)   1.10

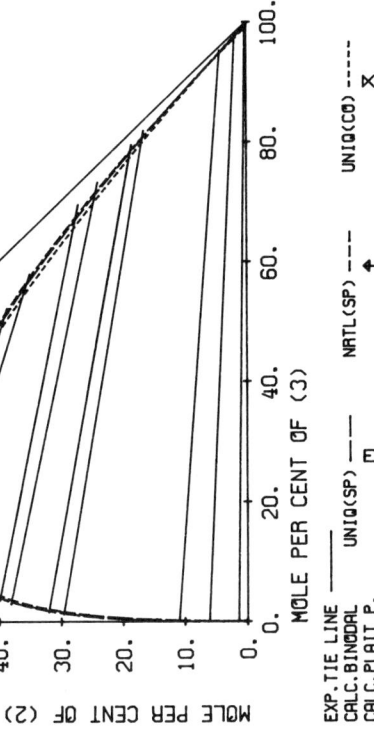

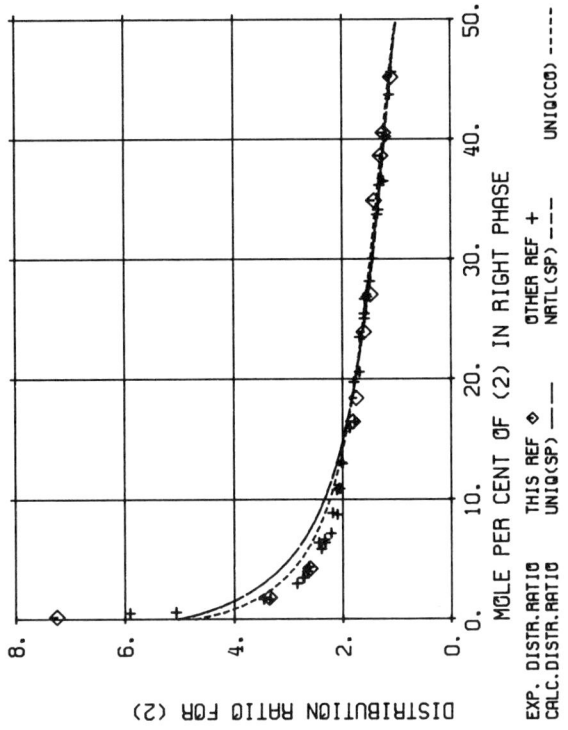

C₂H₄O₂-C₆H₆

(1) H2O   WATER
(2) C2H4O2   ACETIC ACID
(3) C6H6   BENZENE

GARNER F.H., ELLIS S.R.M., ROY U.N.G.
CHEM.ENG.SCI. 2(1953)14

TEMPERATURE = 30.0 DEG C   TYPE OF SYSTEM = 1

EXPERIMENTAL TIE LINES IN MOLE PCT (GRAPH.INTERPOL.)

	LEFT PHASE			RIGHT PHASE	
(1)	(2)	(3)	(1)	(2)	(3)
91.474	8.424	0.102	1.025	2.943	96.026
76.464	22.790	0.747	2.933	10.780	86.287
57.270	39.358	3.372	6.192	23.482	70.326

SPECIFIC MODEL PARAMETERS IN KELVIN

		UNIQUAC		NRTL(ALPHA=.2)	
I	J	AIJ	AJI	AIJ	AJI
1	2	-266.09	-129.95	-88.738	-325.86
1	3	596.58	703.31	1804.8	924.96
2	3	-69.726	0.82389E-01	188.14	-154.94

R1 = 0.9200   R2 = 2.2024   R3 = 3.1878
Q1 = 1.400   Q2 = 2.072   Q3 = 2.400

MEAN DEV. BETWEEN CALC. AND EXP. CONC. IN MOLE PCT

UNIQUAC (SPECIFIC PARAMETERS)   0.75
NRTL   (SPECIFIC PARAMETERS)   0.85
UNIQUAC (COMMON PARAMETERS)   0.43

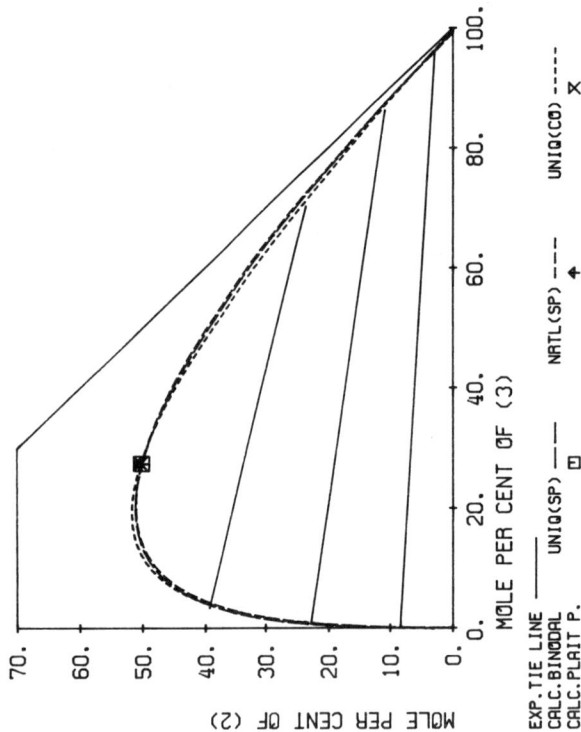

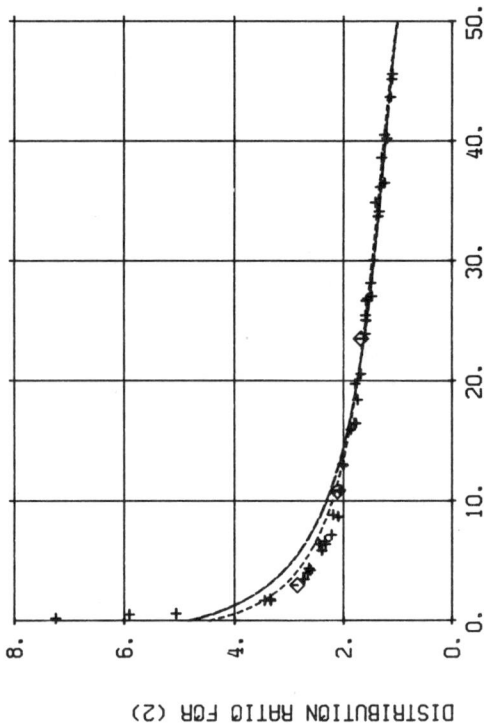

MOLE PER CENT OF (3)

MOLE PER CENT OF (2)

EXP.TIE LINE

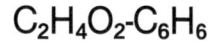

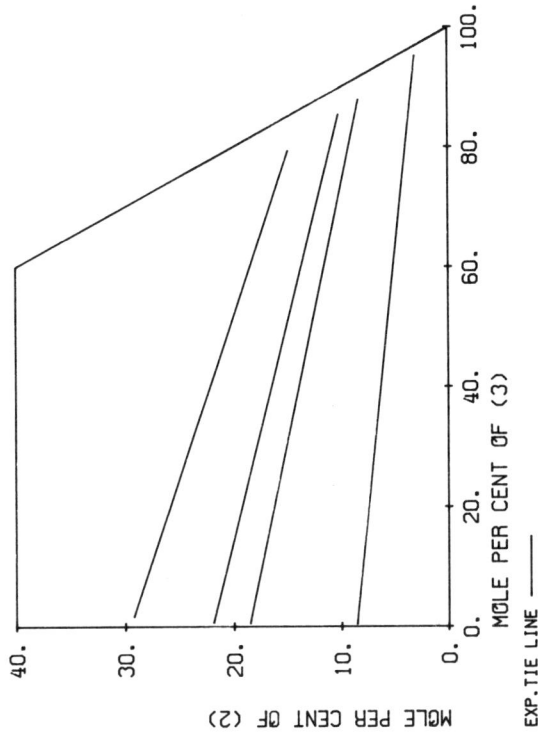

(1) H2O   WATER
(2) C2H4O2   ACETIC ACID
(3) C6H6   BENZENE

GARNER F.H., ELLIS S.R.M., ROY U.N.G.
CHEM.ENG.SCI. 2(1953)14

TEMPERATURE = 40.0 DEG C   TYPE OF SYSTEM = 1

EXPERIMENTAL TIE LINES IN MOLE PCT (GRAPH.INTERPOL.)

LEFT PHASE			RIGHT PHASE		
(1)	(2)	(3)	(1)	(2)	(3)
91.024	8.556	0.420	1.741	3.058	95.200
80.910	18.504	0.586	3.836	8.292	87.872
77.372	21.880	0.747	4.493	10.170	85.337
69.054	29.222	1.725	5.791	14.858	79.350

$C_2H_4O_2$-C_6H_6

MOLE PER CENT OF (3)

EXP.TIE LINE ⎯

MOLE PER CENT OF (2)

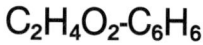

(1) H2O   WATER
(2) C2H4O2   ACETIC ACID
(3) C6H6   BENZENE

GARNER F.H., ELLIS S.R.M., ROY U.N.G.
CHEM.ENG.SCI. 2(1953)14

TEMPERATURE = 50.0 DEG C   TYPE OF SYSTEM = 1

EXPERIMENTAL TIE LINES IN MOLE PCT (GRAPH.INTERPOL.)

LEFT PHASE			RIGHT PHASE		
(1)	(2)	(3)	(1)	(2)	(3)
96.013	3.651	0.336	0.903	0.799	98.298
89.834	9.549	0.616	2.446	3.670	93.884
76.418	22.687	0.895	6.340	11.195	82.465
66.670	30.920	2.410	8.900	16.948	74.152

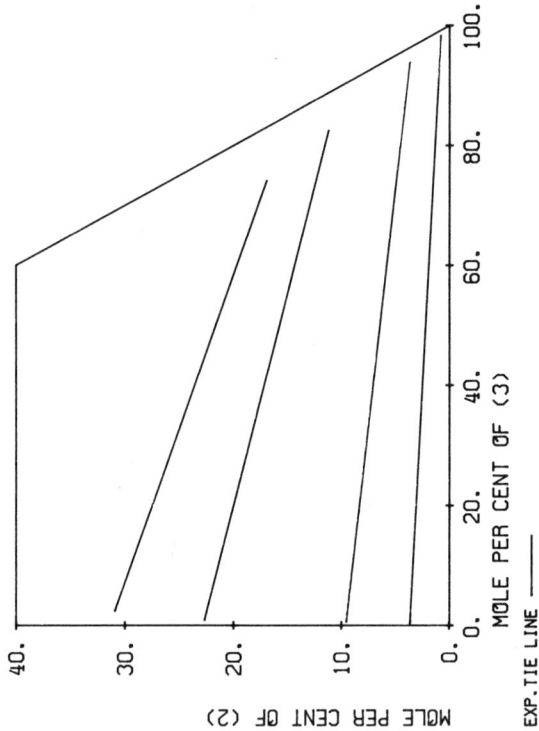

$C_2H_4O_2-C_6H_6$

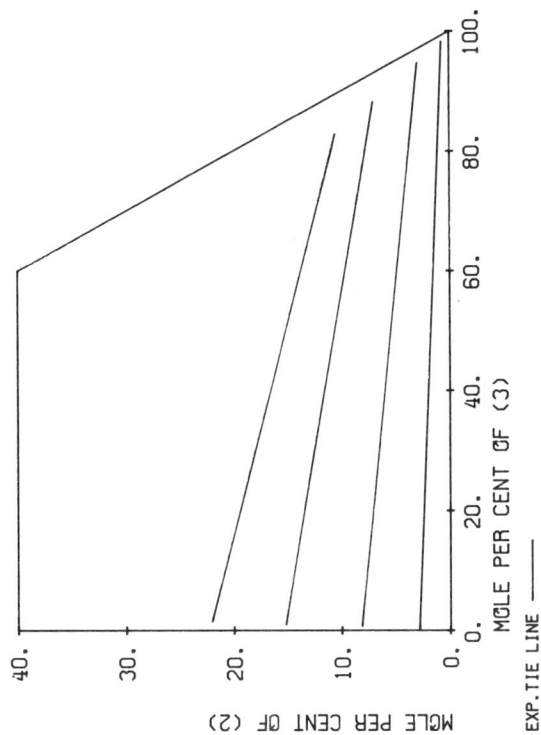

MOLE PER CENT OF (3)

MOLE PER CENT OF (2)

EXP.TIE LINE ———

(1) H2O   WATER
(2) C2H4O2   ACETIC ACID
(3) C6H6   BENZENE

GARNER F.H., ELLIS S.R.M., ROY U.N.G.
CHEM.ENG.SCI. 2(1953)14

TEMPERATURE = 60.0 DEG C   TYPE OF SYSTEM = 1

EXPERIMENTAL TIE LINES IN MOLE PCT (GRAPH.INTERPOL.)

LEFT PHASE			RIGHT PHASE		
(1)	(2)	(3)	(1)	(2)	(3)
96.806	2.846	0.348	0.988	0.722	98.290
91.043	8.177	0.780	2.326	2.969	94.705
83.759	15.213	1.028	4.764	7.023	88.213
76.504	22.037	1.460	6.655	10.523	82.822

$C_2H_4O_2-C_6H_6$

(1) H2O   WATER
(2) C2H4O2   ACETIC ACID
(3) C6H6   BENZENE

JODRA L.G., OTERO J.L., SOLE J.
AN.R.SOC.ESP.FIS.QUIM. 51B(1955)741

TEMPERATURE = 25.0 DEG C   TYPE OF SYSTEM = 1

EXPERIMENTAL TIE LINES IN MOLE PCT

LEFT PHASE			RIGHT PHASE		
(1)	(2)	(3)	(1)	(2)	(3)
96.865	3.061	0.074	0.432	0.518	99.050
94.062	5.781	0.158	0.358	1.673	97.468
88.283	11.482	0.235	0.353	4.350	94.798
85.641	14.050	0.309	1.271	5.845	92.885
81.272	18.329	0.399	1.262	8.708	90.030
76.841	22.588	0.570	1.669	10.888	87.444
73.202	25.931	0.867	1.660	12.951	85.388
69.133	29.628	1.239	2.054	15.899	82.046
63.286	34.592	2.122	2.825	20.582	76.593
56.474	39.737	3.789	3.953	25.025	71.021
47.044	45.937	7.020	7.144	33.725	59.131
41.784	48.297	9.919	11.372	40.193	48.434

SPECIFIC MODEL PARAMETERS IN KELVIN

		UNIQUAC		NRTL(ALPHA=.2)	
I	J	AIJ	AJI	AIJ	AJI
1	2	-266.09	-129.95	-88.738	-325.86
1	3	596.58	703.31	1804.8	924.96
2	3	-69.726	0.82389E-01	188.14	-154.94

R1 = 0.9200   R2 = 2.024   R3 = 3.1878
Q1 = 1.400   Q2 = 2.072   Q3 = 2.400

MEAN DEV. BETWEEN CALC. AND EXP. CONC. IN MOLE PCT

UNIQUAC (SPECIFIC PARAMETERS)   0.66
NRTL   (SPECIFIC PARAMETERS)   0.68
UNIQUAC (COMMON PARAMETERS)   0.71

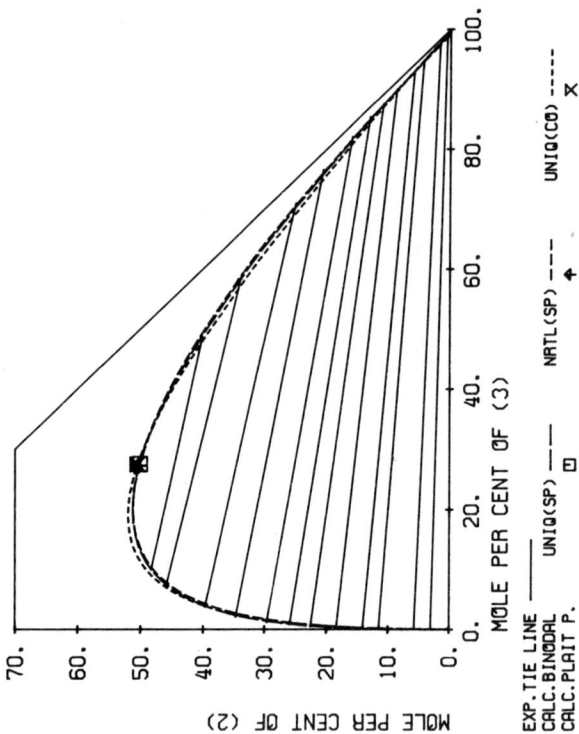

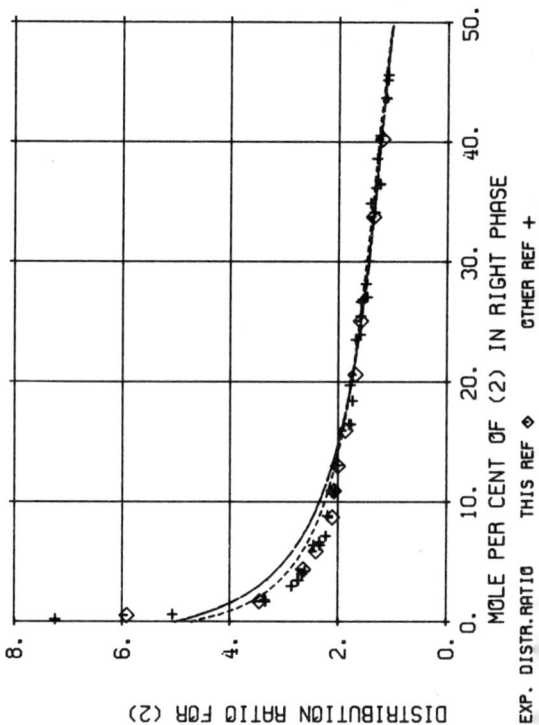

EXP.TIE LINE ———
CALC.BINODAL - - - -
CALC.PLAIT P.

UNIQ(SP) ⊟   NRTL(SP) ✦   UNIQ(CO) - - - - ✕

MOLE PER CENT OF (3)

MOLE PER CENT OF (2)

DISTRIBUTION RATIO FOR (2)

MOLE PER CENT OF (2) IN RIGHT PHASE

EXP. DISTR.RATIO   THIS REF ◇   OTHER REF +

$C_2H_4O_2\text{-}C_6H_6$

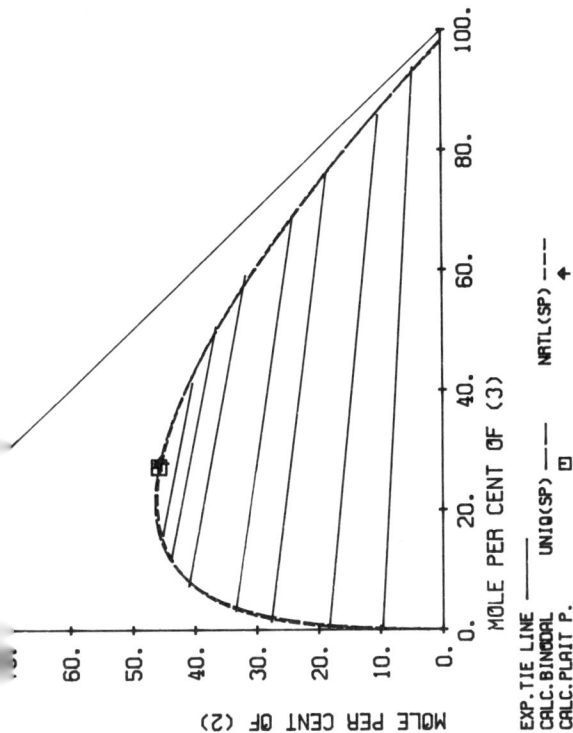

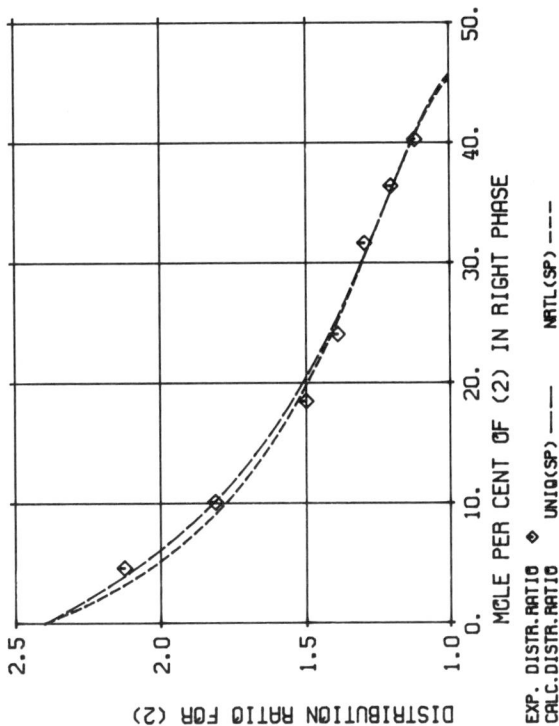

| (2) C2H4O2 | ACETIC ACID |
| (3) C6H6 | BENZENE |

TAGLIAVINI G., ARICH G., BIANCANI M.
ANN.CHIM.(ROME) 45(1955)292

TEMPERATURE = 60.0 DEG C   TYPE OF SYSTEM = 1

EXPERIMENTAL TIE LINES IN MOLE PCT

	LEFT PHASE			RIGHT PHASE	
(1)	(2)	(3)	(1)	(2)	(3)
90.142	9.716	0.142	1.694	4.573	93.734
81.479	18.290	0.232	4.098	10.082	85.819
70.921	27.674	1.405	5.553	18.443	76.005
63.380	33.444	3.175	7.317	24.031	68.651
51.704	40.989	7.306	9.275	31.608	59.117
44.276	43.840	11.884	13.080	36.376	50.544
38.994	45.166	15.841	18.556	40.259	41.186

SPECIFIC MODEL PARAMETERS IN KELVIN

		UNIQUAC		NRTL(ALPHA=.2)	
I J	AIJ	AJI		AIJ	AJI
1 2	-166.24	-105.19		40.441	-221.33
1 3	-447.58	558.39		2517.2	814.26
2 3	-94.433	99.974		71.044	56.883

R1 = 0.9200   R2 = 2.2024   R3 = 3.1878
Q1 = 1.400   Q2 = 2.072   Q3 = 2.400

MEAN DEV. BETWEEN CALC. AND EXP. CONC. IN MOLE PCT

UNIQUAC (SPECIFIC PARAMETERS)   0.62
NRTL (SPECIFIC PARAMETERS)   0.54

$C_2H_4O_2$-C_6H_6

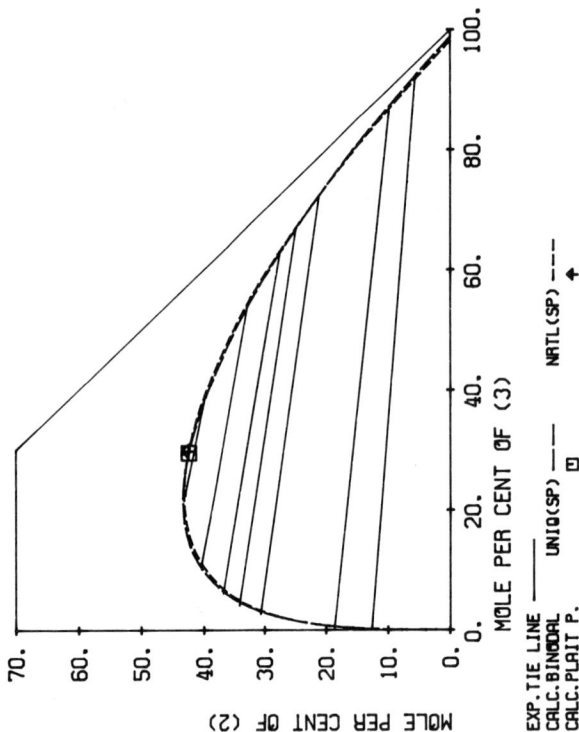

MOLE PER CENT OF (3)

MOLE PER CENT OF (2)

EXP.TIE LINE ———   UNIQ(SP) — ▣
CALC.BINODAL
CALC.PLAIT P.   NRTL(SP) - - - ◆

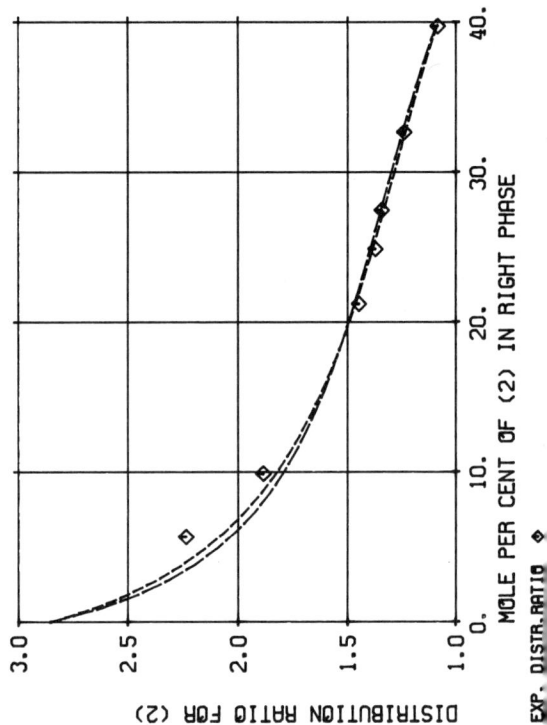

MOLE PER CENT OF (2) IN RIGHT PHASE

DISTRIBUTION RATIO FOR (2)

EXP. DISTR.RATIO ◆

(1) H2O	WATER	
(2) C2H4O2	ACETIC ACID	
(3) C6H6	BENZENE	

TAGLIAVINI G., ARICH G., BIANCANI M.
ANN.CHIM.(ROME) 45(1955)292

TEMPERATURE = 70.0 DEG C   TYPE OF SYSTEM = 1

EXPERIMENTAL TIE LINES IN MOLE PCT

	LEFT PHASE			RIGHT PHASE	
(1)	(2)	(3)	(1)	(2)	(3)
87.130	12.659	0.210	2.517	5.664	91.819
81.014	18.619	0.367	3.702	9.873	86.424
66.632	30.661	2.707	6.534	21.191	72.175
61.822	34.079	4.099	8.021	24.864	67.115
57.111	36.803	6.086	9.371	27.439	63.190
48.813	40.409	10.779	12.885	32.641	54.474
34.637	43.049	22.314	21.036	39.709	39.255

SPECIFIC MODEL PARAMETERS IN KELVIN

		UNIQUAC		NRTL(ALPHA=.2)	
I	J	AIJ	AJI	AIJ	AJI
1	2	-754.90	-144.43	-399.95	-199.58
1	3	825.64	581.32	2904.3	885.89
2	3	-196.71	-351.26	-8.9633	-195.11

R1 = 0.9200 R2 = 2.2024 R3 = 3.1878
Q1 = 1.400   Q2 = 2.072   Q3 = 2.400

MEAN DEV. BETWEEN CALC. AND EXP. CONC. IN MOLE PCT

UNIQUAC (SPECIFIC PARAMETERS)   0.42
NRTL   (SPECIFIC PARAMETERS)   0.36

(2) C2H4O2   ACETIC ACID

(3) C6H6   BENZENE

TAGLIAVINI G., ARICH G., BIANCANI M.
ANN.CHIM.(ROME) 45(1955)292

TEMPERATURE = 80.0 DEG C   TYPE OF SYSTEM = 1

EXPERIMENTAL TIE LINES IN MOLE PCT

	LEFT PHASE			RIGHT PHASE	
(1)	(2)	(3)	(1)	(2)	(3)
90.293	9.508	0.198	2.937	4.279	92.784
80.844	18.452	0.704	4.879	10.734	84.387
76.309	22.282	1.409	6.009	12.979	81.013
66.127	30.250	3.623	7.733	21.112	71.155
66.261	34.254	5.485	10.791	25.452	63.757
54.936	37.306	7.758	13.634	29.277	57.089
45.984	40.595	13.421	19.916	35.152	44.932

SPECIFIC MODEL PARAMETERS IN KELVIN

		UNIQUAC		NRTL(ALPHA=.2)	
I	J	AIJ	AJI	AIJ	AJI
1	2	-103.27	-177.80	550.11	-586.49
1	3	325.26	546.15	2129.4	725.37
2	3	-83.302	95.250	129.76	43.847

R1 = 0.9200   R2 = 2.2024   R3 = 3.1878
Q1 = 1.400   Q2 = 2.072   Q3 = 2.400

MEAN DEV. BETWEEN CALC. AND EXP. CONC. IN MOLE PCT

UNIQUAC (SPECIFIC PARAMETERS)   0.43
NRTL   (SPECIFIC PARAMETERS)   0.28

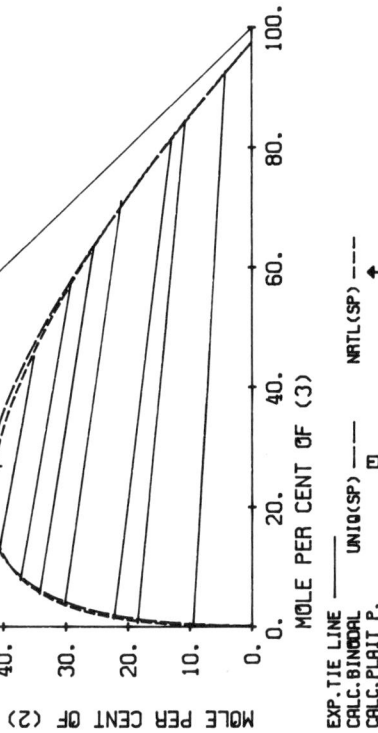

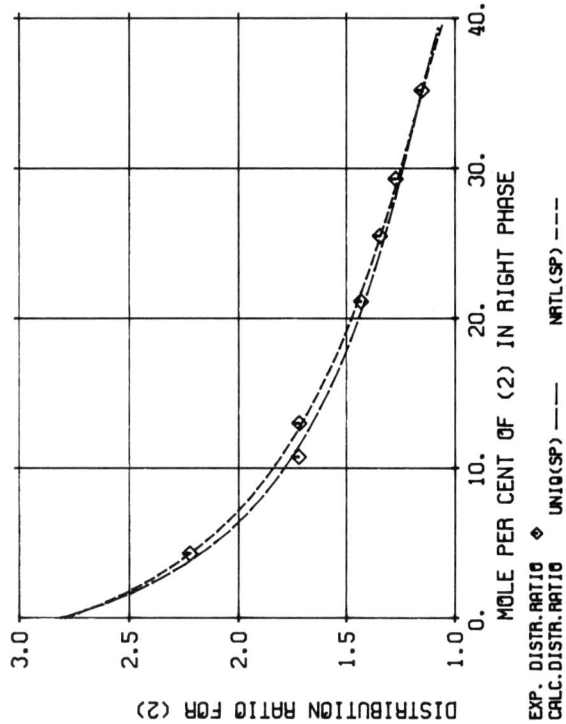

$C_2H_4O_2$-C_6H_6

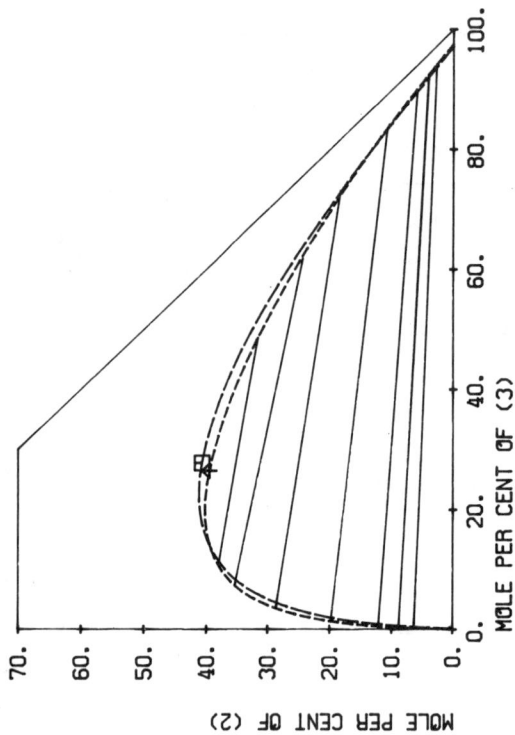

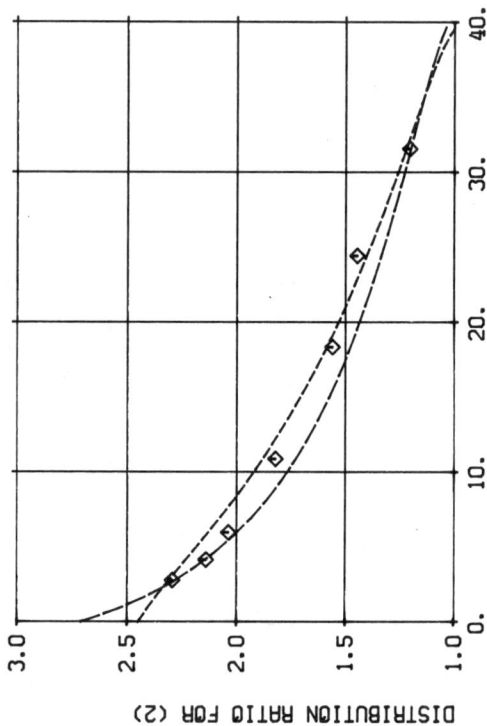

EXP.TIE LINE ———   UNIQ(SP) ⊡   NRTL(SP) ---- ♦
CALC.BINODAL
CALC.PLAIT P.

MOLE PER CENT OF (3)

MOLE PER CENT OF (2)

MOLE PER CENT OF (2) IN RIGHT PHASE

DISTRIBUTION RATIO FOR (2)

EXP. D STR.RATIO ♦

(1) H2O	WATER	
(2) C2H4O2	ACETIC ACID	
(3) C6H6	BENZENE	

TAGLIAVINI G., ARICH G., BIANCANI M.
ANN.CHIM.(ROME) 45(1955)292

TEMPERATURE = 90.0 DEG C   TYPE OF SYSTEM = 1

EXPERIMENTAL TIE LINES IN MOLE PCT

	LEFT PHASE			RIGHT PHASE	
(1)	(2)	(3)	(1)	(2)	(3)
93.438	6.375	0.186	2.947	2.779	94.274
90.890	8.858	0.252	3.347	4.141	92.512
87.438	12.143	0.419	4.138	5.959	89.902
78.837	19.807	1.356	6.039	10.870	83.091
67.613	28.623	3.764	8.511	18.337	73.153
57.333	35.269	7.398	13.799	24.402	61.799
51.217	38.008	10.775	19.817	31.540	48.643

SPECIFIC MODEL PARAMETERS IN KELVIN

		UNIQUAC		NRTL(ALPHA=.2)	
I J	AIJ	AJI		AIJ	AJI
1 2	-45.563	-195.33		1141.3	-847.36
1 3	256.07	562.68		2028.9	742.40
2 3	-59.285	104.05		508.12	-183.01

R1 = 0.9200   R2 = 2.2024   R3 = 3.1878
Q1 = 1.400   Q2 = 2.072   Q3 = 2.400

MEAN DEV. BETWEEN CALC. AND EXP. CONC. IN MOLE PCT

UNIQUAC (SPECIFIC PARAMETERS)   0.79
NRTL   (SPECIFIC PARAMETERS)   0.40

$C_2H_4O_2\text{-}C_6H_6$

(2) C2H4O2   ACETIC ACID

(3) C6H6   BENZENE

TAGLIAVINI G., ARICH G., BIANCANI M.
ANN.CHIM.(ROME) 45(1955)292

TEMPERATURE = 100.0 DEG C   TYPE OF SYSTEM = 1

EXPERIMENTAL TIE LINES IN MOLE PCT

LEFT PHASE		RIGHT PHASE	
(1)	(2)	(3)	(1)	(2)	(3)
80.168	18.470	1.362	6.430	10.249	83.321
75.371	22.420	2.209	7.877	13.470	78.653
70.799	25.920	3.281	9.259	16.898	73.843
66.959	28.689	4.352	10.594	19.864	69.541
58.233	34.202	7.566	15.308	26.486	58.206
53.778	36.382	9.841	18.489	29.479	52.032
48.998	37.828	13.174	23.021	32.489	44.490

SPECIFIC MODEL PARAMETERS IN KELVIN

		UNIQUAC		NRTL(ALPHA=.2)	
I J		AIJ	AJI	AIJ	AJI
1 2		-80.571	-195.76	952.47	-792.83
1 3		268.73	501.72	1974.1	668.80
2 3		-74.366	95.626	341.24	-105.93

R1 = 0.9200   R2 = 2.2024   R3 = 3.1878
Q1 = 1.400   Q2 = 2.072   Q3 = 2.400

MEAN DEV. BETWEEN CALC. AND EXP. CONC. IN MOLE PCT

UNIQUAC (SPECIFIC PARAMETERS)   0.65
NRTL   (SPECIFIC PARAMETERS)   0.18

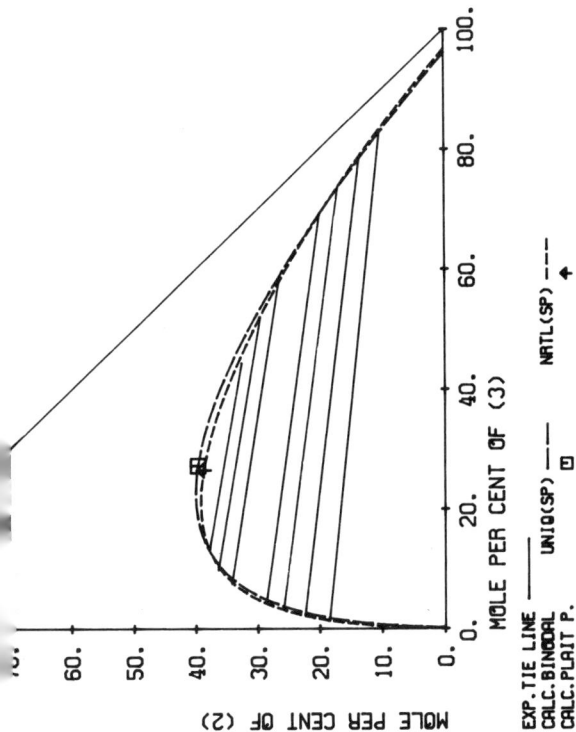

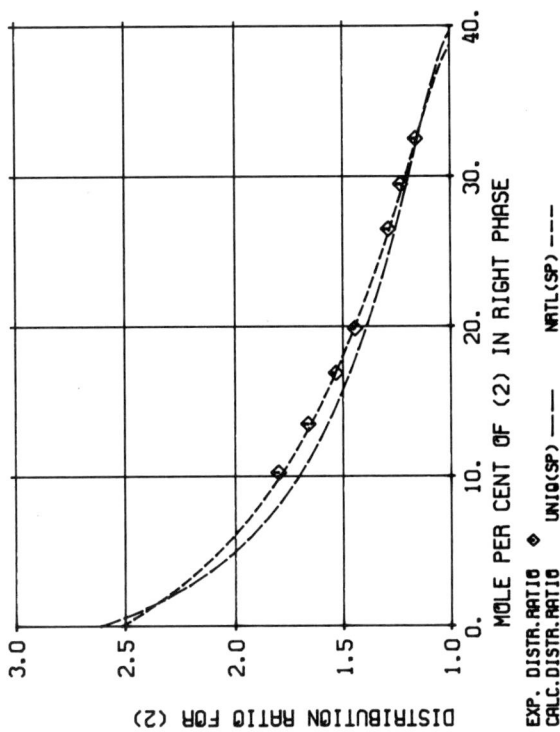

$C_2H_4O_2$-C_6H_6

(1) H2O   WATER
(2) C2H4O2   ACETIC ACID
(3) C6H6   BENZENE

TAGLIAVINI G., ARICH G., BIANCANI M.
ANN.CHIM.(ROME) 45(1955)292

TEMPERATURE = 110.0 DEG C   TYPE OF SYSTEM = 1

EXPERIMENTAL TIE LINES IN MOLE PCT

	LEFT PHASE			RIGHT PHASE	
(1)	(2)	(3)	(1)	(2)	(3)
89.830	9.654	0.516	4.546	5.207	90.247
79.843	18.511	1.646	7.188	10.303	82.510
74.991	22.449	2.561	9.568	14.619	75.714
65.820	29.064	5.116	12.952	20.316	66.733
62.875	30.898	6.227	14.506	22.521	62.973
58.122	33.513	8.365	18.001	27.314	54.685
52.789	35.600	11.611	23.115	30.863	46.022

SPECIFIC MODEL PARAMETERS IN KELVIN

		UNIQUAC		NRTL(ALPHA=.2)	
I	J	AIJ	AJI	AIJ	AJI
1	2	-45.366	-212.52	1142.5	-883.75
1	3	228.44	528.36	2017.7	668.34
2	3	-54.183	78.784	459.18	-192.59

R1 = 0.9200   R2 = 2.2024   R3 = 3.1878
Q1 = 1.400   Q2 = 2.072   Q3 = 2.400

MEAN DEV. BETWEEN CALC. AND EXP. CONC. IN MOLE PCT

UNIQUAC (SPECIFIC PARAMETERS)   0.79
NRTL   (SPECIFIC PARAMETERS)   0.31

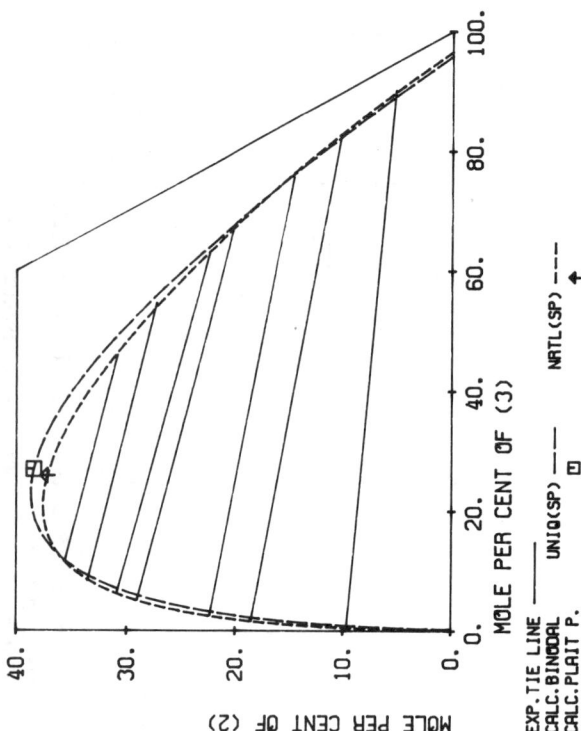

MOLE PER CENT OF (2)
MOLE PER CENT OF (3)

EXP.TIE LINE ———   UNIQ(SP) ——— ▣   NRTL(SP) –––– ✦
CALC.BINODAL
CALC.PLAIT P.

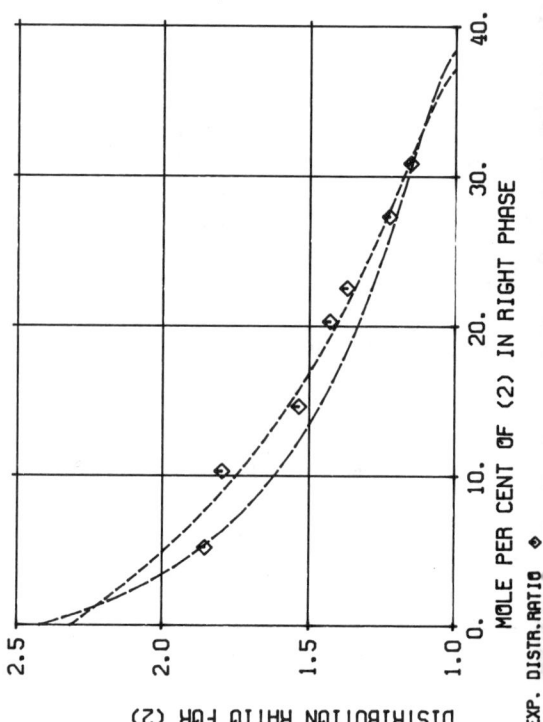

DISTRIBUTION RATIO FOR (2)
MOLE PER CENT OF (2) IN RIGHT PHASE

EXP. DISTR.RATIO ◇

(2) C2H4O2   ACETIC ACID
(3) C6H6   BENZENE

TAGLIAVINI G., ARICH G., BIANCANI M.
ANN.CHIM.(ROME) 45(1955)292

TEMPERATURE = 120.0 DEG C   TYPE OF SYSTEM = 1

EXPERIMENTAL TIE LINES IN MOLE PCT

	LEFT PHASE			RIGHT PHASE	
(1)	(2)	(3)	(1)	(2)	(3)
89.815	9.554	0.631	5.335	5.417	89.247
79.662	18.339	1.999	9.011	11.754	79.235
75.513	21.646	2.841	11.062	15.220	73.717
67.278	27.745	4.976	15.499	21.411	63.090
64.071	29.659	6.270	17.275	23.374	59.350
56.886	33.057	10.057	20.599	27.150	52.252
49.533	34.626	15.841	29.250	31.261	39.488

SPECIFIC MODEL PARAMETERS IN KELVIN

		UNIQUAC		NRTL(ALPHA=.2)	
I J	AIJ	AJI		AIJ	AJI
1 2	-101.08	-197.16		1025.7	-827.30
1 3	278.51	502.35		2144.1	595.18
2 3	-181.39	211.44		94.297	93.817

R1 = 0.9200   R2 = 2.2024   R3 = 3.1878
Q1 = 1.400   Q2 = 2.072   Q3 = 2.400

MEAN DEV. BETWEEN CALC. AND EXP. CONC. IN MOLE PCT

UNIQUAC (SPECIFIC PARAMETERS)   0.68
NRTL   (SPECIFIC PARAMETERS)   0.45

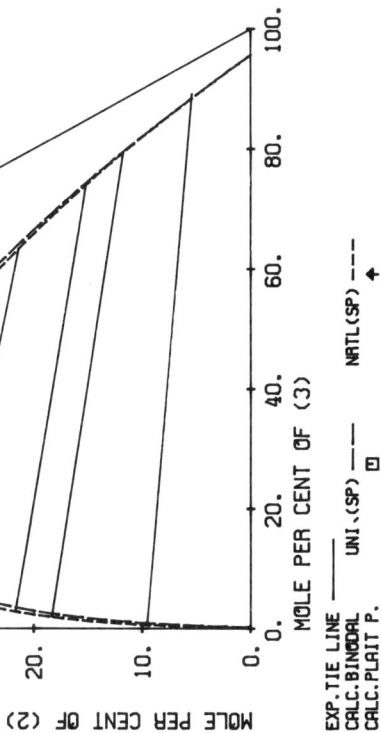

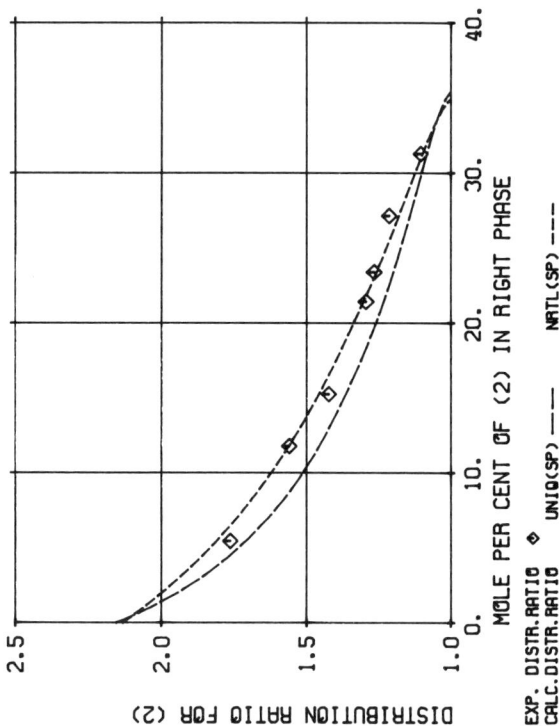

C$_2$H$_4$O$_2$-C$_6$H$_6$

(1) H2O   WATER
(2) C2H4O2   ACETIC ACID
(3) C6H6   BENZENE

PRINCE R.G.H., HUNTER T.G.
CHEM.ENG.SCI. 6(1957)245

TEMPERATURE = 25.0 DEG C   TYPE OF SYSTEM = 1

EXPERIMENTAL TIE LINES IN MOLE PCT

	LEFT PHASE		RIGHT PHASE	
(1)	(2)	(3)	(1)	(2)	(3)
90.450	9.389	0.161	0.514	3.419	96.066
90.423	9.416	0.161	0.514	3.445	96.041
84.864	14.845	0.291	0.722	6.360	92.918
83.791	15.878	0.331	0.805	7.133	92.062
52.941	42.106	4.953	4.603	28.142	67.255
50.793	43.490	5.717	5.323	30.031	64.646

SPECIFIC MODEL PARAMETERS IN KELVIN

		UNIQUAC		NRTL(ALPHA=.2)	
I	J	AIJ	AJI	AIJ	AJI
1	2	-266.09	-129.95	-88.738	-325.86
1	3	596.58	703.31	1804.8	924.96
2	3	-69.726	0.82389E-01	189.14	-154.94

R1 = 0.9200   R2 = 2.2024   R3 = 3.1878
Q1 = 1.400   Q2 = 2.072   Q3 = 2.400

MEAN DEV. BETWEEN CALC. AND EXP. CONC. IN MOLE PCT

UNIQUAC (SPECIFIC PARAMETERS)   0.53
NRTL   (SPECIFIC PARAMETERS)   0.51
UNIQUAC (COMMON PARAMETERS)   0.58

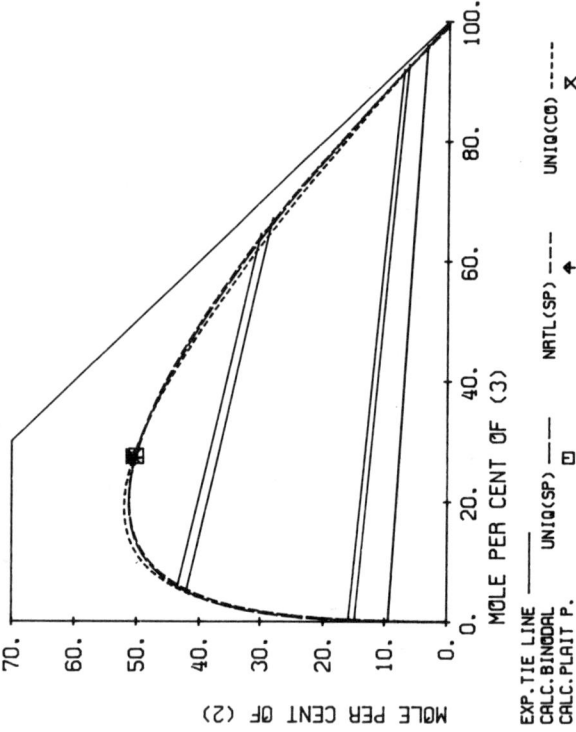

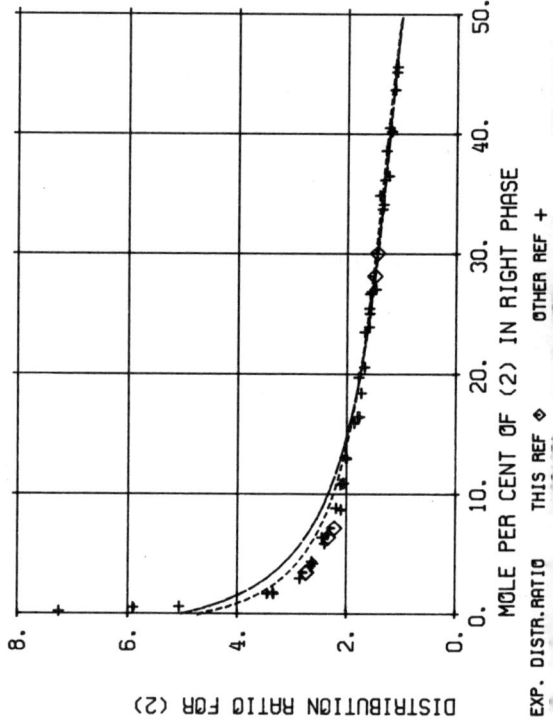

$C_2H_4O_2-C_6H_6$

FUSE K., IGUCHI A.,
KAGAKU KOGAKU 35(1971)107

(2) C2H4O2   ACETIC ACID
(3) C6H6   BENZENE

TEMPERATURE = 25.0 DEG C   TYPE OF SYSTEM = 1

EXPERIMENTAL TIE LINES IN MOLE PCT (GRAPH.INTERPOL.)

	LEFT PHASE			RIGHT PHASE	
(1)	(2)	(3)	(1)	(2)	(3)
84.177	15.539	0.285	0.849	6.341	92.810
69.576	29.073	1.351	2.052	16.421	81.527
53.050	42.256	4.694	3.555	26.665	69.780
47.727	45.684	6.589	6.387	36.497	57.115
37.551	49.942	12.507	10.631	43.672	45.697

SPECIFIC MODEL PARAMETERS IN KELVIN

		UNIQUAC		NRTL(ALPHA=.2)	
I J	AIJ	AJI		AIJ	AJI
1 2	-266.09	-129.95		-88.738	-325.86
1 3	-596.58	-703.31		1804.8	-924.96
2 3	-69.726	0.82389E-01		188.14	-154.94

R1 = 0.9200   R2 = 2.2024   R3 = 3.1878
Q1 = 1.400   Q2 = 2.072   Q3 = 2.400

MEAN DEV. BETWEEN CALC. AND EXP. CONC. IN MOLE PCT

UNIQUAC (SPECIFIC PARAMETERS)   1.11
NRTL   (SPECIFIC PARAMETERS)   1.05
UNIQUAC (COMMON PARAMETERS)   1.35

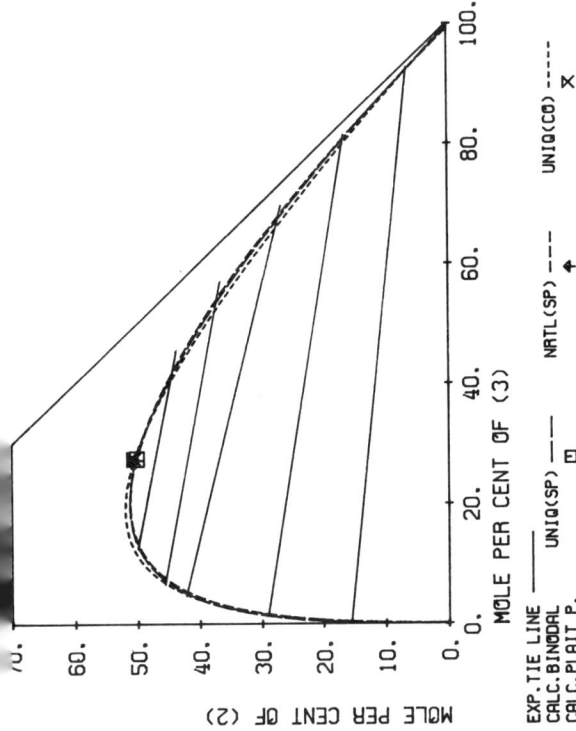

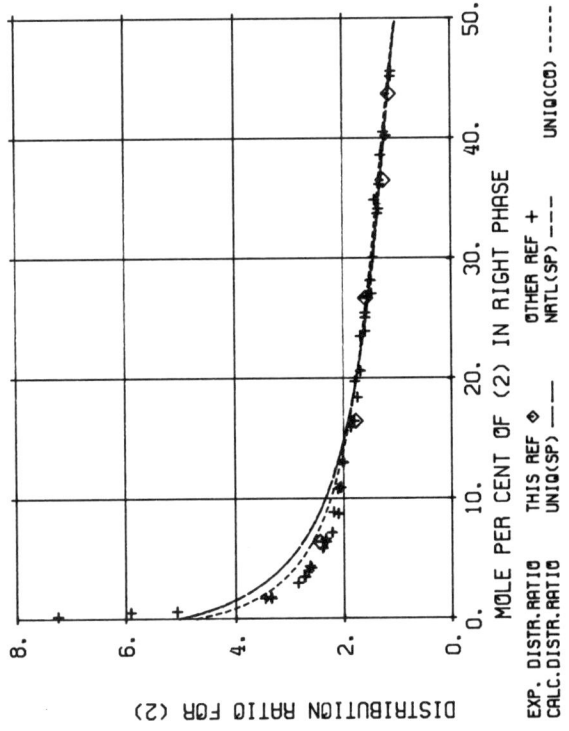

$C_2H_4O_2$-$C_6H_{10}O_4$

(1) C6H1004   ACETALDEHYDE,DIACETATE
----
(2) C2H4O2   ACETIC ACID
----
(3) H2O   WATER

SMITH J.C.
J.PHYS.CHEM. 46(1942)229

TEMPERATURE = 25.0 DEG C   TYPE OF SYSTEM = 1

EXPERIMENTAL TIE LINES IN MOLE PCT (GRAPH.INTERPOL.)

       LEFT PHASE   RIGHT PHASE
    (1)   (2)   (3)   (1)   (2)   (3)

 61.404   5.608   32.988   0.748   1.300   97.952
 56.058   8.882   35.060   0.871   2.557   96.572
 47.866   13.446   38.688   1.052   4.033   94.915
 41.731   16.307   41.962   1.322   5.624   93.055
 37.525   17.762   44.712   1.710   6.974   91.316
 30.153   19.059   50.788   2.410   8.604   88.986
 21.570   18.716   59.713   3.637   10.405   85.958

SPECIFIC MODEL PARAMETERS IN KELVIN

                UNIQUAC   NRTL(ALPHA=.2)
 I J   AIJ   AJI   AIJ   AJI

 1 2 -129.46   109.27   357.15   -377.74
 1 3   351.69   36.746   -75.473   1586.2
 2 3 -125.10   125.00   -2.1213   190.88

 R1 = 5.1542   R2 = 2.2024   R3 = 0.9200
 Q1 = 4.532   Q2 = 2.072   Q3 = 1.400

MEAN DEV. BETWEEN CALC. AND EXP. CONC. IN MOLE PCT

UNIQUAC (SPECIFIC PARAMETERS)   0.60
NRTL (SPECIFIC PARAMETERS)   0.62
UNIQUAC (COMMON PARAMETERS)   1.20

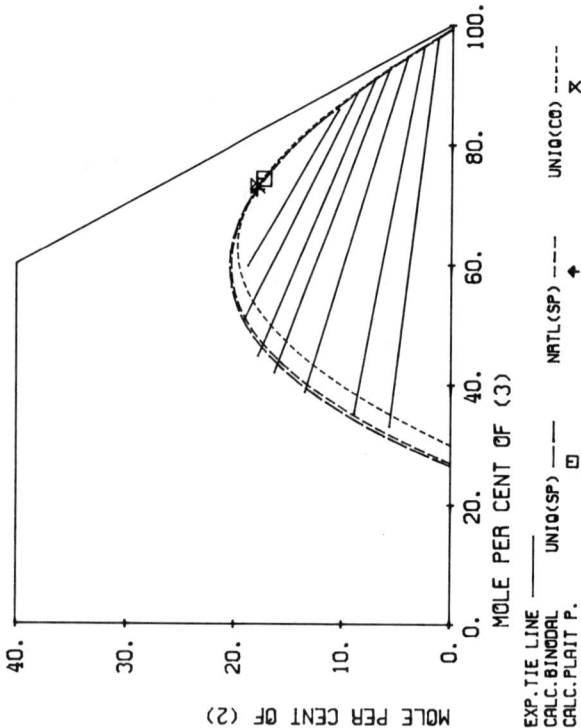

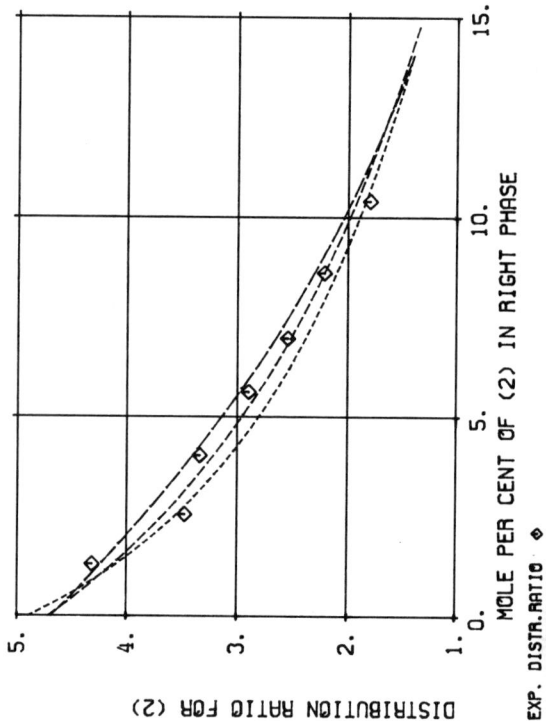

$C_2H_4O_2-C_6H_{12}$

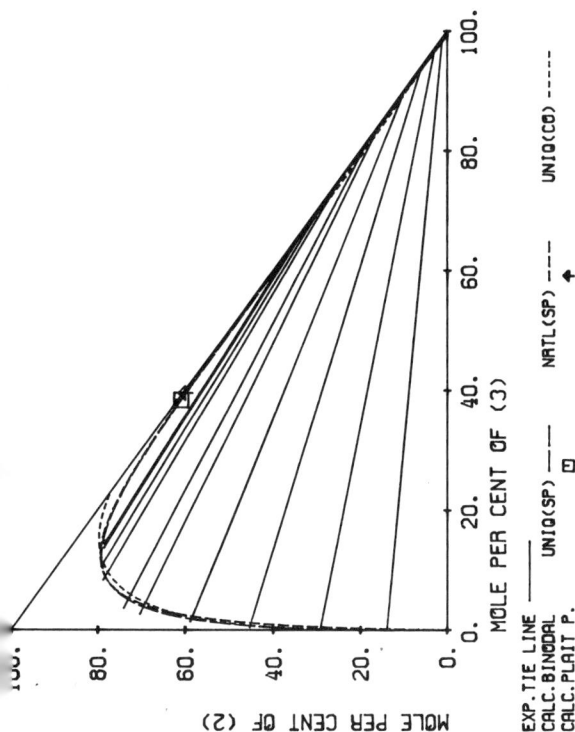

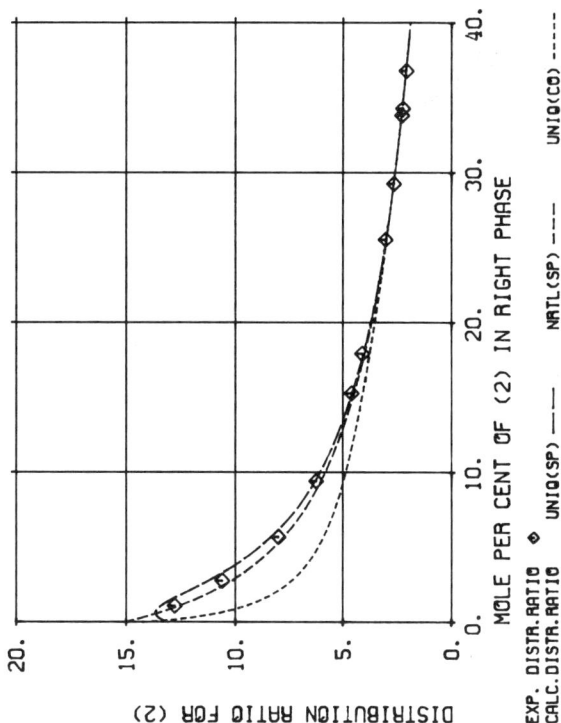

(2) C2H4O2   ACETIC ACID
(3) C6H12   CYCLOHEXANE

FUSE K., IGUCHI A.
KAGAKU KOGAKU 34(1970)1226

TEMPERATURE = 25.0 DEG C   TYPE OF SYSTEM = 1

EXPERIMENTAL TIE LINES IN MOLE PCT (GRAPH.INTERPOL.)

	LEFT PHASE			RIGHT PHASE	
(1)	(2)	(3)	(1)	(2)	(3)
86.007	13.936	0.057	0.0	1.090	98.910
70.671	29.184	0.144	0.0	2.753	97.247
53.993	45.473	0.534	0.0	5.680	94.320
39.750	58.796	1.454	0.0	9.410	90.590
27.009	70.294	2.697	0.0	15.252	84.748
22.273	74.050	3.677	0.0	17.922	82.073
12.920	78.736	8.344	0.431	25.535	74.034
10.067	78.812	11.122	0.426	29.258	70.315
7.708	78.577	13.715	0.420	33.811	65.769
7.186	78.437	14.376	0.420	34.277	65.303
7.069	78.160	14.771	0.416	36.800	62.783

SPECIFIC MODEL PARAMETERS IN KELVIN

		UNIQUAC		NRTL(ALPHA=.2)	
I J	AIJ	AJI		AIJ	AJI
1 2	-360.25	118.84		-632.67	467.82
1 3	487.29	722.96		1580.6	1289.1
2 3	47.981	208.12		459.33	186.26

R1 = 0.9200   R2 = 2.2024   R3 = 4.0464
Q1 = 1.400   Q2 = 2.072   Q3 = 3.240

MEAN DEV. BETWEEN CALC. AND EXP. CONC. IN MOLE PCT

UNIQUAC (SPECIFIC PARAMETERS)   0.45
NRTL   (SPECIFIC PARAMETERS)   0.55
UNIQUAC (COMMON PARAMETERS)   1.69

$C_2H_4O_2$-$C_6H_{12}O$

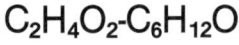

(1) C6H12O   CYCLOHEXANOL
(2) C2H4O2   ACETIC ACID
(3) H2O   WATER

SKRZEC A.E., MURPHY N.F.
IND.ENG.CHEM. 46(1954)2245

TEMPERATURE = 26.7 DEG C   TYPE OF SYSTEM = 1

EXPERIMENTAL TIE LINES IN MOLE PCT (GRAPH.INTERPOL.)

LEFT PHASE			RIGHT PHASE		
(1)	(2)	(3)	(1)	(2)	(3)
52.491	0.0	47.509	0.667	0.0	99.333
50.400	2.836	46.764	0.736	0.667	98.597
36.333	8.940	54.727	0.983	2.780	96.231
34.635	9.586	55.779	1.072	3.257	95.671
31.745	10.590	57.665	1.173	3.878	94.949
15.757	14.280	69.963	4.517	9.815	85.668

SPECIFIC MODEL PARAMETERS IN KELVIN

		UNIQUAC		NRTL(ALPHA=.2)	
I	J	AIJ	AJI	AIJ	AJI
1	2	143.30	-230.04	-26.344	-269.64
1	3	-17.053	230.49	-308.40	1691.2
2	3	-16.603	-167.24	24.180	-23.108

R1 = 4.3489   R2 = 2.2024   R3 = 0.9200
Q1 = 3.512   Q2 = 2.072   Q3 = 1.400

MEAN DEV. BETWEEN CALC. AND EXP. CONC. IN MOLE PCT

UNIQUAC (SPECIFIC PARAMETERS)   0.35
NRTL   (SPECIFIC PARAMETERS)   0.45
UNIQUAC (COMMON PARAMETERS)   0.58

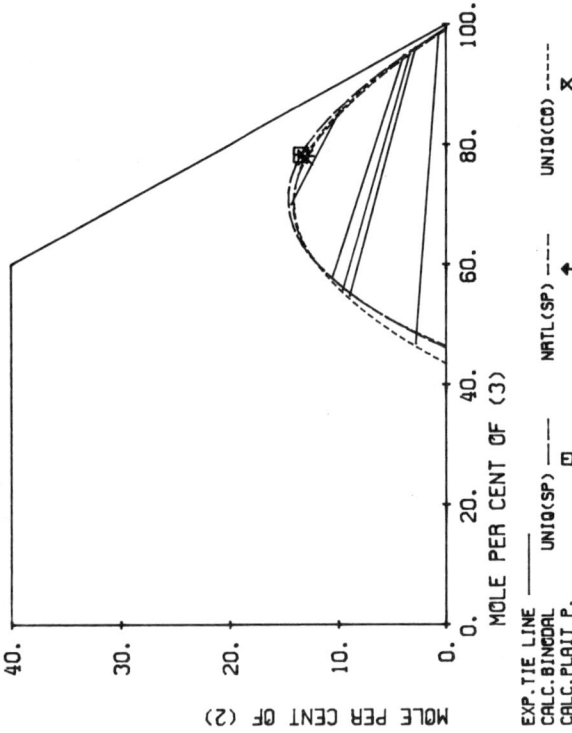

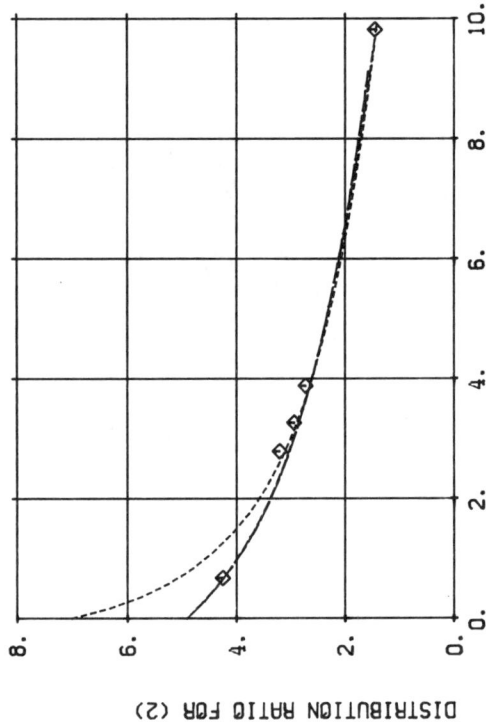

EXP.TIE LINE ——
CALC.BINODAL
CALC.PLAIT P.

UNIQ(SP) ⊡   NRTL(SP) ——— ✦   UNIQ(CO) ----- ✕

C$_2$H$_4$O$_2$-C$_6$H$_{12}$O

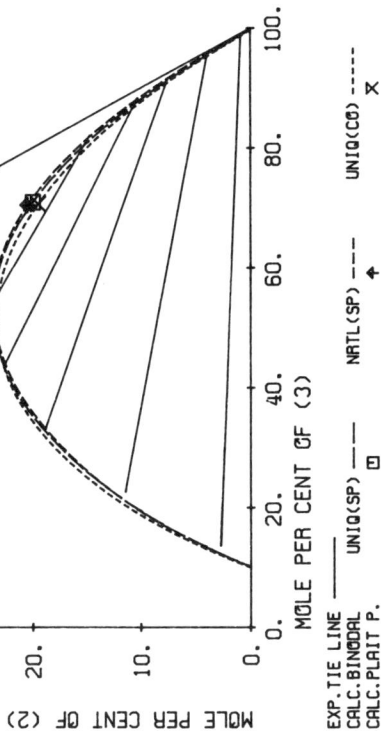

MOLE PER CENT OF (3)

MOLE PER CENT OF (2)

EXP.TIE LINE ——— UNIQ(SP) ——— ▢   NRTL(SP) –·– ✦   UNIQ(CC) –––– ✗
CALC.BINODAL
CALC.PLAIT P.

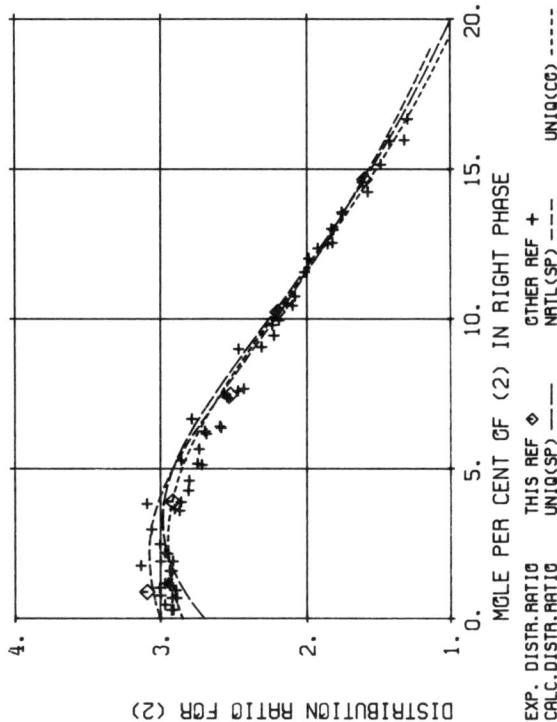

MOLE PER CENT OF (2) IN RIGHT PHASE

DISTRIBUTION RATIO FOR (2)

EXP. DISTR.RATIO   THIS REF ◇   OTHER REF +   UNIQ(CC) ––––
CALC.DISTR.RATIO   UNIQ(SP) ———   NRTL(SP) –·–

---

(2) C2H4O2   ACETIC ACID

(3) H20   WATER

SHERWOOD T.K., EVANS J.E., LONGCOR J.V.A.
IND.ENG.CHEM. 31(1939)1144

TEMPERATURE = 25.0 DEG C   TYPE OF SYSTEM = 1

EXPERIMENTAL TIE LINES IN MOLE PCT (GRAPH.INTERPOL.)

	LEFT PHASE			RIGHT PHASE	
(1)	(2)	(3)	(1)	(2)	(3)
83.611	2.736	13.654	0.322	0.885	98.793
55.943	11.409	22.647	0.503	3.911	95.586
48.089	18.854	33.057	0.855	7.460	91.684
33.499	22.533	43.967	1.369	10.226	88.405
21.638	23.402	54.960	3.227	14.652	82.121

SPECIFIC MODEL PARAMETERS IN KELVIN

		UNIQUAC		NRTL(ALPHA=.2)	
I J	AIJ	AJI		AIJ	AJI
1 2	-225.65	-13.128		696.81	-565.15
1 3	437.77	107.98		228.70	1827.0
2 3	-278.01	128.06		-11.779	44.146

R1 = 4.5959   R2 = 2.2024   R3 = 0.9200
Q1 = 3.952   Q2 = 2.072   Q3 = 1.400

MEAN DEV. BETWEEN CALC. AND EXP. CONC. IN MOLE PCT

UNIQUAC (SPECIFIC PARAMETERS)	0.39
NRTL (SPECIFIC PARAMETERS)	0.32
UNIQUAC (COMMON PARAMETERS)	0.62

C$_2$H$_4$O$_2$-C$_6$H$_{12}$O

(1) C6H12O   2-PENTANONE, 4-METHYL

(2) C2H4O2   ACETIC ACID

(3) H2O   WATER

OTHMER D.F., WHITE R.E., TRUEGER E.
IND.ENG.CHEM. 33(1941)1240

TEMPERATURE = 22.0 DEG C   TYPE OF SYSTEM = 1

EXPERIMENTAL TIE LINES IN MOLE PCT (GRAPH.INTERPOL.)

	LEFT PHASE			RIGHT PHASE	
(1)	(2)	(3)	(1)	(2)	(3)
70.529	9.100	20.370	0.481	2.973	96.546
57.790	15.096	27.114	0.643	5.280	94.078
36.537	21.857	41.606	1.374	9.957	88.669
26.926	22.832	50.242	2.124	12.538	85.338
19.920	22.442	57.638	2.956	14.248	82.796
14.922	21.159	63.919	4.416	15.963	79.621

SPECIFIC MODEL PARAMETERS IN KELVIN

		UNIQUAC		NRTL(ALPHA=.2)	
I J	AIJ	AJI		AIJ	AJI
1 2	-225.65	-13.128		696.81	-565.15
1 3	437.77	107.98		228.70	1827.0
2 3	-278.01	128.06		-11.779	44.146

R1 = 4.5959   R2 = 2.2024   R3 = 0.9200
Q1 = 3.952   Q2 = 2.072   Q3 = 1.400

MEAN DEV. BETWEEN CALC. AND EXP. CONC. IN MOLE PCT

UNIQUAC (SPECIFIC PARAMETERS)   0.69
NRTL (SPECIFIC PARAMETERS)   0.67
UNIQUAC (COMMON PARAMETERS)   0.94

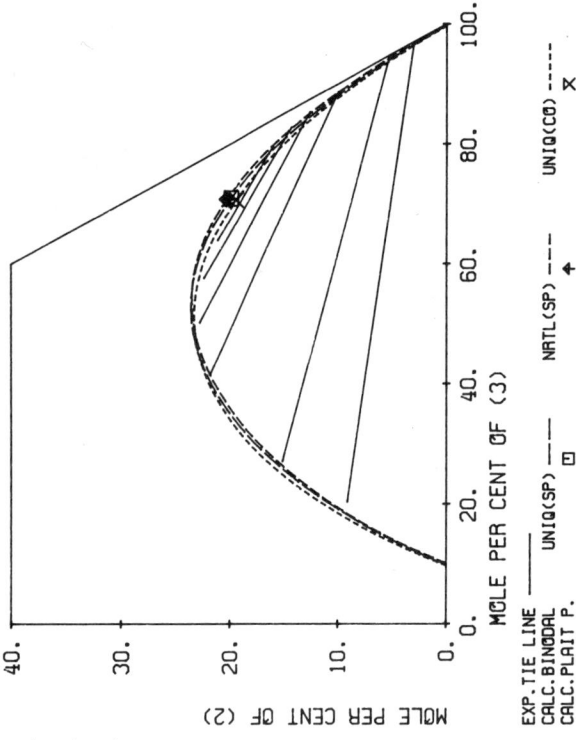

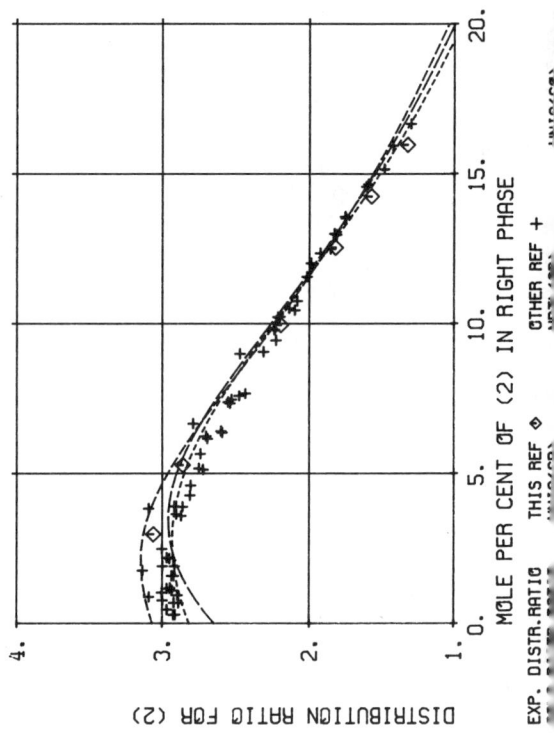

(2) C2H4O2   ACETIC ACID
(3) H2O   WATER

IGUCHI A., FUSE K.
KAGAKU KOGAKU 35(1971)477

TEMPERATURE = 25.0 DEG C   TYPE OF SYSTEM = 1

EXPERIMENTAL TIE LINES IN MOLE PCT (GRAPH.INTERPOL.)

	LEFT PHASE			RIGHT PHASE	
(1)	(2)	(3)	(1)	(2)	(3)
86.360	5.533	8.107	0.420	1.765	97.815
72.554	11.857	15.589	0.541	3.834	95.625
56.227	18.589	25.185	0.727	6.665	92.608
44.697	22.226	33.077	0.947	8.996	90.058
30.801	23.884	45.315	1.607	12.011	86.382
24.833	23.823	51.345	2.246	13.572	84.181
21.433	23.505	55.062	2.856	14.573	82.571
16.720	22.743	60.537	3.965	15.935	80.100

SPECIFIC MODEL PARAMETERS IN KELVIN

		UNIQUAC		NRTL(ALPHA=.2)	
I	J	AIJ	AJI	AIJ	AJI
1	2	-225.65	-13.128	696.81	-565.15
1	3	437.77	107.98	223.70	1827.0
2	3	-278.01	128.06	-11.779	44.146

R1 = 4.5959   R2 = 2.2024   R3 = 0.9200
Q1 = 3.952   Q2 = 2.072   Q3 = 1.400

MEAN DEV. BETWEEN CALC. AND EXP. CONC. IN MOLE PCT

UNIQUAC (SPECIFIC PARAMETERS)   1.28
NRTL   (SPECIFIC PARAMETERS)   1.29
UNIQUAC (COMMON PARAMETERS)   1.22

C₂H₄O₂-C₆H₁₂O

(1) C6H12O   2-PENTANONE, 4-METHYL
(2) C2H4O2   ACETIC ACID
(3) H2O   WATER

HLAVATY K.; LINEK J.
COLLECT.CZECH.CHEM.COMMUN. 38(1973)374

TEMPERATURE = 24.6 DEG C   TYPE OF SYSTEM = 1

EXPERIMENTAL TIE LINES IN MOLE PCT

	LEFT PHASE			RIGHT PHASE	
(1)	(2)	(3)	(1)	(2)	(3)
87.973	0.0	12.027	0.347	0.0	99.653
64.377	12.015	23.608	0.484	4.274	95.242
47.383	18.786	33.830	0.853	7.587	91.560
35.629	22.486	41.885	1.364	10.512	88.124
26.485	23.787	49.728	2.088	13.008	84.904
25.233	23.679	51.088	2.327	13.519	84.155

SPECIFIC MODEL PARAMETERS IN KELVIN

		UNIQUAC		NRTL(ALPHA=.2)	
I J	AIJ	AJI		AIJ	AJI
1 2	-225.65	-13.128		696.81	-565.15
1 3	437.77	-107.98		228.70	1827.0
2 3	-278.01	128.06		-11.779	44.146

R1 = 4.5959   R2 = 2.2024   R3 = 0.9200
Q1 = 3.952   Q2 = 2.072   Q3 = 1.400

MEAN DEV. BETWEEN CALC. AND EXP. CONC. IN MOLE PCT

UNIQUAC (SPECIFIC PARAMETERS)   0.50
NRTL   (SPECIFIC PARAMETERS)   0.41
UNIQUAC (COMMON PARAMETERS)   0.66

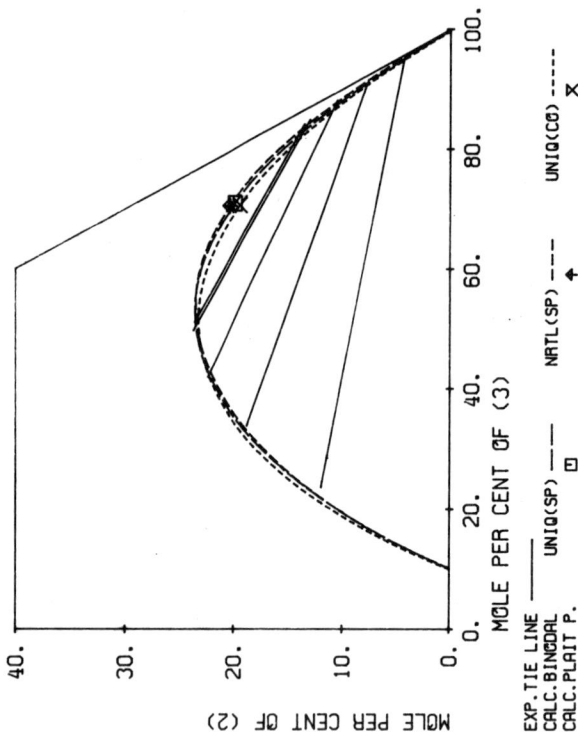

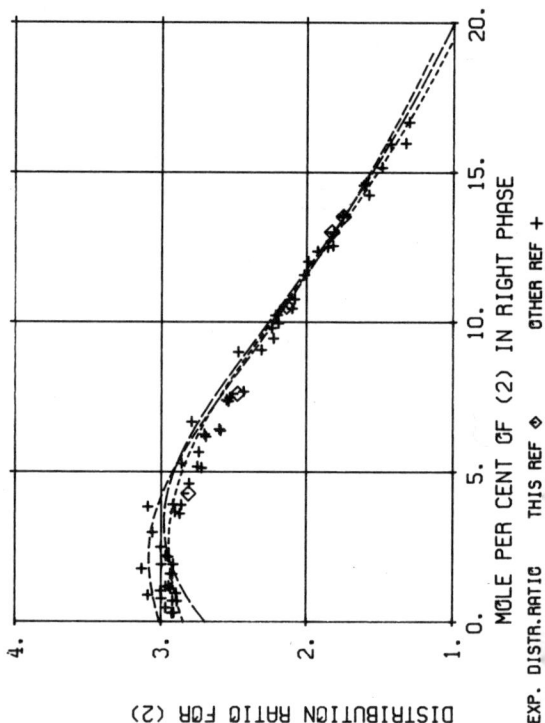

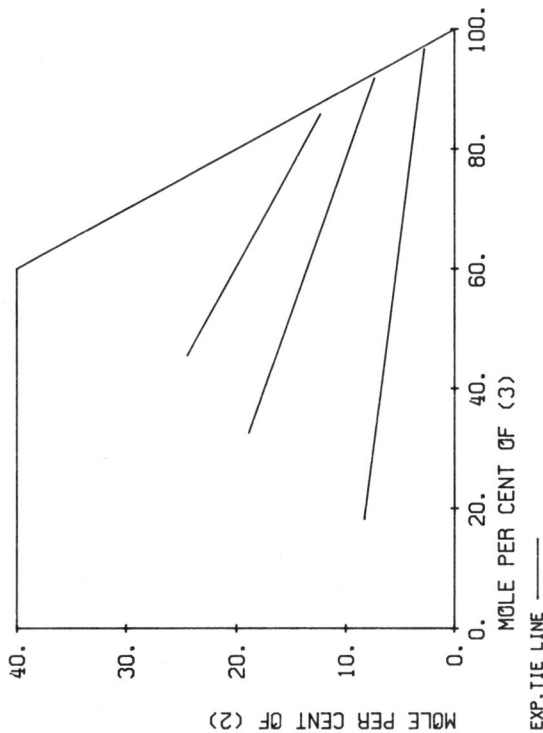

(1) C6H12O   2-PENTANONE, 4-METHYL
(2) C2H4O2   ACETIC ACID
(3) H2O   WATER

GONDO S., JOSE H., KUSUNOKI K., YUTANI K., SODA T.
KAGAKU KOGAKU 37(1973)849

TEMPERATURE = 15.0 DEG C   TYPE OF SYSTEM = 1

EXPERIMENTAL TIE LINES IN MOLE PCT

	LEFT PHASE			RIGHT PHASE	
(1)	(2)	(3)	(1)	(2)	(3)
73.476	8.302	18.222	0.489	2.810	96.701
48.515	18.869	32.616	0.793	7.364	91.843
29.980	24.434	45.585	1.864	12.314	85.823

MOLE PER CENT OF (3)

MOLE PER CENT OF (2)

EXP.TIE LINE ———

$C_2H_4O_2$-$C_6H_{12}O$

(1) C6H12O 2-PENTANONE, 4-METHYL

(2) C2H4O2 ACETIC ACID

(3) H2O WATER

GONDO S., JOSE H., KUSUNOKI K., YUTANI K., SODA T.
KAGAKU KOGAKU 37(1973)849

TEMPERATURE = 25.0 DEG C   TYPE OF SYSTEM = 1

EXPERIMENTAL TIE LINES IN MOLE PCT

	LEFT PHASE			RIGHT PHASE	
(1)	(2)	(3)	(1)	(2)	(3)
83.004	3.104	13.892	0.356	1.033	98.611
73.828	7.465	18.707	0.436	2.487	97.077
66.531	11.142	22.327	0.521	3.896	95.583
57.690	15.484	26.825	0.701	5.653	93.646
49.152	18.613	32.235	0.882	7.334	91.784
41.038	20.958	38.005	1.148	9.061	89.791
34.101	22.410	43.490	1.520	10.753	87.727
27.684	23.197	49.119	2.069	12.499	85.432
14.960	21.716	63.324	5.138	16.662	78.200

SPECIFIC MODEL PARAMETERS IN KELVIN

		UNIQUAC		NRTL(ALPHA=.2)	
I J	AIJ	AJI		AIJ	AJI
1 2	-225.65	-13.128		696.81	-565.15
1 3	437.77	107.98		228.70	1827.0
2 3	-278.01	128.06		-11.779	44.146

R1 = 4.5959 R2 = 2.2024 R3 = 0.9200
Q1 = 3.952   Q2 = 2.072   Q3 = 1.400

MEAN DEV. BETWEEN CALC. AND EXP. CONC. IN MOLE PCT

UNIQUAC (SPECIFIC PARAMETERS)   0.40
NRTL   (SPECIFIC PARAMETERS)   0.40
UNIQUAC (COMMON PARAMETERS)   0.57

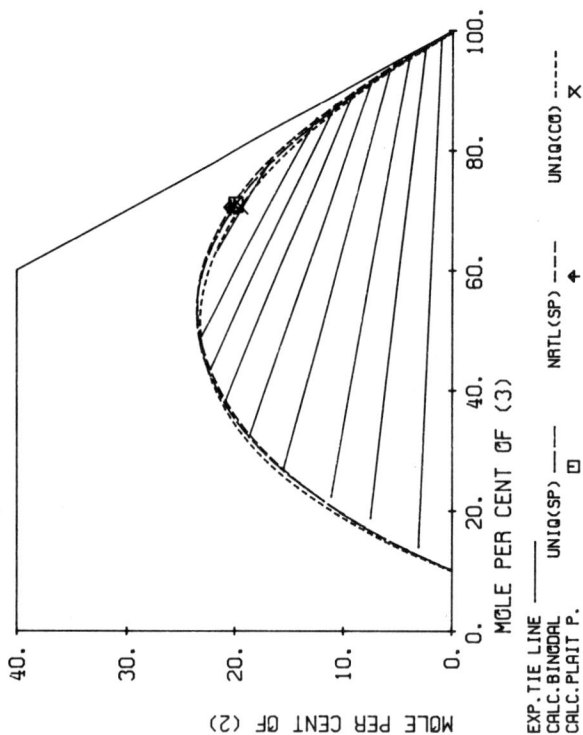

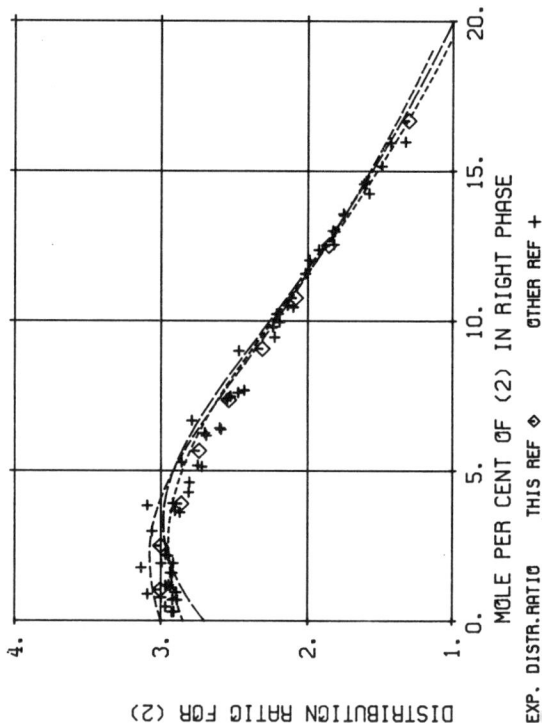

MOLE PER CENT OF (3)

MOLE PER CENT OF (2)

EXP.TIE LINE ⎯⎯

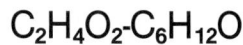

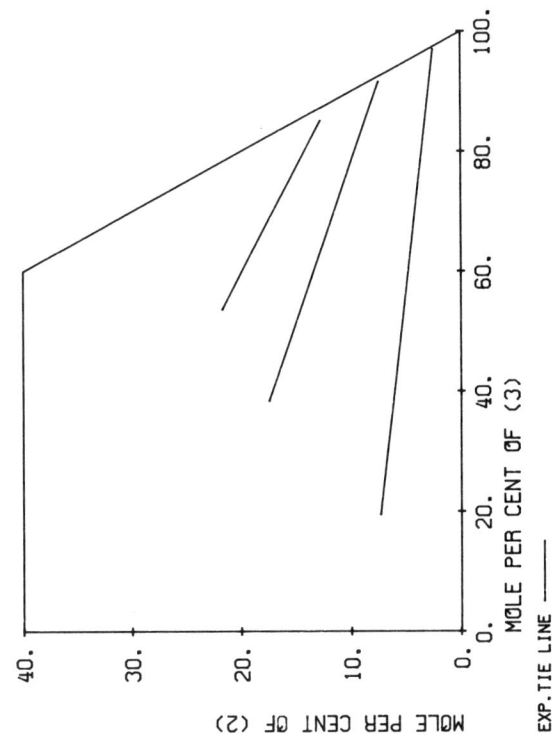

(1) C6H12O   2-PENTANONE, 4-METHYL
(2) C2H4O2   ACETIC ACID
(3) H2O   WATER

GONDO S., JOSE H., KUSUNOKI K., YUTANI K., SODA T.
KAGAKU KOGAKU 37(1973)849

TEMPERATURE = 35.0 DEG C   TYPE OF SYSTEM = 1

EXPERIMENTAL TIE LINES IN MOLE PCT

LEFT PHASE			RIGHT PHASE		
(1)	(2)	(3)	(1)	(2)	(3)
73.180	7.344	19.476	0.417	2.526	97.058
44.173	17.414	38.413	0.873	7.454	91.673
24.642	21.687	53.671	2.132	12.738	85.130

C₂H₄O₂-C₆H₁₂O

(1) C6H12O   2-PENTANONE, 4-METHYL

(2) C2H4O2   ACETIC ACID

(3) H2O   WATER

ALDERS L., KOS J., LEUR J.
EUR.FED.CHEM.ENG.,RECOMM.SYST.LIQ.EXTR.STUD.,EDITOR:T.MISEK
(1978)

TEMPERATURE = 10.0 DEG C   TYPE OF SYSTEM = 1

EXPERIMENTAL TIE LINES IN MOLE PCT (GRAPH.INTERPOL.)

	LEFT PHASE			RIGHT PHASE	
(1)	(2)	(3)	(1)	(2)	(3)
90.144	0.630	9.225	0.330	0.214	99.456
89.002	1.328	9.670	0.351	0.450	99.200
84.768	3.357	11.865	0.375	1.137	98.488
82.207	4.678	13.116	0.398	1.573	98.029
73.749	6.546	14.705	0.424	2.200	97.375
69.619	10.895	19.486	0.478	3.704	95.818
56.037	16.738	27.225	0.590	6.176	93.234
47.625	19.995	32.380	0.677	7.853	91.470
37.169	23.064	39.766	0.969	10.230	88.801
26.773	24.400	48.827	1.630	12.813	85.557

SPECIFIC MODEL PARAMETERS IN KELVIN

		UNIQUAC		NRTL(ALPHA=.2)	
I	J	AIJ	AJI	AIJ	AJI
1	2	-182.75	-50.925	694.44	-524.22
1	3	-425.11	113.65	246.85	1745.0
2	3	-272.66	135.60	86.502	-10.269

R1 = 4.5959   R2 = 2.2024   R3 = 0.9200
Q1 = 3.952   Q2 = 2.072   Q3 = 1.400

MEAN DEV. BETWEEN CALC. AND EXP. CONC. IN MOLE PCT

UNIQUAC (SPECIFIC PARAMETERS)   0.22
NRTL   (SPECIFIC PARAMETERS)   0.22

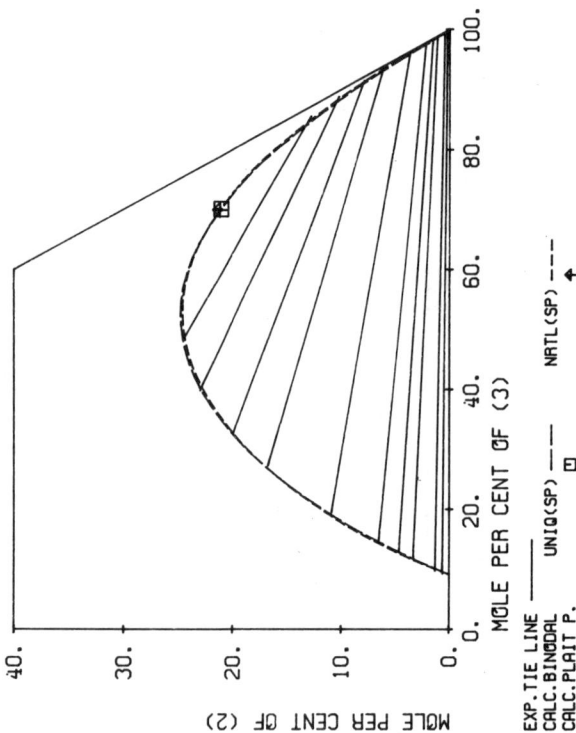

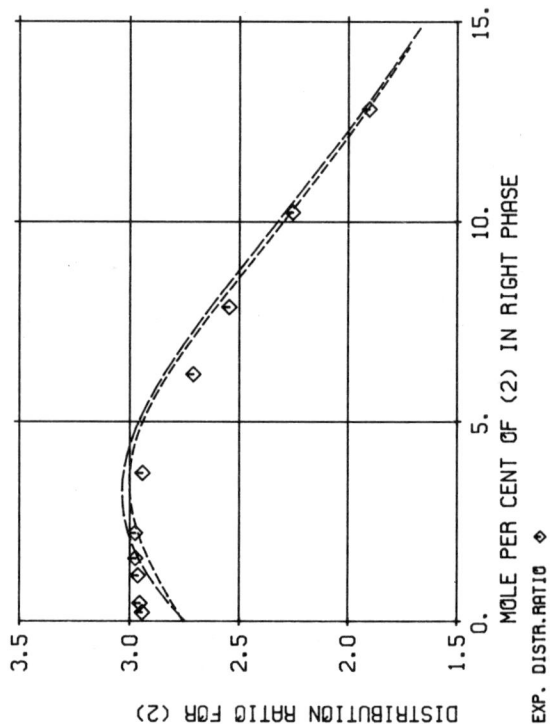

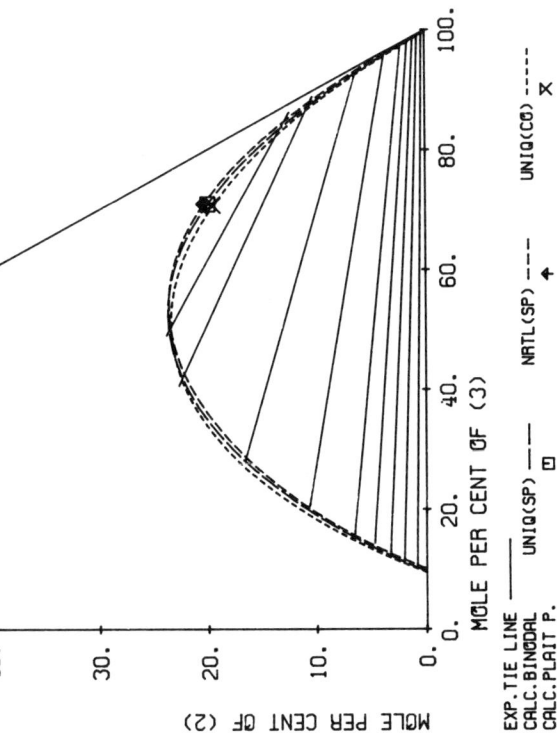

MOLE PER CENT OF (3)

MOLE PER CENT OF (2)

EXP.TIE LINE ——— UNIQ(SP) — — □ NRTL(SP) — — ✦ UNIQ(CO) - - - - ✕
CALC.BINODAL
CALC.PLAIT P.

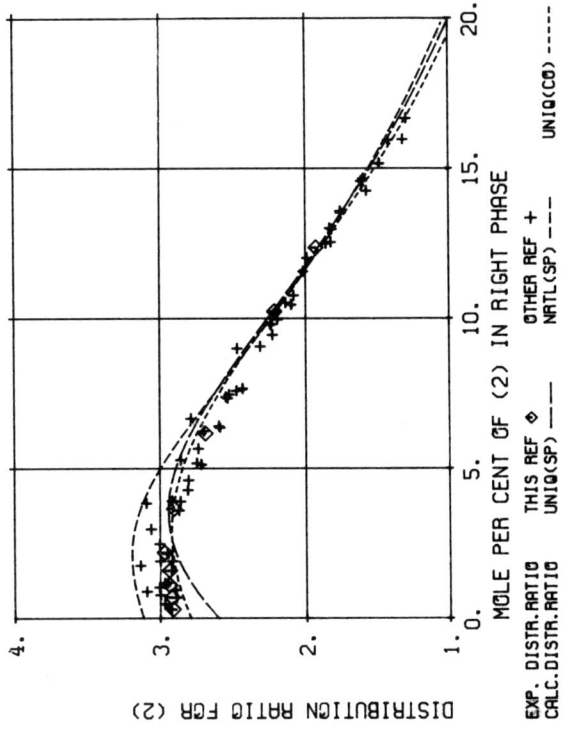

MOLE PER CENT OF (2) IN RIGHT PHASE

DISTRIBUTION RATIO (2)

EXP. DISTR.RATIO ◇ THIS REF ◇ OTHER REF + UNIQ(CO) - - - - -
CALC.DISTR.RATIO UNIQ(SP) ——— NRTL(SP) — — —

---

(2) C2H4O2   ACETIC ACID

(3) H2O   WATER

ALDERS L., KOS J., LEUR J.
EUR.FED.CHEM.ENG.,RECOMM.SYST.LIQ.EXTR.STUD.,EDITOR:T.MISEK
(1978)

TEMPERATURE = 20.0 DEG C   TYPE OF SYSTEM = 1

EXPERIMENTAL TIE LINES IN MOLE PCT (GRAPH.INTERPOL.)

	LEFT PHASE			RIGHT PHASE	
(1)	(2)	(3)	(1)	(2)	(3)
88.517	0.865	10.618	0.294	0.297	99.409
86.498	2.021	11.480	0.296	0.692	99.012
83.992	3.253	12.755	0.318	1.108	98.574
80.910	4.693	14.397	0.341	1.598	98.061
77.536	6.525	15.938	0.365	2.198	97.437
68.782	10.614	20.605	0.397	3.654	95.949
55.287	16.577	28.136	0.525	6.162	93.313
36.685	22.657	40.658	0.921	10.222	88.856
27.099	23.785	49.116	1.385	12.353	86.261

SPECIFIC MODEL PARAMETERS IN KELVIN

I J	UNIQUAC AIJ	AJI	NRTL(ALPHA=.2) AIJ	AJI
1 2	-225.65	-13.128	696.81	-565.15
1 3	437.77	107.98	228.70	1827.0
2 3	-278.01	128.06	-11.779	44.146

R1 = 4.5959   R2 = 2.2024   R3 = 0.9200
Q1 = 3.952   Q2 = 2.072   Q3 = 1.400

MEAN DEV. BETWEEN CALC. AND EXP. CONC. IN MOLE PCT

UNIQUAC (SPECIFIC PARAMETERS)   0.33
NRTL (SPECIFIC PARAMETERS)   0.31
UNIQUAC (COMMON PARAMETERS)   0.54

C$_2$H$_4$O$_2$-C$_6$H$_{12}$O

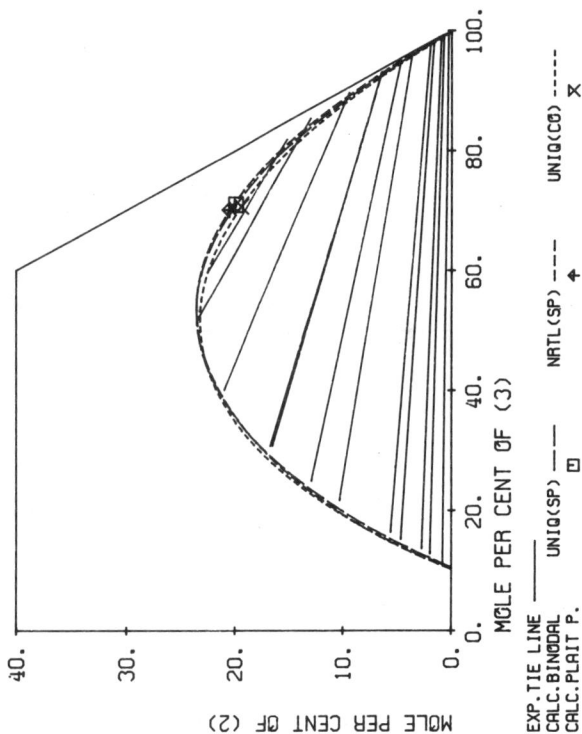

MOLE PER CENT OF (3)

MOLE PER CENT OF (2)

EXP.TIE LINE	——		UNIQ(CO) -----
CALC.BINODAL	UNIQ(SP) -----	☐	✕
CALC.PLAIT P.	NRTL(SP) -----	✦	

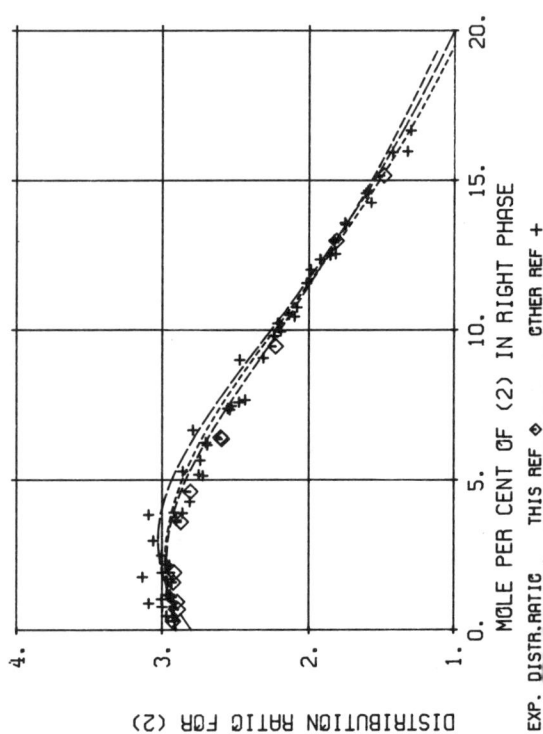

DISTRIBUTION RATIO FOR (2)

MOLE PER CENT OF (2) IN RIGHT PHASE

EXP. DISTR.RATIO   THIS REF ◇   OTHER REF +

---

(1) C6H120   2-PENTANONE, 4-METHYL
(2) C2H4O2   ACETIC ACID
(3) H2O   WATER

ALDERS L., KOS J., LEUR J.
EUR.FED.CHEM.ENG.,RECOMM.SYST.LIQ.EXTR.STUD.,EDITOR:T.MISEK
(1978)

TEMPERATURE = 30.0 DEG C   TYPE OF SYSTEM = 1

EXPERIMENTAL TIE LINES IN MOLE PCT (GRAPH.INTERPOL.)

LEFT PHASE			RIGHT PHASE		
(1)	(2)	(3)	(1)	(2)	(3)
88.090	0.831	11.079	0.256	0.284	99.459
85.183	1.997	12.820	0.278	0.691	99.031
83.611	2.736	13.654	0.298	0.944	98.757
80.133	4.629	15.238	0.322	1.584	98.095
78.024	5.569	16.407	0.324	1.910	97.766
67.973	10.326	21.702	0.376	3.597	96.026
62.246	12.916	24.838	0.426	4.604	94.970
52.799	16.485	30.717	0.527	6.363	93.110
52.633	16.676	30.691	0.528	6.416	93.056
38.824	21.057	40.118	0.908	9.648	89.448
24.762	23.544	51.694	1.609	12.975	85.416
17.609	22.539	59.852	2.944	15.153	81.903

SPECIFIC MODEL PARAMETERS IN KELVIN

		UNIQUAC		NRTL(ALPHA=.2)	
I	J	AIJ	AJI	AIJ	AJI
1	2	-225.65	-13.128	696.81	-565.15
1	3	437.77	107.98	228.70	1827.0
2	3	-278.01	128.06	-11.779	44.146

R1 = 4.5959   R2 = 2.2024   R3 = 0.9200
Q1 = 3.952   Q2 = 2.072   Q3 = 1.400

MEAN DEV. BETWEEN CALC. AND EXP. CONC. IN MOLE PCT

UNIQUAC (SPECIFIC PARAMETERS)   0.48
NRTL   (SPECIFIC PARAMETERS)   0.61
UNIQUAC (COMMON PARAMETERS)   0.73

$C_2H_4O_2$-$C_6H_{12}O$

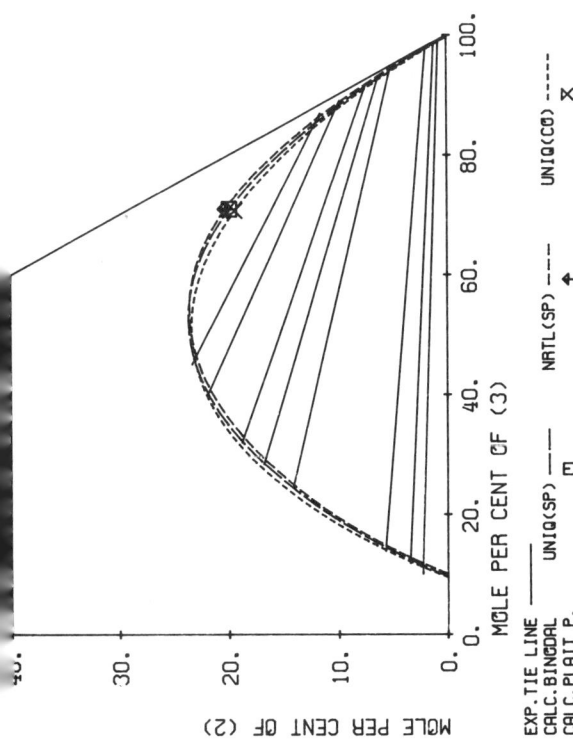

MOLE PER CENT OF (3)

MOLE PER CENT OF (2)

EXP.TIE LINE	────	UNIQ(SP)	▢	UNIQ(CO)	─────
CALC.BINODAL	───	NRTL(SP)	✦		✕
CALC.PLAIT P.					

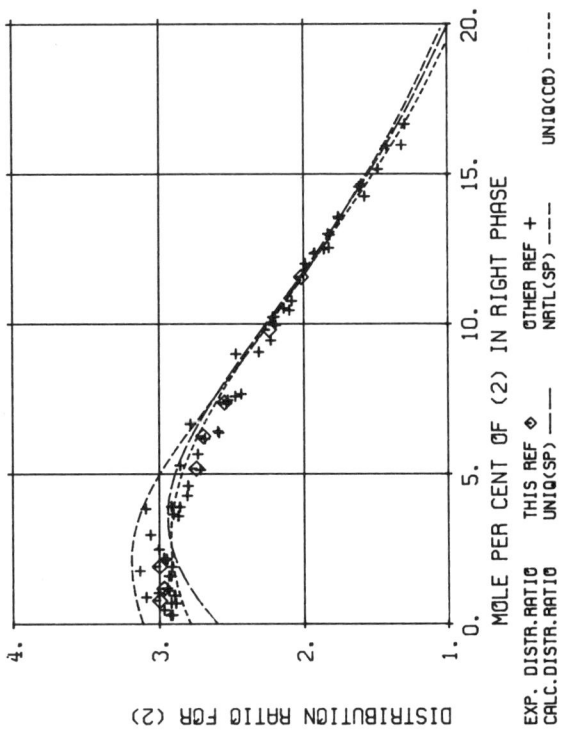

MOLE PER CENT OF (2) IN RIGHT PHASE

DISTRIBUTION RATIO FOR (2)

	THIS REF ◇	OTHER REF +	
EXP. DISTR.RATIO	UNIQ(SP) ────	NRTL(SP) ───	UNIQ(CO) ─────
CALC.DISTR.RATIO			

| (2) C2H4O2 | ACETIC ACID |
| (3) H2O | WATER |

WOLFERSDORFF W.D.B., SOIKA M.
EUR.FED.CHEM.ENG.,RECOMM.SYST.LIQ.EXTR.STUD.,EDITOR:T.MISEK
(1978)

TEMPERATURE = 20.0 DEG C   TYPE OF SYSTEM = 1

EXPERIMENTAL TIE LINES IN MOLE PCT (GRAPH.INTERPOL.)

| | LEFT PHASE | | | RIGHT PHASE | |
(1)	(2)	(3)	(1)	(2)	(3)
87.587	2.317	10.096	0.353	0.773	98.874
84.671	3.469	11.860	0.376	1.169	98.456
80.344	5.743	13.913	0.421	1.915	97.664
61.406	14.217	24.377	0.620	5.170	94.210
54.744	16.862	28.394	0.677	6.244	93.079
49.162	18.816	32.022	0.782	7.377	91.841
38.293	21.955	39.752	1.104	9.795	89.101
31.603	23.320	45.076	1.466	11.560	86.973

SPECIFIC MODEL PARAMETERS IN KELVIN

| | | UNIQUAC | | NRTL(ALPHA=.2) | |
I	J	AIJ	AJI	AIJ	AJI
1	2	-225.65	-13.128	696.81	-565.15
1	3	437.77	107.98	228.70	1827.0
2	3	-278.01	128.06	-11.779	44.146

R1 = 4.5959   R2 = 2.2024   R3 = 0.9200
Q1 = 3.952   Q2 = 2.072   Q3 = 1.400

MEAN DEV. BETWEEN CALC. AND EXP. CONC. IN MOLE PCT

UNIQUAC (SPECIFIC PARAMETERS)	0.31
NRTL (SPECIFIC PARAMETERS)	0.49
UNIQUAC (COMMON PARAMETERS)	0.40

$C_2H_4O_2$-$C_6H_{12}O$

MOLE PER CENT OF (3)

MOLE PER CENT OF (2)

MOLE PER CENT OF (2) IN RIGHT PHASE

DISTRIBUTION RATIO FOR (2)

EXP.TIE LINE ———
CALC.BINODAL
CALC.PLAIT P.

UNIQ(SP) —— ⊡   NRTL(SP) —— +   UNIQ(CO) -----⊠

EXP. DISTR.RATIO ◇   THIS REF ◇   OTHER REF +

(1) C6H12O   2-PENTANONE, 4-METHYL
(2) C2H4O2   ACETIC ACID
(3) H2O   WATER

WOLFERSDORFF W.D.B., SOIKA M.
EUR.FED.CHEM.ENG.,RECOMM.SYST.LIQ.EXTR.STUD.,EDITOR:T.MISEK
(1978)

TEMPERATURE = 30.0 DEG C   TYPE OF SYSTEM = 1

EXPERIMENTAL TIE LINES IN MOLE PCT (GRAPH.INTERPOL.)

LEFT PHASE			RIGHT PHASE		
(1)	(2)	(3)	(1)	(2)	(3)
87.567	1.378	11.055	0.370	0.464	99.166
82.888	3.500	13.611	0.395	1.187	98.418
76.465	6.383	17.152	0.423	2.163	97.414
59.110	13.950	26.939	0.598	5.129	94.273
46.758	18.641	34.600	0.854	7.661	91.485
35.689	21.933	42.378	1.338	10.442	88.220

SPECIFIC MODEL PARAMETERS IN KELVIN

		UNIQUAC		NRTL(ALPHA=.2)	
I J	AIJ	AJI		AIJ	AJI
1 2	-225.65	-13.128		696.81	-565.15
1 3	437.77	-107.98		228.70	1827.0
2 3	-278.01	128.06		-11.779	44.146

R1 = 4.5959   R2 = 2.2024   R3 = 0.9200
Q1 = 3.952   Q2 = 2.072   Q3 = 1.400

MEAN DEV. BETWEEN CALC. AND EXP. CONC. IN MOLE PCT

UNIQUAC (SPECIFIC PARAMETERS)   0.36
NRTL   (SPECIFIC PARAMETERS)   0.37
UNIQUAC (COMMON PARAMETERS)   0.52

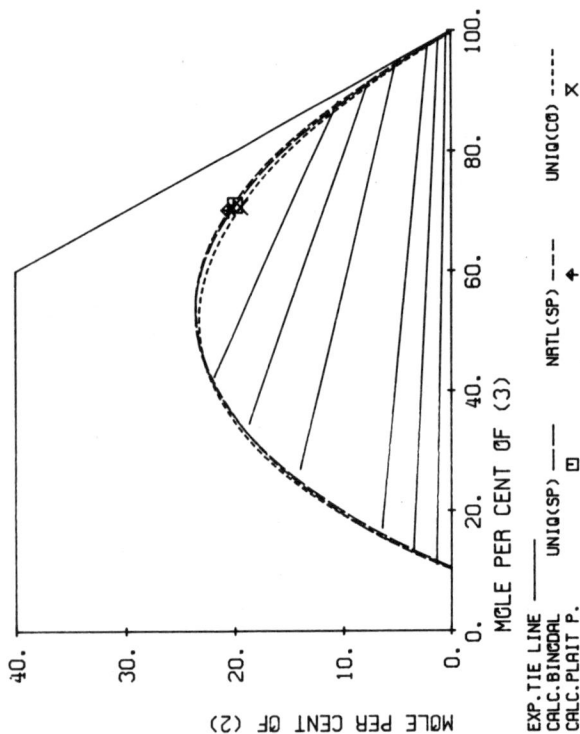

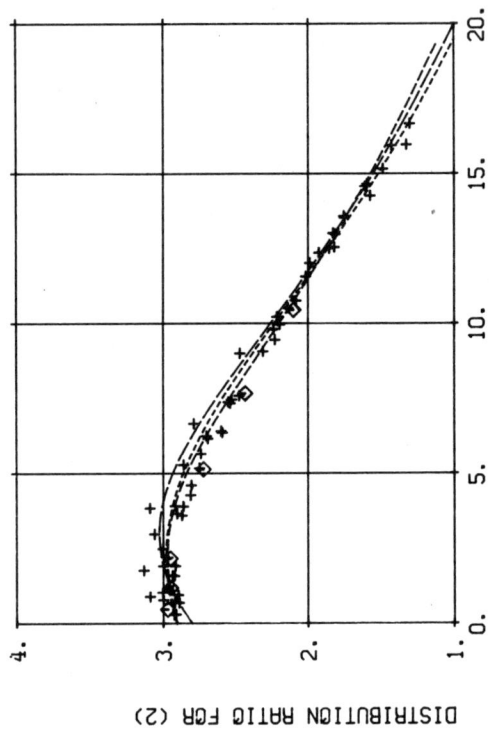

$C_2H_4O_2\text{-}C_6H_{12}O_2$

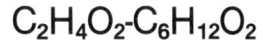

(2) C2H4O2   ACETIC ACID
(3) H2O   WATER

HEYBERGER A., HORACEK J., BULICKA J., PROCHAZKA J.
COLLECT.CZECH.CHEM.COMMUN. 42(1977)3355

TEMPERATURE = 25.0 DEG C   TYPE OF SYSTEM = 1

EXPERIMENTAL TIE LINES IN MOLE PCT (GRAPH.INTERPOL.)

	LEFT PHASE			RIGHT PHASE	
(1)	(2)	(3)	(1)	(2)	(3)
90.464	1.197	8.339	0.095	0.556	99.349
89.245	1.896	8.859	0.095	0.871	99.034
88.046	2.583	9.371	0.112	1.181	98.706
86.818	3.310	9.872	0.113	1.496	98.391
84.779	4.902	10.320	0.131	2.121	97.748
82.479	6.242	11.278	0.150	2.699	97.151
79.759	8.068	12.174	0.169	3.442	96.339
74.670	10.956	14.374	0.191	4.705	95.104
72.055	12.307	15.638	0.212	5.356	94.432
65.780	15.889	18.331	0.238	7.116	92.646
57.514	20.365	22.121	0.346	9.447	90.207
51.882	22.834	25.284	0.500	11.238	88.263
37.974	27.211	34.815	1.048	15.014	83.938
33.649	28.000	38.351	1.383	16.552	82.065
31.346	28.252	40.403	1.518	17.168	81.314

SPECIFIC MODEL PARAMETERS IN KELVIN

		UNIQUAC		NRTL(ALPHA=.2)	
I J	AIJ	AJI		AIJ	AJI
1 2	-114.20	3.1591		356.30	-319.24
1 3	494.86	124.58		350.10	2408.7
2 3	-227.48	118.70		-204.91	318.66

R1 = 4.8266   R2 = 2.2024   R3 = 0.9200
Q1 = 4.192   Q2 = 2.072   Q3 = 1.400

MEAN DEV. BETWEEN CALC. AND EXP. CONC. IN MOLE PCT

UNIQUAC (SPECIFIC PARAMETERS)   0.32
NRTL   (SPECIFIC PARAMETERS)   0.19
UNIQUAC (COMMON PARAMETERS)   0.34

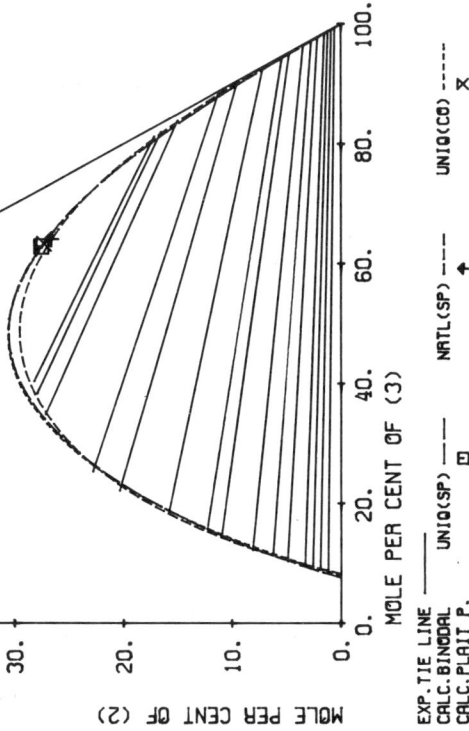

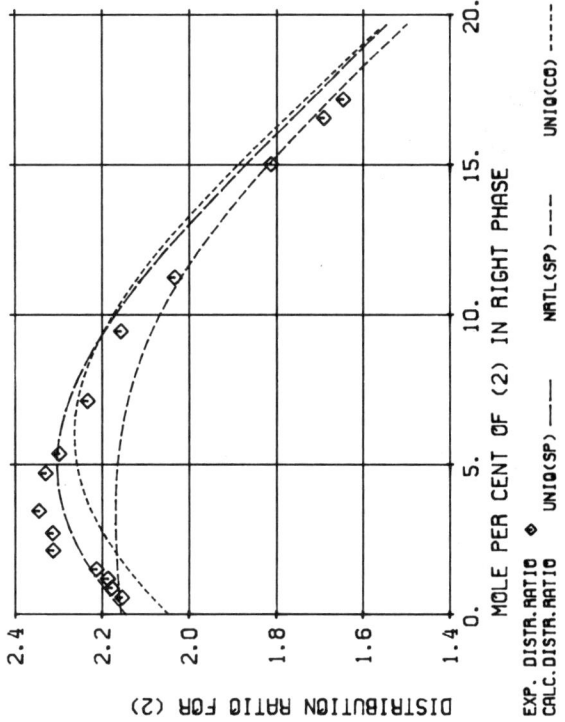

C$_2$H$_4$O$_2$-C$_6$H$_{12}$O$_2$

(1) C6H12O2   ACETIC ACID,ISOBUTYL ESTER
(2) C2H4O2   ACETIC ACID
(3) H2O   WATER

HEYBERGER A., HORACEK J., BULICKA J., PROCHAZKA J.
COLLECT.CZECH.CHEM.COMMUN. 42(1977)3355

TEMPERATURE = 50.0 DEG C   TYPE OF SYSTEM = 1

EXPERIMENTAL TIE LINES IN MOLE PCT (GRAPH.INTERPOL.)

LEFT PHASE		RIGHT PHASE	
(1)	(2)	(3)	(1)	(2)	(3)

(1)	(2)	(3)	(1)	(2)	(3)
78.665	1.147	20.187	0.111	0.563	99.327
77.666	1.772	20.563	0.111	0.862	99.027
76.084	3.051	20.865	0.113	1.448	98.440
74.495	4.350	21.156	0.114	2.066	97.820
71.195	7.113	21.692	0.134	3.242	96.623
66.832	10.233	22.935	0.156	4.651	95.193
64.631	11.860	23.509	0.176	5.471	94.352
57.871	16.291	25.838	0.222	7.765	92.013
54.898	18.171	26.930	0.284	8.819	90.897
47.698	21.194	31.108	0.497	10.927	88.576
47.606	21.304	31.089	0.519	11.084	88.397
39.404	24.220	36.376	0.932	13.501	85.567
34.015	25.611	40.374	1.255	15.079	83.666

SPECIFIC MODEL PARAMETERS IN KELVIN

		UNIQUAC		NRTL(ALPHA=.2)	
I	J	AIJ	AJI	AIJ	AJI
1	2	-230.06	167.68	289.01	-108.17
1	3	270.71	226.93	67.593	2311.5
2	3	-234.94	118.71	-125.35	406.64

R1 = 4.8266   R2 = 2.2024   R3 = 0.9200
Q1 = 4.192   Q2 = 2.072   Q3 = 1.400

MEAN DEV. BETWEEN CALC. AND EXP. CONC. IN MOLE PCT

UNIQUAC (SPECIFIC PARAMETERS)   0.49
NRTL (SPECIFIC PARAMETERS)   0.46

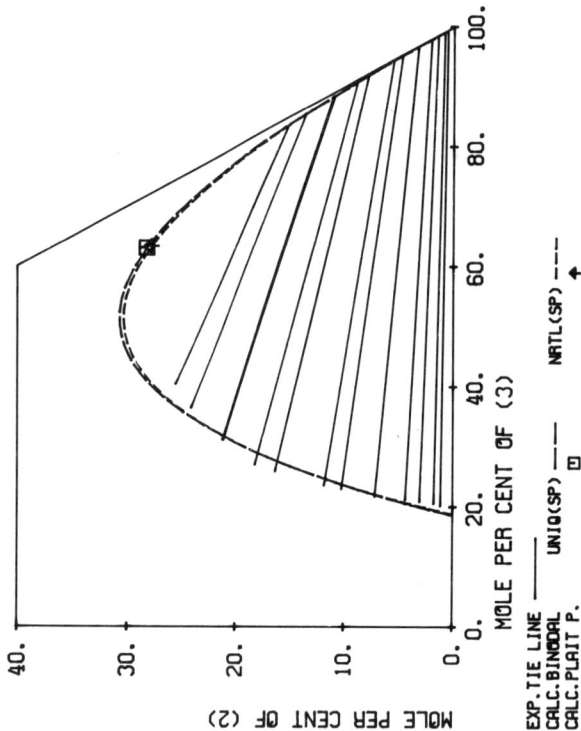

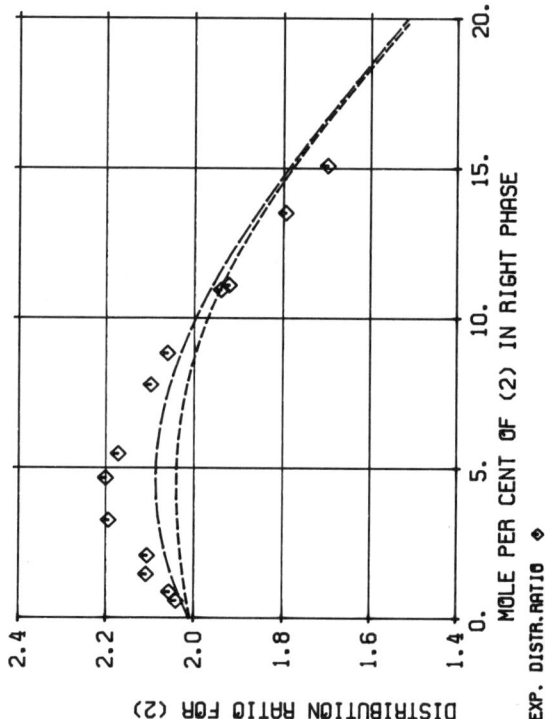

$C_2H_4O_2-C_6H_{12}O_2$

MOLE PER CENT OF (3)

MOLE PER CENT OF (2)

EXP.TIE LINE ——— UNIQ(CC) -------
CALC.BINODAL
CALC.PLAIT P. ⊼

MOLE PER CENT OF (2) IN RIGHT PHASE

DISTRIBUTION RATIO FOR (2)

EXP. DISTR.RATIO ◇ UNIQ(CC) -------
CALC.DISTR.RATIO

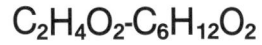

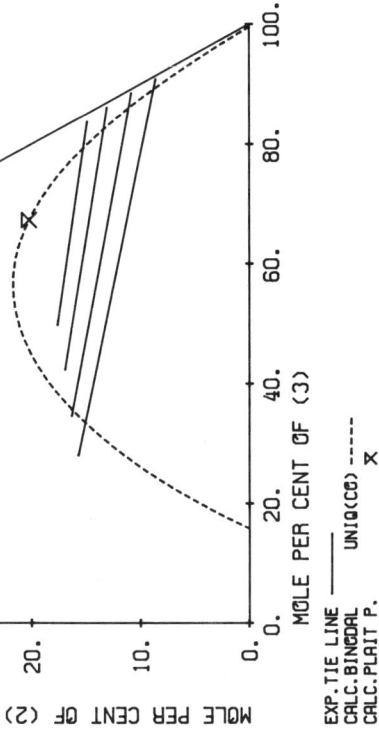

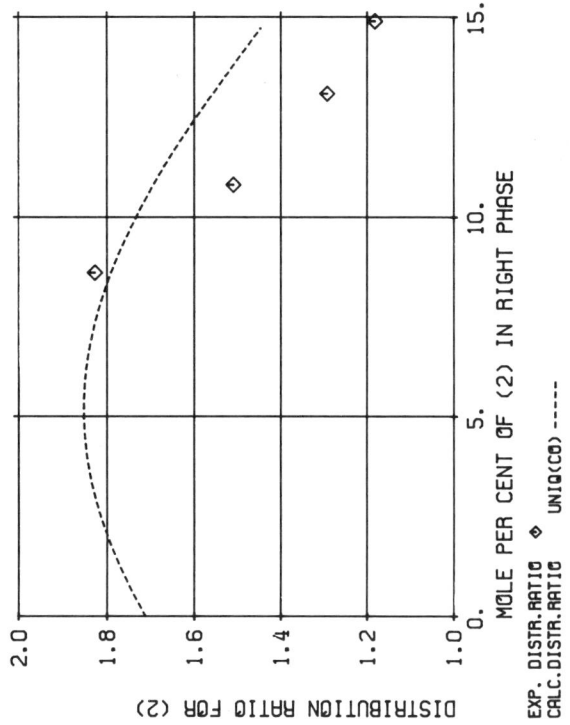

(2) C2H4O2   ACETIC ACID
(3) H2O   WATER

OTHMER D.F., SERRANO J.
IND.ENG.CHEM. 41(1949)1030

TEMPERATURE = 25.0 DEG C   TYPE OF SYSTEM = 1

EXPERIMENTAL TIE LINES IN MOLE PCT (GRAPH.INTERPOL.)

	LEFT PHASE			RIGHT PHASE	
(1)	(2)	(3)	(1)	(2)	(3)
56.348	15.704	27.948	0.536	8.596	90.868
49.173	16.320	34.507	0.628	10.805	88.566
40.770	16.923	42.307	0.752	13.086	86.162
32.644	17.604	49.752	1.298	14.890	83.812

MEAN DEV. BETWEEN CALC. AND EXP. CONC. IN MOLE PCT

UNIQUAC (COMMON PARAMETERS)   1.63

C$_2$H$_4$O$_2$-C$_6$H$_{12}$O$_2$

(1) C6H12O2   BUTANOIC ACID,ETHYL ESTER
(2) C2H4O2   ACETIC ACID
(3) H2O   WATER

RAMANA RAO M.V., DAKSHINAMURTY P.
J.CHEM.ENG.DATA 10(1965)248

TEMPERATURE = 28.0 DEG C   TYPE OF SYSTEM = 1

EXPERIMENTAL TIE LINES IN MOLE PCT

	LEFT PHASE			RIGHT PHASE	
(1)	(2)	(3)	(1)	(2)	(3)
71.635	11.315	17.050	0.177	5.573	94.250
64.239	16.246	19.515	0.280	8.091	91.629
56.670	20.093	23.237	0.332	10.402	89.266
51.174	22.515	26.311	0.583	13.282	86.135
45.240	25.875	28.886	1.243	17.857	80.900

SPECIFIC MODEL PARAMETERS IN KELVIN

		UNIQUAC		NRTL(ALPHA=.2)	
I	J	AIJ	AJI	AIJ	AJI
1	2	-69.230	12.948	368.32	17.135
1	3	457.54	137.31	255.60	2149.5
2	3	-208.20	124.34	-460.45	1120.4

R1 = 4.8274   R2 = 2.2024   R3 = 0.9200
Q1 = 4.196   Q2 = 2.072   Q3 = 1.400

MEAN DEV. BETWEEN CALC. AND EXP. CONC. IN MOLE PCT

UNIQUAC (SPECIFIC PARAMETERS)   0.94
NRTL (SPECIFIC PARAMETERS)   0.80
UNIQUAC (COMMON PARAMETERS)   1.19

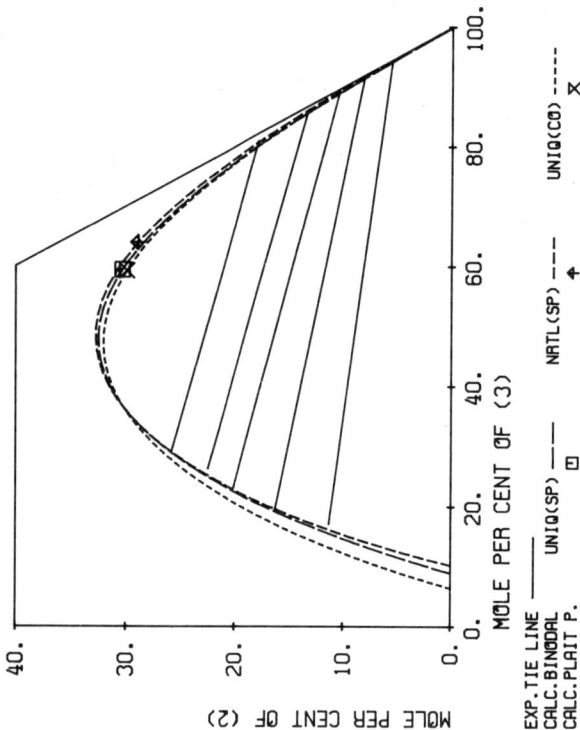

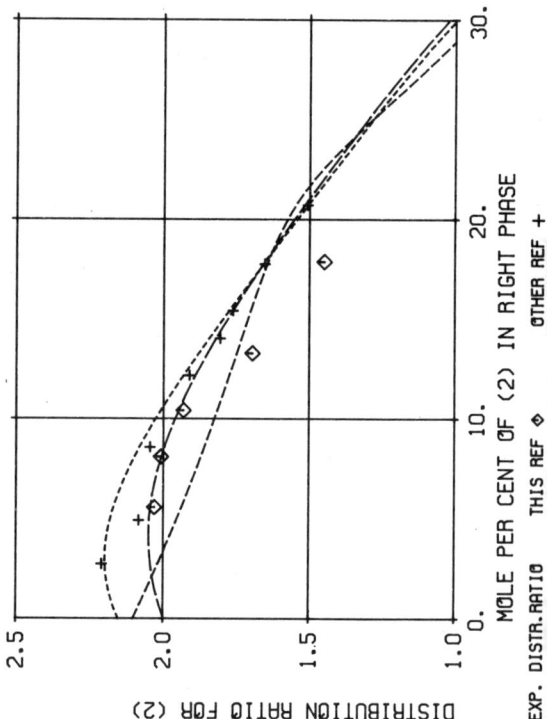

$C_2H_4O_2-C_6H_{12}O_2$

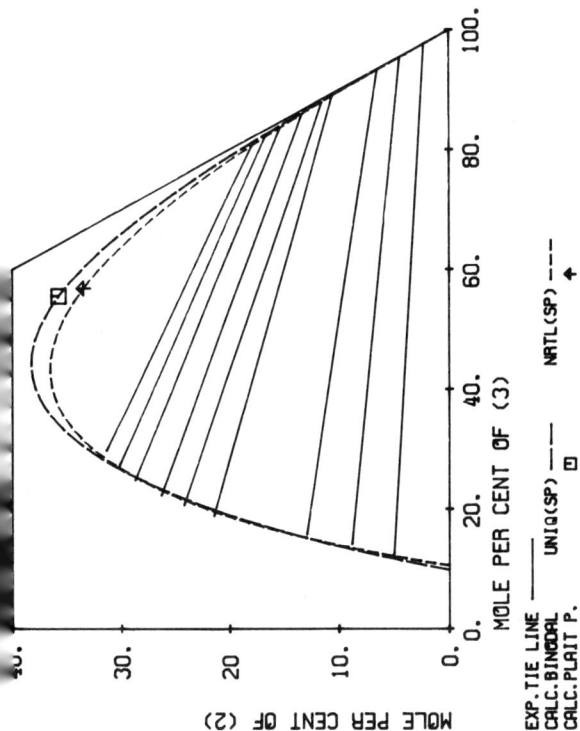

MOLE PER CENT OF (3)

MOLE PER CENT OF (2)

EXP.TIE LINE ———   UNIQ(SP) ⊡   NRTL(SP) – – –
CALC.BINODAL
CALC.PLAIT P.   ✦

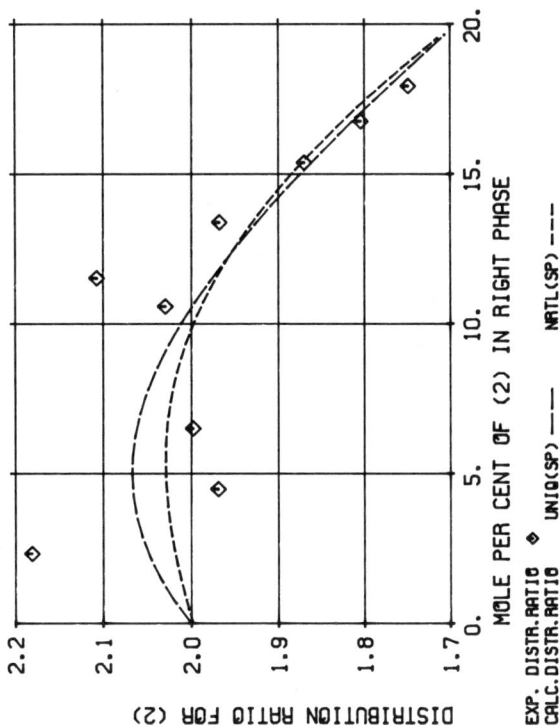

MOLE PER CENT OF (2) IN RIGHT PHASE

DISTRIBUTION RATIO FOR (2)

EXP. DISTR.RATIO ◆   UNIQ(SP) ◇   NRTL(SP) – – –
CALC.DISTR.RATIO

---

(2) C2H4O2   ACETIC ACID

(3) H2O   WATER

DAKSHINAMURTY P., ET AL.
J.CHEM.ENG.DATA 17(1972)379

TEMPERATURE = 10.0 DEG C   TYPE OF SYSTEM = 1

EXPERIMENTAL TIE LINES IN MOLE PCT

	LEFT PHASE			RIGHT PHASE	
(1)	(2)	(3)	(1)	(2)	(3)
83.211	5.035	11.754	0.140	2.309	97.551
77.073	8.821	14.106	0.173	4.478	95.349
71.211	12.982	15.807	0.198	6.497	93.305
59.723	21.451	18.827	0.433	10.571	88.995
55.138	24.252	20.610	0.523	11.510	87.967
51.367	24.347	22.287	0.691	13.385	85.924
46.466	28.742	24.792	0.918	15.369	83.713
42.701	30.210	27.090	1.102	16.743	82.155
38.927	31.368	29.704	1.293	17.928	80.780

SPECIFIC MODEL PARAMETERS IN KELVIN

		UNIQUAC		NRTL(ALPHA=.2)	
I	J	AIJ	AJI	AIJ	AJI
1	2	-155.28	2.3778	219.71	-81.017
1	3	363.12	198.47	230.99	2018.4
2	3	-167.23	-25.428	1.1484	256.67

R1 = 4.8274   R2 = 2.2024   R3 = 0.9200
Q1 = 4.196   Q2 = 2.072   Q3 = 1.400

MEAN DEV. BETWEEN CALC. AND EXP. CONC. IN MOLE PCT

UNIQUAC (SPECIFIC PARAMETERS)   0.34
NRTL   (SPECIFIC PARAMETERS)   0.28

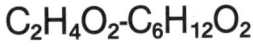

C₂H₄O₂-C₆H₁₂O₂

(1) C6H12O2   BUTANOIC ACID,ETHYL ESTER

(2) C2H4O2   ACETIC ACID

(3) H2O   WATER

DAKSHINAMURTY P., ET AL.:
J.CHEM.ENG.DATA 17(1972)379

TEMPERATURE = 28.0 DEG C   TYPE OF SYSTEM = 1

EXPERIMENTAL TIE LINES IN MOLE PCT

	LEFT PHASE		RIGHT PHASE		
(1)	(2)	(3)	(1)	(2)	(3)
82.696	6.013	11.291	0.158	2.721	97.121
74.924	10.231	14.844	0.201	4.909	94.890
62.972	17.524	19.504	0.331	8.568	91.101
52.227	23.246	24.527	0.571	12.161	87.268
47.245	25.312	27.444	0.731	14.016	85.253
42.366	27.164	30.471	0.907	15.406	83.687
36.529	29.334	34.137	1.337	17.739	80.924
26.123	31.208	42.669	2.106	20.703	77.191

SPECIFIC MODEL PARAMETERS IN KELVIN

		UNIQUAC		NRTL(ALPHA=.2)	
I	J	AIJ	AJI	AIJ	AJI
1	2	-69.230	12.943	368.32	17.136
1	3	457.54	137.31	255.60	2149.5
2	3	-208.20	124.34	-460.45	1120.4

R1 = 4.8274   R2 = 2.2024   R3 = 0.9200
Q1 = 4.196   Q2 = 2.072   Q3 = 1.400

MEAN DEV. BETWEEN CALC. AND EXP. CONC. IN MOLE PCT

UNIQUAC (SPECIFIC PARAMETERS)   0.50
NRTL (SPECIFIC PARAMETERS)   0.44
UNIQUAC (COMMON PARAMETERS)   0.67

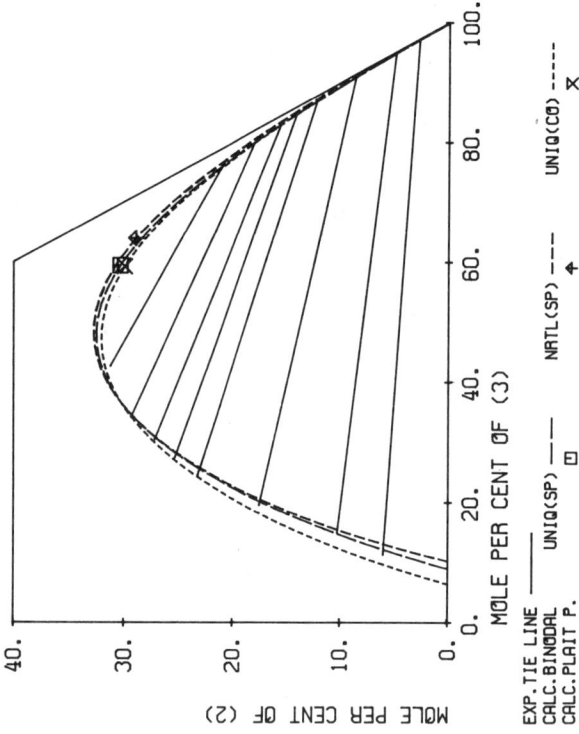

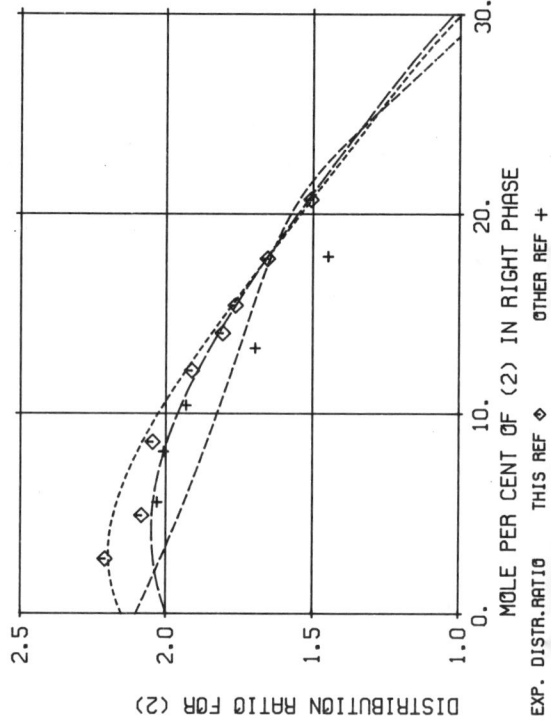

MOLE PER CENT OF (3)

MOLE PER CENT OF (2)

EXP.TIE LINE ———   UNIQ(SP) ⊡   NRTL(SP) ——— ▲   UNIQ(CO) -----
CALC.BINODAL -----
CALC.PLAIT P. ✕   ✕

MOLE PER CENT OF (2) IN RIGHT PHASE

DISTRIBUTION RATIO FOR (2)

EXP. DISTR.RATIO   THIS REF ◇   OTHER REF +

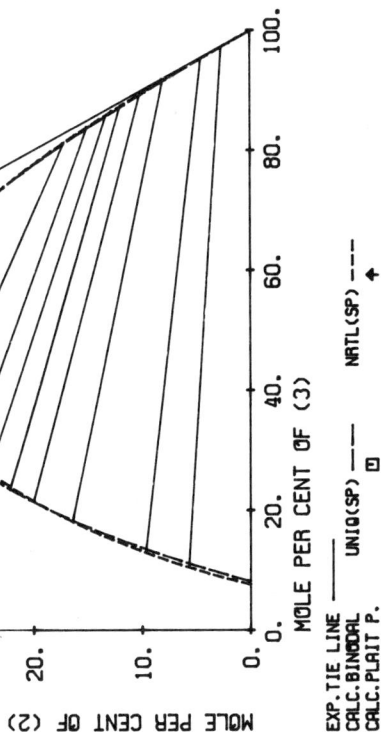

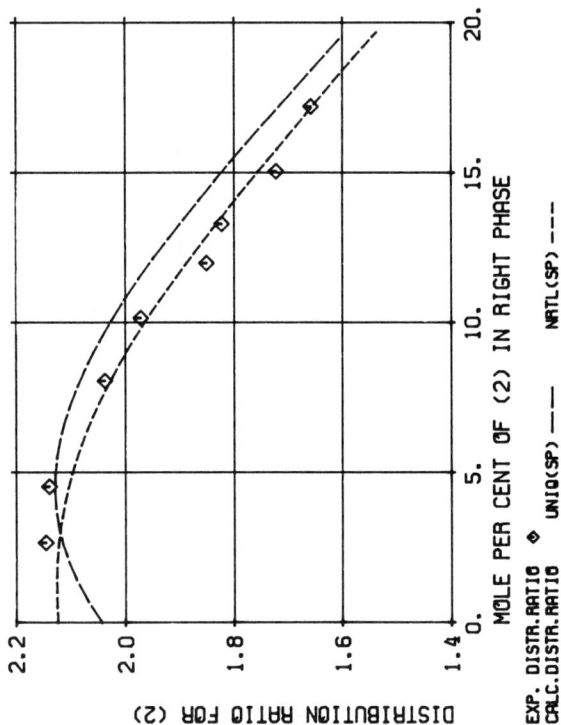

$C_2H_4O_2-C_6H_{12}O_2$

(2) C2H4O2   ACETIC ACID

(3) H2O   WATER

DAKSHINAMURTY P., ET AL.
J.CHEM.ENG.DATA 17(1972)379

TEMPERATURE = 50.0 DEG C   TYPE OF SYSTEM = 1

EXPERIMENTAL TIE LINES IN MOLE PCT

	LEFT PHASE			RIGHT PHASE	
(1)	(2)	(3)	(1)	(2)	(3)
83.706	5.705	10.590	0.149	2.658	97.192
77.415	9.684	12.901	0.173	4.526	95.301
65.076	16.395	18.529	0.261	8.044	91.695
60.608	20.004	21.388	0.330	10.148	89.472
54.437	22.163	23.400	0.507	11.970	87.523
50.064	24.223	25.713	0.711	13.288	86.000
45.182	25.909	28.909	1.003	15.036	83.962
36.856	28.507	34.637	1.519	17.186	81.295

SPECIFIC MODEL PARAMETERS IN KELVIN

		UNIQUAC		NRTL(ALPHA=.2)	
I J	AIJ	AJI		AIJ	AJI
1 2	-63.808	-64.084		-388.14	36.245
1 3	513.40	144.92		348.59	2119.8
2 3	-221.42	60.403		-614.53	667.85

R1 = 4.8274   R2 = 2.2024   R3 = 0.9200
Q1 = 4.196   Q2 = 2.072   Q3 = 1.400

MEAN DEV. BETWEEN CALC. AND EXP. CONC. IN MOLE PCT

UNIQUAC (SPECIFIC PARAMETERS)   0.37
NRTL   (SPECIFIC PARAMETERS)   0.23

$C_2H_4O_2$-$C_6H_{12}O_2$

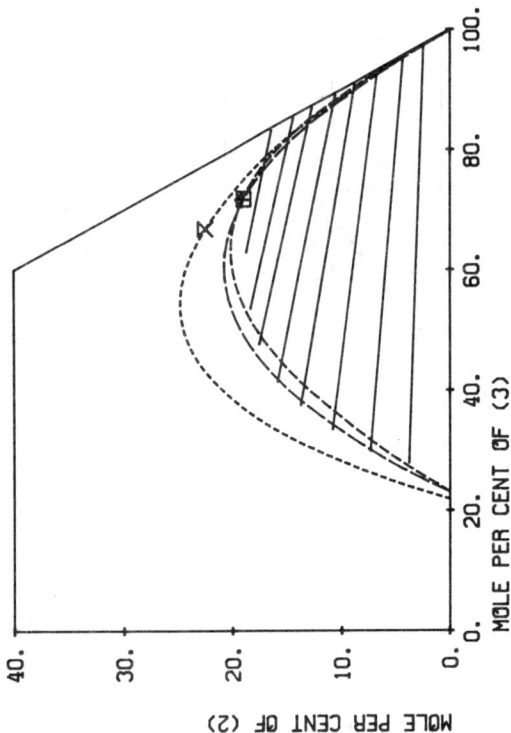

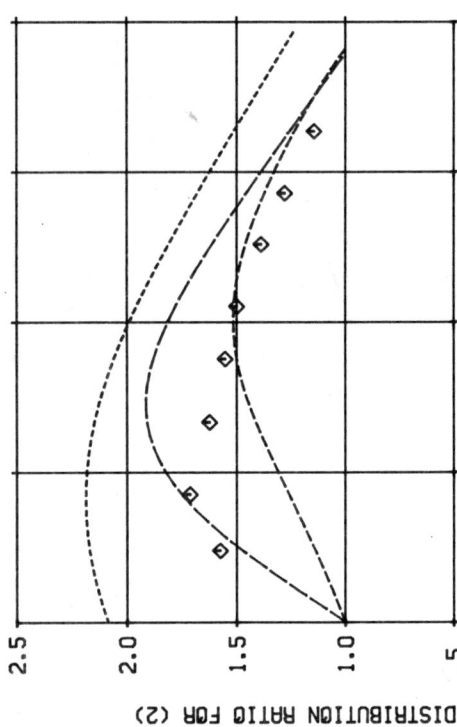

(1) C6H1202   HEXANOIC ACID
(2) C2H4O2   ACETIC ACID
(3) H2O   WATER

OTHMER D.F., SERRANO J.
IND.ENG.CHEM. 41(1949)1030

TEMPERATURE = 25.0 DEG C   TYPE OF SYSTEM = 1

EXPERIMENTAL TIE LINES IN MOLE PCT (GRAPH.INTERPOL.)

	LEFT PHASE			RIGHT PHASE	
(1)	(2)	(3)	(1)	(2)	(3)
68.344	3.753	27.903	0.186	2.382	97.432
62.837	7.304	29.859	0.200	4.264	95.537
55.739	10.808	33.453	0.223	6.654	93.123
48.930	13.624	37.446	0.252	9.773	90.975
42.803	15.761	41.436	0.281	10.514	89.205
34.890	17.468	47.641	0.319	12.586	87.094
28.138	18.291	53.571	0.362	14.289	85.349
18.376	18.724	62.900	0.427	16.352	83.221

SPECIFIC MODEL PARAMETERS IN KELVIN

	UNIQUAC		NRTL(ALPHA=.2)	
I J	AIJ	AJI	AIJ	AJI
1 2	-195.40	-96.576	648.67	-334.62
1 3	228.27	170.70	2.2011	2288.4
2 3	-361.77	116.83	-114.39	-25.577

R1 = 4.9000   R2 = 2.2024   R3 = 0.9200
Q1 = 4.232   Q2 = 2.072   Q3 = 1.400

MEAN DEV. BETWEEN CALC. AND EXP. CONC. IN MOLE PCT

UNIQUAC (SPECIFIC PARAMETERS)   1.24
NRTL   (SPECIFIC PARAMETERS)   1.07
UNIQUAC (COMMON PARAMETERS)   2.79

$C_2H_4O_2$-$C_6H_{12}O_2$

MOLE PER CENT OF (3)

MOLE PER CENT OF (2)

EXP.TIE LINE ——
CALC.BINODAL UNIQ(CO) -----
CALC.PLAIT P. ✕

MOLE PER CENT OF (2) IN RIGHT PHASE

DISTRIBUTION RATIO FOR (2)

EXP. DISTR.RATIO ◇ UNIQ(CO) -----
CALC.DISTR.RATIO -----

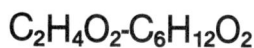

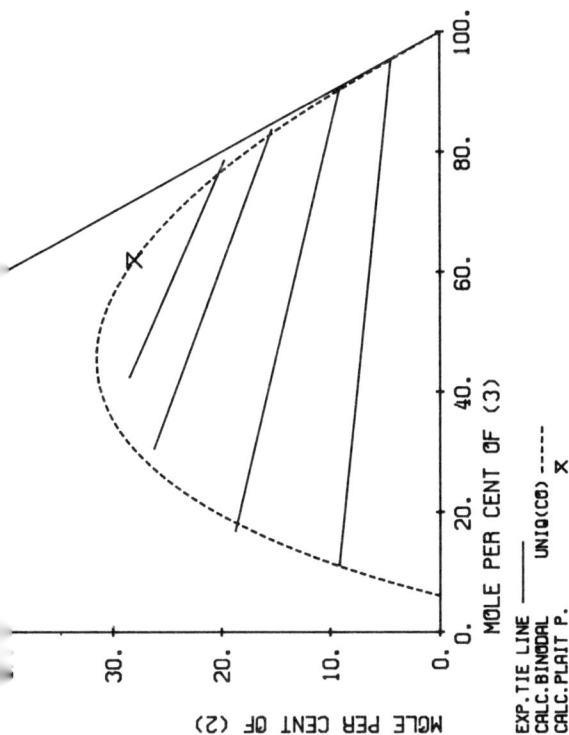

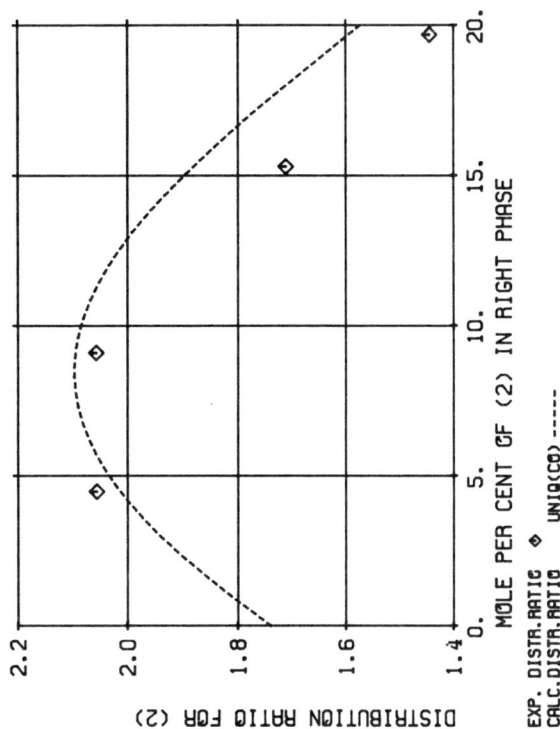

(2) C2H4O2   ACETIC ACID

(3) H2O   WATER

BOMSHTEIN A.L., TROFIMOV A.N., SERAFIMOV L.A.
ZH.PRIKL.KHIM.(LENINGRAD) 51(1978)1280

TEMPERATURE = 20.0 DEG C   TYPE OF SYSTEM = 1

EXPERIMENTAL TIE LINES IN MOLE PCT

LEFT PHASE			RIGHT PHASE		
(1)	(2)	(3)	(1)	(2)	(3)
79.730	9.170	11.100	0.240	4.460	95.300
64.480	18.700	16.820	0.360	9.090	90.550
43.270	26.180	30.550	1.010	15.290	83.700
29.060	28.430	42.510	1.740	19.670	78.590

MEAN DEV. BETWEEN CALC. AND EXP. CONC. IN MOLE PCT

UNIQUAC (COMMON PARAMETERS)   0.92

$C_2H_4O_2$-C_6H_{14}

(1) H2O   WATER

(2) C2H4O2   ACETIC ACID

(3) C6H14   HEXANE

FUSE K., IGUCHI A.
KAGAKU KOGAKU 34(1970)1226

TEMPERATURE = 25.0 DEG C   TYPE OF SYSTEM = 1

EXPERIMENTAL TIE LINES IN MOLE PCT (GRAPH.INTERPOL.)

	LEFT PHASE			RIGHT PHASE	
(1)	(2)	(3)	(1)	(2)	(3)
85.161	14.810	0.028	0.0	1.614	98.386
70.025	29.904	0.071	0.0	4.026	95.974
54.481	44.869	0.650	0.465	8.075	91.460
42.508	56.110	1.382	0.458	13.045	86.497
40.408	58.079	1.513	0.457	13.256	86.285
33.824	64.240	1.936	0.453	16.450	83.097
27.653	69.841	2.506	0.891	20.392	78.717
19.654	75.299	4.047	0.873	26.680	72.447
19.381	76.186	4.433	1.270	34.565	64.164
17.845	76.835	5.320	1.261	36.777	61.962
9.746	80.407	9.847	1.248	39.705	59.047
6.858	80.141	13.001	1.624	45.590	52.786

SPECIFIC MODEL PARAMETERS IN KELVIN

I J	UNIQUAC AIJ	AJI	NRTL(ALPHA=.2) AIJ	AJI
1 2	-331.11	57.720	-657.45	458.96
1 3	452.74	780.59	2202.2	1218.5
2 3	51.890	169.84	576.36	12.474

R1 = 0.9200   R2 = 2.2024   R3 = 4.4998
Q1 = 1.400   Q2 = 2.072   Q3 = 3.856

MEAN DEV. BETWEEN CALC. AND EXP. CONC. IN MOLE PCT

UNIQUAC (SPECIFIC PARAMETERS)	0.94
NRTL (SPECIFIC PARAMETERS)	0.97
UNIQUAC (COMMON PARAMETERS)	1.98

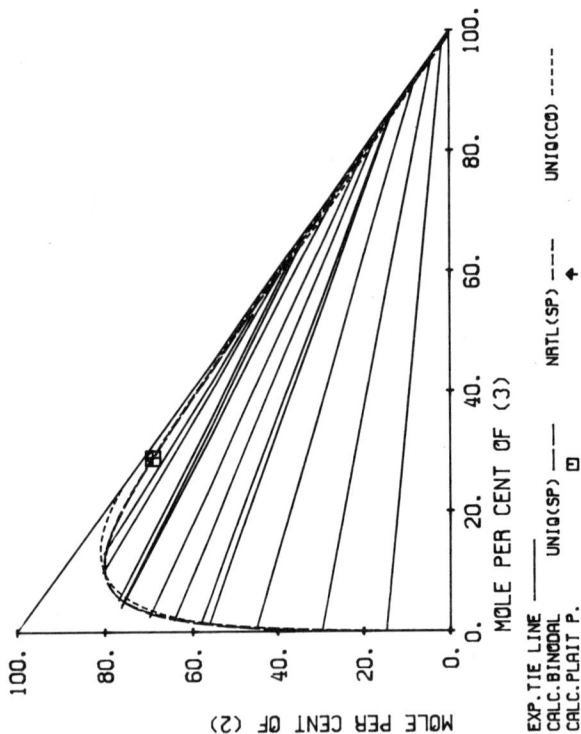

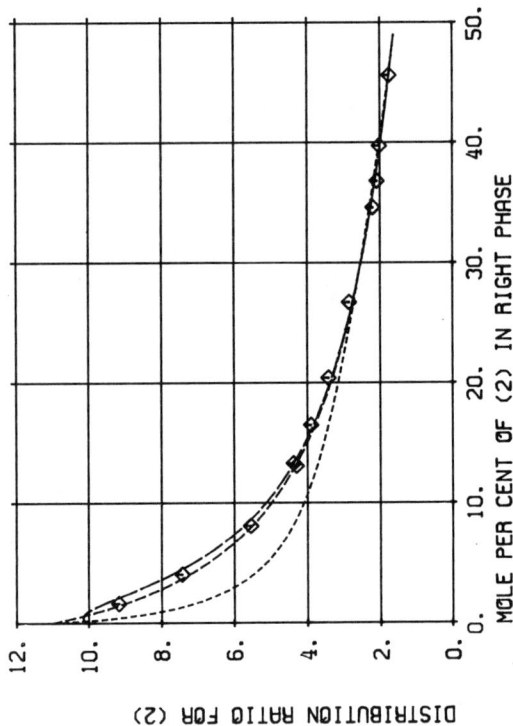

(2) C2H4O2   ACETIC ACID
(3) H2O   WATER

OTHMER D.F., WHITE R.E., TRUEGER E.
IND.ENG.CHEM. 33(1941)1240

TEMPERATURE = 23.5 DEG C   TYPE OF SYSTEM = 1

EXPERIMENTAL TIE LINES IN MOLE PCT (GRAPH.INTERPOL.)

	LEFT PHASE			RIGHT PHASE	
	(1)	(2)	(3)	(1)	(2)	(3)
83.028	9.327	7.645	0.292	5.518	94.190
76.027	13.741	10.232	0.360	7.890	91.750
62.407	22.289	15.303	0.677	13.544	85.779
37.500	31.617	30.884	2.328	22.786	74.886

SPECIFIC MODEL PARAMETERS IN KELVIN

		UNIQUAC		NRTL(ALPHA=.2)	
I	J	AIJ	AJI	AIJ	AJI
1	2	46.858	-161.23	-386.51	26.660
1	3	571.50	-154.62	451.42	1811.3
2	3	-203.29	-37.785	-545.17	444.89

R1 = 4.7421   R2 = 2.2024   R3 = 0.9200
Q1 = 4.088   Q2 = 2.072   Q3 = 1.400

MEAN DEV. BETWEEN CALC. AND EXP. CONC. IN MOLE PCT

UNIQUAC (SPECIFIC PARAMETERS)   0.52
NRTL   (SPECIFIC PARAMETERS)   0.47
UNIQUAC (COMMON PARAMETERS)   2.05

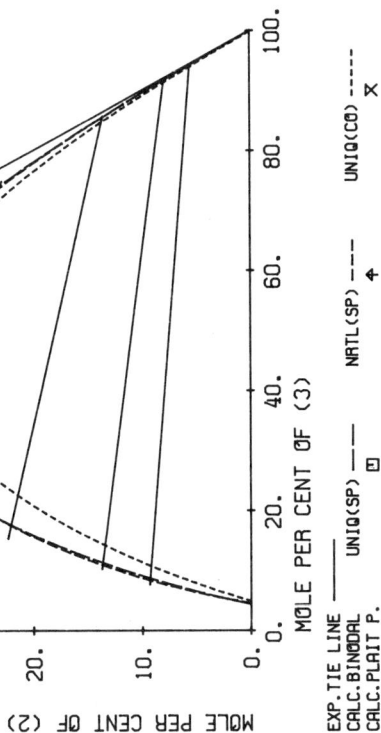

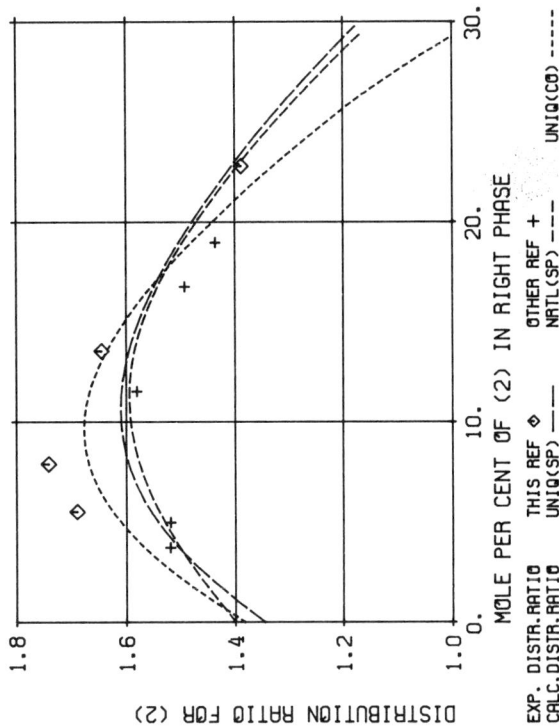

$C_2H_4O_2$-$C_6H_{14}O$

(1) C6H14O   ETHER,DIISOPROPYL

(2) C2H4O2   ACETIC ACID

(3) H2O   WATER

HLAVATY K., LINEK J.
COLLECT.CZECH.CHEM.COMMUN. 38(1973)374

TEMPERATURE = 24.6 DEG C   TYPE OF SYSTEM = 1

EXPERIMENTAL TIE LINES IN MOLE PCT

	LEFT PHASE			RIGHT PHASE	
(1)	(2)	(3)	(1)	(2)	(3)
95.101	0.0	4.899	0.250	0.0	99.750
87.570	5.641	6.790	0.390	3.712	95.898
84.748	7.549	7.703	0.400	4.970	94.630
67.186	18.210	14.604	0.712	11.524	87.764
52.730	25.011	22.259	1.359	16.755	81.886
47.357	27.245	25.398	1.803	18.961	79.235

SPECIFIC MODEL PARAMETERS IN KELVIN

		UNIQUAC		NRTL(ALPHA=.2)	
I J	AIJ	AJI	AIJ	AJI	
1 2	46.858	-161.23	-386.51	26.660	
1 3	571.50	-154.62	451.42	1811.3	
2 3	-203.29	-37.785	-545.17	444.89	

R1 = 4.7421   R2 = 2.2024   R3 = 0.9200
Q1 = 4.088   Q2 = 2.072   Q3 = 1.400

MEAN DEV. BETWEEN CALC. AND EXP. CONC. IN MOLE PCT

UNIQUAC (SPECIFIC PARAMETERS)   0.33
NRTL   (SPECIFIC PARAMETERS)   0.28
UNIQUAC (COMMON PARAMETERS)   1.41

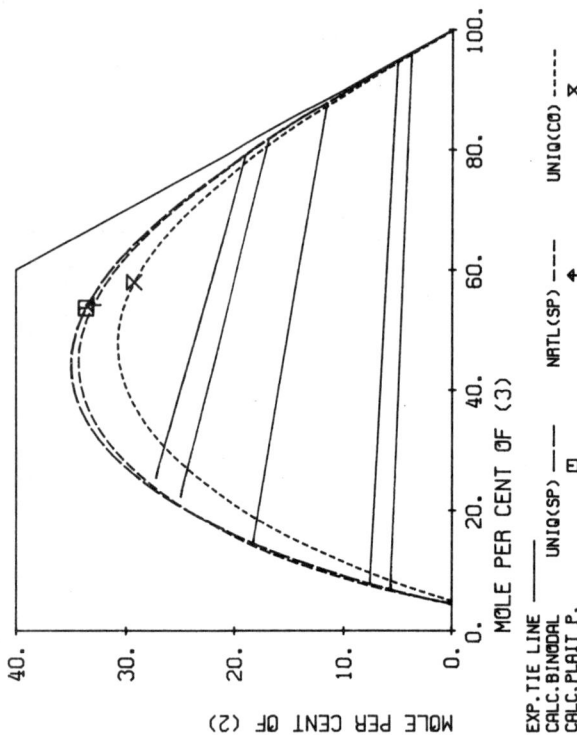

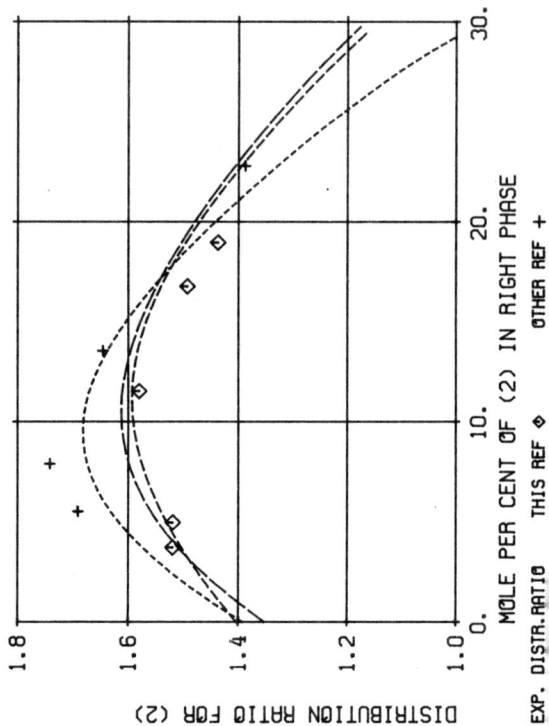

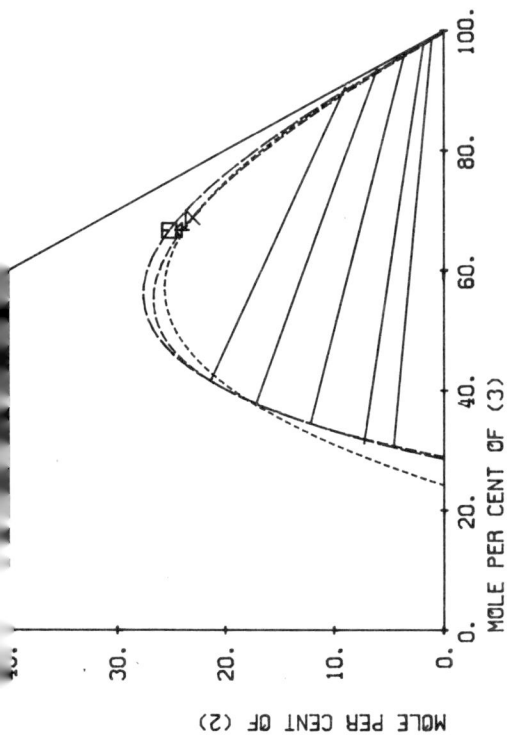

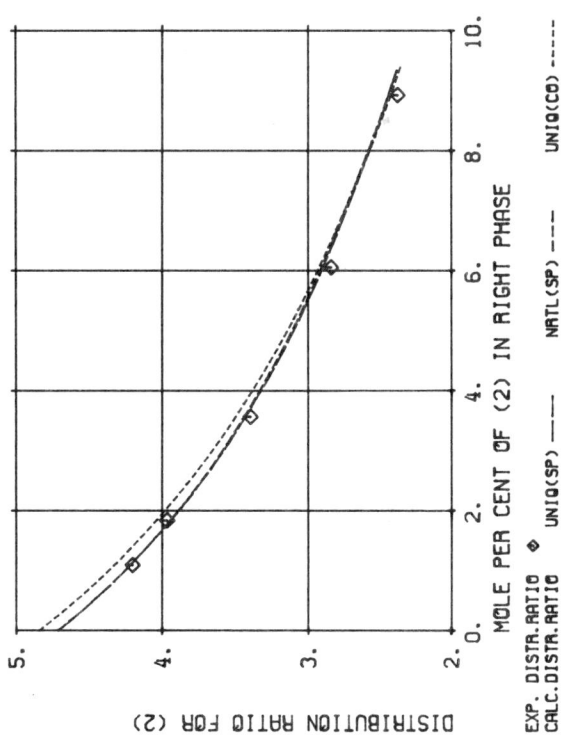

(2) C2H4O2   ACETIC ACID

(3) H2O   WATER

RAJA RAO M.; RAMAMURTY M., VENKATA RAO C.
CHEM.ENG.SCI. 8(1958)265

TEMPERATURE = 30.0 DEG C   TYPE OF SYSTEM = 1

EXPERIMENTAL TIE LINES IN MOLE PCT

	LEFT PHASE			RIGHT PHASE	
(1)	(2)	(3)	(1)	(2)	(3)
64.518	4.574	30.907	0.238	1.088	98.674
61.562	7.283	31.155	0.260	1.836	97.904
53.227	12.099	34.675	0.329	3.559	96.112
45.222	17.182	37.596	0.388	6.052	93.560
36.896	21.249	41.855	0.566	8.927	90.507

SPECIFIC MODEL PARAMETERS IN KELVIN

		UNIQUAC		NRTL(ALPHA=.2)	
I	J	AIJ	AJI	AIJ	AJI
1	2	-40.585	-117.17	-76.118	-115.48
1	3	186.84	164.93	-110.34	1739.2
2	3	-147.25	36.804	78.907	143.05

R1 = 4.8015   R2 = 2.2024   R3 = 0.9200
Q1 = 4.124   Q2 = 2.072   Q3 = 1.400

MEAN DEV. BETWEEN CALC. AND EXP. CONC. IN MOLE PCT

UNIQUAC (SPECIFIC PARAMETERS)   0.20
NRTL   (SPECIFIC PARAMETERS)   0.32
UNIQUAC (COMMON PARAMETERS)   0.79

$C_2H_4O_2$-C_7H_6O

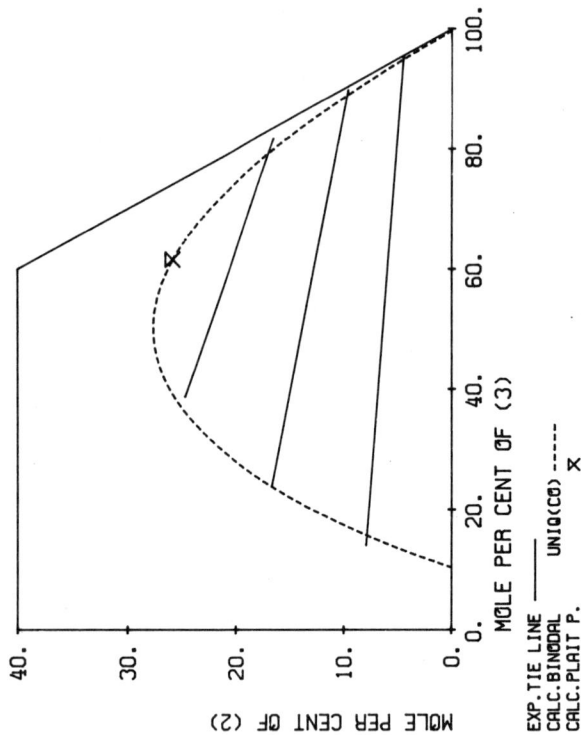

MOLE PER CENT OF (2)

MOLE PER CENT OF (3)

EXP.TIE LINE ——
CALC.BINODAL ----
CALC.PLAIT P.   ✗

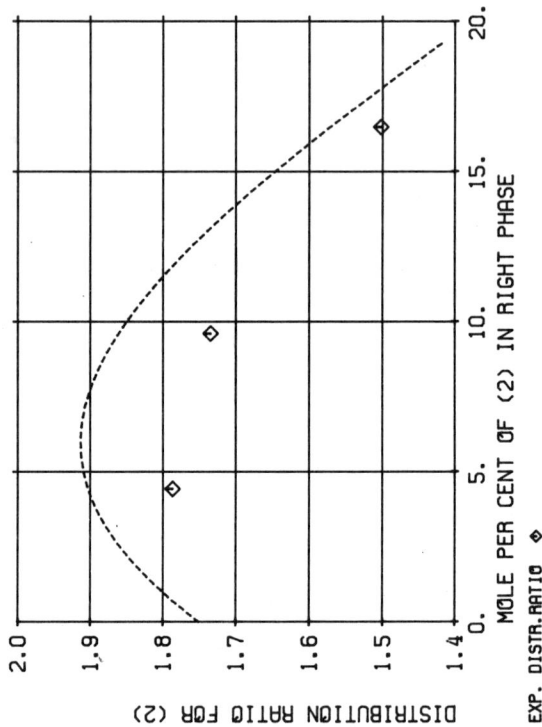

DISTRIBUTION RATIO FOR (2)

MOLE PER CENT OF (2) IN RIGHT PHASE

EXP. DISTR.RATIO ◇

(1) C7H6O   BENZALDEHYDE
(2) C2H4O2   ACETIC ACID
(3) H2O   WATER

SEHRT B.
CHEM.TECH.(BERLIN) 15(1963)624

TEMPERATURE = 20.0 DEG C   TYPE OF SYSTEM = 1

EXPERIMENTAL TIE LINES IN MOLE PCT (GRAPH.INTERPOL.)

LEFT PHASE			RIGHT PHASE		
(1)	(2)	(3)	(1)	(2)	(3)
78.093	7.896	14.011	0.267	4.420	95.313
59.506	16.636	23.858	0.615	9.592	39.792
36.561	24.736	38.703	1.786	16.467	31.747

MEAN DEV. BETWEEN CALC. AND EXP. CONC. IN MOLE PCT
UNIQUAC (COMMON PARAMETERS)   0.70

$C_2H_4O_2-C_7H_8$

(2) C2H4O2   ACETIC ACID

(3) C7H8   TOLUENE

WOODMAN R.M.
J.PHYS.CHEM. 30(1926)1283

TEMPERATURE = 25.0 DEG C   TYPE OF SYSTEM = 1

EXPERIMENTAL TIE LINES IN MOLE PCT (GRAPH.INTERPOL.)

	LEFT PHASE			RIGHT PHASE	
(1)	(2)	(3)	(1)	(2)	(3)
92.415	7.553	0.032	0.380	2.087	97.534
88.090	11.845	0.065	0.502	4.212	95.286
84.436	15.439	0.125	0.608	6.086	93.307
78.123	21.596	0.281	0.801	9.382	89.817
69.270	30.098	0.632	1.195	14.905	83.899
64.321	34.702	0.976	1.475	18.405	80.120
61.220	37.507	1.274	1.728	20.987	77.285
56.647	41.519	1.834	2.230	25.433	72.335

SPECIFIC MODEL PARAMETERS IN KELVIN

		UNIQUAC		NRTL(ALPHA=.2)	
I J	AIJ	AJI	AIJ	AJI	
1 2	-323.46	-60.465	-108.27	-237.84	
1 3	433.38	797.83	2173.5	1160.3	
2 3	-35.149	-44.159	401.73	-274.26	

R1 = 0.9200	R2 = 2.2024	R3 = 3.9228
Q1 = 1.400	Q2 = 2.072	Q3 = 2.968

MEAN DEV. BETWEEN CALC. AND EXP. CONC. IN MOLE PCT

UNIQUAC (SPECIFIC PARAMETERS)	0.65
NRTL (SPECIFIC PARAMETERS)	0.60
UNIQUAC (COMMON PARAMETERS)	0.92

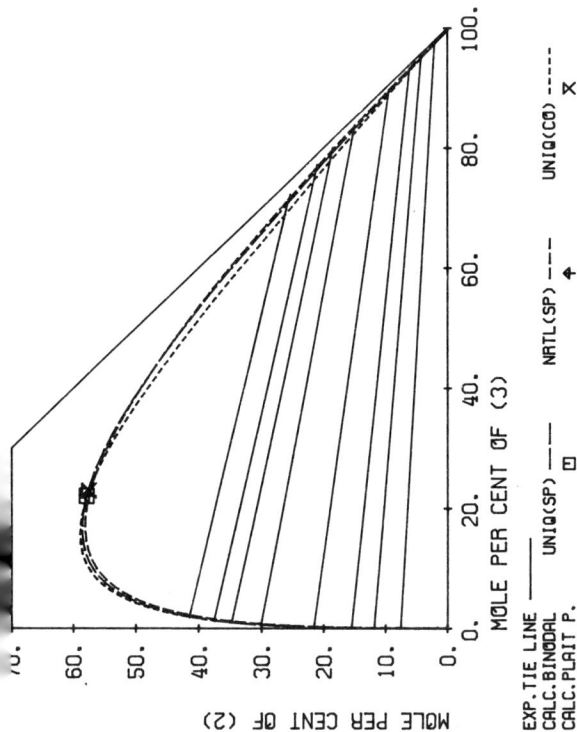

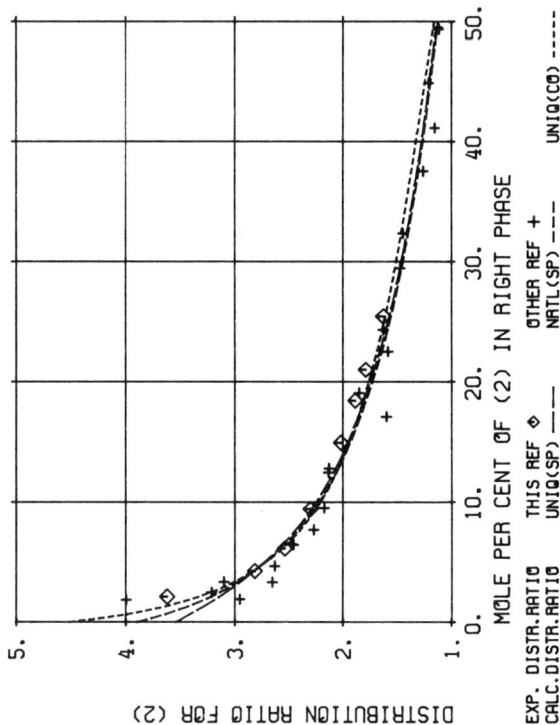

C₂H₄O₂-C₇H₈

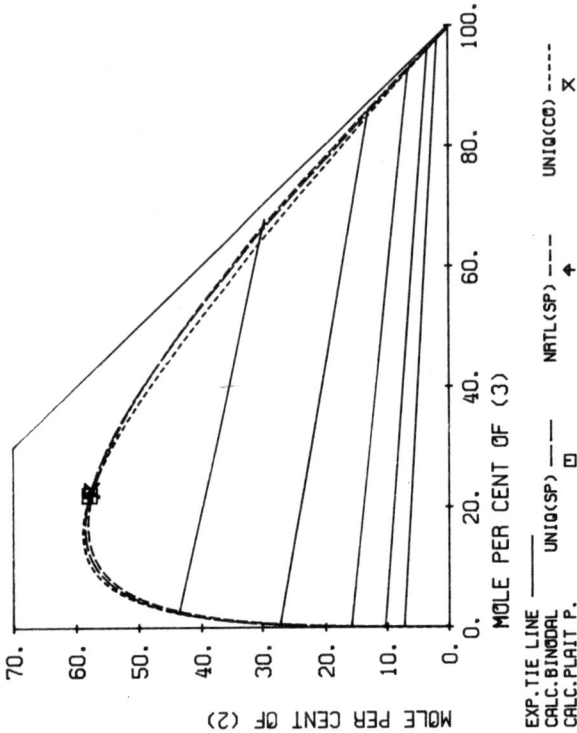

(1) H2O   WATER

(2) C2H4O2   ACETIC ACID

(3) C7H8   TOLUENE

RIUS MIRO A., OTERO DE LA GRANADA J.L.
AN.R.SOC.ESP.FIS.QUIM. 48B(1952)569

TEMPERATURE = 25.0 DEG C   TYPE OF SYSTEM = 1

EXPERIMENTAL TIE LINES IN MOLE PCT (GRAPH.INTERPOL.)

	LEFT PHASE			RIGHT PHASE	
(1)	(2)	(3)	(1)	(2)	(3)
92.620	7.289	0.091	0.395	1.824	97.781
89.557	10.316	0.127	0.478	3.324	96.198
84.021	15.775	0.205	0.696	5.414	92.890
72.290	27.214	0.495	1.138	12.779	86.083
54.071	43.658	2.271	2.732	29.424	67.845

SPECIFIC MODEL PARAMETERS IN KELVIN

		UNIQUAC		NRTL(ALPHA=.2)	
I J	AIJ	AJI		AIJ	AJI
1 2	-323.46	-60.466		-108.27	-237.84
1 3	433.38	797.83		2173.5	1160.3
2 3	-35.149	-44.159		401.73	-274.26

R1 = 0.9200   R2 = 2.2024   R3 = 3.9228
Q1 = 1.400   Q2 = 2.072   Q3 = 2.968

MEAN DEV. BETWEEN CALC. AND EXP. CONC. IN MOLE PCT

UNIQUAC (SPECIFIC PARAMETERS)   0.42
NRTL   (SPECIFIC PARAMETERS)   0.34
UNIQUAC (COMMON PARAMETERS)   0.70

MOLE PER CENT OF (2)

MOLE PER CENT OF (3)

EXP.TIE LINE   UNIQ(SP) ⊟   NRTL(SP) ✦   UNIQ(CO) ✕
CALC.BINODAL
CALC.PLAIT P.

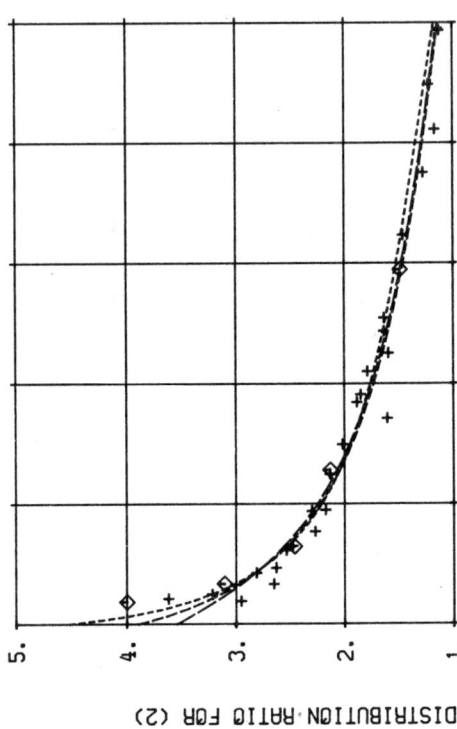

DISTRIBUTION RATIO FOR (2)

MOLE PER CENT OF (2) IN RIGHT PHASE

EXP. DISTR.RATIO   THIS REF ◇   OTHER REF +

$C_2H_4O_2$-C_7H_8

MOLE PER CENT OF (3)

MOLE PER CENT OF (2)

EXP.TIE LINE ——
CALC.BINODAL   UNIQ(SP) □ ——   NRTL(SP) ● ——
CALC.PLAIT P.

MOLE PER CENT OF (2) IN RIGHT PHASE

DISTRIBUTION RATIO FOR (2)

EXP. DISTR.RATIO ◇   UNIQ(SP) ◇ ——
CALC.DISTR.RATIO   NRTL(SP) ——

---

(2) C2H4O2   ACETIC ACID
(3) C7H8   TOLUENE

RIUS MIRO A., OTERO DE LA GRANADA J.L.
AN.R.SOC.ESP.FIS.QUIM. 48B(1952)569

TEMPERATURE = 40.0 DEG C   TYPE OF SYSTEM = 1

EXPERIMENTAL TIE LINES IN MOLE PCT (GRAPH.INTERPOL.)

LEFT PHASE			RIGHT PHASE		
(1)	(2)	(3)	(1)	(2)	(3)
96.108	3.833	0.060	0.508	0.914	98.578
89.984	9.866	0.150	0.852	3.908	95.240
80.714	18.956	0.330	1.397	8.678	89.924
73.679	25.772	0.549	1.945	13.108	84.947
64.163	34.363	1.475	2.834	19.792	77.374

SPECIFIC MODEL PARAMETERS IN KELVIN

		UNIQUAC		NRTL(ALPHA=.2)	
I J	AIJ	AJI		AIJ	AJI
1 2	-162.60	-127.84		-94.072	-274.93
1 3	388.83	720.22		1660.5	939.20
2 3	37.507	-14.202		621.12	-398.95

R1 = 0.9200   R2 = 2.2024   R3 = 3.9228
Q1 = 1.400   Q2 = 2.072   Q3 = 2.968

MEAN DEV. BETWEEN CALC. AND EXP. CONC. IN MOLE PCT

UNIQUAC (SPECIFIC PARAMETERS)   0.31
NRTL   (SPECIFIC PARAMETERS)   0.25

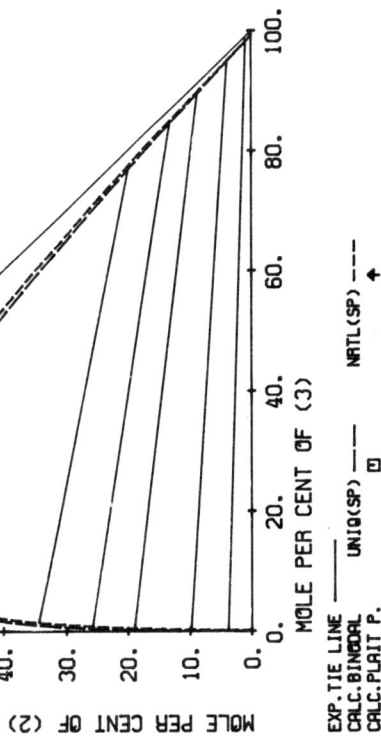

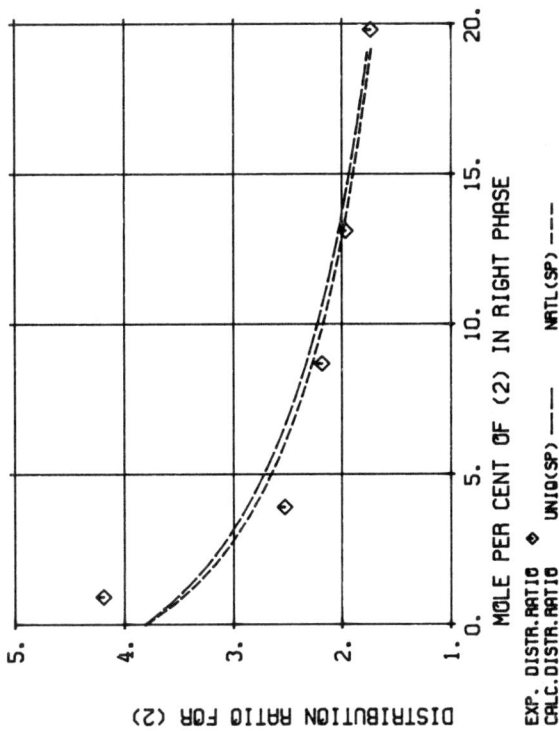

$C_2H_4O_2$-C_7H_8

(1) H2O   WATER
(2) C2H4O2   ACETIC ACID
(3) C7H8   TOLUENE

RIUS MIRO A., OTERO DE LA GRANADA J.L.
AN.R.SOC.ESP.FIS.QUIM. 48B(1952)569

TEMPERATURE = 55.0 DEG C   TYPE OF SYSTEM = 1

EXPERIMENTAL TIE LINES IN MOLE PCT (GRAPH.INTERPOL.)

LEFT PHASE			RIGHT PHASE		
(1)	(2)	(3)	(1)	(2)	(3)
95.751	4.161	0.088	0.709	1.215	98.076
91.507	8.324	0.169	1.028	3.160	95.811
87.092	12.653	0.255	1.389	5.506	93.106
74.494	24.906	0.600	2.694	13.301	84.006
64.730	33.579	1.692	3.986	20.754	75.261

SPECIFIC MODEL PARAMETERS IN KELVIN

		UNIQUAC			NRTL(ALPHA=.2)	
I J		AIJ	AJI		AIJ	AJI
1 2		-141.19	-145.31		15.932	-348.44
1 3		363.26	682.77		1698.0	915.46
2 3		29.713	-3.5008		591.44	-387.18

R1 = 0.9200   R2 = 2.2024   R3 = 3.9228
Q1 = 1.400   Q2 = 2.072   Q3 = 2.968

MEAN DEV. BETWEEN CALC. AND EXP. CONC. IN MOLE PCT

UNIQUAC (SPECIFIC PARAMETERS)   0.29
NRTL   (SPECIFIC PARAMETERS)   0.23

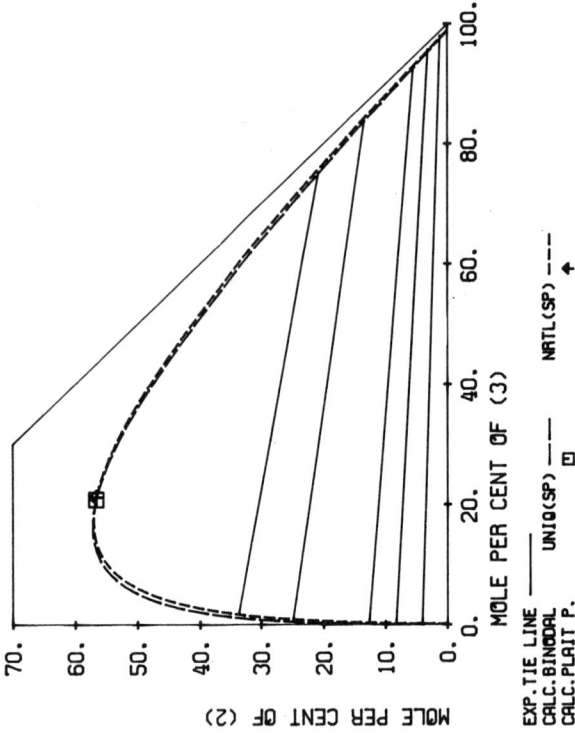

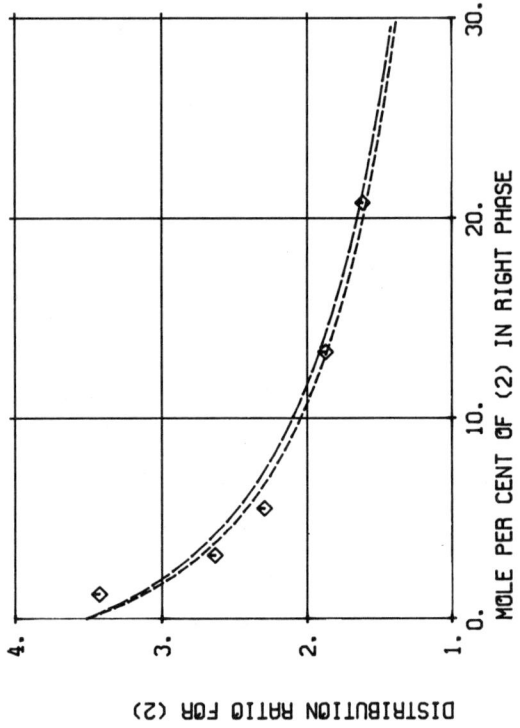

EXP.TIE LINE ——————   UNIQ(SP) ————— ⊟   NRTL(SP) ----- ✦
CALC.BINODAL
CALC.PLAIT P.

MOLE PER CENT OF (3)

MOLE PER CENT OF (2)

DISTRIBUTION RATIO FOR (2)

MOLE PER CENT OF (2) IN RIGHT PHASE

EXP. DISTR.RATIO ◆

$C_2H_4O_2\text{-}C_7H_8$

MOLE PER CENT OF (3)

MOLE PER CENT OF (2)

EXP.TIE LINE ———

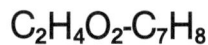

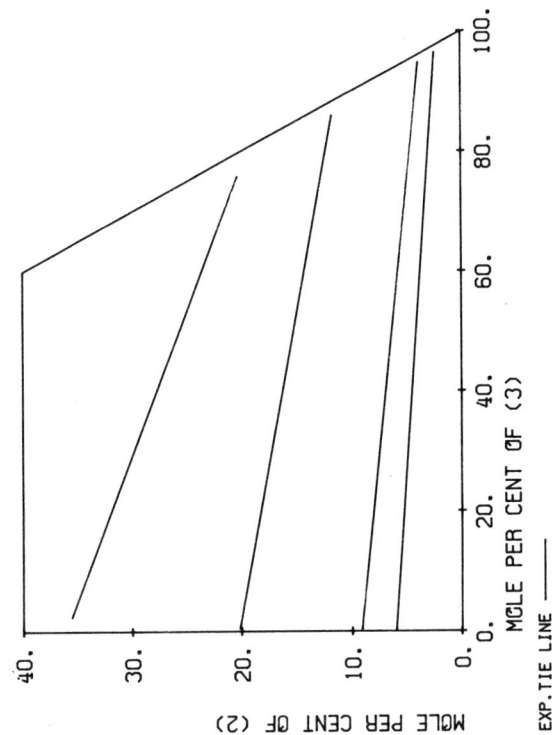

(1) H2O   WATER
(2) C2H4O2   ACETIC ACID
(3) C7H8   TOLUENE

RIUS MIRO A., OTERO DE LA GRANADA J.L.
AN.R.SOC.ESP.FIS.QUIM. 48B(1952)569

TEMPERATURE = 75.0 DEG C   TYPE OF SYSTEM = 1

EXPERIMENTAL TIE LINES IN MOLE PCT (GRAPH.INTERPOL.)

| | LEFT PHASE | | | RIGHT PHASE | |
(1)	(2)	(3)	(1)	(2)	(3)
93.833	6.051	0.117	1.106	2.413	96.481
90.708	9.113	0.179	1.298	3.894	94.808
79.325	20.141	0.534	2.310	11.693	85.998
62.029	35.539	2.433	3.905	20.258	75.833

C$_2$H$_4$O$_2$-C$_7$H$_8$

(1)	H2O	WATER
(2)	C2H4O2	ACETIC ACID
(3)	C7H8	TOLUENE

RIUS MIRO A., OTERO DE LA GRANADA J.L.
AN.R.SOC.ESP.FIS.QUIM. 48B(1952)569

TEMPERATURE = 65.0 DEG C   TYPE OF SYSTEM = 1

EXPERIMENTAL TIE LINES IN MOLE PCT (GRAPH.INTERPOL.)

LEFT PHASE			RIGHT PHASE		
(1)	(2)	(3)	(1)	(2)	(3)
92.605	7.229	0.165	1.277	3.005	95.717
85.251	14.384	0.364	2.061	6.919	91.020
83.061	16.500	0.439	2.153	7.487	90.360
74.265	24.755	0.981	3.257	12.696	84.047

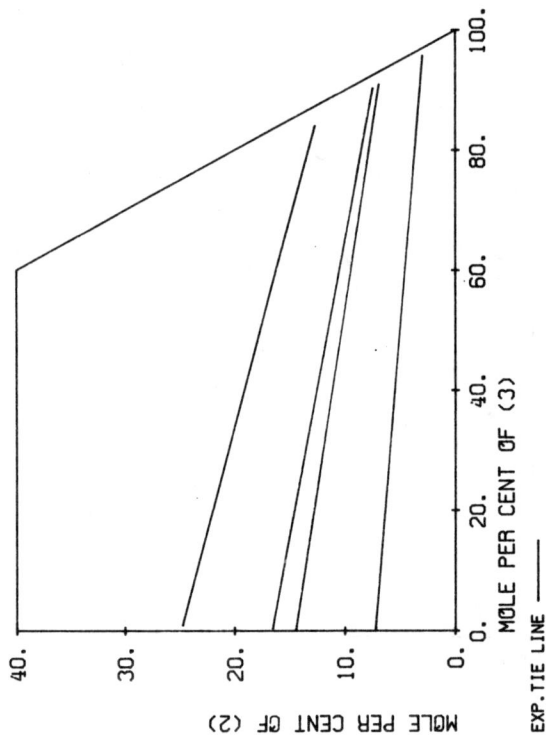

$C_2H_4O_2\text{-}C_7H_8$

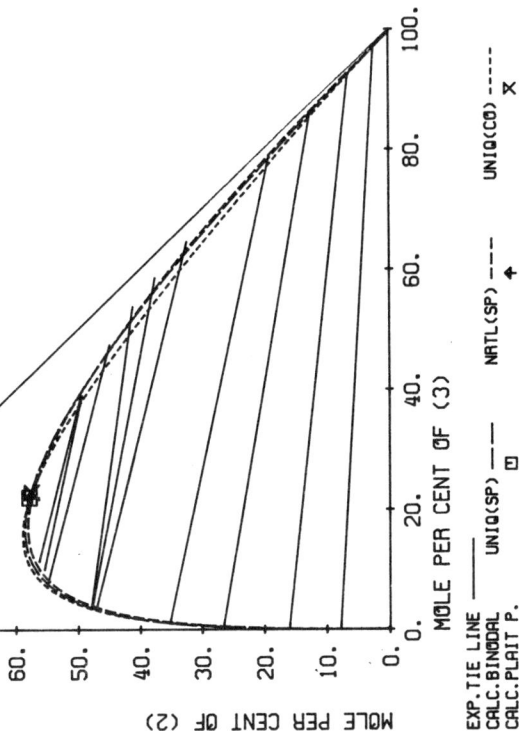

MOLE PER CENT OF (2)

MOLE PER CENT OF (3)

EXP.TIE LINE ———
CALC.BINODAL
CALC.PLAIT P.

UNIQ(SP) ——— ▣
NRTL(SP) ——— ✦
UNIQ(CO) –––– ✕

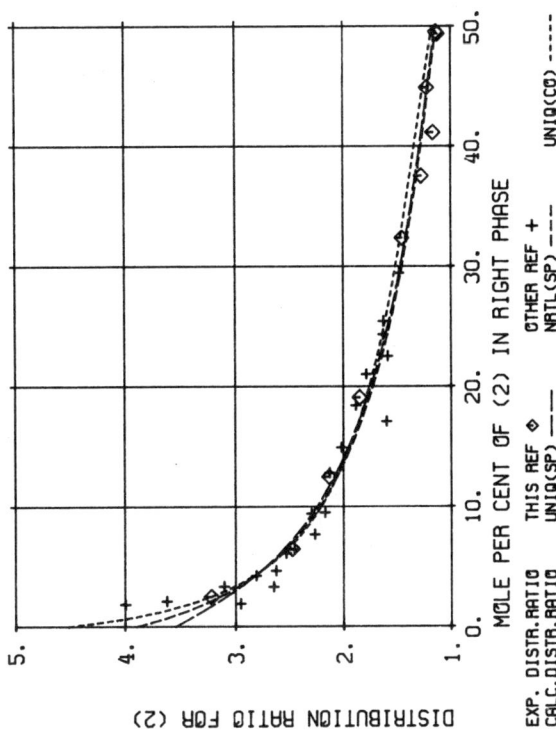

DISTRIBUTION RATIO FOR (2)

MOLE PER CENT OF (2) IN RIGHT PHASE

EXP. DISTR.RATIO   THIS REF ◇
CALC.DISTR.RATIO   UNIQ(SP) ———
                     OTHER REF +
                     NRTL(SP) ––––
                     UNIQ(CO) –––––

(2) C2H4O2   ACETIC ACID

(3) C7H8   TOLUENE

FUSE K., IGUCHI A.
KAGAKU KOGAKU 35(1971)107

TEMPERATURE = 25.0 DEG C   TYPE OF SYSTEM = 1

EXPERIMENTAL TIE LINES IN MOLE PCT (GRAPH.INTERPOL.)

	LEFT PHASE			RIGHT PHASE	
(1)	(2)	(3)	(1)	(2)	(3)
92.104	7.850	0.046	0.505	2.440	97.055
83.803	16.008	0.189	0.992	6.457	92.551
72.899	26.587	0.513	1.450	12.456	86.094
63.642	35.226	1.132	1.879	19.043	79.078
49.587	47.136	3.277	3.089	32.323	64.588
48.657	47.807	3.536	3.859	37.511	58.630
48.538	47.922	3.541	5.012	41.119	53.869
37.268	54.840	7.892	7.605	44.863	47.531
34.495	55.447	10.057	11.646	49.341	39.013
32.289	56.283	11.428	11.638	49.509	38.853

SPECIFIC MODEL PARAMETERS IN KELVIN

		UNIQUAC		NRTL (ALPHA=.2)	
I	J	AIJ	AJI	AIJ	AJI
1	2	-323.46	-60.466	-108.27	-237.84
1	3	433.38	797.83	2173.5	1160.3
2	3	-35.149	-44.159	401.73	-274.26

R1 = 0.9200   R2 = 2.2024   R3 = 3.9228
Q1 = 1.400   Q2 = 2.072   Q3 = 2.968

MEAN DEV. BETWEEN CALC. AND EXP. CONC. IN MOLE PCT

UNIQUAC (SPECIFIC PARAMETERS)   1.02
NRTL   (SPECIFIC PARAMETERS)   0.95
UNIQUAC (COMMON PARAMETERS)   1.55

$C_2H_4O_2\text{-}C_7H_8$

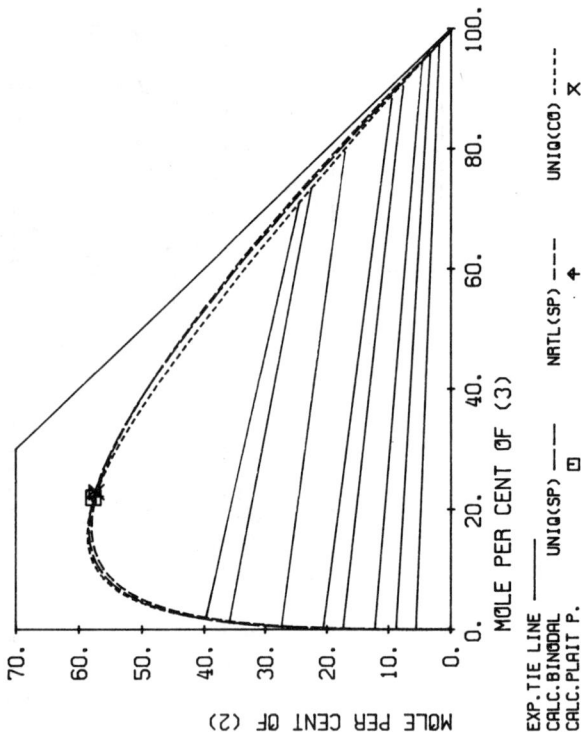

MOLE PER CENT OF (3)

MOLE PER CENT OF (2)

EXP.TIE LINE —————— UNIQ(SP) ——— ⊡ NRTL(SP) ——— ✦ UNIQ(C0) ————— ✕
CALC.BINODAL
CALC.PLAIT P.

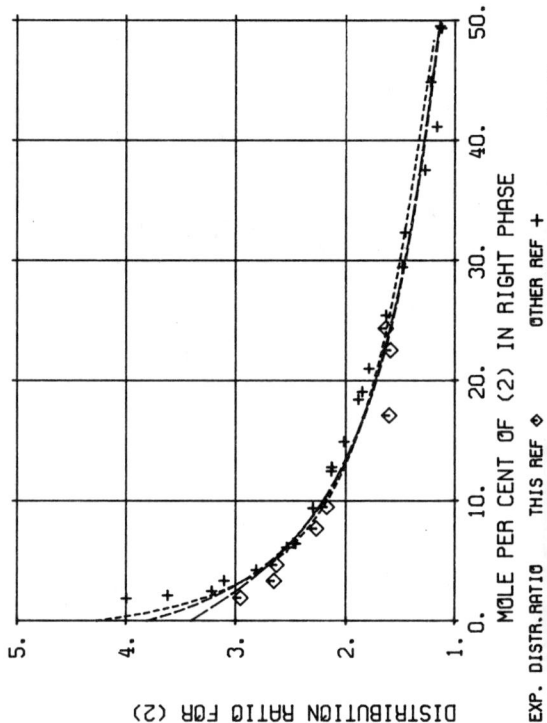

MOLE PER CENT OF (2) IN RIGHT PHASE

DISTRIBUTION RATIO FOR (2)

EXP. DISTR.RATIO ◇ THIS REF + OTHER REF

(1) H2O   WATER
(2) C2H4O2   ACETIC ACID
(3) C7H8   TOLUENE

DAKSHINAMURTY P., SUBRAHMANYAM V., NARSIMHA RAO M.
J.APPL.CHEM.BIOTECHNOL. 23(1973)323

TEMPERATURE = 30.0 DEG C   TYPE OF SYSTEM = 1

EXPERIMENTAL TIE LINES IN MOLE PCT

	LEFT PHASE		RIGHT PHASE	
(1)	(2)	(3)	(2)	(3)
94.404	5.596	0.0	0.758	1.894	97.349
91.141	8.788	0.071	0.903	3.312	95.784
87.612	12.211	0.177	0.998	4.642	94.360
82.406	17.358	0.235	1.863	7.648	90.490
79.137	20.586	0.277	2.185	9.470	88.344
72.204	27.375	0.421	3.500	17.080	79.419
63.045	35.851	1.104	4.091	22.502	73.407
58.448	39.752	1.800	4.280	24.330	71.390

SPECIFIC MODEL PARAMETERS IN KELVIN

I J	UNIQUAC		NRTL(ALPHA=.2)	
	AIJ	AJI	AIJ	AJI
1 2	-323.46	-50.466	-108.27	-237.84
1 3	43.38	797.83	2173.5	1160.3
2 3	-35.149	-44.159	401.73	-274.26

R1 = 0.9200 R2 = 2.2024 R3 = 3.9228
Q1 = 1.400   Q2 = 2.072   Q3 = 2.968

MEAN DEV. BETWEEN CALC. AND EXP. CONC. IN MOLE PCT

UNIQUAC (SPECIFIC PARAMETERS)	0.88
NRTL (SPECIFIC PARAMETERS)	1.01
UNIQUAC (COMMON PARAMETERS)	0.70

$C_2H_4O_2\text{-}C_7H_8$

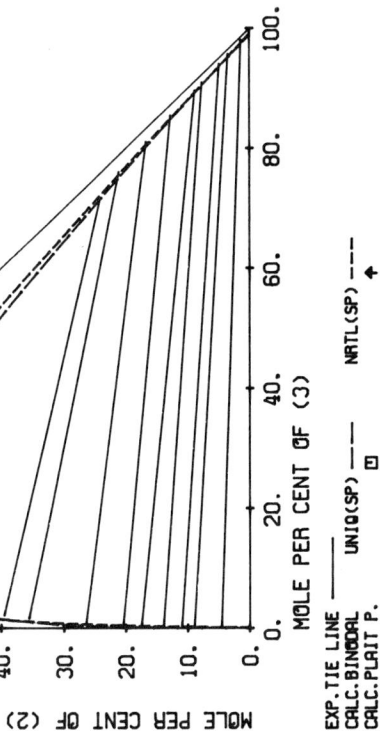

MOLE PER CENT OF (3)

MOLE PER CENT OF (2)

EXP.TIE LINE —— UNIQ(SP) ——⊡ NRTL(SP) ----
CALC.BINODAL
CALC.PLAIT P.

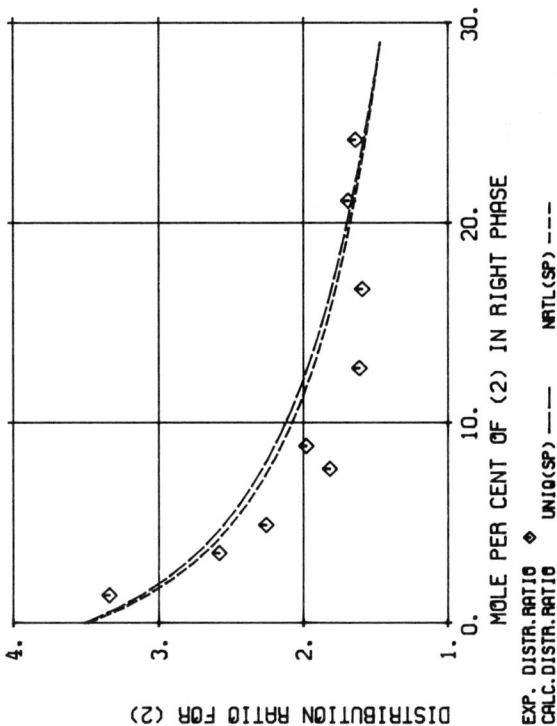

DISTRIBUTION RATIO FOR (2)

MOLE PER CENT OF (2) IN RIGHT PHASE

EXP. DISTR.RATIO ◇ UNIQ(SP) —— NRTL(SP) ----
CALC.DISTR.RATIO

(2) C2H4O2   ACETIC ACID
(3) C7H8   TOLUENE

DAKSHINAMURTY P., SUBRAHMANYAM V., NARSIMHA RAO M.
J.APPL.CHEM.BIOTECHNOL. 23(1973)323

TEMPERATURE = 40.0 DEG C   TYPE OF SYSTEM = 1

EXPERIMENTAL TIE LINES IN MOLE PCT

	LEFT PHASE			RIGHT PHASE	
(1)	(2)	(3)	(1)	(2)	(3)
95.371	4.574	0.054	0.507	1.369	98.124
90.923	8.959	0.119	0.753	3.465	95.781
88.869	10.982	0.148	0.997	4.863	94.140
85.859	13.959	0.182	1.232	7.687	91.081
82.310	17.468	0.222	1.469	8.816	89.715
79.243	20.466	0.291	1.687	12.722	85.591
72.854	26.519	0.627	2.129	16.675	81.196
62.796	35.572	1.632	2.775	21.091	76.134
58.320	39.517	2.162	4.066	24.126	71.808

SPECIFIC MODEL PARAMETERS IN KELVIN

		UNIQUAC		NRTL(ALPHA=.2)	
I	J	AIJ	AJI	AIJ	AJI
1	2	-66.855	-158.19	54.548	-354.05
1	3	399.66	686.41	1613.7	-854.44
2	3	55.870	6.7477	964.11	-489.99

R1 = 0.9200   R2 = 2.2024   R3 = 3.9228
Q1 = 1.400   Q2 = 2.072   Q3 = 2.968

MEAN DEV. BETWEEN CALC. AND EXP. CONC. IN MOLE PCT

UNIQUAC (SPECIFIC PARAMETERS)   0.85
NRTL   (SPECIFIC PARAMETERS)   0.79

C$_2$H$_4$O$_2$-C$_7$H$_8$

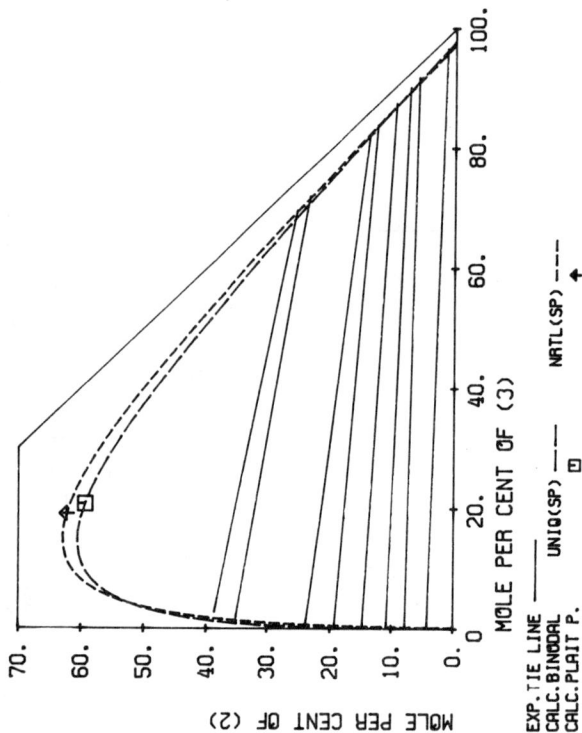

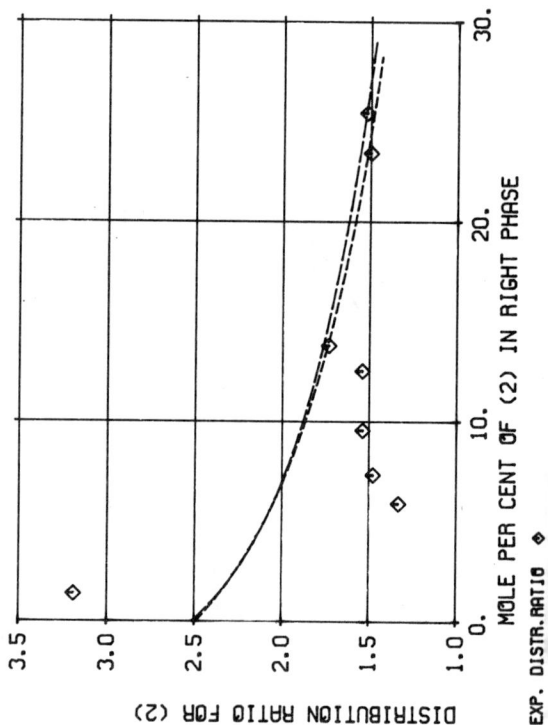

(1) H2O	WATER
(2) C2H4O2	ACETIC ACID
(3) C7H8	TOLUENE

DAKSHINAMURTY P., SUBRAHMANYAM V., NARSIMHA RAO M.
J.APPL.CHEM.BIOTECHNOL. 23(1973)323

TEMPERATURE = 60.0 DEG C   TYPE OF SYSTEM = 1

EXPERIMENTAL TIE LINES IN MOLE PCT

LEFT PHASE			RIGHT PHASE		
(1)	(2)	(3)	(1)	(2)	(3)
95.570	4.311	0.119	2.003	1.352	96.644
91.983	7.854	0.163	2.214	5.903	91.884
88.970	10.820	0.210	2.442	7.326	90.232
85.031	14.704	0.265	2.895	9.555	87.550
80.323	19.304	0.373	3.559	12.528	83.913
75.596	23.910	0.494	3.771	13.789	82.439
63.164	35.062	1.775	4.512	23.418	72.069
58.612	38.672	2.717	4.906	25.423	69.670

SPECIFIC MODEL PARAMETERS IN KELVIN

		UNIQUAC		NRTL(ALPHA=.2)	
I J		AIJ	AJI	AIJ	AJI
1 2		58.624	-193.77	-111.72	-164.03
1 3		462.76	523.85	1767.9	643.54
2 3		134.28	-9.1953	783.45	-438.58

R1 = 0.9200	R2 = 2.2024	R3 = 3.9228	
Q1 = 1.400	Q2 = 2.072	Q3 = 2.968	

MEAN DEV. BETWEEN CALC. AND EXP. CONC. IN MOLE PCT

UNIQUAC (SPECIFIC PARAMETERS)	0.86
NRTL (SPECIFIC PARAMETERS)	0.82

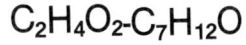

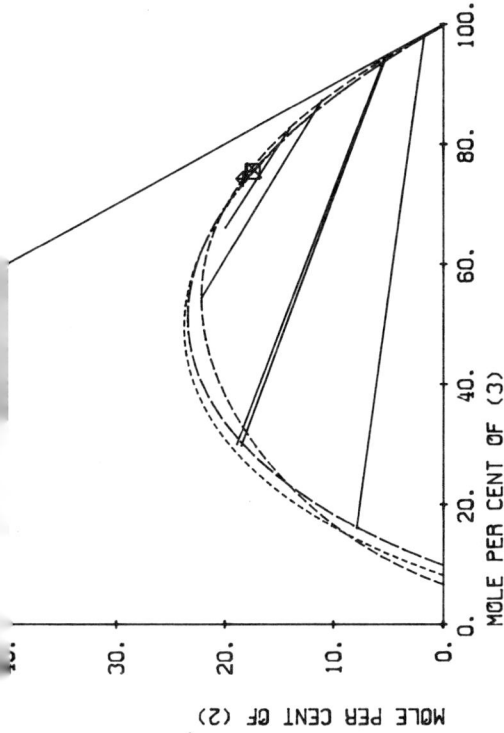

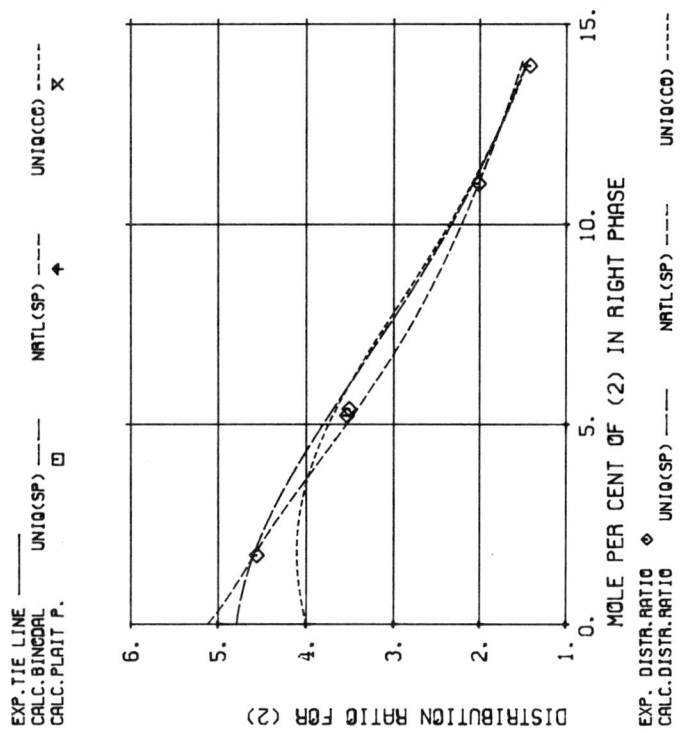

$C_2H_4O_2$-$C_7H_{12}O$

(2) C2H4O2   ACETIC ACID

(3) H2O   WATER

OTHMER D.F., WHITE R.E., TRUEGER E.
IND.ENG.CHEM. 33(1941)1240

TEMPERATURE = 23.5 DEG C   TYPE OF SYSTEM = 1

EXPERIMENTAL TIE LINES IN MOLE PCT (GRAPH.INTERPOL.)

	LEFT PHASE			RIGHT PHASE	
(1)	(2)	(3)	(1)	(2)	(3)
76.143	7.853	16.005	0.271	1.724	98.005
51.768	18.451	29.781	0.498	5.216	94.286
51.112	18.873	30.015	0.518	5.377	94.105
23.498	22.152	54.350	1.593	11.018	87.390
13.982	19.926	66.091	3.350	13.957	82.693

SPECIFIC MODEL PARAMETERS IN KELVIN

		UNIQUAC		NRTL(ALPHA=.2)	
I J	AIJ	AJI		AIJ	AJI
1 2	-250.15	158.39		817.16	-753.79
1 3	449.35	51.902		357.13	-1952.7
2 3	-220.44	211.89		-64.599	-42.357

R1 = 4.8169   R2 = 2.2024   R3 = 0.9200
Q1 = 3.876   Q2 = 2.072   Q3 = 1.400

MEAN DEV. BETWEEN CALC. AND EXP. CONC. IN MOLE PCT

UNIQUAC (SPECIFIC PARAMETERS)   0.54
NRTL   (SPECIFIC PARAMETERS)   0.60
UNIQUAC (COMMON PARAMETERS)   0.63

C$_2$H$_4$O$_2$-C$_7$H$_{14}$O$_2$

MOLE PER CENT OF (3)

MOLE PER CENT OF (2)

EXP.TIE LINE ———— UNIQ(SP) ⊟ UNIQ(CO) ----- ⊠
CALC.BINODAL -------- NRTL(SP) ----- ✦
CALC.PLAIT P.

DISTRIBUTION RATIO FOR (2)

MOLE PER CENT OF (2) IN RIGHT PHASE

EXP. DISTR.RATIO ◇

(1) C7H14O2   ACETIC ACID,3-METHYLBUTYL ESTER
(2) C2H4O2   ACETIC ACID
(3) H2O   WATER

OTHMER D.F., WHITE R.E., TRUEGER E.
IND.ENG.CHEM. 33(1941)1240

TEMPERATURE = 23.5 DEG C   TYPE OF SYSTEM = 1

EXPERIMENTAL TIE LINES IN MOLE PCT (GRAPH.INTERPOL.)

	LEFT PHASE			RIGHT PHASE	
(1)	(2)	(3)	(1)	(2)	(3)
80.631	2.186	17.183	0.030	1.073	98.892
77.964	5.016	17.019	0.043	2.448	97.509
75.398	7.574	17.028	0.051	3.271	96.678
73.361	9.555	17.083	0.059	4.017	95.924
70.590	12.093	17.317	0.071	4.975	94.954
67.644	14.485	17.871	0.085	5.859	94.055
65.242	16.155	18.603	0.089	6.095	93.817
59.907	19.269	20.823	0.118	7.607	92.275
56.222	21.290	22.489	0.133	8.206	91.661
52.451	23.222	24.328	0.152	9.044	90.804
44.489	27.381	28.130	0.167	9.590	90.243
42.634	28.359	29.007	0.181	10.054	89.765

SPECIFIC MODEL PARAMETERS IN KELVIN

		UNIQUAC		NRTL(ALPHA=.2)	
I	J	AIJ	AJI	AIJ	AJI
1	2	472.20	-167.75	460.57	8.5970
1	3	325.33	190.84	184.48	2391.9
2	3	2.0428	124.70	-80.526	658.34

R1 = 5.5010   R2 = 2.2024   R3 = 0.9200
Q1 = 4.732   Q2 = 2.072   Q3 = 1.400

MEAN DEV. BETWEEN CALC. AND EXP. CONC. IN MOLE PCT

UNIQUAC (SPECIFIC PARAMETERS)   0.55
NRTL   (SPECIFIC PARAMETERS)   0.53
UNIQUAC (COMMON PARAMETERS)   2.25

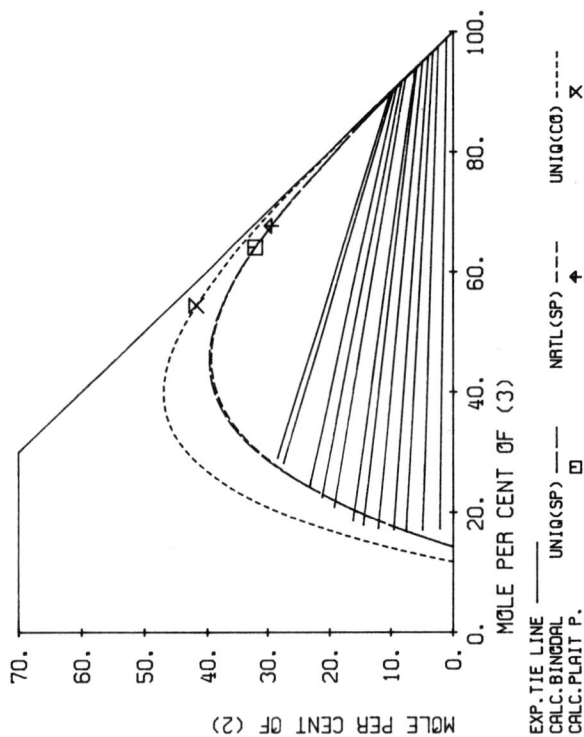

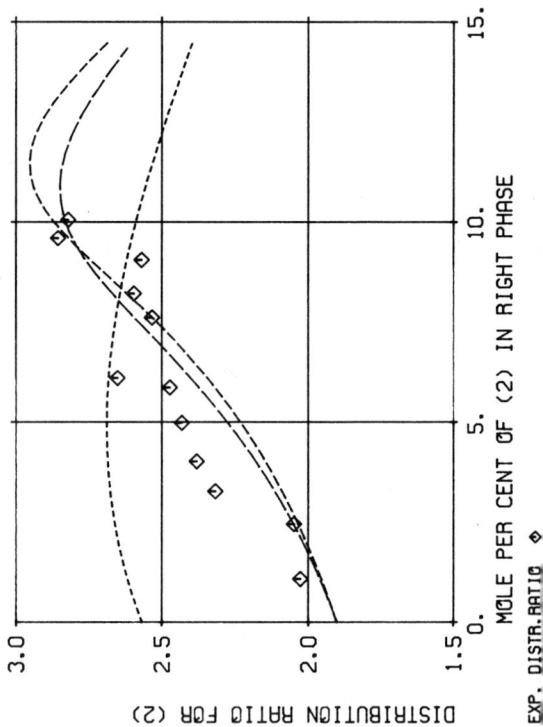

$C_2H_4O_2$-$C_7H_{14}O_2$

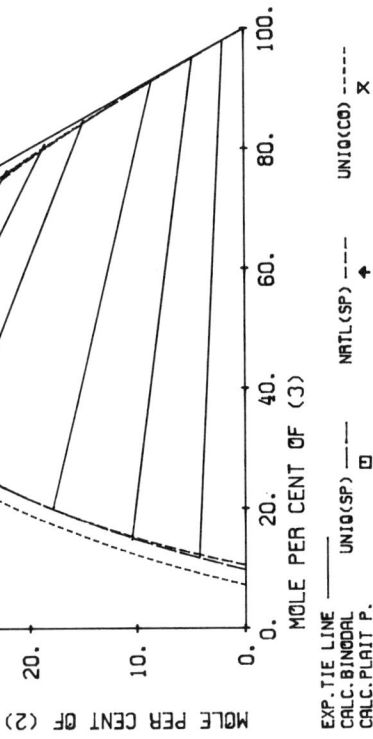

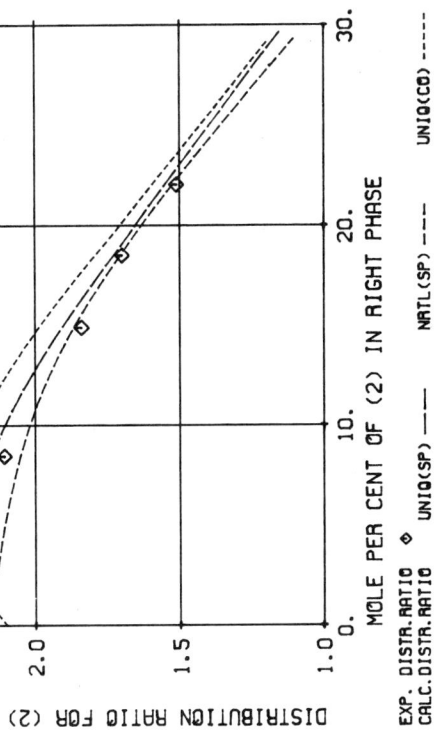

(2) C2H4O2   ACETIC ACID

(3) H2O   WATER

KRISHNAMURTY R., JAYARAMA RAO G., VENKATA RAO C.
J.SCI.IND.RES. 21D(1962)282

TEMPERATURE =   28.0 DEG C   TYPE OF SYSTEM = 1

EXPERIMENTAL TIE LINES IN MOLE PCT

| | LEFT PHASE | | | RIGHT PHASE | |
(1)	(2)	(3)	(1)	(2)	(3)
83.940	4.232	11.829	0.057	1.922	98.022
74.801	10.446	14.754	0.096	4.714	95.190
61.882	17.831	20.287	0.164	8.455	91.382
44.958	27.390	27.652	0.437	14.881	84.681
33.861	31.444	34.695	0.736	18.476	80.788
25.607	33.270	41.123	1.560	22.018	76.422

SPECIFIC MODEL PARAMETERS IN KELVIN

| | | UNIQUAC | | NRTL(ALPHA=.2) | |
I	J	AIJ	AJI	AIJ	AJI
1	2	-76.780	-30.705	214.55	-37.943
1	3	456.47	129.92	254.47	2221.5
2	3	-203.23	65.127	-352.11	750.27

R1 = 5.5018   R2 = 2.2024   R3 = 0.9200
Q1 = 4.736   Q2 = 2.072   Q3 = 1.400

MEAN DEV. BETWEEN CALC. AND EXP. CONC. IN MOLE PCT

UNIQUAC (SPECIFIC PARAMETERS)   0.35
NRTL   (SPECIFIC PARAMETERS)   0.33
UNIQUAC (COMMON PARAMETERS)   0.88

C₂H₄O₂-C₇H₁₄O₂

(1) C7H1402   PENTANOIC ACID,ETHYL ESTER
(2) C2H402   ACETIC ACID
(3) H2O   WATER

BOMSHTEIN A.L., TROFIMOV A.N., SERAFIMOV L.A.
ZH.PRIKL.KHIM.(LENINGRAD) 51(1978)1280

TEMPERATURE = 20.0 DEG C   TYPE OF SYSTEM = 1

EXPERIMENTAL TIE LINES IN MOLE PCT

	LEFT PHASE			RIGHT PHASE	
(1)	(2)	(3)	(1)	(2)	(3)
81.870	8.690	9.440	0.090	5.020	94.890
62.890	23.970	13.140	0.190	10.480	89.330
55.250	26.550	18.200	0.430	16.410	83.160
47.030	32.790	20.180	0.760	20.640	78.600

MEAN DEV. BETWEEN CALC. AND EXP. CONC. IN MOLE PCT

UNIQUAC (COMMON PARAMETERS)   1.73

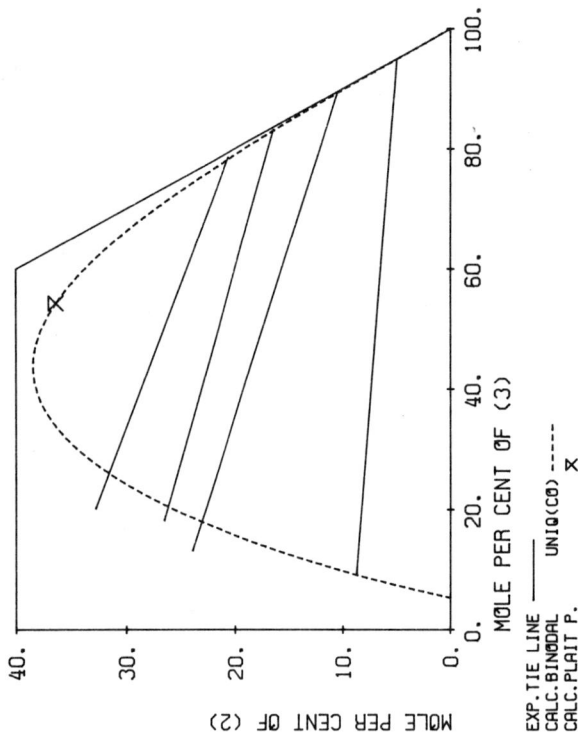

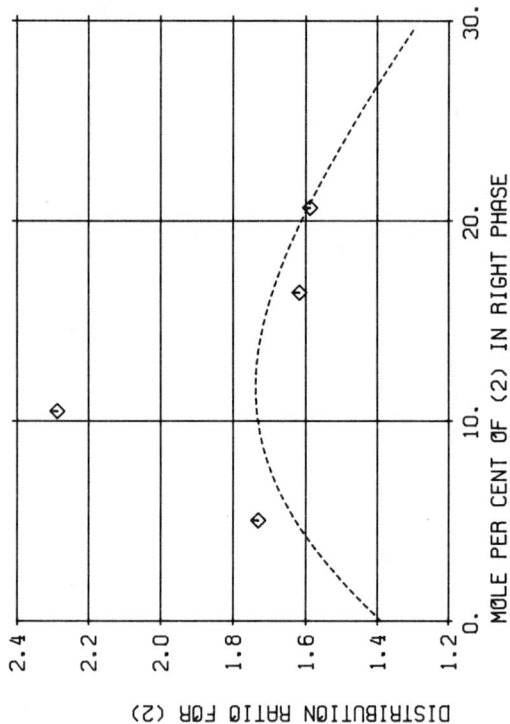

C$_2$H$_4$O$_2$-C$_8$H$_{11}$N

MOLE PER CENT OF (3)

MOLE PER CENT OF (2)

EXP.TIE LINE ———
CALC.BINODAL
CALC.PLAIT P.

UNIQ(SP) ——— ⊟
NRTL(SP) ——— ✦
UNIQ(CO) ----- ✗

MOLE PER CENT OF (2) IN RIGHT PHASE

DISTRIBUTION RATIO FOR (2)

EXP. DISTR.RATIO ◇
CALC.DISTR.RATIO

UNIQ(SP) ◆
NRTL(SP) -----
UNIQ(CO) -----

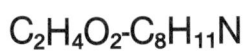

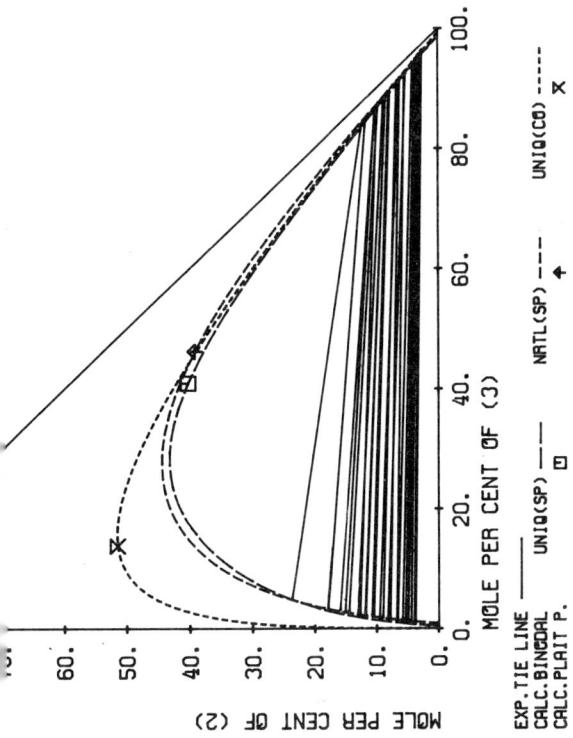

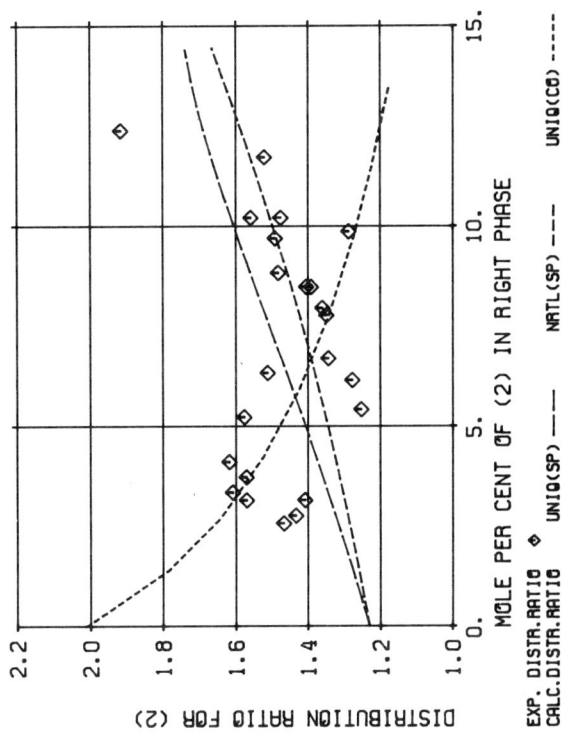

(2) C2H4O2   ACETIC ACID
(3) C8H11N   ANILINE,N,N-DIMETHYL

GARWIN L., HADDAD P.O.
ANAL.CHEM. 25(1953)435

TEMPERATURE = 25.0 DEG C   TYPE OF SYSTEM = 1

EXPERIMENTAL TIE LINES IN MOLE PCT (GRAPH.INTERPOL.)

	LEFT PHASE			RIGHT PHASE	
(1)	(2)	(3)	(1)	(2)	(3)
95.455	3.771	0.774	0.857	2.570	96.573
95.234	3.962	0.804	0.921	2.764	96.315
93.981	4.943	1.076	1.050	3.149	95.802
94.674	4.434	0.893	1.050	3.149	95.800
93.558	5.375	1.067	1.113	3.340	95.546
93.004	5.844	1.153	1.241	3.722	95.038
92.065	6.638	1.297	1.367	4.101	94.532
90.181	8.236	1.533	1.741	5.224	93.034
91.907	6.773	1.320	1.803	5.409	92.787
90.644	7.846	1.510	2.043	6.145	91.807
88.615	9.559	1.826	2.109	6.327	91.564
89.297	8.986	1.717	2.230	6.690	91.079
87.539	10.472	1.989	2.601	7.766	89.632
87.151	10.803	2.046	2.661	7.944	89.395
85.840	11.903	2.257	2.843	8.473	83.684
85.978	11.786	2.236	2.843	8.473	88.684
84.434	13.073	2.492	2.960	8.823	88.217
84.806	12.755	2.439	3.261	9.689	87.050
84.899	12.692	2.409	3.318	9.861	86.821
81.054	15.902	3.044	3.432	10.203	86.365
82.076	15.051	2.873	3.432	10.203	86.365
78.734	17.812	3.454	3.954	11.717	84.329
71.368	23.722	4.909	4.186	12.377	83.437

SPECIFIC MODEL PARAMETERS IN KELVIN

		UNIQUAC		NRTL(ALPHA=.2)	
I J	AIJ	AJI		AIJ	AJI
1 2	-132.62	198.68		631.71	-212.74
1 3	-101.73	1033.9		950.38	811.06
2 3	-30.321	225.39		50.658	409.82

R1 = 0.9200   R2 = 2.2024   R3 = 5.1094
Q1 = 1.400   Q2 = 2.072   Q3 = 3.920

MEAN DEV. BETWEEN CALC. AND EXP. CONC. IN MOLE PCT

UNIQUAC (SPECIFIC PARAMETERS)   0.50
NRTL   (SPECIFIC PARAMETERS)   0.47
UNIQUAC (COMMON PARAMETERS)   1.73

$C_2H_4O_2$-$C_8H_{14}O_2$

```
(1) C8H14O2    ACETIC ACID,CYCLOHEXYL ESTER
(2) C2H4O2     ACETIC ACID
(3) H2O        WATER

OTHMER D.F., WHITE R.E., TRUEGER E.
IND.ENG.CHEM. 33(1941)1240

TEMPERATURE = 23.5 DEG C      TYPE OF SYSTEM = 1

EXPERIMENTAL TIE LINES IN MOLE PCT (GRAPH.INTERPOL.)
```

	LEFT PHASE			RIGHT PHASE	
(1)	(2)	(3)	(1)	(2)	(3)
82.741	5.639	11.620	0.053	1.657	98.290
73.597	12.708	13.695	0.094	5.555	94.351
67.430	17.147	15.423	0.136	7.748	92.117
46.087	29.142	24.771	0.476	15.330	84.194
27.358	33.714	38.929	1.579	22.318	76.103

```
SPECIFIC MODEL PARAMETERS IN KELVIN
```

		UNIQUAC		NRTL(ALPHA=.2)	
I J	AIJ	AJI		AIJ	AJI
1 2	-120.38	76.466		147.03	-117.21
1 3	-491.37	83.146		311.10	2205.0
2 3	-168.49	88.008		-232.66	470.72

```
R1 = 5.7220   R2 = 2.2024   R3 = 0.9200
Q1 = 4.656    Q2 = 2.072    Q3 = 1.400

MEAN DEV. BETWEEN CALC. AND EXP. CONC. IN MOLE PCT

UNIQUAC (SPECIFIC PARAMETERS)     0.65
NRTL   (SPECIFIC PARAMETERS)      0.54
UNIQUAC (COMMON PARAMETERS)       1.05
```

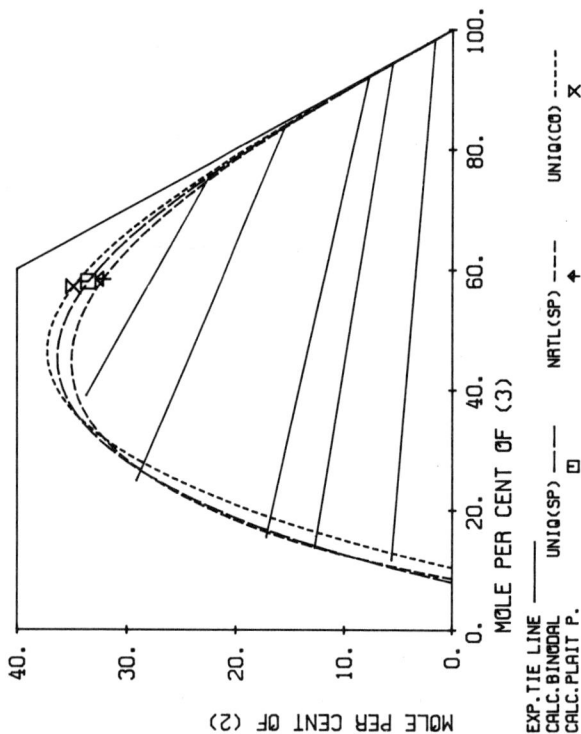

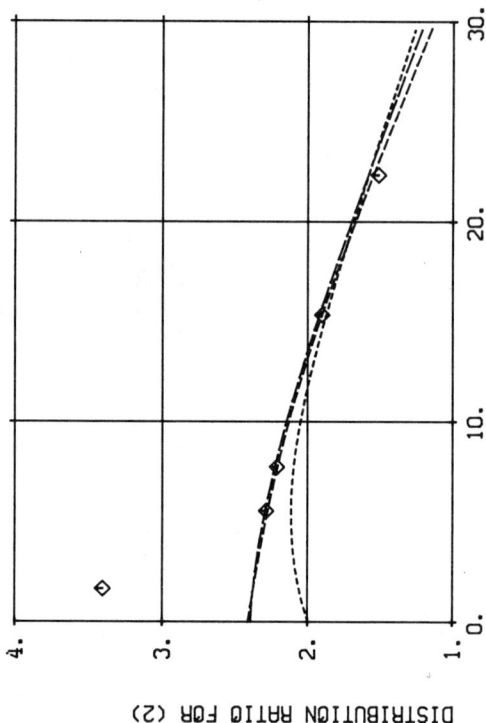

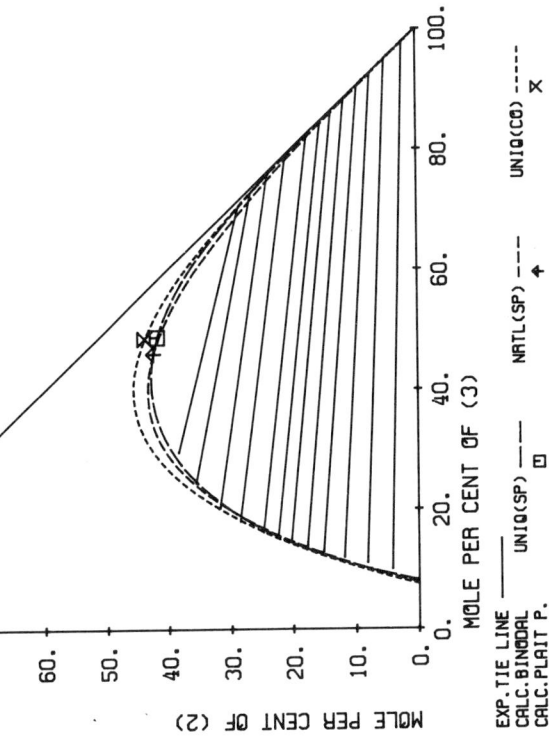

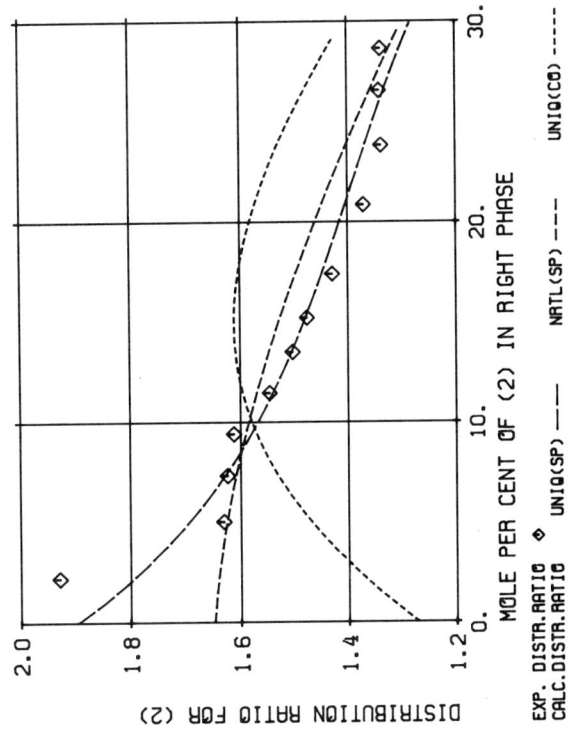

(2) C2H4O2   ACETIC ACID

(3) H2O   WATER

OTHMER D.F., SERRANO J.
IND.ENG.CHEM. 41(1949)1030

TEMPERATURE = 25.0 DEG C   TYPE OF SYSTEM = 1

EXPERIMENTAL TIE LINES IN MOLE PCT (GRAPH.INTERPOL.)

LEFT PHASE			RIGHT PHASE		
(1)	(2)	(3)	(1)	(2)	(3)
85.782	4.265	9.953	0.026	2.212	97.761
80.811	8.224	10.965	0.042	5.043	94.915
76.186	11.907	11.907	0.059	7.333	92.608
72.031	15.168	12.800	0.080	9.409	90.512
68.639	17.734	13.627	0.104	11.475	88.421
65.244	20.310	14.447	0.134	13.512	86.354
62.351	22.448	15.201	0.164	15.207	84.629
58.987	24.848	16.165	0.215	17.391	82.394
53.459	28.574	17.967	0.347	20.849	78.804
47.955	31.839	20.206	0.543	23.808	75.648
40.049	35.609	24.342	0.860	26.562	72.578
32.174	38.331	29.495	1.193	28.644	70.164

SPECIFIC MODEL PARAMETERS IN KELVIN

		UNIQUAC		NRTL(ALPHA=.2)	
I J	AIJ	AJI	AIJ	AJI	
1 2	216.29	-197.32	-468.10	401.99	
1 3	368.20	-373.36	-317.48	1992.7	
2 3	121.42	-283.15	-348.36	354.23	

R1 = 6.2430   R2 = 2.2024   R3 = 0.9200
Q1 = 5.308   Q2 = 2.072   Q3 = 1.400

MEAN DEV. BETWEEN CALC. AND EXP. CONC. IN MOLE PCT

UNIQUAC (SPECIFIC PARAMETERS)   0.48
NRTL   (SPECIFIC PARAMETERS)   0.71
UNIQUAC (COMMON PARAMETERS)   1.08

$C_2H_4O_2$-$C_8H_{18}O$

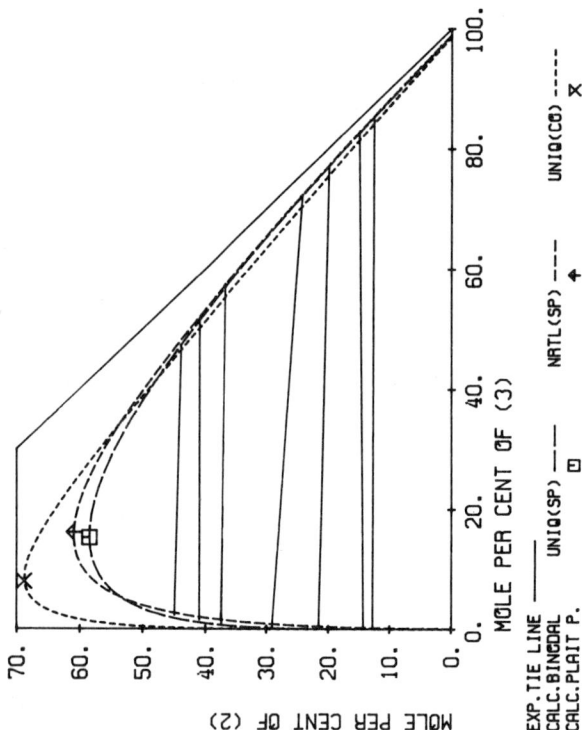

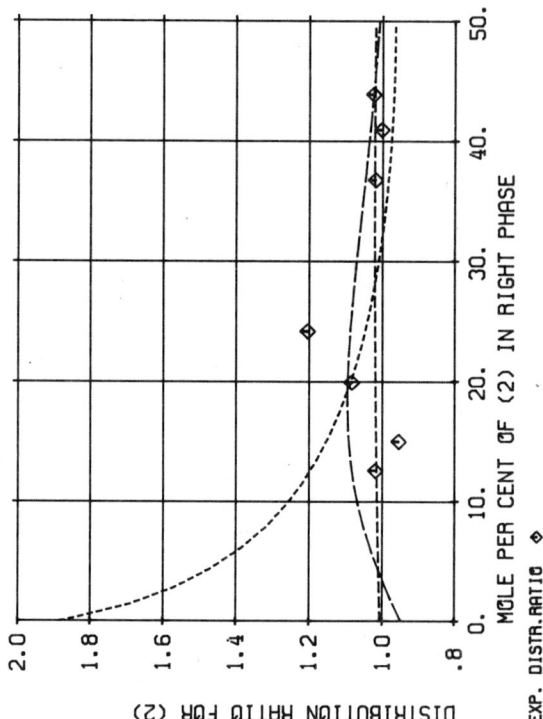

(1) H2O   WATER

(2) C2H4O2   ACETIC ACID

(3) C8H18O   ETHER,DIBUTYL

OTHMER D.F., WHITE R.E., TRUEGER E.
IND.ENG.CHEM. 33(1941)1240

TEMPERATURE = 24.8 DEG C   TYPE OF SYSTEM = 1

EXPERIMENTAL TIE LINES IN MOLE PCT (GRAPH.INTERPOL.)

	LEFT PHASE			RIGHT PHASE	
(1)	(2)	(3)	(1)	(2)	(3)
87.165	12.772	0.063	2.502	12.525	84.973
85.650	14.284	0.067	2.594	14.960	82.446
78.291	21.583	0.125	3.008	19.907	77.085
70.590	29.095	0.315	3.403	24.153	72.445
61.536	37.468	0.996	5.714	36.734	57.552
57.389	41.021	1.590	7.208	40.917	51.874
52.683	44.978	2.338	9.023	43.875	47.101

SPECIFIC MODEL PARAMETERS IN KELVIN

		UNIQUAC		NRTL(ALPHA=.2)	
I J	AIJ	AJI		AIJ	AJI
1 2	-201.97	80.231		128.71	-125.40
1 3	400.73	761.06		1590.9	845.90
2 3	-122.65	198.77		711.62	-429.46

R1 = 0.9200   R2 = 2.2024   R3 = 6.0925
Q1 = 1.400   Q2 = 2.072   Q3 = 5.176

MEAN DEV. BETWEEN CALC. AND EXP. CONC. IN MOLE PCT

UNIQUAC (SPECIFIC PARAMETERS)   0.65
NRTL   (SPECIFIC PARAMETERS)   0.82
UNIQUAC (COMMON PARAMETERS)   1.63

C$_2$H$_4$O$_2$-C$_8$H$_{18}$O

(2) C2H4O2   ACETIC ACID

(3) H2O   WATER

HLAVATY K., LINEK J.
COLLECT.CZECH.CHEM.COMMUN. 38(1973)374

TEMPERATURE = 24.6 DEG C   TYPE OF SYSTEM = 1

EXPERIMENTAL TIE LINES IN MOLE PCT

	LEFT PHASE			RIGHT PHASE	
	(1)	(2)	(3)	(1)	(2)	(3)
	87.144	0.0	12.856	0.0	0.0	100.000
	66.759	15.571	17.669	0.016	5.679	94.306
	57.953	21.126	20.921	0.017	9.056	90.927
	52.538	24.239	23.223	0.035	11.321	88.644
	43.558	28.813	27.629	0.153	16.076	83.771
	33.361	33.455	33.184	0.491	21.932	77.578

SPECIFIC MODEL PARAMETERS IN KELVIN

I J		UNIQUAC			NRTL(ALPHA=.2)	
		AIJ	AJI		AIJ	AJI
1 2		83.029	-210.22		-444.93	34.718
1 3		353.02	192.13		176.44	1913.5
2 3		-37.128	-210.43		-213.35	148.46

R1 = 6.1511   R2 = 2.2024   R3 = 0.9200
Q1 = 5.208   Q2 = 2.072   Q3 = 1.400

MEAN DEV. BETWEEN CALC. AND EXP. CONC. IN MOLE PCT

UNIQUAC (SPECIFIC PARAMETERS)   0.27
NRTL   (SPECIFIC PARAMETERS)   0.47
UNIQUAC (COMMON PARAMETERS)   0.79

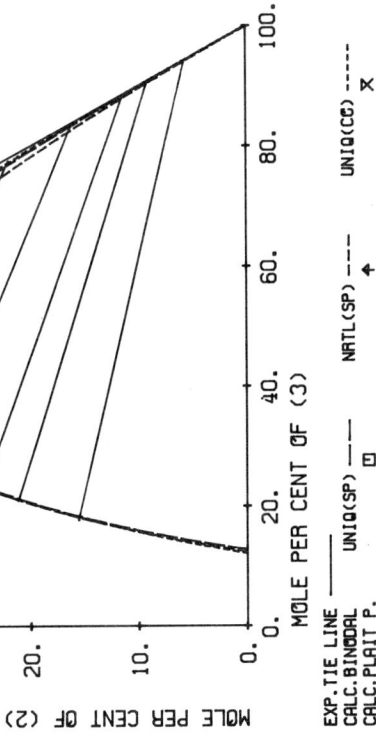

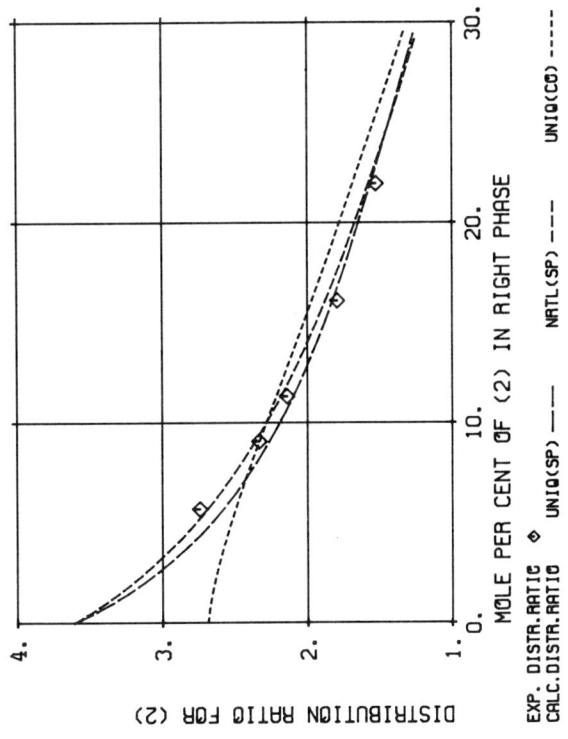

C$_2$H$_4$O$_2$-C$_9$H$_{10}$O$_2$

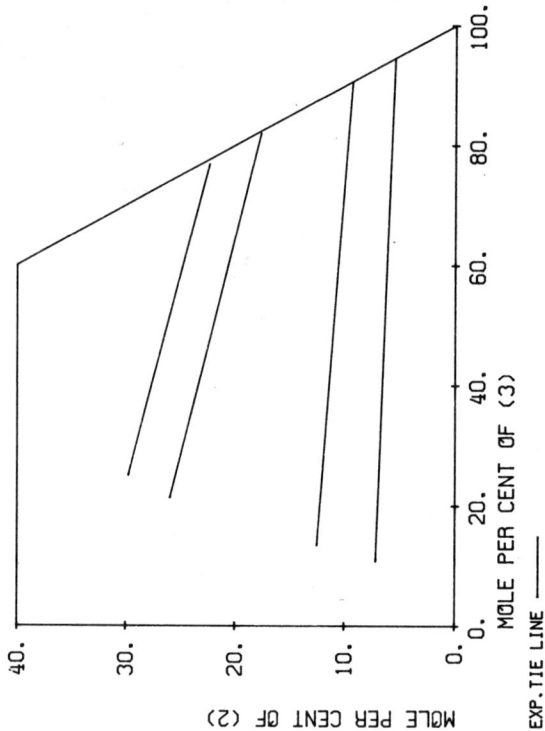

(1) C9H10O2   BENZOIC ACID,ETHYL ESTER
----
(2) C2H4O2   ACETIC ACID
----
(3) H2O   WATER

JAYA RAMA RAO G., VEKKATA RAO C.
J.SCI.IND.RES. 16B(1957)102

TEMPERATURE = 31.0 DEG C   TYPE OF SYSTEM = 1

EXPERIMENTAL TIE LINES IN MOLE PCT

LEFT PHASE			RIGHT PHASE		
(1)	(2)	(3)	(1)	(2)	(3)
82.099	7.117	10.784	0.068	5.432	94.500
74.045	12.507	13.448	0.088	9.328	90.584
52.716	25.941	21.344	0.345	17.627	82.029
45.192	29.758	25.049	0.757	22.392	75.850

C$_2$H$_4$O$_2$-C$_9$H$_{14}$O

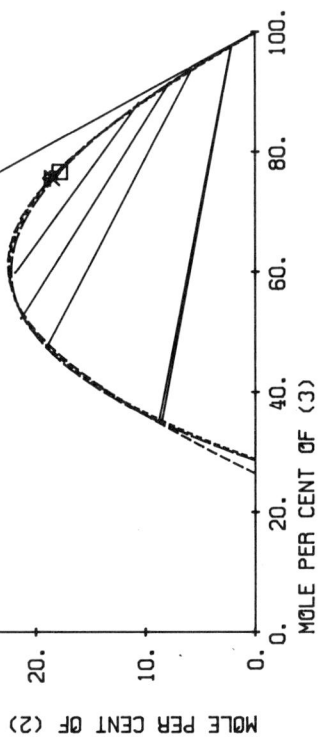

MOLE PER CENT OF (3)

MOLE PER CENT OF (2)

EXP.TIE LINE ——— UNIQ(SP) ☐ NRTL(SP) ——— ✦ UNIQ(CO) ·—·—·—
CALC.BINODAL
CALC.PLAIT P.   ✗

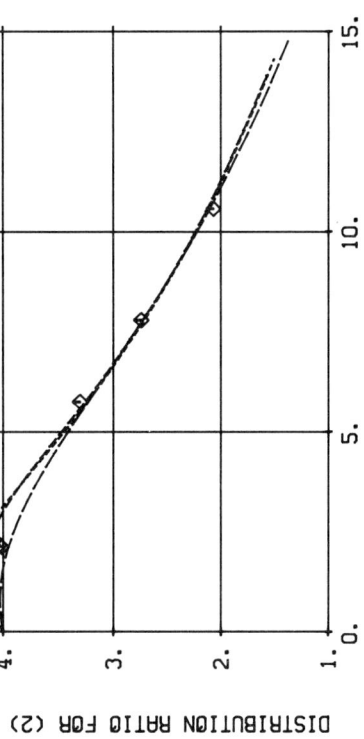

MOLE PER CENT OF (2) IN RIGHT PHASE

DISTRIBUTION RATIO FOR (2)

EXP. DISTR.RATIO ◇ UNIQ(SP) ——— NRTL(SP) ——— UNIQ(CO) ·—·—·—
CALC.DISTR.RATIO

---

(2) C2H4O2   ACETIC ACID

(3) H2O   WATER

OTHMER D.F., WHITE R.E., TRUEGER E.
IND.CHEM. 33(1941)1240

TEMPERATURE = 24.0 DEG C   TYPE OF SYSTEM = 1

EXPERIMENTAL TIE LINES IN MOLE PCT (GRAPH.INTERPOL.)

	LEFT PHASE			RIGHT PHASE	
(1)	(2)	(3)	(1)	(2)	(3)
56.643	8.512	34.845	0.208	2.121	97.671
56.111	8.758	35.131	0.212	2.172	97.615
33.532	18.991	47.477	0.409	5.749	93.842
26.454	21.339	52.207	0.630	7.782	91.588
18.300	21.879	59.820	1.165	10.568	88.267

SPECIFIC MODEL PARAMETERS IN KELVIN

		UNIQUAC		NRTL(ALPHA=.2)	
I J	AIJ	AJI		AIJ	AJI
1 2	-280.16	1.1642		710.02	-407.84
1 3	189.24	151.96		-71.981	2014.8
2 3	-296.11	180.57		-4.6892	339.67

R1 = 5.9315   R2 = 2.2024   R3 = 0.9200
Q1 = 4.940   Q2 = 2.072   Q3 = 1.400

MEAN DEV. BETWEEN CALC. AND EXP. CONC. IN MOLE PCT

UNIQUAC (SPECIFIC PARAMETERS)   0.16
NRTL   (SPECIFIC PARAMETERS)   0.22
UNIQUAC (COMMON PARAMETERS)   0.26

$C_2H_4O_2$-$C_9H_{18}O$

(1) C9H18O   4-HEPTANONE,2,6-DIMETHYL
(2) C2H4O2   ACETIC ACID
(3) H2O   WATER

OTHMER D.F., WHITE R.E., TRUEGER E.
IND.ENG.CHEM. 33(1941)1240

TEMPERATURE = 23.5 DEG C   TYPE OF SYSTEM = 1

EXPERIMENTAL TIE LINES IN MOLE PCT (GRAPH.INTERPOL.)

	LEFT PHASE			RIGHT PHASE	
(1)	(2)	(3)	(1)	(2)	(3)
88.330	6.741	4.929	0.041	3.550	96.409
77.845	14.379	7.776	0.070	9.223	90.707
57.237	28.350	14.413	0.260	20.520	79.221
43.583	35.533	20.885	0.595	27.519	71.886

SPECIFIC MODEL PARAMETERS IN KELVIN

		UNIQUAC		NRTL(ALPHA=.2)	
I	J	AIJ	AJI	AIJ	AJI
1	2	182.95	-217.96	-295.83	312.05
1	3	608.90	274.88	766.35	2795.2
2	3	-80.562	-209.42	-342.65	419.13

R1 = 6.6183   R2 = 2.2024   R3 = 0.9200
Q1 = 5.568   Q2 = 2.072   Q3 = 1.400

MEAN DEV. BETWEEN CALC. AND EXP. CONC. IN MOLE PCT

UNIQUAC (SPECIFIC PARAMETERS)   1.15
NRTL   (SPECIFIC PARAMETERS)   1.02
UNIQUAC (COMMON PARAMETERS)   1.31

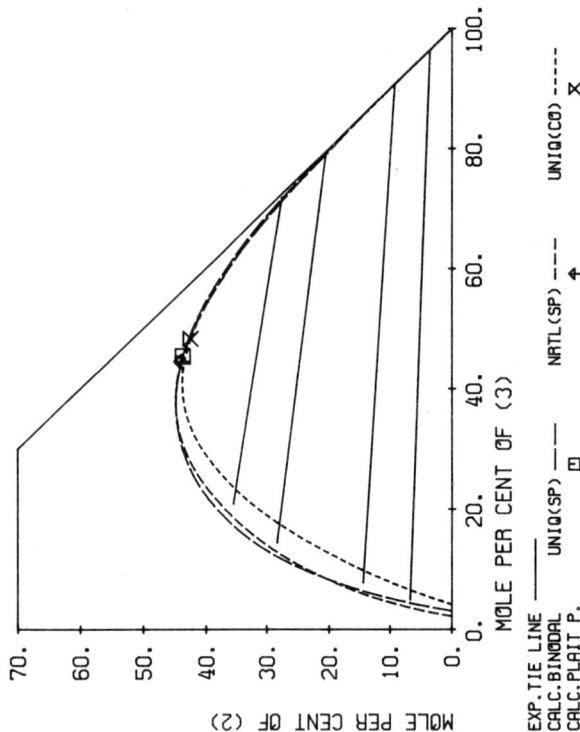

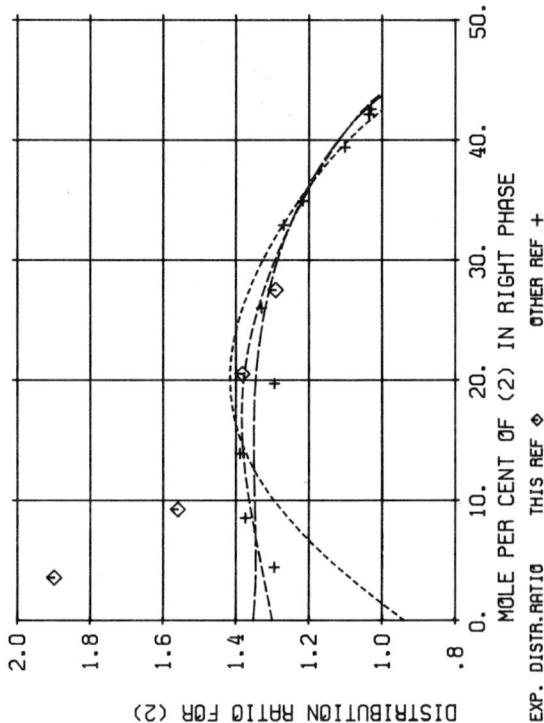

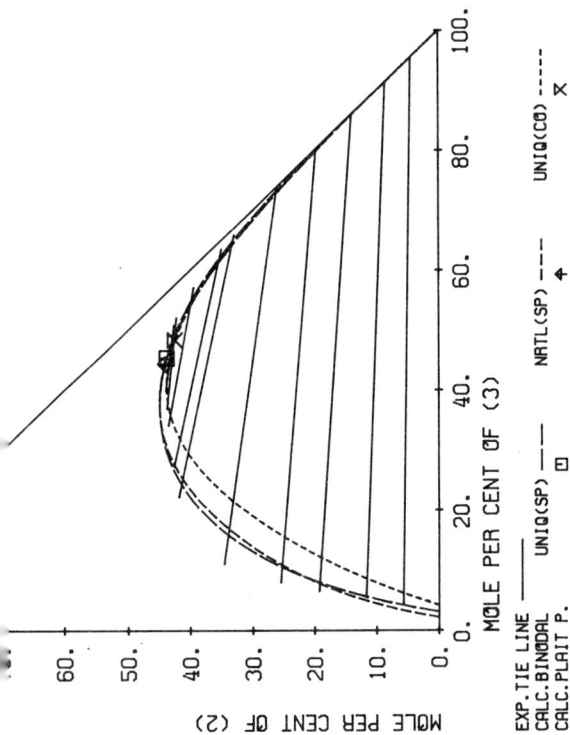

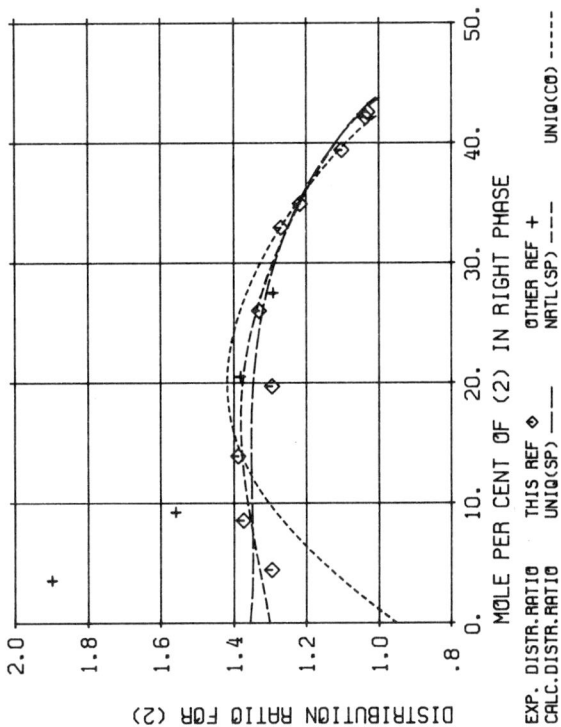

(2) C2H4O2   ACETIC ACID

(3) H2O   WATER

IGUCHI A., FUSE K.
KAGAKU KOGAKU 35(1971)477

TEMPERATURE = 25.0 DEG C   TYPE OF SYSTEM = 1

EXPERIMENTAL TIE LINES IN MOLE PCT (GRAPH.INTERPOL.)

| | LEFT PHASE | | | RIGHT PHASE | |
(1)	(2)	(3)	(1)	(2)	(3)
89.903	5.697	4.399	0.070	4.399	95.531
82.718	11.700	5.582	0.092	8.517	91.391
74.159	19.277	6.564	0.118	13.888	85.994
65.479	25.488	8.033	0.168	19.691	80.141
54.304	34.593	11.104	0.309	26.019	73.671
35.932	41.802	22.266	1.122	32.920	65.958
30.188	42.518	27.294	1.507	34.934	63.559
22.409	43.429	34.162	3.411	39.392	57.197
19.416	43.653	36.931	5.721	42.121	52.158
16.226	43.865	39.909	6.566	42.544	50.890

SPECIFIC MODEL PARAMETERS IN KELVIN

| | | UNIQUAC | | NRTL(ALPHA=.2) | |
I	J	AIJ	AJI	AIJ	AJI
1	2	182.95	-217.96	-295.83	312.05
1	3	608.90	274.88	766.35	2785.2
2	3	-80.562	-209.42	-342.65	419.13

R1 = 6.6183   R2 = 2.2024   R3 = 0.9200
Q1 = 5.568   Q2 = 2.072   Q3 = 1.400

MEAN DEV. BETWEEN CALC. AND EXP. CONC. IN MOLE PCT

UNIQUAC (SPECIFIC PARAMETERS)   0.98
NRTL   (SPECIFIC PARAMETERS)   1.19
UNIQUAC (COMMON PARAMETERS)   2.87

C₂H₄O₂-C₁₀H₁₂

(1) H2O   WATER
(2) C2H4O2   ACETIC ACID
(3) C10H12   TETRALIN

GOTO S., MATSUBARA M., WASHINO K.
KAGAKU KOGAKU 38(1974)369

TEMPERATURE = 90.0 DEG C   TYPE OF SYSTEM = 1

EXPERIMENTAL TIE LINES IN MOLE PCT

	LEFT PHASE			RIGHT PHASE	
(1)	(2)	(3)	(1)	(2)	(3)
96.795	3.194	0.012	0.073	0.701	99.226
93.516	6.461	0.024	0.145	1.570	98.285
90.032	9.932	0.035	0.145	2.473	97.382
87.984	11.971	0.045	0.216	3.174	96.610
76.801	23.068	0.131	0.560	7.480	91.960
56.577	42.359	1.064	0.984	17.818	81.198
31.762	63.090	5.148	3.640	33.197	63.163
30.291	64.073	5.635	4.219	36.112	59.669
25.394	62.882	11.724	9.717	46.552	43.731

SPECIFIC MODEL PARAMETERS IN KELVIN

		UNIQUAC		NRTL(ALPHA=.2)	
I	J	AIJ	AJI	AIJ	AJI
1	2	-62.538	-118.07	524.65	-598.15
1	3	458.53	909.12	2561.0	1731.8
2	3	-120.13	374.37	372.68	-168.90

R1 = 0.9200   R2 = 2.2024   R3 = 5.5532
Q1 = 1.400   Q2 = 2.072   Q3 = 4.000

MEAN DEV. BETWEEN CALC. AND EXP. CONC. IN MOLE PCT

UNIQUAC (SPECIFIC PARAMETERS)   1.03
NRTL   (SPECIFIC PARAMETERS)   1.10

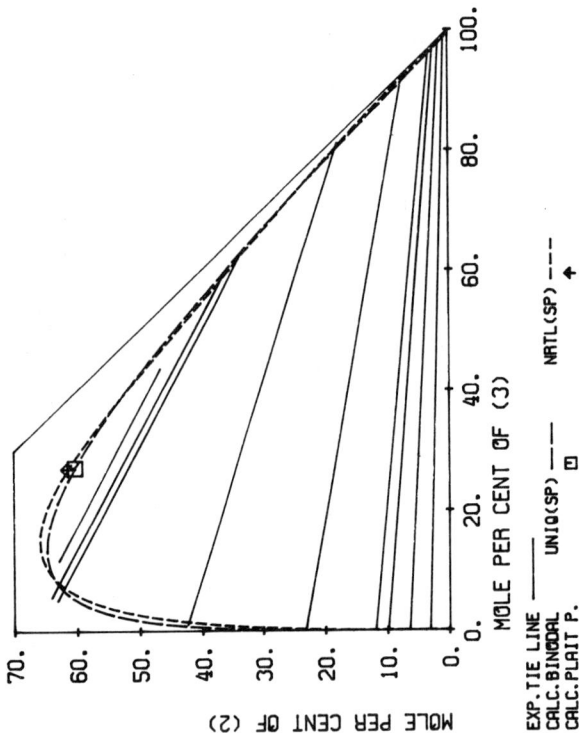

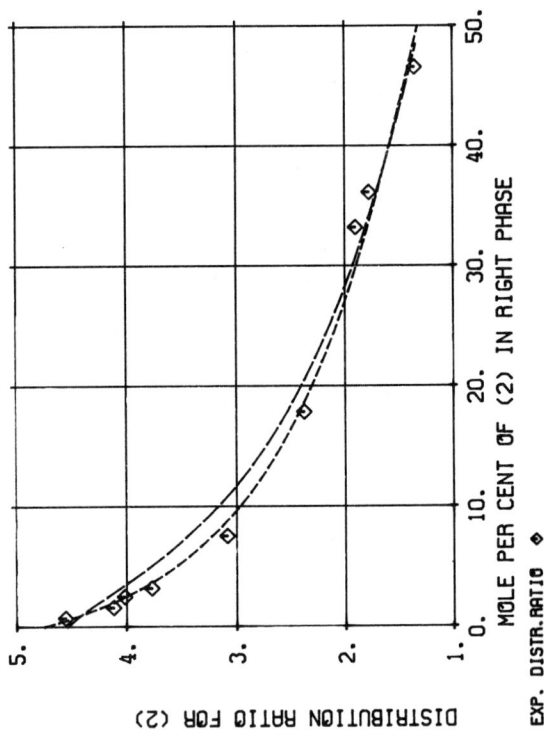

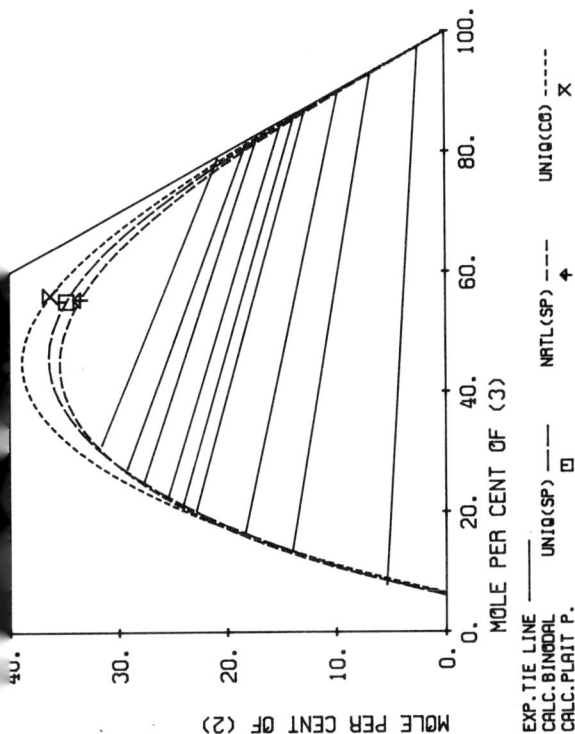

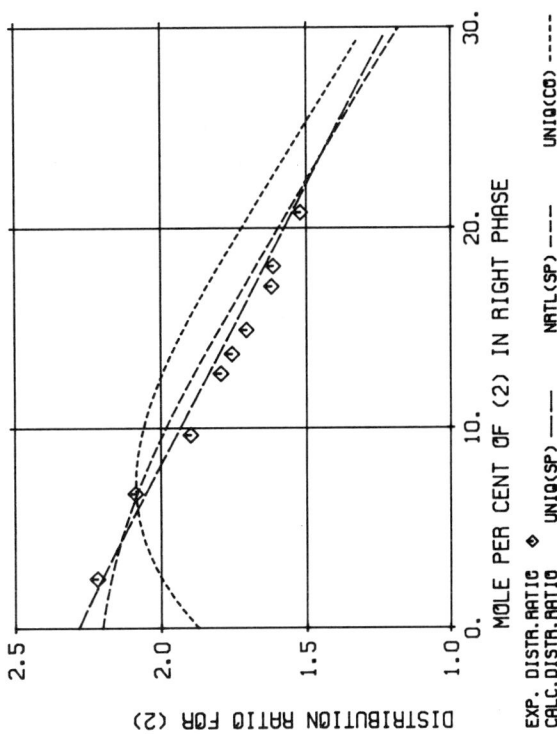

$C_2H_4O_2$-$C_{10}H_{16}O$

(2) C2H4O2   ACETIC ACID

(3) H2O   WATER

OTHMER D.F., WHITE R.E., TRUEGER E.
IND.ENG.CHEM. 33(1941)1240

TEMPERATURE = 23.3 DEG C   TYPE OF SYSTEM = 1

EXPERIMENTAL TIE LINES IN MOLE PCT (GRAPH.INTERPOL.)

	LEFT PHASE			RIGHT PHASE	
(1)	(2)	(3)	(1)	(2)	(3)
86.741	5.442	7.817	0.046	2.453	97.500
72.458	14.023	13.519	0.096	6.718	93.185
64.999	18.322	16.679	0.146	9.648	90.205
56.934	22.853	20.213	0.199	12.722	87.080
54.712	24.088	21.201	0.218	13.693	86.089
51.973	25.451	22.576	0.242	14.891	34.867
47.267	27.671	25.062	0.294	17.055	82.651
43.484	29.203	27.313	0.316	18.065	81.619
37.196	31.564	31.240	0.394	20.751	78.855

SPECIFIC MODEL PARAMETERS IN KELVIN

		UNIQUAC		NRTL(ALPHA=.2)	
I J	AIJ	AJI	AIJ	AJI	
1 2	-13.033	59.794	-417.60	135.66	
1 3	604.20	19.178	354.14	1817.6	
2 3	-152.25	159.53	-485.25	526.63	

R1 = 6.3837   R2 = 2.2024   R3 = 0.9200
Q1 = 5.032   Q2 = 2.072   Q3 = 1.400

MEAN DEV. BETWEEN CALC. AND EXP. CONC. IN MOLE PCT

UNIQUAC (SPECIFIC PARAMETERS)   0.33
NRTL   (SPECIFIC PARAMETERS)   0.49
UNIQUAC (COMMON PARAMETERS)   0.90

C₂H₄O₂-C₁₀H₂₀O₂

(1) C10H20O2   ACETIC ACID,OCTYL ESTER
(2) C2H4O2   ACETIC ACID
(3) H2O   WATER

OTHMER D.F., WHITE R.E., TRUEGER E.
IND.ENG.CHEM. 33(1941)1240

TEMPERATURE = 23.0 DEG C   TYPE OF SYSTEM = 1

EXPERIMENTAL TIE LINES IN MOLE PCT (GRAPH.INTERPOL.)

	LEFT PHASE			RIGHT PHASE	
(1)	(2)	(3)	(1)	(2)	(3)
94.845	3.590	1.565	0.003	2.275	97.722
90.474	5.677	2.849	0.006	4.696	95.299
85.074	10.491	4.435	0.009	7.557	92.435
78.863	14.820	6.317	0.010	10.364	89.625
72.694	19.099	8.207	0.014	13.609	86.377
66.009	23.549	10.442	0.032	17.418	82.550
52.110	32.466	15.424	0.178	25.817	74.005
49.505	34.025	16.470	0.235	27.531	72.235
40.280	39.081	20.639	0.451	32.417	67.132
30.836	43.595	25.570	0.971	37.673	61.356

SPECIFIC MODEL PARAMETERS IN KELVIN

		UNIQUAC		NRTL(ALPHA=.2)	
I	J	AIJ	AJI	AIJ	AJI
1	2	93.573	-6.7773	-573.32	902.77
1	3	828.89	100.58	786.07	2295.4
2	3	-169.23	145.79	-531.58	751.80

R1 = 7.5250   R2 = 2.2024   R3 = 0.9200
Q1 = 6.356   Q2 = 2.072   Q3 = 1.400

MEAN DEV. BETWEEN CALC. AND EXP. CONC. IN MOLE PCT

UNIQUAC (SPECIFIC PARAMETERS)   0.45
NRTL   (SPECIFIC PARAMETERS)   0.40
UNIQUAC (COMMON PARAMETERS)   1.23

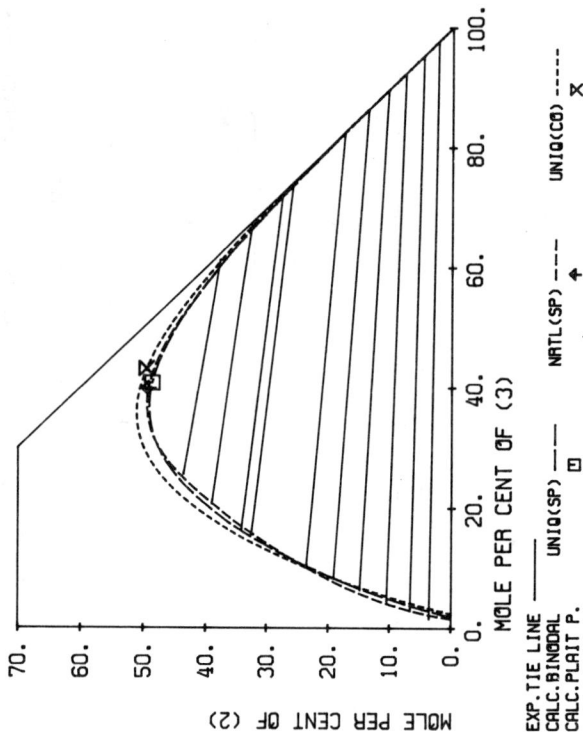

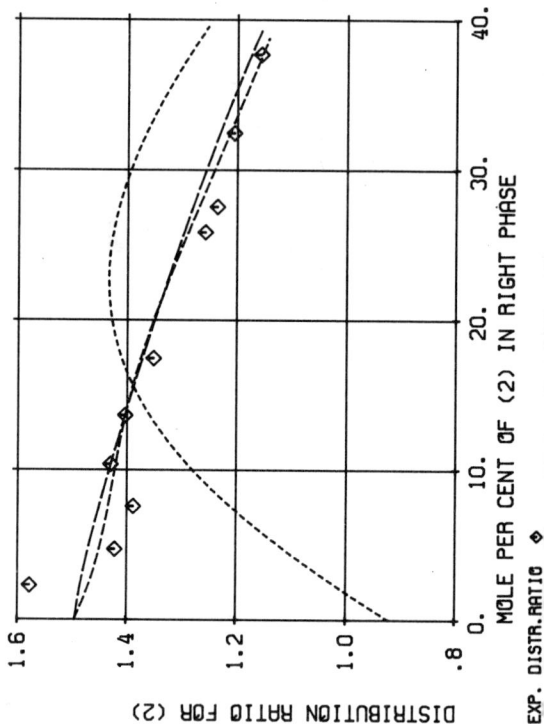

$C_2H_4O_2-C_{12}H_{10}O$

MOLE PER CENT OF (2)

MOLE PER CENT OF (3)

EXP.TIE LINE   UNIQ(SP)   NRTL(SP) -----   UNIQ(CD) -----
CALC.BINODAL
CALC.PLAIT P.

DISTRIBUTION RATIO (2)

MOLE PER CENT OF (2) IN RIGHT PHASE

EXP. DISTR.RATIO   UNIQ(SP)   NRTL(SP) -----   UNIQ(CD) -----
CALC.DISTR.RATIO

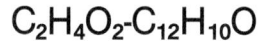

(2) C2H4O2   ACETIC ACID

(3) C12H10O   ETHER,DIPHENYL

PURNELL J.H., BOWDEN S.T.
J.CHEM.SOC. (1954)539

TEMPERATURE = 25.0 DEG C   TYPE OF SYSTEM = 1

EXPERIMENTAL TIE LINES IN MOLE PCT

| | LEFT PHASE | | | RIGHT PHASE | |
(1)	(2)	(3)	(1)	(2)	(3)
91.129	8.871	0.0	0.0	1.682	98.318
83.356	16.629	0.015	0.906	5.162	93.932
79.009	20.975	0.016	0.893	7.234	91.873
74.333	25.616	0.051	1.757	8.433	89.810
69.081	30.827	0.091	2.525	13.382	84.093
63.091	36.731	0.178	3.278	15.979	80.744
54.513	44.910	0.577	3.866	22.731	73.403
48.091	50.972	0.936	4.448	27.133	68.419
36.791	60.706	2.503	4.686	38.560	56.754

SPECIFIC MODEL PARAMETERS IN KELVIN

| | | UNIQUAC | | NRTL(ALPHA=.2) | |
I	J	AIJ	AJI	AIJ	AJI
1	2	-9.5328	-210.21	170.34	-487.77
1	3	259.82	662.44	1566.5	323.43
2	3	17.465	104.02	1212.1	-428.67

R1 = 0.9200   R2 = 2.2024   R3 = 6.2873
Q1 = 1.400   Q2 = 2.072   Q3 = 4.480

MEAN DEV. BETWEEN CALC. AND EXP. CONC. IN MOLE PCT

UNIQUAC (SPECIFIC PARAMETERS)	0.49
NRTL (SPECIFIC PARAMETERS)	0.50
UNIQUAC (COMMON PARAMETERS)	0.71

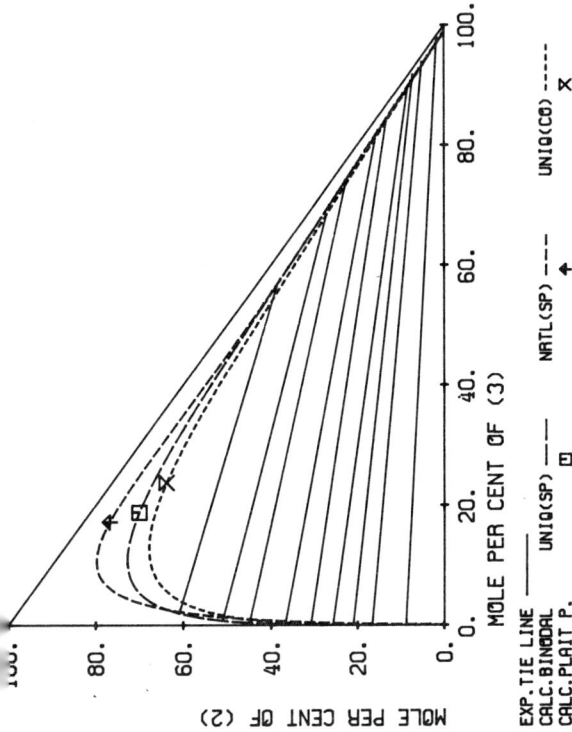

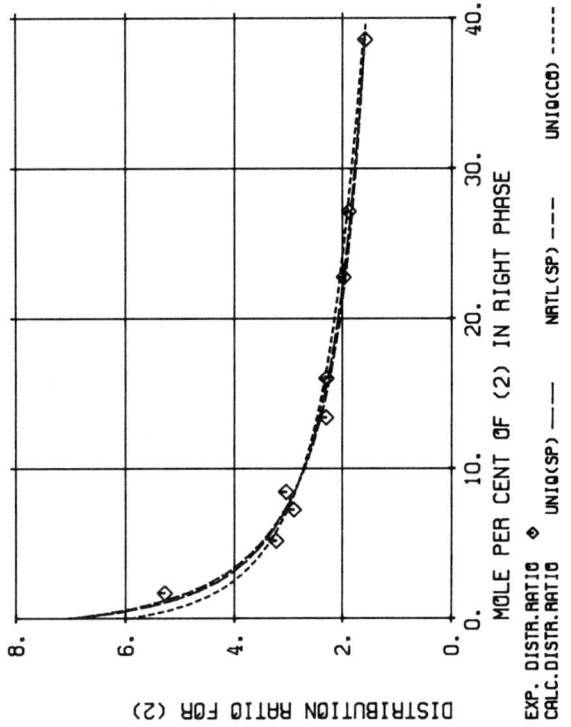

$C_2H_4O_2$-$C_{17}H_{36}O$

(1) C17H36O   1-HEPTADECANOL
(2) C2H4O2   ACETIC ACID
(3) H2O   WATER

UPCHURCH J.C., VAN WINKLE M.
IND.ENG.CHEM. 44(1952)518

TEMPERATURE = 25.0 DEG C   TYPE OF SYSTEM = 1

EXPERIMENTAL TIE LINES IN MOLE PCT (GRAPH.INTERPOL.)

	LEFT PHASE			RIGHT PHASE	
(1)	(2)	(3)	(1)	(2)	(3)
87.588	9.223	3.189	0.008	2.444	97.547
78.387	16.951	4.662	0.010	6.187	93.803
69.347	24.513	6.139	0.013	12.046	87.941
57.951	34.078	7.971	0.019	22.015	77.967
43.587	46.135	10.278	0.027	36.418	63.556
33.877	54.276	11.843	0.089	46.342	53.569

SPECIFIC MODEL PARAMETERS IN KELVIN

		UNIQUAC		NRTL(ALPHA=.2)	
I	J	AIJ	AJI	AIJ	AJI
1	2	166.48	-31.046	-846.97	1817.3
1	3	806.45	299.14	699.54	1792.4
2	3	-158.00	284.91	-255.00	389.31

R1 = 12.2215   R2 = 2.2024   R3 = 0.9200
Q1 = 10.072   Q2 = 2.072   Q3 = 1.400

MEAN DEV. BETWEEN CALC. AND EXP. CONC. IN MOLE PCT

UNIQUAC (SPECIFIC PARAMETERS)   0.32
NRTL   (SPECIFIC PARAMETERS)   0.66
UNIQUAC (COMMON PARAMETERS)   2.14

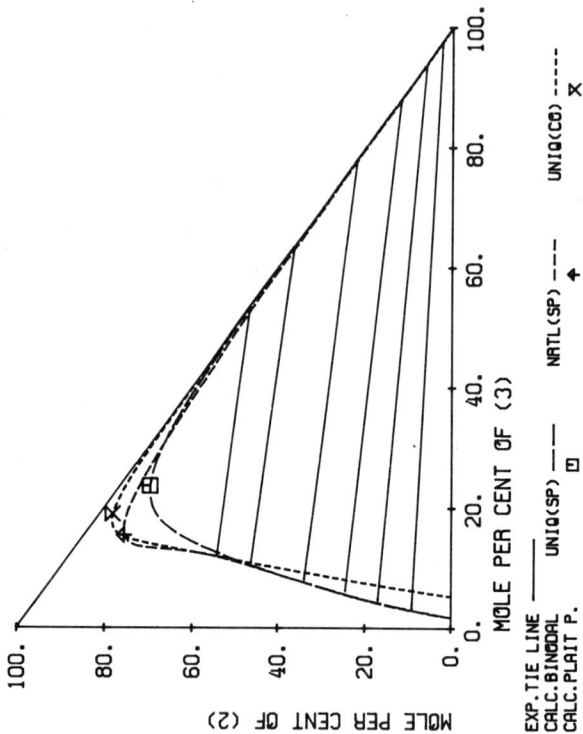

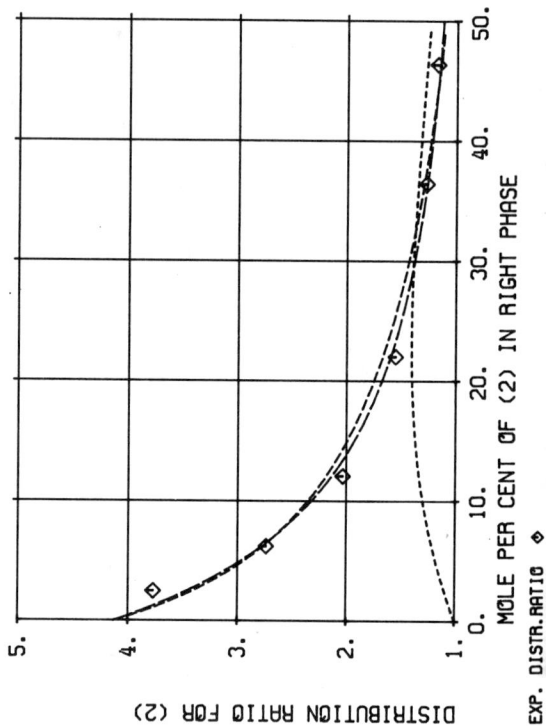

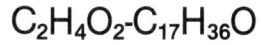

(2) C2H4O2   ACETIC ACID

(3) H2O   WATER

UPCHURCH J.C., VAN WINKLE M.
IND.ENG.CHEM. 44(1952)618

TEMPERATURE = 50.0 DEG C   TYPE OF SYSTEM = 1

EXPERIMENTAL TIE LINES IN MOLE PCT (GRAPH.INTERPOL.)

	LEFT PHASE			RIGHT PHASE	
(1)	(2)	(3)	(1)	(2)	(3)
84.710	4.620	10.670	0.007	2.444	97.549
70.791	17.460	11.748	0.008	6.107	93.885
63.257	24.462	12.281	0.009	11.990	88.001
53.090	33.860	13.050	0.011	21.922	78.067
41.382	44.669	13.950	0.030	36.533	63.437
28.265	56.781	14.954	0.092	45.967	53.941

SPECIFIC MODEL PARAMETERS IN KELVIN

		UNIQUAC		NRTL(ALPHA=.2)	
I	J	AIJ	AJI	AIJ	AJI
1	2	189.76	-61.456	-934.69	1428.9
1	3	462.81	310.61	375.03	3081.7
2	3	20.103	-12.369	-117.91	81.891

R1 =12.2215   R2 = 2.2024   R3 = 0.9200
Q1 =10.072   Q2 = 2.072   Q3 = 1.400

MEAN DEV. BETWEEN CALC. AND EXP. CONC. IN MOLE PCT

UNIQUAC (SPECIFIC PARAMETERS)   0.78
NRTL   (SPECIFIC PARAMETERS)   0.87

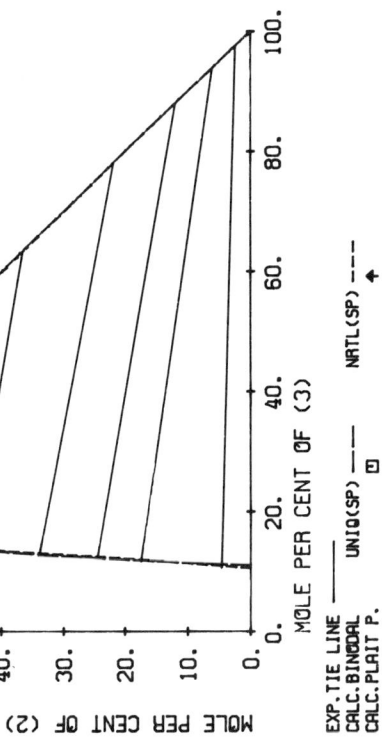

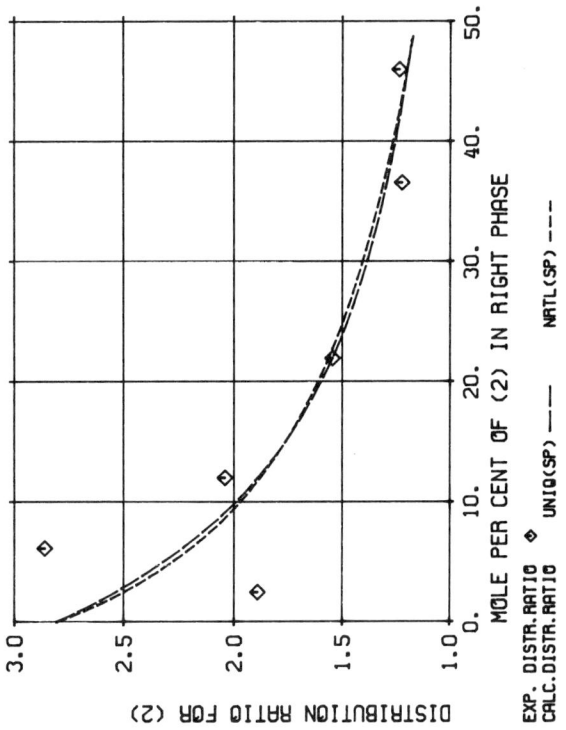

C$_2$H$_4$O$_2$-C$_4$H$_9$NO

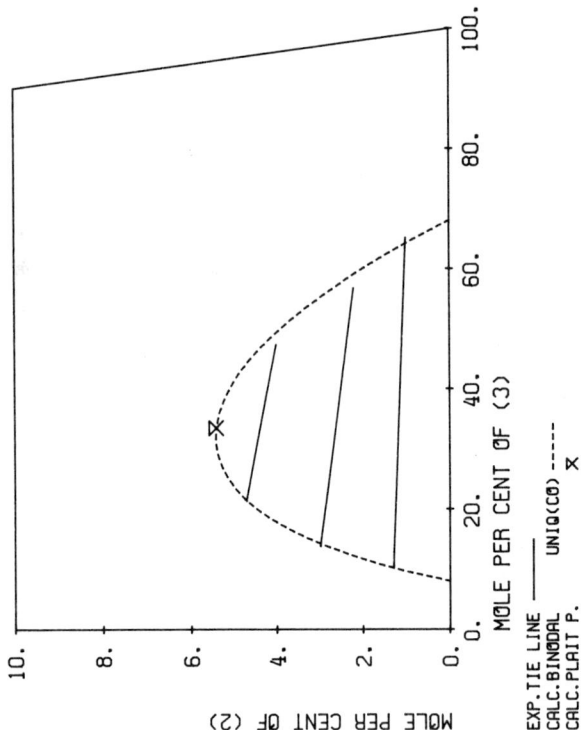

MOLE PER CENT OF (3)

MOLE PER CENT OF (2)

EXP.TIE LINE ——— UNIQ(CC) ———
CALC.BINODAL - - - - -
CALC.PLAIT P. X

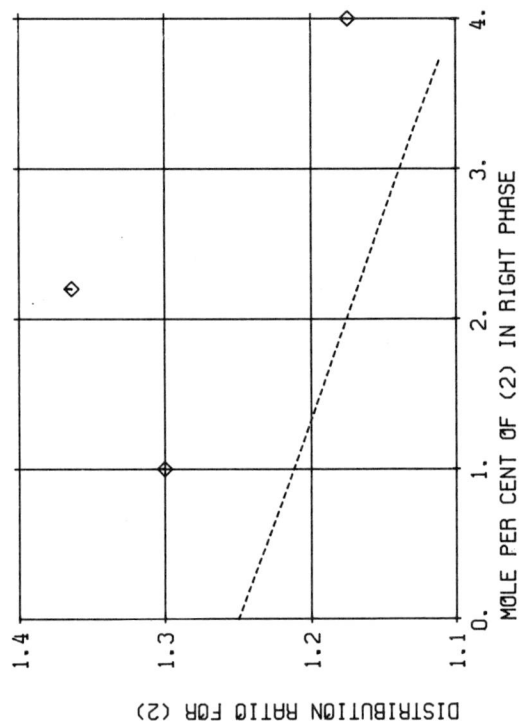

DISTRIBUTION RATIO FOR (2)

MOLE PER CENT OF (2) IN RIGHT PHASE

EXP. DISTR.RATIO ◇ UNIQ(CC) ———
CALC. DISTR.RATIO - - - - -

(1) H2O   WATER
(2) C4H9NO   ACETIC ACID,AMIDE,N,N-DIMETHYL
(3) C2H4O2   FORMIC ACID,METHYL ESTER

KRUPATKIN I.L., AVTONOMOVA E.N.
IZV.VYSSH.UCHEBN.ZAVED.KHIM.KHIM.TEKHNOL. 15(1972)1480

TEMPERATURE = 25.0 DEG C   TYPE OF SYSTEM = 1

EXPERIMENTAL TIE LINES IN MOLE PCT

	LEFT PHASE			RIGHT PHASE	
(1)	(2)	(3)	(1)	(2)	(3)
88.400	1.300	10.300	33.800	1.000	65.200
83.200	3.000	13.800	41.000	2.200	56.800
73.900	4.700	21.400	48.600	4.000	47.400

MEAN DEV. BETWEEN CALC. AND EXP. CONC. IN MOLE PCT
UNIQUAC (COMMON PARAMETERS)   0.41

$C_2H_4O_3S-C_7H_8$

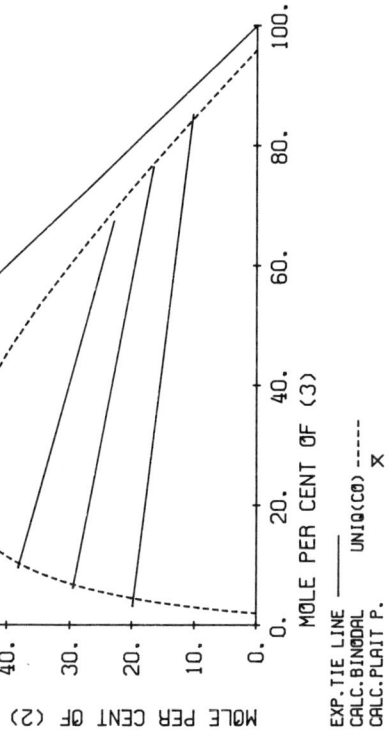

MOLE PER CENT OF (3)

MOLE PER CENT OF (2)

EXP.TIE LINE ————
CALC.BINODAL   UNIQ(C0) - - - -
CALC.PLAIT P.   ⊠

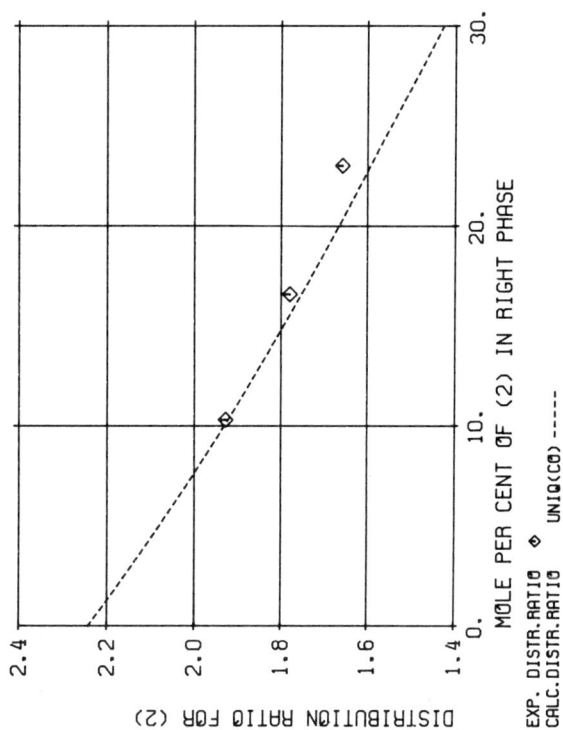

MOLE PER CENT OF (2) IN RIGHT PHASE

DISTRIBUTION RATIO FOR (2)

EXP. DISTR.RATIO ◇   UNIQ(C0) - - - -
CALC.DISTR.RATIO - - - -

(2) C7H8   TOLUENE

(3) C2H4O3S   SULFUROUS ACID,ETHYLENE ESTER

RAWAT B.S., GULATI I.B.
J.APPL.CHEM.BIOTECHNOL. 26(1976)425

TEMPERATURE = 25.0 DEG C   TYPE OF SYSTEM = 1

EXPERIMENTAL TIE LINES IN MOLE PCT

	LEFT PHASE			RIGHT PHASE	
(1)	(2)	(3)	(1)	(2)	(3)
76.933	19.848	3.219	4.480	10.295	85.225
64.286	29.502	6.211	6.890	16.579	76.531
52.224	38.169	9.606	9.592	23.010	67.398

MEAN DEV. BETWEEN CALC. AND EXP. CONC. IN MOLE PCT

UNIQUAC (COMMON PARAMETERS)   0.66

C$_2$H$_5$ClO-C$_6$H$_5$Cl

(1) C6H5CL   BENZENE,CHLORO
(2) C2H5CLO   ETHANOL,2-CHLORO
(3) H2O   WATER

ABABI V., POPA A., MIHAILA GH.
AN.STIINT.UNIV.AL.I.CUZA IASI. 9(1963)233

TEMPERATURE = 20.5 DEG C   TYPE OF SYSTEM = 1

EXPERIMENTAL TIE LINES IN MOLE PCT

	LEFT PHASE			RIGHT PHASE	
(1)	(2)	(3)	(1)	(2)	(3)
94.496	5.504	0.0	0.0	4.263	95.737
91.808	8.192	0.0	0.0	6.231	93.769
86.039	13.961	0.0	0.043	9.667	90.291
77.542	20.721	1.737	0.188	13.155	86.657
69.436	26.665	3.898	0.433	16.427	83.139
61.158	31.423	7.420	0.995	20.861	78.144
55.861	34.401	9.738	1.732	24.255	74.013
50.230	36.980	12.790	2.742	27.256	70.002
44.959	39.016	16.025	4.437	30.493	65.065

SPECIFIC MODEL PARAMETERS IN KELVIN

		UNIQUAC		NRTL(ALPHA=.2)	
I	J	AIJ	AJI	AIJ	AJI
1	2	441.03	-230.15	-1660.0	6.1839
1	3	901.37	493.25	1494.5	1636.4
2	3	407.66	-250.89	-361.90	-1166.3

R1 = 3.8127   R2 = 2.6698   R3 = 0.9200
Q1 = 2.844   Q2 = 2.392   Q3 = 1.400

MEAN DEV. BETWEEN CALC. AND EXP. CONC. IN MOLE PCT

UNIQUAC (SPECIFIC PARAMETERS)   0.70
NRTL   (SPECIFIC PARAMETERS)   0.55
UNIQUAC (COMMON PARAMETERS)   1.21

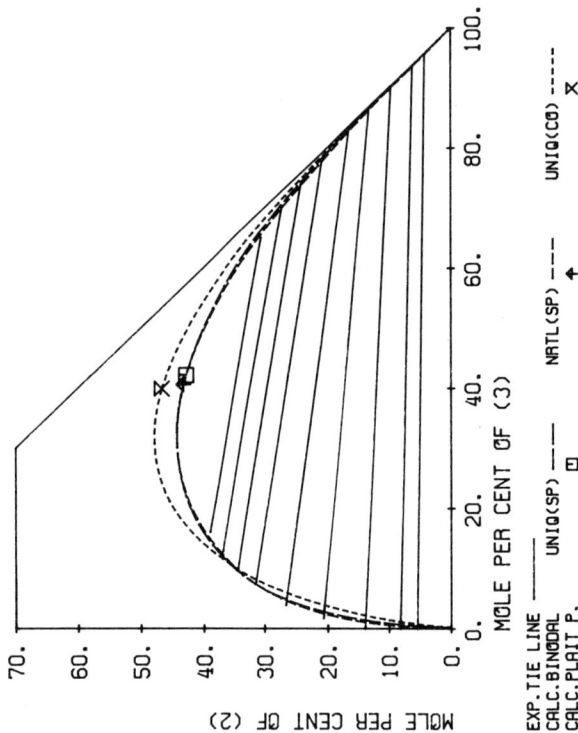

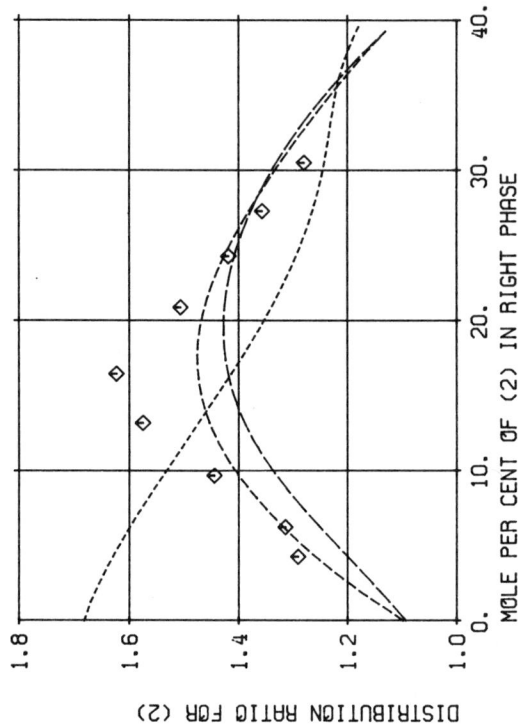

MOLE PER CENT OF (3)
MOLE PER CENT OF (2)

EXP.TIE LINE —— UNIQ(SP) —·— NRTL(SP) ---- UNIQ(C0) -----
CALC.BINODAL
CALC.PLAIT P.

UNIQ(SP) ⊡   NRTL(SP) ◆   UNIQ(C0) ✕

DISTRIBUTION RATIO FOR (2)
MOLE PER CENT OF (2) IN RIGHT PHASE

EXP. DISTR.RATIO ◇

C$_2$H$_5$ClO-C$_6$H$_{14}$O

MOLE PER CENT OF (3)

MOLE PER CENT OF (2)

EXP.TIE LINE ———   UNIQ(SP) ⊟   NRTL(SP) ╈   UNIQ(CO) ----
CALC.BINODAL
CALC.PLAIT P. ☒

MOLE PER CENT OF (2) IN RIGHT PHASE

DISTRIBUTION RATIO FOR (2)

EXP. DISTR.RATIO ◇   UNIQ(SP) ———   NRTL(SP) ----   UNIQ(CO) ----
CALC.DISTR.RATIO

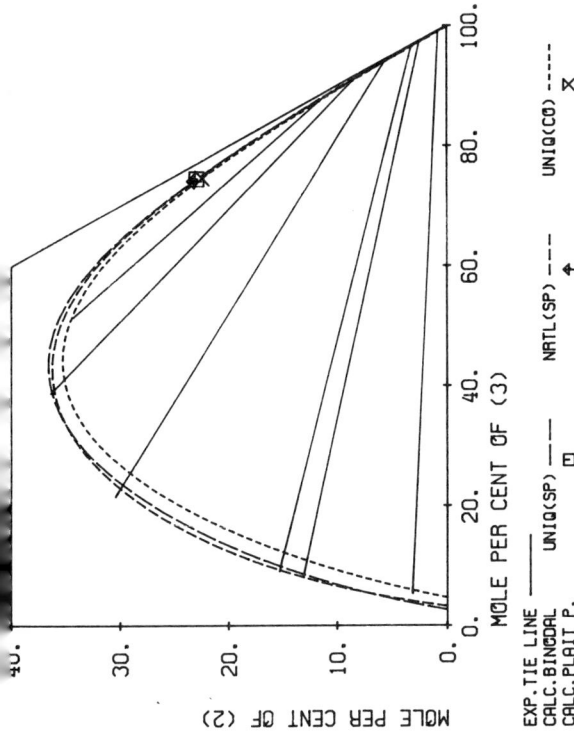

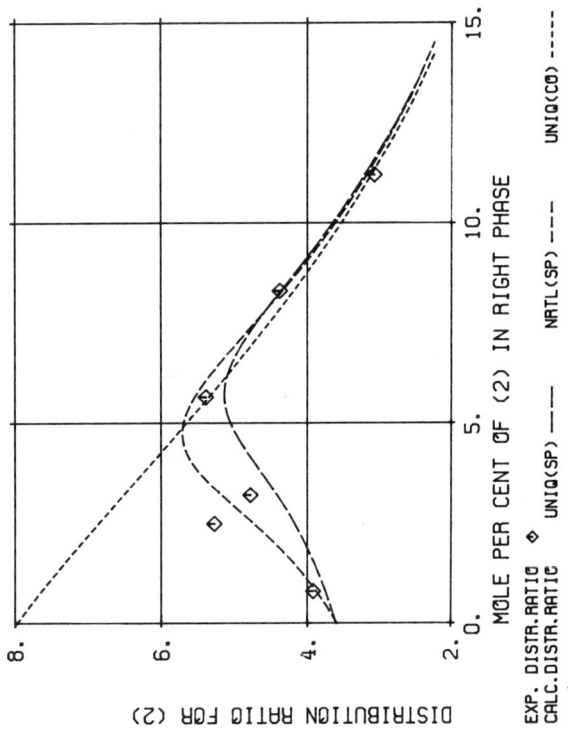

---
(2) C2H5CLO   ETHANOL,2-CHLORO
---
(3) H2O   WATER

OGORODNIKOV S.K., KOGAN V.B.
KHIM.PROM-ST.(MOSCOW) (1963)270

TEMPERATURE = 20.0 DEG C   TYPE OF SYSTEM = 1

EXPERIMENTAL TIE LINES IN MOLE PCT (GRAPH.INTERPOL.)

	LEFT PHASE			RIGHT PHASE	
(1)	(2)	(3)	(1)	(2)	(3)
91.486	3.131	5.383	0.098	0.799	99.103
78.703	13.087	8.210	0.211	2.479	97.311
75.597	15.290	9.112	0.251	3.198	96.551
48.081	30.411	21.507	0.340	5.641	94.019
24.948	36.335	38.717	0.424	8.298	91.278
14.306	34.264	51.430	0.699	11.195	88.106

SPECIFIC MODEL PARAMETERS IN KELVIN

		UNIQUAC		NRTL(ALPHA=.2)	
I J	AIJ	AJI		AIJ	AJI
1 2	356.48	-147.47		817.71	-426.31
1 3	703.84	93.084		525.49	1738.3
2 3	-81.202	224.79		-347.11	1006.1

R1 = 4.7421   R2 = 2.6698   R3 = 0.9200
Q1 = 4.088   Q2 = 2.392   Q3 = 1.400

MEAN DEV. BETWEEN CALC. AND EXP. CONC. IN MOLE PCT

UNIQUAC (SPECIFIC PARAMETERS)   0.52
NRTL (SPECIFIC PARAMETERS)   0.38
UNIQUAC (COMMON PARAMETERS)   1.02

$C_2H_5NO-C_6H_6$

(1) C6H14   HEXANE

(2) C6H6   BENZENE

(3) C2H5NO   FORMIC ACID,AMIDE,N-METHYL

STEIB H.
J.PRAKT.CHEM. 4,31(1966)179

TEMPERATURE = 20.0 DEG C   TYPE OF SYSTEM = 1

EXPERIMENTAL TIE LINES IN MOLE PCT

	LEFT PHASE			RIGHT PHASE	
(1)	(2)	(3)	(1)	(2)	(3)
86.289	13.610	0.101	3.400	5.548	91.051
71.530	27.196	1.274	4.198	11.419	84.382
58.906	40.814	0.280	4.520	18.720	75.759
58.580	40.999	0.420	4.447	18.887	76.666
43.554	55.343	1.103	5.670	28.486	65.844

SPECIFIC MODEL PARAMETERS IN KELVIN

		UNIQUAC		NRTL(ALPHA=.2)	
I J		AIJ	AJI	AIJ	AJI
1 2		36.701	-48.960	13.908	-83.588
1 3		780.44	7.7798	1472.7	494.69
2 3		535.54	-146.57	1024.9	-268.10

R1 = 4.4998   R2 = 3.1878   R3 = 2.4317
Q1 = 3.856   Q2 = 2.400   Q3 = 2.192

MEAN DEV. BETWEEN CALC. AND EXP. CONC. IN MOLE PCT

UNIQUAC (SPECIFIC PARAMETERS)   0.51
NRTL (SPECIFIC PARAMETERS)   0.55
UNIQUAC (COMMON PARAMETERS)   1.12

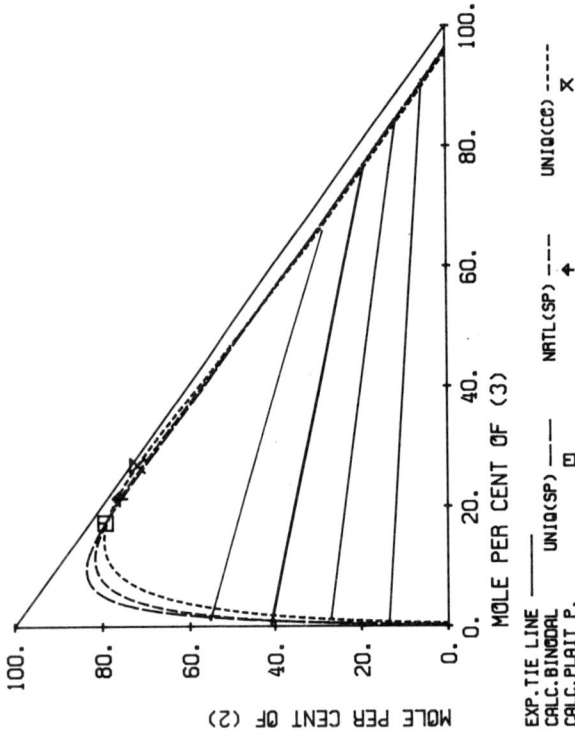

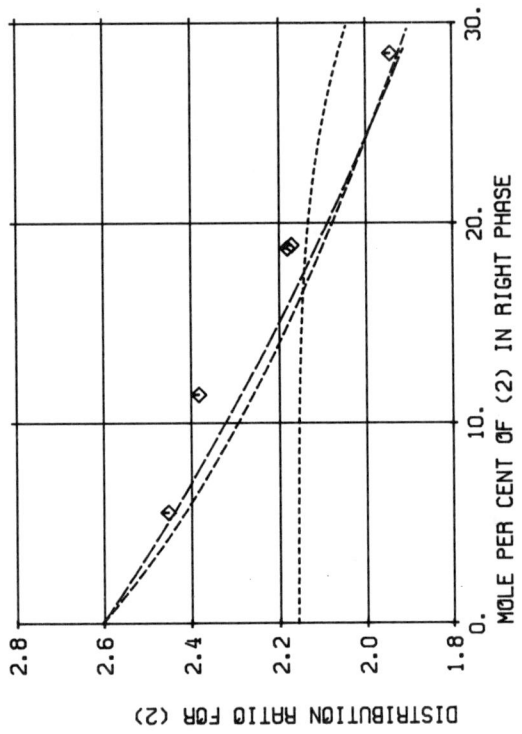

MOLE PER CENT OF (3)

MOLE PER CENT OF (2)

EXP.TIE LINE ———

(1) C7H16   HEPTANE

(2) C6H6   BENZENE

(3) C2H5NO   FORMIC ACID,AMIDE,N-METHYL

STEIB H.
J.PRAKT.CHEM. 4,30(1965)39

TEMPERATURE = 20.0 DEG C   TYPE OF SYSTEM = 1

EXPERIMENTAL TIE LINES IN MOLE PCT

| | LEFT PHASE | | | RIGHT PHASE | |
(1)	(2)	(3)	(1)	(2)	(3)
70.527	28.998	0.475	2.731	10.509	86.760
53.208	45.426	1.365	3.602	20.460	75.938
49.933	49.464	0.603	3.758	22.605	73.638
36.105	62.008	1.887	4.619	32.119	63.262

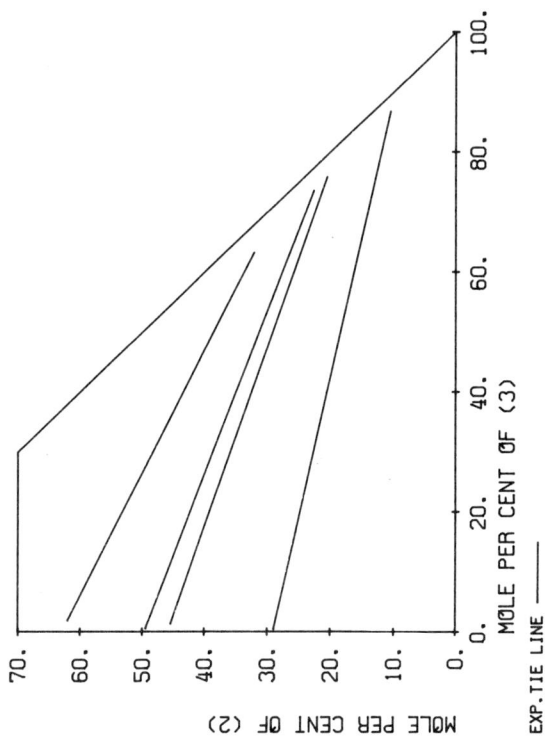

$C_2H_5NO-C_6H_6$

(1) H2O   WATER

(2) C2H5NO   FORMIC ACID,AMIDE,N-METHYL

(3) C6H6   BENZENE

STEIB H.
J.PRAKT.CHEM. 4,30(1965)39

TEMPERATURE = 20.0 DEG C   TYPE OF SYSTEM = 1

EXPERIMENTAL TIE LINES IN MOLE PCT

	LEFT PHASE		RIGHT PHASE	
(1)	(2)	(3)	(1)	(2)	(3)
86.664	13.215	0.120	0.432	0.132	99.436
71.720	27.518	0.762	0.431	1.445	98.125
63.351	35.026	1.623	0.429	2.749	96.821
39.511	51.890	8.599	0.429	3.009	96.562
31.073	54.282	14.645	0.428	3.917	95.655
31.292	54.024	14.683	0.423	4.046	95.526

SPECIFIC MODEL PARAMETERS IN KELVIN

	UNIQUAC		NRTL(ALPHA=.2)	
I J	AIJ	AJI	AIJ	AJI
1 2	-147.88	-194.60	364.20	-621.17
1 3	350.56	492.36	2200.2	959.94
2 3	-49.309	425.32	-8.2538	817.93

R1 = 0.9200   R2 = 2.4317   R3 = 3.1878
Q1 = 1.400   Q2 = 2.192   Q3 = 2.400

MEAN DEV. BETWEEN CALC. AND EXP. CONC. IN MOLE PCT

UNIQUAC (SPECIFIC PARAMETERS)   0.71
NRTL   (SPECIFIC PARAMETERS)   0.61
UNIQUAC (COMMON PARAMETERS)   0.66

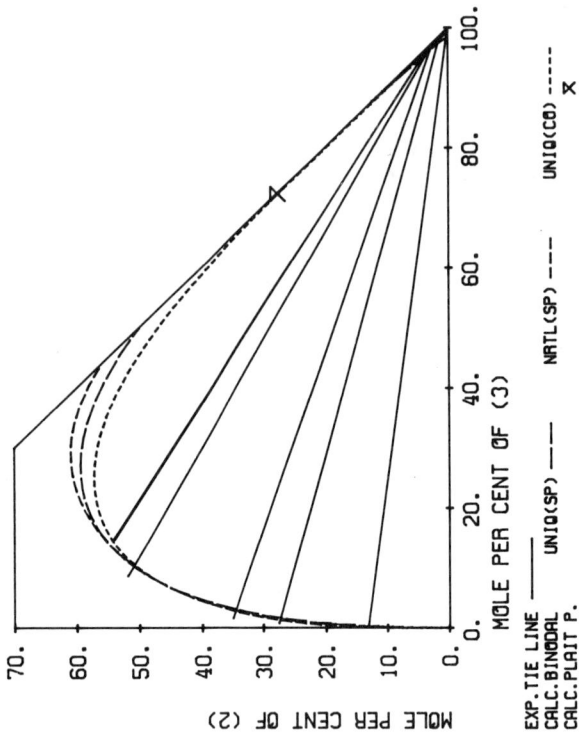

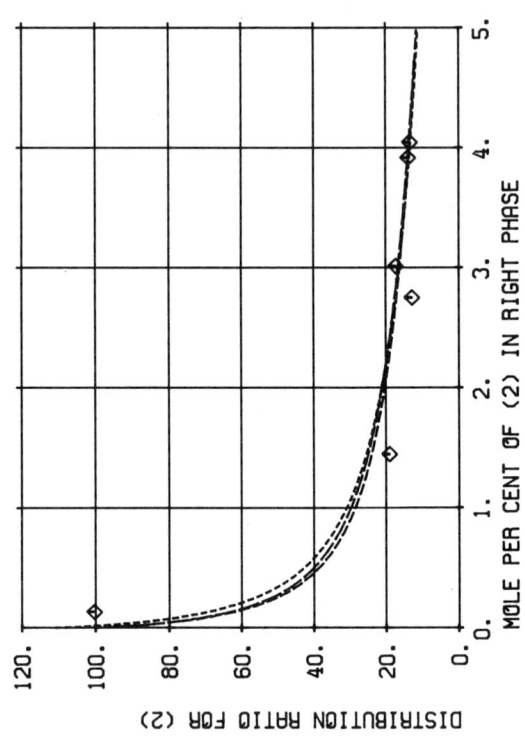

$C_2H_5NO-C_6H_{12}$

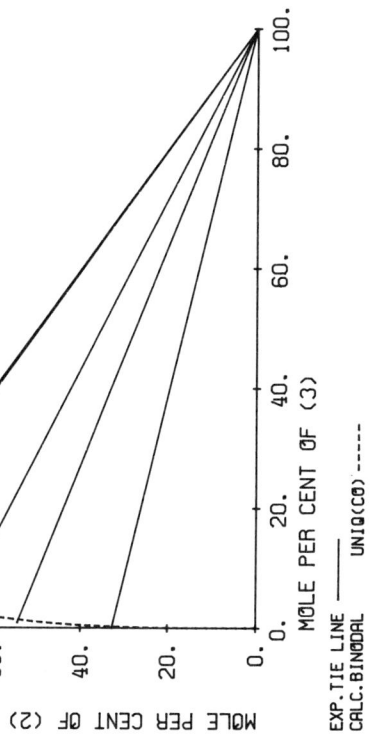

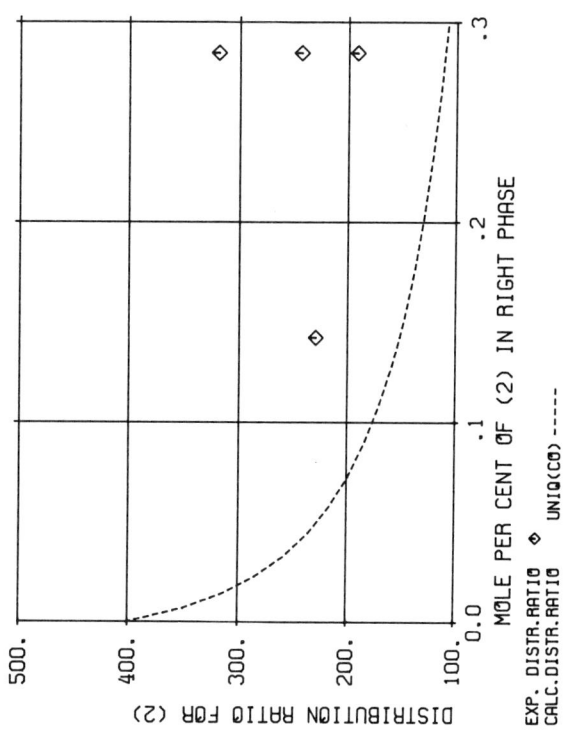

(2) C2H5NO   FORMIC ACID,AMIDE,N-METHYL

(3) C6H12   1-HEXENE

STEIB H.
J.PRAKT.CHEM. 4,31(1966)179

TEMPERATURE = 20.0 DEG C   TYPE OF SYSTEM = 2

EXPERIMENTAL TIE LINES IN MOLE PCT

	LEFT PHASE			RIGHT PHASE	
(1)	(2)	(3)	(1)	(2)	(3)
67.252	32.710	0.037	0.047	0.142	99.811
44.590	54.534	0.877	0.047	0.285	99.669
28.849	69.338	1.813	0.047	0.285	99.669
0.677	90.780	8.544	0.047	0.285	99.669

MEAN DEV. BETWEEN CALC. AND EXP. CONC. IN MOLE PCT

UNIQUAC (COMMON PARAMETERS)   0.82

$C_2H_5NO_2$-C_3H_8O

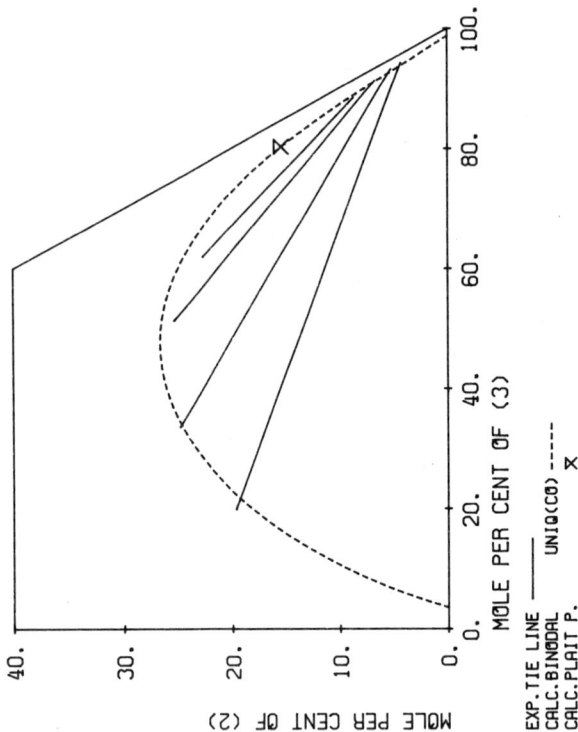

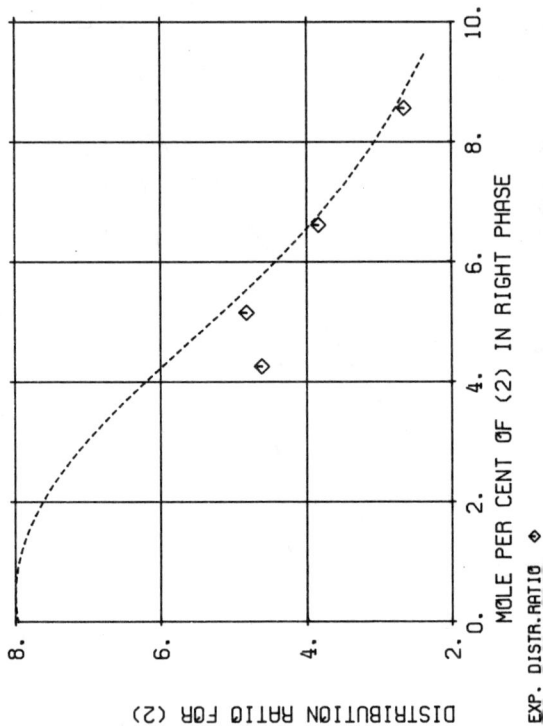

(1) C2H5NO2   ETHANE,NITRO

(2) C3H8O   1-PROPANOL

(3) H2O   WATER

MALONE J.W., VINING R.W.
J.CHEM.ENG.DATA 12(1967)387

TEMPERATURE = 25.0 DEG C   TYPE OF SYSTEM = 1

EXPERIMENTAL TIE LINES IN MOLE PCT

	LEFT PHASE			RIGHT PHASE	
(1)	(2)	(3)	(1)	(2)	(3)
60.479	19.616	19.905	1.400	4.250	94.350
41.569	24.829	33.602	1.543	5.151	93.306
23.334	25.367	51.299	2.017	6.609	91.374
15.345	22.730	61.925	2.915	8.560	88.525

MEAN DEV. BETWEEN CALC. AND EXP. CONC. IN MOLE PCT

UNIQUAC (COMMON PARAMETERS)   0.90

$C_2H_5NO_2$-C_8H_{16}

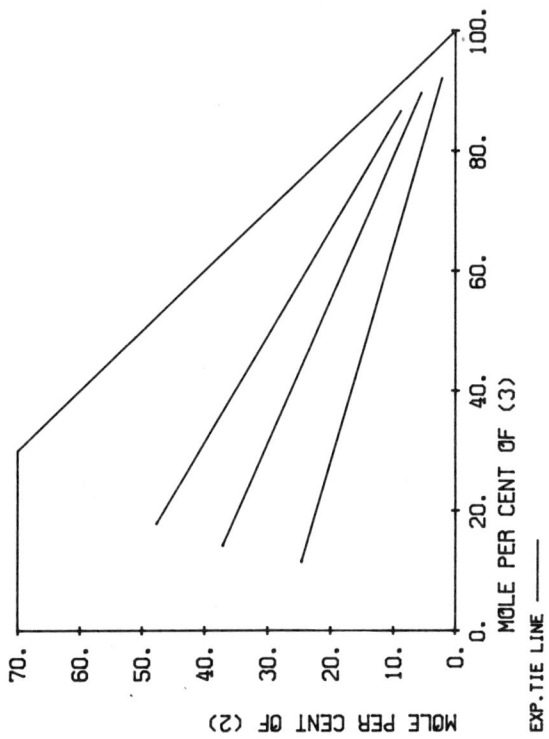

MOLE PER CENT OF (3)

EXP.TIE LINE ——

MOLE PER CENT OF (2)

(1) C8H18   OCTANE
(2) C8H16   1-OCTENE
(3) C2H5NO2   ETHANE,NITRO

HWA S.C.P., TECHO R., ZIEGLER W.T.
J.CHEM.ENG.DATA 8(1963)409

TEMPERATURE = 0.0 DEG C   TYPE OF SYSTEM = 1

EXPERIMENTAL TIE LINES IN MOLE PCT

	LEFT PHASE			RIGHT PHASE	
(1)	(2)	(3)	(1)	(2)	(3)
63.878	24.627	11.495	5.674	2.157	92.168
48.651	37.118	14.231	4.841	5.491	89.667
34.520	47.640	17.839	4.559	8.781	86.660

$C_2H_5NO_2$-C_8H_{18}

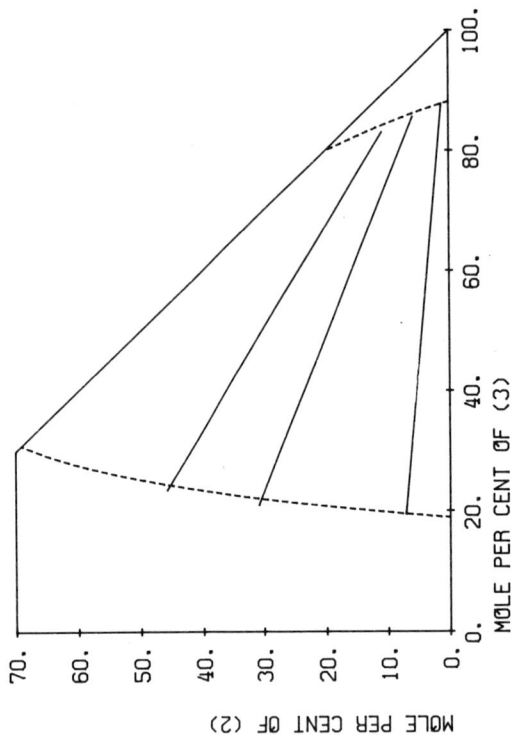

MOLE PER CENT OF (3)

MOLE PER CENT OF (2)

EXP.TIE LINE ——— UNIQ(C0)
CALC.BINODAL --------

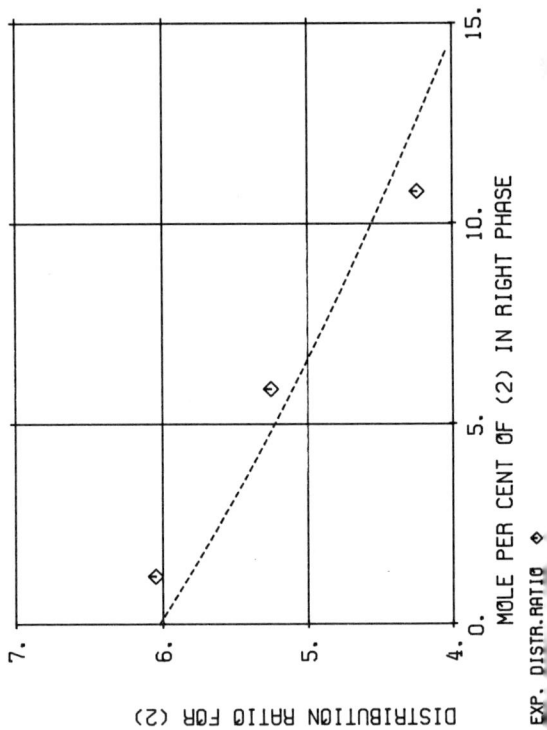

MOLE PER CENT OF (2) IN RIGHT PHASE

DISTRIBUTION RATIO FOR (2)

EXP. DISTR.RATIO ◇

(1) C8H18   OCTANE

(2) C8H18   PENTANE,2,2,4-TRIMETHYL

(3) C2H5NO2   ETHANE,NITRO

HWA S.C.P., TECHO R., ZIEGLER W.T.
J.CHEM.ENG.DATA 8(1963)409

TEMPERATURE = 25.0 DEG C   TYPE OF SYSTEM = 2

EXPERIMENTAL TIE LINES IN MOLE PCT

	LEFT PHASE			RIGHT PHASE	
(1)	(2)	(3)	(1)	(2)	(3)
73.357	7.186	19.456	10.972	1.188	87.840
48.157	30.806	21.037	8.476	5.863	85.661
30.709	45.787	23.504	6.078	10.797	93.125

MEAN DEV. BETWEEN CALC. AND EXP. CONC. IN MOLE PCT

UNIQUAC (COMMON PARAMETERS)   0.58

$C_2H_5NO_2-C_8H_{18}$

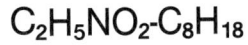

(1) C8H18   OCTANE
(2) C8H18   PENTANE,2,2,4-TRIMETHYL
(3) C2H5NO2   ETHANE,NITRO

HWA S.C.P., TECHO R., ZIEGLER W.T.
J.CHEM.ENG.DATA 8(1963)409

TEMPERATURE = 35.0 DEG C   TYPE OF SYSTEM = 1

EXPERIMENTAL TIE LINES IN MOLE PCT

	LEFT PHASE			RIGHT PHASE	
(1)	(2)	(3)	(1)	(2)	(3)
54.507	10.655	34.838	16.932	3.343	79.725
46.430	18.379	35.191	15.902	6.155	77.943
29.991	30.423	39.586	13.811	13.436	72.754
27.349	31.036	41.615	12.992	15.635	71.373

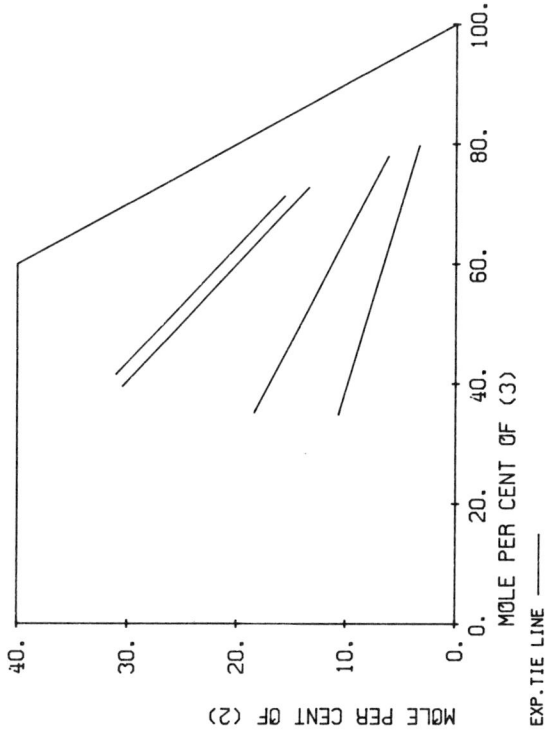

C_2H_6O-C_3H_7Br

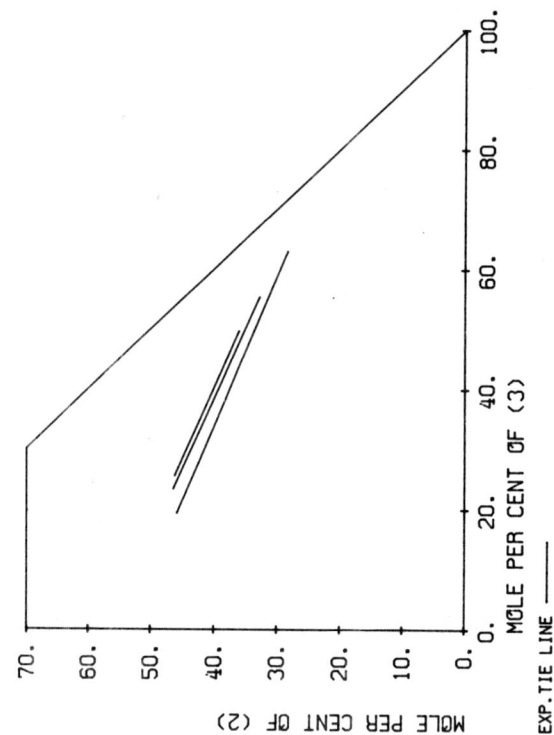

```
(1) -- H2O          WATER
(2) -- C2H6O        ETHANOL
(3) -- C3H7BR       PROPANE,1-BROMO
```

```
BONNER W.D.
J.PHYS.CHEM. 14(1910)738

TEMPERATURE =   0.0 DEG C       TYPE OF SYSTEM = 1

EXPERIMENTAL TIE LINES IN MOLE PCT (GRAPH.INTERPOL.)

        LEFT PHASE               RIGHT PHASE
     (1)     (2)     (3)      (1)     (2)     (3)

  34.898  45.821  19.282    8.687  28.373  62.940
  30.222  46.427  23.351   11.843  32.738  55.368
  28.243  46.185  25.572   14.260  36.071  49.669
```

MOLE PER CENT OF (2)

MOLE PER CENT OF (3)

EXP.TIE LINE ───

$C_2H_6O-C_3H_8O_2$

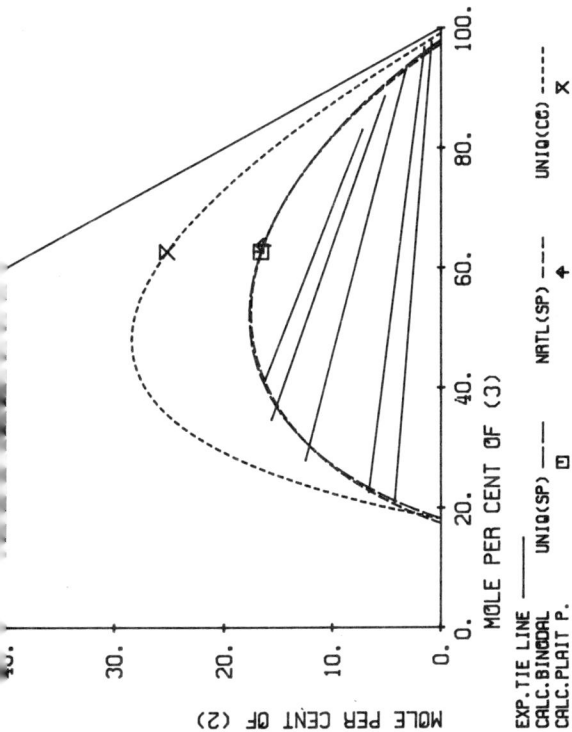

MOLE PER CENT OF (2)

MOLE PER CENT OF (3)

EXP.TIE LINE ———   UNIQ(SP) ——— ☐   NRTL(SP) ——— ✦   UNIQ(CC) ——— ✕
CALC.BINODAL ———
CALC.PLAIT P.

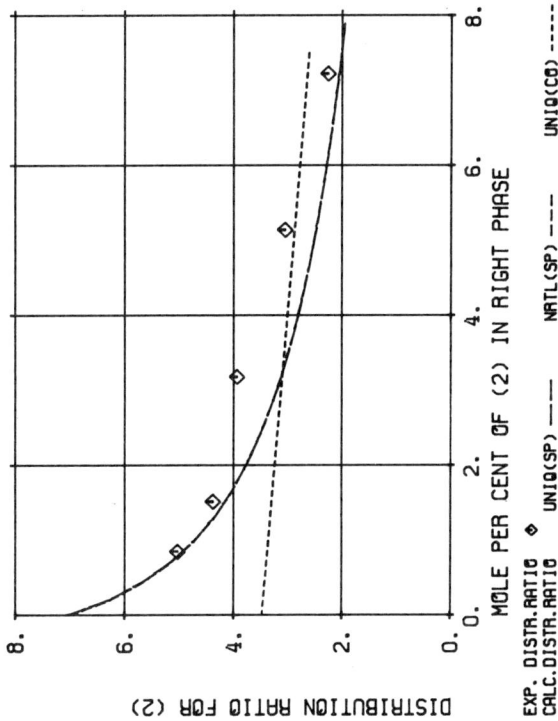

DISTRIBUTION RATIO FOR (2)

MOLE PER CENT OF (2) IN RIGHT PHASE

EXP. DISTR.RATIO ◆   UNIQ(SP) ———   NRTL(SP) ———   UNIQ(CC) ———
CALC.DISTR.RATIO ———

(2) C2H6O	ETHANOL
(3) C6H6	BENZENE

DEGALEESAN T.E., LADDHA G.S.
J.APPL.CHEM. 12(1962)111

TEMPERATURE = 30.0 DEG C   TYPE OF SYSTEM = 1

EXPERIMENTAL TIE LINES IN MOLE PCT

	LEFT PHASE			RIGHT PHASE	
	(1)	(2)	(3)	(1)	(2)	(3)
	74.781	4.247	20.972	1.401	0.845	97.755
	70.740	6.636	22.623	1.631	1.516	96.853
	59.589	12.503	27.908	3.239	3.177	93.584
	49.699	15.651	34.649	6.099	5.137	88.763
	42.592	16.309	41.099	9.637	7.221	83.142

SPECIFIC MODEL PARAMETERS IN KELVIN

		UNIQUAC		NRTL(ALPHA=.2)	
I J		AIJ	AJI	AIJ	AJI
1 2		32.959	-302.35	-187.08	-567.46
1 3		42.453	395.69	147.64	1030.2
2 3		-117.91	-67.578	-177.20	-229.99

R1 = 3.2824   R2 = 2.1055   R3 = 3.1878
Q1 = 2.784   Q2 = 1.972   Q3 = 2.400

MEAN DEV. BETWEEN CALC. AND EXP. CONC. IN MOLE PCT

UNIQUAC (SPECIFIC PARAMETERS)	0.90
NRTL (SPECIFIC PARAMETERS)	1.01
UNIQUAC (COMMON PARAMETERS)	3.06

C_2H_6O-$C_3H_8O_2$

DEGALEESAN T.E., LADDHA G.S.
J.APPL.CHEM. 12(1962)111

(1) C3H8O2   1,2-PROPANEDIOL

(2) C2H6O   ETHANOL

(3) C6H12   CYCLOHEXANE

TEMPERATURE = 30.0 DEG C   TYPE OF SYSTEM = 1

EXPERIMENTAL TIE LINES IN MOLE PCT

	LEFT PHASE			RIGHT PHASE	
(1)	(2)	(3)	(1)	(2)	(3)
78.655	16.106	5.239	1.156	0.637	98.207
67.746	25.661	6.593	1.153	1.360	97.487
47.342	41.418	11.240	1.196	3.413	95.391
34.411	48.549	17.040	1.298	4.609	94.093
27.743	50.102	22.155	1.453	5.600	92.947

SPECIFIC MODEL PARAMETERS IN KELVIN

		UNIQUAC			NRTL(ALPHA=.2)	
I	J	AIJ	AJI		AIJ	AJI
1	2	-238.22	-14.535		-85.198	-251.55
1	3	156.71	338.10		756.14	823.16
2	3	-129.11	494.51		-64.087	817.81

R1 = 3.2824   R2 = 2.1055   R3 = 4.0464
Q1 = 2.784   Q2 = 1.972   Q3 = 3.240

MEAN DEV. BETWEEN CALC. AND EXP. CONC. IN MOLE PCT

UNIQUAC (SPECIFIC PARAMETERS)   0.38
NRTL   (SPECIFIC PARAMETERS)   0.48
UNIQUAC (COMMON PARAMETERS)   0.96

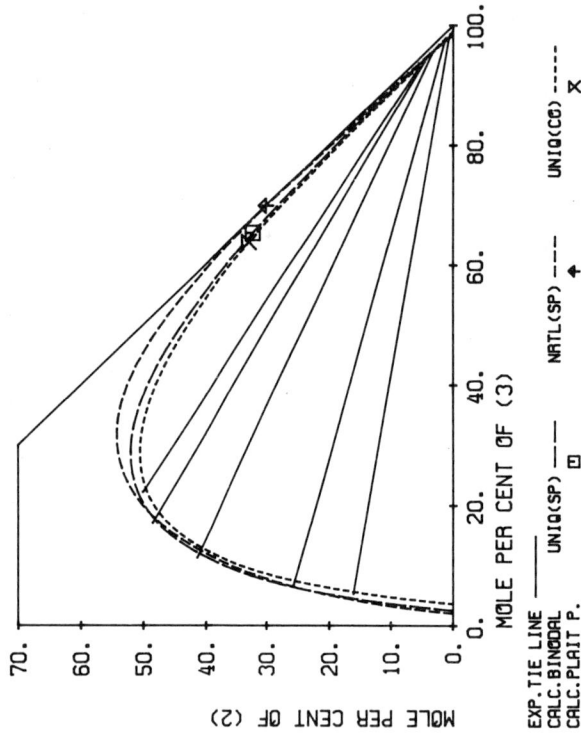

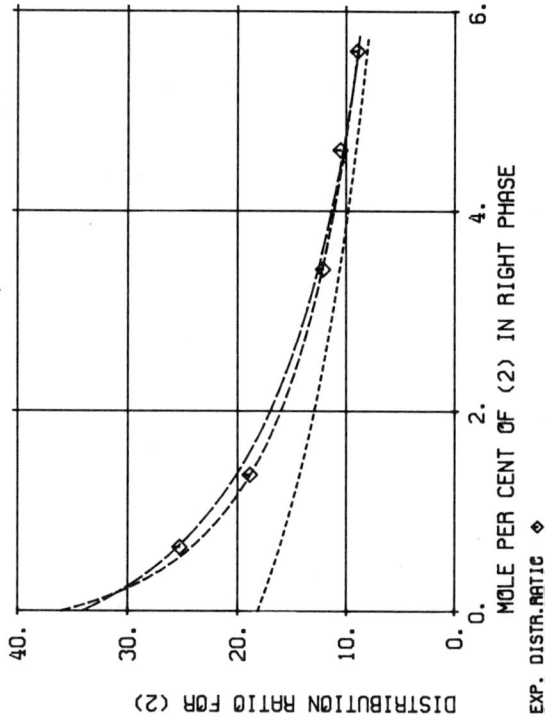

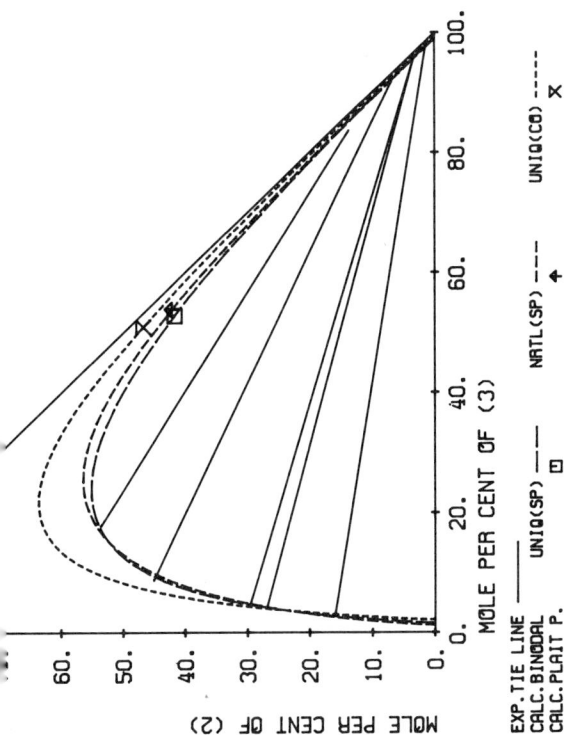

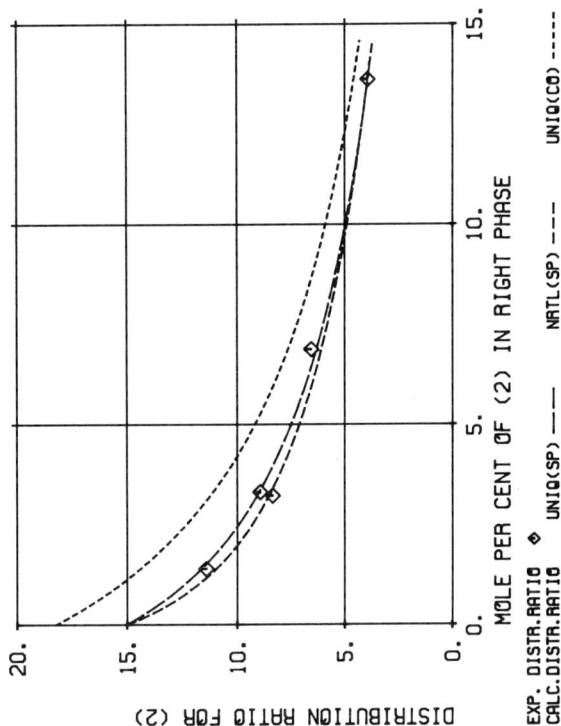

$C_2H_6O-C_3H_8O_2$

(2) C2H6O   ETHANOL

(3) C6H14   HEXANE

DEGALEESAN T.E., LADDHA G.S.
J.APPL.CHEM. 12(1962)111

TEMPERATURE = 30.0 DEG C   TYPE OF SYSTEM = 1

EXPERIMENTAL TIE LINES IN MOLE PCT

	LEFT PHASE			RIGHT PHASE	
(1)	(2)	(3)	(1)	(2)	(3)
80.889	15.868	3.243	0.506	1.393	98.101
68.813	26.897	4.290	0.502	3.223	96.276
65.653	29.570	4.777	0.891	3.312	95.797
46.178	45.098	8.723	1.204	6.871	91.925
28.654	53.640	17.707	2.643	13.621	83.737

SPECIFIC MODEL PARAMETERS IN KELVIN

		UNIQUAC		NRTL (ALPHA=.2)	
I	J	AIJ	AJI	AIJ	AJI
1	2	-202.26	16.873	-94.339	-168.39
1	3	116.04	415.91	366.35	968.00
2	3	-128.74	428.69	70.778	488.79

R1 = 3.2824	R2 = 2.1055	R3 = 4.4998	
Q1 = 2.784	Q2 = 1.972	Q3 = 3.856	

MEAN DEV. BETWEEN CALC. AND EXP. CONC. IN MOLE PCT

UNIQUAC (SPECIFIC PARAMETERS)	0.40
NRTL (SPECIFIC PARAMETERS)	0.52
UNIQUAC (COMMON PARAMETERS)	2.01

C$_2$H$_6$O-C$_3$H$_8$O$_3$

(1) C3H8O3   GLYCEROL
(2) C2H6O   ETHANOL
(3) C6H6   BENZENE

MCDONALD H.J.
J.AM.CHEM.SOC. 62(1940)3183

TEMPERATURE = 25.0 DEG C   TYPE OF SYSTEM = 1

EXPERIMENTAL TIE LINES IN MOLE PCT

	LEFT PHASE			RIGHT PHASE	
(1)	(2)	(3)	(1)	(2)	(3)
82.351	16.135	1.514	0.168	2.352	97.480
70.932	25.801	3.268	0.244	10.239	89.517
55.710	38.353	5.937	0.840	24.588	74.572
48.499	43.319	8.182	1.626	32.063	66.311
40.631	47.611	11.758	3.308	38.530	58.162
27.495	51.563	20.942	7.296	45.306	47.398

SPECIFIC MODEL PARAMETERS IN KELVIN

		UNIQUAC		NRTL(ALPHA=.2)	
I	J	AIJ	AJI	AIJ	AJI
1	2	-30.266	26.356	-569.14	1142.4
1	3	583.49	721.57	1595.2	1318.8
2	3	-246.08	635.16	-244.93	665.75

R1 = 3.5857   R2 = 2.1055   R3 = 3.1878
Q1 = 3.060   Q2 = 1.972   Q3 = 2.400

MEAN DEV. BETWEEN CALC. AND EXP. CONC. IN MOLE PCT

UNIQUAC (SPECIFIC PARAMETERS)   0.56
NRTL   (SPECIFIC PARAMETERS)   1.25
UNIQUAC (COMMON PARAMETERS)   1.43

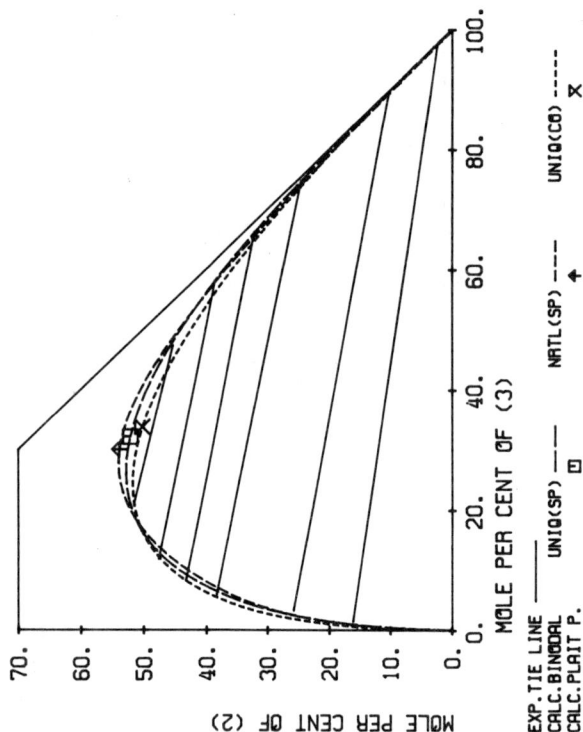

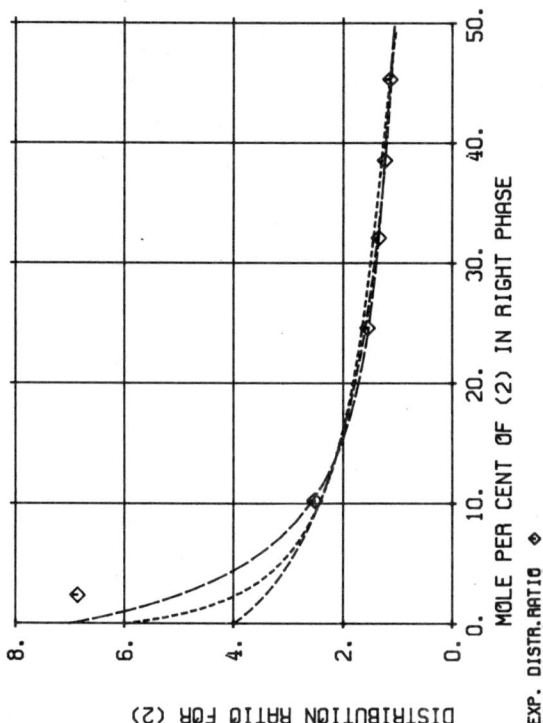

MOLE PER CENT OF (3)

MOLE PER CENT OF (2)

EXP.TIE LINE

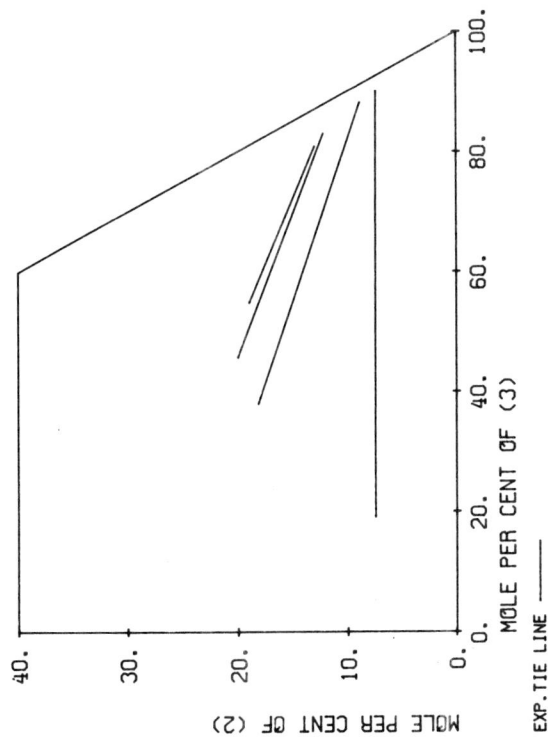

(1) C4H8O2   ACETIC ACID,ETHYL ESTER
(2) C2H6O   ETHANOL
(3) H2O   WATER

BONNER W.D.
J.PHYS.CHEM. 14(1910)738

TEMPERATURE =   0.0 DEG C   TYPE OF SYSTEM = 1

EXPERIMENTAL TIE LINES IN MOLE PCT (GRAPH.INTERPOL.)

| LEFT PHASE | | | RIGHT PHASE | | |
(1)	(2)	(3)	(1)	(2)	(3)
73.460	7.461	19.079	2.502	7.344	90.154
43.896	18.125	37.979	2.943	8.810	88.247
34.282	19.961	45.757	4.780	12.154	83.066
26.218	18.919	54.863	6.123	12.951	80.926

C$_2$H$_6$O-C$_4$H$_8$O$_2$

(1) C4H8O2   ACETIC ACID,ETHYL ESTER
----
(2) C2H6O   ETHANOL
----
(3) H2O   WATER

BEECH D.G., GLASSTONE S.
J.CHEM.SOC. (1938)67

TEMPERATURE =   0.0 DEG C   TYPE OF SYSTEM = 1

EXPERIMENTAL TIE LINES IN MOLE PCT

        LEFT PHASE   RIGHT PHASE
    (1)   (2)   (3)   (1)   (2)   (3)

 89.097   0.175   10.728   2.106   1.646   96.248
 84.861   2.547   12.592   2.098   2.954   94.949
 79.558   5.394   15.048   2.133   4.896   92.971
 72.730   9.005   18.264   2.233   6.454   91.313
 62.542   13.086   24.372   2.411   7.846   89.743
 49.328   16.657   34.015   2.926   9.807   87.267

SPECIFIC MODEL PARAMETERS IN KELVIN

                                      NRTL(ALPHA=.2)
          UNIQUAC   AIJ   AJI
I J   AIJ   AJI
1 2   -233.93   -90.992   1040.4   -427.20
1 3   361.56   183.42   250.96   1116.9
2 3   -367.51   52.664   236.42   -31.704

R1 = 3.4786   R2 = 2.1055   R3 = 0.9200
Q1 = 3.116   Q2 = 1.972   Q3 = 1.400

MEAN DEV. BETWEEN CALC. AND EXP. CONC. IN MOLE PCT

UNIQUAC (SPECIFIC PARAMETERS)   0.66
NRTL   (SPECIFIC PARAMETERS)   0.48

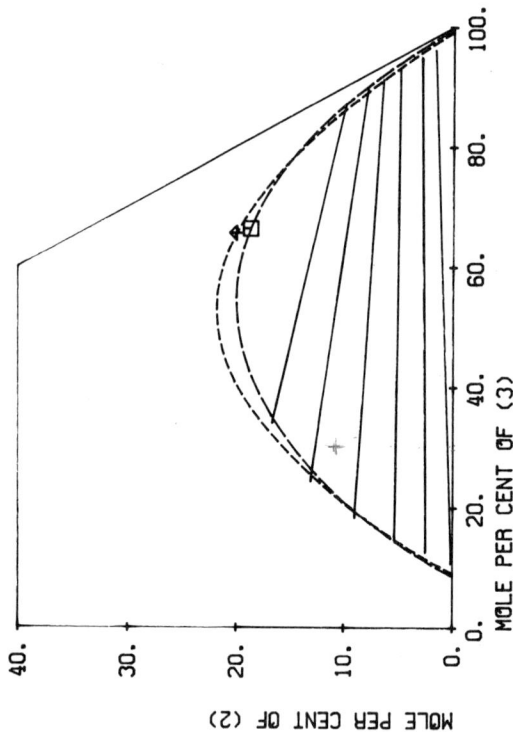

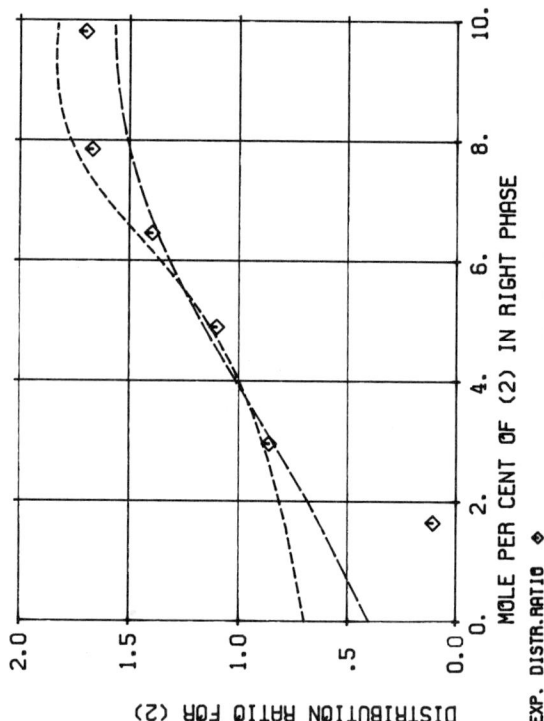

$C_2H_6O-C_4H_8O_2$

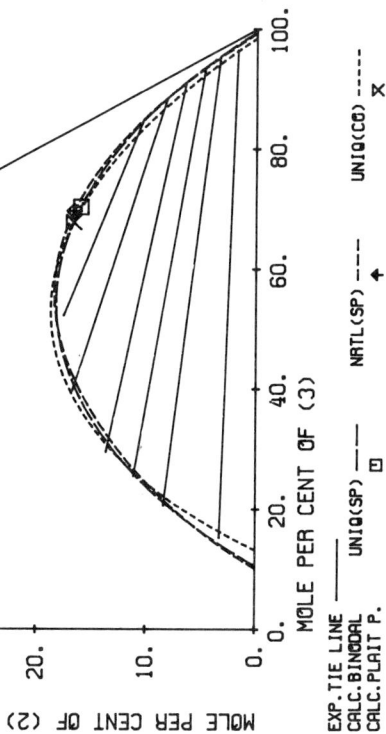

MOLE PER CENT OF (3)

MOLE PER CENT OF (2)

EXP.TIE LINE —— UNIQ(SP) —— ▣ NRTL(SP) ---- ✦ UNIQ(CB) -----
CALC.BINODAL
CALC.PLAIT P. ✕

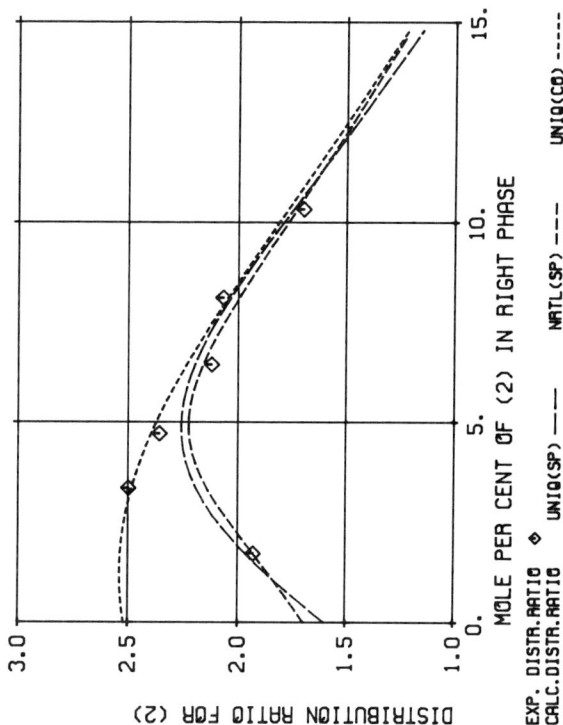

MOLE PER CENT OF (2) IN RIGHT PHASE

DISTRIBUTION RATIO FOR (2)

EXP. DISTR.RATIO ◇ UNIQ(SP) —— NRTL(SP) ---- UNIQ(CB) -----
CALC.DISTR.RATIO

(2) C2H6O   ETHANOL

(3) H2O   WATER

BEECH D.G., GLASSTONE S.
J.CHEM.SOC. (1938)67

TEMPERATURE = 20.0 DEG C   TYPE OF SYSTEM = 1

EXPERIMENTAL TIE LINES IN MOLE PCT

	LEFT PHASE		RIGHT PHASE	
(1)	(2)	(3)	(1)	(2)	(3)
81.495	3.303	15.202	1.769	1.714	96.517
70.994	8.373	20.633	1.936	3.351	94.713
63.412	11.104	25.484	2.178	4.710	93.112
56.872	13.665	29.463	2.631	6.443	90.926
43.596	16.770	39.633	3.338	8.105	88.557
30.262	17.537	52.201	4.953	10.320	84.727

SPECIFIC MODEL PARAMETERS IN KELVIN

		UNIQUAC		NRTL(ALPHA=.2)	
I J	AIJ	AJI	AIJ	AJI
1 2	-275.79	-48.995	885.15	-628.21
1 3	365.96	171.06	234.74	1261.3
2 3	-346.45	149.21	38.993	-174.97

R1 = 3.4786   R2 = 2.1055   R3 = 0.9200
Q1 = 3.116   Q2 = 1.972   Q3 = 1.400

MEAN DEV. BETWEEN CALC. AND EXP. CONC. IN MOLE PCT

UNIQUAC (SPECIFIC PARAMETERS)   0.47
NRTL   (SPECIFIC PARAMETERS)   0.54
UNIQUAC (COMMON PARAMETERS)   1.05

$C_2H_6O-C_4H_8O_2$

(1) C4H8O2   ACETIC ACID,ETHYL ESTER

(2) C2H6O   ETHANOL

(3) H2O   WATER

GRISWOLD J.; CHU P.L.; WINSAUER W.O.
IND.ENG.CHEM. 41(1949)2352

TEMPERATURE = 70.0 DEG C   TYPE OF SYSTEM = 1

EXPERIMENTAL TIE LINES IN MOLE PCT

	LEFT PHASE			RIGHT PHASE	
(1)	(2)	(3)	(1)	(2)	(3)
76.854	0.0	23.146	1.356	0.0	98.644
73.292	1.516	25.192	1.491	0.923	97.587
65.831	4.401	29.769	1.680	1.927	96.393
50.689	9.391	39.920	2.173	3.755	94.072
41.536	11.945	45.519	2.990	5.297	91.712
34.494	13.256	52.250	3.823	6.289	89.888
30.184	13.524	56.291	4.544	6.933	88.523

SPECIFIC MODEL PARAMETERS IN KELVIN

		UNIQUAC		NRTL(ALPHA=.2)	
I	J	AIJ	AJI	AIJ	AJI
1	2	-572.76	-184.03	791.02	-601.96
1	3	187.56	306.90	-24.637	1761.2
2	3	-504.36	18.912	1.6432	-484.99

R1 = 3.4786   R2 = 2.1055   R3 = 0.9200
Q1 = 3.116   Q2 = 1.972   Q3 = 1.400

MEAN DEV. BETWEEN CALC. AND EXP. CONC. IN MOLE PCT

UNIQUAC (SPECIFIC PARAMETERS)   0.33
NRTL   (SPECIFIC PARAMETERS)   0.57

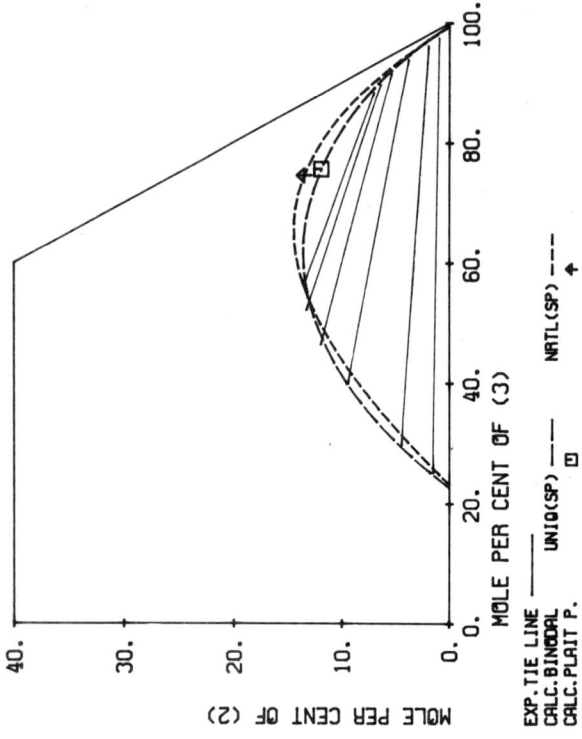

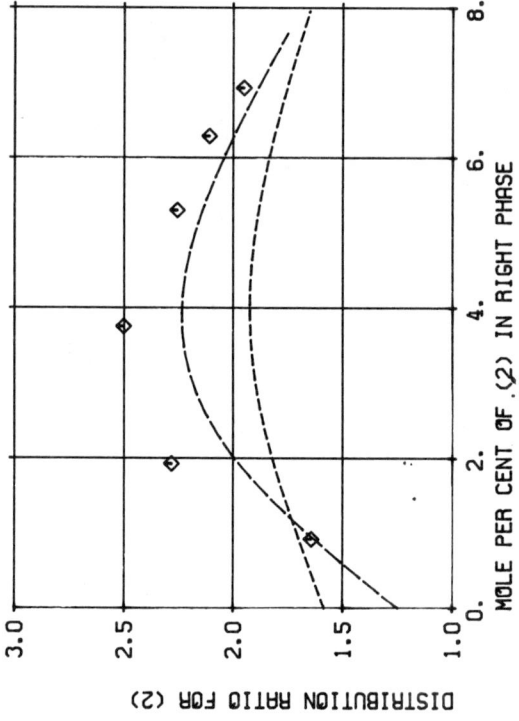

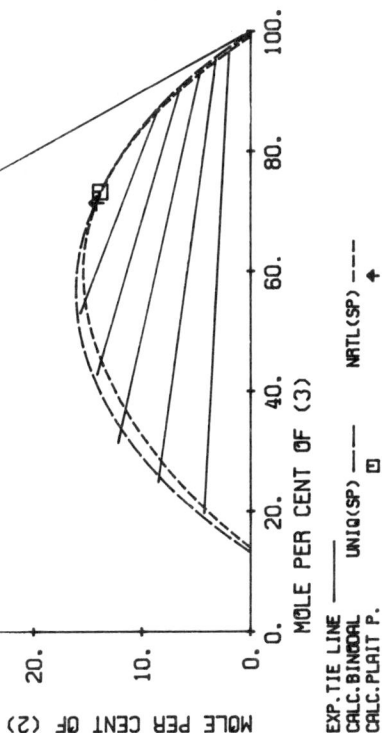

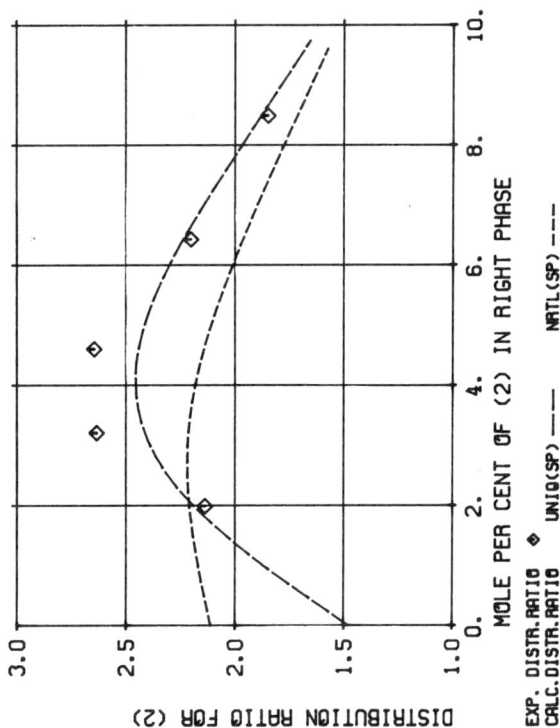

(2) C2H6O   ETHANOL

(3) H2O   WATER

MERTL I.
COLLECT.CZECH.CHEM.COMMUN. 37(1972)366

TEMPERATURE = 40.0 DEG C   TYPE OF SYSTEM = 1

EXPERIMENTAL TIE LINES IN MOLE PCT

	LEFT PHASE			RIGHT PHASE	
(1)	(2)	(3)	(1)	(2)	(3)
76.029	4.249	19.722	1.305	1.988	96.707
66.607	8.445	24.948	1.495	3.207	95.298
56.367	12.171	31.462	1.822	4.602	93.575
43.017	14.147	42.836	2.792	6.428	90.779
31.216	15.672	53.111	4.304	8.487	87.208

SPECIFIC MODEL PARAMETERS IN KELVIN

		UNIQUAC		NRTL(ALPHA=.2)	
I J	AIJ	AJI		AIJ	AJI
1 2	-345.55	-148.34		684.00	-687.50
1 3	301.67	249.15		158.17	1349.4
2 3	-445.68	212.66		-16.719	-426.33

R1 = 3.4786   R2 = 2.1055   R3 = 0.9200
Q1 = 3.116   Q2 = 1.972   Q3 = 1.400

MEAN DEV. BETWEEN CALC. AND EXP. CONC. IN MOLE PCT

UNIQUAC (SPECIFIC PARAMETERS)   0.33
NRTL   (SPECIFIC PARAMETERS)   0.93

MOLE PER CENT OF (2)

MOLE PER CENT OF (3)

EXP.TIE LINE   UNIQ(SP) ---- ▣   NRTL(SP) ---- ↑
CALC.BINODAL
CALC.PLAIT P.

DISTRIBUTION RATIO FOR (2)

MOLE PER CENT OF (2) IN RIGHT PHASE

EXP. DISTR.RATIO ◇   UNIQ(SP) ----   NRTL(SP) ----
CALC.DISTR.RATIO

$C_2H_6O-C_4H_8O_2$

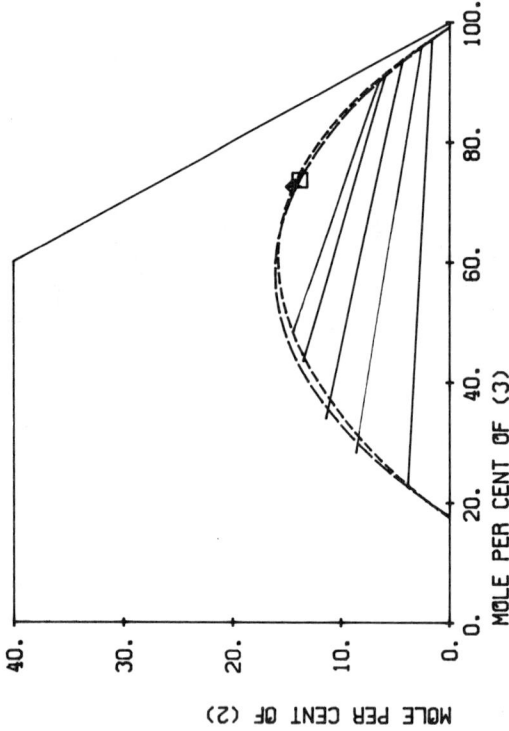

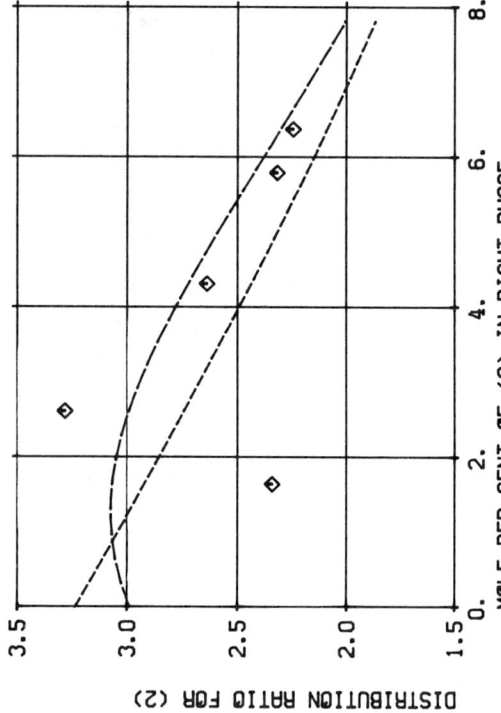

(1) C4H8O2   ACETIC ACID,ETHYL ESTER

(2) C2H6O   ETHANOL

(3) H2O   WATER

MERTL I.
COLLECT.CZECH.CHEM.COMMUN. 37(1972)366

TEMPERATURE = 55.0 DEG C   TYPE OF SYSTEM = 1

EXPERIMENTAL TIE LINES IN MOLE PCT

LEFT PHASE			RIGHT PHASE		
(1)	(2)	(3)	(1)	(2)	(3)
73.447	3.830	22.723	1.206	1.635	97.159
63.091	8.560	28.349	1.364	2.609	96.027
54.620	11.364	34.016	1.765	4.308	93.927
43.037	13.419	43.544	2.421	5.789	91.789
37.146	14.294	48.560	2.981	6.367	90.652

SPECIFIC MODEL PARAMETERS IN KELVIN

		UNIQUAC		NRTL(ALPHA=.2)	
I J		AIJ	AJI	AIJ	AJI
1 2		-274.77	-83.181	541.56	-684.44
1 3		298.79	191.12	86.945	1505.2
2 3		-374.55	230.81	-19.876	-379.88

R1 = 3.4786   R2 = 2.1055   R3 = 0.9200
Q1 = 3.116   Q2 = 1.972   Q3 = 1.400

MEAN DEV. BETWEEN CALC. AND EXP. CONC. IN MOLE PCT

UNIQUAC (SPECIFIC PARAMETERS)   0.44
NRTL   (SPECIFIC PARAMETERS)   0.62

$C_2H_6O-C_4H_8O_2$

(2) C2H6O   ETHANOL

(3) H2O   WATER

MERTL I.
COLLECT.CZECH.CHEM.COMMUN. 37(1972)366

TEMPERATURE = 70.0 DEG C   TYPE OF SYSTEM = 1

EXPERIMENTAL TIE LINES IN MOLE PCT

	LEFT PHASE			RIGHT PHASE	
(1)	(2)	(3)	(1)	(2)	(3)
67.750	4.789	27.462	1.291	1.298	97.411
61.333	8.105	30.563	1.791	2.515	95.694
49.954	11.349	38.697	2.082	4.072	93.845
42.126	12.524	45.350	2.341	4.981	92.677
33.403	13.979	52.617	3.305	6.179	90.516

SPECIFIC MODEL PARAMETERS IN KELVIN

		UNIQUAC		NRTL(ALPHA=.2)	
I J	AIJ	AJI		AIJ	AJI
1 2	-265.30	-51.025		482.41	-737.26
1 3	-285.60	177.58		47.832	1506.5
2 3	-348.59	242.51		-40.115	-364.76

R1 = 3.4786   R2 = 2.1055   R3 = 0.9200
Q1 = 3.116   Q2 = 1.972   Q3 = 1.400

MEAN DEV. BETWEEN CALC. AND EXP. CONC. IN MOLE PCT

UNIQUAC (SPECIFIC PARAMETERS)   0.53
NRTL   (SPECIFIC PARAMETERS)   0.66

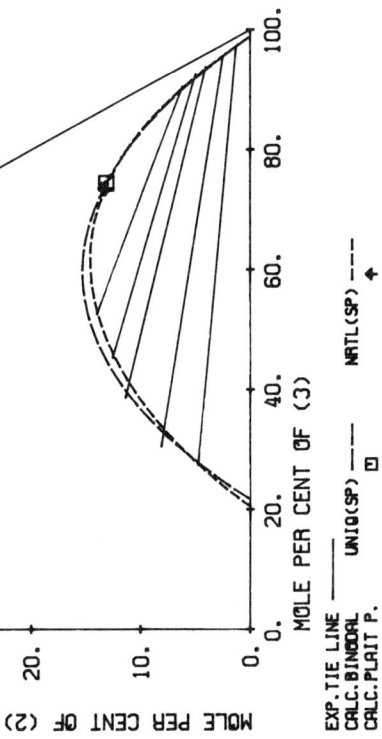

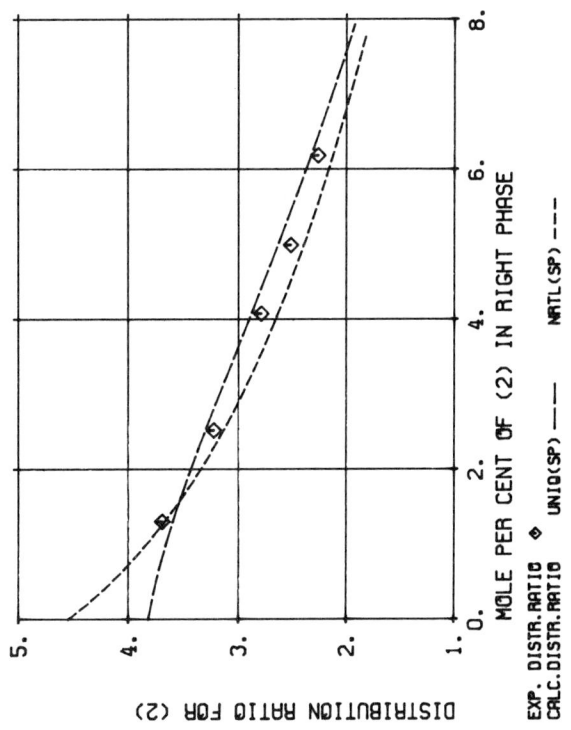

C$_2$H$_6$O-C$_4$H$_{10}$O

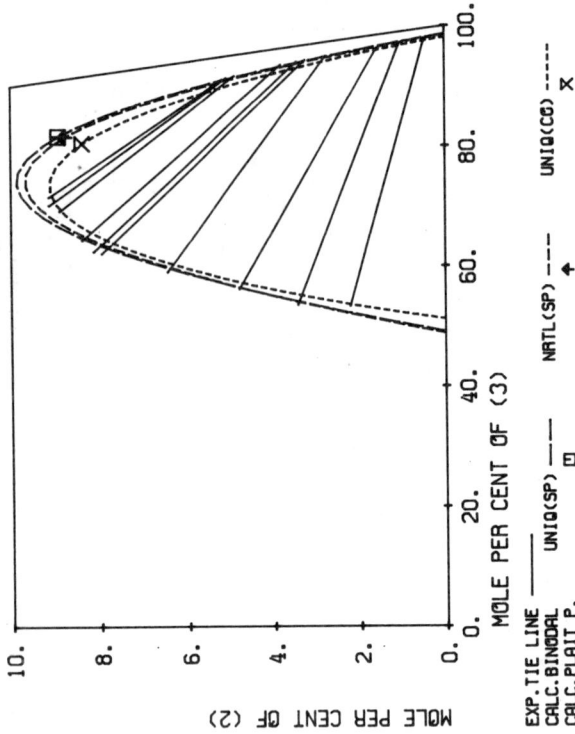

(1) C4H10O   1-BUTANOL -------
(2) C2H6O   ETHANOL   -------
(3) H2O   WATER   -------

SOLOMKO V.P., PANASYUK V.D., ZELENSKAYA A.M.
ZH.PRIKL.KHIM.(LENINGRAD) 35(1962)628

TEMPERATURE = 25.0 DEG C   TYPE OF SYSTEM = 1

EXPERIMENTAL TIE LINES IN MOLE PCT

LEFT PHASE			RIGHT PHASE		
(1)	(2)	(3)	(1)	(2)	(3)
44.529	2.206	53.265	2.038	0.486	97.476
43.086	3.428	53.486	1.967	1.054	96.979
39.082	4.776	56.142	2.120	1.565	96.315
34.580	6.454	58.966	2.312	2.760	94.929
30.004	7.943	62.053	2.465	3.231	94.304
29.497	8.125	62.378	2.541	3.432	94.027
27.270	8.385	64.345	2.695	3.799	93.506
21.907	8.900	69.193	3.451	4.858	91.692
20.708	9.141	70.151	3.619	5.045	91.336
19.330	9.148	71.523	3.731	5.146	91.124

SPECIFIC MODEL PARAMETERS IN KELVIN

I J	UNIQUAC AIJ	AJI	NRTL(ALPHA=.2) AIJ	AJI
1 2	190.76	-243.34	270.88	-450.93
1 3	-23.464	308.83	-311.68	1579.4
2 3	-16.989	-199.44	-35.903	-180.00

R1 = 3.4543   R2 = 2.1055   R3 = 0.9200
Q1 = 3.052   Q2 = 1.972   Q3 = 1.400

MEAN DEV.,BETWEEN CALC. AND EXP. CONC. IN MOLE PCT

UNIQUAC (SPECIFIC PARAMETERS)   0.36
NRTL   (SPECIFIC PARAMETERS)   0.37
UNIQUAC (COMMON PARAMETERS)   0.73

EXP.TIE LINE   UNIQ(SP) ----□   NRTL(SP) ----◆   UNIQ(CO) ----✕
CALC.BINODAL
CALC.PLAIT P.

MOLE PER CENT OF (3)

MOLE PER CENT OF (2)

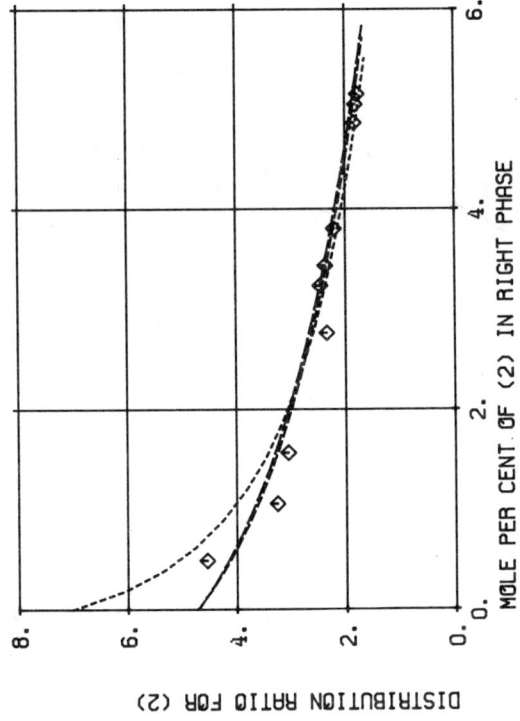

EXP. DISTR.RATIO ◆

DISTRIBUTION RATIO FOR (2)

MOLE PER CENT OF (2) IN RIGHT PHASE

$C_2H_6O-C_4H_{10}O$

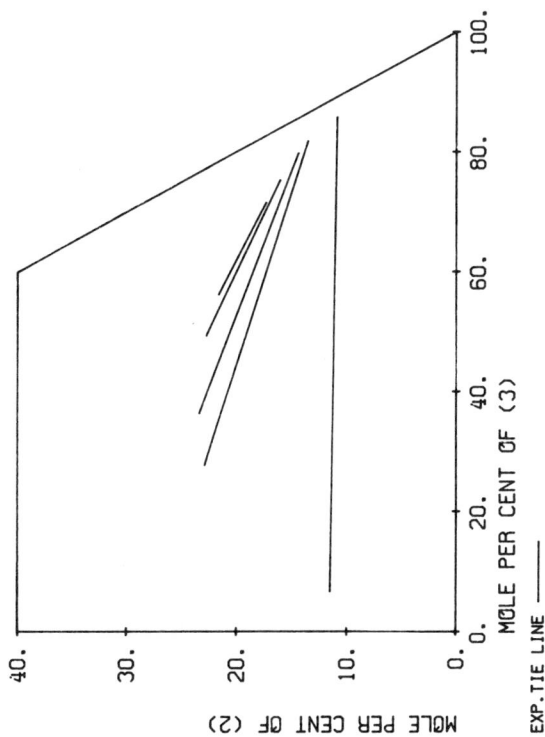

MOLE PER CENT OF (3)

MOLE PER CENT OF (2)

EXP.TIE LINE

(1) C4H1OO   ETHER,DIETHYL
(2) C2H6O   ETHANOL
(3) H2O   WATER

BONNER W.D.
J.PHYS.CHEM. 14(1910)738

TEMPERATURE =   0.0 DEG C   TYPE OF SYSTEM = 1

EXPERIMENTAL TIE LINES IN MOLE PCT (GRAPH.INTERPOL.)

	LEFT PHASE			RIGHT PHASE	
(1)	(2)	(3)	(1)	(2)	(3)
81.782	11.511	6.707	3.376	10.863	85.761
49.368	22.855	27.777	4.681	13.526	81.793
40.275	23.339	36.387	5.805	14.394	79.801
27.941	22.694	49.364	8.638	15.034	75.328
22.236	21.575	56.189	11.113	17.311	71.575

C_2H_6O-$C_4H_{10}O$

MOLE PER CENT OF (3)

MOLE PER CENT OF (2)

EXP.TIE LINE ——— UNIQ(SP) ▣ NRTL(SP) ✦ UNIQ(C0) ---- ✗
CALC.BINODAL
CALC.PLAIT P.

MOLE PER CENT OF (2) IN RIGHT PHASE

DISTRIBUTION RATIO FOR (2)

EXP. DISTR.RATIO ◇ THIS REF ◇ OTHER REF +

(1) C4H10O   ETHER, DIETHYL

(2) C2H6O   ETHANOL

(3) H2O   WATER

HORIBA S.
MEM.COLL.SCI.ENG.KYOTO IMP.UNIV. 3(1911)53

TEMPERATURE = 25.0 DEG C   TYPE OF SYSTEM = 1

EXPERIMENTAL TIE LINES IN MOLE PCT

| | LEFT PHASE | | | RIGHT PHASE | |
(1)	(2)	(3)	(1)	(2)	(3)
94.935	0.0	5.065	1.466	0.0	98.534
90.996	2.422	6.582	1.588	1.382	97.030
86.808	4.877	8.315	1.695	3.462	94.843
83.409	6.955	9.635	1.812	4.681	93.507
76.115	10.793	13.093	2.029	6.165	91.807
74.384	11.625	13.990	2.132	6.632	91.236
69.308	14.137	16.555	2.309	7.431	90.260
63.168	16.518	20.314	2.638	8.539	88.539
55.137	19.490	25.374	2.927	9.662	87.411
47.990	21.459	30.551	3.379	10.921	85.700
37.369	23.435	39.196	4.288	12.398	83.313
30.384	23.402	46.214	5.532	13.838	80.630
21.405	22.094	56.501	8.060	16.018	75.923
18.544	21.182	60.274	9.620	17.007	73.373
15.682	20.156	64.162	11.255	17.980	70.765

SPECIFIC MODEL PARAMETERS IN KELVIN

| | | UNIQUAC | | NRTL(ALPHA=.2) | |
I J	AIJ	AJI		AIJ	AJI
1 2	366.86	-106.48		1118.4	-626.33
1 3	636.29	53.213		448.19	1163.2
2 3	-162.21	363.92		-573.64	1033.1

R1 = 3.3949   R2 = 2.1055   R3 = 0.9200
Q1 = 3.016   Q2 = 1.972   Q3 = 1.400

MEAN DEV. BETWEEN CALC. AND EXP. CONC. IN MOLE PCT

UNIQUAC (SPECIFIC PARAMETERS)   0.57
NRTL   (SPECIFIC PARAMETERS)   0.45
UNIQUAC (COMMON PARAMETERS)   1.58

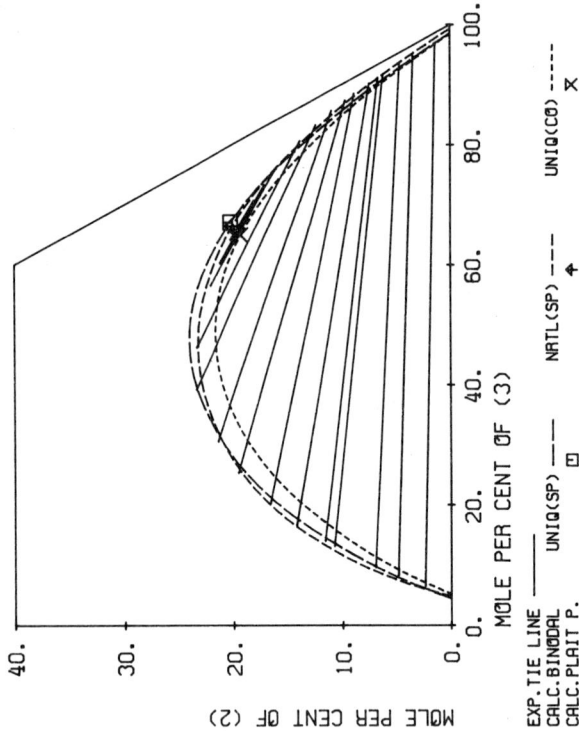

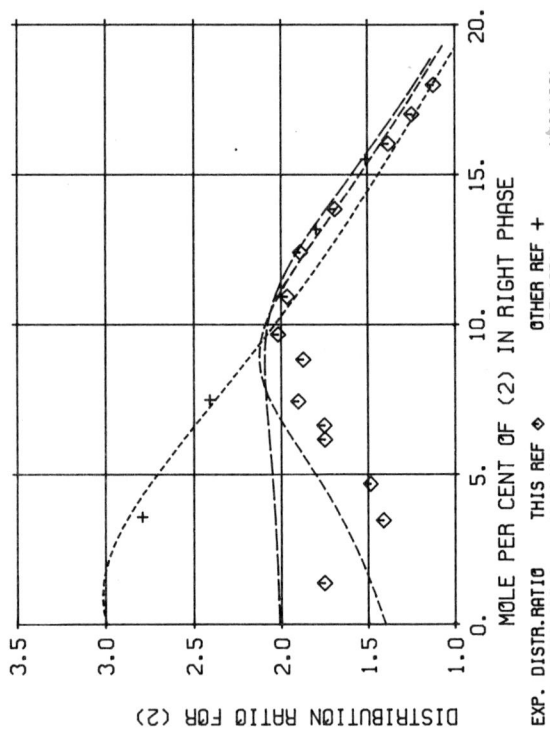

$C_2H_6O\text{-}C_4H_{10}O$

MOLE PER CENT OF (2)

MOLE PER CENT OF (3)

EXP.TIE LINE —— UNIQ(SP) ——— NRTL(SP) ✦ UNIQ(CO) ----- ✕
CALC.BINODAL ———— □
CALC.PLAIT P. ----

DISTRIBUTION RATIO FOR (2)

MOLE PER CENT OF (2) IN RIGHT PHASE

EXP. DISTR.RATIO ◇ THIS REF ◇ OTHER REF + UNIQ(CO) -----
CALC.DISTR.RATIO ———— UNIQ(SP) ——— NRTL(SP) --- +

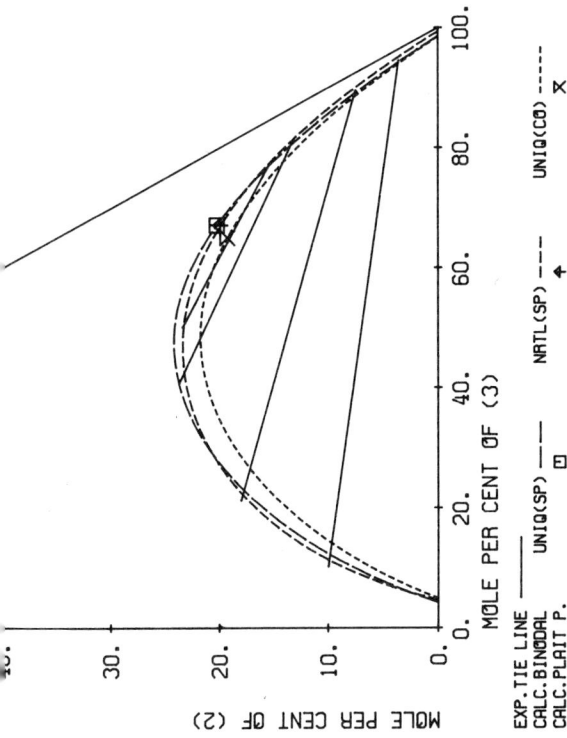

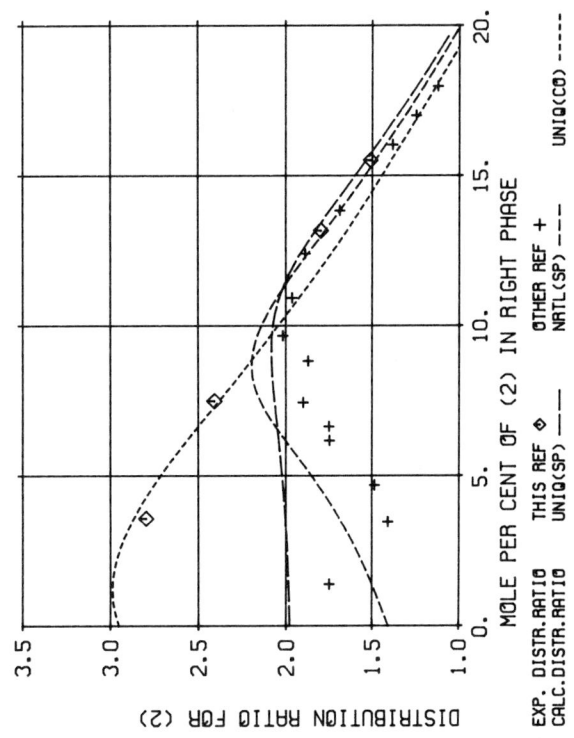

(2) C2H6O   ETHANOL
(3) H2O   WATER

BORISOVA I.A., VATSKOVA V.G., GORBUNOV A.I., SOKOLOV N.M.
ZH.FIZ.KHIM. 52(1978)1563

TEMPERATURE = 20.0 DEG C   TYPE OF SYSTEM = 1

EXPERIMENTAL TIE LINES IN MOLE PCT

LEFT PHASE		RIGHT PHASE	
(1)	(2)	(3)	(1)	(2)	(3)

(1)	(2)	(3)	(1)	(2)	(3)
79.836	9.968	10.196	1.839	3.568	94.592
60.782	18.019	21.199	2.882	7.475	89.643
35.477	23.702	40.821	5.446	13.170	81.383
26.486	23.417	50.097	7.462	15.503	77.036

SPECIFIC MODEL PARAMETERS IN KELVIN

		UNIQUAC		NRTL(ALPHA=.2)	
I	J	AIJ	AJI	AIJ	AJI
1	2	366.86	-106.48	1118.4	-626.33
1	3	636.29	53.213	448.19	1163.2
2	3	-162.21	363.92	-573.64	1033.1

R1 = 3.3949   R2 = 2.1055   R3 = 0.9200
Q1 = 3.016   Q2 = 1.972   Q3 = 1.400

MEAN DEV. BETWEEN CALC. AND EXP. CONC. IN MOLE PCT

UNIQUAC (SPECIFIC PARAMETERS)   0.79
NRTL   (SPECIFIC PARAMETERS)   0.82
UNIQUAC (COMMON PARAMETERS)   1.29

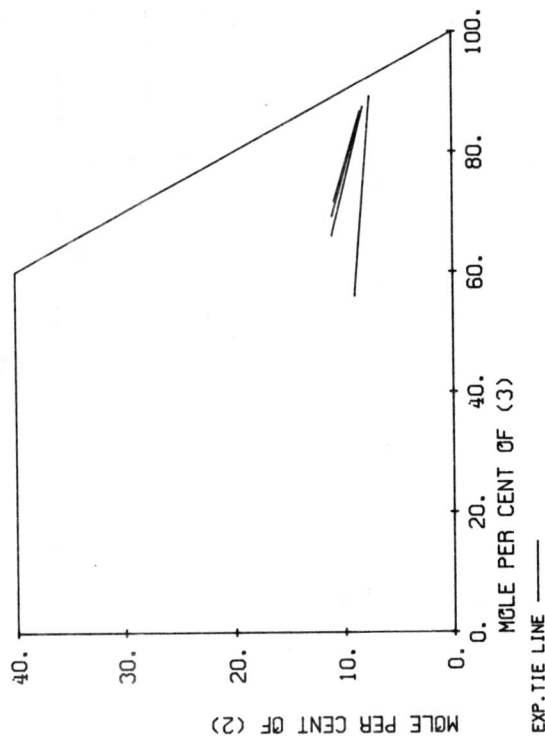

C$_2$H$_6$O-C$_4$H$_{10}$O

(1) C4H10O   1-PROPANOL,2-METHYL

(2) C2H6O   ETHANOL

(3) H2O   WATER

BONNER W.D.
J.PHYS.CHEM. 14(1910)738

TEMPERATURE = 0.0 DEG C   TYPE OF SYSTEM = 1

EXPERIMENTAL TIE LINES IN MOLE PCT (GRAPH.INTERPOL.)

LEFT PHASE			RIGHT PHASE		
(1)	(2)	(3)	(1)	(2)	(3)
35.025	8.985	55.991	3.258	7.530	89.212
23.000	11.079	65.921	4.358	8.146	87.496
19.700	11.030	69.270	4.932	8.386	86.682
17.433	10.844	71.724	5.258	8.460	86.281

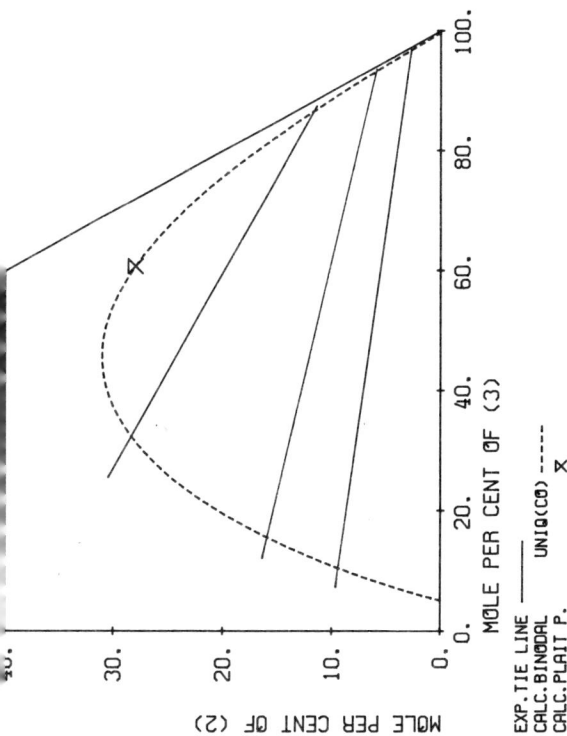

(2) C2H6O   ETHANOL

(3) H2O   WATER

FROLOV A.F., ET AL.
ZH.FIZ.KHIM. 40(1966)100

TEMPERATURE = 20.0 DEG C   TYPE OF SYSTEM = 1

EXPERIMENTAL TIE LINES IN MOLE PCT

	LEFT PHASE			RIGHT PHASE	
(1)	(2)	(3)	(1)	(2)	(3)
94.908	0.0	5.092	0.370	0.0	99.630
82.960	9.642	7.397	0.401	2.695	96.905
71.417	16.404	12.179	0.512	5.886	93.603
43.871	30.425	25.704	1.082	11.421	87.497

MEAN DEV. BETWEEN CALC. AND EXP. CONC. IN MOLE PCT

UNIQUAC (COMMON PARAMETERS)   1.37

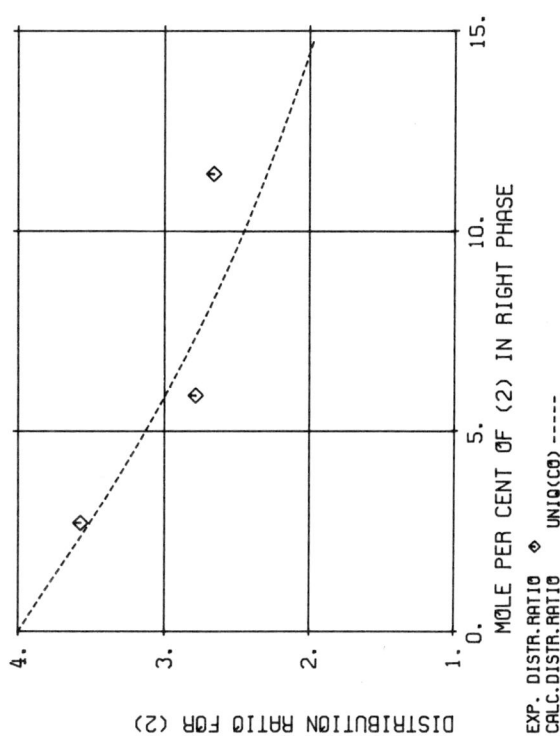

$C_2H_6O-C_5H_{10}O_2$

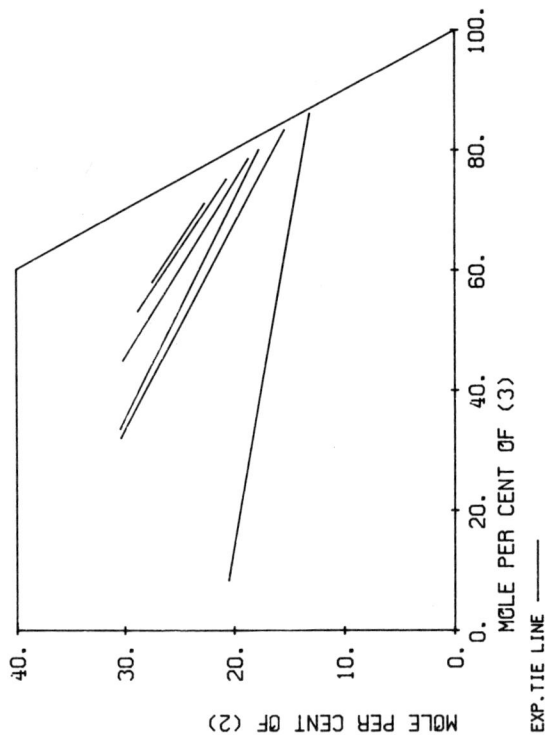

(1) C5H10O2   PROPANOIC ACID,ETHYL ESTER

(2) C2H6O   ETHANOL

(3) H2O   WATER

BONNER W.D.
J.PHYS.CHEM. 14(1910)738

TEMPERATURE =   0.0 DEG C   TYPE OF SYSTEM = 1

EXPERIMENTAL TIE LINES IN MOLE PCT (GRAPH.INTERPOL.)

	LEFT PHASE			RIGHT PHASE	
(1)	(2)	(3)	(1)	(2)	(3)
71.139	20.507	8.354	0.811	13.114	86.076
37.679	30.328	31.993	1.284	15.400	83.316
36.115	30.394	33.492	2.292	17.675	80.033
25.034	30.121	44.845	2.836	18.636	78.529
18.193	28.744	53.063	4.204	20.657	75.140
14.683	27.428	57.889	6.201	22.674	71.125

MOLE PER CENT OF (3)

MOLE PER CENT OF (2)

EXP.TIE LINE

$C_2H_6O-C_5H_{12}O$

MOLE PER CENT OF (3)

MOLE PER CENT OF (2)

EXP.TIE LINE ———

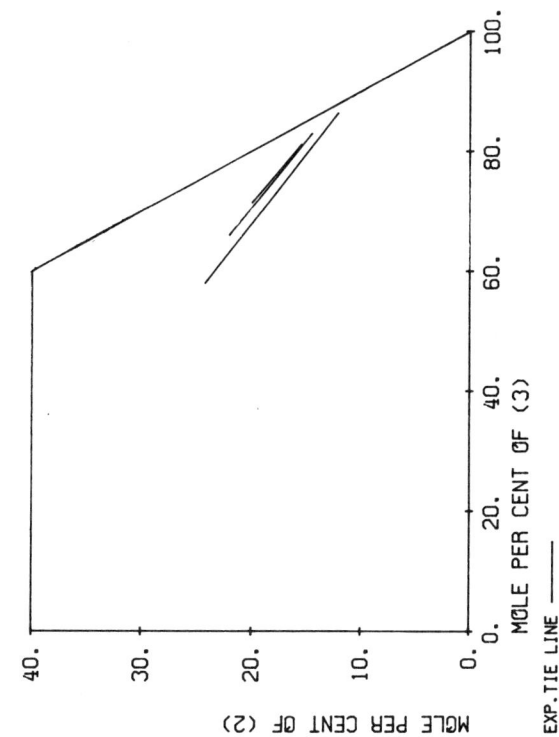

(1) C5H12O   1-BUTANOL,3-METHYL
----
(2) C2H6O   ETHANOL
----
(3) H2O   WATER

BONNER W.D.
J.PHYS.CHEM. 14(1910)738

TEMPERATURE =   0.0 DEG C   TYPE OF SYSTEM = 1

EXPERIMENTAL TIE LINES IN MOLE PCT (GRAPH.INTERPOL.)

         LEFT PHASE   RIGHT PHASE
    (1)   (2)   (3)   (1)   (2)   (3)

  17.815   24.209   57.976   1.530   12.103   86.366
  11.978   21.998   66.024   2.545   14.508   82.947
   8.611   19.953   71.436   3.395   15.462   81.143

C$_2$H$_6$O-C$_5$H$_{12}$O

(1) C5H12O   1-PENTANOL
(2) C2H6O   ETHANOL
(3) H2O   WATER

OTHMER D.F., WHITE R.E., TRUEGER E.
IND.ENG.CHEM. 33(1941)1240

TEMPERATURE = 25.5 DEG C   TYPE OF SYSTEM = 1

EXPERIMENTAL TIE LINES IN MOLE PCT (GRAPH.INTERPOL.)

LEFT PHASE			RIGHT PHASE		
(1)	(2)	(3)	(1)	(2)	(3)
60.560	3.157	36.283	0.164	1.601	98.235
46.247	11.698	42.055	0.182	4.915	94.903
33.312	17.289	49.398	0.195	7.193	92.612
20.370	19.324	60.306	1.178	11.608	87.214

MEAN DEV. BETWEEN CALC. AND EXP. CONC. IN MOLE PCT
UNIQUAC (COMMON PARAMETERS)   1.60

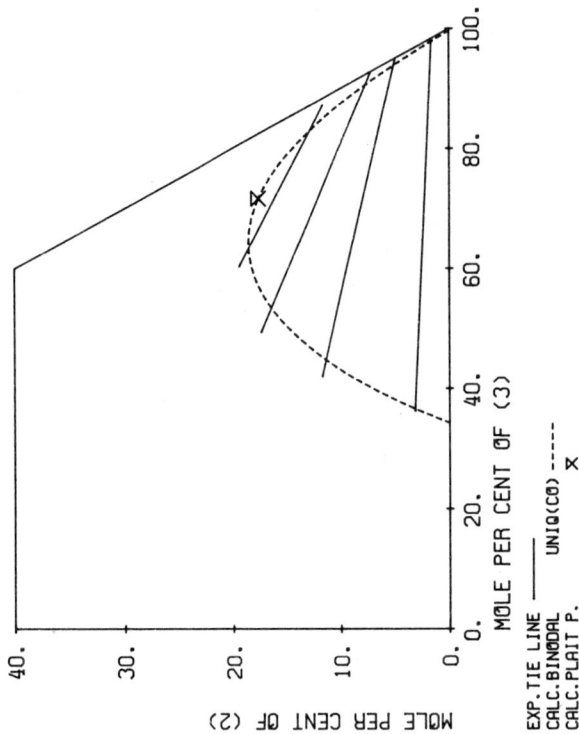

MOLE PER CENT OF (3)

MOLE PER CENT OF (2)

EXP.TIE LINE ———
CALC.BINODAL UNIQ(C0) -----
CALC.PLAIT P. ⊀   X

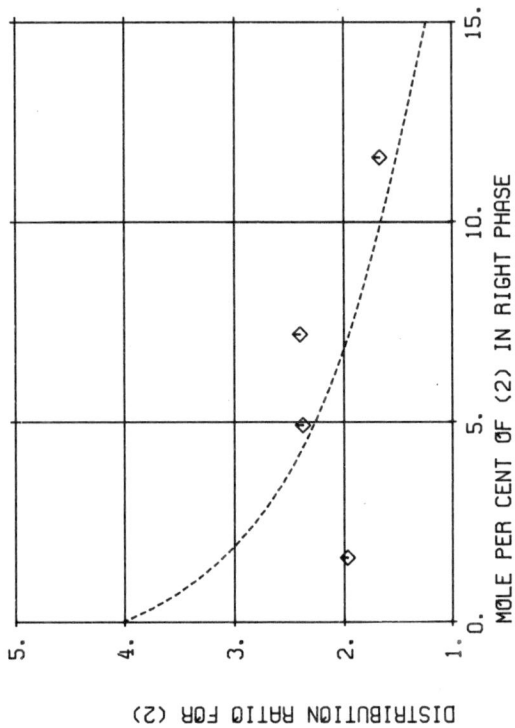

MOLE PER CENT OF (2) IN RIGHT PHASE

DISTRIBUTION RATIO FOR (2)

EXP. DISTR.RATIO ◇

C_2H_6O-$C_6H_5NO_2$

MOLE PER CENT OF (3)

EXP.TIE LINE ──────

MOLE PER CENT OF (2)

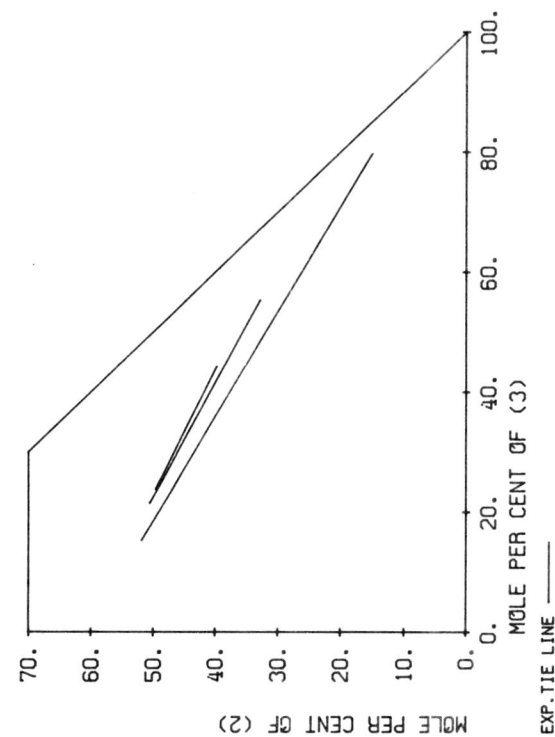

(1) H2O   WATER
----
(2) C2H6O   ETHANOL
----
(3) C6H5NO2   BENZENE,NITRO

BONNER W.D.
J.PHYS.CHEM. 14(1910)738

TEMPERATURE = 15.0 DEG C   TYPE OF SYSTEM = 1

EXPERIMENTAL TIE LINES IN MOLE PCT (GRAPH.INTERPOL.)

     LEFT PHASE   RIGHT PHASE
   (1)   (2)   (3)   (1)   (2)   (3)

 32.917 51.803 15.280   5.297 14.958 79.745
 27.975 50.534 21.491   11.849 32.805 55.346
 26.568 49.596 23.836   15.967 39.762 44.270

C₂H₆O-C₆H₆

(1) H2O   WATER

(2) C2H6O   ETHANOL

(3) C6H6   BENZENE

TAYLOR S.F.
J.PHYS.CHEM. 1(1896-97)461

TEMPERATURE = 25.0 DEG C   TYPE OF SYSTEM = 1

EXPERIMENTAL TIE LINES IN MOLE PCT

	LEFT PHASE			RIGHT PHASE	
(1)	(2)	(3)	(1)	(2)	(3)
76.637	22.489	0.874	2.759	12.669	84.572
61.567	28.379	2.055	4.454	17.237	78.308
61.875	33.908	4.218	7.239	22.408	70.353
51.056	39.619	9.325	11.558	28.502	59.940
38.183	42.335	19.482	17.497	34.869	47.634
31.753	41.892	26.356	20.830	37.401	41.769

SPECIFIC MODEL PARAMETERS IN KELVIN

		UNIQUAC		NRTL(ALPHA=.2)	
I	J	AIJ	AJI	AIJ	AJI
1	2	266.93	-266.09	376.33	-441.74
1	3	249.80	807.82	2797.7	986.99
2	3	-73.352	256.10	87.744	118.04

R1 = 0.9200   R2 = 2.1055   R3 = 3.1878
Q1 = 1.400   Q2 = 1.972   Q3 = 2.400

MEAN DEV. BETWEEN CALC. AND EXP. CONC. IN MOLE PCT

UNIQUAC (SPECIFIC PARAMETERS)   1.27
NRTL   (SPECIFIC PARAMETERS)   1.10
UNIQUAC (COMMON PARAMETERS)   1.46

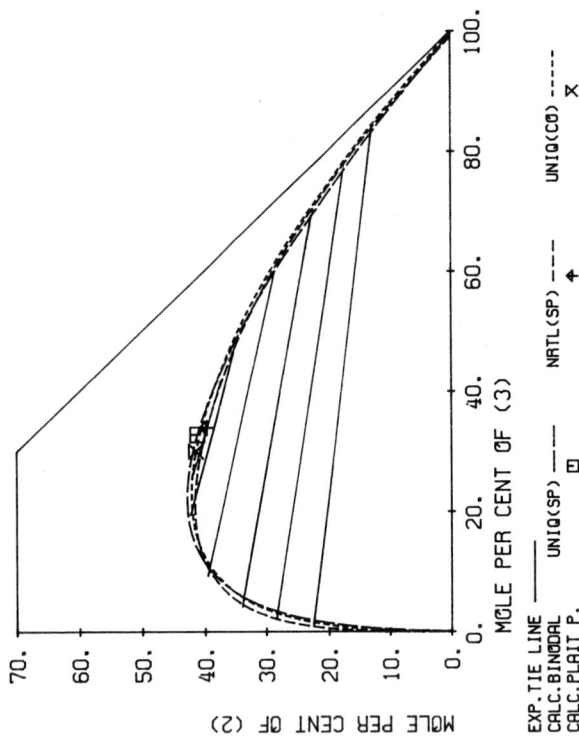

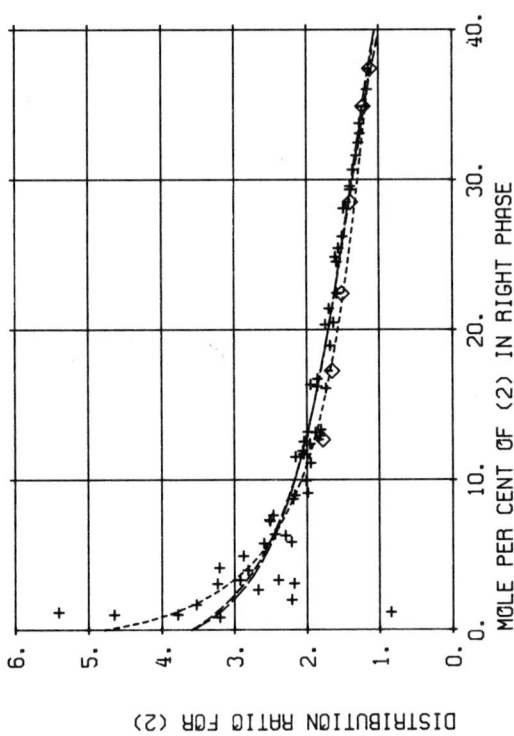

C$_2$H$_6$O-C$_6$H$_6$

(2) C2H6O   ETHANOL

(3) C6H6   BENZENE

WASHBURN E.R., HNIZDA V., VOLD R.
J.AM.CHEM.SOC. 53(1931)3237

TEMPERATURE = 25.0 DEG C   TYPE OF SYSTEM = 1

EXPERIMENTAL TIE LINES IN MOLE PCT (GRAPH.INTERPOL.)

LEFT PHASE			RIGHT PHASE		
(1)	(2)	(3)	(1)	(2)	(3)
98.959	0.994	0.047	0.473	1.177	98.350
95.481	4.442	0.077	0.599	2.009	97.392
92.748	7.149	0.103	0.683	2.669	96.648
88.661	11.203	0.136	0.847	3.977	95.176
84.106	15.582	0.312	1.254	6.376	92.369
80.237	19.094	0.670	1.651	8.713	89.636
77.433	21.659	0.908	2.275	11.120	86.604
73.866	24.788	1.346	2.845	13.163	83.992

SPECIFIC MODEL PARAMETERS IN KELVIN

		UNIQUAC		NRTL(ALPHA=.2)	
I	J	AIJ	AJI	AIJ	AJI
1	2	266.93	-266.09	376.33	-441.74
1	3	249.80	807.82	2797.7	986.99
2	3	-73.352	256.10	87.744	118.04

R1 = 0.9200   R2 = 2.1055   R3 = 3.1878
Q1 = 1.400   Q2 = 1.972   Q3 = 2.400

MEAN DEV. BETWEEN CALC. AND EXP. CONC. IN MOLE PCT

UNIQUAC (SPECIFIC PARAMETERS)   0.54
NRTL   (SPECIFIC PARAMETERS)   0.39
UNIQUAC (COMMON PARAMETERS)   0.37

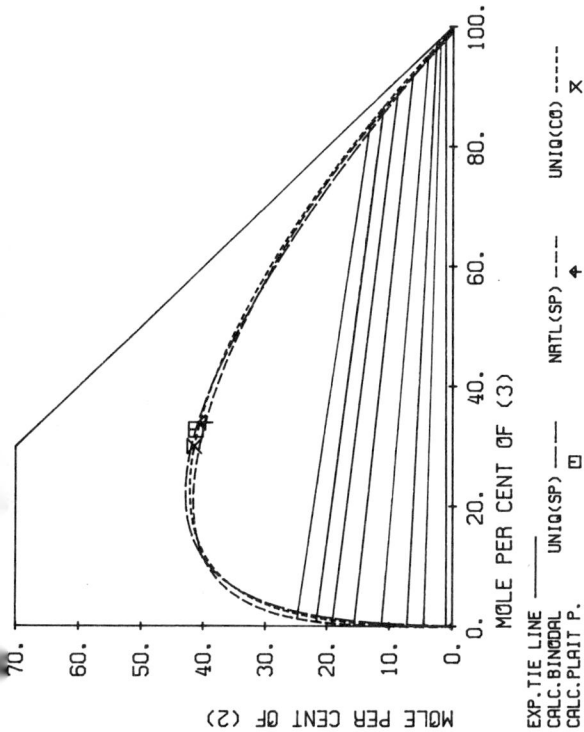

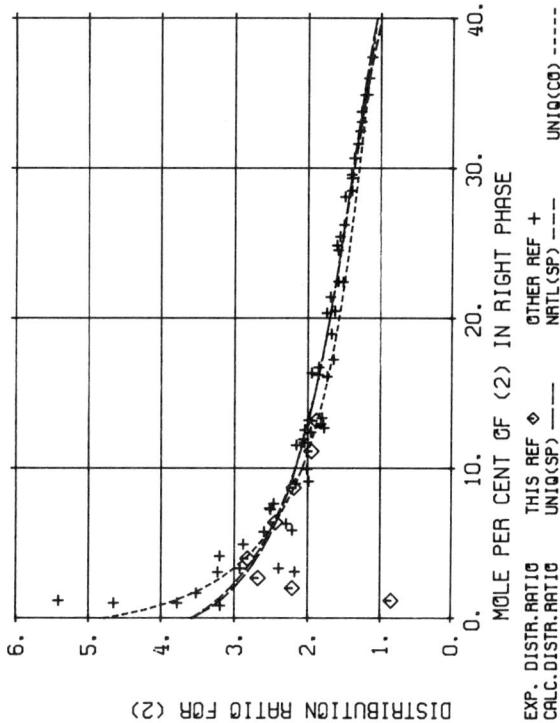

C₂H₆O-C₆H₆

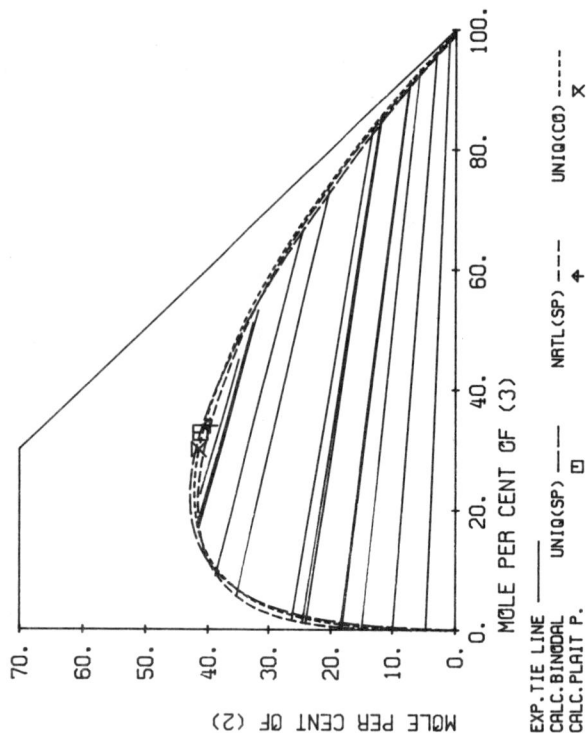

EXP.TIE LINE
CALC.BINODAL   UNIQ(SP) ----- □   NRTL(SP) ----- ♦   UNIQ(CO) ----- X
CALC.PLAIT P.

MOLE PER CENT OF (3)

MOLE PER CENT OF (2)

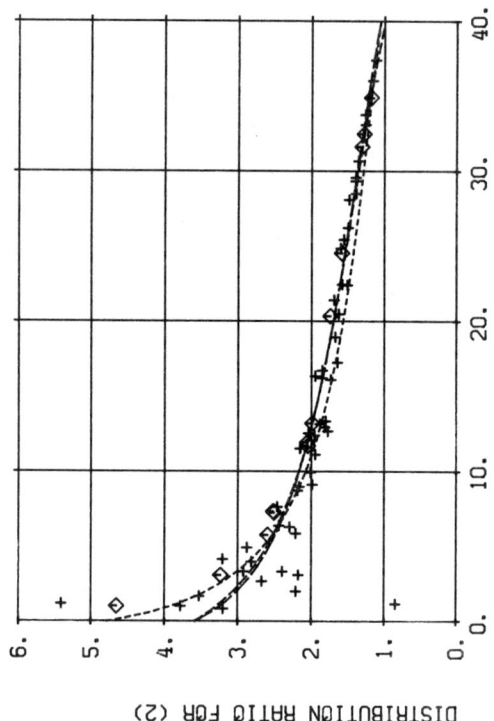

DISTRIBUTION RATIO FOR (2)

MOLE PER CENT OF (2) IN RIGHT PHASE

EXP. DISTR.RATIO ◇   THIS REF ◇   OTHER REF +
CALC.DISTR.RATIO   UNIQ(SP)   NRTL(SP)

(1)	H2O	WATER
(2)	C2H6O	ETHANOL
(3)	C6H6	BENZENE

VARTERESSIAN K.A.; FENSKE M.R.
IND.ENG.CHEM. 28(1936)928

TEMPERATURE = 25.0 DEG C   TYPE OF SYSTEM = 1

EXPERIMENTAL TIE LINES IN MOLE PCT

| LEFT PHASE | | | RIGHT PHASE | | |
(1)	(2)	(3)	(1)	(2)	(3)
95.263	4.705	0.032	0.216	1.011	98.773
89.967	9.958	0.075	0.639	3.082	96.279
84.890	14.959	0.151	0.841	5.756	93.403
81.483	18.213	0.304	1.250	7.249	91.502
81.182	18.488	0.329	1.249	7.328	91.423
74.930	23.981	1.089	2.424	11.711	85.861
74.255	24.531	1.214	2.624	11.918	85.458
72.123	26.258	1.619	3.196	13.200	83.604
59.128	35.394	5.478	6.572	20.338	73.090
52.182	38.837	8.981	9.165	24.525	66.309
41.569	41.565	16.929	15.192	31.619	53.190
40.311	41.565	18.124	16.382	32.469	51.149
36.092	41.261	22.646	20.098	34.889	45.013

SPECIFIC MODEL PARAMETERS IN KELVIN

| | UNIQUAC | | NRTL (ALPHA=.2) | |
I J	AIJ	AJI	AIJ	AJI
1 2	266.93	-266.09	376.33	-441.74
1 3	249.80	807.82	2797.7	986.99
2 3	-73.352	256.10	87.744	118.04

R1 = 0.9200   R2 = 2.1055   R3 = 3.1878
Q1 = 1.400   Q2 = 1.972   Q3 = 2.400

MEAN DEV. BETWEEN CALC. AND EXP. CONC. IN MOLE PCT

UNIQUAC (SPECIFIC PARAMETERS)	0.73
NRTL (SPECIFIC PARAMETERS)	0.55
UNIQUAC (COMMON PARAMETERS)	1.35

(2) C2H6O   ETHANOL

(3) C6H6   BENZENE

BANCROFT W.D., HUBARD S.D.
J.AM.CHEM.SOC. 64(1942)347

TEMPERATURE = 25.0 DEG C   TYPE OF SYSTEM = 1

EXPERIMENTAL TIE LINES IN MOLE PCT

	LEFT PHASE			RIGHT PHASE	
(1)	(2)	(3)	(1)	(2)	(3)
96.134	3.817	0.049	0.430	1.010	98.560
91.980	7.968	0.052	0.850	3.323	95.827
86.884	12.977	0.139	2.081	5.860	92.059
81.478	18.134	0.388	2.455	9.121	88.424
75.459	23.540	1.001	3.588	12.939	83.474
74.824	24.069	1.107	3.967	13.340	82.694
70.281	27.892	1.828	5.046	16.090	78.864
64.904	31.725	3.371	6.434	18.943	74.623
59.095	35.510	5.395	7.727	22.444	69.829
51.033	39.382	9.584	10.233	26.216	63.551
45.629	41.062	13.309	12.562	29.341	58.096
37.176	41.771	21.053	16.607	33.093	50.300

SPECIFIC MODEL PARAMETERS IN KELVIN

		UNIQUAC		NRTL(ALPHA=.2)	
I J		AIJ	AJI	AIJ	AJI
1 2		266.93	-266.09	376.33	-441.74
1 3		249.80	807.82	2797.7	986.99
2 3		-73.352	256.10	87.744	118.04

R1 = 0.9200   R2 = 2.1055   R3 = 3.1878
Q1 = 1.400   Q2 = 1.972   Q3 = 2.400

MEAN DEV. BETWEEN CALC. AND EXP. CONC. IN MOLE PCT

UNIQUAC (SPECIFIC PARAMETERS)	0.93
NRTL (SPECIFIC PARAMETERS)	0.79
UNIQUAC (COMMON PARAMETERS)	1.39

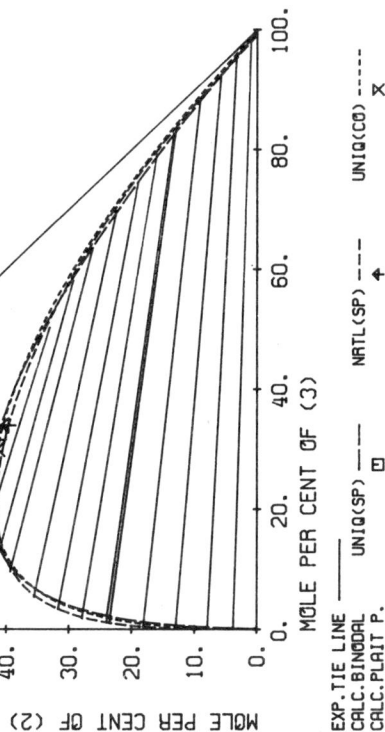

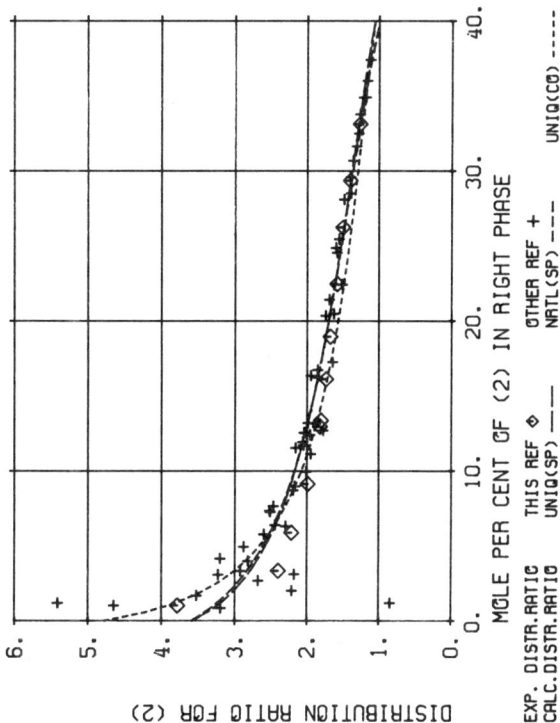

MOLE PER CENT OF (2)

MOLE PER CENT OF (3)

EXP.TIE LINE   UNIQ(SP) ⊟   NRTL(SP) ✦   UNIQ(CO) ✕
CALC.BINODAL
CALC.PLAIT P.

DISTRIBUTION RATIO FOR (2)

MOLE PER CENT OF (2) IN RIGHT PHASE

EXP.DISTR.RATIO   THIS REF ◇   OTHER REF +
CALC.DISTR.RATIO   UNIQ(SP)   NRTL(SP)   UNIQ(CO)

$C_2H_6O\text{-}C_6H_6$

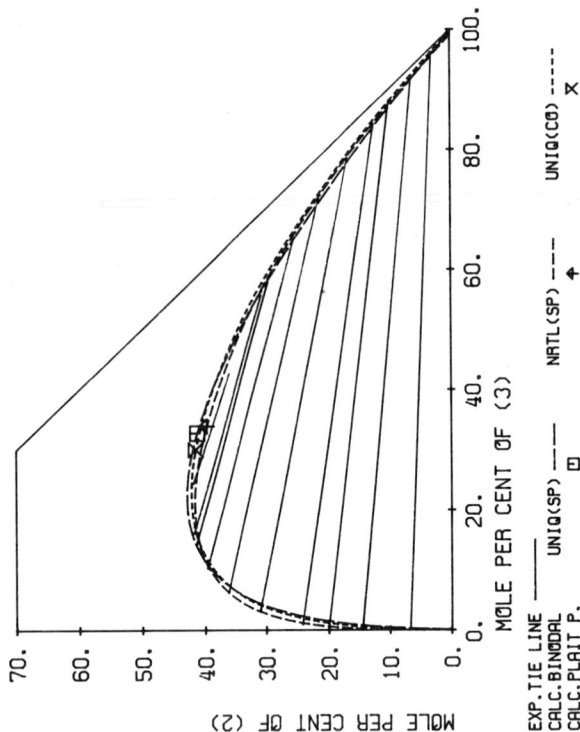

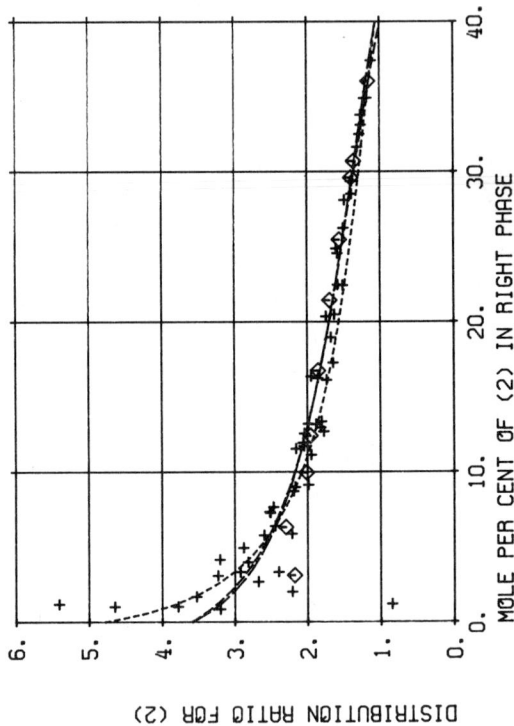

(1) H2O   WATER

(2) C2H6O   ETHANOL

(3) C6H6   BENZENE

YI-CHUNG CHANG, MOULTON R.W.
IND.ENG.CHEM. 45(1953)2350

TEMPERATURE = 25.0 DEG C   TYPE OF SYSTEM = 1

EXPERIMENTAL TIE LINES IN MOLE PCT

	LEFT PHASE			RIGHT PHASE	
(1)	(2)	(3)	(1)	(2)	(3)
93.195	6.756	0.048	0.597	3.099	96.304
85.368	14.448	0.185	1.379	6.291	92.331
79.467	20.009	0.524	1.925	9.945	88.131
74.828	24.237	0.936	3.363	12.384	84.253
65.746	30.969	3.286	4.623	16.710	78.667
57.580	36.212	6.212	6.787	21.407	71.806
50.082	39.643	10.275	9.424	25.419	65.157
44.378	41.395	14.227	12.549	29.575	57.877
42.189	41.756	16.055	13.613	30.673	55.714
33.753	41.861	24.386	21.311	36.007	42.682

SPECIFIC MODEL PARAMETERS IN KELVIN

		UNIQUAC		NRTL(ALPHA=.2)	
I J	AIJ	AJI		AIJ	AJI
1 2	266.93	-266.09		376.33	-441.74
1 3	249.80	807.82		2797.7	986.99
2 3	-73.352	256.10		87.744	118.04

R1 = 0.9200   R2 = 2.1055   R3 = 3.1878
Q1 = 1.400   Q2 = 1.972   Q3 = 2.400

MEAN DEV. BETWEEN CALC. AND EXP. CONC. IN MOLE PCT

UNIQUAC (SPECIFIC PARAMETERS)   0.70
NRTL   (SPECIFIC PARAMETERS)   0.57
UNIQUAC (COMMON PARAMETERS)   1.73

$C_2H_6O\text{-}C_6H_6$

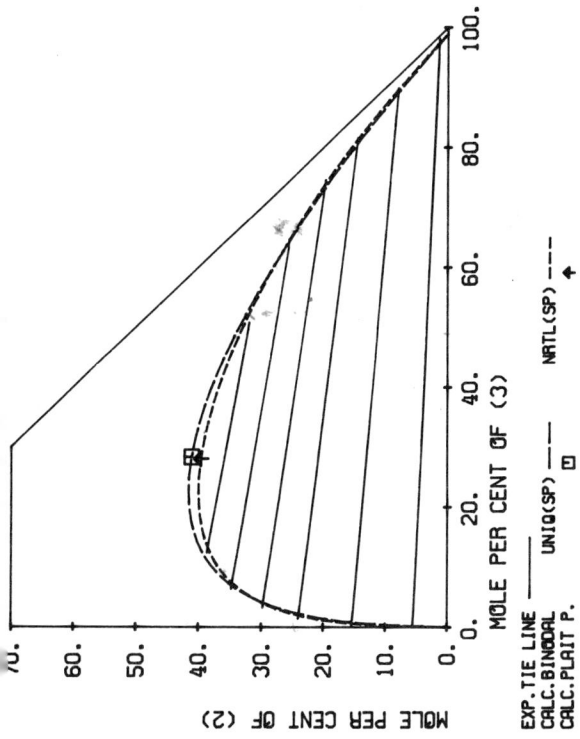

MOLE PER CENT OF (3)

MOLE PER CENT OF (2)

EXP.TIE LINE ⎯⎯⎯⎯ UNIQ(SP) ▣ NRTL(SP) ⎯ ⎯ ⎯
CALC.BINODAL
CALC.PLAIT P. ✦

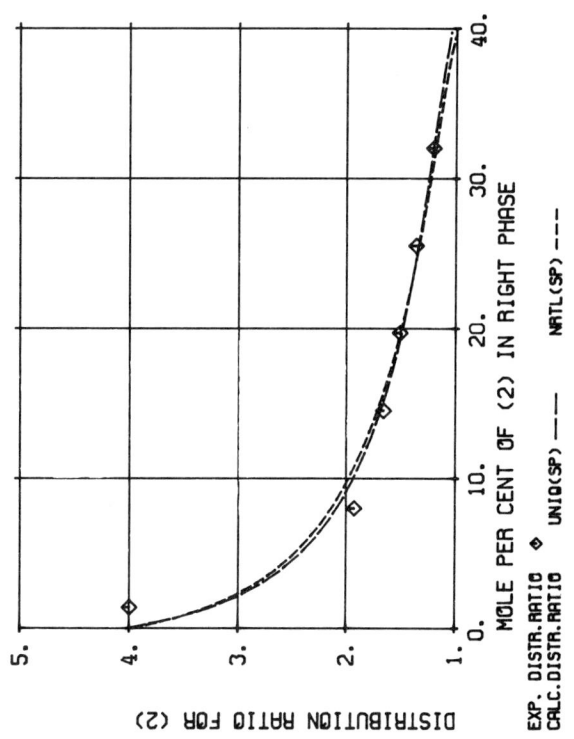

MOLE PER CENT OF (2) IN RIGHT PHASE

DISTRIBUTION RATIO FOR (2)

EXP. DISTR.RATIO ⎯⎯⎯⎯ UNIQ(SP) ◇ NRTL(SP) ⎯ ⎯ ⎯
CALC. DISTR.RATIO

(2) C2H6O   ETHANOL
(3) C6H6   BENZENE

MORACHEVSKII A.G., BELOUSOV V.P.
VESTN.LENINGR.UNIV.FIZ.KHIM. 13,4(1958)117

TEMPERATURE = 35.0 DEG C   TYPE OF SYSTEM = 1

EXPERIMENTAL TIE LINES IN MOLE PCT

	LEFT PHASE			RIGHT PHASE	
(1)	(2)	(3)	(1)	(2)	(3)
94.300	5.600	0.100	0.400	1.400	98.200
84.300	15.400	0.300	2.100	8.000	89.900
74.400	24.100	1.500	3.900	14.500	81.600
66.900	29.800	3.300	5.700	19.700	74.600
58.800	34.800	6.400	9.600	25.500	64.900
49.000	38.600	12.400	17.200	32.000	50.800

SPECIFIC MODEL PARAMETERS IN KELVIN

		UNIQUAC		NRTL(ALPHA=.2)	
I J	AIJ	AJI		AIJ	AJI
1 2	-137.53	-193.32		442.95	-585.19
1 3	-326.15	611.94		1957.1	875.62
2 3	-150.19	153.44		4.4456	48.459

R1 = 0.9200   R2 = 2.1055   R3 = 3.1878
Q1 = 1.400   Q2 = 1.972   Q3 = 2.400

MEAN DEV. BETWEEN CALC. AND EXP. CONC. IN MOLE PCT

UNIQUAC (SPECIFIC PARAMETERS)   0.61
NRTL   (SPECIFIC PARAMETERS)   0.50

C_2H_6O-C_6H_6

(1) H2O   WATER
(2) C2H6O   ETHANOL
(3) C6H6   BENZENE

MORACHEVSKII A.G., BELOUSOV V.P.
VESTN.LENINGR.UNIV.FIZ.KHIM. 13,4(1958)117

TEMPERATURE = 45.0 DEG C   TYPE OF SYSTEM = 1

EXPERIMENTAL TIE-LINES IN MOLE PCT

| | LEFT PHASE | | | RIGHT PHASE | |
(1)	(2)	(3)	(1)	(2)	(3)
94.800	5.100	0.100	0.400	1.700	97.900
85.700	14.000	0.300	2.500	8.600	88.900
77.100	21.700	1.200	4.300	15.600	80.100
70.300	27.000	2.700	6.000	21.300	72.700
62.700	32.000	5.300	11.100	27.000	61.900

SPECIFIC MODEL PARAMETERS IN KELVIN

| | | UNIQUAC | | | NRTL(ALPHA=.2) | |
I J	AIJ	AJI		AIJ	AJI
1 2	-99.331	-226.14		142.11	-549.02
1 3	316.75	631.49		1575.4	855.59
2 3	-62.773	-1.1154		151.46	-335.53

R1 = 0.9200   R2 = 2.1055   R3 = 3.1878
Q1 = 1.400   Q2 = 1.972   Q3 = 2.400

MEAN DEV. BETWEEN CALC. AND EXP. CONC. IN MOLE PCT

UNIQUAC (SPECIFIC PARAMETERS)   0.58
NRTL   (SPECIFIC PARAMETERS)   0.54

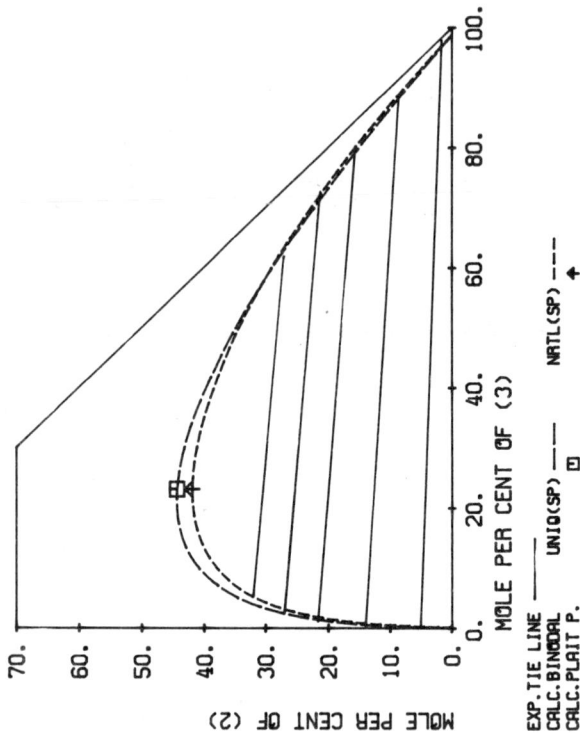

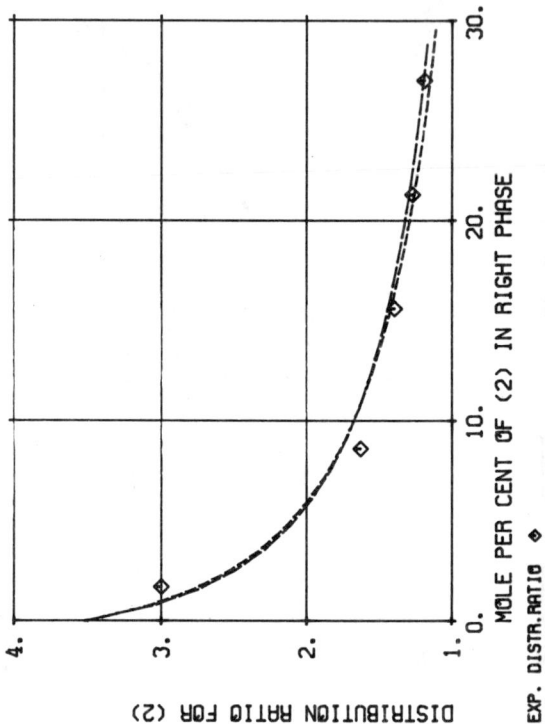

$C_2H_6O-C_6H_6$

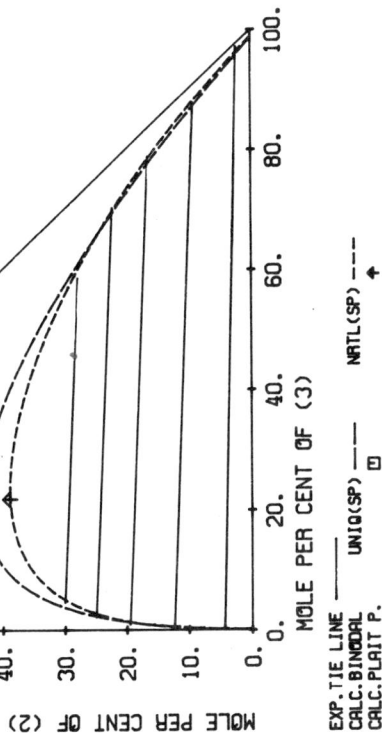

MOLE PER CENT OF (3)

MOLE PER CENT OF (2)

EXP.TIE LINE —— CALC.BINODAL UNIQ(SP) ——☐ NRTL(SP) ----
CALC.PLAIT P. ✦

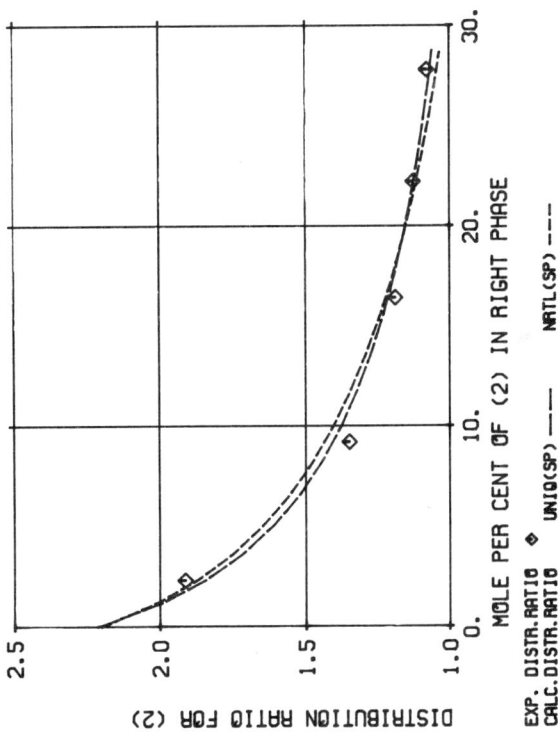

DISTRIBUTION RATIO FOR (2)

MOLE PER CENT OF (2) IN RIGHT PHASE

EXP. DISTR.RATIO ◆ UNIQ(SP) —— NRTL(SP) ----
CALC.DISTR.RATIO

(2) C2H6O   ETHANOL
(3) C6H6   BENZENE

MORACHEVSKII A.G., BELOUSOV V.P.
VESTN.LENINGR.UNIV.FIZ.KHIM. 13,4(1958)117

TEMPERATURE = 55.0 DEG C   TYPE OF SYSTEM = 1

EXPERIMENTAL TIE LINES IN MOLE PCT

	LEFT PHASE			RIGHT PHASE	
(1)	(2)	(3)	(1)	(2)	(3)
95.500	4.400	0.100	0.500	2.300	97.200
87.300	12.400	0.300	2.900	9.200	87.900
79.500	19.500	1.000	4.700	16.400	78.900
72.800	25.000	2.200	7.400	22.200	70.400
65.300	30.000	4.700	13.600	27.800	58.600

SPECIFIC MODEL PARAMETERS IN KELVIN

		UNIQUAC		NRTL(ALPHA=.2)	
I	J	AIJ	AJI	AIJ	AJI
1	2	-66.208	-216.57	353.51	-565.31
1	3	301.54	642.57	1996.1	927.73
2	3	-69.009	-4.4733	-44.688	-100.65

R1 = 0.9200   R2 = 2.1055   R3 = 3.1878
Q1 = 1.400   Q2 = 1.972   Q3 = 2.400

MEAN DEV. BETWEEN CALC. AND EXP. CONC. IN MOLE PCT

UNIQUAC (SPECIFIC PARAMETERS)   0.62
NRTL   (SPECIFIC PARAMETERS)   0.43

$C_2H_6O-C_6H_6$

(1) H2O   WATER

(2) C2H6O   ETHANOL

(3) C6H6   BENZENE

MORACHEVSKII A.G., BELOUSOV V.P.
VESTN.LENINGR.UNIV.FIZ.KHIM. 13,4(1958)117

TEMPERATURE = 64.0 DEG C   TYPE OF SYSTEM = 1

EXPERIMENTAL TIE LINES IN MOLE PCT

	LEFT PHASE			RIGHT PHASE	
(1)	(2)	(3)	(1)	(2)	(3)
96.200	3.700	0.100	0.900	2.000	97.100
89.000	10.800	0.200	2.800	9.900	87.300
82.000	17.200	0.800	5.000	17.400	77.600
74.500	23.500	2.000	8.700	22.800	68.500
67.800	28.000	4.200	15.300	28.400	56.300

SPECIFIC MODEL PARAMETERS IN KELVIN

		UNIQUAC			NRTL(ALPHA=.2)	
I	J	AIJ	AJI		AIJ	AJI
1	2	-42.947	-237.79		259.98	-584.98
1	3	-309.88	646.08		1788.6	900.30
2	3	-81.437	-24.511		4.2625	-314.65

R1 = 0.9200   R2 = 2.1055   R3 = 3.1878
Q1 = 1.400   Q2 = 1.972   Q3 = 2.400

MEAN DEV. BETWEEN CALC. AND EXP. CONC. IN MOLE PCT

UNIQUAC (SPECIFIC PARAMETERS)   0.71
NRTL   (SPECIFIC PARAMETERS)   0.59

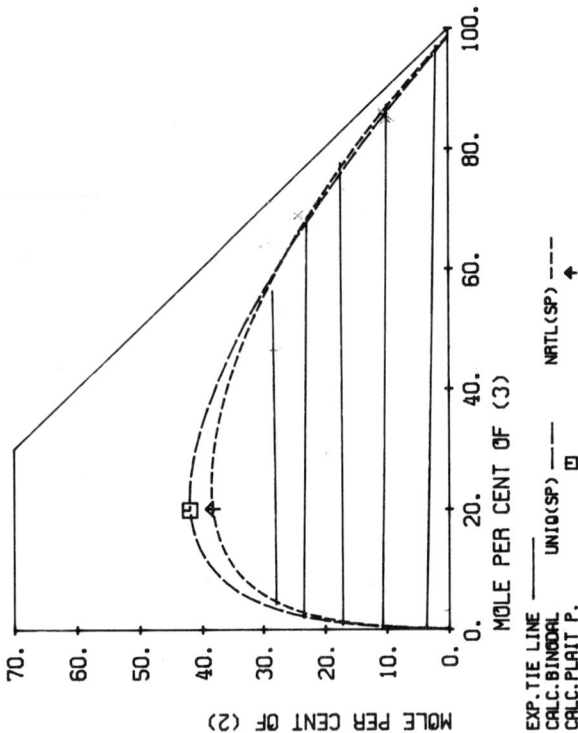

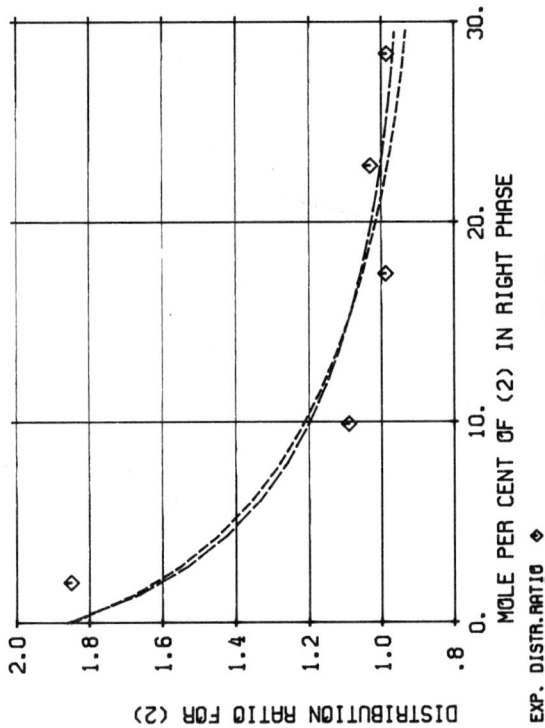

$C_2H_6O\text{-}C_6H_6$

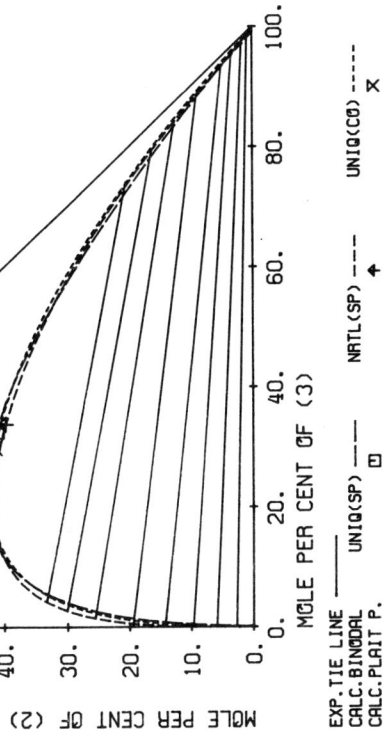

MOLE PER CENT OF (2)

MOLE PER CENT OF (3)

EXP.TIE LINE ——— UNIQ(SP) —— NRTL(SP) —— UNIQ(CO) ----×
CALC.BINODAL ⊡
CALC.PLAIT P.

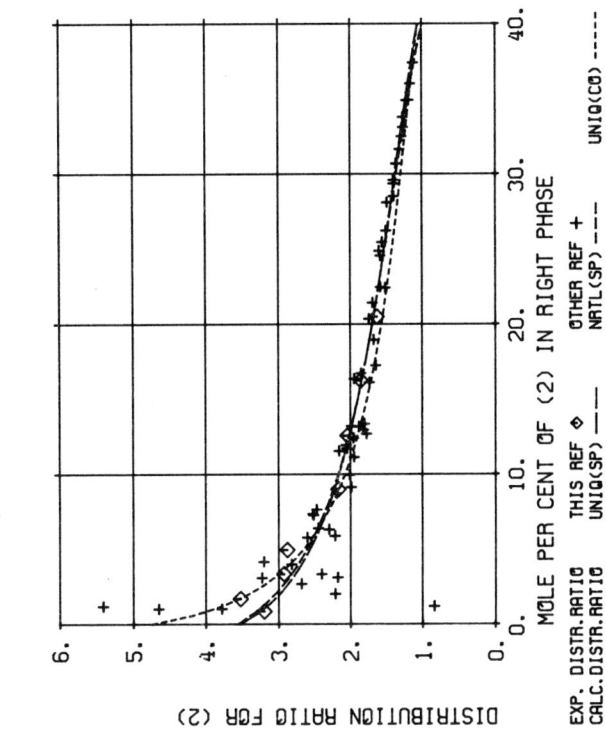

DISTRIBUTION RATIO FOR (2)

MOLE PER CENT OF (2) IN RIGHT PHASE

EXP. DISTR.RATIO THIS REF ◇ OTHER REF +
CALC.DISTR.RATIO UNIQ(SP) —— NRTL(SP) --- UNIQ(CO) -----

(2) C2H6O   ETHANOL
(3) C6H6   BENZENE

MERTSLIN R.V., NIKURASHINA N.I., KAMAEVSKAYA L.A.
ZH.FIZ.KHIM. 35(1961)2628

TEMPERATURE = 26.0 DEG C   TYPE OF SYSTEM = 1

EXPERIMENTAL TIE LINES IN MOLE PCT (GRAPH.INTERPOL.)

	LEFT PHASE			RIGHT PHASE	
(1)	(2)	(3)	(1)	(2)	(3)
97.246	2.694	0.060	0.431	0.842	98.727
94.018	5.906	0.076	0.600	1.676	97.724
90.185	9.708	0.107	0.976	3.320	95.704
85.622	14.202	0.175	1.345	4.931	93.723
80.152	19.454	0.395	2.296	8.978	88.726
73.027	25.590	1.383	3.127	12.541	84.332
67.169	30.079	2.753	4.107	16.211	79.683
62.438	33.311	4.251	5.313	20.478	74.209

SPECIFIC MODEL PARAMETERS IN KELVIN

I J	UNIQUAC			NRTL(ALPHA=.2)	
	AIJ	AJI		AIJ	AJI
1 2	266.93	-266.09		376.33	-441.74
1 3	249.80	807.82		2797.7	986.99
2 3	-73.352	256.10		87.744	118.04

R1 = 0.9200   R2 = 2.1055   R3 = 3.1878
Q1 = 1.400   Q2 = 1.972   Q3 = 2.400

MEAN DEV. BETWEEN CALC. AND EXP. CONC. IN MOLE PCT

UNIQUAC (SPECIFIC PARAMETERS)   0.58
NRTL   (SPECIFIC PARAMETERS)   0.34
UNIQUAC (COMMON PARAMETERS)   0.53

C₂H₆O-C₆H₆

(1) H2O   WATER
(2) C2H6O   ETHANOL
(3) C6H6   BENZENE

ROSS S., PATTERSON R.E.
J.CHEM.ENG.DATA 24(1979)111

TEMPERATURE = 20.0 DEG C   TYPE OF SYSTEM = 1

EXPERIMENTAL TIE LINES IN MOLE PCT

	LEFT PHASE			RIGHT PHASE	
(1)	(2)	(3)	(1)	(2)	(3)
93.578	6.371	0.051	0.430	1.177	98.393
86.559	13.274	0.168	0.847	4.140	95.014
80.771	18.807	0.422	1.653	7.619	90.723
73.926	24.890	1.185	2.826	11.524	85.650
64.679	31.788	3.534	4.293	16.329	79.378
50.837	39.758	9.405	7.302	24.840	67.858
44.857	41.802	13.341	10.463	28.093	51.444
37.618	42.677	19.706	16.824	33.763	49.412

SPECIFIC MODEL PARAMETERS IN KELVIN

		UNIQUAC		NRTL(ALPHA=.2)	
I J		AIJ	AJI	AIJ	AJI
1 2	266.93	-266.09	376.33	-441.74
1 3	249.80	807.82	2797.7	986.99
2 3	-73.352	256.10	87.744	118.04

R1 = 0.9200   R2 = 2.1055   R3 = 3.1878
Q1 = 1.400   Q2 = 1.972   Q3 = 2.400

MEAN DEV. BETWEEN CALC. AND EXP. CONC. IN MOLE PCT

UNIQUAC (SPECIFIC PARAMETERS)   0.83
NRTL   (SPECIFIC PARAMETERS)   0.83
UNIQUAC (COMMON PARAMETERS)   1.87

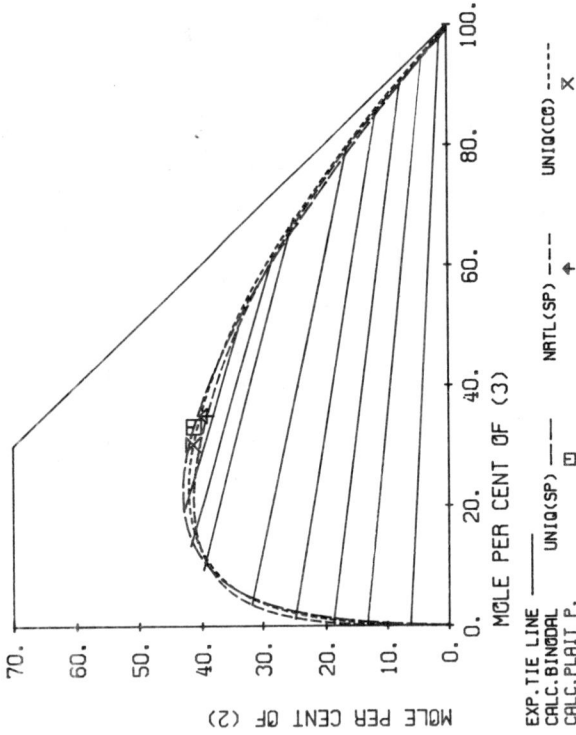

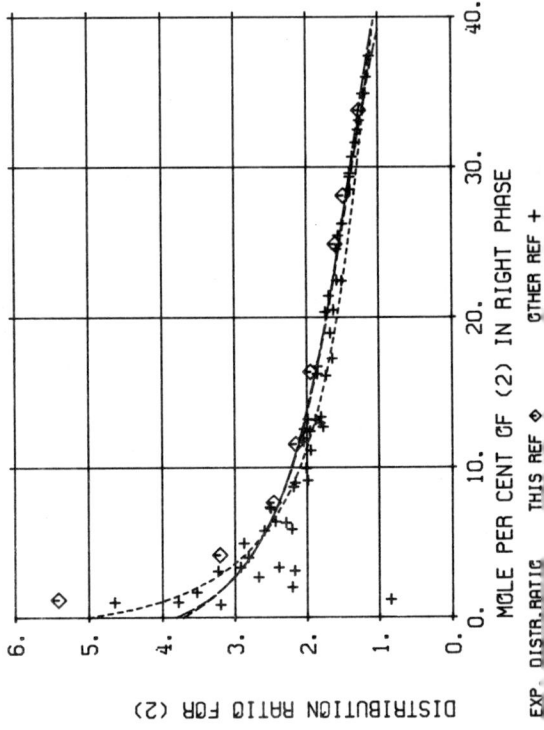

$C_2H_6O-C_6H_{10}$

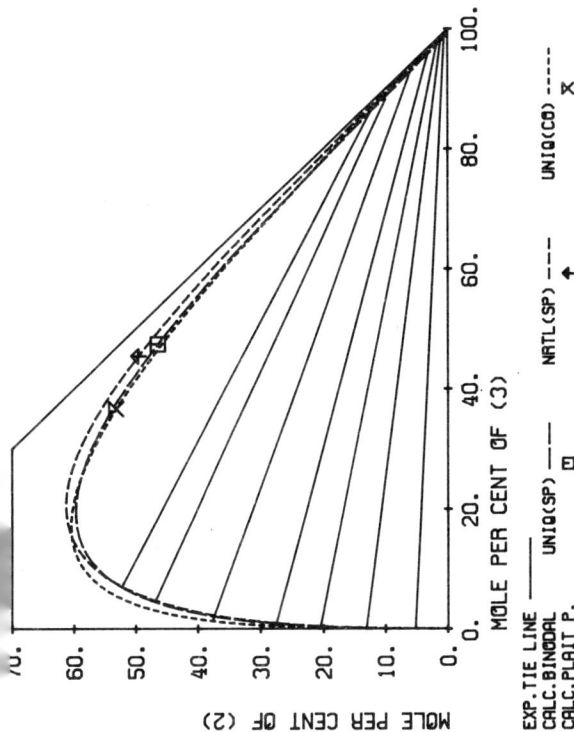

MOLE PER CENT OF (3)

MOLE PER CENT OF (2)

EXP.TIE LINE ——— UNIQ(SP) ——— ⊟ NRTL(SP) — — — ✦ UNIQ(C6) ----- ✕
CALC.BINODAL
CALC.PLAIT P.

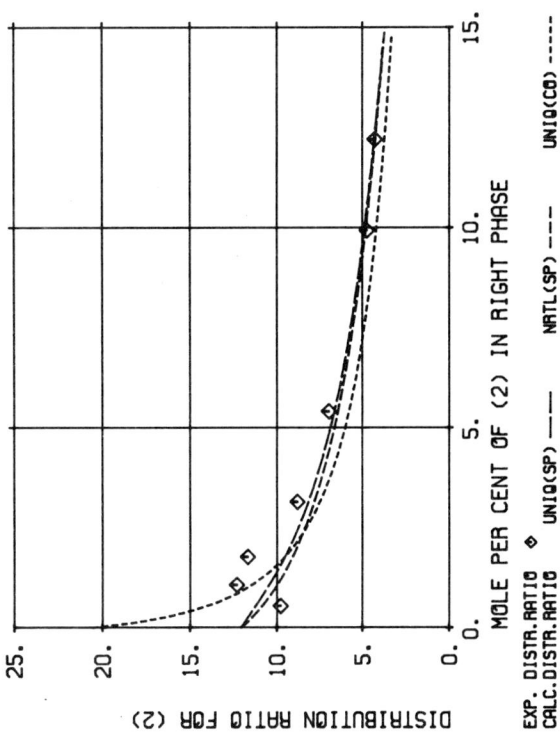

MOLE PER CENT OF (2) IN RIGHT PHASE

DISTRIBUTION RATIO FOR (2)

EXP. DISTR.RATIO ——— UNIQ(SP) ◇ NRTL(SP) — — — UNIQ(C6) -----
CALC.DISTR.RATIO

---

(2) C2H6O ———— ETHANOL
(3) C6H10 ———— CYCLOHEXENE

WASHBURN E.R., GRAHAM C.L., ARNOLD G.B., TRANSUE L.F.
J.AM.CHEM.SOC. 62(1940)1454

TEMPERATURE = 25.0 DEG C   TYPE OF SYSTEM = 1

EXPERIMENTAL TIE LINES IN MOLE PCT (GRAPH.INTERPOL.)

| | LEFT PHASE | | | RIGHT PHASE | |
(1)	(2)	(3)	(1)	(2)	(3)
	94.716	5.212	0.071	0.0	0.534	99.466
	86.914	13.086	0.0	0.452	1.061	98.487
	79.376	20.537	0.087	0.899	1.757	97.345
	71.803	27.624	0.573	0.893	3.143	95.964
	60.810	37.532	1.658	0.0	5.397	94.603
	48.497	46.892	4.611	1.720	9.921	88.359
	40.601	52.085	7.314	1.281	12.190	86.529

SPECIFIC MODEL PARAMETERS IN KELVIN

| | | UNIQUAC | | NRTL(ALPHA=.2) | |
I J	AIJ	AJI		AIJ	AJI
1 2	-186.18	-78.926		-311.58	16.398
1 3	-301.72	722.06		2047.8	1115.9
2 3	-51.248	318.76		113.43	357.32

R1 = 0.9200   R2 = 2.1055   R3 = 3.8143
Q1 = 1.400   Q2 = 1.972   Q3 = 3.027

MEAN DEV. BETWEEN CALC. AND EXP. CONC. IN MOLE PCT

UNIQUAC (SPECIFIC PARAMETERS)   0.48
NRTL   (SPECIFIC PARAMETERS)   0.46
UNIQUAC (COMMON PARAMETERS)   1.60

C₂H₆O-C₆H₁₂

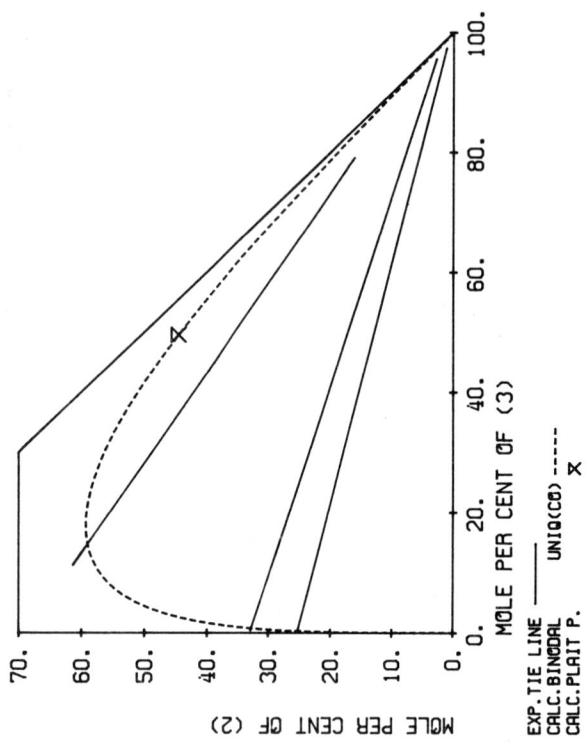

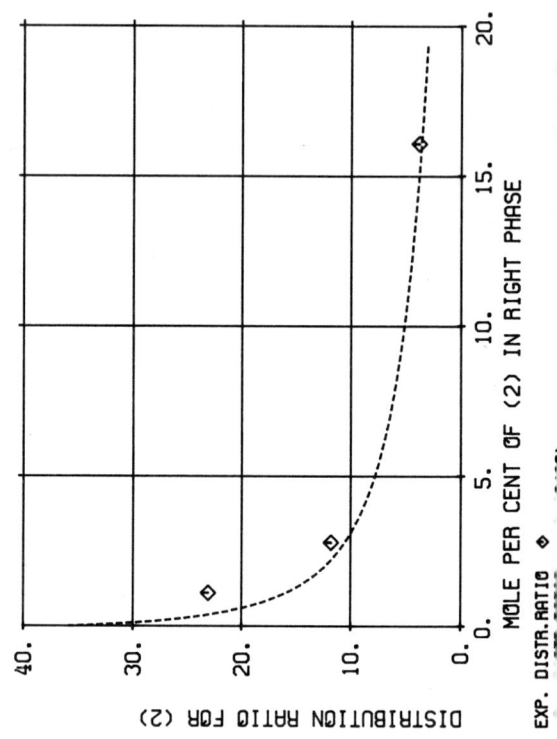

(1) H2O   WATER

(2) C2H6O   ETHANOL

(3) C6H12   CYCLOHEXANE

TARASENKOV D.N., PAULSEN I.A.
ZH.OBSHCH.KHIM. 7(1937)2143

TEMPERATURE = 25.0 DEG C   TYPE OF SYSTEM = 1

EXPERIMENTAL TIE LINES IN MOLE PCT

	LEFT PHASE			RIGHT PHASE	
(1)	(2)	(3)	(1)	(2)	(3)
74.341	25.301	0.358	1.470	1.096	97.434
66.714	32.844	0.441	1.594	2.778	95.628
27.403	61.347	11.250	4.780	16.074	79.146

MEAN DEV. BETWEEN CALC. AND EXP. CONC. IN MOLE PCT

UNIQUAC (COMMON PARAMETERS)   1.44

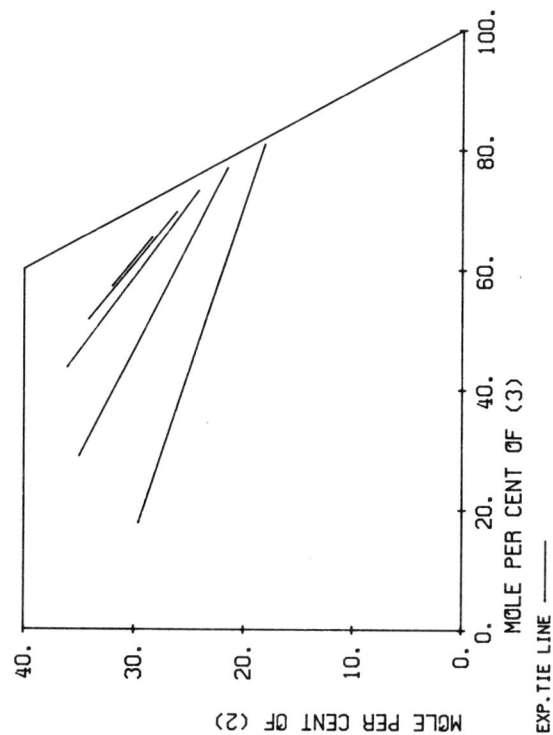

(1) C6H12O2   BUTANOIC ACID,ETHYL ESTER
(2) C2H6O   ETHANOL
(3) H2O   WATER

BONNER W.D.
J.PHYS.CHEM. 14(1910)738

TEMPERATURE = 0.0 DEG C   TYPE OF SYSTEM = 1

EXPERIMENTAL TIE LINES IN MOLE PCT (GRAPH.INTERPOL.)

LEFT PHASE			RIGHT PHASE		
(1)	(2)	(3)	(1)	(2)	(3)
52.631	29.621	17.747	0.951	18.096	80.952
36.222	34.936	28.843	1.471	21.530	76.999
20.249	36.052	43.699	2.666	24.154	73.180
14.238	34.094	51.668	4.185	26.153	69.662
10.812	31.996	57.192	6.215	28.417	65.368

$C_2H_6O\text{-}C_6H_{14}$

(1) H2O   WATER

(2) C2H6O   ETHANOL

(3) C6H14   HEXANE

VOROBEVA A.I., KARAPETYANTS M.KH.
ZH.FIZ.KHIM. 40(1966)3018

TEMPERATURE = 25.0 DEG C   TYPE OF SYSTEM = 1

EXPERIMENTAL TIE LINES IN MOLE PCT

LEFT PHASE			RIGHT PHASE		
(1)	(2)	(3)	(1)	(2)	(3)
69.423	30.111	0.466	0.474	1.297	98.230
40.227	56.157	3.616	0.921	6.482	92.597
26.643	64.612	8.745	1.336	12.540	86.124
19.805	65.678	14.517	2.539	20.515	76.946
15.284	61.759	22.957	3.959	30.339	65.702
12.879	58.444	28.676	4.940	35.808	59.253
11.732	56.258	32.010	5.908	38.983	55.109
11.271	55.091	33.639	6.529	40.849	52.622

SPECIFIC MODEL PARAMETERS IN KELVIN

		UNIQUAC		NRTL(ALPHA=.2)	
I J	AIJ	AJI		AIJ	AJI
1 2	-380.36	-72.446		154.75	-514.60
1 3	495.06	649.69		2054.0	1267.6
2 3	-80.747	337.78		300.20	248.20

R1 = 0.9200   R2 = 2.1055   R3 = 4.4998
Q1 = 1.400   Q2 = 1.972   Q3 = 3.856

MEAN DEV. BETWEEN CALC. AND EXP. CONC. IN MOLE PCT

UNIQUAC (SPECIFIC PARAMETERS)   1.82
NRTL   (SPECIFIC PARAMETERS)   2.49
UNIQUAC (COMMON PARAMETERS)   2.35

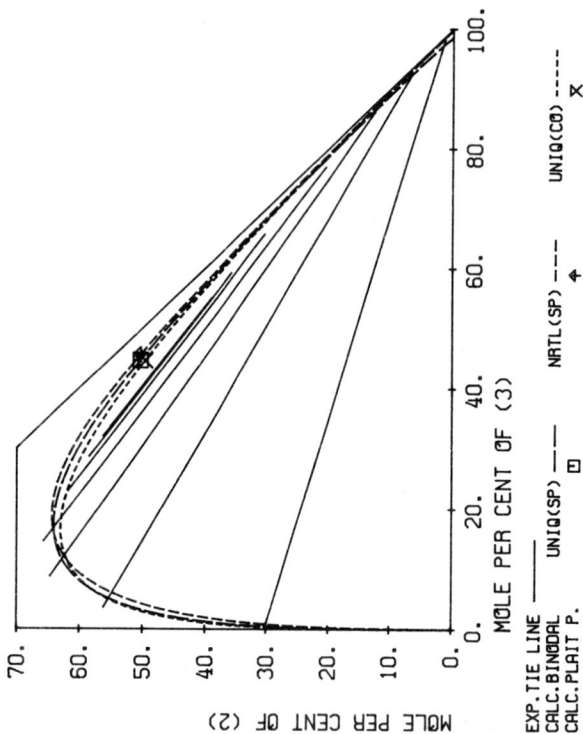

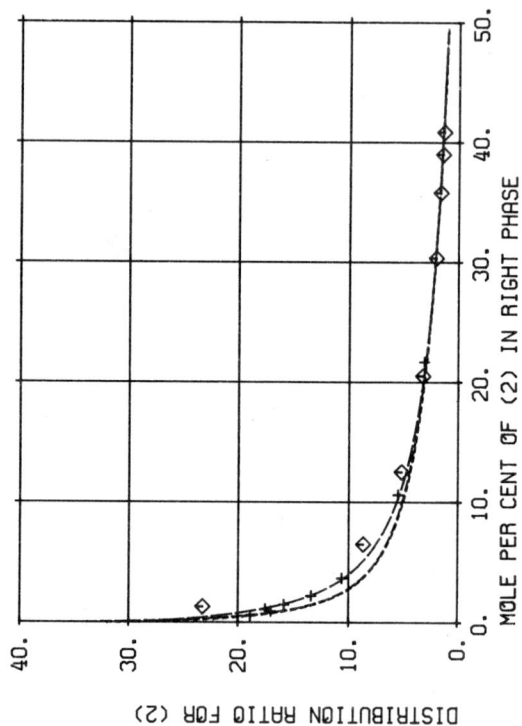

MOLE PER CENT OF (2)
MOLE PER CENT OF (3)

EXP.TIE LINE ———   UNIQ(SP) ---□   NRTL(SP) ---✦   UNIQ(C0) ----✕
CALC.BINODAL
CALC.PLAIT P.

DISTRIBUTION RATIO FOR (2)
MOLE PER CENT OF (2) IN RIGHT PHASE

EXP. DISTR.RATIO ◇   THIS REF ◇   OTHER REF +

$C_2H_6O-C_6H_{14}$

MOLE PER CENT OF (3)

MOLE PER CENT OF (2)

EXP.TIE LINE —— UNIQ(SP) —— NRTL(SP) —— UNIQ(CO) ——
CALC.BINODAL   □   +   X
CALC.PLAIT P.

DISTRIBUTION RATIO FOR (2)

MOLE PER CENT OF (2) IN RIGHT PHASE

EXP. DISTR.RATIO ◇   THIS REF +   OTHER REF +   UNIQ(CO) ——
CALC.DISTR.RATIO   UNIQ(SP) —— NRTL(SP) ——

---

(2) C2H6O   ETHANOL

(3) C6H14   HEXANE

ROSS S., PATTERSON R.E.
J.CHEM.ENG.DATA 24(1979)111

TEMPERATURE = 20.0 DEG C   TYPE OF SYSTEM = 1

EXPERIMENTAL TIE LINES IN MOLE PCT

	LEFT PHASE		RIGHT PHASE	
(1)	(2)	(3)	(2)	(3)
89.435	10.565	0.0	0.0	0.560	99.440
84.114	15.860	0.026	0.0	0.931	99.069
80.373	19.573	0.055	0.0	1.117	98.883
76.335	23.580	0.086	0.0	1.486	98.514
70.020	29.796	0.184	0.0	2.222	97.778
60.180	39.136	0.684	0.0	3.677	96.323
38.321	58.185	3.494	0.453	10.628	88.919
16.087	67.099	16.814	2.523	21.705	75.772

SPECIFIC MODEL PARAMETERS IN KELVIN

		UNIQUAC		NRTL(ALPHA=.2)	
I J	AIJ	AJI	AIJ	AJI
1 2	-380.36	-72.446	154.75	-514.60
1 3	495.06	649.69	2054.0	1267.6
2 3	-80.747	337.78	300.20	248.20

R1 = 0.9200   R2 = 2.1055   R3 = 4.4998
Q1 = 1.400   Q2 = 1.972   Q3 = 3.856

MEAN DEV. BETWEEN CALC. AND EXP. CONC. IN MOLE PCT

UNIQUAC (SPECIFIC PARAMETERS)   0.92
NRTL   (SPECIFIC PARAMETERS)   1.02
UNIQUAC (COMMON PARAMETERS)   1.36

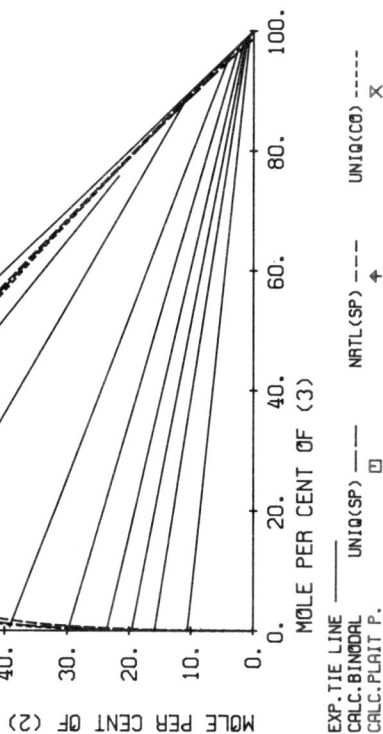

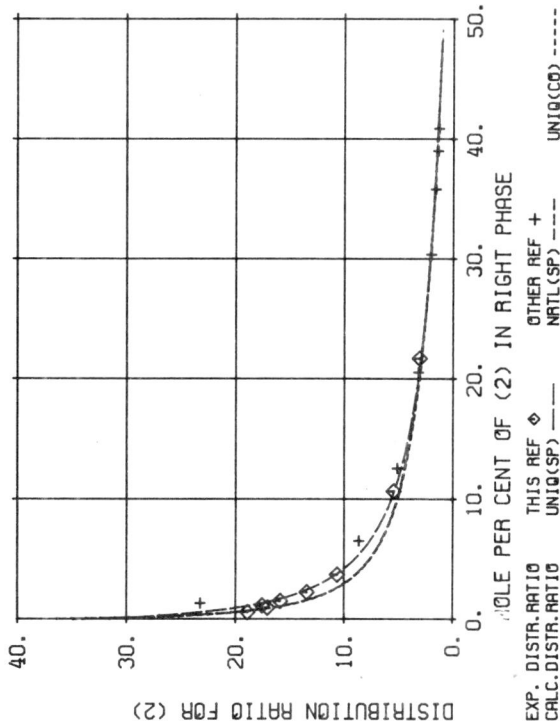

C$_2$H$_6$O-C$_6$H$_{14}$O

(1) C6H14O   2-PENTANOL, 4-METHYL
----
(2) C2H6O   ETHANOL
----
(3) H2O   WATER

DAKSHINAMURTY P., ET AL.
J.CHEM.ENG.DATA 17(1972)379

TEMPERATURE = 30.0 DEG C   TYPE OF SYSTEM = 1

EXPERIMENTAL TIE LINES IN MOLE PCT

	LEFT PHASE			RIGHT PHASE	
(1)	(2)	(3)	(1)	(2)	(3)
66.404	10.163	23.433	0.258	2.041	97.702
59.082	15.324	25.593	0.265	4.036	95.698
49.885	20.870	29.245	0.271	5.404	94.325
43.870	23.957	32.173	0.257	7.015	92.728
36.918	26.925	36.157	0.282	8.310	91.408
30.027	28.893	41.080	0.603	9.734	89.663
26.448	29.383	44.169	0.505	10.923	88.572
18.211	28.877	52.912	0.947	13.195	85.858
15.487	28.258	56.254	1.586	14.731	83.684
14.880	28.002	57.118	3.086	17.165	79.749
15.487	28.258	56.254	3.441	17.622	78.936

SPECIFIC MODEL PARAMETERS IN KELVIN

I J	UNIQUAC AIJ	AJI	NRTL(ALPHA=.2) AIJ	AJI
1 2	414.92	-233.74	345.87	-278.39
1 3	426.92	60.307	95.464	1781.4
2 3	-140.94	254.90	-300.58	783.08

R1 = 4.8015   R2 = 2.1055   R3 = 0.9200
Q1 = 4.124   Q2 = 1.972   Q3 = 1.400

MEAN DEV. BETWEEN CALC. AND EXP. CONC. IN MOLE PCT

UNIQUAC (SPECIFIC PARAMETERS)   0.84
NRTL   (SPECIFIC PARAMETERS)   0.75
UNIQUAC (COMMON PARAMETERS)   1.99

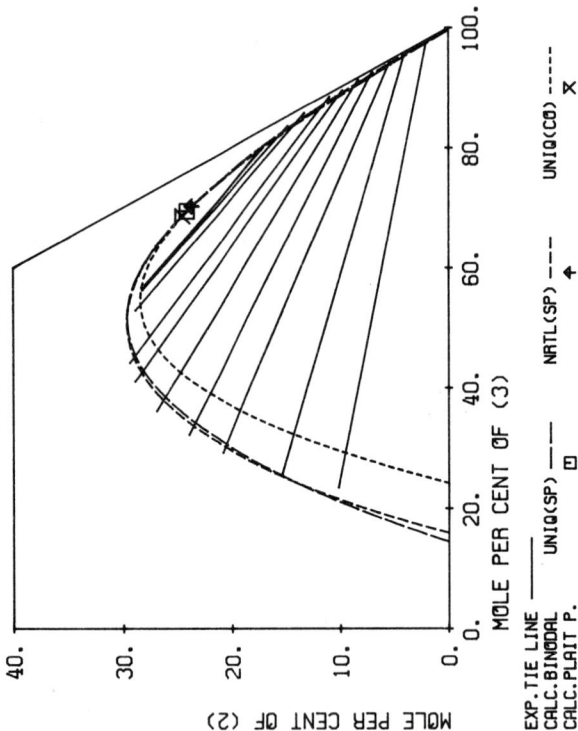

MOLE PER CENT OF (2)

MOLE PER CENT OF (3)

EXP.TIE LINE ———   UNIQ(SP) ▣   NRTL(SP) ——— ♦   UNIQ(CO) ---- ✕
CALC.BINODAL
CALC.PLAIT P.

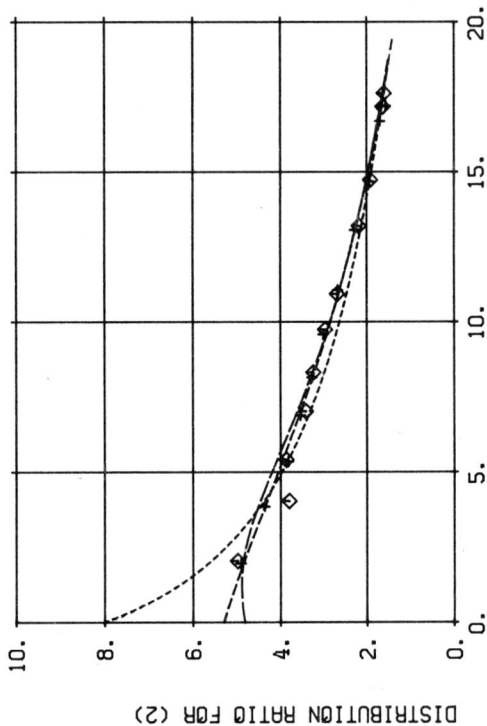

DISTRIBUTION RATIO FOR (2)

MOLE PER CENT OF (2) IN RIGHT PHASE

EXP. DISTR.RATIO ◇   THIS REF ◇   OTHER REF +

$C_2H_6O-C_6H_{14}O$

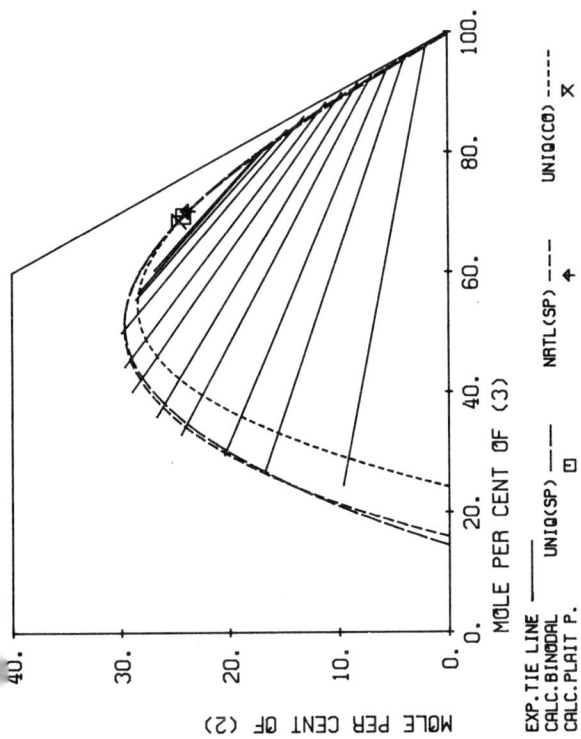

MOLE PER CENT OF (3)

MOLE PER CENT OF (2)

EXP.TIE LINE ———
CALC.BINODAL
CALC.PLAIT P.

UNIQ(SP) ——— ☐
NRTL(SP) ——— ✦
UNIQ(CO) - - - - - ✗

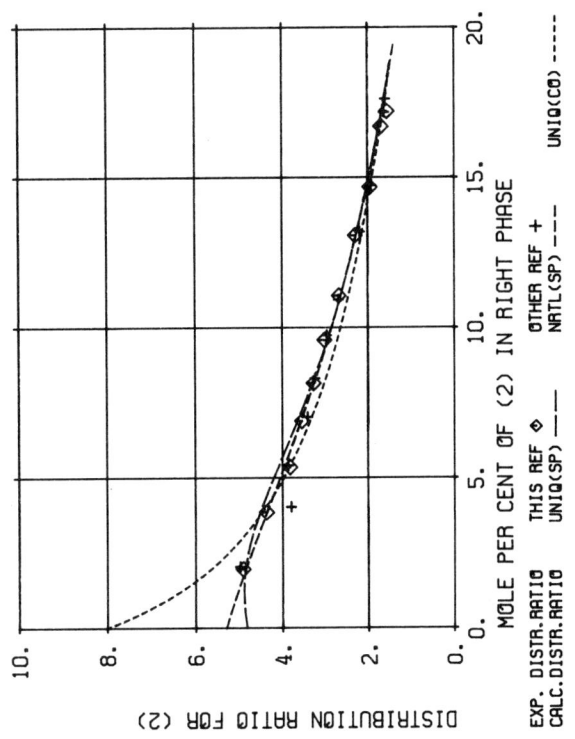

MOLE PER CENT OF (2) IN RIGHT PHASE

DISTRIBUTION RATIO FOR (2)

EXP. DISTR.RATIO ◇
CALC.DISTR.RATIO

THIS REF ◇
UNIQ(SP) ———
OTHER REF +
NRTL(SP) - - - -
UNIQ(CO) - - - - -

(1) C6H14O   2-PENTANOL, 4-METHYL
(2) C2H6O   ETHANOL
(3) H2O   WATER

DAKSHINAMURTY P., ET AL.
J.CHEM.ENG.DATA 17(1972)379

TEMPERATURE = 30.0 DEG C   TYPE OF SYSTEM = 1

EXPERIMENTAL TIE LINES IN MOLE PCT

	LEFT PHASE			RIGHT PHASE	
(1)	(2)	(3)	(1)	(2)	(3)
65.891	9.588	24.521	0.220	1.953	97.826
56.625	16.806	26.569	0.227	3.851	95.922
49.981	20.428	29.591	0.231	5.348	94.421
42.598	24.378	33.024	0.296	6.882	92.821
37.312	26.679	36.009	0.363	8.140	91.497
30.849	28.951	40.200	0.496	9.582	89.922
25.988	29.604	44.403	0.657	11.053	88.289
20.093	29.898	50.010	1.014	13.059	85.927
15.886	28.564	55.550	1.584	14.685	83.748
14.983	28.292	56.725	2.687	16.685	80.628
12.745	26.849	60.406	3.032	17.195	79.773

SPECIFIC MODEL PARAMETERS IN KELVIN

I J	UNIQUAC AIJ	AJI		NRTL(ALPHA=.2) AIJ	AJI
1 2	414.92	-233.74		345.87	-278.39
1 3	426.92	60.307		95.464	1781.4
2 3	-140.94	254.90		-300.58	783.08

R1 = 4.8015   R2 = 2.1055   R3 = 0.9200
Q1 = 4.124   Q2 = 1.972   Q3 = 1.400

MEAN DEV. BETWEEN CALC. AND EXP. CONC. IN MOLE PCT

UNIQUAC (SPECIFIC PARAMETERS)   0.80
NRTL   (SPECIFIC PARAMETERS)   0.63
UNIQUAC (COMMON PARAMETERS)   1.90

C$_2$H$_6$O-C$_7$H$_6$O

MOLE PER CENT OF (3)

EXP.TIE LINE ──────

MOLE PER CENT OF (2)

40. 30. 20. 10. 0.

0. 20. 40. 60. 80. 100.

(1) C7H60   BENZALDEHYDE
(2) C2H60   ETHANOL
(3) H20   WATER

BONNER W.D.
J.PHYS.CHEM. 14(1910)738

TEMPERATURE =   0.0 DEG C   TYPE OF SYSTEM = 1

EXPERIMENTAL TIE LINES IN MOLE PCT (GRAPH.INTERPOL.)

	LEFT PHASE			RIGHT PHASE	
(1)	(2)	(3)	(1)	(2)	(3)
62.282	22.338	15.380	0.310	13.113	86.577
44.745	29.369	25.886	0.591	16.679	82.730
42.171	30.264	27.565	2.728	23.671	73.601
33.183	32.412	34.406	3.811	25.351	70.838
21.798	33.242	44.960	6.525	28.004	65.471
18.379	32.910	48.711	8.086	28.952	62.962

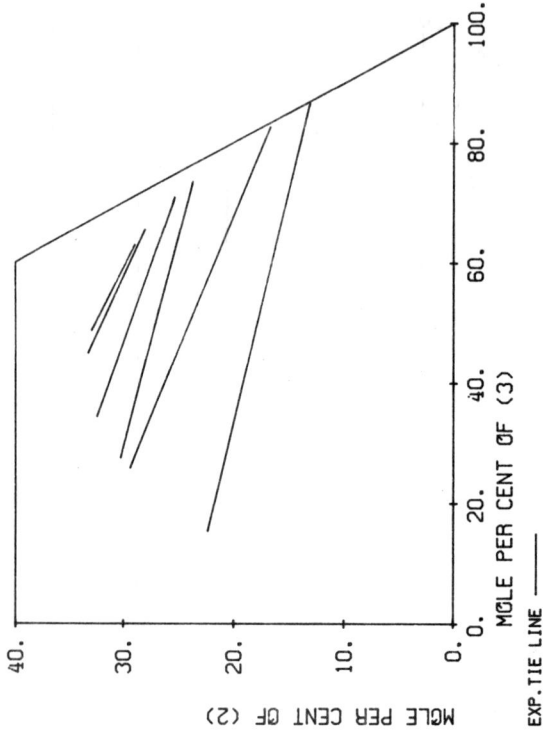

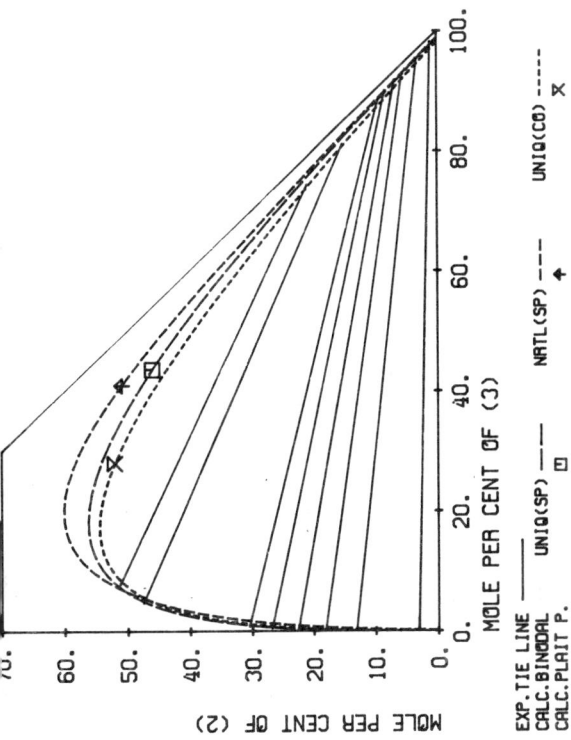

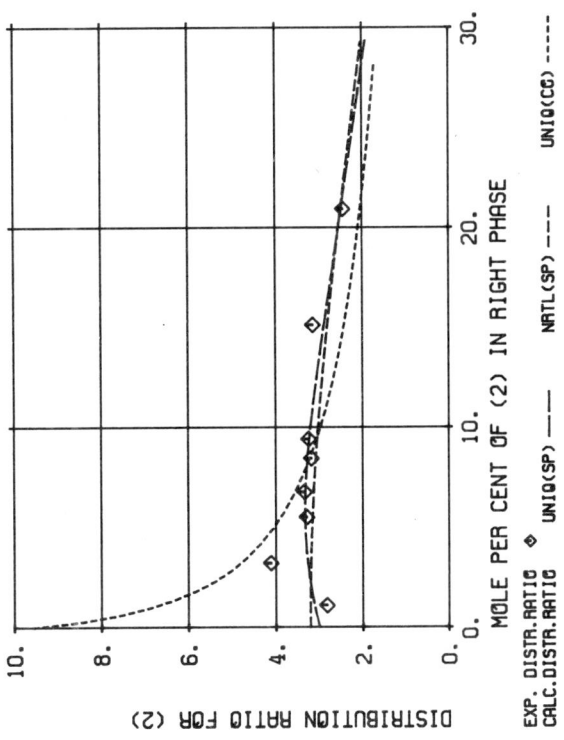

(1) WATER

(2) C2H6O   ETHANOL

(3) C7H7CL   TOLUENE,2-CHLORO

SEJONG NAM,TOYOHIKO HAYAKAWA, SHIGEFUMI FUJITA
J.CHEM.ENG.JPN. 5(1972)327

TEMPERATURE = 25.0 DEG C   TYPE OF SYSTEM = 1

EXPERIMENTAL TIE LINES IN MOLE PCT

| | LEFT PHASE | | | RIGHT PHASE | |
(1)	(2)	(3)	(1)	(2)	(3)
96.891	3.079	0.030	0.694	1.085	98.221
86.822	13.127	0.052	1.360	3.192	95.448
81.817	18.109	0.073	1.340	5.502	93.158
77.174	22.593	0.233	1.329	6.756	91.915
72.673	26.833	0.494	1.960	8.429	89.611
68.700	30.498	0.802	1.947	9.390	88.663
47.777	47.498	4.725	3.678	15.102	81.220
40.890	51.255	7.856	4.650	20.909	74.441

SPECIFIC MODEL PARAMETERS IN KELVIN

| | | UNIQUAC | | NRTL(ALPHA=.2) | |
I	J	AIJ	AJI	AIJ	AJI
1	2	-62.614	8.0261	347.43	-222.42
1	3	-210.18	692.86	1488.9	822.47
2	3	-96.017	381.84	317.40	204.12

R1 = 0.9200   R2 = 2.1055   R3 = 4.5477
Q1 = 1.400   Q2 = 1.972   Q3 = 3.412

MEAN DEV. BETWEEN CALC. AND EXP. CONC. IN MOLE PCT

UNIQUAC (SPECIFIC PARAMETERS)   0.53
NRTL   (SPECIFIC PARAMETERS)   0.71
UNIQUAC (COMMON PARAMETERS)   2.56

C₂H₆O-C₇H₇Cl

(1) H2O   WATER

(2) C2H6O   ETHANOL

(3) C7H7CL   TOLUENE,3-CHLORO

SEJONG NAM,TOYOHIKO HAYAKAWA, SHIGEFUMI FUJITA
J.CHEM.ENG.JPN. 5(1972)327

TEMPERATURE = 25.0 DEG C   TYPE OF SYSTEM = 1

EXPERIMENTAL TIE LINES IN MOLE PCT

LEFT PHASE			RIGHT PHASE		
(1)	(2)	(3)	(1)	(2)	(3)
95.280	4.704	0.015	0.696	0.544	98.760
92.708	7.277	0.016	0.695	0.815	98.490
92.708	7.277	0.016	0.695	0.815	98.490
85.932	14.033	0.035	2.699	2.638	94.663
81.817	18.109	0.073	2.676	3.924	93.400
76.883	22.883	0.234	3.288	5.657	91.055
53.425	43.529	3.046	3.100	14.306	82.594
47.824	47.581	4.595	4.288	14.373	81.339

SPECIFIC MODEL PARAMETERS IN KELVIN

		UNIQUAC			NRTL(ALPHA=.2)	
I	J	AIJ	AJI		AIJ	AJI
1	2	100.98	-252.52		483.51	-586.24
1	3	137.50	667.05		1660.9	658.96
2	3	-10.270	258.53		485.90	38.256

R1 = 0.9200   R2 = 2.1055   R3 = 4.5477
Q1 = 1.400   Q2 = 1.972   Q3 = 3.412

MEAN DEV. BETWEEN CALC. AND EXP. CONC. IN MOLE PCT

UNIQUAC (SPECIFIC PARAMETERS)   0.64
NRTL   (SPECIFIC PARAMETERS)   0.82
UNIQUAC (COMMON PARAMETERS)   1.30

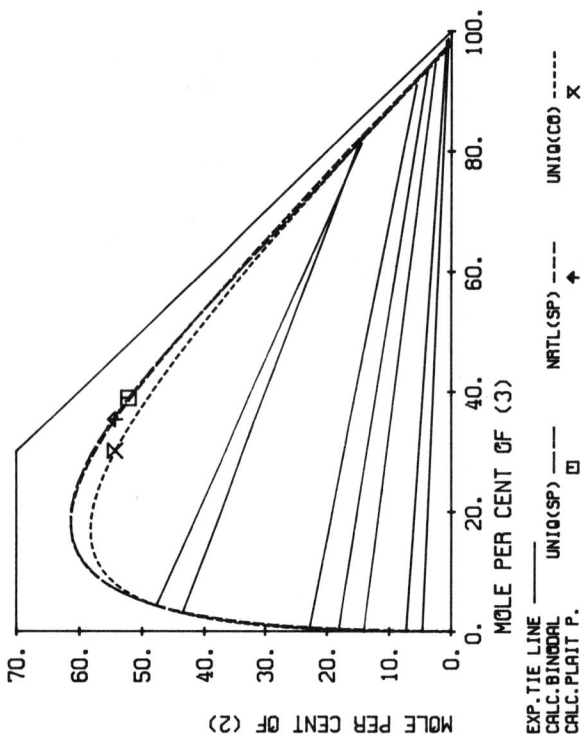

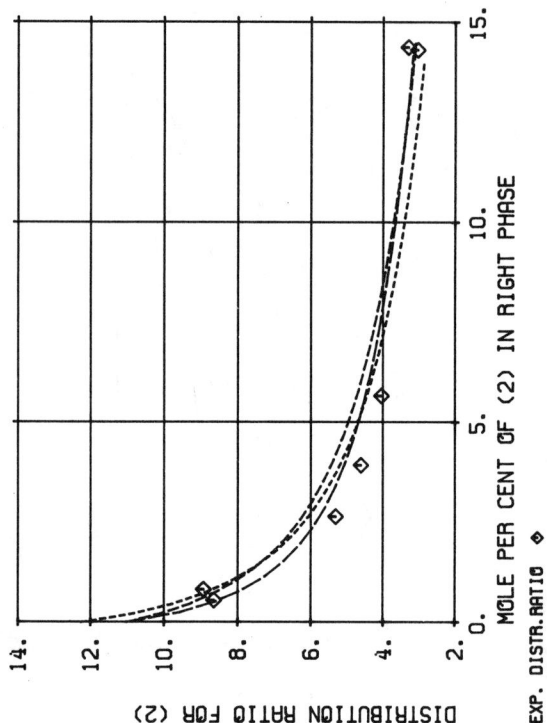

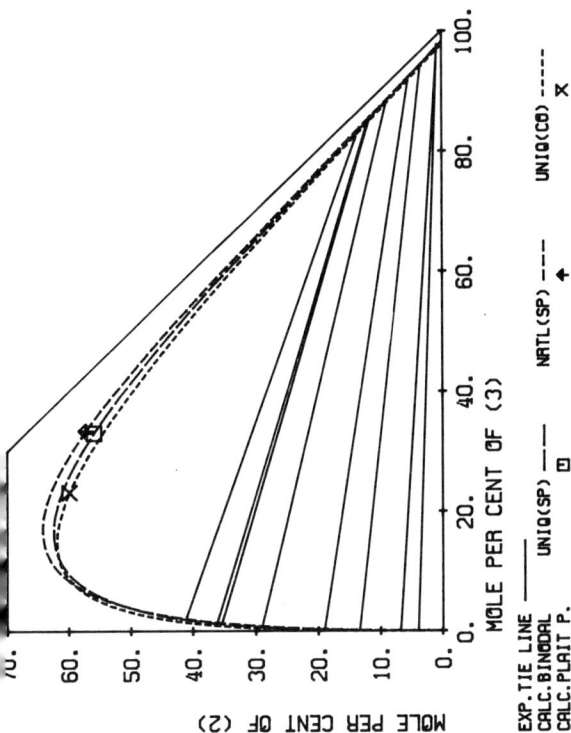

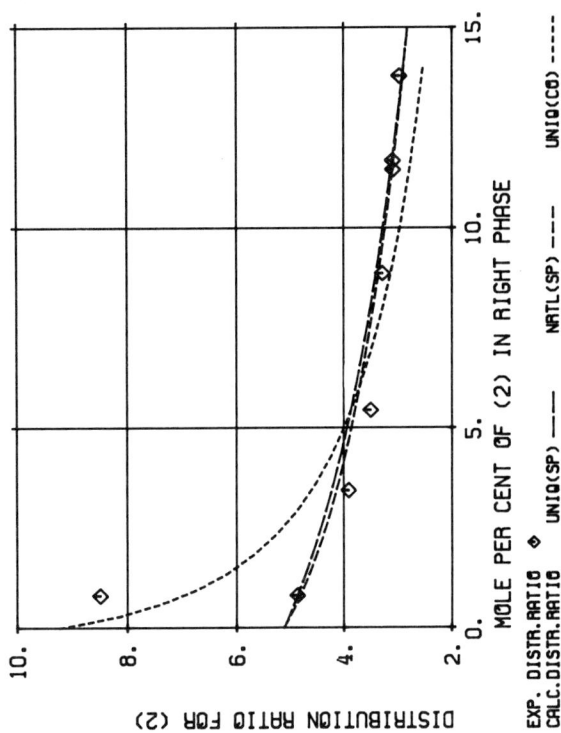

$C_2H_6O-C_7H_7Cl$

(2) C2H6O   ETHANOL

(3) C7H7CL   TOLUENE, 4-CHLORO

SEJONG NAM,TOYOHIKO HAYAKAWA, SHIGEFUMI FUJITA
J.CHEM.ENG.JPN. 5(1972)327

TEMPERATURE = 25.0 DEG C   TYPE OF SYSTEM = 1

EXPERIMENTAL TIE LINES IN MOLE PCT

	LEFT PHASE			RIGHT PHASE	
(1)	(2)	(3)	(1)	(2)	(3)
93.142	6.843	0.016	2.060	0.806	97.134
96.038	3.947	0.015	1.382	0.810	97.808
86.536	13.412	0.052	2.025	3.432	94.542
80.810	19.041	0.149	2.650	5.440	91.911
70.316	29.069	0.615	2.590	8.862	88.549
63.281	35.396	1.323	2.545	11.444	86.011
62.397	36.147	1.456	2.541	11.674	85.785
56.562	41.026	2.412	3.712	13.790	82.498

SPECIFIC MODEL PARAMETERS IN KELVIN

		UNIQUAC		NRTL(ALPHA=.2)	
I	J	AIJ	AJI	AIJ	AJI
1	2	-57.238	-124.55	333.06	-370.32
1	3	173.85	626.90	1765.3	681.00
2	3	-9.6977	211.78	511.08	25.322

R1 = 0.9200   R2 = 2.1055   R3 = 4.5477
Q1 = 1.400   Q2 = 1.972   Q3 = 3.412

MEAN DEV. BETWEEN CALC. AND EXP. CONC. IN MOLE PCT

UNIQUAC (SPECIFIC PARAMETERS)   0.31
NRTL   (SPECIFIC PARAMETERS)   0.31
UNIQUAC (COMMON PARAMETERS)   1.23

C₂H₆O-C₇H₈

(1) H2O   WATER
(2) C2H6O   ETHANOL
(3) C7H8   TOLUENE

WASHBURN R.E., BEGUIN A.E., BECKORD O.C.
J.AM.CHEM.SOC. 61(1939)1694

TEMPERATURE = 25.0 DEG C   TYPE OF SYSTEM = 1

EXPERIMENTAL TIE LINES IN MOLE PCT (GRAPH.INTERPOL.)

	LEFT PHASE			RIGHT PHASE	
(1)	(2)	(3)	(1)	(2)	(3)
95.320	4.621	0.059	0.356	0.398	99.246
91.360	8.538	0.102	0.506	1.385	98.109
87.856	11.990	0.154	0.703	2.552	96.745
82.021	17.712	0.267	1.150	5.210	93.640
77.323	22.245	0.432	1.796	8.849	89.355
72.623	26.547	0.829	2.007	9.905	88.088
69.166	29.572	1.263	2.425	11.970	85.605
64.420	33.500	2.080	3.111	14.779	82.110
59.615	37.191	3.194	3.719	16.672	79.609

SPECIFIC MODEL PARAMETERS IN KELVIN

		UNIQUAC		NRTL(ALPHA=.2)	
I J		AIJ	AJI	AIJ	AJI
1 2		-107.71	-235.67	103.47	-567.47
1 3		272.30	678.16	1544.2	808.51
2 3		-32.362	103.87	495.88	-302.16

R1 = 0.9200   R2 = 2.1055   R3 = 3.9228
Q1 = 1.400   Q2 = 1.972   Q3 = 2.968

MEAN DEV. BETWEEN CALC. AND EXP. CONC. IN MOLE PCT

UNIQUAC (SPECIFIC PARAMETERS)   0.57
NRTL   (SPECIFIC PARAMETERS)   0.65
UNIQUAC (COMMON PARAMETERS)   1.21

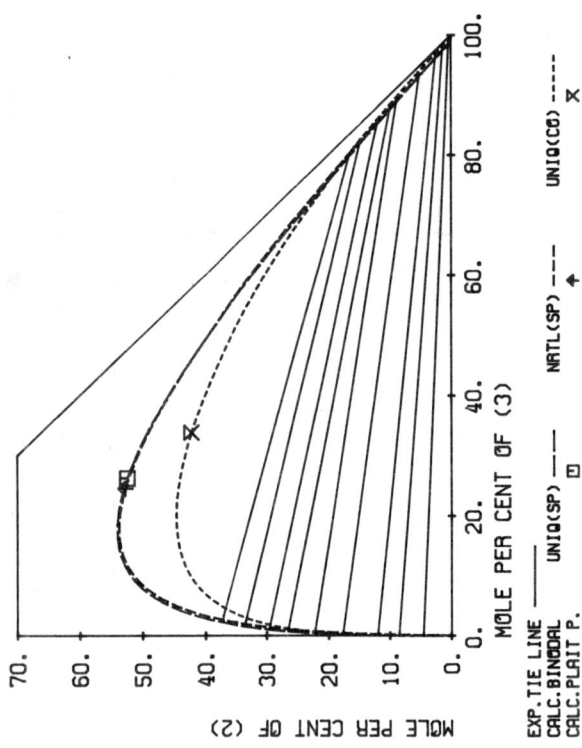

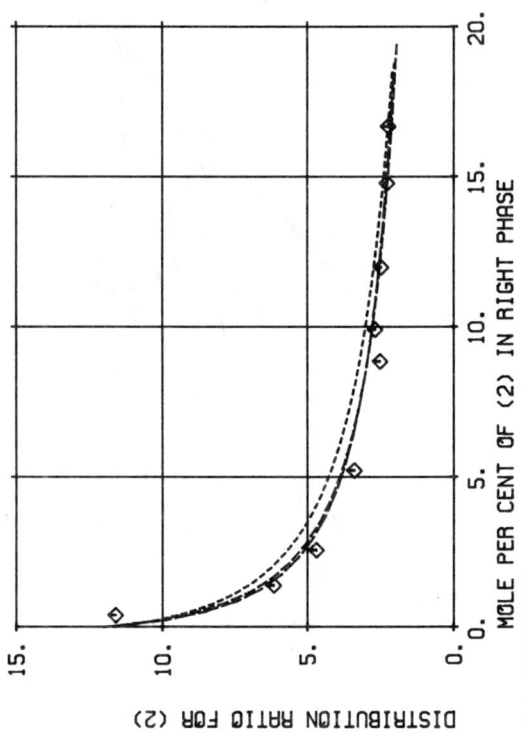

$C_2H_6O-C_7H_9N$

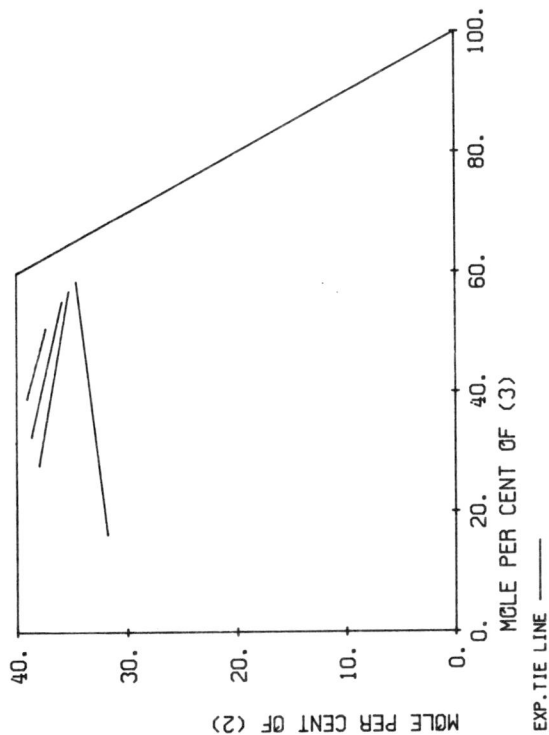

MOLE PER CENT OF (3)

EXP.TIE LINE ———

MOLE PER CENT OF (2)

(1) C7H9N   ANILINE,N-METHYL

(2) C2H6O   ETHANOL

(3) H2O   WATER

BONNER W.D.
J.PHYS.CHEM. 14(1910)738

TEMPERATURE =   0.0 DEG C   TYPE OF SYSTEM = 1

EXPERIMENTAL TIE LINES IN MOLE PCT (GRAPH.INTERPOL.)

	LEFT PHASE			RIGHT PHASE	
(1)	(2)	(3)	(1)	(2)	(3)
51.943	31.795	16.262	6.849	34.577	58.574
34.140	37.974	27.886	7.847	35.215	56.938
28.709	38.648	32.643	8.947	35.832	55.221
21.906	39.057	39.037	11.892	37.334	50.775

C₂H₆O-C₇H₁₄O

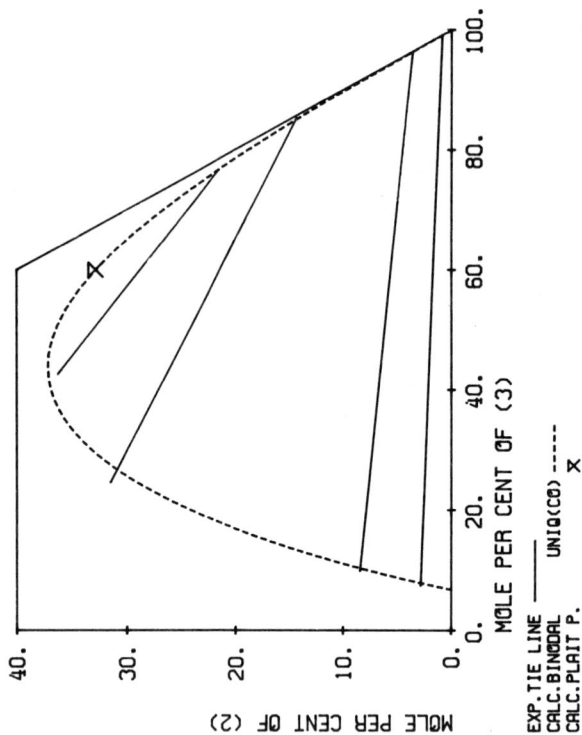

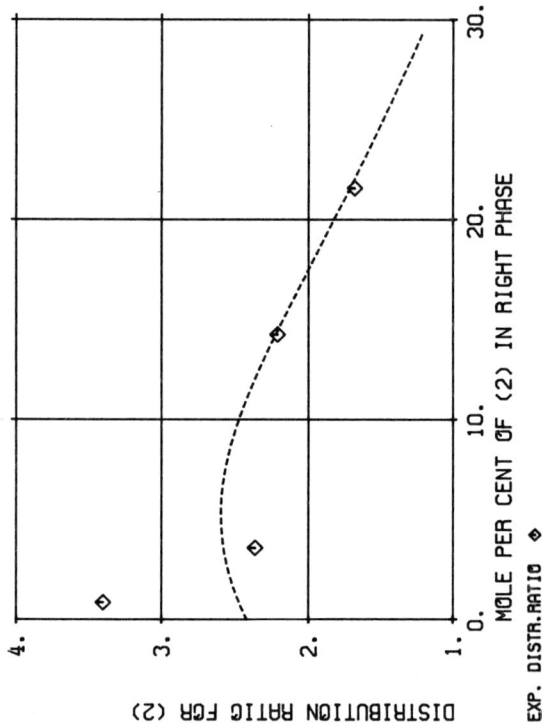

(1) C7H14O   4-HEPTANONE
(2) C2H6O   ETHANOL
(3) H2O   WATER

OTHMER D.F., WHITE R.E., TRUEGER E.
IND.ENG.CHEM. 33(1941)1240

TEMPERATURE = 25.5 DEG C   TYPE OF SYSTEM = 1

EXPERIMENTAL TIE LINES IN MOLE PCT (GRAPH.INTERPOL.)

	LEFT PHASE			RIGHT PHASE	
(1)	(2)	(3)	(1)	(2)	(3)
89.684	2.851	7.465	0.064	0.839	99.097
81.675	8.440	9.885	0.084	3.573	96.343
43.922	31.494	24.584	0.402	14.245	85.353
21.071	36.291	42.638	1.808	21.563	76.629

MEAN DEV. BETWEEN CALC. AND EXP. CONC. IN MOLE PCT

UNIQUAC (COMMON PARAMETERS)   0.62

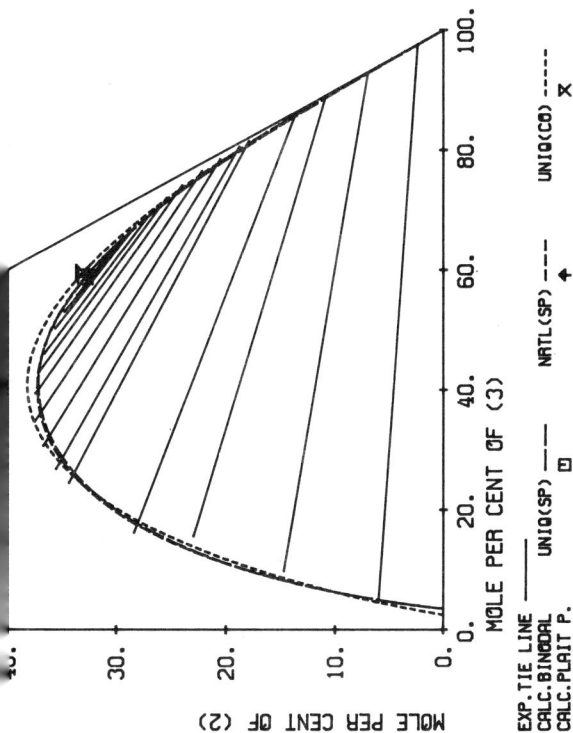

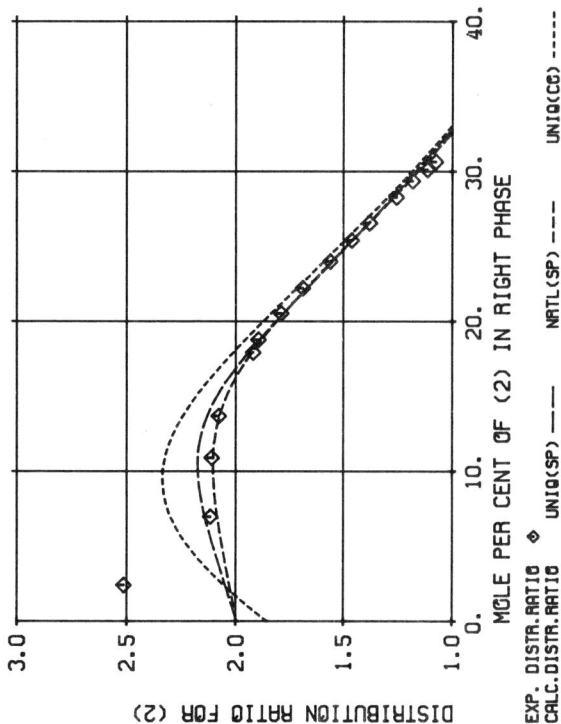

(2) C2H6O   ETHANOL

(3) H2O   WATER

YI-CHUNG CHANG, MOULTON R.W.
IND.ENG.CHEM. 45(1953)2350

TEMPERATURE = 25.0 DEG C   TYPE OF SYSTEM = 1

EXPERIMENTAL TIE LINES IN MOLE PCT

	LEFT PHASE			RIGHT PHASE	
(1)	(2)	(3)	(1)	(2)	(3)
89.450	6.019	4.531	0.036	2.394	97.570
75.585	14.740	9.676	0.094	6.963	92.943
61.570	22.959	15.470	0.155	10.881	88.964
55.531	28.396	16.072	0.212	13.663	86.125
41.250	34.374	24.377	0.526	17.905	81.569
37.717	35.553	26.730	0.649	18.765	80.586
32.714	36.719	30.567	0.954	20.503	78.544
27.245	37.532	35.615	1.265	22.197	76.537
23.245	37.448	39.307	1.859	23.980	74.161
19.875	37.128	42.997	2.549	25.398	72.053
17.368	36.665	45.967	3.267	26.565	70.168
14.184	35.639	50.178	4.366	28.265	67.369
12.294	34.779	52.926	5.214	29.335	65.452
10.703	33.610	55.688	5.972	30.090	63.938
9.648	33.002	57.350	6.518	30.619	62.863

SPECIFIC MODEL PARAMETERS IN KELVIN

		UNIQUAC		NRTL(ALPHA=.2)	
I	J	AIJ	AJI	AIJ	AJI
1	2	417.94	-231.81	332.60	-182.18
1	3	681.87	-99.139	600.20	2498.6
2	3	112.94	-133.43	-196.40	501.17

R1 = 5.5010   R2 = 2.1055   R3 = 0.9200
Q1 = 4.732   Q2 = 1.972   Q3 = 1.400

MEAN DEV. BETWEEN CALC. AND EXP. CONC. IN MOLE PCT

UNIQUAC (SPECIFIC PARAMETERS)   0.65
NRTL   (SPECIFIC PARAMETERS)   0.52
UNIQUAC (COMMON PARAMETERS)   0.87

C₂H₆O-C₇H₁₆

(1) H2O   WATER
(2) C2H6O   ETHANOL
(3) C7H16   HEPTANE

SCHWEPPE J.L., LORAH J.R.
IND.ENG.CHEM. 46(1954)2391

TEMPERATURE = 30.0 DEG C   TYPE OF SYSTEM = 1

EXPERIMENTAL TIE LINES IN MOLE PCT (GRAPH.INTERPOL.)

	LEFT PHASE			RIGHT PHASE	
(1)	(2)	(3)	(1)	(2)	(3)
95.541	4.436	0.023	0.236	6.088	93.675
86.195	13.785	0.020	0.251	6.694	93.054
74.927	25.033	0.040	0.341	10.244	89.416
79.813	20.160	0.027	0.341	10.244	89.416

SPECIFIC MODEL PARAMETERS IN KELVIN

		UNIQUAC		NRTL(ALPHA=.2)	
I	J	AIJ	AJI	AIJ	AJI
1	2	-247.73	208.72	-208.55	440.77
1	3	-292.78	649.98	2336.7	841.13
2	3	-59.141	381.49	332.88	294.82

R1 = 0.9200   R2 = 2.1055   R3 = 5.1742
Q1 = 1.400   Q2 = 1.972   Q3 = 4.396

MEAN DEV. BETWEEN CALC. AND EXP. CONC. IN MOLE PCT

UNIQUAC (SPECIFIC PARAMETERS)   2.48
NRTL   (SPECIFIC PARAMETERS)   2.04
UNIQUAC (COMMON PARAMETERS)   3.69

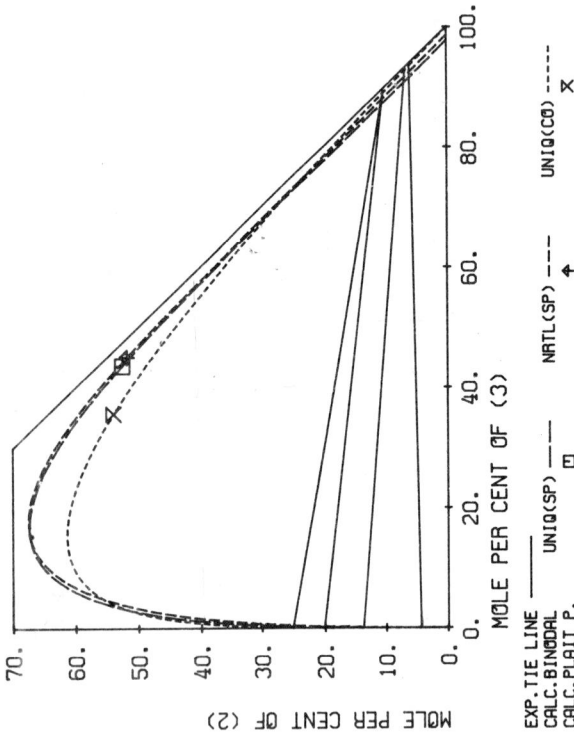

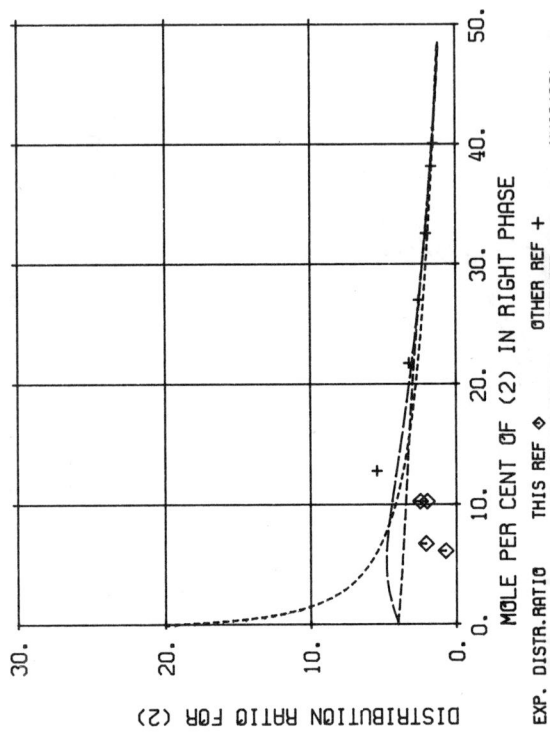

MOLE PER CENT OF (3)

MOLE PER CENT OF (2)

EXP.TIE LINE —— UNIQ(SP) —— ⊡ NRTL(SP) —— ✦ UNIQ(CO) ---- ✗
CALC.BINODAL
CALC.PLAIT P.

MOLE PER CENT OF (2) IN RIGHT PHASE

DISTRIBUTION RATIO FOR (2)

EXP. DISTR.RATIO THIS REF ◇ OTHER REF +
CALC.DISTR.RATIO UNIQ(SP) —— NRTL(SP) --- UNIQ(CO) ----

(2) C2H6O   ETHANOL

(3) C7H16   HEPTANE

VOROBEVA A.I., KARAPETYANTS M.KH.
ZH.FIZ.KHIM. 40(1966)3018

TEMPERATURE = 25.0 DEG C   TYPE OF SYSTEM = 1

EXPERIMENTAL TIE LINES IN MOLE PCT

	LEFT PHASE			RIGHT PHASE	
(1)	(2)	(3)	(1)	(2)	(3)
22.230	69.156	8.614	1.533	12.785	85.683
16.996	69.521	13.483	2.399	21.768	75.833
14.124	67.740	18.136	3.223	27.005	69.772
12.143	65.645	22.212	4.384	32.570	63.046
10.838	62.573	26.589	5.024	38.147	56.828
10.395	61.225	28.381	5.736	40.053	54.211

SPECIFIC MODEL PARAMETERS IN KELVIN

		UNIQUAC		NRTL(ALPHA=.2)	
I J	AIJ	AJI		AIJ	AJI
1 2	-247.73	208.72		-208.55	440.77
1 3	292.78	649.98		2336.7	841.13
2 3	-59.141	381.49		332.88	294.82

R1 = 0.9200   R2 = 2.1055   R3 = 5.1742
Q1 = 1.400   Q2 = 1.972   Q3 = 4.396

MEAN DEV. BETWEEN CALC. AND EXP. CONC. IN MOLE PCT

UNIQUAC (SPECIFIC PARAMETERS)   2.72
NRTL   (SPECIFIC PARAMETERS)   3.44
UNIQUAC (COMMON PARAMETERS)   6.80

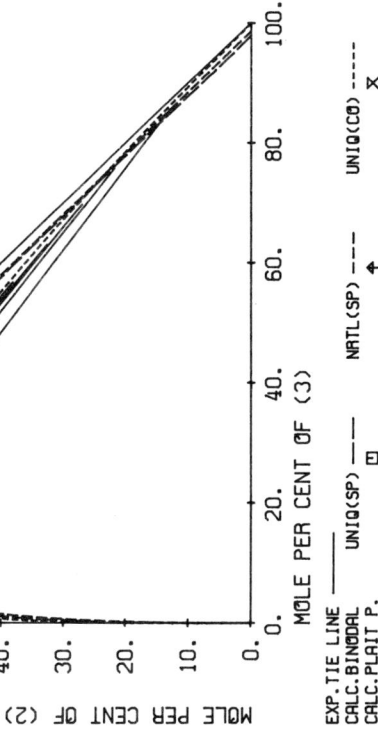

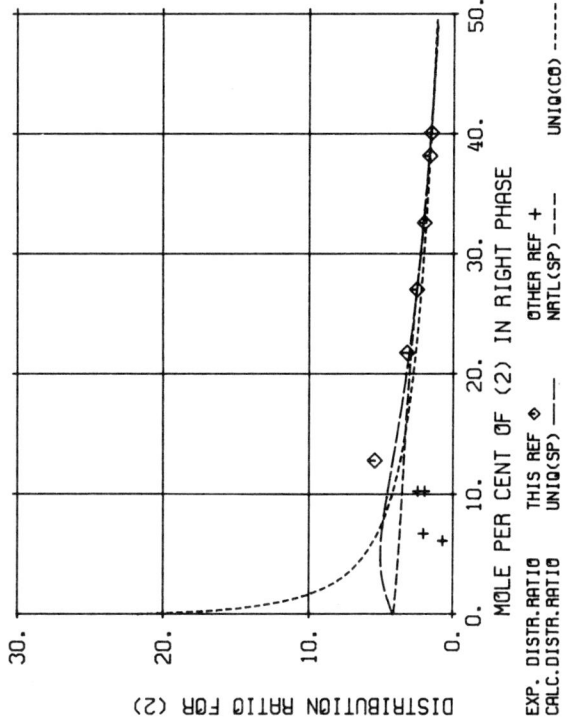

(1) H2O   WATER

(2) C2H6O   ETHANOL

(3) C8H10   BENZENE,1,2-DIMETHYL

SEJONG NAM,TOYOHIKO HAYAKAWA, SHIGEFUMI FUJITA
J.CHEM.ENG.JPN. 5(1972)327

TEMPERATURE =   25.0 DEG C   TYPE OF SYSTEM = 1

EXPERIMENTAL TIE LINES IN MOLE PCT

| | LEFT PHASE | | | RIGHT PHASE | |
(1)	(2)	(3)	(1)	(2)	(3)
80.853	18.970	0.177	1.682	6.138	92.180
76.776	22.945	0.279	1.663	8.022	90.315
73.129	26.409	0.462	1.649	9.458	88.893
73.105	26.457	0.438	1.649	9.458	88.893
68.610	30.619	0.771	2.178	10.222	87.600
68.233	31.019	0.748	2.160	11.614	86.226
62.403	36.156	1.441	2.142	12.984	84.874
57.327	40.337	2.336	3.657	14.707	81.636
51.569	44.804	3.627	3.611	16.741	79.648

SPECIFIC MODEL PARAMETERS IN KELVIN

| | | UNIQUAC | | NRTL(ALPHA=.2) | |
I	J	AIJ	AJI	AIJ	AJI
1	2	-60.089	-43.928	245.81	-164.80
1	3	194.07	690.47	1903.9	763.13
2	3	-62.052	297.42	361.05	155.91

R1 = 0.9200   R2 = 2.1055   R3 = 4.6578
Q1 = 1.400   Q2 = 1.972   Q3 = 3.536

MEAN DEV. BETWEEN CALC. AND EXP. CONC. IN MOLE PCT

UNIQUAC (SPECIFIC PARAMETERS)   0.38
NRTL   (SPECIFIC PARAMETERS)   0.35
UNIQUAC (COMMON PARAMETERS)   3.74

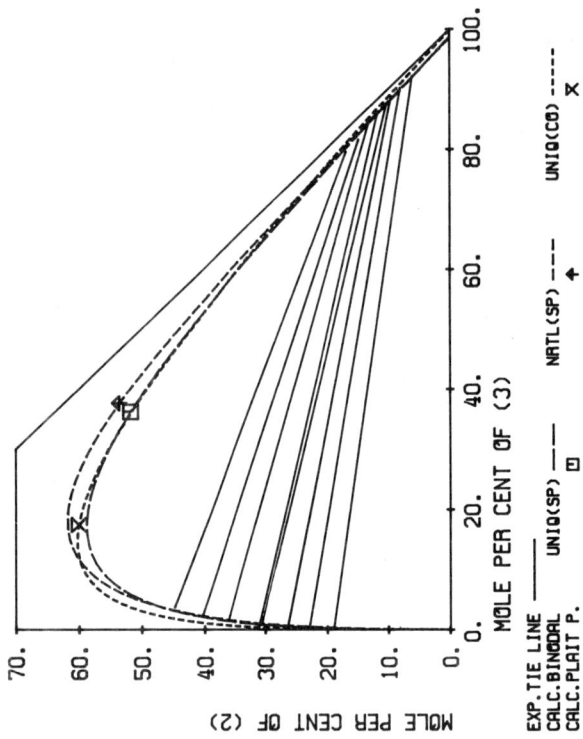

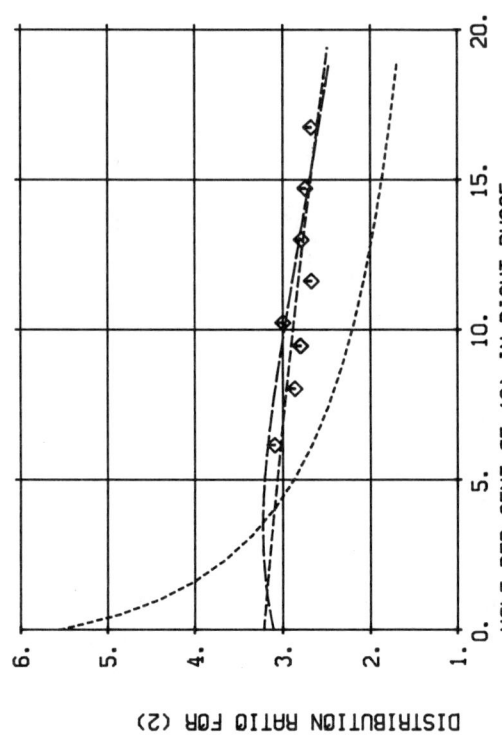

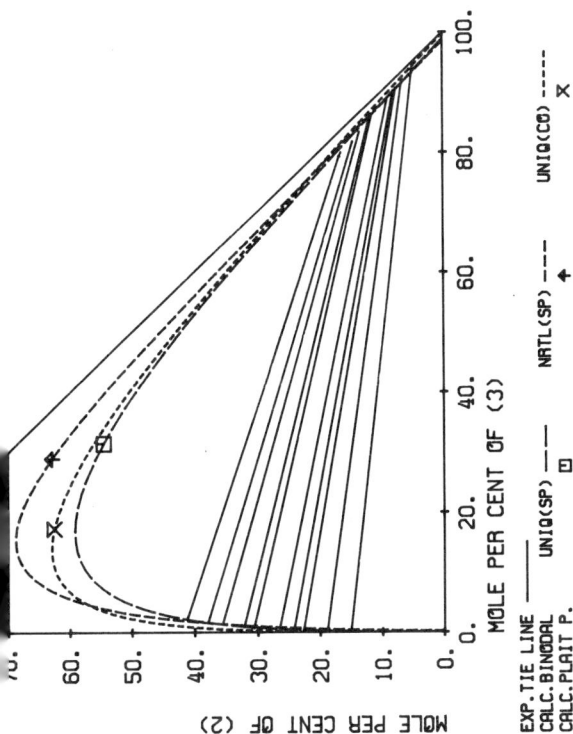

MOLE PER CENT OF (2)

MOLE PER CENT OF (3)

EXP.TIE LINE   UNIQ(SP) ⊡   NRTL(SP) --- ♠   UNIQ(CD) -----
CALC.BINODAL
CALC.PLAIT P.   ✕

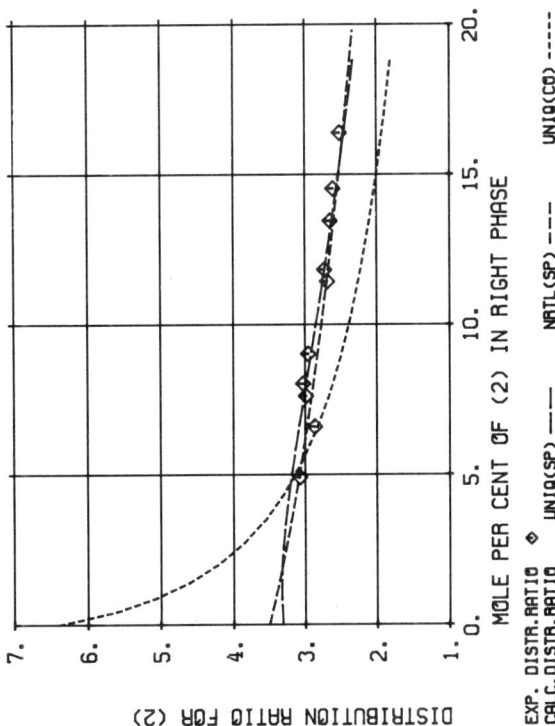

DISTRIBUTION RATIO FOR (2)

MOLE PER CENT OF (2) IN RIGHT PHASE

EXP. DISTR.RATIO ◇   UNIQ(SP) ◇   NRTL(SP) ---   UNIQ(CD) -----
CALC.DISTR.RATIO

---

(2) C2H6O   ETHANOL

(3) C8H10   BENZENE,1,3-DIMETHYL

SEJONG NAM,TOYOHIKO HAYAKAWA, SHIGEFUMI FUJITA
J.CHEM.ENG.JPN. 5(1972)327

TEMPERATURE = 25.0 DEG C   TYPE OF SYSTEM = 1

EXPERIMENTAL TIE LINES IN MOLE PCT

	LEFT PHASE			RIGHT PHASE	
(1)	(2)	(3)	(1)	(2)	(3)
84.852	15.064	0.084	0.570	4.905	94.525
80.962	18.883	0.155	1.124	6.591	92.285
77.069	22.723	0.208	1.667	7.607	90.726
75.389	24.303	0.307	1.663	8.022	90.315
72.946	26.615	0.439	2.194	9.009	88.797
68.629	30.679	0.693	2.163	11.417	86.421
66.950	32.212	0.838	2.158	11.811	86.031
63.094	35.616	1.290	3.174	13.445	83.382
60.470	37.886	1.644	3.661	14.520	81.819
56.434	41.179	2.386	3.619	16.375	80.006

SPECIFIC MODEL PARAMETERS IN KELVIN

		UNIQUAC		NRTL(ALPHA=.2)	
I J	AIJ	AJI		AIJ	AJI
1 2	-106.74	-48.208		87.009	-136.32
1 3	-195.28	722.34		1464.2	745.65
2 3	-53.862	244.50		623.79	-73.710

R1 = 0.9200   R2 = 2.1055   R3 = 4.6578
Q1 = 1.400   Q2 = 1.972   Q3 = 3.536

MEAN DEV. BETWEEN CALC. AND EXP. CONC. IN MOLE PCT

UNIQUAC (SPECIFIC PARAMETERS)   0.32
NRTL   (SPECIFIC PARAMETERS)   0.46
UNIQUAC (COMMON PARAMETERS)   2.24

$C_2H_6O-C_8H_{10}$

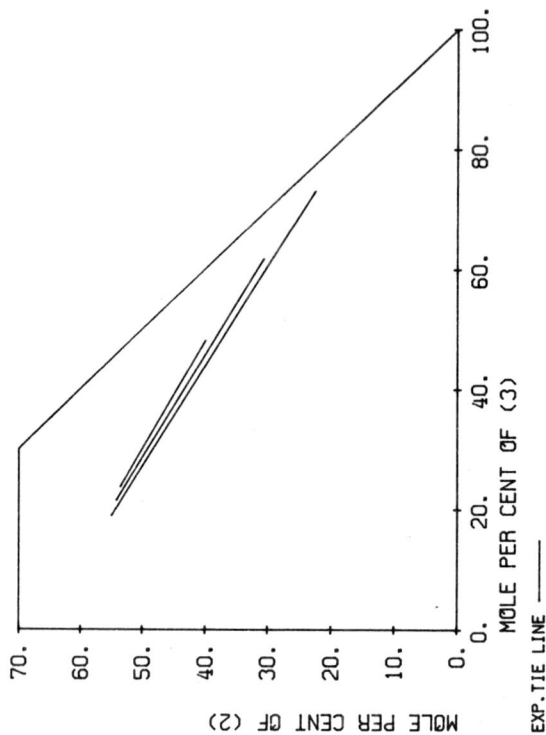

MOLE PER CENT OF (3) ——

EXP.TIE LINE ———

MOLE PER CENT OF (2)

(1) H2O   WATER

(2) C2H6O   ETHANOL

(3) C8H10   BENZENE,1,4-DIMETHYL

BONNER W.D.
J.PHYS.CHEM. 14(1910)738

TEMPERATURE = 15.0 DEG C   TYPE OF SYSTEM = 1

EXPERIMENTAL TIE LINES IN MOLE PCT (GRAPH.INTERPOL.)

LEFT PHASE			RIGHT PHASE		
(1)	(2)	(3)	(1)	(2)	(3)
26.146	54.928	18.926	4.432	22.532	73.036
24.331	54.141	21.527	7.647	30.606	61.747
22.758	53.513	23.729	11.929	39.962	48.108

$C_2H_6O-C_8H_{10}$

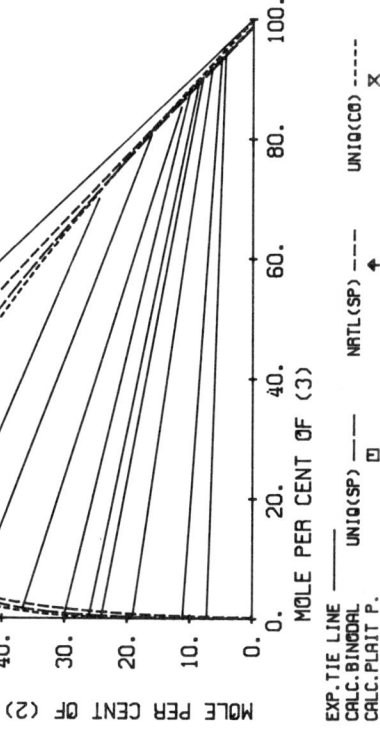

MOLE PER CENT OF (2)

MOLE PER CENT OF (3)

EXP.TIE LINE
CALC.BINODAL
CALC.PLAIT P.

UNIQ(SP) ———   NRTL(SP) ———   UNIQ(C6) ------

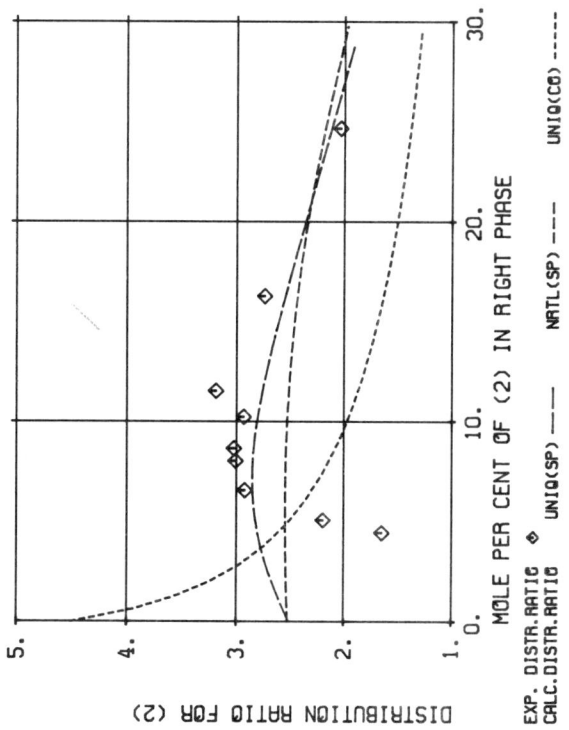

DISTRIBUTION RATIO FOR (2)

MOLE PER CENT OF (2) IN RIGHT PHASE

EXP. DISTR.RATIO ◇   UNIQ(SP) ———   NRTL(SP) ———   UNIQ(C6) ------
CALC. DISTR.RATIO

(2) C2H6O   ETHANOL
(3) C8H10   BENZENE, 1,4-DIMETHYL

SEJONG NAM, TOYOHIKO HAYAKAWA, SHIGEFUMI FUJITA
J.CHEM.ENG.JPN. 5(1972)327

TEMPERATURE = 25.0 DEG C   TYPE OF SYSTEM = 1

EXPERIMENTAL TIE LINES IN MOLE PCT

LEFT PHASE			RIGHT PHASE		
(1)	(2)	(3)	(1)	(2)	(3)
92.679	7.283	0.038	1.699	4.429	93.873
88.809	11.131	0.060	1.692	5.074	93.234
80.717	19.172	0.111	1.678	6.561	91.762
75.592	24.125	0.283	1.663	8.022	90.315
73.510	26.127	0.363	1.657	8.640	89.702
69.368	29.996	0.636	2.178	10.222	87.600
61.757	36.706	1.537	3.211	11.511	85.278
51.740	44.570	3.690	3.119	16.262	80.619
42.526	50.222	7.252	5.293	24.650	70.057

SPECIFIC MODEL PARAMETERS IN KELVIN

	UNIQUAC		NRTL(ALPHA=.2)	
I J	AIJ	AJI	AIJ	AJI
1 2	-155.63	111.33	203.44	-48.234
1 3	207.17	679.14	1409.3	809.54
2 3	-93.552	346.88	331.92	183.00

R1 = 0.9200   R2 = 2.1055   R3 = 4.6578
Q1 = 1.400   Q2 = 1.972   Q3 = 3.536

MEAN DEV. BETWEEN CALC. AND EXP. CONC. IN MOLE PCT

UNIQUAC (SPECIFIC PARAMETERS)   0.79
NRTL   (SPECIFIC PARAMETERS)   1.16
UNIQUAC (COMMON PARAMETERS)   5.12

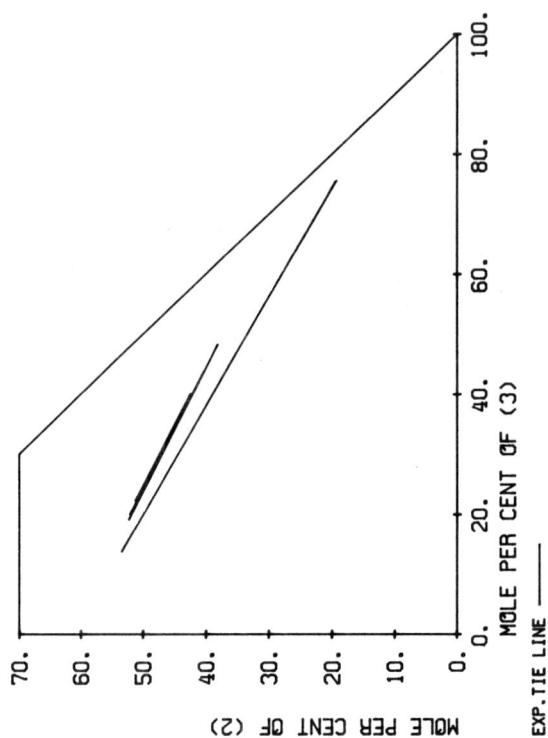

$C_2H_6O-C_8H_{10}O$

(1) H2O   WATER

(2) C2H6O   ETHANOL

(3) C8H100   BENZENE,ETHOXY

BONNER W.D.
J.PHYS.CHEM. 14(1910)738

TEMPERATURE =   0.0 DEG C   TYPE OF SYSTEM = 1

EXPERIMENTAL TIE LINES IN MOLE PCT.(GRAPH.INTERPOL.)

LEFT PHASE			RIGHT PHASE		
(1)	(2)	(3)	(1)	(2)	(3)
32.856	53.345	13.798	5.104	19.294	75.601
28.628	52.203	19.169	13.599	38.081	48.320
27.916	52.027	20.057	15.816	40.363	43.821
26.539	51.164	22.297	17.497	42.452	40.051

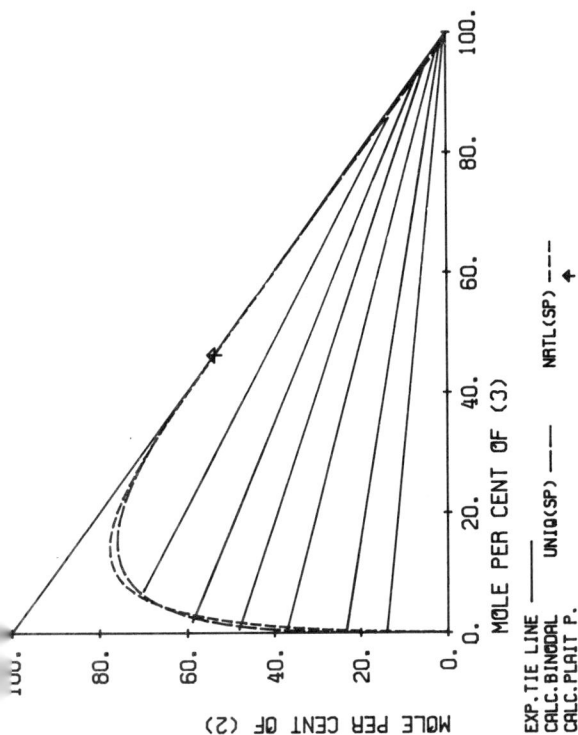

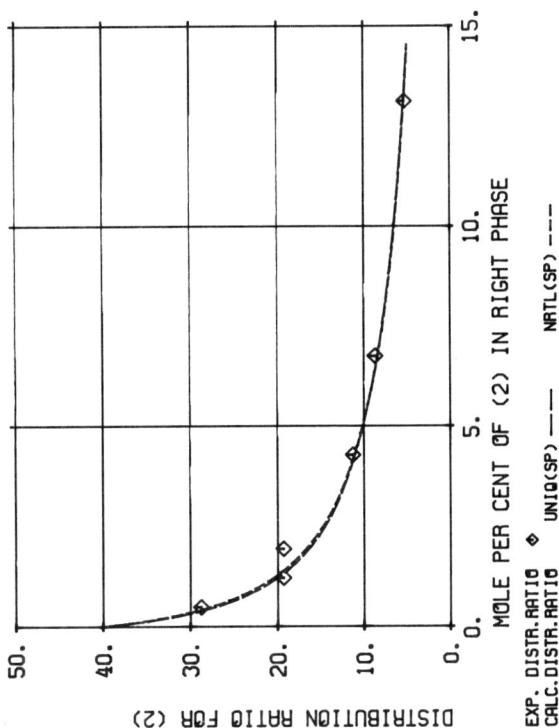

$C_2H_6O-C_8H_{16}$

(2) C2H6O   ETHANOL

(3) C8H16   1-OCTENE

NOWAKOWSKA J., KRETSCHMER C.B., WIEBE R.
J.CHEM.ENG.DATA 1(1956)42

TEMPERATURE =   0.0 DEG C   TYPE OF SYSTEM = 1

EXPERIMENTAL TIE LINES IN MOLE PCT

	LEFT PHASE			RIGHT PHASE	
(1)	(2)	(3)	(1)	(2)	(3)
86.021	13.960	0.020	0.0	0.486	99.514
76.618	23.338	0.044	0.0	1.209	98.791
62.679	37.168	0.153	0.000	1.927	98.073
51.305	48.066	0.629	0.000	4.274	95.726
39.282	58.843	1.876	0.595	6.748	92.657
23.069	69.958	6.973	1.138	13.123	85.740

SPECIFIC MODEL PARAMETERS IN KELVIN

		UNIQUAC		NRTL(ALPHA=.2)	
I J	AIJ	AJI		AIJ	AJI
1 2	-232.00	-115.35		-416.99	-67.082
1 3	291.86	654.06		1512.9	1133.8
2 3	-54.996	394.56		360.51	255.02

R1 = 0.9200   R2 = 2.1055   R3 = 5.6185
Q1 = 1.400   Q2 = 1.972   Q3 = 4.724

MEAN DEV. BETWEEN CALC. AND EXP. CONC. IN MOLE PCT

UNIQUAC (SPECIFIC PARAMETERS)   0.66
NRTL   (SPECIFIC PARAMETERS)   0.51

$C_2H_6O-C_8H_{16}$

(1) H2O   WATER

(2) C2H6O   ETHANOL

(3) C8H16   1-OCTENE

NOWAKOWSKA J., KRETSCHMER C.B., WIEBE R.
J.CHEM.ENG.DATA 1(1956)42

TEMPERATURE = 25.0 DEG C   TYPE OF SYSTEM = 1

EXPERIMENTAL TIE LINES IN MOLE PCT

	LEFT PHASE			RIGHT PHASE	
(1)	(2)	(3)	(1)	(2)	(3)
94.341	5.659	0.0	0.0	0.0	100.000
86.914	13.086	0.0	0.616	0.964	98.420
76.424	23.576	0.0	0.613	1.679	97.707
69.687	30.194	0.119	1.209	3.309	95.482
52.305	46.869	0.825	1.772	6.237	91.991
38.754	58.612	2.633	2.289	10.519	87.192
32.413	63.025	4.562	2.786	13.944	83.271
23.334	67.342	9.324	4.653	22.236	73.112

SPECIFIC MODEL PARAMETERS IN KELVIN

I J		UNIQUAC			NRTL(ALPHA=.2)	
	AIJ	AJI			AIJ	AJI
1 2	-295.53	-77.397			-661.68	144.04
1 3	332.16	734.25			2582.3	889.88
2 3	-72.250	360.45			359.84	115.75

R1 = 0.9200   R2 = 2.1055   R3 = 5.6185
Q1 = 1.400   Q2 = 1.972   Q3 = 4.724

MEAN DEV. BETWEEN CALC. AND EXP. CONC. IN MOLE PCT

UNIQUAC (SPECIFIC PARAMETERS)	0.89
NRTL (SPECIFIC PARAMETERS)	1.00
UNIQUAC (COMMON PARAMETERS)	1.31

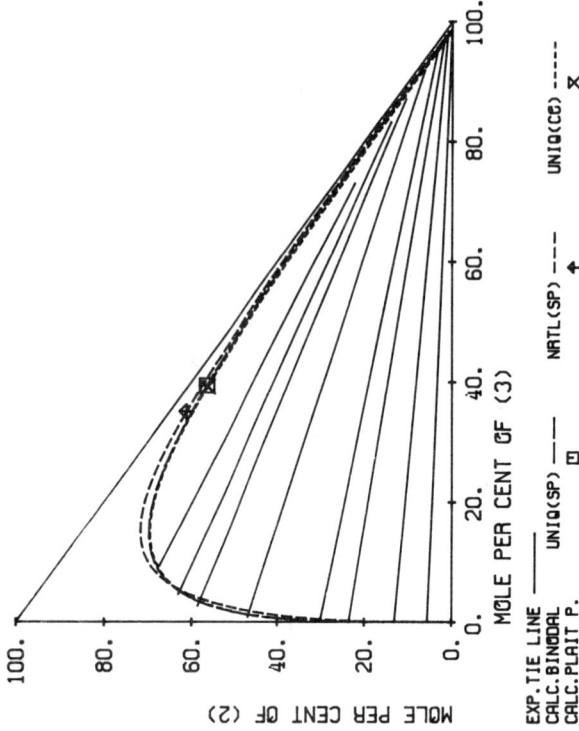

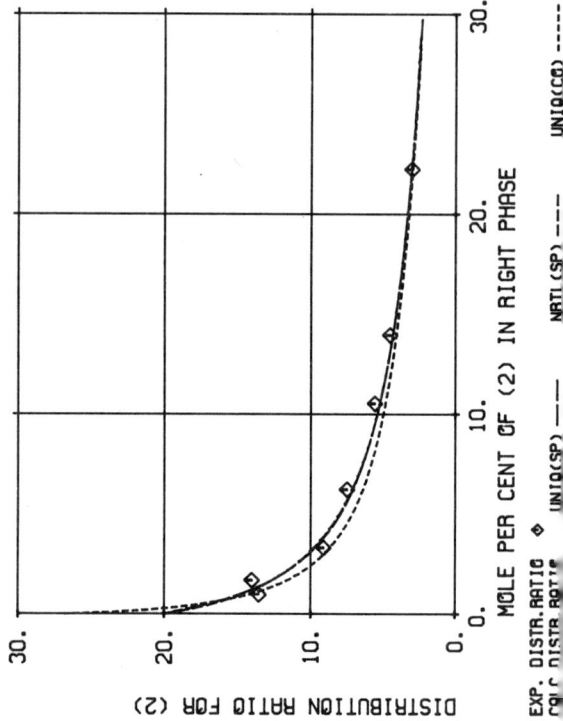

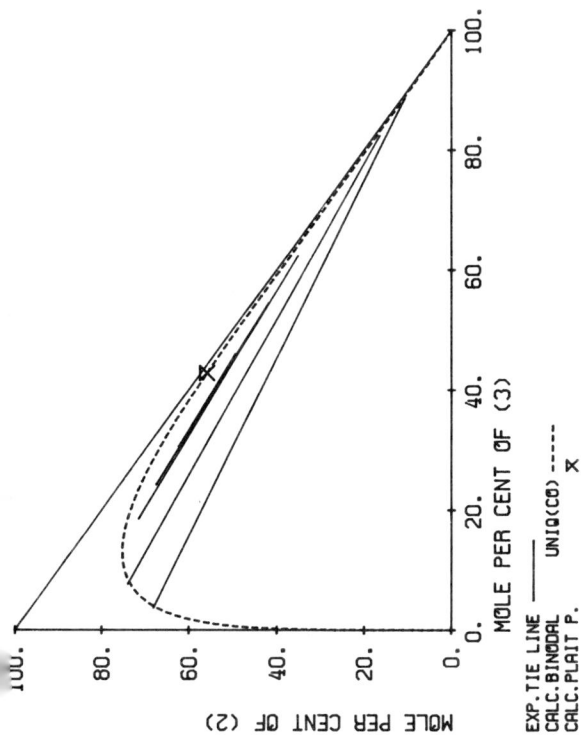

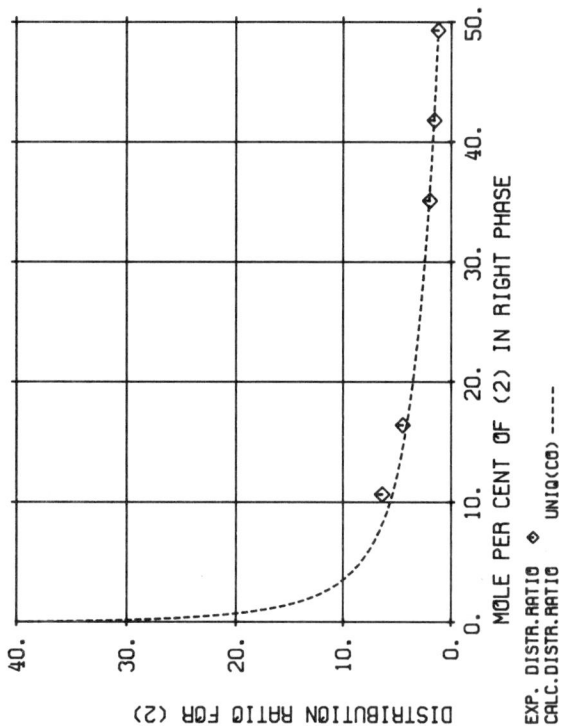

$C_2H_6O-C_8H_{18}$

(2) C2H6O   ETHANOL
(3) C8H18   OCTANE

VOROBEVA A.I., KARAPETYANTS M.KH.
ZH.FIZ.KHIM. 40(1966)3018

TEMPERATURE = 25.0 DEG C   TYPE OF SYSTEM = 1

EXPERIMENTAL TIE LINES IN MOLE PCT

	LEFT PHASE			RIGHT PHASE	
(1)	(2)	(3)	(1)	(2)	(3)
28.239	68.021	3.741	0.591	10.626	88.783
18.436	73.892	7.672	1.132	16.380	82.488
9.942	71.438	18.620	2.441	35.123	62.436
8.363	67.368	24.269	3.652	41.773	54.574
7.202	62.242	30.555	4.651	49.271	46.078

MEAN DEV. BETWEEN CALC. AND EXP. CONC. IN MOLE PCT
UNIQUAC (COMMON PARAMETERS)   2.80

C₂H₆O-C₈H₁₈

(1) H2O   WATER
(2) C2H6O   ETHANOL
(3) C8H18   PENTANE,2,2,4-TRIMETHYL

NOWAKOWSKA J.; KRETSCHMER C.B., WIEBE R.
J.CHEM.ENG.DATA 1(1956)42

TEMPERATURE =   0.0 DEG C   TYPE OF SYSTEM = 1

EXPERIMENTAL TIE LINES IN MOLE PCT

	LEFT PHASE			RIGHT PHASE	
(1)	(2)	(3)	(1)	(2)	(3)
76.667	23.311	0.022	0.0	0.741	99.259
62.112	37.712	0.176	0.0	2.202	97.798
52.097	47.373	0.529	0.0	2.923	97.077
39.263	59.220	1.516	0.609	5.717	93.674
24.271	70.437	5.291	0.588	11.272	88.140
18.878	73.519	7.604	1.143	14.969	83.888
11.829	72.735	15.437	1.624	22.226	76.150

SPECIFIC MODEL PARAMETERS IN KELVIN

		UNIQUAC		NRTL(ALPHA=.2)	
I J	AIJ	AJI	AIJ	AJI
1 2	-328.29	-76.777	-551.97	20.806
1 3	371.02	600.05	1574.2	1013.8
2 3	-70.947	401.34	287.51	294.88

R1 = 0.9200   R2 = 2.1055   R3 = 5.8463
Q1 = 1.400   Q2 = 1.972   Q3 = 5.008

MEAN DEV. BETWEEN CALC. AND EXP. CONC. IN MOLE PCT

UNIQUAC (SPECIFIC PARAMETERS)   0.83
NRTL   (SPECIFIC PARAMETERS)   0.94

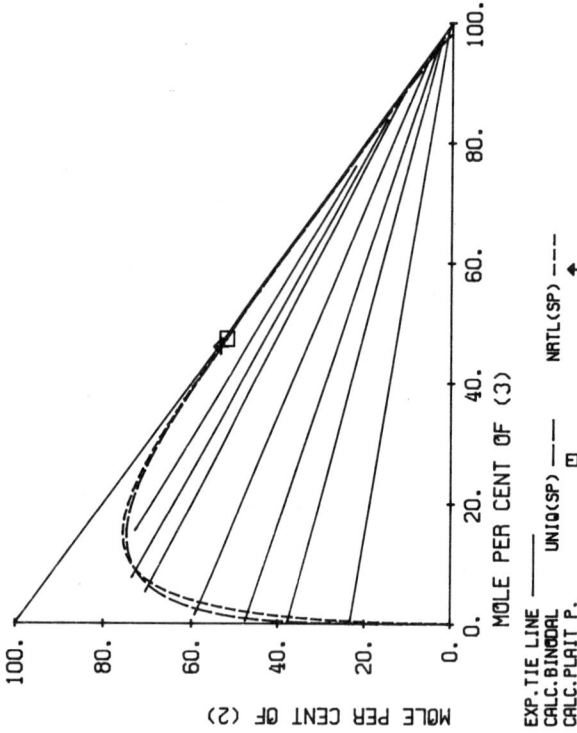

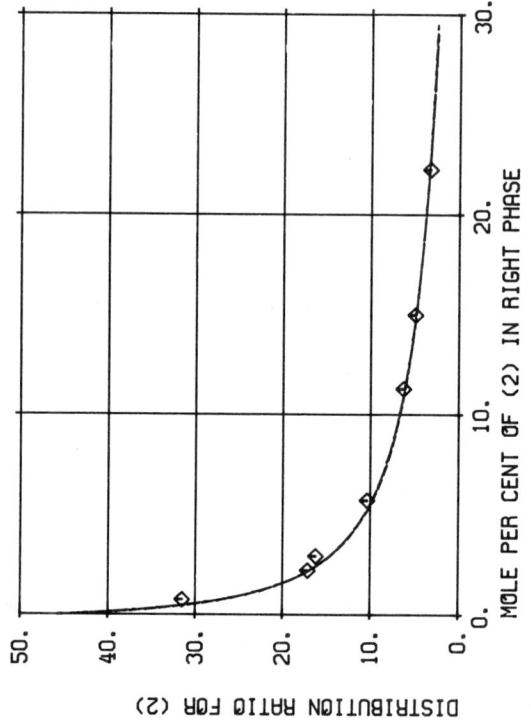

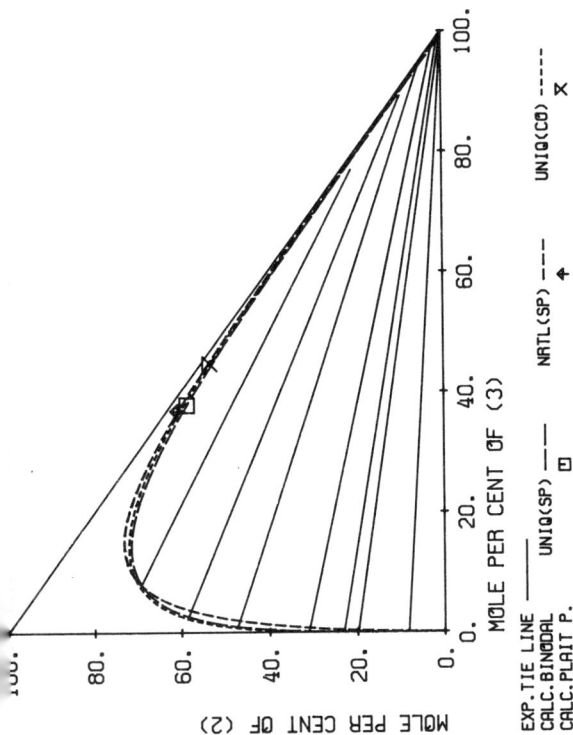

MOLE PER CENT OF (3)

MOLE PER CENT OF (2)

EXP.TIE LINE ——   UNIQ(SP) —— ☐   NRTL(SP) —— ✦   UNIQ(CO) ---- ✕
CALC.BINODAL
CALC.PLAIT P.

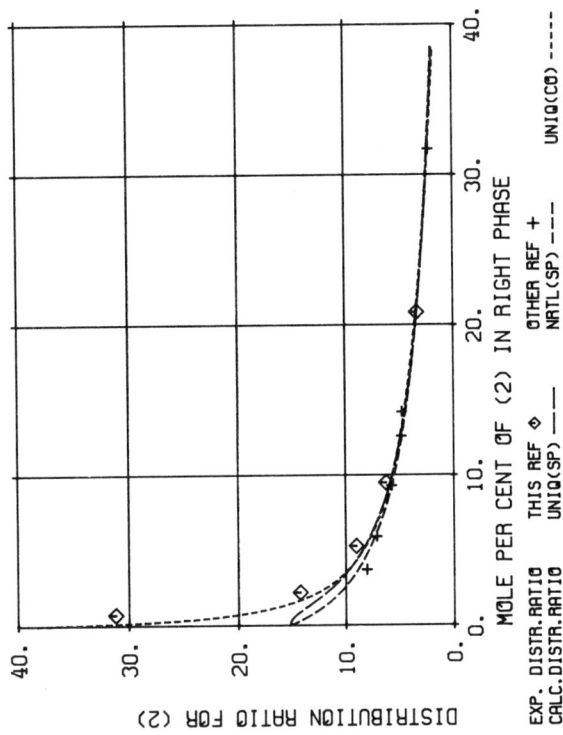

MOLE PER CENT OF (2) IN RIGHT PHASE

DISTRIBUTION RATIO FOR (2)

EXP. DISTR.RATIO   THIS REF ◇   OTHER REF +
CALC.DISTR.RATIO   UNIQ(SP) ——   NRTL(SP) ---   UNIQ(CO) -----

(2) C2H6O   ETHANOL
(3) C8H18   PENTANE,2,2,4-TRIMETHYL

NOWAKOWSKA J., KRETSCHMER C.B., WIEBE R.
J.CHEM.ENG.DATA 1(1956)42

TEMPERATURE = 25.0 DEG C   TYPE OF SYSTEM = 1

EXPERIMENTAL TIE LINES IN MOLE PCT

	LEFT PHASE			RIGHT PHASE	
(1)	(2)	(3)	(1)	(2)	(3)
80.091	19.888	0.021	0.0	0.0	100.000
91.748	8.252	0.0	0.0	0.0	100.000
76.813	23.101	0.086	0.0	0.741	99.259
63.744	31.138	0.118	0.622	2.191	97.187
51.928	47.287	0.785	0.611	5.256	94.133
39.073	58.649	2.279	1.184	9.489	89.327
22.666	69.033	8.301	2.175	20.833	76.992

SPECIFIC MODEL PARAMETERS IN KELVIN

I J	UNIQUAC AIJ	AJI	NRTL(ALPHA=.2) AIJ	AJI
1 2	-385.26	41.721	-604.94	304.48
1 3	407.57	836.57	1750.6	1218.3
2 3	-81.031	363.16	326.39	201.15

R1 = 0.9200   R2 = 2.1055   R3 = 5.8463
Q1 = 1.400   Q2 = 1.972   Q3 = 5.003

MEAN DEV. BETWEEN CALC. AND EXP. CONC. IN MOLE PCT

UNIQUAC (SPECIFIC PARAMETERS)   0.80
NRTL   (SPECIFIC PARAMETERS)   0.96
UNIQUAC (COMMON PARAMETERS)   0.72

$C_2H_6O-C_8H_{18}$

HUBER J.F.K., MEIJERS C.A.M., HULSMAN J.A.R.J.
ANAL.CHEM. 44(1972)111

(1) H2O   WATER

(2) C2H6O   ETHANOL

(3) C8H18   PENTANE,2,2,4-TRIMETHYL

TEMPERATURE = 25.0 DEG C   TYPE OF SYSTEM = 1

EXPERIMENTAL TIE LINES IN MOLE PCT

	LEFT PHASE			RIGHT PHASE	
(1)	(2)	(3)	(1)	(2)	(3)
70.200	29.600	0.200	0.200	3.700	96.100
58.100	41.600	0.300	0.600	5.900	93.500
46.200	52.900	0.900	0.700	9.300	90.000
37.800	60.100	2.100	0.900	12.600	86.500
27.800	67.100	5.100	1.100	14.200	84.700
16.700	70.000	13.300	2.500	31.700	65.800

SPECIFIC MODEL PARAMETERS IN KELVIN

		UNIQUAC		NRTL(ALPHA=.2)	
I J	AIJ	AJI		AIJ	AJI
1 2	-385.26	41.721		-604.94	304.48
1 3	407.57	836.57		1750.6	1218.3
2 3	-81.031	363.16		326.39	201.15

R1 = 0.9200   R2 = 2.1055   R3 = 5.8463
Q1 = 1.400   Q2 = 1.972   Q3 = 5.008

MEAN DEV. BETWEEN CALC. AND EXP. CONC. IN MOLE PCT

UNIQUAC (SPECIFIC PARAMETERS)   0.98
NRTL   (SPECIFIC PARAMETERS)   1.26
UNIQUAC (COMMON PARAMETERS)   1.57

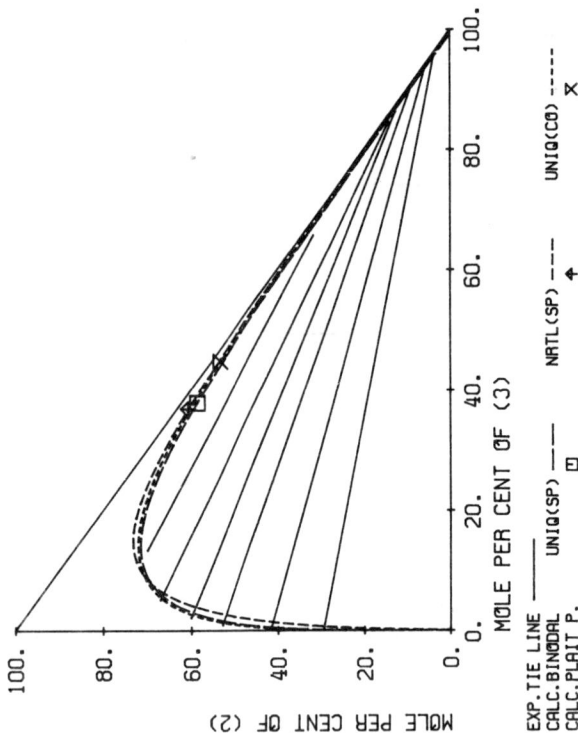

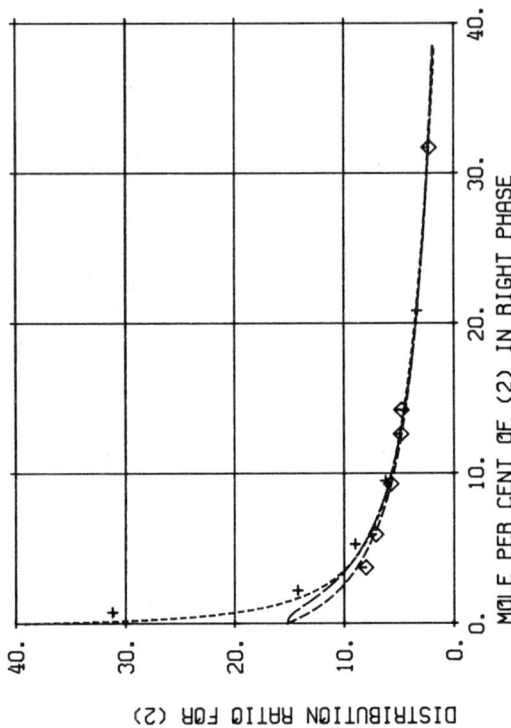

$C_2H_6O-C_8H_{18}O$

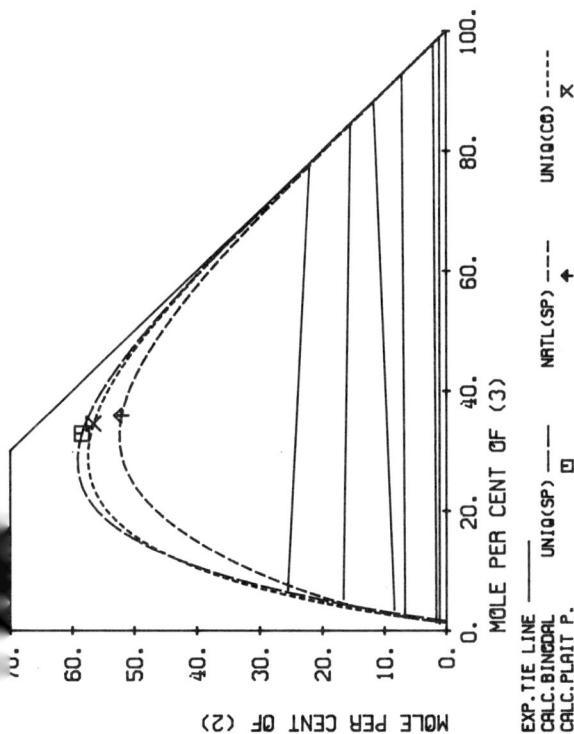

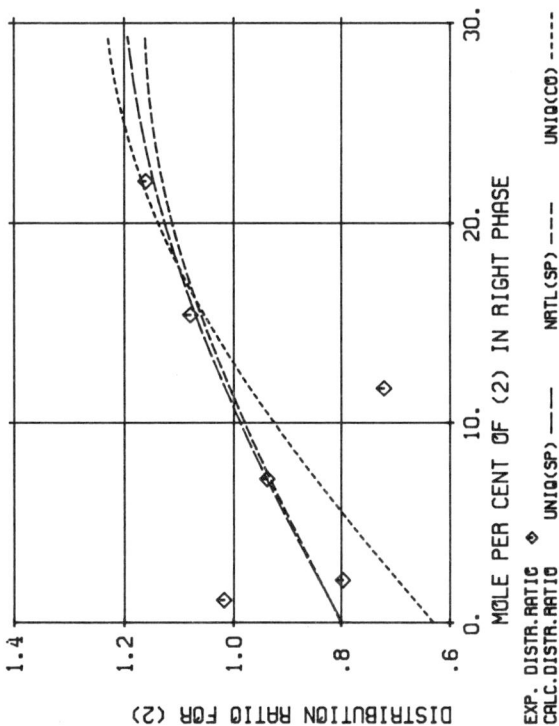

(2) C2H6O	ETHANOL	
(3) H2O	WATER	

OTHMER D.F., WHITE R.E., TRUEGER E.
IND.ENG.CHEM. 33(1941)1240

TEMPERATURE = 25.5 DEG C   TYPE OF SYSTEM = 1

EXPERIMENTAL TIE LINES IN MOLE PCT (GRAPH.INTERPOL.)

LEFT PHASE			RIGHT PHASE		
(1)	(2)	(3)	(1)	(2)	(3)
97.689	1.139	1.172	0.013	1.119	98.869
96.937	1.685	1.378	0.013	2.114	97.873
90.271	6.735	2.994	0.021	7.182	92.797
88.064	8.461	3.474	0.041	11.725	88.234
78.108	16.595	5.297	0.080	15.377	84.542
67.573	25.613	6.814	0.208	22.069	77.723

SPECIFIC MODEL PARAMETERS IN KELVIN

		UNIQUAC		NRTL(ALPHA=.2)	
I	J	AIJ	AJI	AIJ	AJI
1	2	-35.312	-3.2762	-616.26	732.81
1	3	710.88	335.18	809.55	2066.2
2	3	-160.08	-76.818	-440.10	320.85

R1 = 6.0925   R2 = 2.1055   R3 = 0.9200
Q1 = 5.176   Q2 = 1.972   Q3 = 1.400

MEAN DEV. BETWEEN CALC. AND EXP. CONC. IN MOLE PCT

UNIQUAC (SPECIFIC PARAMETERS)   0.58
NRTL   (SPECIFIC PARAMETERS)   0.56
UNIQUAC (COMMON PARAMETERS)   0.57

$C_2H_6O-C_9H_{10}O_2$

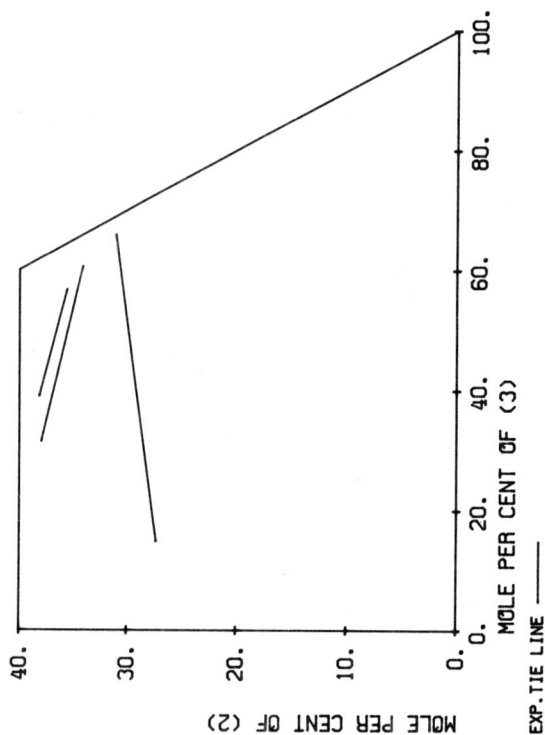

MOLE PER CENT OF (3)

MOLE PER CENT OF (2)

EXP.TIE LINE ———

(1) C9H10O2   ACETIC ACID,BENZYL ESTER
(2) C2H6O   ETHANOL
(3) H2O   WATER

BONNER W.D.
J.PHYS.CHEM. 14(1910)738

TEMPERATURE =   0.0 DEG C   TYPE OF SYSTEM = 1

EXPERIMENTAL TIE LINES IN MOLE PCT (GRAPH.INTERPOL.)

LEFT PHASE			RIGHT PHASE		
(1)	(2)	(3)	(1)	(2)	(3)
57.942	27.300	14.758	2.936	31.109	65.955
30.598	37.948	31.453	5.269	34.124	60.607
22.895	38.148	38.957	7.669	35.596	56.735

$C_2H_6O\text{-}C_9H_{20}$

MOLE PER CENT OF (3)

EXP.TIE LINE ——— UNIQ(CO) ----
CALC.BINODAL
CALC.PLAIT P. ⟡

MOLE PER CENT OF (2)

MOLE PER CENT OF (2) IN RIGHT PHASE

EXP. DISTR.RATIO ⟡ UNIQ(CO) -----
CALC.DISTR.RATIO

DISTRIBUTION RATIO FOR (2)

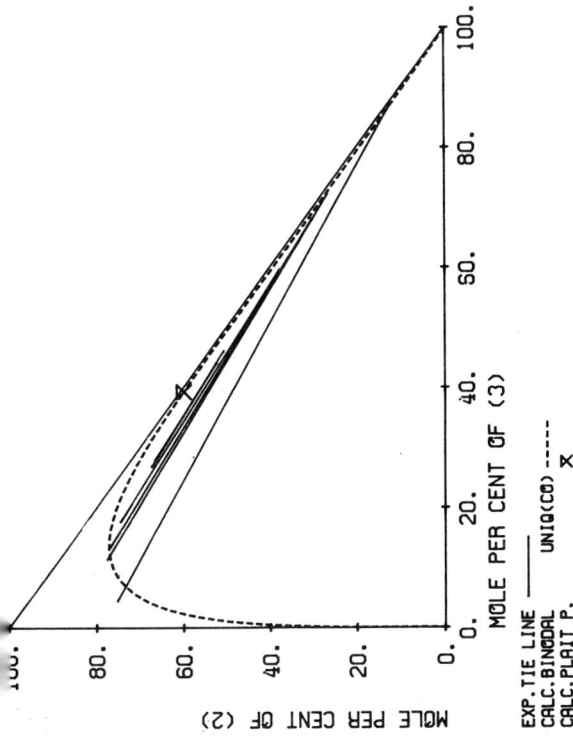

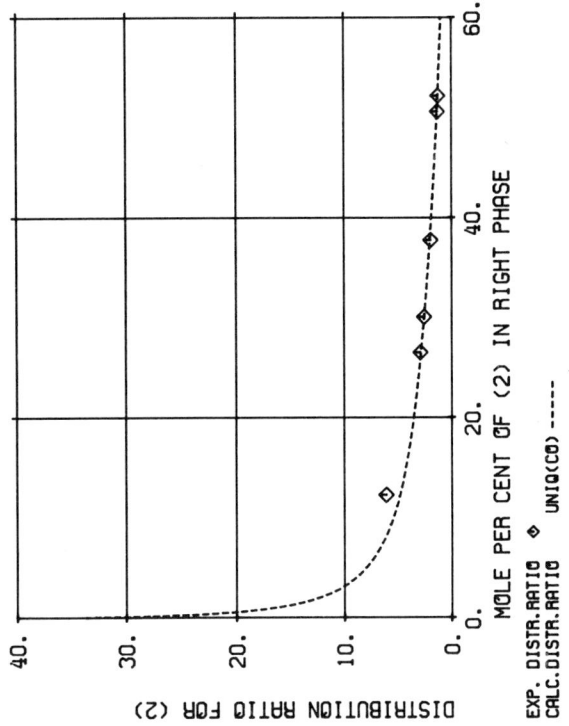

(2) C2H6O   ETHANOL

(3) C9H20   NONANE

VOROBEVA A.I., KARAPETYANTS M.KH.
ZH.FIZ.KHIM. 40(1966)3018

TEMPERATURE = 25.0 DEG C   TYPE OF SYSTEM = 1

EXPERIMENTAL TIE LINES IN MOLE PCT

LEFT PHASE			RIGHT PHASE		
(1)	(2)	(3)	(1)	(2)	(3)
20.306	75.193	4.501	0.652	12.240	87.108
11.046	77.522	11.432	1.168	26.489	72.343
9.672	76.827	13.501	1.694	30.026	68.280
7.782	74.549	17.669	2.620	37.698	59.682
5.905	67.246	26.850	3.231	50.541	46.228
5.616	66.463	27.922	3.616	52.137	44.247

MEAN DEV. BETWEEN CALC. AND EXP. CONC. IN MOLE PCT

UNIQUAC (COMMON PARAMETERS)   2.72

C_2H_6O-$C_{12}H_{10}O$

PURNELL J.H., BOWDEN S.T.
J.CHEM.SOC.(1954)539

(1) H2O   WATER
(2) C2H6O   ETHANOL
(3) C12H10O   ETHER,DIPHENYL

TEMPERATURE = 25.0 DEG C   TYPE OF SYSTEM = 1

EXPERIMENTAL TIE LINES IN MOLE PCT

	LEFT PHASE			RIGHT PHASE	
(1)	(2)	(3)	(1)	(2)	(3)
91.175	8.813	0.012	0.0	1.462	98.538
84.366	15.568	0.066	0.0	3.593	96.402
76.000	23.883	0.117	0.0	4.984	95.016
63.400	36.382	0.218	0.0	8.000	91.998
58.264	41.454	0.283	0.865	10.488	88.646
48.582	50.117	1.300	0.859	11.419	87.722
44.104	54.080	1.816	1.664	14.317	84.019
29.550	65.732	4.717	3.085	21.412	75.503

SPECIFIC MODEL PARAMETERS IN KELVIN

		UNIQUAC		NRTL(ALPHA=.2)	
I	J	AIJ	AJI	AIJ	AJI
1	2	-95.056	-29.371	-109.89	56.891
1	3	285.44	790.35	1977.0	1149.6
2	3	-101.68	411.47	524.26	110.48

R1 = 0.9200   R2 = 2.1055   R3 = 6.2873
Q1 = 1.400   Q2 = 1.972   Q3 = 4.480

MEAN DEV. BETWEEN CALC. AND EXP. CONC. IN MOLE PCT

UNIQUAC (SPECIFIC PARAMETERS)   0.56
NRTL   (SPECIFIC PARAMETERS)   0.66
UNIQUAC (COMMON PARAMETERS)   2.58

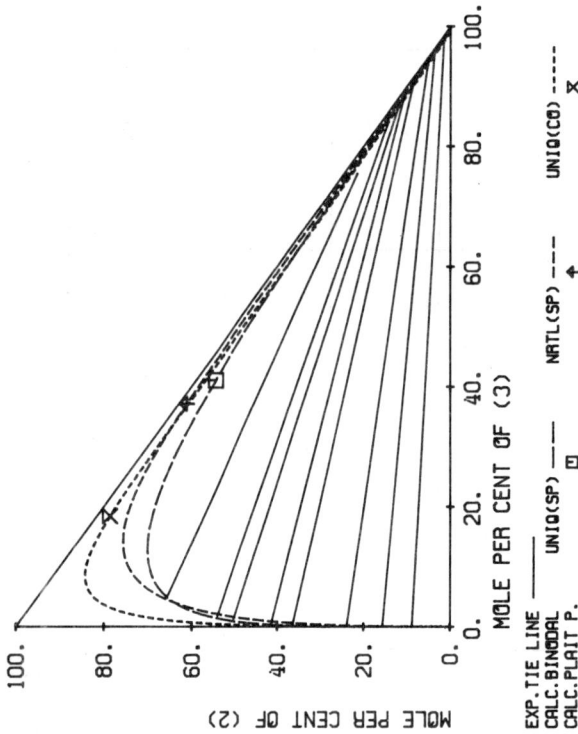

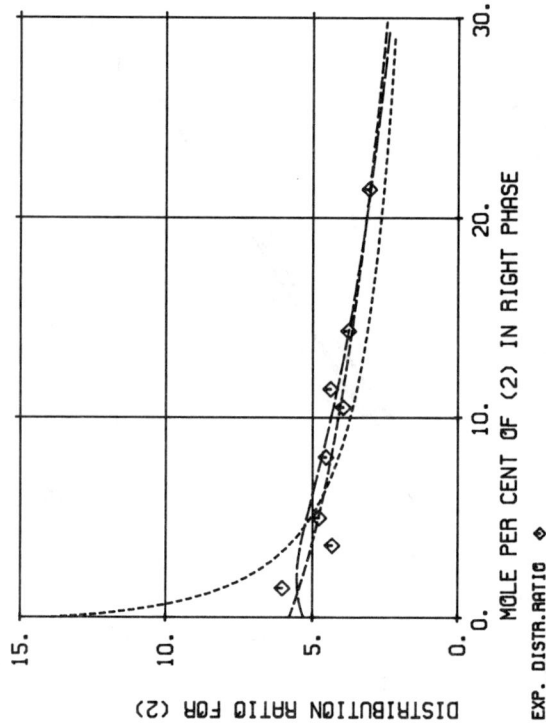

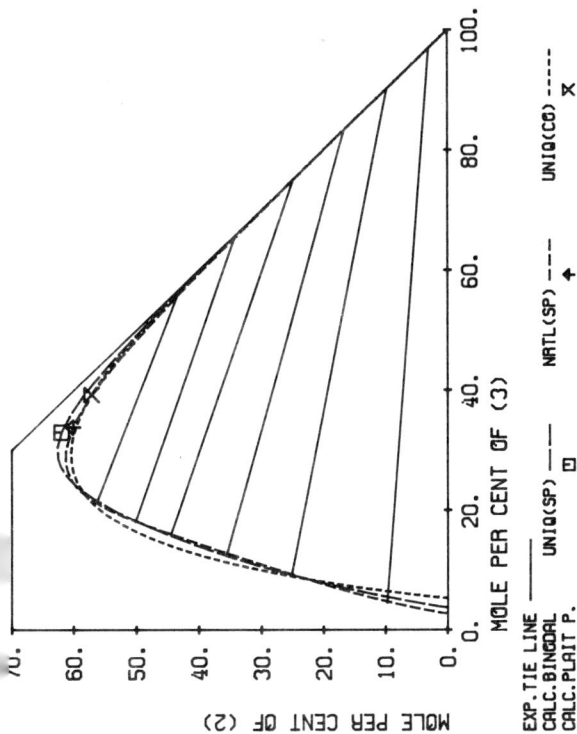

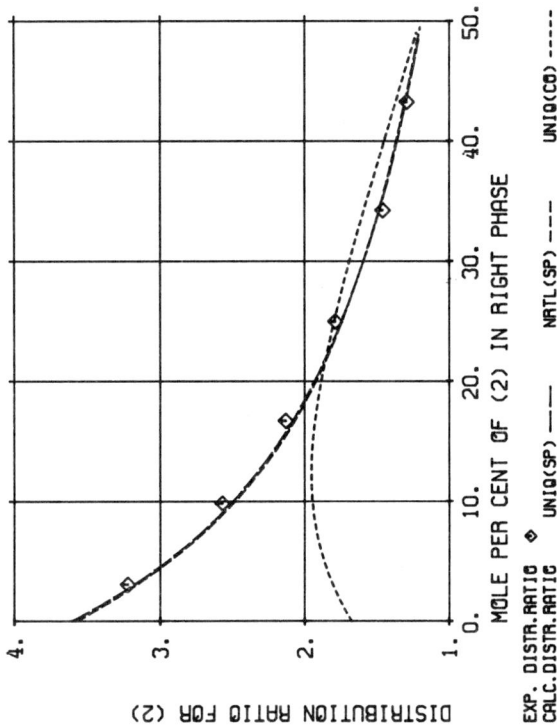

MOLE PER CENT OF (3)

MOLE PER CENT OF (2)

EXP.TIE LINE ——— UNIQ(SP) ······ ⊡ NRTL(SP) — — · + UNIQ(CO) – – – –
CALC.BINODAL   ⋈
CALC.PLAIT P.

DISTRIBUTION RATIO FOR (2)

MOLE PER CENT OF (2) IN RIGHT PHASE

EXP. DISTR.RATIO ◇ UNIQ(SP) ——— ◇ NRTL(SP) — — · UNIQ(CO) – – – –
CALC.DISTR.RATIO

(2) C2H6O   ETHANOL

(3) H2O   WATER

UPCHURCH J.C., VAN WINKLE M.
IND.ENG.CHEM. 44(1952)618

TEMPERATURE = 25.0 DEG C   TYPE OF SYSTEM = 1

EXPERIMENTAL TIE LINES IN MOLE PCT (GRAPH.INTERPOL.)

	LEFT PHASE			RIGHT PHASE	
(1)	(2)	(3)	(1)	(2)	(3)
85.725	9.776	4.500	0.009	3.034	96.957
65.541	25.197	9.263	0.012	9.793	90.195
51.884	35.567	12.549	0.016	16.677	83.308
39.713	44.577	15.710	0.038	24.962	75.000
31.838	50.174	17.988	0.207	34.234	65.559
22.247	56.161	21.591	0.433	43.247	56.320

SPECIFIC MODEL PARAMETERS IN KELVIN

		UNIQUAC		NRTL(ALPHA=.2)	
I	J	AIJ	AJI	AIJ	AJI
1	2	-30.085	117.60	-1020.7	1311.1
1	3	755.80	119.23	696.77	2703.4
2	3	-134.64	139.71	-337.73	281.16

R1 =12.2215   R2 = 2.1055   R3 = 0.9200
Q1 =10.072   Q2 = 1.972   Q3 = 1.400

MEAN DEV. BETWEEN CALC. AND EXP. CONC. IN MOLE PCT

UNIQUAC (SPECIFIC PARAMETERS)   0.39
NRTL   (SPECIFIC PARAMETERS)   0.37
UNIQUAC (COMMON PARAMETERS)   1.66

C₂H₆OS-C₇H₈

(1) C7H16   HEPTANE
------
(2) C7H8   TOLUENE
------
(3) C2H6OS   ETHANOL,2-MERCAPTO

RAWAT B.S., GULATI I.B.
J.APPL.CHEM.BIOTECHNOL. 26(1976)425

TEMPERATURE = 25.0 DEG C   TYPE OF SYSTEM = 1

EXPERIMENTAL TIE LINES IN MOLE PCT

	LEFT PHASE			RIGHT PHASE	
(1)	(2)	(3)	(1)	(2)	(3)
71.942	25.249	2.809	2.279	6.780	90.941
61.423	34.561	4.016	2.652	9.512	87.836
54.165	40.710	5.125	3.102	12.324	84.574
38.656	54.106	7.238	3.978	16.775	79.246

MEAN DEV. BETWEEN CALC. AND EXP. CONC. IN MOLE PCT

UNIQUAC (COMMON PARAMETERS)   0.55

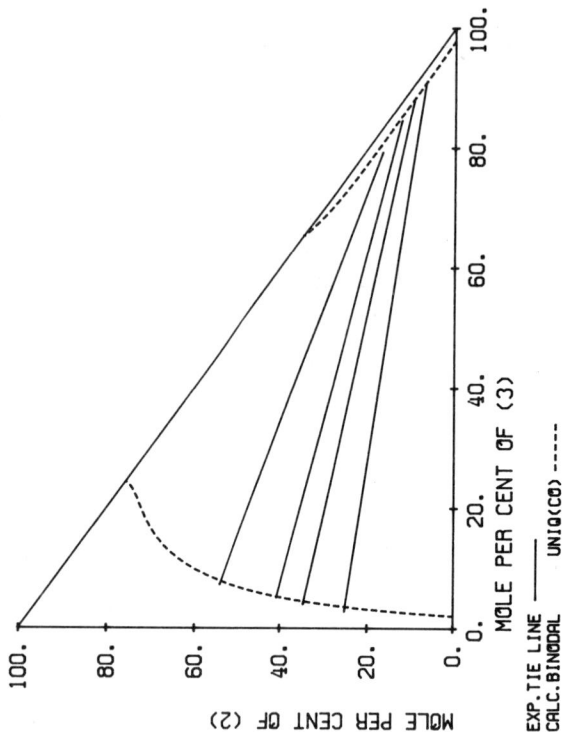

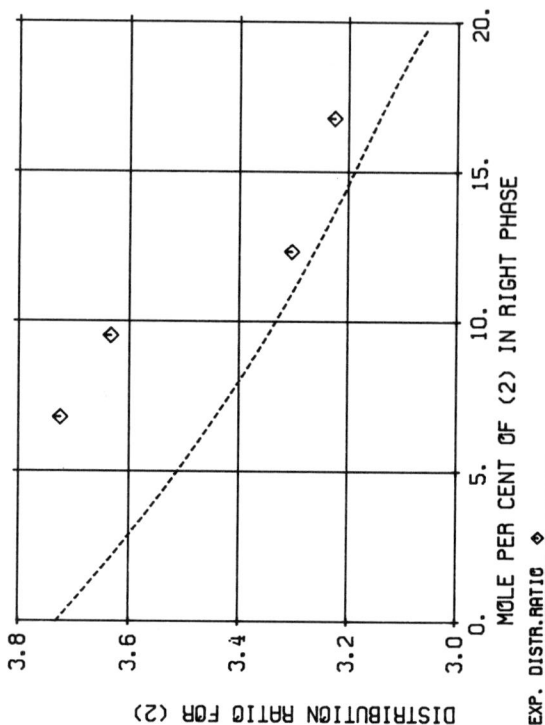

C$_2$H$_6$OS-C$_4$H$_4$S

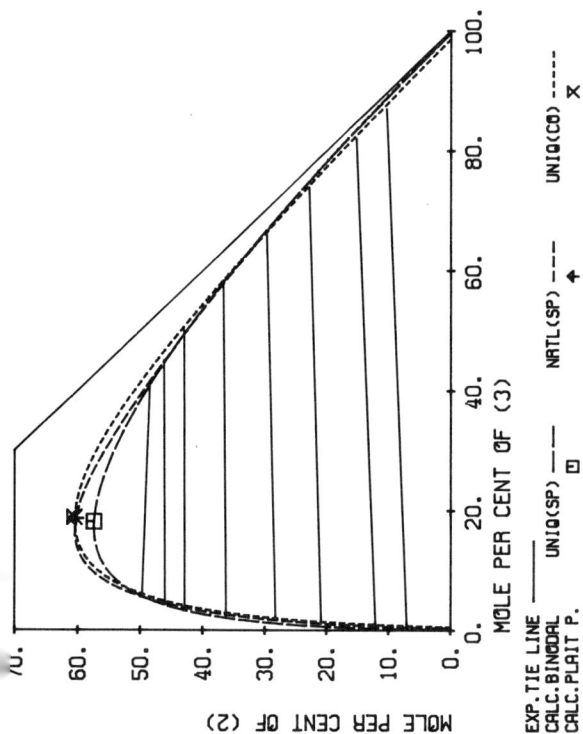

MOLE PER CENT OF (3)

MOLE PER CENT OF (2)

EXP.TIE LINE —— UNIQ(SP) ⊡ NRTL(SP) ——✦ UNIQ(CO) ----✕
CALC.BINODAL ---
CALC.PLAIT P.

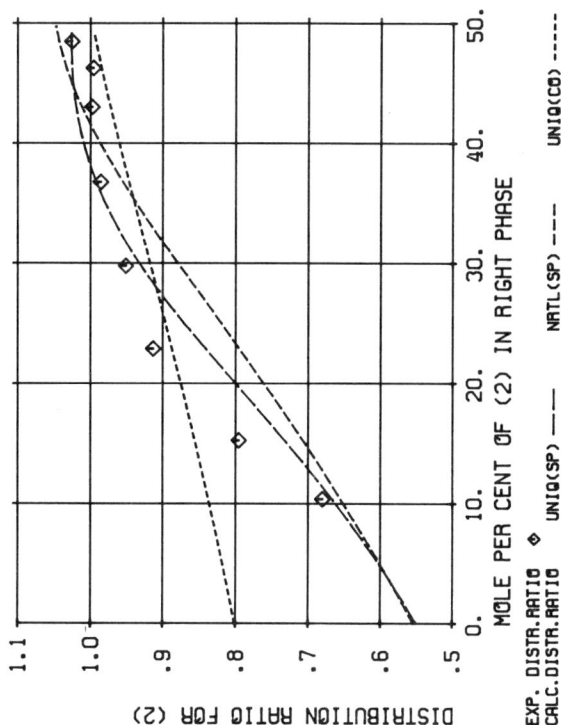

DISTRIBUTION RATIO FOR (2)

MOLE PER CENT OF (2) IN RIGHT PHASE

EXP. DISTR.RATIO ◇ UNIQ(SP) ◇ NRTL(SP) —— UNIQ(CO) ----
CALC.DISTR.RATIO ---

(2) C4H4S   THIOPHENE

(3) C2H6OS   SULFOXIDE,DIMETHYL

HANSON C.; PATEL A.N.; CHANG-KAKOTI D.K.
J.APPL.CHEM. 19(1969)320

TEMPERATURE = 20.5 DEG C   TYPE OF SYSTEM = 1

EXPERIMENTAL TIE LINES IN MOLE PCT

| | LEFT PHASE | | | RIGHT PHASE | |
	(1)	(2)	(3)	(1)	(2)	(3)
	92.187	7.053	0.760	2.612	10.371	87.016
	87.000	12.121	0.879	2.702	15.239	82.058
	77.873	20.891	1.236	3.202	22.880	73.919
	69.868	28.303	1.829	3.869	29.759	66.372
	61.345	36.252	2.403	4.875	36.770	58.355
	53.769	42.913	3.318	6.977	43.015	50.008
	49.690	46.077	4.233	8.679	46.266	45.055
	44.465	49.715	5.820	9.968	48.479	41.553

SPECIFIC MODEL PARAMETERS IN KELVIN

| I J | UNIQUAC | | NRTL(ALPHA=.2) | |
	AIJ	AJI	AIJ	AJI
1 2	442.71	-222.57	383.55	-184.30
1 3	733.39	111.34	1127.6	1224.7
2 3	443.71	-224.02	1141.2	-521.12

R1 = 5.1742   R2 = 2.8569   R3 = 2.8266
Q1 = 4.396   Q2 = 2.140   Q3 = 2.472

MEAN DEV. BETWEEN CALC. AND EXP. CONC. IN MOLE PCT

UNIQUAC (SPECIFIC PARAMETERS)   0.56
NRTL   (SPECIFIC PARAMETERS)   0.73
UNIQUAC (COMMON PARAMETERS)   0.62

C$_2$H$_6$OS-C$_4$H$_8$O

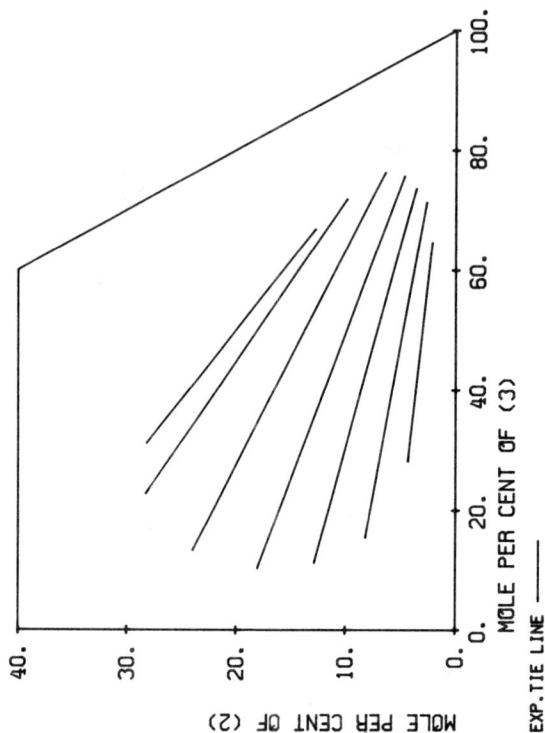

MOLE PER CENT OF (3)

MOLE PER CENT OF (2)

EXP.TIE LINE

(1) H2O   WATER
(2) C2H6OS   SULFOXIDE,DIMETHYL
(3) C4H8O   FURAN,TETRAHYDRO

WOLSKI T.
ROCZ.CHEM. 44(1970)2237

TEMPERATURE = 20.0 DEG C   TYPE OF SYSTEM = 0

EXPERIMENTAL TIE LINES IN MOLE PCT

LEFT PHASE			RIGHT PHASE		
(1)	(2)	(3)	(1)	(2)	(3)
67.504	4.299	28.198	33.189	2.113	64.697
76.412	8.148	15.440	26.030	2.634	71.336
75.960	12.824	11.216	22.779	3.556	73.665
71.684	18.019	10.296	19.643	4.619	75.739
62.761	23.980	13.259	17.319	6.372	76.309
49.000	28.233	22.767	18.227	9.866	71.907
40.730	28.160	31.110	20.382	12.744	66.874

$C_2H_6OS-C_5H_8$

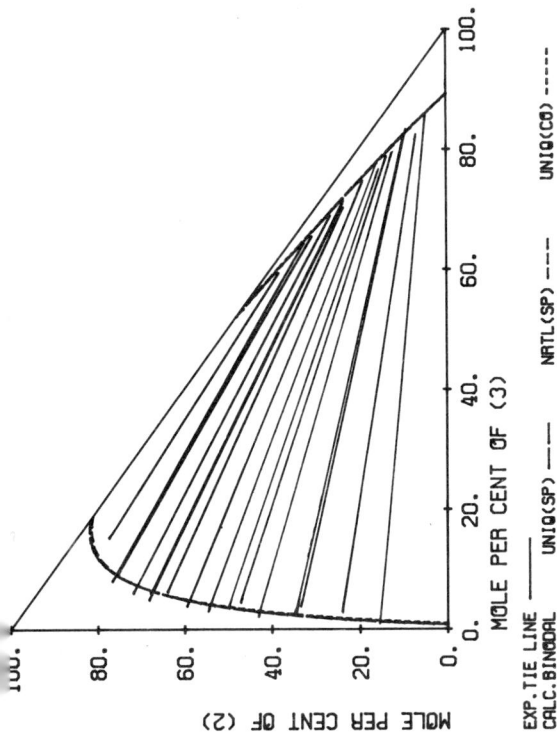

MOLE PER CENT OF (3)

MOLE PER CENT OF (2)

EXP.TIE LINE ——   UNIQ(SP) ——   NRTL(SP) ——   UNIQ(CO) ----
CALC.BINODAL

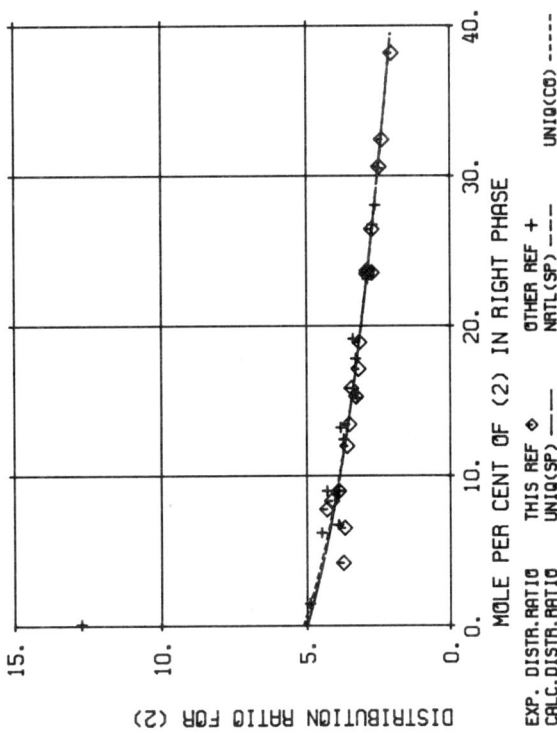

MOLE PER CENT OF (2) IN RIGHT PHASE

DISTRIBUTION RATIO FOR (2)

EXP. DISTR.RATIO ◇   THIS REF ◇   OTHER REF +   UNIQ(CO) ----
CALC.DISTR.RATIO   UNIQ(SP) ——   NRTL(SP) ——

---

(2) C5H8   1,3-BUTADIENE,2-METHYL

(3) C2H6OS   SULFOXIDE,DIMETHYL

ENOMOTO M.; UCHIDA M.
KAGAKU KOGAKU 35(1971)1035

TEMPERATURE =   20.0 DEG C   TYPE OF SYSTEM = 1

EXPERIMENTAL TIE LINES IN MOLE PCT

	LEFT PHASE			RIGHT PHASE	
(1)	(2)	(3)	(1)	(2)	(3)
99.641	0.0	0.359	9.814	0.0	90.186
99.191	0.0	0.809	7.187	0.0	92.813
83.329	15.598	1.074	8.566	4.183	87.251
73.134	24.003	2.862	10.598	6.524	82.879
62.671	33.489	3.840	7.550	7.773	84.677
62.775	34.461	2.764	8.629	8.321	83.051
62.641	34.953	2.406	7.647	8.997	83.356
54.758	43.019	2.222	8.264	11.979	79.757
48.294	47.166	4.540	7.603	13.418	78.979
46.329	49.938	3.733	7.907	15.277	76.817
42.382	54.442	3.549	6.611	15.845	77.545
42.382	54.514	3.104	7.459	17.138	75.403
36.633	59.467	3.900	6.156	18.901	74.943
26.167	67.805	6.028	5.051	23.456	71.493
29.765	63.932	6.303	6.118	23.538	70.344
26.822	68.221	4.957	5.263	23.666	71.071
22.095	71.795	6.110	4.498	26.464	69.038
15.521	76.516	8.963	3.837	30.616	65.547
15.302	76.633	8.065	3.086	32.429	64.485
7.365	77.357	15.278	2.430	38.185	59.385

SPECIFIC MODEL PARAMETERS IN KELVIN

	UNIQUAC		NRTL(ALPHA=.2)	
I J	AIJ	AJI	AIJ	AJI
1 2	-63.931	43.672	-208.44	141.52
1 3	602.47	-16.091	1237.0	222.48
2 3	358.10	-41.071	701.21	47.635

R1 = 3.5919   R2 = 3.3638   R3 = 2.8266
Q1 = 3.220   Q2 = 3.012   Q3 = 2.472

MEAN DEV. BETWEEN CALC. AND EXP. CONC. IN MOLE PCT

UNIQUAC (SPECIFIC PARAMETERS)   0.99
NRTL   (SPECIFIC PARAMETERS)   1.02
UNIQUAC (COMMON PARAMETERS)   1.01

C$_2$H$_6$OS-C$_5$H$_8$

(1) C5H10   2-BUTENE,2-METHYL

(2) C5H8   1,3-BUTADIENE,2-METHYL

(3) C2H6OS   SULFOXIDE,DIMETHYL

ENOMOTO M., UCHIDA M.
KAGAKU KOGAKU 35(1971)1035

TEMPERATURE = 30.0 DEG C   TYPE OF SYSTEM = 1

EXPERIMENTAL TIE LINES IN MOLE PCT

	LEFT PHASE			RIGHT PHASE	
(1)	(2)	(3)	(1)	(2)	(3)
99.011	0.0	0.989	12.210	0.0	87.790
99.011	0.0	0.989	11.558	0.0	88.442
96.306	1.444	2.249	10.794	0.113	89.092
91.904	7.199	0.897	12.512	1.469	86.019
68.993	27.879	3.128	8.869	6.200	84.931
70.703	26.347	2.950	10.595	6.747	82.659
62.641	34.953	2.406	8.837	8.986	82.177
57.504	38.751	3.745	7.321	9.000	83.679
50.147	46.117	3.737	7.936	12.424	79.640
46.207	50.490	2.840	7.066	13.205	79.729
46.207	48.897	4.896	7.801	15.169	77.031
41.967	51.887	6.146	7.475	15.502	77.023
37.156	58.497	4.347	5.519	17.826	76.654
27.533	64.643	7.824	4.972	19.141	75.887
26.995	66.170	6.835	5.588	23.121	71.291
19.000	72.027	8.973	4.603	26.780	68.617
17.623	72.777	9.599	4.702	28.052	67.246
14.635	76.045	9.321	4.156	30.500	65.343

SPECIFIC MODEL PARAMETERS IN KELVIN

		UNIQUAC		NRTL(ALPHA=.2)	
I	J	AIJ	AJI	AIJ	AJI
1	2	-63.931	43.672	-208.44	141.52
1	3	602.47	-16.091	1237.0	222.48
2	3	358.10	-41.071	701.21	47.635

R1 = 3.5919   R2 = 3.3638   R3 = 2.8266
Q1 = 3.220   Q2 = 3.012   Q3 = 2.472

MEAN DEV. BETWEEN CALC. AND EXP. CONC. IN MOLE PCT

UNIQUAC (SPECIFIC PARAMETERS)   0.98
NRTL   (SPECIFIC PARAMETERS)   1.02
UNIQUAC (COMMON PARAMETERS)   0.96

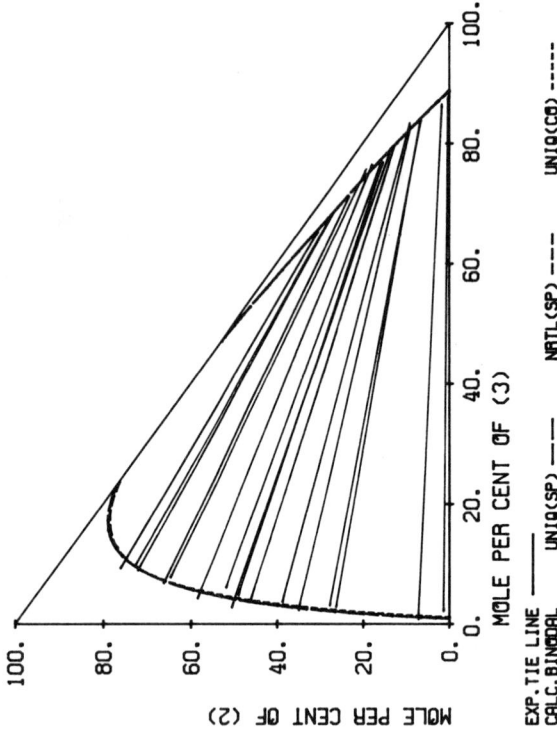

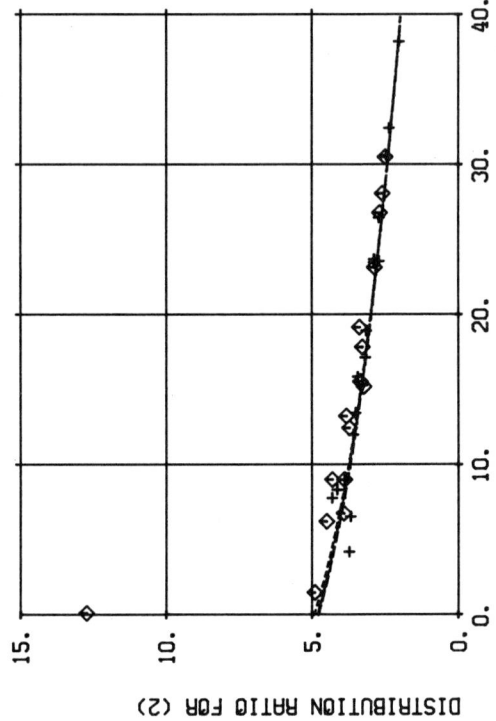

$C_2H_6OS-C_5H_8$

(2) C5H8   1,3-BUTADIENE, 2-METHYL
(3) C2H6OS   SULFOXIDE, DIMETHYL

ENOMOTO M., UCHIDA M.
KAGAKU KOGAKU 35(1971)1035

TEMPERATURE = 40.0 DEG C   TYPE OF SYSTEM = 1

EXPERIMENTAL TIE LINES IN MOLE PCT

LEFT PHASE			RIGHT PHASE		
(1)	(2)	(3)	(1)	(2)	(3)
98.471	0.0	1.529	13.945	0.0	86.055
98.741	0.0	1.259	13.295	0.0	86.705
89.210	8.546	2.245	13.909	2.029	84.062
61.564	32.255	6.181	11.838	10.063	78.099
62.031	34.489	3.480	11.510	10.508	77.982
52.913	42.720	4.367	10.177	13.821	76.003
49.761	44.889	5.350	10.164	14.805	75.032
48.030	46.893	5.078	9.621	15.359	75.020
46.408	48.696	4.896	8.970	16.132	74.898
40.403	54.441	5.156	9.240	20.352	70.409
31.681	59.403	8.916	7.733	21.783	70.484
31.589	59.315	9.096	9.075	25.171	65.753
21.937	67.548	10.515	6.500	28.741	64.760
25.921	64.004	10.075	7.976	29.013	63.011
23.051	62.974	13.975	7.017	29.990	62.993
23.588	66.266	12.146	6.906	30.519	62.574
15.479	63.948	20.573	8.152	41.747	50.102

SPECIFIC MODEL PARAMETERS IN KELVIN

	UNIQUAC		NRTL(ALPHA=.2)	
I J	AIJ	AJI	AIJ	AJI
1 2	-80.432	41.172	-330.20	227.12
1 3	701.63	-48.545	1449.3	144.28
2 3	315.59	-35.117	588.94	73.978

R1 = 3.5919   R2 = 3.3638   R3 = 2.8266
Q1 = 3.220   Q2 = 3.012   Q3 = 2.472

MEAN DEV. BETWEEN CALC. AND EXP. CONC. IN MOLE PCT

UNIQUAC (SPECIFIC PARAMETERS)   1.07
NRTL   (SPECIFIC PARAMETERS)   1.07

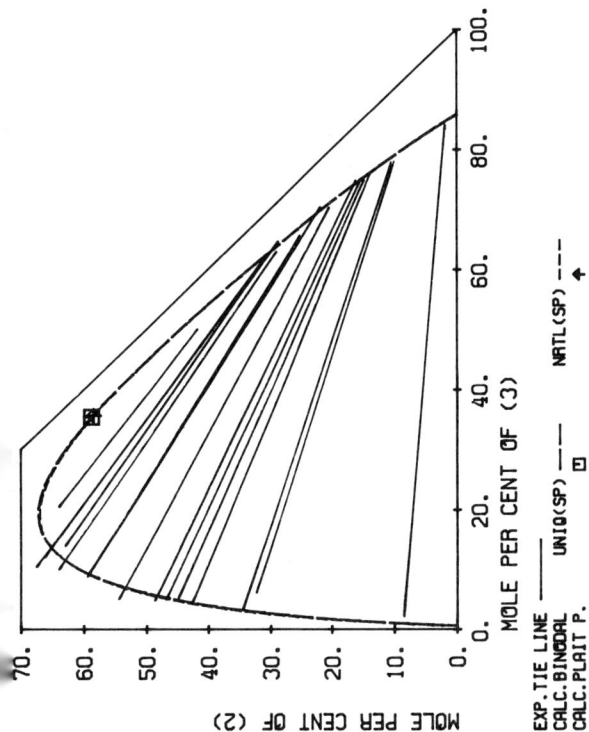

MOLE PER CENT OF (2) vs MOLE PER CENT OF (3)

EXP.TIE LINE ——
CALC.BINODAL UNIQ(SP) ———
CALC.PLAIT P. NRTL(SP) ———

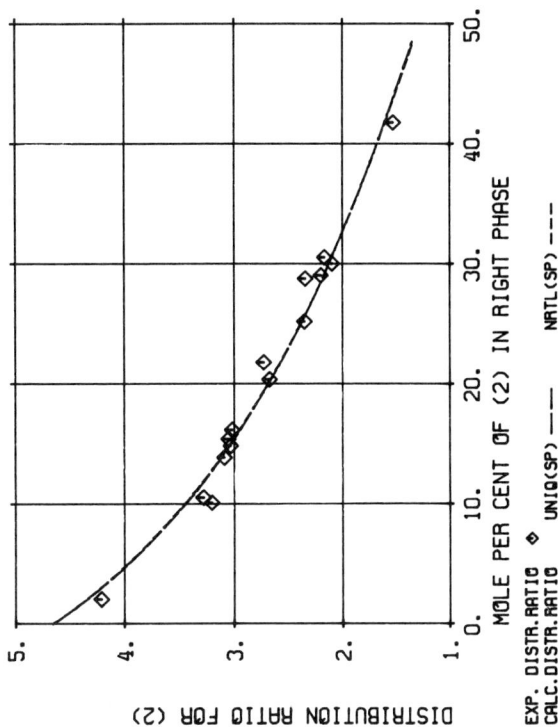

DISTRIBUTION RATIO FOR (2) vs MOLE PER CENT OF (2) IN RIGHT PHASE

EXP. DISTR.RATIO ◇
CALC.DISTR.RATIO UNIQ(SP) ———   NRTL(SP) ———

C$_2$H$_6$OS-C$_6$H$_6$

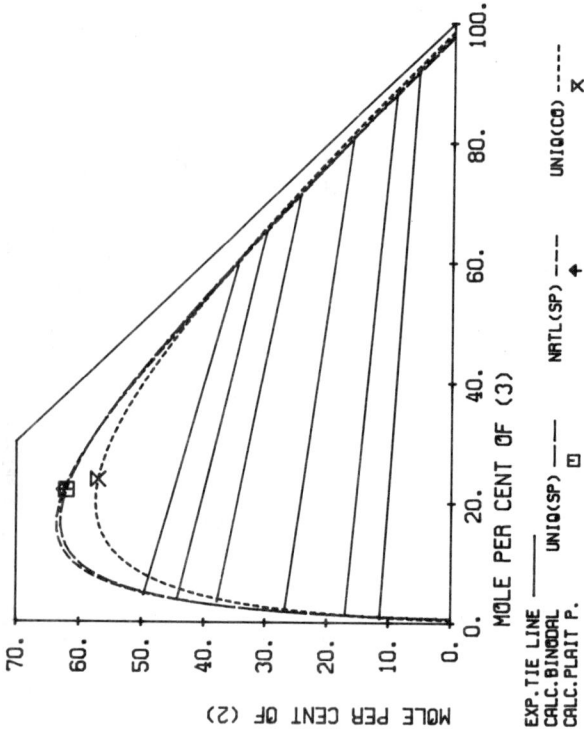

EXP.TIE LINE
CALC.BINODAL
CALC.PLAIT P.

UNIQ(SP) —·— ⊟   NRTL(SP) —— ♦   UNIQ(CO) ----- ⊼

MOLE PER CENT OF (3)

MOLE PER CENT OF (2)

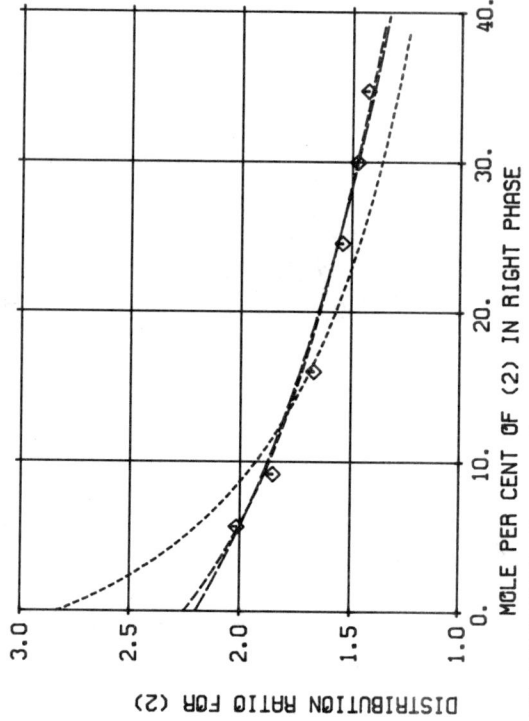

MOLE PER CENT OF (2) IN RIGHT PHASE

DISTRIBUTION RATIO FOR (2)

EXP. DISTR.RATIO ◇

(1) C7H16   HEPTANE

(2) C6H6   BENZENE

(3) C2H6OS   SULFOXIDE,DIMETHYL

HANSON C., PATEL A.N., CHANG-KAKOTI D.K.
J.APPL.CHEM. 19(1969)320

TEMPERATURE = 20.5 DEG C   TYPE OF SYSTEM = 1

EXPERIMENTAL TIE LINES IN MOLE PCT

	LEFT PHASE		RIGHT PHASE	
(1)	(2)	(3)	(1)	(2)	(3)
87.889	11.362	0.749	2.355	5.638	92.008
81.778	16.991	1.231	2.592	9.168	88.241
70.833	26.767	2.400	2.829	16.029	81.142
58.616	37.887	3.497	3.862	24.568	71.571
51.843	44.259	3.898	4.740	29.999	65.260
45.786	49.697	4.517	5.704	34.754	59.542

SPECIFIC MODEL PARAMETERS IN KELVIN

		UNIQUAC		NRTL(ALPHA=.2)	
I	J	AIJ	AJI	AIJ	AJI
1	2	33.541	-74.889	-126.73	-43.366
1	3	593.54	-22.173	1024.0	-694.30
2	3	241.80	-129.91	785.72	-371.08

R1 = 5.1742   R2 = 3.1878   R3 = 2.8266
Q1 = 4.396   Q2 = 2.400   Q3 = 2.472

MEAN DEV. BETWEEN CALC. AND EXP. CONC. IN MOLE PCT

UNIQUAC (SPECIFIC PARAMETERS)   0.27
NRTL   (SPECIFIC PARAMETERS)   0.22
UNIQUAC (COMMON PARAMETERS)   1.09

$C_2H_6OS-C_6H_{12}$

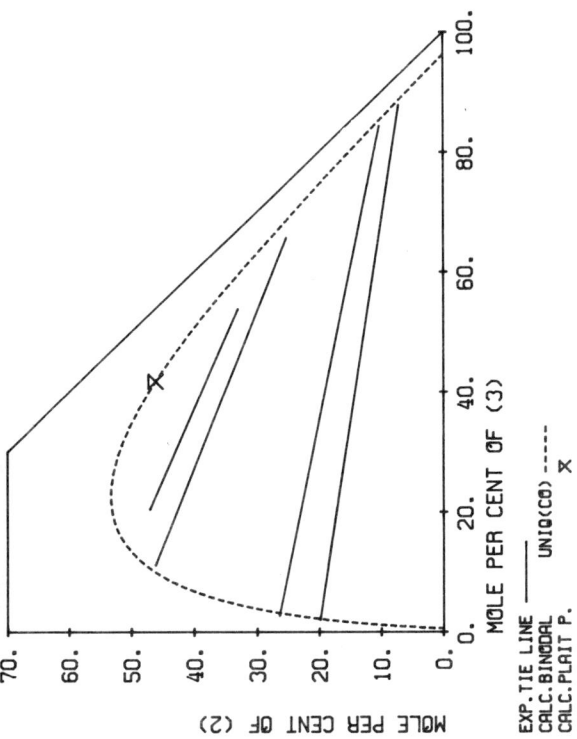

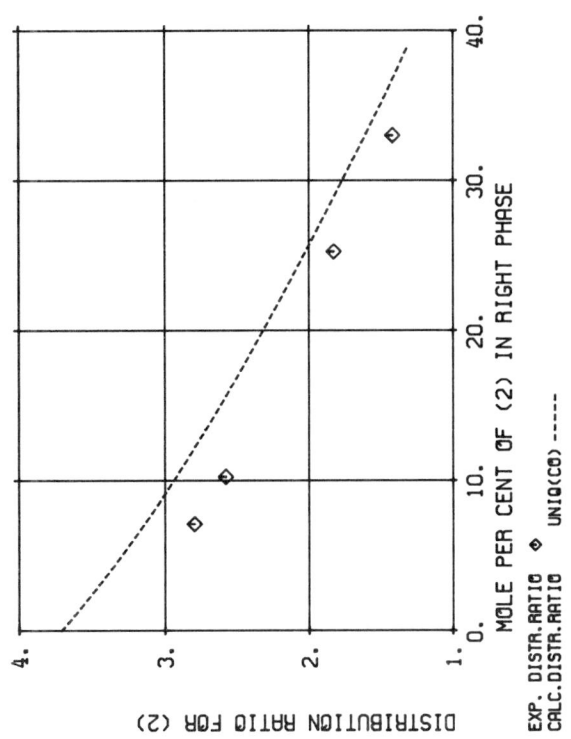

(2) C7H8   TOLUENE

(3) C2H6OS   SULFOXIDE,DIMETHYL

NISSEMA A.; SARKELA S.
SUOM.KEMISTIL. B46(1973)58

TEMPERATURE = 20.0 DEG C   TYPE OF SYSTEM = 1

EXPERIMENTAL TIE LINES IN MOLE PCT

	LEFT PHASE			RIGHT PHASE	
(1)	(2)	(3)	(1)	(2)	(3)
78.100	19.850	2.050	4.950	7.100	87.950
70.800	26.450	2.750	5.350	10.250	84.400
42.750	46.150	11.100	8.950	25.250	65.800
32.500	47.050	20.450	13.150	33.000	53.850

MEAN DEV. BETWEEN CALC. AND EXP. CONC. IN MOLE PCT

UNIQUAC (COMMON PARAMETERS)   3.42

C₂H₆OS-C₇H₈

(1) C7H16   HEPTANE
(2) C7H8   TOLUENE
(3) C2H6OS   SULFOXIDE,DIMETHYL

RAWAT B.S., GULATI I.B.
J.APPL.CHEM.BIOTECHNOL. 26(1975)425

TEMPERATURE = 25.0 DEG C   TYPE OF SYSTEM = 1

EXPERIMENTAL TIE LINES IN MOLE PCT

	LEFT PHASE		RIGHT PHASE	
(1)	(2)	(3)	(1)	(2)	(3)
86.851	11.377	1.772	1.182	4.027	94.791
84.332	14.088	1.579	1.421	4.895	93.684
77.458	19.782	2.760	1.578	7.124	91.298
67.000	31.256	1.744	1.968	11.396	86.636
57.315	37.522	5.163	2.298	15.450	82.251
41.697	54.413	3.890	3.118	24.269	72.613

SPECIFIC MODEL PARAMETERS IN KELVIN

		UNIQUAC		NRTL(ALPHA=.2)	
I	J	AIJ	AJI	AIJ	AJI
1	2	-4.4578	2.8134	-144.58	140.36
1	3	459.72	75.823	796.54	857.79
2	3	386.02	-106.39	824.33	-147.79

R1 = 5.1742   R2 = 3.9228   R3 = 2.8266
Q1 = 4.396   Q2 = 2.968   Q3 = 2.472

MEAN DEV. BETWEEN CALC. AND EXP. CONC. IN MOLE PCT

UNIQUAC (SPECIFIC PARAMETERS)   0.56
NRTL   (SPECIFIC PARAMETERS)   0.54
UNIQUAC (COMMON PARAMETERS)   2.83

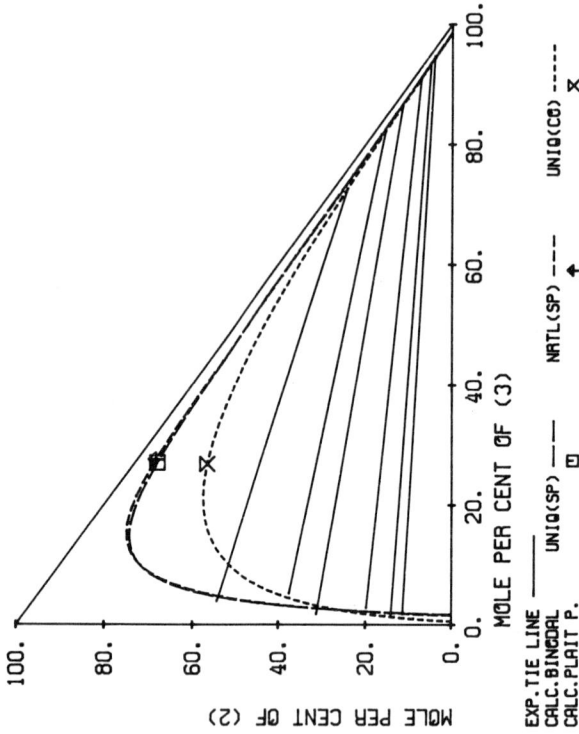

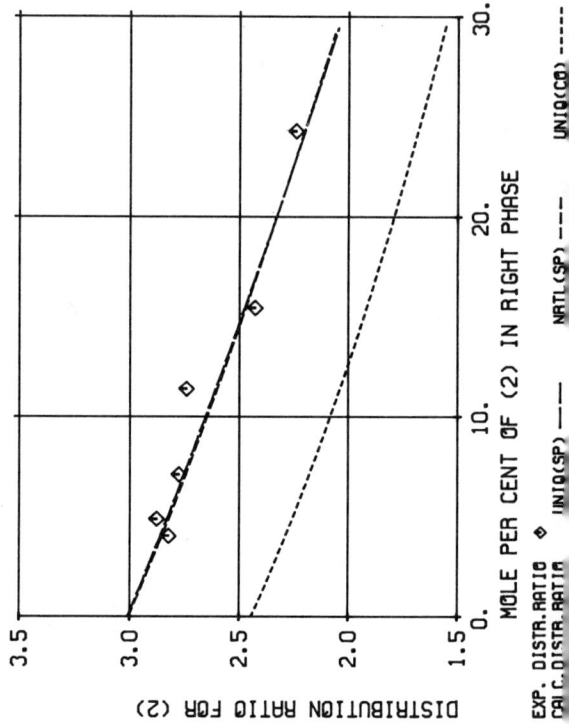

$C_2H_6O_2-C_3H_6O$

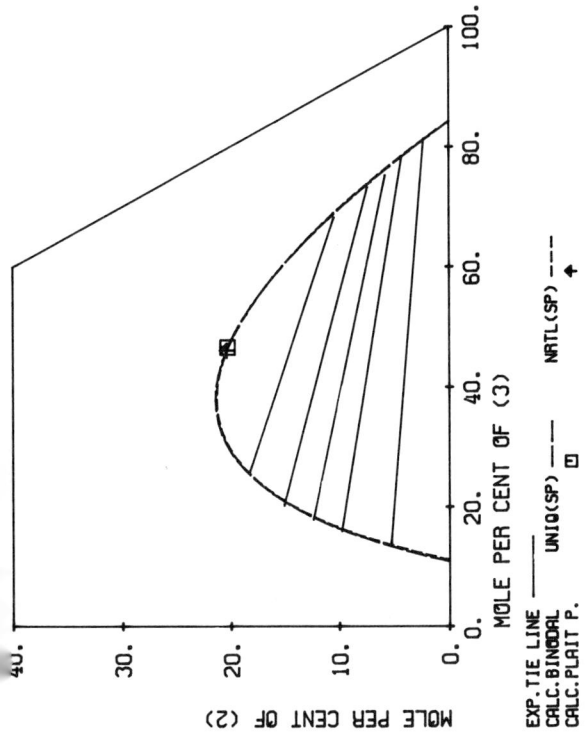

MOLE PER CENT OF (3)

MOLE PER CENT OF (2)

EXP.TIE LINE ——   UNIQ(SP) —— ☐   NRTL(SP) ——
CALC.BINODAL
CALC.PLAIT P.

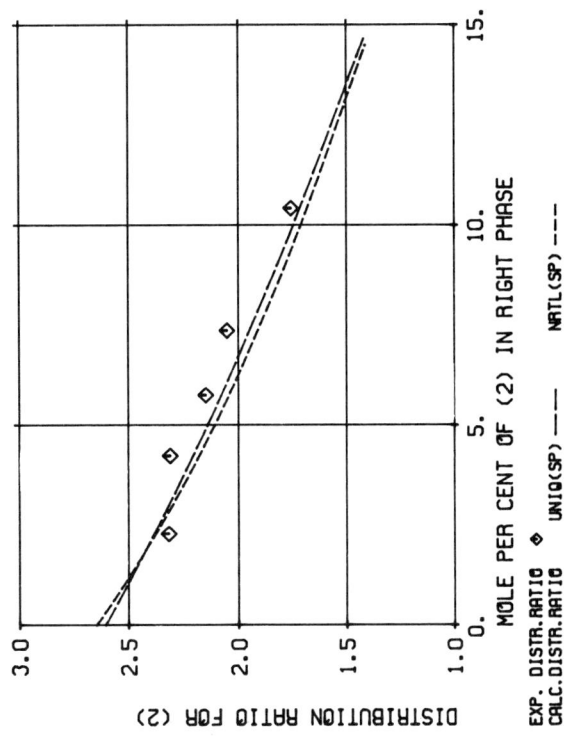

MOLE PER CENT OF (2) IN RIGHT PHASE

DISTRIBUTION RATIO FOR (2)

EXP. DISTR.RATIO ◇   UNIQ(SP) ——   NRTL(SP) ——
CALC. DISTR.RATIO

(2) C3H6O   2-PROPANONE
(3) C2H6O2   1,2-ETHANEDIOL

RAJA RAO M., VENTAKA RAO C.
J.SCI.IND.RES. 14B(1955)204

TEMPERATURE = 31.0 DEG C   TYPE OF SYSTEM = 1

EXPERIMENTAL TIE LINES IN MOLE PCT

	LEFT PHASE			RIGHT PHASE	
(1)	(2)	(3)	(1)	(2)	(3)
80.825	5.282	13.893	16.309	2.281	81.411
74.305	9.767	15.928	17.254	4.230	78.516
69.781	12.359	17.860	18.944	5.749	75.307
64.707	15.104	20.190	19.334	7.362	73.304
56.221	18.316	25.463	21.275	10.413	68.312

SPECIFIC MODEL PARAMETERS IN KELVIN

		UNIQUAC			NRTL(ALPHA=.2)	
I J	AIJ	AJI			AIJ	AJI
1 2	-54.773	2.8889			-83.440	-93.533
1 3	348.92	40.202			553.20	362.79
2 3	34.573	93.152			26.213	220.72

R1 = 3.4786   R2 = 2.5735   R3 = 2.4088
Q1 = 3.116   Q2 = 2.336   Q3 = 2.248

MEAN DEV. BETWEEN CALC. AND EXP. CONC. IN MOLE PCT

UNIQUAC (SPECIFIC PARAMETERS)   0.28
NRTL   (SPECIFIC PARAMETERS)   0.33

$C_2H_6O_2$-C_3H_6O

(1) C5H1002   PROPANOIC ACID,ETHYL ESTER

(2) C3H6O   2-PROPANONE

(3) C2H6O2   1,2-ETHANEDIOL

RAJA RAO M., VENTAKA RAO C.
J.SCI.IND.RES. 14B(1955)204

TEMPERATURE = 31.0 DEG C   TYPE OF SYSTEM = 1

EXPERIMENTAL TIE LINES IN MOLE PCT

	LEFT PHASE			RIGHT PHASE	
(1)	(2)	(3)	(1)	(2)	(3)
80.782	11.649	7.569	6.641	3.448	89.912
70.573	20.195	9.232	7.621	6.589	85.790
61.441	26.973	11.585	8.675	10.094	81.231
51.912	32.581	15.506	9.942	13.648	76.410
41.431	36.427	22.143	12.606	18.967	68.427
36.906	36.861	26.233	15.136	23.144	61.720

SPECIFIC MODEL PARAMETERS IN KELVIN

		UNIQUAC		NRTL(ALPHA=.2)	
I	J	AIJ	AJI	AIJ	AJI
1	2	-56.729	-32.735	539.50	-485.97
1	3	480.65	24.363	714.71	507.72
2	3	45.687	63.909	349.57	89.593

R1 = 4.1530   R2 = 2.5735   R3 = 2.4088
Q1 = 3.656   Q2 = 2.336   Q3 = 2.248

MEAN DEV. BETWEEN CALC. AND EXP. CONC. IN MOLE PCT

UNIQUAC (SPECIFIC PARAMETERS)   0.78
NRTL   (SPECIFIC PARAMETERS)   0.66

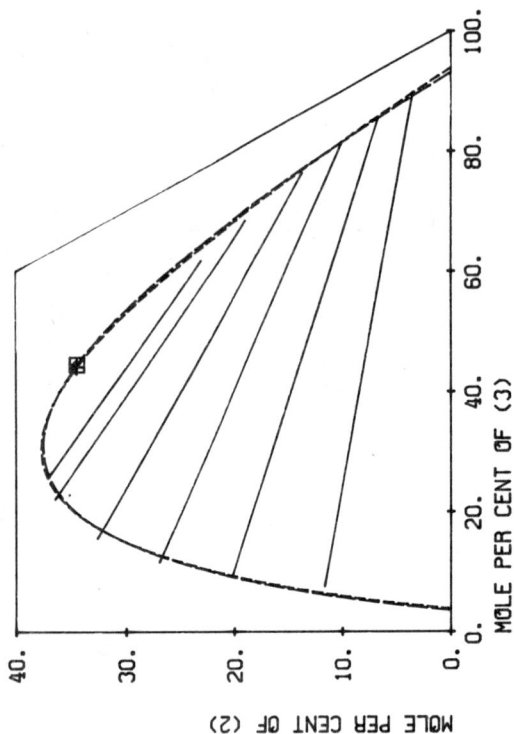

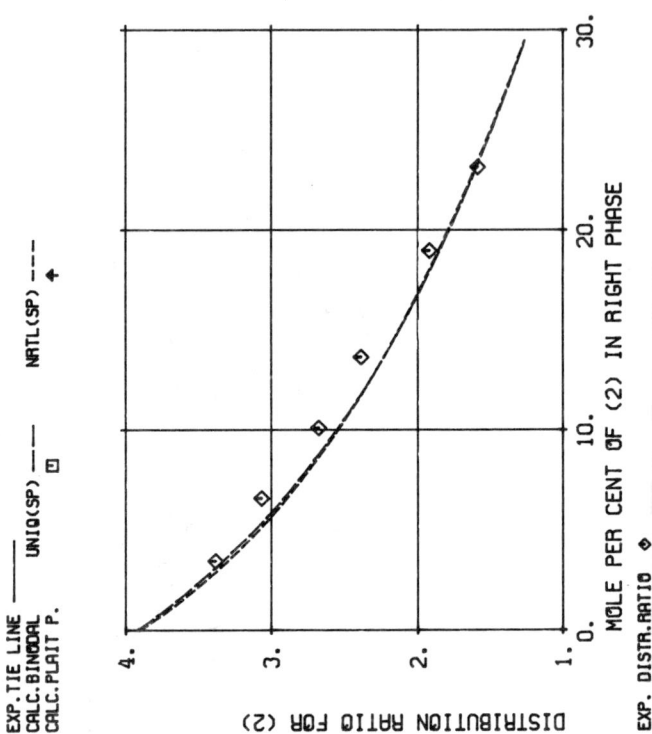

MOLE PER CENT OF (3)

MOLE PER CENT OF (2)

EXP.TIE LINE ——   UNIQ(SP) ——   NRTL(SP) ----
CALC.BINODAL   ⊡
CALC.PLAIT P.   ✦

MOLE PER CENT OF (2) IN RIGHT PHASE

DISTRIBUTION RATIO FOR (2)

EXP. DISTR.RATIO ◇

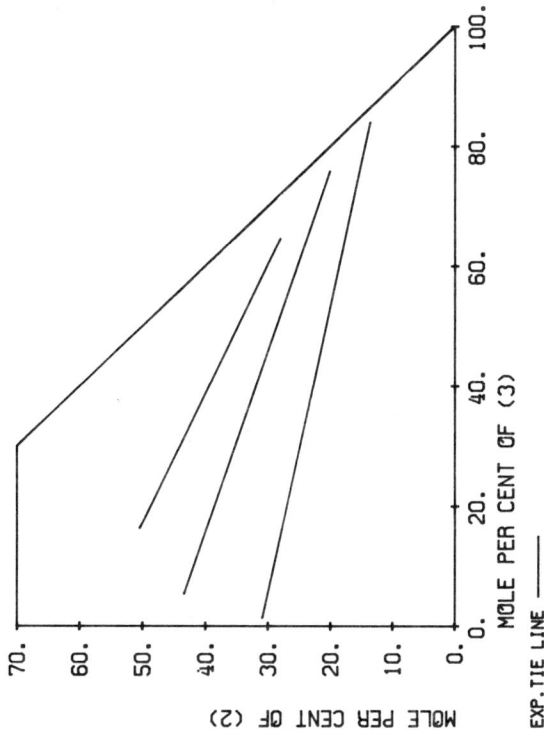

MOLE PER CENT OF (3)

MOLE PER CENT OF (2)

EXP.TIE LINE

(1) C6H5BR   BENZENE,BROMO

(2) C3H6O   2-PROPANONE

(3) C2H6O2   1,2-ETHANEDIOL

TRIMBLE H.M., FRAZER G.E.
IND.ENG.CHEM. 21(1929)1063

TEMPERATURE = 25.0 DEG C   TYPE OF SYSTEM = 1

EXPERIMENTAL TIE LINES IN MOLE PCT (GRAPH.INTERPOL.)

	LEFT PHASE			RIGHT PHASE	
(1)	(2)	(3)	(1)	(2)	(3)
67.758	30.890	1.352	2.381	13.603	84.017
51.181	43.431	5.388	4.039	20.068	75.893
33.259	50.433	16.307	7.307	28.066	64.628

C₂H₆O₂-C₃H₆O

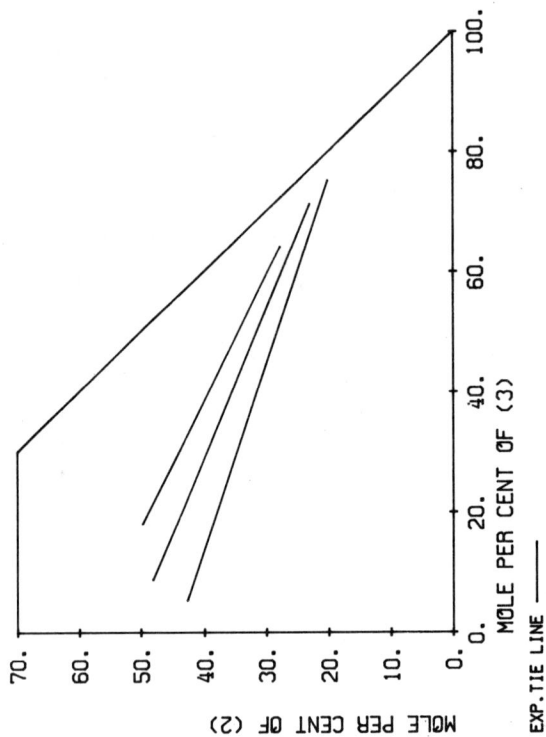

(1) C6H5CL   BENZENE,CHLORO

(2) C3H6O   2-PROPANONE

(3) C2H6O2   1,2-ETHANEDIOL

TRIMBLE H.M., FRAZER G.E.
IND.ENG.CHEM. 21(1929)1063

TEMPERATURE = 23.0 DEG C   TYPE OF SYSTEM = 1

EXPERIMENTAL TIE LINES IN MOLE PCT (GRAPH.INTERPOL.)

LEFT PHASE			RIGHT PHASE		
(1)	(2)	(3)	(1)	(2)	(3)
51.964	42.735	5.301	4.600	20.112	75.288
43.168	48.093	8.738	5.658	22.989	71.352
32.247	49.722	18.031	8.052	27.735	64.213

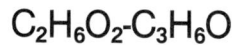

$C_2H_6O_2$-C_3H_6O

(1) C6H5NO2   BENZENE,NITRO

(2) C3H6O   2-PROPANONE

(3) C2H6O2   1,2-ETHANEDIOL

TRIMBLE H.M., FRAZER G.E.
IND.ENG.CHEM. 21(1929)1063

TEMPERATURE = 22.0 DEG C   TYPE OF SYSTEM = 1

EXPERIMENTAL TIE LINES IN MOLE PCT (GRAPH.INTERPOL.)

LEFT PHASE			RIGHT PHASE		
(1)	(2)	(3)	(1)	(2)	(3)
68.907	28.819	2.275	5.274	14.424	80.302
55.932	37.277	6.792	5.957	15.987	78.056
44.304	41.844	13.853	7.766	19.718	72.517

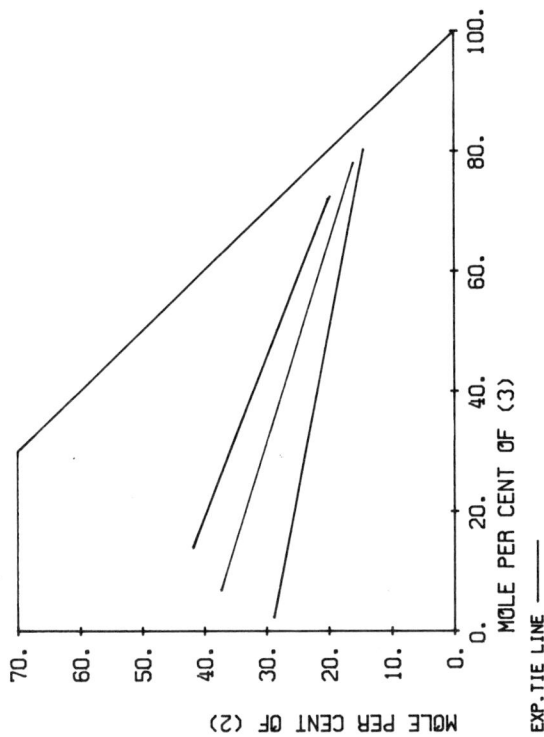

C₂H₆O₂-C₃H₆O

(1) C6H6   BENZENE

(2) C3H6O   2-PROPANONE

(3) C2H6O2   1,2-ETHANEDIOL

TRIMBLE H.M., FRAZER G.E.
IND.ENG.CHEM. 21(1929)1063

TEMPERATURE = 27.0 DEG C   TYPE OF SYSTEM = 1

EXPERIMENTAL TIE LINES IN MOLE PCT (GRAPH.INTERPOL.)

	LEFT PHASE			RIGHT PHASE	
(1)	(2)	(3)	(1)	(2)	(3)
91.665	8.100	0.235	3.390	8.118	88.492
69.652	28.534	1.814	4.709	13.016	82.275
51.779	41.267	6.955	6.736	18.378	74.886
37.112	45.600	17.288	12.443	28.500	59.057

MEAN DEV. BETWEEN CALC. AND EXP. CONC. IN MOLE PCT

UNIQUAC (COMMON PARAMETERS)   1.85

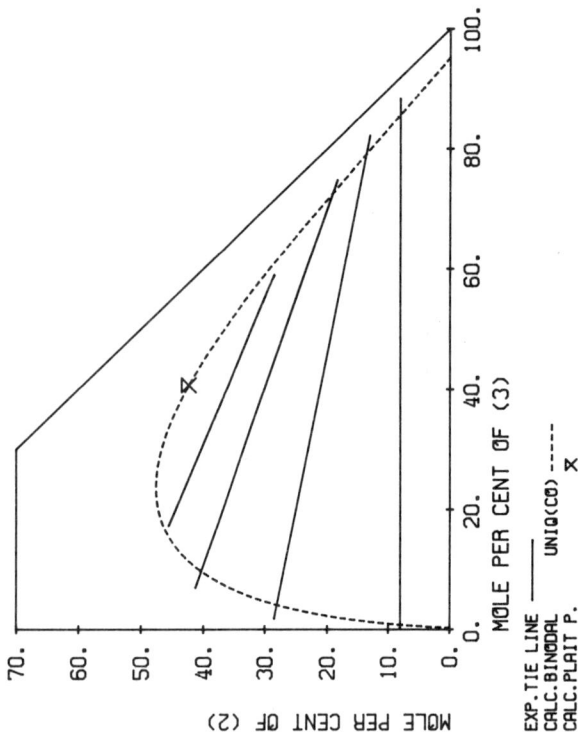

MOLE PER CENT OF (3)

MOLE PER CENT OF (2)

EXP.TIE LINE ——————
CALC.BINODAL UNIQ(C0) ------
CALC.PLAIT P.   ✕

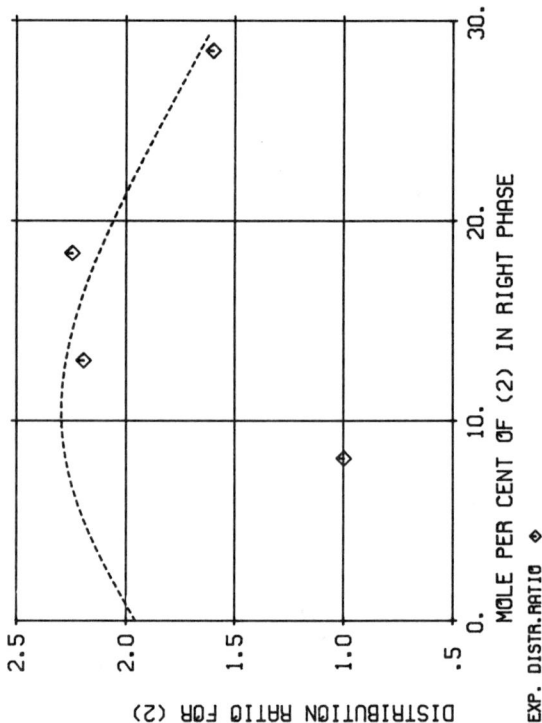

MOLE PER CENT OF (2) IN RIGHT PHASE

DISTRIBUTION RATIO FOR (2)

EXP. DISTR.RATIO ◇

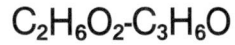

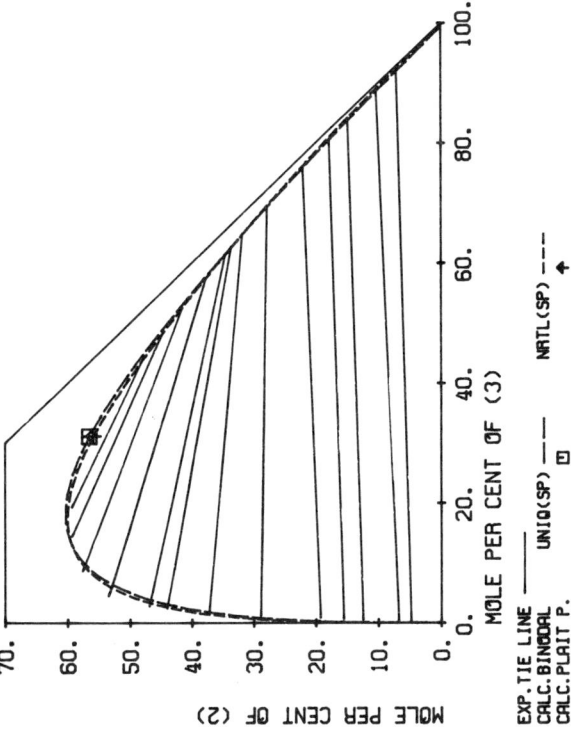

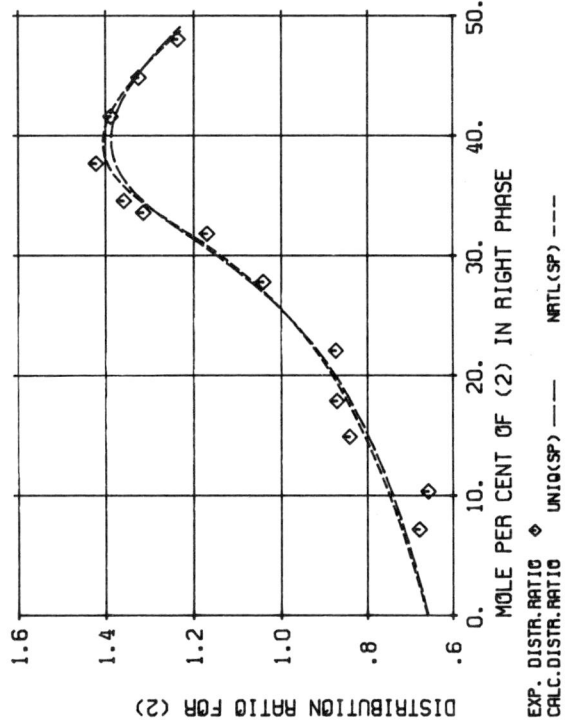

$C_2H_6O_2$-C_3H_6O

MOLE PER CENT OF (3)

MOLE PER CENT OF (2)

EXP.TIE LINE —— UNIQ(SP) ▣ NRTL(SP) ——⭑
CALC.BINODAL ——
CALC.PLAIT P.

MOLE PER CENT OF (2) IN RIGHT PHASE

DISTRIBUTION RATIO FOR (2)

EXP. DISTR.RATIO ◆ UNIQ(SP) ◆ NRTL(SP) ———
CALC.DISTR.RATIO

---

(2) C3H6O   2-PROPANONE

(3) C2H6O2   1,2-ETHANEDIOL

RAJA RAO M., VENTAKA RAO C.
J.SCI.IND.RES. 14B(1955)204

TEMPERATURE = 31.0 DEG C   TYPE OF SYSTEM = 1

EXPERIMENTAL TIE LINES IN MOLE PCT

	LEFT PHASE			RIGHT PHASE	
(1)	(2)	(3)	(1)	(2)	(3)
94.950	4.850	0.200	0.294	7.135	92.571
92.931	6.804	0.265	0.661	10.322	89.017
86.962	12.518	0.521	0.879	14.866	84.255
83.799	15.557	0.644	1.392	17.837	80.771
79.812	19.298	0.890	1.830	22.058	76.112
69.851	28.919	1.230	2.780	27.777	69.443
60.872	37.220	1.908	3.512	31.806	64.682
53.476	44.084	2.440	3.582	33.578	62.840
50.336	46.907	2.758	3.950	34.555	61.494
52.028	53.511	4.461	4.685	37.653	57.662
33.704	57.633	8.663	5.866	41.543	52.592
26.227	59.317	14.456	6.979	44.814	48.208
21.356	59.340	19.304	8.172	48.006	43.822

SPECIFIC MODEL PARAMETERS IN KELVIN

		UNIQUAC		NRTL(ALPHA=.2)	
I J	AIJ	AJI		AIJ	AJI
1 2	360.05	-106.67		578.04	-69.427
1 3	1338.8	181.01		2417.1	1387.0
2 3	132.33	75.555		443.73	55.921

R1 = 4.0464   R2 = 2.5735   R3 = 2.4088
Q1 = 3.240   Q2 = 2.336   Q3 = 2.248

MEAN DEV. BETWEEN CALC. AND EXP. CONC. IN MOLE PCT

UNIQUAC (SPECIFIC PARAMETERS)   0.50
NRTL   (SPECIFIC PARAMETERS)   0.39

$C_2H_6O_2$-C_3H_6O

(1) C6H12O2   ACETIC ACID, BUTYL ESTER

(2) C3H6O   2-PROPANONE

(3) C2H6O2   1,2-ETHANEDIOL

RAJA RAO M., VENTAKA RAO C.
J.SCI.IND.RES. 14B(1955)204

TEMPERATURE = 31.0 DEG C   TYPE OF SYSTEM = 1

EXPERIMENTAL TIE LINES IN MOLE PCT

	LEFT PHASE			RIGHT PHASE	
(1)	(2)	(3)	(1)	(2)	(3)
82.321	11.695	5.984	3.573	3.628	92.798
71.703	21.692	6.605	4.135	6.726	89.140
64.273	28.016	7.710	4.987	9.975	85.038
55.288	34.935	9.778	5.789	13.692	80.519
47.135	40.402	12.463	6.953	17.383	75.663
39.317	44.653	16.030	8.314	21.831	69.856
34.433	45.231	20.336	10.321	26.261	63.418

SPECIFIC MODEL PARAMETERS IN KELVIN

		UNIQUAC		NRTL(ALPHA=.2)	
I	J	AIJ	AJI	AIJ	AJI
1	2	21.621	-19.693	294.42	-281.71
1	3	502.49	24.415	699.86	658.67
2	3	106.97	70.166	432.34	54.197

R1 = 4.8274   R2 = 2.5735   R3 = 2.4083
Q1 = 4.196   Q2 = 2.336   Q3 = 2.248

MEAN DEV. BETWEEN CALC. AND EXP. CONC. IN MOLE PCT

UNIQUAC (SPECIFIC PARAMETERS)   0.85
NRTL   (SPECIFIC PARAMETERS)   0.75

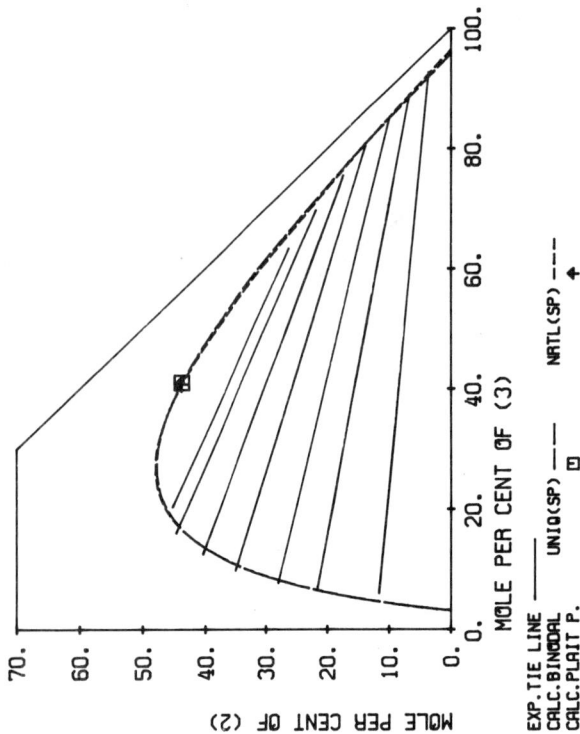

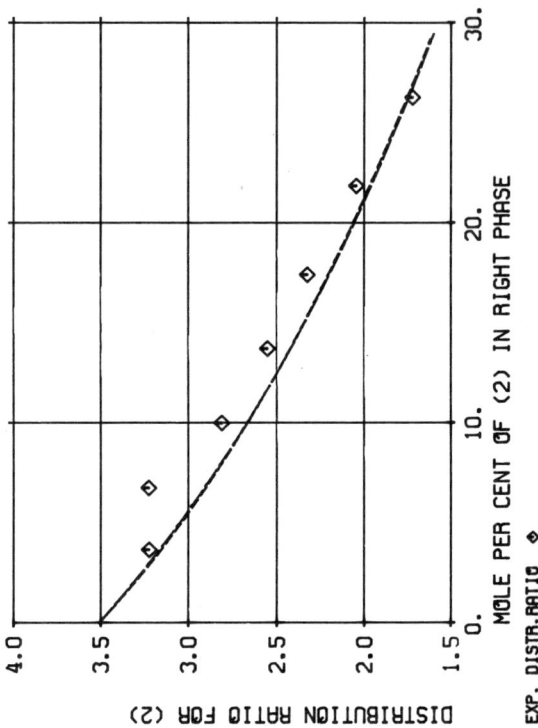

$C_2H_6O_2\text{-}C_3H_6O$

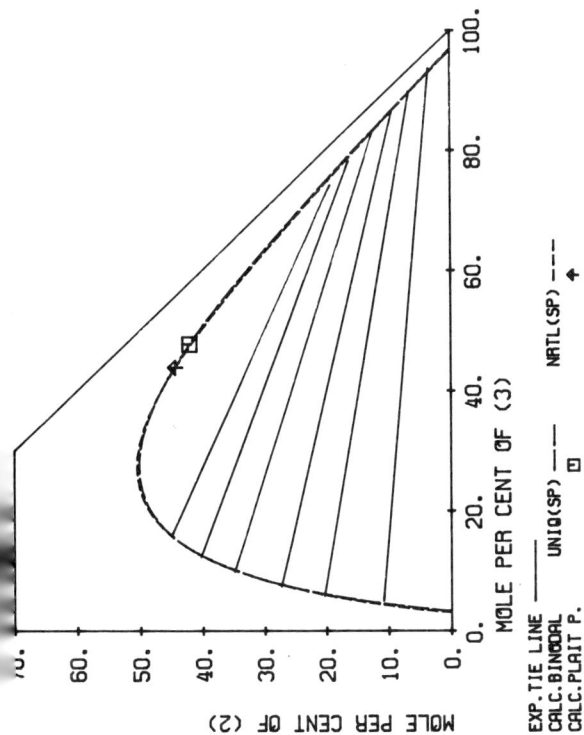

MOLE PER CENT OF (3)

MOLE PER CENT OF (2)

EXP.TIE LINE ——   UNIQ(SP) □   NRTL(SP) ——
CALC.BINODAL
CALC.PLAIT P.

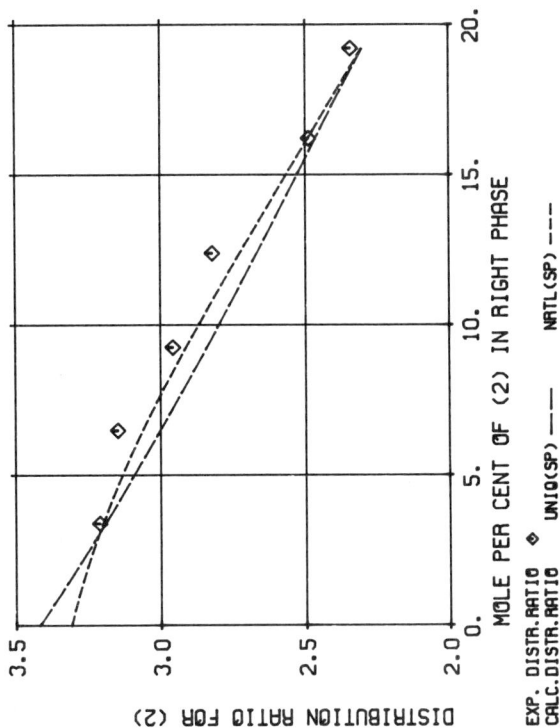

MOLE PER CENT OF (2) IN RIGHT PHASE

DISTRIBUTION RATIO FOR (2)

EXP. DISTR.RATIO ◇   UNIQ(SP) ◇   NRTL(SP) ——
CALC.DISTR.RATIO

(2) C3H6O   2-PROPANONE

(3) C2H6O2   1,2-ETHANEDIOL

RAJA RAO M., VENTAKA RAO C.
J.SCI.IND.RES. 14B(1955)204

TEMPERATURE = 31.0 DEG C   TYPE OF SYSTEM = 1

EXPERIMENTAL TIE LINES IN MOLE PCT

	LEFT PHASE		RIGHT PHASE	
(1)	(2)	(3)	(1)	(2)	(3)
84.112	10.883	5.005	2.897	3.389	93.713
73.757	20.377	5.867	3.567	6.475	89.958
65.212	27.344	7.443	4.128	9.247	86.625
55.172	34.911	9.917	4.749	12.370	82.881
47.349	40.310	12.342	5.546	16.195	78.259
39.024	44.980	15.996	6.530	19.201	74.269

SPECIFIC MODEL PARAMETERS IN KELVIN

		UNIQUAC		NRTL(ALPHA=.2)	
I	J	AIJ	AJI	AIJ	AJI
1	2	41.157	58.556	286.32	-197.59
1	3	478.43	43.664	689.66	-698.68
2	3	112.93	151.82	352.48	178.78

R1 = 4.8274   R2 = 2.5735   R3 = 2.4088
Q1 = 4.196   Q2 = 2.336   Q3 = 2.248

MEAN DEV. BETWEEN CALC. AND EXP. CONC. IN MOLE PCT

UNIQUAC (SPECIFIC PARAMETERS)   0.43
NRTL   (SPECIFIC PARAMETERS)   0.35

$C_2H_6O_2-C_3H_6O$

(1) C7H8   TOLUENE

(2) C3H6O   2-PROPANONE

(3) C2H6O2   1,2-ETHANEDIOL

TRIMBLE H.M., FRAZER G.E.
IND.ENG.CHEM. 21(1929)1063

TEMPERATURE = 27.0 DEG C   TYPE OF SYSTEM = 1

EXPERIMENTAL TIE LINES IN MOLE PCT (GRAPH.INTERPOL.)

	LEFT PHASE			RIGHT PHASE	
(1)	(2)	(3)	(1)	(2)	(3)
64.937	30.696	4.368	3.182	13.027	83.791
49.637	44.684	5.679	4.156	19.220	76.624
35.862	51.641	12.497	5.845	25.894	68.261

SPECIFIC MODEL PARAMETERS IN KELVIN

		UNIQUAC		NRTL(ALPHA=.2)	
I J		AIJ	AJI	AIJ	AJI
1 2		39.400	12.749	505.07	-349.73
1 3		685.80	119.01	1555.7	702.16
2 3		63.339	160.12	221.83	238.34

R1 = 3.9228   R2 = 2.5735   R3 = 2.4088
Q1 = 2.968   Q2 = 2.336   Q3 = 2.248

MEAN DEV. BETWEEN CALC. AND EXP. CONC. IN MOLE PCT

UNIQUAC (SPECIFIC PARAMETERS)   0.81
NRTL   (SPECIFIC PARAMETERS)   0.84
UNIQUAC (COMMON PARAMETERS)   0.87

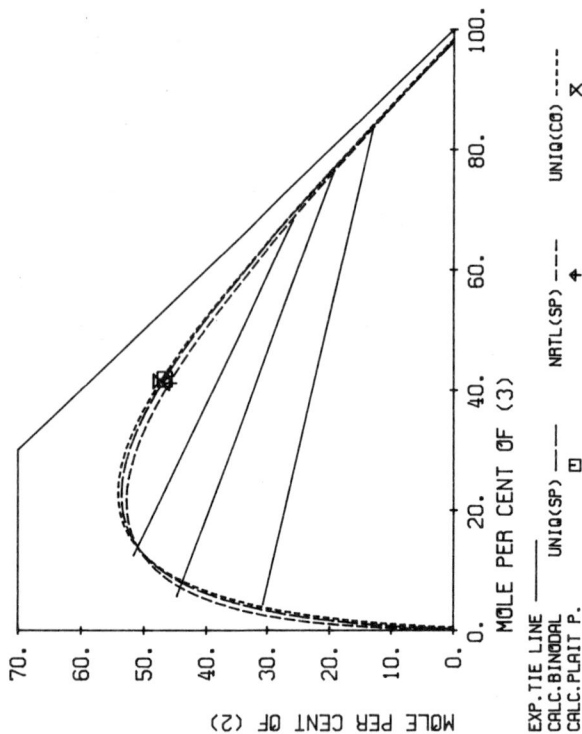

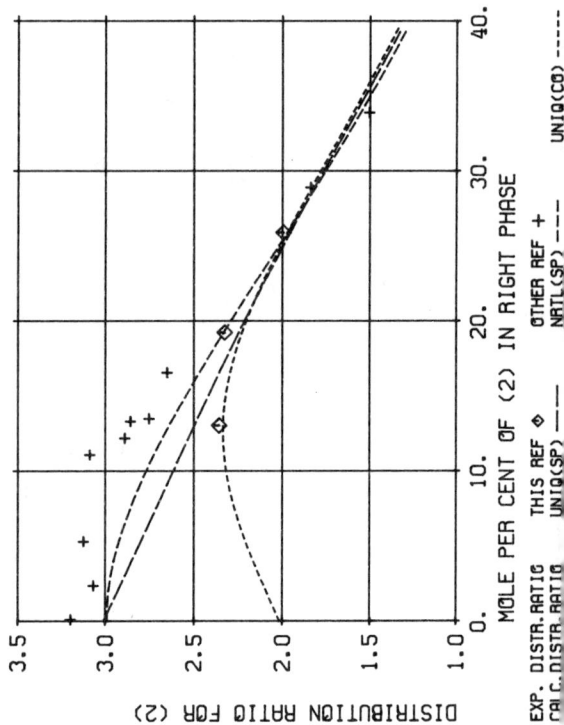

$C_2H_6O_2 - C_3H_6O$

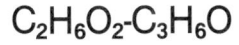

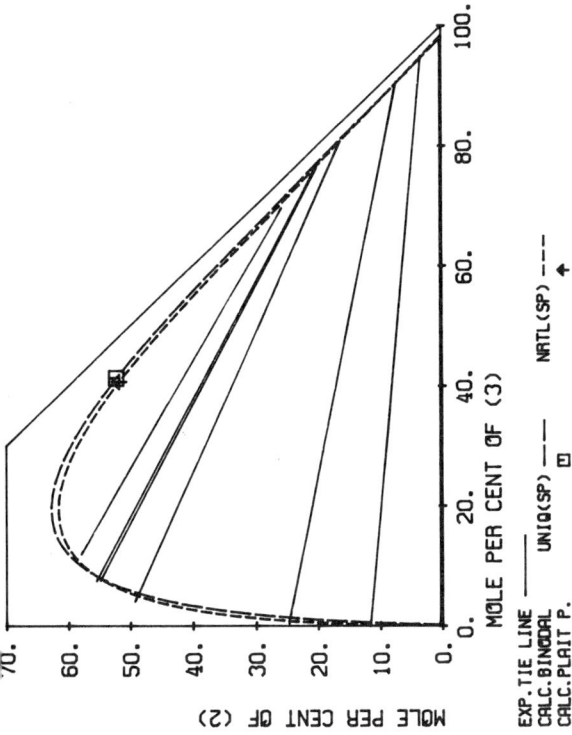

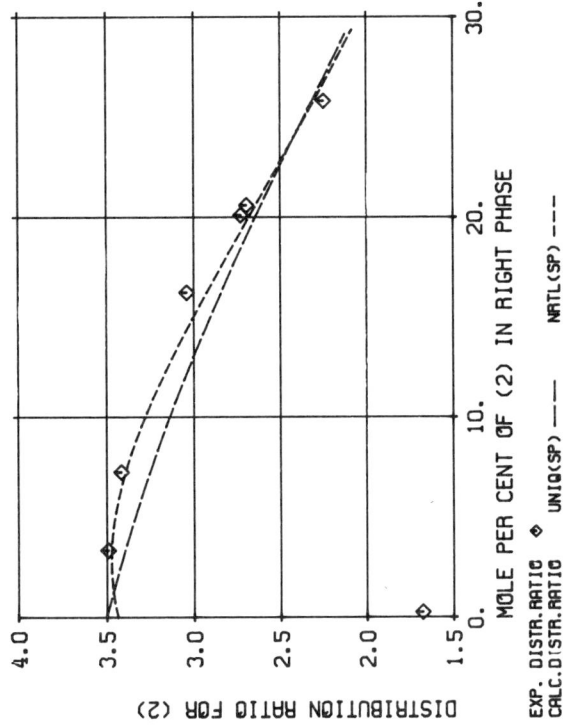

(2) C3H6O   2-PROPANONE
(3) C2H6O2   1,2-ETHANEDIOL

SIMS L.L., BOLME D.W.
J.CHEM.ENG.DATA 10(1965)111

TEMPERATURE =   0.0 DEG C   TYPE OF SYSTEM = 1

EXPERIMENTAL TIE LINES IN MOLE PCT

	LEFT PHASE			RIGHT PHASE	
(1)	(2)	(3)	(1)	(2)	(3)
99.530	0.420	0.050	1.660	0.250	98.090
88.300	11.700	0.0	1.910	3.350	94.740
74.730	24.800	0.470	1.980	7.260	90.760
46.610	49.300	4.090	3.020	16.220	80.760
37.600	54.840	7.560	3.560	20.100	76.340
37.000	55.510	7.490	3.560	20.600	75.840
30.100	57.980	11.920	4.530	25.800	69.670

SPECIFIC MODEL PARAMETERS IN KELVIN

		UNIQUAC		NRTL(ALPHA=.2)	
I	J	AIJ	AJI	AIJ	AJI
1	2	82.764	-49.526	463.19	-316.47
1	3	638.05	111.16	1604.8	672.57
2	3	139.61	87.365	350.10	163.63

R1 = 3.9228   R2 = 2.5735   R3 = 2.4088
Q1 = 2.968   Q2 = 2.336   Q3 = 2.248

MEAN DEV. BETWEEN CALC. AND EXP. CONC. IN MOLE PCT

UNIQUAC (SPECIFIC PARAMETERS)   0.66
NRTL   (SPECIFIC PARAMETERS)   0.45

$C_2H_6O_2$-C_3H_6O

(1) C7H8   TOLUENE

(2) C3H6O   2-PROPANONE

(3) C2H6O2   1,2-ETHANEDIOL

SIMS L.L., BOLME D.W.
J.CHEM.ENG.DATA 10(1965)111

TEMPERATURE = 24.0 DEG C   TYPE OF SYSTEM = 1

EXPERIMENTAL TIE LINES IN MOLE PCT

	LEFT PHASE			RIGHT PHASE	
(1)	(2)	(3)	(1)	(2)	(3)
99.470	0.320	0.210	1.990	0.100	97.910
92.270	7.210	0.520	2.100	2.350	95.550
82.640	16.600	0.760	2.190	5.310	92.500
62.960	34.200	2.840	3.170	11.070	85.760
61.970	35.200	2.830	3.070	12.180	84.750
58.400	38.000	3.600	3.160	13.480	83.540
59.390	37.100	3.510	3.510	13.480	83.010
50.230	43.900	5.870	3.880	16.560	79.560
28.800	53.180	18.020	7.210	28.900	63.890
23.300	51.000	25.700	9.360	33.900	56.740

SPECIFIC MODEL PARAMETERS IN KELVIN

		UNIQUAC		NRTL(ALPHA=.2)	
I J	AIJ	AJI	AIJ	AIJ	AJI
1 2	39.400	12.749	505.07	505.07	-349.73
1 3	685.80	119.01	1555.7	1555.7	702.16
2 3	63.339	160.12	221.83	221.83	238.34

R1 = 3.9228   R2 = 2.5735   R3 = 2.4088
Q1 = 2.968   Q2 = 2.336   Q3 = 2.248

MEAN DEV. BETWEEN CALC. AND EXP. CONC. IN MOLE PCT

UNIQUAC (SPECIFIC PARAMETERS)   1.07
NRTL (SPECIFIC PARAMETERS)   0.64
UNIQUAC (COMMON PARAMETERS)   1.35

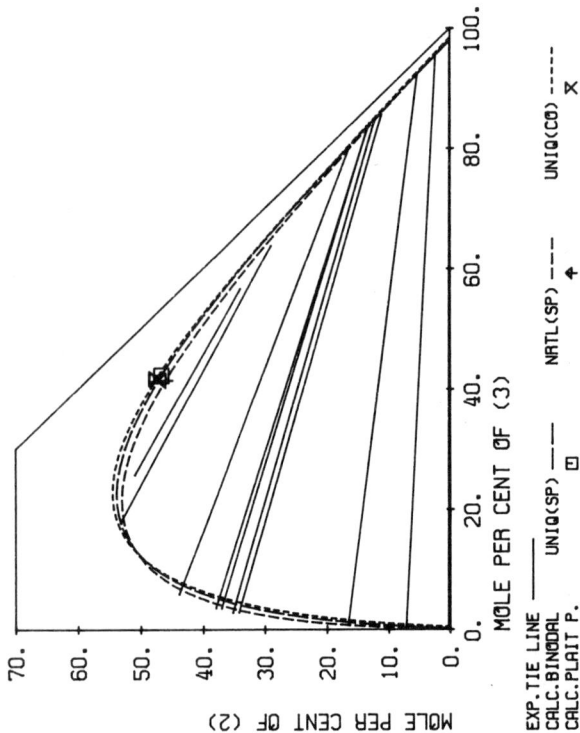

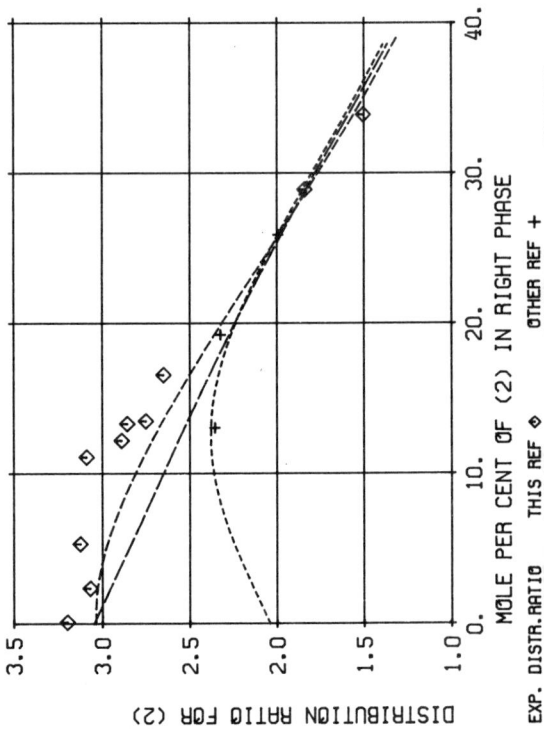

$C_2H_6O_2$-C_3H_6O

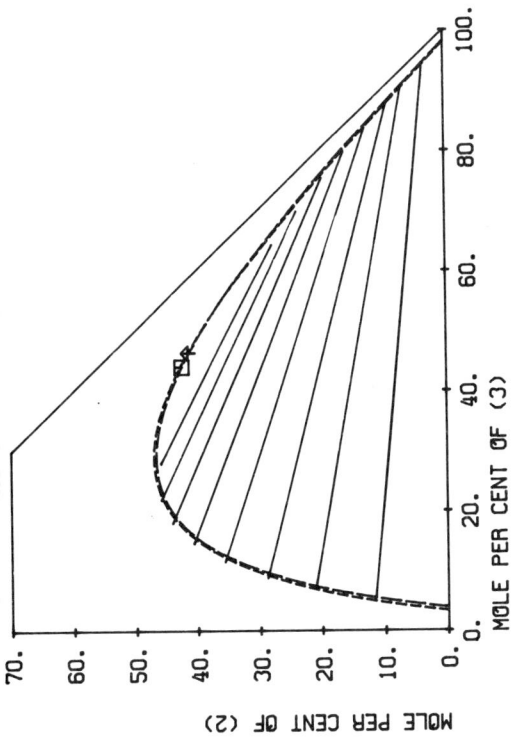

EXP.TIE LINE ——— UNIQ(SP) ——— □ NRTL(SP) ———
CALC.BINODAL
CALC.PLAIT P.   ✛

MOLE PER CENT OF (3)

MOLE PER CENT OF (2)

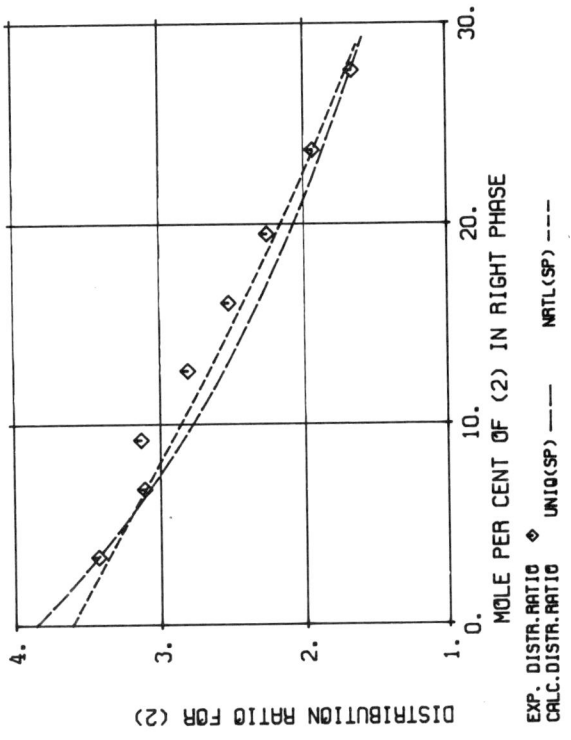

EXP. DISTR.RATIO ◇ UNIQ(SP) ——— NRTL(SP) ———
CALC.DISTR.RATIO

MOLE PER CENT OF (2) IN RIGHT PHASE

DISTRIBUTION RATIO FOR (2)

(1) C3H4O2   ACETIC ACID, PENYL ESTER
(2) C3H6O   2-PROPANONE
(3) C2H6O2   1,2-ETHANEDIOL

RAJA RAO M., VENTAKA RAO C.
J.SCI.IND.RES. 14B(1955)204

TEMPERATURE = 31.0 DEG C   TYPE OF SYSTEM = 1

EXPERIMENTAL TIE LINES IN MOLE PCT

	LEFT PHASE			RIGHT PHASE	
(1)	(2)	(3)	(1)	(2)	(3)
82.722	11.577	5.701	1.993	3.378	94.629
71.845	21.055	7.100	2.437	6.774	90.789
62.336	28.836	8.828	2.936	9.213	87.851
52.862	35.615	11.524	3.538	12.667	83.795
44.784	40.562	14.654	4.197	16.050	79.753
37.911	44.048	18.041	5.020	19.498	75.482
32.273	45.869	21.858	6.387	23.674	69.939
26.215	45.853	27.933	7.955	27.660	64.385

SPECIFIC MODEL PARAMETERS IN KELVIN

		UNIQUAC		NRTL(ALPHA=.2)	
I	J	AIJ	AJI	AIJ	AJI
1	2	-103.52	117.16	169.32	-132.06
1	3	-452.69	57.647	626.51	-832.73
2	3	55.606	95.900	192.50	263.12

R1 = 5.5018   R2 = 2.5735   R3 = 2.4088
Q1 = 4.736   Q2 = 2.336   Q3 = 2.248

MEAN DEV. BETWEEN CALC. AND EXP. CONC. IN MOLE PCT

UNIQUAC (SPECIFIC PARAMETERS)   0.95
NRTL   (SPECIFIC PARAMETERS)   0.56

$C_2H_6O_2\text{-}C_3H_6O_2$

Upper plot axes: MOLE PER CENT OF (2) [vertical] vs MOLE PER CENT OF (3) [horizontal, 0. to 100.]

EXP.TIE LINE —— NRTL(SP) — — ★ UNIQ(CO) — — X
CALC.BINODAL   UNIQ(SP) — — ▢
CALC.PLAIT P.

Lower plot axes: DISTRIBUTION RATIO FOR (2) [vertical, 2. to 6.] vs MOLE PER CENT OF (2) IN RIGHT PHASE [horizontal, 0. to 15.]

EXP. DISTR.RATIO ◇ UNIQ(SP) ▢ NRTL(SP) — — — UNIQ(CO) — — —
CALC.DISTR.RATIO

(1) H2O   WATER
(2) C2H6O2   1,2-ETHANEDIOL
(3) C3H6O2   ACETIC ACID,METHYL ESTER

SAMUEL T., LADDHA G.S.
J.MADRAS UNIV. B28(1958)147

TEMPERATURE = 30.0 DEG C   TYPE OF SYSTEM = 1

EXPERIMENTAL TIE LINES IN MOLE PCT

	LEFT PHASE			RIGHT PHASE	
(1)	(2)	(3)	(1)	(2)	(3)
86.404	5.096	8.499	27.530	1.338	71.132
80.286	9.985	9.728	23.927	3.157	72.915
70.450	17.681	11.870	20.892	5.566	73.542
61.629	24.027	14.344	18.339	7.249	74.412
49.716	31.633	18.651	14.299	11.178	74.524
39.292	37.840	22.868	13.731	13.541	72.727

SPECIFIC MODEL PARAMETERS IN KELVIN

		UNIQUAC		NRTL(ALPHA=.2)	
I J	AIJ	AJI	AIJ	AJI	
1 2	-123.56	-89.715	-148.22	-222.95	
1 3	91.819	288.63	835.70	104.67	
2 3	107.49	174.63	172.93	337.06	

R1 = 0.9200   R2 = 2.4088   R3 = 2.8042
Q1 = 1.400   Q2 = 2.248   Q3 = 2.576

MEAN DEV. BETWEEN CALC. AND EXP. CONC. IN MOLE PCT

UNIQUAC (SPECIFIC PARAMETERS)   1.08
NRTL   (SPECIFIC PARAMETERS)   1.39
UNIQUAC (COMMON PARAMETERS)   1.21

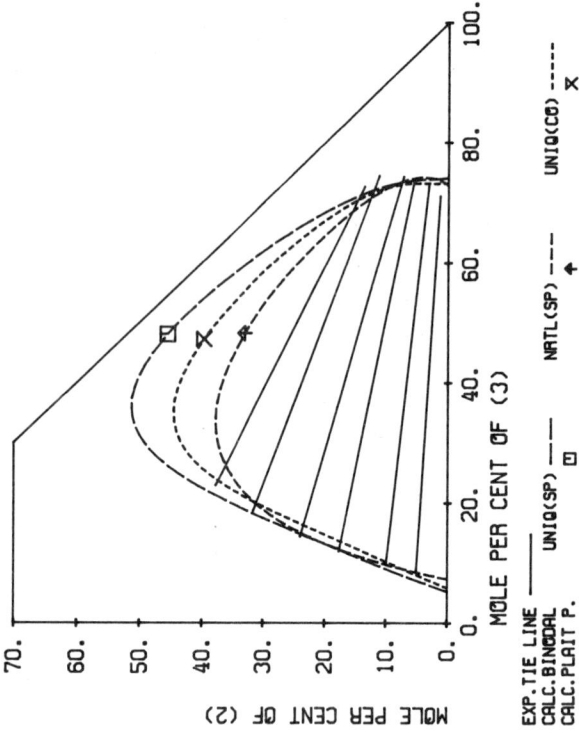

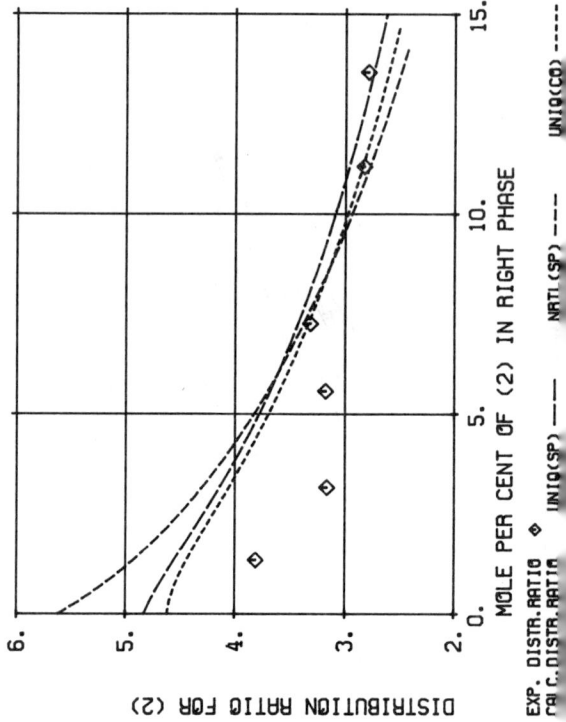

$$C_2H_6O_2\text{-}C_3H_6O_2$$

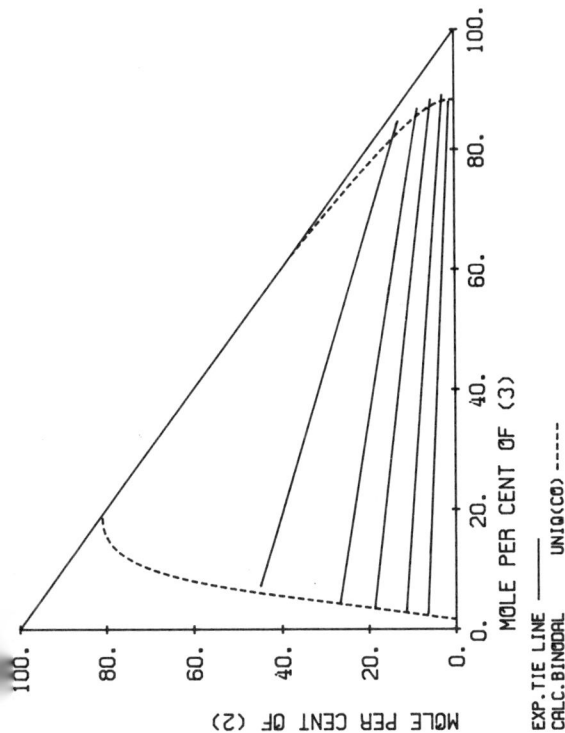

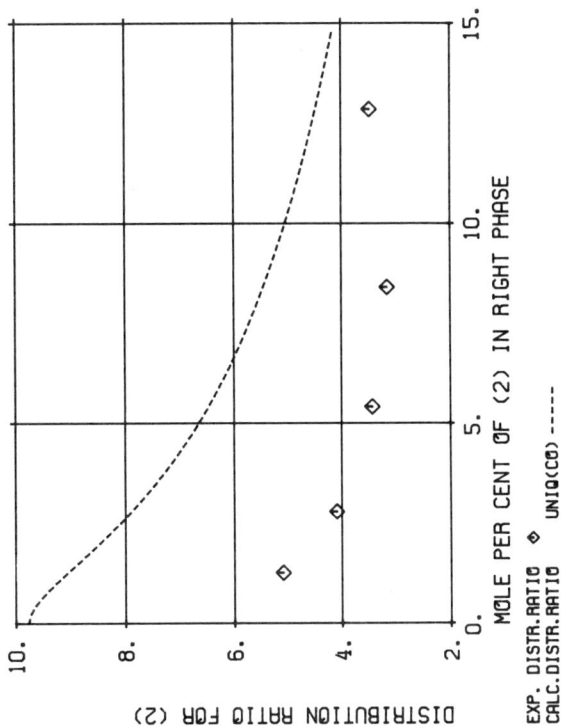

```
------|----------------------------------
 (2)  C2H6O2    1,2-ETHANEDIOL
------|----------------------------------
 (3)  C3H6O2    FORMIC ACID,ETHYL ESTER
------|----------------------------------

SAMUEL T., LADDHA G.S.
J.MADRAS UNIV. B28(1958)147

TEMPERATURE =  30.0 DEG C      TYPE OF SYSTEM = 2

EXPERIMENTAL TIE LINES IN MOLE PCT
---------------------------------------------
        LEFT PHASE              RIGHT PHASE
     (1)      (2)      (3)       (1)      (2)      (3)
---------------------------------------------
   90.908    6.407    2.684     10.743    1.258   87.999
   85.417   11.424    3.159      8.069    2.788   89.143
   77.766   18.612    3.622      6.212    5.409   88.378
   68.996   26.614    4.390      4.692    8.398   86.910
   47.865   44.846    7.289      2.372   12.849   84.780

MEAN DEV. BETWEEN CALC. AND EXP. CONC. IN MOLE PCT
UNIQUAC (COMMON PARAMETERS)           1.84
```

C$_2$H$_6$O$_2$-C$_4$H$_8$O

(1) H2O   WATER

(2) C2H6O2   1,2-ETHANEDIOL

(3) C4H8O   2-BUTANONE

RAJA RAO M., VENKATA RAO C.
J.APPL.CHEM. 7(1957)659

TEMPERATURE = 30.0 DEG C   TYPE OF SYSTEM = 1

EXPERIMENTAL TIE LINES IN MOLE PCT

	LEFT PHASE			RIGHT PHASE	
(1)	(2)	(3)	(1)	(2)	(3)
89.758	3.551	6.691	33.307	0.435	66.258
87.777	5.204	7.019	33.483	1.475	65.042
83.874	8.184	7.942	33.621	3.281	63.098
80.293	10.674	9.033	34.206	4.878	60.916
75.962	13.295	10.743	34.565	6.461	58.974

SPECIFIC MODEL PARAMETERS IN KELVIN

		UNIQUAC		NRTL(ALPHA=.2)	
I J		AIJ	AJI	AIJ	AJI
1 2	-75.212		-281.23	-162.51	-659.29
1 3	82.602		219.54	1115.5	-62.498
2 3	47.055		13.686	-105.17	-126.70

R1 = 0.9200   R2 = 2.4088   R3 = 3.2479
Q1 = 1.400   Q2 = 2.248   Q3 = 2.875

MEAN DEV. BETWEEN CALC. AND EXP. CONC. IN MOLE PCT

UNIQUAC (SPECIFIC PARAMETERS)   0.48
NRTL   (SPECIFIC PARAMETERS)   0.50
UNIQUAC (COMMON PARAMETERS)   2.50

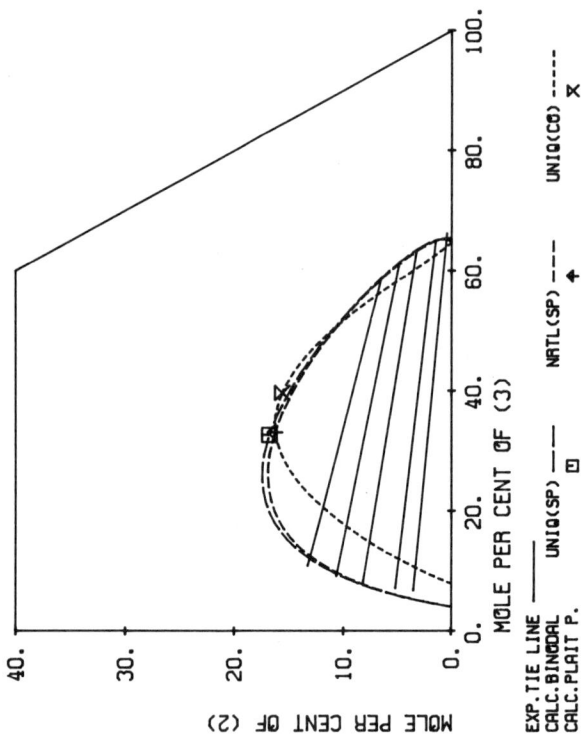

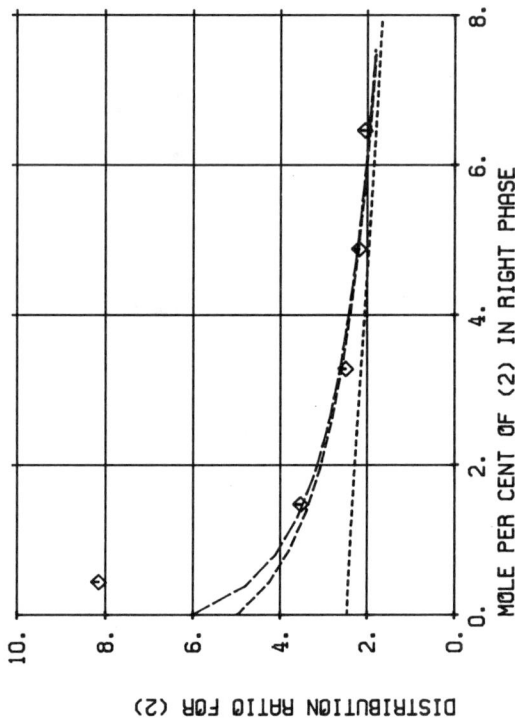

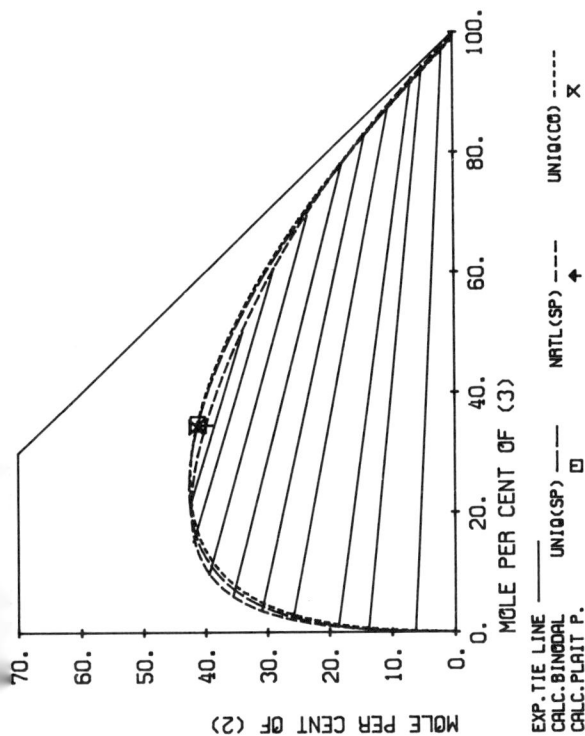

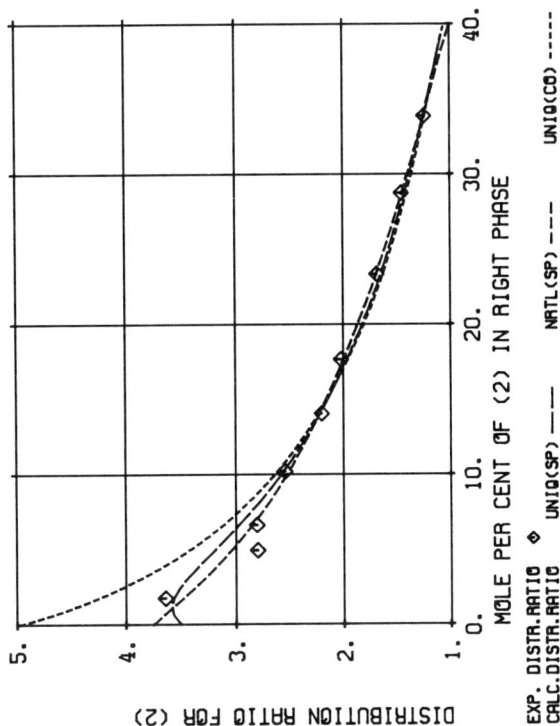

$C_2H_6O_2$-$C_4H_{10}O$

(2) C4H10O   1-BUTANOL

(3) C7H16   HEPTANE

KOGAN V.B., FRIDMAN V.M., ROMANOVA T.G.
ZH.PRIKL.KHIM.(LENINGRAD) 32(1959)847

TEMPERATURE = 20.0 DEG C   TYPE OF SYSTEM = 1

EXPERIMENTAL TIE LINES IN MOLE PCT

| | LEFT PHASE | | | RIGHT PHASE | |
(1)	(2)	(3)	(1)	(2)	(3)
93.413	6.367	0.220	0.161	1.749	98.091
85.690	13.800	0.510	0.636	4.926	94.438
80.409	18.591	1.001	0.870	6.621	92.509
71.819	25.973	2.208	1.795	10.194	88.012
65.354	30.924	3.723	2.997	14.030	82.972
58.345	35.660	5.995	3.942	17.647	78.411
50.887	39.475	9.639	7.214	23.301	69.485
42.870	41.998	15.132	10.436	28.732	60.832
35.680	42.219	22.100	15.942	33.835	50.223

SPECIFIC MODEL PARAMETERS IN KELVIN

| I J | UNIQUAC | | NRTL(ALPHA=.2) | |
	AIJ	AJI	AIJ	AJI
1 2	340.57	-224.40	907.30	-692.25
1 3	325.98	457.50	1680.5	1088.4
2 3	-181.00	386.65	315.87	-47.916

R1 = 2.4088   R2 = 3.4543   R3 = 5.1742
Q1 = 2.248   Q2 = 3.052   Q3 = 4.396

MEAN DEV. BETWEEN CALC. AND EXP. CONC. IN MOLE PCT

UNIQUAC (SPECIFIC PARAMETERS)   0.54
NRTL   (SPECIFIC PARAMETERS)   0.22
UNIQUAC (COMMON PARAMETERS)   0.74

MOLE PER CENT OF (3)

MOLE PER CENT OF (2)

EXP.TIE LINE ———   UNIQ(SP) ⊡ ----   NRTL(SP) ---- ✦   UNIQ(CO) ---- ✕
CALC.BINODAL
CALC.PLAIT P.

MOLE PER CENT OF (2) IN RIGHT PHASE

DISTRIBUTION RATIO FOR (2)

EXP. DISTR.RATIO ◇   UNIQ(SP) ———   NRTL(SP) ---   UNIQ(CO) ----
CALC.DISTR.RATIO

$C_2H_6O_2$-$C_4H_{10}O$

(1) H2O   WATER
----
(2) C2H6O2   1,2-ETHANEDIOL
----
(3) C4H10O   1-BUTANOL

RAJA RAO M., VENKATA RAO C.
J.APPL.CHEM. 7(1957)659

TEMPERATURE = 27.0 DEG C   TYPE OF SYSTEM = 1

EXPERIMENTAL TIE LINES IN MOLE PCT

	LEFT PHASE			RIGHT PHASE	
(1)	(2)	(3)	(1)	(2)	(3)
95.236	2.787	1.977	51.571	2.531	45.899
93.621	4.246	2.133	51.957	3.303	44.740
92.305	5.414	2.281	52.884	4.527	42.589
90.927	6.544	2.529	54.870	6.041	39.089
89.378	7.793	2.828	56.798	6.874	35.328
88.232	8.587	3.182	58.346	7.946	33.708
85.938	10.064	3.998	60.977	9.068	29.955
83.503	11.073	5.424	65.147	10.244	24.609

SPECIFIC MODEL PARAMETERS IN KELVIN

		UNIQUAC		NRTL(ALPHA=.2)	
I J	AIJ	AJI		AIJ	AJI
1 2	-251.87	22.178		-337.95	23.810
1 3	489.01	-81.612		1801.0	-340.88
2 3	-68.654	304.64		-56.161	512.09

R1 = 0.9200   R2 = 2.4083   R3 = 3.4543
Q1 = 1.400   Q2 = 2.248   Q3 = 3.052

MEAN DEV. BETWEEN CALC. AND EXP. CONC. IN MOLE PCT

UNIQUAC (SPECIFIC PARAMETERS)   0.41
NRTL   (SPECIFIC PARAMETERS)   0.41
UNIQUAC (COMMON PARAMETERS)   1.49

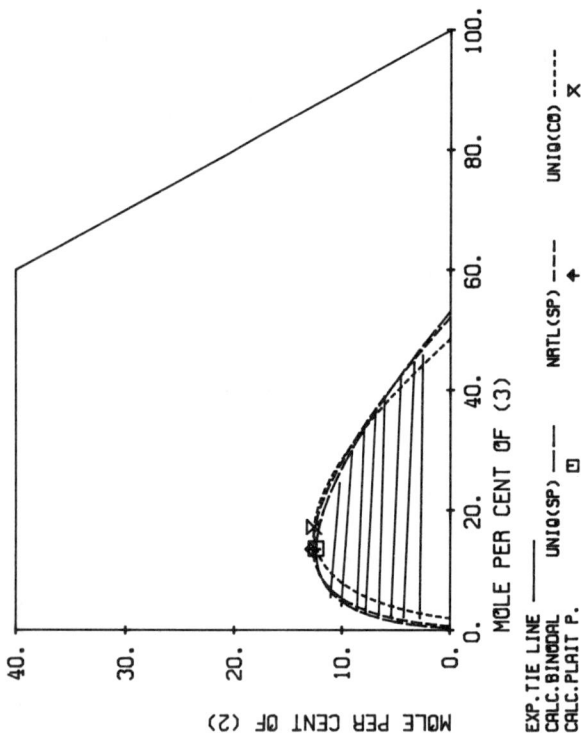

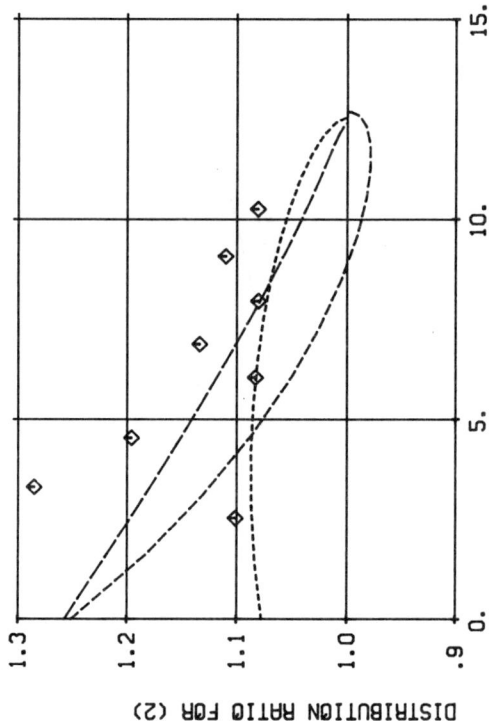

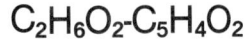

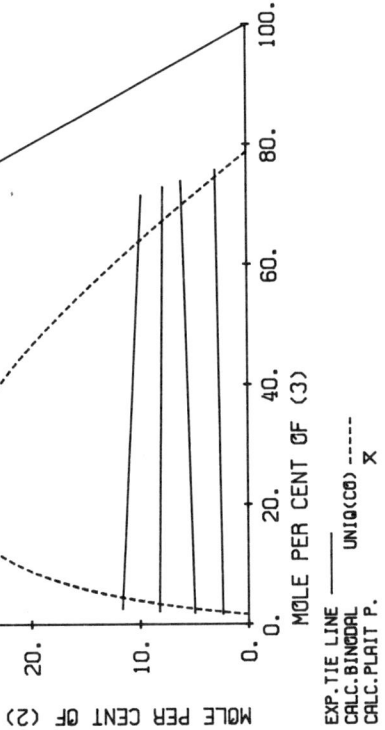

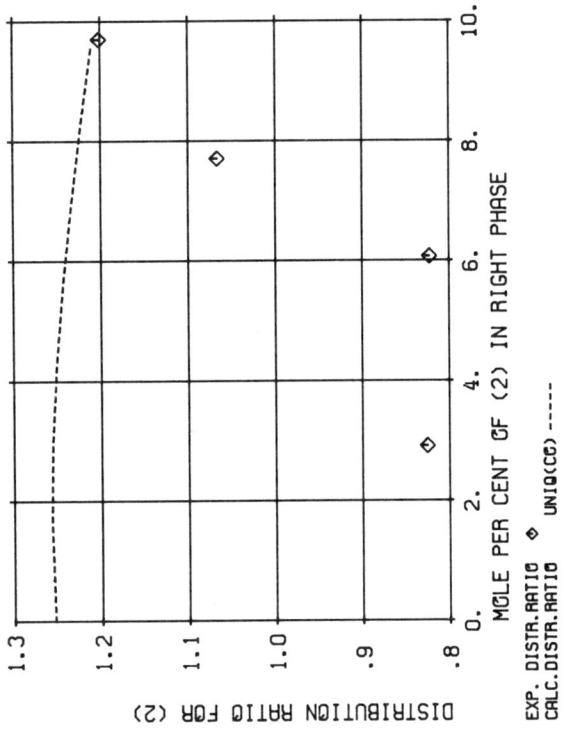

(2) C2H6O2   1,2-ETHANEDIOL

(3) C5H4O2   FURFURAL

CONWAY J.B., NORTON J.J.
IND.ENG.CHEM. 43(1951)1433

TEMPERATURE = 25.0 DEG C   TYPE OF SYSTEM = 1

EXPERIMENTAL TIE LINES IN MOLE PCT (GRAPH.INTERPOL.)

	LEFT PHASE			RIGHT PHASE	
(1)	(2)	(3)	(1)	(2)	(3)
95.893	2.400	1.707	21.336	2.907	75.757
93.148	4.992	1.860	19.982	6.065	73.953
89.675	8.214	2.110	19.356	7.701	72.943
85.839	11.650	2.511	18.742	9.696	71.562

MEAN DEV. BETWEEN CALC. AND EXP. CONC. IN MOLE PCT

UNIQUAC (COMMON PARAMETERS)   2.21

C₂H₆O₂-C₅H₁₂O

```
(1) C7H16      HEPTANE
(2) C5H12O     1-PENTANOL
(3) C2H6O2     1,2-ETHANEDIOL

ALEKSANDROVA M.V., ET AL.
IZV.VYSSH.UCHEBN.ZAVED.KHIM.KHIM.TEKHNOL. 10(1967)678

TEMPERATURE = 20.0 DEG C      TYPE OF SYSTEM = 1

EXPERIMENTAL TIE LINES IN MOLE PCT

        LEFT PHASE              RIGHT PHASE
   (1)     (2)     (3)      (1)     (2)     (3)

 97.121   2.718   0.161    0.383   6.824  92.793
 89.834   9.529   0.637    0.988  13.706  85.306
 84.623  14.586   0.791    1.537  16.564  81.899
 72.517  22.099   5.385    3.659  21.894  74.447
 66.152  24.784   9.064    5.424  24.745  69.830
 62.227  26.278  11.494    5.805  25.510  68.685

SPECIFIC MODEL PARAMETERS IN KELVIN

              UNIQUAC              NRTL(ALPHA=.2)
 I  J    AIJ       AJI          AIJ        AJI

 1  2   505.33   -287.04      -502.99    -64.331
 1  3   627.86    329.15      1425.6     1307.3
 2  3   -26.170   -47.868     -626.62    36.399

 R1 = 5.1742   R2 = 4.1287   R3 = 2.4088
 Q1 = 4.396    Q2 = 3.592    Q3 = 2.248

MEAN DEV. BETWEEN CALC. AND EXP. CONC. IN MOLE PCT

UNIQUAC (SPECIFIC PARAMETERS)    0.48
NRTL   (SPECIFIC PARAMETERS)     0.43
UNIQUAC (COMMON PARAMETERS)      1.28
```

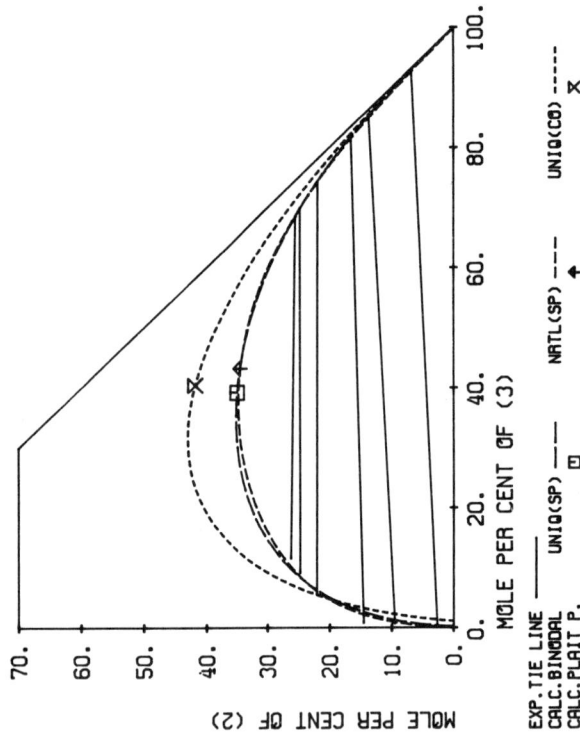

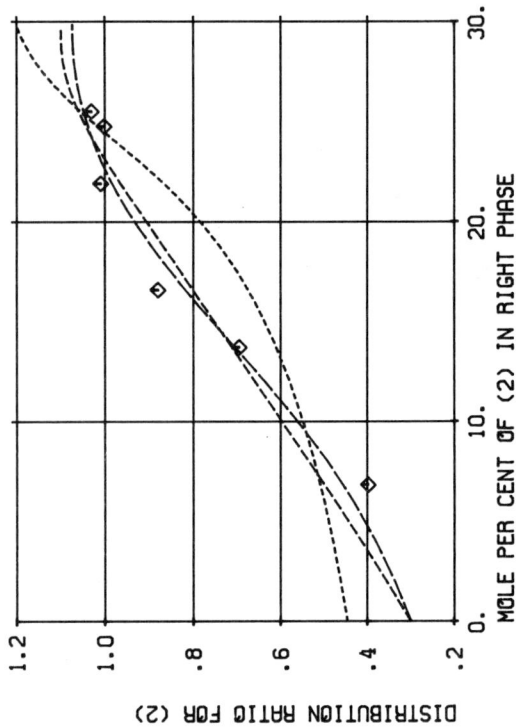

$C_2H_6O_2$-$C_5H_{12}O$

EXP.TIE LINE ——— UNIQ(SP) ——— ☐ NRTL(SP) ----- ♦ UNIQ(CC) -----
CALC.BINODAL
CALC.PLAIT P.

MOLE PER CENT OF (3)

MOLE PER CENT OF (2)

MOLE PER CENT OF (2) IN RIGHT PHASE

DISTRIBUTION RATIO FOR (2)

EXP. DISTR.RATIO ◇ UNIQ(SP) ——— NRTL(SP) —— UNIQ(CC) -----
CALC.DISTR.RATIO

(2) C2H6O2   1,2-ETHANEDIOL
(3) C5H12O   1-PENTANOL

LADDHA G.S., SMITH J.M.
IND.ENG.CHEM. 40(1948)494

TEMPERATURE = 20.0 DEG C   TYPE OF SYSTEM = 1

EXPERIMENTAL TIE LINES IN MOLE PCT

	LEFT PHASE			RIGHT PHASE	
(1)	(2)	(3)	(1)	(2)	(3)
94.188	5.461	0.352	32.420	1.986	65.593
90.996	8.578	0.426	33.033	3.814	63.153
87.221	12.294	0.485	33.341	5.908	60.751
82.656	16.756	0.586	33.622	8.149	58.229
79.636	19.655	0.709	34.475	9.510	56.015
74.475	24.656	0.870	36.398	13.610	49.993
68.679	30.081	1.240	38.280	16.527	45.193

SPECIFIC MODEL PARAMETERS IN KELVIN

I J	UNIQUAC AIJ	AJI	NRTL(ALPHA=.2) AIJ	AJI
1 2	-179.30	-119.62	-475.37	433.34
1 3	684.12	-46.268	2212.8	-165.98
2 3	-52.267	138.11	353.38	207.80

R1 = 0.9200   R2 = 2.4088   R3 = 4.1287
Q1 = 1.400   Q2 = 2.248   Q3 = 3.592

MEAN DEV. BETWEEN CALC. AND EXP. CONC. IN MOLE PCT

UNIQUAC (SPECIFIC PARAMETERS)   1.01
NRTL   (SPECIFIC PARAMETERS)   1.11
UNIQUAC (COMMON PARAMETERS)   1.75

C₂H₆O₂-C₆H₆

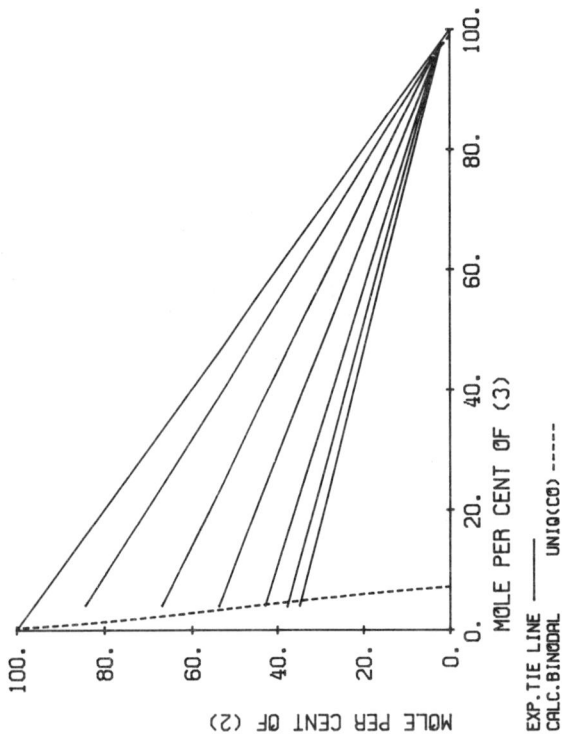

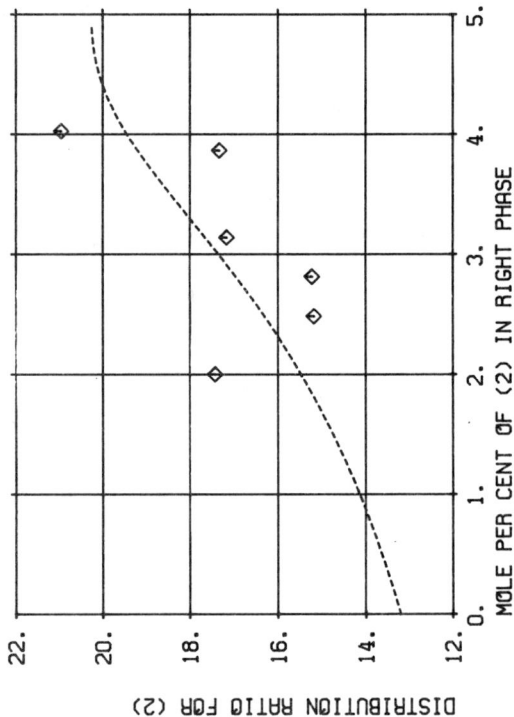

(1) C6H14   HEXANE

(2) C6H6   BENZENE

(3) C2H6O2   1,2-ETHANEDIOL

KUGO M., FUJIKAWA M., TAMAHORI T.
MEM.FAC.ENG.HOKKAIDO UNIV. 11(1960)41

TEMPERATURE = 30.0 DEG C   TYPE OF SYSTEM = 2

EXPERIMENTAL TIE LINES IN MOLE PCT

	LEFT PHASE			RIGHT PHASE	
(1)	(2)	(3)	(1)	(2)	(3)
61.203	34.814	3.983	0.217	1.998	97.784
53.481	37.678	3.841	0.290	2.482	97.228
53.119	42.798	4.083	0.947	2.811	96.242
42.241	53.849	3.910	1.167	3.138	95.695
29.084	66.929	3.987	0.729	3.863	95.408
11.646	84.312	4.042	0.657	4.024	95.319

MEAN DEV. BETWEEN CALC. AND EXP. CONC. IN MOLE PCT

UNIQUAC (COMMON PARAMETERS)   0.82

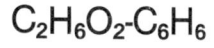

MOLE PER CENT OF (3)

MOLE PER CENT OF (2)

EXP.TIE LINE

(1) C6H14   HEXANE

(2) C6H6   BENZENE

(3) C2H6O2   1,2-ETHANEDIOL

KUGO M., FUJIKAWA M., TAMAHORI T.
MEM.FAC.ENG.HOKKAIDO UNIV. 11(1960)41

TEMPERATURE = 50.0 DEG C   TYPE OF SYSTEM = 2

EXPERIMENTAL TIE LINES IN MOLE PCT

	LEFT PHASE			RIGHT PHASE	
(1)	(2)	(3)	(1)	(2)	(3)
60.950	34.673	4.377	1.528	1.605	96.867
59.417	35.295	5.288	1.090	1.924	96.985
57.106	40.243	2.652	1.163	1.925	96.912
43.831	53.552	2.617	1.164	2.006	96.831
59.093	36.275	4.632	1.457	2.250	96.293
30.784	67.921	1.295	0.733	5.495	93.772

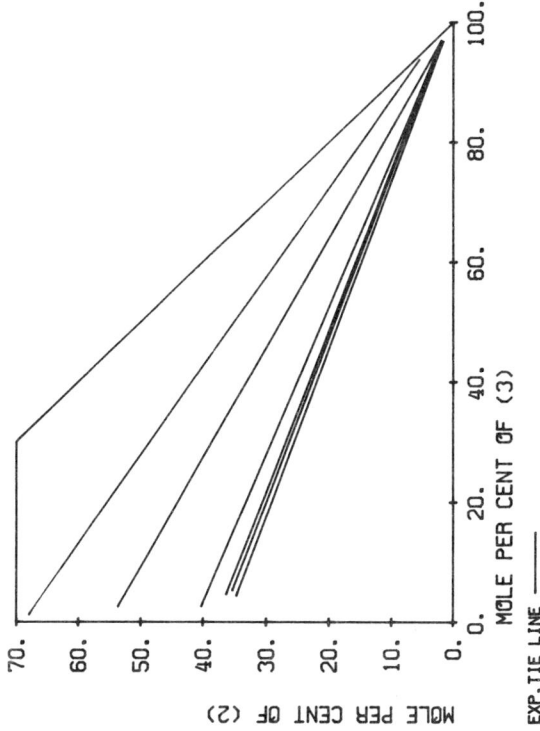

$C_2H_6O_2$-C_6H_7N

(1) C7H9N   ANILINE, N-METHYL

(2) C6H7N   ANILINE

(3) C2H6O2   1,2-ETHANEDIOL

CRUETZEN J.L., JOST W., SIEG L.
Z.ELEKTROCHEM. 61(1957)230

TEMPERATURE = 20.0 DEG C   TYPE OF SYSTEM = 1

EXPERIMENTAL TIE LINES IN MOLE PCT

	LEFT PHASE			RIGHT PHASE	
(1)	(2)	(3)	(1)	(2)	(3)
85.200	1.400	13.400	18.000	0.750	81.250
83.300	2.800	13.900	19.000	1.600	79.400
78.000	5.500	16.500	21.400	3.000	75.600
72.500	7.500	20.200	25.500	4.500	70.000
65.000	8.400	26.600	34.600	6.100	59.300

SPECIFIC MODEL PARAMETERS IN KELVIN

		UNIQUAC		NRTL(ALPHA=.2)	
I J	AIJ	AJI		AIJ	AJI
1 2	-137.38	171.13		125.67	-723.36
1 3	391.79	-16.970		565.99	316.56
2 3	-90.032	255.81		-240.67	-404.31

R1 = 4.4555   R2 = 3.7165   R3 = 2.4088
Q1 = 3.364   Q2 = 2.815   Q3 = 2.248

MEAN DEV. BETWEEN CALC. AND EXP. CONC. IN MOLE PCT

UNIQUAC (SPECIFIC PARAMETERS)   0.73
NRTL   (SPECIFIC PARAMETERS)   0.67
UNIQUAC (COMMON PARAMETERS)   1.49

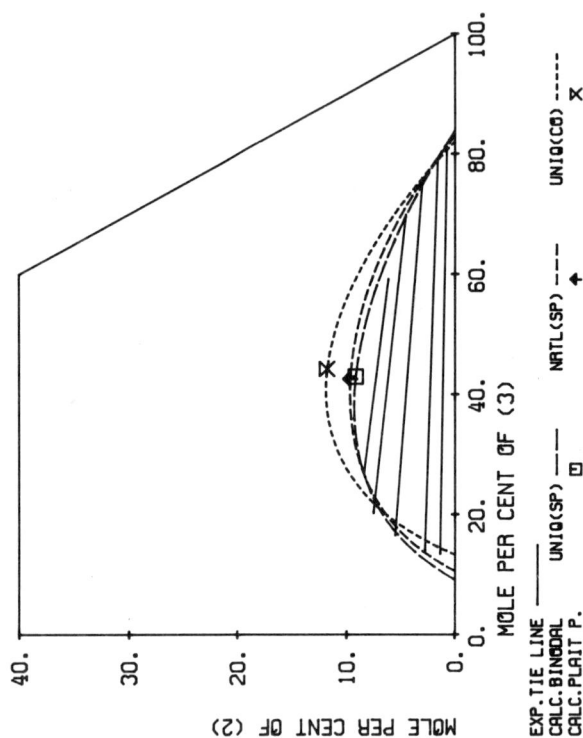

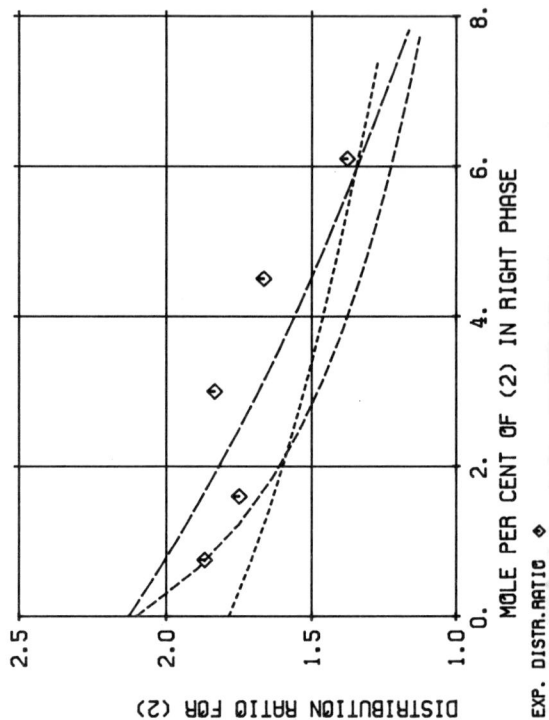

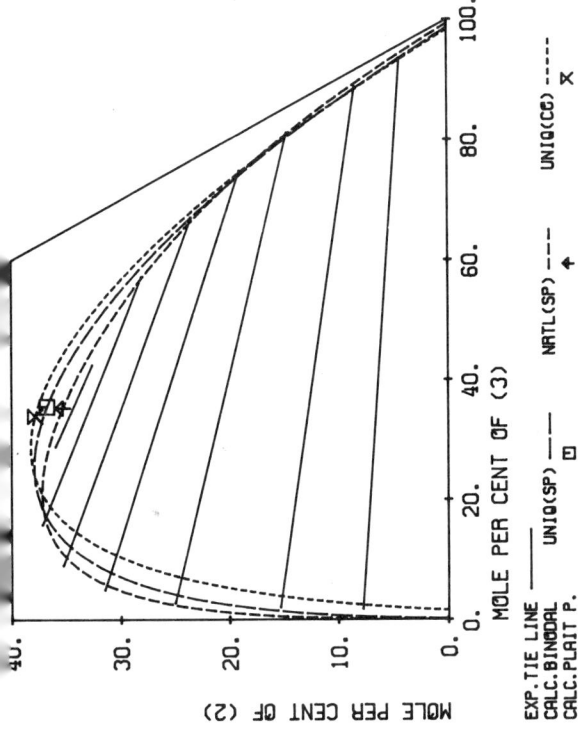

MOLE PER CENT OF (3)

MOLE PER CENT OF (2)

EXP.TIE LINE ———   UNIQ(SP) ⊟   NRTL(SP) ⊹   UNIQ(CC) - - - -
CALC.BINODAL - - -
CALC.PLAIT P.   ⋈

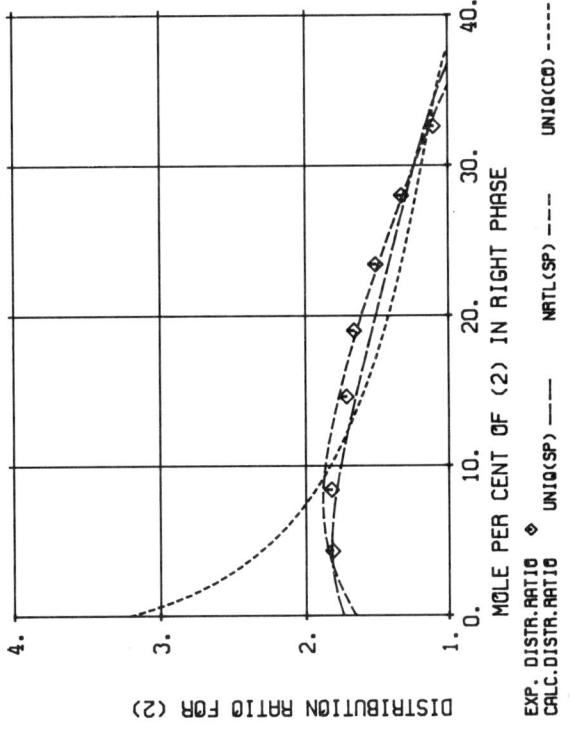

MOLE PER CENT OF (2) IN RIGHT PHASE

DISTRIBUTION RATIO FOR (2)

EXP. DISTR.RATIO ◇   UNIQ(SP) ———   NRTL(SP) - - -   UNIQ(CC) - - - -
CALC.DISTR.RATIO ———

(2) C6H7N   ANILINE

(3) C2H6O2   1,2-ETHANEDIOL

CRUETZEN J.L., JOST W.; SIEG L.
Z.ELEKTROCHEM. 61(1957)230

TEMPERATURE = 20.0 DEG C   TYPE OF SYSTEM = 1

EXPERIMENTAL TIE LINES IN MOLE PCT

| | LEFT PHASE | | | RIGHT PHASE | |
(1)	(2)	(3)	(1)	(2)	(3)
90.500	7.800	1.700	2.100	4.300	93.600
82.700	15.300	2.000	2.600	8.400	89.000
72.300	25.000	2.700	4.100	14.600	81.300
63.600	31.500	4.900	6.000	19.000	75.000
55.700	35.300	9.000	9.400	23.400	67.200
47.000	37.200	15.800	14.700	28.000	57.300
35.200	36.000	28.800	24.800	32.600	42.600

SPECIFIC MODEL PARAMETERS IN KELVIN

| | | UNIQUAC | | NRTL(ALPHA=.2) | |
I	J	AIJ	AJI	AIJ	AJI
1	2	429.17	-278.68	1045.6	-655.00
1	3	806.06	52.769	1982.6	963.06
2	3	179.91	-127.27	100.81	107.27

R1 = 5.1094   R2 = 3.7165   R3 = 2.4088
Q1 = 3.920   Q2 = 2.816   Q3 = 2.248

MEAN DEV. BETWEEN CALC. AND EXP. CONC. IN MOLE PCT

UNIQUAC (SPECIFIC PARAMETERS)   1.01
NRTL   (SPECIFIC PARAMETERS)   0.54
UNIQUAC (COMMON PARAMETERS)   1.71

$C_2H_6O_2$-$C_6H_{10}O$

(1). H2O   WATER

(2) C2H6O2   1,2-ETHANEDIOL

(3) C6H10O   CYCLOHEXANONE

SAMUEL T., LADDHA G.S.
J.MADRAS UNIV. B28(1958)147

TEMPERATURE = 30.0 DEG C   TYPE OF SYSTEM = 1

EXPERIMENTAL TIE LINES IN MOLE PCT

	LEFT PHASE			RIGHT PHASE	
(1)	(2)	(3)	(1)	(2)	(3)
92.946	5.132	1.923	31.073	1.523	67.404
86.923	10.628	2.450	31.235	3.984	64.780
81.409	15.385	3.206	31.323	5.985	62.692
74.047	20.988	4.965	31.537	9.019	59.445
64.600	27.122	8.278	31.451	12.978	55.571

SPECIFIC MODEL PARAMETERS IN KELVIN

		UNIQUAC		NRTL(ALPHA=.2)	
I J	AIJ	AJI	AIJ	AJI	
1 2	-133.89	-122.17	-300.56	46.911	
1 3	161.98	99.226	1617.1	-138.38	
2 3	-38.264	197.18	255.69	223.73	

R1 = 0.9200   R2 = 2.4088   R3 = 4.1433
Q1 = 1.400   Q2 = 2.248   Q3 = 3.340

MEAN DEV. BETWEEN CALC. AND EXP. CONC. IN MOLE PCT

UNIQUAC (SPECIFIC PARAMETERS)   0.45
NRTL   (SPECIFIC PARAMETERS)   0.30
UNIQUAC (COMMON PARAMETERS)   3.07

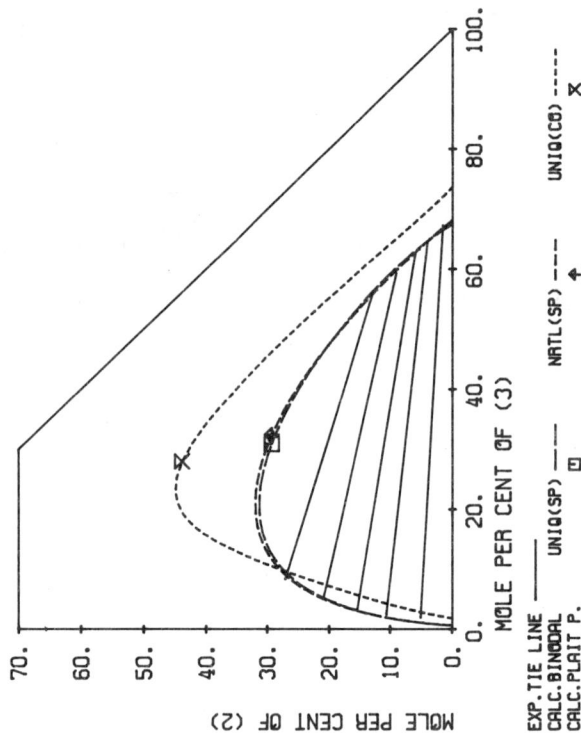

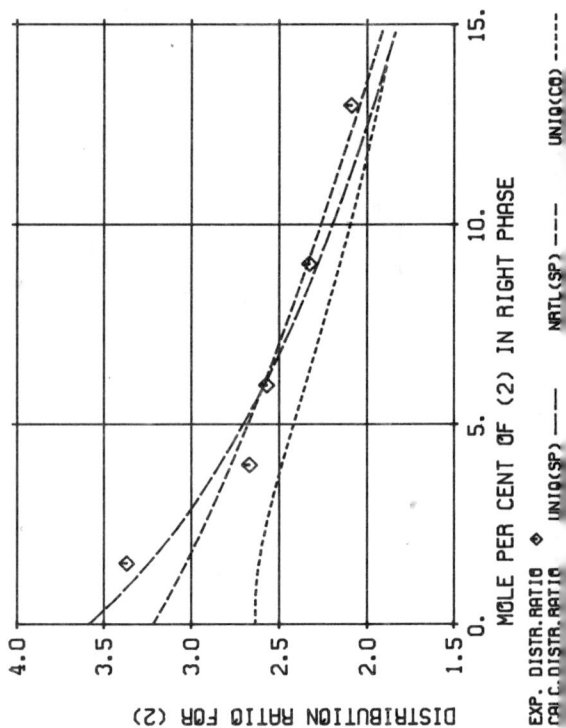

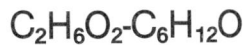

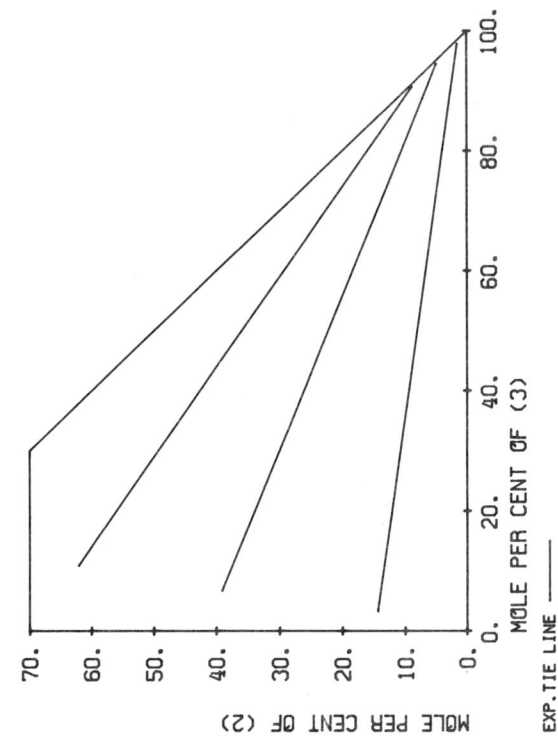

$C_2H_6O_2-C_6H_{12}O$

MOLE PER CENT OF (3)

MOLE PER CENT OF (2)

EXP.TIE LINE ——

(1) C6H14   HEXANE

(2) C6H12O   FURAN,2,5-DIMETHYL,TETRAHYDRO

(3) C2H6O2   1,2-ETHANEDIOL

HUTTON D.G., JONES J.H.
J.CHEM.ENG.DATA 8(1963)617

TEMPERATURE = 120.0 DEG C   TYPE OF SYSTEM = 2

EXPERIMENTAL TIE LINES IN MOLE PCT

	LEFT PHASE			RIGHT PHASE	
(1)	(2)	(3)	(1)	(2)	(3)
82.361	14.400	3.239	0.583	1.504	97.913
54.223	39.104	6.673	0.565	4.887	94.548
27.080	62.072	10.849	0.426	8.792	90.782

430

$C_2H_6O_2-C_6H_{12}O$

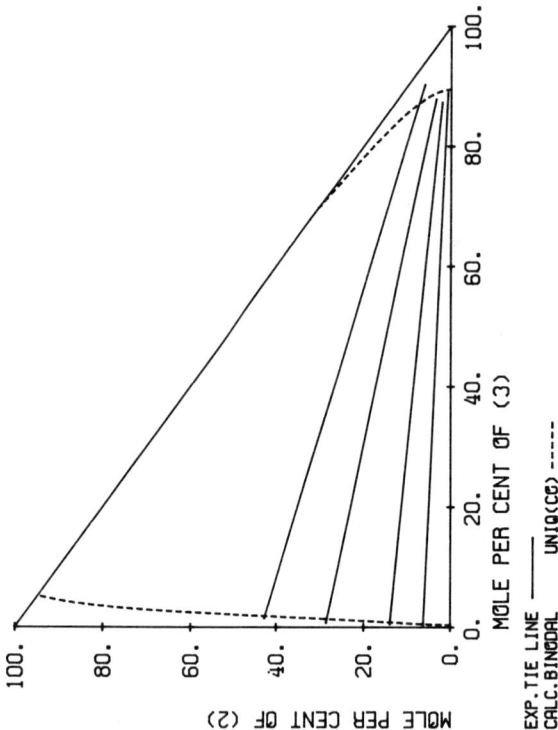

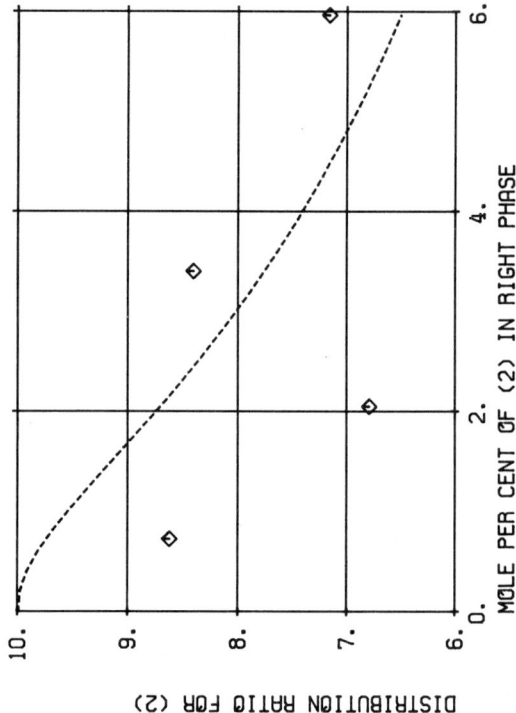

(1) H2O   WATER
----
(2) C2H6O2   1,2-ETHANEDIOL
----
(3) C6H12O   2-PENTANONE, 4-METHYL

SAMUEL T., LADDHA G.S.
J.MADRAS UNIV. B28(1958)147

TEMPERATURE = 30.0 DEG C   TYPE OF SYSTEM = 2

EXPERIMENTAL TIE LINES IN MOLE PCT

	LEFT PHASE		RIGHT PHASE		
(1)	(2)	(3)	(1)	(2)	(3)
93.480	6.228	0.291	10.209	0.722	89.069
85.597	13.893	0.510	10.572	2.046	87.382
70.684	28.580	0.735	8.658	3.400	87.942
55.888	42.682	1.430	3.686	5.960	90.354

MEAN DEV. BETWEEN CALC. AND EXP. CONC. IN MOLE PCT

UNIQUAC (COMMON PARAMETERS)   1.00

$C_2H_6O_2-C_6H_{14}O$

(2) C6H14O   1-HEXANOL

(3) C2H6O2   1,2-ETHANEDIOL

ALEKSANDROVA M.V., ET AL.
IZV.VYSSH.UCHEBN.ZAVED.KHIM.KHIM.TEKHNOL. 10(1967)678

TEMPERATURE = 20.0 DEG C   TYPE OF SYSTEM = 1

EXPERIMENTAL TIE LINES IN MOLE PCT

	LEFT PHASE			RIGHT PHASE		
(1)	(2)	(3)		(1)	(2)	(3)
97.289	2.550	0.161		0.190	3.102	96.709
94.480	5.197	0.323		0.255	4.315	95.430
86.451	12.742	0.807		0.453	6.671	92.875
80.533	17.378	2.089		0.589	8.216	91.195
73.070	22.467	4.464		0.858	9.320	89.822

SPECIFIC MODEL PARAMETERS IN KELVIN

		UNIQUAC			NRTL(ALPHA=.2)	
I	J	AIJ	AJI		AIJ	AJI
1	2	324.63	-153.18		855.78	-197.83
1	3	607.17	174.30		929.28	1195.1
2	3	126.64	48.812		259.95	409.12

R1 = 5.1742   R2 = 4.8031   R3 = 2.4088
Q1 = 4.396   Q2 = 4.132   Q3 = 2.248

MEAN DEV. BETWEEN CALC. AND EXP. CONC. IN MOLE PCT

UNIQUAC (SPECIFIC PARAMETERS)   0.45
NRTL   (SPECIFIC PARAMETERS)   0.55
UNIQUAC (COMMON PARAMETERS)   1.07

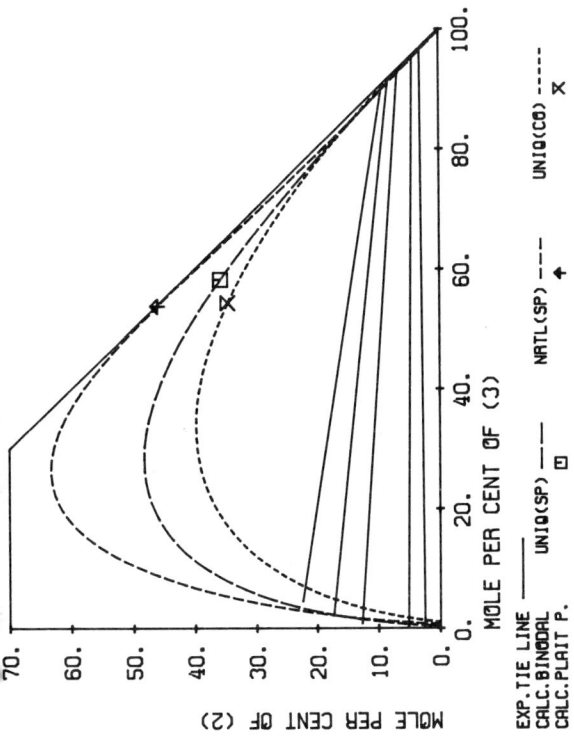

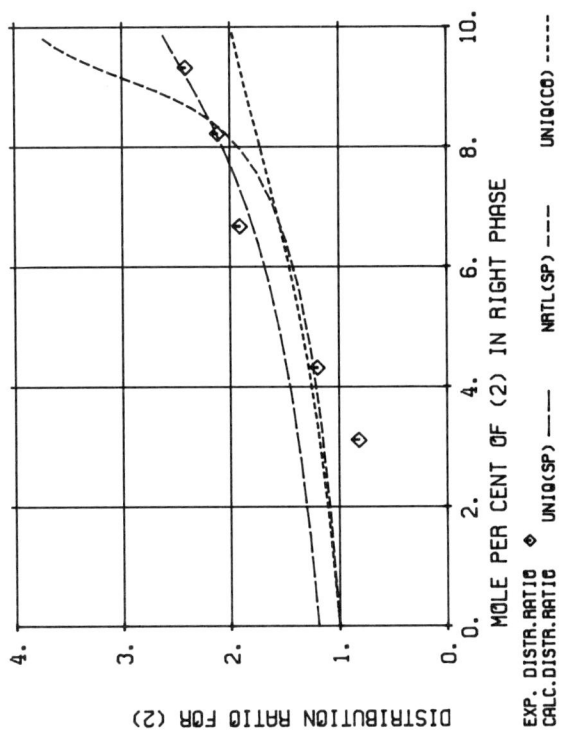

$C_2H_6O_2$-$C_6H_{14}O$

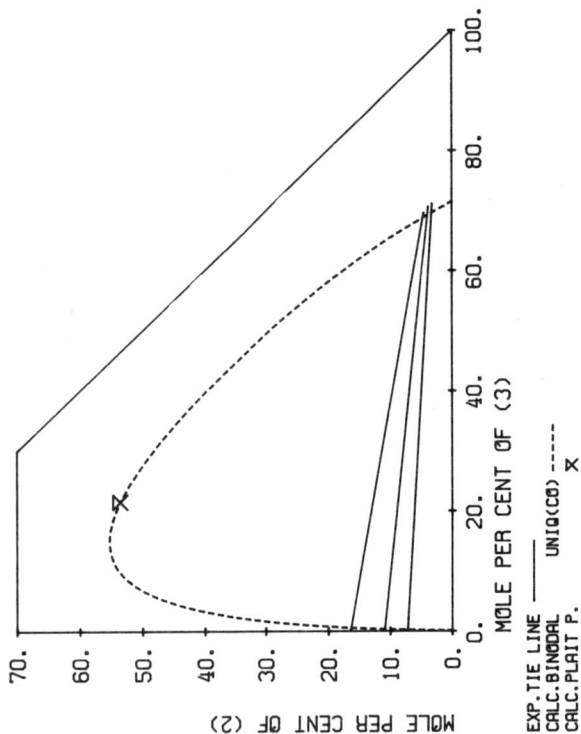

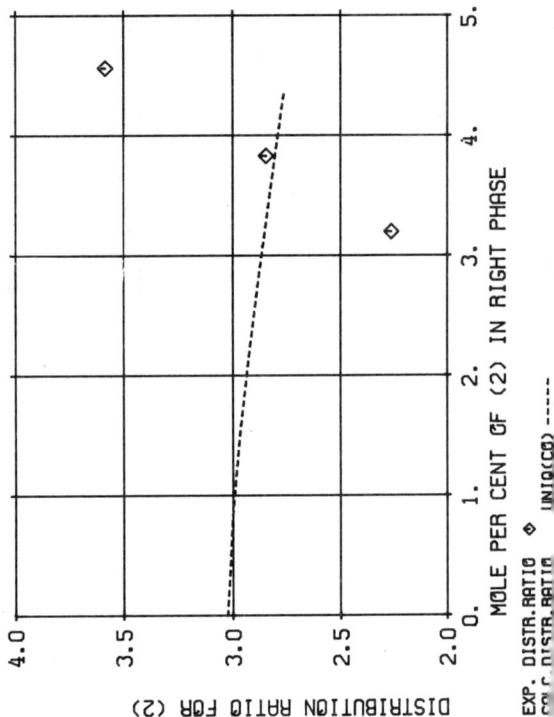

(1) H2O   WATER

(2) C2H6O2   1,2-ETHANEDIOL

(3) C6H14O   1-HEXANOL

LADDHA G.S., SMITH J.M.
IND.ENG.CHEM. 40(1948)494

TEMPERATURE = 20.0 DEG C   TYPE OF SYSTEM = 1

EXPERIMENTAL TIE LINES IN MOLE PCT

LEFT PHASE			RIGHT PHASE		
(1)	(2)	(3)	(1)	(2)	(3)
92.658	7.238	0.104	25.559	3.197	71.244
88.968	10.898	0.135	25.495	3.827	70.678
83.482	16.344	0.174	25.764	4.563	69.674

MEAN DEV. BETWEEN CALC. AND EXP. CONC. IN MOLE PCT

UNIQUAC (COMMON PARAMETERS)   0.90

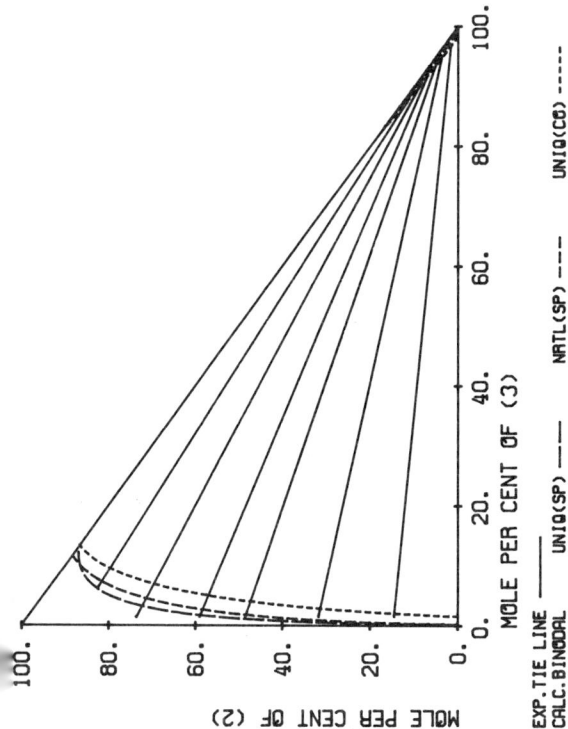

MOLE PER CENT OF (2)

MOLE PER CENT OF (3)

EXP.TIE LINE ——— UNIQ(SP) ◇
CALC.BINODAL ——— NRTL(SP) ———— UNIQ(C6) -----

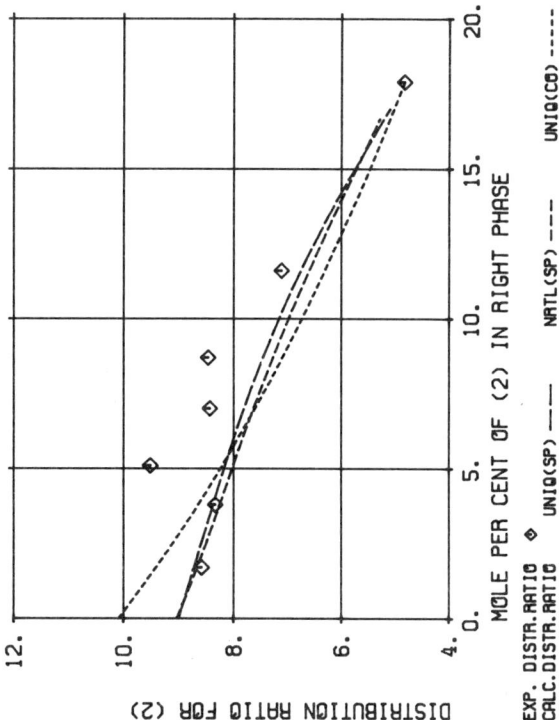

DISTRIBUTION RATIO FOR (2)

MOLE PER CENT OF (2) IN RIGHT PHASE

EXP. DISTR.RATIO ◇ UNIQ(SP) ◇ NRTL(SP) ——— UNIQ(C6) -----
CALC.DISTR.RATIO

(2) C7H9N   ANILINE,N-METHYL

(3) C2H6O2   1,2-ETHANEDIOL

CRUETZEN J.L., JOST W., SIEG L.
Z.ELEKTROCHEM. 61(1957)230

TEMPERATURE = 20.0 DEG C   TYPE OF SYSTEM = 2

EXPERIMENTAL TIE LINES IN MOLE PCT

	LEFT PHASE			RIGHT PHASE	
(1)	(2)	(3)	(1)	(2)	(3)
98.450	0.0	1.550	1.720	0.0	98.280
83.860	14.600	1.540	1.500	1.700	96.800
66.800	31.700	1.500	1.400	3.800	94.800
50.000	48.550	1.450	1.100	5.100	93.800
39.500	59.100	1.400	1.000	7.000	92.000
25.000	73.700	1.300	1.000	8.700	90.300
11.200	82.600	6.200	0.800	11.600	87.600
0.000	86.700	13.300	0.000	17.900	82.100

SPECIFIC MODEL PARAMETERS IN KELVIN

		UNIQUAC			NRTL(ALPHA=.2)	
I J	AIJ	AJI			AIJ	AJI
1 2	-47.255	80.640			-207.60	319.09
1 3	614.31	432.32			1321.3	887.83
2 3	342.79	-3.2916			529.18	340.68

R1 = 5.1094   R2 = 4.4555   R3 = 2.4038
Q1 = 3.920   Q2 = 3.364   Q3 = 2.248

MEAN DEV. BETWEEN CALC. AND EXP. CONC. IN MOLE PCT

UNIQUAC (SPECIFIC PARAMETERS)   0.88
NRTL   (SPECIFIC PARAMETERS)   1.00
UNIQUAC (COMMON PARAMETERS)   1.59

$C_2H_6O_2$-C_7H_{16}

MOLE PER CENT OF (3)

MOLE PER CENT OF (2)

EXP. TIE LINE ———
CALC. BINODAL   UNIQ(CO) ------

MOLE PER CENT OF (2) IN RIGHT PHASE

DISTRIBUTION RATIO FOR (2)

EXP. DISTR. RATIO ◇
CALC. DISTR. RATIO   UNIQ(CO)

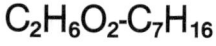

(1) C7H16   HEPTANE
(2) C7H16O   1-HEPTANOL
(3) C2H6O2   1,2-ETHANEDIOL

ALEKSANDROVA M.V., ET AL.
IZV.VYSSH.UCHEBN.ZAVED.KHIM.KHIM.TEKHNOL. 10(1967)678

TEMPERATURE = 20.0 DEG C   TYPE OF SYSTEM = 1

EXPERIMENTAL TIE LINES IN MOLE PCT

	LEFT PHASE			RIGHT PHASE	
(1)	(2)	(3)	(1)	(2)	(3)
97.416	2.422	0.162	0.050	0.118	99.833
88.180	10.838	0.982	0.156	1.052	98.791
71.611	24.260	4.129	0.284	1.959	97.757
46.767	38.165	15.068	0.413	2.797	96.789

MEAN DEV. BETWEEN CALC. AND EXP. CONC. IN MOLE PCT
UNIQUAC (COMMON PARAMETERS)   0.69

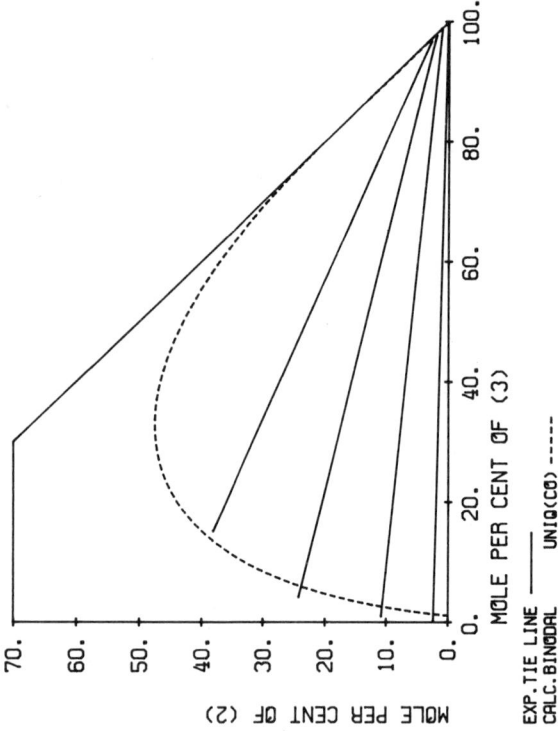

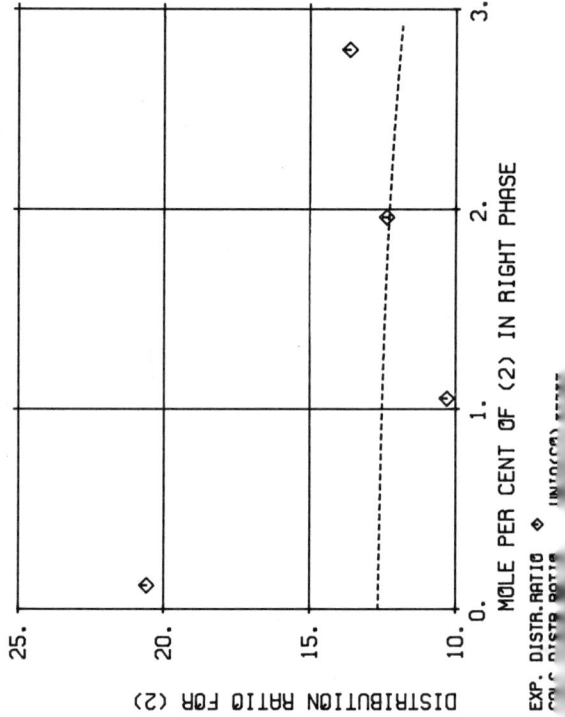

$C_2H_6O_4S-C_6H_6$

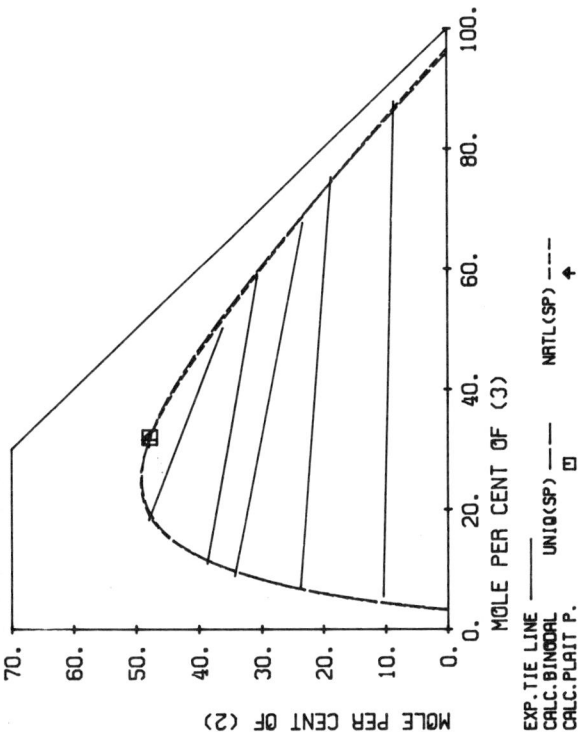

MOLE PER CENT OF (2)

MOLE PER CENT OF (3)

EXP.TIE LINE ———   UNIQ(SP) ———   NRTL(SP) — — —
CALC.BINODAL
CALC.PLAIT P.   □   ✦

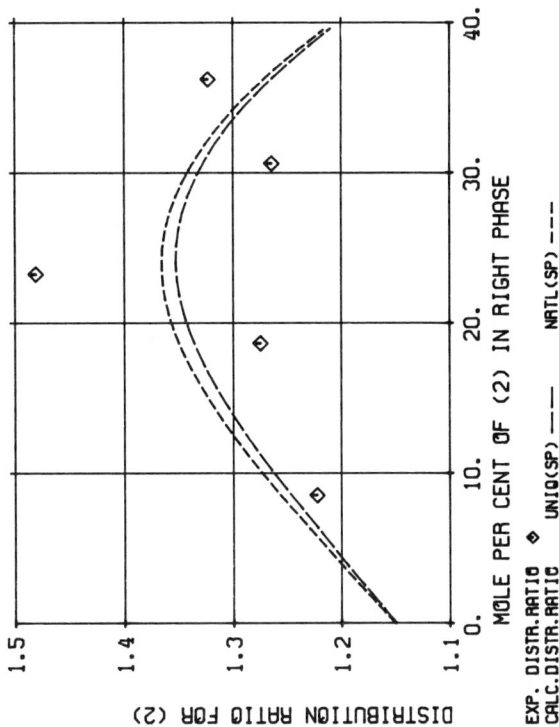

DISTRIBUTION RATIO FOR (2)

MOLE PER CENT OF (2) IN RIGHT PHASE

EXP. DISTR.RATIO ———   UNIQ(SP) ———   NRTL(SP) — — —
CALC.DISTR.RATIO   ◆

(2) C6H6   BENZENE

(3) C2H6O4S   SULFURIC ACID,DIMETHYL ESTER

PASCAL P.; QUINET M.-L.
ANN.CHIM.ANAL.CHIM.APPL. 23(1941)5

TEMPERATURE = 17.0 DEG C   TYPE OF SYSTEM = 1

EXPERIMENTAL TIE LINES IN MOLE PCT

	LEFT PHASE			RIGHT PHASE	
(1)	(2)	(3)	(1)	(2)	(3)
84.060	10.427	5.513	3.505	8.528	87.967
69.401	23.773	6.825	6.000	18.649	75.352
56.736	34.385	8.880	8.993	23.217	67.790
50.354	38.708	10.938	10.522	30.613	58.864
33.800	47.948	18.252	13.545	36.233	50.222

SPECIFIC MODEL PARAMETERS IN KELVIN

		UNIQUAC		NRTL(ALPHA=.2)	
I	J	AIJ	AJI	AIJ	AJI
1	2	273.75	-141.15	259.86	-64.086
1	3	306.16	29.482	653.86	646.60
2	3	157.62	-6.3830	752.22	-200.66

R1 = 5.1742   R2 = 3.1878   R3 = 4.1160
Q1 = 4.396   Q2 = 2.400   Q3 = 3.056

MEAN DEV. BETWEEN CALC. AND EXP. CONC. IN MOLE PCT

UNIQUAC (SPECIFIC PARAMETERS)   1.01
NRTL   (SPECIFIC PARAMETERS)   0.93

$C_2H_6O_4S-C_7H_8$

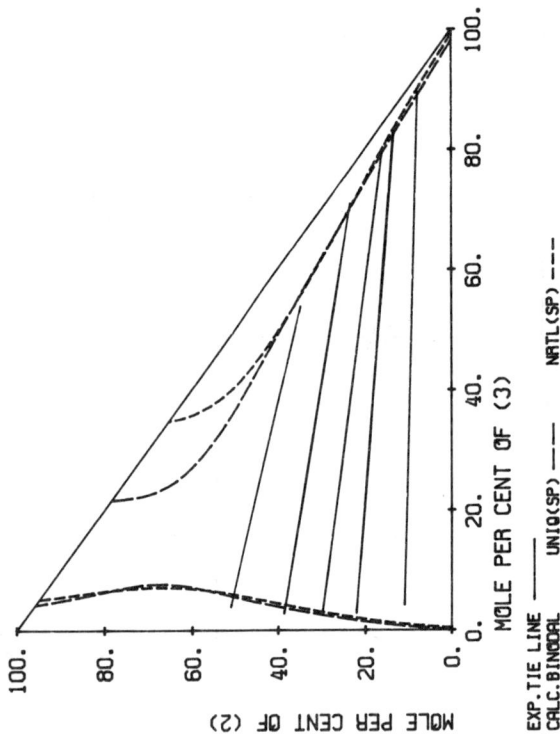

MOLE PER CENT OF (2)

MOLE PER CENT OF (3)

EXP.TIE LINE ———   UNIQ(SP) ———   NRTL(SP) ----

CALC.BINODAL

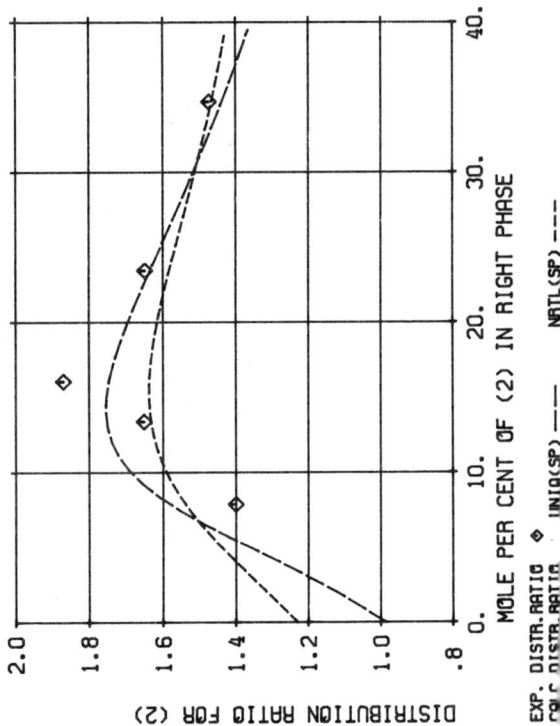

DISTRIBUTION RATIO FOR (2)

MOLE PER CENT OF (2) IN RIGHT PHASE

EXP. DISTR.RATIO ◇   UNIQ(SP) ———   NRTL(SP) ----

CALC. DISTR.RATIO

(1) C7H16   HEPTANE

(2) C7H8   TOLUENE

(3) C2H6O4S   SULFURIC ACID,DIMETHYL ESTER

PASCAL P., QUINET M.-L.
ANN.CHIM.ANAL.CHIM.APPL. 23(1941)5

TEMPERATURE = 17.0 DEG C   TYPE OF SYSTEM = 1

EXPERIMENTAL TIE LINES IN MOLE PCT

	LEFT PHASE			RIGHT PHASE	
(1)	(2)	(3)	(1)	(2)	(3)
84.857	11.004	4.139	2.573	7.862	89.565
75.109	22.061	2.830	3.493	13.359	83.149
67.316	30.030	2.655	5.003	16.063	78.933
58.654	38.632	2.714	5.591	23.432	70.977
45.052	51.099	3.849	11.443	34.688	53.864

SPECIFIC MODEL PARAMETERS IN KELVIN

		UNIQUAC		NRTL(ALPHA=.2)	
I J	AIJ	AJI		AIJ	AJI
1 2	488.07	-272.57		852.27	-537.05
1 3	584.36	6.8918		1132.1	887.97
2 3	601.67	-201.97		1371.6	-378.48

R1 = 5.1742   R2 = 3.9228   R3 = 4.1160
Q1 = 4.396   Q2 = 2.968   Q3 = 3.056

MEAN DEV. BETWEEN CALC. AND EXP. CONC. IN MOLE PCT

UNIQUAC (SPECIFIC PARAMETERS)   1.06
NRTL   (SPECIFIC PARAMETERS)   1.17

C$_2$H$_7$NO-C$_5$H$_4$O$_2$

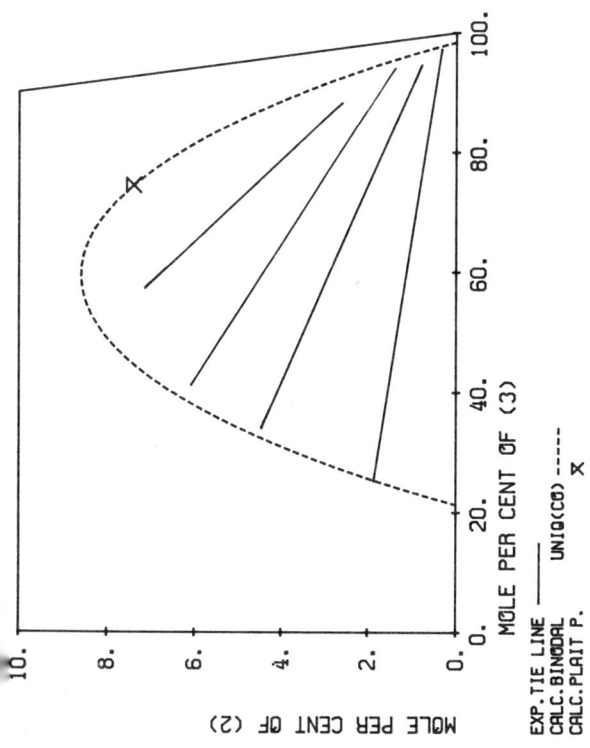

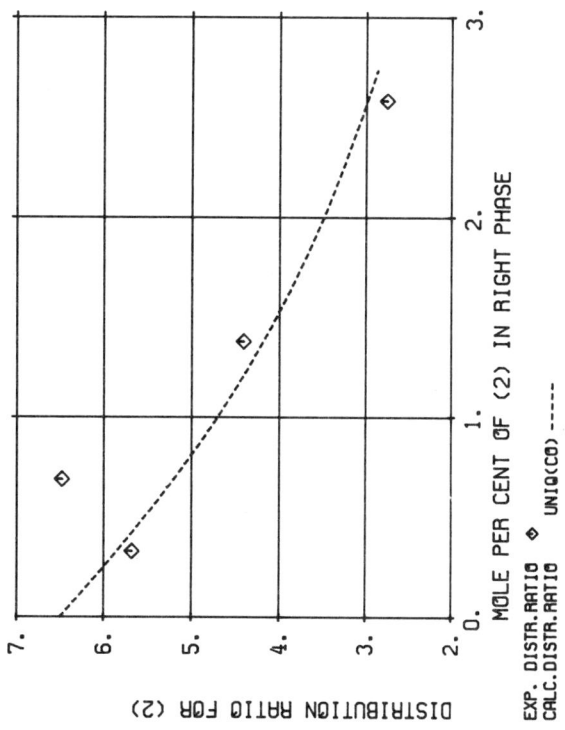

(2) C2H7NO   ETHANOL, 2-AMINO

(3) H2O   WATER

MASKHULIYA V.P., KRUPATKIN I.L.,
FAZOVYE RAVNOVESIYA,NO.2,KALININ,EDITOR:I.KRUPATKIN
(1975)66

TEMPERATURE = 25.0 DEG C   TYPE OF SYSTEM = 1

EXPERIMENTAL TIE LINES IN MOLE PCT

	LEFT PHASE			RIGHT PHASE	
(1)	(2)	(3)	(1)	(2)	(3)
72.914	1.860	25.227	2.397	0.328	97.275
61.573	4.453	33.975	3.454	0.688	95.858
52.752	6.071	41.177	4.542	1.378	94.080
35.615	7.123	57.262	9.162	2.581	88.257

MEAN DEV. BETWEEN CALC. AND EXP. CONC. IN MOLE PCT

UNIQUAC (COMMON PARAMETERS)   0.64

C₂H₇NO-C₅H₉NO

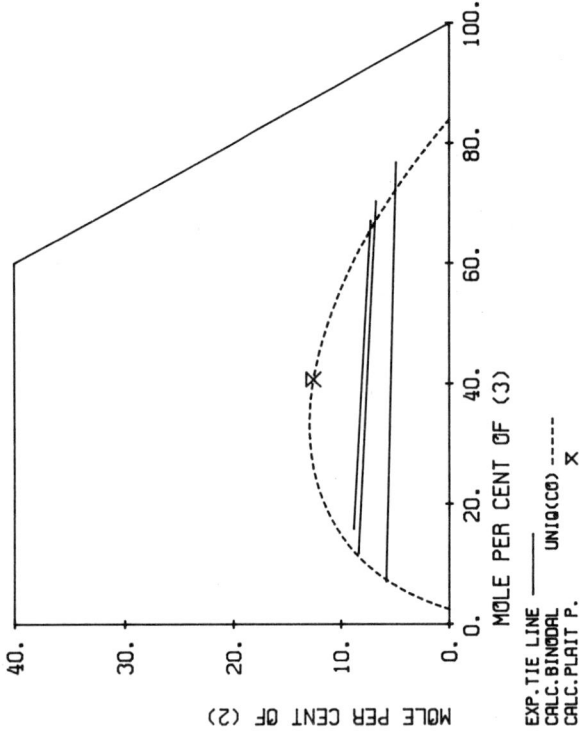

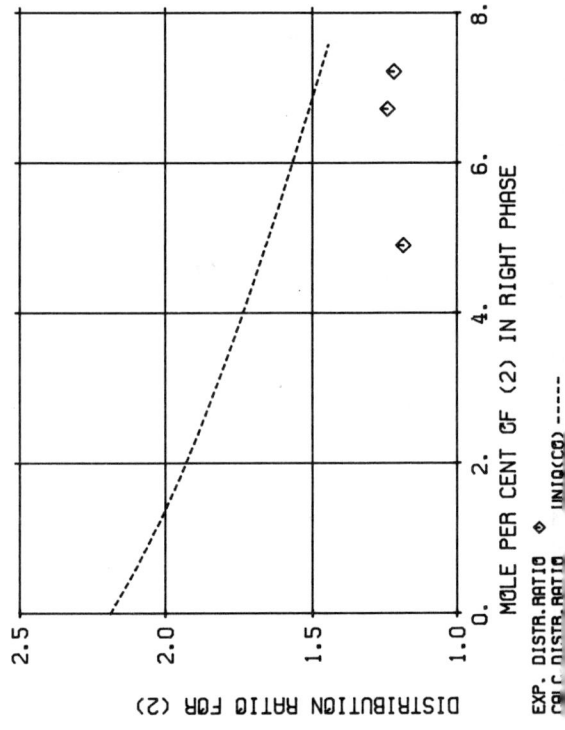

(1) C6H6   BENZENE
----
(2) C5H9NO   2-PYRROLIDONE,1-METHYL
----
(3) C2H7NO   ETHANOL,2-AMINO

FABRIES J.-F., GUSTIN J.-L., RENON H.
J.CHEM.ENG.DATA 22(1977)303

TEMPERATURE = 25.0 DEG C   TYPE OF SYSTEM = 1

EXPERIMENTAL TIE LINES IN MOLE PCT

          LEFT PHASE   RIGHT PHASE
     (1)   (2)   (3)   (1)   (2)   (3)

   87.110   5.810   7.080   18.260   4.900   76.840
   79.950   8.340   11.710   22.890   6.720   70.390
   75.430   8.800   15.770   25.630   7.220   67.150

MEAN DEV. BETWEEN CALC. AND EXP. CONC. IN MOLE PCT

UNIQUAC (COMMON PARAMETERS)   0.83

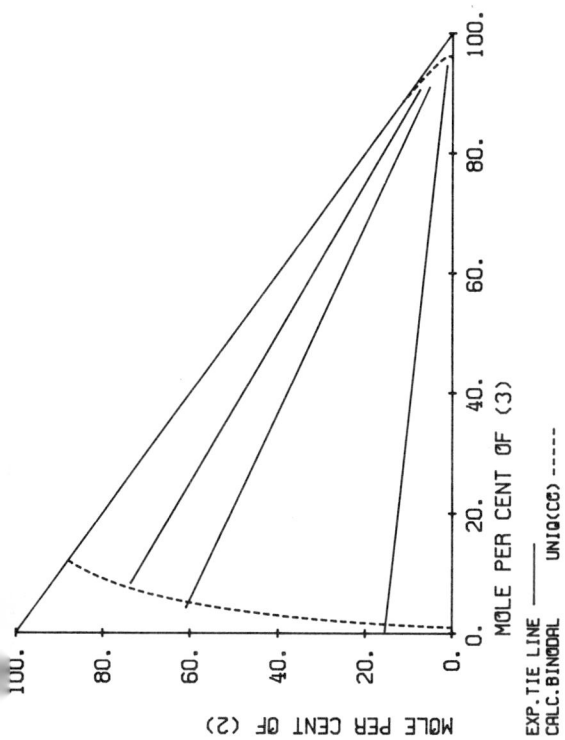

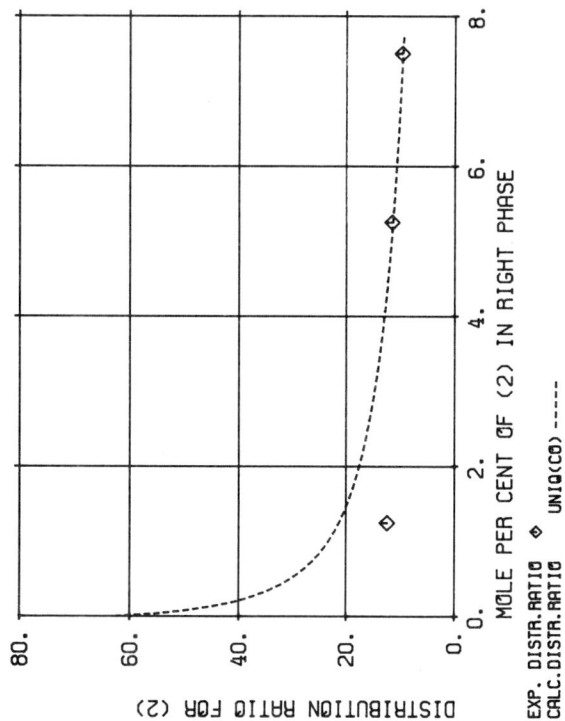

C₂H₇NO-C₅H₉NO

(2) C5H9NO   2-PYRROLIDONE,1-METHYL

(3) C7H16   HEPTANE

FABRIES J.-F.; GUSTIN J.-L., RENON H.
J.CHEM.ENG.DATA 22(1977)303

TEMPERATURE = 25.0 DEG C   TYPE OF SYSTEM = 2

EXPERIMENTAL TIE LINES IN MOLE PCT

	LEFT PHASE			RIGHT PHASE	
(1)	(2)	(3)	(1)	(2)	(3)
84.170	15.410	0.420	4.100	1.240	94.660
34.970	60.810	4.220	3.680	5.250	91.070
18.300	73.420	8.280	1.860	7.500	90.640

MEAN DEV. BETWEEN CALC. AND EXP. CONC. IN MOLE PCT

UNIQUAC (COMMON PARAMETERS)   1.11

MOLE PER CENT OF (3)

MOLE PER CENT OF (2)

EXP.TIE LINE ——— UNIQ(CC) -----
CALC.BINODAL

MOLE PER CENT OF (2) IN RIGHT PHASE

DISTRIBUTION RATIO FOR (2)

EXP. DISTR.RATIO ◇ UNIQ(CC) -----
CALC.DISTR.RATIO

$C_2H_7NO\text{-}C_6H_6$

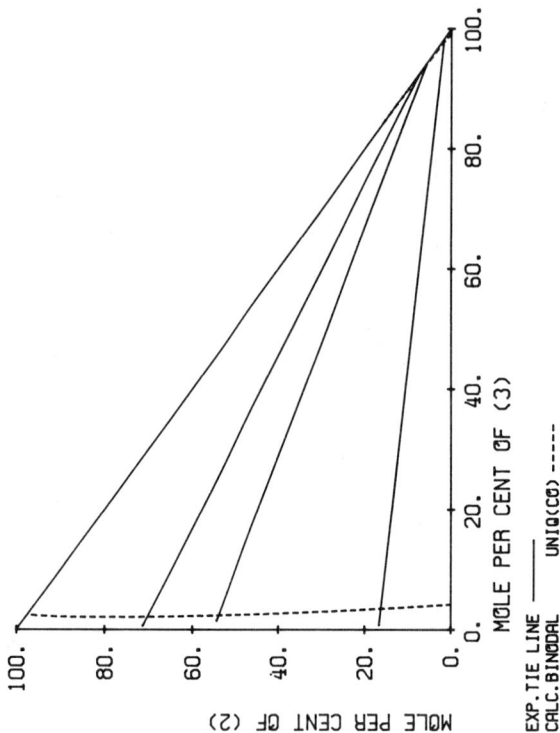

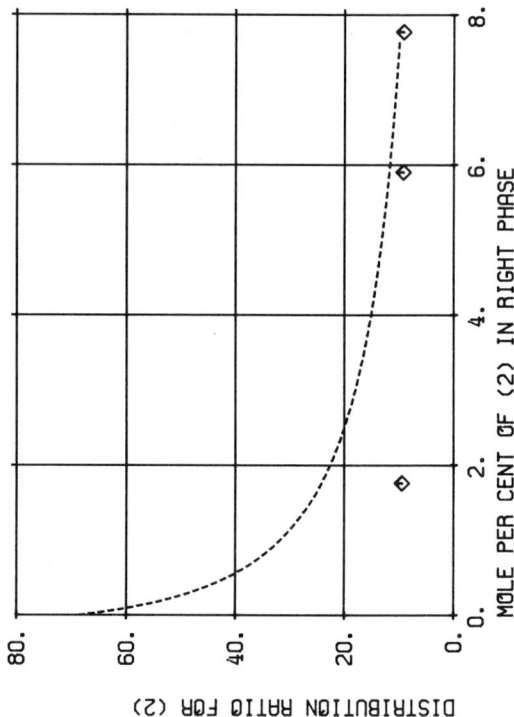

(1) C7H16   HEPTANE
(2) C6H6   BENZENE
(3) C2H7NO   ETHANOL,2-AMINO

FABRIES J.-F., GUSTIN J.-L., RENON H.
J.CHEM.ENG.DATA 22(1977)303

TEMPERATURE = 25.0 DEG C   TYPE OF SYSTEM = 2

EXPERIMENTAL TIE LINES IN MOLE PCT

LEFT PHASE			RIGHT PHASE		
(1)	(2)	(3)	(1)	(2)	(3)
82.580	16.630	0.790	0.140	1.760	98.100
44.150	54.410	1.440	0.150	5.900	93.950
28.170	71.230	0.600	0.140	7.770	92.090

MEAN DEV. BETWEEN CALC. AND EXP. CONC. IN MOLE PCT

UNIQUAC (COMMON PARAMETERS)   1.47

C$_2$H$_7$NO-C$_6$H$_7$N

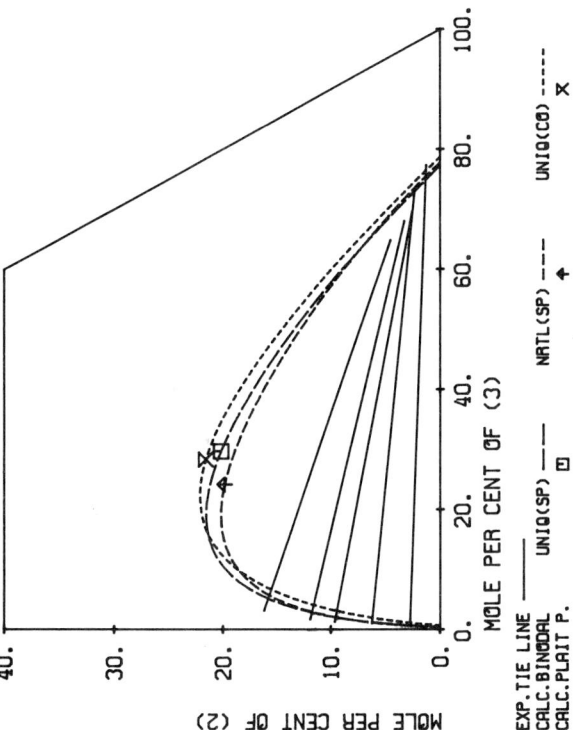

MOLE PER CENT OF (3)

MOLE PER CENT OF (2)

EXP.TIE LINE ——— UNIQ(SP) ☐ NRTL(SP) ✛ UNIQ(CO) ✕
CALC.BINODAL ----
CALC.PLAIT P.

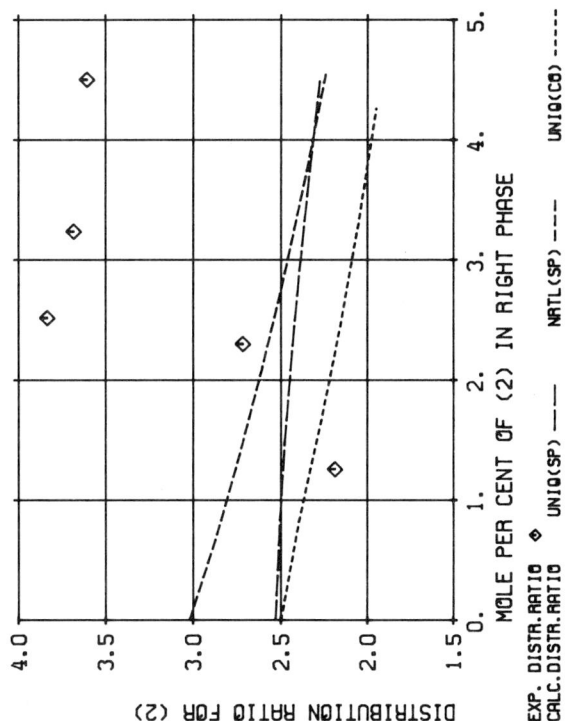

MOLE PER CENT OF (2) IN RIGHT PHASE

DISTRIBUTION RATIO FOR (2)

EXP. DISTR.RATIO ◇ UNIQ(SP) ◇ NRTL(SP) ◇ UNIQ(CO) -----
CALC.DISTR.RATIO

---

(2) C2H7NO   ETHANOL,2-AMINO
---
(3) C6H7N   ANILINE

ZHURAVLEVA I.K., MASLOVSKAYA N.V., ZHURAVLEV E.F.,
IZV.VYSSH.UCHEBN.ZAVED.KHIM.KHIM.TEKHNOL. 20(1977)791

TEMPERATURE = 20.0 DEG C   TYPE OF SYSTEM = 1

EXPERIMENTAL TIE LINES IN MOLE PCT

	LEFT PHASE			RIGHT PHASE	
(1)	(2)	(3)	(1)	(2)	(3)
96.509	2.749	0.742	21.296	1.256	77.448
92.789	6.244	0.966	24.614	2.299	73.087
89.117	9.642	1.240	27.769	2.515	69.716
86.490	11.916	1.595	28.610	3.236	68.154
80.685	16.235	3.080	30.532	4.502	64.966

SPECIFIC MODEL PARAMETERS IN KELVIN

		UNIQUAC		NRTL(ALPHA=.2)	
I J	AIJ	AJI	AIJ	AJI	
1 2	-201.78	12.016	-124.64	-3.2262	
1 3	214.76	70.819	1631.8	-13.893	
2 3	-124.09	375.89	-172.91	839.51	

R1 = 0.9200   R2 = 2.5736   R3 = 3.7165
Q1 = 1.400   Q2 = 2.360   Q3 = 2.816

MEAN DEV. BETWEEN CALC. AND EXP. CONC. IN MOLE PCT

UNIQUAC (SPECIFIC PARAMETERS)   1.09
NRTL   (SPECIFIC PARAMETERS)   1.15
UNIQUAC (COMMON PARAMETERS)   1.55

C$_2$H$_8$N$_2$-C$_6$H$_6$

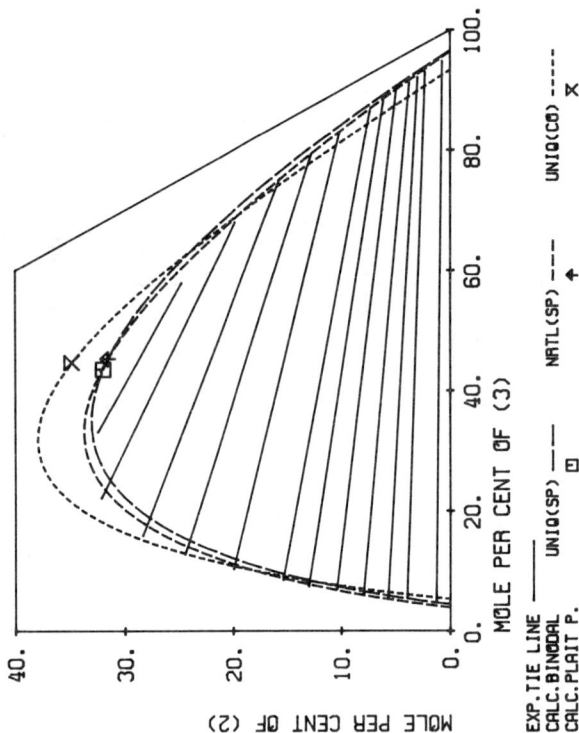

MOLE PER CENT OF (3)

MOLE PER CENT OF (2)

EXP.TIE LINE   UNIQ(SP) ⊡   NRTL(SP) — ✛   UNIQ(C6) ----- ✗
CALC.BINODAL
CALC.PLAIT P.

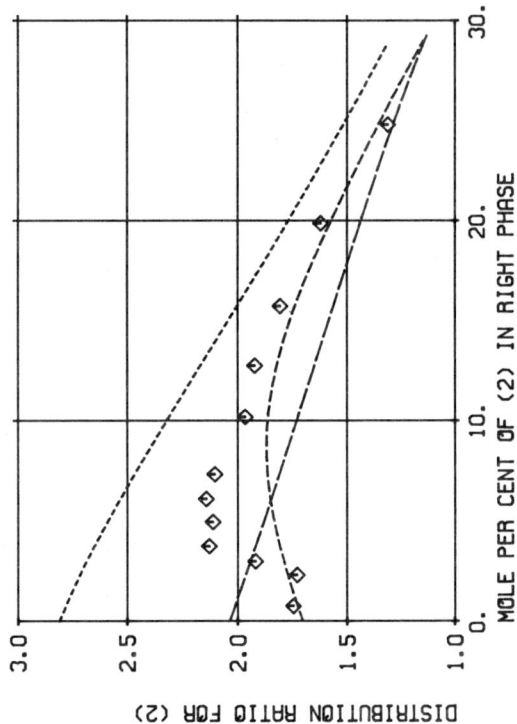

MOLE PER CENT OF (2) IN RIGHT PHASE

DISTRIBUTION RATIO FOR (2)

EXP. DISTR.RATIO ◆

(1) C6H12   CYCLOHEXANE

(2) C6H6   BENZENE

(3) C2H8N2   ETHANE,1,2-DIAMINO

KOMAROVA E.G., KOGAN V.B.
ZH.PRIKL.KHIM.(LENINGRAD) 37(1964)1570

TEMPERATURE = 20.0 DEG C   TYPE OF SYSTEM = 1

EXPERIMENTAL TIE LINES IN MOLE PCT

	LEFT PHASE			RIGHT PHASE	
(1)	(2)	(3)	(1)	(2)	(3)
95.280	0.0	4.720	4.210	0.0	95.790
93.770	1.310	4.920	4.370	0.750	94.880
90.540	3.960	5.500	4.610	2.290	93.100
88.770	5.700	5.530	4.810	2.970	92.220
85.840	7.940	6.220	4.900	3.730	91.370
82.690	10.430	6.880	5.250	4.940	89.810
79.620	13.040	7.340	5.570	6.090	88.340
76.230	15.390	8.380	5.980	7.320	86.700
69.830	19.990	10.180	6.800	10.170	83.030
62.830	24.520	12.650	7.790	12.750	79.460
55.780	28.390	15.830	9.010	15.710	75.280
45.790	32.200	22.010	12.020	19.860	68.120
34.500	32.510	32.990	17.230	24.800	57.970

SPECIFIC MODEL PARAMETERS IN KELVIN

		UNIQUAC		NRTL(ALPHA=.2)	
I J	AIJ	AJI		AIJ	AJI
1 2	-122.78	81.108		633.07	-374.69
1 3	289.97	141.56		615.98	646.18
2 3	-85.389	112.41		292.56	87.486

R1 = 4.0464   R2 = 3.1878   R3 = 2.7384
Q1 = 3.240   Q2 = 2.400   Q3 = 2.472

MEAN DEV. BETWEEN CALC. AND EXP. CONC. IN MOLE PCT

UNIQUAC (SPECIFIC PARAMETERS)   0.90
NRTL   (SPECIFIC PARAMETERS)   0.50
UNIQUAC (COMMON PARAMETERS)   1.43

$C_2H_8N_2\text{-}C_6H_6$

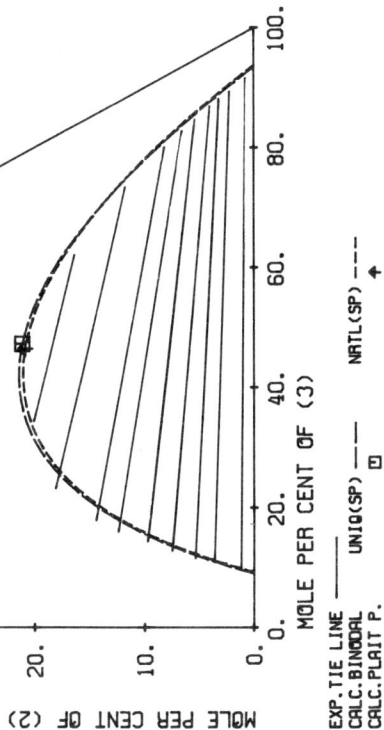

MOLE PER CENT OF (3)

MOLE PER CENT OF (2)

EXP.TIE LINE ———   UNIQ(SP) □   NRTL(SP) ---
CALC.BINODAL
CALC.PLAIT P.

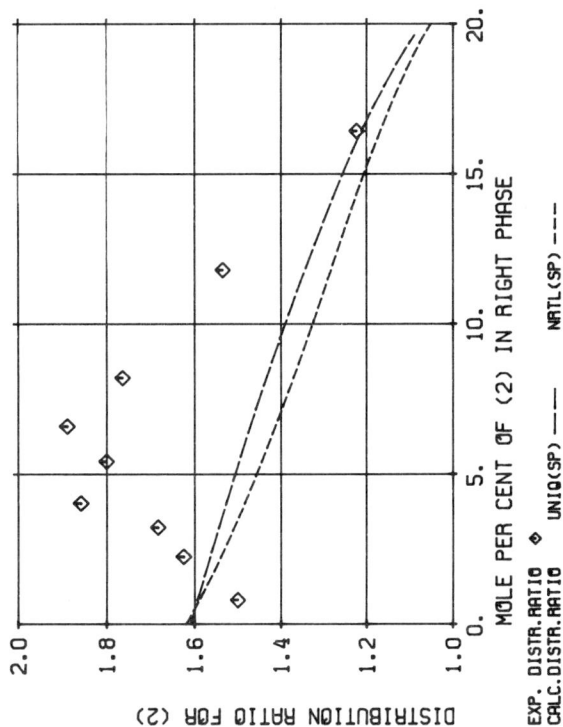

MOLE PER CENT OF (2) IN RIGHT PHASE

DISTRIBUTION RATIO FOR (2)

EXP. DISTR.RATIO ◇   UNIQ(SP) ———   NRTL(SP) ---
CALC.DISTR.RATIO

(2) C6H6   BENZENE
(3) C2H8N2   ETHANE,1,2-DIAMINO

KOMAROVA E.G., KOGAN V.B.
ZH.PRIKL.KHIM.(LENINGRAD) 37(1964)1570

TEMPERATURE = 40.0 DEG C   TYPE OF SYSTEM = 1

EXPERIMENTAL TIE LINES IN MOLE PCT

	LEFT PHASE			RIGHT PHASE	
(1)	(2)	(3)	(1)	(2)	(3)
90.960	0.0	9.040	7.050	0.0	92.950
89.200	1.200	9.600	7.520	0.800	91.680
85.560	3.640	10.800	8.320	2.240	89.440
83.100	5.400	11.500	8.570	3.210	88.220
79.770	7.490	12.740	9.060	4.030	86.910
75.930	9.740	14.330	9.910	5.410	84.680
71.680	12.430	15.890	10.620	6.580	82.800
67.730	14.460	17.810	11.770	8.200	80.090
58.750	18.100	23.150	14.770	11.790	73.440
45.620	20.080	34.300	21.480	16.420	62.100

SPECIFIC MODEL PARAMETERS IN KELVIN

		UNIQUAC		NRTL(ALPHA=.2)	
I J	AIJ	AJI		AIJ	AJI
1 2	-68.716	-89.186		-15.837	-238.98
1 3	257.56	12.36		456.95	649.19
2 3	-155.49	66.225		-262.21	172.56

R1 = 4.0464   R2 = 3.1878   R3 = 2.7384
Q1 = 3.240   Q2 = 2.400   Q3 = 2.472

MEAN DEV. BETWEEN CALC. AND EXP. CONC. IN MOLE PCT

UNIQUAC (SPECIFIC PARAMETERS)   0.61
NRTL   (SPECIFIC PARAMETERS)   0.82

$C_2H_8N_2$-C_6H_6

(1) C6H12   CYCLOHEXANE

(2) C6H6   BENZENE

(3) C2H8N2   ETHANE,1,2-DIAMINO

KOMAROVA E.G., KOGAN V.B.
ZH.PRIKL.KHIM.(LENINGRAD) 37(1964)1570

TEMPERATURE = 60.0 DEG C   TYPE OF SYSTEM = 1

EXPERIMENTAL TIE LINES IN MOLE PCT

	LEFT PHASE			RIGHT PHASE	
(1)	(2)	(3)	(1)	(2)	(3)
80.680	0.0	19.320	13.310	0.0	86.690
78.700	1.300	20.000	13.850	0.800	85.350
73.050	4.100	22.850	15.600	2.100	82.300
69.050	5.600	25.350	17.580	3.100	79.320
63.060	7.250	29.690	19.210	4.110	76.680
56.280	8.570	35.150	24.480	6.270	69.250

SPECIFIC MODEL PARAMETERS IN KELVIN

		UNIQUAC		NRTL(ALPHA=.2)	
I	J	AIJ	AJI	AIJ	AJI
1	2	-167.20	5.2046	99.983	-512.81
1	3	223.54	107.30	360.08	596.51
2	3	-197.05	183.67	-184.81	-158.97

R1 = 4.0464   R2 = 3.1878   R3 = 2.7384
Q1 = 3.240   Q2 = 2.400   Q3 = 2.472

MEAN DEV. BETWEEN CALC. AND EXP. CONC. IN MOLE PCT

UNIQUAC (SPECIFIC PARAMETERS)   0.39
NRTL   (SPECIFIC PARAMETERS)   0.45

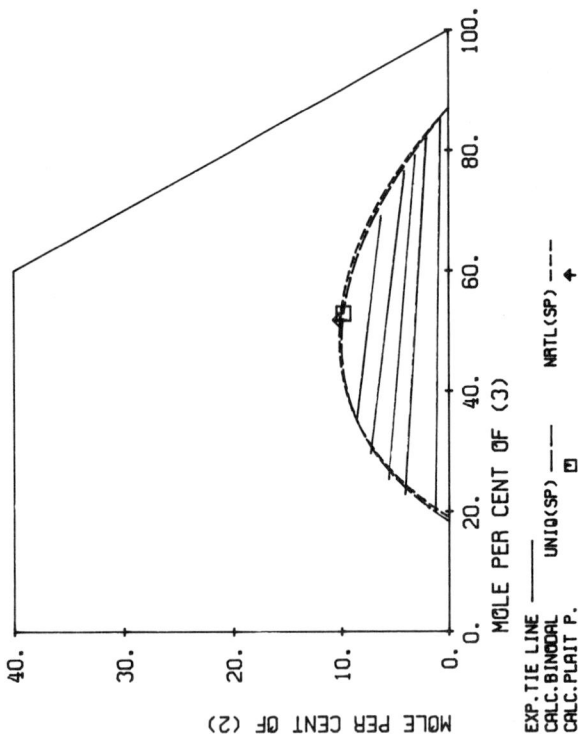

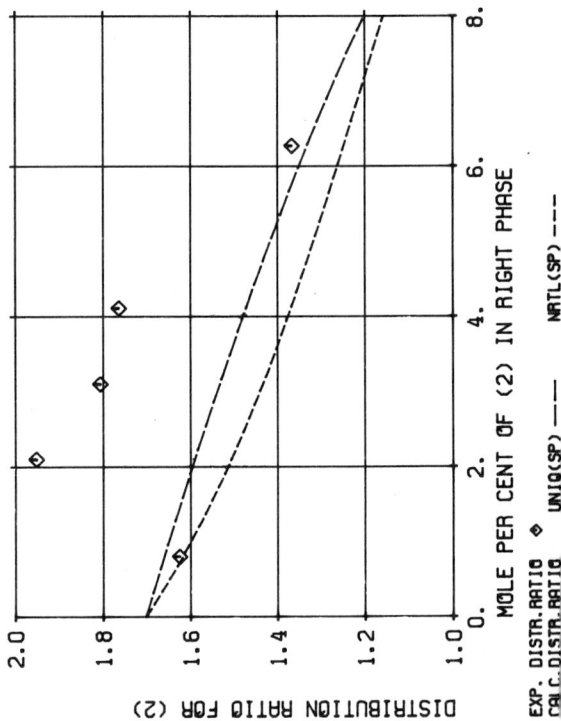

$C_2H_8N_2$-C_6H_6

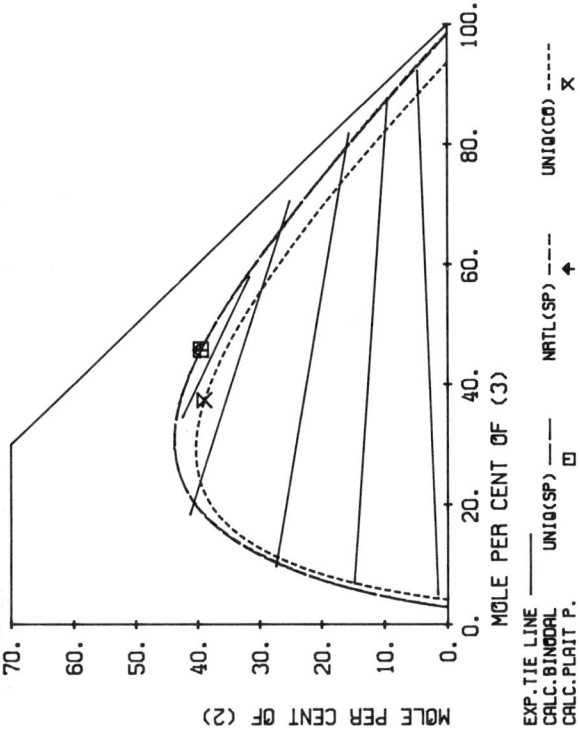

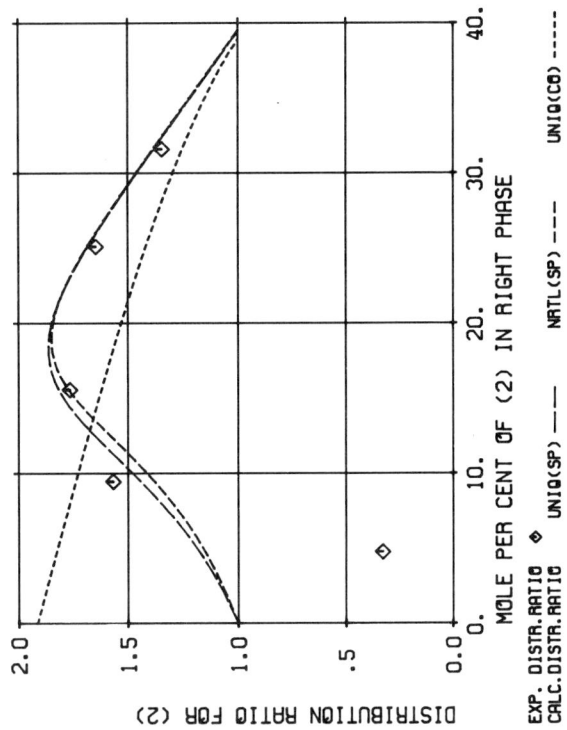

(2) C6H6   BENZENE
(3) C2H8N2   ETHANE,1,2-DIAMINO

CUMMING A.P.C., MORTON F.
J.APPL.CHEM. 3(1953)358

TEMPERATURE = 20.0 DEG C   TYPE OF SYSTEM = 1

EXPERIMENTAL TIE LINES IN MOLE PCT

	LEFT PHASE			RIGHT PHASE	
(1)	(2)	(3)	(1)	(2)	(3)
95.894	0.0	4.106	6.379	0.0	93.621
93.491	1.573	4.936	2.864	4.740	92.396
78.409	14.807	6.784	2.757	9.445	87.798
62.953	27.425	9.622	2.508	15.541	81.952
40.407	41.278	18.315	4.272	25.078	70.651
23.025	42.492	34.484	10.497	31.583	57.920

SPECIFIC MODEL PARAMETERS IN KELVIN

		UNIQUAC		NRTL(ALPHA=.2)	
I J	AIJ	AJI		AIJ	AJI
1 2	426.05	-184.26		684.36	-218.34
1 3	355.58	126.44		619.82	882.71
2 3	85.970	74.032		362.26	140.19

R1 = 4.4998   R2 = 3.1878   R3 = 2.7384
Q1 = 3.856   Q2 = 2.400   Q3 = 2.472

MEAN DEV. BETWEEN CALC. AND EXP. CONC. IN MOLE PCT

UNIQUAC (SPECIFIC PARAMETERS)   1.46
NRTL   (SPECIFIC PARAMETERS)   1.52
UNIQUAC (COMMON PARAMETERS)   4.33

$C_3H_3N-C_3H_5N$

MOLE PER CENT OF (2)

MOLE PER CENT OF (3)

EXP.TIE LINE ——— UNIQ(SP) ——— NRTL(SP) ——— UNIQ(CO) -----
CALC.BINODAL -----

DISTRIBUTION RATIO FOR (2)

MOLE PER CENT OF (2) IN RIGHT PHASE

EXP. DISTR.RATIO ◇ UNIQ(SP) ——— NRTL(SP) ——— UNIQ(CO) -----
CALC.DISTR.RATIO -----

(1) C3H3N	PROPENOIC ACID,NITRILE
(2) C3H5N	PROPANOIC ACID,NITRILE
(3) H2O	WATER

PROKHOROVA V.V., ET AL.
ZH.FIZ.KHIM. 38(1964)1488

TEMPERATURE = 25.0 DEG C   TYPE OF SYSTEM = 2

EXPERIMENTAL TIE LINES IN MOLE PCT

LEFT PHASE			RIGHT PHASE		
(1)	(2)	(3)	(1)	(2)	(3)
77.066	11.322	11.612	2.182	0.531	97.288
64.341	23.992	11.667	1.990	0.945	97.065
40.044	46.971	12.986	1.637	1.566	96.797
26.202	58.326	15.473	1.254	2.226	96.520
18.675	65.955	15.370	0.584	3.083	96.333
11.404	72.698	15.898	0.432	3.453	96.115

SPECIFIC MODEL PARAMETERS IN KELVIN

	UNIQUAC		NRTL(ALPHA=.2)	
I J	AIJ	AJI	AIJ	AJI
1 2	27.070	38.361	79.547	54.918
1 3	316.84	211.78	299.68	936.43
2 3	303.09	123.77	192.68	892.20

R1 = 2.3144   R2 = 2.5445   R3 = 0.9200
Q1 = 2.052   Q2 = 2.264   Q3 = 1.400

MEAN DEV. BETWEEN CALC. AND EXP. CONC. IN MOLE PCT

UNIQUAC (SPECIFIC PARAMETERS)	0.33
NRTL (SPECIFIC PARAMETERS)	0.31
UNIQUAC (COMMON PARAMETERS)	0.46

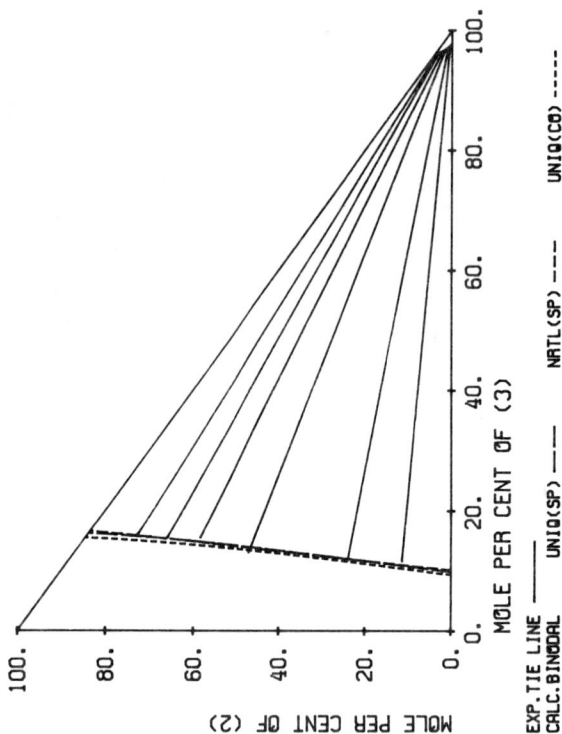

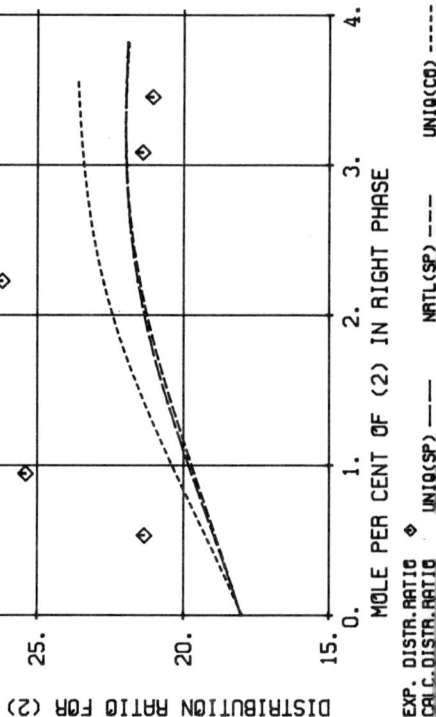

$C_3H_3N\text{-}C_3H_5N$

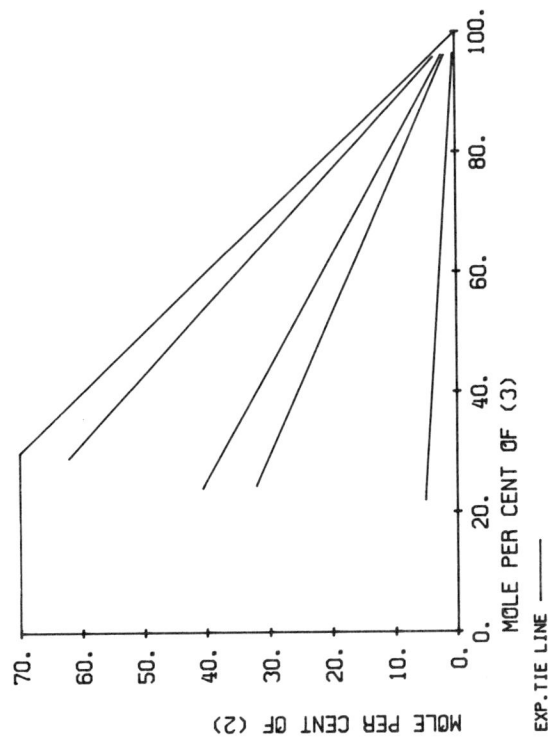

MOLE PER CENT OF (3)

EXP.TIE LINE ———

MOLE PER CENT OF (2)

(1) C3H3N   PROPENOIC ACID,NITRILE

(2) C3H5N   PROPANOIC ACID,NITRILE

(3) H2O   WATER

PROKHOROVA V.V., ET AL.
ZH.FIZ.KHIM. 38(1964)1488

TEMPERATURE = 70.0 DEG C   TYPE OF SYSTEM = 2

EXPERIMENTAL TIE LINES IN MOLE PCT

	LEFT PHASE			RIGHT PHASE	
(1)	(2)	(3)	(1)	(2)	(3)
72.870	5.150	21.980	3.237	0.346	96.416
43.496	32.050	24.454	2.156	1.760	96.084
35.421	40.534	24.044	1.699	2.197	96.104
8.746	62.140	29.114	0.716	3.529	95.756

C₃H₃N-C₄H₅Cl

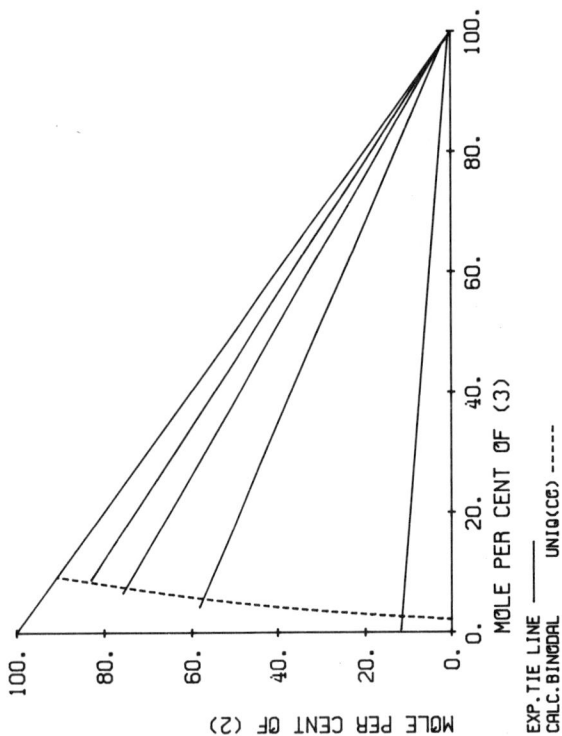

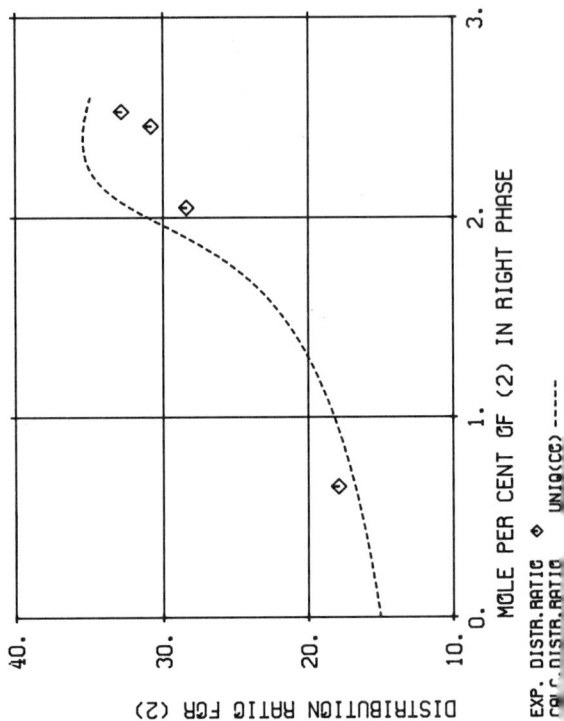

(1) C4H5CL   1,3-BUTADIENE, 2-CHLORO
(2) C3H3N   PROPENOIC ACID,NITRILE
(3) H2O   WATER

NOVOTNY M., RYCHTR L.
COLLECT.CZECH.CHEM.COMMUN. 29(1964)2558

TEMPERATURE = 24.0 DEG C   TYPE OF SYSTEM = 2

EXPERIMENTAL TIE LINES IN MOLE PCT

	LEFT PHASE			RIGHT PHASE	
(1)	(2)	(3)	(1)	(2)	(3)
87.811	11.722	0.467	0.041	0.654	99.304
37.502	58.178	4.320	0.017	2.049	97.934
17.708	75.646	6.646	0.011	2.456	97.534
8.245	82.946	8.809	0.006	2.530	97.464

MEAN DEV. BETWEEN CALC. AND EXP. CONC. IN MOLE PCT

UNIQUAC (COMMON PARAMETERS)   0.69

$C_3H_4O_2-C_5H_{10}O_2$

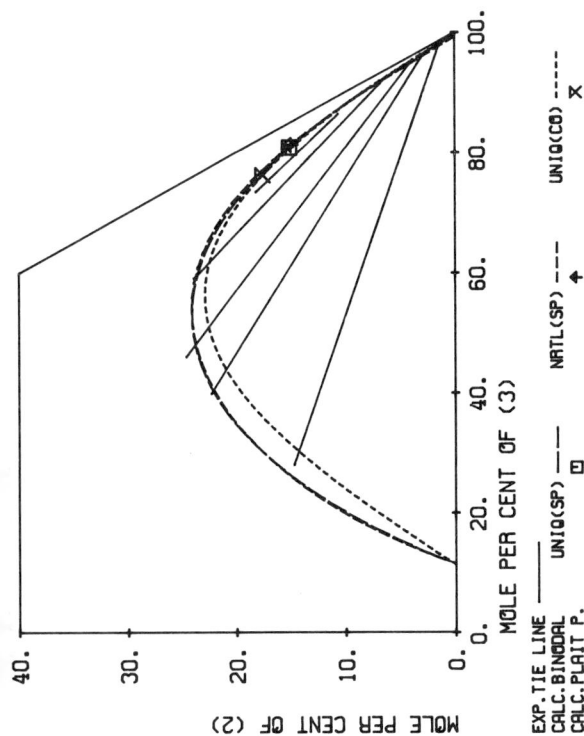

MOLE PER CENT OF (3)

MOLE PER CENT OF (2)

EXP.TIE LINE ——— UNIQ(SP) ⊡ NRTL(SP) ✦ UNIQ(CD) -----
CALC.BINODAL ✗
CALC.PLAIT P.

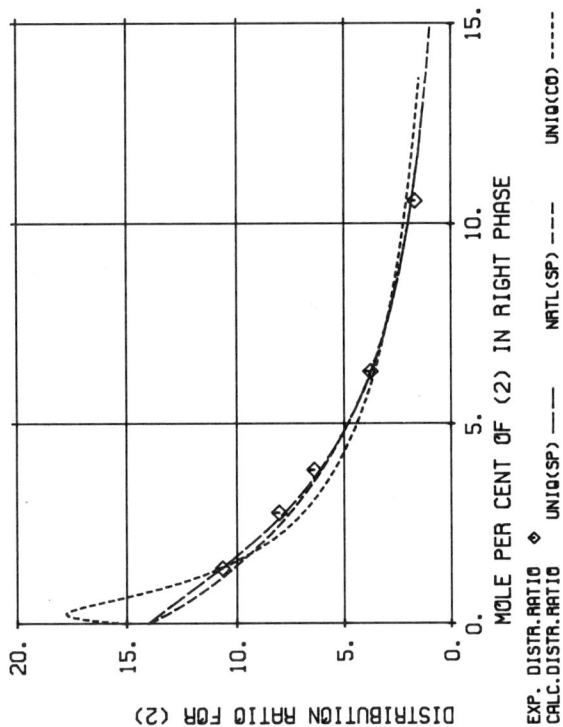

MOLE PER CENT OF (2) IN RIGHT PHASE

DISTRIBUTION RATIO FOR (2)

EXP. DISTR.RATIO ◇ UNIQ(SP) ——— NRTL(SP) ——— UNIQ(CD) -----
CALC.DISTR.RATIO

(2) C3H4O2   PROPENOIC ACID

(3) H2O   WATER

LINEK J., HLAVATY K., WICHTERLE I.
COLLECT.CZECH.CHEM.COMMUN. 38(1973)1840

TEMPERATURE = 24.6 DEG C   TYPE OF SYSTEM = 1

EXPERIMENTAL TIE LINES IN MOLE PCT

	LEFT PHASE			RIGHT PHASE	
(1)	(2)	(3)	(1)	(2)	(3)
88.690	0.0	11.310	0.395	0.0	99.605
57.291	14.717	27.992	0.431	1.380	98.189
37.815	22.331	39.853	0.548	2.772	96.681
29.342	24.641	46.017	0.667	3.840	95.492
17.019	23.933	59.048	1.140	6.305	92.555
8.407	18.164	73.429	2.894	10.564	86.542

SPECIFIC MODEL PARAMETERS IN KELVIN

		UNIQUAC			NRTL(ALPHA=.2)	
I J	AIJ	AJI			AIJ	AJI
1 2	-165.19	110.89			439.63	-467.40
1 3	-406.33	119.24			202.73	1290.0
2 3	-177.31	301.85			-599.20	1391.6

R1 = 4.1522   R2 = 2.6467   R3 = 0.9200
Q1 = 3.652   Q2 = 2.400   Q3 = 1.400

MEAN DEV. BETWEEN CALC. AND EXP. CONC. IN MOLE PCT

UNIQUAC (SPECIFIC PARAMETERS)   0.42
NRTL   (SPECIFIC PARAMETERS)   0.55
UNIQUAC (COMMON PARAMETERS)   1.66

C$_3$H$_4$O$_2$-C$_6$H$_{12}$O

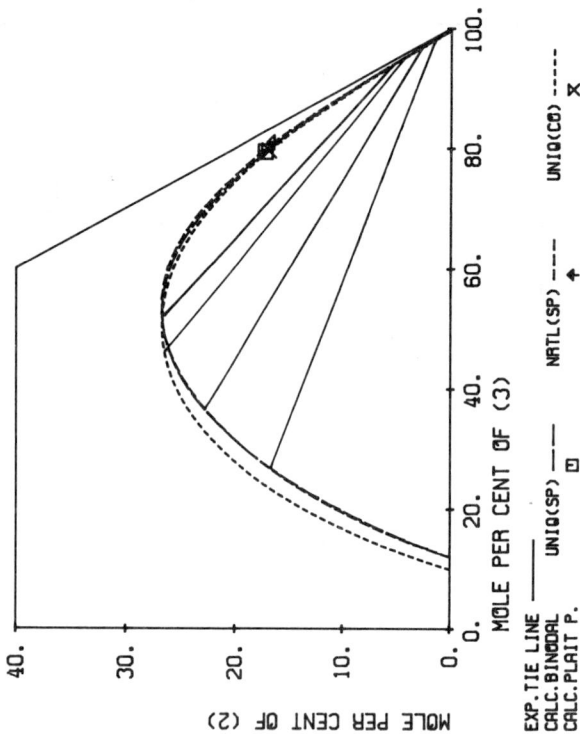

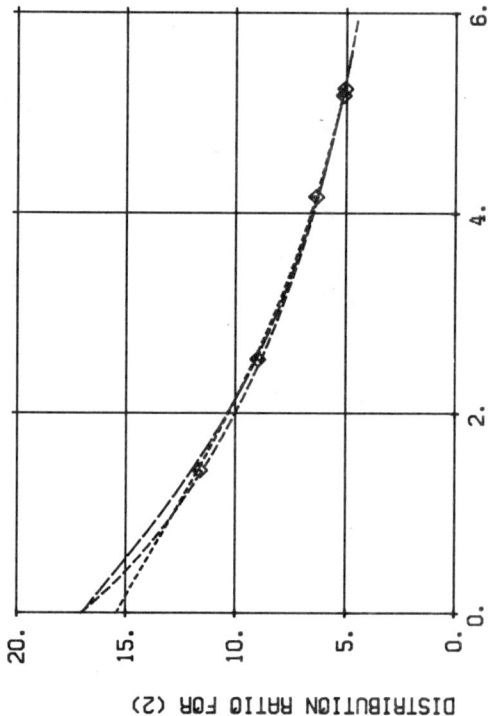

(1) C6H12O   2-PENTANONE, 4-METHYL
(2) C3H4O2   PROPENOIC ACID
(3) H2O   WATER

LINEK J.,HLAVATY K.,WICHTERLE I.
COLLECT.CZECH.CHEM.COMMUN. 38(1973)1840

TEMPERATURE = 24.6 DEG C   TYPE OF SYSTEM = 1

EXPERIMENTAL TIE LINES IN MOLE PCT

	LEFT PHASE			RIGHT PHASE	
(1)	(2)	(3)	(1)	(2)	(3)
87.973	0.0	12.027	0.347	0.0	99.653
56.504	16.604	26.892	0.362	1.430	98.208
40.694	22.837	36.469	0.374	2.542	97.084
27.340	26.449	46.210	0.516	4.163	95.320
21.379	26.530	52.090	0.617	5.178	94.205
21.227	26.534	52.239	0.618	5.246	94.135

SPECIFIC MODEL PARAMETERS IN KELVIN

		UNIQUAC		NRTL(ALPHA=.2)	
I	J	AIJ	AJI	AIJ	AJI
1	2	-162.87	76.867	150.56	-298.74
1	3	-422.68	86.627	177.04	1484.1
2	3	-151.91	258.12	-571.01	1362.3

R1 = 4.5959   R2 = 2.6467   R3 = 0.9200
Q1 = 3.952   Q2 = 2.400   Q3 = 1.400

MEAN DEV. BETWEEN CALC. AND EXP. CONC. IN MOLE PCT

UNIQUAC (SPECIFIC PARAMETERS)   0.10
NRTL   (SPECIFIC PARAMETERS)   0.12
UNIQUAC (COMMON PARAMETERS)   0.57

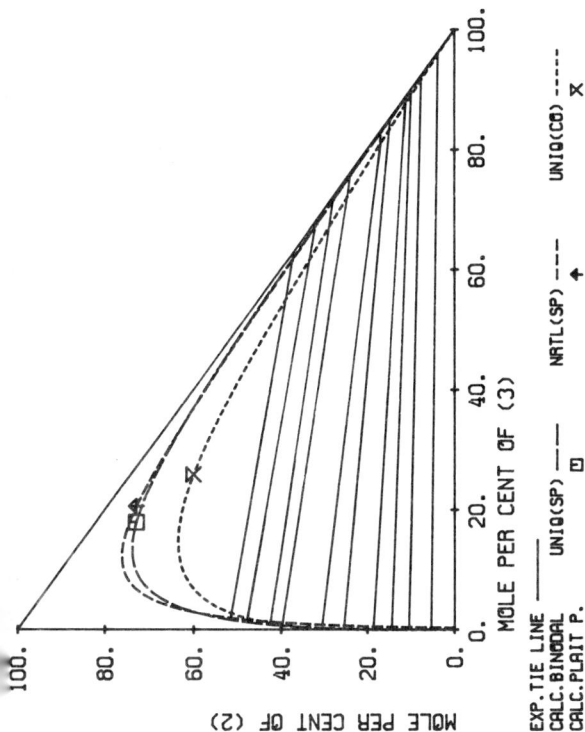

MOLE PER CENT OF (2)
MOLE PER CENT OF (3)

EXP.TIE LINE —— UNIQ(SP) □ NRTL(SP) + UNIQ(CO) -----
CALC.BINODAL
CALC.PLAIT P. ×

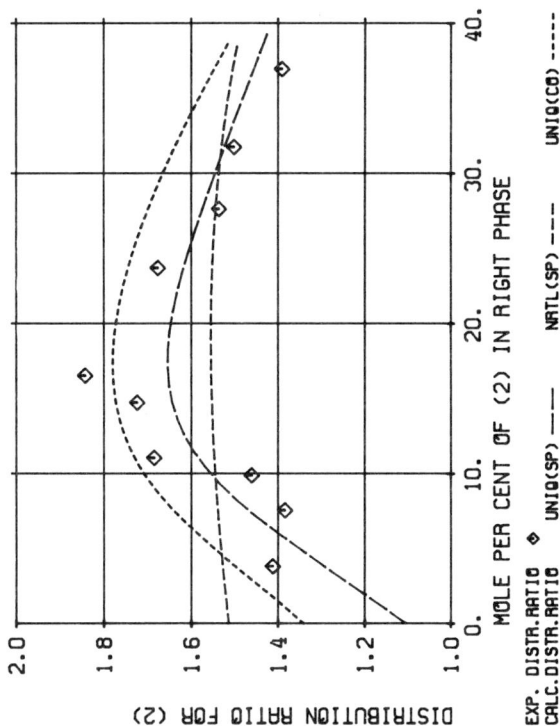

DISTRIBUTION RATIO FOR (2)
MOLE PER CENT OF (2) IN RIGHT PHASE

EXP. DISTR.RATIO ◇ UNIQ(SP) —— NRTL(SP) —— UNIQ(CO) -----
CALC.DISTR.RATIO

(1) H2O   WATER
(2) C3H4O2   PROPENOIC ACID
(3) C6H14   HEXANE

ABABI V., MIHAILA GH.
STUD.UNIV.BABES-BOLYAI.SER.CHEM. (1963)429

TEMPERATURE = 25.0 DEG C   TYPE OF SYSTEM = 1

EXPERIMENTAL TIE LINES IN MOLE PCT

	LEFT PHASE			RIGHT PHASE	
(1)	(2)	(3)	(1)	(2)	(3)
94.630	5.370	0.0	0.0	3.803	96.197
89.483	10.462	0.055	0.0	7.559	92.441
85.385	14.435	0.181	0.0	9.883	90.117
81.113	18.592	0.295	0.0	11.039	88.961
74.102	25.339	0.559	0.0	14.706	85.294
58.775	30.453	0.772	0.0	16.523	83.477
59.296	39.679	1.025	0.0	23.681	76.319
56.351	42.437	1.212	0.0	27.632	72.368
50.636	47.643	1.721	0.0	31.745	68.255
46.282	51.410	2.309	0.0	36.964	63.036

SPECIFIC MODEL PARAMETERS IN KELVIN

		UNIQUAC		NRTL(ALPHA=.2)	
I	J	AIJ	AJI	AIJ	AJI
1	2	-301.50	576.52	-156.25	277.46
1	3	336.68	1241.5	1157.9	1694.3
2	3	40.057	79.525	873.12	-289.66

R1 = 0.9200   R2 = 2.6467   R3 = 4.4998
Q1 = 1.400   Q2 = 2.400   Q3 = 3.856

MEAN DEV. BETWEEN CALC. AND EXP. CONC. IN MOLE PCT

UNIQUAC (SPECIFIC PARAMETERS)   0.58
NRTL (SPECIFIC PARAMETERS)   0.83
UNIQUAC (COMMON PARAMETERS)   1.51

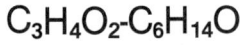

$C_3H_4O_2\text{-}C_6H_{14}O$

(1) C6H14O   ETHER,DIISOPROPYL

(2) C3H4O2   PROPENOIC ACID

(3) H2O   WATER

LINEK J., HLAVATY K., WICHTERLE I.
COLLECT.CZECH.CHEM.COMMUN. 38(1973)1840

TEMPERATURE = 24.6 DEG C   TYPE OF SYSTEM = 1

EXPERIMENTAL TIE LINES IN MOLE PCT

	LEFT PHASE			RIGHT PHASE	
(1)	(2)	(3)	(1)	(2)	(3)
95.101	0.0	4.899	0.250	0.0	99.750
67.010	17.192	15.798	0.169	1.946	97.885
49.543	26.254	24.203	0.175	3.233	96.592
29.691	33.376	36.934	0.359	6.052	93.589
26.111	34.265	39.624	0.481	7.217	92.302
20.158	33.754	46.088	0.766	9.344	89.890

SPECIFIC MODEL PARAMETERS IN KELVIN

		UNIQUAC		NRTL(ALPHA=.2)	
I	J	AIJ	AJI	AIJ	AJI
1	2	-61.138	108.82	-229.71	184.81
1	3	557.90	110.32	384.66	1586.6
2	3	-103.43	251.18	-481.97	1217.5

R1 = 4.7421   R2 = 2.6467   R3 = 0.9200
Q1 = 4.088   Q2 = 2.400   Q3 = 1.400

MEAN DEV. BETWEEN CALC. AND EXP. CONC. IN MOLE PCT

UNIQUAC (SPECIFIC PARAMETERS)   0.38
NRTL (SPECIFIC PARAMETERS)   0.38
UNIQUAC (COMMON PARAMETERS)   0.68

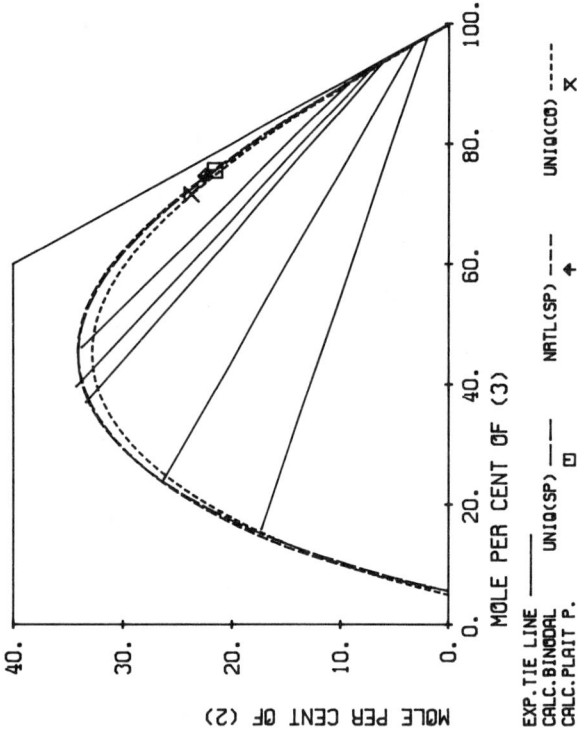

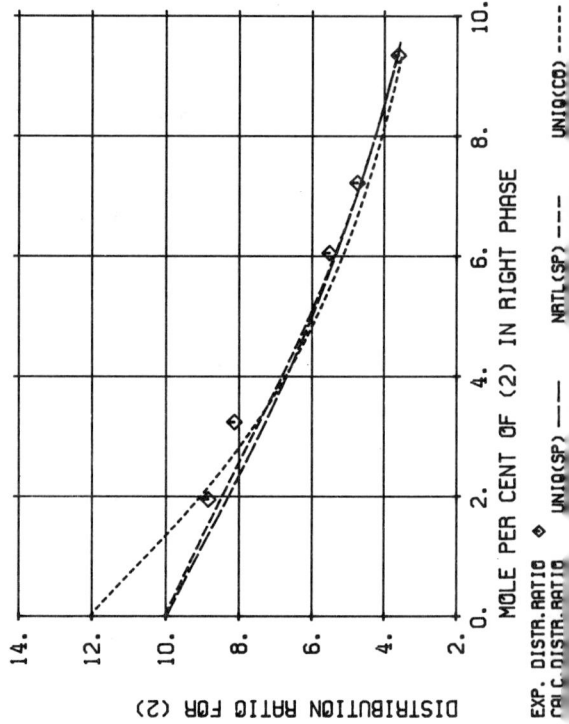

C₃H₄O₂-C₆H₁₄O

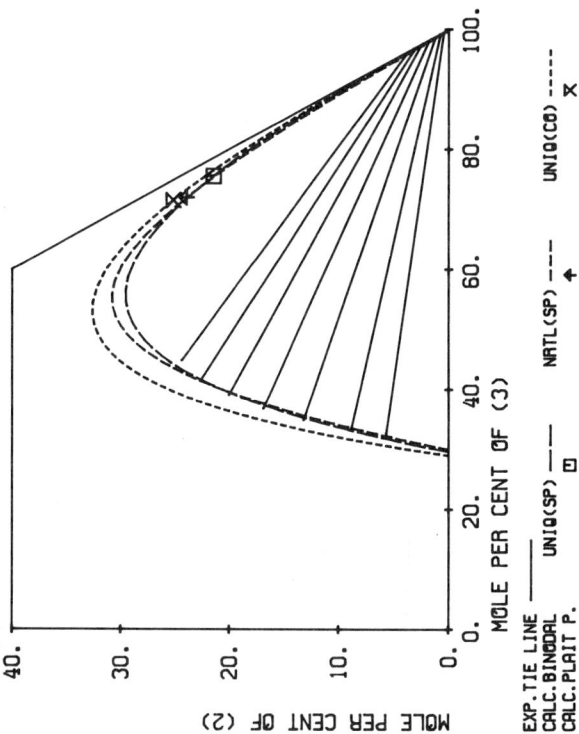

MOLE PER CENT OF (3)

MOLE PER CENT OF (2)

EXP.TIE LINE ———   UNIQ(SP) ---- □   NRTL(SP) ---- ✦   UNIQ(CB) ---- ✕
CALC.BINODAL ------
CALC.PLAIT P.

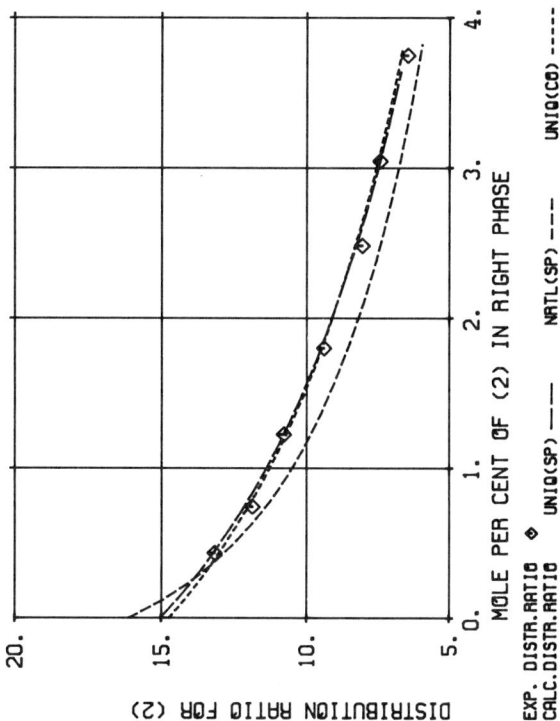

DISTRIBUTION RATIO FOR (2)

MOLE PER CENT OF (2) IN RIGHT PHASE

EXP. DISTR.RATIO ◇   UNIQ(SP) ———   NRTL(SP) ———   UNIQ(CB) -----
CALC.DISTR.RATIO

(2) C3H4O2   PROPENOIC ACID

(3) H2O   WATER

ABABI V.; MIHAILA GH.
STUD.UNIV.BABES-BOLYAI.SER.CHEM. (1963)429

TEMPERATURE = 25.0 DEG C   TYPE OF SYSTEM = 1

EXPERIMENTAL TIE LINES IN MOLE PCT

	LEFT PHASE			RIGHT PHASE	
(1)	(2)	(3)	(1)	(2)	(3)
61.840	5.683	32.477	0.036	0.431	99.533
57.847	8.827	33.326	0.054	0.743	99.203
51.995	13.197	34.808	0.073	1.222	98.705
46.370	16.897	36.733	0.093	1.799	98.108
40.957	20.032	39.011	0.095	2.481	97.423
35.970	22.688	41.342	0.116	3.043	96.841
30.889	24.360	44.752	0.138	3.748	96.114

SPECIFIC MODEL PARAMETERS IN KELVIN

		UNIQUAC		NRTL(ALPHA=.2)	
I J	AIJ	AJI		AIJ	AJI
1 2	5.3670	-60.639		207.57	-360.56
1 3	121.97	244.18		-126.81	1890.3
2 3	-148.20	288.52		35.658	493.16

R1 = 4.8031   R2 = 2.6467   R3 = 0.9200
Q1 = 4.132   Q2 = 2.400   Q3 = 1.400

MEAN DEV. BETWEEN CALC. AND EXP. CONC. IN MOLE PCT

UNIQUAC (SPECIFIC PARAMETERS)   0.24
NRTL   (SPECIFIC PARAMETERS)   0.48
UNIQUAC (COMMON PARAMETERS)   1.06

C₃H₄O₂-C₇H₈

(1)	C7H8	TOLUENE
(2)	C3H4O2	PROPENOIC ACID
(3)	H2O	WATER

ABABI V.; MIHAILA GH.
STUD.UNIV.BABES-BOLYAI.SER.CHEM. (1963)429

TEMPERATURE = 25.0 DEG C   TYPE OF SYSTEM = 1

EXPERIMENTAL TIE LINES IN MOLE PCT

	LEFT PHASE			RIGHT PHASE	
(1)	(2)	(3)	(1)	(2)	(3)
96.070	3.930	0.0	0.063	2.651	97.285
89.993	10.007	0.0	0.089	4.465	95.446
86.231	13.769	0.0	0.115	5.841	94.044
80.092	18.034	0.975	0.145	7.548	92.307
74.689	23.400	1.910	0.185	11.406	88.410
71.348	25.364	3.288	0.277	13.438	86.286
63.850	31.132	5.018	0.962	20.004	79.034
62.172	32.391	5.436	1.200	21.875	76.925
59.497	34.238	6.266	1.492	23.784	74.724

SPECIFIC MODEL PARAMETERS IN KELVIN

			UNIQUAC		NRTL(ALPHA=.2)	
I	J	AIJ	AJI		AIJ	AJI
1	2	243.44	-103.86		59.492	5.5576
1	3	726.70	348.73		941.92	1763.7
2	3	251.66	-105.79		101.09	136.13

R1 = 3.9228   R2 = 2.6467   R3 = 0.9200
Q1 = 2.968   Q2 = 2.400   Q3 = 1.400

MEAN DEV. BETWEEN CALC. AND EXP. CONC. IN MOLE PCT

UNIQUAC (SPECIFIC PARAMETERS)	1.00
NRTL (SPECIFIC PARAMETERS)	0.97
UNIQUAC (COMMON PARAMETERS)	1.24

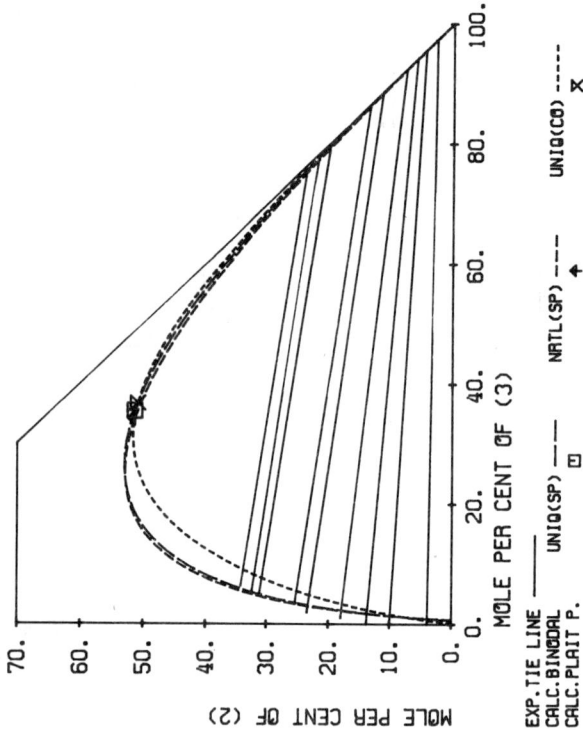

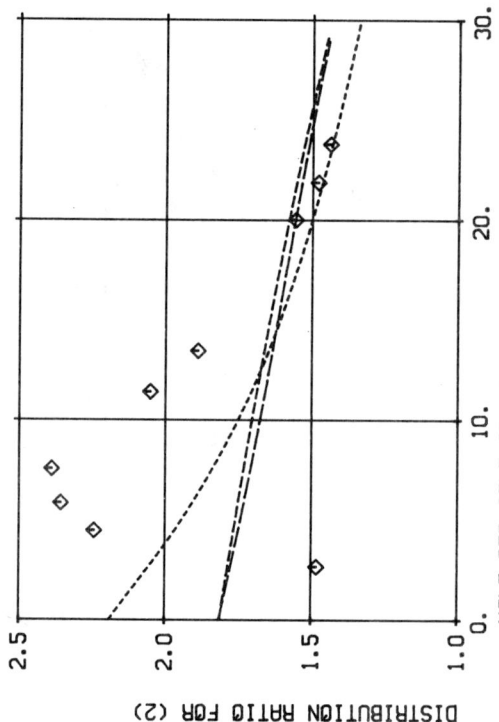

$C_3H_4O_2\text{-}C_8H_{18}O$

(1) C8H18O   1-HEXANOL,2-ETHYL
----
(2) C3H4O2   PROPENOIC ACID
----
(3) H2O   WATER

LINEK J., HLAVATY K., WICHTERLE I.
COLLECT.CZECH.CHEM.COMMUN. 38(1973)1840

TEMPERATURE = 24.6 DEG C   TYPE OF SYSTEM = 1

EXPERIMENTAL TIE LINES IN MOLE PCT

        LEFT PHASE   RIGHT PHASE
    (1)   (2)   (3)   (1)   (2)   (3)

  87.144   0.0   12.856   0.0   0.0   100.000
  61.955   18.549   19.496   0.015   1.682   98.303
  46.215   28.670   25.116   0.015   3.759   96.226
  35.360   33.159   31.480   0.049   5.935   94.016
  29.187   34.875   35.938   0.120   7.712   92.168
  24.395   35.737   39.868   0.218   9.893   89.889

SPECIFIC MODEL PARAMETERS IN KELVIN

             UNIQUAC   NRTL(ALPHA=.2)
  I J   AIJ   AJI   AIJ   AJI
----
  1 2   4.8241   -98.289   60.754   -290.35
  1 3   457.73   80.732   185.45   1912.8
  2 3 -80.053   123.53   -217.61   739.56

  R1 = 6.1511   R2 = 2.6467   R3 = 0.9200
  Q1 = 5.208   Q2 = 2.400   Q3 = 1.400

MEAN DEV. BETWEEN CALC. AND EXP. CONC. IN MOLE PCT

UNIQUAC (SPECIFIC PARAMETERS)   0.44
NRTL   (SPECIFIC PARAMETERS)   0.39
UNIQUAC (COMMON PARAMETERS)   0.90

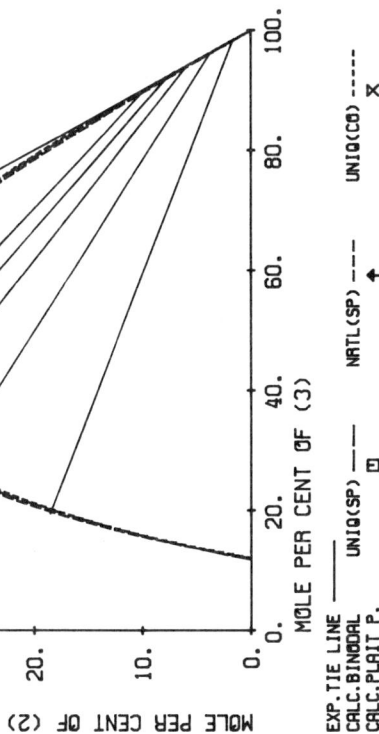

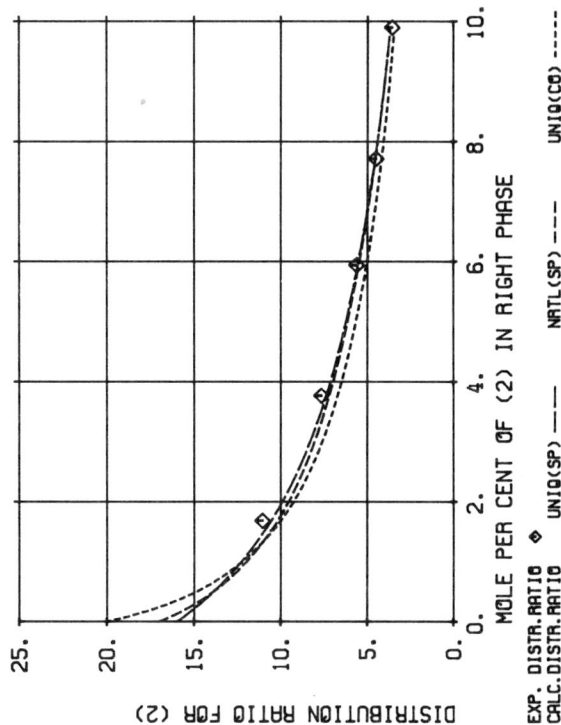

C$_3$H$_4$O$_4$-C$_4$H$_{10}$O

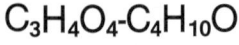

(1) C4H10O   ETHER,DIETHYL
(2) C3H4O4   MALONIC ACID
(3) H2O   WATER

KLOBBIE E.A.
Z.PHYS.CHEM.(LEIPZIG) 24(1897)615

TEMPERATURE =   15.0 DEG C   TYPE OF SYSTEM = 1

EXPERIMENTAL TIE LINES IN MOLE PCT

	LEFT PHASE			RIGHT PHASE	
(1)	(2)	(3)	(1)	(2)	(3)
95.241	0.0	4.759	2.007	0.0	97.993
93.452	0.490	6.058	2.140	0.889	96.971
90.767	1.478	7.756	2.454	2.391	95.155
84.983	3.300	11.717	3.215	4.687	92.098
75.195	5.978	18.826	4.274	6.959	88.766
25.956	12.735	61.309	14.891	12.069	73.040
24.159	12.707	63.135	24.691	12.278	63.031

SPECIFIC MODEL PARAMETERS IN KELVIN

		UNIQUAC		NRTL(ALPHA=.2)	
I J	AIJ	AJI		AIJ	AJI
1 2	-74.286	-76.938		742.06	-835.46
1 3	585.38	46.514		380.25	965.49
2 3	-273.45	41.534		-351.58	-727.36

R1 = 3.3949   R2 = 3.2770   R3 = 0.9200
Q1 = 3.016   Q2 = 2.988   Q3 = 1.400

MEAN DEV. BETWEEN CALC. AND EXP. CONC. IN MOLE PCT

UNIQUAC (SPECIFIC PARAMETERS)   1.27
NRTL   (SPECIFIC PARAMETERS)   1.42

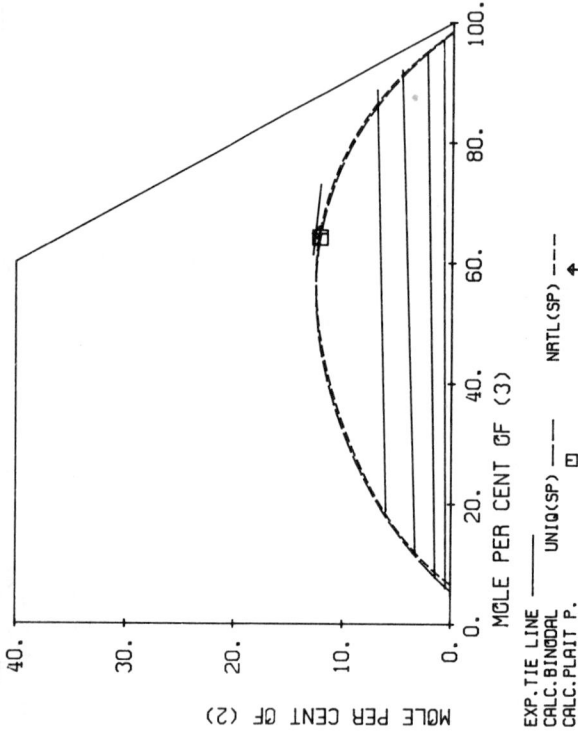

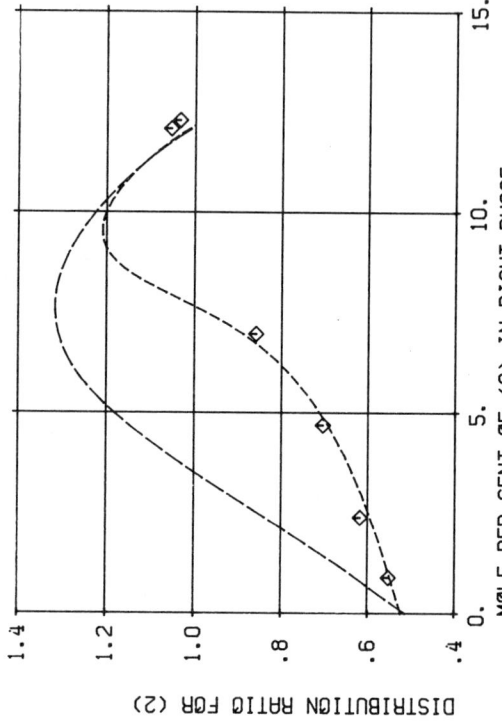

$C_3H_5N-C_7H_8$

(1) C3H5N   PROPANOIC ACID,NITRILE
----
(2) C7H8   TOLUENE
----
(3) C8F16O   OCTANE,1,8-OXY,PERFLUORO

KIKIC I., ALESSI P.
CAN.J.CHEM.ENG. 53(1975)192

TEMPERATURE = 25.0 DEG C   TYPE OF SYSTEM = 2

EXPERIMENTAL TIE LINES IN MOLE PCT

         LEFT PHASE   RIGHT PHASE
      (1)   (2)   (3)   (1)   (2)   (3)

   99.710   0.0   0.290   2.010   0.0   97.990
   94.870   4.690   0.440   2.300   0.590   97.110
   35.100   64.340   0.560   1.270   6.440   92.290
   13.220   86.140   0.640   0.590   8.510   90.900
    4.740   94.580   0.680   0.230   9.410   90.360
    0.0   99.270   0.730   0.0   9.820   90.180

SPECIFIC MODEL PARAMETERS IN KELVIN

              UNIQUAC   NRTL(ALPHA=.2)
  I J   AIJ   AJI   AIJ   AJI

  1 2   13.368   26.265   16.094   143.99
  1 3 142.20   464.15   1269.4   647.70
  2 3   19.700   235.74   1286.6   244.57

  R1 = 2.5445   R2 = 3.9228   R3 = 8.3279
  Q1 = 2.264   Q2 = 2.968   Q3 = 7.600

MEAN DEV. BETWEEN CALC. AND EXP. CONC. IN MOLE PCT

  UNIQUAC (SPECIFIC PARAMETERS)   0.15
  NRTL   (SPECIFIC PARAMETERS)   0.07
  UNIQUAC (COMMON PARAMETERS)   0.24

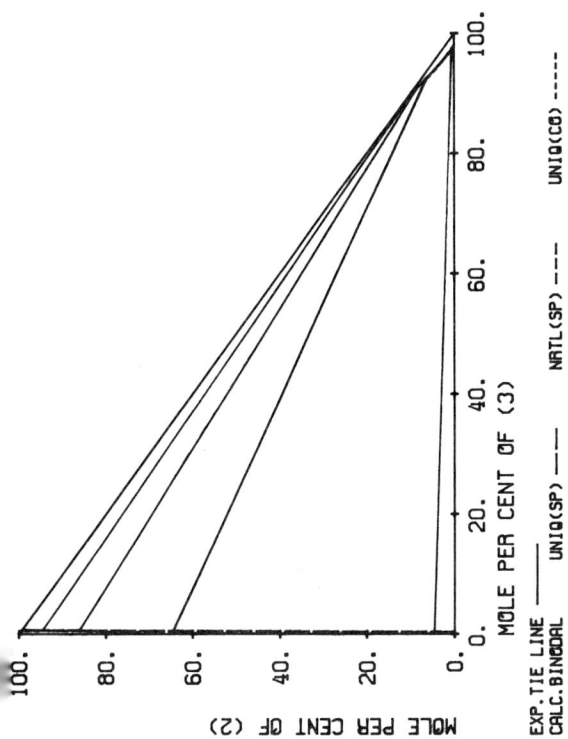

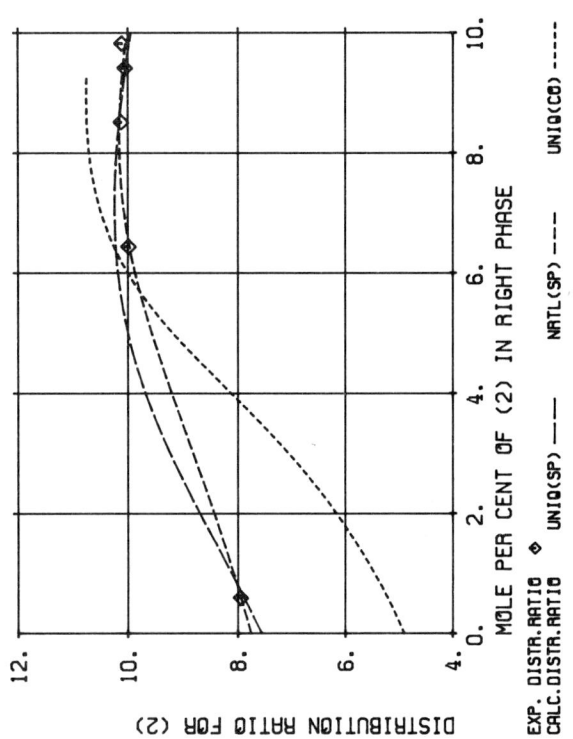

C₃H₅N-C₇H₁₄

KIKIC I., ALESSI P.
CAN.J.CHEM.ENG. 53(1975)192

(1) C3H5N   PROPANOIC ACID,NITRILE
(2) C7H14   CYCLOHEXANE,METHYL
(3) C8F16O   OCTANE,1,8-OXY,PERFLUORO

TEMPERATURE = 25.0 DEG C   TYPE OF SYSTEM = 2

EXPERIMENTAL TIE LINES IN MOLE PCT

	LEFT PHASE			RIGHT PHASE	
(1)	(2)	(3)	(1)	(2)	(3)
99.710	0.0	0.290	2.010	0.0	97.990
96.900	2.820	0.280	2.380	1.510	96.110
72.750	26.850	0.400	2.760	9.880	87.360
68.550	30.830	0.620	2.430	10.280	87.290
42.030	56.720	1.250	2.320	12.000	85.680
13.380	84.880	1.740	1.500	13.040	85.460
16.140	92.010	1.850	0.960	14.540	84.500
0.0	96.810	3.190	0.0	18.200	81.800

SPECIFIC MODEL PARAMETERS IN KELVIN

		UNIQUAC		NRTL(ALPHA=.2)	
I J		AIJ	AJI	AIJ	AJI
1 2	-49.789	295.68	233.71	320.47	
1 3	145.83	460.91	1634.5	640.45	
2 3	29.893	144.27	999.43	162.68	

R1 = 2.5445   R2 = 4.7200   R3 = 8.3279
Q1 = 2.264   Q2 = 3.776   Q3 = 7.600

MEAN DEV. BETWEEN CALC. AND EXP. CONC. IN MOLE PCT

UNIQUAC (SPECIFIC PARAMETERS)   0.55
NRTL   (SPECIFIC PARAMETERS)   0.53
UNIQUAC (COMMON PARAMETERS)   0.80

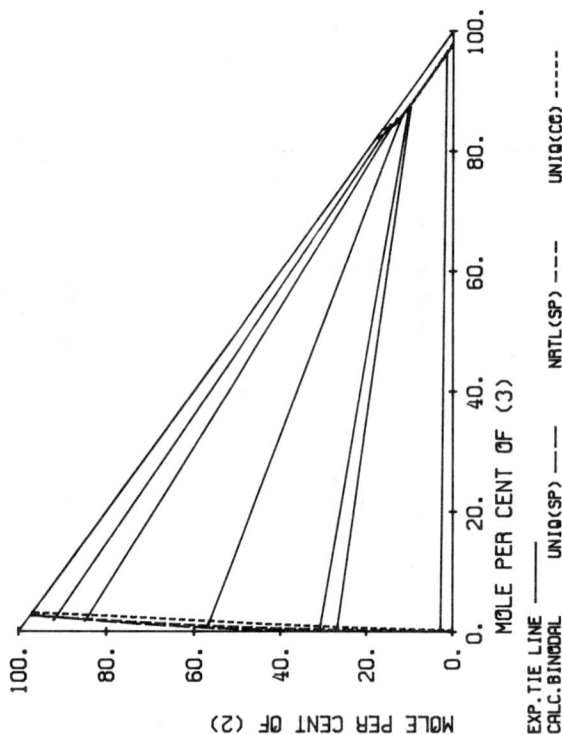

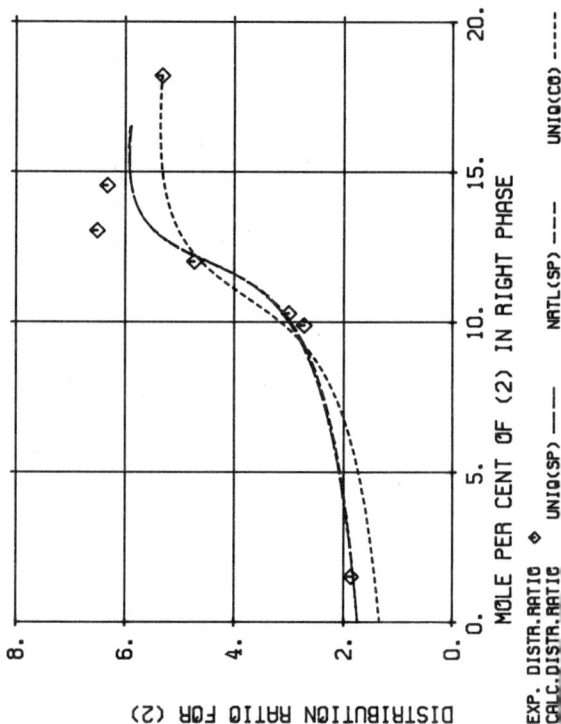

$$C_3H_5N-C_7H_{14}$$

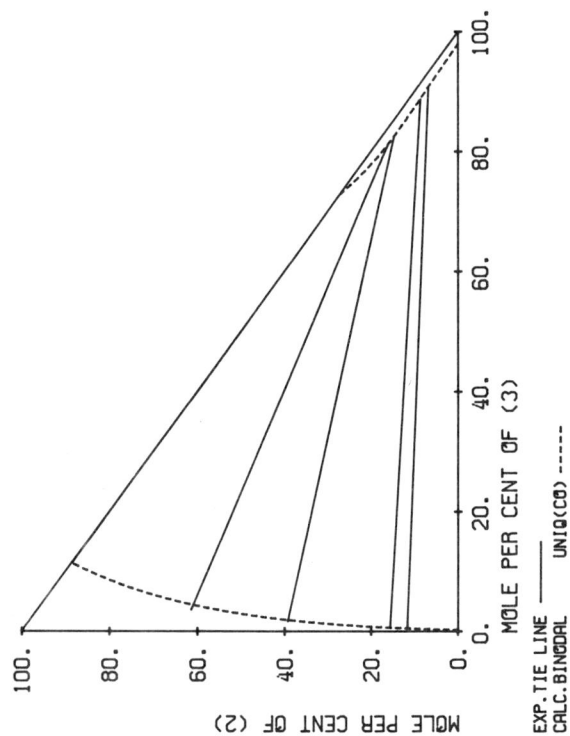

MOLE PER CENT OF (2)

MOLE PER CENT OF (3)

EXP.TIE LINE ———
CALC.BINODAL UNIQ(CO) -----

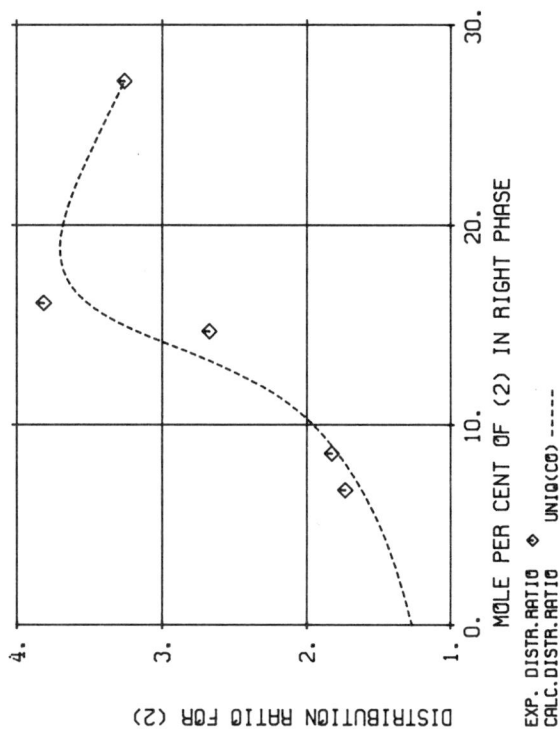

DISTRIBUTION RATIO FOR (2)

MOLE PER CENT OF (2) IN RIGHT PHASE

EXP. DISTR.RATIO ◇   UNIQ(CO) -----
CALC.DISTR.RATIO

(1) C3H5N   PROPANOIC ACID,NITRILE
----
(2) C7H14   1-HEPTENE
----
(3) C8F16O   OCTANE,1,8-OXY,PERFLUORO

KIKIC I., ALESSI P.
CAN.J.CHEM.ENG. 53(1975)192

TEMPERATURE = 25.0 DEG C   TYPE OF SYSTEM = 2

EXPERIMENTAL TIE LINES IN MOLE PCT

       LEFT PHASE   RIGHT PHASE
   (1)   (2)   (3)   (1)   (2)   (3)

 87.680 11.670   0.650   2.560   6.740 90.700
 83.620 15.630   0.750   2.610   8.570 88.820
 59.000 39.270   1.730   2.740 14.680 82.580
 34.970 61.390   3.640   2.680 16.090 81.230
  0.0   88.600 11.400   0.0   27.200 72.800

MEAN DEV. BETWEEN CALC. AND EXP. CONC. IN MOLE PCT

UNIQUAC (COMMON PARAMETERS)   0.36

C$_3$H$_5$N-C$_7$H$_{16}$

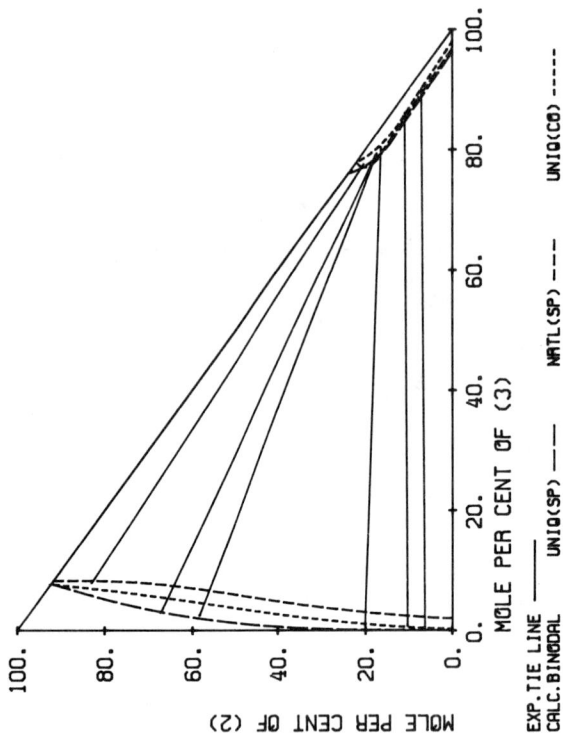

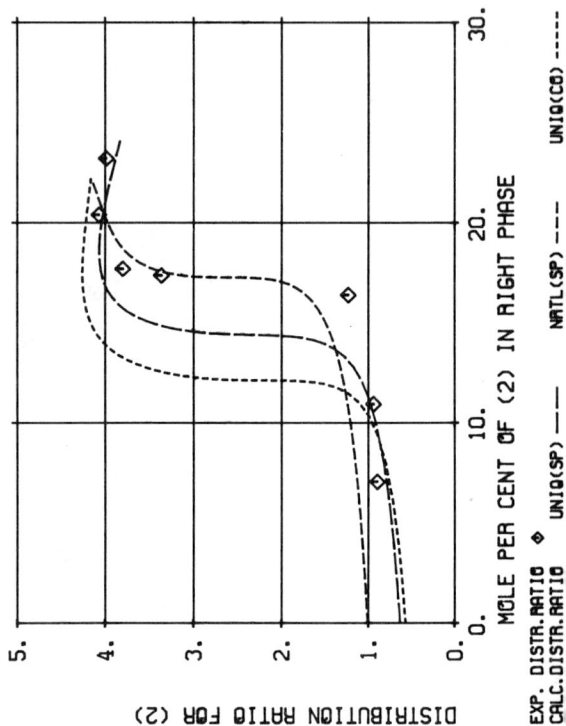

(1) C3H5N   PROPANOIC ACID,NITRILE
(2) C7H16   HEPTANE
(3) C8F16O   OCTANE,1,8-OXY,PERFLUORO

KIKIC I., ALESSI P.
CAN.J.CHEM.ENG. 53(1975)192

TEMPERATURE = 25.0 DEG C   TYPE OF SYSTEM = 2

EXPERIMENTAL TIE LINES IN MOLE PCT

	LEFT PHASE			RIGHT PHASE	
(1)	(2)	(3)	(1)	(2)	(3)
99.710	0.0	0.290	2.010	0.0	97.990
93.380	6.310	0.310	2.670	7.060	90.270
89.020	10.270	0.710	2.870	10.940	86.190
79.100	20.150	0.750	3.170	16.380	80.450
39.130	58.400	2.470	3.260	17.360	79.380
29.820	67.250	2.930	3.200	17.700	79.100
9.110	82.980	7.910	2.000	20.400	77.600
0.0	92.470	7.530	0.0	23.210	76.790

SPECIFIC MODEL PARAMETERS IN KELVIN

		UNIQUAC		NRTL(ALPHA=.2)	
I	J	AIJ	AJI	AIJ	AJI
1	2	56.555	165.38	368.82	319.85
1	3	197.65	366.06	809.92	626.55
2	3	21.338	119.29	723.40	192.26

R1 = 2.5445   R2 = 5.1742   R3 = 3.3279
Q1 = 2.264   Q2 = 4.396   Q3 = 7.600

MEAN DEV. BETWEEN CALC. AND EXP. CONC. IN MOLE PCT

UNIQUAC (SPECIFIC PARAMETERS)   0.96
NRTL   (SPECIFIC PARAMETERS)   1.57
UNIQUAC (COMMON PARAMETERS)   1.34

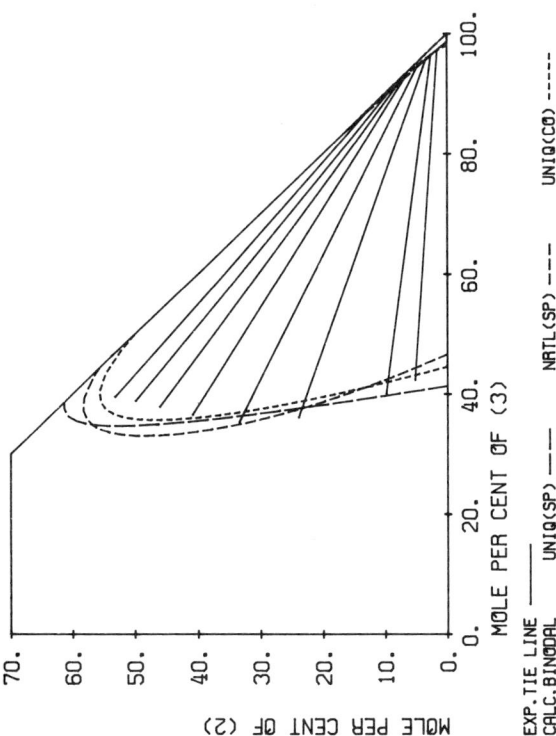

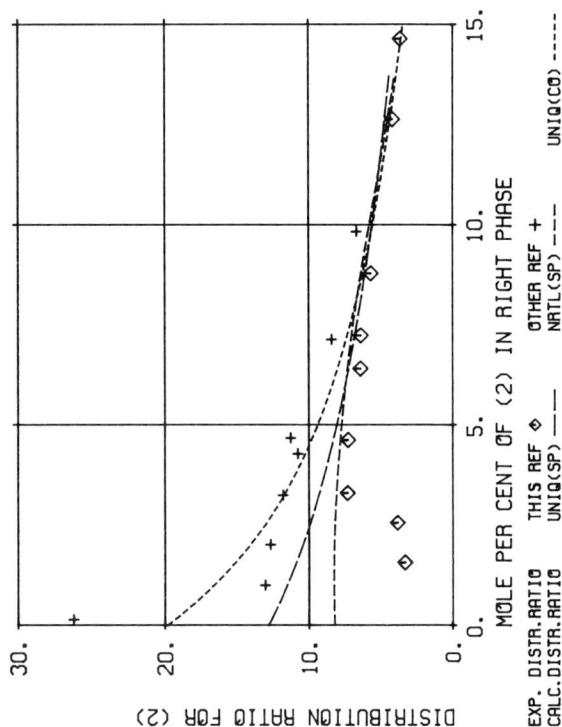

$C_3H_6O-C_4H_{10}O$

(1) C4H10O   1-PROPANOL,2-METHYL

(2) C3H6O   PROPANAL

(3) H2O   WATER

MOZZHUKIN A.S., ET AL.
KHIM.TEKHNOL.TOPL.MASEL (1966)4,11

TEMPERATURE = 20.0 DEG C   TYPE OF SYSTEM = 2

EXPERIMENTAL TIE LINES IN MOLE PCT

	LEFT PHASE			RIGHT PHASE	
(1)	(2)	(3)	(1)	(2)	(3)
55.337	0.0	44.663	2.208	0.0	97.792
52.457	5.202	42.341	1.118	1.559	97.323
50.137	9.830	40.033	0.842	2.553	96.605
39.956	23.945	36.100	0.692	3.294	96.014
31.231	33.452	35.316	0.517	4.615	94.868
22.533	40.906	36.561	0.365	6.406	93.229
16.014	46.120	37.866	0.342	7.231	92.428
11.239	49.897	38.864	0.293	8.779	90.929
7.322	53.152	39.526	0.094	12.621	87.285
0.0	53.700	46.300	0.0	14.637	85.363

SPECIFIC MODEL PARAMETERS IN KELVIN

		UNIQUAC		NRTL(ALPHA=.2)	
I J	AIJ	AJI		AIJ	AJI
1 2	-161.62	167.78		-495.79	788.88
1 3	40.718	248.00		-80.605	3830.7
2 3	318.03	13.194		-8.6013	784.31

R1 = 3.4535   R2 = 2.5735   R3 = 0.9200
Q1 = 3.048   Q2 = 2.336   Q3 = 1.400

MEAN DEV. BETWEEN CALC. AND EXP. CONC. IN MOLE PCT

UNIQUAC (SPECIFIC PARAMETERS)   2.07
NRTL   (SPECIFIC PARAMETERS)   1.89
UNIQUAC (COMMON PARAMETERS)   1.37

$C_3H_6O-C_4H_{10}O$

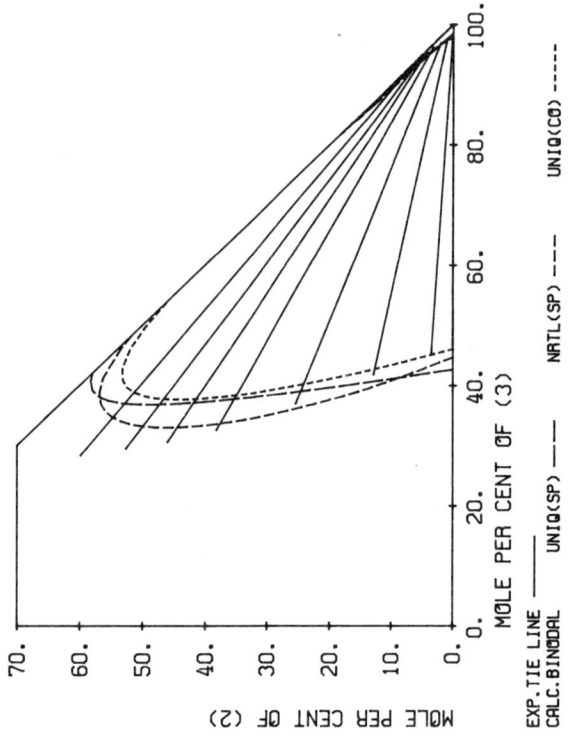

MOLE PER CENT OF (3)

MOLE PER CENT OF (2)

EXP.TIE LINE ———
CALC.BINODAL - - - -
UNIQ(SP) ———
NRTL(SP) - - - -
UNIQ(CO) - - - -

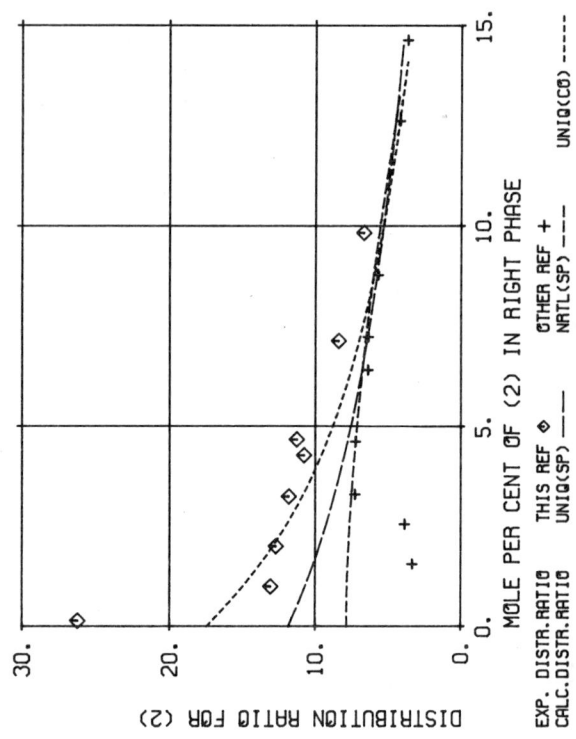

MOLE PER CENT OF (2) IN RIGHT PHASE

DISTRIBUTION RATIO FOR (2)

EXP. DISTR.RATIO THIS REF ◇   OTHER REF +
CALC.DISTR.RATIO UNIQ(SP) ——— NRTL(SP) - - - - UNIQ(CO) - - - -

(1) C4H10O   1-PROPANOL,2-METHYL
(2) C3H6O   PROPANAL
(3) H2O   WATER

MOZZHUKIN A.S., ET AL.
KHIM.TEKHNOL.TOPL.MASEL (1966)4,11   TYPE OF SYSTEM = 2

TEMPERATURE = 30.0 DEG C

EXPERIMENTAL TIE LINES IN MOLE PCT

	LEFT PHASE			RIGHT PHASE	
(1)	(2)	(3)	(1)	(2)	(3)
53.398	0.0	46.602	1.933	0.0	93.067
51.134	3.478	45.388	2.076	0.132	97.792
45.230	12.966	41.803	1.453	0.993	97.554
37.658	25.387	36.955	0.993	2.001	97.005
29.176	38.261	32.563	0.529	3.241	96.229
23.418	46.024	30.557	0.214	4.271	95.514
17.903	52.633	29.464	0.081	4.667	95.253
11.807	59.854	28.338	0.056	7.125	92.818
0.0	65.581	34.419	0.0	9.827	90.173

SPECIFIC MODEL PARAMETERS IN KELVIN

		UNIQUAC		NRTL(ALPHA=.2)	
I	J	AIJ	AJI	AIJ	AJI
1	2	-161.62	167.78	-495.79	788.88
1	3	40.718	248.00	-80.605	3830.7
2	3	318.03	13.194	8.6013	784.31

R1 = 3.4535   R2 = 2.5735   R3 = 0.9200
Q1 = 3.048   Q2 = 2.336   Q3 = 1.400

MEAN DEV. BETWEEN CALC. AND EXP. CONC. IN MOLE PCT

UNIQUAC (SPECIFIC PARAMETERS)   3.22
NRTL   (SPECIFIC PARAMETERS)   2.43
UNIQUAC (COMMON PARAMETERS)   3.55

(1) C3H6O2   ACETIC ACID,METHYL ESTER
(2) C3H6O   2-PROPANONE
(3) H2O   WATER

VENKATARATNAM A., JAGANNADHA RAO R., VENKATA RAO C.
CHEM.ENG.SCI. 7(1957)102

TEMPERATURE = 30.0 DEG C   TYPE OF SYSTEM = 1

EXPERIMENTAL TIE LINES IN MOLE PCT

	LEFT PHASE			RIGHT PHASE	
(1)	(2)	(3)	(1)	(2)	(3)
65.398	0.0	34.602	7.647	0.0	92.353
62.791	1.389	35.820	7.961	0.508	91.532
58.132	3.034	38.834	8.450	1.078	90.472
49.630	6.330	44.040	9.160	2.064	88.776
39.715	7.449	52.836	11.273	3.595	85.132

SPECIFIC MODEL PARAMETERS IN KELVIN

		UNIQUAC		NRTL(ALPHA=.2)	
I J		AIJ	AJI	AIJ	AJI
1 2		-30.816	-21.084	213.08	-511.34
1 3		247.85	84.707	5.3125	942.96
2 3		-96.844	47.294	-63.315	-186.16

R1 = 2.8042   R2 = 2.5735   R3 = 0.9200
Q1 = 2.576   Q2 = 2.336   Q3 = 1.400

MEAN DEV. BETWEEN CALC. AND EXP. CONC. IN MOLE PCT

UNIQUAC (SPECIFIC PARAMETERS)   0.50
NRTL   (SPECIFIC PARAMETERS)   0.62
UNIQUAC (COMMON PARAMETERS)   1.21

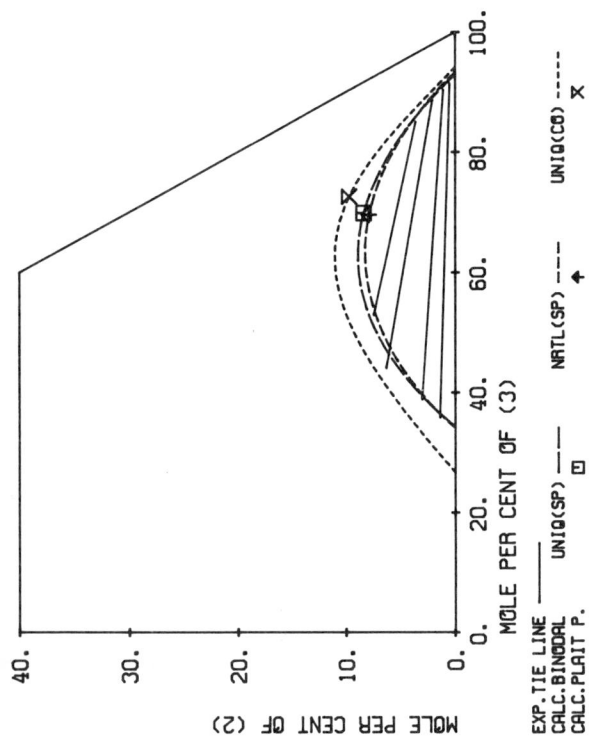

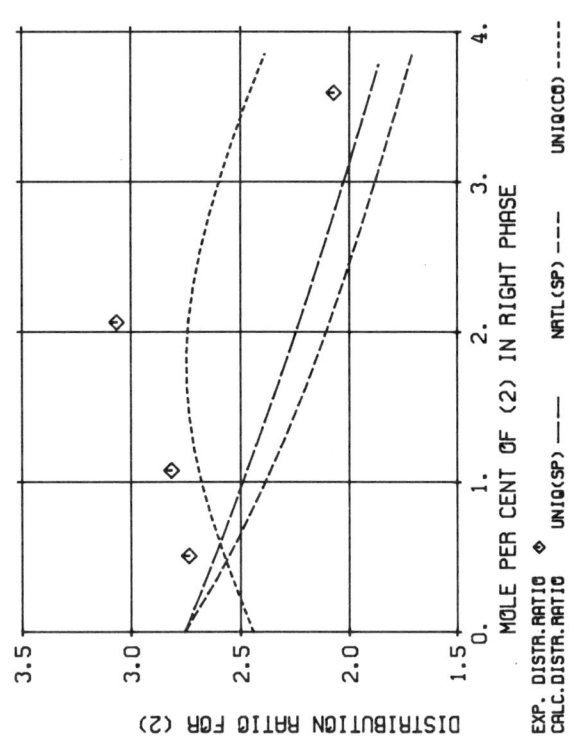

MOLE PER CENT OF (2)

MOLE PER CENT OF (3)

EXP.TIE LINE ——   UNIQ(SP) ⊟   NRTL(SP) ——   UNIQ(CO) - - - -
CALC.BINODAL
CALC.PLAIT P.

DISTRIBUTION RATIO FOR (2)

MOLE PER CENT OF (2) IN RIGHT PHASE

EXP. DISTR.RATIO ◇   UNIQ(SP) ——   NRTL(SP) ———   UNIQ(CO) - - - -
CALC.DISTR.RATIO

$C_3H_6O-C_4H_6O$

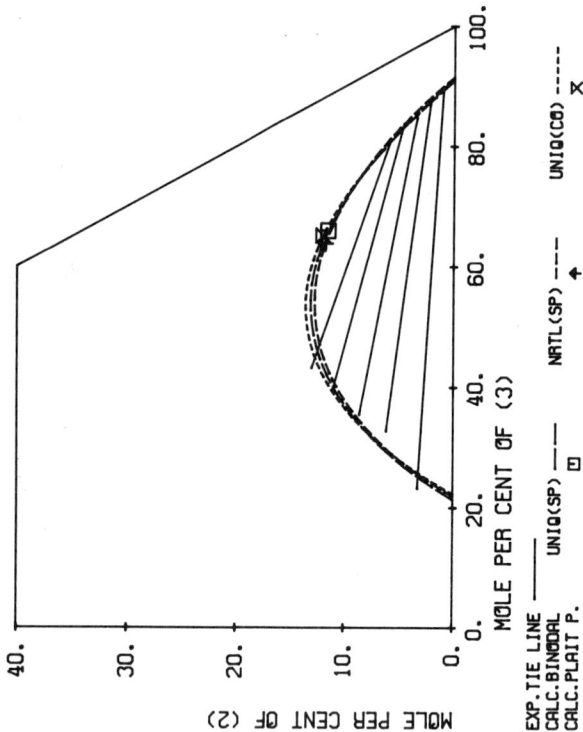

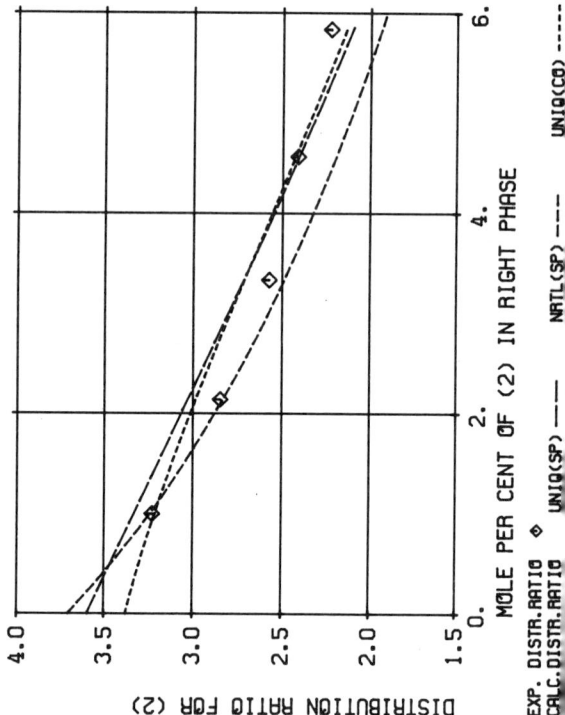

(1) C4H6O	3-BUTEN-2-ONE
(2) C3H6O	2-PROPANONE
(3) H2O	WATER

FROLOV A.F., LOGINOVA M.A., KIRSANOVA G.N.
ZH.PRIKL.KHIM.(LENINGRAD) 42(1969)2095

TEMPERATURE = 25.0 DEG C   TYPE OF SYSTEM = 1

EXPERIMENTAL TIE LINES IN MOLE PCT

| | LEFT PHASE | | RIGHT PHASE | |
	(1)	(2)	(3)	(1)	(2)	(3)
73.584	3.225	23.191	0.998	9.542	89.460
61.245	6.093	32.662	2.141	10.574	87.285
56.009	8.585	35.406	3.335	11.781	84.884
49.238	11.007	39.755	4.563	11.788	83.649
43.811	13.000	43.189	5.837	13.419	80.743

SPECIFIC MODEL PARAMETERS IN KELVIN

| | | UNIQUAC | | NRTL(ALPHA=.2) | |
I J	AIJ	AJI	AIJ	AJI
1 2	41.526	10.568	-384.11	-171.01
1 3	469.67	-28.875	199.99	712.03
2 3	94.519	35.552	-438.52	462.51

R1 = 3.0178   R2 = 2.5735   R3 = 0.9200
Q1 = 2.664   Q2 = 2.336   Q3 = 1.400

MEAN DEV. BETWEEN CALC. AND EXP. CONC. IN MOLE PCT

UNIQUAC (SPECIFIC PARAMETERS)	0.53
NRTL (SPECIFIC PARAMETERS)	0.66
UNIQUAC (COMMON PARAMETERS)	0.56

(1) ~~ACETIC ACID,ETHENYL ESTER~~

(2) C3H6O   2-PROPANONE

(3) H2O   WATER

SMITH J.C.
J.PHYS.CHEM. 46(1942)229

TEMPERATURE = 25.0 DEG C   TYPE OF SYSTEM = 1

EXPERIMENTAL TIE LINES IN MOLE PCT (GRAPH.INTERPOL.)

	LEFT PHASE			RIGHT PHASE	
(1)	(2)	(3)	(1)	(2)	(3)
84.078	10.971	4.952	0.566	0.708	98.726
73.883	18.226	7.891	0.631	2.057	97.312
67.151	23.124	9.724	0.695	3.309	95.996
60.954	27.865	11.181	0.749	4.211	95.040
55.340	32.103	12.557	0.818	5.296	93.832
50.727	35.384	13.889	0.881	6.296	92.822
46.330	38.189	15.480	0.937	7.150	91.912
42.431	40.409	17.160	1.023	8.266	90.710
35.238	43.920	20.842	1.127	9.441	89.432
29.426	45.703	24.871	1.432	12.369	86.199
23.940	45.923	30.137	1.880	15.444	82.677
18.081	43.489	38.431	2.547	18.269	79.184

SPECIFIC MODEL PARAMETERS IN KELVIN

		UNIQUAC		NRTL(ALPHA=.2)	
I J	AIJ	AJI		AIJ	AJI
1 2	270.81	-166.94		115.29	-130.76
1 3	524.66	207.80		467.98	1253.3
2 3	437.12	-103.43		-25.639	560.45

R1 = 3.2485   R2 = 2.5735   R3 = 0.9200
Q1 = 2.904   Q2 = 2.336   Q3 = 1.400

MEAN DEV. BETWEEN CALC. AND EXP. CONC. IN MOLE PCT

UNIQUAC (SPECIFIC PARAMETERS)   0.82
NRTL   (SPECIFIC PARAMETERS)   1.02
UNIQUAC (COMMON PARAMETERS)   0.99

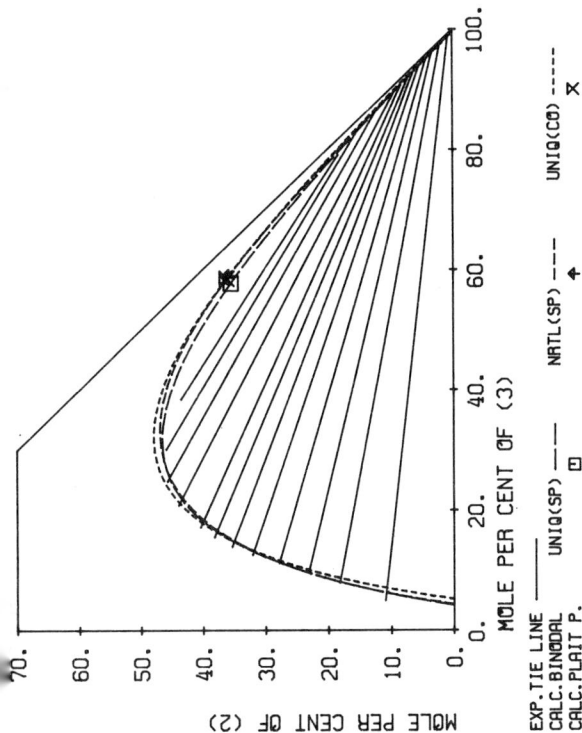

MOLE PER CENT OF (3)

MOLE PER CENT OF (2)

EXP.TIE LINE ——   UNIQ(SP) ⊡   NRTL(SP) ✦   UNIQ(CO) -----
CALC.BINODAL -----   ✕
CALC.PLAIT P.

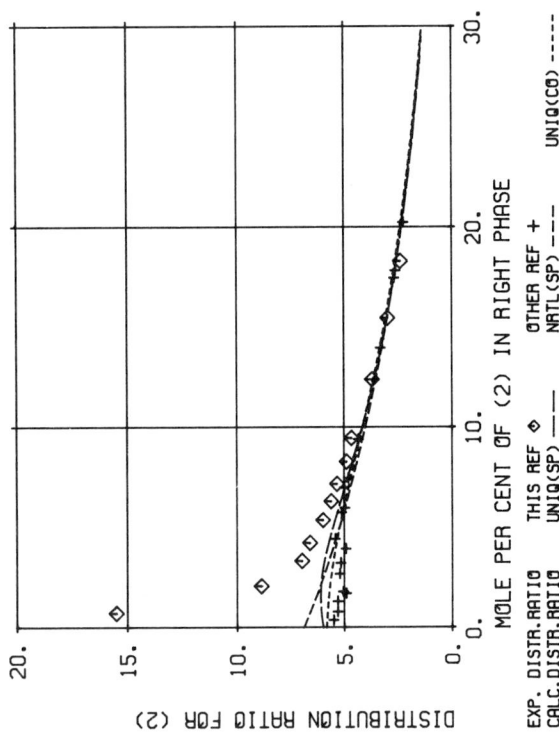

MOLE PER CENT OF (2) IN RIGHT PHASE

DISTRIBUTION RATIO FOR (2)

EXP. DISTR.RATIO ◇   THIS REF ◇ ——   OTHER REF +
CALC.DISTR.RATIO   UNIQ(SP) ——   NRTL(SP) ---   UNIQ(CO) -----

C₃H₆O-C₄H₆O₂

$C_3H_6O\text{-}C_4H_6O_2$

(1) C4H6O2   ACETIC ACID, ETHENYL ESTER
(2) C3H6O   2-PROPANONE
(3) H2O   WATER

PRATT H.R.C., GLOVER S.T.
TRANS.INST.CHEM.ENG. 24(1946)54

TEMPERATURE = 20.0 DEG C   TYPE OF SYSTEM = 1

EXPERIMENTAL TIE LINES IN MOLE PCT (GRAPH.INTERPOL.)

	LEFT PHASE			RIGHT PHASE	
(1)	(2)	(3)	(1)	(2)	(3)
92.350	1.983	5.667	0.256	0.360	99.385
89.772	4.174	6.055	0.291	0.785	98.925
86.708	6.869	6.422	0.327	1.296	98.382
85.075	8.325	6.600	0.363	1.692	97.941
84.209	8.998	6.792	0.375	1.781	97.844
78.398	13.913	7.689	0.428	2.669	96.904
75.372	16.614	8.013	0.479	3.210	96.311
71.981	19.305	8.714	0.533	3.915	95.552
66.368	23.936	9.697	0.574	4.411	95.015
59.608	29.451	10.941	0.676	5.741	93.583
59.397	29.676	10.932	0.691	5.967	93.342
52.194	34.876	12.930	0.768	6.979	92.253
50.777	35.846	13.377	0.814	7.461	91.725
43.570	40.632	15.798	0.978	9.390	89.633
31.843	46.049	22.108	1.398	13.953	84.650
24.461	46.538	29.001	1.926	17.443	80.631
25.424	46.561	28.015	2.013	17.811	80.171
19.839	45.610	34.551	2.817	20.226	75.957

SPECIFIC MODEL PARAMETERS IN KELVIN

		UNIQUAC		NRTL(ALPHA=.2)	
I	J	AIJ	AJI	AIJ	AJI
1	2	270.81	-166.94	115.29	-130.76
1	3	524.66	207.80	467.98	1253.3
2	3	437.12	-103.43	-25.639	560.45

R1 = 3.2485   R2 = 2.5735   R3 = 0.9200
Q1 = 2.904   Q2 = 2.336   Q3 = 1.400

MEAN DEV. BETWEEN CALC. AND EXP. CONC. IN MOLE PCT

UNIQUAC (SPECIFIC PARAMETERS)   0.61
NRTL   (SPECIFIC PARAMETERS)   0.74
UNIQUAC (COMMON PARAMETERS)   0.71

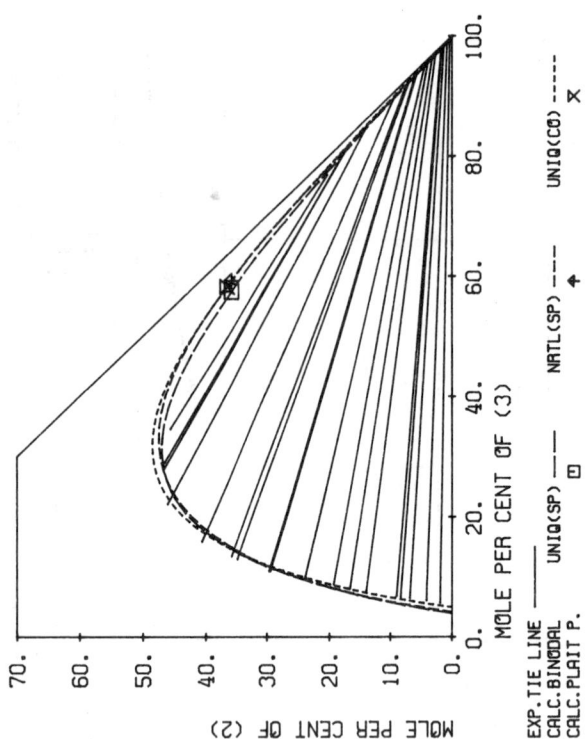

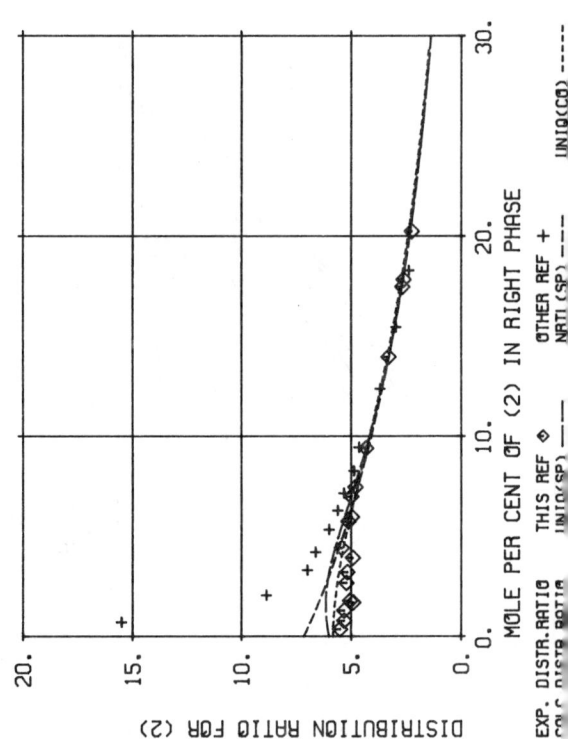

MOLE PER CENT OF (3)

MOLE PER CENT OF (2)

EXP.TIE LINE ————
CALC.BINODAL   UNIQ(SP) —————— ⊡
CALC.PLAIT P.

NRTL(SP) —·—·— ⧫

UNIQ(CO) ------ ✕

MOLE PER CENT OF (2) IN RIGHT PHASE

DISTRIBUTION RATIO FOR (2)

EXP. DISTR.RATIO ◇
CALC. DISTR.RATIO

THIS REF ◇
UNIQ(SP)

OTHER REF +
NRTL(SP) —·—·—

UNIQ(CO) ------

$C_3H_6O-C_4H_8O$

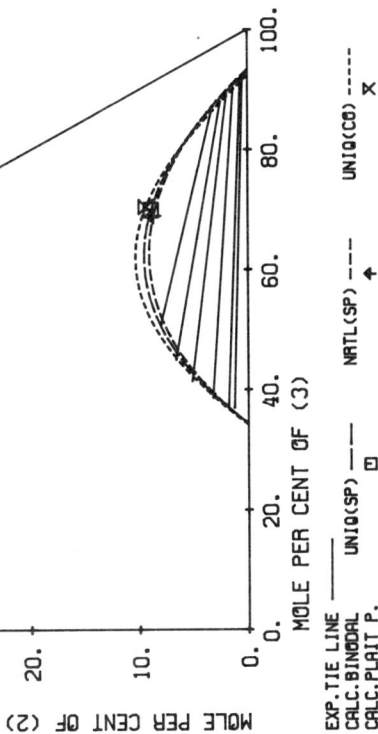

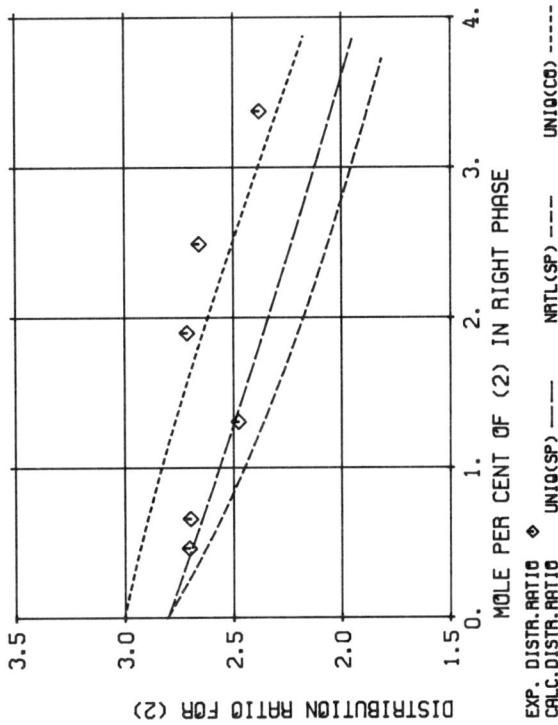

```
(2) C3H6O      2-PROPANONE
(3) H2O        WATER
```

OTHMER D.F., CHUDGAR M.M., LEVY S.L.
IND.ENG.CHEM. 44(1952)1872

TEMPERATURE = 25.0 DEG C   TYPE OF SYSTEM = 1

EXPERIMENTAL TIE LINES IN MOLE PCT

	LEFT PHASE			RIGHT PHASE	
(1)	(2)	(3)	(1)	(2)	(3)
61.824	1.253	36.923	7.804	0.463	91.733
60.920	1.780	37.300	7.876	0.660	91.464
58.153	3.235	38.613	8.224	1.306	90.470
53.492	5.148	41.360	8.651	1.898	89.450
48.396	6.614	44.990	9.386	2.489	88.125
41.230	8.021	50.750	10.620	3.372	86.007

SPECIFIC MODEL PARAMETERS IN KELVIN

I J	UNIQUAC AIJ	AJI	NRTL(ALPHA=.2) AIJ	AJI
1 2	10.588	12.234	169.84	-445.40
1 3	325.54	7.6964	0.43916	933.40
2 3	1.6127	21.310	-45.666	-94.382

```
R1 = 3.2479   R2 = 2.5735   R3 = 0.9200
Q1 = 2.876    Q2 = 2.336    Q3 = 1.400
```

MEAN DEV. BETWEEN CALC. AND EXP. CONC. IN MOLE PCT

```
UNIQUAC (SPECIFIC PARAMETERS)   0.37
NRTL    (SPECIFIC PARAMETERS)   0.54
UNIQUAC (COMMON PARAMETERS)     0.34
```

$C_3H_6O-C_4H_8O_2$

MOLE PER CENT OF (3)

MOLE PER CENT OF (2)

EXP.TIE LINE ———   UNIQ(SP) ◇   NRTL(SP) ———   UNIQ(CG) ------
CALC.BINODAL   UNIQ(SP) □   NRTL(SP) ✦   UNIQ(CG) ✕
CALC.PLAIT P.

MOLE PER CENT OF (2) IN RIGHT PHASE

DISTRIBUTION RATIO FOR (2)

EXP. DISTR.RATIO ◇   UNIQ(SP) ———   NRTL(SP) ———   UNIQ(CG) ------
CALC.DISTR.RATIO

(1) C4H8O2   ACETIC ACID, ETHYL ESTER

(2) C3H6O   2-PROPANONE

(3) H2O   WATER

VENKATARATNAM A., JAGANNADHA RAO R., VENKATA RAO C.
CHEM.ENG.SCI. 7(1957)102

TEMPERATURE = 30.0 DEG C   TYPE OF SYSTEM = 1

EXPERIMENTAL TIE LINES IN MOLE PCT

	LEFT PHASE			RIGHT PHASE	
(1)	(2)	(3)	(1)	(2)	(3)
84.932	0.0	15.068	1.607	0.0	98.393
76.582	6.129	17.289	1.861	1.088	97.051
68.855	11.472	19.673	1.827	2.079	96.093
61.766	15.716	22.519	1.954	3.393	94.653
58.171	18.978	22.852	2.243	4.243	93.022
53.058	22.055	24.887	2.443	5.599	91.958
49.589	24.076	26.335	2.613	6.801	90.586
43.780	26.569	29.651	3.255	8.014	88.732
40.241	27.376	32.383	3.175	8.654	88.172
31.951	29.266	38.783	4.390	11.722	83.888

SPECIFIC MODEL PARAMETERS IN KELVIN

I	J	UNIQUAC		NRTL(ALPHA=.2)	
		AIJ	AJI	AIJ	AJI
1	2	55.689	-78.658	146.07	-249.00
1	3	401.94	76.437	172.18	1120.5
2	3	210.36	-83.086	-73.092	496.65

R1 = 3.4786   R2 = 2.5735   R3 = 0.9200
Q1 = 3.116   Q2 = 2.336   Q3 = 1.400

MEAN DEV. BETWEEN CALC. AND EXP. CONC. IN MOLE PCT

UNIQUAC (SPECIFIC PARAMETERS)   0.33
NRTL (SPECIFIC PARAMETERS)   0.39
UNIQUAC (COMMON PARAMETERS)   0.56

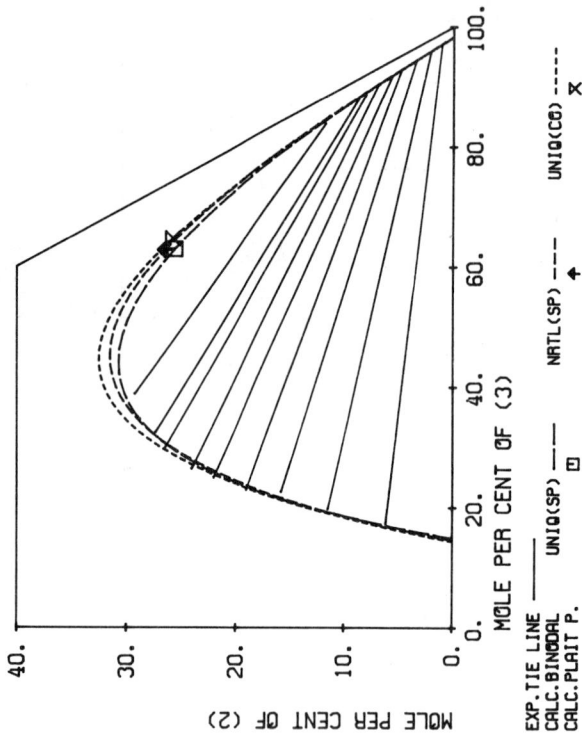

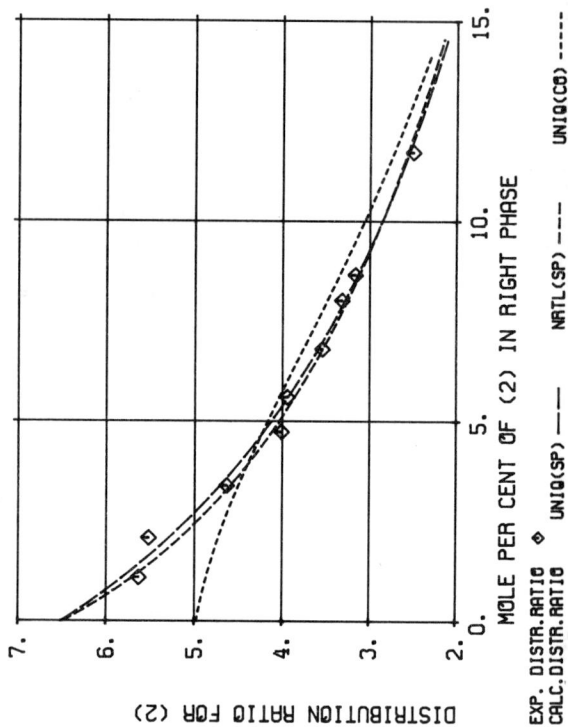

$C_3H_6O-C_4H_{10}O$

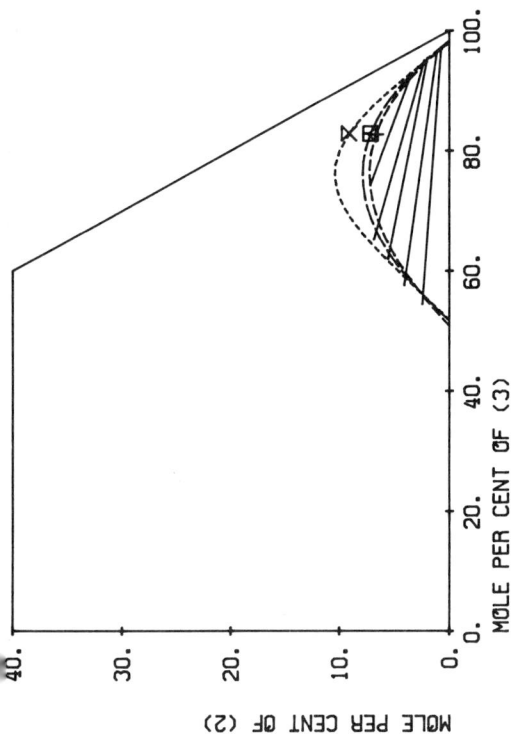

MOLE PER CENT OF (3)

MOLE PER CENT OF (2)

EXP.TIE LINE ———   UNIQ(SP) ▣   NRTL(SP) ——— ◆   UNIQ(CO) ----- ✕
CALC.BINODAL
CALC.PLAIT P.

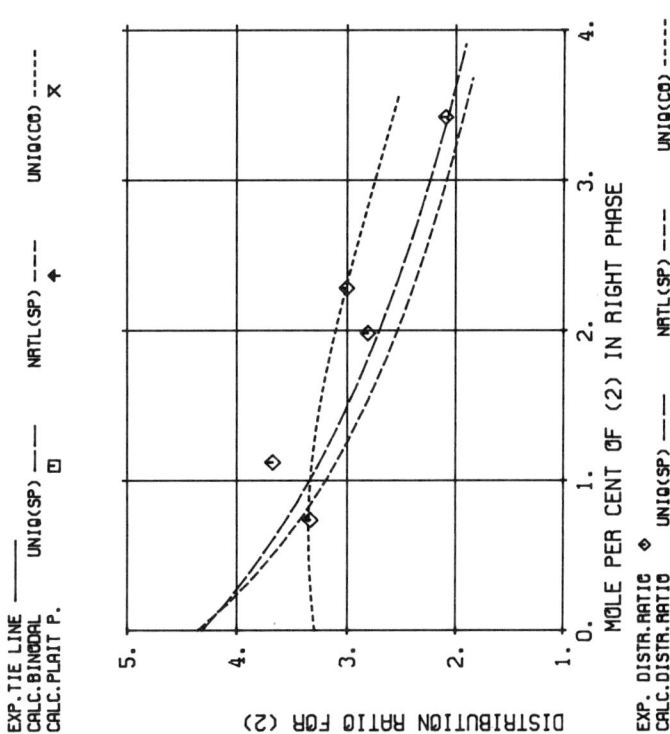

MOLE PER CENT OF (2) IN RIGHT PHASE

DISTRIBUTION RATIO FOR (2)

EXP. DISTR.RATIO ◇   UNIQ(SP) ◇   NRTL(SP) ———   UNIQ(CO) -----
CALC.DISTR.RATIO

(1) C4H10O   1-BUTANOL
(2) C3H6O   2-PROPANONE
(3) H2O   WATER

VENKATARATNAM A., JAGANNADHA RAO R., VENKATA RAO C.
J.SCI.IND.RES. 17B(1958)108

TEMPERATURE = 30.0 DEG C   TYPE OF SYSTEM = 1

EXPERIMENTAL TIE LINES IN MOLE PCT

	LEFT PHASE			RIGHT PHASE	
(1)	(2)	(3)	(1)	(2)	(3)
43.122	2.454	54.424	1.964	0.735	97.300
38.316	4.110	57.574	2.206	1.119	96.674
32.497	5.565	61.939	2.451	1.981	95.568
27.989	6.842	65.168	2.555	2.279	95.166
18.592	7.172	74.236	3.515	3.420	93.065

SPECIFIC MODEL PARAMETERS IN KELVIN

		UNIQUAC		NRTL(ALPHA=.2)	
I J	AIJ	AJI		AIJ	AJI
1 2	167.70	-220.40		284.82	-589.00
1 3	-24.158	293.99		-322.87	1549.3
2 3	-36.425	-178.72		-272.02	-302.97

R1 = 3.4543   R2 = 2.5735   R3 = 0.9200
Q1 = 3.052   Q2 = 2.336   Q3 = 1.400

MEAN DEV. BETWEEN CALC. AND EXP. CONC. IN MOLE PCT

UNIQUAC (SPECIFIC PARAMETERS)   0.73
NRTL (SPECIFIC PARAMETERS)   0.87
UNIQUAC (COMMON PARAMETERS)   0.82

C$_3$H$_6$O-C$_4$H$_{10}$O

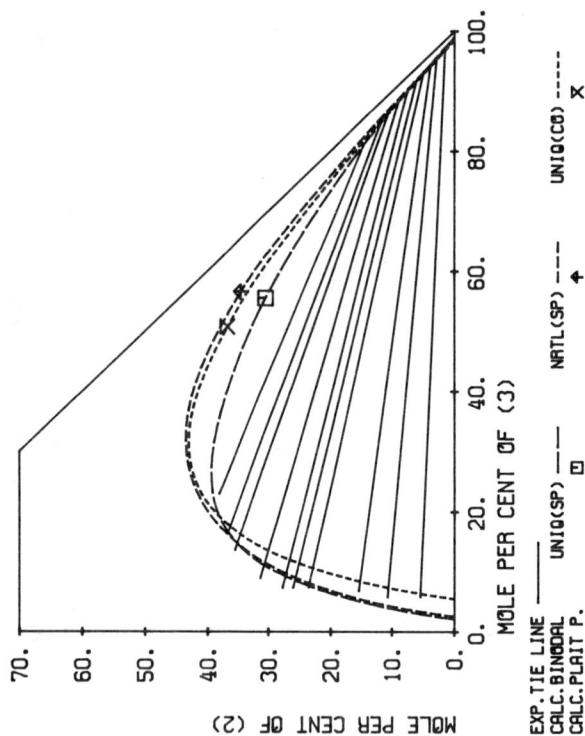

MOLE PER CENT OF (3)

MOLE PER CENT OF (2)

EXP.TIE LINE ——— UNIQ(SP) □ NRTL(SP) + UNIQ(CO) ----- ✗
CALC.BINODAL
CALC.PLAIT P.

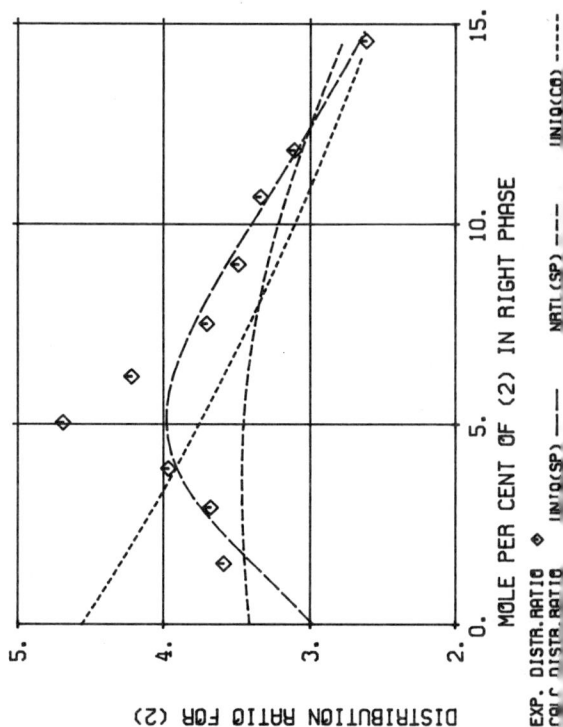

MOLE PER CENT OF (2) IN RIGHT PHASE

DISTRIBUTION RATIO FOR (2)

EXP. DISTR.RATIO ◊ UNIQ(SP) ----- NRTL(SP) ----- UNIQ(CO) -----

(1) C4H10O   ETHER,DIETHYL
(2) C3H6O   2-PROPANONE
(3) H2O   WATER

KRISHNAMURTY V.V.G.; MURTI P.S., VENKATA RAO C.
J.SCI.IND.RES. 12B(1953)583

TEMPERATURE = 30.0 DEG C   TYPE OF SYSTEM = 1

EXPERIMENTAL TIE LINES IN MOLE PCT

	LEFT PHASE			RIGHT PHASE	
(1)	(2)	(3)	(1)	(2)	(3)
88.728	5.446	5.827	1.662	1.519	96.819
83.522	10.718	5.759	1.741	2.916	95.344
77.802	15.419	6.779	1.834	3.888	94.278
69.052	23.598	7.351	2.058	5.034	92.908
66.619	26.072	7.309	2.261	6.182	91.558
64.943	27.777	7.281	2.383	7.507	90.110
59.760	31.344	8.896	2.816	8.986	88.198
50.883	35.610	13.507	3.254	10.671	86.075
46.827	36.790	16.383	3.502	11.832	84.666
38.896	38.154	22.949	4.325	14.569	81.105

SPECIFIC MODEL PARAMETERS IN KELVIN

		UNIQUAC		NRTL(ALPHA=.2)	
I J	AIJ	AJI		AIJ	AJI
1 2	413.57	-204.19		383.90	-194.43
1 3	774.94	69.928		666.11	921.56
2 3	700.41	-149.81		-32.627	572.32

R1 = 3.3949   R2 = 2.5735   R3 = 0.9200
Q1 = 3.016   Q2 = 2.336   Q3 = 1.400

MEAN DEV. BETWEEN CALC. AND EXP. CONC. IN MOLE PCT

UNIQUAC (SPECIFIC PARAMETERS)   0.78
NRTL   (SPECIFIC PARAMETERS)   0.96
UNIQUAC (COMMON PARAMETERS)   1.41

$C_3H_6O\text{-}C_5H_4O_2$

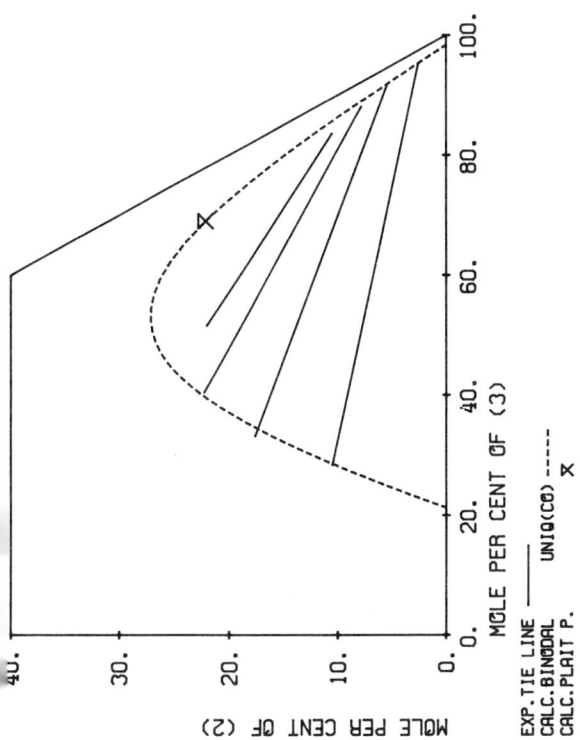

MOLE PER CENT OF (3)

MOLE PER CENT OF (2)

EXP.TIE LINE ——
CALC.BINODAL ----- UNIQ(CO) -----
CALC.PLAIT P. ✕

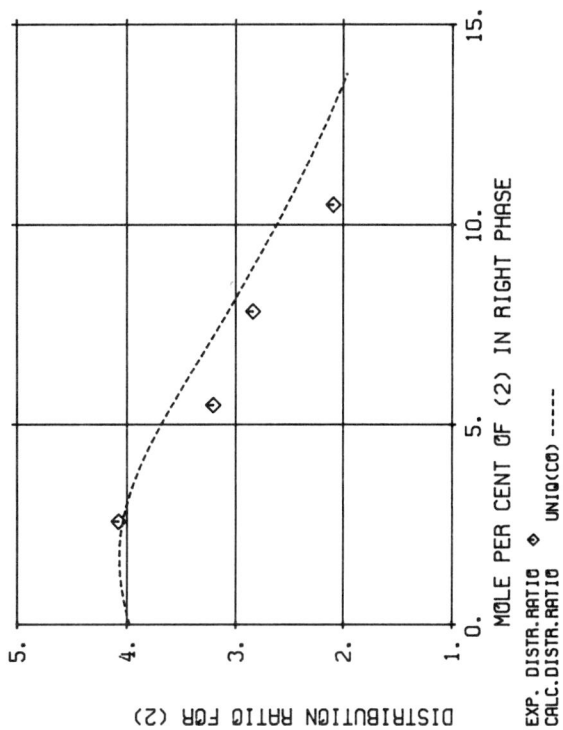

MOLE PER CENT OF (2) IN RIGHT PHASE

DISTRIBUTION RATIO FOR (2)

EXP. DISTR.RATIO ◇ UNIQ(CO) -----
CALC.DISTR.RATIO -----

(1) C3H4O2   FURFURAL
(2) C3H6O   2-PROPANONE
(3) H2O   WATER

LLOYD B.A., THOMPSON S.O., FERGUSON J.B.
CAN.J.RES. B15(1937)98

TEMPERATURE = 25.0 DEG C   TYPE OF SYSTEM = 1

EXPERIMENTAL TIE LINES IN MOLE PCT (GRAPH.INTERPOL.)

	LEFT PHASE			RIGHT PHASE	
(1)	(2)	(3)	(1)	(2)	(3)
61.166	10.481	28.353	2.157	2.570	95.273
49.248	17.608	33.144	2.849	5.486	91.666
37.341	22.202	40.457	4.048	7.828	88.124
26.413	21.972	51.614	5.855	10.494	83.651

MEAN DEV. BETWEEN CALC. AND EXP. CONC. IN MOLE PCT

UNIQUAC (COMMON PARAMETERS)   1.43

C₃H₆O-C₅H₈O₂

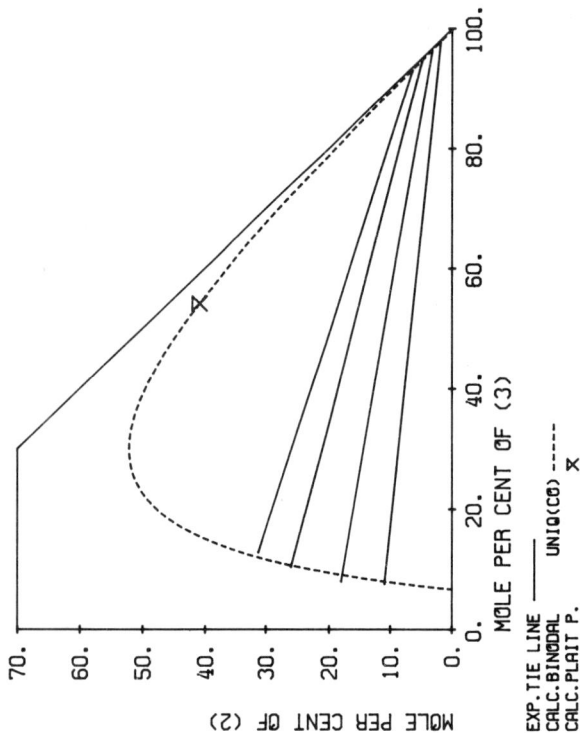

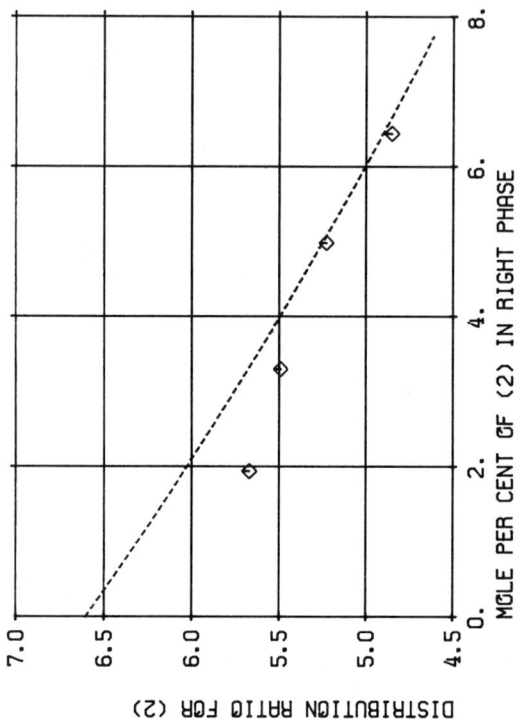

(1) C5H8O2   PROPENOIC ACID,2-METHYL,METHYL ESTER

(2) C3H6O   2-PROPANONE

(3) H2O   WATER

CHUBAROV G.A., DANOV S.M., LOGUTOV V.I., BROVKINA G.V.
ZH.PRIKL.KHIM.(LENINGRAD) 52(1979)1082

TEMPERATURE = 25.0 DEG C   TYPE OF SYSTEM = 1

EXPERIMENTAL TIE LINES IN MOLE PCT

	LEFT PHASE		RIGHT PHASE	
	(1)	(2)	(3)	(1)	(2)	(3)
81.497	10.966	7.537	0.304	1.934	97.762
74.041	18.077	7.882	0.343	3.293	96.364
63.698	25.994	10.308	0.410	4.974	94.615
56.025	31.230	12.745	0.497	6.437	93.066

MEAN DEV. BETWEEN CALC. AND EXP. CONC. IN MOLE PCT

UNIQUAC (COMMON PARAMETERS)   0.36

$C_3H_6O\text{-}C_5H_{10}O_2$

(1) C5H10O2   ACETIC ACID, PROPYL ESTER

(2) C3H6O   2-PROPANONE

(3) H2O   WATER

VENKATARATNAM A., JAGANNADHA RAO R., VENKATA RAO C.
CHEM.ENG.SCI. 7(1957)102

TEMPERATURE = 30.0 DEG C   TYPE OF SYSTEM = 1

EXPERIMENTAL TIE LINES IN MOLE PCT

	LEFT PHASE			RIGHT PHASE	
(1)	(2)	(3)	(1)	(2)	(3)
91.560	0.0	8.440	0.322	0.0	99.678
89.195	1.451	9.354	0.368	1.198	98.434
84.380	5.495	10.124	0.374	1.907	97.719
81.206	8.818	9.976	0.401	2.817	96.782
71.304	16.932	11.765	0.531	4.046	95.424
61.507	24.369	14.124	0.619	6.352	93.029
51.093	31.724	17.183	0.855	7.701	91.444
31.016	40.515	28.469	2.047	15.976	81.976
26.108	41.524	32.368	2.468	17.498	80.034

SPECIFIC MODEL PARAMETERS IN KELVIN

		UNIQUAC		NRTL(ALPHA=.2)	
I	J	AIJ	AJI	AIJ	AJI
1	2	312.05	-120.62	614.21	-221.93
1	3	393.61	276.94	315.61	1707.9
2	3	387.96	-116.39	338.40	232.55

R1 = 4.1530   R2 = 2.5735   R3 = 0.9200
Q1 = 3.656   Q2 = 2.336   Q3 = 1.400

MEAN DEV. BETWEEN CALC. AND EXP. CONC. IN MOLE PCT

UNIQUAC (SPECIFIC PARAMETERS)   0.86
NRTL   (SPECIFIC PARAMETERS)   0.93
UNIQUAC (COMMON PARAMETERS)   0.62

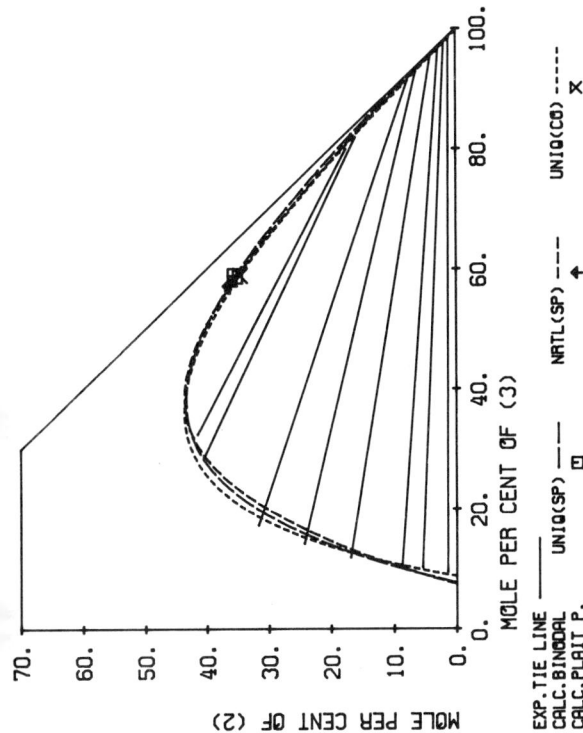

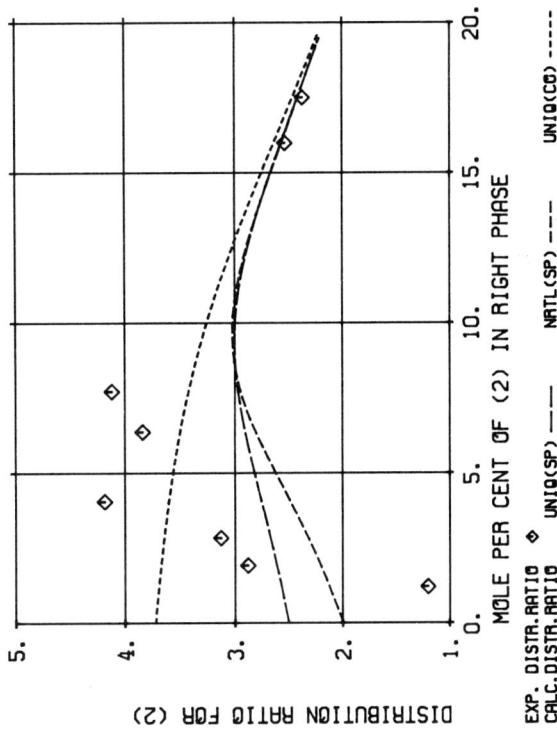

MOLE PER CENT OF (3)

MOLE PER CENT OF (2)

EXP.TIE LINE ———   UNIQ(SP) ▣   NRTL(SP) ✦   UNIQ(CO) -----
CALC.BINODAL -----
CALC.PLAIT P.   ✕

DISTRIBUTION RATIO FOR (2)

MOLE PER CENT OF (2) IN RIGHT PHASE

EXP. DISTR.RATIO ◇   UNIQ(SP) ———   NRTL(SP) ———   UNIQ(CO) -----
CALC.DISTR.RATIO

$C_3H_6O\text{-}C_5H_{10}O_2$

(1) C5H10O2   PROPANOIC ACID, ETHYL ESTER
----
(2) C3H6O   2-PROPANONE
----
(3) H2O   WATER

VENKATARATNAM A.; JAGANNADHA RAO R., VENKATA RAO C.
CHEM.ENG.SCI. 7(1957)102

TEMPERATURE = 30.0 DEG C   TYPE OF SYSTEM = 1

EXPERIMENTAL TIE LINES IN MOLE PCT

	LEFT PHASE			RIGHT PHASE	
(1)	(2)	(3)	(1)	(2)	(3)
90.587	0.0	9.413	0.487	0.0	99.513
77.011	13.679	9.310	0.530	2.163	97.307
66.547	23.720	9.733	0.576	4.472	94.952
54.997	31.354	13.649	0.643	6.566	92.791
38.790	41.762	19.448	0.884	10.961	88.155
34.250	43.620	22.130	0.918	11.626	87.456
26.683	45.606	27.711	1.197	14.771	84.032

SPECIFIC MODEL PARAMETERS IN KELVIN

		UNIQUAC		NRTL(ALPHA=.2)	
I	J	AIJ	AJI	AIJ	AJI
1	2	31.954	-0.65932	-75.016	37.644
1	3	497.65	138.57	351.87	1298.8
2	3	208.67	-26.396	14.351	535.18

R1 = 4.1530   R2 = 2.5735   R3 = 0.9200
Q1 = 3.656   Q2 = 2.336   Q3 = 1.400

MEAN DEV. BETWEEN CALC. AND EXP. CONC. IN MOLE PCT

UNIQUAC (SPECIFIC PARAMETERS)   0.65
NRTL   (SPECIFIC PARAMETERS)   0.68
UNIQUAC (COMMON PARAMETERS)   0.84

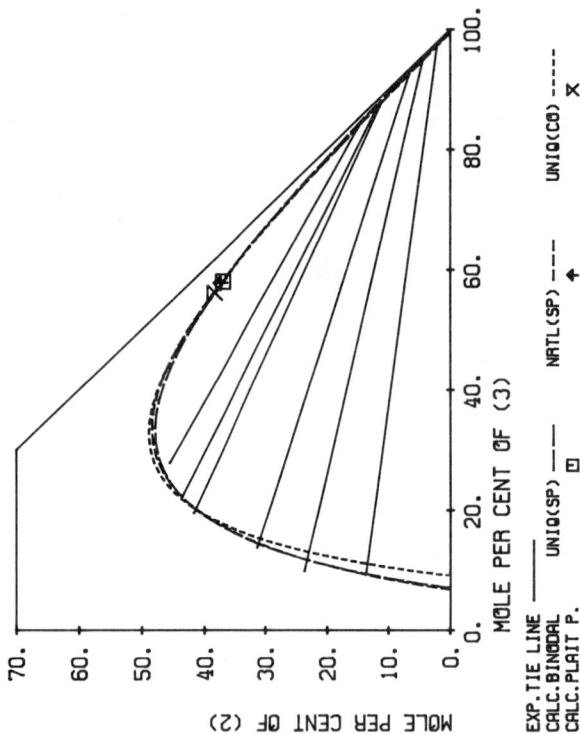

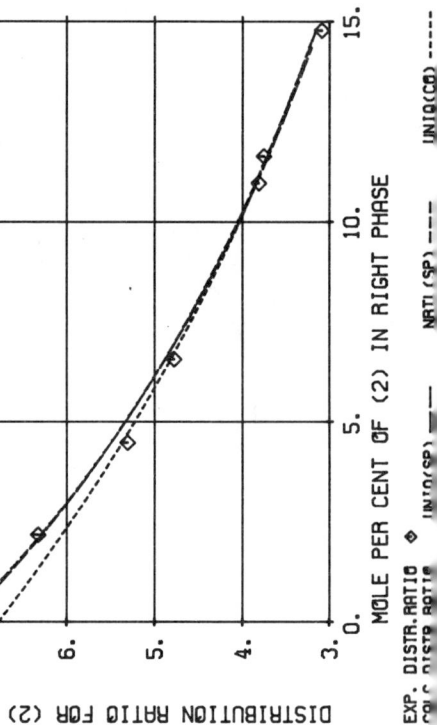

MOLE PER CENT OF (3)

MOLE PER CENT OF (2)

EXP.TIE LINE ——   UNIQ(SP) ⊡   NRTL(SP) ✦   UNIQ(CO) -----
CALC.BINODAL -----
CALC.PLAIT P.   ✕

DISTRIBUTION RATIO FOR (2)

MOLE PER CENT OF (2) IN RIGHT PHASE

EXP. DISTR.RATIO ◇   UNIQ(SP) ◇   NRTL(SP) ——   UNIQ(CO) -----

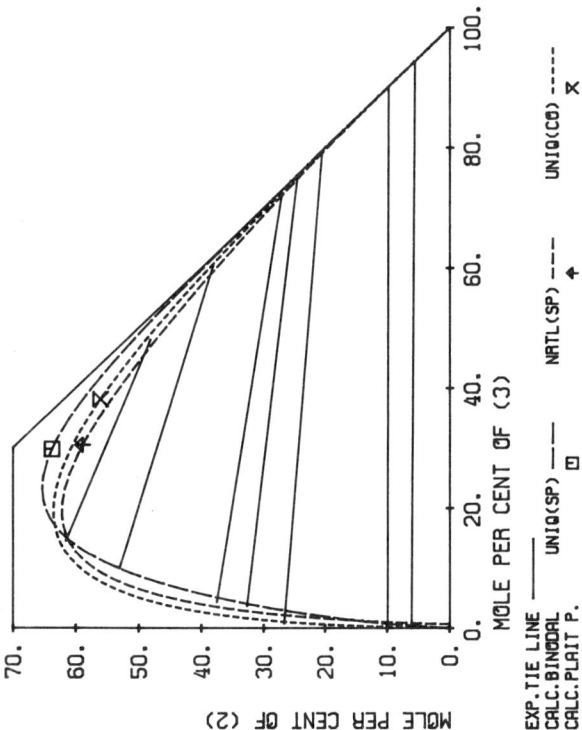

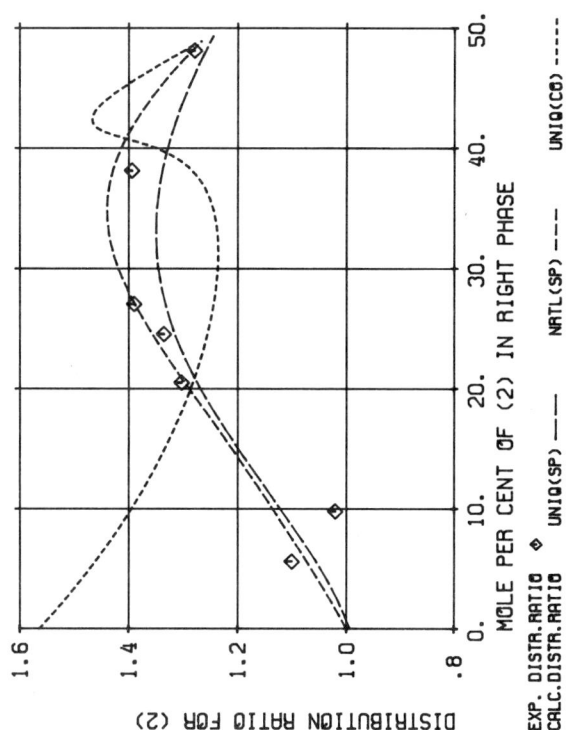

$C_3H_6O-C_5H_{12}$

MOLE PER CENT OF (3)

MOLE PER CENT OF (2)

EXP.TIE LINE ———   UNIQ(SP) —— ☐   NRTL(SP) ——·· ✦   UNIQ(CB) ——·—· ⊠
CALC.BINODAL
CALC.PLAIT P.

DISTRIBUTION RATIO FOR (2)

MOLE PER CENT OF (2) IN RIGHT PHASE

EXP. DISTR.RATIO ◇   UNIQ(SP) ——   NRTL(SP) ———   UNIQ(CB) ----
CALC.DISTR.RATIO

(2) C3H6O   2-PROPANONE

(3) H2O   WATER

KHANINA E.P.,BEREGOVYKH V.V., PAVLENKO T.G., TIMOFEEV V.S.
ZH.FIZ.KHIM. 52(1978)1558

TEMPERATURE = 20.0 DEG C   TYPE OF SYSTEM = 1

EXPERIMENTAL TIE LINES IN MOLE PCT

	LEFT PHASE			RIGHT PHASE	
(1)	(2)	(3)	(1)	(2)	(3)
93.863	6.137	0.0	0.0	5.579	94.421
90.012	9.988	0.0	0.0	9.781	90.219
72.556	26.689	0.755	0.073	20.482	79.445
63.619	32.741	3.640	0.155	24.518	75.327
58.143	37.558	4.299	0.403	27.016	72.582
37.003	53.117	9.880	1.416	38.105	60.480
23.901	61.586	14.513	3.819	48.115	48.066

SPECIFIC MODEL PARAMETERS IN KELVIN

		UNIQUAC		NRTL(ALPHA=.2)	
I J		AIJ	AJI	AIJ	AJI
1 2		-681.20	275.40	248.16	66.707
1 3		1189.5	79.48	1015.8	2366.8
2 3		-19.422	-633.75	577.71	-87.037

R1 = 3.8254   R2 = 2.5735   R3 = 0.9200
Q1 = 3.316   Q2 = 2.336   Q3 = 1.400

MEAN DEV. BETWEEN CALC. AND EXP. CONC. IN MOLE PCT

UNIQUAC (SPECIFIC PARAMETERS)   0.81
NRTL (SPECIFIC PARAMETERS)   0.69
UNIQUAC (COMMON PARAMETERS)   1.97

C₃H₆O-C₆H₅Br

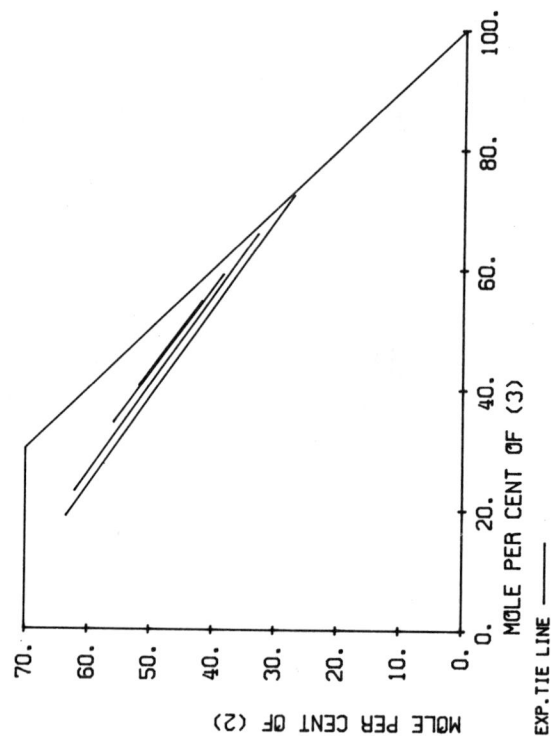

(1) C6H5BR   BENZENE,BROMO

(2) C3H6O   2-PROPANONE

(3) H2O   WATER

BONNER W.D.
J.PHYS.CHEM. 14(1910)738

TEMPERATURE = 0.0 DEG C   TYPE OF SYSTEM = 1

EXPERIMENTAL TIE LINES IN MOLE PCT (GRAPH.INTERPOL.)

LEFT PHASE			RIGHT PHASE		
(1)	(2)	(3)	(1)	(2)	(3)
17.669	63.458	18.873	0.528	27.130	72.342
14.914	62.125	22.961	1.215	32.893	65.892
9.716	55.873	34.411	2.488	38.390	59.122
7.643	51.784	40.573	3.539	41.822	54.639

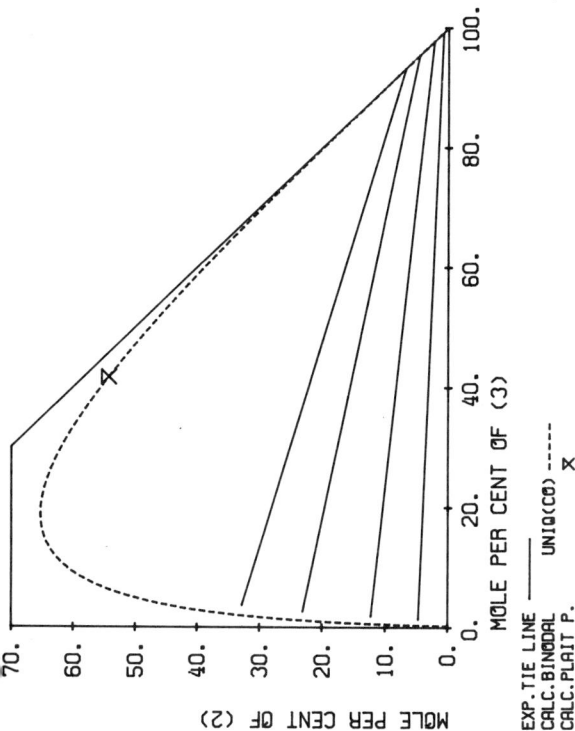

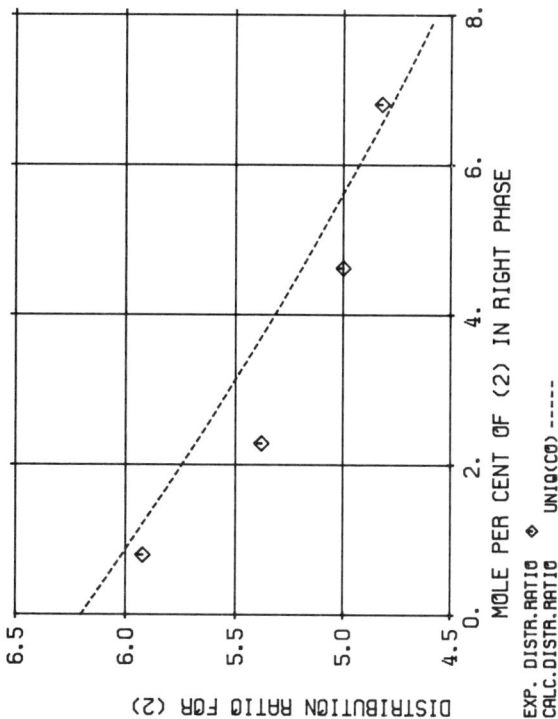

(1) ~~BENZENE, CHLORO~~

(2) C3H6O   2-PROPANONE

(3) H2O   WATER

OTHMER D.F., WHITE R.E., TRUEGER E.
IND.ENG.CHEM. 33(1941)1240

TEMPERATURE = 25.5 DEG C   TYPE OF SYSTEM = 1

EXPERIMENTAL TIE LINES IN MOLE PCT (GRAPH.INTERPOL.)

	LEFT PHASE			RIGHT PHASE	
(1)	(2)	(3)	(1)	(2)	(3)
93.922	4.678	1.400	0.024	0.790	99.185
85.860	12.279	1.861	0.034	2.282	97.684
74.279	23.080	2.641	0.045	4.617	95.339
63.508	32.821	3.671	0.057	6.803	93.141

MEAN DEV. BETWEEN CALC. AND EXP. CONC. IN MOLE PCT

UNIQUAC (COMMON PARAMETERS)   0.68

C₃H₆O-C₆H₆

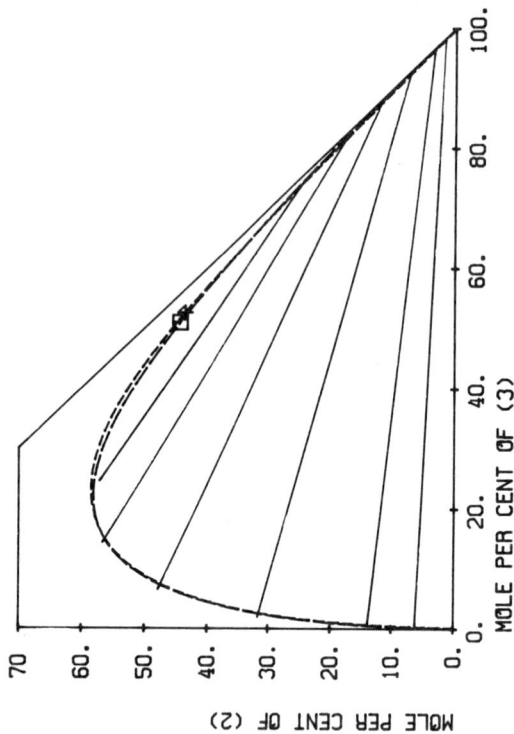

MOLE PER CENT OF (3)

MOLE PER CENT OF (2)

EXP.TIE LINE ——— CALC.BINODAL ——— UNIQ(SP) ——— ▣ NRTL(SP) ---- ◆
CALC.PLAIT P.

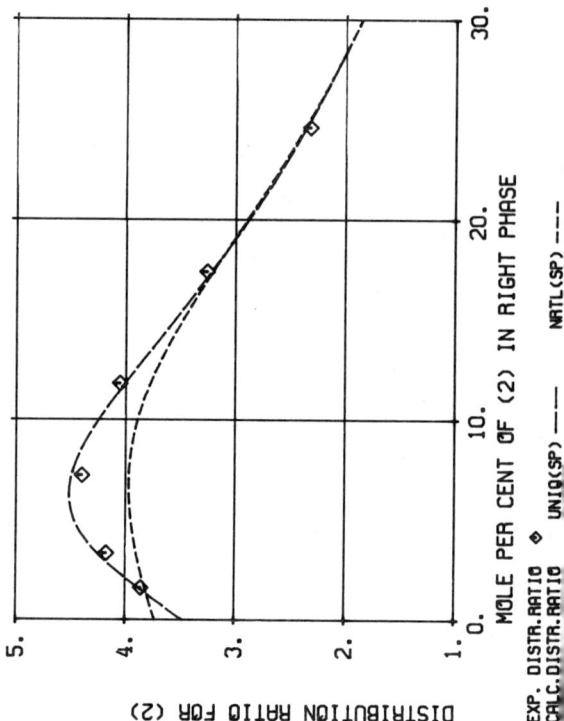

MOLE PER CENT OF (2) IN RIGHT PHASE

DISTRIBUTION RATIO FOR (2)

EXP. DISTR.RATIO ◇ UNIQ(SP) ◆ NRTL(SP) ———
CALC. DISTR.RATIO ———

(1) C6H6   BENZENE
(2) C3H6O   2-PROPANONE
(3) H2O   WATER

BRIGGS S.W., COMINGS E.W.
IND.ENG.CHEM. 35(1943)411

TEMPERATURE = 15.0 DEG C   TYPE OF SYSTEM = 1

EXPERIMENTAL TIE LINES IN MOLE PCT

	LEFT PHASE			RIGHT PHASE	
(1)	(2)	(3)	(1)	(2)	(3)
93.374	6.200	0.425	0.024	1.608	98.369
85.254	13.915	0.831	0.025	3.334	96.641
66.321	31.720	1.959	0.080	7.216	92.704
45.679	47.861	6.459	0.205	11.814	87.981
29.141	56.639	14.220	0.453	17.394	82.154
18.350	57.144	24.506	1.170	24.599	74.231

SPECIFIC MODEL PARAMETERS IN KELVIN

		UNIQUAC		NRTL(ALPHA=.2)	
I	J	AIJ	AJI	AIJ	AJI
1	2	304.49	-163.95	366.35	-188.57
1	3	1018.6	378.79	1246.4	-1332.8
2	3	352.45	-81.433	133.58	410.86

R1 = 3.1878   R2 = 2.5735   R3 = 0.9200
Q1 = 2.400   Q2 = 2.336   Q3 = 1.400

MEAN DEV. BETWEEN CALC. AND EXP. CONC. IN MOLE PCT

UNIQUAC (SPECIFIC PARAMETERS)   0.32
NRTL   (SPECIFIC PARAMETERS)   0.60

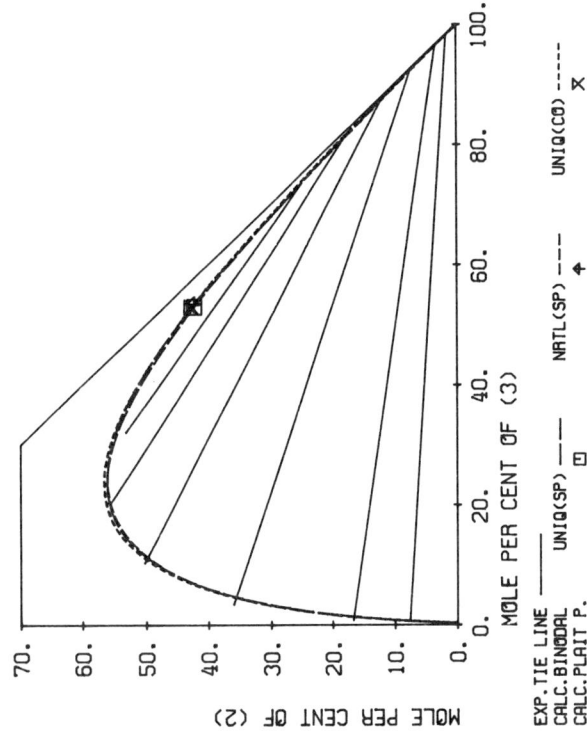

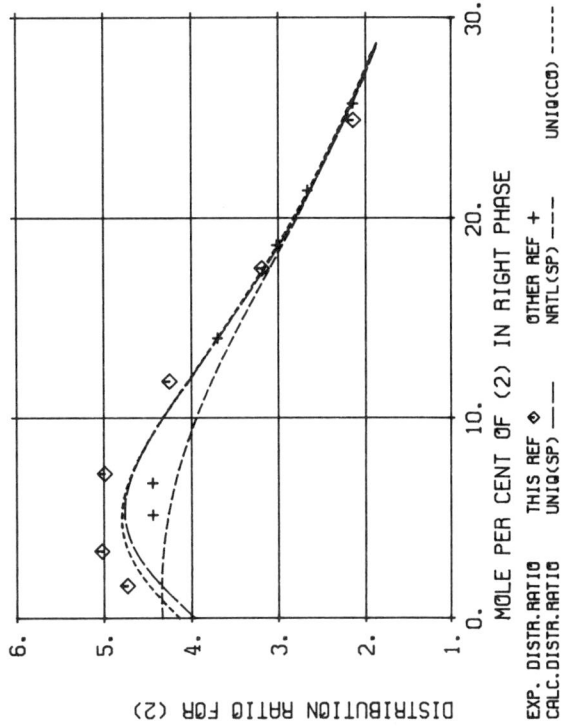

C_3H_6O-C_6H_6

MOLE PER CENT OF (3)

MOLE PER CENT OF (2)

EXP.TIE LINE ——
CALC.BINODAL ——
CALC.PLAIT P.

UNIQ(SP) —— ▣
NRTL(SP) —— ✦
UNIQ(CO) —— ✕

DISTRIBUTION RATIO FOR (2)

MOLE PER CENT OF (2) IN RIGHT PHASE

EXP. DISTR.RATIO ◇
CALC.DISTR.RATIO

THIS REF ◇
UNIQ(SP) ——

OTHER REF +
NRTL(SP) ——

UNIQ(CO) ——

(1) C6H6   BENZENE

(2) C3H6O   2-PROPANONE

(3) H2O   WATER

BRIGGS S.W., COMINGS E.W.
IND.ENG.CHEM. 35(1943)411

TEMPERATURE = 30.0 DEG C   TYPE OF SYSTEM = 1

EXPERIMENTAL TIE LINES IN MOLE PCT

	LEFT PHASE		RIGHT PHASE	
(1)	(2)	(3)	(1) (2)	(3)
91.557	7.598	0.845	0.024 1.608	98.369
82.424	16.751	0.825	0.050 3.337	96.613
60.533	36.028	3.439	0.107 7.222	92.670
39.431	50.268	10.301	0.264 11.837	87.899
24.237	55.656	20.106	0.584 17.469	81.946
14.739	53.170	32.091	1.516 24.872	73.612

SPECIFIC MODEL PARAMETERS IN KELVIN

		UNIQUAC		NRTL(ALPHA=.2)	
I J	AIJ	AJI	AIJ	AJI
1 2	295.28	-165.93	336.01	-201.69
1 3	907.18	-268.16	1167.3	1376.1
2 3	356.30	-78.297	85.020	469.68

R1 = 3.1878   R2 = 2.5735   R3 = 0.9200
Q1 = 2.400   Q2 = 2.336   Q3 = 1.400

MEAN DEV. BETWEEN CALC. AND EXP. CONC. IN MOLE PCT

UNIQUAC (SPECIFIC PARAMETERS)   0.73
NRTL   (SPECIFIC PARAMETERS)   1.10
UNIQUAC (COMMON PARAMETERS)   0.79

C$_3$H$_6$O-C$_6$H$_6$

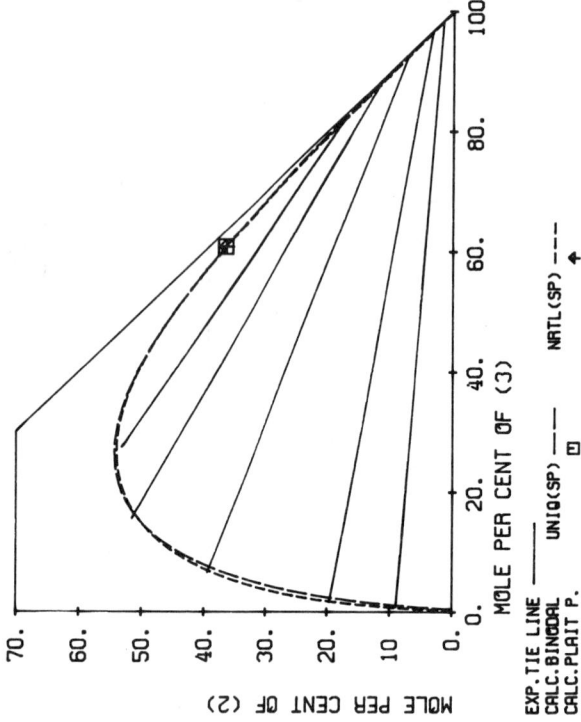

MOLE PER CENT OF (3)

MOLE PER CENT OF (2)

EXP.TIE LINE ——— UNIQ(SP) ——— ⊡ NRTL(SP) – – –
CALC.BINODAL
CALC.PLAIT P.

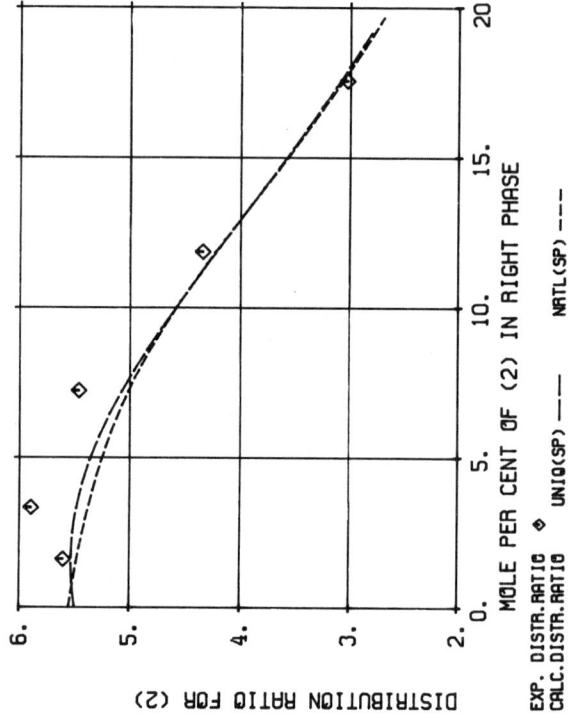

MOLE PER CENT OF (2) IN RIGHT PHASE

DISTRIBUTION RATIO FOR (2)

EXP. DISTR.RATIO ◇ UNIQ(SP) ——— NRTL(SP) – – –
CALC.DISTR.RATIO

(1) C6H6   BENZENE
(2) C3H6O   2-PROPANONE
(3) H2O   WATER

BRIGGS S.W., COMINGS E.W.
IND.ENG.CHEM. 35(1943)411

TEMPERATURE = 45.0 DEG C   TYPE OF SYSTEM = 1

EXPERIMENTAL TIE LINES IN MOLE PCT

	LEFT PHASE			RIGHT PHASE	
(1)	(2)	(3)	(1)	(2)	(3)
90.152	9.006	0.842	0.048	1.609	98.343
78.713	19.662	1.625	0.050	3.337	96.613
53.924	39.458	6.618	0.134	7.229	92.637
32.914	51.515	15.570	0.323	11.860	87.817
19.377	53.179	27.443	0.751	17.564	81.685

SPECIFIC MODEL PARAMETERS IN KELVIN

		UNIQUAC		NRTL(ALPHA=.2)	
I J	AIJ	AJI	AIJ	AJI
1 2	171.18	-101.92	257.31	-139.20
1 3	803.09	361.65	1234.9	1471.1
2 3	179.61	14.224	-48.684	699.41

R1 = 3.1878   R2 = 2.5735   R3 = 0.9200
Q1 = 2.400   Q2 = 2.336   Q3 = 1.400

MEAN DEV. BETWEEN CALC. AND EXP. CONC. IN MOLE PCT

UNIQUAC (SPECIFIC PARAMETERS)   0.47
NRTL   (SPECIFIC PARAMETERS)   0.43

C$_3$H$_6$O-C$_6$H$_6$

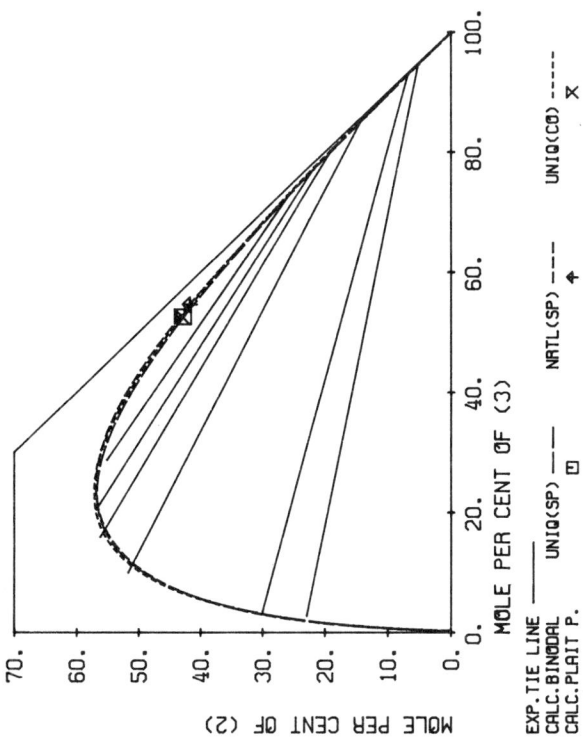

MOLE PER CENT OF (3)

MOLE PER CENT OF (2)

EXP.TIE LINE —— UNIQ(SP) —— ⊡   NRTL(SP) —— ✦   UNIQ(CO) —— ✕
CALC.BINODAL
CALC.PLAIT P.

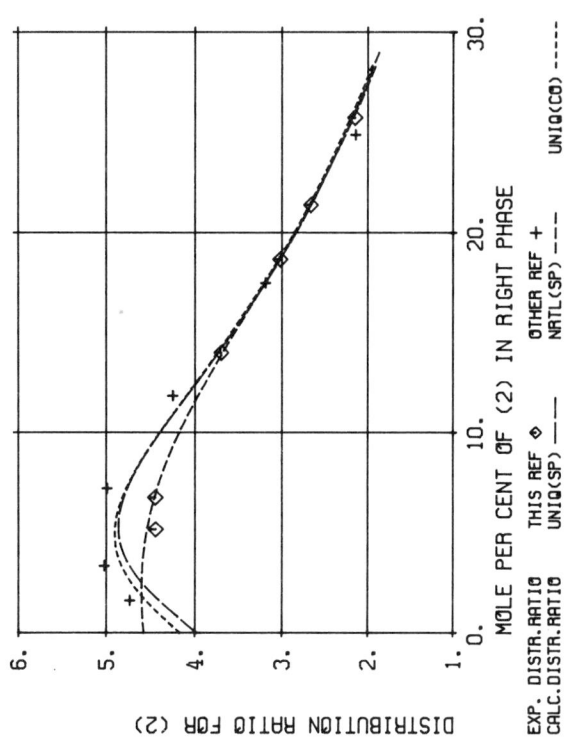

MOLE PER CENT OF (2) IN RIGHT PHASE

DISTRIBUTION RATIO FOR (2)

EXP.DISTR.RATIO   THIS REF ◇   OTHER REF +
CALC.DISTR.RATIO   UNIQ(SP) ——   NRTL(SP) ———   UNIQ(CO) ———

(1) C6H6	BENZENE
(2) C3H6O	2-PROPANONE
(3) H2O	WATER

ARICH G., TAGLIAVINI G.
RIC.SCI., 28(1958)1620

TEMPERATURE = 20.0 DEG C   TYPE OF SYSTEM = 1

EXPERIMENTAL TIE LINES IN MOLE PCT

	LEFT PHASE			RIGHT PHASE	
(1)	(2)	(3)	(1)	(2)	(3)
74.330	22.880	2.790	0.080	5.160	94.760
66.500	30.000	3.500	0.110	6.760	93.130
38.360	51.700	9.940	0.240	14.000	85.760
27.880	56.220	15.900	0.490	18.650	80.860
22.630	56.840	20.530	0.760	21.370	77.870
16.130	55.130	28.740	1.420	25.720	72.860

SPECIFIC MODEL PARAMETERS IN KELVIN

		UNIQUAC		NRTL(ALPHA=.2)	
I	J	AIJ	AJI	AIJ	AJI
1	2	295.28	-165.93	336.01	-201.69
1	3	907.18	-268.16	1167.3	1376.1
2	3	356.30	-78.297	85.020	459.68

R1 = 3.1878   R2 = 2.5735   R3 = 0.9200
Q1 = 2.400   Q2 = 2.336   Q3 = 1.400

MEAN DEV. BETWEEN CALC. AND EXP. CONC. IN MOLE PCT

UNIQUAC (SPECIFIC PARAMETERS)	0.65
NRTL (SPECIFIC PARAMETERS)	0.91
UNIQUAC (COMMON PARAMETERS)	0.66

C$_3$H$_6$O-C$_6$H$_6$

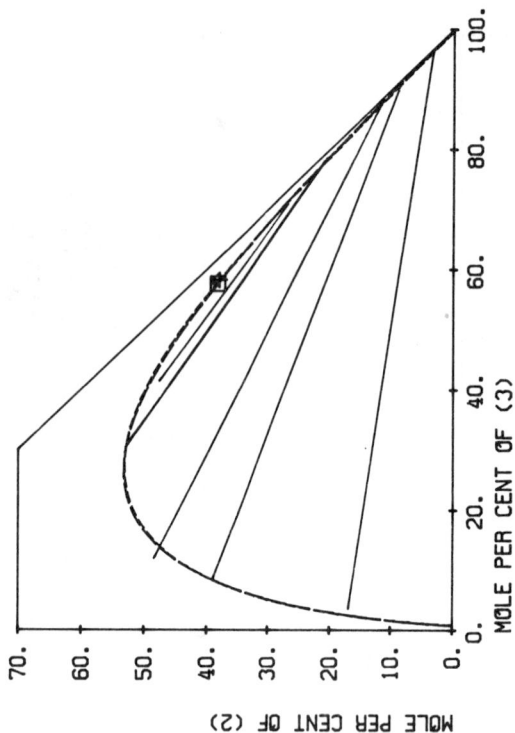

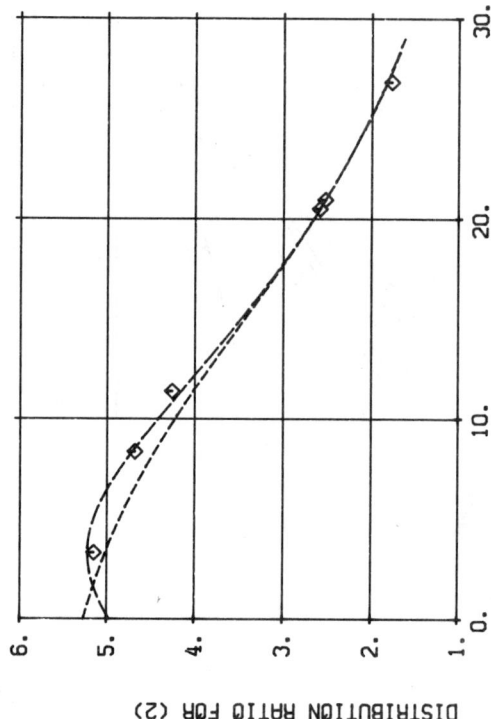

(1) C6H6   BENZENE

(2) C3H6O   2-PROPANONE

(3) H2O   WATER

ARICH G., TAGLIAVINI G.
RIC.SCI. 28(1958)1620

TEMPERATURE = 40.0 DEG C   TYPE OF SYSTEM = 1

EXPERIMENTAL TIE LINES IN MOLE PCT

	LEFT PHASE			RIGHT PHASE	
(1)	(2)	(3)	(1)	(2)	(3)
79.400	16.980	3.620	0.080	3.300	96.620
52.560	39.110	8.330	0.190	8.360	91.450
39.690	48.410	11.900	0.320	11.370	83.310
16.720	52.930	30.350	0.960	20.490	78.550
16.460	52.780	30.760	1.040	20.940	78.020
10.960	47.590	41.450	2.200	26.830	70.970

SPECIFIC MODEL PARAMETERS IN KELVIN

		UNIQUAC		NRTL(ALPHA=.2)	
I	J	AIJ	AJI	AIJ	AJI
1	2	234.85	-142.98	165.85	-66.609
1	3	764.79	214.99	960.22	1274.2
2	3	285.06	-46.300	-21.083	635.64

R1 = 3.1878   R2 = 2.5735   R3 = 0.9200
Q1 = 2.400   Q2 = 2.336   Q3 = 1.400

MEAN DEV. BETWEEN CALC. AND EXP. CONC. IN MOLE PCT

UNIQUAC (SPECIFIC PARAMETERS)   0.51
NRTL   (SPECIFIC PARAMETERS)   0.66

$C_3H_6O-C_6H_6$

(1) C6H6   BENZENE
(2) C3H6O   2-PROPANONE
(3) H2O   WATER

ARICH G., TAGLIAVINI G.
RIC.SCI. 28(1958)1620

TEMPERATURE = 60.0 DEG C   TYPE OF SYSTEM = 1

EXPERIMENTAL TIE LINES IN MOLE PCT

	LEFT PHASE			RIGHT PHASE	
(1)	(2)	(3)	(1)	(2)	(3)
75.790	18.310	5.900	0.100	2.810	97.090
60.290	30.840	8.870	0.180	5.070	94.750
45.500	42.390	12.110	0.270	7.950	91.780
34.950	48.210	16.840	0.380	10.420	89.200
25.960	50.690	23.350	0.480	12.630	86.890
17.980	50.230	31.790	0.800	16.700	82.500
11.870	45.860	42.270	1.770	22.780	75.450

SPECIFIC MODEL PARAMETERS IN KELVIN

		UNIQUAC		NRTL(ALPHA=.2)	
I J	AIJ	AJI		AIJ	AJI
1 2	126.74	-92.436		59.028	-10.550
1 3	568.91	235.22		769.83	1395.4
2 3	211.58	-0.25357		-73.727	759.84

R1 = 3.1878   R2 = 2.5735   R3 = 0.9200
Q1 = 2.400   Q2 = 2.336   Q3 = 1.400

MEAN DEV. BETWEEN CALC. AND EXP. CONC. IN MOLE PCT

UNIQUAC (SPECIFIC PARAMETERS)   0.68
NRTL   (SPECIFIC PARAMETERS)   0.79

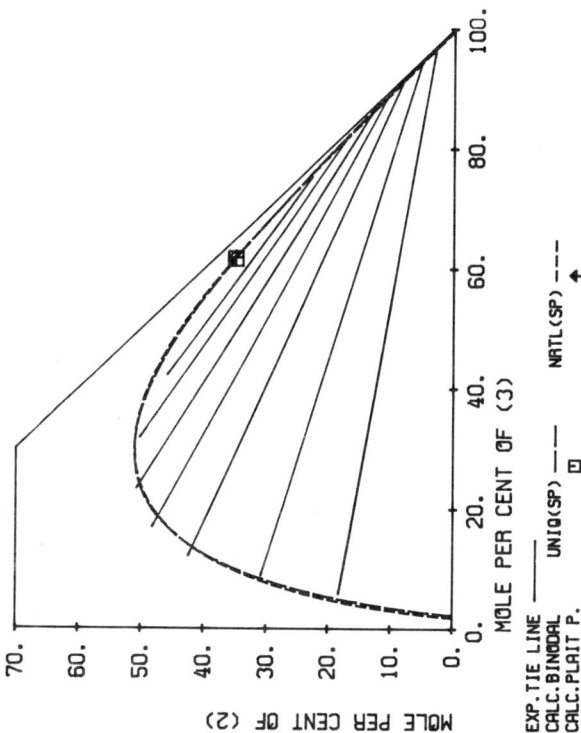

MOLE PER CENT OF (3)
MOLE PER CENT OF (2)

EXP.TIE LINE ———   UNIQ(SP) ⊡   NRTL(SP) -----
CALC.BINODAL ———
CALC.PLAIT P.

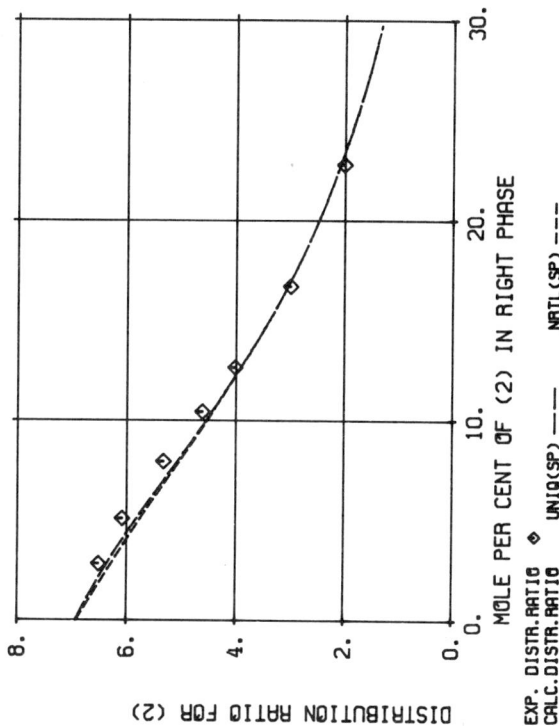

MOLE PER CENT OF (2) IN RIGHT PHASE
DISTRIBUTION RATIO FOR (2)

EXP. DISTR.RATIO ◇   UNIQ(SP) ———   NRTL(SP) -----
CALC.DISTR.RATIO ———

$C_3H_6O-C_6H_6O$

(1) C6H6O   PHENOL

(2) C3H6O   2-PROPANONE

(3) H2O   WATER

SCHREINEMAKERS F.A.H.
Z.PHYS.CHEM.(LEIPZIG) 39(1902)485

TEMPERATURE = 56.5 DEG C   TYPE OF SYSTEM = 1

EXPERIMENTAL TIE LINES IN MOLE PCT

	LEFT PHASE			RIGHT PHASE	
(1)	(2)	(3)	(1)	(2)	(3)
22.308	0.0	77.692	3.144	0.0	96.856
27.772	5.181	67.047	2.102	0.409	97.490
28.303	13.202	58.495	1.462	1.320	97.218
26.174	17.661	56.190	1.277	1.798	96.924
23.304	21.360	55.336	1.275	2.513	96.213
23.507	21.746	55.746	1.286	2.953	95.761
14.515	24.052	61.433	1.645	6.346	92.008
14.293	24.089	61.617	1.729	6.624	91.647
10.679	22.768	66.553	2.212	8.562	89.227

SPECIFIC MODEL PARAMETERS IN KELVIN

		UNIQUAC		NRTL(ALPHA=.2)	
I	J	AIJ	AJI	AIJ	AJI
1	2	-268.76	-248.26	-476.48	-859.84
1	3	-187.88	290.44	-513.66	1604.0
2	3	30.503	48.440	-28.469	597.58

R1 = 3.5517   R2 = 2.5735   R3 = 0.9200
Q1 = 2.680   Q2 = 2.336   Q3 = 1.400

MEAN DEV. BETWEEN CALC. AND EXP. CONC. IN MOLE PCT

UNIQUAC (SPECIFIC PARAMETERS)   1.52
NRTL   (SPECIFIC PARAMETERS)   1.61

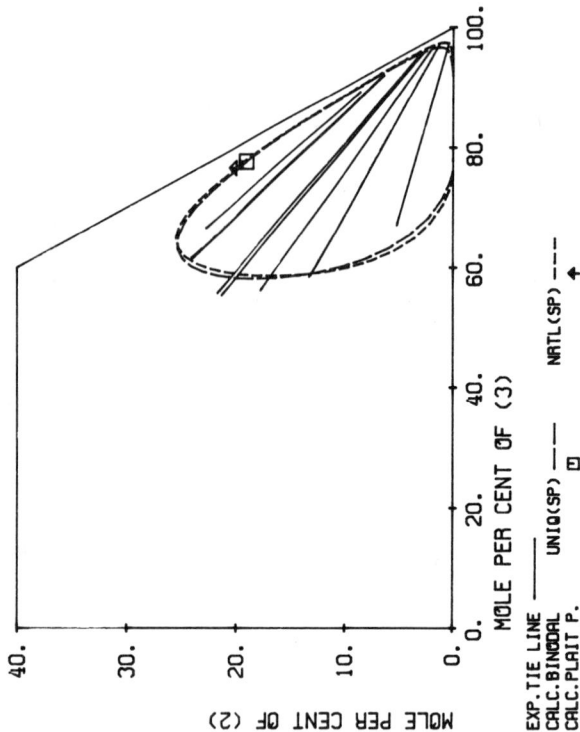

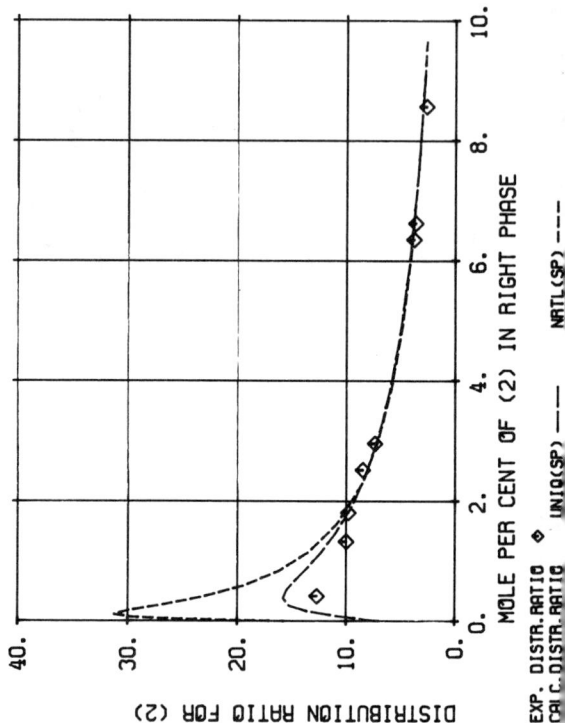

MOLE PER CENT OF (2)
MOLE PER CENT OF (3)

EXP.TIE LINE ——   UNIQ(SP) □   NRTL(SP) ----
CALC.BINODAL ----
CALC.PLAIT P.

DISTRIBUTION RATIO FOR (2)
MOLE PER CENT OF (2) IN RIGHT PHASE

EXP. DISTR.RATIO ◇   UNIQ(SP) ◇   NRTL(SP) ----
CALC.DISTR.RATIO

$C_3H_6O-C_6H_{12}O$

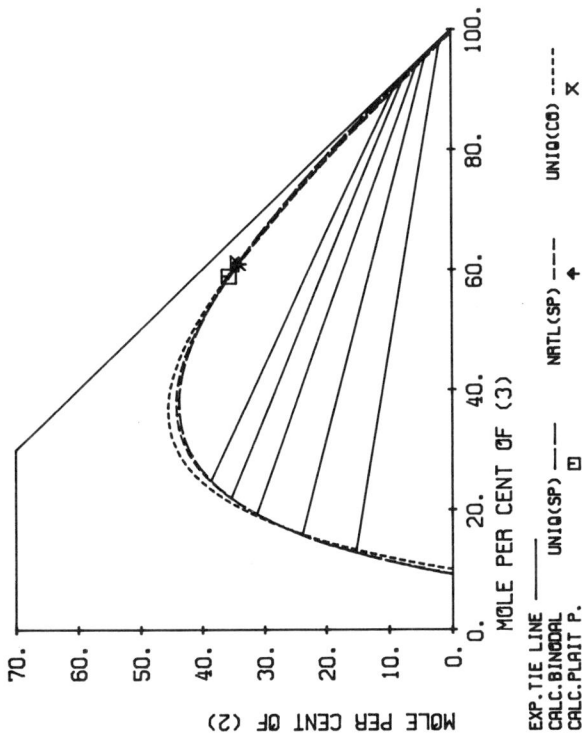

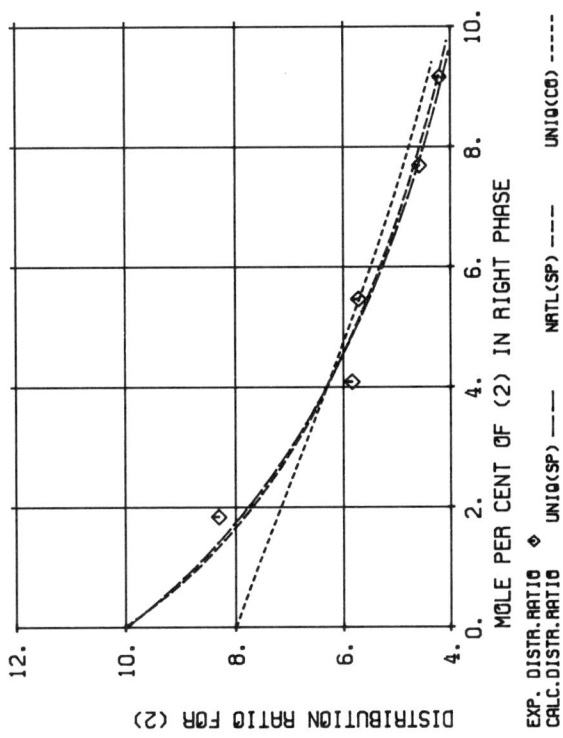

(1) C6H12O   2-PENTANONE, 4-METHYL

(2) C3H6O   2-PROPANONE

(3) H2O   WATER

OTHMER D.F., WHITE R.E., TRUEGER E.
IND.ENG.CHEM. 33(1941)1240

TEMPERATURE = 25.5 DEG C   TYPE OF SYSTEM = 1

EXPERIMENTAL TIE LINES IN MOLE PCT (GRAPH.INTERPOL.)

	LEFT PHASE			RIGHT PHASE	
(1)	(2)	(3)	(1)	(2)	(3)
71.752	15.246	13.002	0.419	1.834	97.746
60.214	23.877	15.909	0.501	4.087	95.413
49.564	31.289	19.147	0.565	5.462	93.972
42.744	35.333	21.923	0.697	7.685	91.619
36.518	38.574	24.908	0.812	9.159	90.028

SPECIFIC MODEL PARAMETERS IN KELVIN

		UNIQUAC		NRTL(ALPHA=.2)	
I J	AIJ	AJI		AIJ	AJI
1 2	-11.928	-35.178		-35.285	-107.08
1 3	428.21	142.82		258.07	1342.3
2 3	75.235	7.1109		-34.930	559.24

R1 = 4.5959   R2 = 2.5735   R3 = 0.9200
Q1 = 3.952   Q2 = 2.336   Q3 = 1.400

MEAN DEV. BETWEEN CALC. AND EXP. CONC. IN MOLE PCT

UNIQUAC (SPECIFIC PARAMETERS)   0.24
NRTL   (SPECIFIC PARAMETERS)   0.19
UNIQUAC (COMMON PARAMETERS)   0.38

$C_3H_6O-C_6H_{12}O_2$

(1) C6H12O2   BUTANOIC ACID, ETHYL ESTER

(2) C3H6O   2-PROPANONE

(3) H2O   WATER

VENKATARATNAM A., JAGANNADHA RAO R., VENKATA RAO C.
CHEM.ENG.SCI. 7(1957)102

TEMPERATURE = 30.0 DEG C   TYPE OF SYSTEM = 1

EXPERIMENTAL TIE LINES IN MOLE PCT

	LEFT PHASE			RIGHT PHASE	
(1)	(2)	(3)	(1)	(2)	(3)
96.254	0.0	3.746	0.125	0.0	99.875
89.203	4.248	6.550	0.111	0.569	99.321
85.940	7.623	6.437	0.144	1.187	98.669
78.194	14.548	7.259	0.181	2.200	97.619
64.620	25.218	10.163	0.277	4.534	95.189
40.700	42.638	16.662	0.391	8.128	91.480
30.373	48.365	21.263	0.442	9.689	89.869
21.209	49.487	29.304	0.642	13.502	85.856

SPECIFIC MODEL PARAMETERS IN KELVIN

		UNIQUAC		NRTL(ALPHA=.2)	
I	J	AIJ	AJI	AIJ	AJI
1	2	158.07	-72.528	26.690	62.538
1	3	552.63	193.00	448.11	1386.8
2	3	128.34	48.075	-29.804	689.46

R1 = 4.8274   R2 = 2.5735   R3 = 0.9200
Q1 = 4.196   Q2 = 2.336   Q3 = 1.400

MEAN DEV. BETWEEN CALC. AND EXP. CONC. IN MOLE PCT

UNIQUAC (SPECIFIC PARAMETERS)   0.47
NRTL   (SPECIFIC PARAMETERS)   0.49
UNIQUAC (COMMON PARAMETERS)   0.92

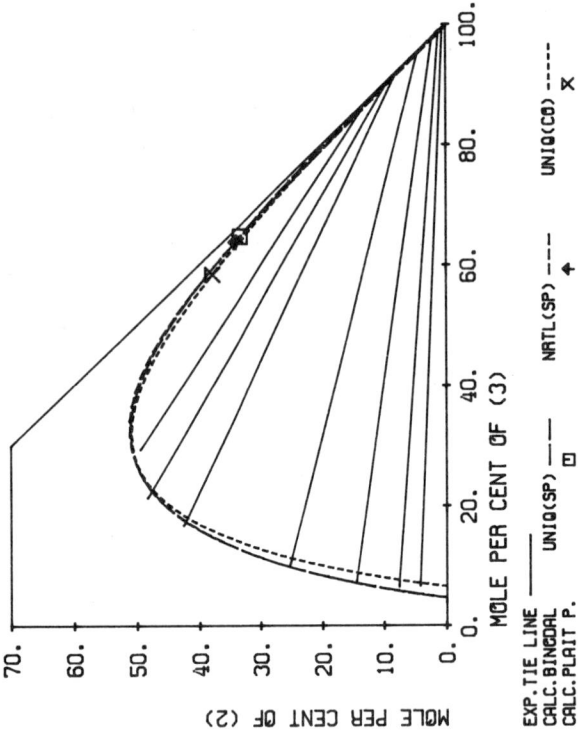

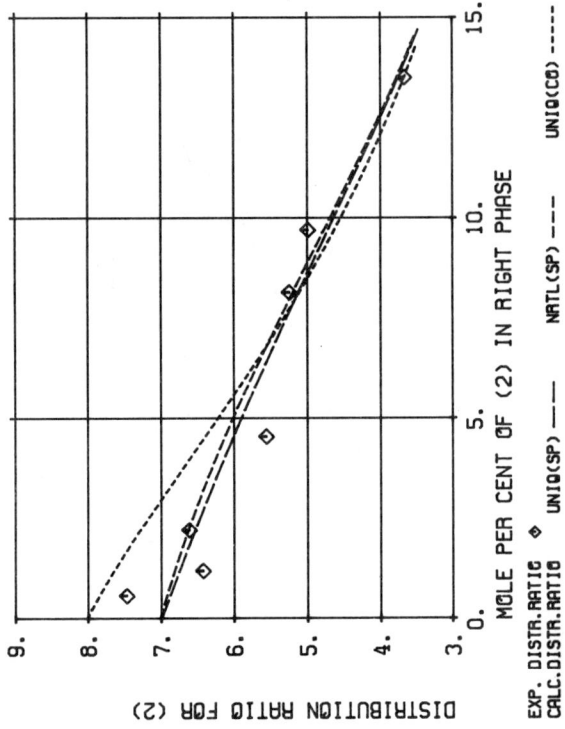

C$_3$H$_6$O-C$_6$H$_{14}$

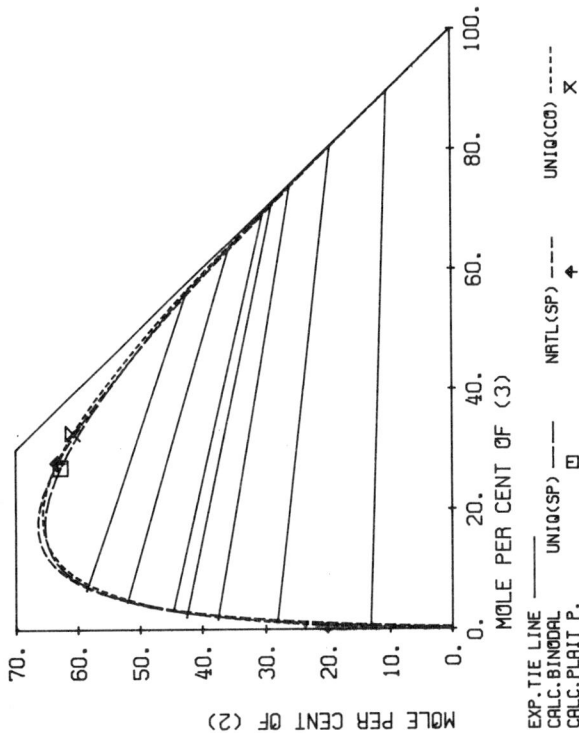

MOLE PER CENT OF (2)

MOLE PER CENT OF (3)

EXP.TIE LINE	UNIQ(SP)	NRTL(SP)	UNIQ(CO)
CALC.BINODAL			
CALC.PLAIT P.			

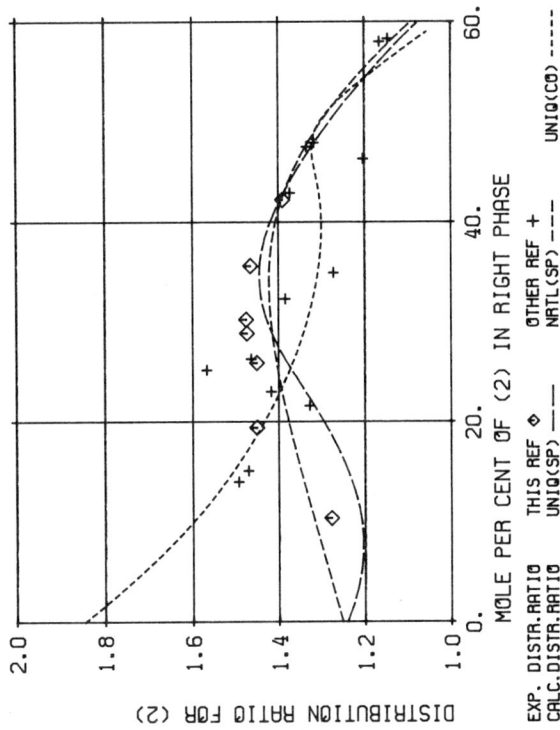

DISTRIBUTION RATIO FOR (2)

MOLE PER CENT OF (2) IN RIGHT PHASE

	EXP. DISTR. RATIO	THIS REF ◇	OTHER REF +
	CALC.DISTR.RATIO	UNIQ(SP)	NRTL(SP)	UNIQ(CO)

(1) C6H14   HEXANE

(2) C3H6O   2-PROPANONE

(3) H2O   WATER

TREYBAL R.E., VONDRAK O.J.
IND.ENG.CHEM. 41(1949)1761

TEMPERATURE = 25.0 DEG C   TYPE OF SYSTEM = 1

EXPERIMENTAL TIE LINES IN MOLE PCT

	LEFT PHASE			RIGHT PHASE	
(1)	(2)	(3)	(1)	(2)	(3)
86.822	13.178	0.0	0.049	10.315	89.636
70.592	28.119	1.289	0.105	19.386	80.509
60.789	37.557	1.654	0.295	25.900	73.804
55.509	42.469	2.022	0.416	28.832	70.752
52.254	44.571	3.175	0.481	30.219	69.300
43.359	52.082	4.559	0.761	35.556	63.684
34.849	58.631	6.521	1.237	42.189	56.574

SPECIFIC MODEL PARAMETERS IN KELVIN

		UNIQUAC		NRTL(ALPHA=.2)	
I J	AIJ	AJI		AIJ	AJI
1 2	332.28	-114.80		48.660	256.45
1 3	941.20	1420.4		1225.0	2646.3
2 3	573.95	-191.57		503.69	-33.946

R1 = 4.4998   R2 = 2.5735   R3 = 0.9200
Q1 = 3.856   Q2 = 2.336   Q3 = 1.400

MEAN DEV. BETWEEN CALC. AND EXP. CONC. IN MOLE PCT

UNIQUAC (SPECIFIC PARAMETERS)   0.53
NRTL   (SPECIFIC PARAMETERS)   0.54
UNIQUAC (COMMON PARAMETERS)   1.33

$C_3H_6O-C_6H_{14}$

```
(1) C6H14      HEXANE
(2) C3H6O      2-PROPANONE
(3) H2O        WATER

RAGAINI V., RAVERDINO V., SANTI S.
QUAD.ING.CHIM.ITAL. 10(1974)19

TEMPERATURE = 25.0 DEG C     TYPE OF SYSTEM = 1

EXPERIMENTAL TIE LINES IN MOLE PCT

        LEFT PHASE                RIGHT PHASE
    (1)      (2)      (3)      (1)      (2)      (3)

  77.800   22.200    0.0     0.300   15.100   84.600
  65.500   32.600    1.900   0.600   23.000   76.400
  59.300   38.500    2.200   0.800   26.300   73.300
  51.900   44.400    3.700   0.800   34.900   64.300
  37.500   55.700    6.800   1.500   46.300   52.200
  21.500   63.100   15.400   2.600   47.900   49.500
  16.700   66.800   16.500   5.400   58.300   36.300

SPECIFIC MODEL PARAMETERS IN KELVIN

              UNIQUAC                 NRTL(ALPHA=.2)
I  J     AIJ        AJI          AIJ          AJI

1  2   332.28    -114.80       48.660       256.45
1  3   941.20    1420.4      1225.0        2645.3
2  3   573.95    -191.57      503.69        -33.946

R1 = 4.4998    R2 = 2.5735    R3 = 0.9200
Q1 = 3.856     Q2 = 2.336     Q3 = 1.400

MEAN DEV. BETWEEN CALC. AND EXP. CONC. IN MOLE PCT

UNIQUAC (SPECIFIC PARAMETERS)    1.75
NRTL    (SPECIFIC PARAMETERS)    1.64
UNIQUAC (COMMON PARAMETERS)      1.84
```

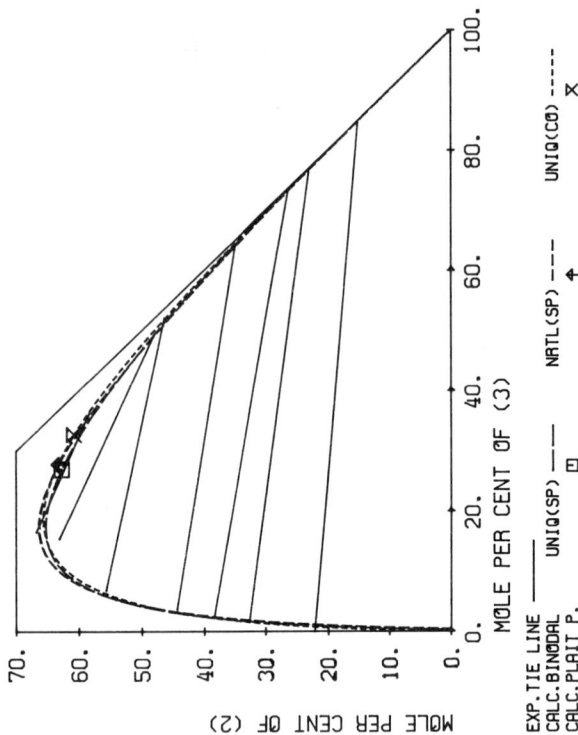

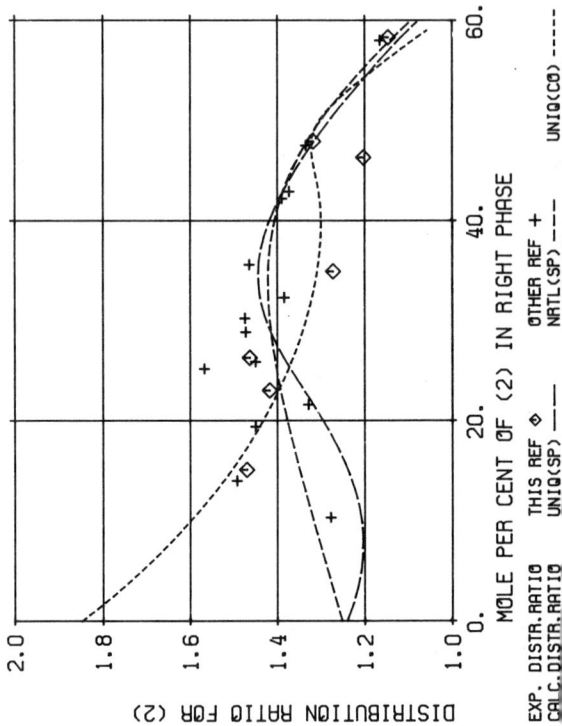

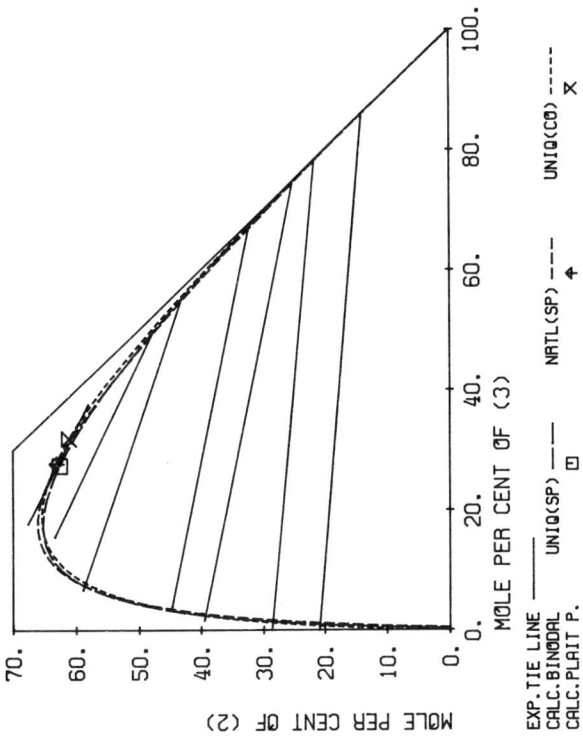

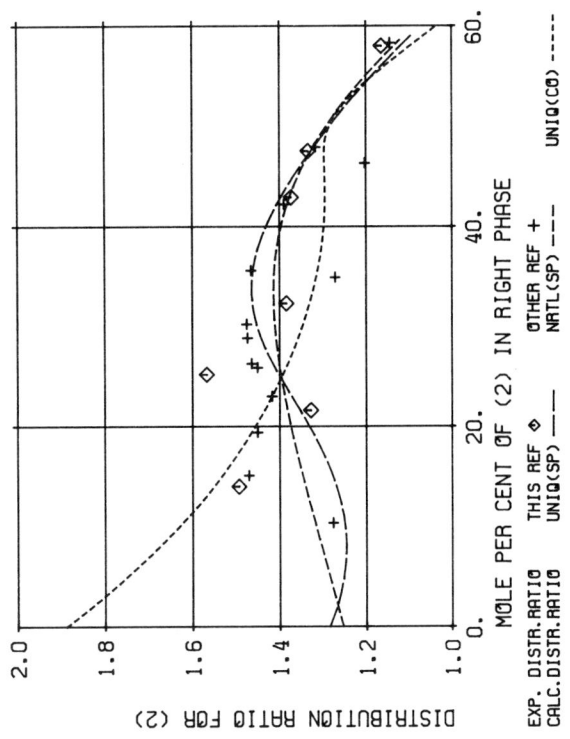

(1) C6H14   HEXANE
(2) C3H6O   2-PROPANONE
(3) H2O   WATER

RAGAINI V., RAVERDINO V., SANTI S.
QUAD.ING.CHIM.ITAL. 10(1974)19

TEMPERATURE = 30.0 DEG C   TYPE OF SYSTEM = 1

EXPERIMENTAL TIE LINES IN MOLE PCT

	LEFT PHASE			RIGHT PHASE	
(1)	(2)	(3)	(1)	(2)	(3)
79.100	20.900	0.0	0.300	14.000	85.700
71.300	28.700	0.0	0.400	21.600	78.000
58.900	39.500	1.600	0.400	25.200	74.400
51.400	44.700	3.900	0.900	32.300	66.800
34.400	58.900	6.700	1.500	42.900	55.600
21.100	63.400	15.500	2.400	47.500	50.100
14.700	67.600	17.700	4.300	58.000	37.700

SPECIFIC MODEL PARAMETERS IN KELVIN

I J	UNIQUAC		NRTL(ALPHA=.2)	
	AIJ	AJI	AIJ	AJI
1 2	332.28	-114.80	48.660	256.45
1 3	941.20	1420.4	1225.0	2645.3
2 3	573.95	-191.57	503.69	-33.946

R1 = 4.4998   R2 = 2.5735   R3 = 0.9200
Q1 = 3.856   Q2 = 2.336   Q3 = 1.400

MEAN DEV. BETWEEN CALC. AND EXP. CONC. IN MOLE PCT

UNIQUAC (SPECIFIC PARAMETERS)   1.27
NRTL   (SPECIFIC PARAMETERS)   1.24
UNIQUAC (COMMON PARAMETERS)   1.57

$C_3H_6O-C_6H_{14}$

(1) C6H14   HEXANE
(2) C3H6O   2-PROPANONE
(3) H2O   WATER

RAGAINI V., RAVERDINO V., SANTI S.
QUAD.ING.CHIM.ITAL. 10(1974)19

TEMPERATURE = 35.0 DEG C   TYPE OF SYSTEM = 1

EXPERIMENTAL TIE LINES IN MOLE PCT

	LEFT PHASE			RIGHT PHASE	
(1)	(2)	(3)	(1)	(2)	(3)
83.000	17.000	0.0	0.300	10.600	89.100
67.000	32.400	0.600	0.400	18.500	81.100
60.600	38.500	0.900	0.800	23.700	76.300
51.100	47.100	1.800	0.800	34.800	64.400
33.800	59.200	7.000	1.200	42.100	56.700
18.900	64.400	16.700	1.900	45.600	52.500
11.300	63.300	25.400	3.500	55.800	40.700

SPECIFIC MODEL PARAMETERS IN KELVIN

		UNIQUAC		NRTL(ALPHA=.2)	
I	J	AIJ	AJI	AIJ	AJI
1	2	364.86	-121.29	43.674	377.90
1	3	1450.2	1363.3	1932.9	2459.6
2	3	435.77	-147.24	276.99	232.94

R1 = 4.4998   R2 = 2.5735   R3 = 0.9200
Q1 = 3.856   Q2 = 2.336   Q3 = 1.400

MEAN DEV. BETWEEN CALC. AND EXP. CONC. IN MOLE PCT

UNIQUAC (SPECIFIC PARAMETERS)   1.75
NRTL   (SPECIFIC PARAMETERS)   1.24

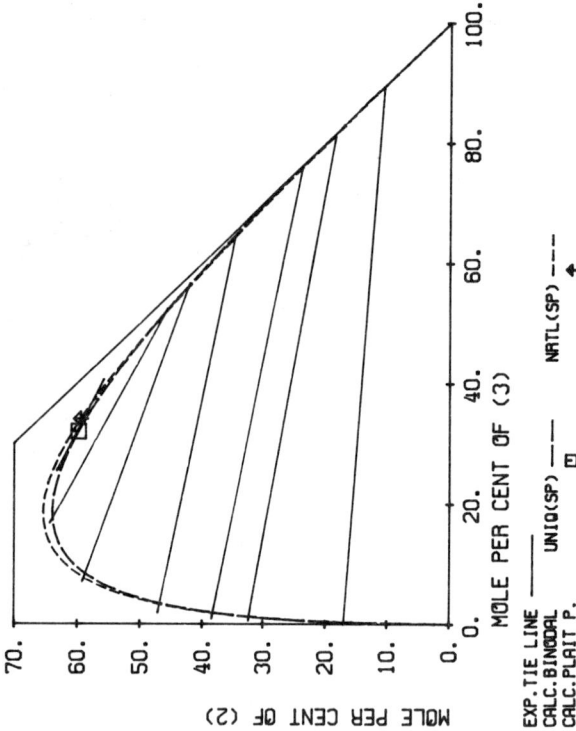

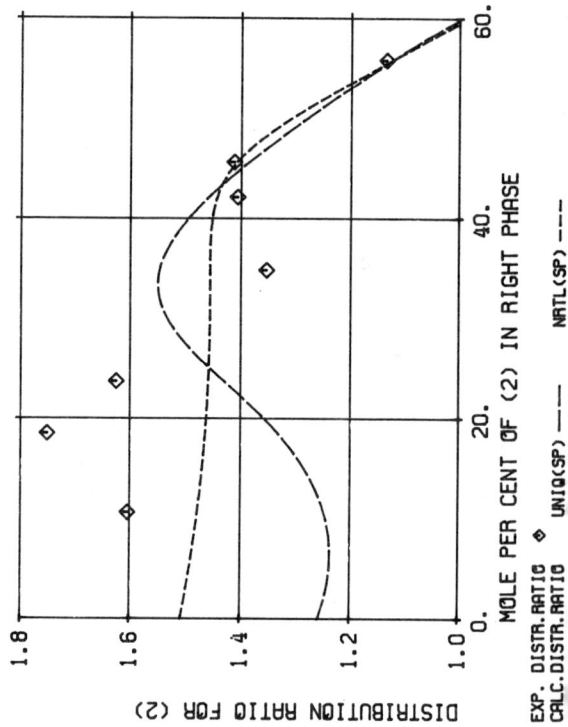

EXP.TIE LINE ———   UNIQ(SP) ⊟   NRTL(SP) - - - -
CALC.BINODAL
CALC.PLAIT P.

EXP. DISTR.RATIO ◇   UNIQ(SP) ———   NRTL(SP) - - - -
CALC.DISTR.RATIO - - - -

$C_3H_6O-C_6H_{14}$

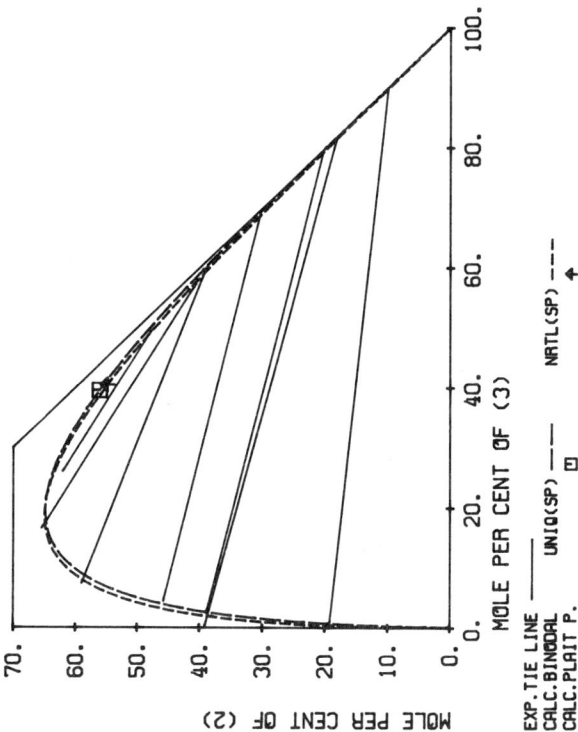

MOLE PER CENT OF (3)

MOLE PER CENT OF (2)

EXP.TIE LINE ———   UNIQ(SP) ---- ☐   NRTL(SP) ---- ♠
CALC.BINODAL
CALC.PLAIT P.

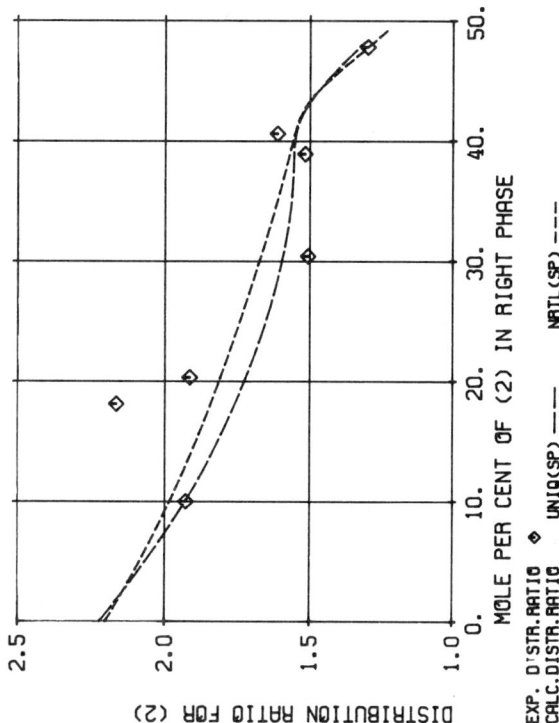

MOLE PER CENT OF (2) IN RIGHT PHASE

DISTRIBUTION RATIO FOR (2)

EXP. D'STR.RATIO ◊   UNIQ(SP) ———   NRTL(SP) ----
CALC.DISTR.RATIO

(1) C6H14   HEXANE
(2) C3H6O   2-PROPANONE
(3) H2O   WATER

RAGAINI V., RAVERDINO V., SANTI S.
QUAD.ING.CHIM.ITAL. 10(1974)19

TEMPERATURE = 40.0 DEG C   TYPE OF SYSTEM = 1

EXPERIMENTAL TIE LINES IN MOLE PCT

| | LEFT PHASE | | | RIGHT PHASE | |
(1)	(2)	(3)	(1)	(2)	(3)
80.700	19.300	0.0	0.0	10.000	90.000
60.700	39.200	0.100	0.000	18.100	81.900
58.700	38.900	2.400	0.000	20.300	79.700
49.800	45.800	4.400	0.500	30.400	69.100
33.700	59.000	7.300	1.000	38.900	60.100
18.500	65.500	16.000	2.000	40.600	57.400
11.900	62.100	26.000	2.700	47.800	49.500

SPECIFIC MODEL PARAMETERS IN KELVIN

| | | UNIQUAC | | NRTL(ALPHA=.2) | |
I	J	AIJ	AJI	AIJ	AJI
1	2	237.79	-33.262	-61.046	520.23
1	3	1287.8	-553.30	1941.2	1587.1
2	3	313.27	-52.562	186.41	395.60

R1 = 4.4998   R2 = 2.5735   R3 = 0.9200
Q1 = 3.856   Q2 = 2.336   Q3 = 1.400

MEAN DEV. BETWEEN CALC. AND EXP. CONC. IN MOLE PCT

UNIQUAC (SPECIFIC PARAMETERS)   1.31
NRTL   (SPECIFIC PARAMETERS)   1.25

C₃H₆O-C₆H₁₄

(1) C6H14   HEXANE

(2) C3H6O   2-PROPANONE

(3) H2O   WATER

RAGAINI V., RAVERDINO V., SANTI S.
QUAD.ING.CHIM.ITAL. 10(1974)19

TEMPERATURE = 45.0 DEG C   TYPE OF SYSTEM = 1

EXPERIMENTAL TIE LINES IN MOLE PCT

	LEFT PHASE			RIGHT PHASE	
(1)	(2)	(3)	(1)	(2)	(3)
79.900	20.100	0.0	0.0	9.300	90.700
65.000	32.300	2.700	0.0	15.300	84.700
55.000	41.400	3.600	0.000	19.600	80.400
48.600	46.400	5.000	0.700	27.900	71.400
32.500	59.300	8.200	1.100	36.700	62.200
16.200	64.700	19.100	1.800	38.400	59.800
10.800	61.100	28.100	3.100	43.200	53.700

SPECIFIC MODEL PARAMETERS IN KELVIN

		UNIQUAC		NRTL(ALPHA=.2)	
I	J	AIJ	AJI	AIJ	AJI
1	2	231.20	-28.927	-96.797	602.34
1	3	1254.8	530.17	1240.9	1454.4
2	3	287.11	-35.245	196.39	419.37

R1 = 4.4998   R2 = 2.5735   R3 = 0.9200
Q1 = 3.856   Q2 = 2.336   Q3 = 1.400

MEAN DEV. BETWEEN CALC. AND EXP. CONC. IN MOLE PCT

UNIQUAC (SPECIFIC PARAMETERS)   1.04
NRTL   (SPECIFIC PARAMETERS)   0.96

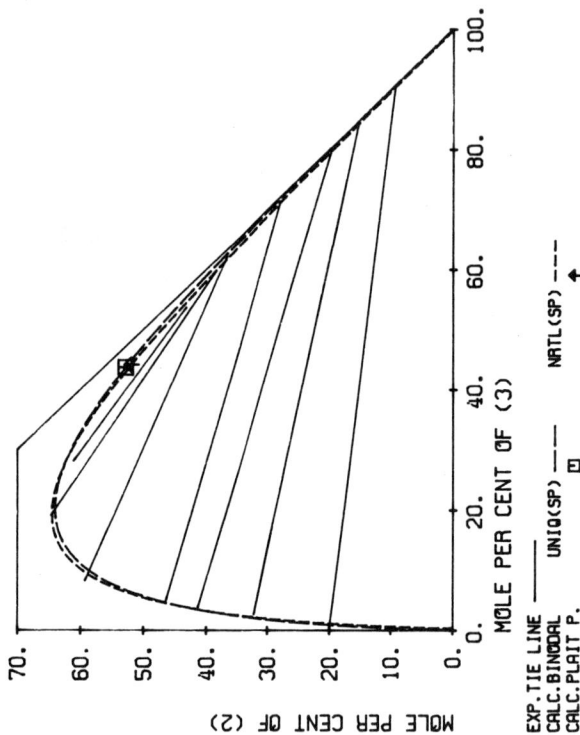

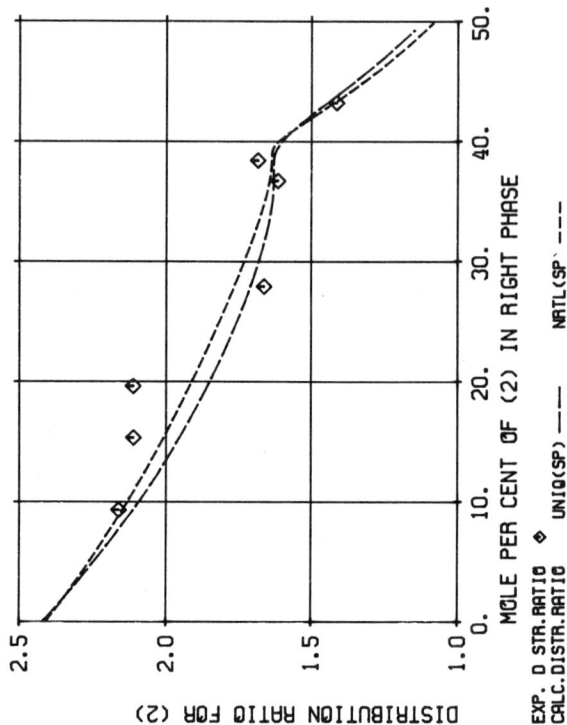

$C_3H_6O\text{-}C_6H_{14}O$

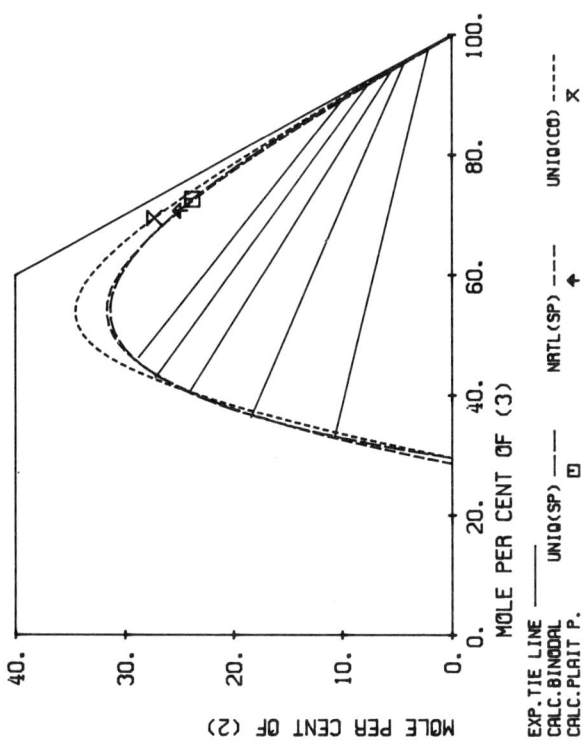

MOLE PER CENT OF (3)

MOLE PER CENT OF (2)

EXP.TIE LINE ——   UNIQ(SP) ——   NRTL(SP) ——   UNIQ(C0) - - - -
CALC.BINODAL - - - -
CALC.PLAIT P. ⊡   ✛   ⚹

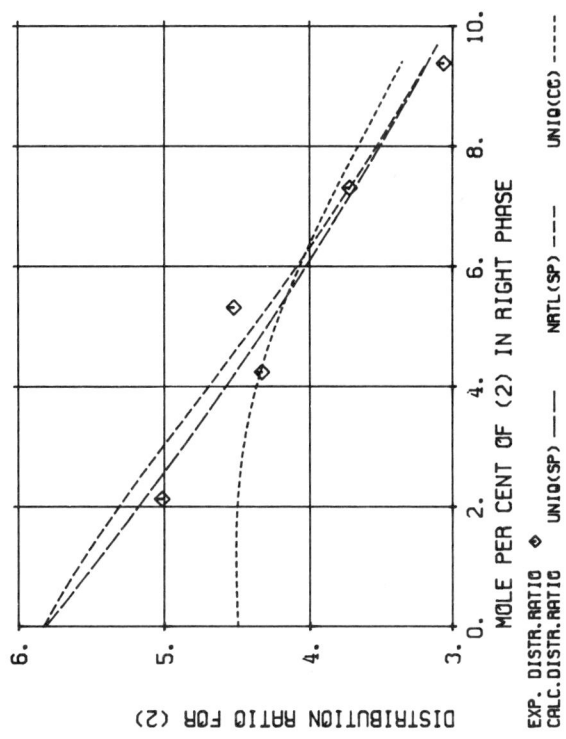

DISTRIBUTION RATIO FOR (2)

MOLE PER CENT OF (2) IN RIGHT PHASE

EXP. DISTR.RATIO ◇   UNIQ(SP) ——   NRTL(SP) - - - -   UNIQ(C0) - - - -
CALC.DISTR.RATIO

(1) C6H14O   1-HEXANOL
----
(2) C3H6O   2-PROPANONE
----
(3) H2O   WATER

VENKATARATNAM A., JAGANNADHA RAO R., VENKATA RAO C.
J.SCI.IND.RES. 17B(1958)108

TEMPERATURE = 30.0 DEG C   TYPE OF SYSTEM = 1

EXPERIMENTAL TIE LINES IN MOLE PCT

        LEFT PHASE   RIGHT PHASE
    (1)   (2)   (3)   (1)   (2)   (3)

  56.147   10.656   33.197   0.149   2.125   97.726
  45.314   18.372   36.314   0.175   4.240   95.585
  35.436   24.046   40.517   0.199   5.316   94.485
  29.897   27.184   42.919   0.207   7.308   92.485
  24.982   28.721   46.297   0.237   9.382   90.381

SPECIFIC MODEL PARAMETERS IN KELVIN

            UNIQUAC   NRTL(ALPHA=.2)
I J   AIJ   AJI   AIJ   AJI

1 2   121.39   -38.438   240.06   -143.54
1 3   140.53   221.68   -109.44   1915.6
2 3   -58.718   191.67   -40.422   596.02

R1 = 4.8031   R2 = 2.5735   R3 = 0.9200
Q1 = 4.132   Q2 = 2.336   Q3 = 1.400

MEAN DEV. BETWEEN CALC. AND EXP. CONC. IN MOLE PCT

UNIQUAC (SPECIFIC PARAMETERS)   0.30
NRTL   (SPECIFIC PARAMETERS)   0.33
UNIQUAC (COMMON PARAMETERS)   0.50

C$_3$H$_6$O-C$_7$H$_8$

MOLE PER CENT OF (3)

NRTL(SP) ----
UNIQ(SP) ---- ☐

EXP.TIE LINE
CALC.BINODAL
CALC.PLAIT P.

MOLE PER CENT OF (2)

MOLE PER CENT OF (2) IN RIGHT PHASE

EXP. DISTR. RATIO ◇ UNIQ(SP) ----
CALC.DISTR.RATIO NRTL(SP) ----

DISTRIBUTION RATIO FOR (2)

(1) C7H8	TOLUENE
(2) C3H6O	2-PROPANONE
(3) H2O	WATER

HACKL A., SOLAR W., ZIEBLAND G.
EUR.FED.CHEM.ENG.,RECOMM.SYST.LIQ.EXTR.STUD.,EDITOR:T.MISEK
(1978)

TEMPERATURE = 10.0 DEG C   TYPE OF SYSTEM = 1

EXPERIMENTAL TIE LINES IN MOLE PCT (GRAPH.INTERPOL.)

LEFT PHASE			RIGHT PHASE		
(1)	(2)	(3)	(1)	(2)	(3)
99.587	0.158	0.255	0.010	0.056	99.934
98.110	1.637	0.254	0.010	0.537	99.453
96.558	2.938	0.504	0.010	0.918	99.072
95.497	4.001	0.502	0.010	1.227	98.763
94.612	4.887	0.500	0.010	1.494	98.506
91.710	7.796	0.495	0.010	2.324	97.666
88.300	10.969	0.732	0.021	3.270	96.709
83.486	15.558	0.956	0.033	4.599	95.380
80.002	18.820	1.178	0.032	5.516	94.451
76.752	21.854	1.393	0.034	6.432	93.534
72.635	25.768	1.597	0.046	7.470	92.484
66.126	31.880	1.994	0.059	9.158	90.783

SPECIFIC MODEL PARAMETERS IN KELVIN

		UNIQUAC		NRTL(ALPHA=.2)	
I	J	AIJ	AJI	AIJ	AJI
1	2	115.14	29.585	211.69	14.071
1	3	814.64	334.88	1057.6	1643.2
2	3	317.21	-22.287	366.44	250.69

R1 = 3.9228   R2 = 2.5735   R3 = 0.9200
Q1 = 2.968   Q2 = 2.336   Q3 = 1.400

MEAN DEV. BETWEEN CALC. AND EXP. CONC. IN MOLE PCT

UNIQUAC (SPECIFIC PARAMETERS)	0.05
NRTL (SPECIFIC PARAMETERS)	0.09

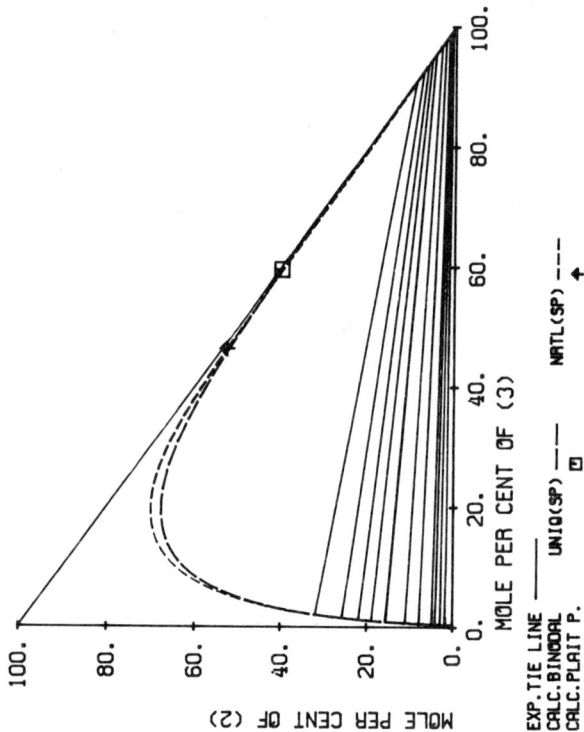

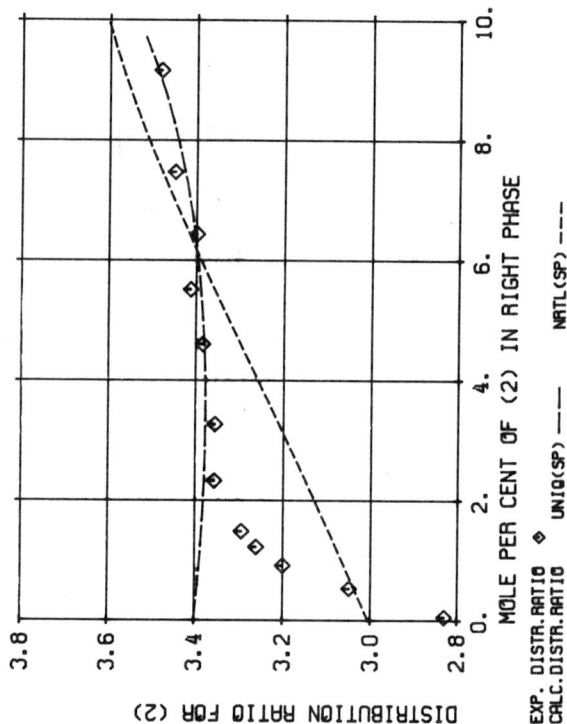

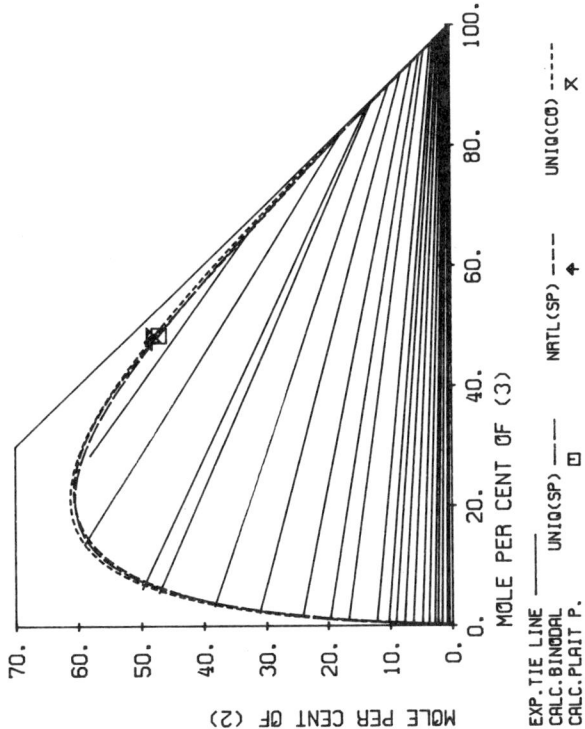

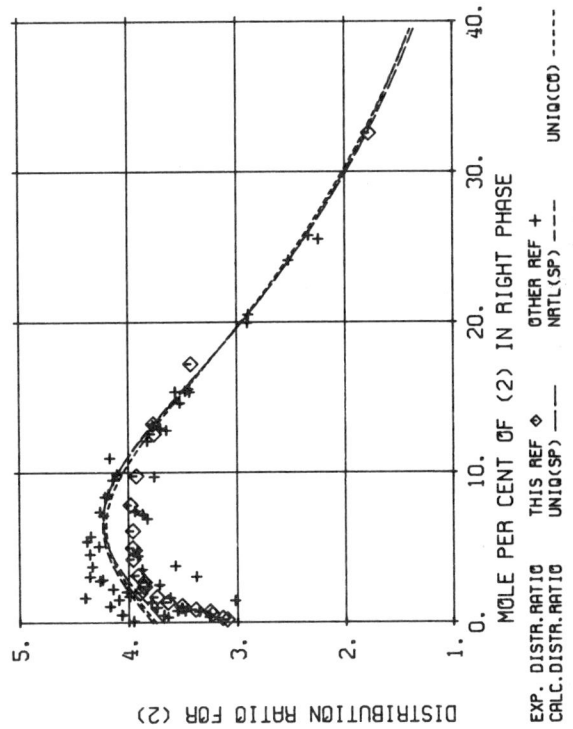

$C_3H_6O-C_7H_8$

(1) C7H8   TOLUENE

(2) C3H6O   2-PROPANONE

(3) H2O   WATER

HACKL A., SOLAR W., ZIEBLAND G.
EUR.FED.CHEM.ENG.,RECOMM.SYST.LIQ.EXTR.STUD.,EDITOR:T.MISEK
(1978)

TEMPERATURE = 20.0 DEG C   TYPE OF SYSTEM = 1

EXPERIMENTAL TIE LINES IN MOLE PCT (GRAPH.INTERPOL.)

	LEFT PHASE			RIGHT PHASE	
(1)	(2)	(3)	(1)	(2)	(3)
99.176	0.569	0.255	0.010	0.184	99.806
98.815	0.931	0.254	0.010	0.297	99.693
98.000	1.746	0.254	0.010	0.540	99.450
97.114	2.381	0.505	0.010	0.732	99.258
96.604	2.892	0.504	0.010	0.854	99.136
95.681	3.817	0.502	0.010	1.087	98.903
94.612	4.887	0.500	0.010	1.342	98.648
93.191	6.312	0.498	0.010	1.687	98.303
91.784	7.721	0.495	0.010	1.983	98.007
90.378	9.130	0.492	0.010	2.368	97.622
88.982	10.285	0.734	0.021	2.667	97.312
87.061	12.211	0.728	0.021	3.121	96.859
82.363	16.685	0.952	0.021	4.217	95.762
79.139	19.687	1.174	0.033	4.970	94.998
74.209	24.183	1.607	0.033	6.110	93.857
66.663	31.119	2.218	0.046	7.819	92.135
58.297	38.303	3.401	0.060	9.759	90.181
47.205	47.206	5.589	0.100	12.540	87.360
43.881	49.855	6.263	0.114	13.221	86.665
27.318	58.942	13.740	0.231	17.203	82.566
13.517	58.017	28.465	1.940	32.569	65.491

SPECIFIC MODEL PARAMETERS IN KELVIN

		UNIQUAC		NRTL(ALPHA=.2)	
I	J	AIJ	AJI	AIJ	AJI
1	2	269.90	-138.80	489.20	-301.51
1	3	987.42	172.79	1318.8	2557.3
2	3	390.94	-86.302	377.45	210.60

R1 = 3.9228   R2 = 2.5735   R3 = 0.9200
Q1 = 2.968   Q2 = 2.336   Q3 = 1.400

MEAN DEV. BETWEEN CALC. AND EXP. CONC. IN MOLE PCT

UNIQUAC (SPECIFIC PARAMETERS)   0.39
NRTL   (SPECIFIC PARAMETERS)   0.40
UNIQUAC (COMMON PARAMETERS)   0.42

C₃H₆O-C₇H₈

(1) C7H8   TOLUENE

(2) C3H6O   2-PROPANONE

(3) H2O   WATER

HACKL A., SOLAR W., ZIEBLAND G.,
EUR.FED.CHEM.ENG.,RECOMM.SYST.LIQ.EXTR.STUD.,EDITOR:T.MISEK
(1978)

TEMPERATURE = 30.0 DEG C   TYPE OF SYSTEM = 1

EXPERIMENTAL TIE LINES IN MOLE PCT (GRAPH.INTERPOL.)

	LEFT PHASE			RIGHT PHASE	
(1)	(2)	(3)	(1)	(2)	(3)
98.501	1.245	0.254	0.010	0.382	99.608
97.052	2.443	0.505	0.010	0.728	99.262
96.111	3.386	0.503	0.010	0.964	99.026
94.369	5.131	0.500	0.010	1.352	98.639
92.283	6.974	0.743	0.010	1.747	98.243
90.078	9.186	0.737	0.010	2.217	97.772
87.230	11.799	0.970	0.010	2.767	97.223
82.559	16.252	1.190	0.021	3.750	96.229
78.757	19.838	1.405	0.022	4.554	95.425
73.195	24.978	1.827	0.022	5.745	94.233
66.116	31.452	2.431	0.034	7.383	92.583
61.221	35.759	3.020	0.058	8.548	91.394
57.081	39.329	3.590	0.071	9.498	90.431
48.562	45.619	5.819	0.098	10.937	88.966

SPECIFIC MODEL PARAMETERS IN KELVIN

		UNIQUAC		NRTL(ALPHA=.2)	
I J	AIJ	AJI		AIJ	AJI
1 2	269.90	-138.80		489.20	-301.51
1 3	987.42	172.79		1318.8	2557.3
2 3	390.94	-86.302		377.45	210.60

R1 = 3.9228   R2 = 2.5735   R3 = 0.9200
Q1 = 2.968   Q2 = 2.336   Q3 = 1.400

MEAN DEV. BETWEEN CALC. AND EXP. CONC. IN MOLE PCT

UNIQUAC (SPECIFIC PARAMETERS)   0.23
NRTL   (SPECIFIC PARAMETERS)   0.30
UNIQUAC (COMMON PARAMETERS)   0.22

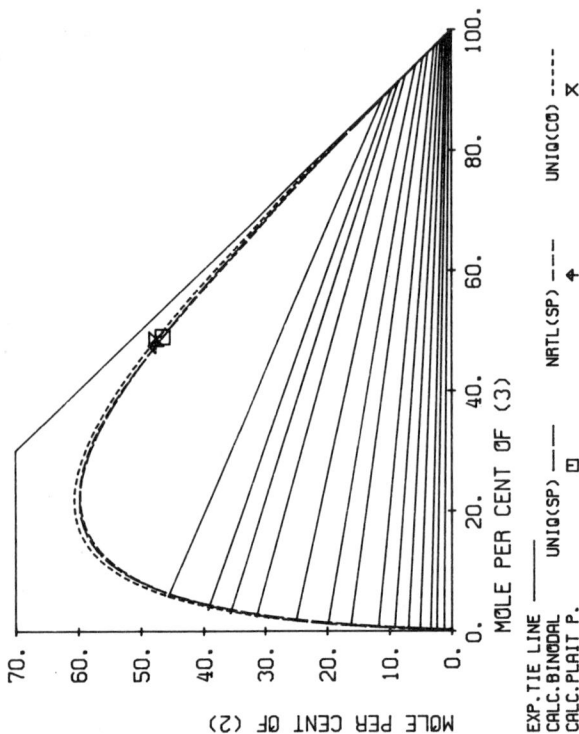

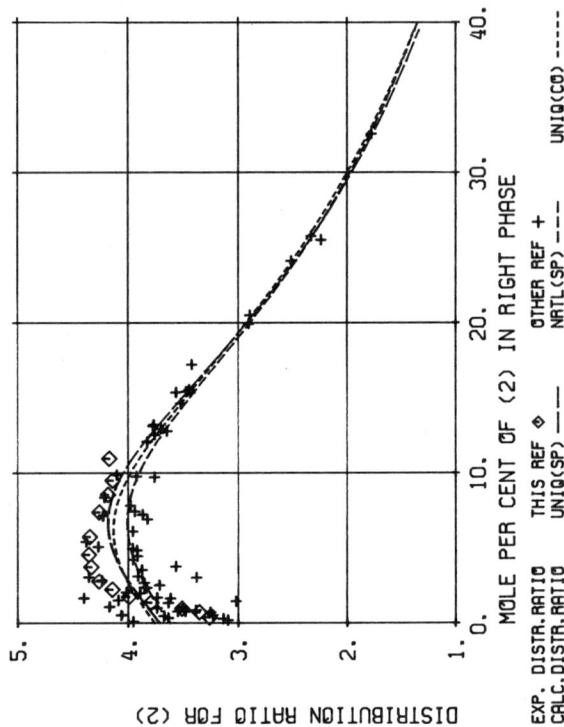

$C_3H_6O-C_7H_8$

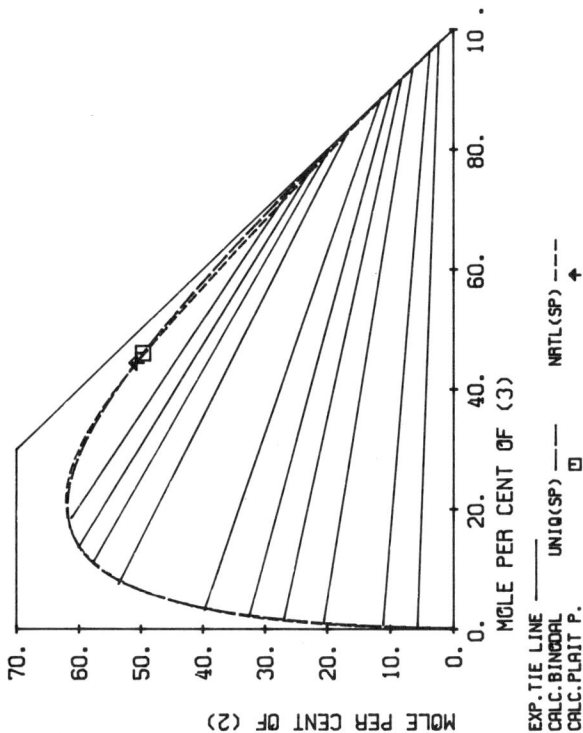

MOLE PER CENT OF (3)

MOLE PER CENT OF (2)

EXP.TIE LINE —— UNIQ(SP) —— $\square$
CALC.BINODAL NRTL(SP) ---
CALC.PLAIT P. $\leftarrow$

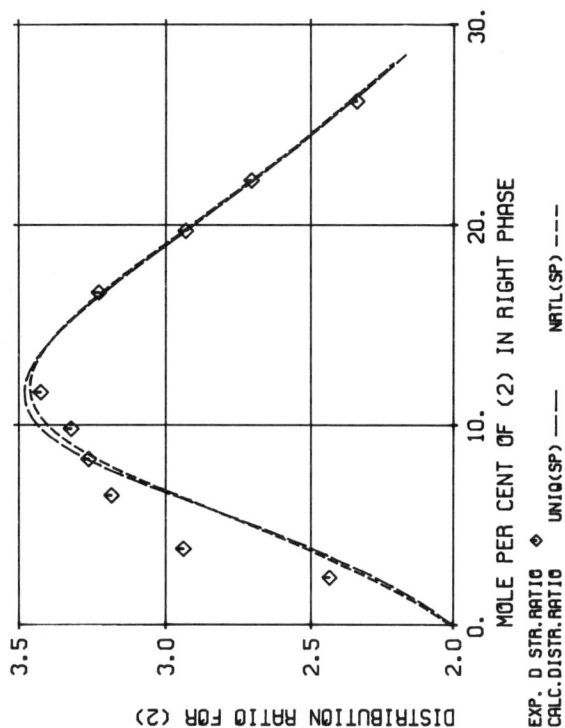

MOLE PER CENT OF (2) IN RIGHT PHASE

DISTRIBUTION RATIO FOR (2)

EXP. D STR.RATIO $\diamond$ UNIQ(SP) ——
CALC.DISTR.RATIO NRTL(SP) ---

(1)	C7H8	TOLUENE
(2)	C3H6O	2-PROPANONE
(3)	H2O	WATER

PAVASOVIC V.
EUR.FED.CHEM.ENG.,RECOMM.SYST.LIQ.EXTR.STUD.,EDITOR:T.MISEK
(1978)

TEMPERATURE = 10.0 DEG C   TYPE OF SYSTEM = 1

EXPERIMENTAL TIE LINES IN MOLE PCT (GRAPH.INTERPOL.)

	LEFT PHASE			RIGHT PHASE	
(1)	(2)	(3)	(1)	(2)	(3)
93.779	5.723	0.499	0.010	2.351	97.639
88.083	11.186	0.731	0.021	3.805	96.174
78.036	20.563	1.401	0.022	6.462	93.516
71.208	26.980	1.812	0.035	8.273	91.692
65.248	32.547	2.204	0.048	9.800	90.152
57.049	39.777	3.174	0.062	11.615	88.323
38.821	53.600	7.579	0.188	16.619	83.193
31.176	57.764	11.060	0.283	19.695	80.022
26.249	60.085	13.666	0.399	22.203	77.398
20.235	61.237	18.529	0.757	26.168	73.075

SPECIFIC MODEL PARAMETERS IN KELVIN

	UNIQUAC		NRTL(ALPHA=.2)	
I J	AIJ	AJI	AIJ	AJI
1 2	357.74	-153.05	664.54	-312.45
1 3	947.23	-413.74	1393.0	2084.4
2 3	391.25	-95.732	416.28	159.00

R1 = 3.9228   R2 = 2.5735   R3 = 0.9200
Q1 = 2.968   Q2 = 2.336   Q3 = 1.400

MEAN DEV. BETWEEN CALC. AND EXP. CONC. IN MOLE PCT

UNIQUAC (SPECIFIC PARAMETERS)   0.20
NRTL   (SPECIFIC PARAMETERS)   0.29

$C_3H_6O-C_7H_8$

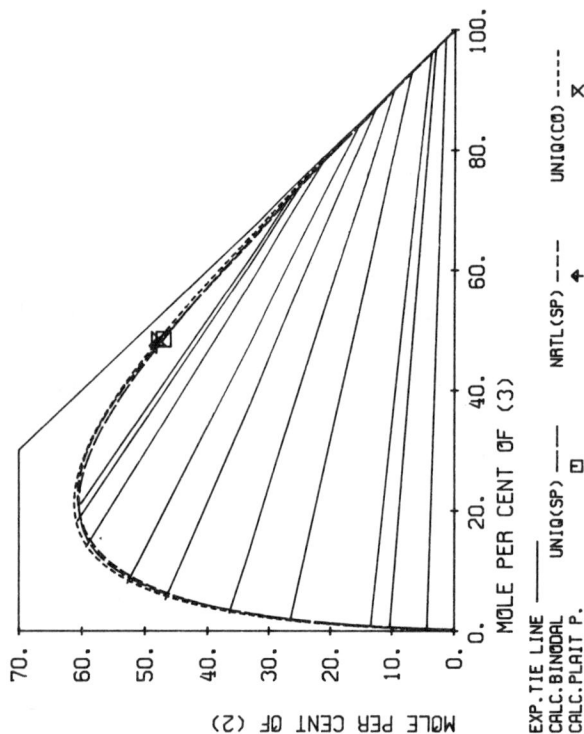

MOLE PER CENT OF (2)

MOLE PER CENT OF (3)

EXP.TIE LINE   UNIQ(SP) ──── □   NRTL(SP) ──── ♦   UNIQ(CO) ──── ✕
CALC.BINODAL
CALC.PLAIT P.

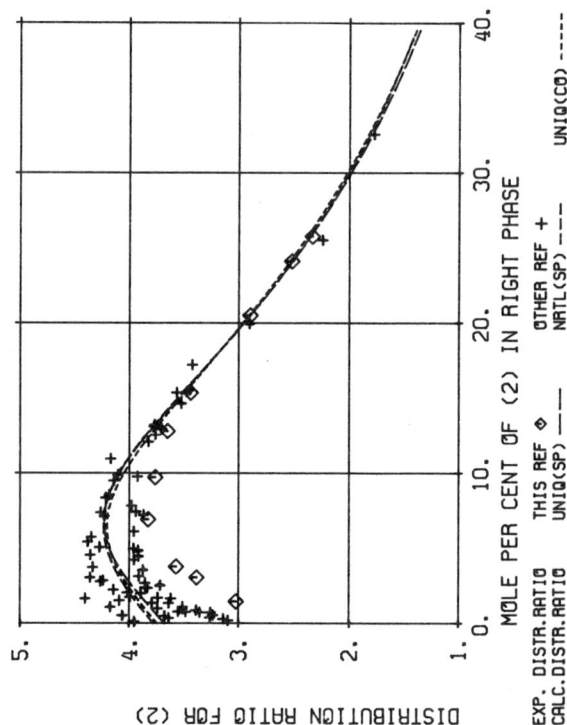

DISTRIBUTION RATIO FOR (2)

MOLE PER CENT OF (2) IN RIGHT PHASE

EXP. DISTR. RATIO ◊   THIS REF ◊   OTHER REF +   UNIQ(CO) ────
CALC.DISTR.RATIO   UNIQ(SP) ────   NRTL(SP) ────

(1) C7H8	TOLUENE
(2) C3H6O	2-PROPANONE
(3) H2O	WATER

PAVASOVIC V.
EUR.FED.CHEM.ENG.,RECOMM.SYST.LIQ.EXTR.STUD.,EDITOR:T.MISEK
(1978)

TEMPERATURE = 20.0 DEG C   TYPE OF SYSTEM = 1

EXPERIMENTAL TIE LINES IN MOLE PCT (GRAPH.INTERPOL.)

	LEFT PHASE			RIGHT PHASE	
(1)	(2)	(3)	(1)	(2)	(3)
95.146	4.353	0.501	0.010	1.441	98.549
88.953	10.314	0.733	0.021	3.050	96.930
85.578	13.458	0.964	0.021	3.768	96.211
71.711	26.473	1.816	0.034	6.915	93.051
60.444	36.546	3.010	0.048	9.708	90.244
48.057	46.721	5.222	0.101	12.790	87.109
39.265	52.777	7.958	0.171	15.350	84.479
26.649	59.343	14.007	0.388	20.499	79.113
20.982	60.701	18.317	0.673	24.107	75.220
18.713	60.153	21.134	0.817	25.763	73.420

SPECIFIC MODEL PARAMETERS IN KELVIN

		UNIQUAC		NRTL(ALPHA=.2)	
I	J	AIJ	AJI	AIJ	AJI
1	2	269.90	-138.80	489.20	-301.51
1	3	987.42	172.79	1318.8	2557.3
2	3	390.94	-86.302	377.45	210.60

R1 = 3.9228   R2 = 2.5735   R3 = 0.9200
Q1 = 2.968   Q2 = 2.336   Q3 = 1.400

MEAN DEV. BETWEEN CALC. AND EXP. CONC. IN MOLE PCT

UNIQUAC (SPECIFIC PARAMETERS)	0.45
NRTL (SPECIFIC PARAMETERS)	0.59
UNIQUAC (COMMON PARAMETERS)	0.38

$C_3H_6O-C_7H_8$

(1) C7H8   TOLUENE

(2) C3H6O   2-PROPANONE

(3) H2O   WATER

PAVASOVIC V.
EUR.FED.CHEM.ENG.,RECOMM.SYST.LIQ.EXTR.STUD.,EDITOR:T.MISEK
(1978)

TEMPERATURE = 30.0 DEG C   TYPE OF SYSTEM = 1

EXPERIMENTAL TIE LINES IN MOLE PCT (GRAPH.INTERPOL.)

	LEFT PHASE			RIGHT PHASE	
(1)	(2)	(3)	(1)	(2)	(3)
93.091	6.164	0.745	0.010	1.507	98.483
87.001	12.029	0.970	0.021	2.838	97.141
74.388	23.776	1.836	0.022	5.427	94.551
67.207	30.132	2.661	0.045	7.126	92.828
55.384	40.437	4.178	0.072	9.856	90.072
43.618	49.562	6.820	0.152	13.127	86.721
37.073	53.631	9.296	0.212	15.544	84.244
25.990	57.896	16.114	0.428	19.941	79.631
15.683	57.186	27.130	0.942	25.504	73.554

SPECIFIC MODEL PARAMETERS IN KELVIN

		UNIQUAC		NRTL(ALPHA=.2)	
I J	AIJ	AJI		AIJ	AJI
1 2	269.90	-138.80		489.20	-301.51
1 3	987.42	-172.79		1318.8	2557.3
2 3	390.94	-86.302		377.45	210.60

R1 = 3.9228   R2 = 2.5735   R3 = 0.9200
Q1 = 2.968   Q2 = 2.336   Q3 = 1.400

MEAN DEV. BETWEEN CALC. AND EXP. CONC. IN MOLE PCT

UNIQUAC (SPECIFIC PARAMETERS)   0.66
NRTL   (SPECIFIC PARAMETERS)   0.79
UNIQUAC (COMMON PARAMETERS)   0.83

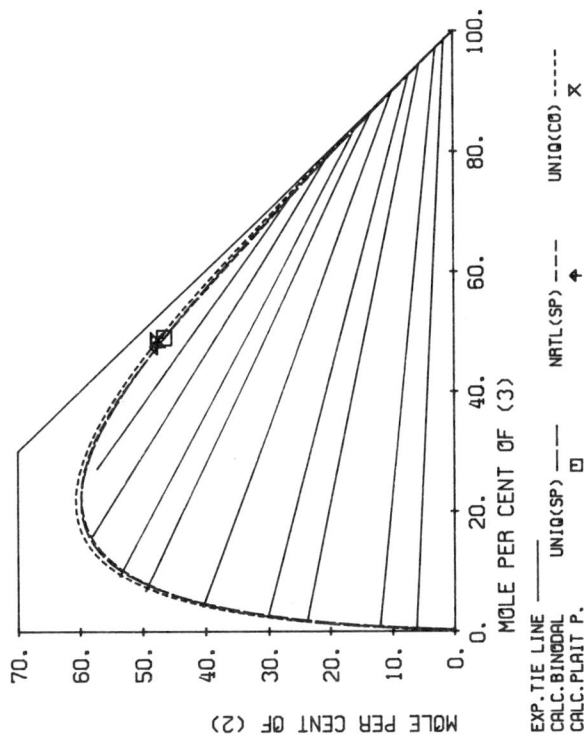

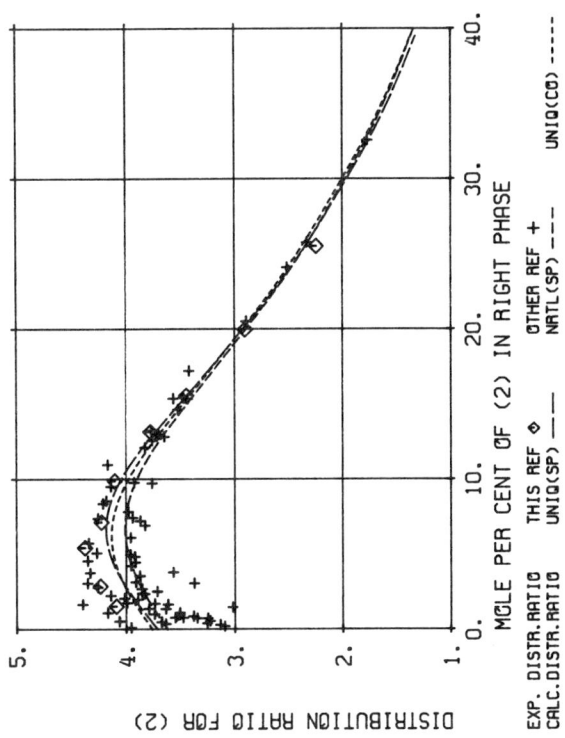

$C_3H_6O-C_7H_8$

(1) C7H8   TOLUENE
(2) C3H6O   2-PROPANONE
(3) H2O   WATER

BRANDT H.W., SCHROETER J., STRAUSS G.
EUR.FED.CHEM.ENG.,RECOMM.SYST.LIQ.EXTR.STUD.,EDITOR:T.MISEK
(1978)

TEMPERATURE = 10.0 DEG C   TYPE OF SYSTEM = 1

EXPERIMENTAL TIE LINES IN MOLE PCT (GRAPH.INTERPOL.)

	LEFT PHASE			RIGHT PHASE	
(1)	(2)	(3)	(1)	(2)	(3)
99.240	0.506	0.255	0.006	0.184	99.810
97.166	2.556	0.278	0.008	0.870	99.122
94.987	4.712	0.301	0.010	1.537	98.453
93.319	6.382	0.299	0.011	2.013	97.975
89.851	9.756	0.393	0.012	2.974	97.013
85.112	14.358	0.530	0.017	4.239	95.744
76.449	21.837	1.714	0.025	6.381	93.594
70.048	27.214	2.738	0.032	7.593	92.375
54.727	39.973	5.301	0.056	11.290	83.654
29.908	58.672	11.420	0.303	20.165	79.532

SPECIFIC MODEL PARAMETERS IN KELVIN

		UNIQUAC			NRTL(ALPHA=.2)	
I	J	AIJ	AJI		AIJ	AJI
1	2	285.44	-130.35		374.49	-142.00
1	3	729.38	379.69		975.96	1454.3
2	3	407.45	-97.166		321.28	245.82

R1 = 3.9228   R2 = 2.5735   R3 = 0.9200
Q1 = 2.968   Q2 = 2.336   Q3 = 1.400

MEAN DEV. BETWEEN CALC. AND EXP. CONC. IN MOLE PCT

UNIQUAC (SPECIFIC PARAMETERS)   0.25
NRTL   (SPECIFIC PARAMETERS)   0.35

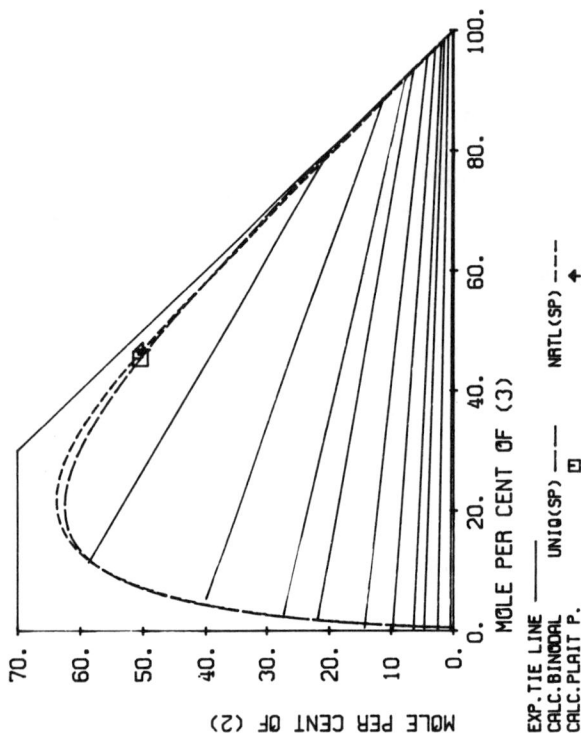

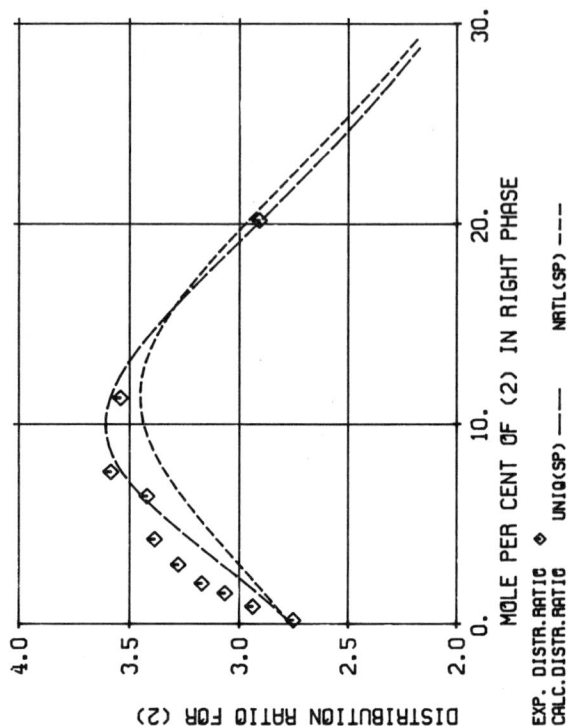

$C_3H_6O\text{-}C_7H_8$

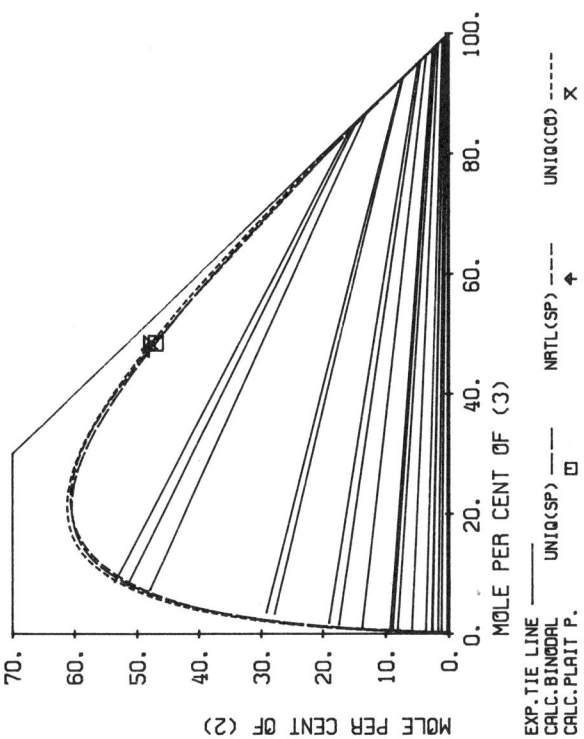

MOLE PER CENT OF (3)

MOLE PER CENT OF (2)

EXP.TIE LINE   UNIQ(SP) ⊟   NRTL(SP) ✦   UNIQ(CB) ✕
CALC.BINODAL
CALC.PLAIT P.

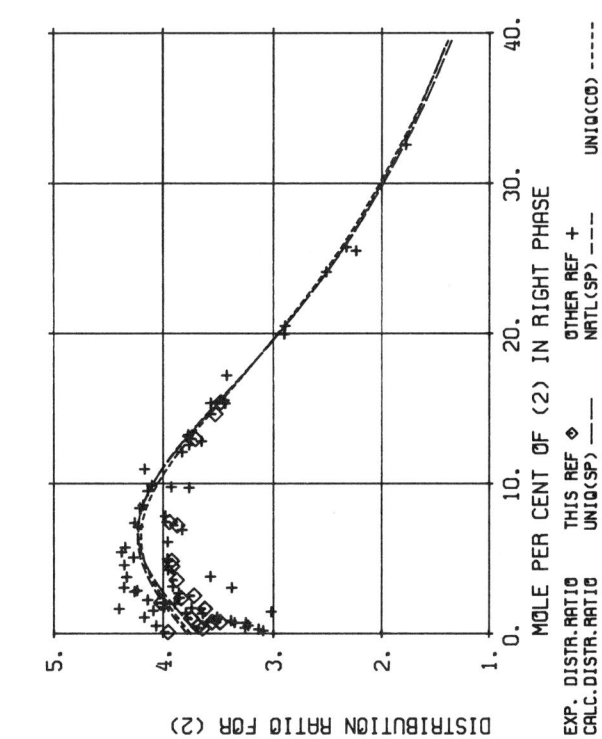

MOLE PER CENT OF (2) IN RIGHT PHASE

DISTRIBUTION RATIO FOR (2)

EXP.DISTR.RATIO ◇   THIS REF ◇   OTHER REF +   UNIQ(CB)
CALC.DISTR.RATIO   UNIQ(SP)   NRTL(SP)

(1) C7H8   TOLUENE

(2) C3H6O   2-PROPANONE

(3) H2O   WATER

BRANDT H.W., SCHROETER J., STRAUSS G.
EUR.FED.CHEM.ENG.,RECOMM.SYST.LIQ.EXTR.STUD.,EDITOR:T.MISEK
(1978)

TEMPERATURE = 20.0 DEG C   TYPE OF SYSTEM = 1

EXPERIMENTAL TIE LINES IN MOLE PCT (GRAPH.INTERPOL.)

	LEFT PHASE			RIGHT PHASE	
(1)	(2)	(3)	(1)	(2)	(3)
99.362	0.332	0.306	0.006	0.084	99.910
98.466	1.229	0.305	0.006	0.338	99.657
97.919	1.777	0.304	0.006	0.483	99.511
97.048	2.649	0.303	0.008	0.760	99.232
96.971	2.726	0.303	0.008	0.767	99.225
95.983	3.715	0.302	0.008	0.993	98.999
93.662	5.939	0.399	0.010	1.640	98.350
91.338	8.168	0.494	0.010	2.034	97.956
90.509	8.949	0.542	0.010	2.334	97.656
90.067	9.344	0.590	0.010	2.513	97.477
85.210	13.731	1.059	0.013	3.542	96.445
81.001	17.441	1.558	0.019	4.452	95.529
79.157	19.018	1.825	0.022	4.849	95.129
68.778	27.885	3.338	0.036	7.207	92.757
67.186	29.238	3.576	0.039	7.418	92.544
44.683	47.974	7.343	0.104	12.946	86.951
40.285	51.572	8.142	0.156	14.621	85.223
37.737	53.612	8.652	0.182	15.413	84.405

SPECIFIC MODEL PARAMETERS IN KELVIN

I J	UNIQUAC AIJ	AJI	NRTL(ALPHA=.2) AIJ	AJI
1 2	269.90	-138.80	489.20	-301.51
1 3	987.42	-172.79	1318.8	2557.3
2 3	390.94	-86.302	377.45	210.60

R1 = 3.9228   R2 = 2.5735   R3 = 0.9200
Q1 = 2.968   Q2 = 2.336   Q3 = 1.400

MEAN DEV. BETWEEN CALC. AND EXP. CONC. IN MOLE PCT

UNIQUAC (SPECIFIC PARAMETERS)	0.26
NRTL (SPECIFIC PARAMETERS)	0.30
UNIQUAC (COMMON PARAMETERS)	0.28

$C_3H_6O-C_7H_8$

(1) C7H8   TOLUENE

(2) C3H6O   2-PROPANONE

(3) H2O   WATER

BRANDT H.W., SCHROETER J., STRAUSS G.
EUR.FED.CHEM.ENG.,RECOMM.SYST.LIQ.EXTR.STUD.,EDITOR:T.MISEK
(1973)

TEMPERATURE = 30.0 DEG C   TYPE OF SYSTEM = 1

EXPERIMENTAL TIE LINES IN MOLE PCT (GRAPH.INTERPOL.)

	LEFT PHASE			RIGHT PHASE	
(1)	(2)	(3)	(1)	(2)	(3)
97.555	2.040	0.405	0.010	0.502	99.483
95.041	4.508	0.451	0.012	1.080	98.908
92.183	7.222	0.594	0.012	1.640	98.347
85.506	13.290	1.204	0.017	3.049	96.934
75.815	21.699	2.485	0.026	5.080	94.894
59.332	35.259	5.409	0.049	8.365	91.586
45.555	46.248	8.197	0.105	12.083	87.813
33.598	54.872	11.530	0.198	15.366	84.436

SPECIFIC MODEL PARAMETERS IN KELVIN

		UNIQUAC		NRTL(ALPHA=.2)	
I J		AIJ	AJI	AIJ	AJI
1 2		269.90	-138.80	489.20	-301.51
1 3		987.42	172.79	1318.8	2557.3
2 3		390.94	-86.302	377.45	210.60

R1 = 3.9228   R2 = 2.5735   R3 = 0.9200
Q1 = 2.968   Q2 = 2.336   Q3 = 1.400

MEAN DEV. BETWEEN CALC. AND EXP. CONC. IN MOLE PCT

UNIQUAC (SPECIFIC PARAMETERS)   0.51
NRTL   (SPECIFIC PARAMETERS)   0.62
UNIQUAC (COMMON PARAMETERS)   0.64

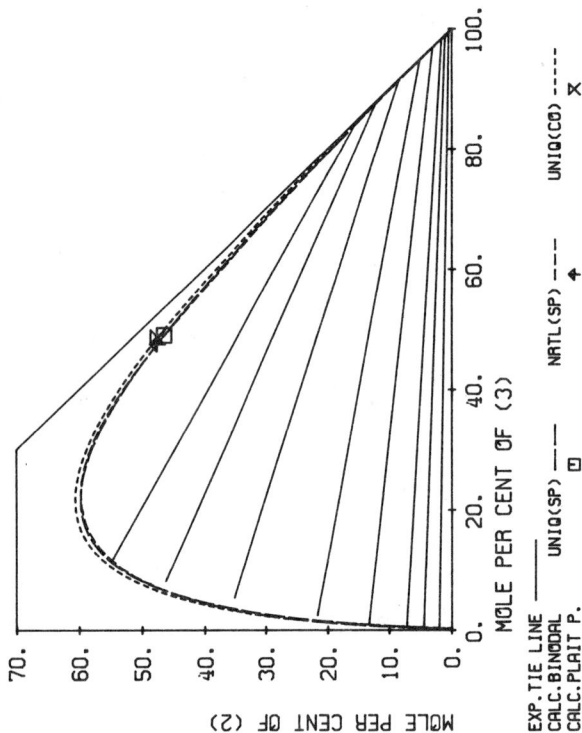

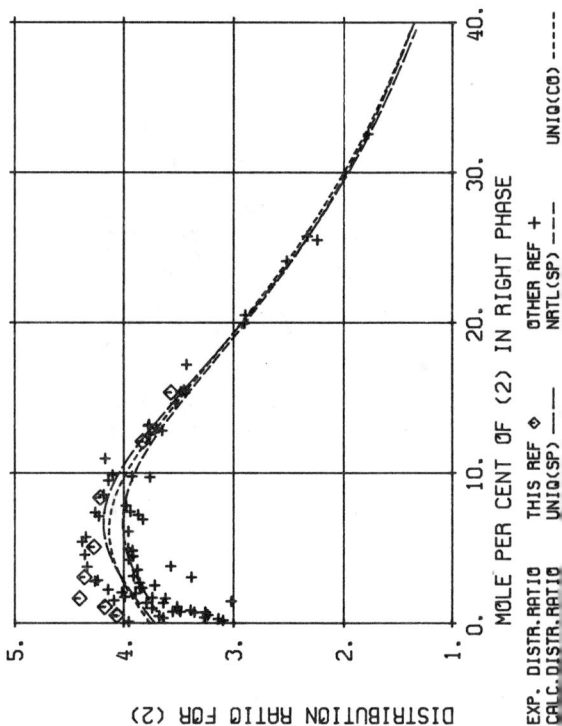

MOLE PER CENT OF (2)

MOLE PER CENT OF (3)

EXP.TIE LINE ———
CALC.BINODAL   UNIQ(SP) ▢   NRTL(SP) ——— ✦   UNIQ(CO) ———— ✕
CALC.PLAIT P.

DISTRIBUTION RATIO FOR (2)

MOLE PER CENT OF (2) IN RIGHT PHASE

EXP. DISTR.RATIO ◇   THIS REF ◇   OTHER REF +   UNIQ(CO) ————
CALC. DISTR.RATIO   UNIQ(SP) ———   NRTL(SP) ———

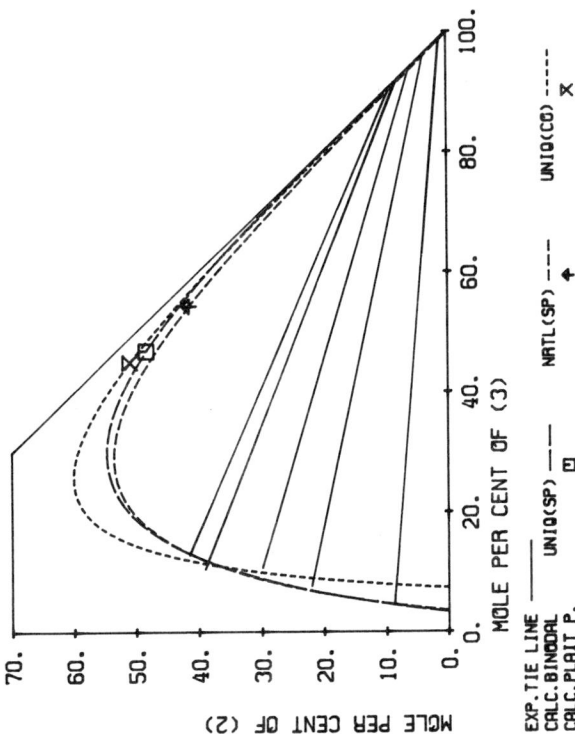

MOLE PER CENT OF (3)

MOLE PER CENT OF (2)

EXP.TIE LINE ——— UNIQ(SP) ---□--- NRTL(SP) ---+--- UNIQ(CO) -------
CALC.BINODAL ---
CALC.PLAIT P.   X

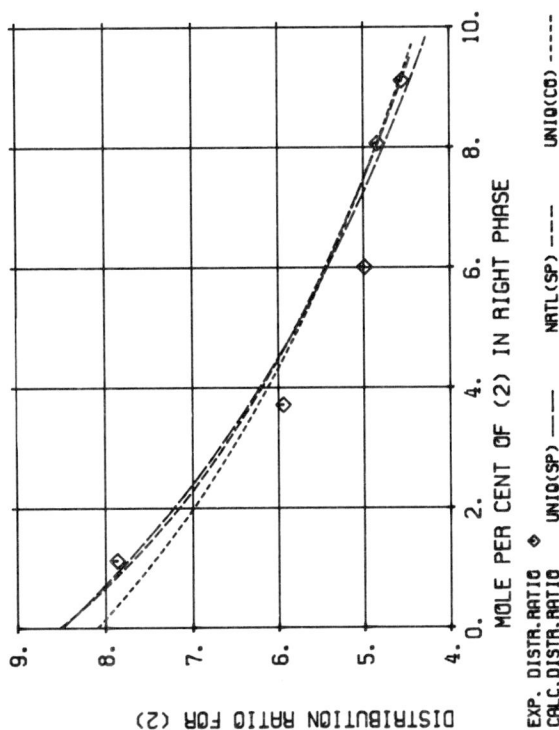

MOLE PER CENT OF (2) IN RIGHT PHASE

DISTRIBUTION RATIO FOR (2)

EXP. DISTR.RATIO ◇ UNIQ(SP) ---◇--- NRTL(SP) ----- UNIQ(CO) -----
CALC.DISTR.RATIO

(1) C7H14O2   ACETIC ACID,PENTYL ESTER

(2) C3H6O   2-PROPANONE

(3) H2O   WATER

VENKATARATNAM A., JAGANNADHA RAO R., VENKATA RAO C.
CHEM.ENG.SCI. 7(1957)102

TEMPERATURE = 30.0 DEG C   TYPE OF SYSTEM = 1

EXPERIMENTAL TIE LINES IN MOLE PCT

	LEFT PHASE			RIGHT PHASE	
(1)	(2)	(3)	(1)	(2)	(3)
97.180	0.0	2.820	0.028	0.0	99.972
86.602	8.787	4.612	0.071	1.117	98.811
70.339	22.030	7.631	0.075	3.709	96.215
59.346	29.931	10.722	0.079	6.003	93.918
50.472	39.007	10.522	0.115	8.059	91.826
45.616	41.521	12.862	0.168	9.105	90.728

SPECIFIC MODEL PARAMETERS IN KELVIN

		UNIQUAC		NRTL(ALPHA=.2)	
I J	AIJ	AJI		AIJ	AJI
1 2	6.2111	24.401		-63.316	-42.344
1 3	594.43	267.20		518.96	1435.5
2 3	49.531	77.909		67.074	465.48

R1 = 5.5018	R2 = 2.5735	R3 = 0.9200	
Q1 = 4.736	Q2 = 2.336	Q3 = 1.400	

MEAN DEV. BETWEEN CALC. AND EXP. CONC. IN MOLE PCT

UNIQUAC (SPECIFIC PARAMETERS)   0.46
NRTL (SPECIFIC PARAMETERS)   0.50
UNIQUAC (COMMON PARAMETERS)   1.32

C$_3$H$_6$O-C$_7$H$_{16}$

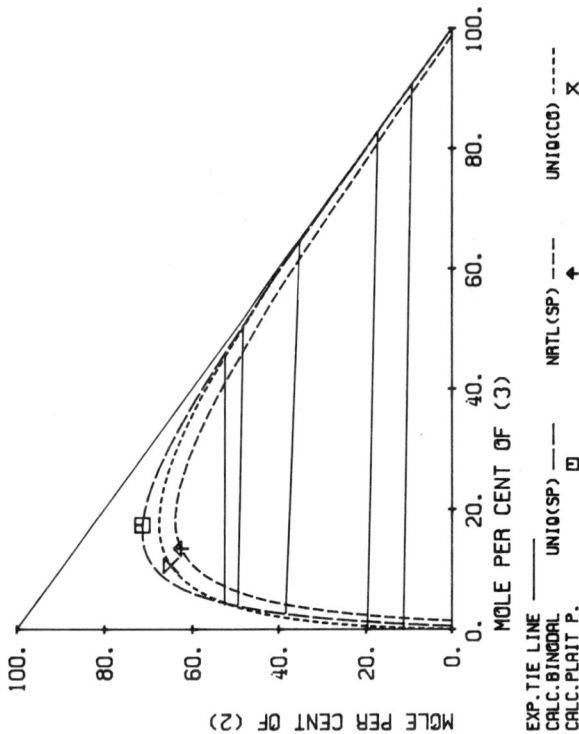

MOLE PER CENT OF (3)

MOLE PER CENT OF (2)

EXP.TIE LINE ---- UNIQ(SP) ---- □ NRTL(SP) ---- ▲ UNIQ(CC) ---- ✕
CALC.BINODAL
CALC.PLAIT P.

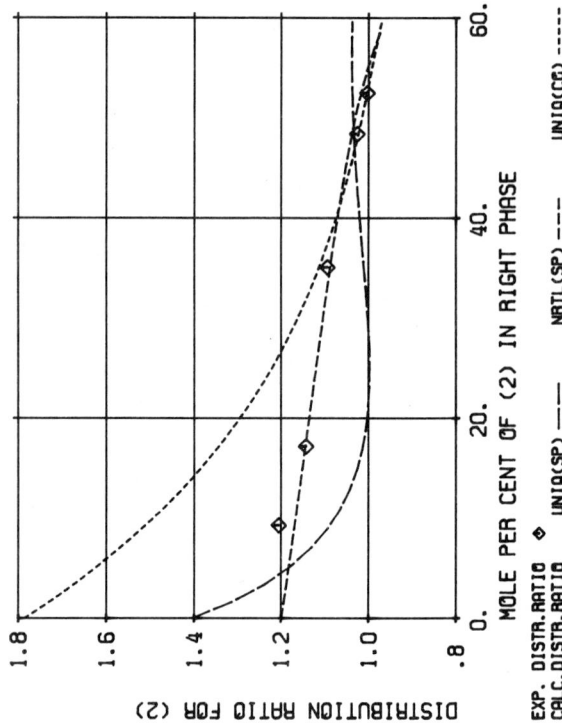

MOLE PER CENT OF (2) IN RIGHT PHASE

DISTRIBUTION RATIO FOR (2)

EXP. DISTR.RATIO ◇ UNIQ(SP) ---- NRTL(SP) ---- UNIQ(CC) ----
CALC.DISTR.RATIO

(1) C7H16	HEPTANE	
(2) C3H6O	2-PROPANONE	
(3) H2O	WATER	

TREYBAL R.E., VONDRAK O.J.:
IND.ENG.CHEM. 41(1949)1761

TEMPERATURE = 25.0 DEG C   TYPE OF SYSTEM = 1

EXPERIMENTAL TIE LINES IN MOLE PCT

	LEFT PHASE			RIGHT PHASE	
(1)	(2)	(3)	(1)	(2)	(3)
88.819	11.181	0.0	0.002	9.276	90.721
80.356	19.644	0.0	0.075	17.194	82.732
58.931	38.344	2.725	0.332	35.055	64.612
46.147	49.643	4.210	1.262	48.384	50.354
43.186	52.672	4.142	1.934	52.498	45.567

SPECIFIC MODEL PARAMETERS IN KELVIN

		UNIQUAC		NRTL(ALPHA=.2)	
I J	AIJ	AJI		AIJ	AJI
1 2	-598.18	73.918		-30.967	590.86
1 3	919.83	254.42		815.10	880.81
2 3	378.80	-869.33		479.67	72.620

R1 = 5.1742   R2 = 2.5735   R3 = 0.9200
Q1 = 4.396   Q2 = 2.336   Q3 = 1.400

MEAN DEV. BETWEEN CALC. AND EXP. CONC. IN MOLE PCT

UNIQUAC (SPECIFIC PARAMETERS)	0.84
NRTL (SPECIFIC PARAMETERS)	1.80
UNIQUAC (COMMON PARAMETERS)	0.87

$C_3H_6O\text{-}C_7H_{16}O$

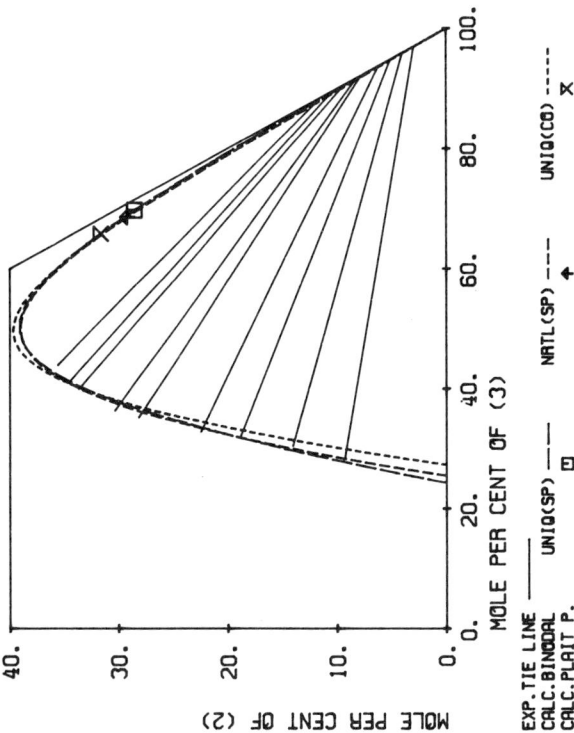

MOLE PER CENT OF (2)

MOLE PER CENT OF (3)

EXP.TIE LINE ——   UNIQ(SP) ⊡   NRTL(SP) ◆   UNIQ(CD) ⊠
CALC.BINODAL ----
CALC.PLAIT P.

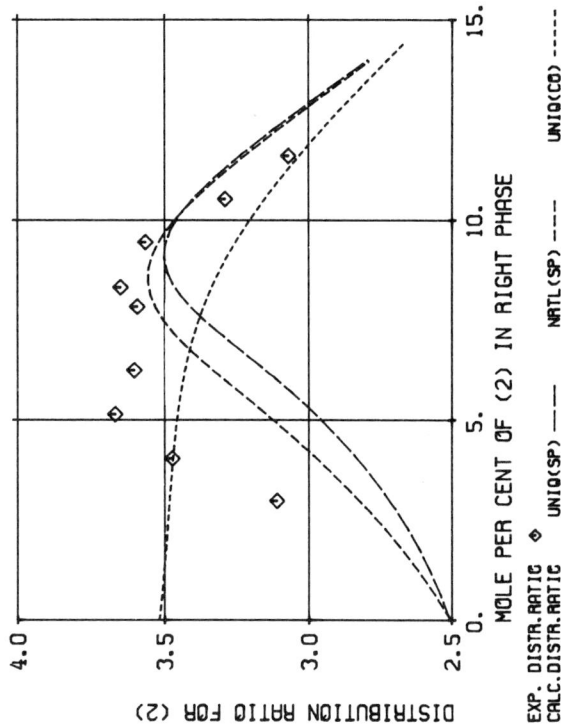

DISTRIBUTION RATIO FOR (2)

MOLE PER CENT OF (2) IN RIGHT PHASE

EXP. DISTR.RATIO ◇   UNIQ(SP) ——   NRTL(SP) ----   UNIQ(CD) -----
CALC.DISTR.RATIO

(1) C7H16O   1-HEPTANOL
(2) C3H6O   2-PROPANONE
(3) H2O   WATER

SOMASUNDARA RAO K., RAMANA RAO M.V., VENKATA RAO C.
J.SCI.IND.RES. 20B(1961)283

TEMPERATURE = 30.0 DEG C   TYPE OF SYSTEM = 1

EXPERIMENTAL TIE LINES IN MOLE PCT

	LEFT PHASE			RIGHT PHASE	
(1)	(2)	(3)	(1)	(2)	(3)
62.224	9.270	28.506	0.017	2.979	97.004
55.599	14.007	30.394	0.051	4.032	95.917
49.196	18.822	31.981	0.069	5.132	94.799
44.789	22.496	32.715	0.089	6.243	93.668
36.707	28.119	35.174	0.091	7.826	92.082
33.349	30.331	36.320	0.111	8.311	91.578
26.464	33.651	39.886	0.132	9.438	90.430
24.425	34.633	40.941	0.154	10.520	89.326
20.465	35.627	43.908	0.177	11.599	88.224

SPECIFIC MODEL PARAMETERS IN KELVIN

		UNIQUAC		NRTL(ALPHA=.2)	
I	J	AIJ	AJI	AIJ	AJI
1	2	359.22	-96.392	405.98	56.667
1	3	140.03	323.70	-46.158	2229.4
2	3	178.46	12.444	14.980	647.38

R1 = 5.4775   R2 = 2.5735   R3 = 0.9200
Q1 = 4.672   Q2 = 2.336   Q3 = 1.400

MEAN DEV. BETWEEN CALC. AND EXP. CONC. IN MOLE PCT

UNIQUAC (SPECIFIC PARAMETERS)   0.49
NRTL   (SPECIFIC PARAMETERS)   0.40
UNIQUAC (COMMON PARAMETERS)   0.66

C$_3$H$_6$O-C$_8$H$_{10}$

(1) C8H10	BENZENE,DIMETHYL(ISOMER NOT SPECIFIED)	
(2) C3H6O	2-PROPANONE	
(3) H2O	WATER	

OTHMER D.F., WHITE R.E., TRUEGER E.
IND.ENG.CHEM. 33(1941)1240

TEMPERATURE = 25.5 DEG C   TYPE OF SYSTEM = 1

EXPERIMENTAL TIE LINES IN MOLE PCT (GRAPH.INTERPOL.)

	LEFT PHASE			RIGHT PHASE	
(1)	(2)	(3)	(1)	(2)	(3)
88.441	10.613	0.946	0.026	3.097	96.877
79.283	19.442	1.275	0.036	5.707	94.257
71.673	26.745	1.582	0.049	8.194	91.757
60.033	37.793	2.174	0.066	10.561	89.373
52.281	44.961	2.758	0.086	12.843	87.071
42.713	52.767	4.519	0.101	14.005	85.894

SPECIFIC MODEL PARAMETERS IN KELVIN

		UNIQUAC		NRTL(ALPHA=.2)	
I J	AIJ	AJI		AIJ	AJI
1 2	182.13	-20.039		-34.520	457.34
1 3	765.77	386.18		1023.2	1677.3
2 3	448.11	-61.198		370.01	356.95

R1 = 4.6578   R2 = 2.5735   R3 = 0.9200
Q1 = 3.536   Q2 = 2.336   Q3 = 1.400

MEAN DEV. BETWEEN CALC. AND EXP. CONC. IN MOLE PCT

UNIQUAC (SPECIFIC PARAMETERS)	0.18
NRTL (SPECIFIC PARAMETERS)	0.18
UNIQUAC (COMMON PARAMETERS)	1.34

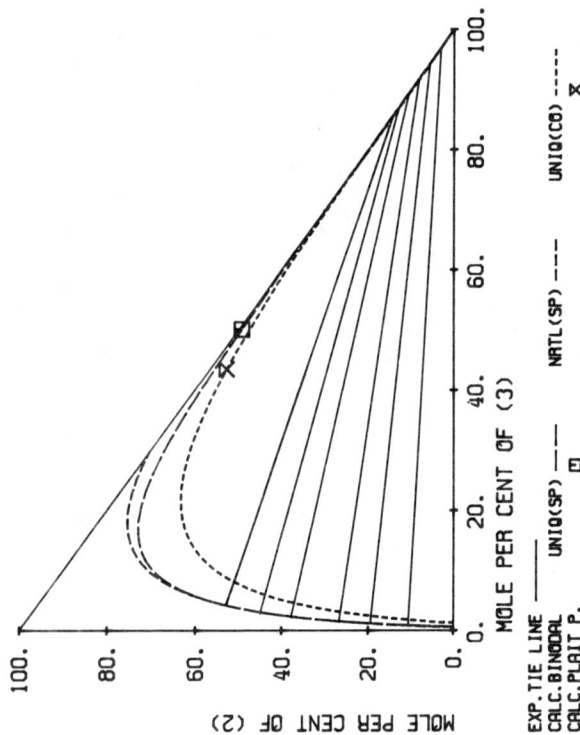

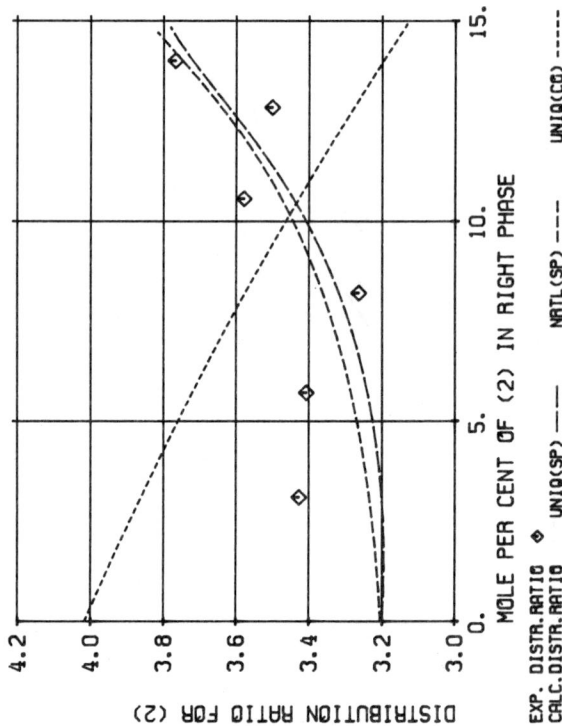

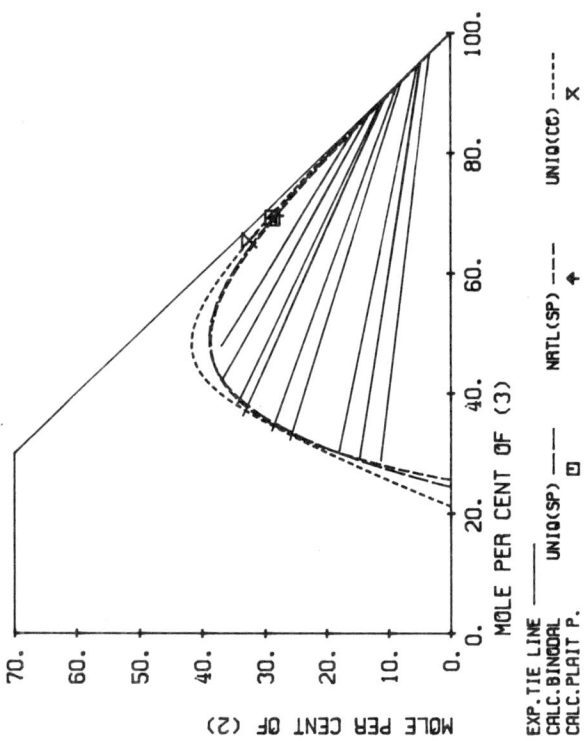

| | EXP.TIE LINE | CALC.BINODAL | CALC.PLAIT P. | UNIQ(SP) □ | NRTL(SP) ♦ | UNIQ(CC) ✕ |

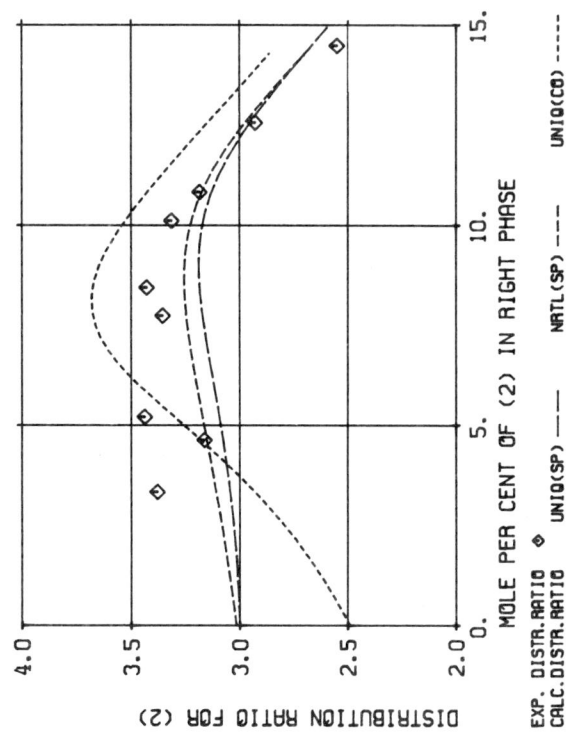

| | EXP. DISTR.RATIO ◇ | UNIQ(SP) | NRTL(SP) | UNIQ(CC) |
| CALC.DISTR.RATIO | | | | |

(1) C8H18O   1-OCTANOL

(2) C3H6O   2-PROPANONE

(3) H2O   WATER

SOMASUNDARA RAO K., RAMANA RAO M.V., VENKATA RAO C.
J.SCI.IND.RES. 20B(1961)283

TEMPERATURE = 30.0 DEG C   TYPE OF SYSTEM = 1

EXPERIMENTAL TIE LINES IN MOLE PCT

| | LEFT PHASE | | | RIGHT PHASE | |
(1)	(2)	(3)	(1)	(2)	(3)
59.850	11.273	28.877	0.015	3.335	96.650
56.003	14.633	29.364	0.031	4.626	95.343
51.647	17.893	30.459	0.046	5.205	94.749
41.838	25.951	32.211	0.065	7.733	92.202
37.334	28.930	33.736	0.083	8.435	91.482
30.204	33.525	36.270	0.085	10.112	89.803
27.368	34.451	38.180	0.086	10.822	89.092
20.869	36.739	42.391	0.089	12.552	87.359
15.200	36.916	47.883	0.184	14.476	85.340

SPECIFIC MODEL PARAMETERS IN KELVIN

| | | UNIQUAC | | NRTL(ALPHA=.2) | |
I	J	AIJ	AJI	AIJ	AJI
1	2	328.40	-103.96	173.25	210.25
1	3	190.26	238.90	-46.259	2254.0
2	3	73.831	71.197	-113.99	764.10

R1 = 6.1519   R2 = 2.5735   R3 = 0.9200
Q1 = 5.212   Q2 = 2.336   Q3 = 1.400

MEAN DEV. BETWEEN CALC. AND EXP. CONC. IN MOLE PCT

UNIQUAC (SPECIFIC PARAMETERS)   0.55
NRTL   (SPECIFIC PARAMETERS)   0.47
UNIQUAC (COMMON PARAMETERS)   1.00

C₃H₆O-C₉H₁₂

(1) C9H12   BENZENE, ISOPROPYL
(2) C3H6O   2-PROPANONE
(3) H2O   WATER

POP A., WEISS G., VANYOLOS A.
STUD.UNIV.BABES-BOLYAI.SER.CHEM. (1967)2,49

TEMPERATURE = 20.0 DEG C   TYPE OF SYSTEM = 1

EXPERIMENTAL TIE LINES IN MOLE PCT

	LEFT PHASE			RIGHT PHASE	
(1)	(2)	(3)	(1)	(2)	(3)
93.336	6.021	0.643	0.016	1.957	98.028
89.212	9.533	1.254	0.022	3.015	96.962
78.876	18.767	2.357	0.027	5.341	94.632
70.630	26.019	3.351	0.035	7.799	92.166
50.868	43.313	5.818	0.059	14.037	85.904
41.597	51.206	7.197	0.082	16.394	83.524
36.459	54.960	8.581	0.172	18.957	80.871
26.685	61.660	11.655	0.430	25.618	73.952
19.239	64.538	16.222	1.022	33.119	65.859
16.954	64.080	18.966	1.405	35.773	62.822
15.760	63.712	20.528	1.677	37.499	60.824

SPECIFIC MODEL PARAMETERS IN KELVIN

		UNIQUAC		NRTL(ALPHA=.2)	
I	J	AIJ	AJI	AIJ	AJI
1	2	98.664	5.8622	-102.80	154.35
1	3	766.92	309.04	778.69	1462.2
2	3	202.66	11.000	325.60	185.95

R1 = 5.2708   R2 = 2.5735   R3 = 0.9200
Q1 = 4.044   Q2 = 2.336   Q3 = 1.400

MEAN DEV. BETWEEN CALC. AND EXP. CONC. IN MOLE PCT

UNIQUAC (SPECIFIC PARAMETERS)   0.56
NRTL   (SPECIFIC PARAMETERS)   0.65
UNIQUAC (COMMON PARAMETERS)   1.20

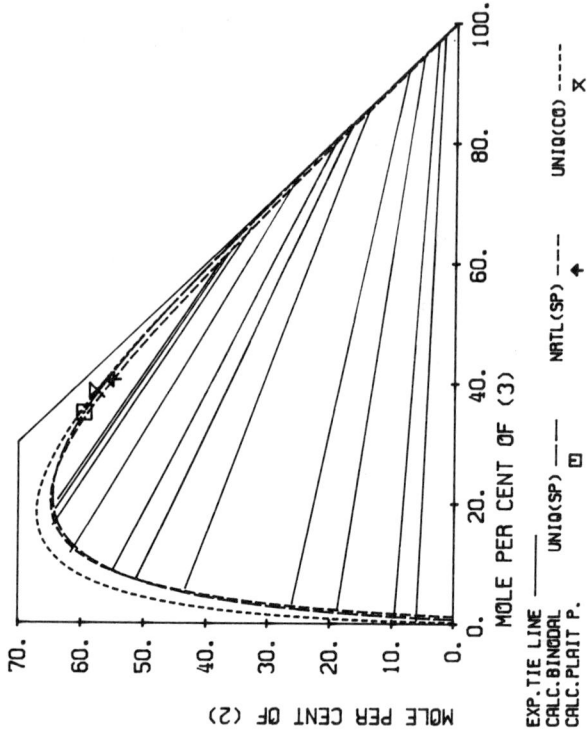

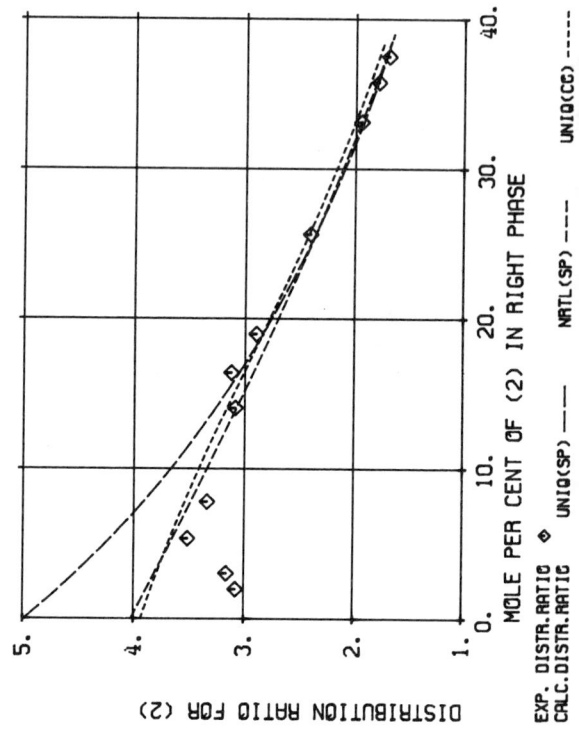

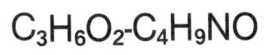

(1) H2O   WATER
----
(2) C4H9NO   ACETIC ACID,AMIDE,N,N-DIMETHYL
----
(3) C3H6O2   FORMIC ACID,ETHYL ESTER

KRUPATKIN I.L., AVTONOMOVA E.N.
IZV.VYSSH.UCHEBN.ZAVED.KHIM.KHIM.TEKHNOL. 15(1972)1480

TEMPERATURE = 25.0 DEG C   TYPE OF SYSTEM = 1

EXPERIMENTAL TIE LINES IN MOLE PCT

	LEFT PHASE			RIGHT PHASE	
(1)	(2)	(3)	(1)	(2)	(3)
94.600	2.200	3.200	20.800	1.100	78.100
91.900	4.200	3.900	23.700	3.200	73.100
88.500	6.700	4.800	25.600	4.800	68.600
83.100	10.100	6.800	34.200	7.700	58.100
78.000	12.500	9.500	42.800	10.300	46.900

SPECIFIC MODEL PARAMETERS IN KELVIN

		UNIQUAC		NRTL(ALPHA=.2)	
I J	AIJ	AJI		AIJ	AJI
1 2	21.762	-279.74		-527.17	183.61
1 3	222.12	-226.09		1163.6	78.532
2 3	23.303	-91.245		-329.29	687.97

R1 = 0.9200   R2 = 3.5332   R3 = 2.8042
Q1 = 1.400   Q2 = 2.968   Q3 = 2.576

MEAN DEV. BETWEEN CALC. AND EXP. CONC. IN MOLE PCT

UNIQUAC (SPECIFIC PARAMETERS)   0.63
NRTL (SPECIFIC PARAMETERS)   0.51
UNIQUAC (COMMON PARAMETERS)   2.27

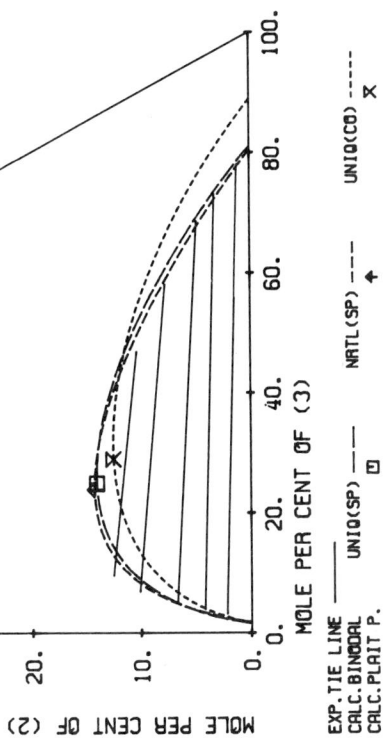

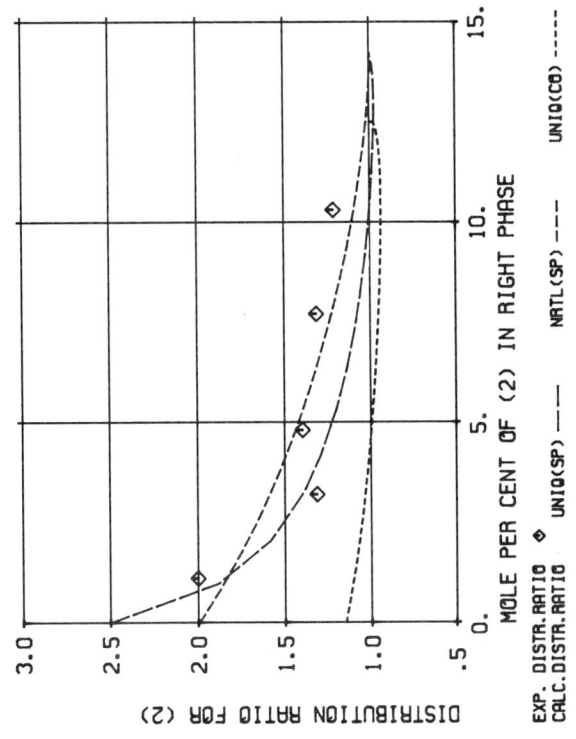

C$_3$H$_6$O$_2$-C$_4$H$_8$O$_2$

(1) C4H8O2   ACETIC ACID,ETHYL ESTER

(2) C3H6O2   PROPANOIC ACID

(3) H2O   WATER

JAYA RAMA RAO G., VENTAKA RAO C.
J.SCI.IND.RES.14B(1955)444

TEMPERATURE = 30.0 DEG C   TYPE OF SYSTEM = 1

EXPERIMENTAL TIE LINES IN MOLE PCT

	LEFT PHASE			RIGHT PHASE	
(1)	(2)	(3)	(1)	(2)	(3)
67.024	8.727	24.249	1.810	0.892	97.298
57.233	12.270	30.497	1.896	1.399	96.705
43.601	15.876	40.523	1.978	2.186	95.836
30.263	17.480	52.257	2.144	3.271	94.584
23.260	17.404	59.335	2.533	4.117	93.350

SPECIFIC MODEL PARAMETERS IN KELVIN

		UNIQUAC		NRTL(ALPHA=.2)	
I	J	AIJ	AJI	AIJ	AJI
1	2	-187.56	74.232	560.66	-864.63
1	3	-421.98	92.058	112.46	1302.3
2	3	-182.28	248.98	-30.207	-279.68

R1 = 3.4786   R2 = 2.8768   R3 = 0.9200
Q1 = 3.116   Q2 = 2.612   Q3 = 1.400

MEAN DEV. BETWEEN CALC. AND EXP. CONC. IN MOLE PCT

UNIQUAC (SPECIFIC PARAMETERS)   0.27
NRTL   (SPECIFIC PARAMETERS)   1.87
UNIQUAC (COMMON PARAMETERS)   1.18

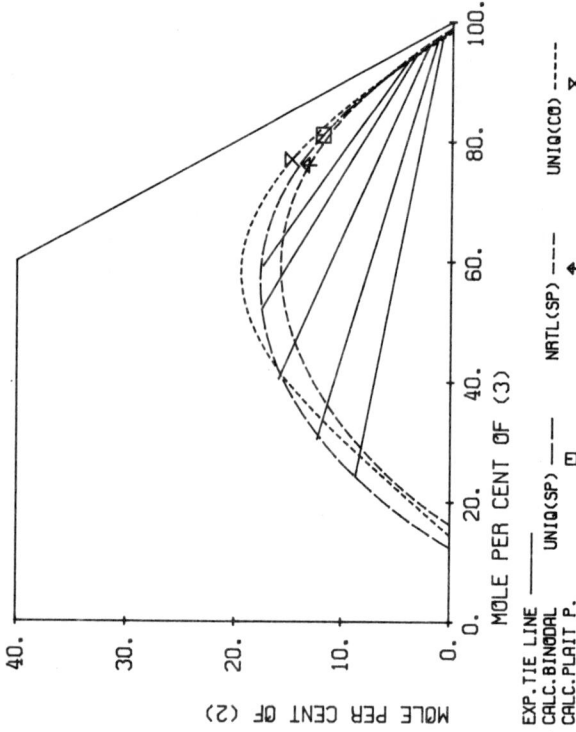

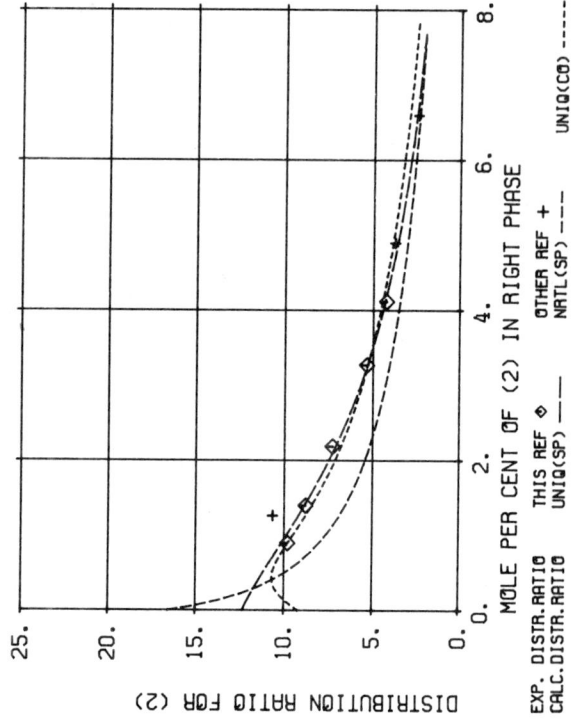

$C_3H_6O_2\text{-}C_4H_8O_2$

```
(1) C4H8O2     ACETIC ACID,ETHYL ESTER
--------------------------------------
(2) C3H6O2     PROPANOIC ACID
--------------------------------------
(3) H2O        WATER
--------------------------------------

MIRADA LILLO R., GONZALES TRIGO G.
AN.R.SOC.ESP.FIS.QUIM. 56B(1960)217

TEMPERATURE = 20.0 DEG C     TYPE OF SYSTEM = 1

EXPERIMENTAL TIE LINES IN MOLE PCT
--------------------------------------
            LEFT PHASE        RIGHT PHASE
          (1)    (2)    (3)    (1)    (2)    (3)
--------------------------------------
        53.892 13.420 32.688  1.647  1.261 97.092
        24.530 18.467 56.953  2.727  4.897 92.376
        15.835 16.256 67.909  3.882  6.597 89.521

SPECIFIC MODEL PARAMETERS IN KELVIN
--------------------------------------
               UNIQUAC            NRTL(ALPHA=.2)
  I  J     AIJ      AJI         AIJ       AJI
--------------------------------------
  1  2  -187.56   74.232      560.66   -864.63
  1  3   421.98   92.058      112.46   1302.3
  2  3  -182.28  248.98      -30.207  -279.68

  R1 = 3.4786   R2 = 2.8768   R3 = 0.9200
  Q1 = 3.116    Q2 = 2.612    Q3 = 1.400

MEAN DEV. BETWEEN CALC. AND EXP. CONC. IN MOLE PCT
--------------------------------------
UNIQUAC (SPECIFIC PARAMETERS)        0.35
NRTL   (SPECIFIC PARAMETERS)         1.99
UNIQUAC (COMMON PARAMETERS)          1.63
```

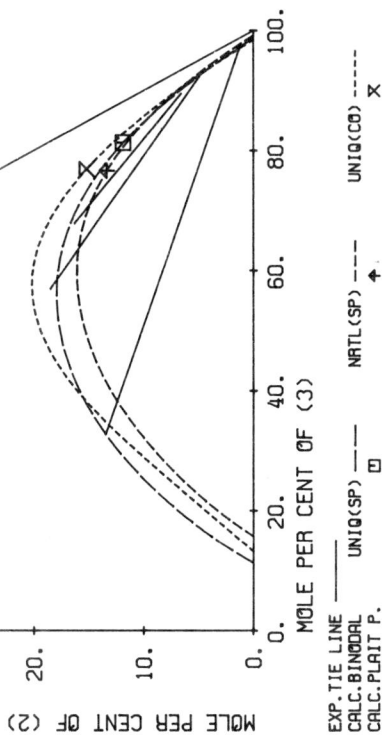

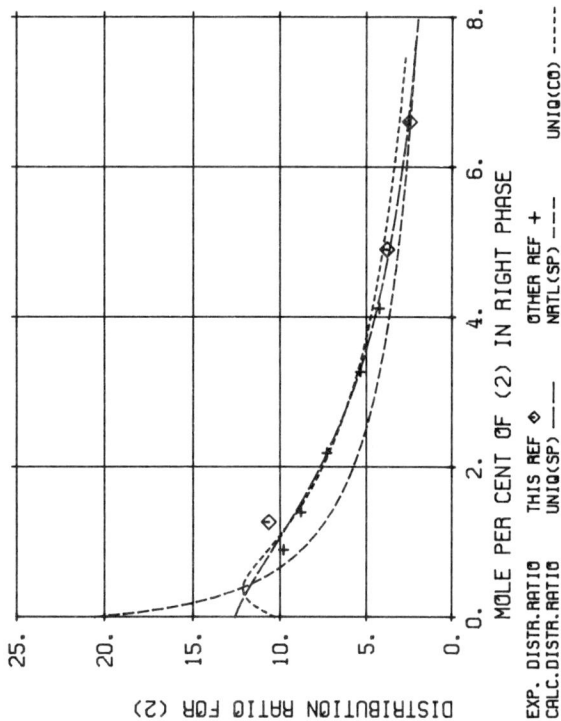

C$_3$H$_6$O$_2$-C$_5$H$_4$O$_2$

(1) C5H4O2   FURFURAL
(2) C3H6O2   PROPANOIC ACID
(3) H2O   WATER

HERIC E.L., RUTLEDGE R.M.
J.CHEM.ENG.DATA 5(1960)272

TEMPERATURE = 25.0 DEG C   TYPE OF SYSTEM = 1

EXPERIMENTAL TIE LINES IN MOLE PCT

	LEFT PHASE			RIGHT PHASE	
(1)	(2)	(3)	(1)	(2)	(3)
67.460	3.834	28.705	1.808	0.560	97.633
56.991	6.900	36.110	1.997	1.118	96.885
45.364	10.046	44.590	2.328	1.919	95.752
35.351	11.852	52.797	2.807	2.750	94.432
25.910	12.527	61.563	3.655	3.993	92.352

SPECIFIC MODEL PARAMETERS IN KELVIN

	UNIQUAC		NRTL(ALPHA=.2)	
I J	AIJ	AJI	AIJ	AJI
1 2	-230.10	93.528	435.49	-803.01
1 3	156.76	106.54	-2.6034	1281.3
2 3	-232.72	282.57	-182.13	-432.61

R1 = 3.1680   R2 = 2.8768   R3 = 0.9200
Q1 = 2.484   Q2 = 2.612   Q3 = 1.400

MEAN DEV. BETWEEN CALC. AND EXP. CONC. IN MOLE PCT

UNIQUAC (SPECIFIC PARAMETERS)   0.46
NRTL   (SPECIFIC PARAMETERS)   1.05
UNIQUAC (COMMON PARAMETERS)   1.61

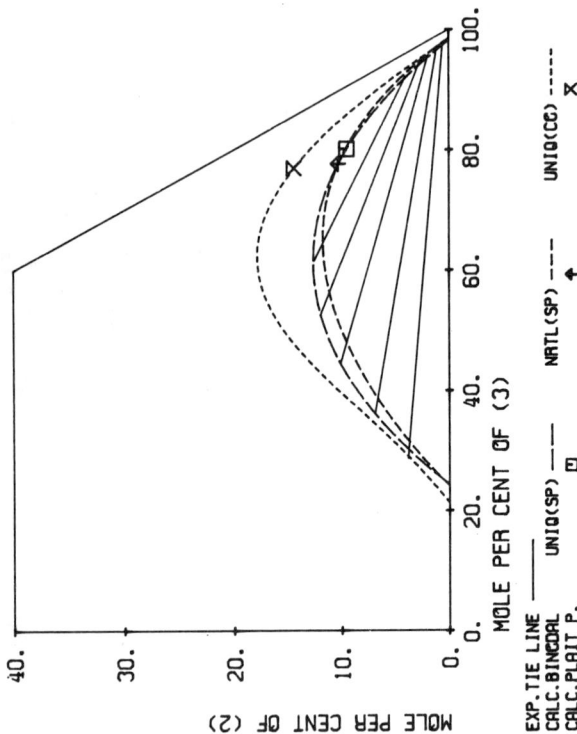

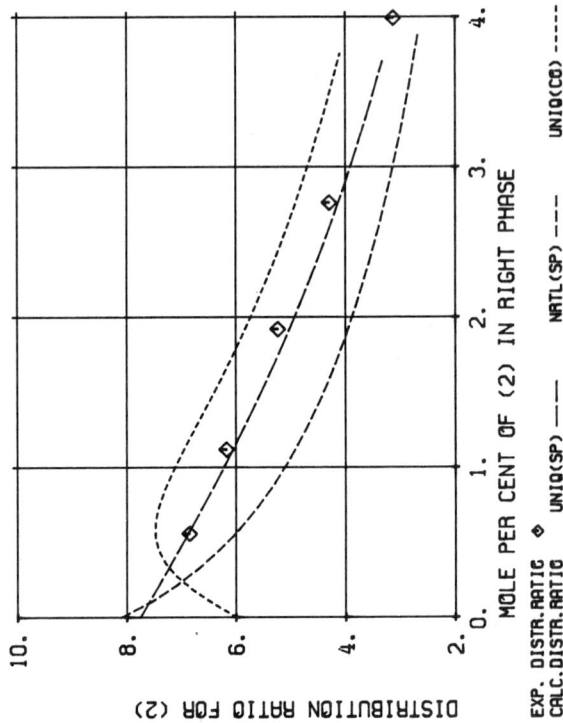

$C_3H_6O_2\text{-}C_5H_4O_2$

(1) C5H4O2   FURFURAL
----
(2) C3H6O2   PROPANOIC ACID
----
(3) H2O   WATER

HERIC E.L., RUTLEDGE R.M.
J.CHEM.ENG.DATA 5(1960)272

TEMPERATURE = 35.0 DEG C   TYPE OF SYSTEM = 1

EXPERIMENTAL TIE LINES IN MOLE PCT

               LEFT PHASE   RIGHT PHASE
          (1)   (3)   (1)   (2)   (3)

        64.216   31.989   1.986   0.564   97.450
        54.438   38.882   2.204   1.127   96.668
        45.359   45.729   2.482   1.722   95.796
        36.983   52.467   2.928   2.396   94.676
        26.522   62.234   3.883   3.545   92.572

SPECIFIC MODEL PARAMETERS IN KELVIN

                UNIQUAC   NRTL(ALPHA=.2)
I J   AIJ   AJI   AIJ   AJI

1 2   -246.81   84.926   382.92   -854.69
1 3   144.34   108.11   -17.455   1241.8
2 3   -246.77   292.74   -244.47   -421.29

R1 = 3.1680   R2 = 2.8768   R3 = 0.9200
Q1 = 2.484   Q2 = 2.612   Q3 = 1.400

MEAN DEV. BETWEEN CALC. AND EXP. CONC. IN MOLE PCT

UNIQUAC (SPECIFIC PARAMETERS)   0.43
NRTL   (SPECIFIC PARAMETERS)   1.15

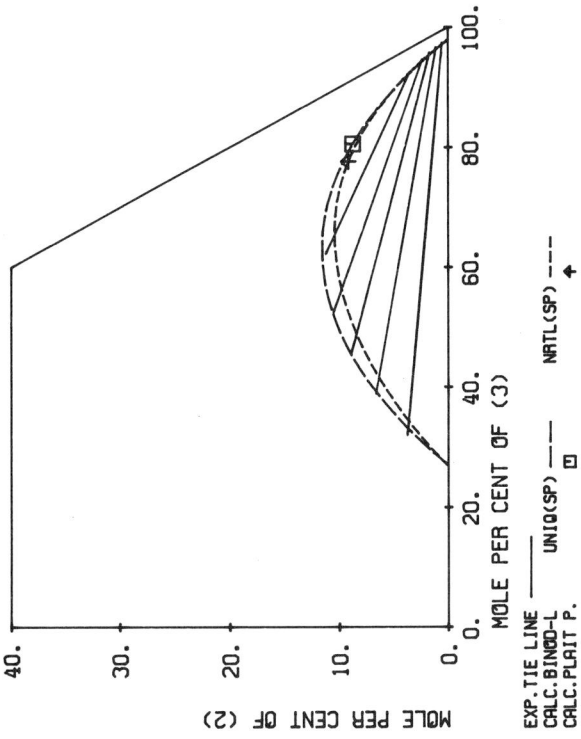

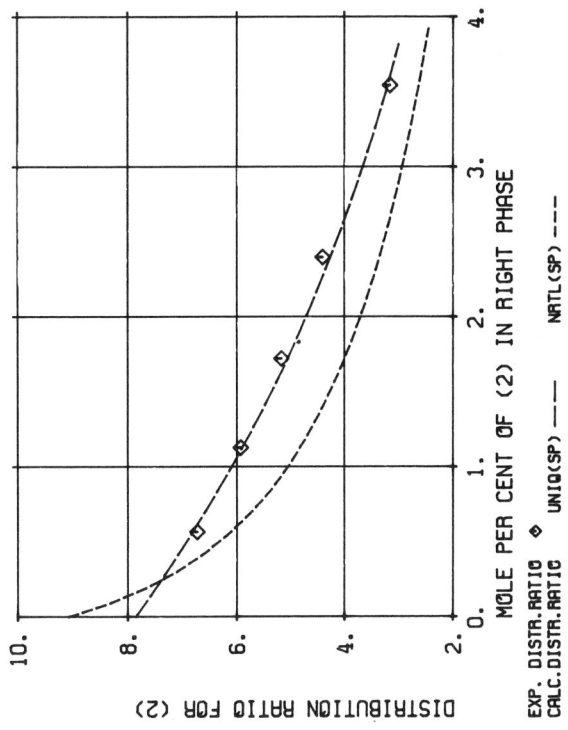

MOLE PER CENT OF (2)
MOLE PER CENT OF (3)

EXP.TIE LINE ――   UNIQ(SP) □   NRTL(SP) ----
CALC.BINOD-L
CALC.PLAIT P.   NRTL(SP) ---- ✦

DISTRIBUTION RATIO FOR (2)
MOLE PER CENT OF (2) IN RIGHT PHASE

EXP. DISTR.RATIO ◇   UNIQ(SP) ◇   NRTL(SP) ----
CALC.DISTR.RATIO ――

$C_3H_6O_2$-$C_5H_{10}O_2$

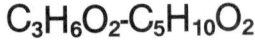

(1) C5H1002   BUTANOIC ACID,METHYL ESTER
(2) C3H602   PROPANOIC ACID
(3) H20   WATER

SITARAMA MURTY N.; SUBRAHMANYAM V., DAKSHINA MURTY P.
J.CHEM.ENG.DATA 11(1966)335

TEMPERATURE = 30.0 DEG C   TYPE OF SYSTEM = 1

EXPERIMENTAL TIE LINES IN MOLE PCT

	LEFT PHASE			RIGHT PHASE	
(1)	(2)	(3)	(1)	(2)	(3)
63.949	14.335	21.717	0.473	1.617	97.910
56.637	18.826	24.536	0.521	2.234	97.245
49.961	21.276	28.763	0.570	2.789	96.642
43.429	23.492	33.078	0.597	3.239	96.164
40.222	24.754	35.025	0.604	3.667	95.729
36.591	25.564	37.844	0.654	4.113	95.233

SPECIFIC MODEL PARAMETERS IN KELVIN

		UNIQUAC		NRTL(ALPHA=.2)	
I	J	AIJ	AJI	AIJ	AJI
1	2	-31.720	88.804	179.44	-249.17
1	3	409.64	125.73	125.19	1539.6
2	3	-94.462	254.19	-401.18	1056.1

R1 = 4.1530   R2 = 2.8768   R3 = 0.9200
Q1 = 3.656   Q2 = 2.612   Q3 = 1.400

MEAN DEV. BETWEEN CALC. AND EXP. CONC. IN MOLE PCT

UNIQUAC (SPECIFIC PARAMETERS)   0.24
NRTL   (SPECIFIC PARAMETERS)   0.55
UNIQUAC (COMMON PARAMETERS)   0.43

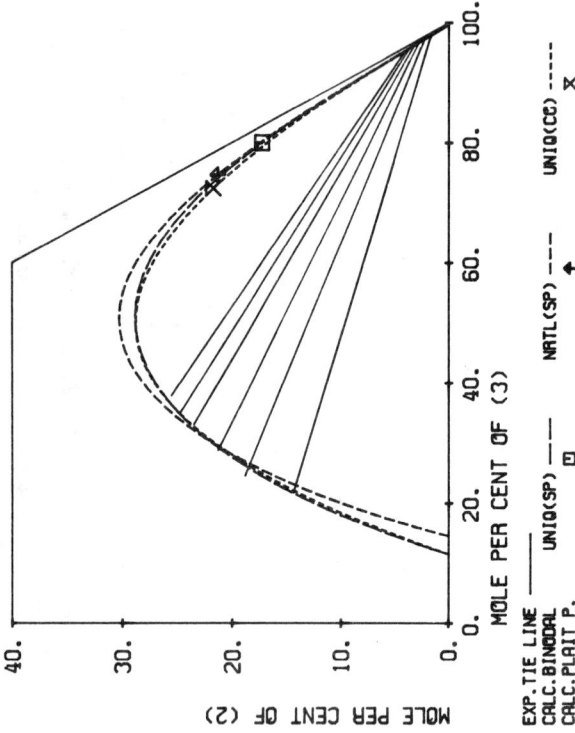

MOLE PER CENT OF (3)

MOLE PER CENT OF (2)

EXP.TIE LINE ——— UNIQ(SP) ⊡   NRTL(SP) ✦   UNIQ(CC) ⤬
CALC.BINODAL
CALC.PLAIT P.

DISTRIBUTION RATIO FOR (2)

MOLE PER CENT OF (2) IN RIGHT PHASE

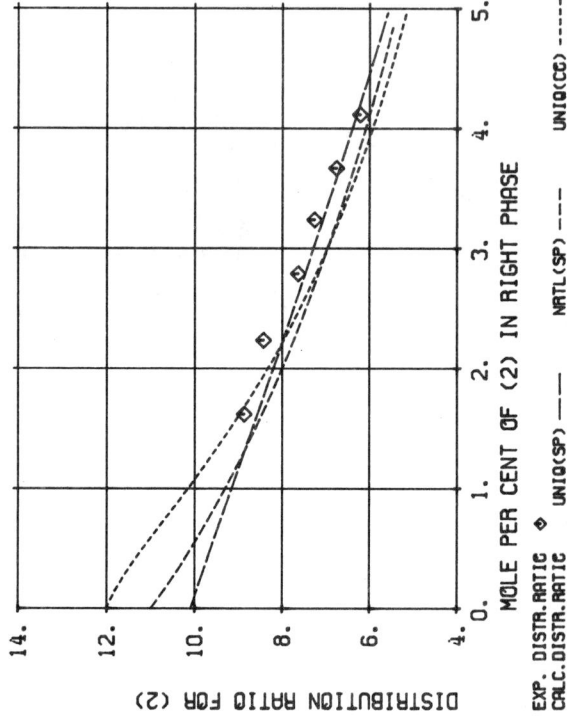

EXP. DISTR.RATIO ◇   UNIQ(SP) ———   NRTL(SP) ———   UNIQ(CC) ———
CALC.DISTR.RATIO

(1) C5H10O2   PROPANOIC ACID,ETHYL ESTER
----
(2) C3H6O2   PROPANOIC ACID
----
(3) H2O   WATER

JAYA RAMA RAO G., VENTAKA RAO C.
J.SCI.IND.RES. 14B(1955)444

TEMPERATURE = 30.0 DEG C   TYPE OF SYSTEM = 1

EXPERIMENTAL TIE LINES IN MOLE PCT

        LEFT PHASE   RIGHT PHASE
      (1)   (2)   (3)   (1)   (2)   (3)

    79.904   6.527 13.568   0.385   0.711 98.904
    65.718 14.607 19.675   0.444   1.601 97.955
    54.234 20.006 25.760   0.533   2.364 97.103
    42.335 24.120 33.546   0.619   3.312 96.069
    32.991 26.227 40.783   0.688   4.270 95.042
    25.874 26.835 47.292   0.848   5.221 93.931
    19.057 26.062 54.881   0.998   6.608 92.393

SPECIFIC MODEL PARAMETERS IN KELVIN

              UNIQUAC   NRTL(ALPHA=.2)
 I J   AIJ   AJI   AIJ   AJI

 1 2 -75.854   199.12   0.38050E-01 -16.634
 1 3   455.12   131.88   277.76   1413.6
 2 3 -120.88   302.15   -582.28   1420.1

 R1 = 4.1530   R2 = 2.8768   R3 = 0.9200
 Q1 = 3.656   Q2 = 2.612   Q3 = 1.400

MEAN DEV. BETWEEN CALC. AND EXP. CONC. IN MOLE PCT

UNIQUAC (SPECIFIC PARAMETERS)   0.30
NRTL   (SPECIFIC PARAMETERS)   0.30
UNIQUAC (COMMON PARAMETERS)   1.06

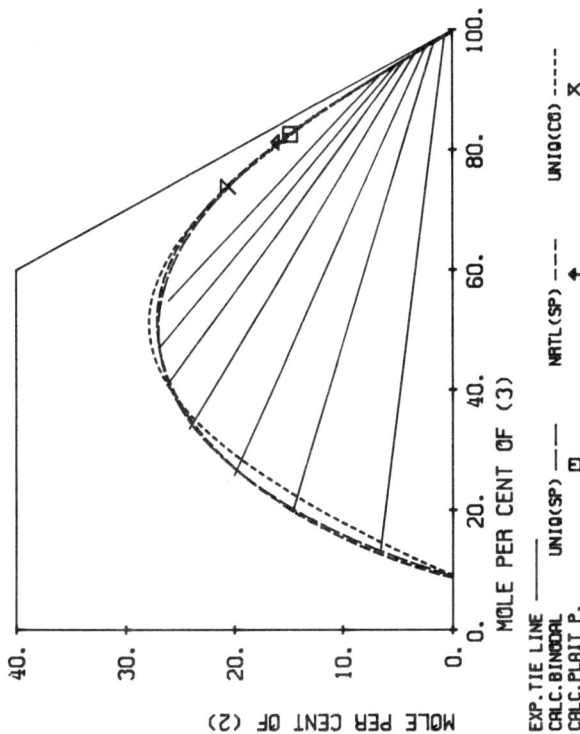

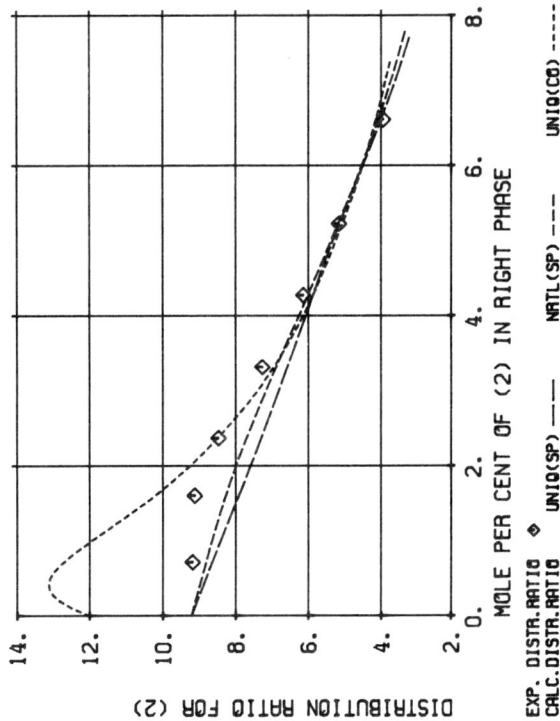

C₃H₆O₂-C₆H₆

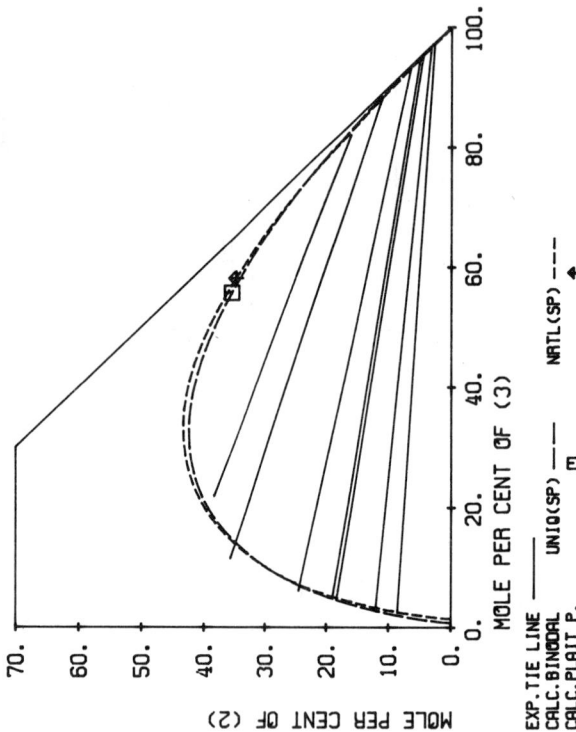

MOLE PER CENT OF (3)

MOLE PER CENT OF (2)

EXP.TIE LINE ——— UNIQ(SP) ——— NRTL(SP) - - - -
CALC.BINODAL
CALC.PLAIT P.

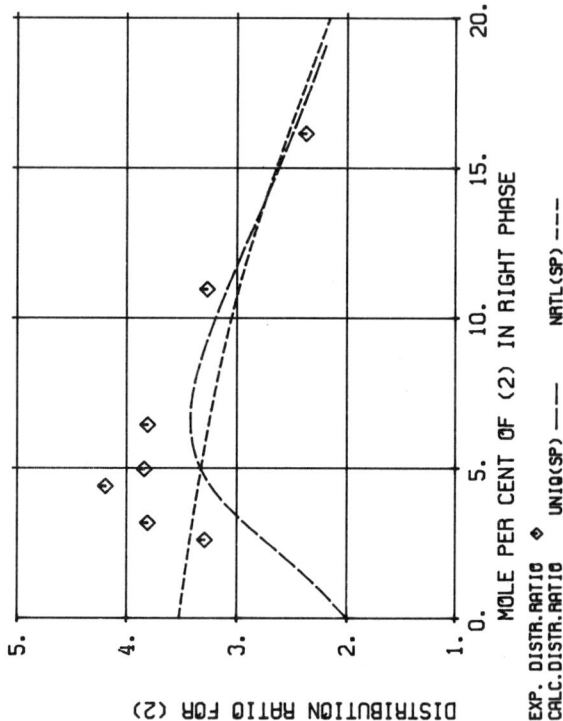

MOLE PER CENT OF (2) IN RIGHT PHASE

DISTRIBUTION RATIO FOR (2)

EXP. DISTR.RATIO ◇ UNIQ(SP) ——— NRTL(SP) - - - -
CALC.DISTR.RATIO

(1) C6H6   BENZENE
----
(2) C3H6O2   PROPANOIC ACID
----
(3) H2O   WATER

BIANCANI M., DE FILIPPO D.
GAZZ.CHIM.ITAL. 88(1958)1202

TEMPERATURE = 40.0 DEG C   TYPE OF SYSTEM = 1

EXPERIMENTAL TIE LINES IN MOLE PCT

	LEFT PHASE			RIGHT PHASE	
(1)	(2)	(3)	(1)	(2)	(3)
88.877	8.584	2.539	0.100	2.608	97.292
84.557	12.085	3.358	0.166	3.174	96.660
76.985	18.457	4.557	0.185	4.404	95.412
75.961	19.085	4.954	0.228	4.972	94.800
69.392	24.491	6.116	0.251	6.434	93.315
52.763	35.806	11.430	0.469	10.956	88.574
39.754	38.411	21.835	1.764	16.150	82.086

SPECIFIC MODEL PARAMETERS IN KELVIN

		UNIQUAC		NRTL(ALPHA=.2)	
I	J	AIJ	AJI	AIJ	AJI
1	2	457.76	-239.76	273.88	-128.45
1	3	672.86	453.55	790.21	1495.0
2	3	348.52	-146.71	-231.93	777.59

R1 = 3.1878   R2 = 2.8758   R3 = 0.9200
Q1 = 2.400   Q2 = 2.612   Q3 = 1.400

MEAN DEV. BETWEEN CALC. AND EXP. CONC. IN MOLE PCT

UNIQUAC (SPECIFIC PARAMETERS)   0.67
NRTL   (SPECIFIC PARAMETERS)   0.75

$C_3H_6O_2\text{-}C_6H_6$

(1) C6H6   BENZENE

(2) C3H6O2   PROPANOIC ACID

(3) H2O   WATER

BIANCANI M., DE FILIPPO D.
GAZZ.CHIM.ITAL. 88(1958)1202

TEMPERATURE = 60.0 DEG C   TYPE OF SYSTEM = 1

EXPERIMENTAL TIE LINES IN MOLE PCT

	LEFT PHASE			RIGHT PHASE	
(1)	(2)	(3)	(1)	(2)	(3)
80.367	12.316	7.316	0.228	2.873	96.899
74.136	16.681	9.183	0.349	4.901	94.751
55.165	29.434	15.402	0.868	10.913	88.219
45.516	33.779	20.705	2.122	15.711	82.167
36.292	34.997	28.710	5.078	21.145	73.776

SPECIFIC MODEL PARAMETERS IN KELVIN

		UNIQUAC		NRTL(ALPHA=.2)	
I J		AIJ	AJI	AIJ	AJI
1 2		7.9077	-76.043	199.80	-435.64
1 3		501.24	152.37	510.95	1566.1
2 3		131.55	-94.896	98.376	73.260

R1 = 3.1878   R2 = 2.8768   R3 = 0.9200
Q1 = 2.400   Q2 = 2.612   Q3 = 1.400

MEAN DEV. BETWEEN CALC. AND EXP. CONC. IN MOLE PCT

UNIQUAC (SPECIFIC PARAMETERS)   0.53
NRTL   (SPECIFIC PARAMETERS)   0.59

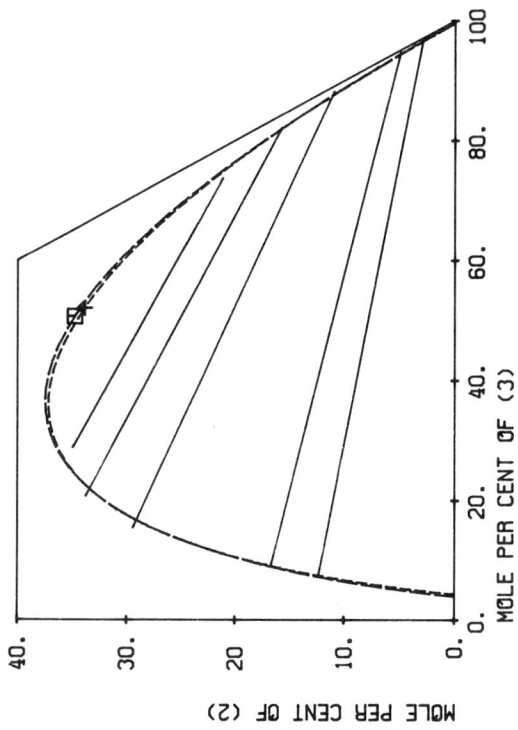

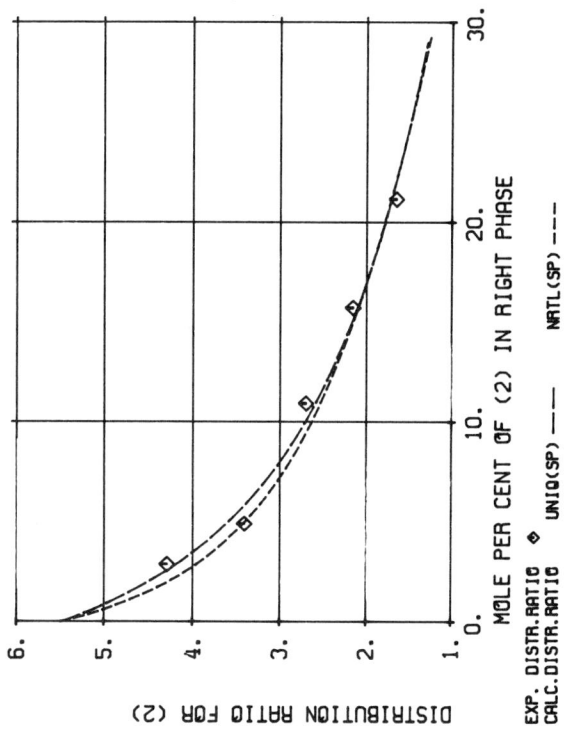

C$_3$H$_6$O$_2$-C$_6$H$_6$

(1) C6H6   BENZENE
(2) C3H6O2   PROPANOIC ACID
(3) H2O   WATER

IGUCHI A., FUSE K.
KAGAKU KOGAKU 36(1972)439

TEMPERATURE = 25.0 DEG C   TYPE OF SYSTEM = 1

EXPERIMENTAL TIE LINES IN MOLE PCT (GRAPH.INTERPOL.)

	LEFT PHASE			RIGHT PHASE	
(1)	(2)	(3)	(1)	(2)	(3)
96.243	3.326	0.431	0.048	1.582	98.370
89.937	9.206	0.857	0.075	2.729	97.196
76.904	21.823	1.274	0.135	5.308	94.557
64.731	33.173	2.096	0.206	8.629	91.166
54.473	40.629	4.898	0.662	13.277	86.061
46.124	43.019	10.858	1.720	18.134	80.146
45.013	43.076	11.911	1.895	18.638	79.467
39.552	43.224	17.224	3.313	22.026	74.661
36.731	42.962	20.306	4.617	24.186	71.197
30.991	41.691	27.318	6.684	27.229	65.087
22.044	38.396	39.560	9.334	30.113	60.554

SPECIFIC MODEL PARAMETERS IN KELVIN

		UNIQUAC		NRTL(ALPHA=.2)	
I J	AIJ	AJI		AIJ	AJI
1 2	608.34	-262.91		1329.4	-651.10
1 3	1010.7	198.16		1767.6	1850.2
2 3	935.03	-200.14		544.28	-53.179

R1 = 3.1878   R2 = 2.8768   R3 = 0.9200
Q1 = 2.400   Q2 = 2.612   Q3 = 1.400

MEAN DEV. BETWEEN CALC. AND EXP. CONC. IN MOLE PCT

UNIQUAC (SPECIFIC PARAMETERS)   0.83
NRTL   (SPECIFIC PARAMETERS)   0.97
UNIQUAC (COMMON PARAMETERS)   1.93

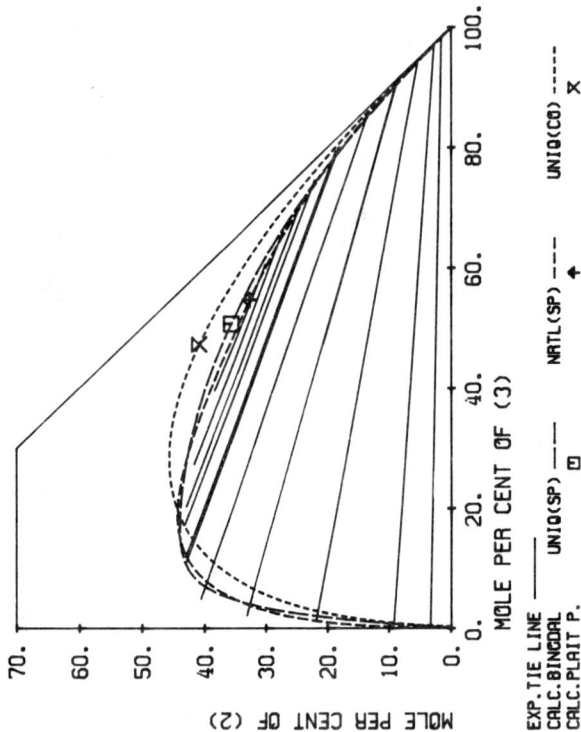

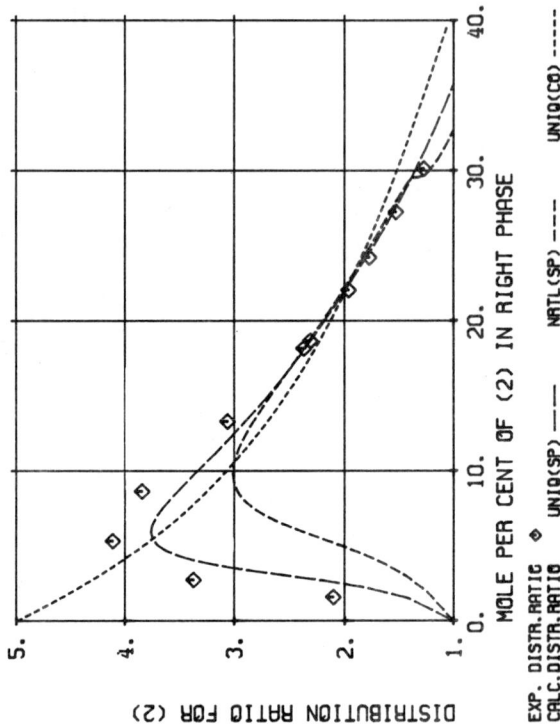

EXP.TIE LINE ——   UNIQ(SP) ☐   NRTL(SP) ----   UNIQ(CO) -----
CALC.BINODAL   UNIQ(SP) ☐   NRTL(SP) ◆   UNIQ(CO) ✕
CALC.PLAIT P.

EXP. DISTR.RATIO ◇   UNIQ(SP) ◇   NRTL(SP) ----   UNIQ(CO) -----
CALC.DISTR.RATIO

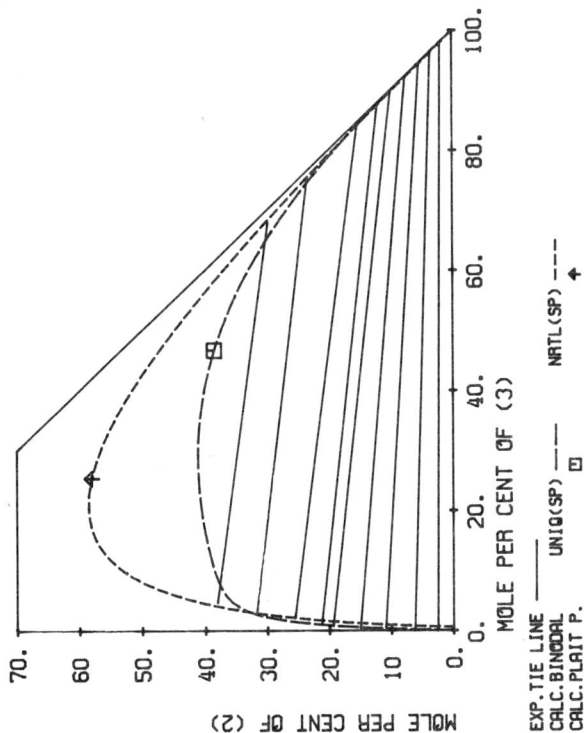

MOLE PER CENT OF (3)

MOLE PER CENT OF (2)

EXP.TIE LINE ——— UNIQ(SP) —🞴— NRTL(SP) ----
CALC.BINODAL ----
CALC.PLAIT P. 🞴

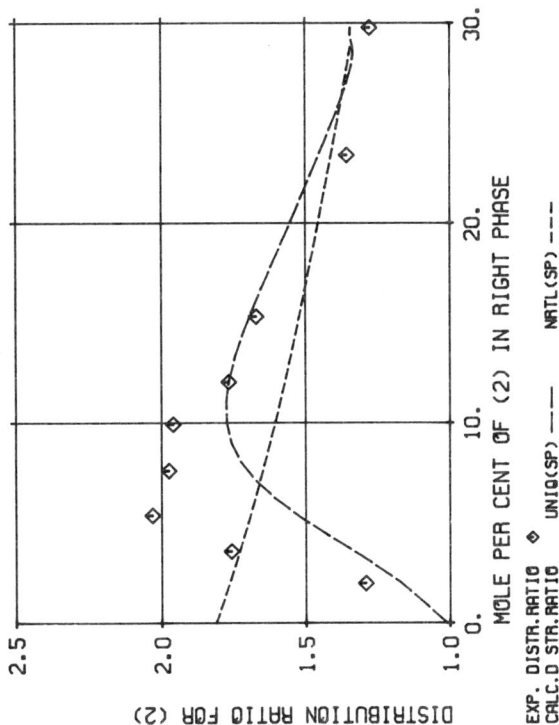

MOLE PER CENT OF (2) IN RIGHT PHASE

DISTRIBUTION RATIO FOR (2)

EXP. DISTR.RATIO ◇ UNIQ(SP) ——— NRTL(SP) ----
CALC.D STR.RATIO

(1) C6H10   CYCLOHEXENE
(2) C3H6O2   PROPANOIC ACID
(3) H2O   WATER

RAJA RAO M., VENKATA RAO C.
J.APPL.CHEM. 6(1956)269

TEMPERATURE = 31.0 DEG C   TYPE OF SYSTEM = 1

EXPERIMENTAL TIE LINES IN MOLE PCT

	LEFT PHASE			RIGHT PHASE	
(1)	(2)	(3)	(1)	(2)	(3)
97.456	2.544	0.0	0.035	1.963	98.002
93.718	6.282	0.0	0.061	3.574	96.365
88.619	10.931	0.450	0.090	5.380	94.530
84.528	15.024	0.448	0.136	7.609	92.255
79.671	19.441	0.888	0.202	9.920	89.878
77.404	21.271	1.325	0.243	12.045	87.713
72.219	25.598	2.184	0.425	15.306	84.269
65.222	31.761	3.017	1.204	23.378	75.419
57.245	38.109	4.646	2.243	29.756	68.001

SPECIFIC MODEL PARAMETERS IN KELVIN

		UNIQUAC		NRTL(ALPHA=.2)	
I J	AIJ	AJI		AIJ	AJI
1 2	551.07	-273.44		-168.47	53.574
1 3	1179.0	678.21		1027.4	1625.1
2 3	1178.0	-309.20		345.50	-216.87

R1 = 3.8143   R2 = 2.8768   R3 = 0.9200
Q1 = 3.027   Q2 = 2.612   Q3 = 1.400

MEAN DEV. BETWEEN CALC. AND EXP. CONC. IN MOLE PCT

UNIQUAC (SPECIFIC PARAMETERS)   0.70
NRTL   (SPECIFIC PARAMETERS)   0.74

$C_3H_6O_2$-C_6H_{12}

(1) C6H12   CYCLOHEXANE

(2) C3H6O2   PROPANOIC ACID

(3) H2O   WATER

RAJA RAO M., VENKATA RAO C.
J.APPL.CHEM. 6(1956)269

TEMPERATURE = 31.0 DEG C   TYPE OF SYSTEM = 1

EXPERIMENTAL TIE LINES IN MOLE PCT

	LEFT PHASE			RIGHT PHASE	
(1)	(2)	(3)	(1)	(2)	(3)
98.073	1.927	0.0	0.0	2.265	97.735
95.368	4.632	0.0	0.024	4.057	95.919
92.123	7.877	0.0	0.051	6.361	93.588
86.834	12.707	0.458	0.142	10.322	89.536
83.052	16.039	0.910	0.217	14.061	85.722
78.980	19.890	1.130	0.339	18.379	81.282
73.798	24.857	1.345	0.699	25.313	73.988
65.733	32.497	1.770	1.645	34.524	63.831
57.396	39.567	3.037	2.466	39.003	58.531
37.709	51.157	11.134	7.090	48.935	43.975

SPECIFIC MODEL PARAMETERS IN KELVIN

		UNIQUAC		NRTL(ALPHA=.2)	
I J	AIJ	AJI	AIJ	AJI
1 2	301.48	-131.07	-96.591	242.13
1 3	920.56	-851.56	1871.1	2363.3
2 3	528.60	-237.24	16.128	92.463

R1 = 4.0464   R2 = 2.8768   R3 = 2.265
Q1 = 3.240   Q2 = 2.612   Q3 = 1.400

MEAN DEV. BETWEEN CALC. AND EXP. CONC. IN MOLE PCT

UNIQUAC (SPECIFIC PARAMETERS)   0.92
NRTL   (SPECIFIC PARAMETERS)   0.74

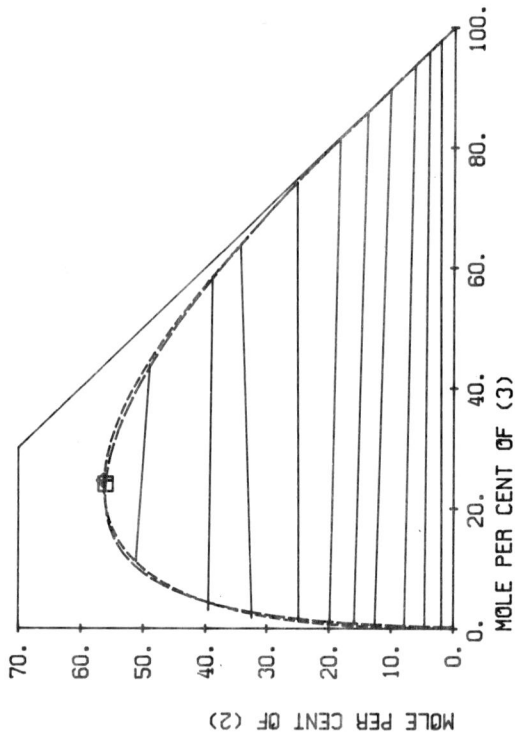

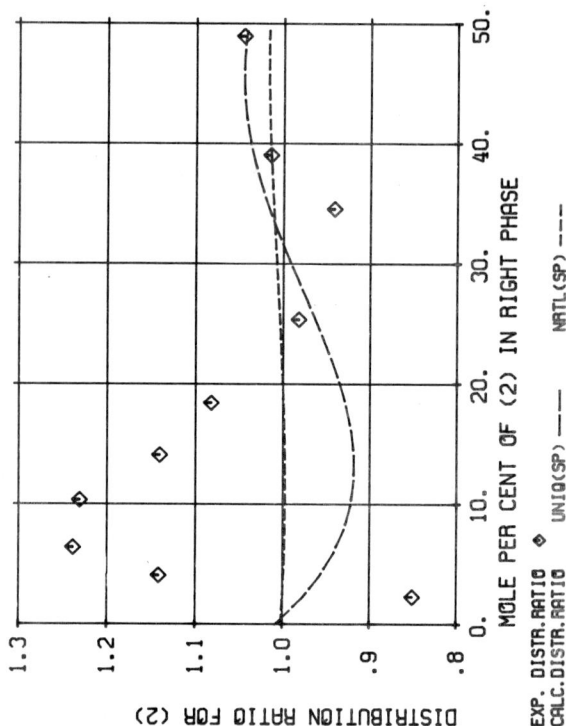

MOLE PER CENT OF (3)

MOLE PER CENT OF (2)

EXP.TIE LINE ——   UNIQ(SP) —□—   NRTL(SP) —◆—
CALC.BINODAL
CALC.PLAIT P.

MOLE PER CENT OF (2) IN RIGHT PHASE

DISTRIBUTION RATIO FOR (2)

EXP. DISTR.RATIO ◇   UNIQ(SP)   NRTL(SP) ----
CALC.DISTR.RATIO

$C_3H_6O_2$-C_6H_{12}

MOLE PER CENT OF (3)

MOLE PER CENT OF (2)

EXP.TIE LINE ———   UNIQ(SP) □   NRTL(SP) ———   UNIQ(CC) -----
CALC.BINODAL
CALC.PLAIT P.   ✕

MOLE PER CENT OF (2) IN RIGHT PHASE

DISTRIBUTION RATIO FOR (2)

EXP. DISTR.RATIO ◇   UNIQ(SP) ———   NRTL(SP) ———   UNIQ(CC) -----
CALC.DISTR.RATIO

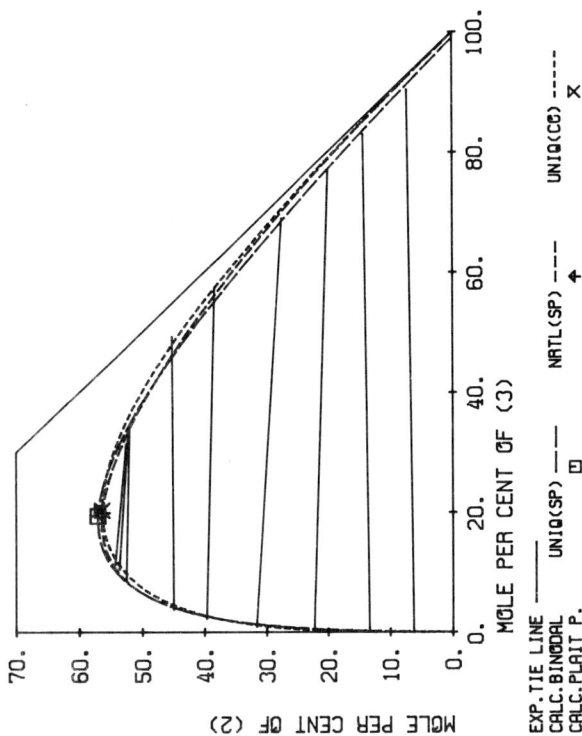

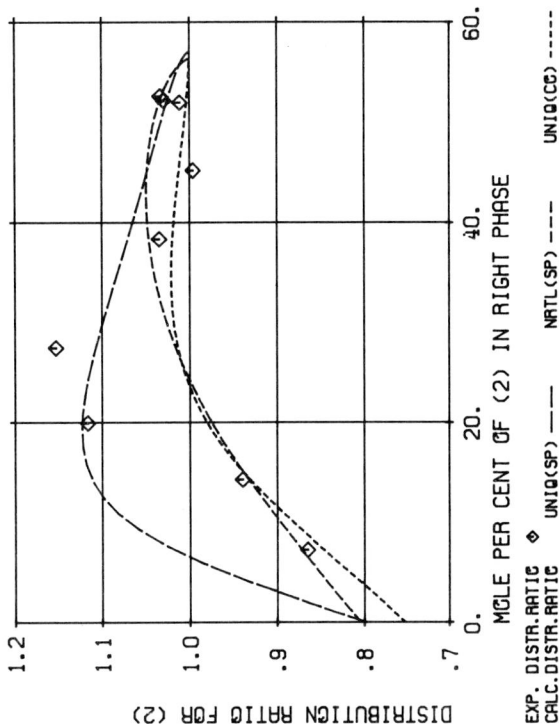

(1)	H2O	WATER
(2)	C3H6O2	PROPANOIC ACID
(3)	C6H12	CYCLOHEXANE

IGUCHI A., FUSE K.
KAGAKU KOGAKU 36(1972)320

TEMPERATURE = 25.0 DEG C   TYPE OF SYSTEM = 1

EXPERIMENTAL TIE LINES IN MOLE PCT (GRAPH.INTERPOL.)

| | LEFT PHASE | | | RIGHT PHASE | |
(1)	(2)	(3)	(1)	(2)	(3)
93.710	6.264	0.026	2.274	7.244	90.483
86.496	13.413	0.091	2.696	14.280	83.024
77.567	22.251	0.182	3.112	19.926	76.962
67.362	31.686	0.952	3.511	27.473	69.016
58.037	39.680	2.283	4.299	38.349	57.352
51.189	45.005	3.806	5.483	45.189	49.328
38.540	52.582	8.878	13.242	51.995	34.763
35.794	53.737	10.469	13.583	52.191	34.226
34.309	54.383	11.309	15.928	52.615	31.457

SPECIFIC MODEL PARAMETERS IN KELVIN

| | | UNIQUAC | | NRTL(ALPHA=.2) | |
I	J	AIJ	AJI	AIJ	AJI
1	2	-667.34	254.33	449.99	-188.16
1	3	31.62	712.71	1710.5	1572.4
2	3	15.883	-410.85	487.28	-177.20

R1 = 0.9200   R2 = 2.8768   R3 = 4.0464
Q1 = 1.400   Q2 = 2.612   Q3 = 3.240

MEAN DEV. BETWEEN CALC. AND EXP. CONC. IN MOLE PCT

UNIQUAC (SPECIFIC PARAMETERS)	0.83
NRTL (SPECIFIC PARAMETERS)	0.92
UNIQUAC (COMMON PARAMETERS)	0.92

$C_3H_6O_2-C_6H_{12}O_2$

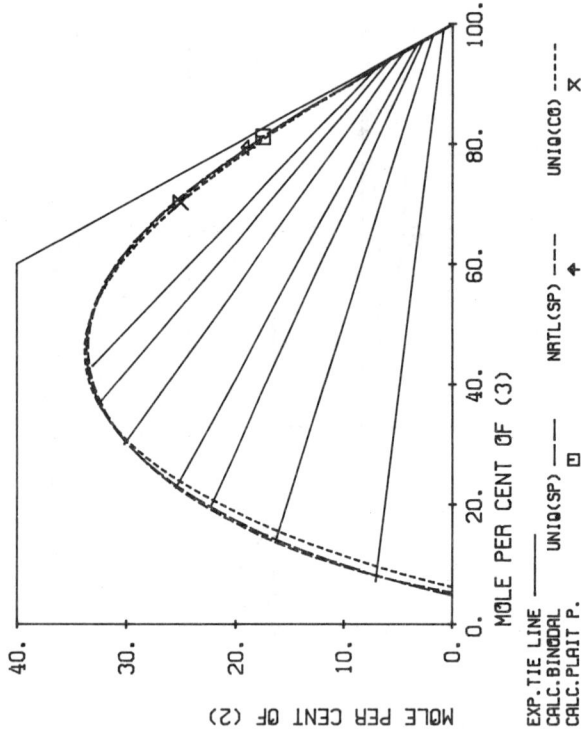

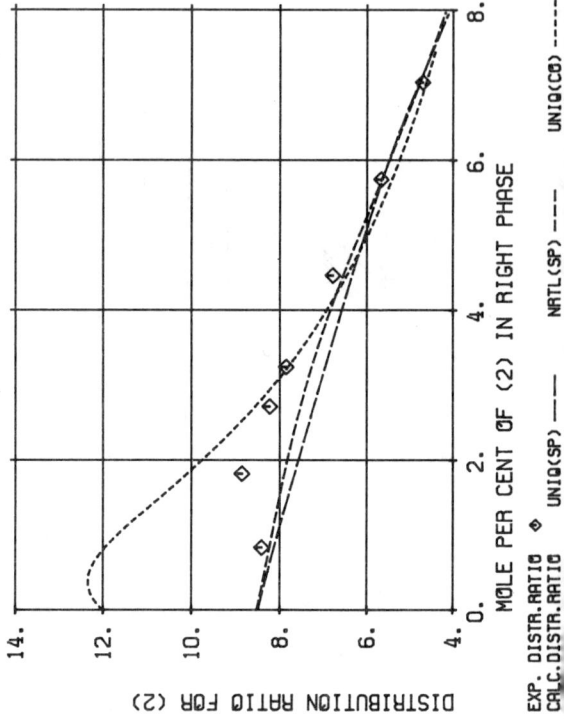

(1) C6H1202   BUTANOIC ACID,ETHYL ESTER
(2) C3H602   PROPANOIC ACID
(3) H20   WATER

JAYA RAMA RAO G., VENTAKA RAO C.
J.SCI.IND.RES. 14B(1955)444

TEMPERATURE = 26.0 DEG C   TYPE OF SYSTEM = 1

EXPERIMENTAL TIE LINES IN MOLE PCT

	LEFT PHASE			RIGHT PHASE	
(1)	(2)	(3)	(1)	(2)	(3)
85.856	7.011	7.134	0.109	0.833	99.053
69.476	16.094	14.430	0.145	1.817	98.038
58.134	22.322	19.544	0.166	2.712	97.122
51.830	25.459	22.711	0.170	3.242	96.587
39.812	30.194	29.993	0.180	4.458	95.362
30.840	32.559	36.602	0.222	5.742	94.036
23.898	33.089	43.013	0.355	7.038	92.607

SPECIFIC MODEL PARAMETERS IN KELVIN

		UNIQUAC		NRTL(ALPHA=.2)	
I J	AIJ	AJI		AIJ	AJI
1 2	36.612	61.651		48.325	1.9422
1 3	602.93	101.71		399.79	1594.1
2 3	-88.603	269.14		-489.56	1301.1

R1 = 4.8274   R2 = 2.8768   R3 = 0.9200
Q1 = 4.196   Q2 = 2.612   Q3 = 1.400

MEAN DEV. BETWEEN CALC. AND EXP. CONC. IN MOLE PCT

UNIQUAC (SPECIFIC PARAMETERS)   0.33
NRTL   (SPECIFIC PARAMETERS)   0.28
UNIQUAC (COMMON PARAMETERS)   0.79

MOLE PER CENT OF (3)

MOLE PER CENT OF (2)

EXP.TIE LINE ———   UNIQ(SP) ⊡   NRTL(SP) ✦   UNIQ(CO) -----
CALC.BINODAL
CALC.PLAIT P. ✕

DISTRIBUTION RATIO FOR (2)

MOLE PER CENT OF (2) IN RIGHT PHASE

EXP. DISTR.RATIO ◇   UNIQ(SP) ———   NRTL(SP) ———   UNIQ(CO) -----
CALC.DISTR.RATIO

$C_3H_6O_2-C_6H_{14}$

(1) C6H14   HEXANE

(2) C3H6O2   PROPANOIC ACID

(3) H2O   WATER

IGUCHI A., FUSE K.
KAGAKU KOGAKU 36(1972)320

TEMPERATURE = 25.0 DEG C   TYPE OF SYSTEM = 1

EXPERIMENTAL TIE LINES IN MOLE PCT (GRAPH.INTERPOL.)

	LEFT PHASE			RIGHT PHASE	
(1)	(2)	(3)	(1)	(2)	(3)
90.119	8.479	1.402	0.075	6.175	93.750
83.865	14.745	1.390	0.144	11.887	87.969
77.140	21.031	1.829	0.239	20.025	79.736
69.039	28.708	2.253	0.615	30.100	69.286
56.247	40.249	3.505	1.677	41.647	56.676
42.387	51.312	6.301	3.631	49.330	47.038
41.486	51.822	6.692	3.961	50.025	46.014
26.840	57.552	15.607	7.364	53.241	39.395
38.446	54.105	7.449	6.056	53.590	40.354
32.648	56.196	11.156	7.069	53.762	39.169

SPECIFIC MODEL PARAMETERS IN KELVIN

		UNIQUAC		NRTL(ALPHA=.2)	
I J		AIJ	AJI	AIJ	AJI
1 2		-185.09	41.426	-336.86	620.48
1 3		1176.8	244.17	1511.3	1459.8
2 3		266.96	-409.04	-59.559	228.54

R1 = 4.4998   R2 = 2.8768   R3 = 0.9200
Q1 = 3.856   Q2 = 2.612   Q3 = 1.400

MEAN DEV. BETWEEN CALC. AND EXP. CONC. IN MOLE PCT

UNIQUAC (SPECIFIC PARAMETERS)	1.18
NRTL (SPECIFIC PARAMETERS)	1.38
UNIQUAC (COMMON PARAMETERS)	1.33

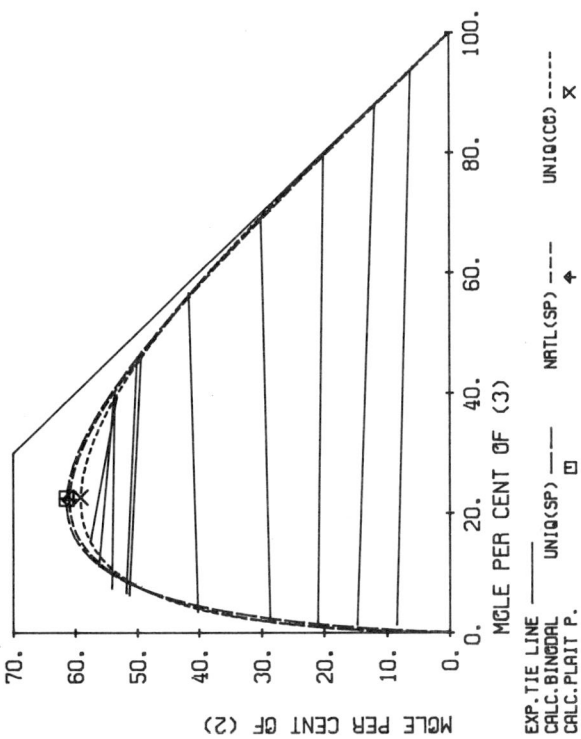

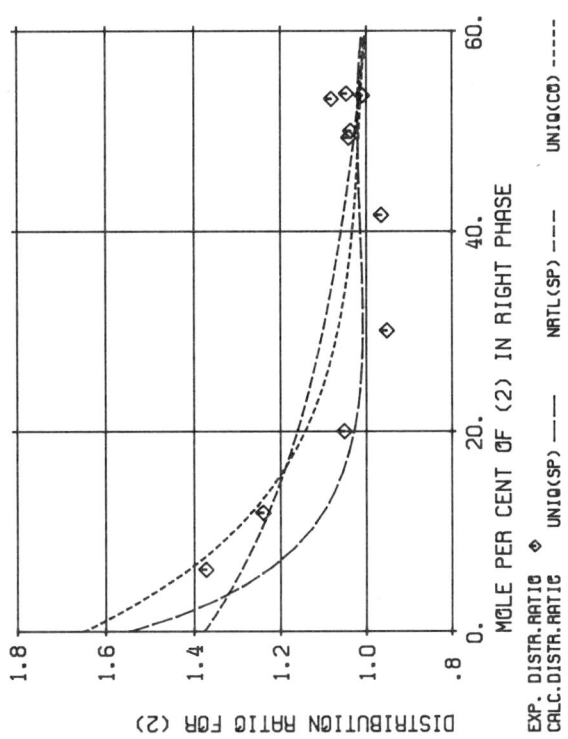

MOLE PER CENT OF (2)

MOLE PER CENT OF (3)

EXP.TIE LINE —— UNIQ(SP) ---□   NRTL(SP) ---+   UNIQ(CC) ---✕
CALC.BINODAL
CALC.PLAIT P.

DISTRIBUTION RATIO FOR (2)

MOLE PER CENT OF (2) IN RIGHT PHASE

EXP. DISTR.RATIO ◇ UNIQ(SP) ——   NRTL(SP) ———   UNIQ(CC) -----
CALC.DISTR.RATIO

C$_3$H$_6$O$_2$-C$_6$H$_{14}$O

(1) C6H14O   1-HEXANOL
(2) C3H6O2   PROPANOIC ACID
(3) H2O   WATER

CHANDY C.A., RAJA RAO M.
J.CHEM.ENG.DATA 7(1962)473

TEMPERATURE = 30.0 DEG C   TYPE OF SYSTEM = 1

EXPERIMENTAL TIE LINES IN MOLE PCT

	LEFT PHASE			RIGHT PHASE	
(1)	(2)	(3)	(1)	(2)	(3)
58.057	7.280	34.663	0.117	0.623	99.259
51.104	12.900	35.996	0.129	1.219	98.653
44.790	17.412	37.798	0.150	1.920	97.930
27.660	24.420	47.920	0.199	3.874	95.927
15.559	24.613	59.829	0.450	6.270	93.280

SPECIFIC MODEL PARAMETERS IN KELVIN

		UNIQUAC		NRTL(ALPHA=.2)	
I J	AIJ	AJI	AIJ	AJI	
1 2	-93.330	91.381	558.53	-471.67	
1 3	201.82	150.10	-87.632	1956.1	
2 3	-113.00	257.49	-371.87	1028.4	

R1 = 4.8031   R2 = 2.8768   R3 = 0.9200
Q1 = 4.132   Q2 = 2.612   Q3 = 1.400

MEAN DEV. BETWEEN CALC. AND EXP. CONC. IN MOLE PCT

UNIQUAC (SPECIFIC PARAMETERS)   0.63
NRTL   (SPECIFIC PARAMETERS)   0.77
UNIQUAC (COMMON PARAMETERS)   1.19

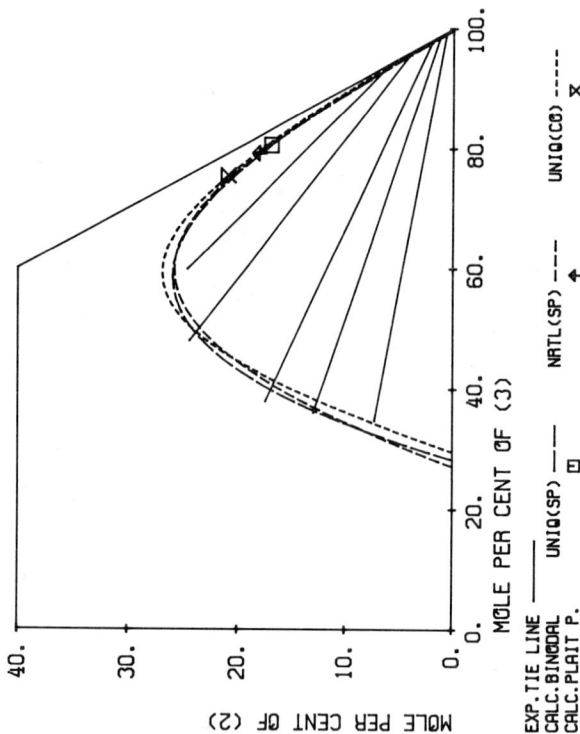

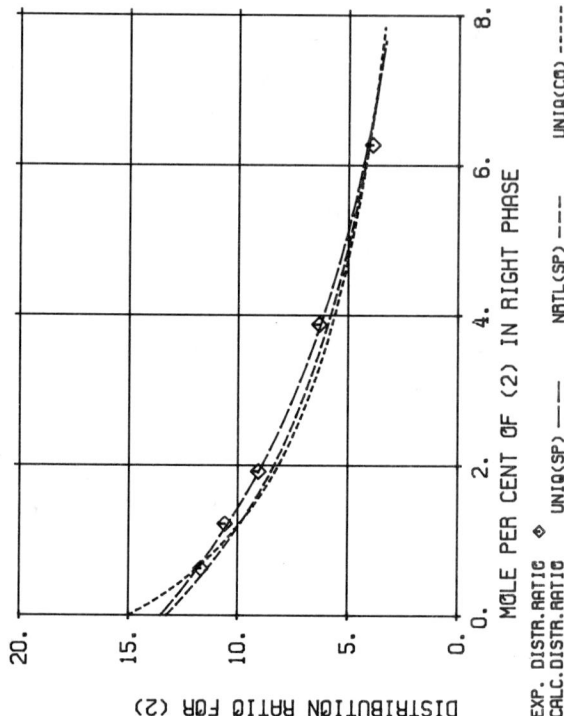

MOLE PER CENT OF (2)

MOLE PER CENT OF (3)

EXP.TIE LINE   UNIQ(SP) ─ ▣   NRTL(SP) ─ ✦   UNIQ(CO) ----- ✕
CALC.BINODAL
CALC.PLAIT P.

DISTRIBUTION RATIO FOR (2)

MOLE PER CENT OF (2) IN RIGHT PHASE

EXP. DISTR.RATIO ◇   UNIQ(SP) ◇   NRTL(SP) -----   UNIQ(CO) -----
CALC.DISTR.RATIO

$C_3H_6O_2-C_6H_{14}O$

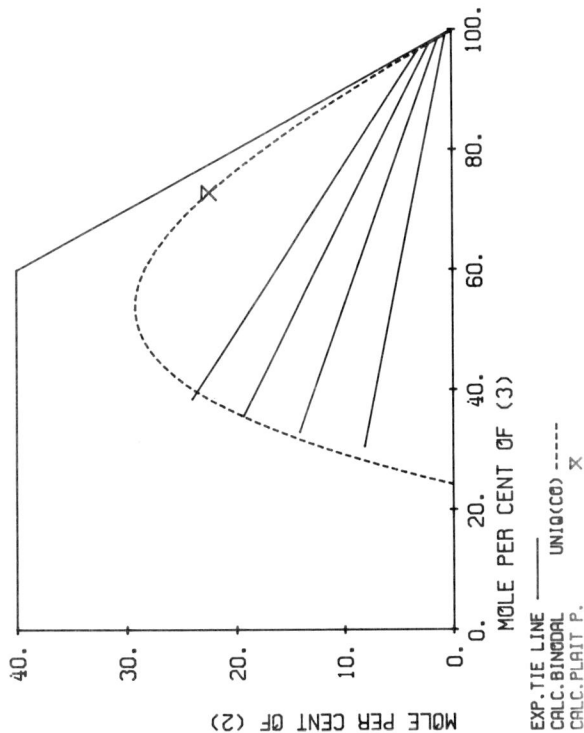

MOLE PER CENT OF (3)

MOLE PER CENT OF (2)

EXP.TIE LINE ———
CALC.BINODAL ------ UNIQ(CO) ------
CALC.PLAIT P.   x

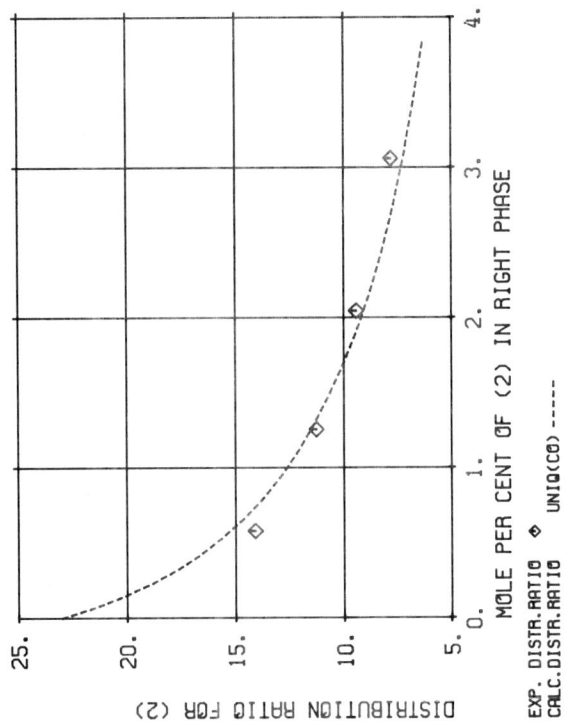

MOLE PER CENT OF (2) IN RIGHT PHASE

DISTRIBUTION RATIO FOR (2)

EXP. DISTR.RATIO ◇ UNIQ(CO) ------
CALC.DISTR.RATIO ------

(1) C6H14O   2-PENTANOL,4-METHYL
(2) C3H6O2   PROPANOIC ACID
(3) H2O   WATER

RAJA RAO M.; RAMAMURTY M., VENKATA RAO C.
CHEM.ENG.SCI. 8(1958)265

TEMPERATURE = 30.0 DEG C   TYPE OF SYSTEM = 1

EXPERIMENTAL TIE LINES IN MOLE PCT

| | LEFT PHASE | | | RIGHT PHASE | |
(1)	(2)	(3)	(1)	(2)	(3)
61.390	8.117	30.493	0.236	0.575	99.189
53.017	14.092	32.890	0.259	1.252	98.488
45.107	19.296	35.596	0.285	2.043	97.673
37.510	23.997	38.492	0.373	3.057	96.570

MEAN DEV. BETWEEN CALC. AND EXP. CONC. IN MOLE PCT

UNIQUAC (COMMON PARAMETERS)   0.55

$C_3H_6O_2$-C_7H_8

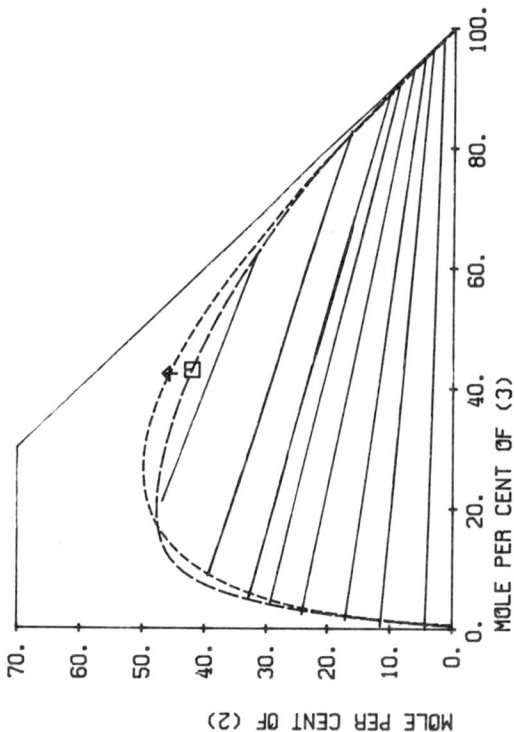

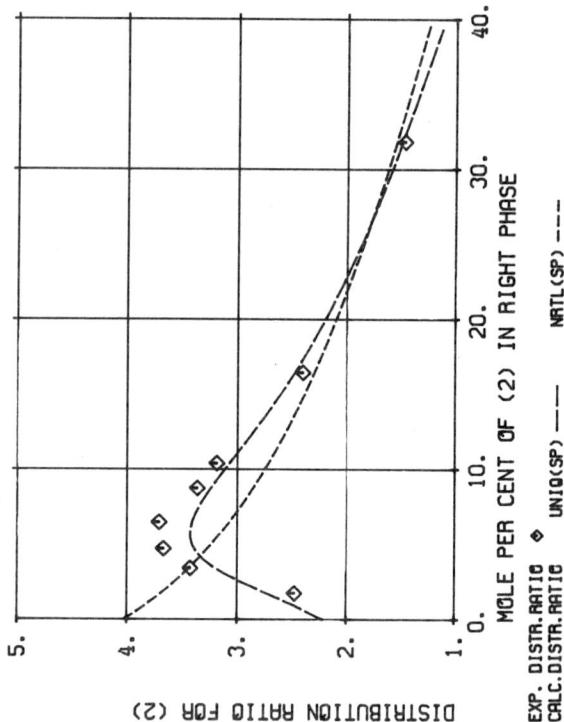

(1) C7H8   TOLUENE

(2) C3H6O2   PROPANOIC ACID

(3) H2O   WATER

RAJA RAO M., VENKATA RAO C.
J.APPL.CHEM. 6(1956)269

TEMPERATURE = 31.0 DEG C   TYPE OF SYSTEM = 1

EXPERIMENTAL TIE LINES IN MOLE PCT

LEFT PHASE			RIGHT PHASE		
(1)	(2)	(3)	(1)	(2)	(3)
95.683	4.317	0.0	0.0	1.743	98.257
87.880	11.622	0.498	0.022	3.390	96.588
81.320	17.215	1.465	0.056	4.691	95.252
73.571	24.041	2.388	0.094	6.479	93.427
66.427	29.390	4.183	0.150	8.733	91.117
61.912	33.054	5.034	0.234	10.368	89.398
51.735	39.546	8.719	0.693	16.443	82.864
32.033	46.824	21.143	5.794	31.810	62.396

SPECIFIC MODEL PARAMETERS IN KELVIN

		UNIQUAC		NRTL(ALPHA=.2)	
I J		AIJ	AJI	AIJ	AJI
1 2		477.94	-251.19	-36.527	-91.570
1 3		877.83	180.45	996.01	1317.7
2 3		940.40	-223.91	-34.860	321.57

R1 = 3.9228   R2 = 2.8768   R3 = 0.9200
Q1 = 2.968   Q2 = 2.612   Q3 = 1.400

MEAN DEV. BETWEEN CALC. AND EXP. CONC. IN MOLE PCT

UNIQUAC (SPECIFIC PARAMETERS)   0.49
NRTL   (SPECIFIC PARAMETERS)   1.15

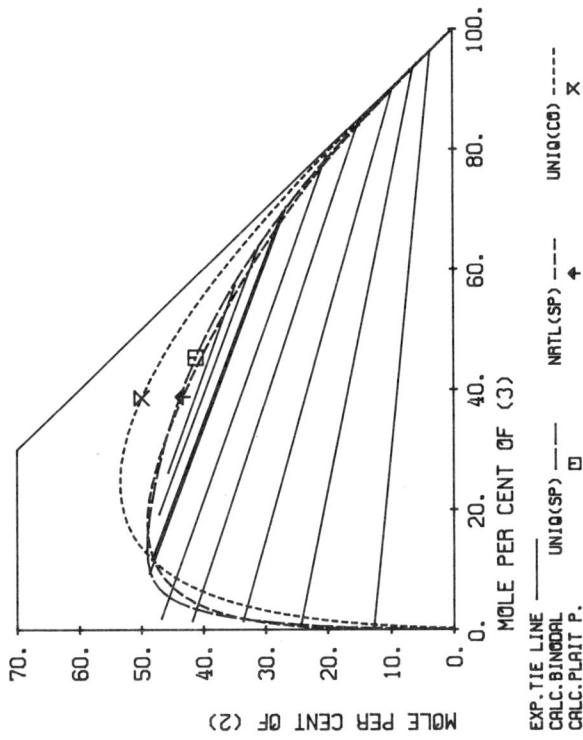

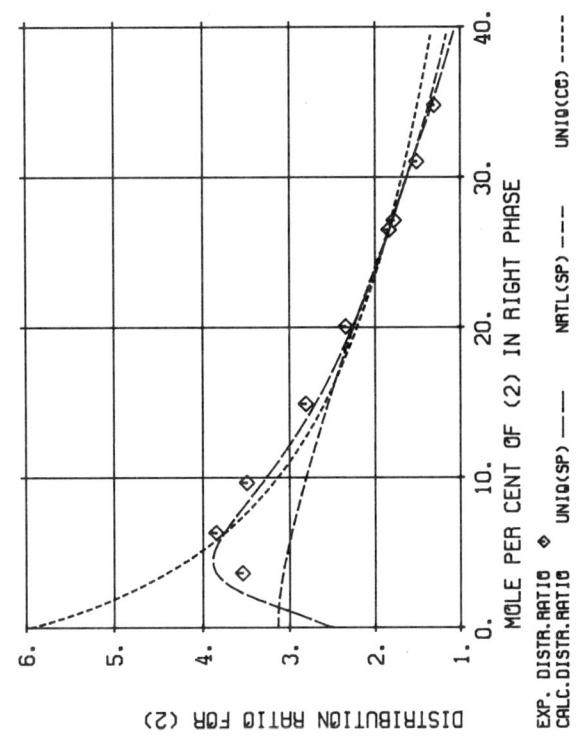

(1) C7H8   TOLUENE
(2) C3H6O2   PROPANOIC ACID
(3) H2O   WATER

IGUCHI A., FUSE K.
KAGAKU KOGAKU 36(1972)439

TEMPERATURE = 25.0 DEG C   TYPE OF SYSTEM = 1

EXPERIMENTAL TIE LINES IN MOLE PCT (GRAPH.INTERPOL.)

	LEFT PHASE			RIGHT PHASE	
(1)	(2)	(3)	(1)	(2)	(3)
86.749	12.755	0.497	0.022	3.606	96.373
74.757	24.276	0.966	0.047	6.312	93.641
64.945	33.639	1.416	0.128	9.644	90.228
56.783	41.825	1.392	0.230	14.879	84.891
51.378	46.794	1.828	0.943	20.047	79.010
41.914	48.754	9.333	2.928	26.491	70.581
40.023	48.352	11.625	3.165	27.131	69.704
33.599	47.150	19.250	5.104	31.023	63.873
28.172	45.686	26.142	7.746	34.788	57.456

SPECIFIC MODEL PARAMETERS IN KELVIN

		UNIQUAC		NRTL(ALPHA=.2)	
I J		AIJ	AJI	AIJ	AJI
1 2		516.39	-277.76	617.12	-450.67
1 3		1349.3	217.03	2581.5	2045.9
2 3		824.38	-223.45	355.73	70.317

R1 = 3.9228   R2 = 2.8768   R3 = 0.9200
Q1 = 2.968   Q2 = 2.612   Q3 = 1.400

MEAN DEV. BETWEEN CALC. AND EXP. CONC. IN MOLE PCT

UNIQUAC (SPECIFIC PARAMETERS)   0.90
NRTL   (SPECIFIC PARAMETERS)   1.47
UNIQUAC (COMMON PARAMETERS)   2.34

C$_3$H$_6$O$_2$-C$_7$H$_{14}$O$_2$

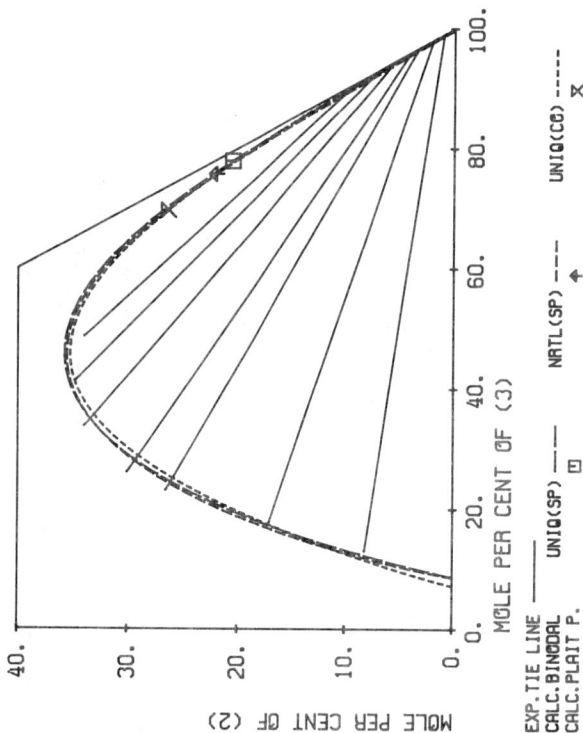

MOLE PER CENT OF (3)

MOLE PER CENT OF (2)

EXP.TIE LINE   UNIQ(SP) —— ☐   NRTL(SP) —— ✦   UNIQ(CC) - - - - -
CALC.BINODAL
CALC.PLAIT P.   UNIQ(CC) - - - - - ✗

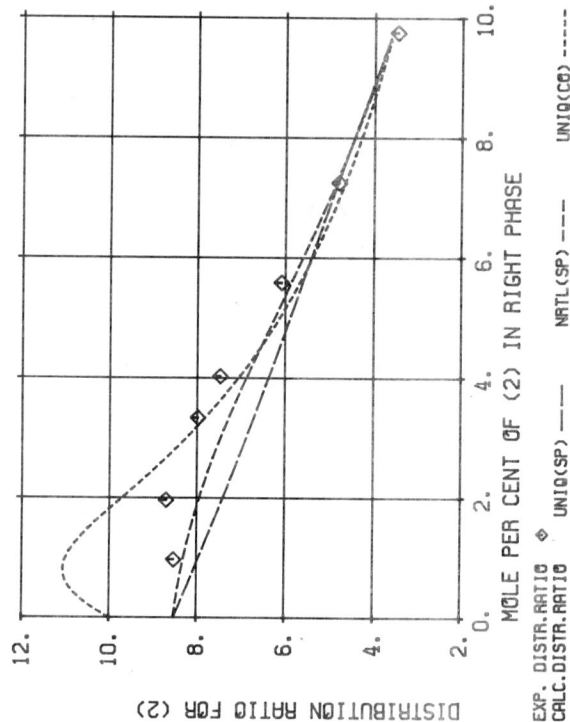

MOLE PER CENT OF (2) IN RIGHT PHASE

DISTRIBUTION RATIO FOR (2)

EXP. DISTR.RATIO ◇   UNIQ(SP) ◇   NRTL(SP)   UNIQ(CC) - - - - -
CALC.DISTR.RATIO

(1)	C7H14O2 ACETIC ACID,PENTYL ESTER
(2)	C3H6O2 PROPANOIC ACID
(3)	H2O WATER

KRISHNAMURTY R., JAYARAMA RAO G., VENKATA RAO C.
J.SCI.IND.RES. 21D(1962)282

TEMPERATURE = 28.0 DEG C   TYPE OF SYSTEM = 1

EXPERIMENTAL TIE LINES IN MOLE PCT

LEFT PHASE			RIGHT PHASE		
(1)	(2)	(3)	(1)	(2)	(3)
78.866	8.187	12.948	0.054	0.960	98.986
65.339	17.077	17.584	0.066	1.961	97.973
50.372	26.551	23.077	0.094	3.326	96.580
43.888	30.062	26.050	0.108	4.025	95.867
32.328	33.970	33.701	0.131	5.581	94.288
23.850	34.718	41.432	0.207	7.250	92.543
17.375	33.970	48.655	0.461	9.751	89.788

SPECIFIC MODEL PARAMETERS IN KELVIN

	UNIQUAC		NRTL(ALPHA=.2)	
I J	AIJ	AJI	AIJ	AJI
1 2	48.354	36.278	323.66	-226.21
1 3	533.26	70.865	273.47	1507.1
2 3	-29.684	185.67	-339.38	1071.1

R1 = 5.5018   R2 = 2.8768   R3 = 0.9200
Q1 = 4.736   Q2 = 2.612   Q3 = 1.400

MEAN DEV. BETWEEN CALC. AND EXP. CONC. IN MOLE PCT

UNIQUAC (SPECIFIC PARAMETERS)	0.66
NRTL (SPECIFIC PARAMETERS)	0.59
UNIQUAC (COMMON PARAMETERS)	0.94

$C_3H_6O_2$-C_7H_{16}

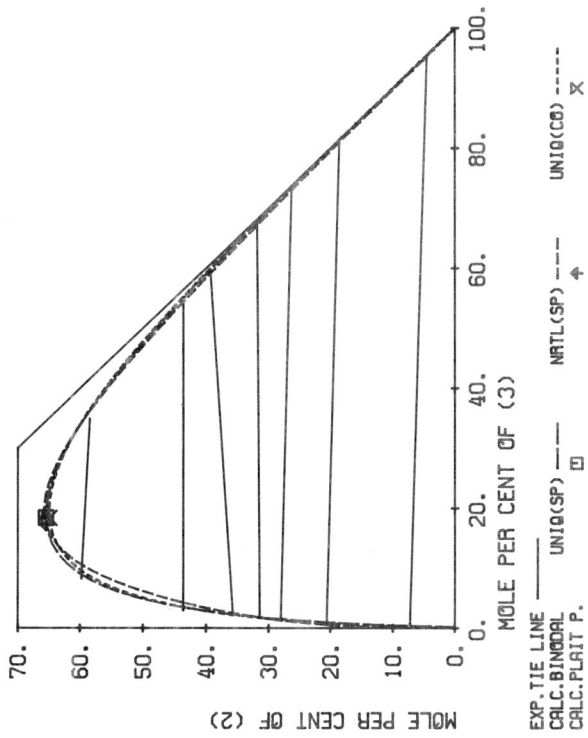

MOLE PER CENT OF (3)

MOLE PER CENT OF (2)

EXP.TIE LINE ——   UNIQ(SP) ⊟   NRTL(SP) ♦   UNIQ(CO) ----
CALC.BINODAL ----
CALC.PLAIT P.   ✕

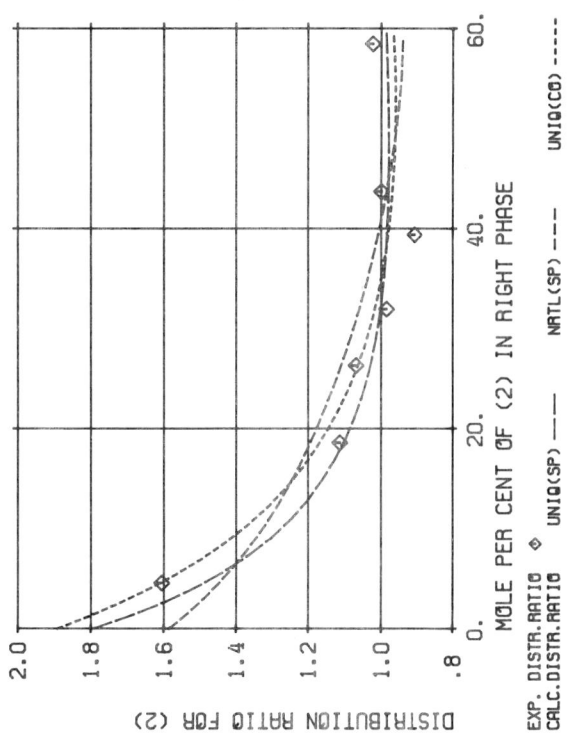

MOLE PER CENT OF (2) IN RIGHT PHASE

DISTRIBUTION RATIO FOR (2)

EXP. DISTR.RATIO ◇   UNIQ(SP) ——   NRTL(SP) — —   UNIQ(CO) ----
CALC.DISTR.RATIO

(1) C7H16   HEPTANE
(2) C3H6O2   PROPANOIC ACID
(3) H2O   WATER

KOVALEVA A.G.
UKR.KHIM.ZH.(RUSS.ED.) 35(1969)1137

TEMPERATURE = 25.0 DEG C   TYPE OF SYSTEM = 1

EXPERIMENTAL TIE LINES IN MOLE PCT

	LEFT PHASE			RIGHT PHASE	
(1)	(2)	(3)	(1)	(2)	(3)
92.191	7.266	0.543	0.021	4.526	95.454
78.836	20.641	0.524	0.057	18.561	81.382
70.400	28.075	1.526	0.098	26.279	73.622
66.531	31.463	2.006	0.361	31.913	67.726
61.795	35.739	2.466	0.855	39.370	59.776
53.481	43.640	2.879	1.260	43.654	55.086
32.003	59.788	8.210	6.620	58.431	34.949

SPECIFIC MODEL PARAMETERS IN KELVIN

		UNIQUAC		NRTL(ALPHA=.2)	
I	J	AIJ	AJI	AIJ	AJI
1	2	70.794	-27.464	-439.04	956.14
1	3	1113.5	352.88	2145.8	1438.1
2	3	369.43	-243.83	-55.498	261.88

R1 = 5.1742   R2 = 2.8768   R3 = 0.9200
Q1 = 4.396   Q2 = 2.612   Q3 = 1.400

MEAN DEV. BETWEEN CALC. AND EXP. CONC. IN MOLE PCT

UNIQUAC (SPECIFIC PARAMETERS)   0.63
NRTL   (SPECIFIC PARAMETERS)   1.14
UNIQUAC (COMMON PARAMETERS)   0.75

C₃H₆O₂-C₉H₁₀O₂

(1) C9H10O2   BENZOIC ACID,ETHYL ESTER

(2) C3H6O2   PROPANOIC ACID

(3) H2O   WATER

JAYA RAMA RAO G., VEKKATA RAO C.
J.SCI.IND.RES. 16B(1957)102

TEMPERATURE = 31.0 DEG C   TYPE OF SYSTEM = 1

EXPERIMENTAL TIE LINES IN MOLE PCT

LEFT PHASE			RIGHT PHASE		
(1)	(2)	(3)	(1)	(2)	(3)
71.319	14.637	14.044	0.064	2.220	97.716
55.727	25.455	18.818	0.068	3.980	95.952
27.134	35.576	37.290	0.169	8.668	91.163
22.291	35.460	42.249	0.321	10.167	89.512

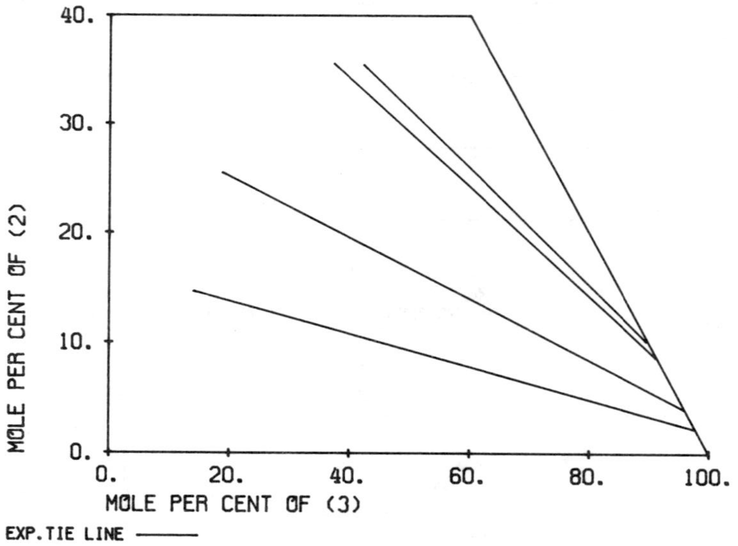

$C_3H_6O_2$-$C_{12}H_{10}O$

(1) C12H10O   ETHER,DIPHENYL

(2) C3H6O2   PROPANOIC ACID

(3) H2O   WATER

PURNELL J.H., BOWDEN S.T.
J.CHEM.SOC.;(1954)539

TEMPERATURE = 25.0 DEG C   TYPE OF SYSTEM = 1

EXPERIMENTAL TIE LINES IN MOLE PCT

	LEFT PHASE			RIGHT PHASE	
(1)	(2)	(3)	(1)	(2)	(3)
74.023	22.008	3.969	0.059	12.325	87.616
68.531	26.893	4.576	0.079	15.659	84.262
55.625	36.184	8.191	0.451	23.712	75.837
44.245	43.568	12.188	1.446	31.492	57.062
35.453	48.074	16.474	2.583	36.994	60.423

SPECIFIC MODEL PARAMETERS IN KELVIN

		UNIQUAC		NRTL(ALPHA=.2)	
I	J	AIJ	AJI	AIJ	AJI
1	2	-34.227	-139.40	-210.87	366.40
1	3	-738.62	-356.79	872.00	1906.4
2	3	-88.059	-147.98	-153.67	448.42

R1 = 6.2873   R2 = 2.8768   R3 = 0.9200
Q1 = 4.480   Q2 = 2.612   Q3 = 1.400

MEAN DEV. BETWEEN CALC. AND EXP. CONC. IN MOLE PCT

UNIQUAC (SPECIFIC PARAMETERS)   0.59
NRTL   (SPECIFIC PARAMETERS)   0.19
UNIQUAC (COMMON PARAMETERS)   0.90

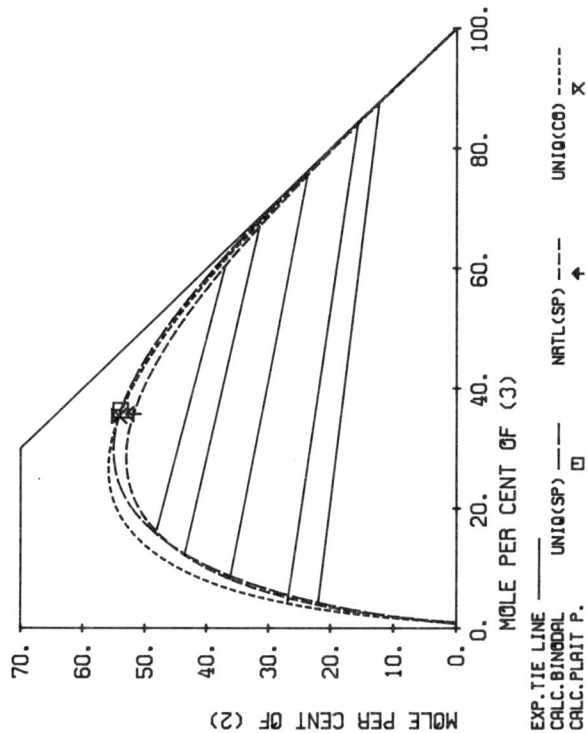

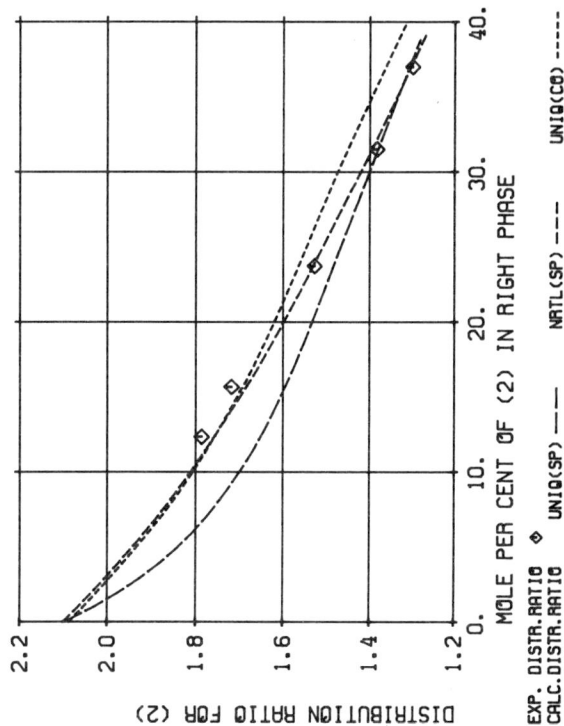

C$_3$H$_6$O$_3$-C$_4$H$_8$O

(1) C4H8O   2-BUTANONE
(2) C3H6O3   PROPANOIC ACID,2-HYDROXY
(3) H2O   WATER

ABABI V., POPA A.
AN.STIINT.UNIV.AL.I.CUZA IASI. 6(1960)1

TEMPERATURE = 25.0 DEG C   TYPE OF SYSTEM = 1

EXPERIMENTAL TIE LINES IN MOLE PCT

| | LEFT PHASE | | | RIGHT PHASE | |
(1)	(2)	(3)	(1)	(2)	(3)
60.520	0.568	38.912	8.118	0.276	91.605
59.665	0.620	39.716	8.221	0.353	91.425
55.355	1.082	43.563	8.734	0.697	90.570
48.676	1.721	49.603	10.219	1.192	88.589
45.401	1.903	52.696	11.310	1.337	87.352
41.452	2.144	56.403	12.943	1.567	85.489
41.322	2.140	56.538	13.212	1.607	85.181
37.069	2.293	60.638	15.141	1.802	83.058

SPECIFIC MODEL PARAMETERS IN KELVIN

		UNIQUAC		NRTL(ALPHA=.2)	
I J	AIJ	AJI	AIJ	AJI	
1 2	-133.42	-24.127	458.89	-850.65	
1 3	-305.77	6.1981	-37.746	-937.90	
2 3	-279.31	131.67	-578.98	-501.56	

R1 = 3.2479   R2 = 3.1793   R3 = 0.9200
Q1 = 2.876   Q2 = 2.884   Q3 = 1.400

MEAN DEV. BETWEEN CALC. AND EXP. CONC. IN MOLE PCT

UNIQUAC (SPECIFIC PARAMETERS)   0.26
NRTL (SPECIFIC PARAMETERS)   0.42
UNIQUAC (COMMON PARAMETERS)   0.34

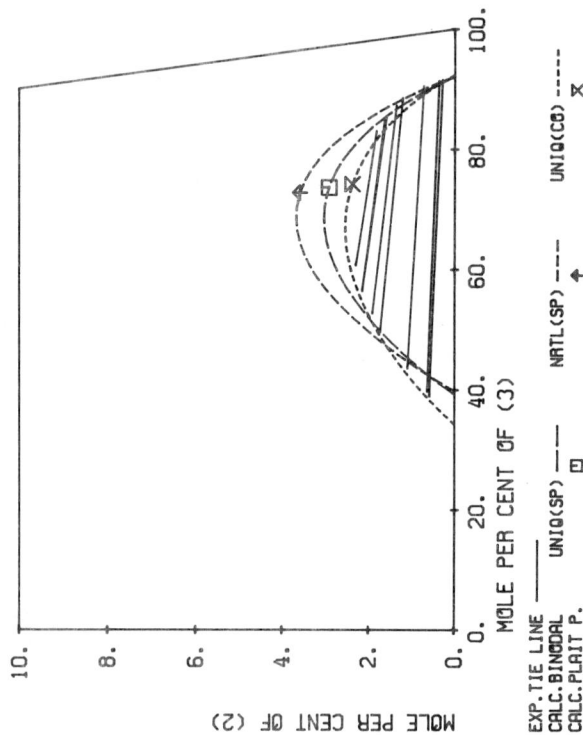

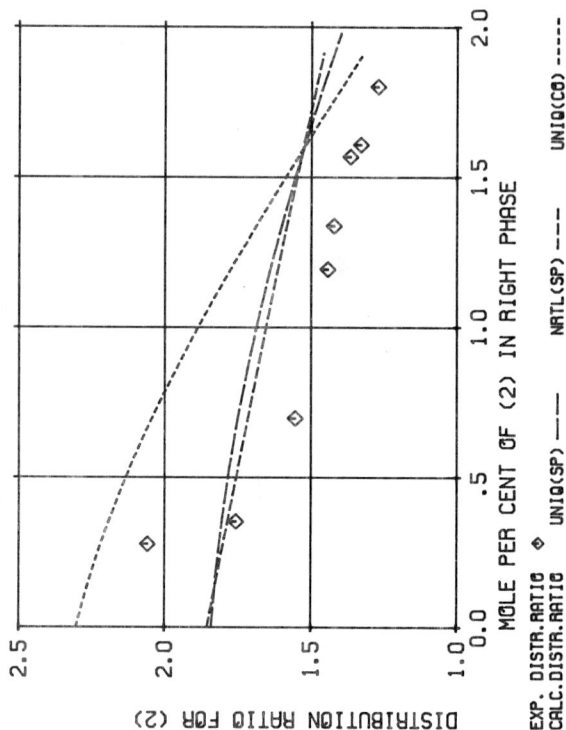

$C_3H_6O_3\text{-}C_4H_{10}O$

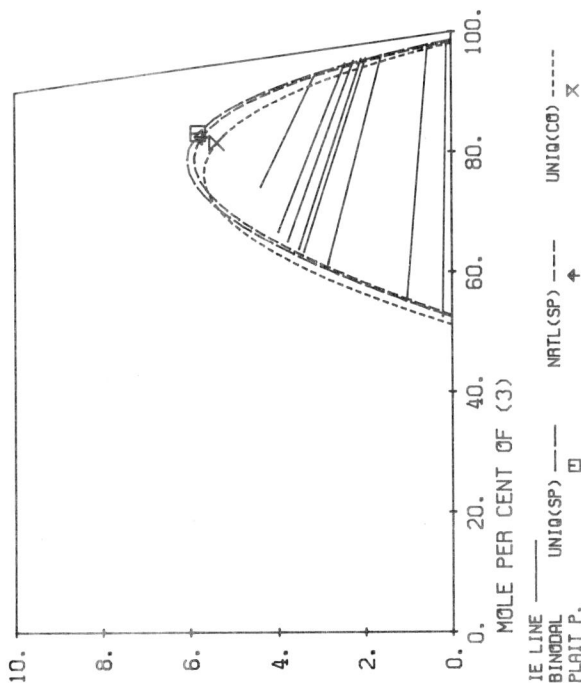

MOLE PER CENT OF (3)

MOLE PER CENT OF (2)

EXP.TIE LINE ———   UNIQ(SP) --- ---   NRTL(SP) --- ---   UNIQ(CD) --- ---
CALC.BINODAL ———   ⊡   ✦   ✗
CALC.PLAIT P.

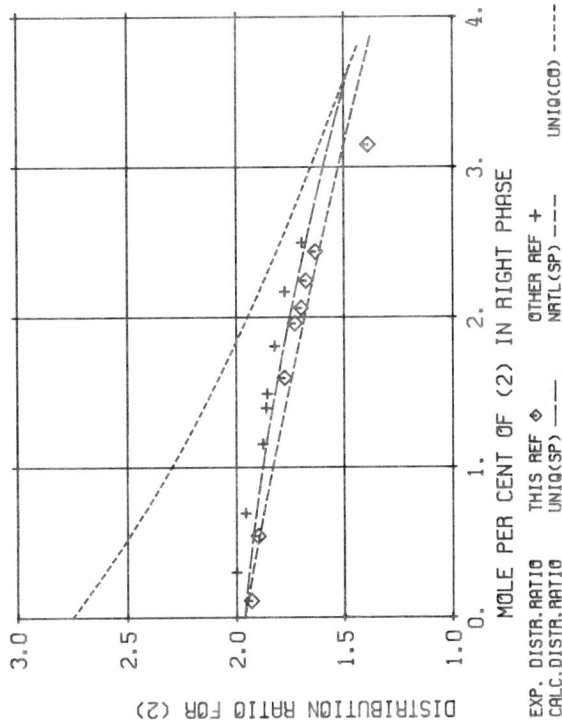

DISTRIBUTION RATIO FOR (2)

MOLE PER CENT OF (2) IN RIGHT PHASE

EXP. DISTR.RATIO ◇   THIS REF ◇   OTHER REF +   UNIQ(CD) --- ---
CALC.DISTR.RATIO ———   UNIQ(SP) --- ---   NRTL(SP) --- ---

(1) C4H10O   1-BUTANOL
(2) C3H6O3   PROPANOIC ACID,2-HYDROXY
(3) H2O   WATER

PETRITIS V.E., GEANKOPLIS C.J.
J.CHEM.ENG.DATA 4(1959)197

TEMPERATURE = 25.0 DEG C   TYPE OF SYSTEM = 1

EXPERIMENTAL TIE LINES IN MOLE PCT (GRAPH.INTERPOL.)

	LEFT PHASE			RIGHT PHASE	
(1)	(2)	(3)	(1)	(2)	(3)
48.216	0.0	51.784	1.797	0.0	98.203
47.226	0.213	52.561	1.832	0.110	98.058
43.714	1.028	55.258	1.947	0.541	97.512
35.845	2.836	61.319	2.231	1.596	96.173
33.235	3.386	63.379	2.351	1.957	95.692
32.651	3.503	63.846	2.390	2.061	95.548
31.143	3.761	65.095	2.528	2.243	95.229
29.308	3.978	66.714	2.730	2.436	94.834
21.397	4.333	74.220	4.285	3.148	92.566

SPECIFIC MODEL PARAMETERS IN KELVIN

		UNIQUAC		NRTL(ALPHA=.2)	
I J	AIJ	AJI		AIJ	AJI
1 2	228.78	-174.22		408.13	-631.85
1 3	-49.640	335.63		-354.66	1506.7
2 3	-48.392	-221.18		-587.83	-466.20

R1 = 3.4543   R2 = 3.1793   R3 = 0.9200
Q1 = 3.052   Q2 = 2.884   Q3 = 1.400

MEAN DEV. BETWEEN CALC. AND EXP. CONC. IN MOLE PCT

UNIQUAC (SPECIFIC PARAMETERS)   0.53
NRTL (SPECIFIC PARAMETERS)   0.68
UNIQUAC (COMMON PARAMETERS)   1.06

C₃H₆O₃-C₄H₁₀O

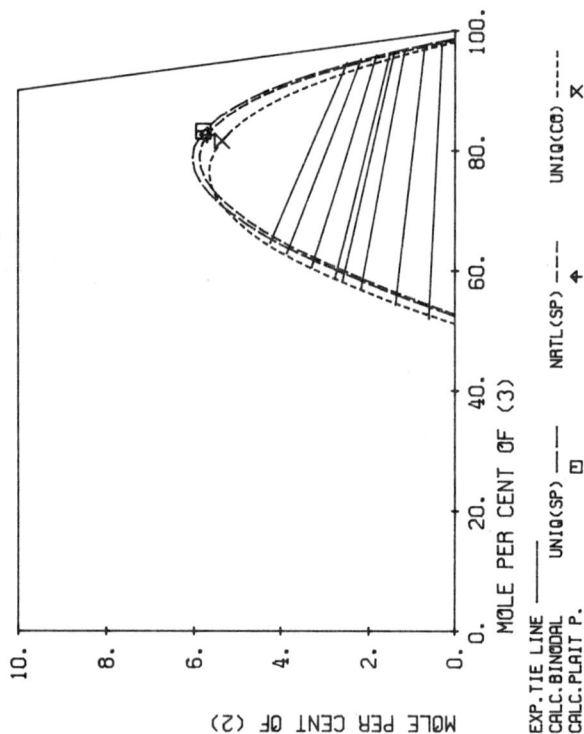

MOLE PER CENT OF (3)

MOLE PER CENT OF (2)

EXP.TIE LINE ——— UNIQ(SP) ⊟ NRTL(SP) ✦ UNIQ(CO) ✗
CALC.BINODAL ----
CALC.PLAIT P.

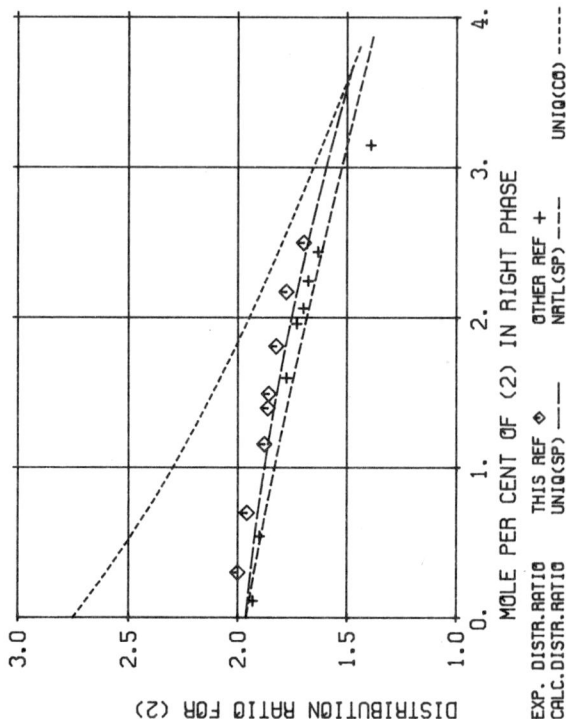

MOLE PER CENT OF (2) IN RIGHT PHASE

DISTRIBUTION RATIO FOR (2)

EXP. DISTR.RATIO ◇ THIS REF + OTHER REF + UNIQ(CO) ----
CALC.DISTR.RATIO UNIQ(SP) ——— NRTL(SP) ----

(1) C4H10O   1-BUTANOL
(2) C3H6O3   PROPANOIC ACID,2-HYDROXY
(3) H2O   WATER

ABABI V., POPA A.
AN.STIINT.UNIV.AL.I.CUZA IASI. 6(1960)1

TEMPERATURE = 25.0 DEG C   TYPE OF SYSTEM = 1

EXPERIMENTAL TIE LINES IN MOLE PCT

	LEFT PHASE			RIGHT PHASE	
(1)	(2)	(3)	(1)	(2)	(3)
47.397	0.600	52.003	1.901	0.300	97.799
44.332	1.364	54.305	1.930	0.696	97.373
40.790	2.168	57.042	2.050	1.154	96.795
38.988	2.596	58.416	2.185	1.393	96.421
38.407	2.768	58.825	2.222	1.490	96.288
36.175	3.297	60.528	2.308	1.808	95.884
33.275	3.855	62.869	2.551	2.169	95.280
31.149	4.236	64.615	2.861	2.497	94.642

SPECIFIC MODEL PARAMETERS IN KELVIN

		UNIQUAC		NRTL(ALPHA=.2)	
I J	AIJ	AJI		AIJ	AJI
1 2	228.78	-174.22		408.13	-631.85
1 3	-49.640	335.63		-354.66	1606.7
2 3	-48.392	-221.18		-587.88	-406.20

R1 = 3.4543   R2 = 3.1793   R3 = 0.9200
Q1 = 3.052   Q2 = 2.884   Q3 = 1.400

MEAN DEV. BETWEEN CALC. AND EXP. CONC. IN MOLE PCT

UNIQUAC (SPECIFIC PARAMETERS)   0.29
NRTL   (SPECIFIC PARAMETERS)   0.34
UNIQUAC (COMMON PARAMETERS)   0.50

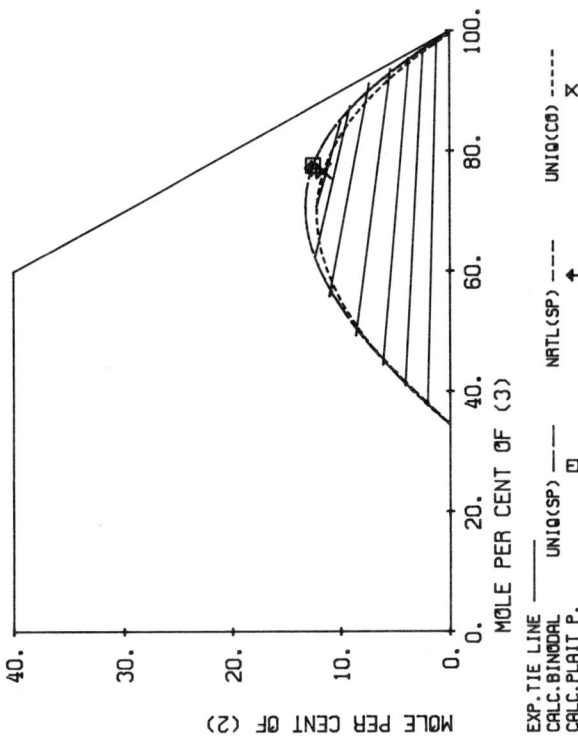

MOLE PER CENT OF (3)

MOLE PER CENT OF (2)

EXP.TIE LINE	———	UNIQ(SP)	———	☐	NRTL(SP)	———	↟	UNIQ(CO)	-----	✗
CALC.BINODAL										
CALC.PLAIT P.										

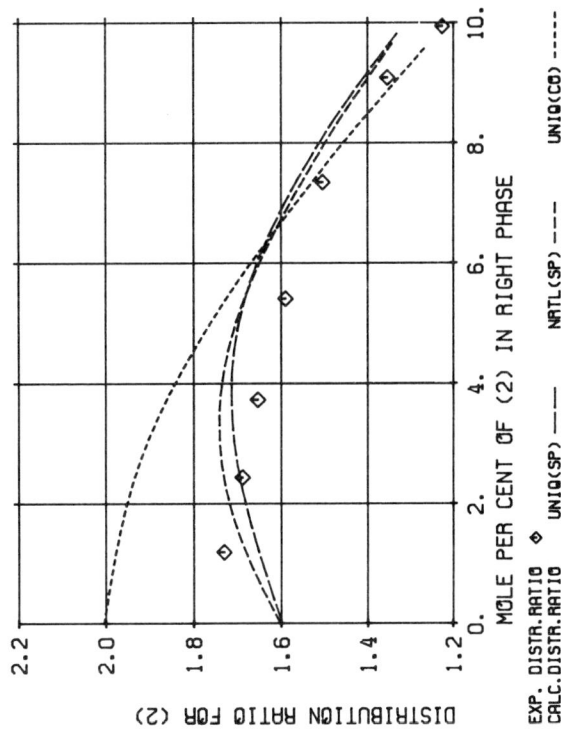

MOLE PER CENT OF (2) IN RIGHT PHASE

DISTRIBUTION RATIO FOR (2)

| EXP. DISTR.RATIO | ——— | ◇ | UNIQ(SP) | ——— | NRTL(SP) | ——— | UNIQ(CO) | ----- |
| CALC.DISTR.RATIO | | | | | | | | |

(1) C5H12O   1-BUTANOL,3-METHYL

(2) C3H6O3   PROPANOIC ACID,2-HYDROXY

(3) H2O   WATER

WEISER R.B., GEANKOPLIS C.J.
IND.ENG.CHEM. 47(1955)858

TEMPERATURE = 25.0 DEG C   TYPE OF SYSTEM = 1

EXPERIMENTAL TIE LINES IN MOLE PCT (GRAPH.INTERPOL.)

	LEFT PHASE			RIGHT PHASE	
(1)	(2)	(3)	(1)	(2)	(3)
65.418	0.0	34.582	0.517	0.0	99.483
60.007	2.051	37.942	0.579	1.185	98.235
54.941	4.102	40.957	0.645	2.431	96.925
49.450	6.152	44.398	0.739	3.721	95.540
42.153	8.596	49.251	0.908	5.407	93.685
33.187	11.044	55.769	1.320	7.344	91.336
25.156	12.309	62.535	3.359	9.084	87.557
17.130	12.195	70.675	5.143	9.937	84.920

SPECIFIC MODEL PARAMETERS IN KELVIN

		UNIQUAC		NRTL(ALPHA=.2)	
I	J	AIJ	AJI	AIJ	AJI
1	2	315.54	-181.27	546.90	-443.58
1	3	58.500	290.76	-185.09	1823.8
2	3	18.424	-189.31	-168.81	-347.82

R1 = 4.1279   R2 = 3.1793   R3 = 0.9200
Q1 = 3.588   Q2 = 2.884   Q3 = 1.400

MEAN DEV. BETWEEN CALC. AND EXP. CONC. IN MOLE PCT

UNIQUAC (SPECIFIC PARAMETERS)	0.37
NRTL (SPECIFIC PARAMETERS)	0.43
UNIQUAC (COMMON PARAMETERS)	0.72

$C_3H_6O_3$-$C_6H_{10}O$

(1) C6H10O   CYCLOHEXANONE

(2) C3H6O3   PROPANOIC ACID,2-HYDROXY

(3) H2O   WATER

ABABI V., POPA A.
AN.STIINT.UNIV.AL.I.CUZA IASI. 6(1960)1

TEMPERATURE = 25.0 DEG C   TYPE OF SYSTEM = 1

EXPERIMENTAL TIE LINES IN MOLE PCT

LEFT PHASE			RIGHT PHASE		
(1)	(2)	(3)	(1)	(2)	(3)
66.562	1.938	31.499	1.871	0.918	97.211
65.302	2.318	32.380	1.974	1.109	96.915
62.087	3.270	34.644	2.103	1.597	96.300
59.363	4.109	36.528	2.256	2.032	95.712
55.270	4.897	39.833	2.439	2.488	95.073
53.936	5.270	40.793	2.505	2.681	94.814
51.236	5.834	42.880	2.792	3.057	94.141
47.790	6.281	45.928	3.030	3.378	93.592
40.517	7.234	52.249	4.187	4.291	91.522
38.565	7.476	53.960	4.674	4.592	90.734

SPECIFIC MODEL PARAMETERS IN KELVIN

	UNIQUAC		NRTL(ALPHA=.2)	
I J	AIJ	AJI	AIJ	AJI
1 2	-152.74	11.128	464.00	-716.02
1 3	306.14	-0.14729E-01	-51.652	1309.9
2 3	-274.61	136.64	-377.04	-498.31

R1 = 4.1433   R2 = 3.1793   R3 = 0.9200
Q1 = 3.340   Q2 = 2.884   Q3 = 1.400

MEAN DEV. BETWEEN CALC. AND EXP. CONC. IN MOLE PCT

UNIQUAC (SPECIFIC PARAMETERS)   0.19
NRTL   (SPECIFIC PARAMETERS)   0.30
UNIQUAC (COMMON PARAMETERS)   0.58

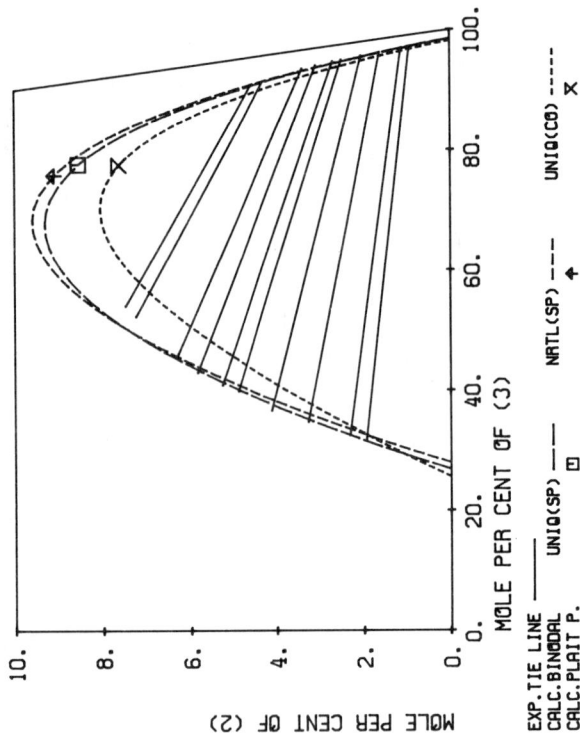

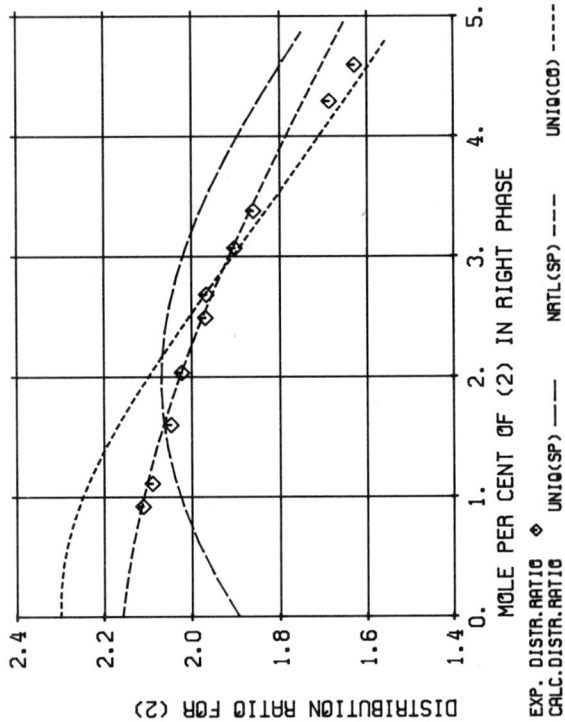

(1) C6H14O   1-HEXANOL

(2) C3H6O3   PROPANOIC ACID,2-HYDROXY

(3) H2O   WATER

ABABI V., POPA A.
AN.STIINT.UNIV.AL.I.CUZA IASI. 6(1960)1

TEMPERATURE = 25.0 DEG C   TYPE OF SYSTEM = 1

EXPERIMENTAL TIE LINES IN MOLE PCT

	LEFT PHASE			RIGHT PHASE	
(1)	(2)	(3)	(1)	(2)	(3)
64.746	1.387	33.867	0.092	0.894	99.014
59.651	3.454	36.895	0.136	2.307	97.557
57.945	4.673	37.383	0.139	3.021	96.840
53.980	6.101	39.919	0.165	4.101	95.734
52.565	6.542	40.893	0.188	4.408	95.403
48.264	8.323	43.413	0.218	5.694	94.087
42.131	10.892	46.976	0.233	7.806	91.960
40.450	11.303	47.771	0.235	8.016	91.749
37.450	12.705	49.845	0.265	8.892	90.842
36.992	12.991	50.018	0.269	9.420	90.311
33.699	13.974	52.327	0.299	9.975	89.726
31.950	14.731	53.318	0.384	10.823	88.793
31.154	15.040	53.806	0.411	10.961	88.623

SPECIFIC MODEL PARAMETERS IN KELVIN

		UNIQUAC		NRTL(ALPHA=.2)	
I	J	AIJ	AJI	AIJ	AJI
1	2	249.29	-149.99	39.882	-274.96
1	3	26.951	409.36	-91.513	2744.9
2	3	43.231	-196.47	-54.994	-596.69

R1 = 4.8031   R2 = 3.1793   R3 = 0.9200
Q1 = 4.132   Q2 = 2.884   Q3 = 1.400

MEAN DEV. BETWEEN CALC. AND EXP. CONC. IN MOLE PCT

UNIQUAC (SPECIFIC PARAMETERS)   0.30
NRTL   (SPECIFIC PARAMETERS)   0.21
UNIQUAC (COMMON PARAMETERS)   0.87

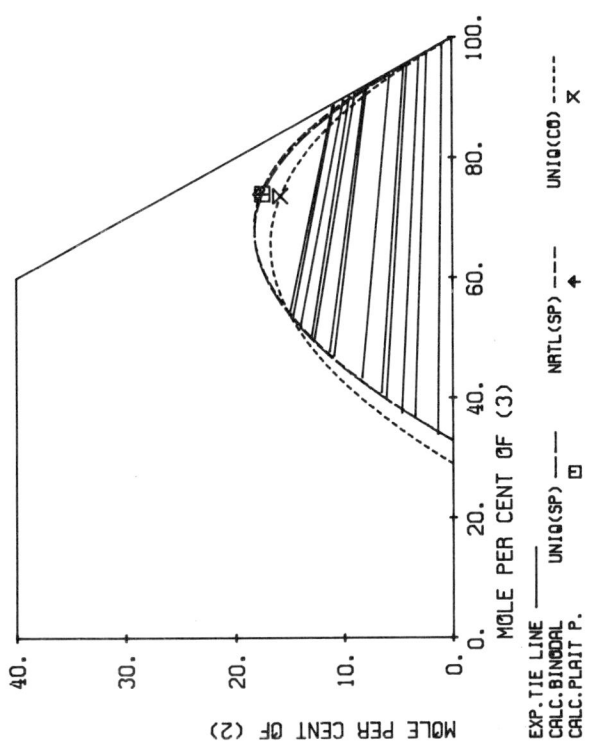

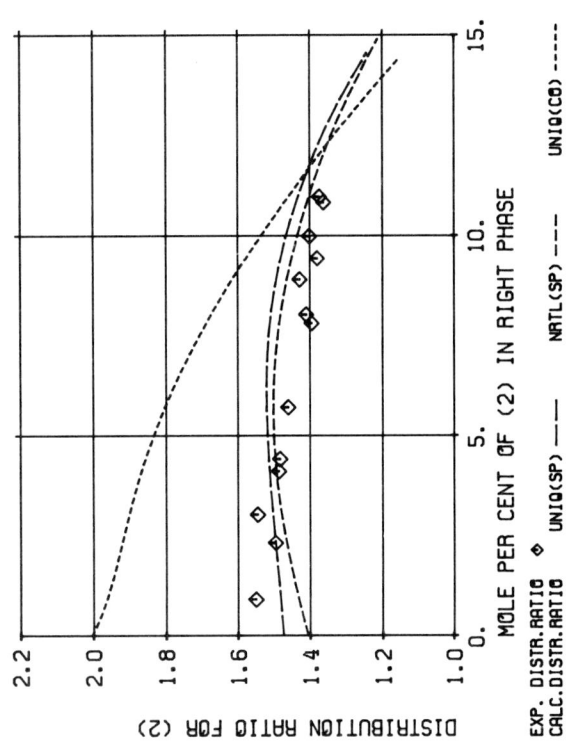

C₃H₆O₃-C₈H₁₈O

$C_3H_6O_3-C_8H_{18}O$

Upper plot axes:
MOLE PER CENT OF (2) (vertical, values 0., 10., 20., 30., 40.)
MOLE PER CENT OF (3) (horizontal, values 0., 20., 40., 60., 80., 100.)

EXP.TIE LINE ——— UNIQ(SP) □ ——— NRTL(SP) ✦ ——— UNIQ(CO) ✗ ─────
CALC.BINODAL
CALC.PLAIT P.

Lower plot axes:
DISTRIBUTION RATIO FOR (2) (vertical, values 1.0, 1.2, 1.4, 1.6)
MOLE PER CENT OF (2) IN RIGHT PHASE (horizontal, values 0., 5., 10., 15., 20.)

EXP. DISTR.RATIO ◇ UNIQ(SP) ◇ ——— NRTL(SP) ——— UNIQ(CO) ─────
CALC.DISTR.RATIO

(1) C8H18O   1-OCTANOL
(2) C3H6O3   PROPANOIC ACID,2-HYDROXY
(3) H2O   WATER

ABABI V., POPA A.
AN.STIINT.UNIV.AL.I.CUZA IASI. 6(1960)1

TEMPERATURE = 25.0 DEG C   TYPE OF SYSTEM = 1

EXPERIMENTAL TIE LINES IN MOLE PCT

	LEFT PHASE			RIGHT PHASE	
(1)	(2)	(3)	(1)	(2)	(3)
67.769	3.319	28.912	0.0	2.704	97.296
53.265	10.312	35.424	0.0	9.041	90.959
50.575	11.939	37.486	0.000	10.390	89.610
45.475	13.762	40.763	0.0	12.469	87.531
40.206	16.489	43.306	0.0	14.910	85.090
37.135	17.788	45.077	0.115	16.397	83.488
32.644	19.626	47.731	0.144	18.086	81.771
31.180	20.360	48.460	0.196	18.934	80.871

SPECIFIC MODEL PARAMETERS IN KELVIN

		UNIQUAC		NRTL(ALPHA=.2)	
I J	AIJ	AJI		AIJ	AJI
1 2	148.86	-159.59		288.16	-51.652
1 3	190.00	172.57		-59.645	-2307.2
2 3	-142.85	-59.119		413.17	-368.73

R1 = 6.1519   R2 = 3.1793   R3 = 0.9200
Q1 = 5.212   Q2 = 2.884   Q3 = 1.400

MEAN DEV. BETWEEN CALC. AND EXP. CONC. IN MOLE PCT

UNIQUAC (SPECIFIC PARAMETERS)   0.94
NRTL   (SPECIFIC PARAMETERS)   0.93
UNIQUAC (COMMON PARAMETERS)   1.71

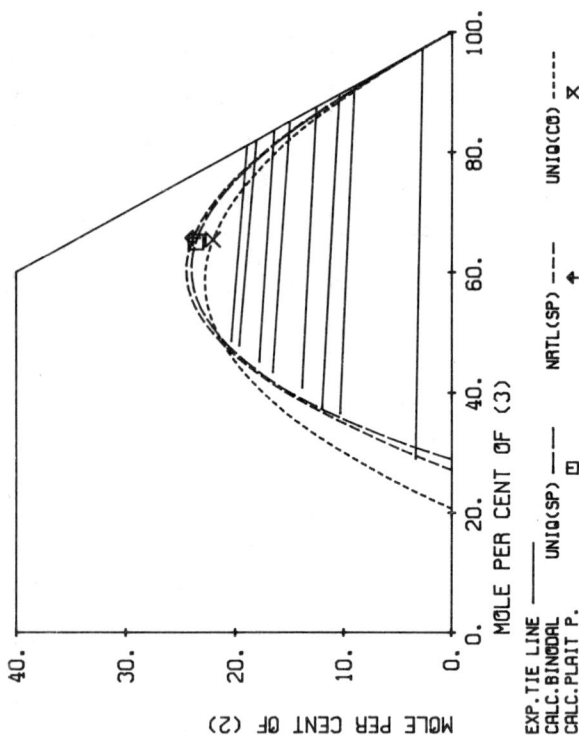

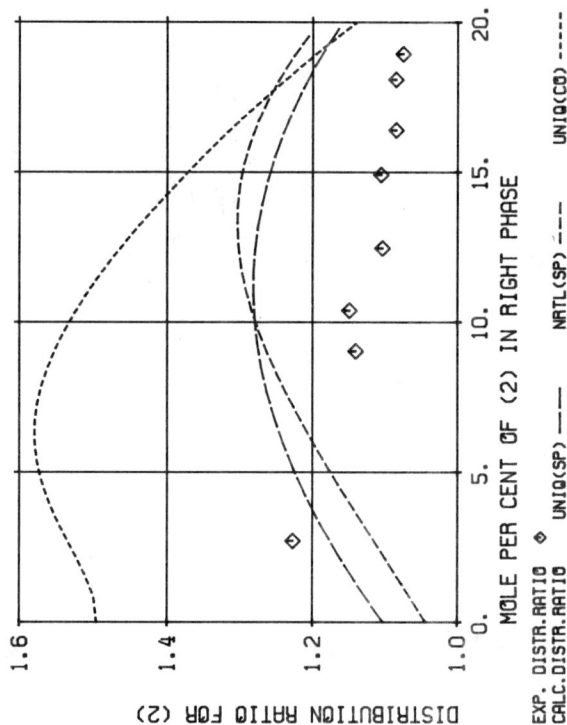

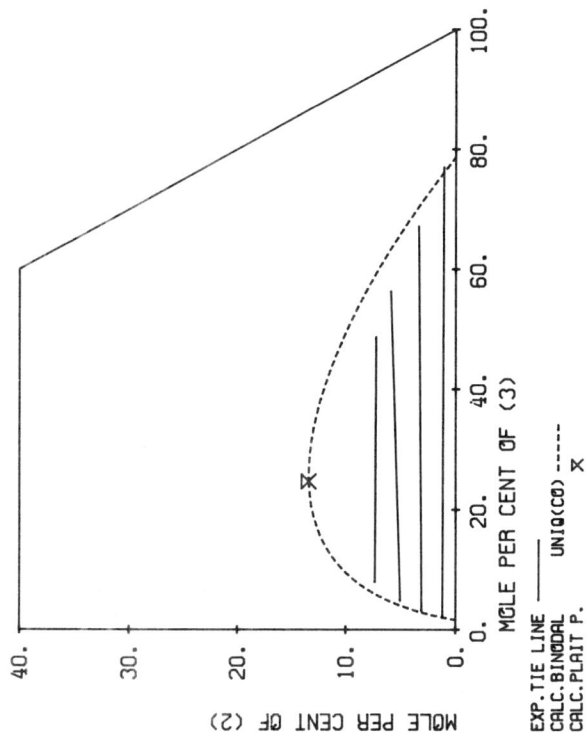

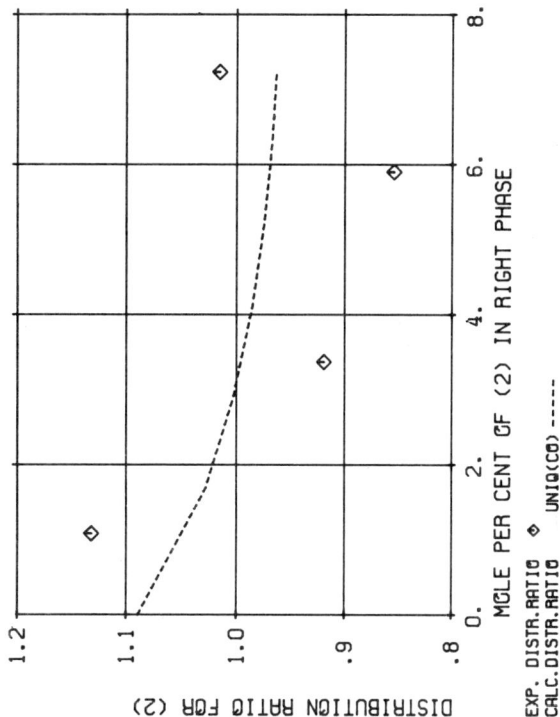

$C_3H_7NO-C_5H_4O_2$

(1) H2O   WATER

(2) C3H7NO   FORMIC ACID,AMIDE,N,N-DIMETHYL

(3) C5H4O2   FURFURAL

MASKHULIYA V.P., KRUPATKIN I.L.,
FAZOVYE RAVNOVESIYA,NO.2,KALININ,EDITOR:I.KRUPATKIN
(1975)25

TEMPERATURE = 25.0 DEG C   TYPE OF SYSTEM = 1

EXPERIMENTAL TIE LINES IN MOLE PCT

	LEFT PHASE			RIGHT PHASE	
(1)	(2)	(3)	(1)	(2)	(3)
96.755	1.220	2.025	21.863	1.078	77.059
93.739	3.097	3.164	29.341	3.368	67.291
90.127	5.041	4.831	37.698	5.897	56.405
84.712	7.347	7.941	44.011	7.232	48.757

MEAN DEV. BETWEEN CALC. AND EXP. CONC. IN MOLE PCT

UNIQUAC (COMMON PARAMETERS)   1.81

C₃H₇NO-C₆H₆

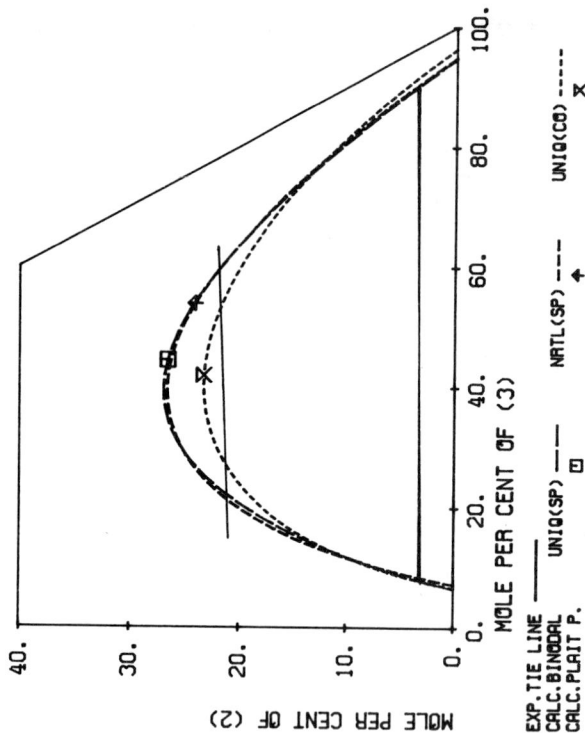

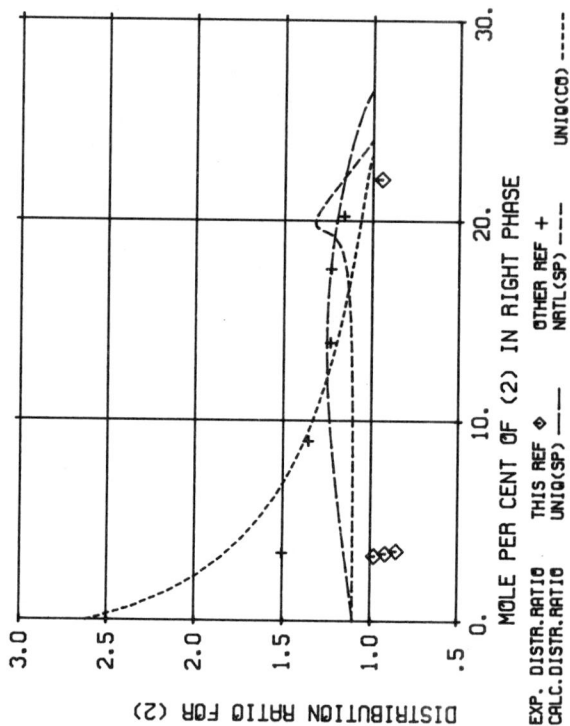

(1) C3H7NO   FORMIC ACID,AMIDE,N,N-DIMETHYL
(2) C6H6   BENZENE
(3) C7H16   HEPTANE

STEIB H.
J.PRAKT.CHEM. 4,28(1965)252

TEMPERATURE = 20.0 DEG C   TYPE OF SYSTEM = 1

EXPERIMENTAL TIE LINES IN MOLE PCT

	LEFT PHASE			RIGHT PHASE	
	(1)	(2)	(3)	(1)	(2)	(3)
93.348	0.0	6.652	6.598	0.0	93.402
92.965	0.0	7.035	6.333	0.0	93.667
93.042	0.0	6.958	5.935	0.0	94.064
89.501	2.987	7.512	6.678	3.499	89.822
89.013	3.088	7.899	6.811	3.374	89.815
88.837	3.186	7.977	6.813	3.250	89.937
64.181	20.917	14.902	15.104	22.074	62.822

SPECIFIC MODEL PARAMETERS IN KELVIN

		UNIQUAC		NRTL(ALPHA=.2)	
I	J	AIJ	AJI	AIJ	AJI
1	2	163.82	-107.08	690.69	-99.229
1	3	9.7438	-353.67	472.93	631.13
2	3	247.89	-110.30	-142.15	759.45

R1 = 3.0856   R2 = 3.1873   R3 = 5.1742
Q1 = 2.736   Q2 = 2.400   Q3 = 4.396

MEAN DEV. BETWEEN CALC. AND EXP. CONC. IN MOLE PCT

UNIQUAC (SPECIFIC PARAMETERS)   2.07
NRTL   (SPECIFIC PARAMETERS)   1.68
UNIQUAC (COMMON PARAMETERS)   2.42

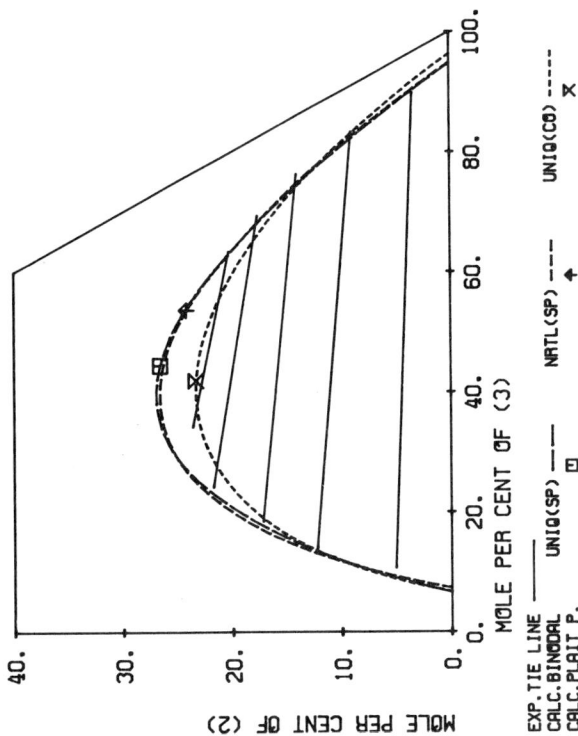

MOLE PER CENT OF (3)

MOLE PER CENT OF (2)

EXP.TIE LINE —— UNIQ(SP) —Ⓑ— NRTL(SP) —⬥— UNIQ(CO) —✕—
CALC.BINODAL
CALC.PLAIT P.

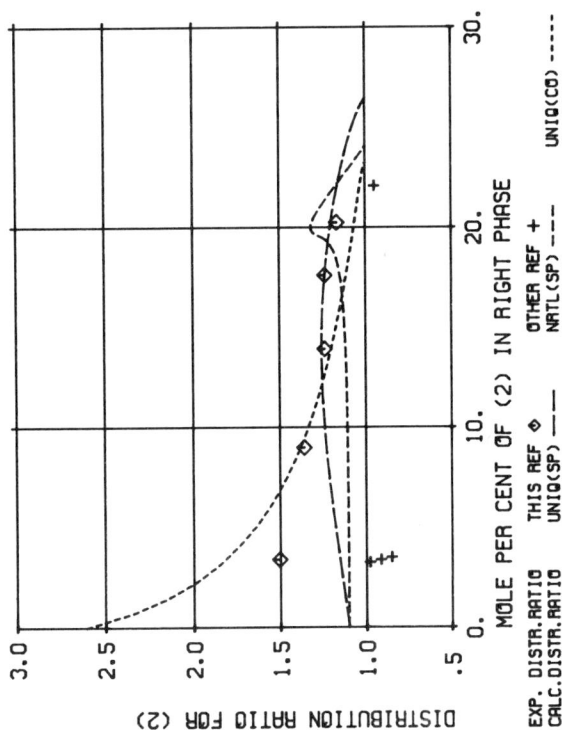

DISTRIBUTION RATIO FOR (2)

MOLE PER CENT OF (2) IN RIGHT PHASE

EXP. DISTR.RATIO   THIS REF ⬥   OTHER REF +   UNIQ(CO) -----
CALC.DISTR.RATIO   UNIQ(SP) ——   NRTL(SP) —— —

(1) C3H7NO   FORMIC ACID,AMIDE,N,N-DIMETHYL
(2) C6H6   BENZENE
(3) C7H16   HEPTANE

HANSON C., PATEL A.N., CHANG-KAKOTI D.K.
J.APPL.CHEM. 19(1969)320

TEMPERATURE = 20.5 DEG C   TYPE OF SYSTEM = 1

EXPERIMENTAL TIE LINES IN MOLE PCT

| | LEFT PHASE | | | RIGHT PHASE | |
(1)	(2)	(3)	(1)	(2)	(3)
84.275	5.074	10.650	6.680	3.375	89.945
73.884	12.199	13.917	7.630	8.985	83.384
64.293	17.186	18.521	9.698	13.912	76.390
54.126	21.703	24.170	12.952	17.584	69.464
42.292	23.527	34.181	16.245	20.227	63.528

SPECIFIC MODEL PARAMETERS IN KELVIN

| | | UNIQUAC | | | NRTL(ALPHA=.2) | |
I	J	AIJ	AJI		AIJ	AJI
1	2	163.82	-107.08		690.69	-99.229
1	3	9.7438	-353.67		472.93	631.13
2	3	247.89	-110.30		-142.15	759.45

R1 = 3.0856   R2 = 3.1873   R3 = 5.1742
Q1 = 2.736   Q2 = 2.400   Q3 = 4.396

MEAN DEV. BETWEEN CALC. AND EXP. CONC. IN MOLE PCT

UNIQUAC (SPECIFIC PARAMETERS)   1.56
NRTL   (SPECIFIC PARAMETERS)   1.65
UNIQUAC (COMMON PARAMETERS)   2.63

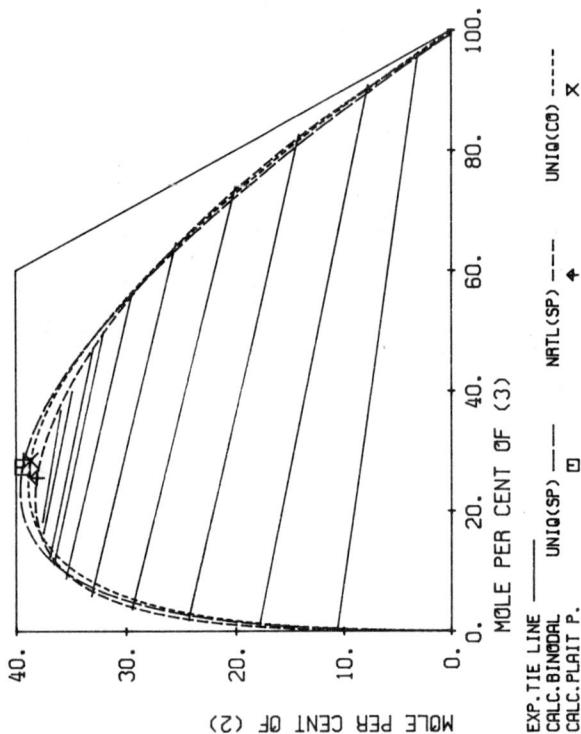

MOLE PER CENT OF (3)

MOLE PER CENT OF (2)

EXP.TIE LINE ──────   UNIQ(SP) ─────□   NRTL(SP) ─────+   UNIQ(CO) ─────✕
CALC.BINODAL
CALC.PLAIT P.

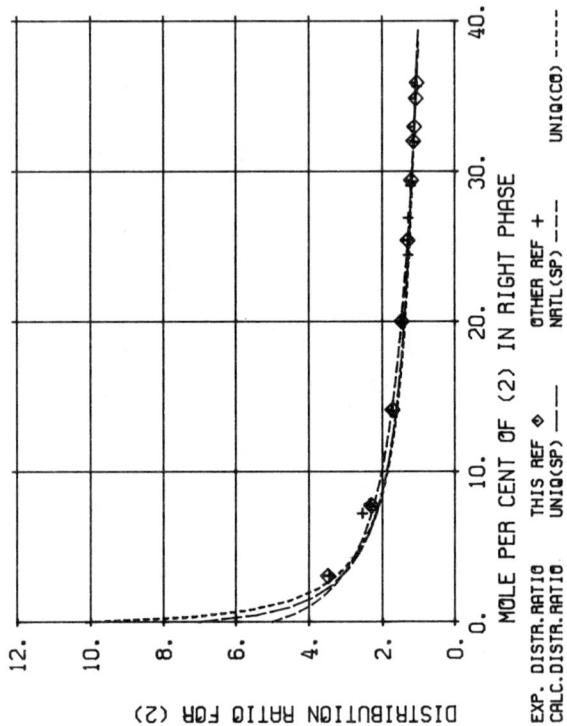

MOLE PER CENT OF (2) IN RIGHT PHASE

DISTRIBUTION RATIO FOR (2)

EXP. DISTR.RATIO ◇   THIS REF ◇   OTHER REF +   UNIQ(CO) ─────
CALC.DISTR.RATIO   UNIQ(SP)   NRTL(SP)

(1) H2O   WATER

(2) C3H7NO   FORMIC ACID,AMIDE,N,N-DIMETHYL

(3) C6H6   BENZENE

ROETHLIN S., CRUETZEN J.L., SCHULTZE G.R.
CHEM.ING.TECH. 29(1957)211

TEMPERATURE = 20.0 DEG C   TYPE OF SYSTEM = 1

EXPERIMENTAL TIE LINES IN MOLE PCT

	LEFT PHASE			RIGHT PHASE	
(1)	(2)	(3)	(1)	(2)	(3)
89.100	10.600	0.300	0.650	3.050	96.300
81.350	17.850	0.800	1.350	7.750	90.900
73.850	24.300	1.850	3.300	14.100	82.600
66.900	29.450	3.650	6.000	20.000	74.000
61.000	33.150	5.850	9.900	25.400	64.700
55.750	35.450	8.800	14.250	29.400	56.350
52.300	36.550	11.150	18.100	32.000	49.900
50.700	36.900	12.400	19.950	32.950	47.100
46.200	37.500	16.300	25.250	34.850	39.900
44.150	37.600	18.250	27.350	35.900	36.750

SPECIFIC MODEL PARAMETERS IN KELVIN

		UNIQUAC		NRTL(ALPHA=.2)	
I	J	AIJ	AJI	AIJ	AJI
1	2	-75.260	-212.92	425.37	-614.21
1	3	381.60	665.25	2101.6	954.30
2	3	-30.674	-7.1154	-4.1328	2.4822

R1 = 0.9200   R2 = 3.0856   R3 = 3.1878
Q1 = 1.400   Q2 = 2.736   Q3 = 2.400

MEAN DEV. BETWEEN CALC. AND EXP. CONC. IN MOLE PCT

UNIQUAC (SPECIFIC PARAMETERS)   1.05
NRTL (SPECIFIC PARAMETERS)   0.34
UNIQUAC (COMMON PARAMETERS)   1.13

$C_3H_7NO\text{-}C_6H_6$

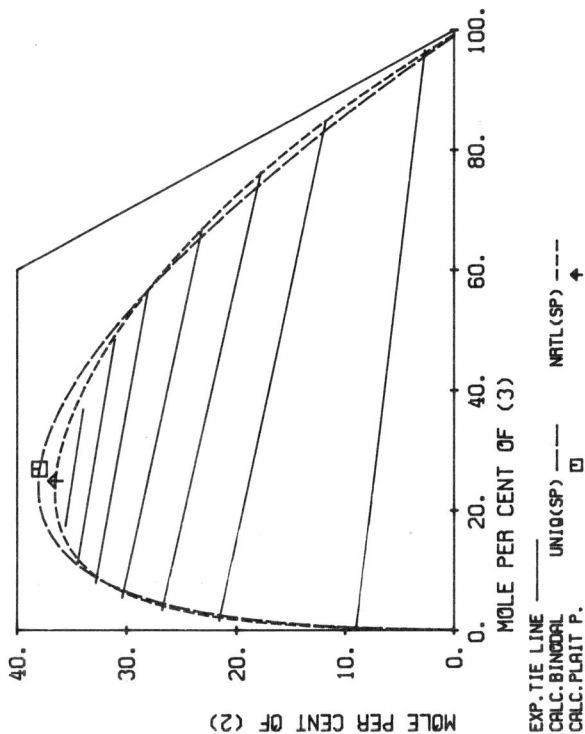

MOLE PER CENT OF (3)

MOLE PER CENT OF (2)

EXP.TIE LINE ———
CALC.BINODAL   UNIQ(SP) ———   NRTL(SP) – – –
CALC.PLAIT P.   ⊟   ✦

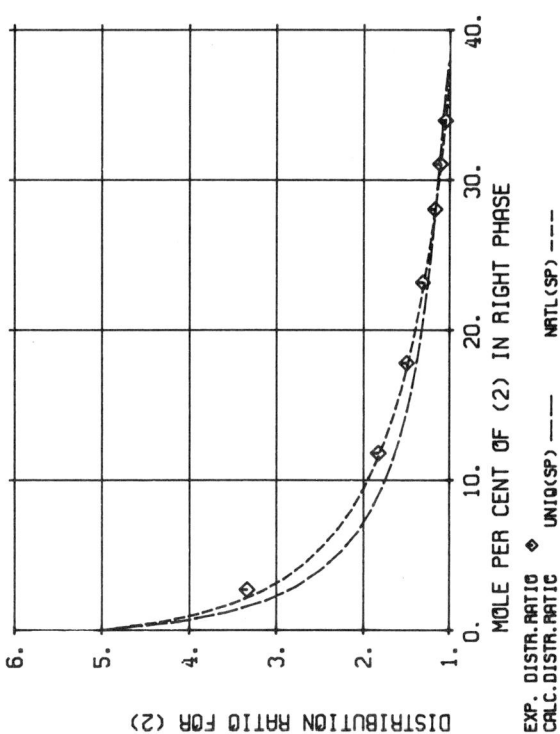

MOLE PER CENT OF (2) IN RIGHT PHASE

DISTRIBUTION RATIO FOR (2)

EXP. DISTR.RATIO ◇   UNIQ(SP) ———   NRTL(SP) – – –
CALC.DISTR.RATIO ———

(1) H2O   WATER
(2) C3H7NO   FORMIC ACID,AMIDE,N,N-DIMETHYL
(3) C6H6   BENZENE

ROETHLIN S., CRUETZEN J.L., SCHULTZE G.R.
CHEM.ING.TECH. 29(1957)211

TEMPERATURE = 35.0 DEG C   TYPE OF SYSTEM = 1

EXPERIMENTAL TIE LINES IN MOLE PCT

	LEFT PHASE			RIGHT PHASE	
(1)	(2)	(3)	(1)	(2)	(3)
90.600	9.000	0.400	0.700	2.700	96.600
76.750	21.600	1.650	3.250	11.800	84.950
69.850	26.800	3.350	5.900	17.800	76.300
64.150	30.400	5.450	9.850	23.150	67.000
59.300	32.800	7.900	15.150	28.050	56.800
54.450	34.450	11.100	20.450	31.050	48.500
47.000	35.600	17.400	29.200	33.950	36.850

SPECIFIC MODEL PARAMETERS IN KELVIN

		UNIQUAC		NRTL(ALPHA=.2)	
I J	AIJ	AJI		AIJ	AJI
1 2	-97.163	-203.13		660.27	-751.59
1 3	291.96	655.04		1664.1	947.75
2 3	3.6436	-56.032		115.80	-139.17

R1 = 0.9200   R2 = 3.0856   R3 = 3.1878
Q1 = 1.400   Q2 = 2.736   Q3 = 2.400

MEAN DEV. BETWEEN CALC. AND EXP. CONC. IN MOLE PCT

UNIQUAC (SPECIFIC PARAMETERS)   1.28
NRTL   (SPECIFIC PARAMETERS)   0.38

$C_3H_7NO-C_6H_6$

(1) H2O   WATER

(2) C3H7NO   FORMIC ACID, AMIDE, N,N-DIMETHYL

(3) C6H6   BENZENE

ROETHLIN S., CRUETZEN J.L., SCHULTZE G.R.
CHEM.ING.TECH. 29(1957)211

TEMPERATURE = 50.0 DEG C   TYPE OF SYSTEM = 1

EXPERIMENTAL TIE LINES IN MOLE PCT

	LEFT PHASE			RIGHT PHASE	
(1)	(2)	(3)	(1)	(2)	(3)
92.250	7.400	0.350	1.650	4.550	93.800
86.000	13.200	0.800	2.900	8.000	89.100
79.950	18.450	1.600	4.500	11.900	83.600
67.600	27.400	5.000	11.150	22.350	66.500
61.900	30.300	7.800	16.000	26.450	57.550
58.050	31.700	10.250	20.800	29.100	50.100
55.100	32.500	12.400	24.200	30.500	45.300
50.450	33.250	16.300	29.700	32.000	38.300

SPECIFIC MODEL PARAMETERS IN KELVIN

	UNIQUAC		NRTL(ALPHA=.2)	
I J	AIJ	AJI	AIJ	AJI
1 2	-47.165	-161.09	1118.0	-768.14
1 3	266.56	666.21	1702.1	919.94
2 3	-59.597	36.108	246.31	-137.08

R1 = 0.9200   R2 = 3.0856   R3 = 3.1878
Q1 = 1.400   Q2 = 2.736   Q3 = 2.400

MEAN DEV. BETWEEN CALC. AND EXP. CONC. IN MOLE PCT
UNIQUAC (SPECIFIC PARAMETERS)   1.09
NRTL (SPECIFIC PARAMETERS)   0.30

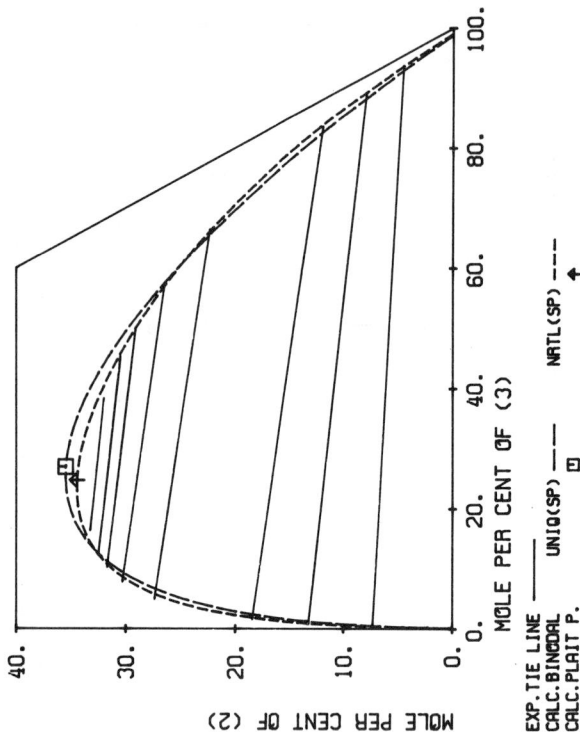

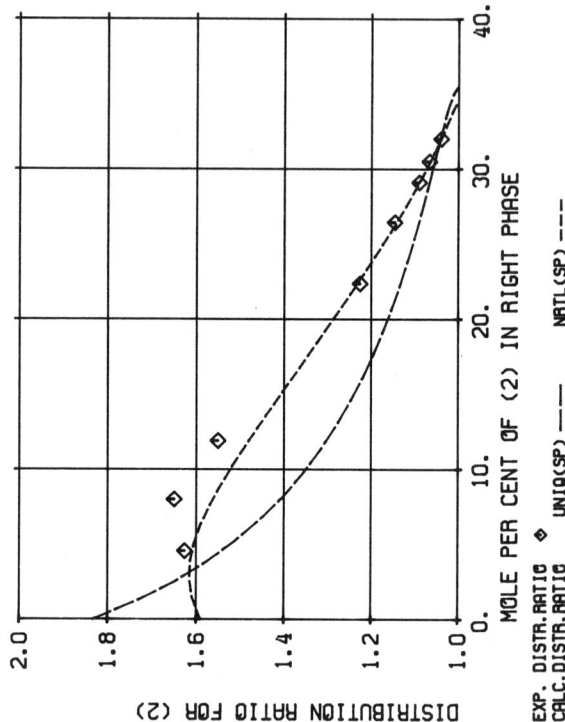

C$_3$H$_7$NO-C$_6$H$_6$

(1) H2O   WATER
(2) C3H7NO   FORMIC ACID,AMIDE,N,N-DIMETHYL
(3) C6H6   BENZENE

STEIB H.
J.PRAKT.CHEM. 4,28(1965)252

TEMPERATURE = 20.0 DEG C   TYPE OF SYSTEM = 1

EXPERIMENTAL TIE LINES IN MOLE PCT

	LEFT PHASE			RIGHT PHASE	
(1)	(2)	(3)	(1)	(2)	(3)
89.106	10.587	0.308	0.860	3.073	96.067
80.418	18.400	1.182	2.123	7.220	90.657
81.328	17.874	0.798	1.282	7.687	91.032
80.886	18.529	0.585	3.770	7.846	88.384
73.833	24.294	1.873	3.348	14.030	82.621
66.911	29.546	3.543	5.115	19.896	73.989
64.016	31.453	4.531	9.117	24.426	66.457
60.978	33.163	5.859	9.843	25.424	64.733
58.415	34.612	6.973	13.012	26.884	60.103
56.077	35.575	8.348	14.348	29.041	52.413
55.817	35.419	8.764	14.348	29.316	56.336
52.288	36.515	11.197	18.200	31.939	49.861
44.142	37.536	18.323	27.539	35.657	36.804

SPECIFIC MODEL PARAMETERS IN KELVIN

	UNIQUAC		NRTL(ALPHA=.2)	
I J	AIJ	AJI	AIJ	AJI
1 2	-75.260	-212.92	425.37	-614.21
1 3	381.60	665.25	2101.6	954.30
2 3	-30.674	-7.1154	-4.1328	2.4822

R1 = 0.9200   R2 = 3.0856   R3 = 3.1878
Q1 = 1.400   Q2 = 2.736   Q3 = 2.400

MEAN DEV. BETWEEN CALC. AND EXP. CONC. IN MOLE PCT

UNIQUAC (SPECIFIC PARAMETERS)   1.07
NRTL   (SPECIFIC PARAMETERS)   0.66
UNIQUAC (COMMON PARAMETERS)   1.14

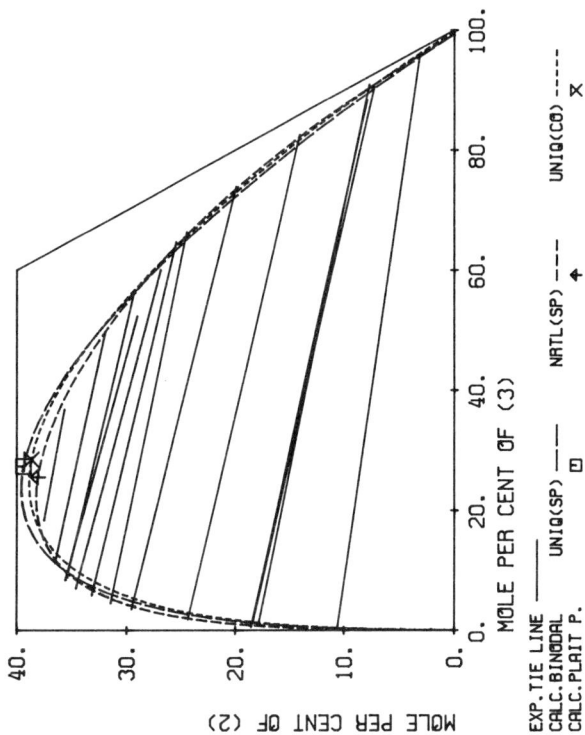

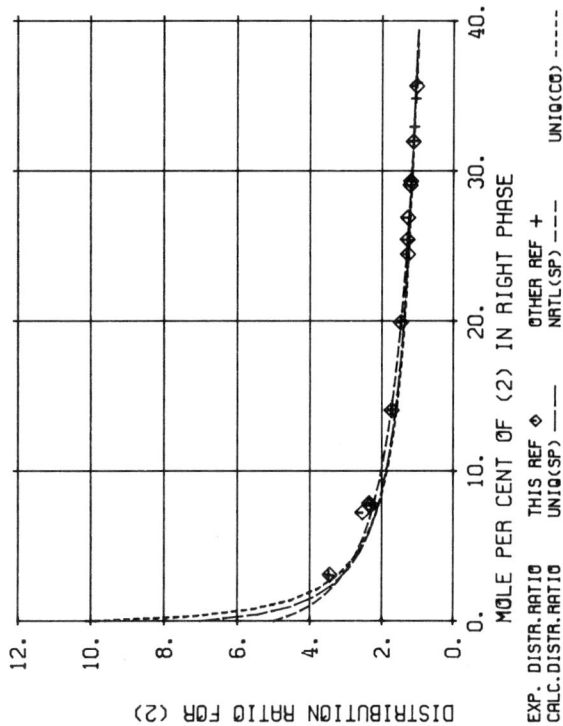

MOLE PER CENT OF (2)
MOLE PER CENT OF (3)

EXP.TIE LINE ——   UNIQ(SP) —— ⊟   NRTL(SP) —— ✦   UNIQ(CO) ----- ✗
CALC.BINODAL
CALC.PLAIT P.

DISTRIBUTION RATIO FOR (2)
MOLE PER CENT OF (2) IN RIGHT PHASE

EXP. DISTR.RATIO ◇   THIS REF +   OTHER REF +
CALC.DISTR.RATIO   UNIQ(SP) ——   NRTL(SP) ---   UNIQ(CO) -----

C₃H₇NO-C₆H₁₂

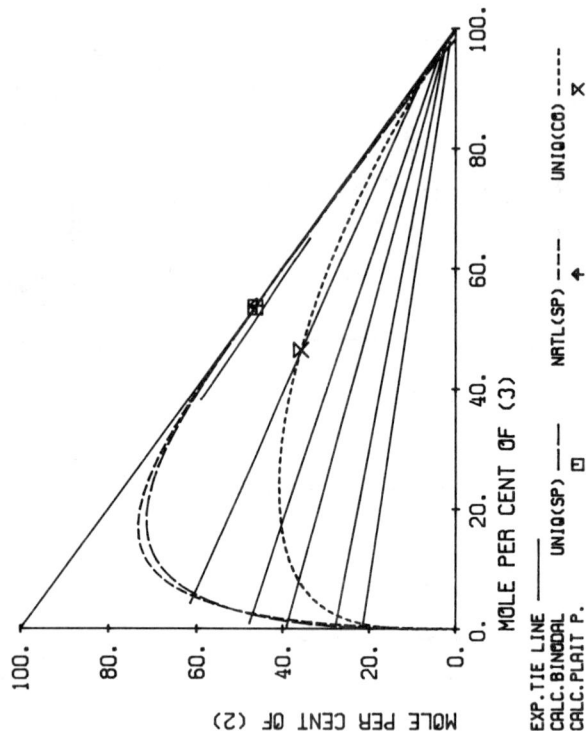

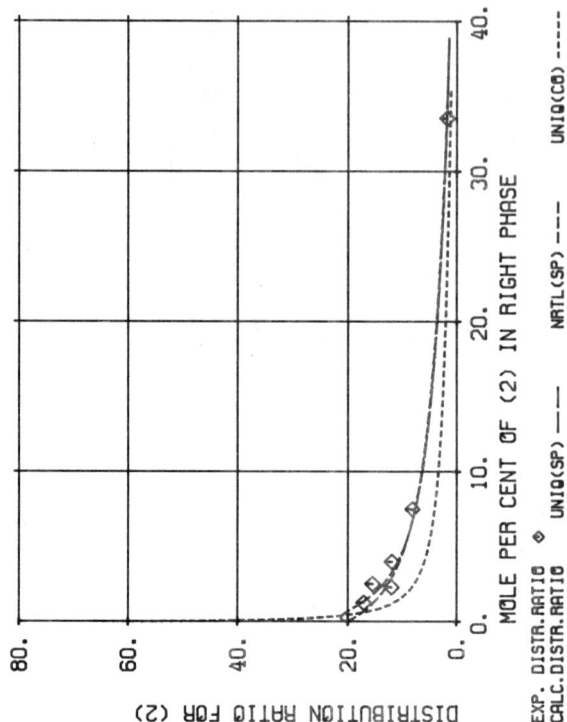

(1) H2O   WATER
(2) C3H7NO   FORMIC ACID,AMIDE,N,N-DIMETHYL
(3) C6H12   1-HEXENE

STEIB H.
J.PRAKT.CHEM. 4,31(1966)179

TEMPERATURE = 20.0 DEG C   TYPE OF SYSTEM = 1

EXPERIMENTAL TIE LINES IN MOLE PCT

	LEFT PHASE			RIGHT PHASE	
(1)	(2)	(3)	(1)	(2)	(3)
78.336	21.522	0.142	0.926	1.255	97.819
77.879	27.722	0.399	0.464	2.288	97.243
60.532	39.042	0.426	0.464	2.516	97.020
51.199	47.892	0.909	0.923	3.980	95.098
34.293	61.485	4.222	0.461	7.497	92.042
2.941	58.925	38.134	1.325	33.531	65.144

SPECIFIC MODEL PARAMETERS IN KELVIN

I J	UNIQUAC AIJ	AJI	NRTL(ALPHA=.2) AIJ	AJI
1 2	-255.12	4.5006	-450.66	215.19
1 3	455.97	583.88	1614.2	1093.5
2 3	-1.8446	235.44	252.07	416.32

R1 = 0.9200   R2 = 3.0856   R3 = 4.2697
Q1 = 1.400   Q2 = 2.736   Q3 = 3.644

MEAN DEV. BETWEEN CALC. AND EXP. CONC. IN MOLE PCT

UNIQUAC (SPECIFIC PARAMETERS)   0.87
NRTL (SPECIFIC PARAMETERS)   0.80
UNIQUAC (COMMON PARAMETERS)   8.66

(1) H2O   WATER
(2) C3H7NO   FORMIC ACID,AMIDE,N,N-DIMETHYL
(3) C6H14   HEXANE

STEIB H.
J.PRAKT.CHEM. 4,31(1966)179

TEMPERATURE = 20.0 DEG C   TYPE OF SYSTEM = 2

EXPERIMENTAL TIE LINES IN MOLE PCT

| | LEFT PHASE | | | RIGHT PHASE | |
(1)	(2)	(3)	(1)	(2)	(3)
40.825	53.292	0.883	0.048	2.349	97.603
25.638	72.507	1.855	0.048	2.700	97.253
0.411	90.830	8.760	0.043	4.332	95.620

MEAN DEV. BETWEEN CALC. AND EXP. CONC. IN MOLE PCT
UNIQUAC (COMMON PARAMETERS)   0.80

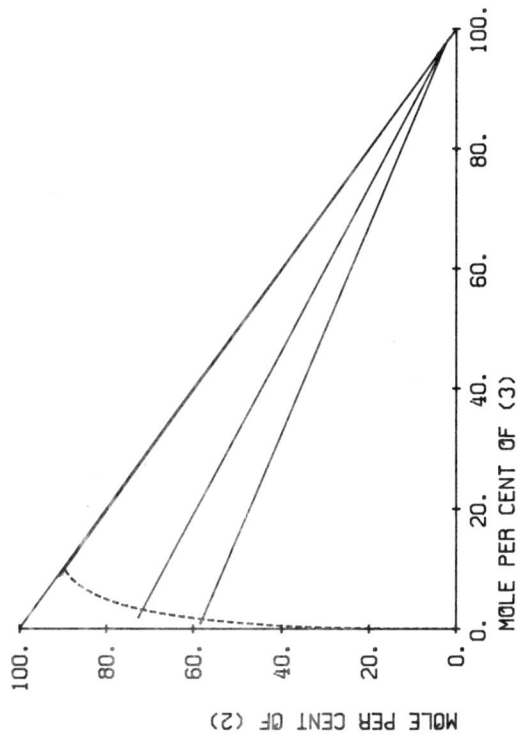

MOLE PER CENT OF (3)

MOLE PER CENT OF (2)

EXP.TIE LINE ——
CALC.BINODAL -----
UNIQ(C0) ——

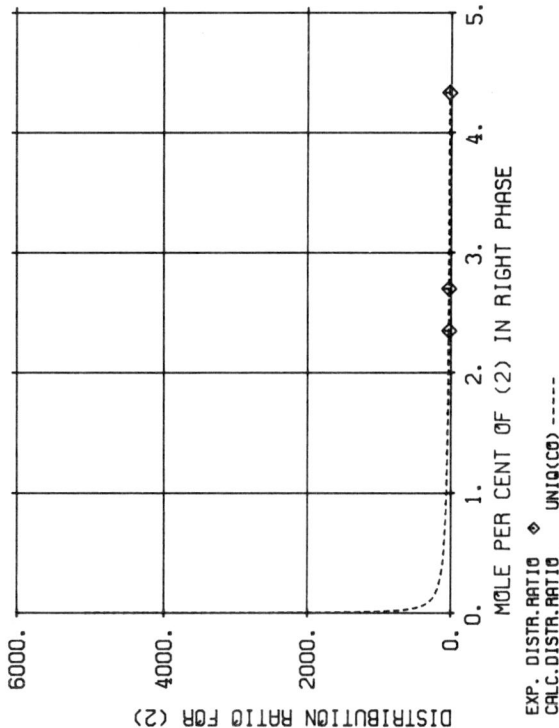

DISTRIBUTION RATIO FOR (2)

MOLE PER CENT OF (2) IN RIGHT PHASE

EXP. DISTR.RATIO   ◆
CALC.DISTR.RATIO   UNIQ(C0) -----

C₃H₇NO-C₇H₁₄

(1) H2O   WATER
(2) C3H7NO   FORMIC ACID,AMIDE,N,N-DIMETHYL
(3) C7H14   1-HEPTENE

STEIB H.
J.PRAKT.CHEM. 4,31(1966)179

TEMPERATURE = 20.0 DEG C   TYPE OF SYSTEM = 1

EXPERIMENTAL TIE LINES IN MOLE PCT

| | LEFT PHASE | | | RIGHT PHASE | |
(1)	(2)	(3)	(1)	(2)	(3)
78.122	21.847	0.031	0.542	0.401	99.057
72.099	27.867	0.034	0.540	1.598	97.861
61.409	38.190	0.401	0.539	2.525	96.936
34.322	63.560	2.118	0.533	6.833	92.634
1.775	67.044	31.181	0.500	31.031	68.470

SPECIFIC MODEL PARAMETERS IN KELVIN

| | | UNIQUAC | | NRTL(ALPHA=.2) | |
I	J	AIJ	AJI	AIJ	AJI
1	2	-281.82	-66.445	-538.39	-1.5127
1	3	382.16	570.45	1890.8	1025.6
2	3	-14.651	253.53	328.06	366.65

R1 = 0.9200   R2 = 3.0856   R3 = 4.9441
Q1 = 1.400   Q2 = 2.736   Q3 = 4.184

MEAN DEV. BETWEEN CALC. AND EXP. CONC. IN MOLE PCT

UNIQUAC (SPECIFIC PARAMETERS)   0.85
NRTL   (SPECIFIC PARAMETERS)   0.63
UNIQUAC (COMMON PARAMETERS)   8.35

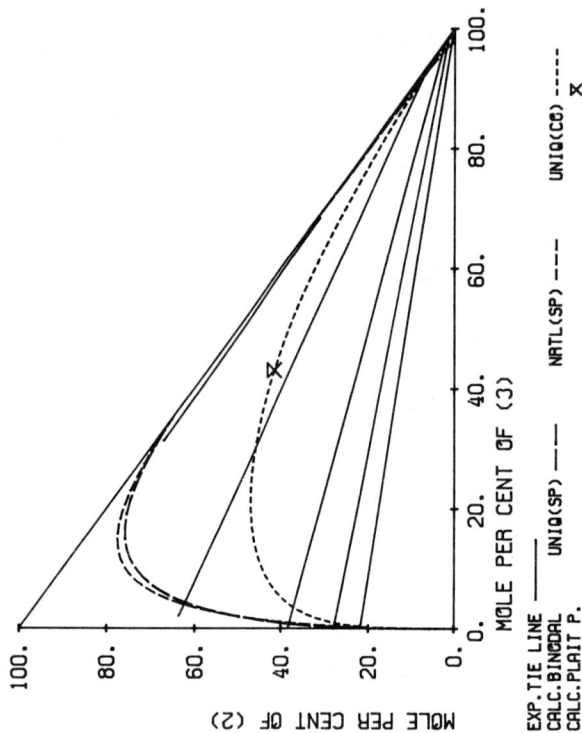

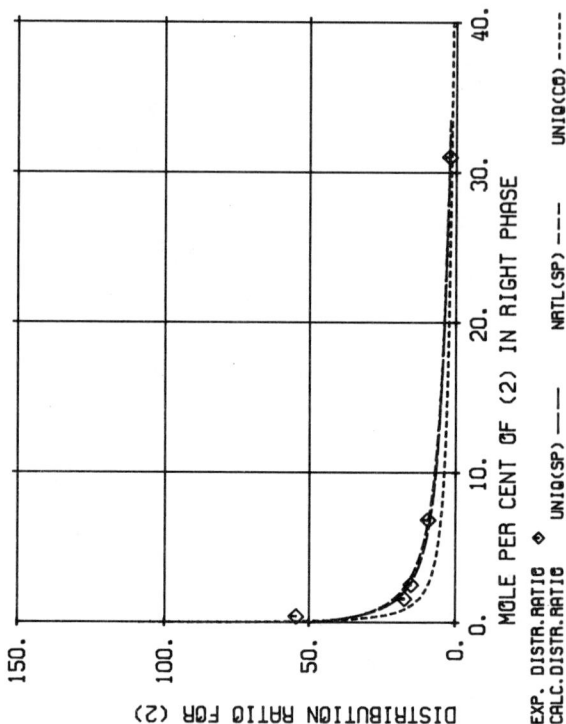

C₃H₈O-C₄H₈O₂

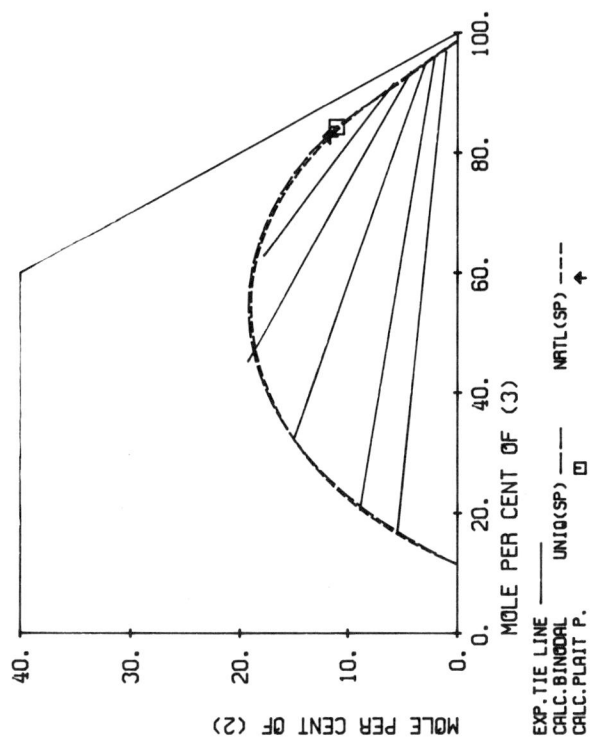

MOLE PER CENT OF (3)

MOLE PER CENT OF (2)

EXP.TIE LINE —— UNIQ(SP) — — — NRTL(SP) — — —
CALC.BINODAL ⊡
CALC.PLAIT P. ✦

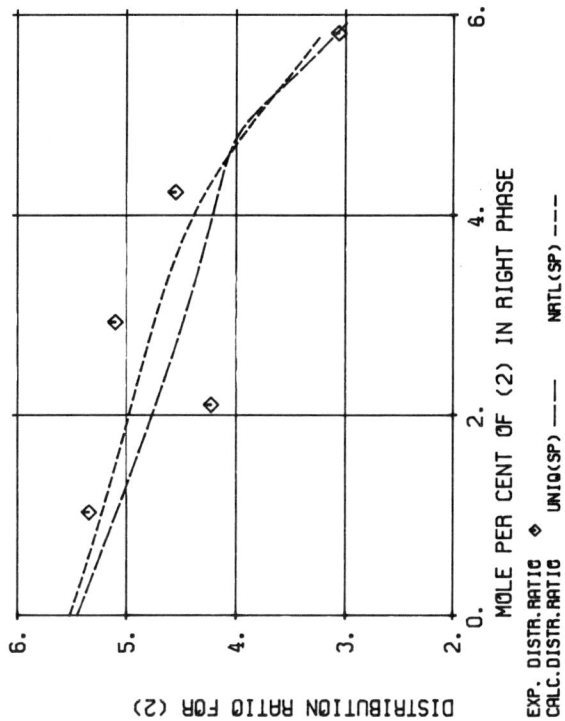

MOLE PER CENT OF (2) IN RIGHT PHASE

DISTRIBUTION RATIO FOR (2)

EXP. DISTR.RATIO ◇ UNIQ(SP) —— NRTL(SP) — — —
CALC.DISTR.RATIO

(1) C4H8O2   ACETIC ACID,ETHYL ESTER
(2) C3H8O   1-PROPANOL
(3) H2O   WATER

BEECH D.G., GLASSTONE S.
J.CHEM.SOC. (1938)67

TEMPERATURE =   0.0 DEG C   TYPE OF SYSTEM = 1

EXPERIMENTAL TIE LINES IN MOLE PCT

| | LEFT PHASE | | | RIGHT PHASE | |
(1)	(2)	(3)	(1)	(2)	(3)
77.880	5.484	16.635	2.078	1.026	96.895
70.557	8.886	20.557	2.104	2.102	95.794
52.494	14.944	32.563	2.092	2.929	94.979
35.466	19.253	45.281	2.230	4.228	93.542
19.324	17.803	62.872	2.578	5.817	91.605

SPECIFIC MODEL PARAMETERS IN KELVIN

| | | UNIQUAC | | | NRTL(ALPHA=.2) | |
I J	AIJ	AJI			AIJ	AJI
1 2	37.674	171.72			-66.299	161.67
1 3	399.47	77.134			-207.65	1034.7
2 3	-116.77	333.47			-597.91	1385.2

R1 = 3.4786   R2 = 2.7799   R3 = 0.9200
Q1 = 3.116   Q2 = 2.512   Q3 = 1.400

MEAN DEV. BETWEEN CALC. AND EXP. CONC. IN MOLE PCT

UNIQUAC (SPECIFIC PARAMETERS)   0.43
NRTL   (SPECIFIC PARAMETERS)   0.37

$C_3H_8O-C_4H_8O_2$

(1) C4H8O2   ACETIC ACID, ETHYL ESTER
----
(2) C3H8O   1-PROPANOL
----
(3) H2O   WATER

BEECH D.G., GLASSTONE S.
J.CHEM.SOC. (1938)67

TEMPERATURE = 20.0 DEG C   TYPE OF SYSTEM = 1

EXPERIMENTAL TIE LINES IN MOLE PCT

	LEFT PHASE			RIGHT PHASE	
(1)	(2)	(3)	(1)	(2)	(3)
76.056	6.660	17.284	1.746	1.148	97.106
59.482	13.015	27.503	1.830	2.046	96.124
47.043	16.590	36.367	1.928	2.622	95.450
32.319	19.227	48.454	2.138	3.378	94.483
24.234	19.124	56.642	2.350	4.014	93.636
14.691	16.993	68.316	3.131	5.719	91.149

SPECIFIC MODEL PARAMETERS IN KELVIN

I J	UNIQUAC AIJ	AJI	NRTL(ALPHA=.2) AIJ	AJI
1 2	-37.019	199.75	-187.31	117.74
1 3	413.07	90.746	251.27	1112.7
2 3	-131.31	329.88	-677.72	1481.1

R1 = 3.4786   R2 = 2.7799   R3 = 0.9200
Q1 = 3.116   Q2 = 2.512   Q3 = 1.400

MEAN DEV. BETWEEN CALC. AND EXP. CONC. IN MOLE PCT

UNIQUAC (SPECIFIC PARAMETERS)   0.50
NRTL (SPECIFIC PARAMETERS)   0.30
UNIQUAC (COMMON PARAMETERS)   0.70

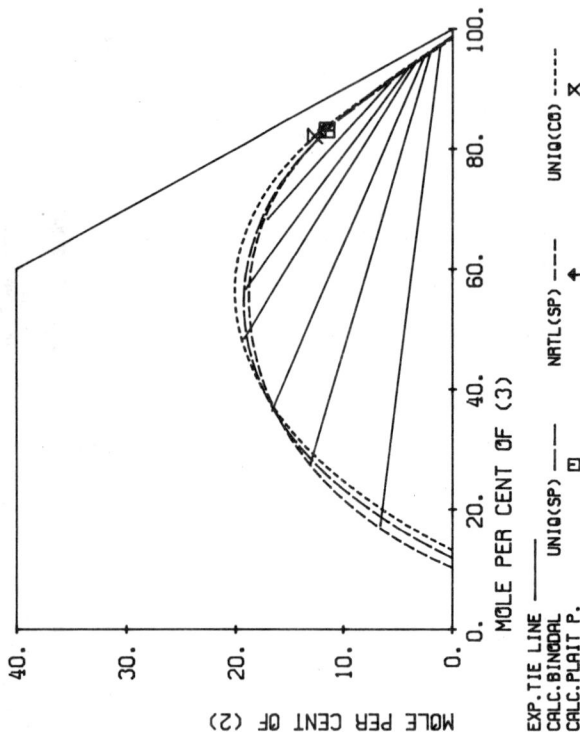

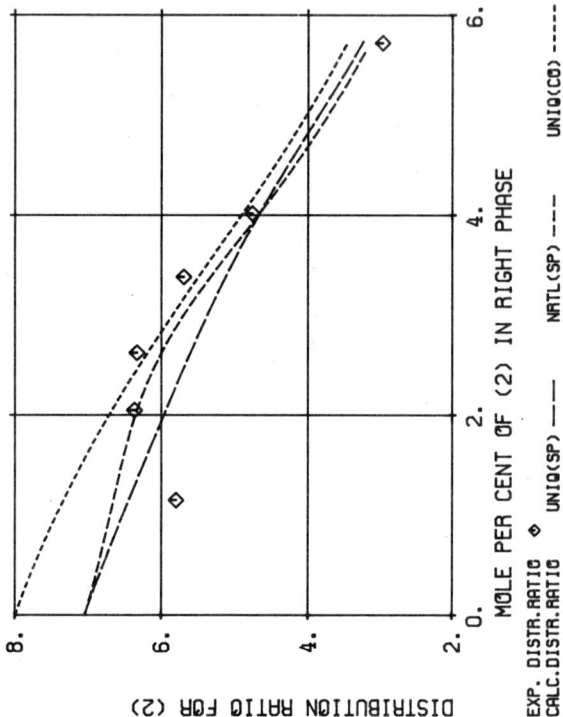

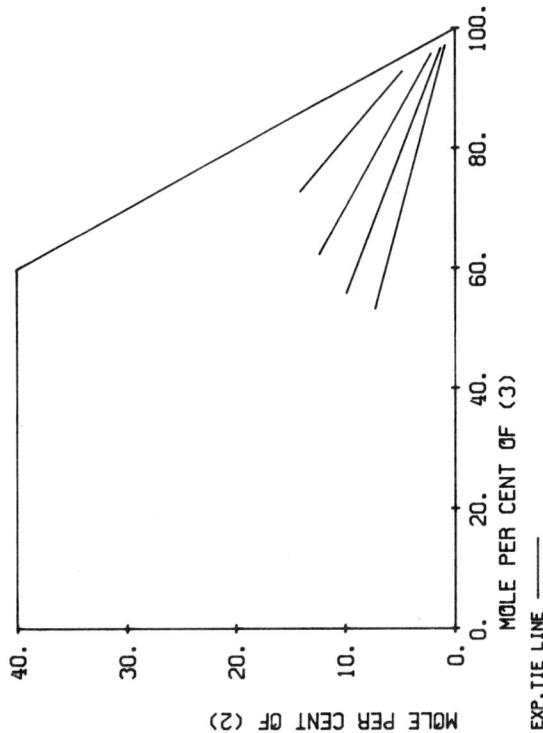

(1) C4H10O   1-BUTANOL

(2) C3H8O   1-PROPANOL

(3) H2O   WATER

MCCANTS J.F., JONES J.H., HOPSON W.H.
IND.ENG.CHEM. 45(1953)454

TEMPERATURE = 37.8 DEG C   TYPE OF SYSTEM = 1

EXPERIMENTAL TIE LINES IN MOLE PCT

	LEFT PHASE			RIGHT PHASE	
(1)	(2)	(3)	(1)	(2)	(3)
39.477	7.262	53.261	1.946	0.908	97.146
34.299	9.850	55.850	1.992	1.310	96.698
25.348	12.328	62.324	2.059	2.172	95.769
13.134	14.060	72.805	2.394	4.802	92.804

MOLE PER CENT OF (2)

MOLE PER CENT OF (3)

EXP.TIE LINE

C₃H₈O-C₄H₁₀O

(1) C4H10O   1-PROPANOL,2-METHYL
(2) C3H8O   1-PROPANOL
(3) H2O   WATER

MOZZHUKIN A.S., ET AL.
KHIM.TEKHNOL.TOPL.MASEL (1966)4,11

TEMPERATURE = 20.0 DEG C   TYPE OF SYSTEM = 1

EXPERIMENTAL TIE LINES IN MOLE PCT

	LEFT PHASE			RIGHT PHASE	
(1)	(2)	(3)	(1)	(2)	(3)
47.740	4.912	47.348	1.707	1.166	97.127
41.258	8.890	49.853	1.509	1.795	96.696
31.863	14.211	53.926	1.505	2.519	95.976
23.501	17.838	58.660	1.569	2.735	95.697
16.648	18.769	64.583	1.521	3.015	95.463

SPECIFIC MODEL PARAMETERS IN KELVIN

		UNIQUAC		NRTL(ALPHA=.2)	
I J	AIJ	AJI		AIJ	AJI
1 2	201.25	-16.768		249.07	-96.446
1 3	0.34576	310.25		-263.43	1593.2
2 3	-147.94	409.66		-498.24	1331.4

R1 = 3.4535	R2 = 2.7799	R3 = 0.9200		
Q1 = 3.048	Q2 = 2.512	Q3 = 1.400		

MEAN DEV. BETWEEN CALC. AND EXP. CONC. IN MOLE PCT

UNIQUAC (SPECIFIC PARAMETERS)   0.49
NRTL   (SPECIFIC PARAMETERS)   0.62
UNIQUAC (COMMON PARAMETERS)   1.14

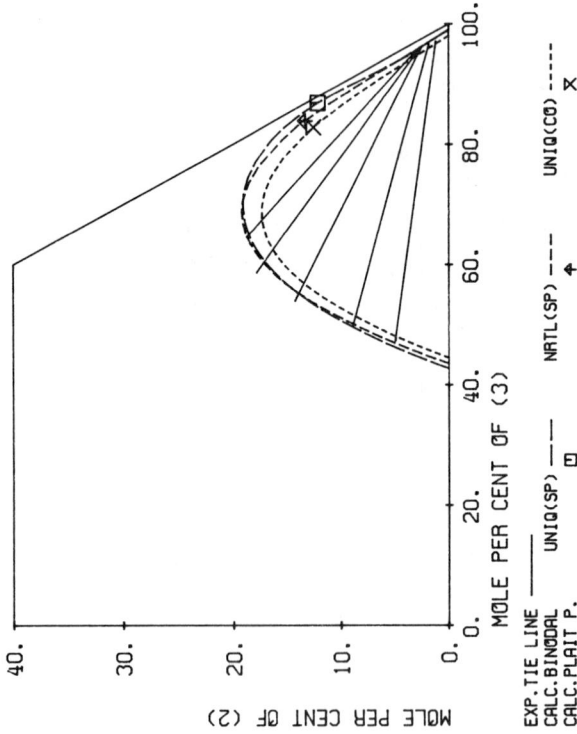

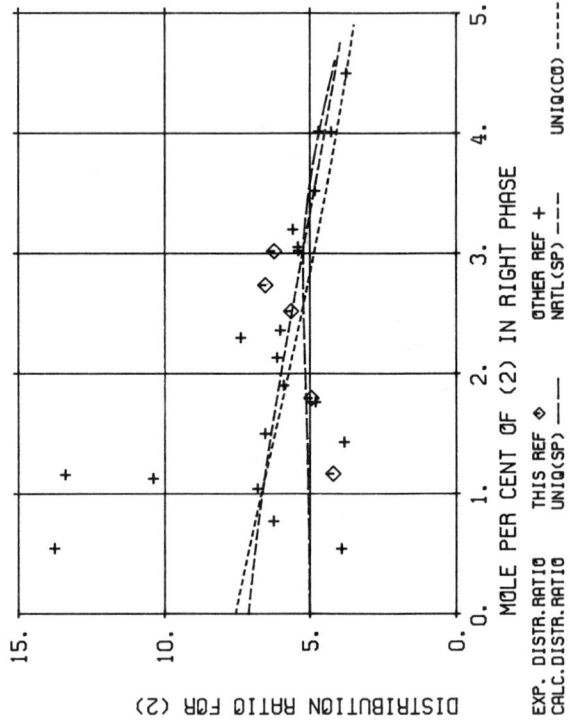

(1) C4H10O   1-PROPANOL,2-METHYL
----
(2) C3H8O   1-PROPANOL
----
(3) H2O   WATER

MOZZHUKIN A.S., ET AL.
KHIM.TEKHNOL.TOPL.MASEL (1966)4,11

TEMPERATURE = 20.0 DEG C   TYPE OF SYSTEM = 1

EXPERIMENTAL TIE LINES IN MOLE PCT

	LEFT PHASE			RIGHT PHASE	
(1)	(2)	(3)	(1)	(2)	(3)
47.562	5.474	46.965	1.579	1.427	96.994
42.712	8.487	48.800	1.535	1.762	96.702
26.231	16.503	57.266	1.925	3.051	95.024
17.012	18.909	64.079	2.264	4.013	93.723

SPECIFIC MODEL PARAMETERS IN KELVIN

		UNIQUAC		NRTL(ALPHA=.2)	
I	J	AIJ	AJI	AIJ	AJI
1	2	201.25	-16.768	249.07	-96.446
1	3	0.34576	310.25	-263.43	1593.2
2	3	-147.94	409.66	-498.24	1331.4

R1 = 3.4535   R2 = 2.7799   R3 = 0.9200
Q1 = 3.048   Q2 = 2.512   Q3 = 1.400

MEAN DEV. BETWEEN CALC. AND EXP. CONC. IN MOLE PCT

UNIQUAC (SPECIFIC PARAMETERS)   0.45
NRTL   (SPECIFIC PARAMETERS)   0.51
UNIQUAC (COMMON PARAMETERS)   0.92

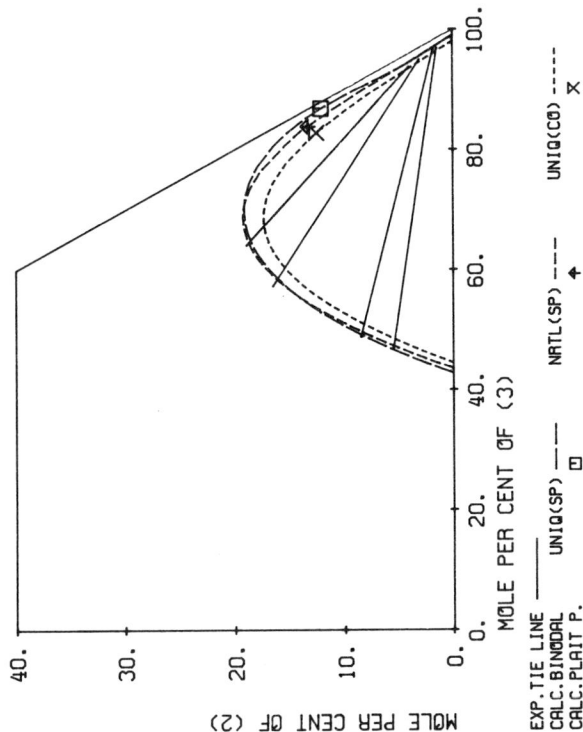

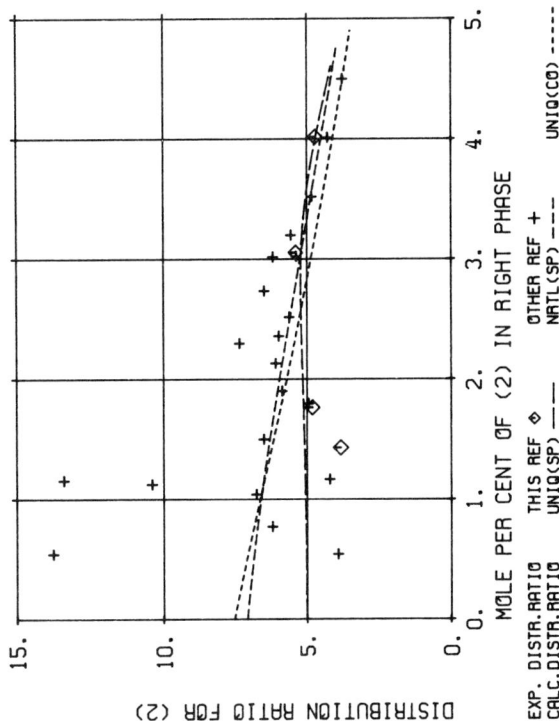

C₃H₈O-C₄H₁₀O

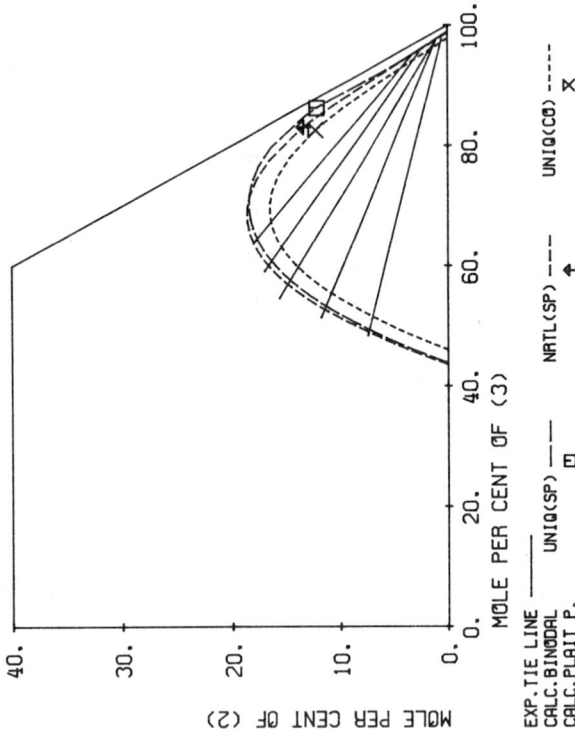

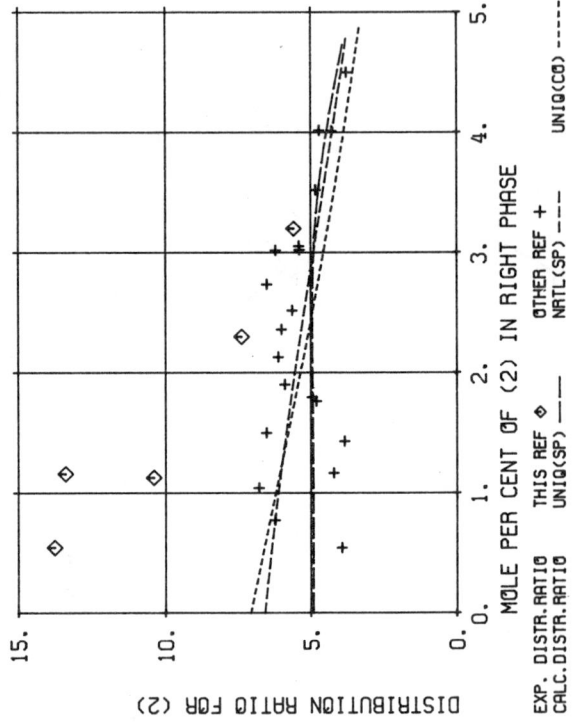

(1) C4H10O   1-PROPANOL,2-METHYL
(2) C3H8O   1-PROPANOL
(3) H2O   WATER

MOZZHUKIN A.S., ET AL.
KHIM.TEKHNOL.TOPL.MASEL (1966)4,11

TEMPERATURE = 30.0 DEG C   TYPE OF SYSTEM = 1

EXPERIMENTAL TIE LINES IN MOLE PCT

	LEFT PHASE			RIGHT PHASE	
(1)	(2)	(3)	(1)	(2)	(3)
44.099	7.484	48.417	1.710	0.543	97.747
36.931	11.693	51.375	1.486	1.125	97.389
29.879	15.535	54.586	1.487	1.158	97.355
23.945	16.945	59.110	1.359	2.299	96.342
18.111	17.878	64.011	1.556	3.198	95.245

SPECIFIC MODEL PARAMETERS IN KELVIN

		UNIQUAC		NRTL(ALPHA=.2)	
I	J	AIJ	AJI	AIJ	AJI
1	2	201.25	-16.768	249.07	-96.446
1	3	0.34576	310.25	-263.43	1593.2
2	3	-147.94	409.66	-498.24	1331.4

R1 = 3.4535   R2 = 2.7799   R3 = 0.9200
Q1 = 3.048   Q2 = 2.512   Q3 = 1.400

MEAN DEV. BETWEEN CALC. AND EXP. CONC. IN MOLE PCT

UNIQUAC (SPECIFIC PARAMETERS)   0.90
NRTL (SPECIFIC PARAMETERS)   0.90
UNIQUAC (COMMON PARAMETERS)   1.69

$C_3H_8O-C_4H_{10}O$

(1) C4H10O   1-PROPANOL, 2-METHYL
----
(2) C3H8O   1-PROPANOL
----
(3) H2O   WATER

PERELYGIN V.M., SHADENKOVA N.S.
IZV.VYSSH.UCHEBN.ZAVED.PISHCH.TEKHNOL. (1979)40

TEMPERATURE = 20.0 DEG C   TYPE OF SYSTEM = 1

EXPERIMENTAL TIE LINES IN MOLE PCT

         LEFT PHASE   RIGHT PHASE
      (1)   (2)   (3)   (1)   (2)   (3)

    51.580   2.120   46.300   2.140   0.540   97.320
    47.670   4.810   47.520   2.070   0.770   97.160
    43.710   7.070   49.220   2.010   1.040   96.950
    39.210   9.810   50.980   1.910   1.500   96.590
    36.510 11.220   52.270   1.870   1.900   96.230
    32.960 13.060   53.980   1.860   2.130   96.010
    29.940 14.210   55.850   1.840   2.360   95.800
    24.200 16.290   59.510   1.810   3.020   95.170
    20.550 17.020   62.430   1.880   3.520   94.600
    17.640 17.160   65.200   1.970   4.010   94.020
    14.940 16.960   68.100   2.010   4.500   93.490

SPECIFIC MODEL PARAMETERS IN KELVIN

          UNIQUAC   NRTL(ALPHA=.2)
 I J   AIJ   AJI   AIJ   AJI
 1 2   201.25   -16.763   249.07   -96.446
 1 3   0.34576   310.25   -263.43   1593.2
 2 3 -147.94   409.66   -498.24   1331.4

 R1 = 3.4535   R2 = 2.7799   R3 = 0.9200
 Q1 = 3.048   Q2 = 2.512   Q3 = 1.400

MEAN DEV. BETWEEN CALC. AND EXP. CONC. IN MOLE PCT

 UNIQUAC (SPECIFIC PARAMETERS)   0.59
 NRTL   (SPECIFIC PARAMETERS)   0.60
 UNIQUAC (COMMON PARAMETERS)   0.56

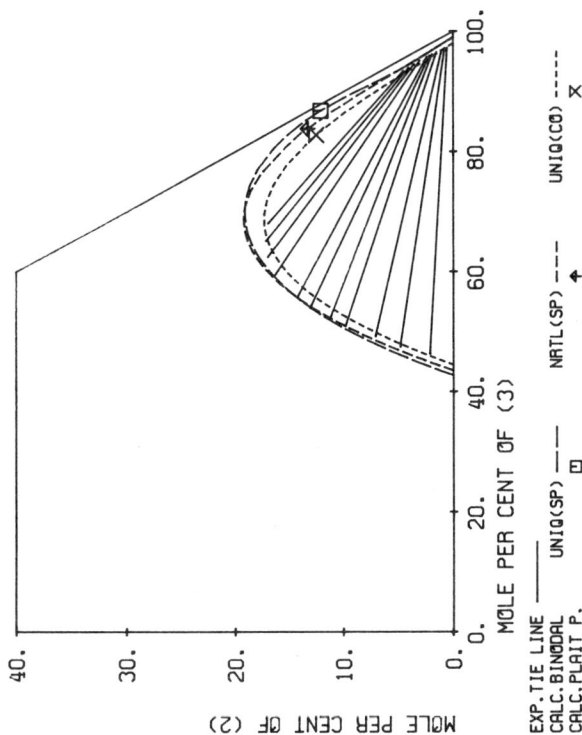

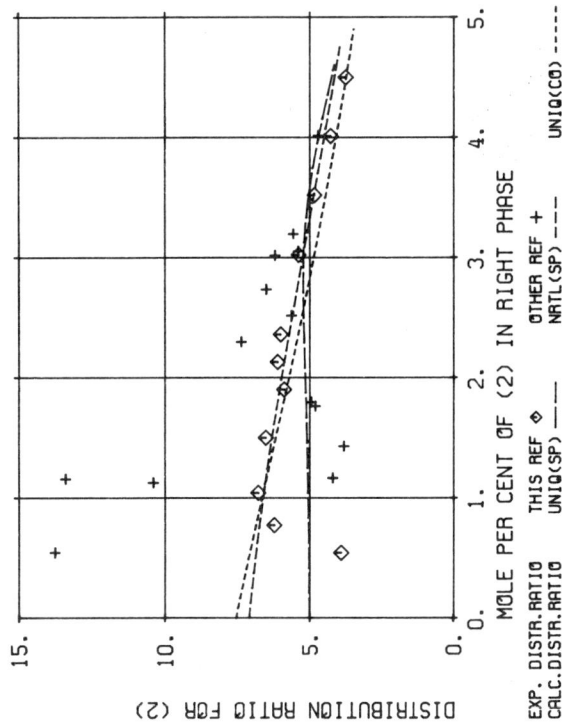

C₃H₈O-C₄H₁₀O

(1) C4H10O   1-PROPANOL,2-METHYL

(2) C3H8O   1-PROPANOL

(3) H2O   WATER

PERELYGIN V.M., SHADENKOVA N.S.
IZV.VYSSH.UCHEBN.ZAVED.PISHCH.TEKHNOL. (1979)40

TEMPERATURE = 40.0 DEG C   TYPE OF SYSTEM = 1

EXPERIMENTAL TIE LINES IN MOLE PCT

	LEFT PHASE			RIGHT PHASE	
(1)	(2)	(3)	(1)	(2)	(3)
47.590	3.210	49.200	1.910	0.300	97.790
43.220	5.560	51.220	1.850	0.550	97.600
38.870	8.280	52.850	1.780	0.880	97.340
34.990	10.080	54.930	1.740	1.210	97.050
32.680	11.020	56.300	1.720	1.390	96.890
29.440	12.890	57.670	1.680	1.830	96.490
23.780	14.590	61.630	1.710	2.340	95.950
18.030	15.640	66.330	1.800	3.160	95.040
15.540	15.500	68.960	1.990	3.710	94.300
11.920	14.430	73.650	2.280	4.470	93.250

SPECIFIC MODEL PARAMETERS IN KELVIN

		UNIQUAC		NRTL(ALPHA=.2)	
I	J	AIJ	AJI	AIJ	AJI
1	2	-104.66	110.83	594.02	-654.45
1	3	16.614	274.24	-269.62	1658.1
2	3	-176.74	404.58	-65.808	296.41

R1 = 3.4535   R2 = 2.7799   R3 = 0.9200
Q1 = 3.048   Q2 = 2.512   Q3 = 1.400

MEAN DEV. BETWEEN CALC. AND EXP. CONC. IN MOLE PCT

UNIQUAC (SPECIFIC PARAMETERS)   0.29
NRTL   (SPECIFIC PARAMETERS)   1.29

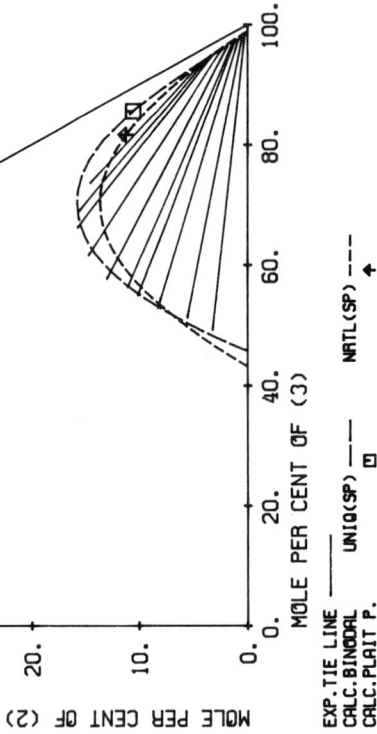

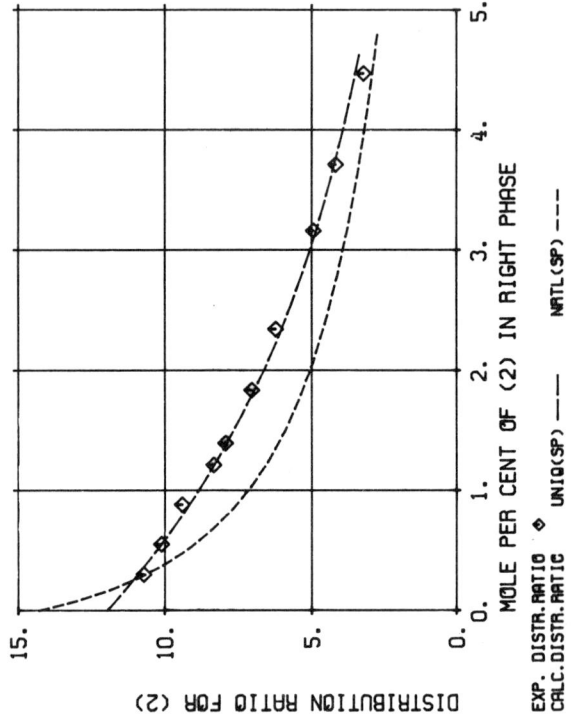

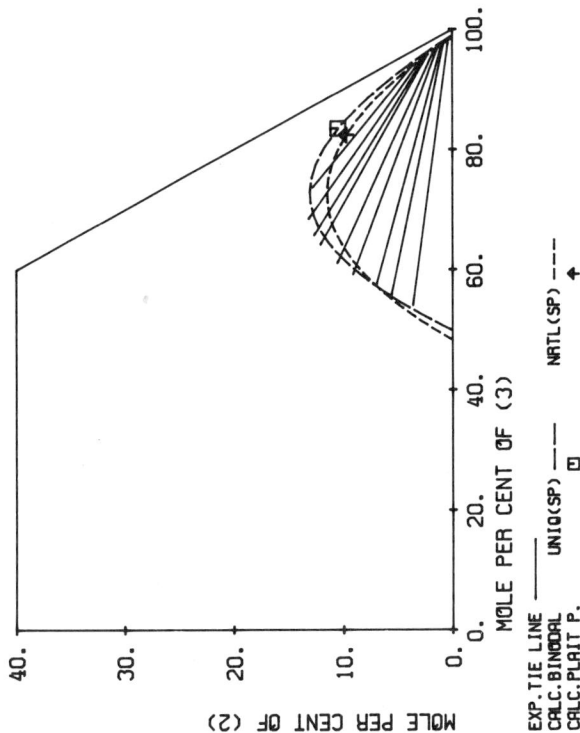

MOLE PER CENT OF (3)

MOLE PER CENT OF (2)

EXP.TIE LINE —— UNIQ(SP) □ NRTL(SP) -·-· ✦
CALC.BINODAL
CALC.PLAIT P.

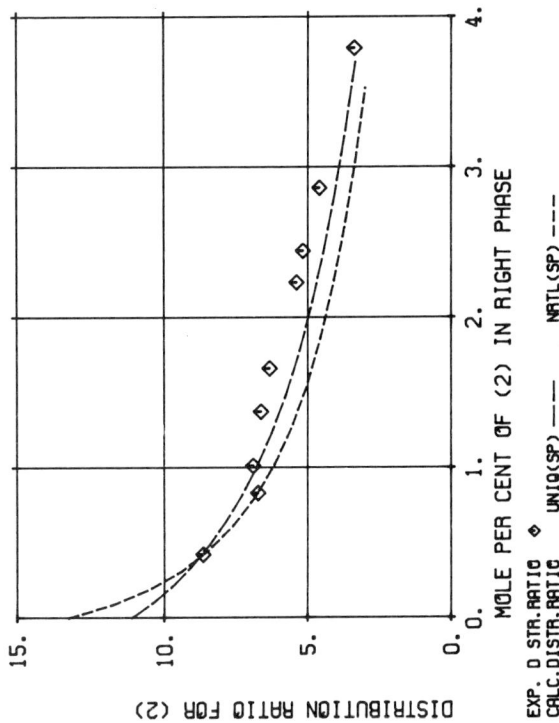

MOLE PER CENT OF (2) IN RIGHT PHASE

DISTRIBUTION RATIO FOR (2)

EXP. D STR.RATIO ◇ UNIQ(SP) —— NRTL(SP) ---
CALC.DISTR.RATIO

(1) C4H10O   1-PROPANOL,2-METHYL

(2) C3H8O   1-PROPANOL

(3) H2O   WATER

PERELYGIN V.M., SHADENKOVA N.S.
IZV.VYSSH.UCHEBN.ZAVED.PISHCH.TEKHNOL. (1979)40

TEMPERATURE = 60.0 DEG C   TYPE OF SYSTEM = 1

EXPERIMENTAL TIE LINES IN MOLE PCT

	LEFT PHASE			RIGHT PHASE	
(1)	(2)	(3)	(1)	(2)	(3)
42.290	3.630	54.080	1.770	0.420	97.810
38.690	5.600	55.710	1.620	0.830	97.550
35.710	6.980	57.310	1.700	1.010	97.290
31.640	9.110	59.250	1.680	1.370	96.950
28.380	10.530	61.090	1.700	1.660	96.640
23.740	12.060	64.200	1.790	2.230	95.980
21.660	12.630	65.710	1.840	2.440	95.720
18.360	13.120	68.520	2.020	2.860	95.120
13.550	12.770	73.680	2.490	3.790	93.720

SPECIFIC MODEL PARAMETERS IN KELVIN

I J	UNIQUAC		NRTL(ALPHA=.2)	
	AIJ	AJI	AIJ	AJI
1 2	170.02	-206.53	311.94	-655.23
1 3	-55.016	397.85	-342.75	1763.8
2 3	78.121	-47.473	15.492	-4.7642

R1 = 3.4535   R2 = 2.7799   R3 = 0.9200
Q1 = 3.048   Q2 = 2.512   Q3 = 1.400

MEAN DEV. BETWEEN CALC. AND EXP. CONC. IN MOLE PCT

UNIQUAC (SPECIFIC PARAMETERS)   0.58
NRTL   (SPECIFIC PARAMETERS)   1.12

C₃H₈O-C₅H₄O₂

(1) C5H4O2   FURFURAL
----
(2) C3H8O   1-PROPANOL
----
(3) H2O   WATER

KRUPATKIN I.L.,GLAGOLEVA M.F.
ZH.OBSHCH.KHIM. 40(1970)12

TEMPERATURE = 25.0 DEG C   TYPE OF SYSTEM = 1

EXPERIMENTAL TIE LINES IN MOLE PCT

	LEFT PHASE			RIGHT PHASE	
(1)	(2)	(3)	(1)	(2)	(3)
62.529	6.240	31.231	1.832	1.331	96.837
51.418	9.847	38.735	2.256	2.531	95.213
45.656	11.483	42.861	2.466	2.886	94.649
35.436	13.118	51.445	3.055	3.965	92.980
31.168	13.596	55.237	3.524	4.650	91.827
22.530	13.377	64.092	5.489	6.685	87.827

SPECIFIC MODEL PARAMETERS IN KELVIN

		UNIQUAC		NRTL(ALPHA=.2)	
I	J	AIJ	AJI	AIJ	AJI
1	2	-213.52	100.07	446.06	-720.14
1	3	143.69	125.42	11.598	1302.8
2	3	-228.88	254.47	-93.964	-337.49

R1 = 3.1680   R2 = 2.7799   R3 = 0.9200
Q1 = 2.484   Q2 = 2.512   Q3 = 1.400

MEAN DEV. BETWEEN CALC. AND EXP. CONC. IN MOLE PCT

UNIQUAC (SPECIFIC PARAMETERS)   0.48
NRTL   (SPECIFIC PARAMETERS)   0.89
UNIQUAC (COMMON PARAMETERS)   1.83

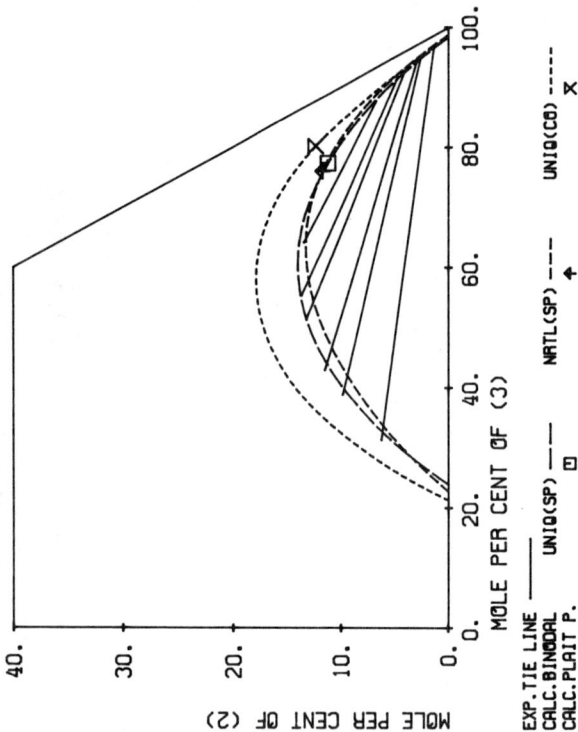

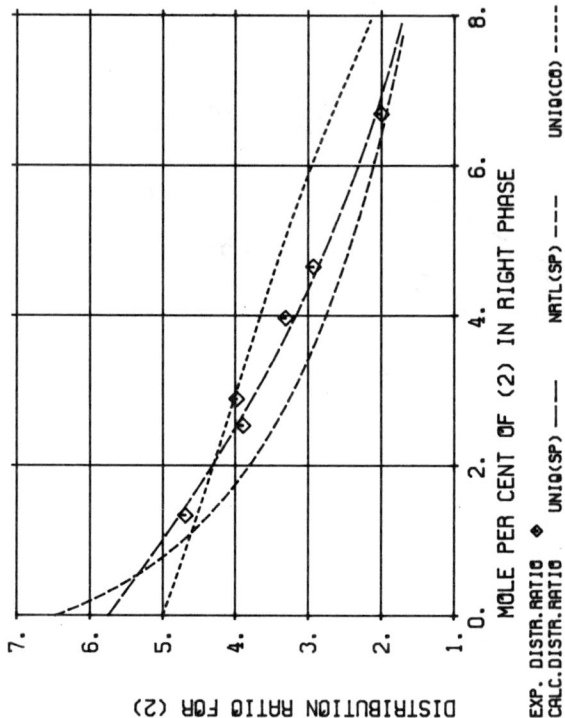

MOLE PER CENT OF (2)

MOLE PER CENT OF (3)

EXP.TIE LINE ——   UNIQ(SP) ⊡   NRTL(SP) ✦   UNIQ(CO) ----
CALC.BINODAL ——
CALC.PLAIT P.

DISTRIBUTION RATIO FOR (2)

MOLE PER CENT OF (2) IN RIGHT PHASE

EXP. DISTR.RATIO ◆   UNIQ(SP) ◇   NRTL(SP) ———   UNIQ(CO) ----
CALC.DISTR.RATIO

MOLE PER CENT OF (3)

MOLE PER CENT OF (2)

EXP.TIE LINE —— UNIQ(SP) ⊟ NRTL(SP) ✦ UNIQ(CO) ⚹
CALC.BINODAL ----
CALC.PLAIT P.

DISTRIBUTION RATIO FOR (2)

MOLE PER CENT OF (2) IN RIGHT PHASE

EXP. DISTR.RATIO   THIS REF ◇   OTHER REF +   UNIQ(CO) ----
CALC.DISTR.RATIO   UNIQ(SP) ——   NRTL(SP) - - -

(1) C5H10O2   ACETIC ACID,PROPYL ESTER

(2) C3H8O   1-PROPANOL

(3) H2O   WATER

SMITH T.E., BONNER R.F.
IND.ENG.CHEM. 42(1950)396

TEMPERATURE = 20.0 DEG C   TYPE OF SYSTEM = 1

EXPERIMENTAL TIE LINES IN MOLE PCT

	LEFT PHASE			RIGHT PHASE	
(1)	(2)	(3)	(1)	(2)	(3)
91.071	0.0	8.929	0.395	0.0	99.605
76.080	9.591	14.329	0.427	1.355	98.218
70.183	12.458	17.359	0.450	1.750	97.800
58.900	19.259	21.841	0.454	2.185	97.361
42.872	25.177	31.951	0.443	3.043	96.514
38.742	26.886	34.372	0.485	3.296	96.219
26.162	29.141	44.698	0.493	4.053	95.454
20.202	29.511	50.287	0.519	4.577	94.905
14.742	28.213	57.045	0.588	5.238	94.174
9.045	24.853	66.101	0.799	6.607	92.594

SPECIFIC MODEL PARAMETERS IN KELVIN

		UNIQUAC		NRTL(ALPHA=.2)	
I	J	AIJ	AJI	AIJ	AJI
1	2	82.901	69.206	235.12	-43.362
1	3	461.89	141.36	303.35	1521.7
2	3	-138.92	384.55	-554.64	1484.1

R1 = 4.1530   R2 = 2.7799   R3 = 0.9200
Q1 = 3.656   Q2 = 2.512   Q3 = 1.400

MEAN DEV. BETWEEN CALC. AND EXP. CONC. IN MOLE PCT

UNIQUAC (SPECIFIC PARAMETERS)   0.49
NRTL   (SPECIFIC PARAMETERS)   0.48
UNIQUAC (COMMON PARAMETERS)   0.55

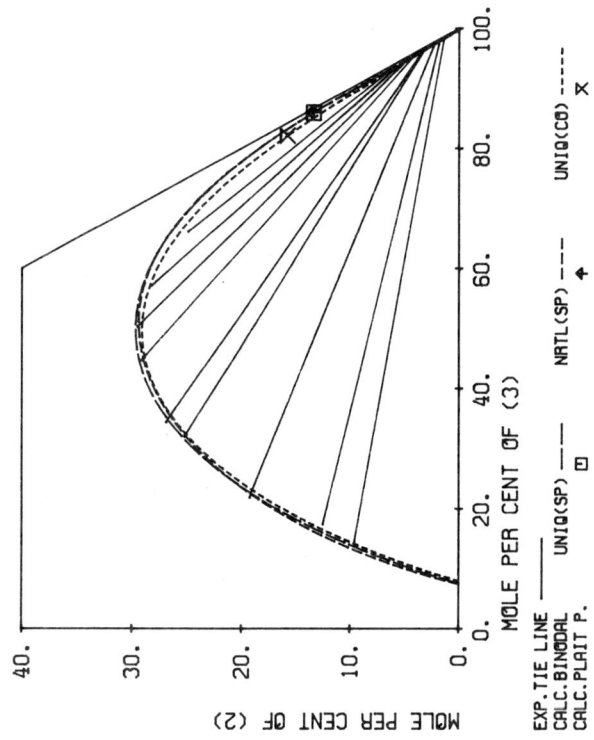

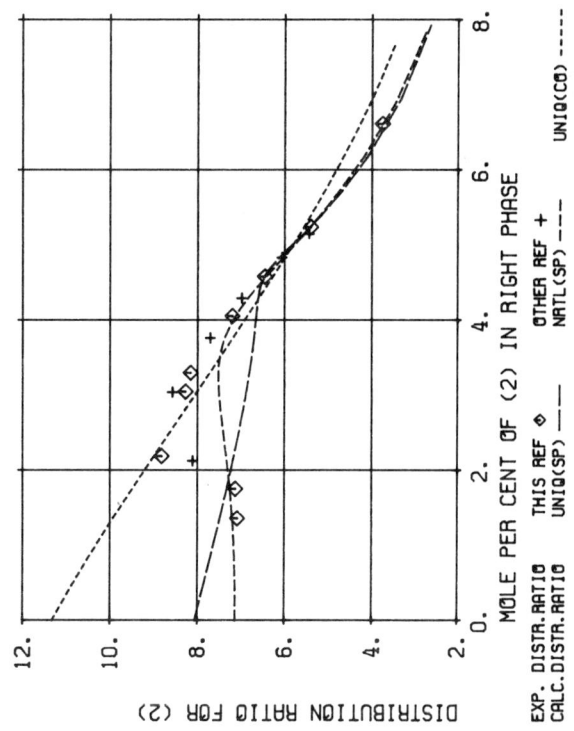

C₃H₈O-C₅H₁₀O₂

(1) C5H1002   ACETIC ACID,PROPYL ESTER
(2) C3H8O   1-PROPANOL
(3) H2O   WATER

SMITH T.E., BONNER R.F.
IND.ENG.CHEM. 42(1950)896

TEMPERATURE = 35.0 DEG C   TYPE OF SYSTEM = 1

EXPERIMENTAL TIE LINES IN MOLE PCT

	LEFT PHASE			RIGHT PHASE	
(1)	(2)	(3)	(1)	(2)	(3)
88.690	0.0	11.310	0.322	0.0	99.678
71.988	10.931	17.082	0.368	1.158	98.474
53.338	19.783	26.878	0.375	1.943	97.682
37.423	26.544	36.033	0.420	2.693	96.887
24.797	29.195	46.003	0.428	3.536	96.036
15.695	28.756	55.549	0.518	4.464	95.018
8.983	25.397	65.620	0.724	5.939	93.337

SPECIFIC MODEL PARAMETERS IN KELVIN

		UNIQUAC		NRTL(ALPHA=.2)	
I J		AIJ	AJI	AIJ	AJI
1 2		-4.8654	167.70	136.82	13.832
1 3		396.00	165.99	228.29	1496.0
2 3		-133.47	385.67	-544.38	1508.5

R1 = 4.1530   R2 = 2.7799   R3 = 0.9200
Q1 = 3.656   Q2 = 2.512   Q3 = 1.400

MEAN DEV. BETWEEN CALC. AND EXP. CONC. IN MOLE PCT

UNIQUAC (SPECIFIC PARAMETERS)   0.42
NRTL   (SPECIFIC PARAMETERS)   0.43

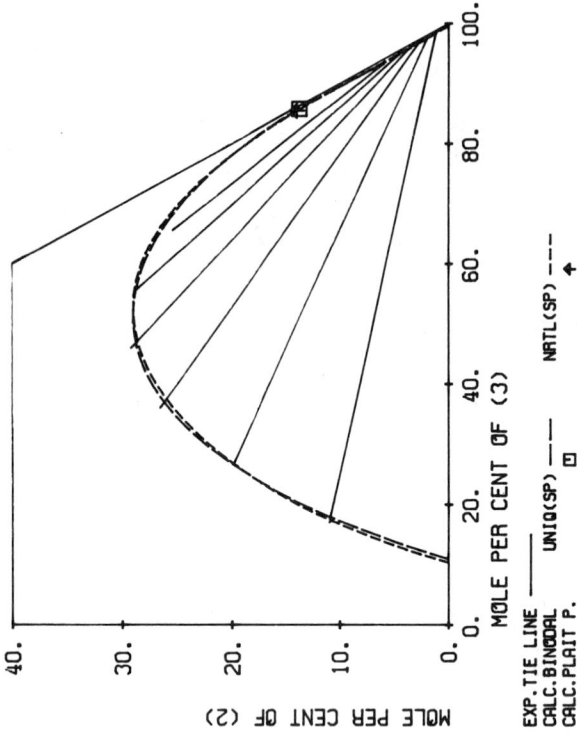

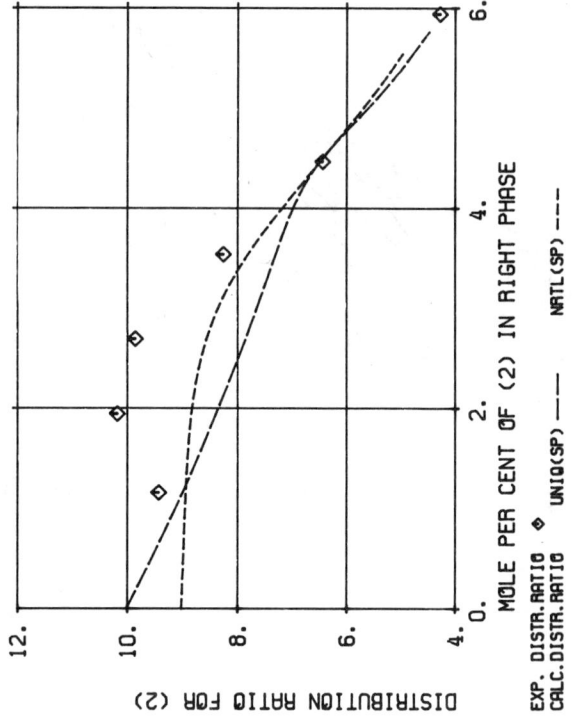

MOLE PER CENT OF (3)

MOLE PER CENT OF (2)

EXP.TIE LINE ——   UNIQ(SP) —— 🔲   NRTL(SP) ----
CALC.BINODAL
CALC.PLAIT P.

DISTRIBUTION RATIO FOR (2)

MOLE PER CENT OF (2) IN RIGHT PHASE

EXP. DISTR.RATIO ◇   UNIQ(SP) ◇   NRTL(SP) ——
CALC.DISTR.RATIO

$C_3H_8O-C_5H_{10}O_2$

(1) C5H1002   ACETIC ACID,PROPYL ESTER

(2) C3H80   1-PROPANOL

(3) H2O   WATER

JAGANNADHA RAO R., VENKATA RAO C.
J.APPL.CHEM. 9(1959)69

TEMPERATURE = 30.0 DEG C   TYPE OF SYSTEM = 1

EXPERIMENTAL TIE LINES IN MOLE PCT

	LEFT PHASE			RIGHT PHASE	
(1)	(2)	(3)	(1)	(2)	(3)
66.152	17.230	16.618	0.511	2.123	97.365
42.404	26.022	31.574	0.641	3.035	96.324
29.270	29.018	41.712	0.712	3.763	95.525
21.433	29.949	48.614	0.761	4.289	94.950
16.244	29.283	54.473	0.812	4.828	94.360
12.407	27.973	59.620	0.860	5.147	93.993

SPECIFIC MODEL PARAMETERS IN KELVIN

		UNIQUAC		NRTL(ALPHA=.2)	
I J	AIJ	AJI		AIJ	AJI
1 2	82.901	69.206		235.12	-43.362
1 3	461.89	141.36		303.35	1521.7
2 3	-138.92	384.55		-554.64	1484.1

R1 = 4.1530   R2 = 2.7799   R3 = 0.9200
Q1 = 3.656   Q2 = 2.512   Q3 = 1.400

MEAN DEV. BETWEEN CALC. AND EXP. CONC. IN MOLE PCT

UNIQUAC (SPECIFIC PARAMETERS)   0.69
NRTL   (SPECIFIC PARAMETERS)   0.56
UNIQUAC (COMMON PARAMETERS)   1.05

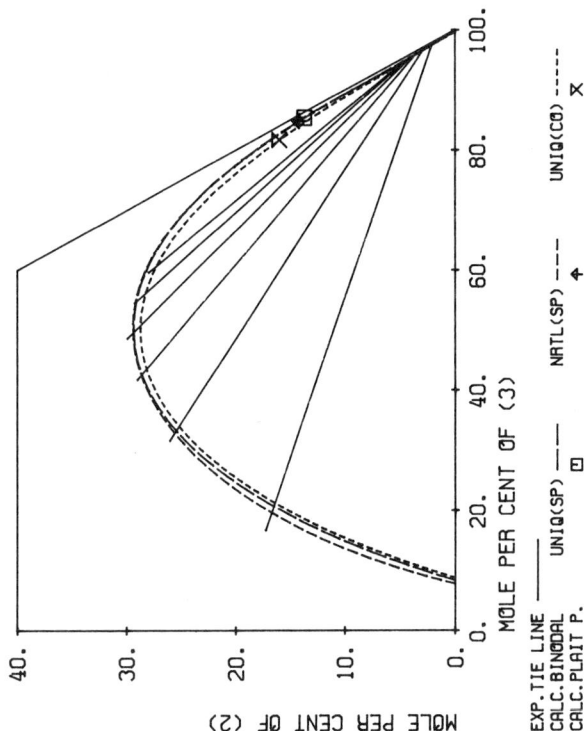

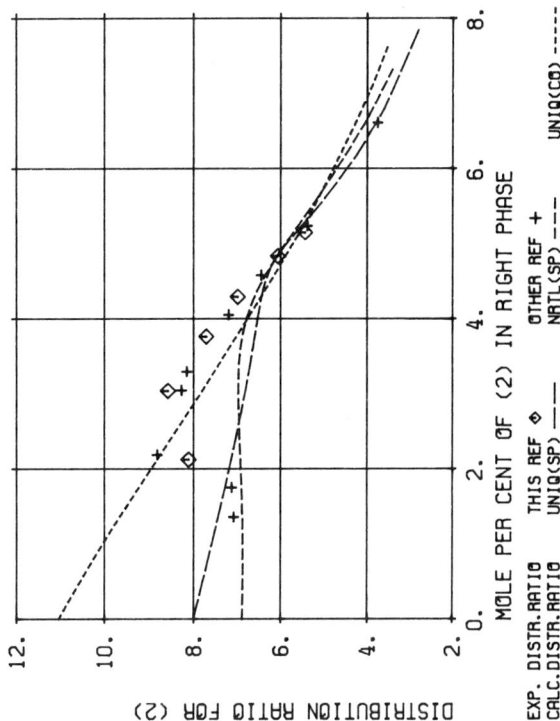

$C_3H_8O-C_5H_{10}O_2$

MOLE PER CENT OF (3)

MOLE PER CENT OF (2)

EXP.TIE LINE   UNIQ(SP) —— $\boxminus$ NRTL(SP) ----
CALC.BINODAL
CALC.PLAIT P.   ✦

MOLE PER CENT OF (2) IN RIGHT PHASE

DISTRIBUTION RATIO FOR (2)

EXP. DISTR.RATIO ◇   UNIQ(SP) ◇   NRTL(SP) ----
CALC.DISTR.RATIO

(1) C5H10O2   ACETIC ACID,PROPYL ESTER

(2) C3H8O   1-PROPANOL

(3) H2O   WATER

SMIRNOVA N.A., MORACHEVSKII A.G., STORONKIN A.V.
VESTN.LENINGR.UNIV.FIZ.KHIM. 18,22(1963)97

TEMPERATURE = 50.0 DEG C   TYPE OF SYSTEM = 1

EXPERIMENTAL TIE LINES IN MOLE PCT

	LEFT PHASE			RIGHT PHASE	
(1)	(2)	(3)	(1)	(2)	(3)
89.630	0.0	10.370	0.322	0.0	99.678
73.485	8.462	18.053	0.366	0.871	98.763
58.637	15.886	25.476	0.390	1.516	98.094
45.565	21.420	33.015	0.376	2.077	97.547
34.122	25.406	40.473	0.419	2.624	96.957
24.069	27.318	48.614	0.445	3.290	96.265
15.733	27.014	57.253	0.575	4.142	95.283
10.855	25.093	64.052	0.712	5.152	94.136

SPECIFIC MODEL PARAMETERS IN KELVIN

I J	UNIQUAC AIJ	AJI	NRTL(ALPHA=.2) AIJ	AJI
1 2	35.614	126.94	46.144	98.263
1 3	442.37	138.13	224.61	1535.1
2 3	-150.41	427.65	-612.43	1654.7

R1 = 4.1530   R2 = 2.7799   R3 = 0.9200
Q1 = 3.656   Q2 = 2.512   Q3 = 1.400

MEAN DEV. BETWEEN CALC. AND EXP. CONC. IN MOLE PCT

UNIQUAC (SPECIFIC PARAMETERS)   0.37
NRTL   (SPECIFIC PARAMETERS)   0.35

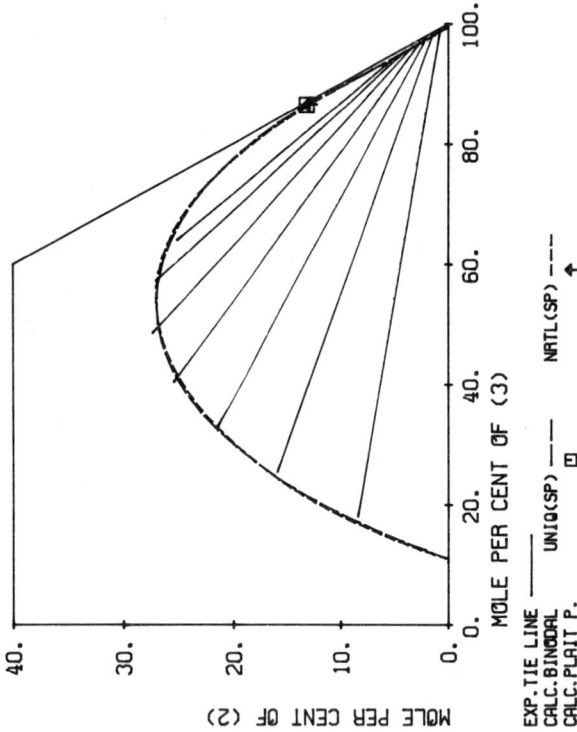

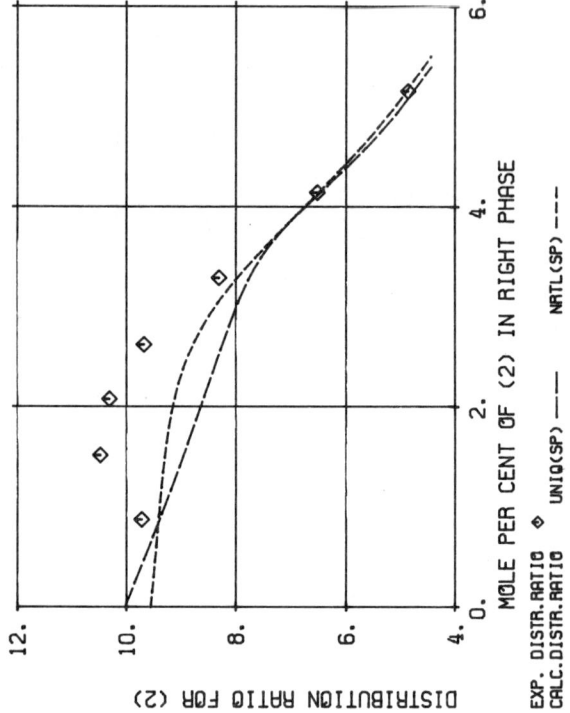

$C_3H_8O-C_5H_{10}O_2$

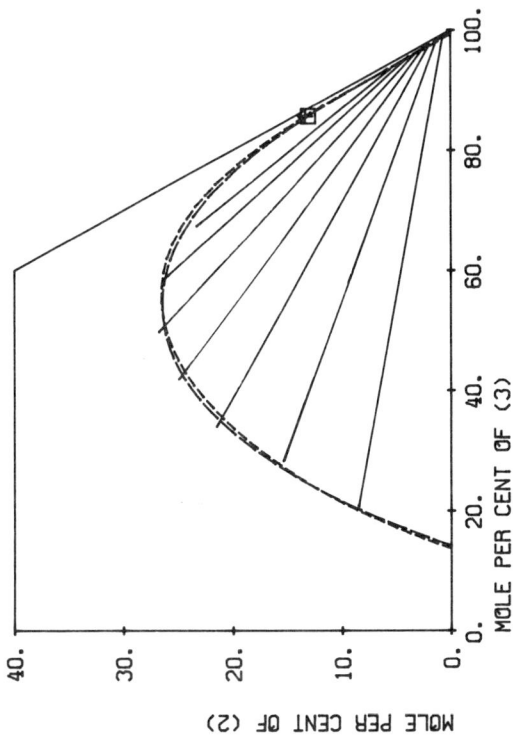

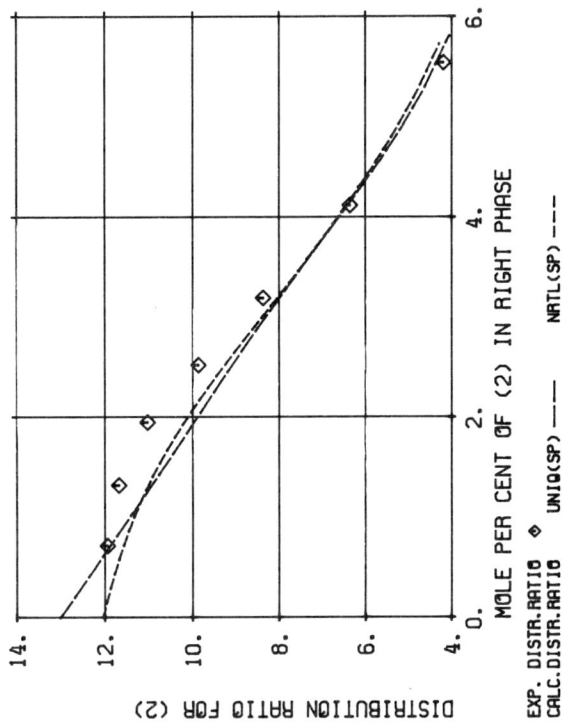

(1) C5H10O2   ACETIC ACID,PROPYL ESTER
----
(2) C3H8O   1-PROPANOL
----
(3) H2O   WATER

SMIRNOVA N.A., MORACHEVSKII A.G., STORONKIN A.V.
VESTN.LENINGR.UNIV.FIZ.KHIM. 18,22(1963)97

TEMPERATURE = 65.0 DEG C   TYPE OF SYSTEM = 1

EXPERIMENTAL TIE LINES IN MOLE PCT

| | LEFT PHASE | | | RIGHT PHASE | |
(1)	(2)	(3)	(1)	(2)	(3)
85.962	0.0	14.038	0.322	0.0	99.678
70.780	8.503	20.716	0.365	0.713	98.923
56.271	15.417	28.312	0.389	1.320	98.291
44.525	21.461	34.014	0.413	1.947	97.640
33.188	24.863	41.949	0.419	2.521	97.060
23.523	26.672	49.805	0.484	3.189	96.327
15.179	26.138	58.683	0.636	4.117	95.248
9.315	23.237	67.447	0.867	5.540	93.593

SPECIFIC MODEL PARAMETERS IN KELVIN

| | | UNIQUAC | | NRTL(ALPHA=.2) | |
I	J	AIJ	AJI	AIJ	AJI
1	2	-51.896	175.69	111.31	-62.878
1	3	383.98	171.90	166.49	1579.9
2	3	-163.67	432.20	-630.11	1687.8

R1 = 4.1530   R2 = 2.7799   R3 = 0.9200
Q1 = 3.656   Q2 = 2.512   Q3 = 1.400

MEAN DEV. BETWEEN CALC. AND EXP. CONC. IN MOLE PCT

UNIQUAC (SPECIFIC PARAMETERS)   0.31
NRTL   (SPECIFIC PARAMETERS)   0.44

MOLE PER CENT OF (3)

EXP.TIE LINE ———   UNIQ(SP) ———   NRTL(SP) ----
CALC.BINODAL   □
CALC.PLAIT P.   ↑

MOLE PER CENT OF (2)

MOLE PER CENT OF (2) IN RIGHT PHASE

EXP. DISTR.RATIO ◇   UNIQ(SP) ———   NRTL(SP) ----
CALC.DISTR.RATIO

DISTRIBUTION RATIO FOR (2)

$C_3H_8O-C_5H_{10}O_2$

(1) C5H10O2   ACETIC ACID,PROPYL ESTER
(2) C3H8O   1-PROPANOL
(3) H2O   WATER

SMIRNOVA N.A., MORACHEVSKII A.G., STORONKIN A.V.
VESTN.LENINGR.UNIV.FIZ.KHIM. 18,22(1963)97

TEMPERATURE = 80.0 DEG C   TYPE OF SYSTEM = 1

EXPERIMENTAL TIE LINES IN MOLE PCT

	LEFT PHASE			RIGHT PHASE	
(1)	(2)	(3)	(1)	(2)	(3)
81.703	0.0	18.297	0.340	0.0	99.660
67.463	8.606	23.931	0.382	0.619	98.999
54.241	15.399	30.361	0.406	1.160	98.434
42.716	20.751	36.533	0.412	1.780	97.808
31.879	23.931	44.191	0.456	2.389	97.155
22.467	25.633	51.900	0.522	3.089	96.388
13.832	24.365	61.803	0.695	4.017	95.288
8.637	20.850	70.514	1.156	5.893	92.951

SPECIFIC MODEL PARAMETERS IN KELVIN

		UNIQUAC		NRTL(ALPHA=.2)	
I	J	AIJ	AJI	AIJ	AJI
1	2	-130.56	222.37	245.91	-286.58
1	3	-304.98	242.53	108.39	1714.2
2	3	-206.48	484.39	-689.85	1772.3

R1 = 4.1530   R2 = 2.7799   R3 = 0.9200
Q1 = 3.656   Q2 = 2.512   Q3 = 1.400

MEAN DEV. BETWEEN CALC. AND EXP. CONC. IN MOLE PCT

UNIQUAC (SPECIFIC PARAMETERS)   0.33
NRTL   (SPECIFIC PARAMETERS)   0.60

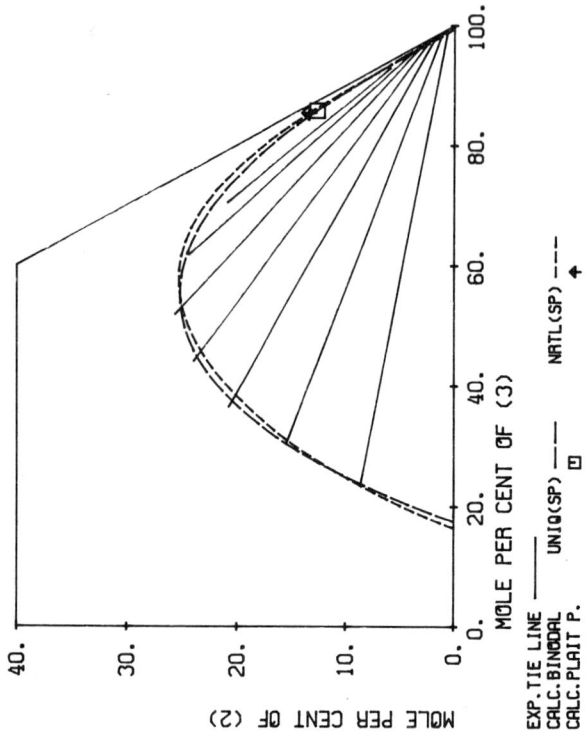

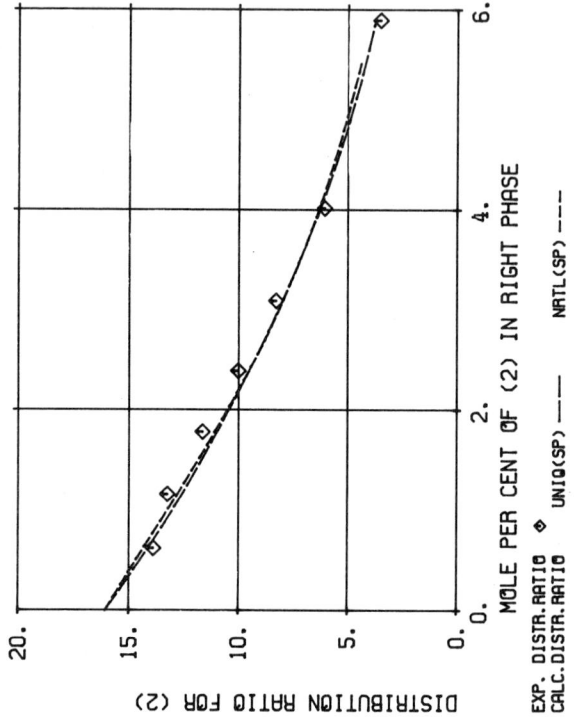

(1) C5H10O2   PROPANOIC ACID,ETHYL ESTER

(2) C3H8O   1-PROPANOL

(3) H2O   WATER

JAGANNADHA RAO R., VENKATA RAO C.
J.APPL.CHEM. 9(1959)69

TEMPERATURE =   30.0 DEG C   TYPE OF SYSTEM = 1

EXPERIMENTAL TIE LINES IN MOLE PCT

	LEFT PHASE			RIGHT PHASE	
(1)	(2)	(3)	(1)	(2)	(3)
64.562	16.917	18.521	0.415	2.114	97.471
44.118	26.456	29.425	0.462	2.976	96.562
30.708	30.083	39.209	0.509	3.693	95.798
22.145	30.777	47.078	0.534	4.134	95.331
16.713	29.958	53.329	0.601	4.631	94.767
12.399	28.538	59.063	0.608	5.166	94.226

SPECIFIC MODEL PARAMETERS IN KELVIN

		UNIQUAC			NRTL(ALPHA=.2)	
I J		AIJ	AJI		AIJ	AJI
1 2		-9.7267	210.74		126.32	61.851
1 3		473.68	143.57		355.35	1350.0
2 3		-103.97	350.47		-512.77	1469.8

R1 = 4.1530   R2 = 2.7799   R3 = 0.9200
Q1 = 3.656 . Q2 = 2.512   Q3 = 1.400

MEAN DEV. BETWEEN CALC. AND EXP. CONC. IN MOLE PCT

UNIQUAC (SPECIFIC PARAMETERS)   0.39
NRTL   (SPECIFIC PARAMETERS)   0.33
UNIQUAC (COMMON PARAMETERS)   1.06

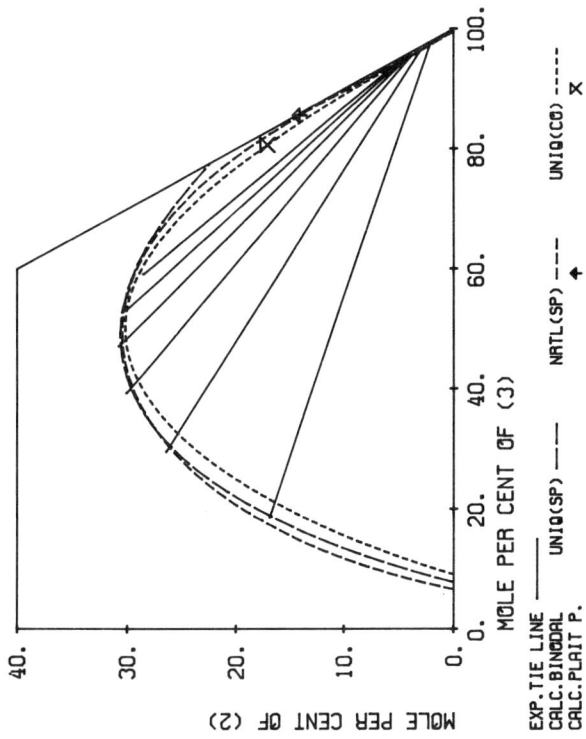

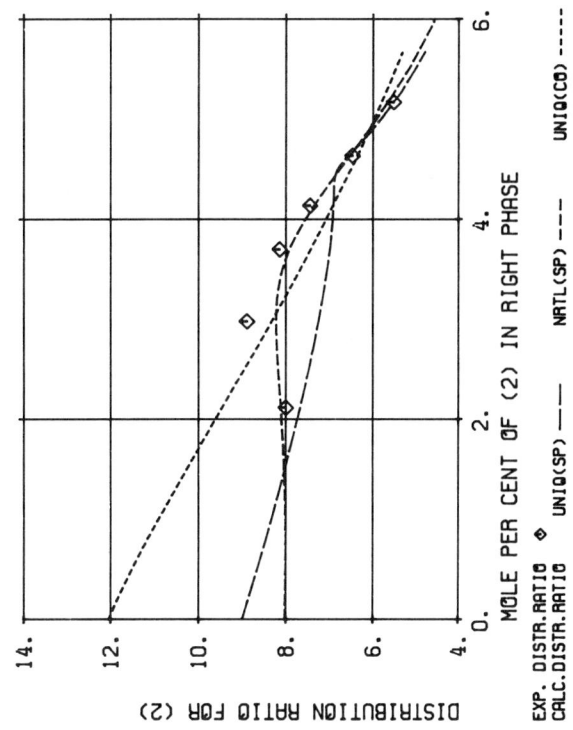

$C_3H_8O-C_5H_{12}O$

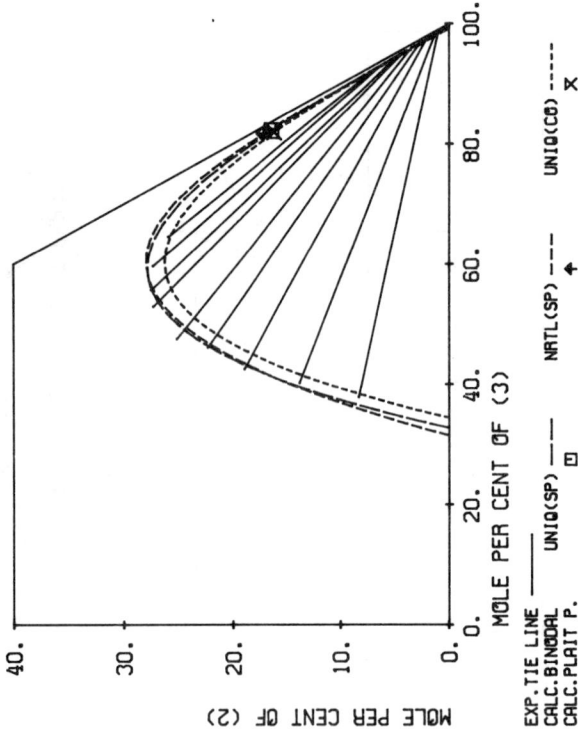

MOLE PER CENT OF (3)

MOLE PER CENT OF (2)

EXP.TIE LINE ——   UNIQ(SP) ⊡   NRTL(SP) ✦   UNIQ(C6) ─ ─ ─ ✕
CALC.BINODAL
CALC.PLAIT P.

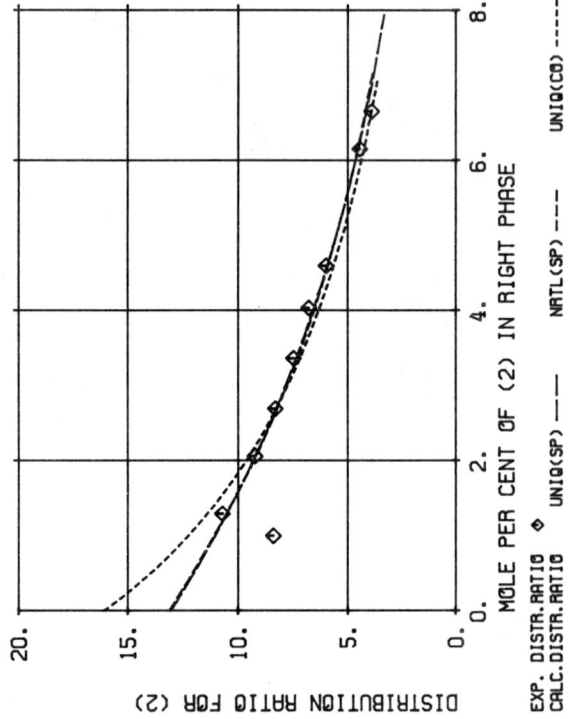

MOLE PER CENT OF (2) IN RIGHT PHASE

DISTRIBUTION RATIO FOR (2)

EXP. DISTR.RATIO ◇   UNIQ(SP) ◇   NRTL(SP) ——   UNIQ(C6) ─ ─ ─
CALC.DISTR.RATIO

(1) C5H12O   1-BUTANOL,3-METHYL
(2) C3H8O   1-PROPANOL
(3) H2O   WATER

COULL J., HOPE H.B.
J.PHYS.CHEM. 39(1935)967

TEMPERATURE = 25.0 DEG C   TYPE OF SYSTEM = 1

EXPERIMENTAL TIE LINES IN MOLE PCT

	LEFT PHASE			RIGHT PHASE	
(1)	(2)	(3)	(1)	(2)	(3)
53.907	8.349	37.744	0.455	0.993	98.553
46.099	13.770	40.131	0.495	1.287	98.219
38.599	18.963	42.438	0.506	2.053	97.440
31.601	22.333	46.066	0.553	2.686	96.760
27.358	25.156	47.486	0.507	3.360	96.133
19.849	27.316	52.835	0.477	4.029	95.493
16.873	27.518	55.609	0.564	4.587	94.849
13.058	27.336	59.606	0.609	6.148	93.243
9.556	25.946	64.498	0.617	6.648	92.735

SPECIFIC MODEL PARAMETERS IN KELVIN

		UNIQUAC		NRTL(ALPHA=.2)	
I J	AIJ	AJI		AIJ	AJI
1 2	4.0813	20.496		129.37	-119.38
1 3	114.16	212.61		-125.42	1533.3
2 3	-97.289	289.96		-395.46	1233.5

R1 = 4.1279   R2 = 2.7799   R3 = 0.9200
Q1 = 3.588   Q2 = 2.512   Q3 = 1.400

MEAN DEV. BETWEEN CALC. AND EXP. CONC. IN MOLE PCT

UNIQUAC (SPECIFIC PARAMETERS)   0.41
NRTL (SPECIFIC PARAMETERS)   0.55
UNIQUAC (COMMON PARAMETERS)   0.86

$C_3H_8O-C_6H_6$

(1) C6H6   BENZENE

(2) C3H8O   1-PROPANOL

(3) H2O   WATER

DENZLER C.G.
J.PHYS.CHEM. 49(1945)358

TEMPERATURE = 20.0 DEG C   TYPE OF SYSTEM = 1

EXPERIMENTAL TIE LINES IN MOLE PCT

	LEFT PHASE		RIGHT PHASE	
(1)	(2)	(3)	(1)	(2)	(3)
42.173	39.806	18.021	0.104	5.534	94.362
34.819	44.388	20.793	0.185	5.897	93.918
28.066	44.789	27.145	0.213	6.256	93.531
16.752	44.713	38.535	0.269	6.708	93.023
9.881	39.605	50.514	0.392	8.584	91.024
7.570	36.453	55.976	0.484	9.364	90.152

SPECIFIC MODEL PARAMETERS IN KELVIN

		UNIQUAC		NRTL(ALPHA=.2)	
I J	AIJ	AJI	AIJ	AJI	
1 2	214.70	-72.359	359.58	-50.553	
1 3	773.44	401.86	1074.2	1362.0	
2 3	-116.96	341.40	-354.00	1228.9	

R1 = 3.1878   R2 = 2.7799   R3 = 0.9200
Q1 = 2.400   Q2 = 2.512   Q3 = 1.400

MEAN DEV. BETWEEN CALC. AND EXP. CONC. IN MOLE PCT

UNIQUAC (SPECIFIC PARAMETERS)   1.18
NRTL   (SPECIFIC PARAMETERS)   1.45
UNIQUAC (COMMON PARAMETERS)   2.08

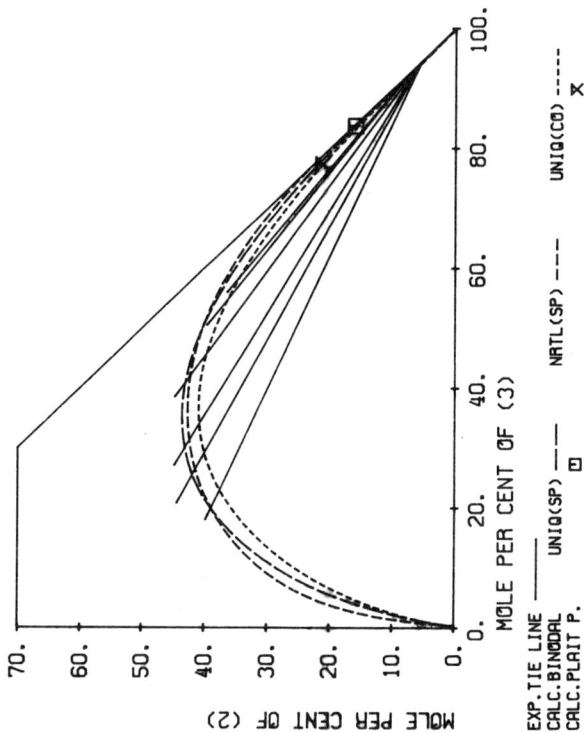

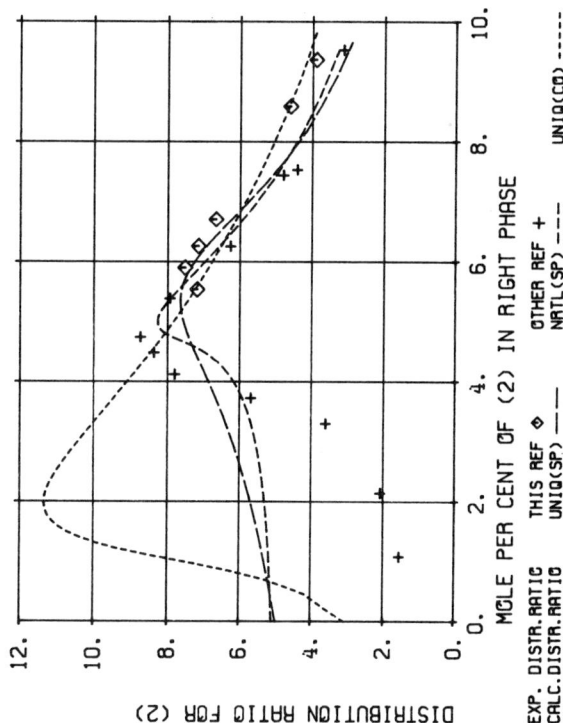

C$_3$H$_8$O-C$_6$H$_6$

(1) C6H6   BENZENE

(2) C3H8O   1-PROPANOL

(3) H2O   WATER

MCCANTS J.F., JONES J.H., HOPSON W.H.
IND.ENG.CHEM. 45(1953)454

TEMPERATURE = 37.8 DEG C   TYPE OF SYSTEM = 1

EXPERIMENTAL TIE LINES IN MOLE PCT

| | LEFT PHASE | | | RIGHT PHASE | |
(1)	(2)	(3)	(1)	(2)	(3)
93.766	4.961	1.273	0.072	1.882	98.046
85.123	11.186	3.691	0.147	2.519	97.334
86.803	9.896	3.302	0.173	2.895	96.931
76.581	17.858	5.561	0.175	3.381	96.444
48.878	32.780	18.342	0.258	4.692	95.050
34.584	38.533	26.878	0.263	5.640	94.097

SPECIFIC MODEL PARAMETERS IN KELVIN

		UNIQUAC		NRTL(ALPHA=.2)	
I	J	AIJ	AJI	AIJ	AJI
1	2	315.16	-105.01	711.99	-188.92
1	3	736.76	204.70	769.13	1392.7
2	3	-70.783	272.64	-367.17	1282.9

R1 = 3.1878   R2 = 2.7799   R3 = 0.9200
Q1 = 2.400   Q2 = 2.512   Q3 = 1.400

MEAN DEV. BETWEEN CALC. AND EXP. CONC. IN MOLE PCT

UNIQUAC (SPECIFIC PARAMETERS)   0.30
NRTL   (SPECIFIC PARAMETERS)   0.42

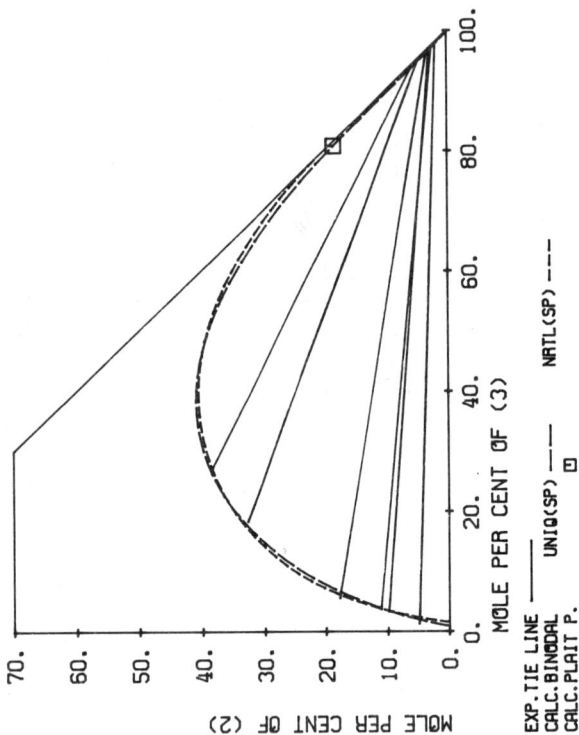

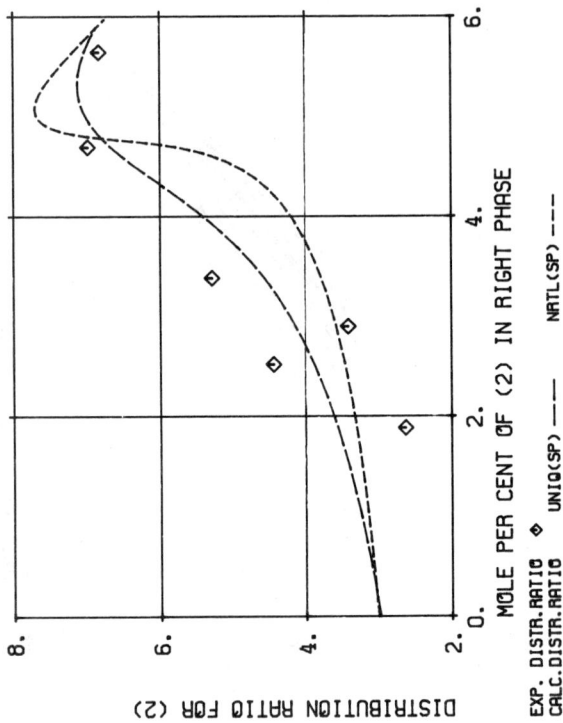

(1) C6H6   BENZENE
----
(2) C3H8O   1-PROPANOL
----
(3) H2O   WATER

UDOVENKO V.V., MAZANKO T.F.
ZH.FIZ.KHIM. 37(1963)1151

TEMPERATURE = 30.0 DEG C   TYPE OF SYSTEM = 1

EXPERIMENTAL TIE LINES IN MOLE PCT

	LEFT PHASE			RIGHT PHASE	
(1)	(2)	(3)	(1)	(2)	(3)
97.892	1.677	0.431	0.024	1.076	98.900
94.675	4.472	0.853	0.036	2.142	97.821
86.112	11.813	2.074	0.050	3.298	96.653
72.918	21.190	5.892	0.075	3.722	96.202
53.666	32.166	14.168	0.089	4.118	95.793
39.988	37.533	22.478	0.102	4.484	95.414
29.163	41.427	29.410	0.116	4.744	95.140
18.142	42.783	39.075	0.130	5.336	94.484
10.133	39.168	50.699	0.186	6.251	93.563
7.178	35.759	57.064	0.245	7.439	92.315
5.928	33.365	60.707	0.273	7.529	92.197
4.329	29.785	65.886	0.428	9.527	90.045

SPECIFIC MODEL PARAMETERS IN KELVIN

		UNIQUAC		NRTL(ALPHA=.2)	
I J	AIJ	AJI		AIJ	AJI
1 2	214.70	-72.359		359.58	-50.553
1 3	773.44	-401.86		1074.2	1362.0
2 3	-116.96	341.40		-354.00	1228.9

R1 = 3.1878   R2 = 2.7799   R3 = 0.9200
Q1 = 2.400   Q2 = 2.512   Q3 = 1.400

MEAN DEV. BETWEEN CALC. AND EXP. CONC. IN MOLE PCT

UNIQUAC (SPECIFIC PARAMETERS)   0.79
NRTL   (SPECIFIC PARAMETERS)   0.92
UNIQUAC (COMMON PARAMETERS)   1.87

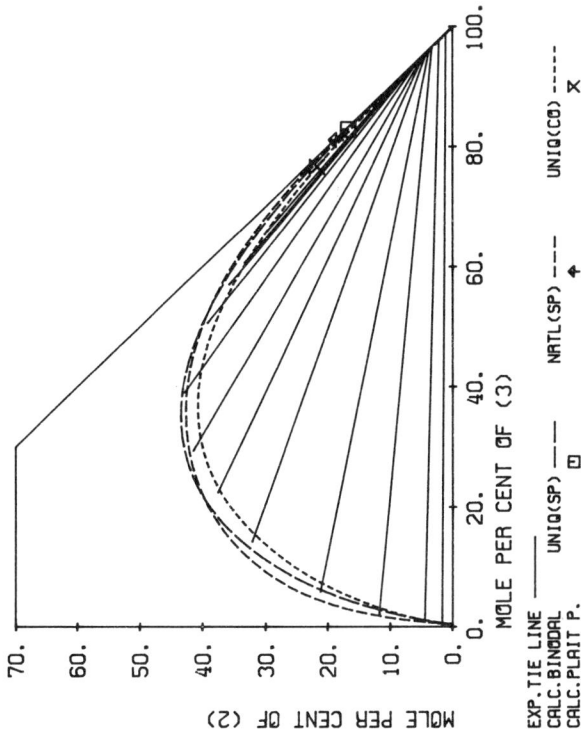

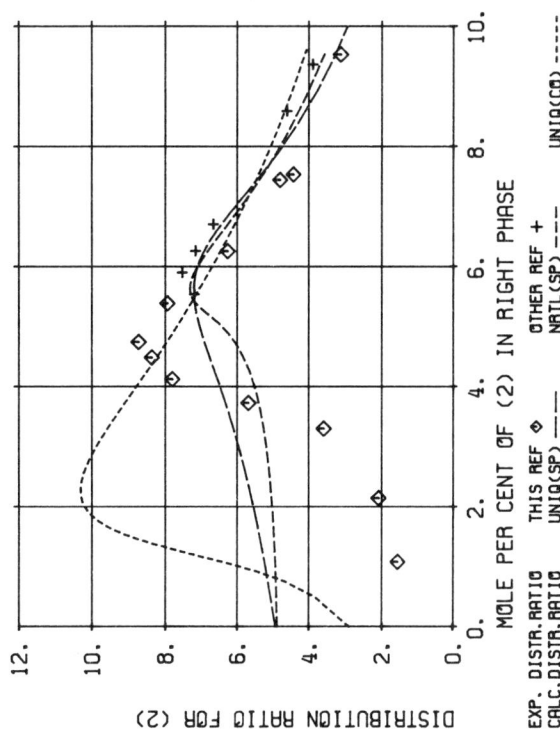

C₃H₈O-C₆H₆

(1) C6H6   BENZENE
(2) C3H8O   1-PROPANOL
(3) H2O   WATER

UDOVENKO V.V., MAZANKO T.F.
ZH.FIZ.KHIM. 37(1963)1151

TEMPERATURE = 45.0 DEG C   TYPE OF SYSTEM = 1

EXPERIMENTAL TIE LINES IN MOLE PCT

	LEFT PHASE		RIGHT PHASE	
(1)	(2)	(3)	(1)	(2)	(3)
98.664	0.905	0.431	0.047	0.826	99.127
96.450	2.694	0.856	0.073	2.046	97.881
82.410	13.525	4.066	0.099	2.957	96.944
70.389	22.261	7.350	0.125	3.341	96.534
53.817	31.686	14.497	0.151	3.696	96.153
40.736	36.741	22.523	0.178	4.130	95.692
27.209	41.212	31.579	0.205	4.352	95.443
16.022	40.898	43.080	0.234	5.100	94.666
8.391	35.759	55.850	0.293	6.193	93.514
5.380	30.698	63.922	0.414	7.860	91.726
3.166	24.337	72.497	0.741	11.176	88.083

SPECIFIC MODEL PARAMETERS IN KELVIN

		UNIQUAC		NRTL(ALPHA=.2)	
I J		AIJ	AJI	AIJ	AJI
1 2		310.90	-113.09	852.26	-371.76
1 3		743.61	304.23	1039.7	1519.8
2 3		-123.69	364.06	-431.16	1374.9

R1 = 3.1878   R2 = 2.7799   R3 = 0.9200
Q1 = 2.400   Q2 = 2.512   Q3 = 1.400

MEAN DEV. BETWEEN CALC. AND EXP. CONC. IN MOLE PCT

UNIQUAC (SPECIFIC PARAMETERS)   0.60
NRTL   (SPECIFIC PARAMETERS)   0.64

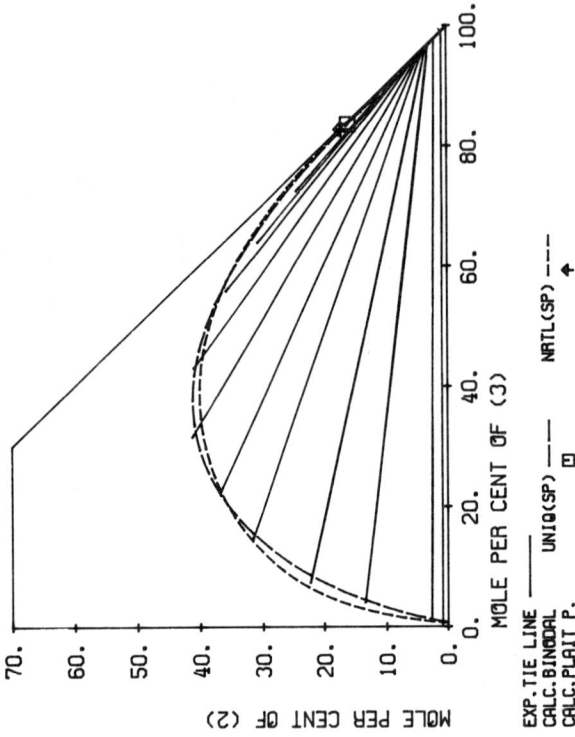

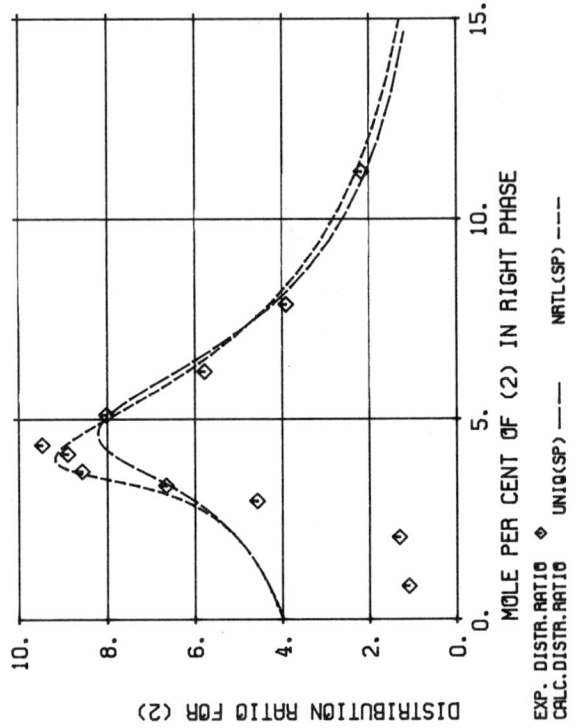

C_3H_8O-C_6H_6

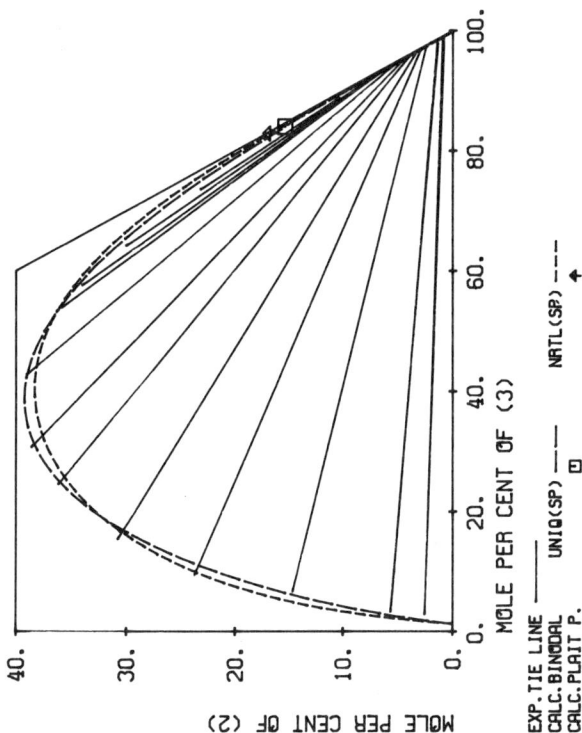

MOLE PER CENT OF (3)

MOLE PER CENT OF (2)

EXP.TIE LINE ——   UNIQ(SP) ⊟   NRTL(SP) ✦
CALC.BINODAL
CALC.PLAIT P.

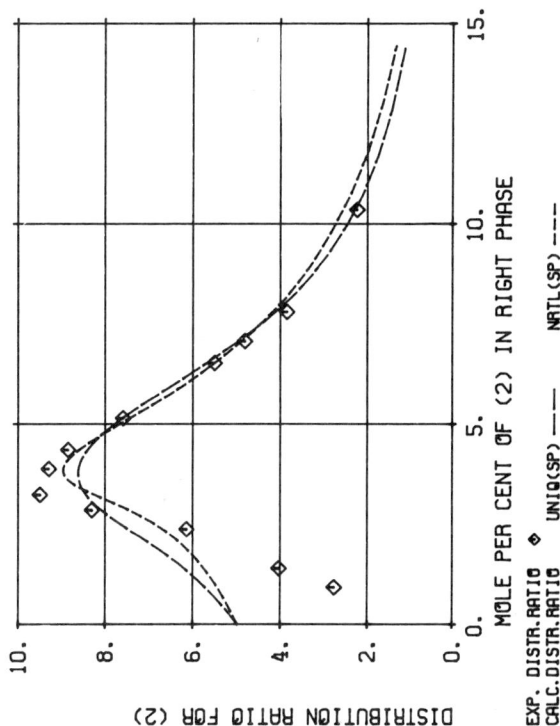

MOLE PER CENT OF (2) IN RIGHT PHASE

DISTRIBUTION RATIO FOR (2)

EXP. DISTR.RATIO ◇   UNIQ(SP) ——   NRTL(SP) ----
CALC.DISTR.RATIO

(1) C6H6   BENZENE
(2) C3H8O   1-PROPANOL
(3) H2O   WATER

UDOVENKO V.V., MAZANKO T.F.
ZH.FIZ.KHIM. 37(1963)1151

TEMPERATURE = 60.0 DEG C   TYPE OF SYSTEM = 1

EXPERIMENTAL TIE LINES IN MOLE PCT

	LEFT PHASE			RIGHT PHASE	
(1)	(2)	(3)	(1)	(2)	(3)
94.526	2.525	2.949	0.024	0.919	99.057
91.042	5.622	3.335	0.048	1.395	98.558
78.641	14.618	6.741	0.098	2.380	97.522
66.839	23.701	9.460	0.099	2.854	97.047
53.796	30.745	15.458	0.125	3.236	96.639
39.282	36.114	24.605	0.202	3.881	95.916
30.699	38.581	30.720	0.230	4.356	95.414
18.176	39.017	42.808	0.260	5.142	94.597
10.191	35.819	53.989	0.349	6.524	93.127
8.417	33.934	57.650	0.408	7.065	92.527
5.810	29.975	64.215	0.498	7.796	91.706
3.272	23.142	73.586	0.819	10.345	88.835

SPECIFIC MODEL PARAMETERS IN KELVIN

		UNIQUAC		NRTL(ALPHA=.2)	
I J	AIJ	AJI		AIJ	AJI
1 2	356.64	-157.59		798.58	-375.94
1 3	648.42	352.51		872.35	1551.6
2 3	-166.09	434.95		-468.79	1458.0

R1 = 3.1878   R2 = 2.7799   R3 = 0.9200
Q1 = 2.400   Q2 = 2.512   Q3 = 1.400

MEAN DEV. BETWEEN CALC. AND EXP. CONC. IN MOLE PCT

UNIQUAC (SPECIFIC PARAMETERS)   0.43
NRTL   (SPECIFIC PARAMETERS)   0.60

C₃H₈O-C₆H₁₂

(1) C6H12   CYCLOHEXANE
(2) C3H8O   1-PROPANOL
(3) H2O   WATER

WASHBURN E.R., BROCKWAY C.E., GRAHAM C.L., DEMING P.
J.AM.CHEM.SOC. 64(1942)1886

TEMPERATURE = 25.0 DEG C   TYPE OF SYSTEM = 1

EXPERIMENTAL TIE LINES IN MOLE PCT

	LEFT PHASE			RIGHT PHASE	
(1)	(2)	(3)	(1)	(2)	(3)
98.744	1.256	0.0	0.022	1.813	98.164
98.051	1.949	0.0	0.068	2.749	97.182
93.946	6.054	0.0	0.119	4.635	95.246
81.183	15.772	3.045	0.147	6.010	93.843
67.094	25.986	6.920	0.199	6.737	93.063
36.603	41.236	22.160	0.177	7.344	92.480
21.012	44.212	34.777	0.204	7.810	91.986
12.405	42.358	45.237	0.284	8.302	91.414
1.706	21.404	76.889	0.982	17.014	82.003

SPECIFIC MODEL PARAMETERS IN KELVIN

		UNIQUAC		NRTL(ALPHA=.2)	
I J	AIJ	AJI		AIJ	AJI
1 2	206.00	-57.559		282.58	-33.686
1 3	720.70	281.46		1254.2	1691.3
2 3	-77.346	271.52		-366.11	1192.4

R1 = 4.0464   R2 = 2.7799   R3 = 0.9200
Q1 = 3.240   Q2 = 2.512   Q3 = 1.400

MEAN DEV. BETWEEN CALC. AND EXP. CONC. IN MOLE PCT

UNIQUAC (SPECIFIC PARAMETERS)   1.34
NRTL (SPECIFIC PARAMETERS)   1.16
UNIQUAC (COMMON PARAMETERS)   2.99

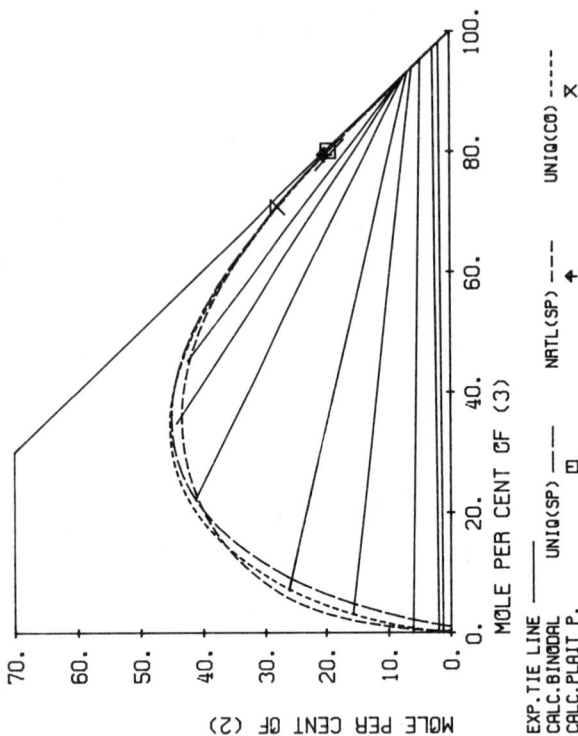

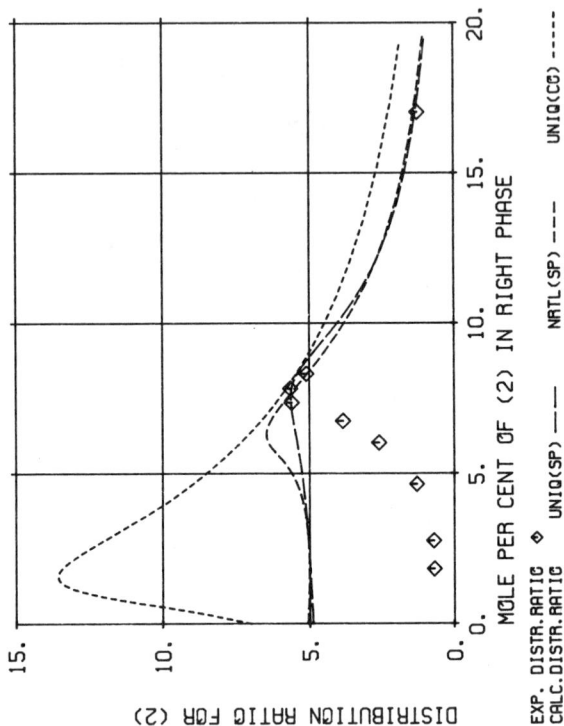

$C_3H_8O-C_6H_{12}$

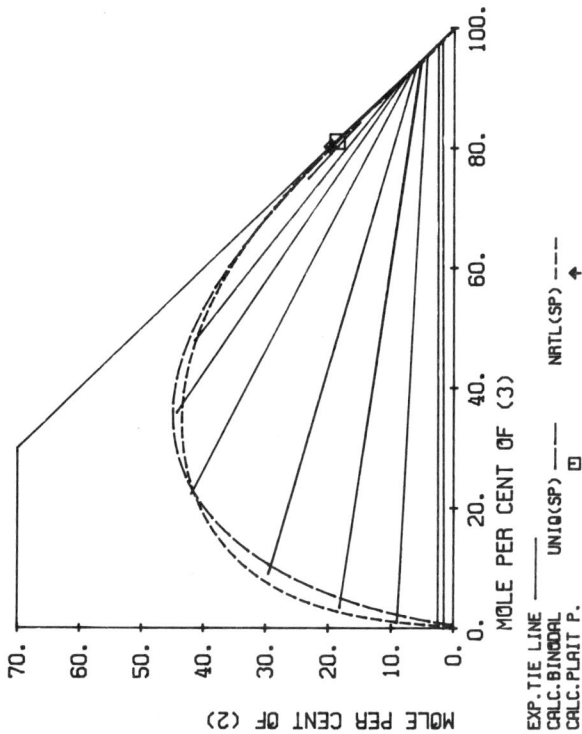

MOLE PER CENT OF (3)

MOLE PER CENT OF (2)

EXP.TIE LINE ——— UNIQ(SP) ⊟ NRTL(SP) ——
CALC.BINODAL
CALC.PLAIT P.

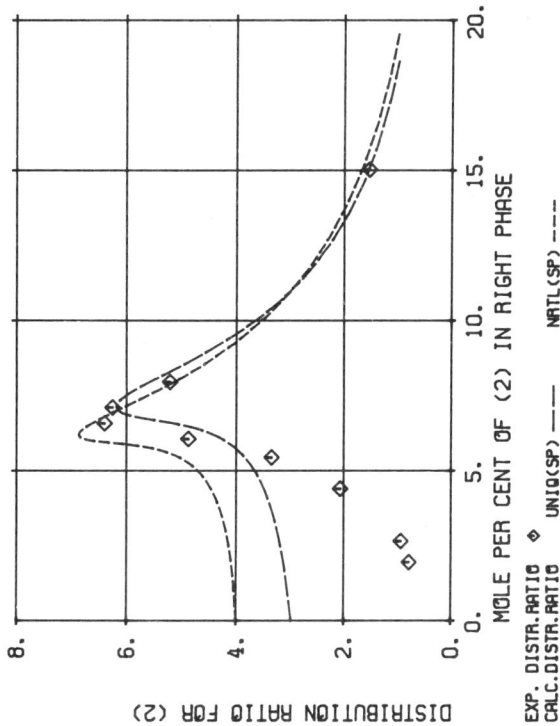

MOLE PER CENT OF (2) IN RIGHT PHASE

DISTRIBUTION RATIO FOR (2)

EXP. DISTR.RATIO ◇ UNIQ(SP) ——— NRTL(SP) ——
CALC.DISTR.RATIO

(1) C6H12   CYCLOHEXANE

(2) C3H8O   1-PROPANOL

(3) H2O   WATER

WASHBURN E.R., BROCKWAY C.E., GRAHAM C.L., DEMING P.
J.AM.CHEM.SOC. 64(1942)1886

TEMPERATURE = 35.0 DEG C   TYPE OF SYSTEM = 1

EXPERIMENTAL TIE LINES IN MOLE PCT

| | LEFT PHASE | | | RIGHT PHASE | |
(1)	(2)	(3)	(1)	(2)	(3)
98.467	1.533	0.0	0.022	1.944	98.034
97.045	2.493	0.462	0.023	2.643	97.334
90.026	9.071	0.903	0.0	4.395	95.605
78.376	18.182	3.442	0.0	5.438	94.562
61.377	29.551	9.071	0.0	6.055	93.945
35.631	42.000	22.370	0.0	6.568	93.432
19.743	44.359	35.898	0.025	7.099	92.876
10.658	41.441	47.901	0.076	7.942	91.981
1.897	23.179	74.924	0.707	15.022	84.271

SPECIFIC MODEL PARAMETERS IN KELVIN

| | | UNIQUAC | | NRTL(ALPHA=.2) | |
I	J	AIJ	AJI	AIJ	AJI
1	2	323.13	-95.237	425.16	-54.936
1	3	872.58	286.68	1219.4	1321.4
2	3	-79.580	286.01	-333.34	1207.8

R1 = 4.0464   R2 = 2.7799   R3 = 0.9200
Q1 = 3.240   Q2 = 2.512   Q3 = 1.400

MEAN DEV. BETWEEN CALC. AND EXP. CONC. IN MOLE PCT

UNIQUAC (SPECIFIC PARAMETERS)   0.79
NRTL   (SPECIFIC PARAMETERS)   0.76

C₃H₈O-C₆H₁₂O₂

$C_3H_8O\text{-}C_6H_{12}O_2$

(1) C6H12O2   ACETIC ACID,BUTYL ESTER

(2) C3H8O   1-PROPANOL

(3) H2O   WATER

JAGANNADHA RAO R., VENKATA RAO C.
J.APPL.CHEM. 9(1959)69

TEMPERATURE = 30.0 DEG C   TYPE OF SYSTEM = 1

EXPERIMENTAL TIE LINES IN MOLE PCT

	LEFT PHASE			RIGHT PHASE	
(1)	(2)	(3)	(1)	(2)	(3)
69.984	16.011	14.004	0.163	1.762	98.075
49.697	27.663	22.640	0.199	2.432	97.369
27.977	34.413	37.610	0.253	3.232	96.514
18.047	35.127	46.826	0.274	3.840	95.885

MEAN DEV. BETWEEN CALC. AND EXP. CONC. IN MOLE PCT
UNIQUAC (COMMON PARAMETERS)   1.40

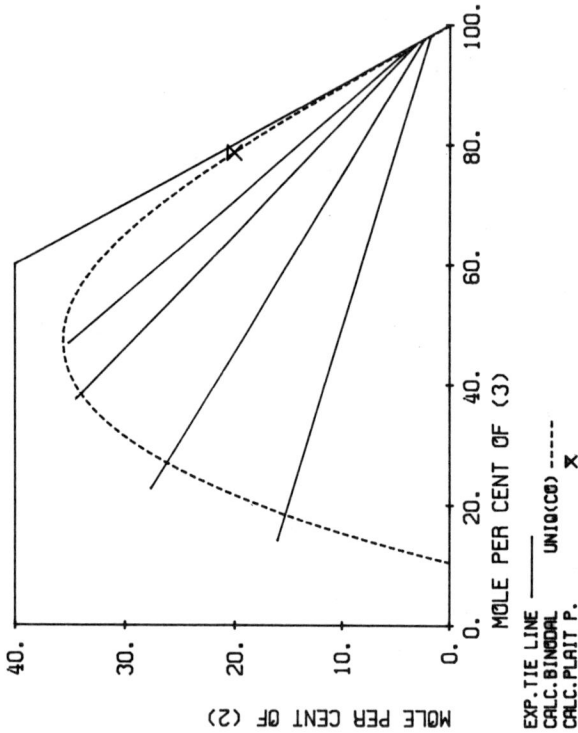

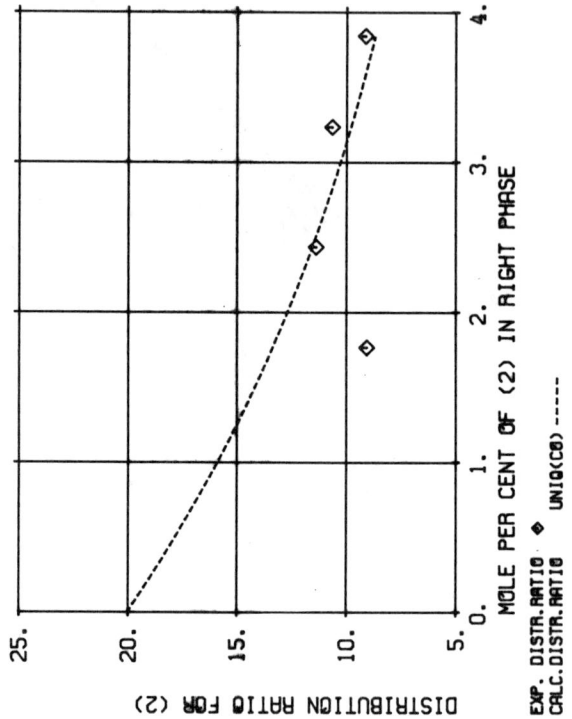

Graph 1 axis labels:
MOLE PER CENT OF (2) [vertical axis, values 0., 10., 20., 30., 40.]
MOLE PER CENT OF (3) [horizontal axis, values 0., 20., 40., 60., 80., 100.]

EXP.TIE LINE ———
CALC.BINODAL   UNIQ(C6) -----
CALC.PLAIT P. ⋊

Graph 2 axis labels:
DISTRIBUTION RATIO FOR (2) [vertical axis, values 5., 10., 15., 20., 25.]
MOLE PER CENT OF (2) IN RIGHT PHASE [horizontal axis, values 1., 2., 3., 4.]

EXP. DISTR.RATIO ◆ UNIQ(C6) -----
CALC.DISTR.RATIO

$C_3H_8O-C_6H_{12}O_2$

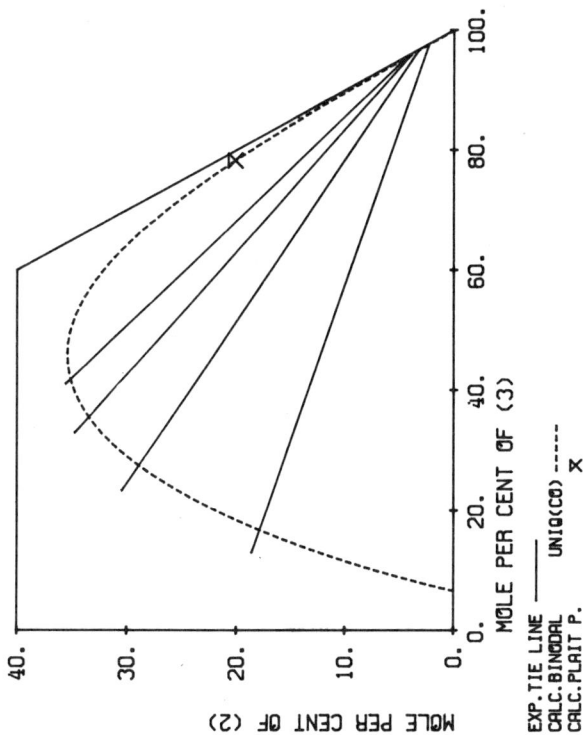

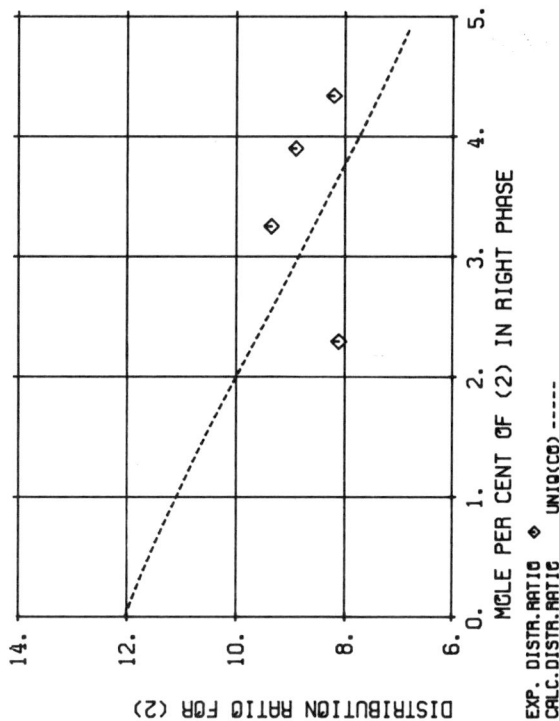

(1) C6H12O2   BUTANOIC ACID,ETHYL ESTER
----
(2) C3H8O   1-PROPANOL
----
(3) H2O   WATER

JAGANNADHA RAO R., VENKATA RAO C.
J.APPL.CHEM. 9(1959)69

TEMPERATURE = 30.0 DEG C   TYPE OF SYSTEM = 1

EXPERIMENTAL TIE LINES IN MOLE PCT

         LEFT PHASE   RIGHT PHASE
     (1)   (2)   (3)   (1)   (2)   (3)

   68.504   18.580   12.916   0.165   2.293   97.542
   46.277   30.444   23.279   0.168   3.252   96.579
   32.385   34.720   32.895   0.205   3.898   95.897
   23.418   35.570   41.012   0.225   4.339   95.437

MEAN DEV. BETWEEN CALC. AND EXP. CONC. IN MOLE PCT
UNIQUAC (COMMON PARAMETERS)   1.29

C$_3$H$_8$O-C$_6$H$_{12}$O$_2$

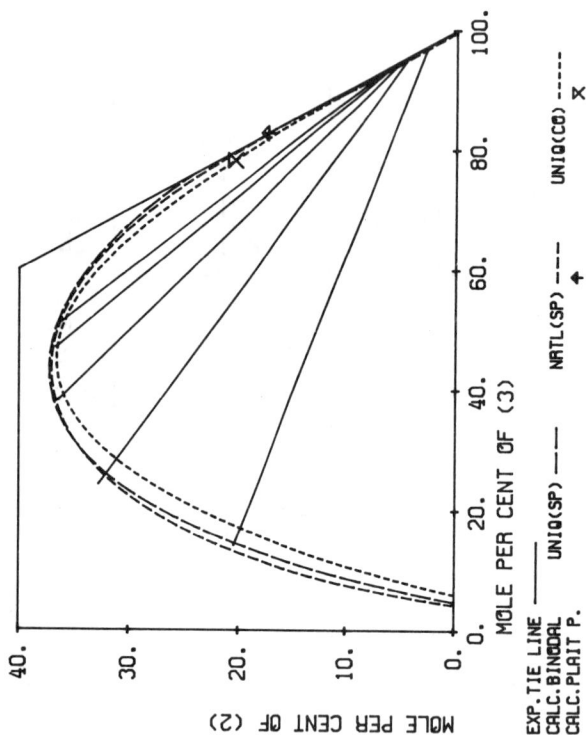

MOLE PER CENT OF (2)

MOLE PER CENT OF (3)

EXP.TIE LINE ------   UNIQ(SP) ------   NRTL(SP) ------   UNIQ(C0) ------
CALC.BINODAL
CALC.PLAIT P.   X

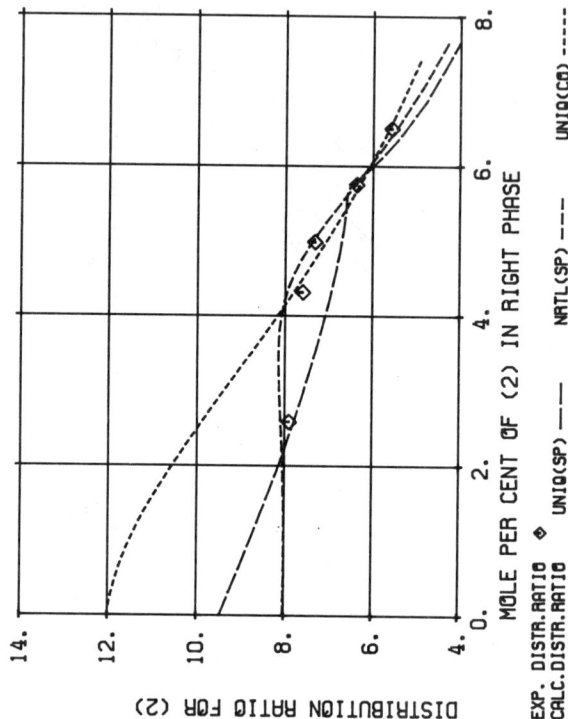

DISTRIBUTION RATIO FOR (2)

MOLE PER CENT OF (2) IN RIGHT PHASE

EXP. DISTR.RATIO ◊   UNIQ(SP) ------   NRTL(SP) ------   UNIQ(C0) ------
CALC.DISTR.RATIO

(1) C6H12O2   PROPANOIC ACID, PROPYL ESTER

(2) C3H8O   1-PROPANOL

(3) H2O   WATER

MOZZHUKIN A.S., ET AL.
ZH.FIZ.KHIM. 41(1967)1687

TEMPERATURE = 20.0 DEG C   TYPE OF SYSTEM = 1

EXPERIMENTAL TIE LINES IN MOLE PCT

	LEFT PHASE			RIGHT PHASE	
(1)	(2)	(3)	(1)	(2)	(3)
65.471	20.406	14.122	0.066	2.583	97.351
43.008	32.752	24.239	0.103	4.310	95.587
25.383	36.449	39.168	0.122	4.981	94.897
16.364	36.647	46.988	0.142	5.750	94.108
12.612	36.216	51.173	0.144	6.499	93.357

SPECIFIC MODEL PARAMETERS IN KELVIN

		UNIQUAC		NRTL (ALPHA=.2)	
I J	AIJ	AJI		AIJ	AJI
1 2	-20.669	190.73		79.049	120.52
1 3	571.78	127.62		456.08	1297.5
2 3	-60.482	275.16		-379.75	1252.9

R1 = 4.8274 R2 = 2.7799 R3 = 0.9200
Q1 = 4.196 Q2 = 2.512 Q3 = 1.400

MEAN DEV. BETWEEN CALC. AND EXP. CONC. IN MOLE PCT

UNIQUAC (SPECIFIC PARAMETERS)   0.42
NRTL (SPECIFIC PARAMETERS)   0.38
UNIQUAC (COMMON PARAMETERS)   1.11

$C_3H_8O-C_6H_{14}$

(1) C6H14   HEXANE

(2) C3H8O   1-PROPANOL

(3) H2O   WATER

MCCANTS J.F., JONES J.H., HOPSON W.H.
IND.ENG.CHEM. 45(1953)454

TEMPERATURE = 37.8 DEG C   TYPE OF SYSTEM = 1

EXPERIMENTAL TIE LINES IN MOLE PCT

	LEFT PHASE			RIGHT PHASE	
(1)	(2)	(3)	(1)	(2)	(3)
91.484	6.213	2.303	0.070	4.590	95.340
69.340	22.368	8.293	0.123	7.289	92.588
40.436	37.729	21.834	0.177	8.791	91.031
23.695	42.270	34.035	0.205	9.331	90.464
17.047	42.179	40.774	0.263	10.542	89.195

SPECIFIC MODEL PARAMETERS IN KELVIN

		UNIQUAC		NRTL(ALPHA=.2)	
I	J	AIJ	AJI	AIJ	AJI
1	2	399.28	-126.27	885.17	-241.39
1	3	823.76	184.93	771.92	1758.1
2	3	-5.1997	165.09	-287.82	1044.1

R1 = 4.4998   R2 = 2.7799   R3 = 0.9200
Q1 = 3.856   Q2 = 2.512   Q3 = 1.400

MEAN DEV. BETWEEN CALC. AND EXP. CONC. IN MOLE PCT

UNIQUAC (SPECIFIC PARAMETERS)   0.46
NRTL   (SPECIFIC PARAMETERS)   0.34

C₃H₈O-C₆H₁₄

```
(1)  C6H14     HEXANE

(2)  C3H8O     1-PROPANOL

(3)  H2O       WATER
```

VOROBEVA A.I., KARAPETYANTS M.KH.
ZH.FIZ.KHIM. 41(1967)1144

TEMPERATURE =   25.0 DEG C   TYPE OF SYSTEM = 1

EXPERIMENTAL TIE LINES IN MOLE PCT

	LEFT PHASE			RIGHT PHASE	
(1)	(2)	(3)	(1)	(2)	(3)
83.093	14.221	2.686	0.049	6.781	93.171
55.222	32.106	12.671	0.154	9.492	90.354
51.774	33.965	14.260	0.182	10.001	89.817
41.794	38.074	20.132	0.182	10.232	89.586
39.009	38.977	22.014	0.210	10.521	89.269
27.743	41.955	30.301	0.210	10.615	89.175
17.930	42.523	39.547	0.239	11.148	88.613
4.714	32.062	63.224	0.478	13.986	85.536
4.061	30.565	65.375	0.570	14.628	84.802
3.378	28.573	68.049	0.699	15.701	83.600
2.374	24.822	72.804	1.106	18.329	80.564

SPECIFIC MODEL PARAMETERS IN KELVIN

		UNIQUAC		NRTL(ALPHA=.2)	
I	J	AIJ	AJI	AIJ	AJI
1	2	20.673	53.282	141.04	164.15
1	3	894.71	286.10	1060.0	1763.6
2	3	-116.97	321.73	-390.96	1192.9

R1 = 4.4998   R2 = 2.7799   R3 = 0.9200
Q1 = 3.856   Q2 = 2.512   Q3 = 1.400

MEAN DEV. BETWEEN CALC. AND EXP. CONC. IN MOLE PCT

```
UNIQUAC (SPECIFIC PARAMETERS)      1.90
NRTL    (SPECIFIC PARAMETERS)      0.56
UNIQUAC (COMMON PARAMETERS)        3.87
```

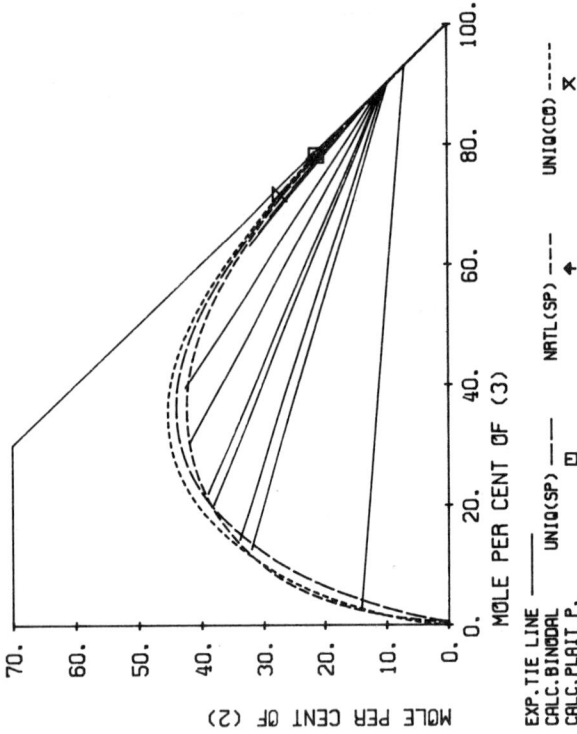

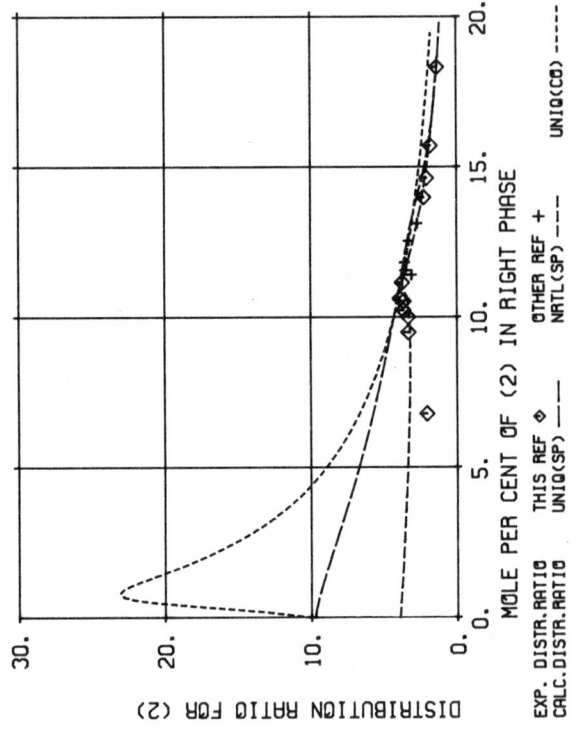

$C_3H_8O-C_6H_{14}$

MOLE PER CENT OF (3)

MOLE PER CENT OF (2)

EXP.TIE LINE ——   UNIQ(SP) --⊟--   NRTL(SP) --◆--   UNIQ(CO) ----   ✕
CALC.BINODAL --------
CALC.PLAIT P.

DISTRIBUTION RATIO FOR (2)

MOLE PER CENT OF (2) IN RIGHT PHASE

EXP. DISTR. RATIO   THIS REF ◇   OTHER REF +
CALC. DISTR. RATIO   UNIQ(SP) ——   NRTL(SP) --◆--   UNIQ(CO) ----

(1) C6H14   HEXANE

(2) C3H8O   1-PROPANOL

(3) H2O   WATER

SUGI H., KATAYAMA T.
J.CHEM.ENG.JPN. 10(1977)400

TEMPERATURE = 25.0 DEG C   TYPE OF SYSTEM = 1

EXPERIMENTAL TIE LINES IN MOLE PCT

	LEFT PHASE			RIGHT PHASE	
(1)	(2)	(3)	(1)	(2)	(3)
47.660	36.030	16.310	0.030	11.400	88.570
33.490	41.700	24.810	0.050	11.550	88.400
22.290	42.720	34.990	0.070	11.810	83.120
11.630	42.000	46.370	0.120	12.530	87.350
7.320	36.060	56.620	0.290	13.110	86.600

SPECIFIC MODEL PARAMETERS IN KELVIN

		UNIQUAC		NRTL(ALPHA=.2)	
I	J	AIJ	AJI	AIJ	AJI
1	2	20.673	53.282	141.04	164.15
1	3	894.71	286.10	1060.0	1763.6
2	3	-116.97	321.73	-390.96	1192.9

R1 = 4.4998   R2 = 2.7799   R3 = 0.9200
Q1 = 3.856   Q2 = 2.512   Q3 = 1.400

MEAN DEV. BETWEEN CALC. AND EXP. CONC. IN MOLE PCT

UNIQUAC (SPECIFIC PARAMETERS)   1.86
NRTL   (SPECIFIC PARAMETERS)   0.54
UNIQUAC (COMMON PARAMETERS)   3.55

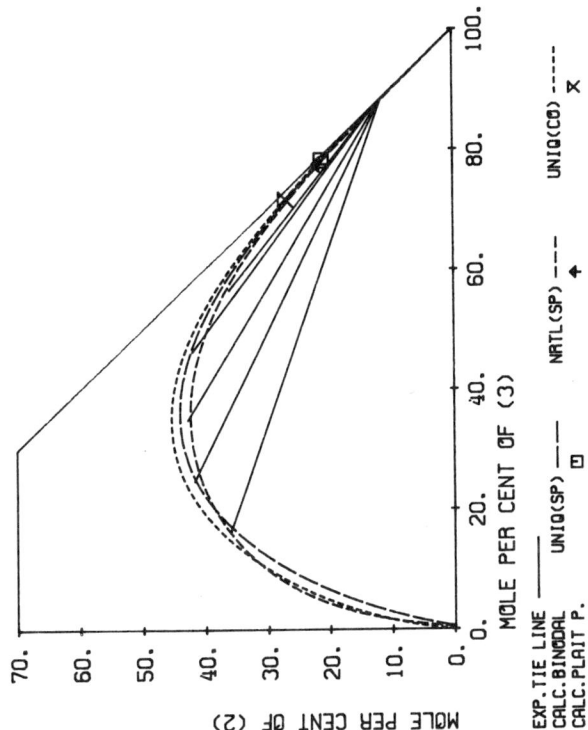

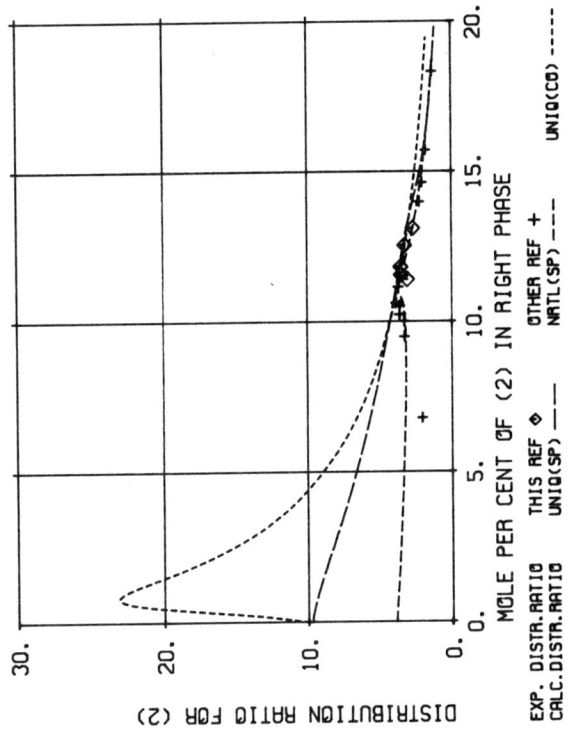

C$_3$H$_8$O-C$_7$H$_8$

(1) C7H8   TOLUENE

(2) C3H8O   1-PROPANOL

(3) H2O   WATER

BAKER E.M.
J.PHYS.CHEM. 59(1955)1182

TEMPERATURE = 25.0 DEG C   TYPE OF SYSTEM = 1

EXPERIMENTAL TIE LINES IN MOLE PCT

	LEFT PHASE			RIGHT PHASE	
(1)	(2)	(3)	(1)	(2)	(3)
96.048	3.700	0.252	0.104	2.618	97.277
87.162	10.423	2.415	0.096	3.870	96.033
78.918	16.450	4.632	0.109	4.673	95.218
64.427	26.193	9.379	0.087	4.929	94.984
57.352	30.815	11.833	0.110	5.235	94.655
48.616	35.315	16.069	0.055	5.374	94.570
36.766	40.266	22.968	0.063	6.071	93.866
23.226	43.388	32.787	0.079	6.630	93.291
21.226	43.401	33.373	0.091	6.713	93.196
12.140	41.285	46.575	0.128	7.920	91.952
7.397	36.710	55.893	0.140	8.008	91.852

SPECIFIC MODEL PARAMETERS IN KELVIN

I J	UNIQUAC AIJ	AJI	NRTL(ALPHA=.2) AIJ	AJI
1 2	364.97	-127.16	622.14	-155.55
1 3	827.95	317.66	936.24	1583.5
2 3	-133.96	372.30	-334.16	1177.9

R1 = 3.9228 R2 = 2.7799 R3 = 0.9200
Q1 = 2.968   Q2 = 2.512   Q3 = 1.400

MEAN DEV. BETWEEN CALC. AND EXP. CONC. IN MOLE PCT

UNIQUAC (SPECIFIC PARAMETERS)   0.43
NRTL   (SPECIFIC PARAMETERS)   0.55
UNIQUAC (COMMON PARAMETERS)   3.88

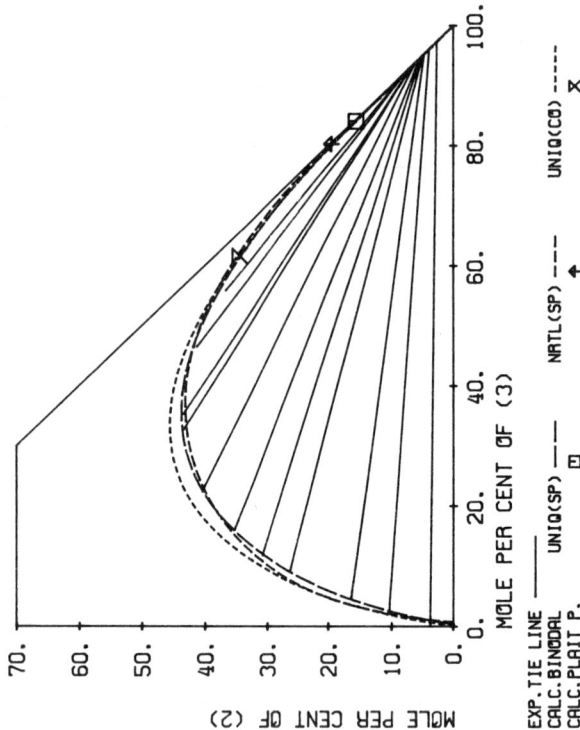

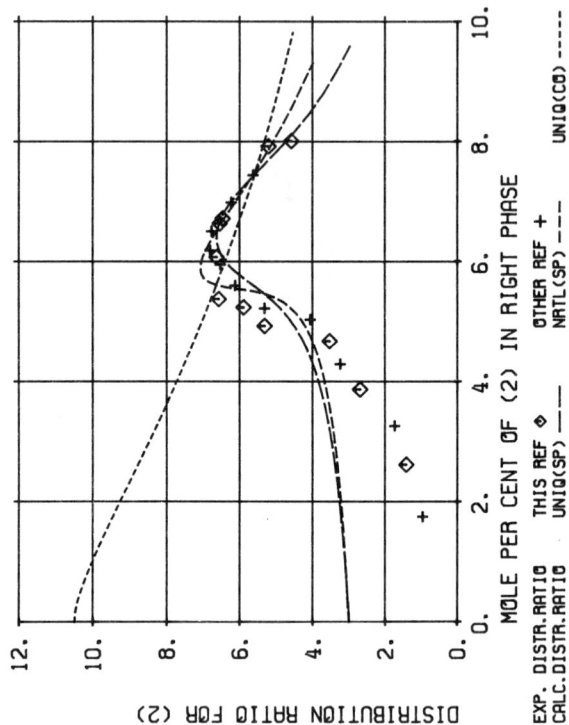

C$_3$H$_8$O-C$_7$H$_8$

(1) C7H8   TOLUENE
----
(2) C3H8O   1-PROPANOL
----
(3) H2O   WATER

NIKURASHINA N.I., ILIN K.K.,
ZH.OBSHCH.KHIM. 42(1972)1657

TEMPERATURE = 25.0 DEG C   TYPE OF SYSTEM = 1

EXPERIMENTAL TIE LINES IN MOLE PCT (GRAPH.INTERPOL.)

	LEFT PHASE			RIGHT PHASE	
(1)	(2)	(3)	(1)	(2)	(3)
97.824	1.670	0.506	0.015	1.748	98.237
93.341	5.664	0.995	0.020	3.260	96.720
83.581	13.847	2.572	0.026	4.291	95.683
74.599	20.444	4.957	0.031	5.029	94.940
63.279	27.789	8.932	0.033	5.218	94.750
52.100	34.289	13.611	0.035	5.599	94.365
41.612	38.840	19.548	0.039	5.948	94.013
32.494	42.101	25.405	0.041	6.183	93.776
25.068	44.024	30.909	0.044	6.500	93.455
17.932	43.505	38.563	0.050	6.985	92.965
12.475	41.909	45.616	0.059	7.438	92.503

SPECIFIC MODEL PARAMETERS IN KELVIN

I J	UNIQUAC			NRTL(ALPHA=.2)	
	AIJ	AJI		AIJ	AJI
1 2	364.97	-127.16		622.14	-155.55
1 3	827.95	317.66		936.24	1553.5
2 3	-133.96	372.30		-334.16	1177.9

R1 = 3.9228   R2 = 2.7799   R3 = 0.9200
Q1 = 2.968   Q2 = 2.512   Q3 = 1.400

MEAN DEV. BETWEEN CALC. AND EXP. CONC. IN MOLE PCT

UNIQUAC (SPECIFIC PARAMETERS)   0.46
NRTL   (SPECIFIC PARAMETERS)   0.43
UNIQUAC (COMMON PARAMETERS)   2.74

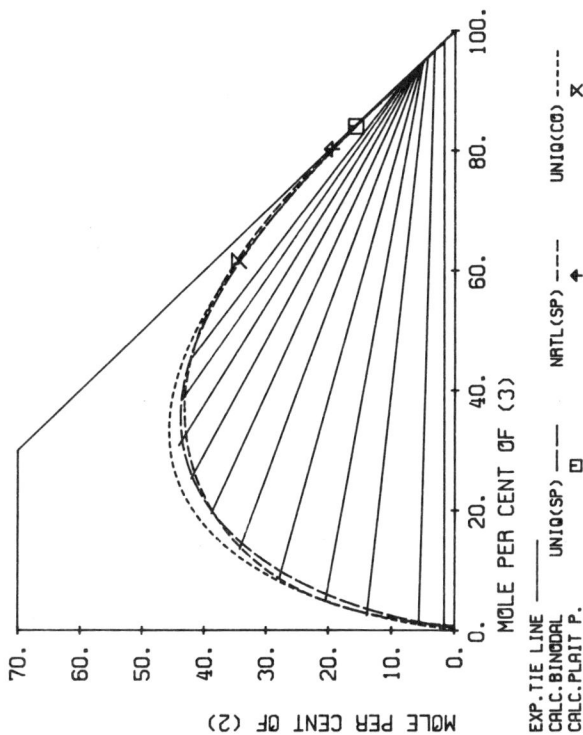

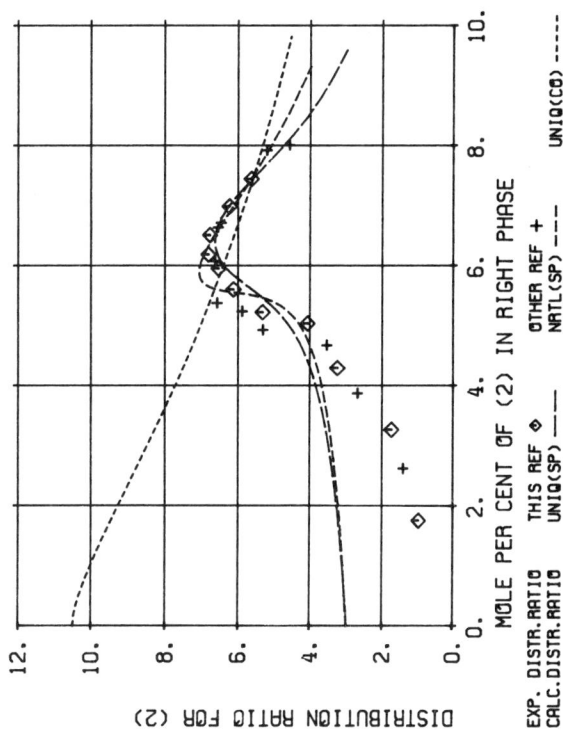

$C_3H_8O-C_7H_{14}O_2$

(1) C7H14O2   ACETIC ACID, PENTYL ESTER

(2) C3H8O   1-PROPANOL

(3) H2O   WATER

JAGANNADHA RAO R., VENKATA RAO C.
J.APPL.CHEM. 9(1959)69

TEMPERATURE = 30.0 DEG C   TYPE OF SYSTEM = 1

EXPERIMENTAL TIE LINES IN MOLE PCT

	LEFT PHASE			RIGHT PHASE	
(1)	(2)	(3)	(1)	(2)	(3)
62.763	21.038	16.199	0.072	1.853	98.075
28.966	36.633	34.401	0.105	3.279	96.616
21.090	37.459	41.451	0.137	3.745	96.119
16.031	36.602	47.367	0.153	4.218	95.629
12.728	35.760	51.512	0.186	4.557	95.257

SPECIFIC MODEL PARAMETERS IN KELVIN

		UNIQUAC		NRTL(ALPHA=.2)	
I J	AIJ	AJI		AIJ	AJI
1 2	-29.945	214.70		183.08	-10.898
1 3	538.44	142.70		434.84	1335.1
2 3	-95.062	354.45		-404.56	1372.0

R1 = 5.5018   R2 = 2.7799   R3 = 0.9200
Q1 = 4.736   Q2 = 2.512   Q3 = 1.400

MEAN DEV. BETWEEN CALC. AND EXP. CONC. IN MOLE PCT

UNIQUAC (SPECIFIC PARAMETERS)   0.42
NRTL   (SPECIFIC PARAMETERS)   0.48
UNIQUAC (COMMON PARAMETERS)   1.46

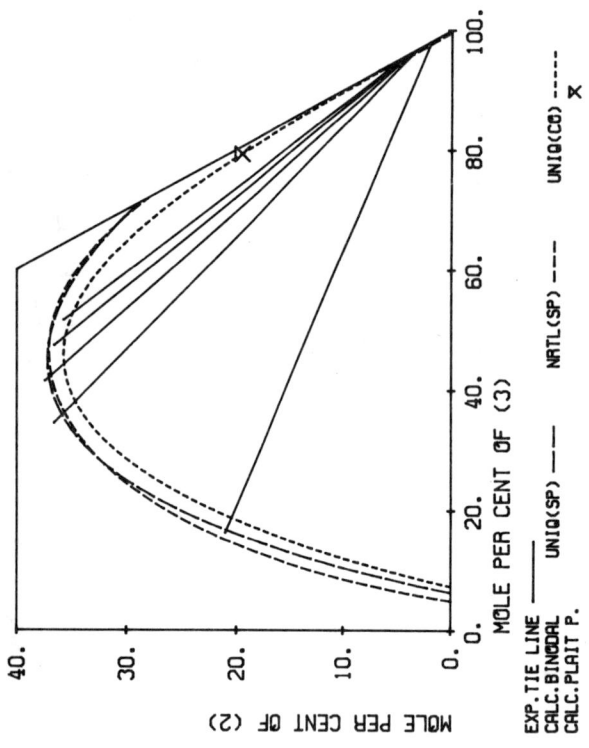

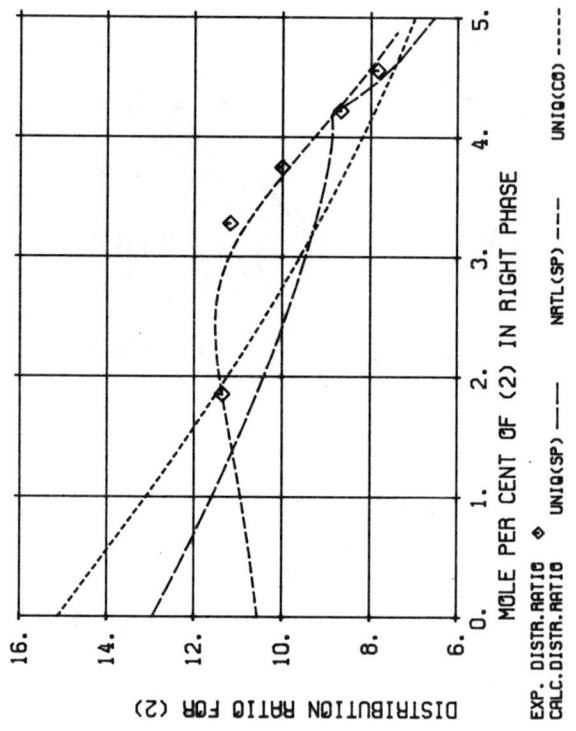

$C_3H_8O-C_7H_{16}$

(1) C7H16   HEPTANE

(2) C3H8O   1-PROPANOL

(3) H2O   WATER

MCCANTS J.F., JONES J.H., HOPSON W.H.
IND.ENG.CHEM. 45(1953)454

TEMPERATURE = 37.8 DEG C   TYPE OF SYSTEM = 1

EXPERIMENTAL TIE LINES IN MOLE PCT

LEFT PHASE			RIGHT PHASE		
(1)	(2)	(3)	(1)	(2)	(3)
75.034	19.121	5.844	0.131	8.829	91.041
70.092	19.220	10.688	0.242	14.424	85.334
55.738	30.633	13.628	1.829	23.848	74.323
52.274	32.400	15.326	2.091	25.259	72.650

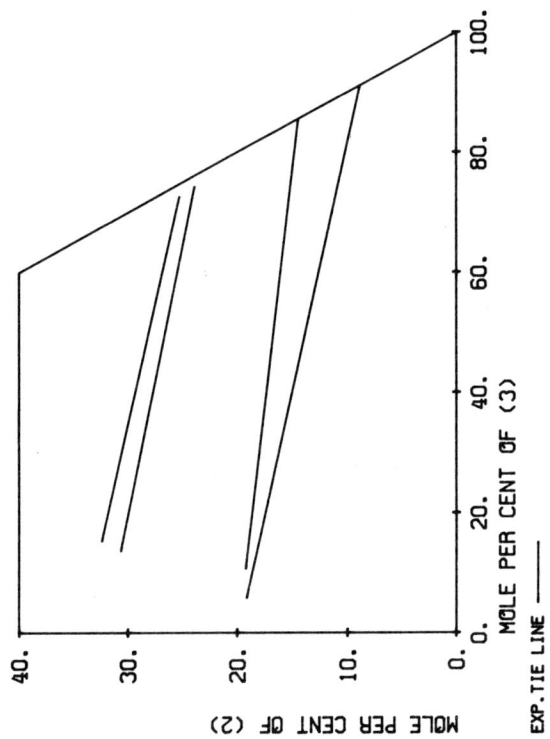

C₃H₈O-C₇H₁₆

(1) H2O   WATER

(2) C3H8O   1-PROPANOL

(3) C7H16   HEPTANE

VOROBEVA A.I., KARAPETYANTS M.KH.
ZH.FIZ.KHIM. 41(1967)1144

TEMPERATURE = 25.0 DEG C   TYPE OF SYSTEM = 1

EXPERIMENTAL TIE LINES IN MOLE PCT

	LEFT PHASE			RIGHT PHASE	
(1)	(2)	(3)	(1)	(2)	(3)
93.483	6.434	0.083	1.066	8.305	90.629
91.407	8.507	0.086	2.570	13.710	83.720
87.375	12.392	0.234	4.889	20.222	74.889
83.828	15.723	0.449	5.770	22.049	72.181
76.854	22.048	1.097	6.622	23.817	69.561
69.732	28.058	2.210	7.048	24.362	68.590
59.883	35.650	4.467	7.870	25.805	66.326
49.894	41.898	8.208	9.855	28.461	61.684
43.410	44.838	11.752	11.691	31.405	56.904
33.386	45.117	21.497	19.811	38.136	42.053
29.263	43.850	26.887	24.086	41.401	34.513

SPECIFIC MODEL PARAMETERS IN KELVIN

		UNIQUAC		NRTL(ALPHA=.2)	
I J	AIJ	AJI		AIJ	AJI
1 2	121.32	-30.623		1023.5	-379.13
1 3	129.11	-906.32		2061.9	775.09
2 3	-43.607	221.42		257.60	143.86

R1 = 0.9200   R2 = 2.7799   R3 = 5.1742
Q1 = 1.400   Q2 = 2.512   Q3 = 4.396

MEAN DEV. BETWEEN CALC. AND EXP. CONC. IN MOLE PCT

UNIQUAC (SPECIFIC PARAMETERS)   1.51
NRTL   (SPECIFIC PARAMETERS)   1.39
UNIQUAC (COMMON PARAMETERS)   3.05

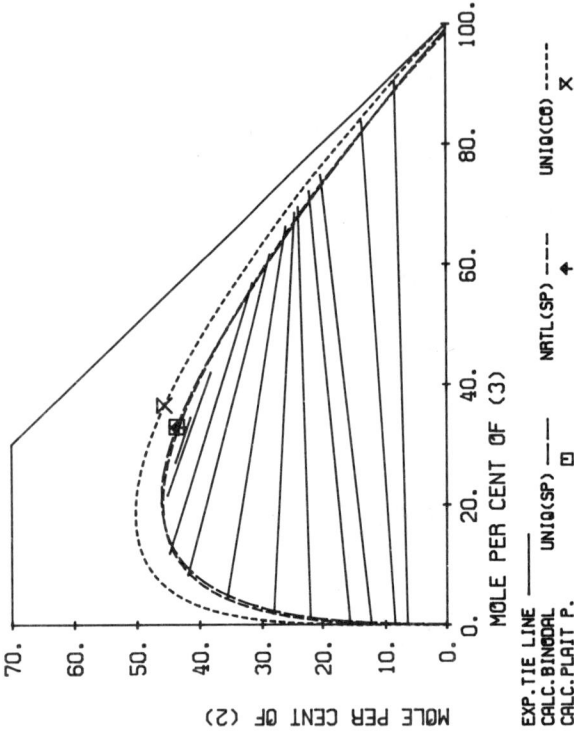

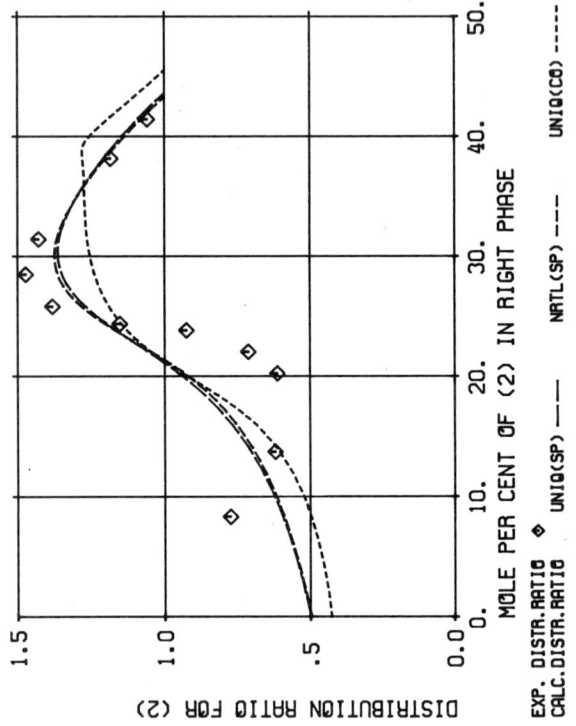

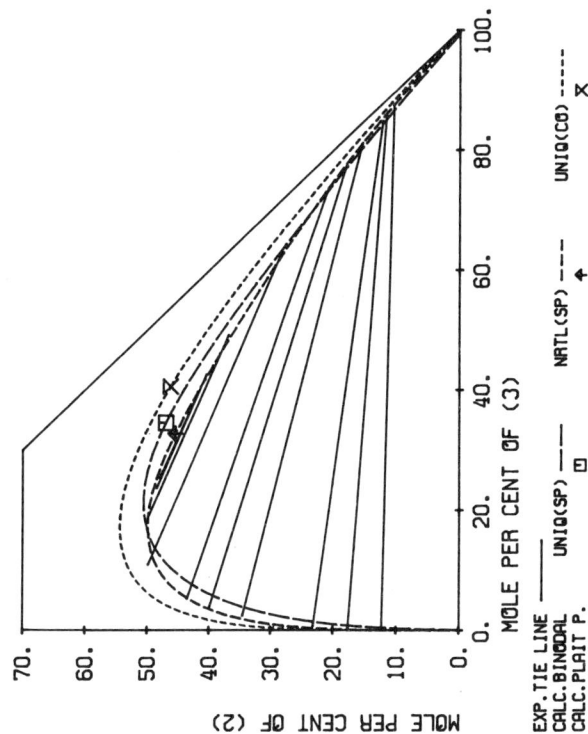

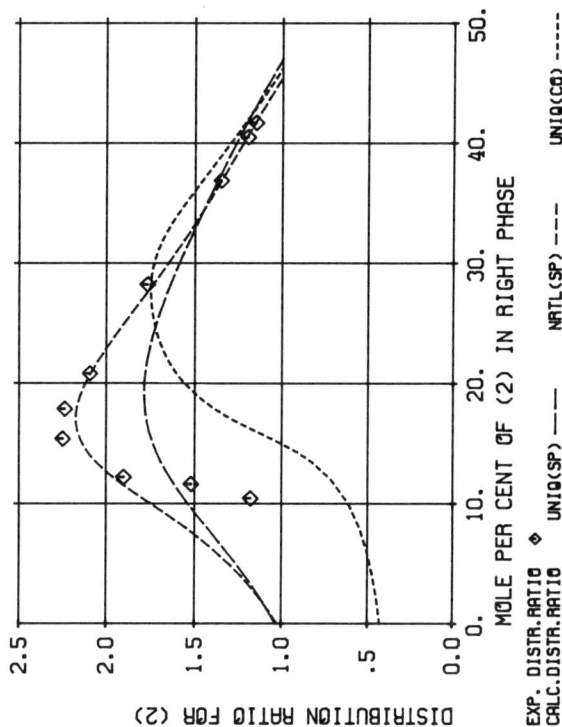

$C_3H_8O\text{-}C_8H_{18}$

(1) H2O   WATER

(2) C3H8O   1-PROPANOL

(3) C8H18   OCTANE

VOROBEVA A.I., KARAPETYANTS M.KH.
ZH.FIZ.KHIM. 41(1967)1144

TEMPERATURE = 25.0 DEG C   TYPE OF SYSTEM = 1

EXPERIMENTAL TIE LINES IN MOLE PCT

	LEFT PHASE			RIGHT PHASE	
(1)	(2)	(3)	(1)	(2)	(3)
87.604	12.314	0.082	2.360	10.435	87.205
82.037	17.647	0.316	2.347	11.605	86.048
76.113	23.190	0.697	2.909	12.207	84.884
63.066	34.555	2.380	3.418	15.366	81.216
56.312	39.954	3.734	4.453	17.851	77.695
51.215	43.569	5.216	5.426	20.814	73.760
39.264	49.867	10.869	8.561	28.222	63.217
31.173	49.764	19.063	13.918	36.870	49.212
27.857	48.463	23.680	16.889	40.494	42.617
26.734	47.969	25.297	18.462	41.682	39.856

SPECIFIC MODEL PARAMETERS IN KELVIN

		UNIQUAC		NRTL(ALPHA=.2)	
I J	AIJ	AJI		AIJ	AJI
1 2	-31.426	24.727		959.71	-422.19
1 3	46.163	1037.8		3488.2	1300.1
2 3	-34.542	205.53		507.10	68.718

R1 = 0.9200   R2 = 2.7799   R3 = 5.8486
Q1 = 1.400   Q2 = 2.512   Q3 = 4.936

MEAN DEV. BETWEEN CALC. AND EXP. CONC. IN MOLE PCT

UNIQUAC (SPECIFIC PARAMETERS)   1.92
NRTL   (SPECIFIC PARAMETERS)   0.69
UNIQUAC (COMMON PARAMETERS)   3.98

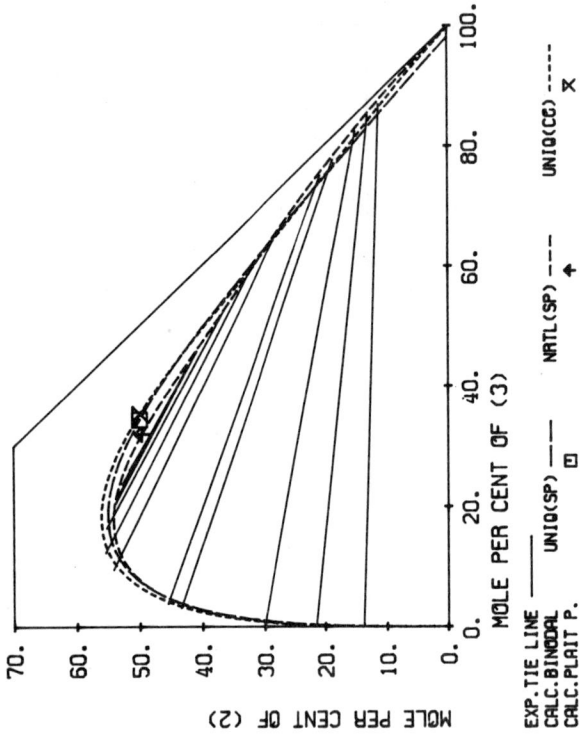

EXP.TIE LINE ———— UNIQ(SP) □ NRTL(SP) ✦ UNIQ(CC) ✕
CALC.BINODAL – – –
CALC.PLAIT P.

MOLE PER CENT OF (3)

MOLE PER CENT OF (2)

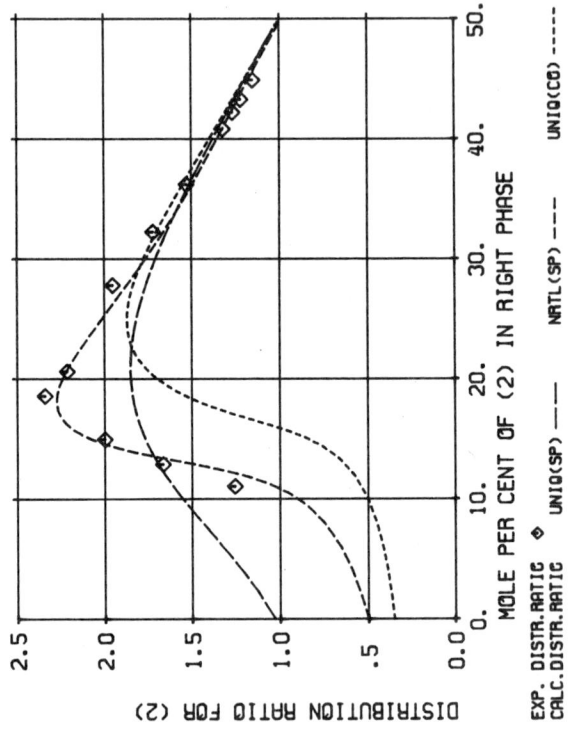

EXP. DISTR.RATIO ◇ UNIQ(SP) ◇ NRTL(SP) – – – UNIQ(CC) – – –
CALC.DISTR.RATIO

DISTRIBUTION RATIO FOR (2)

MOLE PER CENT OF (2) IN RIGHT PHASE

(1) H2O   WATER
(2) C3H8O   1-PROPANOL
(3) C9H20   NONANE

VOROBEVA A.I., KARAPETYANTS M.KH.
ZH.FIZ.KHIM. 41(1967)1144

TEMPERATURE = 25.0 DEG C   TYPE OF SYSTEM = 1

EXPERIMENTAL TIE LINES IN MOLE PCT

	LEFT PHASE			RIGHT PHASE	
(1)	(2)	(3)	(1)	(2)	(3)
86.072	13.853	0.075	1.974	11.045	86.981
78.172	21.464	0.364	1.954	12.882	85.164
69.123	29.888	0.988	2.559	14.955	82.486
53.411	43.458	3.131	3.714	18.553	77.733
50.578	45.612	3.810	4.256	20.591	75.153
36.389	54.188	9.423	6.784	27.787	65.429
32.294	55.424	12.281	8.600	32.216	59.185
28.331	55.289	16.380	10.701	36.193	53.106
25.583	53.891	20.471	13.329	40.803	45.868
24.753	53.328	21.919	14.395	42.167	43.438
23.565	52.748	23.686	15.432	43.257	41.311
22.394	51.644	25.962	17.072	44.868	38.059

SPECIFIC MODEL PARAMETERS IN KELVIN

		UNIQUAC		NRTL(ALPHA=.2)	
I J		AIJ	AJI	AIJ	AJI
1 2	-42.385	54.138	1179.9	-422.91
1 3	142.85	779.60	3024.1	1549.5
2 3	-79.603	281.16	583.67	28.118

R1 = 0.9200   R2 = 2.7799   R3 = 6.5230
Q1 = 1.400   Q2 = 2.512   Q3 = 5.476

MEAN DEV. BETWEEN CALC. AND EXP. CONC. IN MOLE PCT

UNIQUAC (SPECIFIC PARAMETERS)   2.03
NRTL (SPECIFIC PARAMETERS)   0.60
UNIQUAC (COMMON PARAMETERS)   2.64

C$_3$H$_8$O-C$_{12}$H$_{10}$O

(1) H2O   WATER
----
(2) C3H8O   1-PROPANOL
----
(3) C12H100   ETHER,DIPHENYL

PURNELL J.H., BOWDEN S.T.
J.CHEM.SOC. (1954)539

TEMPERATURE = 25.0 DEG C   TYPE OF SYSTEM = 1

EXPERIMENTAL TIE LINES IN MOLE PCT

	LEFT PHASE			RIGHT PHASE	
(1)	(2)	(3)	(1)	(2)	(3)
97.004	2.984	0.011	1.806	4.331	93.863
93.494	6.469	0.037	4.055	16.286	79.659
91.439	8.510	0.051	5.428	20.218	74.354
87.793	12.111	0.096	6.685	23.819	69.496
82.061	17.604	0.335	7.326	24.592	68.082
75.252	23.920	0.827	7.316	24.776	67.908
68.463	29.925	1.612	8.546	26.253	65.201
57.084	39.342	3.573	8.418	28.173	63.409
47.872	45.849	6.278	10.778	29.475	59.747
43.797	48.408	7.795	12.828	31.911	55.261

SPECIFIC MODEL PARAMETERS IN KELVIN

		UNIQUAC		NRTL(ALPHA=.2)	
I J		AIJ	AJI	AIJ	AJI
1 2		267.00	-73.860	1136.7	-343.25
1 3		97.114	755.80	2484.9	897.42
2 3		-98.937	302.02	483.35	28.143

R1 = 0.9200   R2 = 2.7799   R3 = 6.2873
.Q1 = 1.400   Q2 = 2.512   Q3 = 4.480

MEAN DEV. BETWEEN CALC. AND EXP. CONC. IN MOLE PCT

UNIQUAC (SPECIFIC PARAMETERS)   0.61
NRTL   (SPECIFIC PARAMETERS)   0.55
UNIQUAC (COMMON PARAMETERS)   1.92

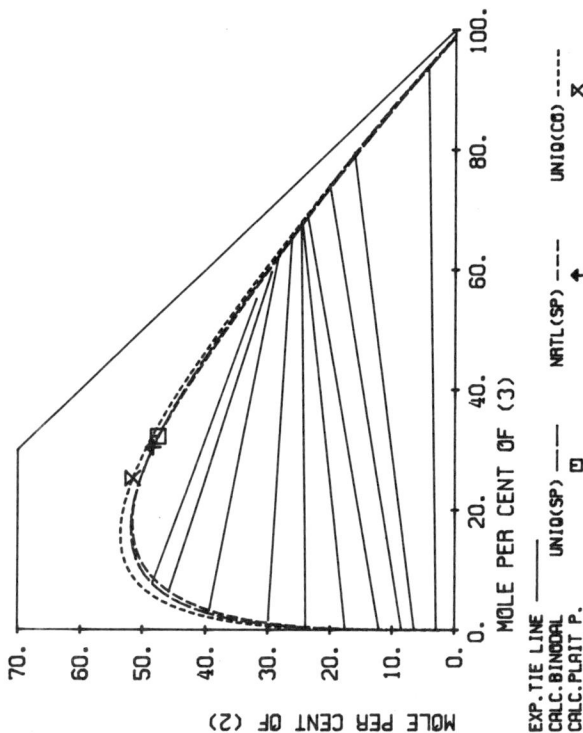

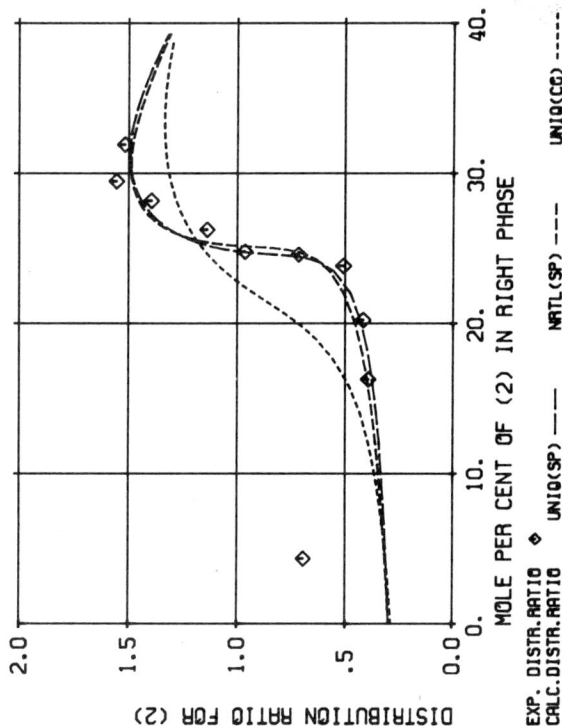

MOLE PER CENT OF (2)
MOLE PER CENT OF (3)

EXP.TIE LINE ———   UNIQ(SP) ❑   NRTL(SP) ✦   UNIQ(CO) ✗
CALC.BINODAL – – –
CALC.PLAIT P.

DISTRIBUTION RATIO FOR (2)
MOLE PER CENT OF (2) IN RIGHT PHASE

EXP. DISTR.RATIO ◇   UNIQ(SP) ———   NRTL(SP) – – –   UNIQ(CO) -----
CALC.DISTR.RATIO

C₃H₈O-C₄H₈O₂

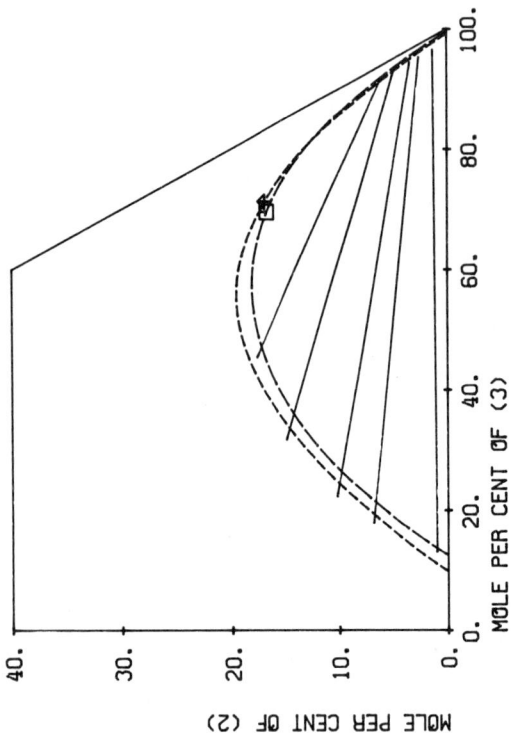

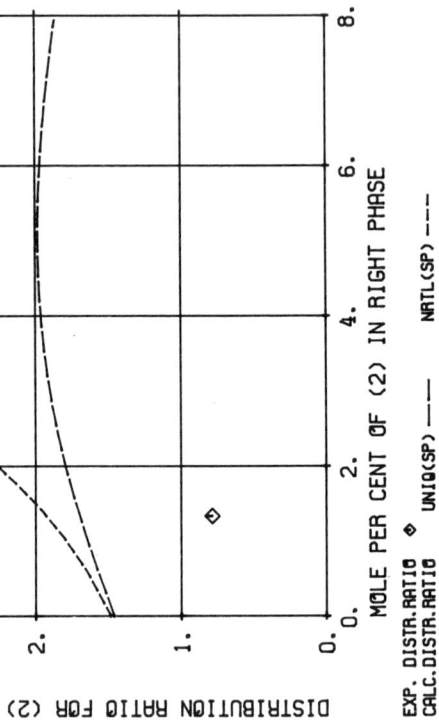

(1) C4H8O2   ACETIC ACID,ETHYL ESTER

(2) C3H8O   2-PROPANOL

(3) H2O   WATER

BEECH D.G., GLASSTONE S.
J.CHEM.SOC. (1938)67

TEMPERATURE = 0.0 DEG C   TYPE OF SYSTEM = 1

EXPERIMENTAL TIE LINES IN MOLE PCT

	LEFT PHASE			RIGHT PHASE		
	(1)	(2)	(3)	(1)	(2)	(3)
85.857	1.047	13.097		2.044	1.331	96.625
75.055	6.976	17.969		2.052	2.598	95.350
67.419	10.184	22.397		2.065	3.444	94.491
53.306	14.870	31.823		2.132	4.958	92.910
36.822	17.652	45.526		2.352	6.229	91.419

SPECIFIC MODEL PARAMETERS IN KELVIN

		UNIQUAC		NRTL(ALPHA=.2)	
I	J	AIJ	AJI	AIJ	AJI
1	2	-49.281	-93.288	1167.5	-614.99
1	3	291.95	187.75	228.35	1157.2
2	3	-295.44	197.07	21.043	65.297

R1 = 3.4786   R2 = 2.7791   R3 = 0.9200
Q1 = 3.116   Q2 = 2.508   Q3 = 1.400

MEAN DEV. BETWEEN CALC. AND EXP. CONC. IN MOLE PCT

UNIQUAC (SPECIFIC PARAMETERS)   1.62
NRTL   (SPECIFIC PARAMETERS)   0.81

(1) C4H8O2   ACETIC ACID,ETHYL ESTER
----
(2) C3H8O   2-PROPANOL
----
(3) H2O   WATER

BEECH D.G., GLASSTONE S.
J.CHEM.SOC. (1938)67

TEMPERATURE = 20.0 DEG C   TYPE OF SYSTEM = 1

EXPERIMENTAL TIE LINES IN MOLE PCT

          LEFT PHASE   RIGHT PHASE
        (1)   (2)   (3)   (1)   (2)   (3)

      77.544   5.083   17.373   1.768   1.116   97.116
      68.669   8.835   22.496   1.963   2.268   95.770
      57.748   12.518   29.734   2.174   3.048   94.778
      45.850   15.460   38.690   2.440   4.221   93.339
      32.403   17.433   50.164   2.888   5.422   91.691
      22.443   17.098   60.459   3.622   6.511   89.868

SPECIFIC MODEL PARAMETERS IN KELVIN

              UNIQUAC   NRTL(ALPHA=.2)
  I J   AIJ   AJI   AIJ   AJI
----
  1 2 -22.131   148.39   863.60   -809.05
  1 3   416.60   72.950   228.96   1391.0
  2 3 -140.87   275.47   14.947   -388.14

  R1 = 3.4786   R2 = 2.7791   R3 = 0.9200
  Q1 = 3.116   Q2 = 2.508   Q3 = 1.400

MEAN DEV. BETWEEN CALC. AND EXP. CONC. IN MOLE PCT

UNIQUAC (SPECIFIC PARAMETERS)   1.08
NRTL   (SPECIFIC PARAMETERS)   1.13
UNIQUAC (COMMON PARAMETERS)   1.12

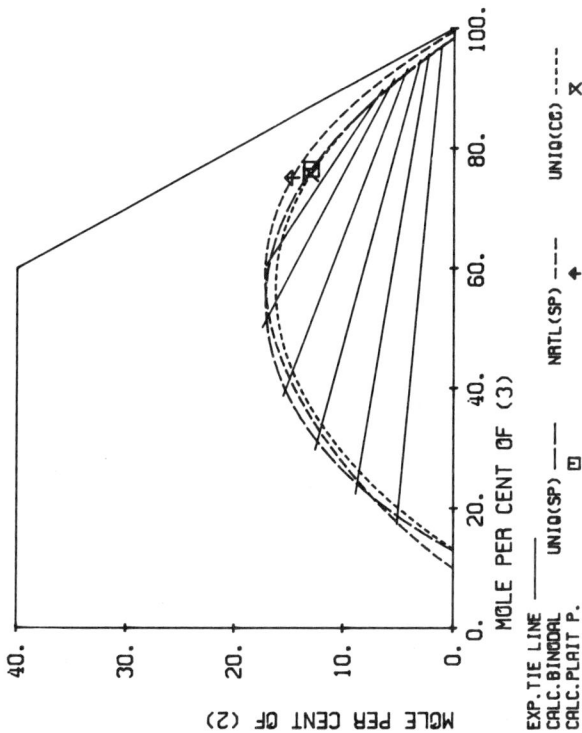

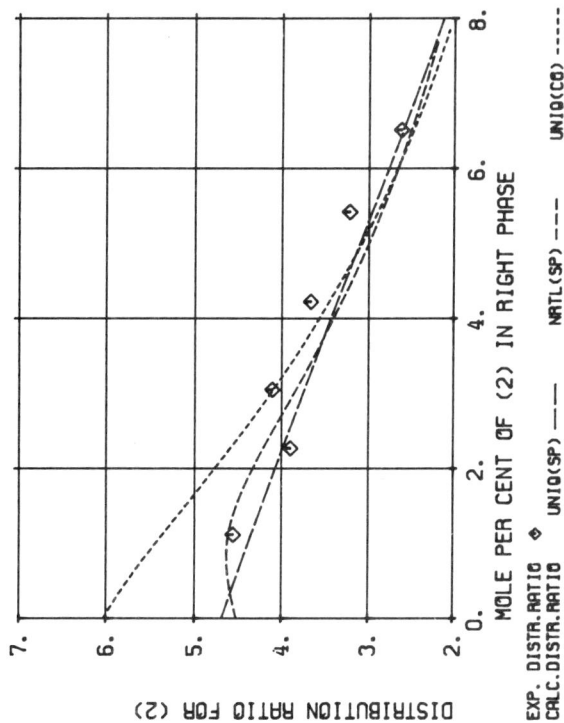

C₃H₈O-C₄H₁₀O

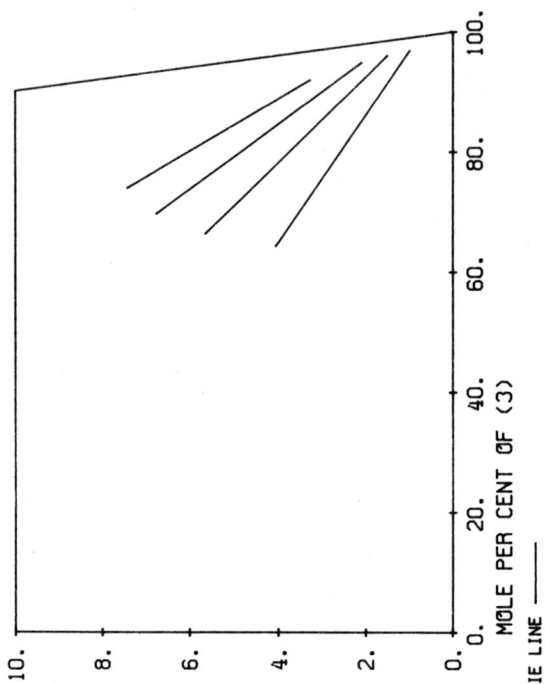

MOLE PER CENT OF (3)

MOLE PER CENT OF (2)

EXP.TIE LINE

(1) C4H10O   1-BUTANOL

(2) C3H8O   2-PROPANOL

(3) H2O   WATER

MOROZOV A.V., SARKISOV A.G., TUROVSKII V.B., ILYASKIN V.I.
ZH.FIZ.KHIM. 52(1978)1821 .

TEMPERATURE = 80.0 DEG C   TYPE OF SYSTEM = 1

EXPERIMENTAL TIE LINES IN MOLE PCT

LEFT PHASE			RIGHT PHASE		
(1)	(2)	(3)	(1)	(2)	(3)
31.656	4.053	64.290	2.117	0.979	96.904
28.011	5.647	66.342	2.459	1.499	96.042
23.599	6.750	69.651	3.059	2.092	94.849
18.686	7.417	73.897	4.797	3.270	91.934

$C_3H_8O\text{-}C_5H_4O_2$

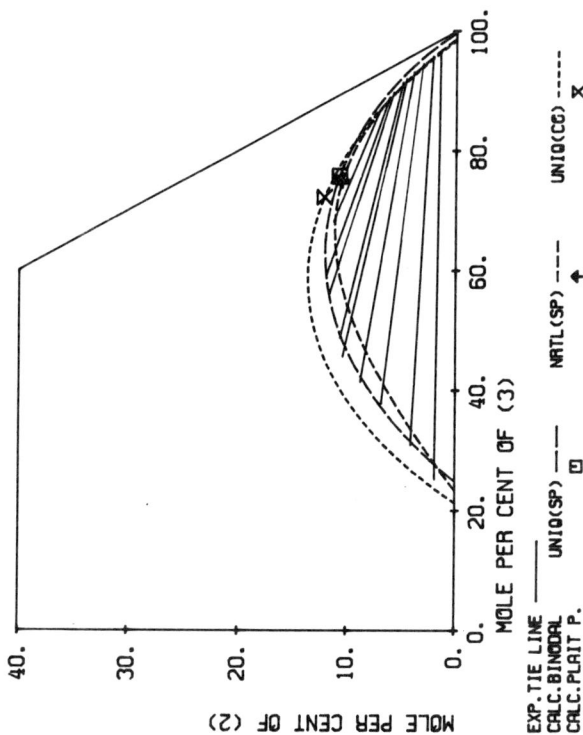

MOLE PER CENT OF (3)

MOLE PER CENT OF (2)

EXP. TIE LINE —— UNIQ(SP) □ —— NRTL(SP) ◆ — — UNIQ(CC) ✗ ·····
CALC. BINODAL
CALC. PLAIT P.

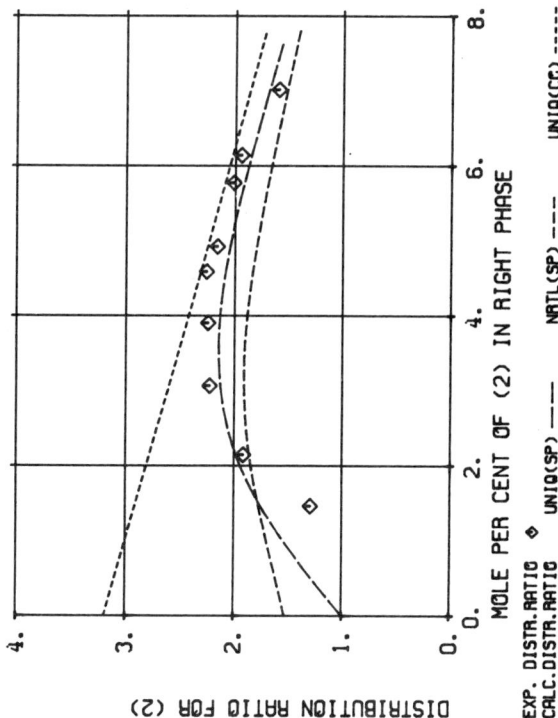

MOLE PER CENT OF (2) IN RIGHT PHASE

DISTRIBUTION RATIO FOR (2)

EXP. DISTR. RATIO ◇ UNIQ(SP) —— NRTL(SP) — — UNIQ(CC) ·····
CALC. DISTR. RATIO

(1) C5H4O2   FURFURAL
(2) C3H8O   2-PROPANOL
(3) H2O   WATER

KRUPATKIN I.L., GLAGOLEVA M.F.
ZH.OBSHCH.KHIM. 40(1970)12

TEMPERATURE = 25.0 DEG C   TYPE OF SYSTEM = 1

EXPERIMENTAL TIE LINES IN MOLE PCT

	LEFT PHASE			RIGHT PHASE	
(1)	(2)	(3)	(1)	(2)	(3)
72.891	1.890	25.219	1.793	1.466	96.741
64.988	4.104	30.908	1.981	2.145	95.875
55.418	6.831	37.751	2.378	3.063	94.559
49.838	8.774	41.389	2.897	3.902	93.201
43.972	10.379	45.649	3.385	4.584	92.031
40.616	10.641	48.744	3.625	4.918	91.457
32.300	11.608	56.093	4.457	5.771	89.772
28.996	11.930	59.075	4.907	6.137	88.957
20.617	11.245	68.138	6.074	7.014	86.912

SPECIFIC MODEL PARAMETERS IN KELVIN

		UNIQUAC		NRTL(ALPHA=.2)	
I	J	AIJ	AJI	AIJ	AJI
1	2	-376.47	-165.32	817.87	-652.61
1	3	-11.493	398.62	6.4841	1288.3
2	3	-453.26	215.05	-152.88	-495.46

R1 = 3.1680   R2 = 2.7791   R3 = 0.9200
Q1 = 2.484   Q2 = 2.508   Q3 = 1.400

MEAN DEV. BETWEEN CALC. AND EXP. CONC. IN MOLE PCT

UNIQUAC (SPECIFIC PARAMETERS)   0.92
NRTL   (SPECIFIC PARAMETERS)   1.22
UNIQUAC (COMMON PARAMETERS)   2.23

C$_3$H$_8$O-C$_5$H$_{12}$O

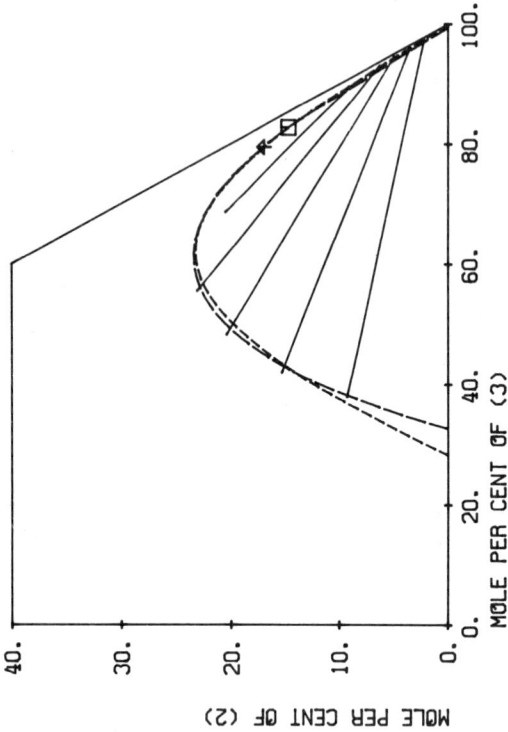

MOLE PER CENT OF (3)

MOLE PER CENT OF (2)

EXP.TIE LINE ——   UNIQ(SP) —□—   NRTL(SP) —◆—
CALC.BINODAL — —
CALC.PLAIT P.

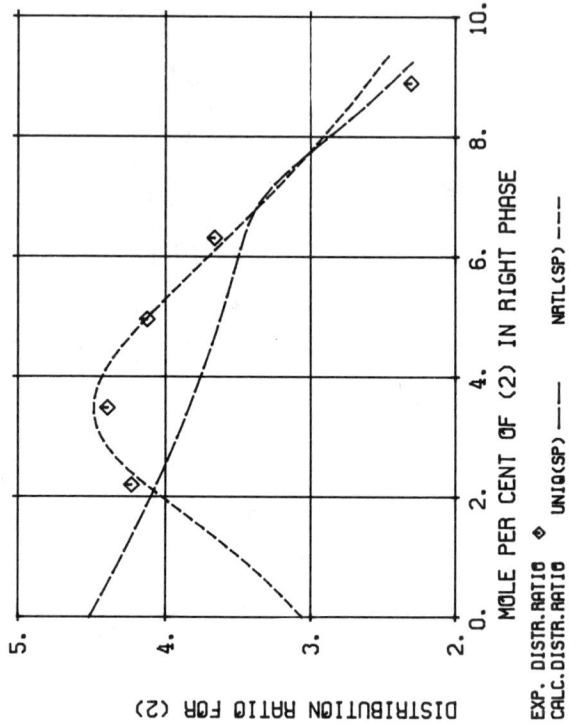

MOLE PER CENT OF (2) IN RIGHT PHASE

DISTRIBUTION RATIO FOR (2)

EXP. DISTR.RATIO ◇   UNIQ(SP) ——   NRTL(SP) — —
CALC.DISTR.RATIO ——

(1) C5H12O   1-BUTANOL, 3-METHYL
(2) C3H8O   2-PROPANOL
(3) H2O   WATER

ARNOLD V.W., WASHBURN E.R.
J.PHYS.CHEM. 62(1958)1088

TEMPERATURE = 10.0 DEG C   TYPE OF SYSTEM = 1

EXPERIMENTAL TIE LINES IN MOLE PCT

	LEFT PHASE			RIGHT PHASE	
(1)	(2)	(3)	(1)	(2)	(3)
52.743	9.313	37.944	0.348	2.202	97.449
42.717	15.299	41.985	0.381	3.482	96.137
31.297	20.443	48.260	0.440	4.957	94.603
21.381	23.107	55.512	0.574	6.311	93.116
10.842	20.515	68.643	1.258	8.888	89.853

SPECIFIC MODEL PARAMETERS IN KELVIN

		UNIQUAC		NRTL(ALPHA=.2)	
I J	AIJ	AJI		AIJ	AJI
1 2	117.08	58.022		860.26	-348.02
1 3	121.54	183.06		-98.521	1770.6
2 3	-91.003	285.01		-356.31	983.03

R1 = 4.1279   R2 = 2.7791   R3 = 0.9200
Q1 = 3.588   Q2 = 2.508   Q3 = 1.400

MEAN DEV. BETWEEN CALC. AND EXP. CONC. IN MOLE PCT

UNIQUAC (SPECIFIC PARAMETERS)   0.73
NRTL   (SPECIFIC PARAMETERS)   0.73

$C_3H_8O\text{-}C_5H_{12}O$

(1) C5H12O   1-BUTANOL,3-METHYL

(2) C3H8O   2-PROPANOL

(3) H2O   WATER

ARNOLD V.W., WASHBURN E.R.
J.PHYS.CHEM. 62(1958)1088

TEMPERATURE = 25.0 DEG C   TYPE OF SYSTEM = 1

EXPERIMENTAL TIE LINES IN MOLE PCT

	LEFT PHASE			RIGHT PHASE	
(1)	(2)	(3)	(1)	(2)	(3)
55.564	6.352	38.084	0.404	1.092	98.503
41.948	14.621	43.431	0.439	2.413	97.148
27.062	20.680	52.258	0.527	4.263	95.210
17.729	21.655	60.616	0.712	5.707	93.581
9.311	18.152	72.537	1.536	8.260	90.203

SPECIFIC MODEL PARAMETERS IN KELVIN

		UNIQUAC		NRTL(ALPHA=.2)	
I J	AIJ	AJI		AIJ	AJI
1 2	52.586	122.32		926.84	-483.50
1 3	112.34	204.18		-145.52	2080.5
2 3	-128.10	354.91		-248.58	741.63

R1 = 4.1279   R2 = 2.7791   R3 = 0.9200
Q1 = 3.588   Q2 = 2.508   Q3 = 1.400

MEAN DEV. BETWEEN CALC. AND EXP. CONC. IN MOLE PCT

UNIQUAC (SPECIFIC PARAMETERS)   0.70
NRTL   (SPECIFIC PARAMETERS)   1.24
UNIQUAC (COMMON PARAMETERS)   1.87

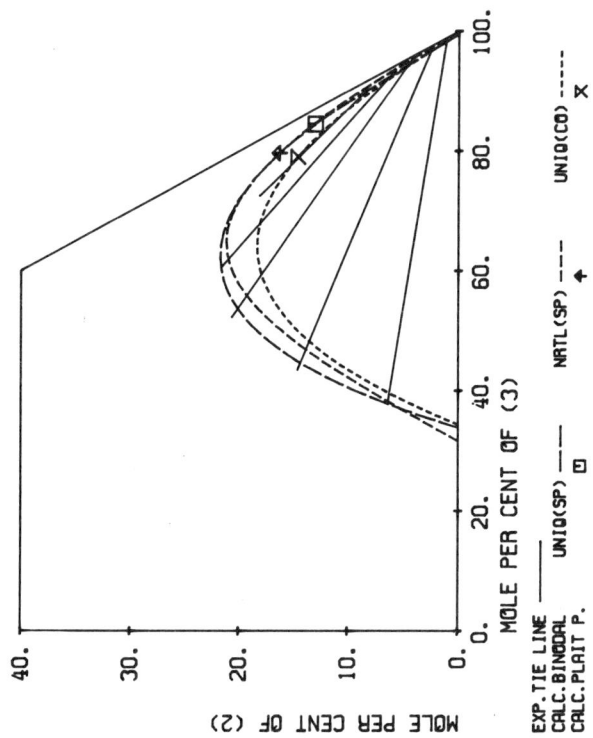

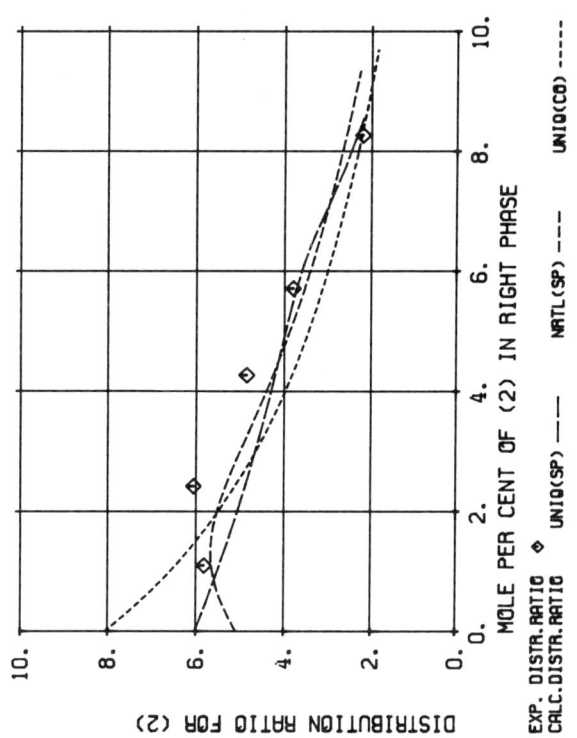

C$_3$H$_8$O-C$_5$H$_{12}$O

(1) C5H12O   1-BUTANOL,3-METHYL

(2) C3H8O   2-PROPANOL

(3) H2O   WATER

ARNOLD V.W., WASHBURN E.R.
J.PHYS.CHEM. 62(1958)1083

TEMPERATURE = 40.0 DEG C   TYPE OF SYSTEM = 1

EXPERIMENTAL TIE LINES IN MOLE PCT

	LEFT PHASE			RIGHT PHASE	
(1)	(2)	(3)	(1)	(2)	(3)
50.897	3.177	40.926	0.405	1.188	93.406
38.291	15.082	46.627	0.414	2.174	97.411
27.784	19.460	52.756	0.472	3.459	96.069
17.096	20.753	62.150	0.989	4.869	94.142
11.599	19.115	69.286	1.097	6.436	92.466

SPECIFIC MODEL PARAMETERS IN KELVIN

		UNIQUAC		NRTL(ALPHA=.2)	
I J	AIJ	AJI		AIJ	AJI
1 2	-12.181	177.09		549.54	-506.20
1 3	-106.10	223.34		-204.76	1847.9
2 3	-151.16	398.58		-106.13	522.38

R1 = 4.1279   R2 = 2.7791   R3 = 0.9200
Q1 = 3.588   Q2 = 2.508   Q3 = 1.400

MEAN DEV. BETWEEN CALC. AND EXP. CONC. IN MOLE PCT

UNIQUAC (SPECIFIC PARAMETERS)   0.44
NRTL   (SPECIFIC PARAMETERS)   1.71

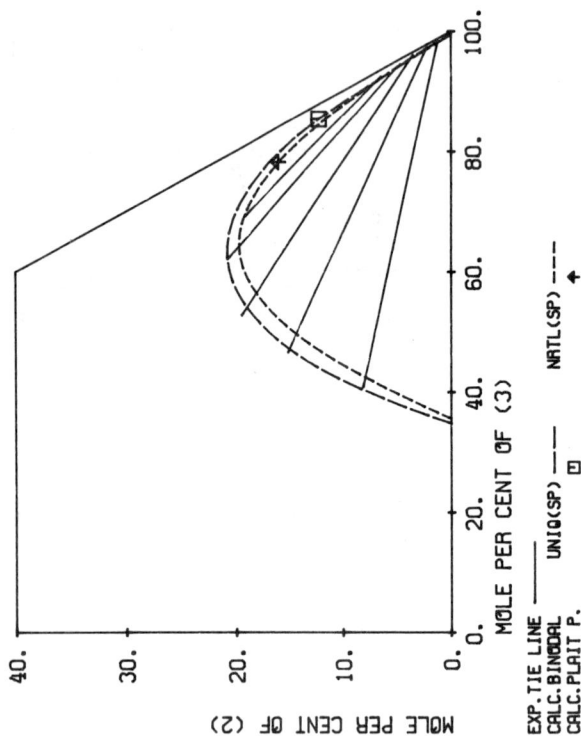

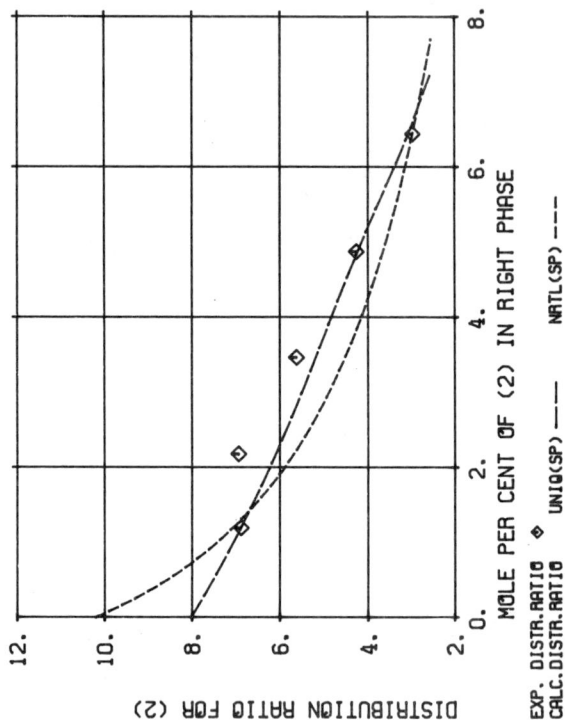

(1) C6H6   BENZENE
(2) C3H8O   2-PROPANOL
(3) H2O   WATER

OLSEN A.L., WASHBURN E.R.
J.AM.CHEM.SOC. 57(1935)303

TEMPERATURE = 25.0 DEG C   TYPE OF SYSTEM = 1

EXPERIMENTAL TIE LINES IN MOLE PCT (GRAPH.INTERPOL.)

LEFT PHASE			RIGHT PHASE		
(1)	(2)	(3)	(1)	(2)	(3)
98.663	1.035	0.302	0.042	0.888	99.069
97.762	1.936	0.301	0.051	1.815	98.135
96.737	2.962	0.301	0.062	2.816	97.122
95.463	4.237	0.300	0.073	3.687	96.241
93.440	6.262	0.298	0.087	4.702	95.210
88.503	11.244	0.253	0.111	6.156	93.733
80.288	18.479	1.233	0.144	7.373	92.483
71.460	23.315	5.226	0.174	8.179	91.647
61.440	29.192	9.367	0.204	8.836	90.960
56.278	32.893	10.829	0.235	9.422	90.343
50.311	36.939	12.750	0.278	9.934	89.789
45.031	39.673	15.295	0.327	10.506	89.167
39.299	41.075	19.626	0.356	10.798	88.845
34.310	41.282	24.408	0.428	11.399	88.173
30.803	41.107	28.090	0.466	11.657	87.877

SPECIFIC MODEL PARAMETERS IN KELVIN

I J	UNIQUAC		NRTL(ALPHA=.2)	
	AIJ	AJI	AIJ	AJI
1 2	439.98	-146.27	978.52	-398.18
1 3	1011.1	195.37	1110.3	1297.9
2 3	-62.339	197.73	-390.53	1082.1

R1 = 3.1878   R2 = 2.7791   R3 = 0.9200
Q1 = 2.400   Q2 = 2.508   Q3 = 1.400

MEAN DEV. BETWEEN CALC. AND EXP. CONC. IN MOLE PCT

UNIQUAC (SPECIFIC PARAMETERS)   1.71
NRTL   (SPECIFIC PARAMETERS)   1.53
UNIQUAC (COMMON PARAMETERS)   2.64

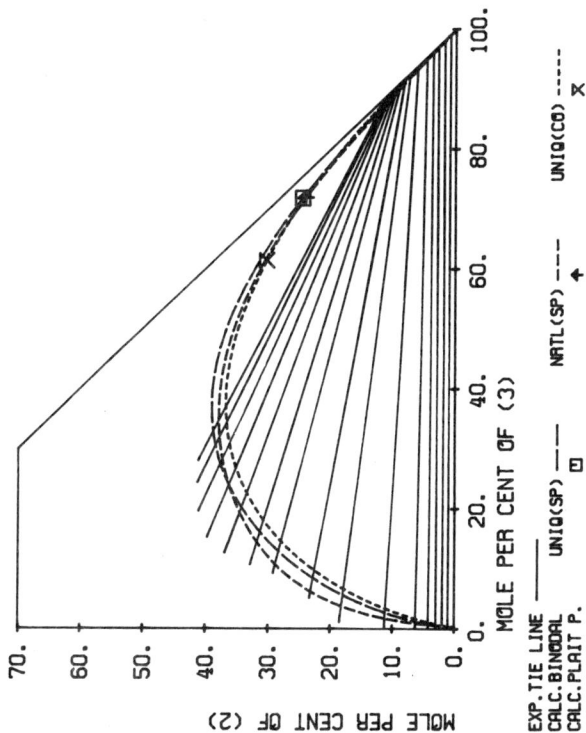

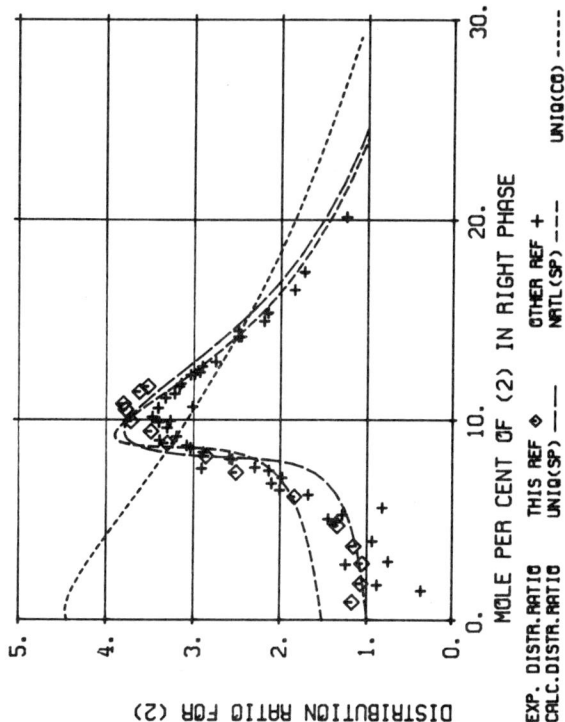

C$_3$H$_8$O-C$_6$H$_6$

(1) C6H6   BENZENE

(2) C3H8O   2-PROPANOL

(3) H2O   WATER

UDOVENKO V.V., MAZANKO T.F.
ZH.FIZ.KHIM. 38(1964)2984

TEMPERATURE = 30.0 DEG C   TYPE OF SYSTEM = 1

EXPERIMENTAL TIE LINES IN MOLE PCT

	LEFT PHASE			RIGHT PHASE	
(1)	(2)	(3)	(1)	(2)	(3)
99.051	0.518	0.432	0.024	1.458	98.518
97.379	2.191	0.430	0.074	2.920	97.006
92.541	6.611	0.848	0.103	4.891	95.006
83.707	13.015	3.278	0.214	6.495	93.291
71.014	21.996	6.990	0.274	7.571	92.155
52.900	30.417	16.683	0.338	8.963	90.699
37.641	35.258	27.101	0.491	10.142	89.366
30.587	36.396	33.016	0.683	11.300	88.018
25.441	36.882	37.677	0.717	11.655	87.628
20.704	36.181	43.115	0.882	12.367	86.751
17.879	35.362	46.760	0.922	12.895	86.184
15.811	35.115	49.074	1.267	14.155	84.578
12.284	32.710	55.006	1.517	14.932	83.551
9.641	30.289	60.070	2.010	16.500	81.491
6.280	24.915	68.803	3.437	20.146	76.417

SPECIFIC MODEL PARAMETERS IN KELVIN

		UNIQUAC		NRTL(ALPHA=.2)	
I	J	AIJ	AJI	AIJ	AJI
1	2	439.98	-146.27	978.52	-398.18
1	3	1011.1	195.37	1110.3	1297.9
2	3	-62.339	197.73	-390.53	1082.1

R1 = 3.1878   R2 = 2.7791   R3 = 0.9200
Q1 = 2.400   Q2 = 2.508   Q3 = 1.400

MEAN DEV. BETWEEN CALC. AND EXP. CONC. IN MOLE PCT

UNIQUAC (SPECIFIC PARAMETERS)	1.06
NRTL (SPECIFIC PARAMETERS)	0.92
UNIQUAC (COMMON PARAMETERS)	4.57

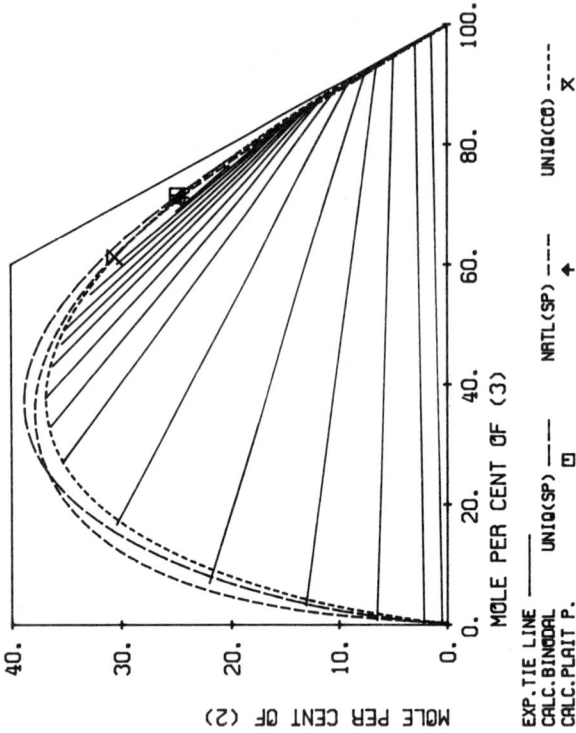

EXP.TIE LINE ——   UNIQ(SP) —— ☐   NRTL(SP) —— +   UNIQ(CO) ----- ✕
CALC.BINODAL ——
CALC.PLAIT P. -----

MOLE PER CENT OF (3)

MOLE PER CENT OF (2)

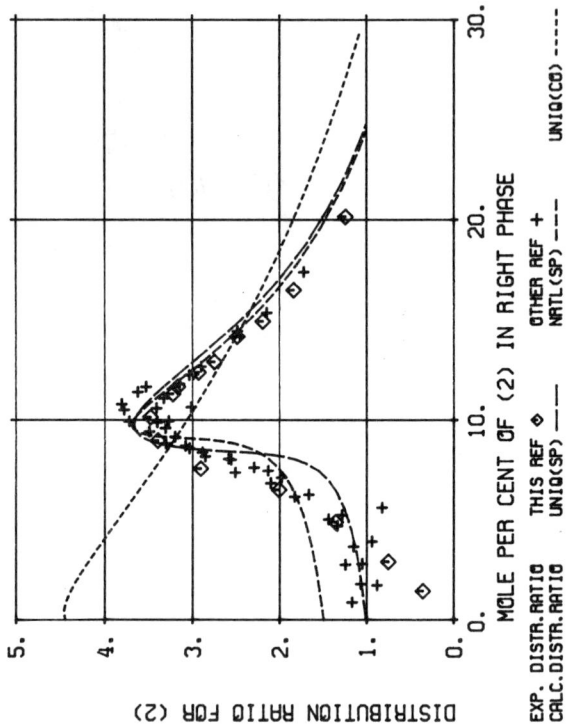

EXP. DISTR.RATIO   THIS REF ◇   OTHER REF +
CALC.DISTR.RATIO   UNIQ(SP) ——   NRTL(SP) -----   UNIQ(CO) -----

MOLE PER CENT OF (2) IN RIGHT PHASE

DISTRIBUTION RATIO FOR (2)

C$_3$H$_8$O-C$_6$H$_6$

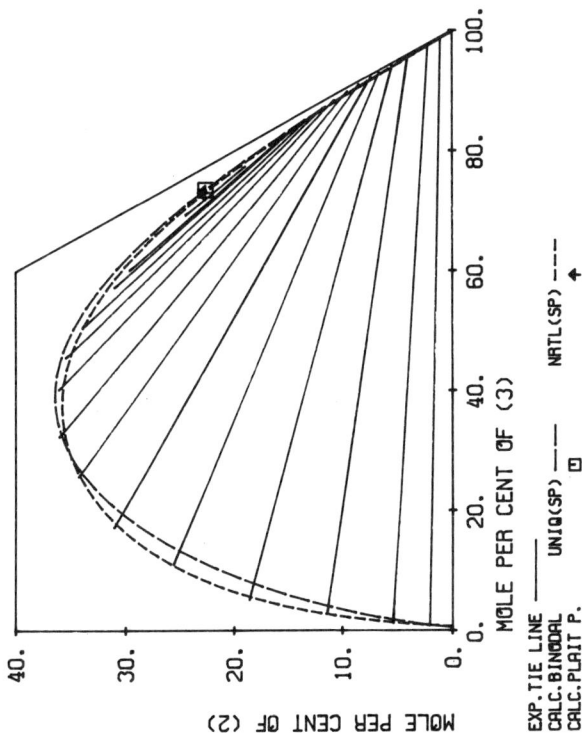

MOLE PER CENT OF (3)

MOLE PER CENT OF (2)

EXP.TIE LINE ——   UNIQ(SP) —E—   NRTL(SP) —+—
CALC.BINODAL
CALC.PLAIT P.

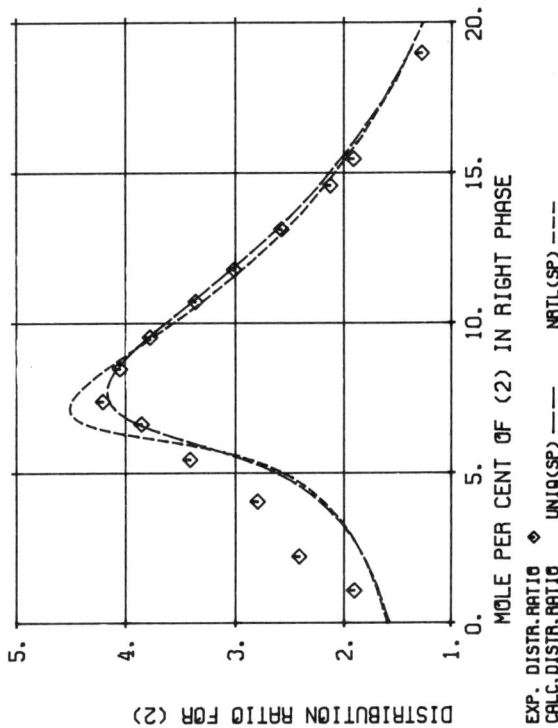

MOLE PER CENT OF (2) IN RIGHT PHASE

DISTRIBUTION RATIO FOR (2)

EXP. DISTR.RATIO ◇   UNIQ(SP) ◇   NRTL(SP) ——
CALC.DISTR.RATIO ——

(1) C6H6   BENZENE
(2) C3H8O   2-PROPANOL
(3) H2O   WATER

UDOVENKO V.V., MAZANKO T.F.
ZH.FIZ.KHIM. 38(1964)2984

TEMPERATURE = 45.0 DEG C   TYPE OF SYSTEM = 1

EXPERIMENTAL TIE LINES IN MOLE PCT

	LEFT PHASE			RIGHT PHASE	
(1)	(2)	(3)	(1)	(2)	(3)
97.087	2.056	0.857	0.071	1.078	98.851
93.390	5.338	1.272	0.097	2.213	97.689
85.368	11.341	3.291	0.127	4.051	95.822
76.231	18.598	5.171	0.262	5.448	94.291
63.302	25.570	11.128	0.295	6.633	93.071
51.723	31.024	17.253	0.327	7.376	92.296
39.987	34.301	25.712	0.448	8.469	91.083
31.575	36.004	32.421	0.573	9.527	89.900
23.674	36.052	40.274	0.795	10.715	88.490
19.023	35.423	45.554	1.027	11.772	87.201
15.419	33.858	50.723	1.469	13.123	85.407
11.703	30.990	57.306	1.843	14.583	83.574
10.219	29.609	60.172	2.148	15.460	82.392
6.192	24.461	69.348	3.489	18.981	77.529

SPECIFIC MODEL PARAMETERS IN KELVIN

		UNIQUAC		NRTL(ALPHA=.2)	
I J		AIJ	AJI	AIJ	AJI
1 2		426.97	-159.28	1188.6	-526.99
1 3		888.60	167.75	1009.8	1377.5
2 3		-108.20	270.04	-453.96	1194.9

R1 = 3.1878   R2 = 2.7791   R3 = 0.9200
Q1 = 2.400   Q2 = 2.508   Q3 = 1.400

MEAN DEV. BETWEEN CALC. AND EXP. CONC. IN MOLE PCT

UNIQUAC (SPECIFIC PARAMETERS)   0.62
NRTL   (SPECIFIC PARAMETERS)   0.42

C₃H₈O-C₆H₆

(1) C6H6   BENZENE

(2) C3H8O   2-PROPANOL

(3) H2O   WATER

UDOVENKO V.V., MAZANKO T.F.
ZH.FIZ.KHIM. 38(1964)2984

TEMPERATURE = 60.0 DEG C   TYPE OF SYSTEM = 1

EXPERIMENTAL TIE LINES IN MOLE PCT

| | LEFT PHASE | | | RIGHT PHASE | |
(1)	(2)	(3)	(1)	(2)	(3)
96.832	2.311	0.857	0.071	0.827	99.103
92.491	5.820	1.689	0.146	2.184	97.670
83.700	12.222	4.078	0.176	3.735	96.089
72.473	19.768	7.760	0.232	4.688	95.080
66.108	24.450	9.442	0.262	5.448	94.291
50.895	30.970	18.135	0.293	6.233	93.474
35.383	35.435	29.182	0.410	7.313	92.277
27.202	35.764	37.034	0.530	8.189	91.281
19.692	35.013	45.295	0.784	9.960	89.256
15.295	33.244	51.460	1.010	10.886	88.104
12.669	31.316	56.015	1.389	12.520	86.091
9.763	28.728	61.508	1.603	13.232	85.166
7.166	25.200	67.634	2.243	15.219	82.538

SPECIFIC MODEL PARAMETERS IN KELVIN

| | | UNIQUAC | | NRTL(ALPHA=.2) | |
I J	AIJ	AJI		AIJ	AJI
1 2	-27.303	150.56		1112.1	-542.27
1 3	944.04	97.088		1042.6	1387.7
2 3	-54.691	225.90		-503.95	1335.0

R1 = 3.1878   R2 = 2.7791   R3 = 0.9200
Q1 = 2.400   Q2 = 2.508   Q3 = 1.400

MEAN DEV. BETWEEN CALC. AND EXP. CONC. IN MOLE PCT

UNIQUAC (SPECIFIC PARAMETERS)   0.70
NRTL (SPECIFIC PARAMETERS)   0.41

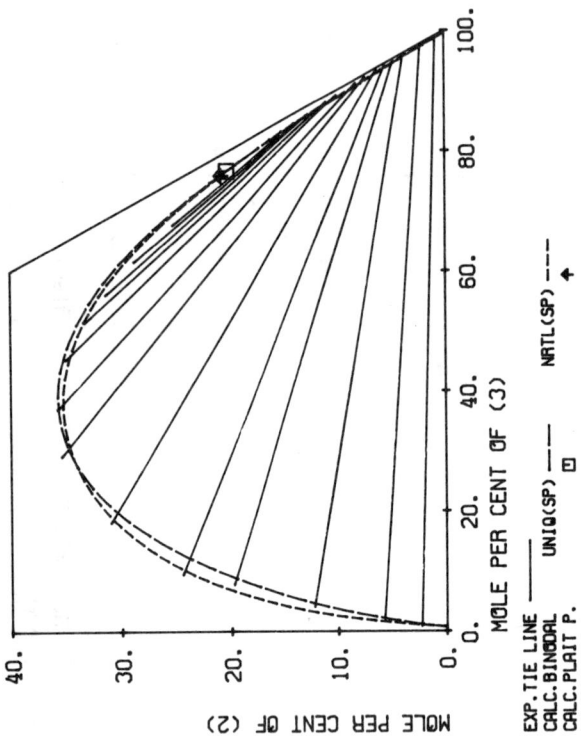

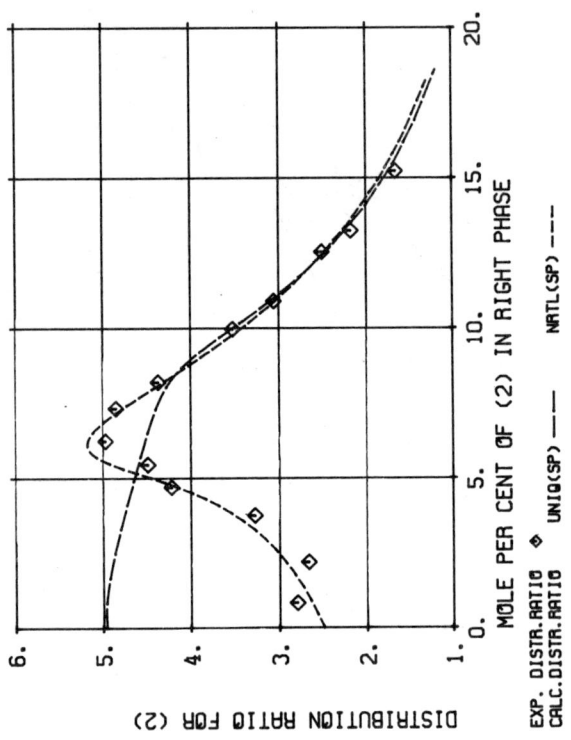

$C_3H_8O-C_6H_6$

(1) C6H6   BENZENE

---

(2) C3H8O   2-PROPANOL

---

(3) H2O   WATER

MORACHEVSKII A.G., LEGOCHKINA L.A.
ZH.PRIKL.KHIM.(LENINGRAD) 38(1965)1789

TEMPERATURE = 20.0 DEG C   TYPE OF SYSTEM = 1

EXPERIMENTAL TIE LINES IN MOLE PCT

	LEFT PHASE			RIGHT PHASE	
(1)	(2)	(3)	(1)	(2)	(3)
94.959	4.613	0.428	0.105	5.611	94.285
74.755	20.478	4.767	0.193	8.013	91.794
50.714	32.078	17.208	0.467	10.648	88.885
33.347	36.664	29.990	0.633	12.268	87.099
19.431	35.830	44.739	1.175	14.440	84.386

SPECIFIC MODEL PARAMETERS IN KELVIN

		UNIQUAC		NRTL(ALPHA=.2)	
I	J	AIJ	AJI	AIJ	AJI
1	2	439.98	-146.27	978.52	-398.18
1	3	1011.1	195.37	1110.3	1297.9
2	3	-62.339	197.73	-390.53	1082.1

R1 = 3.1878   R2 = 2.7791   R3 = 0.9200
Q1 = 2.400   Q2 = 2.508   Q3 = 1.400

MEAN DEV. BETWEEN CALC. AND EXP. CONC. IN MOLE PCT

UNIQUAC (SPECIFIC PARAMETERS)   1.24
NRTL   (SPECIFIC PARAMETERS)   1.34
UNIQUAC (COMMON PARAMETERS)   2.57

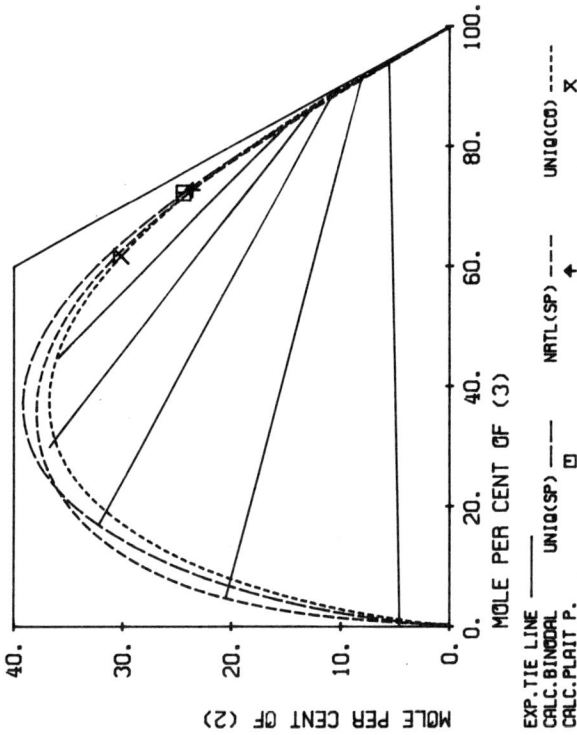

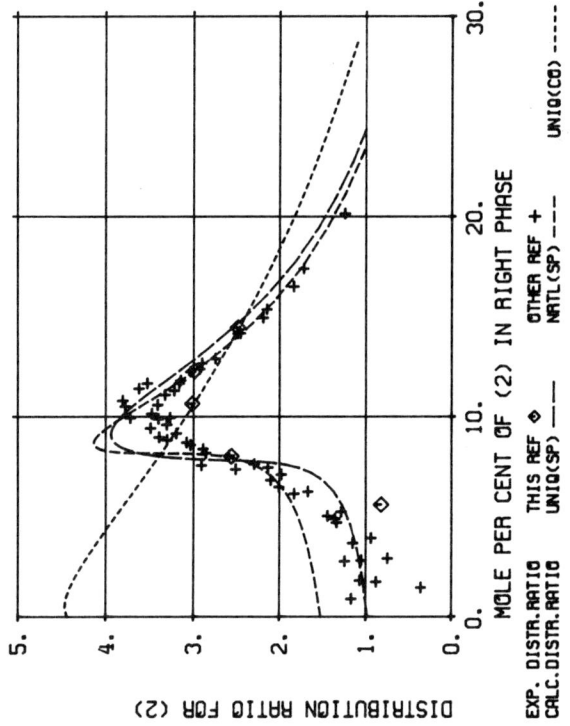

MOLE PER CENT OF (3)

MOLE PER CENT OF (2)

EXP.TIE LINE ——   UNIQ(SP) —— ⊟   NRTL(SP) —— ✦   UNIQ(CO) ---- ✕
CALC.BINODAL
CALC.PLAIT P.

DISTRIBUTION RATIO FOR (2)

MOLE PER CENT OF (2) IN RIGHT PHASE

EXP. DISTR.RATIO   THIS REF ◇   OTHER REF +
CALC.DISTR.RATIO   UNIQ(SP) ——   NRTL(SP) ---   UNIQ(CO) ----

C₃H₈O-C₆H₆

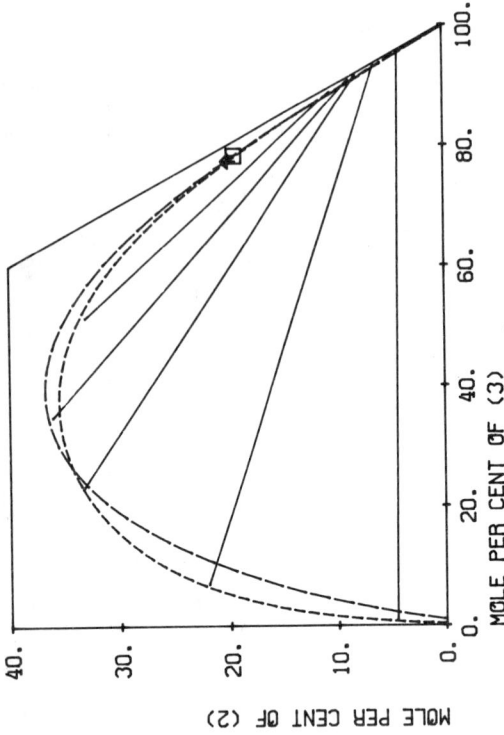

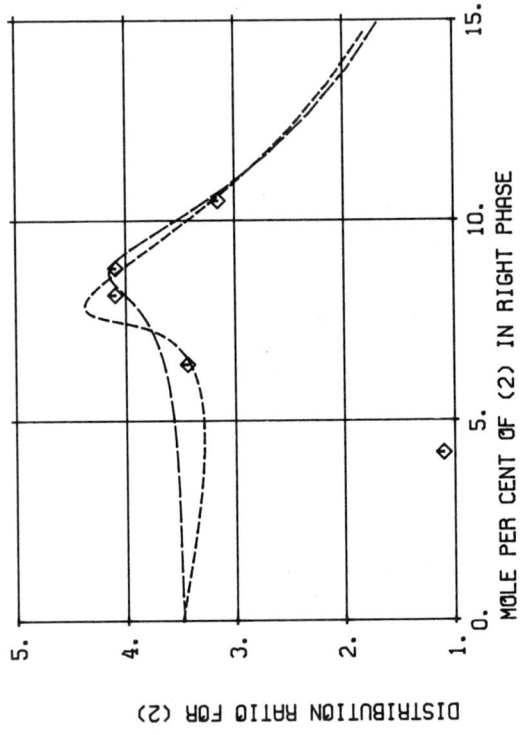

(1) C6H6   BENZENE

(2) C3H8O   2-PROPANOL

(3) H2O   WATER

MORACHEVSKII A.G., LEGOCHKINA L.A.,
ZH.PRIKL.KHIM.(LENINGRAD) 38(1965)1789

TEMPERATURE = 40.0 DEG C   TYPE OF SYSTEM = 1

EXPERIMENTAL TIE LINES IN MOLE PCT

	LEFT PHASE			RIGHT PHASE	
(1)	(2)	(3)	(1)	(2)	(3)
94.549	4.598	0.852	0.076	4.188	95.735
71.014	21.996	6.990	0.160	6.404	93.437
43.658	33.364	22.977	0.221	8.149	91.630
29.054	36.074	34.872	0.280	8.814	90.905
15.596	33.049	51.354	0.702	10.494	88.804

SPECIFIC MODEL PARAMETERS IN KELVIN

		UNIQUAC		NRTL(ALPHA=.2)	
I J	AIJ	AJI		AIJ	AJI
1 2	146.54	4.1662		607.07	-305.63
1 3	741.33	149.53		1205.8	1407.2
2 3	-90.987	274.41		-519.59	1345.9

R1 = 3.1878   R2 = 2.7791   R3 = 0.9200
Q1 = 2.400   Q2 = 2.508   Q3 = 1.400

MEAN DEV. BETWEEN CALC. AND EXP. CONC. IN MOLE PCT

UNIQUAC (SPECIFIC PARAMETERS)   1.16
NRTL   (SPECIFIC PARAMETERS)   0.74

$C_3H_8O-C_6H_6$

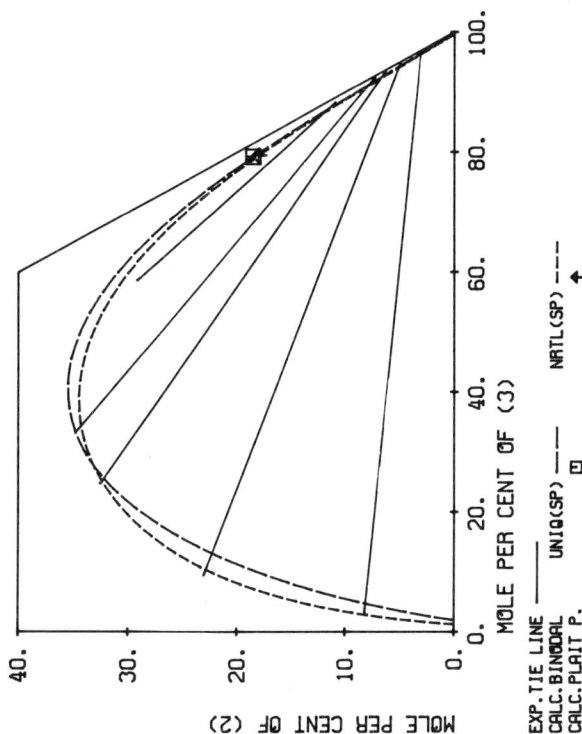

EXP.TIE LINE ——— UNIQ(SP) ⊡ NRTL(SP) ----
CALC.BINODAL
CALC.PLAIT P.

MOLE PER CENT OF (3)

MOLE PER CENT OF (2)

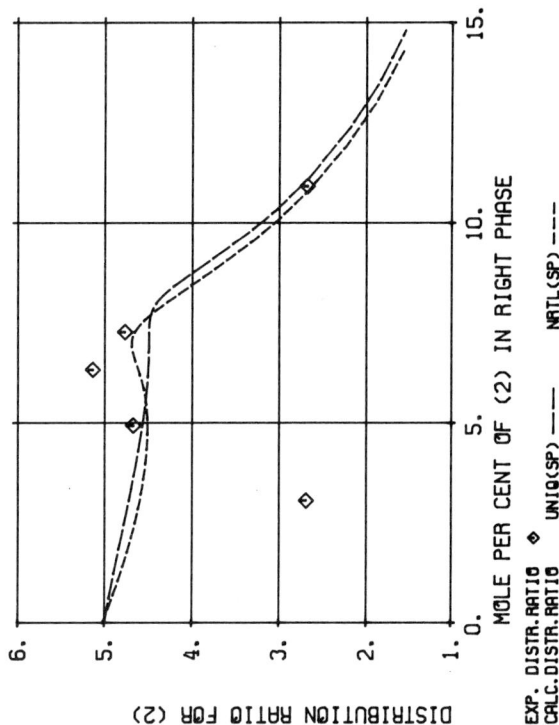

EXP. DISTR.RATIO ◇ UNIQ(SP) ——— NRTL(SP) ----
CALC.DISTR.RATIO

MOLE PER CENT OF (2) IN RIGHT PHASE

DISTRIBUTION RATIO FOR (2)

(1) C6H6   BENZENE
(2) C3H8O   2-PROPANOL
(3) H2O   WATER

MORACHEVSKII A.G., LEGOCHKINA L.A.,
ZH.PRIKL.KHIM.(LENINGRAD) 38(1965)1789

TEMPERATURE = 60.0 DEG C   TYPE OF SYSTEM = 1

EXPERIMENTAL TIE LINES IN MOLE PCT

	LEFT PHASE			RIGHT PHASE	
(1)	(2)	(3)	(1)	(2)	(3)
88.488	8.197	3.315	0.074	3.058	96.868
67.469	23.057	9.474	0.155	4.937	94.908
42.645	32.528	24.826	0.186	6.330	93.484
31.765	34.660	33.575	0.245	7.273	92.482
12.180	29.156	58.663	0.828	10.918	88.254

SPECIFIC MODEL PARAMETERS IN KELVIN

		UNIQUAC		NRTL(ALPHA=.2)	
I	J	AIJ	AJI	AIJ	AJI
1	2	17.653	122.82	214.27	-10.803
1	3	654.12	149.04	895.42	1302.7
2	3	-88.228	285.60	-516.17	1427.9

R1 = 3.1878   R2 = 2.7791   R3 = 0.9200
Q1 = 2.400   Q2 = 2.508   Q3 = 1.400

MEAN DEV. BETWEEN CALC. AND EXP. CONC. IN MOLE PCT

UNIQUAC (SPECIFIC PARAMETERS)   1.09
NRTL   (SPECIFIC PARAMETERS)   0.69

C_3H_8O-C_6H_6

(1) C6H6   BENZENE
(2) C3H8O   2-PROPANOL
(3) H2O   WATER

UDOVENKO V.V., MAZANKO T.F.
ZH.FIZ.KHIM. 41(1967)395

TEMPERATURE = 30.0 DEG C   TYPE OF SYSTEM = 1

EXPERIMENTAL TIE LINES IN MOLE PCT

	LEFT PHASE			RIGHT PHASE	
(1)	(2)	(3)	(1)	(2)	(3)
97.598	1.544	0.858	0.048	1.750	98.202
95.023	3.700	1.277	0.076	3.936	95.988
89.562	6.708	3.730	0.182	5.243	94.575
81.593	14.349	4.057	0.269	6.829	92.901
78.505	15.908	5.587	0.328	7.460	92.212
62.167	26.048	11.785	0.392	8.584	91.024
48.852	32.506	18.642	0.577	9.940	89.483

SPECIFIC MODEL PARAMETERS IN KELVIN

		UNIQUAC		NRTL(ALPHA=.2)	
I J	AIJ	AJI		AIJ	AJI
1 2	439.98	-146.27		978.52	-398.18
1 3	1011.1	195.37		1110.3	1297.9
2 3	-62.339	197.73		-390.53	1082.1

R1 = 3.1878   R2 = 2.7791   R3 = 0.9200
Q1 = 2.400   Q2 = 2.508   Q3 = 1.400

MEAN DEV. BETWEEN CALC. AND EXP. CONC. IN MOLE PCT

UNIQUAC (SPECIFIC PARAMETERS)   0.62
NRTL   (SPECIFIC PARAMETERS)   0.95
UNIQUAC (COMMON PARAMETERS)   1.57

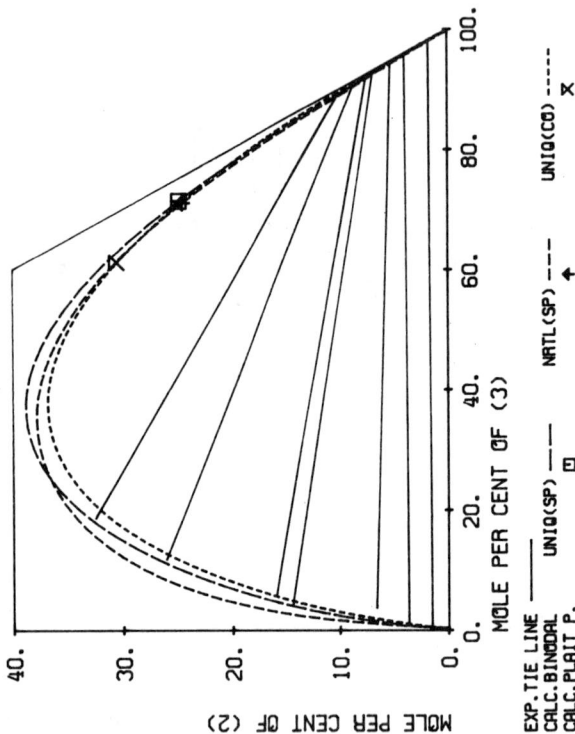

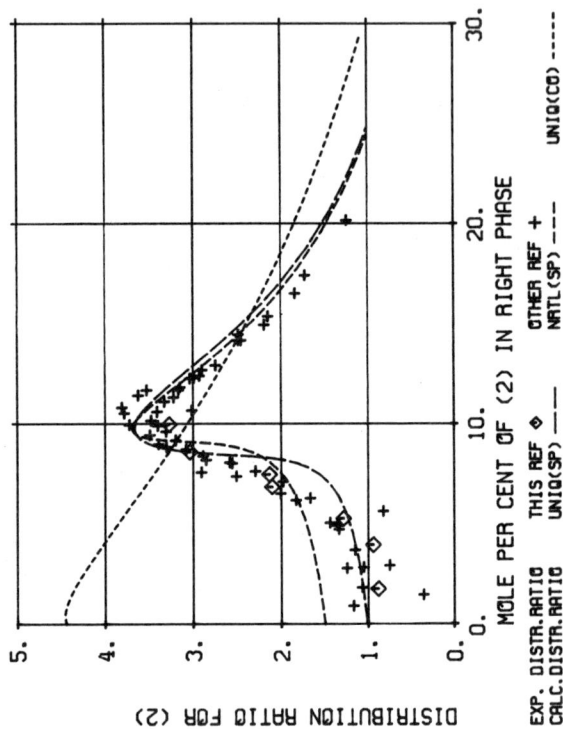

MOLE PER CENT OF (3)
MOLE PER CENT OF (2)

EXP.TIE LINE ——   UNIQ(SP) ——   NRTL(SP) ——+   UNIQ(CO) -----
CALC.BINODAL
CALC.PLAIT P.

DISTRIBUTION RATIO FOR (2)
MOLE PER CENT OF (2) IN RIGHT PHASE

EXP. DISTR.RATIO   THIS REF ◇   OTHER REF +
CALC.DISTR.RATIO   UNIQ(SP) ——   NRTL(SP) ——   UNIQ(CO) -----

$C_3H_8O-C_6H_6$

(1) C6H6   BENZENE

(2) C3H8O   2-PROPANOL

(3) H2O   WATER

UDOVENKO V.V., MAZANKO T.F.
ZH.FIZ.KHIM. 41(1967)395

TEMPERATURE = 45.0 DEG C   TYPE OF SYSTEM = 1

EXPERIMENTAL TIE LINES IN MOLE PCT

	LEFT PHASE			RIGHT PHASE	
(1)	(2)	(3)	(1)	(2)	(3)
96.195	2.949	0.856	0.048	1.685	98.267
86.911	9.382	3.707	0.075	3.687	96.238
82.527	13.407	4.067	0.181	5.054	94.765
75.346	19.109	5.544	0.263	5.640	94.097
56.339	29.395	14.266	0.381	7.224	92.395
34.257	35.439	30.304	0.537	8.889	90.574

SPECIFIC MODEL PARAMETERS IN KELVIN

		UNIQUAC		NRTL(ALPHA=.2)	
I J	AIJ	AJI		AIJ	AJI
1 2	431.18	-158.05		1389.2	-574.87
1 3	661.90	199.66		854.37	1862.4
2 3	-19.125	158.42		-260.98	819.00

R1 = 3.1878   R2 = 2.7791   R3 = 0.9200
Q1 = 2.400   Q2 = 2.508   Q3 = 1.400

MEAN DEV. BETWEEN CALC. AND EXP. CONC. IN MOLE PCT

UNIQUAC (SPECIFIC PARAMETERS)   0.76
NRTL   (SPECIFIC PARAMETERS)   0.60

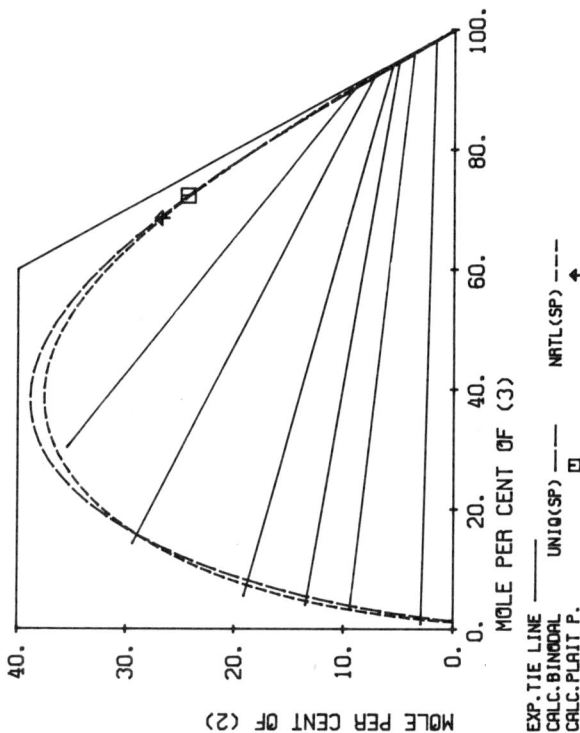

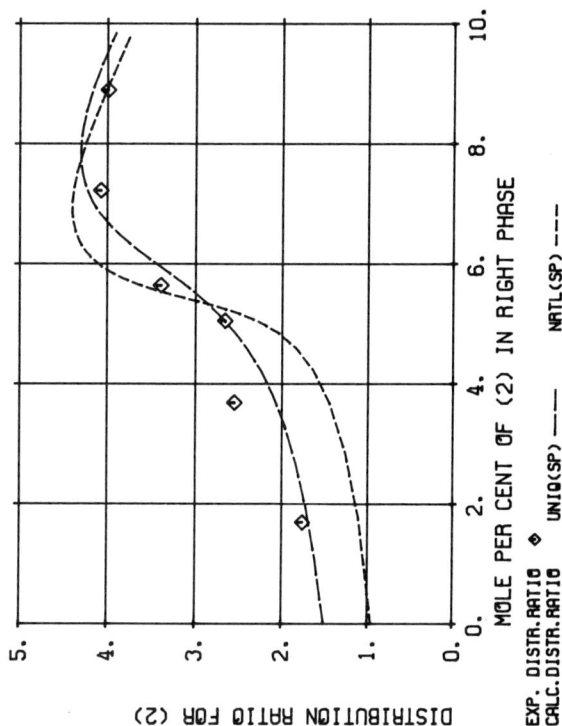

C$_3$H$_8$O-C$_6$H$_6$

(1) C6H6   BENZENE
(2) C3H8O   2-PROPANOL
(3) H2O   WATER

UDOVENKO V.V., MAZANKO T.F.
ZH.FIZ.KHIM. 41(1967)395

TEMPERATURE =   60.0 DEG C   TYPE OF SYSTEM = 1

EXPERIMENTAL TIE LINES IN MOLE PCT

	LEFT PHASE			RIGHT PHASE	
(1)	(2)	(3)	(1)	(2)	(3)
96.832	2.311	0.857	0.095	0.984	98.921
92.294	4.773	2.934	0.145	1.952	97.903
86.552	9.744	3.704	0.175	3.381	96.443
73.784	18.432	7.785	0.208	4.870	94.923
70.787	20.751	8.463	0.208	5.133	94.659
57.787	27.242	14.971	0.316	5.650	94.034
52.644	30.062	17.294	0.347	6.204	93.449
45.031	32.188	22.782	0.407	6.942	92.651
23.587	35.763	40.650	0.592	8.641	90.768
17.960	34.114	47.927	0.785	10.054	89.161

SPECIFIC MODEL PARAMETERS IN KELVIN

		UNIQUAC		NRTL(ALPHA=.2)	
I	J	AIJ	AJI	AIJ	AJI
1	2	271.90	-86.574	1099.1	-502.10
1	3	761.60	145.21	885.76	1621.9
2	3	-84.287	267.57	-488.94	1306.5

R1 = 3.1878   R2 = 2.7791   R3 = 0.9200
Q1 = 2.400   Q2 = 2.508   Q3 = 1.400

MEAN DEV. BETWEEN CALC. AND EXP. CONC. IN MOLE PCT

UNIQUAC (SPECIFIC PARAMETERS)   0.59
NRTL   (SPECIFIC PARAMETERS)   0.29

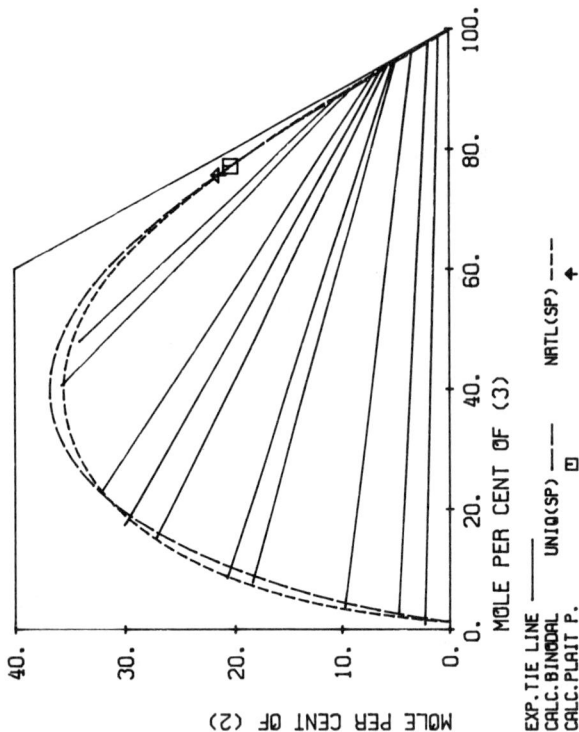

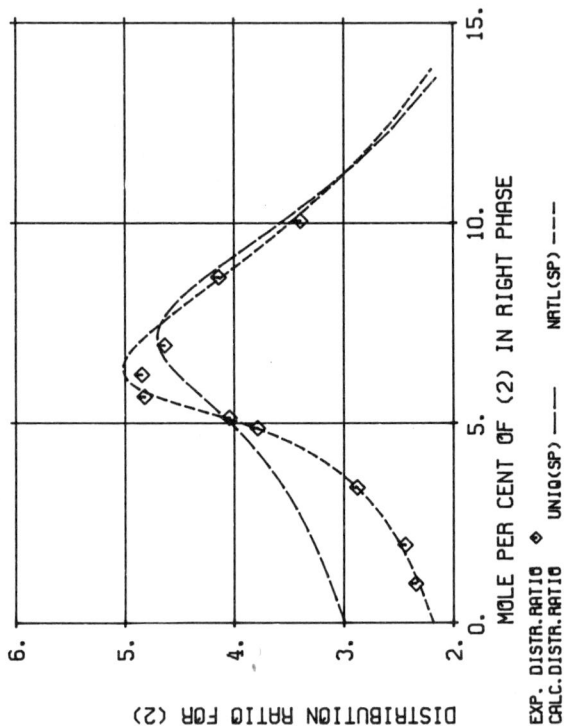

MOLE PER CENT OF (2)

MOLE PER CENT OF (3)

EXP.TIE LINE ─────   UNIQ(SP) ─── ─── □   NRTL(SP) ─ ─ ─ ─
CALC.BINODAL   ◆
CALC.PLAIT P.

DISTRIBUTION RATIO FOR (2)

MOLE PER CENT OF (2) IN RIGHT PHASE

EXP. DISTR.RATIO   ◇   UNIQ(SP) ─── ───   NRTL(SP) ─ ─ ─
CALC.DISTR.RATIO

C$_3$H$_8$O-C$_6$H$_6$

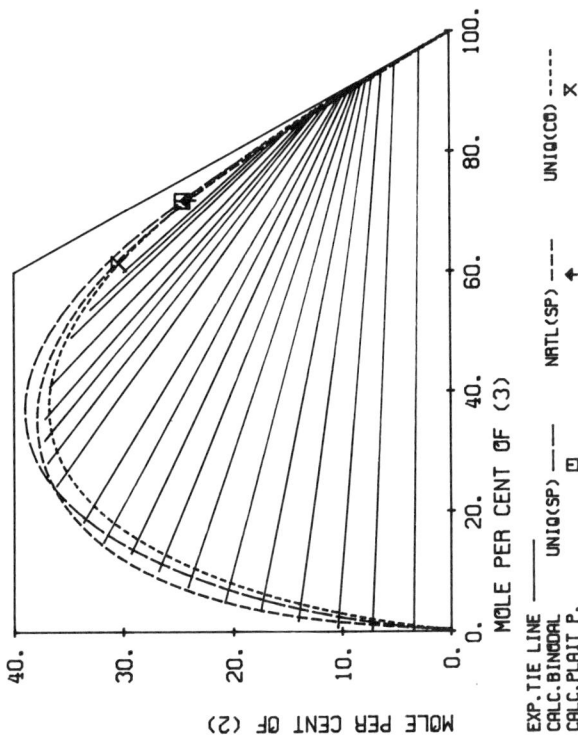

MOLE PER CENT OF (3)

MOLE PER CENT OF (2)

EXP.TIE LINE ———   UNIQ(SP) —□—   NRTL(SP) —+—   UNIQ(CO) —x—
CALC.BINODAL ----
CALC.PLAIT P.

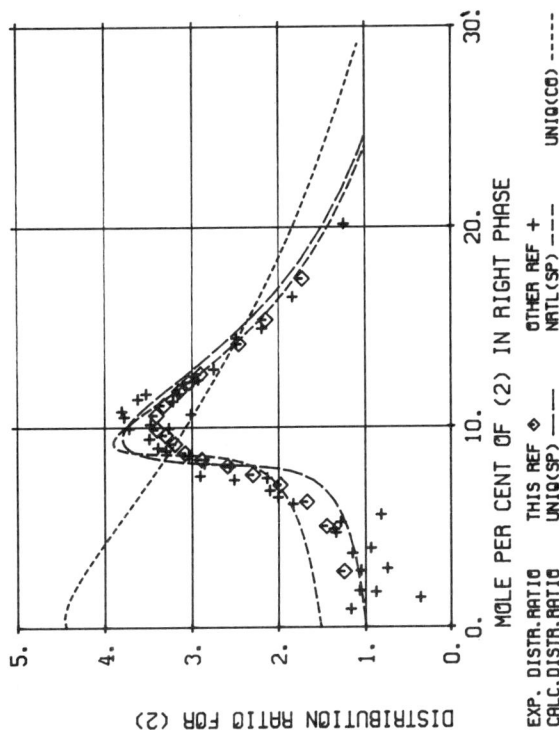

DISTRIBUTION RATIO FOR (2)

MOLE PER CENT OF (2) IN RIGHT PHASE

EXP. DISTR.RATIO ◇   THIS REF ◇   OTHER REF +   UNIQ(CO) -----
CALC.DISTR.RATIO   UNIQ(SP) ———   NRTL(SP) ———

(1) C6H6   BENZENE
(2) C3H8O   2-PROPANOL
(3) H2O   WATER

NIKURASHINA N.I., SINEGUBOVA S.I.
ZH.OBSHCH.KHIM. 43(1973)2100

TEMPERATURE = 25.0 DEG C   TYPE OF SYSTEM = 1

EXPERIMENTAL TIE LINES IN MOLE PCT

LEFT PHASE			RIGHT PHASE	
(1)	(2)	(3)	(1)	(2)	(3)
96.349	3.437	0.215	0.039	2.770	97.191
92.433	7.227	0.340	0.072	5.035	94.892
88.917	10.409	0.674	0.109	6.258	93.633
84.139	14.042	1.820	0.127	7.102	92.771
79.049	17.472	3.479	0.145	7.622	92.233
74.291	20.793	4.916	0.179	8.052	91.769
68.261	24.170	7.569	0.197	8.378	91.425
62.975	26.812	10.214	0.212	8.707	91.081
57.566	29.277	13.157	0.222	9.150	90.627
53.266	31.744	14.991	0.230	9.600	90.170
48.049	33.588	18.363	0.246	9.877	89.877
39.464	36.080	24.456	0.278	10.582	89.140
34.805	36.871	28.324	0.346	11.080	88.574
30.817	37.151	32.033	0.449	11.798	87.753
27.259	37.116	35.625	0.550	12.235	87.215
22.375	36.672	40.952	0.585	12.652	86.762
16.110	34.695	49.195	1.029	14.156	84.815
13.260	32.955	53.785	1.337	15.351	83.313
10.095	29.887	50.017	2.101	17.390	80.510

SPECIFIC MODEL PARAMETERS IN KELVIN

		UNIQUAC			NRTL(ALPHA=.2)	
I J	AIJ	AJI			AIJ	AJI
1 2	439.98	-146.27			978.52	-398.18
1 3	1011.1	195.37			1110.3	1297.9
2 3	-62.339	197.73			-390.53	1082.1

R1 = 3.1878   R2 = 2.7791   R3 = 0.9200
Q1 = 2.400   Q2 = 2.508   Q3 = 1.400

MEAN DEV. BETWEEN CALC. AND EXP. CONC. IN MOLE PCT

UNIQUAC (SPECIFIC PARAMETERS)   0.87
NRTL   (SPECIFIC PARAMETERS)   0.62
UNIQUAC (COMMON PARAMETERS)   3.27

C₃H₈O-C₆H₆O

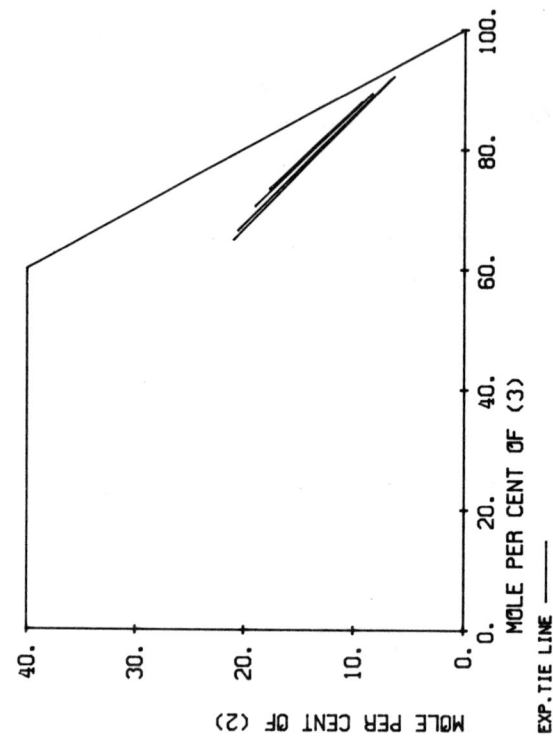

(1) C6H6O   PHENOL
(2) C3H8O   2-PROPANOL
(3) H2O   WATER

RUSANOV A.I.
VESTN.LENINGR.UNIV.FIZ.KHIM. 14,4(1959)132

TEMPERATURE = 15.0 DEG C   TYPE OF SYSTEM = 1

EXPERIMENTAL TIE LINES IN MOLE PCT

LEFT PHASE			RIGHT PHASE		
(1)	(2)	(3)	(1)	(2)	(3)
14.142	20.953	64.905	1.436	6.381	92.183
12.963	20.538	66.500	1.542	6.659	91.799
10.470	18.992	70.538	2.274	8.360	89.366
8.945	17.653	73.403	2.784	9.278	87.938

MOLE PER CENT OF (3)

MOLE PER CENT OF (2)

EXP.TIE LINE ———

$C_3H_8O-C_6H_6O$

(1) C6H6O   PHENOL
(2) C3H8O   2-PROPANOL
(3) H2O   WATER

RUSANOV A.I.
VESTN.LENINGR.UNIV.FIZ.KHIM. 14,4(1959)132

TEMPERATURE = 25.0 DEG C   TYPE OF SYSTEM = 1

EXPERIMENTAL TIE LINES IN MOLE PCT

LEFT PHASE			RIGHT PHASE		
(1)	(2)	(3)	(1)	(2)	(3)
31.911	0.0	68.089	1.747	0.0	98.253
31.905	1.427	66.668	1.667	0.193	98.140
31.001	6.334	62.665	1.493	0.877	97.630
30.035	8.989	60.976	1.420	1.307	97.273
28.086	12.140	59.774	1.306	1.946	96.747
21.950	17.674	60.376	1.251	3.175	95.574
17.581	19.005	63.414	1.325	4.184	94.491
14.346	18.901	66.753	1.499	5.262	93.240
12.347	18.280	69.373	1.799	6.181	92.020
9.499	16.423	74.077	2.455	7.648	89.897
7.504	14.425	78.071	3.333	8.939	87.728

SPECIFIC MODEL PARAMETERS IN KELVIN

		UNIQUAC			NRTL(ALPHA=.2)	
I	J	AIJ	AJI		AIJ	AJI
1	2	-120.76	76.992		-398.83	-576.16
1	3	-204.45	427.00		-459.87	1573.4
2	3	-2.8678	147.83		-31.970	396.66

R1 = 3.5517   R2 = 2.7791   R3 = 0.9200
Q1 = 2.680   Q2 = 2.508   Q3 = 1.400

MEAN DEV. BETWEEN CALC. AND EXP. CONC. IN MOLE PCT

UNIQUAC (SPECIFIC PARAMETERS)   1.86
NRTL   (SPECIFIC PARAMETERS)   1.30
UNIQUAC (COMMON PARAMETERS)   4.69

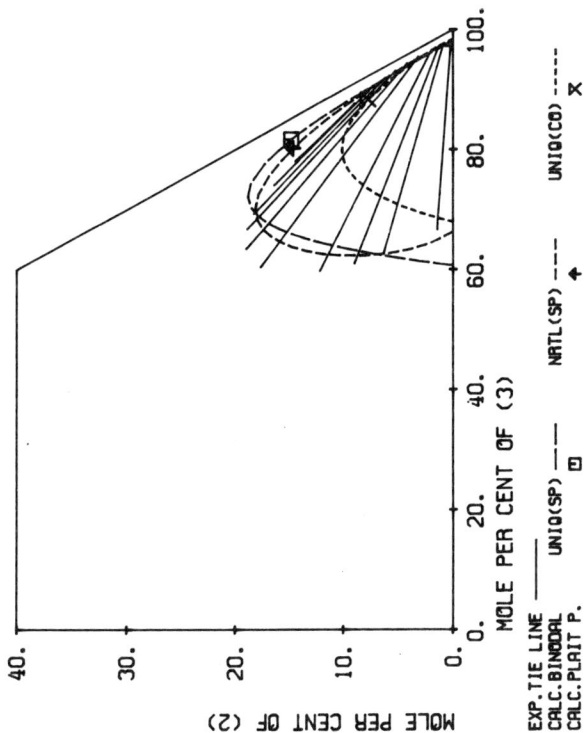

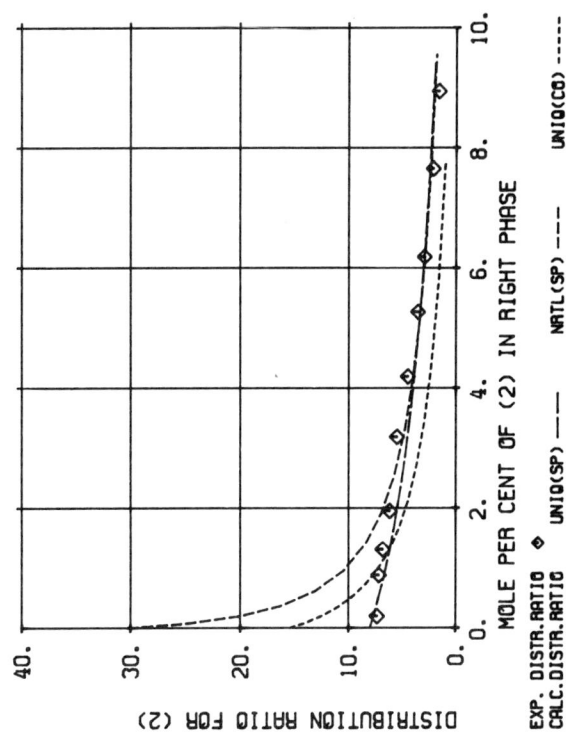

C₃H₈O-C₆H₆O

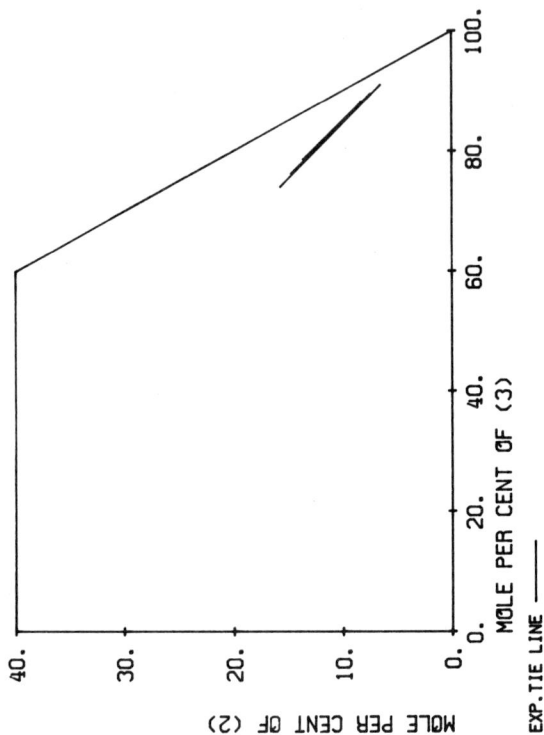

MOLE PER CENT OF (3)

EXP.TIE LINE

MOLE PER CENT OF (2)

(1) C6H6O   PHENOL

(2) C3H8O   2-PROPANOL

(3) H2O   WATER

RUSANOV A.I.
VESTN.LENINGR.UNIV.FIZ.KHIM. 14,4(1959)132

TEMPERATURE = 35.0 DEG C   TYPE OF SYSTEM = 1

EXPERIMENTAL TIE LINES IN MOLE PCT

LEFT PHASE			RIGHT PHASE		
(1)	(2)	(3)	(1)	(2)	(3)
10.360	15.679	73.962	2.503	6.480	91.017
9.176	14.678	76.146	3.068	7.419	89.513
7.897	13.551	78.551	3.606	8.266	88.128

$C_3H_8O-C_6H_6O$

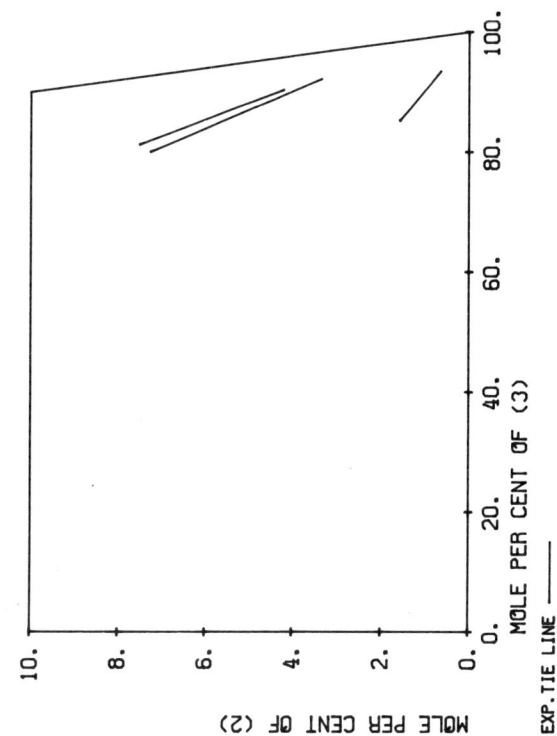

(1) C6H6O   PHENOL
----
(2) C3H8O   2-PROPANOL
----
(3) H2O   WATER

RUSANOV A.I.
VESTN.LENINGR.UNIV.FIZ.KHIM. 14,4(1959)132

TEMPERATURE = 70.0 DEG C   TYPE OF SYSTEM = 0

EXPERIMENTAL TIE LINES IN MOLE PCT

          LEFT PHASE   RIGHT PHASE
      (1)   (2)   (3)   (1)   (2)   (3)

   13.177   1.576   85.247   5.876   0.644   93.480
   12.707   7.262   80.031   4.461   3.340   92.199
   11.251   7.521   81.228   5.380   4.211   90.409

C₃H₈O-C₆H₁₀

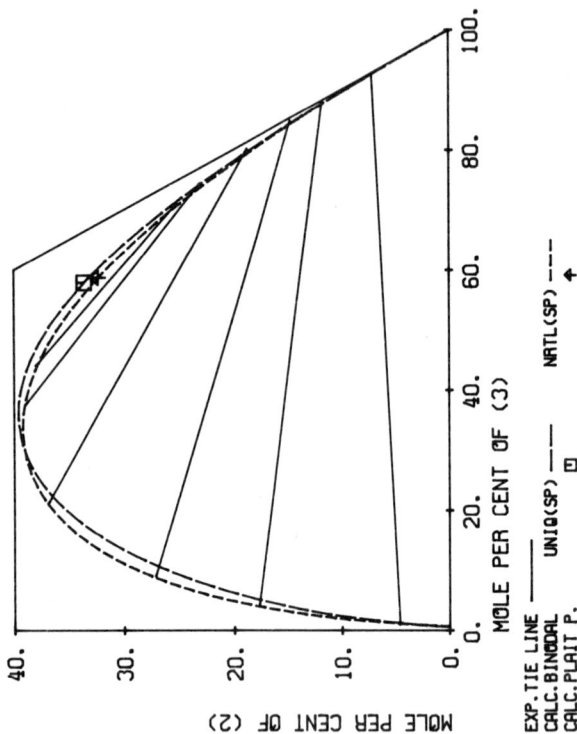

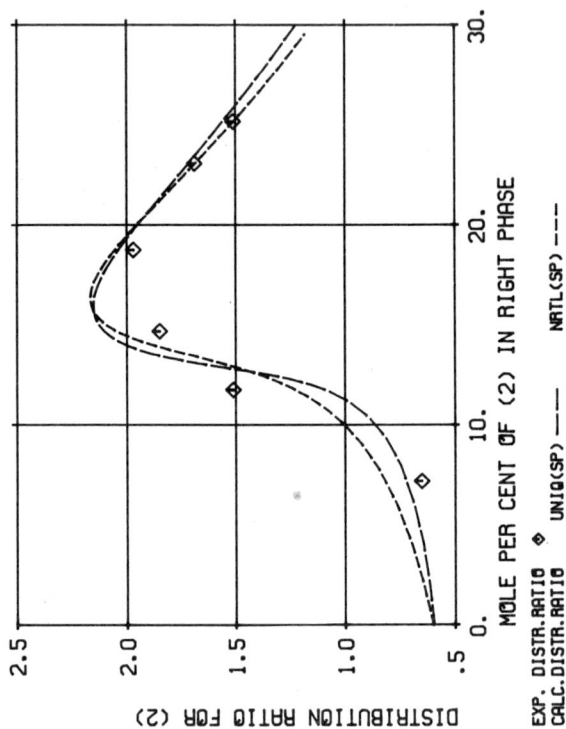

(1) C6H10   CYCLOHEXENE
(2) C3H8O   2-PROPANOL
(3) H2O   WATER

WASHBURN E.R., GRAHAM C.L., ARNOLD G.B., TRANSUE L.F.,
J.AM.CHEM.SOC. 62(1940)1454

TEMPERATURE = 15.0 DEG C   TYPE OF SYSTEM = 1

EXPERIMENTAL TIE LINES IN MOLE PCT (GRAPH.INTERPOL.)

	LEFT PHASE			RIGHT PHASE	
	(1)	(2)	(3)	(1)	(2)	(3)
94.331	4.686	0.983	0.028	7.181	92.791
78.188	17.740	4.072	0.174	11.744	88.031
63.842	27.171	8.988	0.402	14.687	84.911
41.947	36.911	21.142	1.045	18.746	80.209
23.533	38.956	37.511	2.480	23.069	74.451
17.938	38.018	44.045	3.737	25.174	71.089

SPECIFIC MODEL PARAMETERS IN KELVIN

		UNIQUAC		NRTL(ALPHA=.2)	
I J		AIJ	AJI	AIJ	AJI
1 2		508.06	-188.82	1028.8	-408.32
1 3		801.89	-236.36	946.77	1892.6
2 3		127.55	-56.292	56.600	264.20

R1 = 3.8143   R2 = 2.7791   R3 = 0.9200
Q1 = 3.027   Q2 = 2.508   Q3 = 1.400

MEAN DEV. BETWEEN CALC. AND EXP. CONC. IN MOLE PCT

UNIQUAC (SPECIFIC PARAMETERS)   0.71
NRTL   (SPECIFIC PARAMETERS)   0.53

(1) C6H10   CYCLOHEXENE
--------- -------------
(2) C3H8O   2-PROPANOL
--------- -------------
(3) H2O   WATER
--------- -------------

WASHBURN E.R., GRAHAM C.L., ARNOLD G.B., TRANSUE L.F.
J.AM.CHEM.SOC. 62(1940)1454

TEMPERATURE = 25.0 DEG C   TYPE OF SYSTEM = 1

EXPERIMENTAL TIE LINES IN MOLE PCT (GRAPH.INTERPOL.)

        LEFT PHASE   RIGHT PHASE
      (1)   (2)   (3)   (1)   (2)   (3)

    98.731   0.815   0.453   0.023   2.813   97.164
    90.305   7.500   2.195   0.077   6.908   93.015
    72.366   21.516   6.119   0.195   11.269   88.536
    55.760   32.275   11.966   0.442   14.061   85.497
    46.241   35.665   18.094   0.701   15.622   83.677
    34.219   38.629   27.152   1.312   17.674   81.014
    17.073   37.304   45.623   2.425   22.094   75.482
     9.889   32.739   57.372   3.842   24.867   71.291

SPECIFIC MODEL PARAMETERS IN KELVIN

                                     NRTL(ALPHA=.2)
          UNIQUAC   AIJ   AJI
I J   AIJ   AJI

1 2   382.20   -128.62   1004.1   -413.44
1 3   759.33   167.97   1084.5   2008.8
2 3   43.061   59.409   -191.28   592.23

         R1 = 3.8143   R2 = 2.7791   R3 = 0.9200
         Q1 = 3.027   Q2 = 2.508   Q3 = 1.400

MEAN DEV. BETWEEN CALC. AND EXP. CONC. IN MOLE PCT

UNIQUAC (SPECIFIC PARAMETERS)   1.02
NRTL   (SPECIFIC PARAMETERS)   0.84
UNIQUAC (COMMON PARAMETERS)   3.12

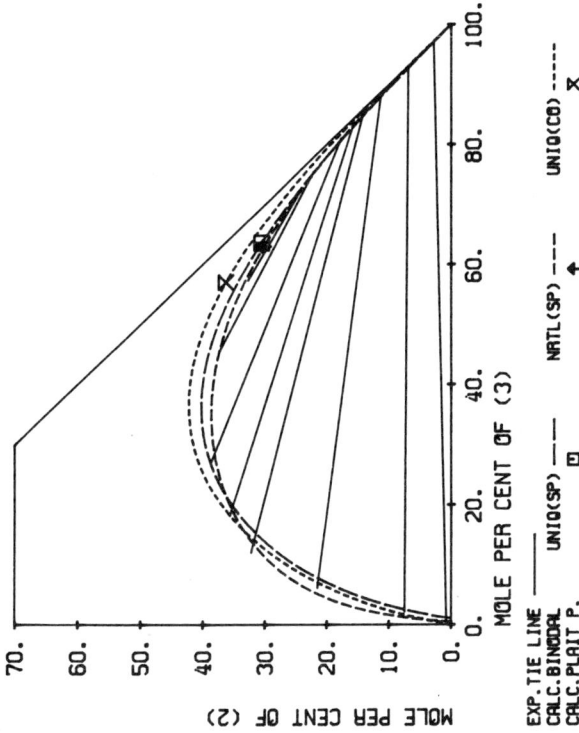

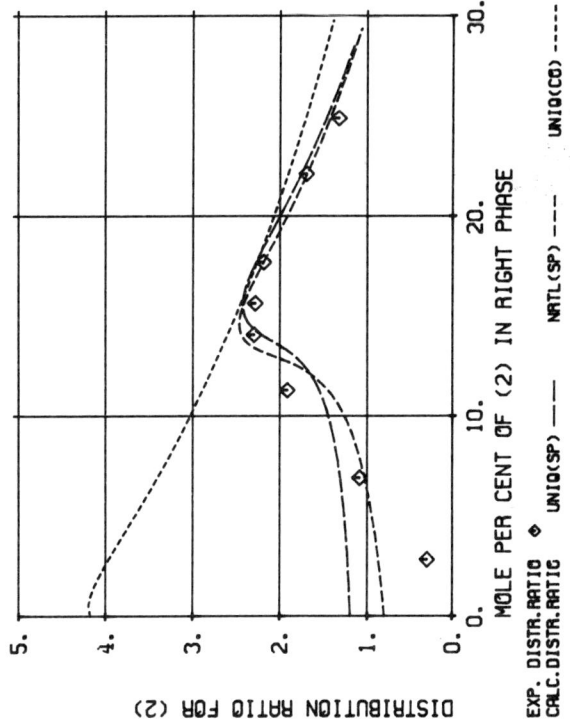

C_3H_8O-C_6H_{10}

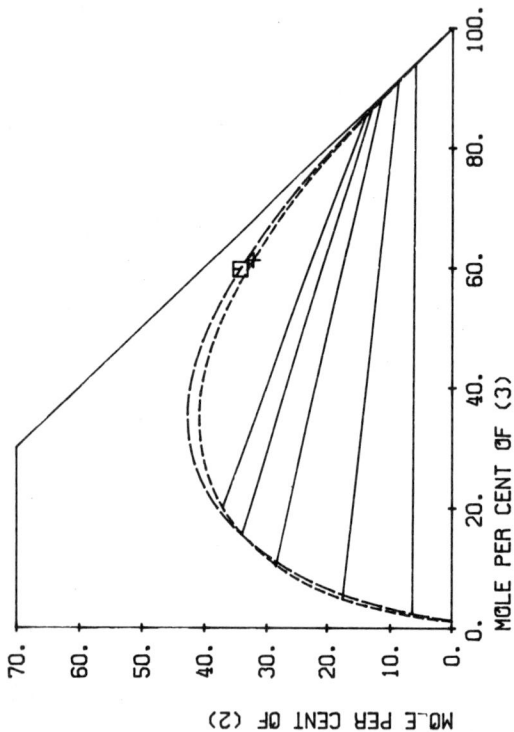

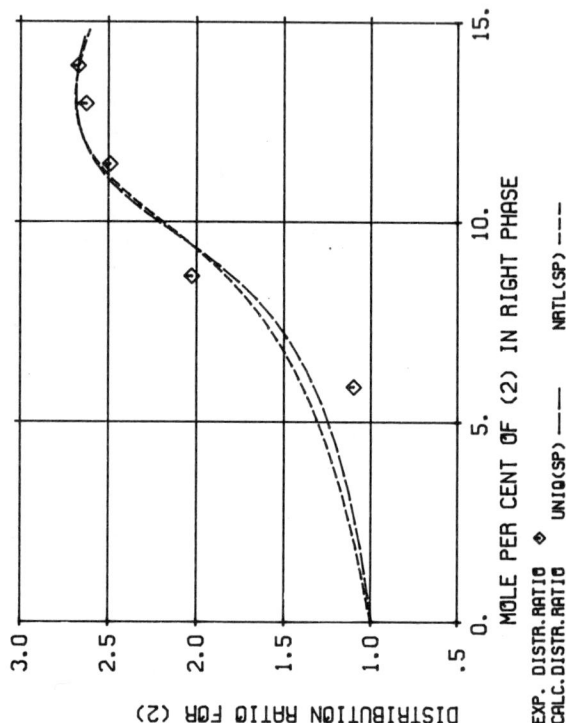

(1) C6H10   CYCLOHEXENE
(2) C3H8O   2-PROPANOL
(3) H2O   WATER

WASHBURN E.R., GRAHAM C.L., ARNOLD G.B., TRANSUE L.F.,
J.AM.CHEM.SOC. 62(1940)1454

TEMPERATURE = 35.0 DEG C   TYPE OF SYSTEM = 1

EXPERIMENTAL TIE LINES IN MOLE PCT (GRAPH.INTERPOL.)

	LEFT PHASE			RIGHT PHASE	
(1)	(2)	(3)	(1)	(2)	(3)
91.047	6.450	2.503	0.062	5.874	94.064
77.682	17.517	4.801	0.122	8.640	91.238
61.313	28.489	10.199	0.269	11.443	88.288
50.426	33.995	15.579	0.410	12.948	86.642
42.861	37.119	20.020	0.536	13.894	85.570

SPECIFIC MODEL PARAMETERS IN KELVIN

		UNIQUAC		NRTL(ALPHA=.2)	
I J		AIJ	AJI	AIJ	AJI
1 2		479.55	-183.72	953.81	-389.40
1 3		656.33	343.46	880.37	2184.3
2 3		127.85	-25.806	14.527	407.78

R1 = 3.8143   R2 = 2.7791   R3 = 0.9200
Q1 = 3.027   Q2 = 2.508   Q3 = 1.400

MEAN DEV. BETWEEN CALC. AND EXP. CONC. IN MOLE PCT

UNIQUAC (SPECIFIC PARAMETERS)   0.33
NRTL   (SPECIFIC PARAMETERS)   0.26

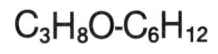

(1) H2O   WATER
(2) C3H8O   2-PROPANOL
(3) C6H12   CYCLOHEXANE

WASHBURN E.R., BROCKWAY C.E., GRAHAM C.L., DEMING P.
J.AM.CHEM.SOC. 64(1942)1886

TEMPERATURE = 25.0 DEG C   TYPE OF SYSTEM = 1

EXPERIMENTAL TIE LINES IN MOLE PCT

	LEFT PHASE			RIGHT PHASE	
(1)	(2)	(3)	(1)	(2)	(3)
98.593	1.363	0.044	0.0	0.0	100.000
97.736	2.241	0.023	0.0	0.280	99.720
94.683	5.196	0.120	0.463	1.804	97.733
90.447	9.500	0.052	0.906	8.013	91.081
86.970	12.778	0.252	1.327	14.852	83.821
81.023	18.136	0.841	4.226	21.783	73.991
69.722	27.285	2.992	10.846	29.952	59.202
56.635	35.282	8.082	18.935	35.205	45.860

SPECIFIC MODEL PARAMETERS IN KELVIN

		UNIQUAC		NRTL(ALPHA=.2)	
I J	AIJ	AJI	AIJ	AJI	
1 2	-726.89	-244.80	-1600.8	-541.64	
1 3	-927.06	1009.2	1669.5	874.98	
2 3	-214.76	-432.15	-9.0112	-1805.8	

R1 = 0.9200   R2 = 2.7791   R3 = 4.0464
Q1 = 1.400   Q2 = 2.508   Q3 = 3.240

MEAN DEV. BETWEEN CALC. AND EXP. CONC. IN MOLE PCT

UNIQUAC (SPECIFIC PARAMETERS)   1.50
NRTL   (SPECIFIC PARAMETERS)   1.99
UNIQUAC (COMMON PARAMETERS)   1.85

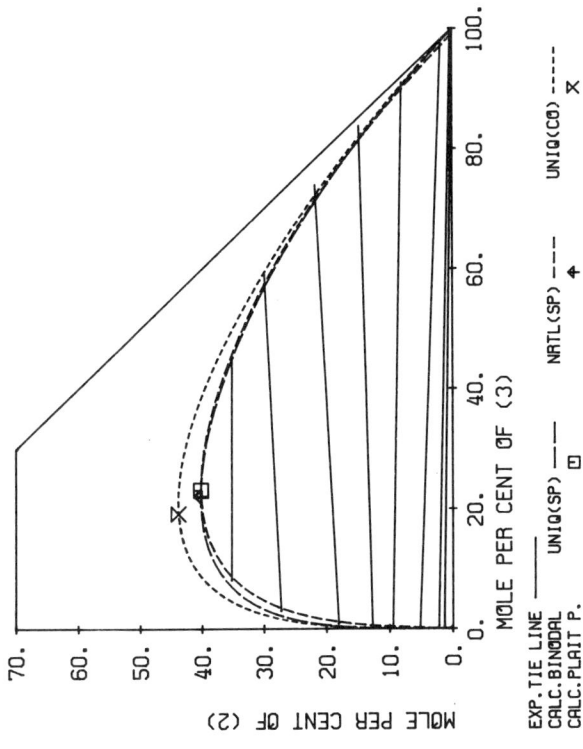

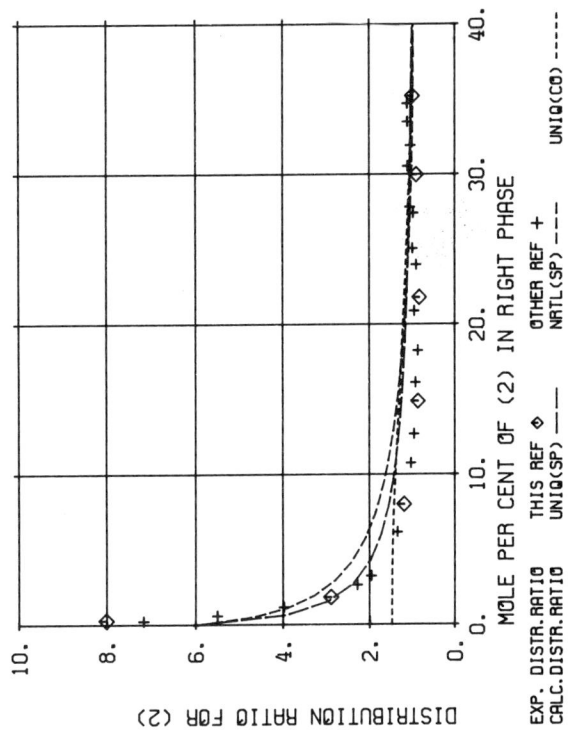

C₃H₈O-C₆H₁₂

(1) H2O   WATER

(2) C3H8O   2-PROPANOL

(3) C6H12   CYCLOHEXANE

VERHOEYE L.A.J.
J.CHEM.ENG.DATA 13(1968)462

TEMPERATURE = 25.0 DEG C   TYPE OF SYSTEM = 1

EXPERIMENTAL TIE LINES IN MOLE PCT

	LEFT PHASE			RIGHT PHASE	
(1)	(2)	(3)	(1)	(2)	(3)
93.903	5.999	0.098	0.462	2.631	96.908
87.677	12.074	0.249	1.336	12.679	85.985
83.905	15.799	0.295	2.174	18.245	79.581
76.401	21.812	1.787	5.676	23.966	70.358
70.379	26.845	2.776	8.022	27.409	64.568
60.317	33.443	6.240	13.168	31.909	54.923

SPECIFIC MODEL PARAMETERS IN KELVIN

		UNIQUAC		NRTL(ALPHA=.2)	
I	J	AIJ	AJI	AIJ	AJI
1	2	-726.89	-244.80	-1600.8	-541.64
1	3	927.06	1009.2	1669.5	874.98
2	3	-214.76	-432.15	-9.0112	-1805.8

R1 = 0.9200   R2 = 2.7791   R3 = 4.0464
Q1 = 1.400   Q2 = 2.508   Q3 = 3.240

MEAN DEV. BETWEEN CALC. AND EXP. CONC. IN MOLE PCT

UNIQUAC (SPECIFIC PARAMETERS)   1.58
NRTL   (SPECIFIC PARAMETERS)   2.12
UNIQUAC (COMMON PARAMETERS)   1.87

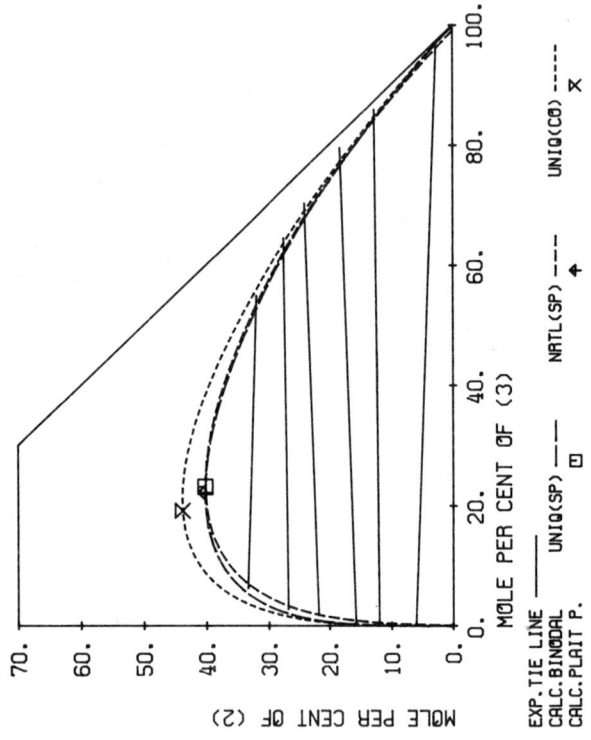

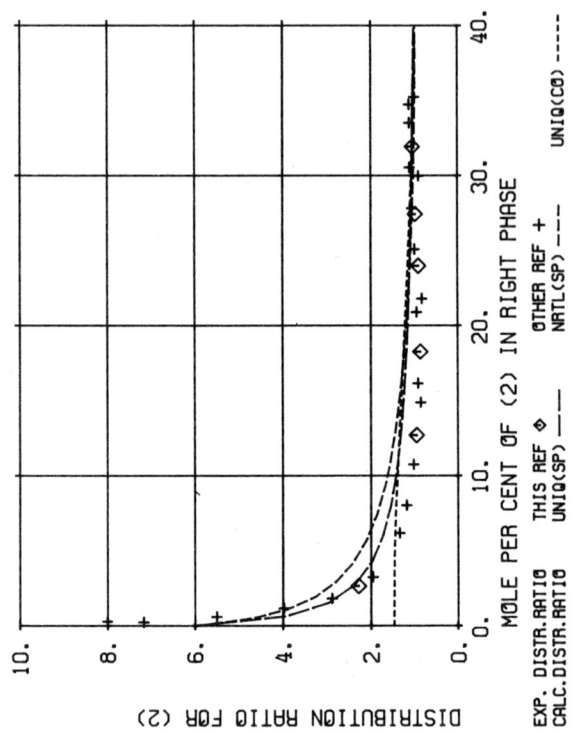

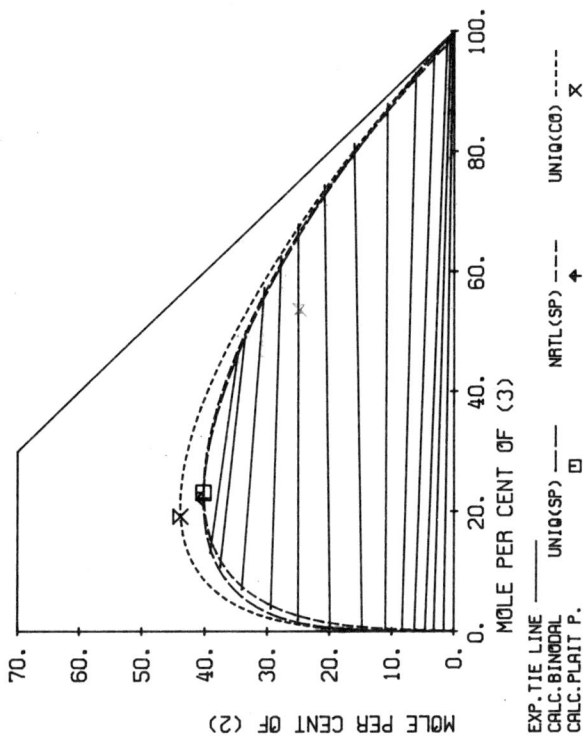

MOLE PER CENT OF (3)

MOLE PER CENT OF (2)

EXP. TIE LINE ———   UNIQ(SP) ———   NRTL(SP) ---+---   UNIQ(CO) ---✗---
CALC. BINODAL   □
CALC. PLAIT P.

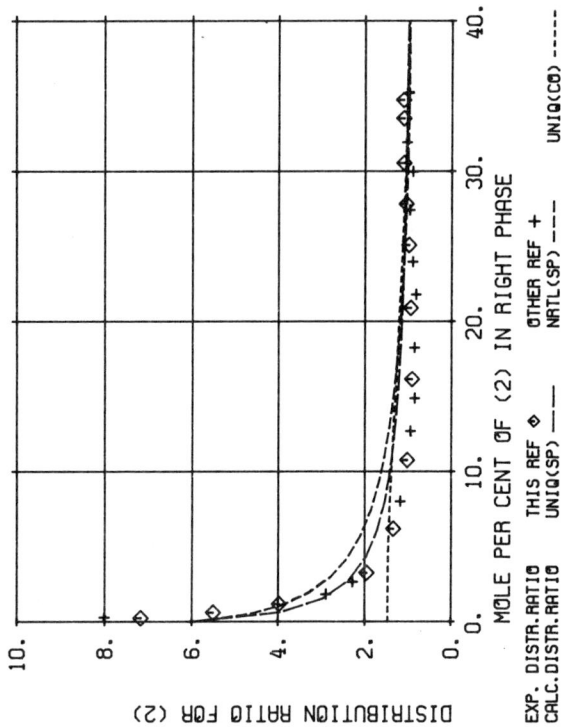

MOLE PER CENT OF (2) IN RIGHT PHASE

DISTRIBUTION RATIO FOR (2)

EXP. DISTR. RATIO ◇   THIS REF ◇   OTHER REF +   UNIQ(CO) ---—---
CALC. DISTR. RATIO ———   UNIQ(SP) ———   NRTL(SP) ---—---

(1) H2O   WATER
(2) C3H8O   2-PROPANOL
(3) C6H12   CYCLOHEXANE

NIKURASHINA N.I., SINEGUBOVA S.I.
ZH.OBSHCH.KHIM. 43(1973)2100

TEMPERATURE = 25.0 DEG C   TYPE OF SYSTEM = 1

EXPERIMENTAL TIE LINES IN MOLE PCT

	LEFT PHASE			RIGHT PHASE	
(1)	(2)	(3)	(1)	(2)	(3)
98.389	1.605	0.007	0.140	0.224	99.636
96.759	3.225	0.016	0.233	0.586	99.181
95.369	4.595	0.036	0.325	1.155	98.520
93.637	6.314	0.049	0.369	3.223	96.408
91.632	8.304	0.064	0.730	6.154	93.116
88.886	11.020	0.094	1.344	10.739	87.918
84.974	14.811	0.214	2.574	16.119	81.308
79.312	19.920	0.767	4.685	20.873	74.442
72.943	25.029	2.028	7.055	25.055	67.889
66.707	29.423	3.869	9.594	27.807	62.599
58.887	34.072	7.041	12.142	30.520	57.337
51.747	37.536	10.717	16.314	33.498	50.187
47.766	39.127	13.106	18.553	34.704	46.743

SPECIFIC MODEL PARAMETERS IN KELVIN

		UNIQUAC		NRTL(ALPHA=.2)	
I	J	AIJ	AJI	AIJ	AJI
1	2	-726.89	-244.80	-1600.8	-541.64
1	3	927.06	1009.2	1669.5	874.98
2	3	-214.76	-432.15	-9.0112	-1805.8

R1 = 0.9200   R2 = 2.7791   R3 = 4.0464
Q1 = 1.400   Q2 = 2.508   Q3 = 3.240

MEAN DEV. BETWEEN CALC. AND EXP. CONC. IN MOLE PCT

UNIQUAC (SPECIFIC PARAMETERS)   1.74
NRTL   (SPECIFIC PARAMETERS)   2.15
UNIQUAC (COMMON PARAMETERS)   2.09

C$_3$H$_8$O-C$_6$H$_{14}$

(1) H2O   WATER

(2) C3H8O   2-PROPANOL

(3) C6H14   HEXANE

VOROBEVA A.I., KARAPETYANTS M.KH.
ZH.FIZ.KHIM. 41(1967)1984

TEMPERATURE = 25.0 DEG C   TYPE OF SYSTEM = 1

EXPERIMENTAL TIE LINES IN MOLE PCT

	LEFT PHASE			RIGHT PHASE	
(1)	(2)	(3)	(1)	(2)	(3)
90.513	9.384	0.102	0.923	9.401	89.677
89.695	10.202	0.104	1.375	10.168	88.457
80.027	19.294	0.679	5.139	21.179	73.682
73.281	25.170	1.549	7.440	25.517	67.044
63.925	32.348	3.727	10.654	30.038	59.308
56.053	37.289	6.658	15.832	34.758	49.410
49.847	40.023	10.129	18.482	36.643	44.875
45.393	41.405	13.202	22.068	38.259	39.673
41.315	41.849	16.836	26.805	39.976	33.219
39.097	42.021	18.882	29.534	40.754	29.712

SPECIFIC MODEL PARAMETERS IN KELVIN

		UNIQUAC		NRTL(ALPHA=.2)	
I J	AIJ	AJI		AIJ	AJI
1 2	-453.04	-15.355		-848.39	-66.744
1 3	474.58	792.57		2628.6	892.99
2 3	-181.03	-67.022		-85.238	-799.14

R1 = 0.9200   R2 = 2.7791   R3 = 4.4998
Q1 = 1.400   Q2 = 2.508   Q3 = 3.856

MEAN DEV. BETWEEN CALC. AND EXP. CONC. IN MOLE PCT

UNIQUAC (SPECIFIC PARAMETERS)   0.78
NRTL   (SPECIFIC PARAMETERS)   0.86
UNIQUAC (COMMON PARAMETERS)   2.67

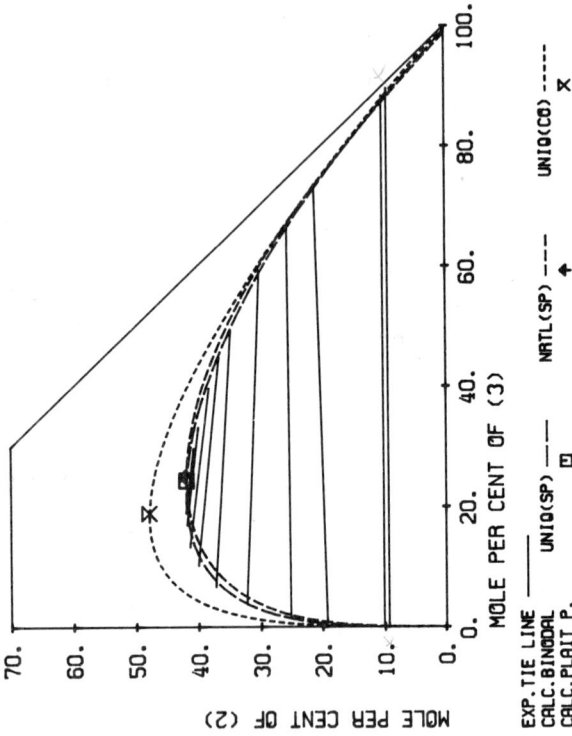

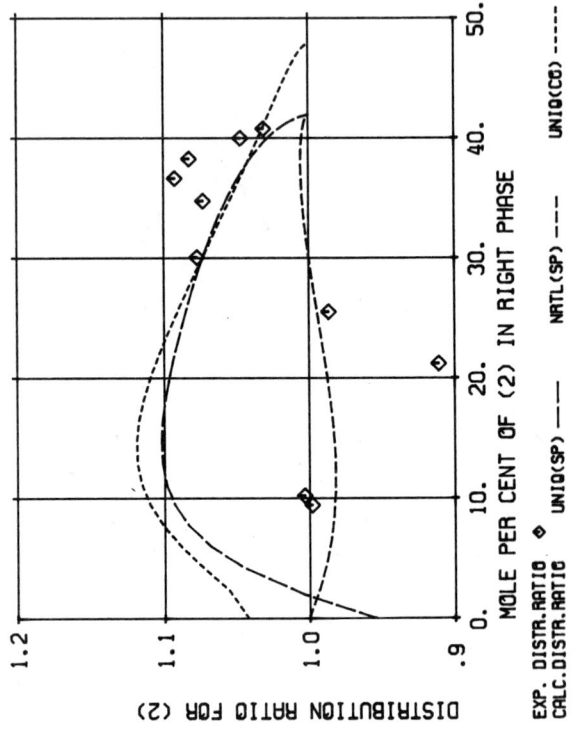

C_3H_8O-C_6H_{14}

(1) C6H14   HEXANE

(2) C3H8O   2-PROPANOL

(3) H2O   WATER

MOROZOV A.V., SARKISOV A.G., TUROVSKII V.B., ILYASKIN V.I.
ZH.FIZ.KHIM. 52(1978)1821

TEMPERATURE = 58.0 DEG C   TYPE OF SYSTEM = 1

EXPERIMENTAL TIE LINES IN MOLE PCT

LEFT PHASE			RIGHT PHASE		
(1)	(2)	(3)	(1)	(2)	(3)
91.752	8.248	0.0	0.0	4.250	95.750
82.871	17.129	0.0	0.039	9.811	90.150
59.808	28.779	11.413	0.350	15.351	83.299
48.718	33.188	18.094	2.552	28.795	68.653
34.060	39.857	26.083	11.135	37.434	51.431

SPECIFIC MODEL PARAMETERS IN KELVIN

		UNIQUAC		NRTL(ALPHA=.2)	
I	J	AIJ	AJI	AIJ	AJI
1	2	284.47	-88.079	95.261	-303.15
1	3	1023.3	165.09	894.44	-2378.4
2	3	5.7246	97.638	32.271	-84.458

R1 = 4.4998   R2 = 2.7791   R3 = 0.9200
Q1 = 3.856   Q2 = 2.508   Q3 = 1.400

MEAN DEV. BETWEEN CALC. AND EXP. CONC. IN MOLE PCT

UNIQUAC (SPECIFIC PARAMETERS)   1.51
NRTL   (SPECIFIC PARAMETERS)   1.83

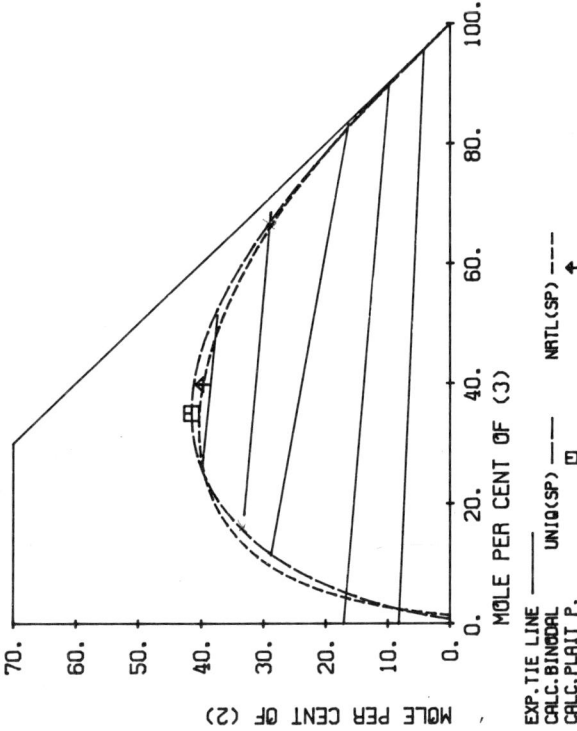

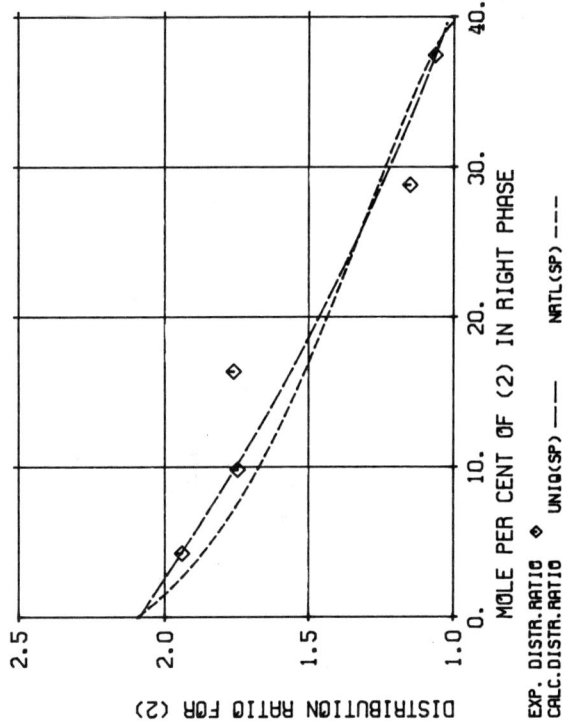

EXP.TIE LINE ———   CALC.BINODAL UNIQ(SP) ——— NRTL(SP) -----
CALC.PLAIT P.

MOLE PER CENT OF (3)

MOLE PER CENT OF (2)

EXP. DISTR.RATIO ◇   UNIQ(SP) ———   NRTL(SP) -----
CALC.DISTR.RATIO ———

MOLE PER CENT OF (2) IN RIGHT PHASE

DISTRIBUTION RATIO FOR (2)

C₃H₈O-C₆H₁₄O

(1) C6H14O	ETHER,DIISOPROPYL
(2) C3H8O	2-PROPANOL
(3) H2O	WATER

FRERE F.J.
IND.ENG.CHEM. 41(1949)2365

TEMPERATURE = 25.0 DEG C   TYPE OF SYSTEM = 1

EXPERIMENTAL TIE LINES IN MOLE PCT

LEFT PHASE			RIGHT PHASE		
(1)	(2)	(3)	(1)	(2)	(3)
91.204	4.503	4.293	0.187	2.192	97.621
84.803	8.990	6.206	0.230	3.187	96.583
79.814	12.649	7.537	0.233	3.933	95.834
73.169	17.191	9.640	0.236	4.481	95.283
67.447	20.078	12.475	0.239	5.079	94.682
57.596	25.278	17.126	0.284	5.582	94.156
51.726	27.928	20.346	0.284	5.860	93.856
41.999	32.027	25.974	0.287	6.375	93.338
31.408	35.067	33.525	0.356	7.293	92.351
23.424	35.607	40.969	0.497	8.670	90.833
16.893	34.552	48.556	0.643	9.767	89.590
13.163	32.321	54.515	0.843	10.850	88.307
9.459	29.044	61.497	1.110	12.329	86.561

SPECIFIC MODEL PARAMETERS IN KELVIN

	UNIQUAC		NRTL(ALPHA=.2)	
I J	AIJ	AJI	AIJ	AJI
1 2	437.93	-154.79	1099.1	-432.16
1 3	730.78	73.536	576.56	1479.8
2 3	-19.791	161.08	-399.76	1134.6

R1 = 4.7421   R2 = 2.7791   R3 = 0.9200
Q1 = 4.088   Q2 = 2.508   Q3 = 1.400

MEAN DEV. BETWEEN CALC. AND EXP. CONC. IN MOLE PCT

UNIQUAC (SPECIFIC PARAMETERS)	0.69
NRTL (SPECIFIC PARAMETERS)	0.42
UNIQUAC (COMMON PARAMETERS)	3.86

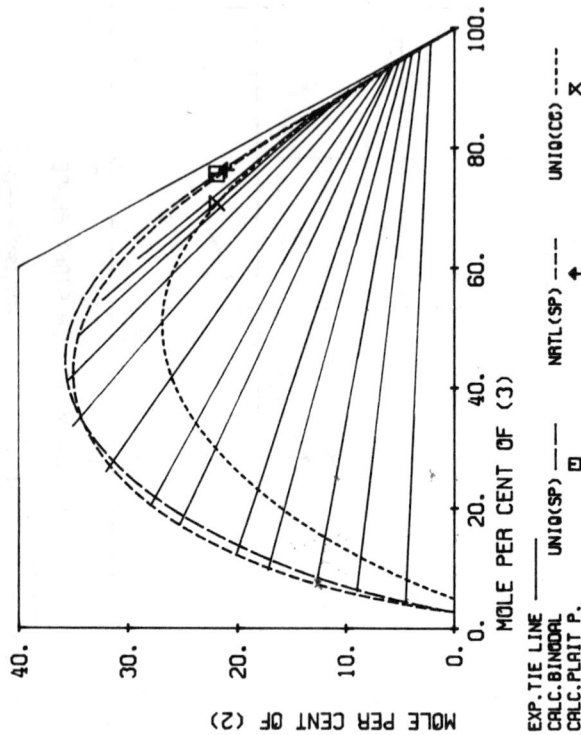

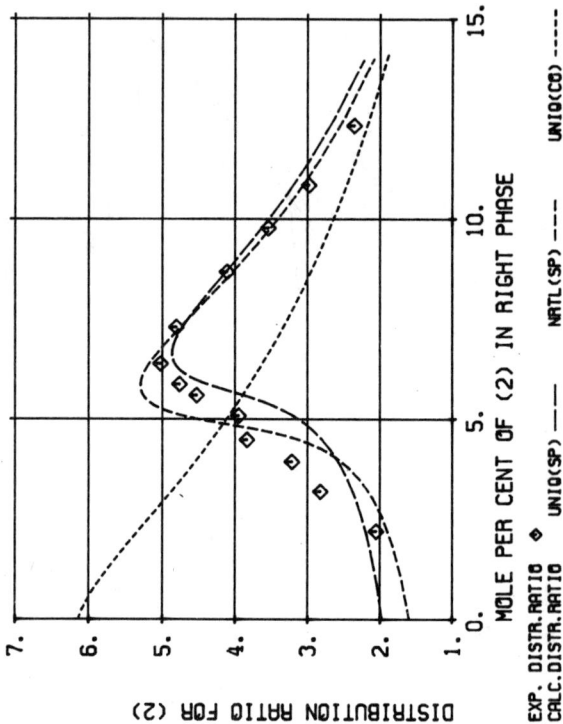

MOLE PER CENT OF (2) — MOLE PER CENT OF (3)

EXP.TIE LINE ——   UNIQ(SP) ☐   NRTL(SP) ✦   UNIQ(CG) ✗
CALC.BINODAL ----   CALC.PLAIT P.

DISTRIBUTION RATIO FOR (2) — MOLE PER CENT OF (2) IN RIGHT PHASE

EXP. DISTR.RATIO ◇   UNIQ(SP)   NRTL(SP)   UNIQ(CG)
CALC.DISTR.RATIO

C₃H₈O-C₇H₈

C_3H_8O-C_7H_8

(1) C7H8   TOLUENE
(2) C3H8O   2-PROPANOL
(3) H2O   WATER

WASHBURN E.R., BEGUIN A.E.
J.AM.CHEM.SOC. 62(1940)579

TEMPERATURE = 25.0 DEG C   TYPE OF SYSTEM = 1

EXPERIMENTAL TIE LINES IN MOLE PCT

	LEFT PHASE			RIGHT PHASE	
(1)	(2)	(3)	(1)	(2)	(3)
96.121	3.426	0.453	0.064	3.994	95.942
85.918	10.685	3.398	0.064	6.895	93.041
74.444	19.413	6.143	0.123	8.751	91.126
56.865	30.190	12.946	0.253	11.020	88.727
46.964	34.259	18.778	0.317	12.221	87.462
35.492	38.047	26.462	0.382	13.763	85.855
26.001	39.752	34.247	0.810	15.693	83.497
19.341	39.617	41.042	1.235	17.450	81.315
16.429	38.589	44.982	1.652	19.109	79.239
12.513	36.020	51.468	2.187	20.864	76.950
12.000	35.572	52.429	2.677	21.670	75.653
10.192	33.955	55.852	3.106	23.307	73.587

SPECIFIC MODEL PARAMETERS IN KELVIN

I J	UNIQUAC		NRTL(ALPHA=.2)	
	AIJ	AJI	AIJ	AJI
1 2	454.74	-161.68	984.99	-393.42
1 3	720.57	115.87	-906.80	1194.0
2 3	54.379	53.732	-210.61	688.75

R1 = 3.9228   R2 = 2.7791   R3 = 0.9200
Q1 = 2.968   Q2 = 2.508   Q3 = 1.400

MEAN DEV. BETWEEN CALC. AND EXP. CONC. IN MOLE PCT

UNIQUAC (SPECIFIC PARAMETERS)   0.62
NRTL   (SPECIFIC PARAMETERS)   0.42
UNIQUAC (COMMON PARAMETERS)   1.73

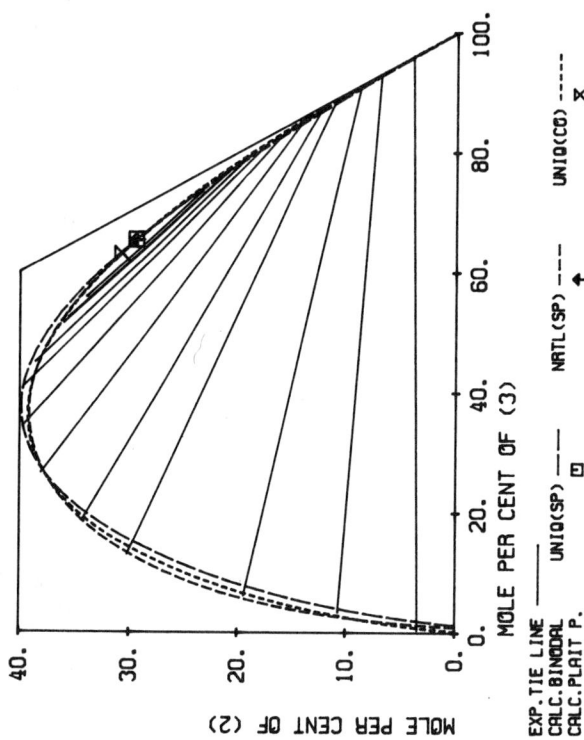

MOLE PER CENT OF (2) / MOLE PER CENT OF (3)

EXP.TIE LINE   UNIQ(SP) ⊞   NRTL(SP) ✦   UNIQ(CO) ⊠
CALC.BINODAL
CALC.PLAIT P.

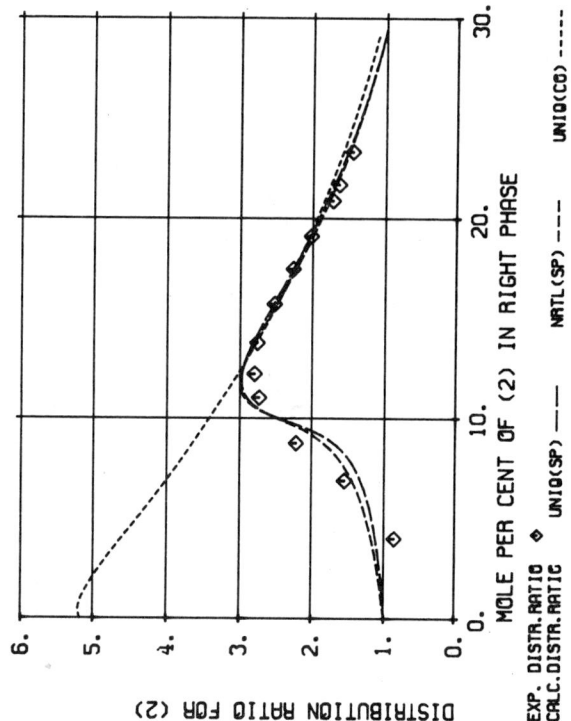

DISTRIBUTION RATIO FOR (2) / MOLE PER CENT OF (2) IN RIGHT PHASE

EXP. DISTR.RATIO ◇   UNIQ(SP)   NRTL(SP)   UNIQ(CO)
CALC.DISTR.RATIO

$C_3H_8O-C_7H_{16}$

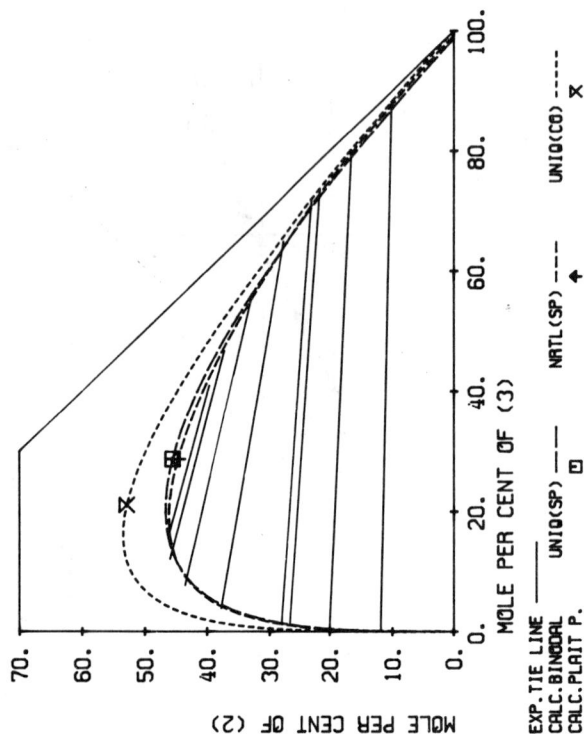

MOLE PER CENT OF (3)

MOLE PER CENT OF (2)

EXP.TIE LINE ——   UNIQ(SP) —— □   NRTL(SP) —— ▲   UNIQ(CO) —— ✕
CALC.BINODAL ——
CALC.PLAIT P.

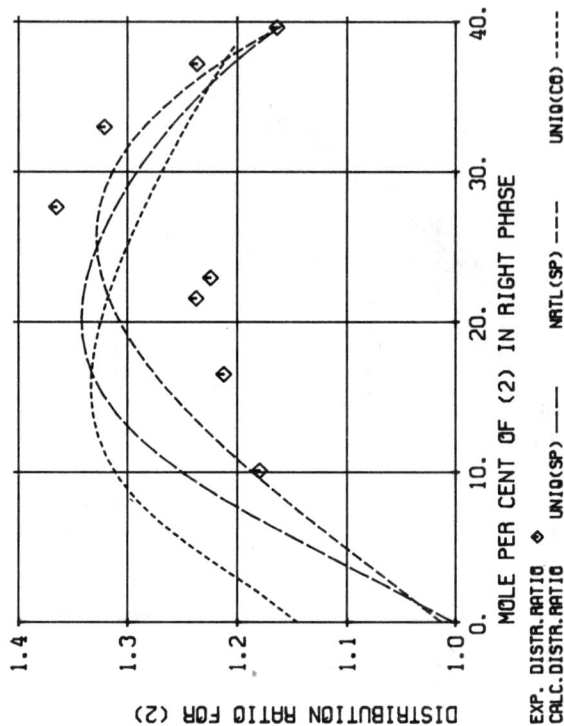

DISTRIBUTION RATIO FOR (2)

MOLE PER CENT OF (2) IN RIGHT PHASE

EXP. DISTR.RATIO ◇   UNIQ(SP) ——   NRTL(SP) ——   UNIQ(CO) ——
CALC.DISTR.RATIO ——

(1) H2O   WATER
(2) C3H8O   2-PROPANOL
(3) C7H16   HEPTANE

VOROBEVA A.I., KARAPETYANTS M.KH.
ZH.FIZ.KHIM. 41(1967)1984

TEMPERATURE = 25.0 DEG C   TYPE OF SYSTEM = 1

EXPERIMENTAL TIE LINES IN MOLE PCT

	LEFT PHASE			RIGHT PHASE	
(1)	(2)	(3)	(1)	(2)	(3)
87.991	11.917	0.092	1.580	10.100	88.320
79.598	20.028	0.374	4.009	16.523	79.468
72.305	26.674	1.021	4.861	21.561	73.578
70.704	28.065	1.231	5.292	22.925	71.783
58.423	37.731	3.846	7.377	27.637	64.985
48.772	43.567	7.661	11.988	32.978	55.034
41.976	45.997	12.027	16.015	37.199	46.786
37.547	46.111	16.342	19.367	39.605	41.028

SPECIFIC MODEL PARAMETERS IN KELVIN

		UNIQUAC		NRTL(ALPHA=.2)	
I J	AIJ	AJI	AIJ	AJI	
1 2	-4.7588	11.946	735.18	-344.69	
1 3	282.38	740.14	2794.1	1036.9	
2 3	-123.22	327.03	265.44	110.29	

R1 = 0.9200   R2 = 2.7791   R3 = 5.1742
Q1 = 1.400   Q2 = 2.508   Q3 = 4.396

MEAN DEV. BETWEEN CALC. AND EXP. CONC. IN MOLE PCT

UNIQUAC (SPECIFIC PARAMETERS)   0.82
NRTL   (SPECIFIC PARAMETERS)   0.59
UNIQUAC (COMMON PARAMETERS)   2.58

$C_3H_8O-C_8H_{10}$

(1) C8H10   BENZENE,ETHYL

(2) C3H8O   2-PROPANOL

(3) H2O   WATER

KNYPL E.T., WOJDYLO S.Z.
J.APPL.CHEM. 17(1967)361

TEMPERATURE = 25.0 DEG C   TYPE OF SYSTEM = 1

EXPERIMENTAL TIE LINES IN MOLE PCT

	LEFT PHASE			RIGHT PHASE	
(1)	(2)	(3)	(1)	(2)	(3)
98.247	1.753	0.0	0.055	3.476	96.469
94.652	5.348	0.0	0.078	6.236	93.686
86.461	11.887	1.652	0.124	8.873	91.003
79.819	17.523	2.658	0.151	11.370	88.479
62.344	28.235	9.421	0.347	14.835	84.818
41.719	39.276	19.004	1.786	22.759	75.455
29.898	39.606	30.496	3.909	27.895	68.195
19.267	39.260	41.473	5.648	31.189	63.163

SPECIFIC MODEL PARAMETERS IN KELVIN

		UNIQUAC		NRTL(ALPHA=.2)	
I	J	AIJ	AJI	AIJ	AJI
1	2	-295.49	152.12	1144.3	-461.32
1	3	900.46	271.85	1069.8	2123.8
2	3	-247.28	-24.085	644.23	-126.63

R1 = 4.5972   R2 = 2.7791   R3 = 0.9200
Q1 = 3.508   Q2 = 2.508   Q3 = 1.400

MEAN DEV. BETWEEN CALC. AND EXP. CONC. IN MOLE PCT

UNIQUAC (SPECIFIC PARAMETERS)   1.28
NRTL   (SPECIFIC PARAMETERS)   0.92
UNIQUAC (COMMON PARAMETERS)   2.53

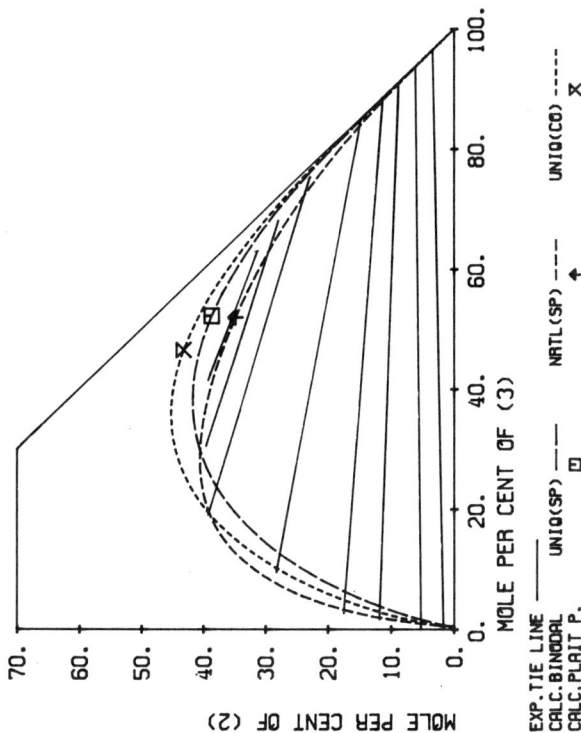

MOLE PER CENT OF (3)

MOLE PER CENT OF (2)

EXP.TIE LINE ———   UNIQ(SP) ☐   NRTL(SP) - - -   UNIQ(CO) - - - -
CALC.BINODAL
CALC.PLAIT P.   ✕   ✦   ✕

DISTRIBUTION RATIO FOR (2)

MOLE PER CENT OF (2) IN RIGHT PHASE

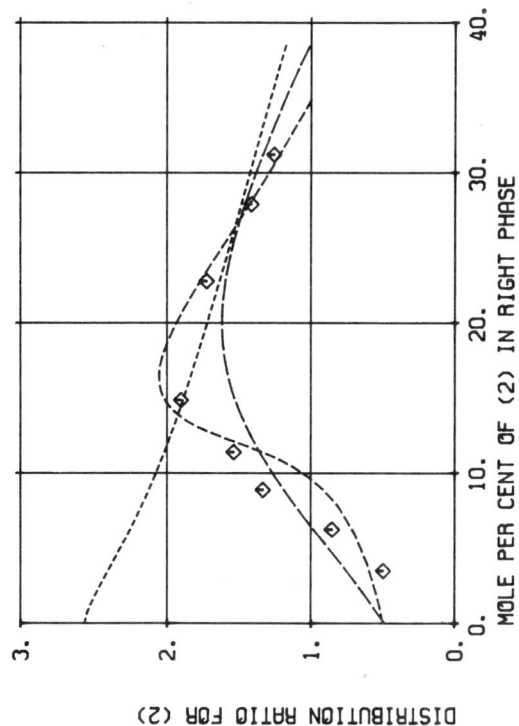

EXP. DISTR.RATIO ◇   UNIQ(SP) - - -   NRTL(SP) - - -   UNIQ(CO) - - - -
CALC.DISTR.RATIO

$C_3H_8O-C_8H_{18}$

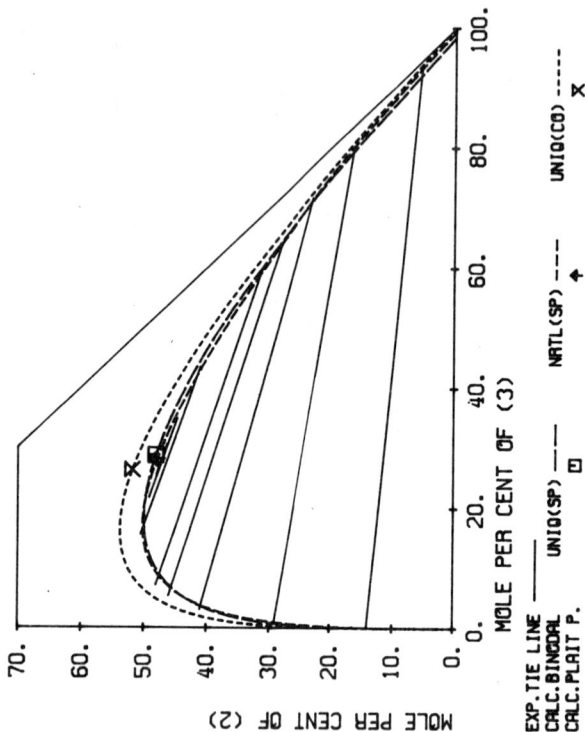

MOLE PER CENT OF (3)

MOLE PER CENT OF (2)

EXP.TIE LINE ——   UNIQ(SP) ◇   NRTL(SP) ----   UNIQ(CO) -----
CALC.BINODAL ----   △
CALC.PLAIT P. ✛   ✕

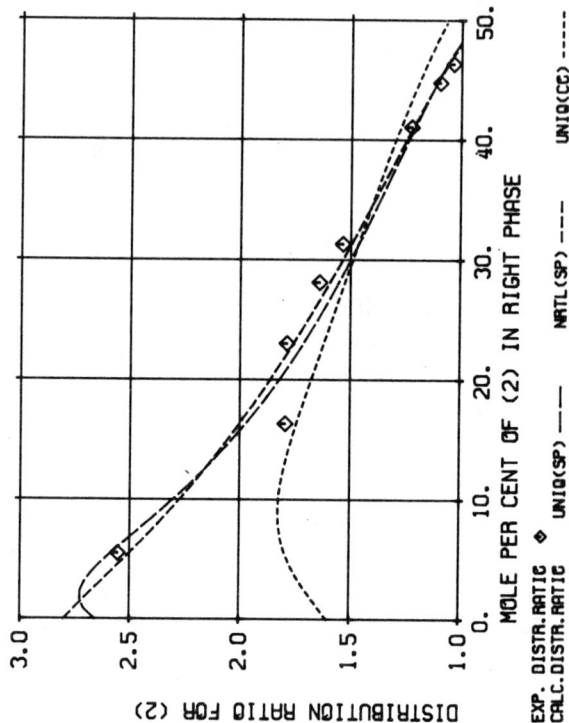

MOLE PER CENT OF (2) IN RIGHT PHASE

DISTRIBUTION RATIO FOR (2)

EXP. DISTR.RATIO ◇   UNIQ(SP) ◇   NRTL(SP) ----   UNIQ(CC) -----
CALC.DISTR.RATIO

(1) H2O   WATER
(2) C3H8O   2-PROPANOL
(3) C8H18   OCTANE

VOROBEVA A.I., KARAPETYANTS M.KH.
ZH.FIZ.KHIM. 41(1967)1984.

TEMPERATURE = 25.0 DEG C   TYPE OF SYSTEM = 1

EXPERIMENTAL TIE LINES IN MOLE PCT

	LEFT PHASE		RIGHT PHASE	
(1)	(2)	(3)	(1)	(2)	(3)
85.926	14.032	0.042	1.222	5.494	93.284
69.994	29.244	0.761	3.402	16.311	80.288
55.860	41.058	3.082	5.364	22.988	71.649
48.624	46.053	5.323	7.635	28.068	64.297
44.807	48.088	7.106	8.868	31.303	59.829
33.995	50.491	15.514	15.420	41.078	43.502
29.114	49.180	21.705	20.333	44.765	34.903
26.477	48.029	25.494	23.048	46.348	30.604

SPECIFIC MODEL PARAMETERS IN KELVIN

		UNIQUAC		NRTL(ALPHA=.2)	
I	J	AIJ	AJI	AIJ	AJI
1	2	-115.89	-42.499	636.26	-491.54
1	3	379.98	677.30	2661.4	1050.6
2	3	-136.89	305.22	399.98	-15.292

R1 = 0.9200   R2 = 2.7791   R3 = 5.8486
Q1 = 1.400   Q2 = 2.508   Q3 = 4.936

MEAN DEV. BETWEEN CALC. AND EXP. CONC. IN MOLE PCT

UNIQUAC (SPECIFIC PARAMETERS)   0.94
NRTL   (SPECIFIC PARAMETERS)   0.63
UNIQUAC (COMMON PARAMETERS)   2.12

C$_3$H$_8$O-C$_9$H$_{20}$

(1) H2O   WATER

(2) C3H8O   2-PROPANOL

(3) C9H2O   NONANE

VOROBEVA A.I., KARAPETYANTS M.KH.
ZH.FIZ.KHIM. 41(1967)1984

TEMPERATURE = 25.0 DEG C   TYPE OF SYSTEM = 1

EXPERIMENTAL TIE LINES IN MOLE PCT

	LEFT PHASE			RIGHT PHASE	
(1)	(2)	(3)	(1)	(2)	(3)
80.529	19.369	0.102	1.995	9.169	88.836
65.689	33.435	0.876	3.175	15.223	81.602
59.606	38.905	1.489	3.733	17.713	78.555
52.473	44.790	2.737	4.797	21.926	73.277
42.074	51.908	6.018	7.249	29.078	63.673
35.665	54.478	9.857	10.329	34.979	54.692
33.413	54.580	12.007	11.063	37.376	51.561
30.157	54.613	15.231	14.070	41.465	44.465
26.795	54.050	19.155	16.320	45.209	38.472
25.666	53.740	20.593	17.229	46.472	36.299
23.676	52.110	24.214	19.845	49.038	31.117

SPECIFIC MODEL PARAMETERS IN KELVIN

		UNIQUAC		NRTL(ALPHA=.2)	
I J	AIJ	AJI		AIJ	AJI
1 2	-186.05	104.60		814.26	-468.11
1 3	-361.91	621.82		3151.0	1367.4
2 3	-126.43	311.70		581.79	-25.910

R1 = 0.9200 R2 = 2.7791 R3 = 6.5230
Q1 = 1.400   Q2 = 2.508   Q3 = 5.476

MEAN DEV. BETWEEN CALC. AND EXP. CONC. IN MOLE PCT

UNIQUAC (SPECIFIC PARAMETERS)   1.40
NRTL   (SPECIFIC PARAMETERS)   0.54
UNIQUAC (COMMON PARAMETERS)   1.68

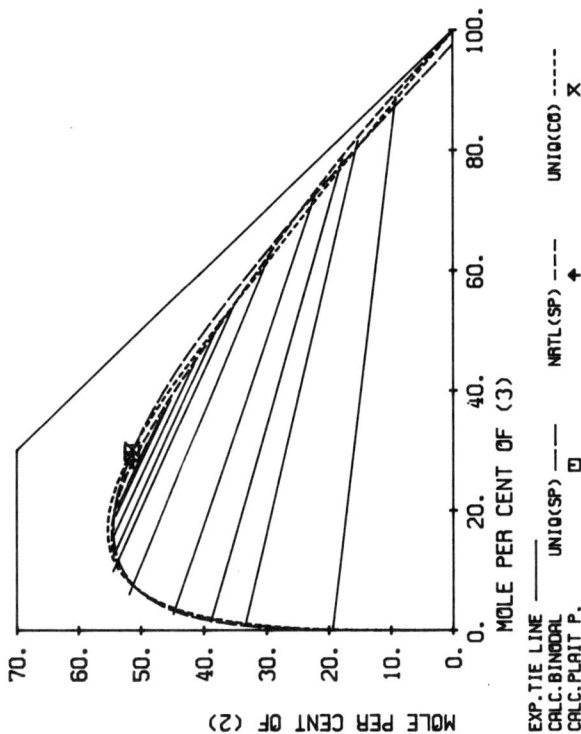

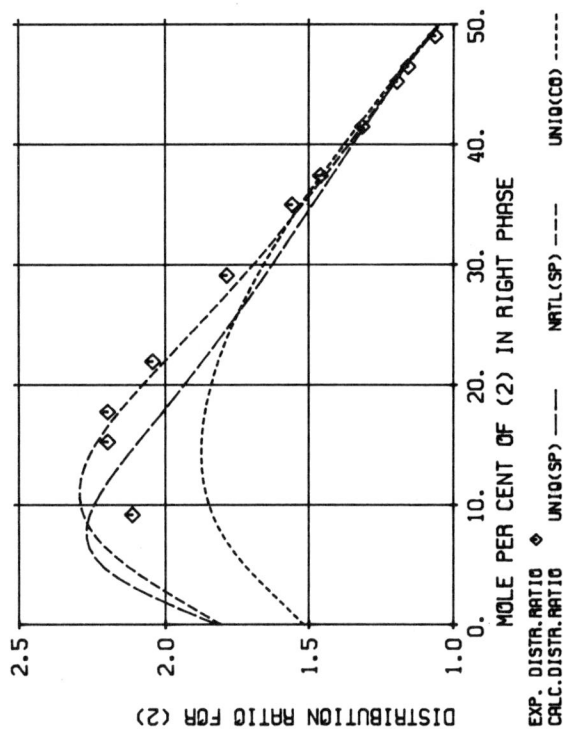

C_3H_8O-$C_{12}H_{10}O$

(1) H2O   WATER
(2) C3H8O   2-PROPANOL
(3) C12H100   ETHER,DIPHENYL

PURNELL J.H., BOWDEN S.T.
J.CHEM.SOC.(1954)1539

TEMPERATURE = 25.0 DEG C   TYPE OF SYSTEM = 1

EXPERIMENTAL TIE LINES IN MOLE PCT

	LEFT PHASE			RIGHT PHASE	
(1)	(2)	(3)	(1)	(2)	(3)
95.589	4.400	0.012	0.914	3.834	95.253
91.418	8.544	0.038	3.531	5.291	91.179
83.735	16.162	0.103	3.448	8.784	87.768
75.843	23.685	0.472	4.214	10.862	84.924
67.259	31.715	1.025	4.885	14.639	80.476
55.227	42.107	2.665	6.945	18.733	74.322
46.568	49.036	4.396	8.191	21.425	70.383
40.327	53.345	6.329	10.482	25.761	63.757
29.346	58.836	11.817	13.633	34.357	52.010

SPECIFIC MODEL PARAMETERS IN KELVIN

		UNIQUAC		NRTL(ALPHA=.2)	
I J		AIJ	AJI	AIJ	AJI
1 2	196.35	-138.33		1139.5	-533.00
1 3	121.11	625.94		3209.0	942.80
2 3	-49.285	227.18		646.68	11.982

R1 = 0.9200   R2 = 2.7791   R3 = 6.2873
Q1 = 1.400   Q2 = 2.508   Q3 = 4.480

MEAN DEV. BETWEEN CALC. AND EXP. CONC. IN MOLE PCT

UNIQUAC (SPECIFIC PARAMETERS)   1.21
NRTL   (SPECIFIC PARAMETERS)   0.67
UNIQUAC (COMMON PARAMETERS)   1.60

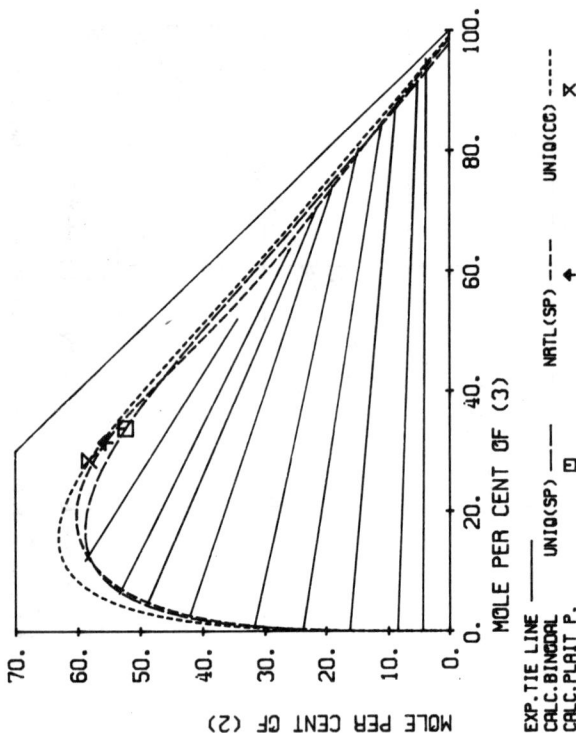

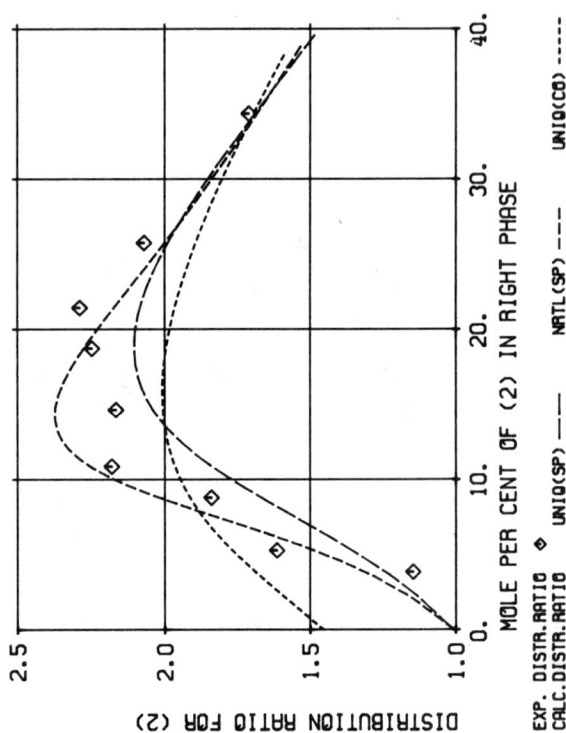

(1) C15H2606   GLYCEROL, TRIBUTANOATE
(2) C14H1806   PHTHALIC ACID, DI(2-METHOXYETHYL) ESTER
(3) C3H802   1,2-PROPANEDIOL

GARY L.H.; CRICHTON J.S.; FEILD R.
J.CHEM.ENG.DATA 3(1958)111

TEMPERATURE = 25.0 DEG C   TYPE OF SYSTEM = 1

EXPERIMENTAL TIE LINES IN MOLE PCT

	LEFT PHASE			RIGHT PHASE	
(1)	(2)	(3)	(1)	(2)	(3)
71.215	8.385	20.400	2.750	0.0	97.250
66.249	12.286	21.465	2.797	0.623	96.581
61.687	15.531	22.782	2.833	1.112	96.055
56.855	18.295	24.850	2.863	1.518	95.619
46.771	24.759	28.469	2.923	2.325	94.751
41.094	28.164	30.742	3.000	2.930	94.070
36.597	30.479	32.924	3.042	3.482	93.477
33.253	32.148	34.599	3.090	3.696	93.214
29.357	33.621	37.022	3.176	4.383	92.441
27.255	34.367	38.378	3.213	4.864	91.923
24.911	35.020	40.070	3.319	5.367	91.314

SPECIFIC MODEL PARAMETERS IN KELVIN

		UNIQUAC		NRTL(ALPHA=.2)	
I J	AIJ	AJI		AIJ	AJI
1 2	-12.428	-40.499		-140.95	-235.93
1 3	362.35	-60.634		191.23	880.53
2 3	158.28	-8.6636		-205.85	897.58

R1 =11.3003   R2 = 9.8472   R3 = 3.2824
Q1 = 9.432   Q2 = 7.936   Q3 = 2.784

MEAN DEV. BETWEEN CALC. AND EXP. CONC. IN MOLE PCT

UNIQUAC (SPECIFIC PARAMETERS)   0.42
NRTL   (SPECIFIC PARAMETERS)   0.43
UNIQUAC (COMMON PARAMETERS)   0.39

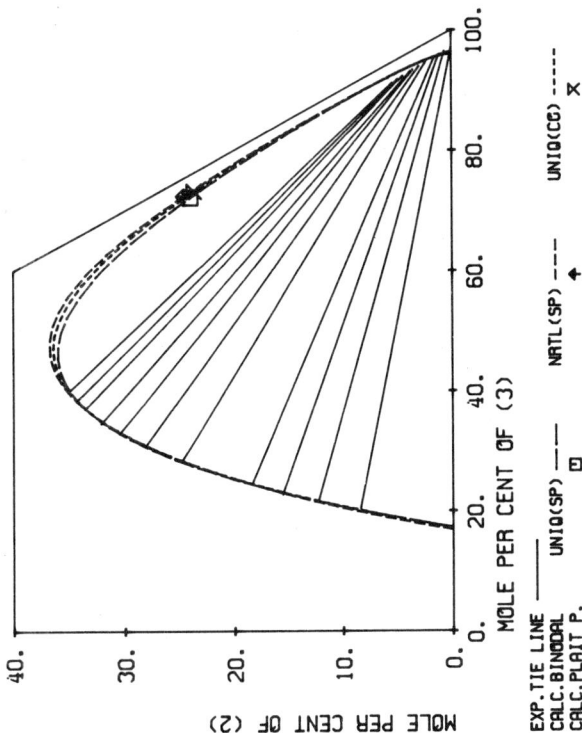

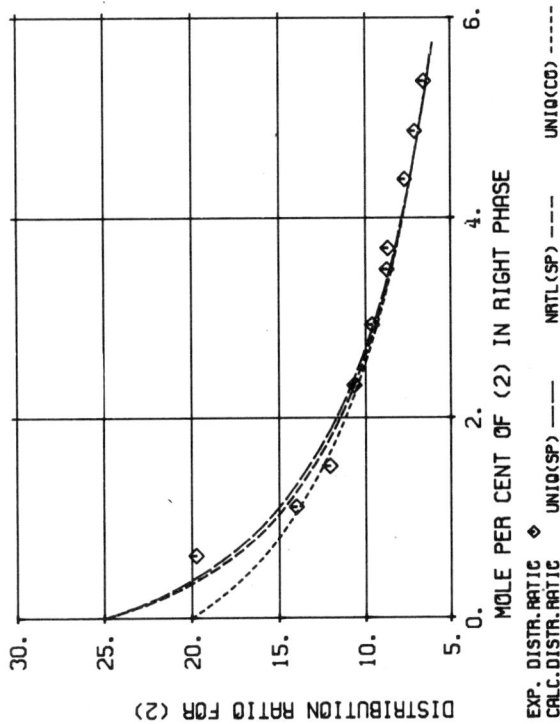

Alphabetical Compound Index
See Part 3